Springer Series in Statistics

For other titles published in this series go to,
http://www.springer.com/series/692

Bertrand Clarke · Ernest Fokoué · Hao Helen Zhang

Principles and Theory for Data Mining and Machine Learning

Springer

Bertrand Clarke
University of Miami
120 NW 14th Street
CRB 1055 (C-213)
Miami, FL, 33136
bclarke2@med.miami.edu

Ernest Fokoué
Center for Quality and Applied Statistics
Rochester Institute of Technology
98 Lomb Memorial Drive
Rochester, NY 14623
ernest.fokoue@gmail.com

Hao Helen Zhang
Department of Statistics
North Carolina State University
Genetics
P.O.Box 8203
Raleigh, NC 27695-8203
USA
hzhang2@stat.ncsu.edu

ISSN 0172-7397
ISBN 978-0-387-98134-5 e-ISBN 978-0-387-98135-2
DOI 10.1007/978-0-387-98135-2
Springer Dordrecht Heidelberg London New York

Library of Congress Control Number: 2009930499

Printed on acid-free paper

Springer is part of Springer Science+Business Media (www.springer.com)

Preface

The idea for this book came from the time the authors spent at the Statistics and Applied Mathematical Sciences Institute (SAMSI) in Research Triangle Park in North Carolina starting in fall 2003. The first author was there for a total of two years, the first year as a Duke/SAMSI Research Fellow. The second author was there for a year as a Post-Doctoral Scholar. The third author has the great fortune to be in RTP permanently. SAMSI was – and remains – an incredibly rich intellectual environment with a general atmosphere of free-wheeling inquiry that cuts across established fields. SAMSI encourages creativity: It is the kind of place where researchers can be found at work in the small hours of the morning – computing, interpreting computations, and developing methodology. Visiting SAMSI is a unique and wonderful experience.

The people most responsible for making SAMSI the great success it is include Jim Berger, Alan Karr, and Steve Marron. We would also like to express our gratitude to Dalene Stangl and all the others from Duke, UNC-Chapel Hill, and NC State, as well as to the visitors (short and long term) who were involved in the SAMSI programs. It was a magical time we remember with ongoing appreciation.

While we were there, we participated most in two groups: Data Mining and Machine Learning, for which Clarke was the group leader, and a General Methods group run by David Banks. We thank David for being a continual source of enthusiasm and inspiration. The first chapter of this book is based on the outline of the first part of his short course on Data Mining and Machine Learning. Moreover, David graciously contributed many of his figures to us. Specifically, we gratefully acknowledge that Figs. 1.1–6, Figs. 2.1,3,4,5,7, Fig. 4.2, Figs. 8.3,6, and Figs. 9.1,2 were either done by him or prepared under his guidance.

On the other side of the pond, the Newton Institute at Cambridge University provided invaluable support and stimulation to Clarke when he visited for three months in 2008. While there, he completed the final versions of Chapters 8 and 9. Like SAMSI, the Newton Institute was an amazing, wonderful, and intense experience.

This work was also partially supported by Clarke's NSERC Operating Grant 2004–2008. In the USA, Zhang's research has been supported over the years by two grants from the National Science Foundation. Some of the research those grants supported is in Chapter 10.

We hope that this book will be of value as a graduate text for a PhD-level course on data mining and machine learning (DMML). However, we have tried to make it comprehensive enough that it can be used as a reference or for independent reading. Our paradigm reader is someone in statistics, computer science, or electrical or computer engineering who has taken advanced calculus and linear algebra, a strong undergraduate probability course, and basic undergraduate mathematical statistics. Someone whose expertise in is one of the topics covered here will likely find that chapter routine, but hopefully find the other chapters are at a comfortable level.

The book roughly separates into three parts. Part I consists of Chapters 1 through 4: This is mostly a treatment of nonparametric regression, assuming a mastery of linear regression. Part II consists of Chapters 5, 6, and 7: This is a mix of classification, recent nonparametric methods, and computational comparisons. Part III consists of Chapters 8 through 11. These focus on high dimensional problems, including clustering, dimension reduction, variable selection, and multiple comparisons. We suggest that a selection of topics from the first two parts would be a good one semester course and a selection of topics from Part III would be a good follow-up course.

There are many topics left out: proper treatments of information theory, VC dimension, PAC learning, Oracle inequalities, hidden Markov models, graphical models, frames, and wavelets are the main absences. We regret this, but no book can be everything.

The main perspective undergirding this work is that DMML is a fusion of large sectors of statistics, computer science, and electrical and computer engineering. The DMML fusion rests on good prediction and a complete assessment of modeling uncertainty as its main organizing principles. The assessment of modeling uncertainty ideally includes all of the contributing factors, including those commonly neglected, in order to be valid. Given this, other aspects of inference – model identification, parameter estimation, hypothesis testing, and so forth – can largely be regarded as a consequence of good prediction. We suggest that the development and analysis of good predictors is the paradigm problem for DMML.

Overall, for students and practitioners alike, DMML is an exciting context in which whole new worlds of reasoning can be productively explored and applied to important problems.

Bertrand Clarke
University of Miami, Miami, FL
Ernest Fokoué
Kettering University, Flint, MI
Hao Helen Zhang
North Carolina State University, Raleigh, NC

Contents

Chapter 1
Variability, Information, and Prediction

Introductory statistics courses often start with summary statistics, then develop a notion of probability, and finally turn to parametric models – mostly the normal – for inference. By the end of the course, the student has seen estimation and hypothesis testing for means, proportions, ANOVA, and maybe linear regression. This is a good approach for a first encounter with statistical thinking. The student who goes on takes a familiar series of courses: survey sampling, regression, Bayesian inference, multivariate analysis, nonparametrics and so forth, up to the crowning glories of decision theory, measure theory, and asymptotics. In aggregate, these courses develop a view of statistics that continues to provide insights and challenges.

All of this was very tidy and cosy, but something changed. Maybe it was computing. All of a sudden, quantities that could only be described could be computed readily and explored. Maybe it was new data sets. Rather than facing small to moderate sample sizes with a reasonable number of parameters, there were 100 data points, 20,000 explanatory variables, and an array of related multitype variables in a time-dependent data set. Maybe it was new applications: bioinformatics, E-commerce, Internet text retrieval. Maybe it was new ideas that just didn't quite fit the existing framework. In a world where model uncertainty is often the limiting aspect of our inferential procedures, the focus became prediction more than testing or estimation. Maybe it was new techniques that were intellectually uncomfortable but extremely effective: What sense can be made of a technique like random forests? It uses randomly generated ensembles of trees for classification, performing better and better as more models are used.

All of this was very exciting. The result of these developments is called data mining and machine earning (DMML).

Data mining refers to the search of large, high-dimensional, multitype data sets, especially those with elaborate dependence structures. These data sets are so unstructured and varied, on the surface, that the search for structure in them is statistical. A famous (possibly apocryphal) example is from department store sales data. Apparently a store found there was an unusually high empirical correlation between diaper sales and beer sales. Investigation revealed that when men buy diapers, they often treat themselves to a six-pack. This might not have surprised the wives, but the marketers would have taken note.

B. Clarke et al., *Principles and Theory for Data Mining and Machine Learning*, Springer Series in Statistics, DOI 10.1007/978-0-387-98135-2_1,

Machine learning refers to the use of formal structures (machines) to do inference (learning). This includes what empirical scientists mean by model building – proposing mathematical expressions that encapsulate the mechanism by which a physical process gives rise to observations – but much else besides. In particular, it includes many techniques that do not correspond to physical modeling, provided they process data into information. Here, information usually means anything that helps reduce uncertainty. So, for instance, a posterior distribution represents "information" or is a "learner" because it reduces the uncertainty about a parameter.

The fusion of statistics, computer science, electrical engineering, and database management with new questions led to a new appreciation of sources of errors. In narrow parametric settings, increasing the sample size gives smaller standard errors. However, if the model is wrong (and they all are), there comes a point in data gathering where it is better to use some of your data to choose a new model rather than just to continue refining an existing estimate. That is, once you admit model uncertainty, you can have a smaller and smaller variance but your bias is constant. This is familiar from decomposing a mean squared error into variance and bias components.

Extensions of this animate DMML. Shrinkage methods (not the classical shrinkage, but the shrinking of parameters to zero as in, say, penalized methods) represent a tradeoff among variable selection, parameter estimation, and sample size. The ideas become trickier when one must select a basis as well. Just as there are well-known sums of squares in ANOVA for quantifying the variability explained by different aspects of the model, so will there be an extra variability corresponding to basis selection. In addition, if one averages models, as in stacking or Bayes model averaging, extra layers of variability (from the model weights and model list) must be addressed. Clearly, good inference requires trade-offs among the biases and variances from each level of modeling. It may be better, for instance, to "stack" a small collection of shrinkage-derived models than to estimate the parameters in a single huge model.

Among the sources of variability that must be balanced – random error, parameter uncertainty and bias, model uncertainty or misspecification, model class uncertainty, generalization error – there is one that stands out: model uncertainty. In the conventional paradigm with fixed parametric models, there is no model uncertainty; only parameter uncertainty remains. In conventional nonparametrics, there is only model uncertainty; there is no parameter, and the model class is so large it is sure to contain the true model. DMML is between these two extremes: The model class is rich beyond parametrization, and may contain the true model in a limiting sense, but the true model cannot be assumed to have the form the model class defines. Thus, there are many parameters, leading to larger standard errors, but when these standard errors are evaluated within the model, they are invalid: The adequacy of the model cannot be assumed, so the standard error of a parameter is about a value that may not be meaningful. It is in these high-variability settings in the mid-range of uncertainty (between parametric and nonparametric) that dealing with model uncertainty carefully usually becomes the dominant issue which can only be tested by predictive criteria.

There are other perspectives on DMML that exist, such as rule mining, fuzzy learning, observational studies, and computational learning theory. To an extent, these can be regarded as elaborations or variations of aspects of the perspective presented here, although advocates of those views might regard that as inadequate. However, no book can cover everything and all perspectives. Details on alternative perspectives to the one perspective presented here can be found in many good texts.

Before turning to an intuitive discussion of several major ideas that will recur throughout this monograph, there is an apparent paradox to note: Despite the novelty ascribed to DMML, many of the topics covered here have been studied for decades. Most of the core ideas and techniques have precedents from before 1990. The slight paradox is resolved by noting that what is at issue is the novel, unexpected way so many ideas, new and old, have been recombined to provide a new, general perspective dramatically extending the conventional framework epitomized by, say, Lehmann's books.

1.0.1 The Curse of Dimensionality

Given that model uncertainty is the key issue, how can it be measured? One crude way is through dimension. The problem is that high model uncertainty, especially of the sort central to DMML, rarely corresponds to a model class that permits a finite-dimensional parametrization. On the other hand, some model classes, such as neural nets, can approximate sets of functions that have an interior in a limiting sense and admit natural finite-dimensional subsets giving arbitrarily good approximations. This is the intermediate tranche between finite-dimensional and genuinely nonparametric models: The members of the model class can be represented as limiting forms of an unusually flexible parametrized family, the elements of which give good, natural approximations. Often the class has a nonvoid interior.

In this context, the real dimension of a model is finite but the dimension of the model space is not bounded. The situation is often summarized by the phrase the Curse of Dimensionality. This phrase was first used by Bellman (1961), in the context of approximation theory, to signify the fact that estimation difficulty not only increases with dimension – which is no surprise – but can increase superlinearly. The result is that difficulty outstrips conventional data gathering even for what one would expect were relatively benign dimensions. A heuristic way to look at this is to think of real functions of x, of y, and of the pair (x,y). Real functions f, g of a single variable represent only a vanishingly small fraction of the functions k of (x,y). Indeed, they can be embedded by writing $k(x,y) = f(x) + g(y)$. Estimating an arbitrary function of two variables is more than twice as hard as estimating two arbitrary functions of one variable.

An extreme case of the Curse of Dimensionality occurs in the "large p, small n" problem in general regression contexts. Here, p customarily denotes the dimension of the space of variables, and n denotes the sample size. A collection of such data is $(\boldsymbol{y}_i, \boldsymbol{x}_{1,i}, \ldots, \boldsymbol{x}_{p,i})$ for $i = 1, \ldots n$. Gathering the explanatory variables, the $\boldsymbol{x}_{i,j}$s, into an $n \times p$ matrix X in which the ith row is $(\boldsymbol{x}_{1,i}, \ldots, \boldsymbol{x}_{p,i})$ means that X is short and fat when $p >> n$. Conventionally, design matrices are tall and skinny, $n >> p$, so there is a relatively high ratio n/p of data to the number of inferences. The short, fat data problem occurs when $n/p << 1$, so that the parameters cannot be estimated directly at all, much less well. These problems need some kind of auxiliary principle, such as shrinkage or other constraints, just to make solutions exist.

The finite-dimensional parametric case and the truly nonparametric case for regression are settings in which it is convenient to discuss some of the recurrent issues in the treatments here. It will be seen that the Curse applies in regression, but the Curse itself is more general, applying to classification, and to nearly all other aspects of multivariate inference. As noted, traditional analysis avoids the issue by making strong model assumptions, such as linearity and normality, to get finite-dimensional behavior or by using distribution-free procedures, and being fully nonparametric. However, the set of practical problems for which these circumventions are appropriate is small, and modern applied statisticians frequently use computer-intensive techniques on the intermediate tranche that are designed to minimize the impact of the Curse.

1.0.2 The Two Extremes

Multiple linear regression starts with n observations of the form $(Y_i, \boldsymbol{X}_i)$ and then makes the strong modeling assumption that the response Y_i is related to the vector of explanatory variables $\boldsymbol{X}_i = (X_{1,i}, ..., X_{p,i})$ by

$$Y_i = \boldsymbol{\beta}^T \boldsymbol{X}_i + \varepsilon_i = \beta_0 + \beta_1 X_{1,i} + \ldots \beta_p X_{p,i} + \varepsilon_i,$$

where each random error ε_i is (usually) an independent draw from a normal distribution with mean zero and fixed but unknown variance. More generally, the ε_is are taken as symmetric, unimodal, and independent. The $\boldsymbol{X}_i$s can be random, or, more commonly, chosen by the experimenter and hence deterministic. In the chapters to follow, instances of this setting will recur several times under various extra conditions.

In contrast, nonparametric regression assumes that the response variable is related to the vector of explanatory variables by

$$Y_i = f(\boldsymbol{X}_i) + \varepsilon_i,$$

where f is some smooth function. The assumptions about the error may be the same as for linear regression, but people tend to put less emphasis on the error structure than on the uncertainty in estimates $\hat{f}$ of f. This is reasonable because, outside of large departures from independent, symmetric, unimodal ε_is, the dominant source of uncertainty comes from estimating f. This setting will recur several times as well; Chapter 2, for instance, is devoted to it.

Smoothness of f is central: For several nonparametric methods, it is the smoothness assumptions that make theorems ensuring good behavior (consistency, for instance) of regression estimators $\hat{f}$ of f possible. For instance, kernel methods often assume f is in a Sobolev space, meaning f and a fixed number, say s, of its derivatives lie in a Hilbert space, say $L_q(\Omega)$, where the open set $\Omega \subset R^p$ is the domain of f.

Other methods, like splines for instance, weaken these conditions by allowing f to be piecewise continuous, so that it is differentiable between prespecified pairs of points, called knots. A third approach penalizes the roughness of the fitted function, so that the data help determine how wiggly the estimate of f should be. Most of these methods include a "bandwidth" parameter, often estimated by cross-validation (to be discussed shortly). The bandwidth parameter is like a resolution defining the scale on which solutions should be valid. A finer-scale, smaller bandwidth suggests high concern with very local behavior of f; a large-scale, higher bandwidth suggests one will have to be satisfied, usually grudgingly, with less information on the detailed behavior of f.

Between these two extremes lies the intermediate tranche, where most of the action in DMML is. The intermediate tranche is where the finite-dimensional methods confront the Curse of Dimensionality on their way to achieving good approximations to the nonparametric setting.

1.1 Perspectives on the Curse

Since almost all finite-dimensional methods break down as the dimension p of $\boldsymbol{X}_i$ increases, it's worth looking at several senses in which the breakdown occurs. This will reveal impediments that methods must overcome. In the context of regression analysis under the squared error loss, the formal statement of the Curse is:

- The mean integrated squared error of fits increases faster than linearly in p.

The central reason is that, as the dimension increases, the amount of extra room in the higher-dimensional space and the flexibility of large function classes is dramatically more than experience with linear models suggests.

For intuition, however, note that there are three nearly equivalent informal descriptions of the Curse of Dimensionality:

- In high dimensions, all data sets are too sparse.
- In high dimensions, the number of possible models to consider increases superexponentially in p.
- In high dimensions, all data sets show multicollinearity (or concurvity , which is the generalization that arises in nonparametric regression).

In addition to these near equivalences, as p increases, the effect of error terms tends to increase and the potential for spurious correlations among the explanatory variables increases. This section discusses these issues in turn.

These issues may not sound very serious, but they are. In fact, scaling up most procedures highlights unforeseen weaknesses in them. To dramatize the effect of scaling from two to three dimensions, recall the high school physics question: What's the first thing that would happen if a spider kept all its proportions the same but was suddenly 10 feet tall? Answer: Its legs would break. The increase in volume in its body (and hence weight) is much greater than the increase in cross-sectional area (and hence strength) of its legs. That's the Curse.

1.1.1 Sparsity

Nonparametric regression uses the data to fit local features of the function f in a flexible way. If there are not enough observations in a neighborhood of some point $\boldsymbol{x}$, then it is hard to decide what $f(\boldsymbol{x})$ should be. It is possible that f has a bump at $\boldsymbol{x}$, or a dip, some kind of saddlepoint feature, or that f is just smoothly increasing or decreasing at $\boldsymbol{x}$. The difficulty is that, as p increases, the amount of local data goes to zero.

This is seen heuristically by noting that the volume of a p-dimensional ball of radius r goes to zero as p increases. This means that the volume of the set centered at $\boldsymbol{x}$ in which a data point $\boldsymbol{x}_i$ must lie in order to provide information about $f(\boldsymbol{x})$ has fewer and fewer points per unit volume as p increases.

This slightly surprising fact follows from a Stirling's approximation argument. Recall the formula for the volume of a ball of radius r in p dimensions:

$$V_r(p) = \frac{\pi^{p/2} r^p}{\Gamma(p/2+1)}. \tag{1.1.1}$$

When p is even, $p = 2k$ for some k. So,

$$\ln V_r(p) = k \ln(\pi r^2) - \ln(k!)$$

since $\Gamma(k+1) = k!$. Stirling's formula gives $k! \approx \sqrt{2\pi} k^{k+1/2} e^{-k}$. So, (1.1.1) becomes

$$\ln V_r(p) = -\frac{1}{2}\ln(2\pi) - \frac{1}{2}\ln k + k[1 + \ln(\pi r^2)] - k \ln k.$$

The last term dominates and goes to $-\infty$ for fixed r. If $p = 2k+1$, one again gets $V_r(p) \to 0$. The argument can be extended by writing $\Gamma(p/2+1) = \Gamma((k+1)+1/2)$ and using bounds to control the extra "$1/2$". As p increases, the volume goes to zero for any r. By contrast, the volume of a cuboid of side length r is r^p, which goes to 0, 1, or ∞ depending on $r < 1$, $r = 1$, or $r > 1$. In addition, the ratio of the volume of the p-dimensional ball of radius r to the volume of the cuboid of side length r typically goes to zero as p gets large.

Therefore, if the $\boldsymbol{x}$ values are uniformly distributed on the unit hypercube, the expected number of observations in any small ball goes to zero. If the data are not uniformly distributed, then the typical density will be even more sparse in most of the domain, if a little less sparse on a specific region. Without extreme concentration in that specific region – concentration on a finite-dimensional hypersurface for instance – the increase in dimension will continue to overwhelm the data that accumulate there, too. Essentially, outside of degenerate cases, for any fixed sample size n, there will be too few data points in regions to allow accurate estimation of f.

To illustrate the speed at which sparsity becomes a problem, consider the best-case scenario for nonparametric regression, in which the $\boldsymbol{x}$ data are uniformly distributed in the p-dimensional unit ball. Figure 1.1 plots r^p on $[0,1]$, the expected proportion of the data contained in a centered ball of radius r for $p = 1, 2, 8$. As p increases, r must grow large rapidly to include a reasonable fraction of the data.

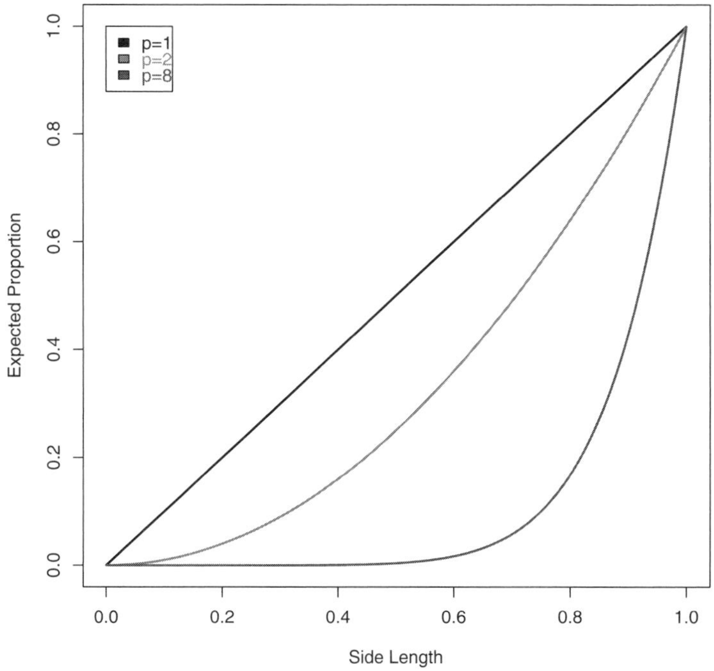

Fig. 1.1 This plots r^p, the expected proportion of the data contained in a centered ball of radius r in the unit ball for $p = 1, 2, 8$. Note that, for large p, the radius needed to capture a reasonable fraction of the data is also large.

To relate this to local estimation of f, suppose one thousand values of are uniformly distributed in the unit ball in $\mathbb{R}^p$. To ensure that at least 10 observations are near $\boldsymbol{x}$ for estimating f near $\boldsymbol{x}$, (1.1.1) implies the expected radius of the requisite ball is $r = \sqrt[p]{.01}$. For $p = 10$, $r = 0.63$ and the value of r grows rapidly to 1 with increasing p. This determines the size of the neighborhood on which the analyst can hope to estimate local features of f. Clearly, the neighborhood size increases with dimension, implying that estimation necessarily gets coarser and coarser. The smoothness assumptions mentioned before – choice of bandwidth, number and size of derivatives – govern how big the class of functions is and so help control how big the neighborhood must be to ensure enough data points are near an $\boldsymbol{x}$ value to permit decent estimation.

Classical linear regression avoids the sparsity issue in the Curse by using the linearity assumption. Linearity ensures that all the points contribute to fitting the estimated surface (i.e., the hyperplane) everywhere on the $\boldsymbol{X}$-space. In other words, linearity permits the estimation of f at any $\boldsymbol{x}$ to borrow strength from all of the $\boldsymbol{x}_i$s, not just the $\boldsymbol{x}_i$s in a small neighborhood of $\boldsymbol{x}$.

More generally, nonlinear models may avoid the Curse when the parametrization does not "pick off" local features. To see the issue, consider the nonlinear model:

$$f(\boldsymbol{x}) = \begin{cases} 17 & \text{if } \boldsymbol{x} \in B_r = \{\boldsymbol{x} : \|\boldsymbol{x} - \boldsymbol{x}_0\| \leq r\} \\ \beta_0 + \sum_{j=1}^{p} \beta_j x_j & \text{if } \boldsymbol{x} \in B_r^c. \end{cases}$$

The ball B_r is a local feature. This nonlinear model borrows strength from the data over most of the space, but even with a large sample it is unlikely that an analyst can estimate f near $\boldsymbol{x}_0$ and the radius r that defines the nonlinear feature. Such cases are not pathological – most nonlinear models have difficulty in some regions; e.g., logistic regression can perform poorly unless observations are concentrated where the sigmoidal function is steep.

1.1.2 Exploding Numbers of Models

The second description of the Curse is that the number of possible models increases superexponentially in dimension. To illustrate the problem, consider a very simple case: polynomial regression with terms of degree 2 or less. Now, count the number of models for different values of p.

For $p = 1$, the seven possible models are:

$$\begin{array}{lll} \mathbb{E}(Y) = \beta_0, & \mathbb{E}(Y) = \beta_1 x_1, & \mathbb{E}(Y) = \beta_2 x_1^2, \\ \mathbb{E}(Y) = \beta_0 + \beta_1 x_1, & \mathbb{E}(Y) = \beta_0 + \beta_2 x_1^2, & \mathbb{E}(Y) = \beta_1 x_1 + \beta_2 x_1^2, \\ \mathbb{E}(Y) = \beta_0 + \beta_1 x_1 + \beta_2 x_1^2. & & \end{array}$$

For $p = 2$, the set of models expands to include terms in x_2 having the form x_2, x_2^2 and $x_1 x_2$. There are 63 such models. In general, the number of polynomial models of order at most 2 in p variables is $2^a - 1$, where $a = 1 + 2p + p(p-1)/2$. (The constant term, which may be included or not, gives 2^1 cases. There are p possible first order terms, and the cardinality of all subsets of p terms is 2^p. There are p second-order terms of the form $\boldsymbol{x}_i^2$, and the cardinality of all subsets is again 2^p. There are $C(p,2) = p(p-1)/2$ distinct subsets of size 2 among p objects. This counts the number of terms of the form $\boldsymbol{x}_i \boldsymbol{x}_j$ for $i \neq j$ and gives $2^{p(p-1)/2}$ terms. Multiplying and subtracting 1 for the disallowed model with no terms gives the result.)

Clearly, the problem worsens if one includes models with more terms, for instance higher powers. The problem remains if polynomial expansions are replaced by more general basis expansions. It may worsen if more basis elements are needed for good approximation or, in the fortunate case, the rate of explosion may decrease somewhat if the basis can express the functions of interest parsimoniously. However, the point remains that an astronomical number of observations are needed to select the best model among so many candidates, even for low-degree polynomial regression.

In addition to fit, consider testing in classical linear regression. Once p is moderately large, one must make a very large number of significance tests, and the family-wise error rate for the collection of inferences will be large or the tests themselves will be conservative to the point of near uselessness. These issues will be examined in detail in Chapter 10, where some resolutions will be presented. However, the practical impossibility of correctly identifying the best model, or even a good one, is a key motivation behind ensemble methods, discussed later.

In DMML, the sheer volume of data and concomitant necessity for flexible regression models forces much harder problems of model selection than arise with low-degree polynomials. As a consequence, the accuracy and precision of inferences for conventional methods in DMML contexts decreases dramatically, which is the Curse.

1.1.3 Multicollinearity and Concurvity

The third description of the Curse relates to instability of fit and was pointed out by Scott and Wand (1991). This complements the two previous descriptions, which focus on sample size and model list complexity. However, all three are different facets of the same issue.

Recall that, in linear regression, multicollinearity occurs when two or more of the explanatory variables are highly correlated. Geometrically, this means that all of the observations lie close to an affine subspace. (An affine subspace is obtained from a linear subspace by adding a constant; it need not contain $\mathbf{0}$.)

Suppose one has response values Y_i associated with observed vectors $\boldsymbol{X}_i$ and does a standard multiple regression analysis. The fitted hyperplane will be very stable in the region where the observations lie, and predictions for similar vectors of explanatory variables will have small variances. But as one moves away from the observed data, the hyperplane fit is unstable and the prediction variance is large. For instance, if the data cluster about a straight line in three dimensions and a plane is fit, then the plane can be rotated about the line without affecting the fit very much. More formally, if the data concentrate close to an affine subspace of the fitted hyperplane, then, essentially, any rotation of the fitted hyperplane around the projection of the affine subspace onto the hyperplane will fit about as well. Informally, one can spin the fitted plane around the affine projection without harming the fit much.

In p-dimensions, there will be p elements in a basis. So, the number of proper subspaces generated by the basis is $2^p - 2$ if $\mathbb{R}^p$ and $\mathbf{0}$ are excluded. So, as p grows, there is an exponential increase in the number of possible affine subspaces. Traditional multicollinearity can occur when, for a finite sample, the explanatory variables concentrate on one of them. This is usually expressed in terms of the design matrix $\boldsymbol{X}$ as $\det \boldsymbol{X}'\boldsymbol{X}$ near zero; i.e., nearly singular. Note that $\boldsymbol{X}$ denotes either a matrix or a vector-valued outcome, the meaning being clear from the context. If needed, a subscript i, as in $\boldsymbol{X}_i$, will indicate the vector case. The chance of multicollinearity happening purely by chance increases with p. That is, as p increases, it is ever more likely that the variables included will be correlated, or seem to be, just by chance. So, reductions to affine subspaces will occur more frequently, decreasing $|\det \boldsymbol{X}'\boldsymbol{X}|$, inflating variances, and giving worse mean squared errors and predictions.

But the problem gets worse. Nonparametric regression fits smooth curves to the data. In analogy with multicollinearity, if the explanatory variables tend to concentrate along a smooth curve that is in the family used for fitting, then the prediction and fit will be good near the projected curve but poor in other regions. This situation is called concurvity . Roughly, it arises when the true curve is not uniquely identifiable, or nearly so. Concurvity is the nonparametric analog of multicollinearity and leads to inflated variances. A more technical discussion will be given in Chapter 4.

1.1.4 The Effect of Noise

The three versions of the Curse so far have been in terms of the model. However, as the number of explanatory variables increases, the error component typically has an ever-larger effect as well.

Suppose one is doing multiple linear regression with $\boldsymbol{Y} = \boldsymbol{X}\boldsymbol{\beta} + \boldsymbol{\varepsilon}$, where $\boldsymbol{\varepsilon} \sim N(\boldsymbol{0}, \sigma^2 \boldsymbol{I})$; i.e., all convenient assumptions hold. Then, from standard linear model theory, the variance in the prediction at a point $\boldsymbol{x}$ given a sample of size n is

$$\mathrm{Var}[\hat{Y}|\boldsymbol{x}] = \sigma^2(1 + \boldsymbol{x}^T(\boldsymbol{X}^T\boldsymbol{X})^{-1}\boldsymbol{x}), \tag{1.1.2}$$

assuming $(\boldsymbol{X}^T\boldsymbol{X})$ is nonsingular so its inverse exists. As $(\boldsymbol{X}^T\boldsymbol{X})$ gets closer to singularity, typically one or more eigenvalues go to 0, so the inverse (roughly speaking) has eigenvalues that go to ∞, inflating the variance. When $p \gg n$, $(\boldsymbol{X}^T\boldsymbol{X})$ is singular, indicating there are directions along which $(\boldsymbol{X}^T\boldsymbol{X})$ cannot be inverted because of zero eigenvalues. If a generalized inverse, such as the Moore-Penrose matrix, is used when $(\boldsymbol{X}^T\boldsymbol{X})$ is singular, a similar formula can be derived (with a limited domain of applicability).

However, consider the case in which the eigenvalues decrease to zero as more and more explanatory variables are included, i.e., as p increases. Then, $(\boldsymbol{X}^T\boldsymbol{X})$ gets ever closer to singularity and so its inverse becomes unbounded in the sense that one or more (usually many) of its eigenvalues go to infinity. Since $\boldsymbol{x}^T(\boldsymbol{X}^T\boldsymbol{X})^{-1}\boldsymbol{x}$ is the norm of $\boldsymbol{x}$ with respect to the inner product defined by $(\boldsymbol{X}^T\boldsymbol{X})^{-1}$, it will usually tend to infinity (as long as the sequence of $\boldsymbol{x}$s used doesn't go to zero). That is, typically, $\mathrm{Var}[\hat{Y}|\boldsymbol{x}]$ tends to infinity as more and more explanatory variables are included. This means the Curse also implies that, for typically occurring values of p and n, the instability of estimates is enormous.

1.2 Coping with the Curse

Data mining, in part, seeks to assess and minimize the effects of model uncertainty to help find useful models and good prediction schemes. Part of this necessitates dealing with the Curse.

In Chapter 4, it will be seen that there is a technical sense in which neural networks can provably avoid the Curse in some cases. There is also evidence (not as clear) that projection pursuit regression can avoid the Curse in some cases. Despite being remarkable intellectual achievements, it is unclear how generally applicable these results are. More typically, other methods rest on other flexible parametric families, nonparametric techniques, or model averaging and so must confront the Curse and other model uncertainty issues directly. In these cases, analysts reduce the impact of the Curse by designing experiments well, extracting low-dimensional features, imposing parsimony, or aggressive variable search and selection.

1.2.1 Selecting Design Points

In some cases (e.g., computer experiments), it is possible to use experimental design principles to minimize the Curse. One selects the $\boldsymbol{x}$s at which responses are to be measured in a smart way. Either one chooses them to be spread as uniformly as possible, to minimize sparsity problems, or one selects them sequentially, to gather information where it is most needed for model selection or to prevent multicollinearity.

There are numerous design criteria that have been extensively studied in a variety of contexts. Mostly, they are criteria on $\boldsymbol{X}^T\boldsymbol{X}$ from (1.1.2). D-optimality, for instance, tries to maximize $\det \boldsymbol{X}^T\boldsymbol{X}$. This is an effort to minimize the variance of the parameter estimates, $\hat{\beta}_i$. A-optimality tries to minimize trace$(\boldsymbol{X}^T\boldsymbol{X})^{-1}$. This is an effort to minimize the average variance of the parameter estimates. G-optimality tries to minimize the maximum prediction variance; i.e., minimize the maximum of $\boldsymbol{x}^T(\boldsymbol{X}^T\boldsymbol{X})^{-1}\boldsymbol{x}$ from (1.1.2) over a fixed range of $\boldsymbol{x}$. In these and many other criteria, the major downside is that the optimality criterion depends on the model chosen. So, the optimum is only optimal for the model and sample size the experimenter specifies. In other words, the uncertainty remaining is conditional on n and the given model. In a fundamental sense, uncertainty in the model and sampling procedure is assumed not to exist.

A fundamental result in this area is the Kiefer and Wolfowitz (1960) equivalence theorem. It states conditions under which D-optimality and G-optimality are the same; see Chernoff (1999) for an easy, more recent introduction. Over the last 50 years, the literature in this general area has become vast. The reader is advised to consult the classic texts of Box et al. (1978), Dodge et al. (1988), or Pukelsheim (1993).

Selection of design points can also be done sequentially; this is very difficult but potentially avoids the model and sample-size dependence of fixed design-point criteria. The full solution uses dynamic programming and a cost function to select the explanatory values for the next response measurement, given all the measurements previously obtained. The cost function penalizes uncertainty in the model fit, especially in regions of particular interest, and perhaps also includes information about different prices for observations at different locations. In general, the solution is intractable, although some approximations (e.g., greedy selection) may be feasible. Unfortunately, many large data sets cannot be collected sequentially.

A separate but related class of design problems is to select points in the domain of integration so that integrals can be evaluated by deterministic algorithms. Traditional Monte Carlo evaluation is based on a Riemann sum approximation,

$$\int_S f(\boldsymbol{x})d\boldsymbol{x} \approx \sum_{i=1}^{n} f(\boldsymbol{X}_i)\Delta(S_i),$$

where the S_i form a partition of $S \subset \mathbb{R}^p$, $\Delta(S_i)$ is the volume of S_i, and the evaluation point $\boldsymbol{X}_i$ is uniformly distributed in S_i. The procedure is often easy to implement, and randomness allows one to make uncertainty statements about the value of the integral. But the procedure suffers from the Curse; error grows faster than linearly in p.

One can sometimes improve the accuracy of the approximation by using nonrandom evaluation points $\boldsymbol{x}_i$. Such sets of points are called quasi-random sequences or low-discrepancy sequences. They are chosen to fill out the region S as evenly as possible and do not depend on f. There are many approaches to choosing quasi-random sequences. The Hammersley points discussed in Note 1.1 were first, but the Halton sequences are also popular (see Niederreiter (1992a)). In general, the grid of points must be fine enough that f looks locally smooth, so a procedure must be capable of generating points at any scale, however fine, and must, in the limit of ever finer scales, reproduce the value of the integral exactly.

1.2.2 Local Dimension

Nearly all DMML methods try to fit the local structure of a function. The problem is that when behavior is local it can change from neighborhood to neighborhood. In particular, an unknown function on a domain may have different low-dimensional functional forms on different regions within its domain. Thus, even though the local low-dimensional expression of a function is easier to uncover, the region on which that form is valid may be difficult to identify.

For the sake of exactitude, define $f : \mathbb{R}^p \to \mathbb{R}$ to have locally low dimension if there exist regions $R_1, R_2, \ldots$ and a set of functions $g_1, g_2, \ldots$ such that $\bigcup R_i \approx \mathbb{R}^p$ and for $\boldsymbol{x} \in R_i$, $f(\boldsymbol{x}) \approx g_i(\boldsymbol{x})$, where g_i depends only on q components of $\boldsymbol{x}$ for $q \ll p$. The sense of approximation and meaning of $\ll$ is vague, but the point is not to make it precise (which can be done easily) so much as to examine the local behavior of functions from a dimensional standpoint.

As examples,

$$f(\boldsymbol{x}) = \begin{cases} 3x_1 & \text{if } x_1 + x_2 < 7 \\ x_2^2 & \text{if } x_1 + x_2 > 7 \\ x_1 + x_2 & \text{if } x_1 = x_2, \end{cases} \quad \text{and} \quad f(\boldsymbol{x}) = \sum_{k=1}^{m} \alpha_k I_{R_k}(\boldsymbol{x})$$

are locally low-dimensional because they reduce to functions of relatively few variables on regions. By contrast,

$$f(\boldsymbol{x}) = \beta_0 + \sum_{j=1}^{p} \beta_j x_j \text{ for } \beta_j \neq 0 \quad \text{and} \quad f(\boldsymbol{x}) = \prod_{j=1}^{p} x_j$$

have high local dimension because they do not reduce anywhere on their domain to functions of fewer than p variables.

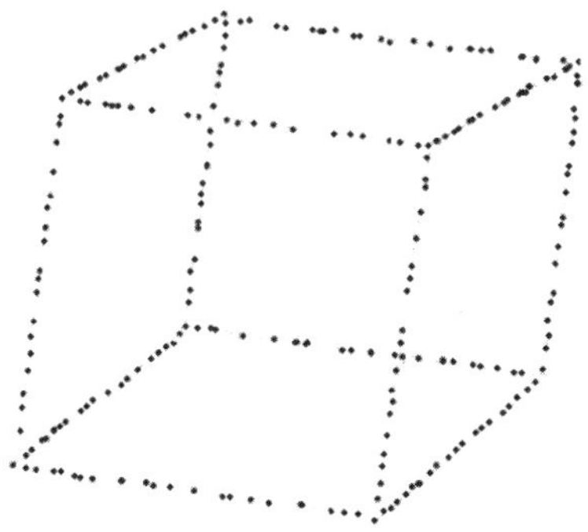

Fig. 1.2 A plot of 200 points uniformly distributed on the 1-cube in $\mathbb{R}^3$, where the plot is tilted 10 degrees from each of the natural axes (otherwise, the image would look like points on the perimeter of a square).

As a pragmatic point, outside of a handful of particularly well-behaved settings, success in multivariate nonparametric regression requires either nonlocal model assumptions or that the regression function have locally low dimension on regions that are not too hard to identify.

Since most DMML methods use local fits (otherwise, they must make global model assumptions), and local fitting succeeds best when the data have locally low dimension, the difficulty is knowing in advance whether the data have simple, low-dimensional structure. There is no standard estimator of average local dimension, and visualization methods are often difficult, especially for large p.

To see how hidden structure, for instance a low-dimensional form, can lurk unsuspected in a scatterplot, consider q-cubes in $\mathbb{R}^p$. These are the q-dimensional boundaries of a p-dimensional cube: A 1-cube in $\mathbb{R}^2$ is the perimeter of a square; a 2-cube in $\mathbb{R}^3$ consists of the faces of a cube; a 3-cube in $\mathbb{R}^3$ is the entire cube. These have simple structure, but it is hard to discern for large p.

Figure 1.2 shows a 1-cube in $\mathbb{R}^3$, tilted 10 degrees from the natural axes in each coordinate. Since $p = 3$ is small, the structure is clear.

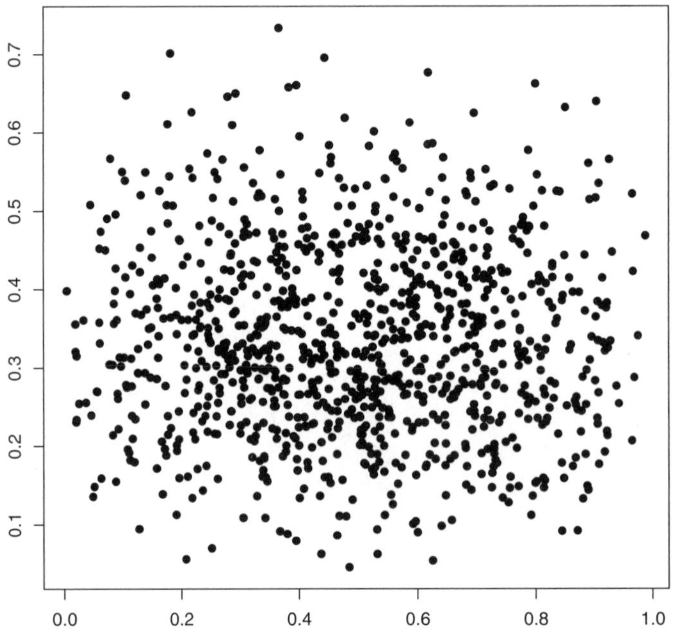

Fig. 1.3 A plot of 200 points uniformly distributed on the 1-cube in $\mathbb{R}^{10}$, where the plot is tilted 10 degrees from each of the natural axes (otherwise, the image would look like points on the perimeter of a square).

In contrast, Fig. 1.3 is a projection of a 1-cube in $\mathbb{R}^{10}$, tilted 10 degrees from the natural axes in each coordinate. This is a visual demonstration that in high dimensions, nearly all projections look Gaussian, see Diaconis and Freedman (1984). This shows that even simple structure can be hard to see in high dimensions.

Although there is no routine estimator for average local dimension and no standard technique for uncovering hidden low-dimensional structures, some template methods are available. A template method is one that links together a sequence of steps but many of the steps could be accomplished by any of a variety of broadly equivalent techniques. For instance, one step in a regression method may involve variable selection and one may use standard testing on the parameters. However, normal-based testing is only one way to do variable selection and one could, in principle, use any other technique that accomplished the same task.

One way to proceed in the search for low local dimension structures is to start by checking if the average local dimension is less than the putative dimension p and, if it is, "grow" sets of data that can be described by low-dimensional models.

To check if the local dimension is lower than the putative dimension, one needs to have a way to decide if data can locally be fit by a lower-dimensional surface. In a perfect mathematical sense, the answer is almost always no, but the dispersal of a portion of a data set in a region may be tight enough about a lower-dimensional surface to justify the approximation. In principle, therefore, one wants to choose a number of points at least as great as p and find that the convex hull it forms really only has $q < p$ dimensions; i.e., in the leftover $p - q$ dimensions, the convex hull is so thin it can be approximated to thickness zero. This means that the solid the data forms can be described by q directions. The question is how to choose q.

Banks and Olszewski (2004) proposed estimating average local dimension in structure discovery problems by obtaining M estimates of the number of vectors required to describe a solid formed by subsets of the data and then averaging the estimates. The subsets are formed by enlarging a randomly chosen sphere to include a certain number of data points, describing them by some dimension reduction technique. We specify principal components, PCs, even though PCs will only be described in detail in Chapter 8, because it is popular. The central idea of PCs needed here is that it is a method that produces vectors from explanatory variable inputs in order of decreasing ability to explain observed variability. Thus, the earlier PCs are more important than later PCs. The parallel is to a factor in an ANOVA: One keeps the factors that explain the biggest portions of the sum of squared errors, and may want to ignore other factors.

The template is as follows.

Let $\{\boldsymbol{X}_i\}$ denote n data points in $\mathbb{R}^p$.

- ☐ Select a random point $\boldsymbol{x}_m^*$ in or near the convex hull of $\boldsymbol{X}_1, \ldots, \boldsymbol{X}_n$ for $m = 1, \ldots, M$.
- ☐ Find a ball centered at $\boldsymbol{x}_m^*$ that contains exactly k points. One must choose $k > p$; $k = 4p$ is one recommended choice.
- ☐ Perform a principal components regression on the k points within the ball.
- ☐ Let c_m be the number of principal components needed to explain a fixed percentage of the variance in the Y_i values; 80% is one recommended choice.

The average $\hat{c} = (1/M)\sum_{m=1}^{M} c_m$ estimates the average local dimension of f. (This assumes a locally linear functional relationship for points within the ball.) If $\hat{c}$ is large relative to p, then the regression relationship is highly multivariate in most of the space; no method has much chance of good prediction. However, if $\hat{c}$ is small, one infers there are substantial regions where the data can be described by lower-dimensional surfaces. It's just a matter of finding them.

Note that this really is a template because one can use any variable reduction technique in place of principal components. In Chapter 4, sliced inverse regression will be introduced and in Chapter 9 partial least squares will be explained, for instance. However, one needn't be so fancy. Throwing out variables with coefficients too close to zero from goodness-of-fit testing is an easily implemented alternative. It is unclear, a priori, which dimension reduction technique is best in a particular setting.

To test the PC-based procedure, Banks and Olszewski (2004) generated $10 * 2^q$ points at random on each of the $2^{p-q}\binom{p}{q}$ sides of a q-cube in $\mathbb{R}^p$. Then independent $N(\mathbf{0}, .25\boldsymbol{I})$ noise was added to each observation. Table 1.1 shows the resulting estimates of the local dimension for given putative dimension p and true lower-dimensional structure dimension q. The estimates are biased down because the principal components regression only uses the number of directions, or linear combinations, required to explain only 80% of the variance. Had 90% been used, the degree of underestimation would have been less.

q	p=1	2	3	4	5	6	7
7							5.03
6						4.25	4.23
5					3.49	3.55	3.69
4				2.75	2.90	3.05	3.18
3			2.04	2.24	2.37	2.50	2.58
2		1.43	1.58	1.71	1.80	1.83	1.87
1	.80	.88	.92	.96	.95	.95	.98

Table 1.1 Estimates of the local dimension of q-cubes in $\mathbb{R}^p$ based on the average of 20 replications per entry. The estimates tend to increase up to the true q as p increases.

Given that one is satisfied that there is a locally low-dimensional structure in the data, one wants to find the regions in terms of the data. However, a locally valid lower-dimensional structure in one region will typically not extend to another. So, the points in a region where a low-dimensional form is valid will fit well (i.e., be good relative to the model), but data outside that region will typically appear to be outliers (i.e., bad relative to the model).

One approach to finding subsamples is as follows. Prespecify the proportion of a sample to be described by a linear model, say 80%. The task is to search for subsets of size $.8n$ of the n data points to find one that fits a prechosen linear model. To begin, select k, the number of subsamples to be constructed, hoping at least one of them matches 80% of the data. (This k can be found as in House and Banks (2004) where this method is described.) So, start with k sets of data, each with $q+2$ data points randomly assigned to them with replacement. This is just enough to permit estimation of q coefficients and assessment of goodness of fit for a model. The q can be chosen near $\hat{c}$ and then nearby values of q tested in refinements. Each of the initial samples can be augmented by randomly chosen data points from the large sample. If including the extra observation improves the goodness of fit, it is retained; otherwise it is discarded. Hopefully, one of the resulting d sets contains all the data well described by the model. These points can be removed and the procedure repeated.

Note that this, too, is a template method, in the sense that various goodness-of-fit measures can be used, various inclusion rules for the addition of data points to a growing "good" subsample can be formulated, and different model classes can be proposed. Linear models are just one good choice because they correspond locally to taking a Taylor expansion of a function on a neighborhood.

1.2.3 Parsimony

One strategy for coping with the Curse is the principle of parsimony. Parsimony is the preference for the simplest explanation that explains the greatest number of observations over more complex explanations. In DMML, this is seen in the fact that simple models often have better predictive accuracy than complex models. This, however, has some qualifications. Let us interpret "simple model" to mean a model that has few parameters, a common notion. Certainly, if two models fit equally well, the one with fewer parameters is preferred because you can get better estimates (smaller standard errors) when there is a higher ratio of data points to number of parameters. Often, however, it is not so clear: The model with more parameters (and hence higher SEs) explains the data better, but is it better enough to warrant the extra complexity?

This question will be addressed further in the context of variance bias decompositions later. From a strictly pragmatic, predictive standpoint, note that:

1. If the true model is complex, one may not be able to make accurate predictions at all.
2. If the true model is simple, then one can probably improve the fit by forcing selection of a simple model.

The inability to make accurate predictions when the true model is complex may be due to n being too small. If n cannot be increased, and this is commonly the case, one is forced to choose oversimple models intelligently.

The most common kind of parsimony arises in variable selection since usually there is at least one parameter per variable included. One wants to choose a model that only includes the covariates that contribute substantially to a good fit. Many data mining methods use stepwise selection to choose variables for the model, but this breaks down for large p – even when a multiple regression model is correct. More generally, as in standard applied statistics contexts, DMML methods try to eliminate explanatory variables that don't explain enough of the variability to be worth including to improve a model that is overcomplex for the available data. One way to do this is to replace a large collection of explanatory variables by a single function of them.

Other kinds of parsimony arise in the context of shrinkage, thresholding, and roughness penalties, as will be discussed in later chapters. Indeed, the effort to find locally low-dimensional representations, as discussed in the last section, is a form of parsimony. Because of data limitations relative to the size of model classes, parsimony is one of the biggest desiderata in DMML.

As a historical note, the principle of parsimony traces back at least to an early logician named William of Ockham (1285–1349?) from Surrey, England. The phrase attributed to him is: "Pluralitas non est ponenda sine neccesitate", which means "entities should not be multiplied unnecessarily". This phrase is not actually found in his writings but the attribution is fitting given his negative stance on papal power. Indeed, William was alive during the Avignon papacy when there were two popes, one in Rome and one in Avignon, France. It is tempting to speculate that William thought this level of theological complexity should be cut down to size.

1.3 Two Techniques

Two of the most important techniques in DMML applications are the bootstrap and cross-validation. The bootstrap estimates uncertainty, and cross-validation assesses model fit. Unfortunately, neither scales up as well as one might want for massive DMML applications – so in many cases one may be back to techniques based on the central limit theorem.

1.3.1 The Bootstrap

The bootstrap was invented by Efron (1979) and was one of the first and most powerful achievements of computer-intensive statistical inference. Very quickly, it became an important method for setting approximate confidence regions on estimates when the underlying distribution is unknown.

The bootstrap uses samples drawn from the empirical distribution function, EDF. For simplicity, consider the univariate case and let $X_1, \ldots, X_n$ be a random sample (i.e., an independent and identically distributed sample, or IID sample) from the distribution F. Then the EDF is

$$\hat{F}_n(x) = \frac{1}{n} \sum_{i=1}^{n} I_{(-\infty, X_i]}(x),$$

where $I_R(x)$ is an indicator function that is one or zero according to whether $x \in \mathbb{R}$ or $x \notin \mathbb{R}$, respectively. The EDF is bounded between 0 and 1 with jumps of size $(1/n)$ at each observation. It is a consistent estimator of F, the true distribution function (DF). Therefore, as n increases, $\hat{F}_n$ converges (in a sense discussed below) to F.

To generalize to the multivariate case, define $\hat{F}_n(\boldsymbol{x})$ as the multivariate DF that for rectangular sets A assigns the probability equal to the proportion of sample points within A. For a random sample $\boldsymbol{X}_1, \ldots, \boldsymbol{X}_n$ in $\mathbb{R}^p$, this multivariate EDF is

$$\hat{F}_n(\boldsymbol{x}) = \frac{1}{n}\sum_{i=1}^{n} I_{R_i}(\boldsymbol{x}),$$

where $R_i = (-\infty, X_{i1}] \times \ldots \times (-\infty, X_{ip}]$ is the set formed by the Cartesian product of all halfspaces determined by the components of $\boldsymbol{X}_i$. For nonrectangular sets, a more careful definition must be given using approximations from rectangular sets.

For univariate data, $\hat{F}$ converges to F in a strong sense. The Glivenko-Cantelli theorem states that, for all $\varepsilon > 0$,

$$\mathbb{P}\left[\limsup_{x} |\hat{F}_n(x) - F(x)| < \varepsilon\right] = 1 \text{ a.s.} \tag{1.3.1}$$

This supremum, sometimes called the Kolmogorov-Smirnov distance, bounds the maximal distance between two distribution functions. Note that the randomness is in the sample defining the EDF. Convergence of EDFs to their limit is fast. Indeed, let $\varepsilon > 0$. Then the Smirnov distributions arise from

$$\lim_{n\to\infty} \mathbb{P}\left(\sqrt{n}\sup_{x\in\mathbb{R}}(F(x) - \hat{F}_n(x)) < \varepsilon\right) = 1 - e^{2\varepsilon^2} \tag{1.3.2}$$

and, from the other side,

$$\lim_{n\to\infty} \mathbb{P}\left(\sqrt{n}\sup_{x\in\mathbb{R}}(\hat{F}_n(x) - F(x)) < \varepsilon\right) = 1 - e^{2\varepsilon^2}. \tag{1.3.3}$$

Moreover, $\hat{F}_n$ also satisfies a large-deviation principle; a large-deviation principle gives conditions under which a class of events has probability decreasing to zero at a rate like $e^{\alpha n}$ for some $\alpha > 0$. Usually, the events have a convergent quantity that is a fixed distance from its limit. For the EDF, it converges to F in Kolmogorov-Smirnov distance and, for $\varepsilon > 0$ bounding that distance away from 0, the Kiefer-Wolfowitz theorem is that $\exists \alpha > 0$ and N so that for $\forall n > N$

$$\mathbb{P}\left(\sup_{x\in\mathbb{R}}|\hat{F}_n(x) - F(x)| > \varepsilon\right) \le e^{-\alpha n}. \tag{1.3.4}$$

Sometimes these results are called Sanov theorems. The earliest version was due to Chernoff (1956), who established an analogous result for the sample mean for distributions with a finite moment generating function on a neighborhood of zero.

Unfortunately, this convergence fails in higher dimensions; Fig. 1.4 illustrates the key problem, namely that the distribution may concentrate on sets that are very badly approximated by rectangles. Suppose the bivariate distribution for (X_1, X_2) is concentrated on the line from (0,1) to (1,0). No finite number of samples $(X_{1,i}, X_{2,i})$, $i = 1, \ldots, n$, covers every point on the line segment. So, consider a point $x = (x_1, x_2)$ on the line segment that is not in the sample. The EDF assigns probability zero to the region $(-\infty, x_1] \times (-\infty, x_2]$, so the limit of the difference is $F(\boldsymbol{x})$, not zero.

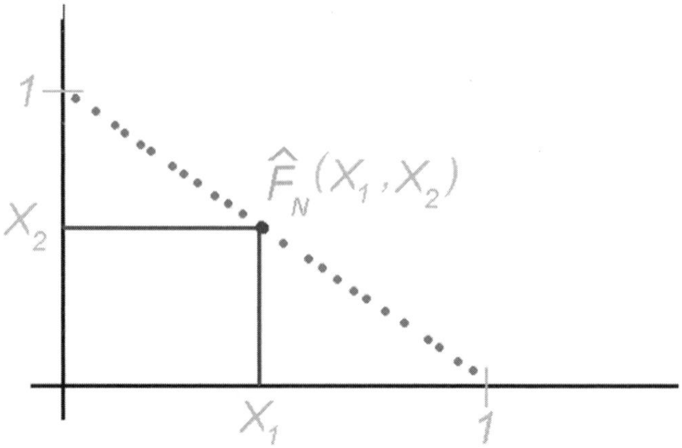

Fig. 1.4 The limsup convergence of the Glivenko-Cantelli theorem does not hold for $p \geq 2$. This figure shows that no finite sample from the (degenerate) bivariate uniform distribution on (0,1) to (1,0) can have the supremal difference going to zero.

Fortunately, for multivariate data, a weaker form of convergence holds, and this is sufficient for bootstrap purposes. The EDF converges in distribution to the true F, which means that, at each point $\boldsymbol{x}$ in $\mathbb{R}^p$ at which F is continuous,

$$\lim_n \hat{F}_n(\boldsymbol{x}) = F(\boldsymbol{x}).$$

Weak convergence, or convergence in distribution, is written as $\hat{F}_n \Rightarrow F$. Convergence in Kolmogorov-Smirnov distance implies weak convergence, but the converse fails. Although weaker, convergence in distribution is enough for the bootstrap because it means that, as data accumulate, the EDF does go to a well-defined limit, the true DF, pointwise, if not uniformly, on its domain. (In fact, the topology of weak convergence is metrizable by the Prohorov metric used in the next proposition.)

Convergence in distribution is also strong enough to ensure that estimates obtained from EDFs converge to their true values. To see this, recognize that many quantities to be estimated can be recognized as functionals of the DF. For instance, the mean is the Lebesgue-Stieltjes integral of x against F. The variance is a function of the first two moments, which are integrals of x^2 and x against F. More exotically, the ratio of the 7th moment to the 5th quantile is another functional. The term functional just means it is a real-valued function whose argument is a function, in this case a DF. Let $T = T(F)$ be a functional of F, and denote the estimate of $T(F)$ based on the sample $\{\boldsymbol{X}_i\}$ by $\hat{T} = T(\{\boldsymbol{X}_i\}) = T(\hat{F}_n)$. Because $\hat{F}_n \Rightarrow F$, we can show $\hat{T} \Rightarrow T$ and the main technical requirement is that T depend smoothly on F.

Proposition: If T is continuous at F, then $\hat{T}$ is consistent for T.

Proof: Recall the definition of the Prohorov metric. For a set A and $\varepsilon > 0$, let

$$A^{\varepsilon} = \{y | d(y,A) < \varepsilon\},$$

where $d(y,A) = \inf_{z \in A} d(y,z)$ and $d(y,z) = |y - z|$. For probabilities G and H, let

$$\nu(G,H) = \inf\{\varepsilon > 0 | \forall A, G(A) < H(A^{\varepsilon}) + \varepsilon\}.$$

Now, the Prohorov metric is $Proh(G,H) = \max[\nu(G,H), \nu(H,G)]$. Prohorov showed that the space of finite measures under $Proh$ is a complete separable metric space and that $Proh(F_n, F) \to 0$ is equivalent to $F_n \to F$ in the sense of weak convergence. (See Billingsley (1968), Appendix III).

Since T is continuous at F, for any $\varepsilon > 0$ there is a $\delta > 0$ such that $Proh(F,G) < \delta$ implies $|T(F) - T(G)| < \varepsilon$. From the consistency of the EDF, we have $Proh(F, \hat{F}_n) \to 0$. So, for any given $\eta > 0$ there is an N_η such that $n > N_\eta$ implies $Proh(F, \hat{F}_n) < \delta$ with probability larger than $1 - \eta$. Now, with probability at least $1 - \eta$, when $n > N_\eta$, $Proh(F, \hat{F}_n) < \delta$ and therefore $|T - \hat{T}| < \varepsilon$. □

Equipped with the EDF, its convergence properties, and how they carry over to functionals of the true DF, we can now describe the bootstrap through one of its simplest incarnations, namely its use in parameter estimation. The intuitive idea underlying the bootstrap method is to use the single available sample as a population and the estimate $\hat{t} = t(x_1, \cdots, x_n)$ as the fixed parameter, and then resample with replacement from the sample to estimate the characteristics of interest. The core idea is to generate bootstrap samples and compute bootstrap replicates as follows:

Given a random sample $\boldsymbol{x} = (x_1, \cdots, x_n)$ and a statistic $\hat{t} = t(x_1, \cdots, x_n)$,

For $b = 1$ to B:

- □ Sample with replacement from $\boldsymbol{x}$ to get $\boldsymbol{x}^{*b} = (x_1^{*b}, \cdots, x_n^{*b})$.
- □ Compute $\hat{\theta}^{*b} = t(x_1^{*b}, \cdots, x_n^{*b})$.

The size of the bootstrap sample could be any number m, but setting $m = n$ is typical. The number of replicates B depends on the problem at hand.

Once the B values $\hat{T}_1, \ldots, \hat{T}_B$ have been computed, they can be used to form a histogram; for instance, to approximate the sampling distribution of $\hat{T}$. In this way, one can evaluate how the sampling variability affects the estimation because the bootstrap is a way to set a confidence region on the functional.

The bootstrap strategy is diagrammed in Fig. 1.5. The top row has the unknown true distribution F. From this one draws the random sample $\boldsymbol{X}_1, \ldots, \boldsymbol{X}_n$, which is used to form the estimate $\hat{T}$ of T and the EDF $\hat{F}_n$. Here, $\hat{T}$ is denoted $T(\{X_i\}, F)$ to emphasize the use of the original sample. Then one draws a series of random samples, the X_i^*s, from the EDF. The fourth row indicates that these bootstrap samples are used to calculate the corresponding estimates, indicated by $T(\{X_i^*\}, F)$, to emphasize the use of the ith bootstrap sample, of the functional for the EDF. Since the EDF is a known

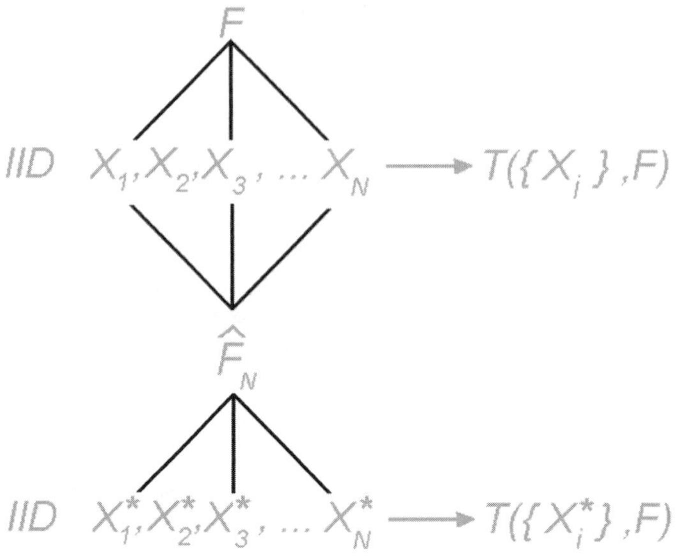

Fig. 1.5 The bootstrap strategy reflects the reflexivity in its name. The relationship between the true distribution, the sample, and the estimate is mirrored by the relationship between the EDF, resamples drawn from the EDF, and estimates based on the resamples. Weak convergence implies that as n increases the sampling distribution for the EDF estimates goes to the sampling distribution of the functional.

function, one knows exactly how much error there is between the functional evaluated for the EDF and its estimate. And since one can draw as many bootstrap samples from the EDF as one wants, repeated resampling produces the sampling distribution for the EDF estimates.

The key point is that, since $\hat{F}_n \Rightarrow F$, the distribution of $T(\{X_i^*\}, \hat{F}_n)$ converges weakly to the distribution of $T(\{X_i\}, F)$, the quantity of interest, as guaranteed by the proposition. That means that a confidence region set from the sampling distribution in the fourth row of Fig. 1.5 converges weakly to the confidence region one would have set in the second row if one could know the true sampling distribution of the functional. The convergence result is, of course, asymptotic, but a great deal of practical experience and simulation studies have shown that bootstrap confidence regions are very reliable, Efron and Tibshirani (1994).

It is important to realize that the effectiveness of the bootstrap does not rest on computing or sampling per se. Foundationally, the bootstrap works because $\hat{F}_n$ is such a good estimator for F. Indeed, (1.3.1) shows that $\hat{F}_n$ is consistent; (1.3.2) and (1.3.3) show that $\hat{F}_n$ has a well-defined asymptotic distribution using a $\sqrt{n}$ rate, and (1.3.4) shows how very unlikely it is for $\hat{F}_n$ to remain a finite distance away from its limit.

1.3.1.1 Bootstrapping an Asymptotic Pivot

As a concrete example to illustrate the power of the bootstrap, suppose $\{X_i\}$ is a random sample and the goal is to find a confidence region for the studentized mean. Then the functional is

$$T(\{X_i\}, F) = \sqrt{n}\left(\frac{\bar{X} - \mu}{s}\right),$$

where $\bar{X}$ and s are the sample mean and standard deviation, respectively, and μ is the mean of F. To set a confidence region, one needs the sampling distribution of $\bar{X}$ in the absence of knowledge of the population standard deviation σ. This is

$$\mathbb{P}_F\left[\sqrt{n}\frac{\bar{X} - \mu}{s} \le t\right]$$

for $t \in \mathbb{R}$. The bootstrap approximation to this sampling distribution is

$$\mathbb{P}_{\hat{F}_n}\left[\sqrt{n}\frac{\bar{X}^* - \bar{X}}{s^*} \le t\right] \tag{1.3.5}$$

for $t \in \mathbb{R}$, where $\bar{X}^*$ and s^* are the mean and standard deviation of a bootstrap sample from $\hat{F}_n$ and $\bar{X}$ is the mean of $\hat{F}_n$. That is, the sample mean $\bar{X}$, from the one available sample, is taken as the population mean under the probability for $\hat{F}_n$. The probability in (1.3.5) can be numerically evaluated by resampling from $\hat{F}_n$.

Aside from the bootstrap, one can use the central limit theorem, CLT, to approximate the distribution of functionals $T(\{X_i\}, F)$ by a normal distribution. However, since the empirical distribution has so many nice properties, it is tempting to conjecture that the sampling distribution will converge faster to its bootstrap approximation than it will to its limiting normal distribution. Tempting – but is it true? That is, as the size n of the actual sample increases, will the actual sampling distribution of T be closer on average to its bootstrap approximation or to its normal limit from the CLT?

To answer this question, recall that a pivot is a function of the data whose distribution is independent of the parameters. For example, the studentized mean

$$T(\{X_i\}, F) = \sqrt{n}\left(\frac{\bar{X} - \mu}{s}\right)$$

is a pivot in the class of normal distributions since this has the Student's-t distribution regardless of the value of μ and σ. In the class of distributions with finite first two moments, $T(\{X_i\}, F)$ is an asymptotic pivot since its asymptotic distribution is the standard normal regardless of the unknown F.

Hall (1992), Chapters 2, 3, and 5, showed that bootstrapping outperforms the CLT when the statistic of interest is an asymptotic pivot but that otherwise the two procedures are asymptotically equivalent.

The reasoning devolves to an Edgeworth expansion argument, which is, perforce, asymptotic. To summarize it, recall little-oh and big-oh notation.

- The little-oh relation written $g(n) = o(h(n))$ means that $g(n)$ gets small faster than $h(n)$ does; i.e., for any $\varepsilon > 0$, there is an M so that for $n > M$

$$g(n)/h(n) \leq \varepsilon.$$

- If little-oh behavior happens in probability, then write $o_p(h(n))$; i.e.,

$$\lim_{n \to \infty} P[|g(n)/h(n)| < \varepsilon] = 1 \;\; \forall \;\; \varepsilon > 0.$$

- The big-oh relation written $g(n) = \mathcal{O}(h(n))$ means that there is an $M > 0$ so that, for some B, $g(n)/h(n) \leq B$ for $n > M$.
- If big-oh behavior happens in probability, then write $\mathcal{O}_p(h(n))$; i.e.,

$$\lim_{B \to \infty} \limsup_{n \to \infty} \mathbb{P}\left[\frac{g(n)}{h(n)} \leq B\right] = 1.$$

Under reasonable technical conditions, the Edgeworth expansion of the sampling distribution of the studentized mean is

$$\mathbb{P}_F\left[\sqrt{n}\left(\frac{\bar{X}-\mu}{s}\right) \leq t\right] = \Phi(t) + n^{-1/2} p_1(t)\phi(t) + \ldots + n^{-j/2} p_j(t)\phi(t) + o(n^{-j/2}),$$

where $\Phi(t)$ is the DF of the standard normal, $\phi(t)$ is its density function, and the $p_j(t)$ functions are related to the Hermite polynomials, involving the jth and lower moments of F. See Note 1.5.2 for details. Note that the -oh notation here and below is used to describe the asymptotic behavior of the error term.

For functionals that are asymptotic pivots with standard normal distributions, the Edgeworth expansion gives

$$\begin{aligned} G(t) &= \mathbb{P}[T(\{X_i\}, F) \leq t] \\ &= \Phi(t) + n^{-1/2} p_1(t)\phi(t) + \mathcal{O}(n^{-1}). \end{aligned}$$

But note that the Edgeworth expansion also applies to the bootstrap estimate of the sampling distribution $G(t)$, giving

$$\begin{aligned} G^*(t) &= \mathbb{P}\left[T(\{X_i^*\}, \hat{F}_n) \leq t \,|\, \{X_i\}\right] \\ &= \Phi(t) + n^{-1/2} \hat{p}_1(t)\phi(t) + \mathcal{O}_p(n^{-1}), \end{aligned}$$

where

$$T(\{X_i^*\}, \hat{F}_n) = \sqrt{n}\left(\frac{\bar{X}^* - \bar{X}}{s^*}\right),$$

and $\hat{p}_1(t)$ is obtained from $p_1(t)$ by replacing the jth and lower moments of F in its coefficients of powers of t by the corresponding moments of the EDF. Consequently, one can show that $\hat{p}_1(t) - p_1(t) = \mathcal{O}_p(n^{-1/2})$; see Note 1.5.3. Thus

$$G^*(t) - G(t) = n^{-1/2}\phi(t)[\hat{p}_1(t) - p_1(t)] + \mathcal{O}_p(n^{-1}) = \mathcal{O}_p(n^{-1}) \qquad (1.3.6)$$

since the first term of the sum is $\mathcal{O}_p(n^{-1})$ and big-oh errors add. This means that using a bootstrap approximation to an asymptotic pivot has error of order n^{-1}.

By contrast, the CLT approximation uses $\Phi(t)$ to estimate $G(t)$, and

$$\begin{aligned} G(t) - \Phi(t) &= n^{-1/2}p_1(t)\phi(t) + \mathcal{O}(n^{-1}) \\ &= \mathcal{O}(n^{-1/2}). \end{aligned}$$

So, the CLT approximation has error of order $n^{-1/2}$ and thus is asymptotically worse than the bootstrap.

The CLT just identifies the first term of the Edgeworth expansion. The bootstrap approximation improves on the CLT approximation by including the extra $p_1\phi/\sqrt{n}$ term in the Edgeworth expansion (1.3.6) for the distribution function of the sampling distribution. The extra term ensures the leading normal terms match and improves the approximation to $\mathcal{O}(1/n)$. (If more terms in the Edgeworth expansion were included in deriving (1.3.6), the result would remain $\mathcal{O}(1/n)$). Having a pivotal quantity is essential because it ensures the leading normal terms cancel, permitting the difference between the $\mathcal{O}(n^{-1/2})$ terms in the Edgeworth expansions of G and $\hat{G}$ to contribute an extra $1/n^{-1/2}$ factor. Without the pivotal quantity, the leading normal terms will not cancel so the error will remain order $\mathcal{O}(1/n^{1/2})$.

Note that the argument here can be applied to functionals other than the studentized mean. As long as T has an Edgeworth expansion and is a pivotal quantity, the derivation will hold. Thus, one can choose T to be a centered and scaled percentile or variances. Both are asymptotically normal and have Edgeworth expansions; see Reiss (1989). U-statistics also have well-known Edgeworth expansions. Bhattacharya and Ranga Rao (1976) treat lattice-valued random variables, and recent work on Edgeworth expansions under censoring can be found in Hwang (2001).

1.3.1.2 Bootstrapping Without Assuming a Pivot

Now suppose the functional of interest $T(\{X_i\}, F)$ is not a pivotal quantity, even asymptotically. It may still be desirable to have an approximation to its sampling distribution. That is, in general we want to replace the sampling distribution

$$\mathbb{P}_F\left[T(\{X_i\}, F) \le t\right]$$

by its bootstrap approximation

$$\mathbb{P}_{\hat{F}_n}\left[T(\{X_i^*\}, \hat{F}_n) \le t\right]$$

for $t \in \mathbb{R}$. The bootstrap procedure is the same as before, of course, but the error decreases as $\mathcal{O}(1/\sqrt{n})$ rather than as $\mathcal{O}(1/\sqrt{n})$. This will be seen from a slightly different Edgeworth expansion argument.

First, to see the mechanics of this argument, take T to be the functional $U(\{X_i\}, F) = \bar{X} - \mu$. The bootstrap takes the sampling distribution of

$$U^* = U(\{X_i^*\}, \hat{F}_n) = \sqrt{n}(\bar{X}^* - \bar{X})$$

as a proxy when making uncertainty statements about $U = \bar{X} - \mu$. Although U is not a pivotal quantity, U/s is. However, for the sake of seeing the argument in a familiar context of a studentized mean, this fact will not be used. That is, the argument below is a template that can be applied anytime a valid Edgeworth expansion for a statistic exists, even though it is written for the mean.

The Edgeworth expansion for the sampling distribution of U is

$$\begin{aligned} H(t) &= \mathbb{P}_F\left[U \le t\right] \\ &= \mathbb{P}_F\left[\sqrt{n}\left(\frac{\bar{X} - \mu}{s}\right) \le t/s\right] \\ &= \Phi(t/s) + n^{-1/2} p_1(t/s) + \mathcal{O}(n^{-1}). \end{aligned}$$

Similarly, the Edgeworth expansion for the sampling distribution of U^* is

$$\begin{aligned} H^*(t) &= \mathbb{P}\left[U^* \le t \mid \{X_i\}\right] \\ &= \Phi(t/s^*) + n^{-1/2} \hat{p}_1(t/s^*)\phi(t/s^*) + \mathcal{O}(n^{-1}). \end{aligned}$$

A careful asymptotic argument (see Note 1.2) shows that

$$\begin{aligned} p_1(y/s) - \hat{p}_1(y/s^*) &= \mathcal{O}_p(n^{-1/2}), \\ s - s^* &= \mathcal{O}_p(n^{-1/2}). \end{aligned}$$

Thus the difference between H and H^* is

$$\begin{aligned} H(t) - H^*(t) &= \Phi(t/s) - \Phi(t/s^*) \\ &\quad + n^{-1/2}[p_1(t/s)\phi(t/s) - \hat{p}_1(t/s^*)\phi(t/s^*)] + \mathcal{O}_p(n^{-1}). \end{aligned} \tag{1.3.7}$$

The second term has order $\mathcal{O}_p(n^{-1})$ but the first has order $\mathcal{O}_p(n^{-1/2})$.

Obviously, if one really wanted the bootstrap for a studentized mean, one would not use U but would use U/s and apply the argument from the previous section. Nevertheless, the point remains that, when the statistic is not an asymptotic pivot, the bootstrap and the CLT have the same asymptotics because estimating a parameter (such as σ) only gives a $\mathcal{O}(1/\sqrt{n})$ rate.

The overall conclusion is that, when the statistic is a pivot, the bootstrap is superior, when it can be implemented, and otherwise the two are roughly equivalent theoretically. This is the main reason that the bootstrap is used so heavily in data mining to make uncertainty statements.

Next, observe that if s did not behave well, U/s would not be an asymptotic pivot. For instance, if F were from a parametric family in which only some of the parameters, say μ, were of interest and the rest, say γ, were nuisance parameters on which σ depended, then while s would remain pivotal under $F_{\mu,\gamma}$, it would not necessarily be pivotal under the mixture $\int F_{\mu,\gamma} w(d\gamma)$. In this case, the data would no longer be IID and other methods would need to be used to assess the variability of $\bar{X}$ as an estimator for μ. For dependent data, the bootstrap and Edgeworth expansions can be applied in principle, but their general behavior is beyond the scope of this monograph. At best, convergence would be at the $\mathscr{O}(1/\sqrt{n})$ rate. More realistically, pivotal quantities are often hard to find for discrete data or for general censoring processes. Thus, whether or not an Edgeworth expansion can be found for these cases, the bootstrap and the CLT will perform comparably.

1.3.2 Cross-Validation

Just as the bootstrap is ubiquitous in assessing uncertainty, cross-validation (CV) has become the standard tool for assessing model fit in a predictive accuracy sense. CV was invented by Stone (1959) in the context of linear regression. He wanted to balance the benefit of using as much data as possible to build the model against the false optimism created when models are tested on the same data that were used to construct them.

The ideal strategy to assess fit is to reserve a random portion of the data, fit the model with the rest, and then use the fitted model to predict the response values in the holdout sample. This approach ensures the estimate of predictive accuracy is unbiased and independent of the model selection and fitting procedures. Realistically, this ideal is nearly impossible to achieve. (The usual exceptions are simulation experiments and large databases of administrative records.) Usually, data are limited, so analysts want to use all the data to build and fit the best possible model – even though it is cheating a little to use the same data for model evaluation as for model building and selection.

In DMML, this problem of sample reuse is exacerbated by the fact that in most problems many models are evaluated for predictive accuracy in an effort to find a good one. Using a fresh holdout sample for each model worth considering would quickly exhaust all available data.

Cross-validation is a compromise between the need to fit and the need to assess a model. Many versions of cross-validation exist; the most common is the K-fold cross-validation algorithm:

Given a random sample $\boldsymbol{x} = (x_1, \cdots, x_n)$:

- ☐ Randomly divide the sample into K equal portions.
- ☐ For $i = 1, \ldots, K$, hold out portion i and fit the model from the rest of the data.
- ☐ For $i = 1, \ldots, K$, use the fitted model to predict the holdout sample.
- ☐ Average the measure of predictive accuracy over the K different fits.

One repeats these steps (including the random division of the sample) for each model to be assessed and looks for the model with the smallest error. The measure of predictive accuracy depends on the situation – for regression it might be predictive mean squared error, while for classification it might be the number of mistakes. In practice, it may not be possible to make the sample sizes of the portions the same; however, one does this as closely as possible. Here, for convenience, set $n = \ell K$, where ℓ is the common size of the portions.

The choice of K requires judgment. If $K = n$, this is called "leave-one-out" or "loo" CV since exactly one data point is predicted by each portion. In this case, there is low bias but possibly high variance in the predictive accuracy, and the computation is lengthy. (The increased variance may be due to the fact that the intersection between the complements of two holdout portions has $n-2$ data points. These data points are used, along with the one extra point, in fitting the model to predict the point left out. Thus, the model is fit twice on almost the same data, giving highly dependent predictions; dependence typically inflates variance.) On the other hand, if K is small, say $K = 4$, then although the dependence from predictor case to predictor case is less than with loo, the bias can be large. Commonly, K is chosen between 5 and 15, depending on n and other aspects of modeling.

One strategy for choosing K, if enough data are available, is to plot the predictive mean squared error as a function of the size of the training sample (see Fig. 1.6). Once the curve levels off, there is no need to increase the size of the portion of the data used for fitting. Thus, the complement gives the size of the holdout portion, and dividing n by this gives an estimate of the optimal K.

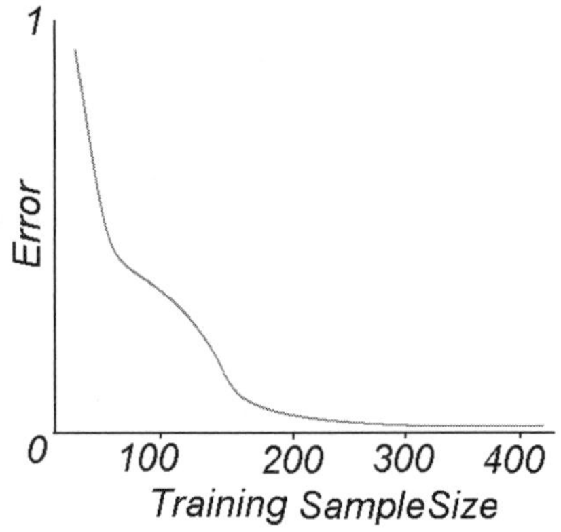

Fig. 1.6 This graph levels off starting around 200, suggesting the gains per additional data point are small after that. Indeed, one can interpret this as suggesting that the remaining error is primarily from reducing the variance in parameter estimation rather than in model selection.

To see the bias–variance trade-off in choosing K, consider regression. Start with the sample $\{Y_i, \boldsymbol{X}_i\}_{i=1}^n$ and randomly partition it into v subsets $S_1, \ldots, S_v$ of size ℓ. Let $f^{(-k)}(\cdot)$ be the regression function fit using all the data except the observations in S_k. The predictive squared error (PSE) for $f^{(-k)}(\cdot)$ on S_k is

$$\mathrm{PSE}_k = \sum_{S_k} \left(\hat{f}^{-(k)}(\boldsymbol{X}_i) - Y_i \right)^2 .$$

Summing over all K subsets gives the CV estimate of the PSE for f:

$$g(v) = \sum_{k=1}^{K} \mathrm{PSE}_k = \sum_{k=1}^{K} \sum_{S_k} \left(\hat{f}^{(-k)}(\boldsymbol{X}_i) - Y_i \right)^2 .$$

Minimizing g over K gives the best K for cross-validation.

The function $g(K)$ has a bias–variance decomposition. By adding and subtracting the terms $(1/n)\sum_{k=1}^K \sum_{S_k} \hat{f}^{(-k)}(\boldsymbol{X}_i)$ and $\bar{Y}$ in the double sum for $g(K)$, one can expand to get

$$\begin{aligned} g(K) = {} & \sum_{k=1}^{K} \sum_{S_k} \left(\hat{f}^{(-k)}(\boldsymbol{X}_i) - (1/n) \sum_{k=1}^{K} \sum_{S_k} \hat{f}^{(-k)}(\boldsymbol{X}_i) \right)^2 \\ & + n \left((1/n) \sum_{k=1}^{K} \sum_{S_k} \hat{f}^{(-k)}(X_i) - \bar{Y} \right)^2 + \sum_{k=1}^{K} \sum_{S_k} (\bar{Y} - Y_i)^2 . \end{aligned}$$

(The three cross-products are zero, as in the usual ANOVA decomposition.) The first term is the empirical variance $\widehat{\mathrm{Var}(f)}$ for f and the covariates together. The second term is the bias between the means of the predictions and the responses. The last term is a variance of the response. Thus, optimizing g over K achieves a trade-off among these three sources of error.

1.3.2.1 Generalized Cross-Validation

Cross-validation is not perfect – some dependency remains in the estimates of predictive error, and the process can absorb a lot of computer time. Many data mining techniques use computational shortcuts to approximate cross-validation.

For example, in many regression models, the estimates are linear functions of the observations; one can write $\hat{\boldsymbol{y}} = \boldsymbol{H}\boldsymbol{y}$, where $\boldsymbol{H} = (h_{i,j})_{n \times n}$. In multiple linear regression, $\boldsymbol{H} = \boldsymbol{X}(\boldsymbol{X}'\boldsymbol{X})^{-1}\boldsymbol{X}'$. Similar forms hold for kernel and spline regressions, as will be seen in Chapters 2, 3, and 10. For such linear estimates, the mean squared error of the cross-validation estimator is

$$n^{-1} \sum_{i=1}^{n} [y_i - \hat{f}^{(-i)}(\boldsymbol{x}_i)]^2 = n^{-1} \sum_{i=1}^{n} \left[\frac{y_i - \hat{f}(\boldsymbol{x}_i)}{1 - h_{ii}} \right]^2 , \tag{1.3.8}$$

where $\hat{f}^{(-i)}(\boldsymbol{x}_i)$ is the estimate of f at $\boldsymbol{x}_i$ based on all the observations except $(y_i, \boldsymbol{x}_i)$ (i.e., the loo cross-validation estimate at $\boldsymbol{x}_i$).

Equation (1.3.8) requires only one calculation of $\hat{f}$, but finding the diagonal elements of $\boldsymbol{H}$ is expensive when n or p is large. Often it is helpful, and not too far wrong, to approximate h_{ii} by $\text{tr}(\boldsymbol{H})/n$. This approximation is generalized cross-validation (GCV); provided not too many of the h_{ii}s are very large or very small this is a computationally convenient and accurate approximation. It is especially useful when doing model selection that necessitates repeated fits. See, for instance, Craven and Wahba (1979).

1.3.2.2 The Twin Problem and SEs

Sometimes a data set can contain cases, say $\{(Y_{i_1}, \boldsymbol{X}_{i_1}\}$ and $\{(Y_{i_2}, \boldsymbol{X}_{i_2}\}$, that are virtually identical in explanatory variable measurements and dependent variable measurements. These are often called twins. If there are a lot of twins relative to n, leave-one-out CV may give an overly optimistic assessment of a model's predictive power because in fitting the near duplication the model does better than it really should. This is particularly a problem in short, fat data settings.

This is the exact opposite of extrapolation, in which the values of the sample are not representative of the region where predictions are to be made. In fact, this is "intrapolation" because the values of the sample are overrepresentative of the region where predictions are to be made. The model cannot avoid overfitting, thereby reducing predictive power.

Two settings where twin data are known to occur regularly are drug discovery and text retrieval. Pharmaceutical companies keep libraries of the compounds they have studied and use them to build data mining models that predict the chemical structure of biologically active molecules. When the company finds a good molecule it promptly makes a number of very similar "twin" molecules (partly to optimize efficacy, partly to ensure an adequately broad patent). Consequently, its library has multiple copies of nearly the same molecule. If cross-validation were applied to this library, then the hold-out sample would usually contain one or more versions of a molecule, while the sample used for fitting contains others. Thus, the predictive accuracy of the fitted model will seem spuriously good; essentially the same data are being used to both fit and assess the model.

In the text retrieval context, the TREC program at the National Institute of Standards and Technology, Voorhees and Harman (2005) makes annual comparisons of search engines on an archive of newspaper articles. These search engines use data mining to build a classification rule that determines whether or not an article is "relevant" to a given search request. But the archive usually contains nearly identical variants of stories distributed by newswire services. Therefore cross-validation can have the same basic text in both the fitting and assessment samples, leading to overestimation of search engine capability.

A related problem is that the data are randomly allocated to the sets S_k. This means that the CV errors are themselves random; a different allocation would give different

CV errors. The implication is that, for model selection, it is not enough to choose the model with the smallest cross-validatory error; the model with the smallest error must have an error so much smaller than that of the model with the second smallest error that it is reasonable to identify the first model as better. Often, it is unclear what the threshold should be. The natural solution would be to find an SE for the CV errors and derive thresholds from it. There are many ways to do this, and several effective ad hoc rules for choosing a model based on CV errors have been proposed. However, none have been universally accepted.

1.3.2.3 CV, GCV, and other model selection procedures

CV and GCV are only two model selection procedures. Many others are available. In general, the asymptotic performance of a model selection procedure (MSP) depends strongly on whether there is a fixed, finite-dimensional model in the set of models the MSP is searching.

Indeed, there is an organized theory that characterizes the behavior of MSPs in a variety of contexts; see Shao (1997) for a thorough treatment. Li (1986, 1987) also provides good background.

The basic quantity that serves as a general criterion for one unified view of model selection is

$$GIC_{\lambda}(m) = \frac{S_n(m)}{n} + \lambda \frac{\hat{\sigma}_n^2 p_n(m)}{n}, \tag{1.3.9}$$

in which m indicates a model ranging over the set A_n of models, $S_n(m) = ||\mathbf{y}_n - \hat{\mu}_n(m)||^2$ is the squared distance between the data vector and the estimate of the mean vector for model m (from n outcomes), $\hat{\sigma}^2$ estimates σ^2, $p_n(m)$ is the dimension of model m, and λ is a constant controlling the trade-off between fit and variability.

Shao (1997) distinguishes three classes of MSPs of the form (1.3.9) in the linear models context. He observes that GIC_2, Mallows' C_p, Akaike's information criterion, leave-one-out CV, and GCV form one class of methods of the form (1.3.9), which are useful when no fixed, finite-dimensional model can be assumed true. A second class of methods of the form (1.3.9) is formed by GIC_{λ_n} when $\lambda_n \to \infty$ and delete-d GCV when $d/n \to 1$. These methods are useful when a true fixed dimension model can be assumed to exist. The third class contains methods that are hybrids between methods in the first two classes, for instance, GIC_{λ} with $\lambda > 2$ and delete-d GCV with $d/n \to \tau \in (0,1)$. The key criteria distinguishing the three classes are expressed in terms of the consistency of model selection or the weaker condition of asymptotic loss efficiency (the loss of the model selected converges to the minimal value of the loss in probability). Along with detailed proofs for a wide variety of settings, Shao (1997) also provides an extensive collection of references.

1.4 Optimization and Search

DMML methods often require searches, and a variety of search procedures are commonly used. Indeed, one can argue that DMML as a whole is a collection of statistically guided search procedures to facilitate good predictive performance. Univariate search is a search for the value of a unidimensional real value, usually but not always assumed to vary continuously over an interval. This arises for instance when finding the best value of a bin width for smoothing or the best K for K-fold CV. Multivariate search is much the same, but multidimensional. The goal is to find the vector that maximizes some function, such as the likelihood or a goodness-of-fit statistic. This is harder because, unlike real numbers, vectors usually have a partial ordering rather than a full ordering. Combinatorial search is the problem of having a finite number of variables each of which can assume one of finitely many values and then seeking the optimal assignment of values to variables. This arises in variable selection when one must decide whether or not to include each variable. More general search procedures do not take account of the specific structure of the problem; these are "uninformed" searches. List and tree searches are general and often arise in model selection.

This section reviews some of the main strategies for each of these cases. In practice, one often creates hybrid techniques that combine more than one strategy. A full discussion of these methods is beyond the scope of this monograph.

1.4.1 Univariate Search

Suppose the goal is to maximize a univariate function $g(\lambda)$ to find

$$\lambda^* = \arg \max_{\lambda} g(\lambda).$$

There are several elementary ways to proceed.

Newton-Raphson iteration: If $g(\lambda)$ is unimodal and not too hard to differentiate, the Newton-Raphson method can be used to find a root; i.e., to solve $g'(\lambda) = 0$. Keeping terms to first order, Taylor expanding gives

$$g(\lambda_0 + \varepsilon) \approx g(\lambda_0) + g'(\lambda_0)\varepsilon.$$

This expression estimates the ε needed to land closer to the root starting from an initial guess λ_0. Setting $g(\lambda_0 + \varepsilon) = 0$ and solving for ε gives $\varepsilon_0 = -(g(\lambda_0))/(g'(\lambda_0))$, which is the first-order adjustment to the root's position. By letting $\lambda_1 = \lambda_0 + \varepsilon_0$, calculating a new ε_1, and so on, the process can be repeated until it converges to a root using $\varepsilon_n = -(g(\lambda_n))/(g'(\lambda_n))$. Unfortunately, this procedure can be unstable near a horizontal asymptote or a local extremum because the derivative is near zero. However, with a good initial choice λ_0 of the root's position, the algorithm can be applied iteratively to obtain a sequence

$$\lambda_{n+1} = \lambda_n - (g(\lambda_n))/(g'(\lambda_n)),$$

which converges.

If $g(\lambda)$ is multimodal, then randomly restarting the procedure is one way to explore the surface the function defines. The idea is to put diffuse distribution on the domain of λ, generate a random starting point from it and "hill-climb" to find a local mode. Hill-climbing means approximating the gradient and taking a step in the direction of function increase. This can be done by a Newton-Raphson procedure that approximates the gradient, by a Fibonacci search (to be described shortly), or by many other methods. Once the top of the "hill" is found, one draws another starting point and repeats. After several runs, the analyst has a good sense of the number and location of the modes. Note that this procedure can be applied to functions that are not easily differentiable, provided the hill-climbing does not require derivatives.

Bracket search: If g is not differentiable but is unimodal, and not too difficult to evaluate, one strategy is to find values to bracket λ^*. Once it is bracketed, the searcher can successively halve the interval, determining on which side of the division λ^* lies, and quickly converge on a very accurate estimate.

Several methods for finding the brackets exist. A popular one with good theoretical properties is Fibonacci search, see Knuth (1988). Start the search at an arbitrary λ_0, and form the sequence of "test" values $\lambda_k = \lambda_0 + F(k)$, where $F(k)$ is the kth Fibonacci number. At some point, one overshoots and $g(\lambda_k)$ is less than a previous value. This means the value of λ^* is bracketed between λ_{k-1} and λ_k. (If the initial λ_0 gives a sequence of evaluations that decreases, then use $\lambda_k = \lambda_0 - F(k)$ instead.)

Diminishing returns: Sometimes the goal is not to find a maximum per se but rather a point at which a trend levels off. For example, one could fit a sequence of regression models using polynomials of successively higher degree. In this case, lack of fit can only decrease as the degree increases, so the task is to find the point of diminishing returns. The standard method is to plot the lack of fit as a function of degree and look for the degree above which improvement is small. Often there is a knee in the curve, indicating where diminishing returns begin. This indicates a useful trade-off between omitting too many terms and including too many terms; it identifies the point at which the benefit of adding one more term, or other entity, abruptly drops in value.

1.4.2 Multivariate Search

In multivariate search for $\boldsymbol{\lambda}^* = \text{argmax}\, g(\boldsymbol{\lambda})$, many of the same techniques apply. If partial derivatives exist, one can find the solution analytically and verify it is an optimum. If the function is multimodal, then random restart can be useful, even when it is hard to differentiate. One can even generalize a Fibonacci search to find hyperrectangles that bracket $\boldsymbol{\lambda}^*$.

However, in multivariate search, the most popular method is the Nelder-Mead algorithm Nelder and Mead (1965). This has a relatively low computational burden and works well whenever $g(\boldsymbol{\lambda})$ is reasonably smooth. Conceptually, Nelder-Mead uses preprocessing to find the right domain on which to apply Newton-Raphson. The basic idea is as follows. To find $\boldsymbol{\lambda}^* \in \mathbb{R}^d$, choose a simplex in $\mathbb{R}^d$ that might contain $\boldsymbol{\lambda}^*$, and evaluate $g(\boldsymbol{\lambda})$ at each of its $d+1$ vertices. Hopefully, one of the vertices $\boldsymbol{v}_i$ will give a smaller value than the others. Reflect $\boldsymbol{v}_i$ through the $d-1$ dimensional hyperplane defined by the other d vertices to give $\boldsymbol{v}_i^*$ and find $g(\boldsymbol{v}_i)$. Then repeat the process. A new worst vertex will be found at each step until the same vertex keeps being reflected back and forth. This suggests (but does not guarantee) that the simplex contains a local mode. At this point, the local mode can be found by Newton-Raphson hill-climbing from any of the vertices. Actual implementation requires the size of the initial simplex and the distance to which the worst vertex is projected on the other side of the hyperflat. These technical details are beyond our present scope.

Some researchers advocate simulated annealing for optimization Kirkpatrick et al. (1983). This is popular, in part, because of a result that guarantees that, with a sufficiently long search, simulated annealing will find the global optimum even for very rough functions with many modes in high dimensions. See, for instance, Andrieu et al. (2001) and Pelletier (1998).

The main idea behind simulated annealing is to start at a value $\boldsymbol{\lambda}_0$ and search randomly in a region D around it. Suppose the search randomly selects a value $\boldsymbol{\lambda}^*$. If $g(\boldsymbol{\lambda}^*) < g(\boldsymbol{\lambda}_0)$, then set $\boldsymbol{\lambda}_1 = \boldsymbol{\lambda}^*$ and relocate the region on the new value. Otherwise, with probability $1-p$, set $\boldsymbol{\lambda}_1 = \boldsymbol{\lambda}_0$ and generate a new $\boldsymbol{\lambda}^*$ that can be tested. This means there is a small probability of leaving a region that contains an optimum. It also means that there is a small probability of jumping to a region that contains a better local minimum. As the search progresses, p is allowed to get smaller, so the current location becomes less and less likely to change by chance rather than discovered improvement. For most applications, simulated annealing is too slow; it is not often used unless the function g is extremely rough, as is the case for neural networks.

1.4.3 General Searches

Searches can be characterized as general and specific or uninformed versus informed. The difference is whether or not there is extra information, unique to the application at hand, available to guide the search. There is some subjectivity in deciding whether a search is informed or not because a search might use generic features of the given problem that are quite narrow. The benefit of an uninformed search is that a single implementation can be used in a wide range of problems. The disadvantage is that the set of objects one must search for a solution, the searchspace, is often extremely large, and an uninformed search may only be computationally feasible for small examples.

The use of one or more specific features of a problem may speed the search. Sometimes this only finds an approximately optimal solution; often the "specific feature" is a heuristic, making the algorithm preferentially examine a region of the search space. Using a good heuristic makes an informed search outperform any uninformed search, but this is very problem-specific so there is little call to treat them generally here.

An important class of searches is called constraint satisfaction. In these cases, the solution is a set of values assigned to a collection of variables. These are usually informed because uninformed methods are typically ineffective. Within this class, combinatorial searches are particularly important for DMML.

1.4.3.1 Uninformed Searches

List search: The simplest search strategy is list search. The goal is to find an element of the searchspace that has a specific property. This is a common problem; there are many solutions whose properties are well known. The simplest algorithm is to examine each element of the list in order. If n is the number of items on the list, the complexity (number of operations that need to be performed) is $\mathscr{O}(n)$ because each item must be examined and tested for the property. Often one speaks of $\mathscr{O}(n)$ as a "running time" assuming the operations are performed at a constant rate. Linear search is very slow; in fact, $\mathscr{O}(n)$ is very high but the algorithm is fully general – no preprocessing of the list is involved.

Binary search: Binary search, by contrast, rules out half the possibilities at each step, usually on the basis of a direction, or ordering on the list. Bracket search is an instance of this. Binary search procedures run in $\mathscr{O}(\log n)$ time, much faster than list searches. Sometimes a very large sorted list can be regarded as nearly continuous. In these cases, it may be possible to use an interpolation procedure rather than a binary criterion.

Note that binary search requires the list be sorted prior to searching. Sorting procedures ensure a list has an order, often numerical but sometimes lexicographical. Other list search procedures perform faster but may require large amounts of memory or have other drawbacks.

Tree search: Less general than list search is tree search; however, it is more typical. The idea is to search the nodes of a tree whether or not the entire tree has been explicitly constructed in full. Often, one starts at the root of the tree and searches downward. Each node may have one or more branches leading to child nodes, and the essence of the algorithm is how to choose a path through child nodes to find a solution. One extreme solution is to search all child nodes from a given node and then systematically search all their child nodes and so forth down to the terminal nodes. This is called breadth first. The opposite extreme solution, depth first, is to start at a node and then follow child nodes from level to level down to the terminal nodes without any backtracking. It is rare that a search is purely depth first or breadth first; trade-offs between the extremes are usually more efficient.

1.4.4 Constraint Satisfaction and Combinatorial Search

The point of constraint satisfaction is to find an assignment of values to a set of variables consistent with the constraint. In the definition of the problem, each variable has an associated range of permissible values. Usually, any assignment of permissible values to variables consistent with the constraints is allowed and there will be many assignments of values to variables that meet the constraint.

Often, one wants to optimize over the set of solutions, not just enumerate them. Tree searches can be used to find solutions, but usually they are inefficient because the order of processing of the variables causes an exponential increase in the size of the searchspace. In such cases, one can attempt a combinatorial search; this is a term that typifies the hardest search problems, involving large searchspaces necessitating efficient search strategies. However, the time required to find a solution can grow exponentially, even factorially, fast with the size of the problem as measured by the number of its most important inputs, often the number of variables, but also the number of values that can be assigned to the variables. For instance, if there are p variables, each of which assumes k values, there are k^p possibilities to examine.

In the general case, the time for solution is intractable, or NP-complete. However, there are many cases where it is easy to determine if a candidate solution meets the constraints. Sometimes these are called NP-problems.

Suppose the goal is to find K solutions from the k^p possibilities. One approach is a branch and bound technique that will recur in Chapter 10. The idea is to organize all the subsets of these k^p possibilities into sets of common size, say i, and then form a lattice based on containment. That is, level i in the lattice corresponds to all sets of cardinality i, and the edges in the (directional) lattice are formed by linking each set to its immediate subsets and supersets; this is the branching part. Once any lattice point is ruled out, so are all of its supersets; this is the bounding part. Now, search algorithms for the K solutions can be visualized as paths through the lattice, usually starting from sets at lower levels and working up to higher levels.

In the variable selection context of Chapter 10, p may be large and one wants to discard variables with little or no predictive power. The lattice of all subsets of variables has 2^p subsets. These can be identified with the 2^p vertices of the unit hypercube, which can be regarded as a directed lattice. A clever search strategy over these vertices would be an attractive way to find a regression model. The Gray code is one procedure for listing the vertices of the hypercube so that there is no repetition, each vertex is one edge away from the previous vertex, and all vertices in a neighborhood are explored before moving on to a new neighborhood. Wilf (1989) describes the mathematical theory and properties of the Gray code system.

In the lattice context, the simulated annealing strategy would move among sets in the lattice that contain a full solution to the problem, attempting to find exact solutions by a series of small changes. If there is no solution in the search space, this kind of search can continue essentially forever. So, one can fail to get a solution and not be able to conclude that no solution exists. A partial correction is to repeat the search from different starting points until either an adequate solution is found or some limit on the number of points in the search space is reached. Again, one can fail to get a solution and still be unable to conclude that no solution exists.

Alternatively, one can seek solutions by building up from smaller sets, at lower levels. The usual procedure is to extend an emerging solution until it is complete or leads to an endpoint past which there can be no solutions. Once an endpoint has been hit, the search returns to one of its earlier decision points, sometimes the first, sometimes the most recent, and tests another sequence of extensions until all paths are exhausted. If this is done so that the whole space is searched and no solution is found, then one can conclude that the problem is unsolvable.

An extension of tree search is graph search. Both of these can be visualized as searches over the lattice of subsets of possibilities. Graph searches, whether on the lattice of possibilities or on other search spaces, rest on the fact that trees are a subclass of graphs and so are also characterized as depth first or breadth first. Many of the problems with graphical search spaces can be solved using efficient search algorithms, such as Dijkstra's or Kruskal's. There are also many classes of search problems that are well studied; the knapsack problem and the traveling salesman problem are merely two classes that are well understood. Both are called NP-complete because they do not admit polynomial time solutions. A standard reference for classes of NP-complete problems and their properties is Garey and Johnson (1979).

1.4.4.1 Search and Selection in Statistics

Bringing the foregoing material back to a more statistical context, consider list search on models and variable selection as a search based on ideas from experimental design.

First, with list search, there is no exploitable structure that links the elements of the list, and the list is usually so long that exhaustive search is infeasible. So, statistically, if one tests entries on the list at random, then one can try some of the following: (1) Estimate the proportion of list entries that give results above some threshold. (2) Use some modeling to estimate the maximum value on the list from a random sample of list entries. (3) Estimate the probability that further search will discover a new maximum within a fixed amount of time. (4) Use the solution to the secretary problem. These are routine, but one may not routinely think of them.

Another strategy, from Maron and Moore (1997), is to "race" the testing. Essentially, this is based on pairwise comparisons of models. At first, one fits only a small random fraction of the data (say a random 1%) to each model on the list. Usually this is sufficient to discover which model is better. If that small fraction does not distinguish the models, then one fits another small fraction. Only very rarely is it necessary to fit all or most of the data to select the better model. Racing can extend one's search by about 100-fold.

Variable selection can be done using ideas from experimental design. One method is due to Clyde (1999). View each explanatory variable as a factor in an experimental design. All factors have two levels, corresponding to whether or not the explanatory variable is included in the model. Now, consider a 2^{p-k} fractional factorial experiment in which one fits a multiple regression model with the included variables and records some measure of goodness of fit. Obviously, k must be sufficiently large that it is possible to perform the computations in a reasonable amount of time and also to limit the effect of multiple testing.

Possible measures of goodness of fit include: (1) adjusted R^2, the proportion of variance in the observations that is explained by the model, but with an adjustment to account for the number of variables in the model; (2) Mallows' C_p, a measure of predictive accuracy that takes account of the number of terms in the model; (3) MISE, the mean integrated squared error of the fitted model over a given region (often the hyperrectangle defined by the minimum and maximum values taken by each explanatory variable used in the model); (4) the square root of the adjusted R^2 since this transformation appears to stabilize the variance and thereby supports use of analysis of variance and response surface methodology in the model search. Weisberg (1985), pp. 185–190 discusses the first three and Scott (1992), Chapter 2.4, discusses MISE.

Treating the goodness-of-fit measure as the response and the presence or absence of each variable as the factor levels, an analysis of variance can be used to examine which factors and factor combinations have a significant influence on the "observations". Significant main effects correspond to explanatory variables that contribute on their own. Significant interaction terms correspond to subsets of variables whose joint inclusion in the model provides explanation. In multiple linear regression, these results are implicit in significance tests on the coefficients. However, this also helps find influential variables for the nonparametric regression techniques popular in data mining (e.g., MARS, PPR, neural nets; see Chapter 4).

1.5 Notes

1.5.1 Hammersley Points

To demonstrate the Hammersley procedure, consider a particular instance. The bivariate Hammersley point set of order k in the unit square starts with the integers from 0 to $2^k - 1$. Write these in binary notation, put a decimal in front, and denote the ith number by a_i for $i = 1, \ldots, 2^k$. From each a_i, generate a b_i by reversing the binary digits of a_i. For example, with $k = 2$, the a_i are .00, .01, .10, .11 (in base 2), or 0, 1/4, 1/2, 3/4. Similarly, the b_i are .00, .10, .01, .11, or 0, 1/2, 1/4, 3/4. Define the Hammersley points as $\boldsymbol{x}_i = (a_i, b_i)$; this gives (0, 0), (1/4, 1/2), (1/2, 1/4), and (3/4, 3/4).

To extend this construction to higher dimensions, represent an integer j between 0 and $b^k - 1$ by its k-digit expansion in base b:

$$j = a_0 + a_1 b + \ldots + a_{k-1} b^{k-1}.$$

The radical inverse of j in base b is

$$\psi_b(j) = a_0 \frac{1}{b} + a_1 \frac{1}{b^2} + \ldots + a_{k-1} \frac{1}{b^k}.$$

The integer radical inverse representation of j is

$$b^k \psi_b(j) = a_{k-1} + \ldots + a_1 b^{k-2} + a_0 b^{k-1}.$$

This is the mirror image of the digits of the usual base b representation of j.

The Hammersley points use a sequence of ψ_bs, where the bs are prime numbers. Let $2 = b_1 < b_2 < \ldots$ be the sequence of all prime numbers in increasing order. The Hammersley sequence with n points in p dimensions contains the points

$$\boldsymbol{x}_i = \left(\frac{i}{n}, \psi_{b_1}(i), \psi_{b_2}(i), \ldots, \psi_{b_{p-1}}(i) \right),$$

where $i = 0, \ldots, n-1$. The points of a Hammersley point set can be pseudorandomized by applying a permutation to the digits of i before finding each coordinate.

It can be verified pictorially that $\{\boldsymbol{x}_1, \ldots, \boldsymbol{x}_n\}$ fills out the space evenly and therefore is a good choice. In particular, the point set is uniform, without clumping or preferred directions. This is accomplished by the Hammersley sequence by using different prime numbers at different stages.

There are a variety of formal ways to measure how well a set of points fills out a space. In general, Hammersley points, see Niederreiter (1992b), are a design that maximizes dispersion and minimizes the discrepancy from the uniform distribution in the Kolmogorov-Smirnov test. Wozniakowski (1991) proved that a modification of Hammersley points avoids the Curse in the context of multivariate integration for smooth functions (those in a Sobolev space), see Traub and Wozniakowski (1992). Moreover, the computations are feasible, see Weisstein (2009). Thus, for high-dimensional integration, one can use Wozniakowski's points and guarantee that the error does not increase faster than linearly in p, at least in some cases. Unfortunately, this result is not directly pertinent to multivariate regression since it does not incorporate errors from model selection and fitting.

1.5.2 Edgeworth Expansions for the Mean

The characteristic function (Fourier transform) of a random sum $S_n = \sum_{j=1}^{n} Y_j$ is

$$\chi_n(t) = \mathbb{E}[e^{itS_n}] = \chi(t/\sqrt{n})^n,$$

where χ is the characteristic function (CS) of Y_1. A Taylor expansion of $\ln \chi(t)$ at $t = 0$ gives

$$\ln \chi(t) = \kappa_1 it + \kappa_2 (it)^2 + \kappa_3 (it)^3 + \ldots. \tag{1.5.1}$$

The coefficients κ_j are the cumulants of Y_i. To simplify the discussion, assume the standardization $Y = (X - \mu)/\sigma$.

Taylor expand the exponential in the integrand of an individual χ, and take logarithms to get another series expansion:

$$\ln \chi(t) = \ln \left[1 + \mathbb{E}(Y)it + \frac{1}{2}\mathbb{E}(Y^2)(it)^2 + \ldots \right]. \tag{1.5.2}$$

By equating the right-hand sides in (1.5.1) and (1.5.2), one sees

$$\sum_{j=1}^{\infty} \frac{\kappa_j}{j!}(it)^j = \ln \left[1 + \sum_{j=1}^{\infty} \frac{1}{j!}\mathbb{E}(Y^j)(it)^j \right]. \tag{1.5.3}$$

Taylor expanding the logarithm shows that the jth cumulant is a sum of products of moments of order j or less and conversely.

When $\mathbb{E}(S_n) = 0$, one has $\kappa_1 = 0$, and when $\mathrm{Var}(S_n) = 1$, one has $\kappa_2 = 1$. Using this and transforming back to χ_n from χ gives

$$\begin{aligned} \chi_n(t) &= \exp\left[-\frac{1}{2}t^2 + \frac{1}{3!n^{1/2}}\kappa_3(it)^3 + \ldots + \frac{1}{j!n^{(j-2)/2}}\kappa_j(it)^j + \ldots \right] \\ &= e^{-t^2/2}\exp\left[\frac{1}{3!n^{1/2}}\kappa_3(it)^3 + \ldots + \frac{1}{j!n^{(j-2)/2}}\kappa_j(it)^j + \ldots \right]. \end{aligned}$$

Taylor expanding the exponential term-by-term and grouping the results in powers of $1/\sqrt{n}$ shows there are polynomials $r_j(it)$ of degree $3j$ with coefficients depending on $\kappa_3, \ldots, \kappa_{j+2}$ such that

$$\chi_n(t) = e^{-t^2/2}\left[1 + \frac{r_1(it)}{n^{1/2}} + \frac{r_2(it)}{n^1} + \frac{r_3(it)}{n^{3/2}} + \ldots \right]. \tag{1.5.4}$$

Next, we set up an application of the inverse Fourier transform (IFT). Write $R_j(x)$ for the IFT of $r_j(it)e^{-t^2/2}$. That is, R_j is the IFT of r_j, weighted by the normal, where r_j is a polynomial with coefficients given by the κ coefficients. Since the IFT of the $N(0,1)$ distribution is $e^{-t^2/2}$ (A.4) gives

$$\mathbb{P}(S_n \le x) = \Phi(x) + \frac{R_1(x)}{n^{1/2}} + \frac{R_2(x)}{n} + \frac{R_3(x)}{n^{3/2}} + \ldots. \tag{1.5.5}$$

This is almost the Edgeworth expansion; it remains to derive an explicit form for the R_js in terms of the Hermite polynomials.

An induction argument on the IFT (with the induction step given by integration by parts) shows

$$\int_{\infty}^{-\infty} e^{itx} d\left[(-D)^j \Phi(x)\right] = (it)^j e^{-t^2/2},$$

where D is the differential operator $\frac{d}{dx}$. By linearity, one can replace the monomial in $-D$ by any polynomial and it will appear on the right-hand side. Take $r_j(-D)$, giving

$$\int_{\infty}^{-\infty} e^{itx} d\left[r_j(-D)\Phi(x)\right] = r_j(it)e^{-t^2/2}.$$

So the (forward) Fourier transform shows

$$R_j(x) = \int_{\infty}^{-\infty} e^{-itx} r_j(it) e^{-t^2/2} = r_j(-D)\Phi(x),$$

which can be used in (1.5.5). Let $H_{j-1}(x)$ denote the Hermite polynomial that arises from the jth signed derivative of the normal distribution:

$$(-D)^j \Phi(x) = H_{j-1}(x)\phi(x).$$

So $r_j(-D)\Phi(x)$ is a sum of Hermite polynomials, weighted by the coefficients of r_j, which depend on the cumulants. To find R_j, one must find the r_js, evaluate the Hermite polynomials and the cumulants, and do the appropriate substitutions. With some work, one finds $R_1(x) = -\kappa_3(x^2-1)\phi(x)$, $R_2(x) = -[\kappa_4 x(x^2-3)/24 + \kappa_3^2(x^4 - 10x + 15)]\phi(x)$ etc. Writing $R_j(x) = q_j(x)\phi(x)$ shows that the Edgeworth expansion for the distribution of the standardized sample mean is

$$\mathbb{P}(S_n \le x) = \Phi(x) + \frac{q_1(x)\phi(x)}{n^{1/2}} \tag{1.5.6}$$

$$+ \frac{q_2(x)\phi(x)}{n} + \ldots + \frac{q_j(x)\phi(x)}{n^{j/2}} + o\left(\frac{1}{n^{j/2}}\right). \tag{1.5.7}$$

Note that in (1.5.7) there is no explicit control on the error term; that requires more care with the Taylor expansion of the CF, see Bhattacharya and Ranga Rao (1976), Petrov (1975). Also, (1.5.7) is pointwise in x as $n \to \infty$. It is a deterministic expansion for probabilities of the random quantity S_n. Thus, for finite n, it is wrong to regard an infinite Edgeworth expansion as necessarily having error zero at every x. The problem is that the limit over n and the limit over j cannot in general be done independently. Nevertheless, Edgeworth expansions tend to be well behaved, so using them cavalierly does not lead to invalid expressions very often.

1.5.3 Bootstrap Asymptotics for the Studentized Mean

Hall (1992), Sections 2.4 and 2.6, details the mathematics needed to find the Edgeworth expansion for the studentized mean. He shows that the expansion

$$\mathbb{P}\left(\frac{(\bar{X}-\mu)}{s} \le x\right) = \Phi(x) + \frac{p_1(x)\phi(x)}{n^{1/2}} \tag{1.5.8}$$

$$+ \frac{p_2(x)\phi(x)}{n} + \ldots + \frac{p_j(x)\phi(x)}{n^{j/2}} + o\left(\frac{1}{n^{j/2}}\right)$$

exists, but the p_j terms are different from the q_j terms derived previously. In particular, the studentized mean is asymptotically equivalent to the standardized mean, but only up to order $\mathcal{O}_p(1/n)$. In fact,

$$\frac{\bar{X}-\mu}{\sigma} - \frac{\bar{X}-\mu}{s} = \frac{(s-\sigma)(\bar{X}-\mu)}{\sigma s}, \tag{1.5.9}$$

and both factors in the numerator are $O_p(1/\sqrt{n})$. So although the first term in their Edgeworth expansions is the same (normal), later terms are not.

For an IID sample of size n from F, let T_n denote the studentized sample mean. Also, for an IID sample of size n from $\hat{F}_n$ (a bootstrap sample), let T_n^* be the studentized mean. Then the Edgeworth expansions are

$$\mathbb{P}(T_n \leq t) = \Phi(t) + \frac{p_1(t)\phi(t)}{\sqrt{n}} + \mathcal{O}\left(\frac{1}{n}\right)$$

$$\mathbb{P}(T_n^* \leq t) = \Phi(t) + \frac{p_1^*(t)\phi(t)}{\sqrt{n}} + \mathcal{O}\left(\frac{1}{n}\right).$$

Note that p_1^* and p_1 are polynomials in the same powers of t with coefficients that are the same function of the cumulants of F and $\hat{F}_n$, respectively. And recall from (1.5.3)) that the cumulants are determined by the moments.

Bootstrap asymptotics for the studentized mean require three technical points about convergence rates:

A. $\hat{p}_1(t) - p_1(t) = \mathcal{O}_p(1/\sqrt{n})$,
B. $p_1(y/s) - \hat{p}_1(y/s^*) = \mathcal{O}_p(1\sqrt{n})$,
C. $s - s^* = \mathcal{O}_p(1\sqrt{n})$.

The first is needed in (1.3.6); B and C are needed in (1.3.8).

If the moments of $\hat{F}_n$ converge to the moments under F at the desired rate, then A follows. So, we need that

$$\sqrt{n}[\mu_k(\hat{F}_n) - \mu_k(F)] = \mathcal{O}_P(1), \tag{1.5.10}$$

where k indicates the order of the moment. Expression (1.5.10) follows from the CLT; the convergence is in F.

Next, we show C. Write

$$\sqrt{n}(s - s^*) = \sqrt{n}\left[\left(\frac{n}{n-1}\bar{X^2} - \bar{X}^2\right)^{1/2}\left(\frac{n}{n-1}\bar{X^2}^* - \bar{X}^{2,*}\right)^{1/2}\right], \tag{1.5.11}$$

in which the superscript $*$ indicates the moment was formed from bootstrap samples rather than the original sample. The convergence is in F, but note that s^* is from a bootstrap sample drawn from $\hat{F}_n$ as determined by the original sample. The conditional structure here matters. The bootstrap samples are taken conditional on the original sample. So, moments from a bootstrap sample converge, conditionally on the original sample, to the original sample, which itself converges unconditionally to the population mean. Another way to see this is to note that, for any function g, $\mathbb{E}_{\hat{F}_n}[g(X)] = g(\bar{X})$ and $\mathbb{E}_F[g(\bar{X})] = \mu_g$. More formally, for $\varepsilon > 0$, we have, conditional on X^n, that for any function Z^* converging in distribution in $\hat{F}_n$ to a constant, say 0,

$$g(X^n) = \mathbb{P}_{\hat{F}_n}(|Z| > \varepsilon) \to 0.$$

Now, $\mathbb{E}_{X^n} g(X^n)$ is bounded by 1 and converges to 0 pointwise, so the dominated convergence theorem gives its convergence to 0 unconditionally. This is convergence in probability, which implies convergence in distribution in the "joint" distribution of $F \times \hat{F}$, which reduces to F. Now, moments such as $\bar{X}^{k,*}$ for $k = 1, 2, \ldots$ are choices of g, and the convergence of functions of moments, like standard deviations, can be handled as well.

Now, s is a function of the first and second moments from the actual sample which converge at a $\mathscr{O}(\sqrt{n})$ rate to $\mu = \mu_1(F)$ and $\mu_2(F)$ by the CLT. The first and second moments in s^* converge at rates $\mathscr{O}(\sqrt{n})$ to $\mu_1(\hat{F}_n)$ and $\mu_2(\hat{F}_n)$ in $\hat{F}_n$ conditional on the original data by a conditional CLT. So, the unconditional CLT gives that the first and second moments in s^* converge at rate $\mathscr{O}(\sqrt{n})$ to $\mu_1(F)$ and $\mu_2(F)$ in F. Since the moments converge, a delta method argument applies to functions of them such as s and s^*. (The delta method is the statement that if $\bar{X}$ is $N(0, 1/n)$, then $g(\bar{X})$ is $N(g(0), (g'(0))^2/n)$, which follows from a Taylor expansion argument.) Now, (1.5.11) is $\mathscr{O}_p(1)$ giving C.

With these results, B follows: The polynomials $p_1(y/s)$ and $\hat{p}_1(y/s)$ have the same powers, and each power has a coefficient that is a function of the cumulants of $\hat{F}_n$ and F, respectively. By (1.5.3), these coefficients are functions of the moments of $\hat{F}_n$ and F, so the multivariate delta method applies. A typical term in the difference $p_1(y/s) - \hat{p}_1(y/s^*)$ has the form

$$\alpha(\mathbb{E}_F(X), \ldots, \mathbb{E}_F(X^k)) \left(\frac{y}{s}\right)^\ell - \alpha(\mathbb{E}_{\hat{F}}(X), \ldots, \mathbb{E}_{\hat{F}}(X^k)) \left(\frac{y}{s^*}\right)^\ell, \qquad (1.5.12)$$

where k is the number of moments and ℓ is the power of the argument. As in the proof of A, for fixed y, the moments under $\hat{F}_n$ converge to the moments under F in probability at a $\sqrt{n}$ rate, as does s^* to s (and as both do to σ). So the delta method gives a $\sqrt{n}$ rate for the term. There are a finite number of terms in the difference $p_1(y/s) - \hat{p}_1(y/s^*)$, so their sum is $\mathscr{O}(1/\sqrt{n})$.

1.6 Exercises

Exercise 1.1. Consider a sphere of radius r in p dimensions. Recall from (1.1.1) that the volume of such a sphere is given by

$$V_r(p) = \frac{\pi^{p/2} r^p}{\Gamma(p/2 + 1)}.$$

1. Write the expression for $V_r(2)$.
2. Let $\varepsilon > 0$ and $r > \varepsilon$ and consider the following game. You throw a coin of radius ε onto a table on which a circle of radius r has been drawn. If the coin lands inside the circle, without touching the boundary, then you win. Otherwise you lose. Show that the probability you win is

$$\mathbb{P}[\text{You Win}] = \left(1 - \frac{\varepsilon}{r}\right)^2.$$

3. Using this show that, in p dimensions, the fraction of the volume $V_r(p)$ for the portion $(r - \varepsilon, r)$ is
$$\delta = 1 - \left(1 - \frac{\varepsilon}{r}\right)^p.$$

Exercise 1.2. The website `http://www.no-free-lunch.org/` is home to a long list of contributions discussing the No-Free-Lunch Theorem (NFLT) introduced in Wolpert and Macready (1995). Applied to the field of combinatorial optimization, the NFLT states that

> *... all algorithms that search for an extremum of a cost function perform exactly the same, when averaged over all possible cost functions. In particular, if algorithm A outperforms algorithm B on some cost functions, then loosely speaking there must exist exactly as many other functions where B outperforms A.*

In other words,

> *over the set of all mathematically possible problems, each search algorithm will do on average as well as any other. This is due to the bias in each search algorithm, because sometimes the assumptions that the algorithm makes are not the correct ones.*

Ho and Pepyne (2002) interpret the No Free Lunch Theorem as meaning that

> *a general-purpose universal optimization strategy is theoretically impossible, and the only way one strategy can outperform another is if it is specialized to the specific problem under consideration.*

In the supervised machine learning context, Wolpert (1992) presents the NFLT through the following assertion:

> *This paper proves that it is impossible to justify a correlation between reproduction of a training set and generalization error off of the training set using only a priori reasoning. As a result, the use in the real world of any generalizer which fits a hypothesis function to a training set (e.g., the use of back-propagation) is implicitly predicated on an assumption about the physical universe.*

1. Give a mathematical formalism for the NFLT.
2. Visit the website and read the contributions on this topic.
3. How large is the range of formalisms for the NFLT? Do some seem more reasonable than others?
4. Construct arguments for or against this result.

Exercise 1.3. Write the Kth order term in a multivariate polynomial in p dimensions as

$$\sum_{i_1=1}^{p} \sum_{i_2=1}^{p} \cdots \sum_{i_K=1}^{p} a_{i_1 i_2 \cdots i_K} x_{i_1} x_{i_2} \cdots x_{i_K}. \tag{1.6.1}$$

Show that (1.6.1) can expressed in the form

$$\sum_{i_1=1}^{p} \sum_{i_2=1}^{i_1} \cdots \sum_{i_K=1}^{i_{K-1}} a_{i_1 i_2 \cdots i_K} x_{i_1} x_{i_2} \cdots x_{i_K}. \tag{1.6.2}$$

Hint: *Use the redundancy in some of the* $a_{i_1 i_2 \cdots i_K}$*s in* (1.6.1) *to obtain* (1.6.2).

Exercise 1.4. Let $\mathscr{D} = \{(X_i, Y_i), i = 1, \cdots, n\}$ be an IID sample of size n arising from an underlying function f. Consider a loss function $\ell(\cdot, \cdot)$ and an estimator $\hat{f}_n$ of f based on $\mathscr{D}$. Explain how the bootstrap can be used the estimate the generalization error

$$\mathbb{E}\left[\ell(Y_{\text{new}}, \hat{f}_n(X_{\text{new}}))\right].$$

Hint: *First clearly identify what aspect of the expression creates the need for techniques like resampling, then explain how bootstrapping helps provide an approximate solution to the problem.*

Exercise 1.5. Let θ be a parameter for which an estimate is sought. The standard two-sided $100(1-\alpha)\%$ confidence interval for θ is given by

$$[\hat{\theta}_n - q_{1-\alpha/2}, \hat{\theta}_n - q_{\alpha/2}],$$

where q_α is the α-quantile of $\hat{\theta}_n - \theta$. Note that calculating the confidence interval rests on knowing compute the quantiles, which requires knowing the distribution of $\hat{\theta}_n - \theta$. In practice, however, the distribution of $\hat{\theta}_n - \theta$ is unknown.

1. Explain how the bootstrap can be used to generate an interval with approximate confidence $1 - \alpha$.
2. Simulate $n = 100$ IID observations from $X_i \sim N(9, 2^2)$, and consider estimating μ from $X_1, \cdots, X_{100}$.

   a. Give an exact 95% confidence interval for μ.

   b. Use the bootstrap to give a 95% confidence interval for μ.

Exercise 1.6. Let $\mathscr{D} = \{(X_i, Y_i), i = 1, \cdots, n\}$ be an IID sample of size n arising from a simple linear regression through the origin,

$$Y_i = \beta \boldsymbol{x}_i + \varepsilon_i.$$

Inferential tasks related to this model require knowledge of the distribution of $\sqrt{n}(\hat{\beta}_n - \beta)$. Often, the noise terms ε_is are taken as IID $N(0, \sigma^2)$ so that $\sqrt{n}(\hat{\beta}_n - \beta)$ is distributed as a $\mathbf{N}\left(0, \sigma^2 / \sum_{i=1}^{n} (x_i - \bar{x})^2\right)$. In practice, however, this distribution is not known, and approximate techniques are used to obtain summary statistics.

1. Describe a bootstrap approach to making inferences about β.
2. Consider $\sqrt{n}(\hat{\beta}_n^* - \hat{\beta}_n)$, the bootstrap approximation of $\sqrt{n}(\hat{\beta}_n - \beta)$. Since the bootstrap is consistent, the consistency of both bias and variance estimators holds. That

is,

$$\frac{\mathbb{E}^*[\hat{\beta}_n^*] - \hat{\beta}_n}{\mathbb{E}[\hat{\beta}_n] - \beta} \xrightarrow{P} 1$$

and

$$\frac{\mathrm{Var}^*(\hat{\beta}_n^*)}{\mathrm{Var}(\hat{\beta}_n)} \xrightarrow{P} 1.$$

a. Simulate from $Y_i = 2\pi \boldsymbol{x}_i + \varepsilon_i$, where $\boldsymbol{x}_i \in [-1, 1]$ and $\varepsilon_i \sim \mathsf{N}(0, .5^2)$.

b. Estimate β.

c. Develop and discuss a computational verification of the consistency of the bootstrap bias and variance.

Exercise 1.7. Let $\mathscr{D} = \{(X_i, Y_i), i = 1, \cdots, n\}$ be an IID sample of size n arising from the nonparametric regression model

$$Y_i = f(\boldsymbol{x}_i) + \varepsilon_i,$$

where f is an underlying real-valued function of a real variable defined on a domain $\mathscr{X}$. Suppose a nonparametric smoothing technique is used to construct an estimate $\hat{f}$ of f. Write $\hat{\boldsymbol{f}}_n = (\hat{f}(\boldsymbol{x}_1), \hat{f}(\boldsymbol{x}_1), \cdots, \hat{f}(\boldsymbol{x}_n))^\mathsf{T} = (\hat{f}_1, \hat{f}_2, \cdots, \hat{f}_n)^\mathsf{T}$ to be the vector of evaluations of the estimator at the design points. Most smoothers that will be studied in Chapters 2 and 3 are linear in the sense that

$$\hat{\boldsymbol{f}}_n = \boldsymbol{H}\boldsymbol{y},$$

where $\boldsymbol{H} = (h_{i,j})_{n \times n}$ is a square matrix whose elements $h_{i,j}$ are functions of both the explanatory variables $\boldsymbol{x}_i$ and the smoothing procedure used. Note that the h_{ii}s are the ith diagonal elements of $\boldsymbol{H}$. Show that for linear smoothers

$$n^{-1} \sum_{i=1}^{n} [y_i - \hat{f}_{n-1}^{(-i)}(\boldsymbol{x}_i)]^2 = n^{-1} \sum_{i=1}^{n} \left[\frac{y_i - \hat{f}_n(\boldsymbol{x}_i)}{1 - h_{ii}} \right]^2.$$

Hint: *Recognize that* $\boldsymbol{y} - \hat{\boldsymbol{f}}_n = (\boldsymbol{I} - \boldsymbol{H})\boldsymbol{y}$ *and that* $\hat{\boldsymbol{f}}_n = \hat{\boldsymbol{f}}_{n-1}^{(-i)} + f_i \boldsymbol{e}_i$, where $\boldsymbol{e}_i$ is the unit vector with only the ith coordinate equal to 1.

Exercise 1.8. The World Wide Web is rich in statistical computing resources. One of the most popular with statisticians and machine learners is the package R. It is free, user-friendly, and can be download from `http://www.r-project.org`. The software package MATLAB also has a wealth of statistical computing resources, however, MATLAB is expensive. A free emulator of MATLAB called OCTAVE can be downloaded from the web. There are also many freely available statistical computing libraries for those who prefer to program in C.

1. Download your favorite statistical computing package, and then install it and get acquainted with its functioning.

2. Implement the cross-validation technique on some simple polynomial regressions to to select the model with the lowest prediction error. For instance, set

$$Y = 1 - 2x + 4x^3 + \varepsilon,$$

where $\varepsilon \sim N(0,1)$, to be a true model. Let $x \in [-1,1]$ and suppose the design points x_i are equally spaced.

a. Generate $n = 200$ data points (x_i, y_i).
b. Perform cross-validation to compute the estimate of the prediction error for each of the following candidate models: $M1 : Y = \beta_0 + \beta_1 x + \varepsilon$, $M2 : Y = \beta_0 + \beta_1 x + \beta_2 x^2 + \varepsilon$, $M3 : Y = \beta_0 + \beta_1 x + \beta_2 x^2 + \beta_3 x^3 + \varepsilon$, $M4 : Y = \beta_0 + \beta_1 x + \beta_2 x^2 + \beta_3 x^3 + \beta_4 x^4 + \varepsilon$.
c. Which model does the technique select? Are you satisfied with the performance? Explain.

Exercise 1.9. Consider a unimodal function g defined on an interval $[a,b]$. Suppose your goal is to find the point x^* in $[a,b]$ where g achieves its maximum. The Fibonacci approach for finding x^* consists of constructing successive subintervals $[a_n, b_n]$ of $[a,b]$ that zero in ever closer on x^*. More specifically, starting from $[a_0, b_0] = [a,b]$, successive subintervals $[a_n, b_n]$ are constructed such that

$$a_{n+1} - a_n = b_n - b_{n+1} = \rho_n (b_n - a_n).$$

The gist of the Fibonacci search technique lies in using the classical Fibonacci sequence as a device for defining the sequence $\{\rho_n\}$. Recall that the Fibonacci sequence is defined as the sequence $F_1, F_2, F_3, \cdots$ such that $\forall\, n \geq 0$

$$F_{n+1} = F_n + F_{n-1}.$$

By convention, $F_{-1} = 0$ and $F_0 = 1$.

1. Show that, for $n \geq 2$,
$$F_{n-2}F_{n+1} - F_{n-1}F_n = (-1)^n.$$
2. Show that
$$F_n = \frac{1}{\sqrt{5}} \left[\left(\frac{1+\sqrt{5}}{2} \right)^{n+1} - \left(\frac{1-\sqrt{5}}{2} \right)^{n+1} \right].$$
3. From the definition of the Fibonacci sequence above, one can define another sequence, $\rho_1, \rho_2, \cdots, \rho_k$, where
$$\rho_1 = 1 - \frac{F_k}{F_{k+1}}, \quad \rho_2 = 1 - \frac{F_{k-1}}{F_k}, \cdots, \quad \rho_n = 1 - \frac{F_{k-n+1}}{F_{k-n+2}}, \cdots, \quad \rho_k = 1 - \frac{F_1}{F_2}.$$

   a. Show that, for each $n = 1, \cdots, k$,
$$0 \leq \rho_n \leq 1/2.$$

b. Show that, for each $n = 1, \cdots, k-1$,

$$\rho_{n+1} = 1 - \frac{\rho_n}{1-\rho_n}.$$

c. Reread the description of the Fibonacci search technique, and explain how this sequence of ρ_n applies to it.

Exercise 1.10. Consider using

$$\theta^{(k+1)} = \theta^{(k)} - \alpha^{(k)} \left[\mathrm{H}(\theta^{(k)})\right]^{-1} \mathbf{g}^{(k)},$$

where

$$\alpha^{(k)} = \arg \min_{\alpha \geq 0} f\left(\theta^{(k)} - \alpha \left[\mathrm{H}(\theta^{(k)})\right]^{-1} \mathbf{g}^{(k)}\right),$$

$\mathbf{g}^{(k)} = \nabla f(\theta^{(k)})$ is the gradient, and $\mathrm{H}(\theta^{(k)})$ is the Hessian matrix of f evaluated at the current point $\theta^{(k)}$ to find the minimum of a twice-differentiable function$f(\theta)$. This is called the modified Newton's algorithm because the updating scheme in the original Newton's algorithm is simply

$$\theta^{(k+1)} = \theta^{(k)} - \left[\mathrm{H}(\theta^{(k)})\right]^{-1} \mathbf{g}^{(k)},$$

which clearly does not have the learning rate $\alpha^{(k)}$. Apply the modified Newton's algorithm to the quadratic function:

$$f(\theta) = \frac{1}{2}\theta^{\mathsf{T}} Q \theta - \theta^{\mathsf{T}} b \quad \text{where} \quad Q = Q^{\mathsf{T}} > 0.$$

Recall that, for quadratic functions, the standard Newton's method reaches the point θ^* such that $\nabla f(\theta^*) = 0$ in just one step starting from any initial point $\theta^{(0)}$.

1. Does the modified Newton's algorithm possess the same property?
2. Justify your answer analytically.

Exercise 1.11. Consider Rosenbrock's famous *banana valley* function

$$f(x_1, x_2) = 100(x_2 - x_1^2)^2 + (1 - x_1)^2.$$

Using your favorite software package (MATLAB, R, or even C):

1. Plot f, and identify its extrema and their main characteristics.
2. Find the numerical value of the extremum of this function. *You may use any of the techniques described earlier, such as Newton-Raphson or modified Newton (Exercise 1.10).*
3. Consider the following widely used optimization technique, called gradient descent/ascent, which iteratively finds the point at which the function $f(\theta)$ reaches

its optimum by updating the vector $\theta = (x_1, x_2)^\top$. The updating formula is

$$\theta^{(k+1)} = \theta^{(k)} - \alpha^{(k)} \nabla f(\theta^{(k)}),$$

where $\nabla f(\theta^{(k)})$ is the gradient and $\alpha^{(k)}$ is a positive scalar known as the *step size* or *learning rate* used to set the magnitude of the move from $\theta^{(k)}$ to $\theta^{(k+1)}$. In the version of gradient descent known as steepest descent, $\alpha^{(k)}$ is chosen to maximize the amount of decrease of the objective function at each iteration,

$$\alpha^{(k)} = \arg \min_{\alpha \geq 0} f(\theta^{(k)} - \alpha \nabla f(\theta^{(k)})).$$

Apply it to the *banana valley* function, and compare your results to those from 1 and 2.

Exercise 1.12. Let $\ell(\cdot,\cdot)$ be a loss function and let $\hat{f}_{n-1}^{(-i)}(X_i)$ be an estimate of a function f using the deleted data; i.e., formed from the data set $\{(x_i, y_i) \mid i = 1, \cdots, n\}$ by deleting the i data point.

1. Show that the variance of the leave-one-out CV error is

$$\mathrm{Var}\left(\frac{1}{n} \sum_{j=1}^{n} \ell(Y_i, \hat{f}_{n-1}^{(-i)}(X_i)) \right) = \frac{1}{n^2} \sum_{i=1}^{n} \sum_{i=1}^{n} \mathrm{Cov}(\ell(Y_i, \hat{f}_{n-1}^{(-i)}(X_i)), \ell(Y_j, \hat{f}_{n-1}^{(-j)}(X_j))).$$

2. Why would you expect

$$\mathrm{Cov}(\ell(Y_i, \hat{f}_{n-1}^{(-i)}(X_i)), \ell(Y_j, \hat{f}_{n-1}^{(-j)}(X_j)))$$

to be typically large? **Hint**: Even though the data points are mutually independent, does it follow that functions of them are?

3. Now consider the bias of leave-one-out CV: Do you expect it to be low or high? Give an intuitive explanation.

Exercise 1.13. One limitation of K-fold CV is that there are many ways to partition n data points into K equal subsets, and K-fold CV only uses one of them. One way around this is to sample k data points at random, use them as a holdout set for testing. Sampling with replacement allows this procedure, called leave-k-out CV, to be repeated many times. As suggested by the last Exercise, leave-one-out CV can be unstable so leave-k-out CV for $k \geq 2$, may give better performance.

1. How would you expect the variance and bias of leave-k-out CV to behave? Use this to suggest why it would be preferred over leave-one-out CV if $k \geq 2$.
2. How many possible choices are there for a random sample of size k from $\mathscr{D}$?
3. Let $\ell(\cdot,\cdot)$ be a loss function and set $q = \binom{n}{k}$. What does the formula

$$\frac{1}{q}\sum_{s=1}^{q}\frac{1}{k}\sum_{i\in\mathscr{D}_s}\ell\left(Y_i,\hat{f}_{n-k}^{(-\mathscr{D}_s)}(X_i)\right) \tag{1.6.3}$$

compute? In (1.6.3), $\mathscr{D}_s$ denotes a sample of size k drawn from $\mathscr{D}$.

4. Briefly explain the main advantage leave-k-out CV has over the K-fold CV.
5. What is the most obvious computational drawback of the leave-k-out CV formula in item 3? Suggest a way to get around it.

Exercise 1.14. It will be seen in Chapter 2 that the best possible mean squared error (MSE) rate of the nonparametric density estimator in p dimensions with a sample of size n is

$$O\left(n^{-4/(4+p)}\right).$$

1. Compute this rate for $d = 1$ and $n = 100$.
2. Construct an entire table of similar MSE rates for $d = 1,2,5,10,15$ and $n = 100,1000,10000,100000$.
3. Compare the rates for $(d = 1, n = 100)$ and $(d = 10, n = 10000)$, and provide an explanation in light of the Curse of Dimensionality.
4. Explain why density estimation is restricted to $d = 2$ in practice.

Exercise 1.15. Let $\mathscr{H} = \{h_1, h_2, \cdots, h_p\}$ be a set of basis functions defined on a domain $\mathscr{X}$. Consider the basis function expansion

$$f(x) = \sum_{j=1}^{p}\beta_j h_j(x)$$

widely used for estimating the functional dependencies underlying a data set $\mathscr{D} = \{(x_1,y_1),\cdots,(x_n,y_n)\}$.

1. Provide a detailed explanation of how the Curse of Dimensionality arises when the set of basis functions is fixed, i.e. the h_is are known prior to collecting the data and remain fixed throughout the learning process.
2. Explain why the use of an adaptive set of basis functions – i.e., possibly rechoosing the list of h_is at each time step – has the potential of evading the Curse of Dimensionality.

Exercise 1.16. It has been found experimentally that leave-one-out CV also referred to as LOOCV is asymptotically suboptimal. In the context of feature selection, for instance, it could select a suboptimal subset of features even if the sample size was infinite.

Explore this fact computationally and provide your own insights as to why this is the case. **Hint**: *Set up a simulation study with p explanatory variables. Consider the case of a truly p-dimensional function and also consider the case where the p dimensions arise from one single variable along with its transforms like x and x^2, and x^p. For* *example, contrast the genuinely three-dimensional setup using x_1, x_2, and x_3 with the setup using x_1, x_1^2, and $x_1^3 + 2x_1^2$. Compare the two and see whether LOOCV does or does not yield a suboptimal set of variables. Imagine, for instance, x_1 and x_1^3 in the interval $[-1,1]$ or $[0,1]$, and note that in this interval the difference may be too small to be picked up by a naive technique.*

Exercise 1.17. It is commonly reported in Machine Learning circles that Ronald Kohavi and Leo Breiman independently found through experimentation that 10 is the best number of "folds" for CV.

1. Consider an interesting problem and explore a variety of "folds" on it, including of course the 10-fold CV. You may want to explore benchmarks problem like the Boston Housing data set in multiple regression or the Piman Indian diabetes data set in classification, since the best performances of learning machines for these tasks can be found on the Web – at the University of California-Irvine data set repository, for example. Is it clear in these benchmark problems that the 10-fold CV yields a better model?
2. Explore the properties of the different folds through a simulation study. Consider regression with orthogonal polynomials, for instance under different sample sizes, and different model complexities. Then perform CV with k smaller than 10, $k = 10$, and k larger than 10 folds.

   a. Do you notice any regularity in the behavior of the estimate of the prediction error as a function of both the number of folds and the complexity of the task?
   b. Is it obvious that smaller folds are less stable than larger ones?
   c. Whether or not you are convinced by the 10-fold CV, could you suggest a way of choosing the number of folds optimally?

Exercise 1.18. Let $\mathscr{H} = \{h_1, h_2, \cdots, h_p\}$ be a set of basis functions defined on a domain $\mathscr{X}$. Consider the basis function expansion

$$f(x) = \sum_{j=1}^{p} \beta_j h_j(x)$$

and the data set $\mathscr{D} = \{(x_1, y_1), \cdots, (x_n, y_n)\}$.

1. Explain what the concept of generalization means in this context. Particularly discuss the interrelationships among model complexity, sample size, model selection bias, bias–variance trade-off, and the choice of $\mathscr{H}$. Are there other aspects of function estimation that should be included?
2. Provide a speculative (or otherwise) discussion on the philosophical and technical difficulties inherent in the goal of generalization.
3. How can one be sure that the estimated function based on a given sample size is getting close to the function that would be obtained in the limit of an infinite sample size?

Exercise 1.19. Cross-validation was discussed earlier as a technique for estimating the prediction error of a learning machine. In essence, the technique functions in two separate steps.

1. Indicate the two steps of cross-validation.
2. Discuss why having two separate steps can be an impediment to the learning task.
3. Cross-validation and generalized cross-validation are often used in regularized function estimation settings as a technique for estimating the tuning parameter. Although GCV provides an improvement over CV in terms of stability, the fact that both are two-step procedures makes them less appealing than any procedure that incorporates the whole analysis into one single sweep. In the Bayesian context, the tuning parameter that requires GCV for its estimation is incorporated into the analysis directly through the prior distribution.

    a. In the regression context, find the literature that introduces the treatment of the tuning parameter as a component of the prior distribution.
    b. Explain why such a treatment helps circumvent the problems inherent in CV and GCV.

Exercise 1.20. Consider once again the task of estimating a function f using the data set $\mathscr{D} = \{(x_1, y_1), \cdots, (x_n, y_n)\}$. Explain in your own words why there isn't any single "one size fits all model" that can solve this problem.

Hint: A deeper understanding of the No Free Lunch Theorem should provide you with solid background knowledge for answering this question. To see how, do the following:

- Find two problems and call the first one A and the second B. For example, let A be a prediction problem and B be a hypothesis-testing problem.
- Our question is whether or not there is a strategy, call it S^*, that does optimally well on both A and B. What one knows in this context is that there is a strategy P that is optimal for A but that performs poorly on B and a strategy T that is optimal for B but performs poorly on A.

    1. Search the literature to find what P and T are. Be sure to provide the authors and details of their findings on the subject.
    2. Is there a way to combine P and T to find some S^* that performs optimally on both A and B?
    3. If the answer to question 2 is no, can you find in the literature or construct yourself two qualitatively different tasks and one single strategy that performs optimally on both of them?

Chapter 2
Local Smoothers

Nonparametric methods in DMML usually refers to the use of finite, possibly small data sets to search large spaces of functions. Large means, in particular, that the elements of the space cannot be indexed by a finite-dimensional parameter. Thus, large spaces are typically infinite-dimensional – and then some. For instance, a Hilbert space of functions may have countably many dimensions, the smallest infinite cardinal number, $\aleph_0$. Other spaces, such as the Banach space of bounded functions on $[0,1]$ under a supremum norm, $L^\infty[0,1]$, have uncountably many dimensions, $\aleph_1$, under the continuum hypothesis. Spaces of functions containing a collection of finite-dimensional parametric families are also called nonparametric when the whole space equals the closure of the set of parametric families and is infinite-dimensional. By construction, it is already complete. Usually, the dimension of the parametric families is unbounded, and it is understood that the whole space is "reasonable" in that it covers a range of behavior believed to contain the true relationship between the explanatory variables and the dependent variables.

This is in contrast to classical nonparametrics including ranking and selection, permutation tests, and measures of location, scale and association, whose goal is to provide good information about a parameter independent of the underlying distributions. These methods are also, typically, for small sample sizes, but they are fundamentally intended for finite-dimensional parametric inference. That is, a sample value of, say, Spearman's ρ or Kendall's τ is an estimator of its population value, which is a real number, rather than a technique for searching a large function space. Moreover, these statistics often satisfy a CLT and so are amenable to conventional inference. Although these statistics regularly occur in DMML, they are conceptually disjoint from the focus here.

Roughly speaking, in DMML, nonparametric methods can be grouped into four categories. Here, by analogy with the terms used in music, they are called Early, Classical, New Wave, and Alternative. Early nonparametrics is more like data summarization than inference. That is, an Early nonparametric function estimator, such as a bin smoother, is better at revealing the picture a scatterplot is trying to express than it is for making inferences about the true function or for making predictions about future outcomes. The central reason is that no optimization has occurred and good properties

B. Clarke et al., *Principles and Theory for Data Mining and Machine Learning*, Springer Series in Statistics, DOI 10.1007/978-0-387-98135-2_2, © Springer Science+Business Media, LLC 2009

cannot necessarily be assumed. So, even if the bin smoother is good for inference and prediction, that cannot be determined without further work.

This central limitation of Early methods is corrected by Classical methods. The key Classical methods in this chapter are LOESS, kernels, and nearest neighbors. LOESS provides a local polynomial fit, generalizing Early smoothers. The idea behind kernel methods is, for regression, to put a bump at each data point, which is achieved by optimization. A parameter is chosen as part of an optimization or inference to ensure the function being fit matches the scatterplot without over- or underfitting: If the match is too close, it is interpolation: if it is too far, one loses details. In kernel methods, the parameter is the bandwidth, or h, used for smoothing. Finite data sets impose severe limits on how to search nonparametric classes of functions. Nearest-neighbor methods try to infer properties of new observations by looking at the data points already accumulated that they resemble most.

LOESS is intermediate between Early and Classical smoothers; kernel methods and nearest neighbors are truly Classical. Another Classical method is splines, which are covered in Chapter 3. Chapter 4 provides an overview of many of the main techniques that realistically fall under the heading of New Wave nonparametrics; some of these – generalized additive models for instance – are transitions between Classical and truly New Wave. The key feature here is that they focus on the intermediate tranche of modeling. Alternative methods, such as the use of ensembles of models, are taken up in Chapter 6; they are Alternative in that they tend to combine the influences of many models or at least do not focus exclusively on a single true model.

While parametric regression imposes a specific form for the approximating function, nonparametric regression implicitly specifies the class the approximand must lie in, usually through desirable properties. One property is smoothness, which is quantified through various senses of continuity and differentiability. The smoothest class is linear. Multiple linear regression (MLR) is one ideal for expressing a response in terms of explanatory variables. Recall that the model class is

$$Y = \beta_0 + \beta_1 X_1 + \ldots + \beta_p X_p + \varepsilon, \tag{2.0.1}$$

where the εs are IID $N(0, \sigma^2)$, independent of $x_1, \ldots, x_p$, and usually have normal distributions. The benefits of multiple regression are well known and include:

- MLR is interpretable – the effect of each explanatory variable is captured by a single coefficient.
- Theory supports inference for the β_is, and prediction is easy.
- Simple interactions between X_i and X_j are easy to include.
- Transformations of the X_is are easy to include, and dummy variables allow the use of categorical information.
- Computation is fast.

The structure of the model makes all the data relevant to estimating all the parameters, no matter where the data points land. However, in general, it may be unreasonable to permit x_is far from some other point x to influence the value of $Y(\boldsymbol{x})$. For instance,

$\bar{\boldsymbol{x}}$ is meaningful for estimating the β_is in (2.0.1) but may not be useful for estimating $Y(\boldsymbol{x})$ if the right-hand side is a general, nonlinear function of $\boldsymbol{X}$, as may be the case in nonparametric settings. Nevertheless, nonparametric regression would like to enjoy as many of the properties of linear regression as possible, and some of the methods are rather successful.

In fact, all the Early and Classical methods presented here are local. That is, rather than allowing all data points to contribute to estimating each function value, as in linear regression, the influence of the data points on the function value depends on the value of $\boldsymbol{x}$. Usually, the influence of a data point $\boldsymbol{x}_i$ is highest for those $\boldsymbol{x}$s close to it and its influence diminishes for $\boldsymbol{x}$s far from it. In fact, many New Wave and Alternative methods re local as well, but the localization is typically more obvious with Early and Classical methods.

To begin, a smoothing algorithm takes the data and returns a function. The output is called a smooth ; it describes the trend in Y as a function of the explanatory variables $X_1,\ldots,X_p$. Essentially, a smooth is an estimate $\hat{f}$ of f in the nonparametric analog of (2.0.1),

$$Y = f(\boldsymbol{x}) + \varepsilon, \tag{2.0.2}$$

in which the error term ε is IID, independent of f and x, with some symmetric, unimodal distribution, sometimes taken as $N(0,\sigma^2)$.

First let $p = 1$, and consider scatterplot smooths. These usually generalize to $p = 2$ and $p = 3$, but the Curse quickly renders them impractical for larger dimensions. As a running example for several techniques, assume one has data generated from the function in Fig. 2.1 by adding $N(0,.25)$ noise. That is, the function graphed in Fig. 2.1 is an instance of the f in (2.0.1), and the scatter of Ys seen in Fig. 2.2 results from choosing an evenly spaced collection of x-values, evaluating $f(x)$ for them, and generating random εs to add to the $f(x)$s. Next, pretend we don't know f and that the ε_is are unavailable as well. The task is to find a technique that will uncover f using only the y_is in Fig. 2.2.

This chapter starts with Early methods to present the basics of descriptive smoothers. Early, simple smoothing algorithms provide the insight, and sometimes the building blocks, for later, more sophisticated procedures. Then, the main Classical techniques, LOESS, kernel and nearest neighbors in this chapter and spline regression in the next, are presented.

2.1 Early Smoothers

Three Early smoothers were bin, running line, and moving average smoothers. For flexibility, they can be combined or modified by adjusting some aspects of their construction. There are many other Early smoothers, many of them variants on those presented here, and there is some ambiguity about names. For instance, sometimes a bin

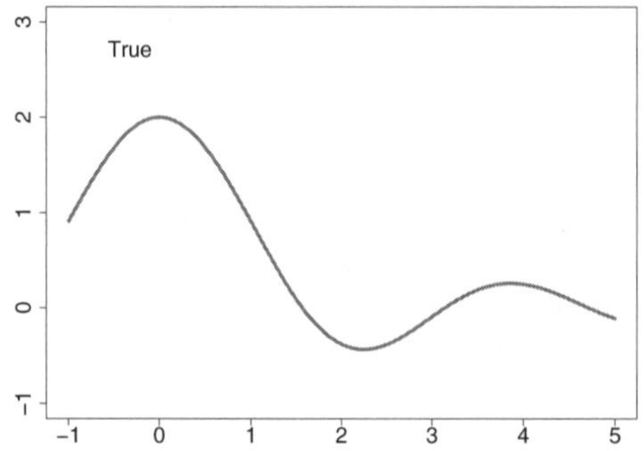

Fig. 2.1 This graph shows the true function $f(x)$ that some of the smoothing techniques presented here will try to find.

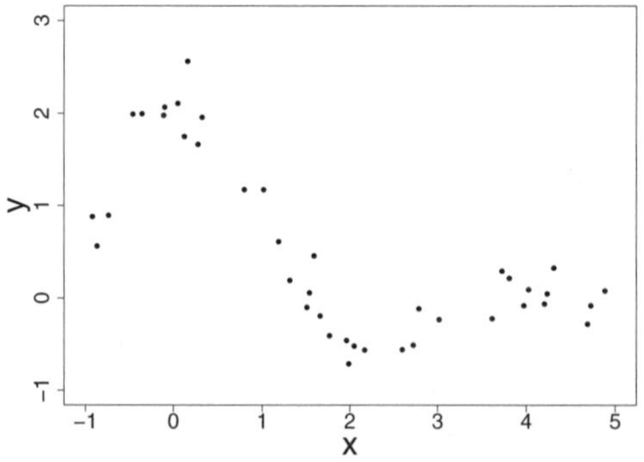

Fig. 2.2 This graph shows the simulated data generated using evenly spaced values of x on $[-1,5]$, the f from Fig. 2.1, and $\varepsilon = N(0,.25)$, IID. For several techniques, the smoothers' performances will be compared using these data as the y_is.

smoother is called a regressogram, see Tukey (1961), and sometimes a moving average smoother is called a nearest neighbors smoother, see Fix and Hodges (1951).

In bin smoothing, one partitions $\mathbb{R}^p$ into prespecified disjoint bins; e.g., for $p = 1$, one might use the integer partition $\{[i, i+1), i \in \mathscr{Z}\}$. The value of the smooth in a bin is the average of the Y-values for the $\boldsymbol{x}$-values inside that bin. For example, Fig. 2.3 shows the results of applying the integer partition above to the data in our running example. Clearly, summarizing the data by their behavior on largish intervals is not very effective in general. One can choose smaller intervals to define the bins and get, hopefully, a closer matching to f. Even so, however, most people consider bin smoothing to be undesirably rough. However, bin smoothers are still often used when one wants to partition the data for some purpose, such as data compression in an information-theoretic context. Also, histogram estimators for densities are bin smoothers; see Yu and Speed (1992).

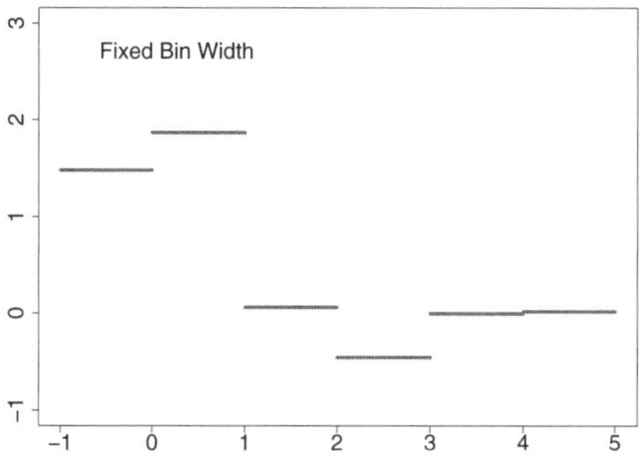

Fig. 2.3 This graph shows the bin smoother for an integer partition of the data in Fig. 2.2. It is always discontinuous, and the steps do not in general track the function values particularly well.

To improve bin smoothing, one might use variable-sized bins containing a fixed number of observations rather than fixed-width bins with a variable number of observations. This smoother is called a moving average; usually the bins are required to contain the nearest x-values – this is called a k-nearest-neighbor smoother, discussed in Section 4. However, other choices for the xs are possible. One could, for instance, take the closest x-values only on one side (making an allowance for the left boundary) or from more distant regions believed to be relevant. Moreover, instead of averaging all the closest data points with equal weights, one can weight the data points closer to x more than to those farther away. These are called weighted average smoothers. If the median is used in place of the mean for the sake of robustness, one gets the running median smoother of Tukey (1977).

Weighted or not, moving average smoothers tend to reflect the local properties of a curve reasonably well. They don't tend to look as coarse as bin smoothers, but they are still relatively rough. Figure 2.4 shows a moving average in which each variable bin contains the three nearest x-values. If one increases the number of observations within the bin above three, the plot becomes smoother.

A further improvement is the running line smoother. This fits a straight line rather than an average to the data in a bin. It can be combined with variable bin widths as in the moving average smoother to give better local matching to the unknown f. As before, one must decide how many observations a bin should contain, and larger numbers give smoother functions. Also, one can fit a more general polynomial than a straight line; see Fan and Gijbels (1996). Figure 2.5 shows the smooth using a linear fit tends to be rough. However, it is typically smoother than the bin smoother for the same choice of bins as in Fig. 2.4.

In all three of these smoothers, there is flexibility in the bin selection. Bins can be chosen by the analyst directly or determined from the data by some rule. Either way, it is of interest to choose the bins to ensure that the function is represented well.

In the context of running line smoothers, Friedman (1984) has used cross-validation (see Chapter 1) to select how many xs to include in variable-length bins.

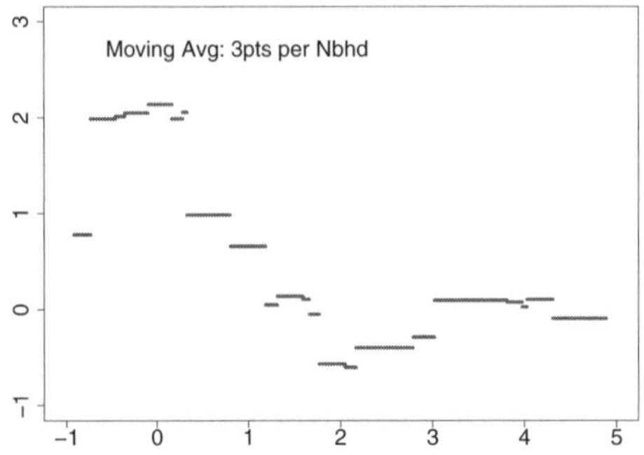

Fig. 2.4 This graph shows the moving average smoother for 3-nearest-neighbor bins and the data in Fig. 2.2. It is always discontinuous, but gives adequate local matching. Note that the values on the bins are often so close as to be indistinguishable in the figure.

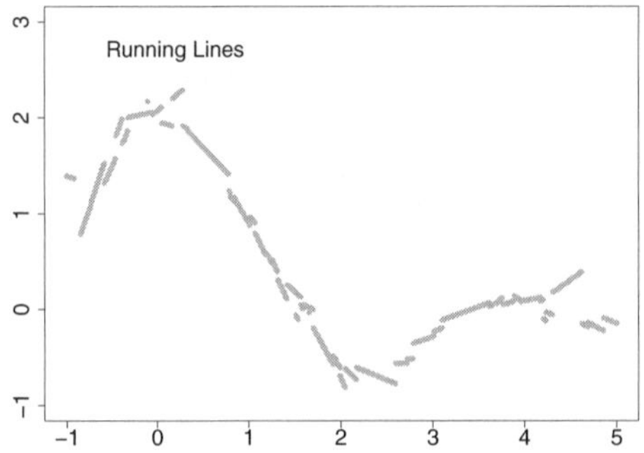

Fig. 2.5 This graph shows the running-line smoother for the 3-nearest-neighbor bins and the data in Fig. 2.2. It is smoother than the curves in Figs. 2.3 and 2.4 because it is a more flexible family, at the cost of less severe summarization inside the bins.

SuperSmoother chooses among three different numbers of observations: $n/2$, $n/5$, and $n/20$. Except near the endpoints of the domain of xs, the values $\hat{f}(x)$ are found using half of the closest observations on each side of x; this forced symmetry is different from merely using, say, the nearest $n/2$ xs, in which more or less than $(1/2)(n/2)$ may be to one side of x. The choice among the three options is made by finding $\hat{f}_1(x)$, $\hat{f}_2(x)$, and $\hat{f}_3(x)$ for the three options and then using leave-one-out cross-validation to determine which has the smallest predictive mean squared error.

Note that the smoothers exhibited so far are linear in an important sense. In squared error, it is well known that the best predictor for Y from $\boldsymbol{X}$ is $f(x) = \mathbb{E}(Y|\boldsymbol{X} = \boldsymbol{x})$. So, the smooths developed so far can be regarded as estimators of $\mathbb{E}(Y|\boldsymbol{X})$ in (2.0.2). Different from the linearity in (2.0.1), a smooth in (2.0.2) is linear if and only if

$$\hat{f}(x) = \sum_{i=1}^{n} W_i(x) y_i = L_n(x)\mathbf{y}, \tag{2.1.1}$$

in which the W_js are weights depending on the whole data set $(y_i, x_i)_{i=1}^n$ and $L_n(x)$ is a linear operator on $\mathbf{y}$ with entries defined by the W_js . In this notation, the choice of bins is implicit in the definition of the W_js. For fixed bins, for instance, $W_j(x_0)$ only depends on the xs in the bin containing x_0. In this form, it is seen that $\hat{f} = L_n\mathbf{y}$ means that $\mathrm{Var}(\hat{f}(x)) = \mathrm{Var}(L_n(x)\mathbf{y}) = L_n(x)\mathrm{Var}(\mathbf{y})L_n(x)^{\mathsf{T}}$, which is $\sigma^2 L_n(x)L_n(x)^{\mathsf{T}}$ for IID $N(0,\sigma^2)$ errors.

Just as there are many other Early smoothers, there are even more linear smoothers; several will be presented in the following sections. Linear smoothers have a common convenient expression for the MSE, bias, and variance. In particular, these expressions will permit the optimizations typical of Classical methods. This will be seen explicitly in the next section for the case of kernel estimators, which are linear. For these and other smoothers, expression (2.1.1) implies that averages of linear smoothers are again linear, so it may be natural to combine linear smoothers as a way to do nonparametric regression better.

2.2 Transition to Classical Smoothers

Early smoothers are good for data summarization; they are easy to understand and provide a good picture. Early smoothers can also be computed quickly, but that has become less important as computational power has improved.

Unfortunately, because their structure is so simplified, Early smoothers are generally inadequate for more precise goals such as estimation, prediction, and inference more generally. Some of the limitations are obvious from looking at the figures. For instance, Early smoothers do not optimize over any parameter, so it is difficult to control the complexity of the smooth they generate. Thus, they do not automatically adapt to the local roughness of the underlying function. In addition, there is no measure of bias or dispersion. Unsurprisingly, Early smoothers don't generalize well to higher dimensions.

A separate point is that Early smoothers lack mathematical theory to support their use, so it is unclear how well they quantify the information in the data. Another way to say this is that often many, many smooths appear equally good and there is no way to compare them to surmise that one curve, or region of curves, is more appropriate than another. Practitioners are often comfortable with this because it reflects the fact that more information – from data or modeling assumptions – is needed to identify the right curve. Indeed, this indeterminacy is just the sort of model uncertainty one anticipates in curve fitting. On the other hand, it is clearly desirable to be able to compare smooths reliably and formally.

At the other end of the curve-fitting spectrum from the Early smoothers of the last section is polynomial interpolation. Motivated perhaps by Taylor expansions, the initial goal was to approximate a function over its domain by a single polynomial. Recall that polynomials are just a basis for a function space; they can be orthogonalized in the L^2 inner product to give the Legendre polynomials. Thus, more generally, the goal was to use basis expansions to get a global representation for a function. The extreme case of this is interpolation, where the approximand equals the function on the available data.

In fact, global polynomial interpolation does not work well because of an unexpected phenomenon: As the degree increases, the oscillations of the approximand around the true function increase without bound. That is, requiring exact matching of function values at points forces ever worse matching away from those points. If one backs off from requiring exact matching, the problem decreases but remains. This is surprising because Taylor expansions often converge uniformly. What seems to be going on is that forcing the error term too small, possibly to zero, at a select number of points in polynomials of high enough degree (necessary for the error term to get small) creates not just a bad fit elsewhere but a bad fit resulting from ever-wilder oscillations. This is another version of bias–variance trade-off. Requiring the bias to be too small forces the variability to increase. In practice, estimating the coefficients in such an expansion will give the same problem.

2.2.1 Global Versus Local Approximations

There exists a vast body of literature in numerical analysis that deals with the approximation of a function from a finite collection of function values at specific points, one of the most important results being the following.

Weierstrass Approximation Theorem: Suppose f is defined and continuous on $[a,b]$. For each $\varepsilon > 0$, there exists a polynomial $g(x)$, defined on $[a,b]$, with the property that

$$|f(x) - g(x)| < \varepsilon, \qquad \forall x \in [a,b]. \quad \square \tag{2.2.1}$$

This theorem simply states that any continuous function on an interval can be approximated to arbitrary precision by a polynomial. However, it says nothing about the properties of g or how to find it.

Unsurprisingly, it turns out that the quality of an approximation deteriorates as the range over which it is used expands. This is the main weakness of global polynomial approximations. To see how global approximation can break down, consider univariate functions. Let $\mathscr{X} = [-1,1]$ and $f(x) : \mathscr{X} \to \mathbb{R}$ be the Runge function

$$f(x) = \frac{1}{1+25x^2}. \tag{2.2.2}$$

Let $x_1, x_2, \cdots, x_n \in \mathscr{X}$, be uniformly spaced points

$$x_i = -1 + (i-1)\frac{2}{n-1}, \qquad i = 1, \cdots, n.$$

C. D. Runge showed that if f is interpolated on the x_is by a polynomial $g_k(x)$ of degree $\leq k$, then as k increases, the interpolant oscillates ever more at the endpoints -1 and 1. This is graphed in Fig. 2.6.

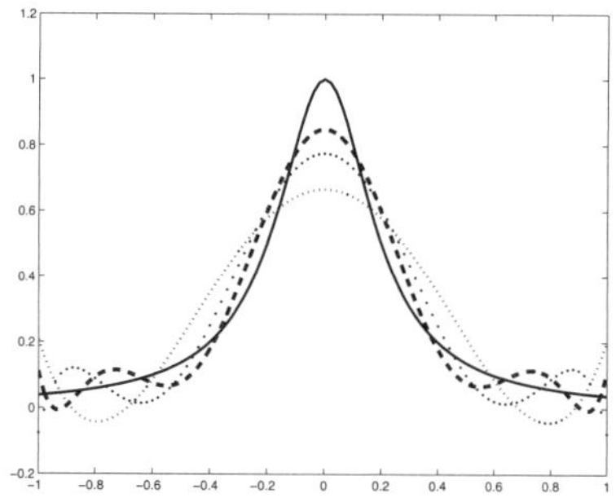

Fig. 2.6 The solid curve is the Runge function. As the dashes in the other curves get smaller, the order of the polynomial gets larger, representing the 5th-, 7th-, and 9th-degree polynomials. As the degree increases, the oscillations of the interpolant at the endpoints are seen to increase as well.

But it's worse than Fig. 2.6 shows: The interpolation error tends to infinity at the endpoints as the degree k increases; i.e.,

$$\lim_{k\to\infty}\left(\max_{-1\leq x\leq 1}\left|f(x)-g_k(x)\right|\right)=\infty. \tag{2.2.3}$$

That is, as k increases, the interpolating polynomial (quickly) gets much bigger than the function it is interpolating; see Exercise 2.3. Although Runge's phenomenon has now been seen only for one function, it clearly demonstrates that high-degree polynomials are generally unsuitable for interpolation. After all, there is nothing unusual about the shape of f.

There is a resolution to Runge's phenomenon. If global approximation won't work well, then patch together a sequence of local approximations that will be more sensitive to the local features of the underlying function, especially near the endpoints. One way to do this is by using splines – a special class of local polynomials – to be discussed in Chapter 3. Comparing Fig. 2.6 to Fig. 2.3, 2.4 and 2.5 suggests that local polynomials (splines) will outperform global polynomial interpolation on the Runge function f. It will be seen later that piecewise fitting the spline does make it a great improvement over global polynomials. (Any decent spline approximation to Runge's function in Fig. 2.6 is indistinguishable from Runge's function up to the resolution of the printer.)

Quantifying the sense in which using low-degree local polynomials with enough pieces does better than high-degree global polynomials requires some definitions. Let $f : \mathscr{X} \longrightarrow \mathbb{R}$, and consider the function space $\mathscr{F} \equiv \mathbb{R}^{\mathscr{X}}$ under the supremum norm; i.e., for $f \in \mathscr{F}$, $\|f\|_\infty \equiv \sup_{x\in\mathscr{X}} |f(x)|$. The space $\mathscr{F}$ is so vast that searching it is simply unreasonable. So, consider a linear subspace $\mathscr{G} \subset \mathscr{F}$ of dimension k with a basis $\mathscr{B}$. The space of polynomials of degree less than or equal to k is a natural choice for $\mathscr{G}$.

Let T be an operator, for instance an interpolant, acting on functions $f \in \mathscr{F}$. If, for $f \in \mathscr{F}$, $TF \in \mathscr{G}$, then a measure of the worst-case scenario is defined by the norm on T inherited from $\|\cdot\|_\infty$, given by

$$\|T\|_\infty \equiv \sup_{f \in \mathscr{F}} \frac{\|Tf\|_\infty}{\|f\|_\infty}. \tag{2.2.4}$$

The operator norm in (2.2.4) is a generalization of the largest absolute eigenvalue for real symmetric matrices; it is the biggest possible ratio of the size of Tf compared with the size of f. The larger $\|T\|_\infty$ is, the bigger the difference between some function f and its interpolant can be. Note that although this is expressed in the supremum norm $\|\cdot\|_\infty$, the properties of the norms used here hold for all norms; later, when a Hilbert space structure is assumed, the norm will be assumed to arise from an inner product. If T is linear in the sense of (2.2.4),

$$\|T\|_\infty = \sup_{\|f\|_\infty = 1} \|Tf\|_\infty.$$

(In general, the norm $\|\cdot\|_\infty$ does not arise from an inner product, so linear spaces equipped with $\|\cdot\|_\infty$ become Banach spaces rather than Hilbert spaces.)

In terms of (2.2.4), if T is the polynomial interpolant, expression (2.2.3) means that $\|T\|_\infty = \infty$. In other words, there is a sequence of functions f_j such that the norm $\|Tf_j\|_\infty$ of the interpolant is getting much larger than the norm $\|f_j\|_\infty$ of the function, as is visible in Fig. 2.6. In fact, there are many such sequences f_j.

One popular interpolant is the nth Lagrange polynomial g_n, seen to be unique in the following.

Theorem (Lagrange interpolation): If $x_0, x_1, \cdots, x_n$ are $n+1$ distinct points and f is a function whose values at these points are $y_0 = f(x_0), y_1 = f(x_1), \cdots, y_n = f(x_n)$, respectively, then there exists a unique polynomial $g(x) = g_n(x)$ of degree at most n with the property that

$$y_i = g(x_i) \qquad \text{for each } i = 0, 1, \cdots, n,$$

where

$$\begin{aligned} g(x) &= y_0\ell_0(x) + y_1\ell_1(x) + \cdots + y_n\ell_n(x) \\ &= \sum_{i=0}^{n} y_i \ell_i(x) \end{aligned} \tag{2.2.5}$$

with

$$\begin{aligned} \ell_i(x) &= \frac{(x-x_0)(x-x_1)\cdots(x-x_{i-1})(x-x_{i+1})\cdots(x-x_n)}{(x_i-x_0)(x_i-x_1)\cdots(x_i-x_{i-1})(x_i-x_{i+1})\cdots(x_i-x_n)} \\ &= \prod_{j=0, j\neq i}^{n} \frac{x-x_j}{x_i-x_j} \end{aligned} \tag{2.2.6}$$

for each $i = 0, 1, \cdots, n$. □

Because of the uniqueness of the interpolating polynomial, one can use a slightly different norm to provide a theoretical quantification of the quality of an interpolation. Specifically, one can represent the interpolating polynomial using the Lagrange interpolation basis $[\ell_0, \ell_1, \cdots, \ell_n]$ so that

$$Tf(x) = \sum_{i=0}^{n} y_i \ell_i(x)$$

and then define a new norm

$$\|T\|_\ell = \sup_{x \in \mathscr{X}} \sum_{i=1}^{n} |\ell_i(x)|.$$

Equipped with this norm, one can easily compare different interpolation operators. In the Runge function case (2.2.2), choosing $n = 16$ uniformly spaced points in $[-1,1]$ in (2.2.5) gives that $\|T\|_\ell$ for interpolating operators can be found straightforwardly: $\|T_{poly}\|_\ell \approx 509.05$ for the polynomial interpolant and $\|T_{spline}\|_\ell \approx 1.97$ for the cubic spline interpolant. The difference is huge. Moreover, it can be shown that, as n increases, $\|T_{poly}\|_\ell = \mathscr{O}(\exp(n/2))$, while $\|T\|_\ell \approx 2$ regardless of n, see Exercise 3.1.

Naturally, it is desirable not only to match a specific polynomial but to ensure that the procedure by which the matching is done will be effective for a class of functions. After all, in practice the function to be "matched" is unknown, apart from being assumed to lie in such a class. So, consider a general class $\mathscr{G}$. Since $Tf \in \mathscr{G}$, the norm $\|f - Tf\|_\infty$ cannot be less than the distance

$$\text{dist}(f, \mathscr{G}) = \inf_{g \in \mathscr{G}} \|f - g\|_\infty$$

from f to $\mathscr{G}$. Also, since $\|T\|_\infty$ is the supremum over all ratios $\|Tf\|_\infty / \|f\|_\infty$, we have

$$\|Tf\|_\infty \leq \|T\|_\infty \|f\|_\infty, \qquad f \in \mathscr{F},$$

when T has a finite norm. By the definition of interpolation, $Tg = g$ for all $g \in \mathscr{G}$, so restricted to $\mathscr{G}$, T would have a finite norm. On the other hand, if T is a linear operator, then T has a finite norm if and only if it is continuous (in the same norm, $\|\cdot\|_\infty$), and it is natural to write

$$f - Tf = f - g + Tg - Tf = (f - g) + T(g - f).$$

Therefore

$$\|f - Tf\|_\infty \leq \|f - g\|_\infty + \|T\|_\infty \|g - f\|_\infty = (1 + \|T\|_\infty)\|f - g\|_\infty$$

for any $g \in \mathscr{G}$. Choosing g to make $\|f - g\|_\infty$ as small as possible gives the Lebesgue inequality: For $f \in \mathscr{F}$,

$$\text{dist}(f, \mathscr{G}) \leq \|f - Tf\|_\infty \leq (1 + \|T\|_\infty)\text{dist}(f, \mathscr{G}).$$

This means that when the norm $\|T\|_\infty$ of an interpolation operator is small, the interpolation error $\|f - Tf\|_\infty$ is within an interpretable factor of the best possible error. From this, it is seen that there are two aspects to good function approximation:

☐ The approximation process T should have a small norm $\|T\|_\infty$.

☐ The distance $\mathrm{dist}(f, \mathscr{G})$ of f from $\mathscr{G}$ should be small.

Note that these two desiderata are stated with the neutral term approximation. Interpolation is one way to approximate, more typical of numerical analysis. In statistics, however, approximation is generally done by function estimation. In statistical terms, these two desiderata correspond to another instance of bias–variance trade-off. Indeed, requiring a "small norm" is akin to asking for a small bias, and requiring a small distance is much like asking for a small variance. As a generality, when randomness is taken into account, local methods such as kernel and cubic spline smoothers will tend to achieve these goals better than polynomials.

2.2.2 LOESS

In the aggregate, the limitations of Early smoothers and the problems with global polynomials (or many other basis expansions) motivated numerous innovations to deal with local behavior better. Arguably, the most important of these are LOESS, kernel smoothing, and spline smoothing. Here, LOESS (pronounced LOW-ESS) will be presented, followed by kernel methods in the next section and splines in the next chapter.

Overall, LOESS is a bridge between Early and Classical smoothers, leaning more to Classical: Like Early smoothers, it is descriptive and lacks optimality, but like Classical smoothers it is locally responsive and permits formal inferences. Locally weighted scatterplot smoothing (i.e., LOESS) was developed by Cleveland (1979) and Cleveland and Devlin (1988). More accurately, LOESS should be called locally weighted polynomial regression. In essence, LOESS extends the running line smooth by using weighted linear regression in the variable-width bins. A key strength of LOESS is its flexibility because it does not fit a closed-form function to the data. A key weakness of LOESS is its flexibility because one does not get a convenient closed-form function.

Informally, the LOESS procedure is as follows. To assign a regression function value to each $\boldsymbol{x}$ in the domain, begin by associating a variable-length bin to it. The bin for $\boldsymbol{x}$ is defined to contain the q observations from the points in the data $\boldsymbol{x}_1, \ldots, \boldsymbol{x}_n$ that are closest to $\boldsymbol{x}$. On this bin, use the q data points to fit a low-degree polynomial by locally weighted least squares. The closer an $\boldsymbol{x}_i$ in the bin is to $\boldsymbol{x}$, the more weight it gets in the optimization. After finding the coefficients in the local polynomial, the value of the regression function $\hat{Y}(\boldsymbol{x})$ is found by evaluating the local polynomial from the bin at $\boldsymbol{x}$. It is seen that LOESS can be computationally demanding; however, it is often satisfactorily smooth, and for reasonable choices of its inputs q and the weights, tracks the unknown curve without overfitting or major departures. As will be seen in later sections, this kind of method suffers the Curse of Dimensionality. However, for low-dimensional problems, local methods work well.

There are three ways in which LOESS is flexible. First, the degree of the polynomial model can be any integer, but using a polynomial of high degree defeats the purpose. Usually, the degree is 1, 2, or 3. In actuality, there is no need to limit the local functions to polynomials, although polynomials are most typical. Any set of functions that provides parsimonious local fits will work well. Second, the choice of how to weight distances between $\boldsymbol{x}$ and $\boldsymbol{x}_i$ is also flexible, but the tri-cube weight (defined below) is relatively standard. Third, the number of data points in the bins ranges from 1 to n, but $n/4 \leq q/n \leq 1/2$ is fairly typical.

More formally, recall that weighted least squares finds

$$\hat{\boldsymbol{\beta}} = \arg \min_{\boldsymbol{\beta}\in\mathbb{R}^p} \sum_{i=1}^{n} w_i(Y_i - \boldsymbol{X}_i^{\mathsf{T}}\boldsymbol{\beta})^2, \tag{2.2.7}$$

in which w_i is a weight function, often derived from the covariance matrix of the ε_is. This $\hat{\boldsymbol{\beta}}$ is used in the model

$$\widehat{\mathbb{E}(Y)}(\boldsymbol{x}) = \boldsymbol{x}^{\mathsf{T}}\hat{\boldsymbol{\beta}} \tag{2.2.8}$$

for inference and prediction.

By contrast, LOESS replaces the w_i with a function $w(\boldsymbol{x})$ derived from the distances between the $\boldsymbol{x}_i$s and $\boldsymbol{x}$ for the q $\boldsymbol{x}_i$s closest to $\boldsymbol{x}$ and set to zero for the $n-q$ $\boldsymbol{x}_i$s furthest from $\boldsymbol{x}$. To see how this works for a value of $\boldsymbol{x}$, say $\boldsymbol{x}_k$ from the data set (for convenience), write $d_i = ||\boldsymbol{x}_k - \boldsymbol{x}_i||$ using the Euclidean norm and sort the d_is into increasing order. Fix α to be the fraction of data included in a bin so that $q = \max(\lfloor \alpha n \rfloor, 1)$. Now, d_q is the qth smallest distance from any $\boldsymbol{x}_i$ to $\boldsymbol{x}_k$. To include just the q closest points in a bin, one can use the tri-cube weight for any $\boldsymbol{x}_i$:

$$w_i(\boldsymbol{x}_k) = \chi_{\{||\boldsymbol{x}_i - \boldsymbol{x}_k|| \leq d_q\}} \left(1 - \left\| \frac{\boldsymbol{x}_i - \boldsymbol{x}_k}{d_q} \right\|^3\right)^3. \tag{2.2.9}$$

Now, given these weights, the weighted polynomial fit using the $\boldsymbol{x}_i$s in the bin of cardinality q around $\boldsymbol{x}_k$ can be found by the usual weighted least squares minimization. The resulting function gives the LOESS fit at $\boldsymbol{x}_k$ and the procedure works for any $\boldsymbol{x} \neq \boldsymbol{x}_k$.

It is seen that the β_is obtained at one $\boldsymbol{x}$ are in general different from the β_is obtained for a different $\boldsymbol{x}$. Moreover, LOESS uses weighted linear regression on polynomials; (2.2.7) is like a special case of LOESS using linear functions and one bin, $q = n$. Thus, the corresponding p for LOESS, say p^*, is not the same as the p in (2.2.7); p^* depends on the choice of local functions and n. It's as if (2.2.7) became

$$\hat{\boldsymbol{\beta}}(\boldsymbol{x}) = \arg \min_{\boldsymbol{\beta}^*\in\mathbb{R}^{p^*}} \sum_{i=1}^{n} w_i(\boldsymbol{x})(y_i - f(\boldsymbol{x}, \boldsymbol{x}_i, \boldsymbol{\beta}^*))^2, \tag{2.2.10}$$

in which $w_i(\boldsymbol{x})$ is the weight function with local dependence and f is the local polynomial, parametrized by $\boldsymbol{\beta}^*$, for the bin containing the $\boldsymbol{x}_i$s associated to $\boldsymbol{x}$. Note that $\hat{\boldsymbol{\beta}}$ depends continuously on $\boldsymbol{x}$, the point the regression is fitting. The resulting $\hat{\boldsymbol{\beta}}(\boldsymbol{x})$ would then be used in the model

$$\widehat{\mathbb{E}(Y)}(\boldsymbol{x}) = f(\boldsymbol{x}, \hat{\boldsymbol{\beta}}(\boldsymbol{x})), \tag{2.2.11}$$

in which the dependence of f on the data $\boldsymbol{x}_i$ is suppressed because it has been used to obtain $\hat{\boldsymbol{\beta}}(\boldsymbol{x})$. The expression in (2.2.11) is awkward because the LOESS "model" can only be expressed locally, differing from (2.2.8), which holds for all $\boldsymbol{x} \in \mathbb{R}^p$. However, the regression function in (2.2.11) is continuous as a function of $\boldsymbol{x}$, unlike earlier smoothers.

The statistical properties of LOESS derive from the fact that LOESS is of the form (2.1.1). Indeed, if we write

$$\hat{f}(x) = \sum_{i=1}^{n} W_i(\boldsymbol{x}) y_i,$$

so that the estimate $\hat{f}(\boldsymbol{x})$ is a linear function L_n of the y_is, with fitted values $\hat{y}_i = \hat{f}(\boldsymbol{x}_i)$, we get

$$\hat{\boldsymbol{y}} = L_n \boldsymbol{y}.$$

The residual vector is $\hat{\boldsymbol{\varepsilon}} = (I_{n\times n} - L_n)\boldsymbol{y}$, so $I_{n\times n} - L_n$ plays the same role as the projection operator in the usual least squares formulation, although it is not in general symmetric or idempotent, Cleveland and Devlin (1988), p. 598.

Theorems that characterize the behavior of LOESS estimators are straightforward to establish. Here, it will be enough to explain them informally. The key assumptions are on the distribution of Y, typically taken as normal, and on the form of the true function f_T, typically assumed to be locally approximable by the polynomial f.

Indeed, for local linear or quadratic fitting, one only gets true consistency when f_T is linear or quadratic. Otherwise, the consistency can only hold in a limiting sense on neighborhoods of a given $\boldsymbol{x}$ on which f_T can be well approximated by the local polynomial. Under the assumption of unbiasedness, the usual normal distribution theory for weighted least squares holds locally. That is, under local consistency and normality, $\hat{\boldsymbol{y}}$ and $\hat{\boldsymbol{\varepsilon}}$ are normally distributed with covariances $\sigma^2 L_n^{\mathsf{T}} L_n$ and $\sigma^2 (I_{n\times n} - L_n)^{\mathsf{T}} (I_{n\times n} - L_n)$. Thus, the expected residual sum of squares is

$$\mathbb{E}(\hat{\boldsymbol{\varepsilon}}^{\mathsf{T}} \hat{\boldsymbol{\varepsilon}}) = \sigma^2 \text{trace}(I_{n\times n} - L_n)^{\mathsf{T}} (I_{n\times n} - L_n),$$

giving the natural estimate $\hat{\sigma} = \hat{\boldsymbol{\varepsilon}}^{\mathsf{T}} \hat{\boldsymbol{\varepsilon}} / \text{trace}(I_{n\times n} - L_n)^{\mathsf{T}} (I_{n\times n} - L_n)$. Using the normality in (2.0.2) gives

$$\widehat{\text{Var}}(\hat{g}(\boldsymbol{x})) = \hat{\sigma}^2 \sum_{i=1}^{n} W_i(\boldsymbol{x}_i)^2.$$

Again, as in the usual normal theory, the distribution of a quadratic form such as $\hat{\boldsymbol{\varepsilon}} \hat{\boldsymbol{\varepsilon}}^{\mathsf{T}}$ can be approximated by a constant times a χ^2 distribution, where the degrees of freedom and the constant are chosen to match the first two moments of the quadratic form.

As noted in Cleveland and Devlin (1988), setting $\delta_1 = \text{trace}(I_{n\times n} - L_n)(I_{n\times n} - L_n)^{\mathsf{T}}$ and $\delta_2 = \text{trace}[(I_{n\times n} - L_n)(I_{n\times n} - L_n)^{\mathsf{T}}]^2$, the distribution of $\delta_1^2\hat{\sigma}^2/(\delta_2\sigma^2)$ is approximately $\chi^2_{\delta_1^2/\delta_2}$ and the distribution of $(\hat{f}(\boldsymbol{x}) - f(\boldsymbol{x}))/\hat{\sigma}(\boldsymbol{x})$ is approximately $t_{\delta_1^2/\delta_2}$. Used together, these give confidence intervals, pointwise in $\boldsymbol{x}$, for $f_T(\boldsymbol{x})$ based on $\hat{f}(\boldsymbol{x})$.

Figure 2.7 gives an indication of how well this approach works in practice. For appropriate choices, LOESS is a locally consistent estimator, but, due to the weighting, may be inefficient at finding even relatively simple structures in the data. Indeed, it is easy to see that, past 4, the LOESS curve misses the downturn in the true curve. If there were more data to the right, LOESS would pick up the downturn, so this can be regarded as an edge effect. However, the fact that it is so strong even for the last sixth of the domain is worrisome. Careful adjustment of q and other inputs can improve the fits, but the point remains that LOESS can be inefficient (i.e., it may need a lot of data to get good fit). Although LOESS was not intended for high-dimensional regression, and data sparsity exacerbates inefficiency as p increases, LOESS is often used because normal theory is easy. Of course, like other methods in this chapter, LOESS works best on large, densely sampled, low-dimensional data sets. These sometimes occur, but are hardly typical.

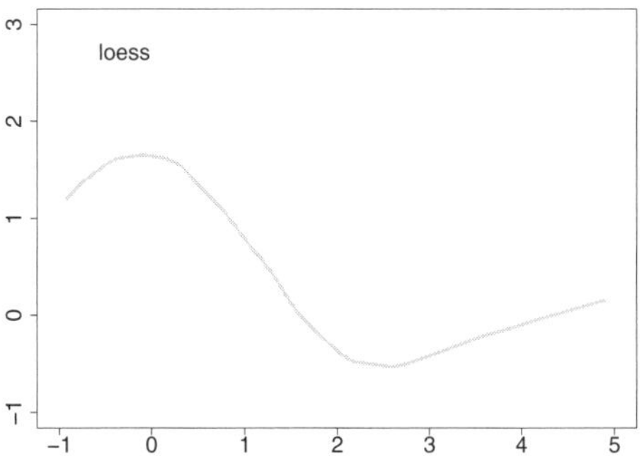

Fig. 2.7 This graph shows the LOESS smoother for the data in Fig. 2.2 for a normal weighting function, polynomials of order 2, and $q = .75n$. The large q makes the graph much smoother than the Early smoothers, but the fit can be poor for unfortunate choices.

2.3 Kernel Smoothers

The first of three truly Classical methods to be presented here is kernel smoothers. However, to do this necessitates some definitions and concepts that run throughout nonparametric function estimation. Although substantial, this is the typical language for nonparametrics.

The problem is to recover an entire function from a random sample of observations $(\boldsymbol{X}_1, Y_1), \cdots, (\boldsymbol{X}_n, Y_n)$, where $Y_i = f(\boldsymbol{X}_i) + \varepsilon_i$ and $\mathbb{E}(\varepsilon_i) = 0$. Under squared error loss, the goal is to find an estimator $\hat{f}(\boldsymbol{x})$ of $f(\boldsymbol{x}) = \mathbb{E}(Y|\boldsymbol{X} = \boldsymbol{x})$. There are a variety of obvious questions: What do we know, or think we know, about the distribution of X? How are the ε_is distributed and how are they related to the Y_is? What is a good choice for $\hat{f}$, and how good is it?

The first subsection introduces how the quality of $\hat{f}$, sometimes denoted $\hat{f}_n$ to emphasize the sample size, is assessed and describes the modes of convergence of $\hat{f}_n$ to f. The second and following subsections explain the kernel methods for forming $\hat{f}$s in several settings, giving their basic properties, including rates of convergence. The discussion will focus on the univariate case, although extensions to low-dimensional $\boldsymbol{X}$s are similar. High-dimensional $\boldsymbol{X}$s suffer the Curse.

2.3.1 Statistical Function Approximation

There are a variety of modes of convergence, some more appropriate than others in some contexts. At the base, there is pointwise convergence. Let $< f_n >$ be a sequence of functions defined on a common domain $\mathscr{X} \subset \mathbb{R}$. The sequence $< f_n >$ converges pointwise to $f(\cdot)$ if

$$\lim_{n \to \infty} f_n(x) = f(x)$$

for each $x \in \mathscr{X}$. This can also be expressed as

$$\forall \varepsilon > 0, \; \exists N, \; \forall n \geq N \text{ such that } |f_n(x) - f(x)| < \varepsilon$$

for each $x \in \mathscr{X}$. Note that this is just the usual notion of convergence for real numbers that happen to be function values at x.

Extending from individual xs to sets of xs is where the complications begin. First, pointwise convergence is not the same as convergence of integrals. This can be seen in standard examples. For instance, consider the sequence of functions $f_n(x) = nx\,(1-x^2)^n$. It can be seen that

$$\lim_{n \to \infty} f_n(x) = \lim_{n \to \infty} nx(1-x^2)^n = 0,$$

but

$$\lim_{n \to \infty} \int_0^1 f_n(x)dx = \lim_{n \to \infty} \frac{n}{2n+2} = \frac{1}{2}.$$

Since the integral of the limiting function over the domain is different from the limit of the sequence of integrals, it follows that pointwise convergence is not a strong mode.

Uniform convergence on a set is clearly stronger than pointwise convergence on that set. Formally, let $< f_n >$ be a sequence of functions all defined on a common domain

$\mathscr{X} \subset \mathbb{R}$. The sequence $< f_n >$ converges uniformly to $f(x)$ if

$$\forall \varepsilon > 0, \ \exists N, \ \forall x \in \mathscr{X} \ \forall n \geq N \ |f_n(x) - f(x)| < \varepsilon.$$

Uniform convergence means that the error between f_n and f can be made arbitrarily small uniformly over $\mathscr{X}$. A fortiori uniform convergence implies pointwise convergence, but the converse fails. (Consider $f_n(x) = x^n$ on $[0,1]$, for example.) Uniform convergence also implies that integrals converge.

Theorem: Let f_n be a sequence of continuous functions defined on a closed interval $[a,b]$. If f_n converges uniformly to $f(x)$ on $[a,b]$, then

$$\lim_{n \to \infty} \int_a^b f_n(x)dx = \int_a^b f(x)dx.$$

In a measure-theoretic context, the monotone convergence theorem, the dominated convergence theorem, and Egorov's theorem together mean that pointwise convergence almost gives convergence of integrals and each of the two modes is almost equivalent to uniform convergence. In this context, behavior at specific points is not important because functions are only defined up to sets of measure zero.

In an inner product space, uniform convergence is expressed in terms of the norm $\|\cdot\|$ derived from the inner product. A sequence f_n in an inner product space converges to f if and only if

$$\forall \varepsilon > 0, \ \exists N, \text{ such that } \forall n \geq N \|f_n - f\| < \varepsilon.$$

The x does not appear in the definition since the norm is on the function as an entity, not necessarily dependent on specific points of its domain.

A sequence $< f_n >$ in an inner product space is Cauchy if

$$\forall \varepsilon > 0, \ \exists N, \text{ such that } \forall n, m \geq N \ \|f_n - f_m\| < \varepsilon.$$

In most common topological spaces, sequences converge if and only if they are Cauchy. A space in which every Cauchy sequence converges in its norm to a member of the space (i.e., the space is closed under Cauchy convergence) is complete. A complete linear space together with a norm defined on it is called a Banach space. A closed Banach space in which the norm arises from an inner product is called a Hilbert space. Finite-dimensional vector spaces $\mathbb{R}^p$ and the space of square-integrable functions L^2 are both Hilbert spaces. Under $\|\cdot\|_\infty$, a linear space such as $\mathscr{C}[0,1]$ is usually Banach but not Hilbert.

Turning to more statistical properties, squared error is used more often than other notions of distance, such as $\|\cdot\|_\infty$ for instance, especially when evaluating error pointwise in $\boldsymbol{x}$. However, different measures of distance have different properties. Euclidean distance is the most widely used because in finite dimensions it corresponds well to our intuitive sense of distance and remains convenient and tractable in higher dimensions. Starting with this, recall that, for good prediction, the MSE of the predictor must be small. If the goal is to predict Y_{new} from $\boldsymbol{X}_{new}$, having already seen $(\boldsymbol{X}_1, Y_1), \ldots, (\boldsymbol{X}_n, Y_n)$, the mean squared error gives the average of the error for each $\boldsymbol{X}$: For function estimation, the mean squared error (MSE) of $\hat{f}$ at any $\boldsymbol{x}$ is

$$\mathsf{MSE}[\hat{f}(x)] = \mathbb{E}\left[\left(\hat{f}(x) - f(x)\right)^2\right].$$

As before, this breaks down into two parts. The bias of $\hat{f}$ at x is

$$\text{Bias}(\hat{f}(x)) = \mathbb{E}(\hat{f}(x)) - f(x);$$

the variance of $\hat{f}$ at x is

$$\text{Var}(\hat{f}(x)) = \mathbb{E}\left[\left(\hat{f}(x) - \mathbb{E}(\hat{f}(x))\right)^2\right];$$

and the MSE can be decomposed:

$$\mathsf{MSE}[\hat{f}(x)] = \text{Var}(\hat{f}(x)) + \text{Bias}(\hat{f}(x))^2.$$

Naively, the minimum-variance unbiased estimator is the most desirable. After all, if $\hat{f}$ is pointwise unbiased (i.e., $\mathsf{Bias}(\hat{f}(x)) = 0$ for each $x \in \mathscr{X}$), then one is certain that enough data will uncover the true function. However, sometimes unbiased estimators don't exist and often there are function estimators with smaller variance and small bias (that goes to zero as n increases) with smaller MSE.

Another measure of distance that is more appropriate for densities is the mean absolute error (MAE). The mean absolute error of $\hat{f}$ at x is

$$\mathsf{MAE}[\hat{f}(x)] = \mathbb{E}[|\hat{f}(x) - f(x)|].$$

Unlike the MSE, the MAE does not allow an obvious decomposition into meaningful quantities such as variance and bias. It also poses severe analytical and computational challenges. However, applied to densities, it corresponds to probability (recall Scheffe's theorem) and is usually equivalent to the total variation distance. Indeed, $\hat{f}$ is weakly pointwise consistent for f when $\hat{f}(\boldsymbol{x})$ converges to $f(\boldsymbol{x})$ in probability (i.e., $\hat{f}(\boldsymbol{x}) \to_P f(\boldsymbol{x})$ for each $\boldsymbol{x}$), and $\hat{f}$ is pointwise consistent for f when

$$\forall \boldsymbol{x} \in \mathscr{X} \quad \mathbb{E}(\hat{f}(x)) \to f(x).$$

For the remainder of this section, the focus will be on the global properties of $\hat{f}$ on the whole domain $\mathscr{X}$ of f rather than on pointwise properties. This means that all the assessments are in terms of $\hat{f}$ and f, with no direct dependence on the values $\boldsymbol{x}$.

A general class of norms comes from the Lebesgue spaces, L^p, given by

$$\|\hat{f} - f\|_p = \left(\int_{\mathscr{X}} |\hat{f}(x) - f(x)|^p dx\right)^{1/p},$$

for $\hat{f} - f$. For the norm to be well defined, there are two key requirements: $\hat{f}$ must be defined on $\mathscr{X}$, and the integral must exist.

Three special cases of L_p norms are $p = 1, 2, \infty$. The L_1 norm, also called integrated absolute error (IAE) is

$$\mathsf{IAE}[\hat{f}] = \int_{\mathscr{X}} |\hat{f}(x) - f(x)| dx.$$

The L_2 norm, also called integrated squared error (ISE) is

$$\mathsf{ISE}[\hat{f}] = \int_{\mathscr{X}} (\hat{f}(x) - f(x))^2 dx.$$

The L_∞ norm, also called supremal absolute error (SAE) is

$$\mathsf{SAE}[\hat{f}] = \sup_{x \in \mathscr{X}} \left| \hat{f}(x) - f(x) \right|.$$

The Csiszar ϕ divergences are another general class of measures of distance. Instead of being defined by expectations of powers, Csiszar ϕ divergences are expectations of convex functions of density ratios. The power divergence family is a subset of Csiszar ϕ divergences. Two of the most important examples are the Kullback-Leibler distance, or relative entropy, given by

$$\mathsf{KL}[\hat{f}, f] = \int_{\mathscr{X}} \hat{f}(x) \log\left(\frac{\hat{f}(x)}{f(x)}\right) dx,$$

and the Hellinger distance, given by

$$\mathsf{H}[\hat{f}, f] = \left(\int_{\mathscr{X}} \left(\hat{f}^{1/p}(x) - f^{1/p}(x)\right)^p dx\right)^{1/p}.$$

These distances are not metrics. However, they do typically have convex neighborhood bases and satisfy some metric-like properties. In addition, the Kullback-Leibler distance represents codelength, and the Hellinger distance represents the closest packing of spheres. (Another ϕ divergence is the χ-squared distance, which represents goodness of fit.) Overall, select members of the Csiszar ϕ divergence class have interpretations that are usually more appropriate to physical modeling than L_p norms have. Whichever distance is chosen, the consistency of $\hat{f}$ is studied from a global perspective by trying to obtain

$$\int_{\mathscr{X}} \mathbb{E}\left(L(\hat{f}(x), f(x))\right) dx \to 0,$$

where $L(\hat{f}(x), f(x))$ indicates the distance chosen as the loss function.

Among these global measures, focus usually is on the ISE. It is more mathematically tractable than the others because the loss function is squared error, giving $L(\hat{f}(x), f(x)) = (\hat{f}(x) - f(x))^2$. Consequently, a full measure of the quality of $\hat{f}$ is often formed by combining the MSE and the ISE into the integrated mean squared error (IMSE) which turns out to equal the mean integrated squared error (MISE). To see this, define the integrated squared bias (ISB),

$$\mathsf{ISB}[\hat{f}] = \int_{\mathscr{X}} \left(\mathbb{E}(\hat{f}(x)) - f(x)\right)^2 dx,$$

and the integrated variance (IV),

$$\mathsf{IV}[\hat{f}] = \int_{\mathscr{X}} \mathrm{Var}(\hat{f}(x))dx = \int_{\mathscr{X}} \mathbb{E}\left[\left(\hat{f}(x) - \mathbb{E}(\hat{f}(x))\right)^2\right]dx.$$

Now, the IMSE is

$$\begin{aligned}\mathsf{IMSE}[\hat{f}] &= \int_{\mathscr{X}} \mathbb{E}\left((\hat{f}(x) - f(x))^2\right) dx \\ &= \mathsf{IV}(\hat{f}) + \mathsf{ISB}(\hat{f}).\end{aligned}$$

Assuming a Fubini theorem, the integrated mean squared error is

$$\begin{aligned}\mathsf{IMSE}[\hat{f}] &= \mathbb{E}\left(\int_{\mathscr{X}} (\hat{f}(x) - f(x))^2 dx\right) \\ &= \mathsf{MISE}(\hat{f}),\end{aligned}$$

the MISE. Unfortunately, as suggested by the Runge function example, global unbiasedness generally does not hold. So, in practice, usually both $\mathsf{IV}(\hat{f})$ and $\mathsf{ISB}(\hat{f})$ must be examined.

Continuing the definitions for squared error, $\hat{f}$ is mean square consistent (or L_2 consistent) for f if the MISE converges to 0. Formally, this is

$$\int_{\mathscr{X}} \mathbb{E}\left((\hat{f}(x) - f(x))^2\right) dx \to 0.$$

The expectation in the integral can be removed, in which case the expression is a function of the data after integrating out the $\boldsymbol{x}$. This reduced expression can still go to zero in probability as n increases, or with probability one, giving familiar notions of weak and strong consistency, respectively.

Next, to respect the fact that X is a random variable, not just a real function, it is important to take into account the stochasticity of X through its density $p(x)$. So, redefine the MISE to be

$$\mathsf{MISE}(\hat{f}) = \int_{\mathscr{X}} \mathbb{E}\left((\hat{f}(x) - f(x))^2\right) p(x)dx.$$

If a weight function $w(x)$ is included in the ISE, then writing

$$d_I(\hat{f}, f) = \int (\hat{f}(x) - f(x))^2 p(x) w(x) dx$$

gives that the MISE is the expectation of d_I with respect to X. That is,

$$\mathsf{MISE}(\hat{f}) = d_M(\hat{f}, f) = \mathbb{E}(d_I(\hat{f}, f)].$$

The notation d_M is a reminder that the MISE is a distance from $\hat{f}$ to f resulting from another distance d_I.

Very often MISE is intractable because it has no closed-form expression. There are two ways around this problem. Theoretically, one can examine the limit of MISE as $n \to \infty$. This gives the asymptotic mean integrated squared error (AMISE). Alternatively, for computational purposes, a discrete approximation of d_I based on a sample $X_1, \cdots, X_n$ can be used. This is the average squared error (ASE),

$$\mathsf{ASE}(\hat{f}, f) = d_A(\hat{f}, f) = \frac{1}{n} \sum_{i=1}^{n} (\hat{f}(X_i) - f(X_i))^2 w(X_i). \tag{2.3.1}$$

The ASE is convenient because, being discrete, d_A avoids numerical integration. Indeed, as a generality, the main quantities appearing in nonparametric reasoning must be discretized to be implemented in practice. (Expressions like (2.3.1) are empirical risks, and there is an established theory for them. However, it will not be presented here in any detail.)

2.3.2 *The Concept of Kernel Methods and the Discrete Case*

In this subsection, the setting for kernel methods is laid out. The basic idea is to smooth the data by associating to each datum a function that looks like a bump at the data point, called a kernel. The kernel spreads out the influence of the observation so that averaging over the bumps gives a smooth. The special case of deterministic choices of the x_is is dealt with here in contrast to (i) the Runge function example, (ii) the stochastic case, which involves an extra normalization, and (iii) the spline setting to be developed later.

2.3.2.1 Key Tools

The central quantity in kernel smoothing is the kernel itself. Technically, a kernel K is a bounded, continuous function on $\mathbb{R}$ satisfying

$$\forall v \;\; K(v) \geq 0 \quad \text{and} \int K(v)dv = 1.$$

To make this more intuitive, K usually is required to satisfy the additional conditions

$$\int vK(v)dv = 0 \quad \text{and} \quad \int v^2 K(v)dv < \infty.$$

For multivariate $\boldsymbol{X}$s, one often takes multiples of p copies of K, one for each x_i in $\boldsymbol{X}$, rapidly making the problem difficult. For notational convenience, define

$$K_h(v) = \frac{1}{h} K\left(\frac{v}{h}\right).$$

Here, K_h is the rescaled kernel and h is the bandwidth or smoothing parameter. It is easy to see that if the support of K is $\text{supp}(K) = [-1, +1]$, then $\text{supp}(K_h) = [-h, +h]$. Also, K_h integrates to 1 over v for each h. It will be seen later that kernel smoothers are linear in the sense of (2.1.1) because K is the basic ingredient for constructing the weights $\{W_i(x)\}_{i=1}^n$. The shape of the weights comes from the shape of K, while their size is determined by h.

Clearly, there are many possible choices for K. Some are better than others, but not usually by much. So, it is enough to restrict attention to a few kernels. The following table shows four of the most popular kernels; graphs of them are in Fig. 2.8.

Kernel name	Equation	Range
Epanechnikov	$K(v) = \frac{3}{4}(1 - v^2)$	$-1 \le v \le 1$
Biweight	$K(v) = \frac{15}{16}(1 - v^2)^2$	$-1 \le v \le 1$
Triangle	$K(v) = (1 - \lvert v \rvert)$	$-1 \le v \le 1$
Gaussian (normal)	$K(v) = \frac{1}{\sqrt{2\pi}} e^{-v^2/2}$	$-\infty < v < \infty$

Three of the kernels in the table above are zero outside a fixed interval. This restriction helps avoid computational numerical underflows resulting from the kernel taking on very small values. In terms of efficiency, the best kernel is the Epanechnikov. The least efficient of the four is the normal.

It turns out that continuity of a function is not a strong enough condition to permit statements and proofs of theorems that characterize the behavior of kernel estimators. A little bit more is needed. This little bit more amounts to continuity with contraction properties akin to uniform continuity but with a rate on ε as a function of δ. Thus, key theorems assume Hölder continuity and Lipschitz continuity of the underlying function f as well as of the other functions (such as kernels) used to estimate it.

Let g be a univariate function with compact domain $\mathscr{X} \subset \mathbb{R}$. Lipschitz continuity asks for a uniform linear rate of contraction of the function values in terms of their arguments. That is, the function g is Lipschitz continuous if

$$\exists \delta > 0 \quad \text{such that} \quad |g(u) - g(v)| \le \delta |u - v|$$

for all $u, v \in \mathscr{X}$. A more general version of this criterion allows upper bounds that are not first order. A univariate function g on a compact domain $\mathscr{X} \subset \mathbb{R}$ is α-Hölder continuous for some $0 < \alpha \le 1$ if

$$\exists \delta_\alpha > 0 \quad \text{such that} \quad |g(u) - g(v)| \le \delta_\alpha |u - v|^\alpha$$

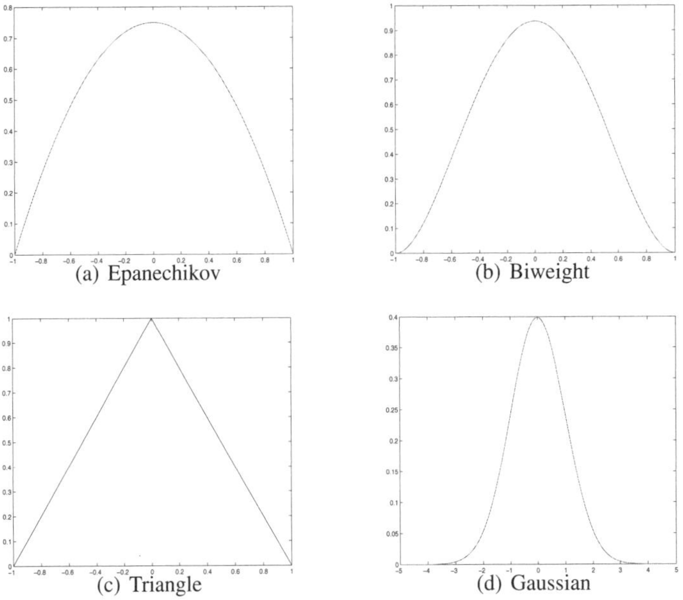

Fig. 2.8 The graphs of the four kernels from the table show how they spread the weight of a data point over a region. Only the normal has noncompact support – and it is least efficient.

for all $u, v \in \mathscr{X}$. Clearly, an α-Hölder continuous function with $\alpha = 1$ is Lipschitz continuous. It is easy to see that on compact sets these two conditions are readily satisfied by most well-behaved functions. Functions that do not satisfy them often have uncontrolled local oscillations.

2.3.2.2 Kernel Smoothing for Deterministic Designs

Assume $(x_1, y_1), \cdots, (x_n, y_n)$ is generated by the model $Y_i = f(x_i) + \varepsilon_i, \quad i = 1, \cdots, n$, where $\varepsilon_1, \cdots, \varepsilon_n$ are IID $(0, \sigma^2)$ and the design points are equidistant in $[0, 1]$; i.e.,

$$x_i = \frac{i-1}{n-1}, \quad i = 1, 2, \cdots, n.$$

Let $f : [0, 1] \longrightarrow \mathbb{R}$ be the underlying function to be estimated, and choose a fixed kernel K symmetric about zero; i.e., $K(-v) = K(v)$.

The Priestley-Chao (PC) kernel estimate of f (see Priestley and Chao (1972)) for a deterministic design is

$$\hat{f}(x) = \frac{1}{n} \sum_{i=1}^{n} K_{h_n}(x - x_i) Y_i = \frac{1}{nh_n} \sum_{i=1}^{n} K\left(\frac{x - x_i}{h_n}\right) Y_i, \tag{2.3.2}$$

where $x \in [0,1]$ and $\{h_n\}$ is a sequence of positive real numbers converging to zero at rate $o(1/n)$; that is, $nh_n \longrightarrow \infty$ as $n \longrightarrow \infty$. In the presence of Lipschitz continuity, the behavior of the AMSE at an arbitrary point x is controlled by the proximity of x to a design point. In this chapter, proofs of theorems are merely sketched since full formal proofs are readily available from the various sources cited.

Theorem (Gasser and Müller, 1984): Suppose K has compact support and is Lipschitz continuous on $\text{supp}(K)$. If f is twice continuously differentiable, then the asymptotic mean squared error at $x \in [0,1]$ is

$$\mathsf{AMSE}(\hat{f}(x)) = \frac{(\mu_2(K)f''(x))^2}{4}h_n^4 + \frac{1}{nh_n}\sigma^2 S(K), \tag{2.3.3}$$

where $S(K) = \int K^2(t)dt$ and $\mu_2(K) = \int t^2 K(t)dt$.

Proof: Recall that the MSE can be decomposed into bias and variance, namely

$$\mathsf{MSE}(\hat{f}(x)) = \text{Var}(\hat{f}(x)) + \text{Bias}^2(\hat{f}(x)).$$

The first ingredient for obtaining the bias is

$$\begin{aligned}\mathbb{E}(\hat{f}(x)) &= \mathbb{E}\left[\frac{1}{nh_n}\sum_{i=1}^{n} K\left(\frac{x-x_i}{h_n}\right)Y_i\right] \\ &= \frac{1}{nh_n}\sum_{i=1}^{n} K\left(\frac{x-x_i}{h_n}\right)\mathbb{E}[Y_i].\end{aligned}$$

Now, with $h_n \longrightarrow 0$ as $n \longrightarrow \infty$, the summation over i can be approximated by an integral over x, namely

$$\mathbb{E}(\hat{f}(x)) = \int \frac{1}{h_n} K\left(\frac{x-v}{h_n}\right) f(v)dv + O\left(\frac{1}{n}\right).$$

The change of variable $t = (x-v)/h_n$ gives $v = x - h_n t$, $dv = -h_n dt$, and

$$\mathbb{E}(\hat{f}(x)) = \int_{(x-1)/h_n}^{x/h_n} K(t) f(x - h_n t)dt + O\left(\frac{1}{n}\right).$$

Taylor expanding $f(\cdot)$ at x gives

$$f(x - h_n t) = f(x) - h_n t f'(x) + \frac{1}{2}h_n^2 t^2 f''(x) + \cdots.$$

Since K is supported on $(-1,1)$, the Taylor expansion can be substituted into the integral to give

$$\begin{aligned}\mathbb{E}(\hat{f}(x)) &= \int_{-1}^{1} K(t)\left[p(x) - h_n t f'(x) + \frac{1}{2}h_n^2 t^2 f''(x) + \cdots\right] dt \\ &= f(x)\int K(t)dt - h_n p'(x)\int tK(t)dt + \frac{1}{2}h_n^2 f''(x)\int t^2 K(t)dt + \cdots.\end{aligned}$$

By definition, $\int K(v)dv = 1$ and $\int vK(v)dv = 0$, so

$$\mathbb{E}(\hat{f}(x)) = f(x) + \frac{1}{2}h_n^2 f''(x)\int t^2 K(t)dt + \cdots .$$

Defining $\mu_2(K) = \int t^2 K(t)dt$, the bias is given by

$$\mathbb{E}(\hat{f}(x)) - f(x) = \frac{1}{2}h^2 \mu_2(K) f''(x) + \mathscr{O}(h_n^2) + \mathscr{O}\left(\frac{1}{n}\right),$$

and the asymptotic squared bias, as claimed, is

$$\mathsf{ASB}[\hat{f}(x)] = \frac{(\mu_2(K)f''(x))^2}{4}h_n^4.$$

For the variance, the same approximation of a sum by an integral and the same change of variable as above leads to

$$\mathrm{Var}(\hat{f}(x)) = \frac{1}{nh_n}\left[\frac{1}{n}\sum_{i=1}^{n}\frac{1}{h_n}K^2\left(\frac{x-x_i}{h_n}\right)\mathrm{Var}(Y_i)\right] = \frac{\sigma^2}{nh_n}\left[\int K^2(t)dt\right] + \mathscr{O}\left(\frac{1}{nh_n}\right)$$

for small enough h_n. With $S(K) = \int K^2(t)dt$, the asymptotic variance is

$$\mathsf{AV}[\hat{f}(x)] = \frac{1}{nh_n}\sigma^2 S(K),$$

also for small h_n, giving the result claimed. □

An immediate consequence of (2.3.3) is the pointwise consistency of $\hat{f}$.

Corollary: If $h_n \longrightarrow 0$ and $nh_n \longrightarrow \infty$ as $n \longrightarrow \infty$, then

$$\mathsf{AMSE}(\hat{f}(x)) \longrightarrow 0.$$

Therefore

$$\hat{f}(x) \underset{p}{\longrightarrow} f(x),$$

and $\hat{f}$ is asymptotically consistent, pointwise in x. □

The expression for $\mathsf{AMSE}(\hat{f}(x))$ provides a way to estimate the optimal bandwidth, along with the corresponding rate of convergence to the true underlying curve. Since this procedure is qualitatively the same for stochastic designs, which are more typical, this estimation is deferred to the discussion in the next section.

2.3.3 Kernels and Stochastic Designs: Density Estimation

For stochastic designs, assume $(X,Y),\cdots,(X_n,Y_n)$ are IID $\mathbb{R}\times\mathbb{R}$-valued random vectors with $\mathbb{E}(|Y|)<\infty$ and that

$$Y_i = f(X_i)+\varepsilon_i, \quad i=1,2,\cdots,n,$$

where $X_1,\cdots,X_n$ have common density $p(x)$ and the $\varepsilon_1,\cdots,\varepsilon_n$ are IID $N(0,\sigma^2)$, independent of $X_1,\cdots,X_n$. The goal, as before, is to estimate the regression function

$$f(x)=\mathbb{E}(Y|X=x)$$

from the data. However, this is different from the Priestley-Chao problem because the design of the x_is is not fixed. Intuitively, the estimator $\hat{f}$ must be responsive to whatever value of X occurs and so the weight assigned to a specific x must be random and generalize the constant nh_n in (2.3.2).

The Nadaraya-Watson (NW) kernel estimate of f is given by

$$\hat{f}(x)=\frac{\sum_{i=1}^n K_h(x-X_i)Y_i}{\sum_{i=1}^n K_h(x-X_i)}. \tag{2.3.4}$$

The denominator is a density estimate, so the NW estimate of f is often expressed in terms of the Parzen-Rosenblatt kernel density estimate $\hat{p}(x)$ of $p(x)$ by writing

$$\hat{f}(x)=\frac{\frac{1}{n}\sum_{i=1}^n K_h(x-X_i)Y_i}{\hat{p}(x)},$$

where

$$\hat{p}(x)=\frac{1}{n}\sum_{i=1}^n K_h(x-X_i). \tag{2.3.5}$$

In effect, the randomness in X makes the estimation of $f(x)$ essentially the same as estimating the numerator and denominator of the conditional expectation of Y given $X=x$:

$$\mathbb{E}(Y|X=x)=\frac{\int y p_{X,Y}(x,y)dy}{p_X(x)}=\frac{\int y p_{X,Y}(x,y)dy}{\int p_{X,Y}(x,y)dy}.$$

The consistency of the NW smoother rests on the consistency of the Parzen-Rosenblatt density estimator.

Expectations of kernel estimators are convolutions of the kernel with $p(x)$; i.e., $\mathbb{E}(\hat{p}(x))=(K_h*p)(x)$. This can be seen by writing the definitions

$$\begin{aligned}\mathbb{E}(\hat{p}(x)) &= \frac{1}{n}\sum_{i=1}^n \mathbb{E}(K_h(x-X_i))=\mathbb{E}(K_h(x-X_1))\\ &= \int K_h(x-v)p(v)dv=(K_h*p)(x).\end{aligned}$$

This last expression shows that the kernel estimator $\hat{p}(x)$ of $p(x)$ is a convolution operator that locally replaces each point by a weighted average of its neighbors.

An extension of the technique of proof of the last theorem gives consistency of the kernel density estimator for stochastic designs. Obtaining the consistency of the NW smoother using this technique is done in the next theorem.

Theorem: Let K be a kernel satisfying $\lim_{|v|\longrightarrow\infty} vK(v) = 0$. Then, for any x at which the density $p(x)$ is defined, we have

$$\hat{p}(x) \xrightarrow{p} p(x)$$

if $h \longrightarrow 0$ and $nh \longrightarrow \infty$ as $n \longrightarrow \infty$. The optimal bandwidth is $h^{\text{opt}} = \mathcal{O}\left(n^{-\frac{1}{5}}\right)$, and the AMISE decreases at rate $n^{-4/5}$.

Proof (sketch): First, we sketch the bias. The change of variable $t = (x-v)/h$ gives $v = x - ht$ and $dv = -hdt$, so the expectation of $\hat{p}(x)$ is

$$\begin{aligned}
\mathbb{E}(\hat{p}(x)) &= \mathbb{E}\left[\frac{1}{h}K\left(\frac{x-X_1}{h}\right)\right] \\
&= \int_a^b \frac{1}{h}K\left(\frac{x-v}{h}\right)p(v)dv = \int_{\frac{x-b}{h}}^{\frac{x-a}{h}} K(t)p(x-ht)dt.
\end{aligned}$$

Taylor expanding $p(\cdot)$ at x gives

$$p(x-ht) = p(x) - ht\,p'(x) + \frac{1}{2}h^2t^2p''(x) + \cdots.$$

As a special case, if the kernel K is supported on $(-\xi, \xi)$, then

$$\begin{aligned}
\mathbb{E}(\hat{p}(x)) &= \int_{-\xi}^{\xi} K(t)\left[p(x) - ht\,p'(x) + \frac{1}{2}h^2t^2p''(x) + \cdots\right]dt \\
&= p(x)\int K(t)dt - hp'(x)\int tK(t)dt + \frac{1}{2}h^2p''(x)\int t^2K(t)dt + \cdots.
\end{aligned}$$

So, using $\int K(v)dv = 1$ and $\int vK(v)dv = 0$ gives

$$\mathbb{E}(\hat{p}(x)) = p(x) + \frac{1}{2}h^2p''(x)\int t^2K(t)dt + \cdots.$$

As a result, setting $\mu_2(K) = \int t^2K(t)dt$ gives an expression for the bias,

$$\mathbb{E}(\hat{p}(x)) - p(x) = \frac{1}{2}h^2\mu_2(K)p''(x) + \mathcal{O}\left(\frac{1}{nh}\right) + \mathcal{O}\left(\frac{1}{n}\right).$$

Now, setting $S(p'') = \int (p''(x))^2dx$, squaring, and ignoring small error terms gives

$$\mathsf{AISB}[\hat{p}] = \frac{(\mu_2(K)S(p''))^2}{4}h^4. \tag{2.3.6}$$

For the variance,

$$\begin{aligned}
\operatorname{Var}[\hat{p}(x)] &= \frac{1}{n}\left[\left(\frac{1}{h}\int K^2(t)p(x-ht)dt\right) - \mathbb{E}(\hat{p}(x))^2\right] \\
&= \frac{1}{nh}\int K^2(t)\left[p(x) - htp'(x) + \frac{1}{2}h^2t^2p''(x) + \cdots\right]dt - \frac{1}{n}[\mathbb{E}(\hat{p}(x))]^2 \\
&= \frac{1}{nh}p(x)\int K^2(t)dt + \mathscr{O}\left(\frac{1}{nh}\right) + \mathscr{O}\left(\frac{1}{n}\right).
\end{aligned}$$

If $h \longrightarrow 0$ and $nh \longrightarrow \infty$ as $n \longrightarrow \infty$, then the asymptotic variance of $\hat{p}(x)$ becomes

$$\mathsf{AV}[\hat{p}(x)] = \frac{1}{nh}p(x)S(K),$$

where $S(K) = \int K^2(t)dt$, and the corresponding asymptotic integrated variance is

$$\mathsf{AIV}[\hat{p}] = \frac{1}{nh}S(K). \tag{2.3.7}$$

Using (2.3.6) and (2.3.7), the expression for the AMISE for $\hat{p}$ is

$$\mathsf{AMISE}(\hat{p}) = \frac{(\mu_2(K)p''(x))^2}{4}h^4 + \frac{1}{nh}S(K),$$

from which one gets the convergence in mean square, and hence in probability, of $\hat{p}(x)$ to $p(x)$. Also, it is easy to see that solving

$$\frac{\partial \mathsf{AMISE}(\hat{p})}{\partial h} = 0$$

yields $h^{\mathsf{opt}} = \mathscr{O}\left(n^{-\frac{1}{5}}\right)$, which in turn corresponds to $\mathsf{AMISE}(\hat{p}) = O\left(n^{-\frac{4}{5}}\right)$. □

In parametric inference, after establishing consistency for an estimator, one tries to show asymptotic normality. This holds here for $\hat{p}(x)$. Indeed, by writing the kernel density estimator $\hat{p}(x)$ in the form of a sum of random variables

$$\hat{p}(x) = \frac{1}{n}\sum_{i=1}^{n}\frac{1}{h}K\left(\frac{x-X_i}{h}\right) = \frac{1}{n}\sum_{i=1}^{n}Z_i,$$

the Lyapunov central limit theorem gives its asymptotic distribution. Thus, if $h \to 0$ and $nh \to \infty$ as $n \to \infty$,

$$\sqrt{nh}\Big(\hat{p}(x) - \mathbb{E}(\hat{p}(x))\Big) \xrightarrow{d} N(0, \sigma_x^2),$$

where $\sigma_x^2 = (nh)\text{Var}[\hat{p}(x)] = p(x)\int K^2(t)dt$. Later, it will be seen that $h = \mathcal{O}(n^{-1/5})$ achieves a good bias-variance trade-off, in which case

$$\sqrt{nh}\left(\hat{p}(x) - p(x)\right) \xrightarrow{d} N\left(\frac{1}{2}\mu_2(K)p''(x), S(K)p(x)\right),$$

where $S(K) = \int K^2(t)dt$ and $\mu_2(K) = \int t^2K(t)dt$. When the bias is of smaller order than the standard deviation, the distribution of $\sqrt{nh}(\hat{p}(x) - p(x))$ coincides with that of $\sqrt{nh}(\hat{p}(x) - \mathbb{E}(\hat{p}(x)))$, which is more appealing because the estimator $\hat{p}$ is available.

2.3.4 Stochastic Designs: Asymptotics for Kernel Smoothers

There are two core results for kernel smoothers. First is consistency, and second is an expression for the AMISE, since variance alone is not enough. Both results are based on analyzing a variance-bias decomposition and extend the result from the last subsection on the consistency of the kernel density estimator. The last theorem will be used for both the numerator and denominator of the NW kernel estimator for f, pulling them together with Slutzky's theorem. Recall that Slutzky's theorem gives the behavior of sequences of variables under convergence in distribution and probability. Thus, let a be a constant, X be a random variable, and $\{X_n\}$ and $\{Y_n\}$ be sequences of random variables satisfying $X_n \xrightarrow{d} X$ and $Y_n \xrightarrow{p} a$. Then (1) $Y_nX_n \xrightarrow{d} aX$ and (2) $X_n + Y_n \xrightarrow{d} X + a$.

To see how this gets used, write the NW estimator as a fraction,

$$\hat{f}(x) = \frac{\hat{q}(x)}{\hat{p}(x)},$$

so that $\hat{q}(x) = \hat{f}(x)\hat{p}(x)$. The content of the last theorem was that when $h \to 0$ and $nh \to \infty$, the denominator $\hat{p}(x)$ of $\hat{f}(x)$ is a consistent estimate of $p(x)$. Similar techniques to deal with $\hat{q}$ are at the core of consistency of the kernel smoother as seen in the proof of the following.

Theorem: Let K be a kernel satisfying $\lim_{|v|\to\infty} vK(v) = 0$, and suppose X gives a stochastic design with $\hat{p}(x)$ consistent for $p(x)$. If $\mathbb{E}(Y_i^2) < \infty$, then for any x at which $p(x)$ and $f(x)$ are continuous and $p(x) > 0$,

$$\hat{f}(x) \xrightarrow{p} f(x)$$

if $h \longrightarrow 0$ and $nh \longrightarrow \infty$ as $n \longrightarrow \infty$.

Proof (sketch): The central task is to verify that, under the same conditions as the last theorem,

$$\hat{q}(x) \underset{p}{\longrightarrow} q(x) \equiv f(x)p(x).$$

To see this, it is enough to show that the MSE of $\hat{q}(x)$ for $q(x)$ goes to zero. Since the MSE is the "squared bias plus variance", it is enough to show that their sum goes to zero under the conditions in the theorem.

First, we address the bias of $\hat{q}(x)$. The change of variable $t = (x-u)/h$ gives

$$\begin{aligned}\mathbb{E}(\hat{q}(x)) &= \mathbb{E}\left[\frac{1}{nh}\sum_{i=1}^{n} K\left(\frac{x-X_i}{h}\right)\cdot Y_i\right] = \mathbb{E}\left[\frac{1}{nh}\sum_{i=1}^{n} K\left(\frac{x-X_i}{h}\right)\cdot f(X_i)\right] \\ &= \int \frac{1}{h} K\left(\frac{x-u}{h}\right) f(u)p(u)du = \int K(t)\, f(x-ht)p(x-ht)dt. \quad (2.3.8)\end{aligned}$$

For convenience, (2.3.8) can be rewritten as

$$\mathbb{E}(\hat{q}(x)) = \int K(t)\, q(x-ht)dt, \tag{2.3.9}$$

which is of the same form as $\mathbb{E}(\hat{p}(x))$. Assuming that $q(x) = f(x)p(x)$ is twice continuously differentiable, and Taylor expanding as before in $\mathbb{E}(\hat{p}(x))$, the bias is

$$\mathbb{E}(\hat{q}(x)) - q(x) = \frac{\mu_2(K)q''(x)}{2}h^2 + o(h^2) = \mathcal{O}\left(h^2\right) + o(h^2) = \mathcal{O}\left(h^2\right)$$

where $\mu_2(K) = \int t^2 K(t)dt$ and $q''(x)$ is

$$q''(x) = (f(x)p(x))'' = f''(x)p(x) + 2f'(x)p'(x) + p''(x)f(x).$$

Using an argument similar to the one above, the variance of $\hat{q}(x)$ is

$$\begin{aligned}\mathrm{Var}(\hat{q}(x)) &= \mathrm{Var}\left[\frac{1}{n}\sum_{i=1}^{n}\frac{1}{h}K\left(\frac{x-X_i}{h}\right)\cdot Y_i\right] = \frac{1}{n}\mathbb{E}\left[\frac{1}{h}K\left(\frac{x-X_i}{h}\right)\cdot Y_i\right]^2 - \frac{1}{n}\left(\mathbb{E}(\hat{q}(x))\right)^2 \\ &= \frac{1}{n}\int \frac{1}{h^2}K^2\left(\frac{x-u}{h}\right)[\sigma^2 + f^2(u)]p(u)du - \frac{1}{n}\left(\mathbb{E}(\hat{q}(x))\right)^2 \\ &= \frac{1}{nh}\int K^2(t)(\sigma^2 + f^2(x-ht))p(x-ht)dt - \frac{1}{n}\left(\mathbb{E}(\hat{q}(x))\right)^2 \\ &= \frac{(\sigma^2 + f^2(x))p(x)}{nh}\int K^2(t)dt + o\left(\frac{1}{nh}\right) \\ &= \mathcal{O}\left(\frac{1}{nh}\right) + o\left(\frac{1}{nh}\right) = \mathcal{O}\left(\frac{1}{nh}\right).\end{aligned}$$

Note that $f(\cdot)$ and $p(\cdot)$ are evaluated at x because, as $h \to 0$, $f(x-ht)$ and $p(x-ht)$ converge to $f(x)$ and $p(x)$. Also, $1/n = o(1/nh)$.

From the expressions for the bias and variance, the MSE of $\hat{q}(x)$ is $[\mathcal{O}(h^2)]^2 + \mathcal{O}(1/(nh))$. As a result, if $h \to 0$ and $nh \to \infty$ as $n \to \infty$, then

$$\hat{q}(x) \xrightarrow{L^2} q(x), \quad \text{implying that} \quad \hat{q}(x) \xrightarrow{p} q(x).$$

Since $\hat{p}(x) \xrightarrow{p} p(x)$, Slutzky's theorem completes the proof:

$$\hat{f}(x) = \frac{\hat{q}(x)}{\hat{p}(x)} \xrightarrow{p} \frac{q(x)}{p(x)} = \frac{f(x)p(x)}{p(x)} = f(x). \quad \square$$

The main step in the proof was consistency of $\hat{q}(x)$ for $q(x)$. As in the last subsection, asymptotic normality for $\hat{q}(x)$ holds for individual xs: The Lyapunov central limit theorem can be applied directly. In this case, if $h \to 0$ and $nh \to \infty$ as $n \to \infty$,

$$\sqrt{nh}\left(\hat{q}(x) - \mathbb{E}(\hat{q}(x))\right) \xrightarrow{d} N\left(0, (\sigma^2 + f^2(x))p(x) \int K^2(t)dt\right).$$

Parallel to $\hat{p}(x)$ and $\hat{q}(x)$, it would be nice to have an asymptotic normality result for $f(x)$. Unfortunately, since the kernel smoother $\hat{f}(x)$ is a ratio of two random variables, direct central limit theorems cannot be used to find its asymptotic distribution. Another, more elaborate technique must be used. Moreover, in general it is not the pointwise behavior in x but the overall behavior for X measured by AMISE that is important. An expression for AMISE will also lead to values for $h = h_n$. Both results – asymptotic normality for $\hat{f}(x)$ and an expression for AMISE – are based on the same bias-variance trade-off reasoning.

For intuition and brevity, consider the following heuristic approach. (Correct mathematics can be found in the standard references.) Start by writing

$$\hat{f}(x) - f(x) \approx \frac{\hat{p}(x)}{p(x)}(\hat{f}(x) - f(x)) = \frac{1}{p(x)}\hat{q}(x) - \frac{f(x)}{p(x)}\hat{p}(x). \qquad (2.3.10)$$

Having established results about both $\hat{q}(x) = \hat{f}(x)\hat{p}(x)$ and $\hat{p}(x)$, asymptotic results for $\hat{f}(x) - f(x)$ can now be obtained using (2.3.10). It is seen that the difference between $\hat{f}(x) - f(x)$ and its "linearized" form $\frac{1}{p(x)}\hat{q}(x) - \frac{f(x)}{p(x)}\hat{p}(x)$ is $o_p\left(1/\sqrt{nh}\right)$.

The bias $\mathbb{E}\hat{f}(x) - f(x)$ is approximately

$$\begin{aligned}\mathbb{E}\left[\frac{\hat{q}(x) - f(x)\hat{p}(x)}{p(x)}\right] &= \frac{\mathbb{E}(\hat{q}(x)) - f(x)\mathbb{E}(\hat{p}(x))}{p(x)} = \left[\frac{\mathbb{E}(\hat{p}(x))}{p(x)}\right]\left[\frac{\mathbb{E}(\hat{q}(x))}{\mathbb{E}(\hat{p}(x))} - f(x)\right] \\ &\approx \frac{\mathbb{E}(\hat{q}(x))}{\mathbb{E}(\hat{p}(x))} - f(x), \qquad \text{assuming that} \quad \mathbb{E}(\hat{p}(x)) \approx p(x).\end{aligned}$$

Adding and subtracting $f(x)p(x)$ and using $\mathbb{E}(\hat{p}(x)) \approx p(x)$ leads to

$$\begin{aligned}
\frac{\mathbb{E}(\hat{q}(x))}{\mathbb{E}(\hat{p}(x))} - f(x) &\approx (p(x))^{-1}\left\{\mathbb{E}(\hat{q}(x)) - f(x)p(x) + f(x)p(x) - f(x)\mathbb{E}(\hat{p}(x))\right\} \\
&= (p(x))^{-1}\left\{\text{Bias}(\hat{q}(x)) - f(x)\text{Bias}(\hat{p}(x))\right\} \\
&\approx (p(x))^{-1}\left\{\frac{h^2}{2}\mu_2(K)q''(x) - f(x)\frac{h^2}{2}\mu_2(K)p''(x)\right\} \\
&= \frac{h^2}{2}\mu_2(K)\left[f''(x) + 2f'(x)(p'(x)/p(x))\right]
\end{aligned}$$

by using $q''(x) = (f(x)p(x))'' = f''(x)p(x) + 2f'(x)p'(x) + p''(x)f(x)$. Next, an approximation for the variance can be found similarly. It can be easily verified that

$$\frac{\hat{q}(x)}{\hat{p}(x)} - \frac{\mathbb{E}(\hat{q}(x))}{\mathbb{E}(\hat{p}(x))} = \left[\frac{\hat{q}(x)}{p(x)} - \frac{\hat{p}(x)}{p(x)}\frac{\mathbb{E}(\hat{q}(x))}{\mathbb{E}(\hat{p}(x))}\right]\left[\frac{p(x)}{\hat{p}(x)}\right].$$

Using $\frac{p(x)}{\hat{p}(x)} \xrightarrow{p} 1$, and pretending $\mathbb{E}\hat{f}(x) = \mathbb{E}\hat{q}(x)/\mathbb{E}\hat{p}(x)$, the desired variance is approximately the same as the variance of

$$G_n(x) = \frac{\hat{q}(x)}{p(x)} - \frac{\hat{p}(x)}{p(x)}\frac{\mathbb{E}(\hat{q}(x))}{\mathbb{E}(\hat{p}(x))}.$$

Now rewrite $G_n(x)$ in terms of $\hat{p}(x)$ and $\hat{q}(x)$ as

$$\begin{aligned}
G_n(x) &= \frac{1}{p(x)}\left[\hat{q}(x) - \mathbb{E}(\hat{q}(x))\right] - \frac{f(x)}{p(x)}\left[\hat{p}(x) - \mathbb{E}(\hat{p}(x))\right] \\
&= \gamma_1\left[\hat{p}(x) - \mathbb{E}(\hat{p}(x))\right] + \gamma_2\left[\hat{q}(x) - \mathbb{E}(\hat{q}(x))\right],
\end{aligned}$$

where $\gamma_1 = -f(x)/p(x)$ and $\gamma_2 = 1/p(x)$. Using the asymptotic normal distributions of $\hat{p}(x) - \mathbb{E}(\hat{p}(x))$ and $\hat{q}(x) - \mathbb{E}(\hat{q}(x))$ stated earlier, the delta method gives that $G_n(x)$ is also asymptotically normally distributed and identifies the variance.

For completeness, the delta method is the content of the following.

Theorem: Let $< Y_n >$ be a sequence of random variables satisfying $\sqrt{n}(Y_n - \theta) \xrightarrow{d} \mathbf{N}(0, \sigma^2)$. Given a differentiable function g and a fixed value of θ with $g'(\theta) \neq 0$,

$$\sqrt{n}\,[g(Y_n) - g(\theta)] \xrightarrow{d} N(0, \sigma^2[g'(\theta)]^2).\ \square$$

Now, it is seen that the variance of $G_n(x)$ is

$$
\begin{aligned}
nh\text{Var}[G_n(x)] &= [\gamma_1^2 + 2\gamma_1\gamma_2 f(x) + \gamma_2^2(f^2(x)+\sigma^2)]p(x)S(K) \\
&= \left[\frac{(f(x))^2}{(p(x))^2} - 2\frac{f(x)}{p(x)}\frac{1}{p(x)}f(x) + \frac{1}{(p(x))^2}(f^2(x)+\sigma^2)\right]p(x)S(K) \\
&= \frac{\sigma^2}{p(x)}S(K),
\end{aligned}
$$

in which the results for the variance of $\hat{p}(x)$, $\hat{q}(x)$ have been used along with the corresponding result for their correlation derived by the same reasoning, which gives the term with $2\gamma_1\gamma_2 f(x)$.

Now, putting together the bias and variance expressions gives the two desired theorems. First, we have the asymptotic normality of the NW estimator.

Theorem: Let K be a bounded, continuous kernel that is symmetric about zero (thus $\int tK(t)dt = 0$). Assume $f(x)$ and $p(x)$ are twice continuously differentiable and $\mathbb{E}(|Y_i|^{2+\delta}|X_i = x) < \infty$ for all x for some $\delta > 0$. Set $h = O(n^{-1/5})$. Then, for all x with $p(x) > 0$, the limiting distribution of $\hat{f}(x)$ is

$$
\sqrt{nh}(\hat{f}(x) - f(x)) \xrightarrow{d} N(B(x), V(x)) \tag{2.3.11}
$$

with asymptotic bias

$$
B(x) = \left(f''(x) + 2f'(x)\frac{p'(x)}{p(x)}\right)\mu_2(K) \tag{2.3.12}
$$

and asymptotic variance

$$
V(x) = \frac{\sigma^2 S(K)}{p(x)}, \tag{2.3.13}
$$

where $\mu_2(K) = \int t^2K(t)dt$ and $S(K) = \int K^2(t)dt$.

Proof: The proof is an application of the Lyapunov central limit theorem since the Lyapunov condition (the $2+\delta$ conditional moment) is satisfied. □

Finally, the key result of this subsection can be stated. The global measure of accuracy of the NW estimator is the $\mathsf{AMISE}(\hat{f})$, and it admits an asymptotic expression as a function of h, n, and several other fixed quantities determined from f and K.

Theorem: Assume the noise ε_i to be homoscedastic with variance σ^2. Then, for $h \to 0$ and $nh \to \infty$ as $n \to \infty$, the $\mathsf{AMISE}(\hat{f})$ of the NW estimator is

$$
\begin{aligned}
\mathsf{AMISE}(\hat{f}) &= \frac{h^4}{4}(\mu_2(K))^2 \int \left(f''(x) + 2f'(x)\frac{p'(x)}{p(x)}\right)^2 dx \\
&\quad + \frac{\sigma^2 S(K)}{nh}\int \frac{1}{p(x)}dx,
\end{aligned} \tag{2.3.14}
$$

where $\mu_2(K) = \int t^2K(t)dt$ and $S(K) = \int K^2(t)dt$. The optimal bandwidth h^{opt} decreases at rate $n^{-\frac{1}{5}}$, which corresponds to an $n^{-\frac{4}{5}}$ rate of decrease of the AMISE.

Proof: The derivation of the expression of $\mathsf{AMISE}(\hat{f})$ follows directly from the previous heuristics. To find h^{opt}, write the AMISE as

$$\mathsf{AMISE}(\hat{f}_h) = C_B^2 \mu_2^2(K) h^4 + C_V S(K) n^{-1} h^{-1},$$

where $C_V = \sigma^2 \int \frac{1}{p(x)} dx$ and $C_B = \frac{1}{2} \int \left(f''(x) + 2f'(x) \frac{p'(x)}{p(x)} \right) dx$ are constants. By solving $\frac{\partial \mathsf{AMISE}(\hat{f}_h)}{\partial h} = 0$, it is straightforward to see that the bandwidth that minimizes the AMISE above is

$$h^{\mathsf{opt}} = \left[\frac{C_V}{4C_B^2} \right]^{1/5} \left[\frac{S(K)}{\mu_2^2(K)} \right]^{1/5} n^{-1/5},$$

with the corresponding optimal AMISE given by

$$\mathsf{AMISE}^{\mathsf{opt}} = C_V^{4/5} C_B^{2/5} [4^{1/5} + 4^{-4/5}] [S(K)]^{4/5} [\mu_2(K)]^{2/5} n^{-4/5}. \square \quad (2.3.15)$$

Note that all the expressions for the measure of accuracy of estimators encountered so far depend on the smoothing parameter (bandwidth) h, hence the central role of estimating h in theory and in practice. Expression (2.3.15) for the optimal AMISE depends on the two kernel constants $\mu_2(K)$ and $S(K)$. This latter fact will be used later in the argument for determining the optimal kernel as a measure of accuracy of the estimator. Extensions of the results here to use derivatives of f of higher order result in faster rates, as will be seen in the next subsection.

2.3.5 Convergence Theorems and Rates for Kernel Smoothers

Although studied in separate subsections, the difference between PC and NW as regression estimators is small in the sense that NW generalizes PC. That is, if X were uniformly distributed to correspond to equispaced points x_i and $K(u) = I_{\{|u| \leq (1/2)\}}(u)$ were used as a kernel, then $\hat{p}$ in (2.3.5) would become h in the limit of large n. In fact, the key difference between PC and NW is that NW is a convex version of the same weights as used in PC by normalization. This is why the two kernel smoothers (PC and NW) along with the kernel density estimator have expressions for certain of their MSEs that are of the form $C_1 h^4 + C_2(1/nh)$, where the constants C_1 and C_2 depend on the local behavior of f, the properties of K, and σ^2; see (2.3.3), (2.3.6), (2.3.7), and (2.3.14).

Looking at the technique of proof of these results for MSE, it is seen that the properties of the kernels and the order of the Taylor expansion are the main hypotheses. Indeed, assuming $\int vK(v)dv = 0$ made the contribution of the first derivative p' zero in (2.3.6) so that the second derivative was needed in the expression. It is possible to generalize the generic form of the MSEs by making assumptions to ensure lower-order terms drop out. Since it is only the terms with an even number of derivatives that contribute, one can obtain a general form $C_1 h^{2d} + C_2(1/nh)$ and therefore an optimal $h^{\text{opt}} = h_n = 1/n^{1/(2d+1)}$, where d is the number of derivatives assumed well behaved and the C_is are new but qualitatively similar.

Rates of this form are generic since they come from a variance-bias decomposition using properties of kernels and Taylor expansions; see Eubank (1988), Chapter 4. To see how they extend to derivatives of f, consider the natural estimator for the kth derivative, $k \le d$, of f in the simplest case, namely the PC kernel smoother. This is the kth derivative $\hat{f}^{(k)}(x)$ of the PC estimator $\hat{f}(x)$,

$$\hat{f}^{(k)}(x) = \frac{1}{nh^{(k+1)}} \sum_{i=1}^{n} K^{(k)}\left(\frac{x - X_i}{h}\right) Y_i; \tag{2.3.16}$$

see Eubank (1988), Chapter 4.8 for more details. The result for the PC estimator is the following; it extends to the NW as well.

Proposition: Consider the deterministic design and the estimate $\hat{f}^{(k)}(x)$ of $f^{(k)}(x)$, where $x \in \mathscr{X} \subset \mathbb{R}$ as defined in (2.3.16). Let $S^{(k)}(K) = \int [K^{(k)}(t)]^2 dt$ and $\mu_2^{(k)}(K) = \int t^{2+k} K^{(k)}(t) dt$, and assume:

1. $K \in \mathscr{C}^k$ with support $[-1, 1]$ and $K^{(j)}(-1) = K^{(j)}(1) = 0$, $j = 0, \cdots, k-1$.
2. $f^{(k)} \in \mathscr{C}^2$; i.e., $f^{(k)}$ is k times continuously differentiable.
3. $\mathbb{V}(\varepsilon_i) = \sigma^2$ for $i = 1, 2, \cdots, n$.
4. $X_i = \frac{i-1}{n-1}$ for $i = 1, 2, \cdots, n$.
5. $h \to 0$ and $nh^{k+1} \to \infty$ as $n \to \infty$.

Then

$$\text{AMSE}(\hat{f}^{(k)}(x)) = \frac{\sigma^2}{nh^{2k+1}} S^{(k)}(K) + \frac{[\mu_2^{(k)}(K) f^{(k+2)}(x)]^2}{[(k+2)!]^2} h^4.$$

Proof: This follows the same reasoning as was used to get (2.3.3). □

Given that all these kernel-based function estimators are so similar in their behavior – in terms of the rates for pointwise AMISE as well as the averaged AMISE – it is possible to state generic rates for the estimators and their sense of errors. Hardle (1990) observes that the rate of convergence depends on four features:

1. Dimension of the covariate X, here p;
2. Object to be estimated; e.g., $f^{(k)}$, the kth derivative of f;
3. Type of estimator used;
4. Smoothness of the function f.

When the dimension of X is $p \ge 2$, it is understood that the appropriate kernel is the product of the univariate kernels for the components X_j, $j = 1, ..., p$ of $\boldsymbol{X}$.

To be more formal, observe that parametric techniques typically produce convergence rates of order $O(n^{-1/2})$, whereas nonparametric estimation is slower, with convergence rates of order n^{-r} for some $r \in (0, 1/2)$ for the function class $\mathscr{F} = \mathscr{C}^{d,\alpha}(\overline{\mathscr{X}})$. This is the smoothness class of d times continuously differentiable functions f on $\mathscr{X}$ such that the dth derivative $f^{(d)}(x)$ of $f(x)$ is globally α-Hölder continuous. The rate is defined by $r = r(p, k, d, \alpha)$; the constant in the rate depends on the form of error used and other aspects of the estimator such as the kernel, σ, and derivatives of f. How large n must be for the rate to kick in is an open question. Clearly, the slower the rate, the more complicated the estimand, so the more data will be needed, pushing the n needed to observe the rate further out. Moreover, under appropriate uniformity criteria on the values of x and n, the pointwise rates can be integrated to give the corresponding AMISE rates.

Hardle (1990) gives an expression for r that establishes its dependence on the four qualitative features above.

Theorem: Let f be d times continuously differentiable. Assume that $f^{(d)}$ is α-Hölder continuous for some α, and let $K \geq 0$ be a nonnegative, continuous kernel satisfying

$$\int K(v)dv = 1, \qquad \int vK(v)dv = 0, \qquad \text{and} \qquad \int |v|^{2+\alpha} K(v)dv < \infty.$$

Then, based on IID samples $(X_1, Y_1), \cdots, (X_n, Y_n) \in \mathbb{R}^p \times \mathbb{R}$, kernel estimates of $f^{(k)}$ have optimal rates of convergence n^{-r}, where

$$r = \frac{2(d - k + \alpha)}{2(d + \alpha) + p}.$$

Proof (sketch): The proof uses the variance-bias decomposition, Taylor expansions, properties of the kernel, and so forth, as before. A detailed presentation of the proof can be found in Stone (1980), see also Stone (1982). Other authors include Ibragimov and Hasminksy (1980), Nussbaum (1985), and Nemirovsky et al. (1985).

For the case $p = 1$, the proposition shows the variance of $\hat{f}^k(x)$ is

$$\mathrm{Var}(\hat{f}^{(k)}(x)) = \frac{\sigma^2}{nh^{2k+1}} S^{(k)}(K)$$

and the bias is

$$\mathbb{E}[\hat{f}^{(k)}(x)] - f^{(k)}(x) = C_{d+\alpha,k} f^{(k+2)}(x) h^{d+\alpha-k}.$$

The leading term of the mean squared error of $\hat{f}^{(k)}$ is such that

$$\mathsf{AMSE}(\hat{f}^{(k)}(x)) = \frac{\sigma^2}{nh^{2k+1}} S^{(k)}(K) + [C_{d+\alpha,k} f^{(k+2)}(x)]^2 h^{2(d+\alpha-k)}.$$

Taking the partial derivative of $\mathsf{AMSE}(\hat{f}^{(k)}(x))$ with respect to h and setting it to zero yields

$$h^{\mathsf{opt}} = \left[\frac{2k+1}{2(d+\alpha-k)} \frac{\sigma^2 S(K)}{n[C_{d+\alpha,k} f^{(k+2)}(x)]^2} \right]^{\frac{1}{2(d+\alpha)+1}}.$$

The mean squared error obtained using h^{opt} is therefore approximately

$$\mathsf{AMSE}_0 \approx C_0[C_{d+\alpha,k}f^{(k+2)}(x)]^{\frac{2(2k+1)}{2(d+\alpha)+1}}\left[\frac{\sigma^2 S(K)}{n}\right]^{\frac{2(d+\alpha-k)}{2(d+\alpha)+1}}. \quad \square$$

Corollary: Suppose IID samples $(X_1,Y_1),\cdots,(X_n,Y_n) \in \mathbb{R}^p \times \mathbb{R}$ are used to form the kernel smoothing estimate $\hat{f}$ of a Lipschitz-continuous function f. Then $\alpha = 1, d = 1$, and $k = 0$, and the rate of convergence is

$$n^{-\frac{4}{4+p}}.$$

For the univariate regression case considered earlier, this corollary gives the rate $n^{-(4/5)}$, as determined in (2.3.14) and (2.3.15). $\square$

The expression provided in the last theorem for finding the rate of convergence of kernel smoothers has the following implications:

- As $(d+\alpha)$ increases, the rate of convergence r increases. Intuitively, this means that smooth functions are easier to estimate.
- As k increases, the rate of convergence r decreases, meaning that derivatives are harder to estimate.
- As p increases, the rate of convergence r decreases, which is simply the Curse of Dimensionality discussed at length in Chapter 1.

One of the most general results on the convergence of the NW estimator is due to Devroye and Wagner (1980). Their theorem is a distribution-free consistency result.

Theorem (Devroye and Wagner, 1980): Let $(X_1,Y_1),\cdots,(X_n,Y_n)$ be an $\mathbb{R}^p \times \mathbb{R}$-valued sample, and consider the NW estimator

$$\hat{f}(x) = \frac{\sum_{i=1}^n K_h(x-X_i)Y_i}{\sum_{i=1}^n K_h(x-X_i)}$$

for $f(x) = \mathbb{E}(Y|\mathbf{X}=\mathbf{x})$. If $\mathbb{E}(|Y|^q) < \infty$, $q \geq 1$, $h_n \longrightarrow_n 0$, and $nh_n^p \longrightarrow_n \infty$, and if (i) K is a nonnegative function on $\mathbb{R}^d$ bounded by $k^* < \infty$; (ii) K has compact support; and (iii) $K(u) \geq \beta I_B(u)$ for some $\beta > 0$ and some closed sphere B centered at the origin with positive radius, then

$$\mathbb{E}\left\{\int |m_n(\boldsymbol{x}) - m(\boldsymbol{x})|^q \mu(d\boldsymbol{x})\right\} \longrightarrow_n 0. \quad \square$$

Although the results of these theorems are highly satisfying, they only hint at a key problem: The risk, as seen in $\mathsf{AMISE}_{\hat{f}_n}(h)$, increases quickly with the dimension of the problem. In other words, kernel methods suffer the Curse of Dimensionality. The following table from Wasserman (2004) shows the sample size required to obtain a relative mean square error less than 0.1 at 0 when the density being estimated is a multivariate normal and the optimal bandwidth has been selected.

Dimension	Sample size
1	4
2	19
3	67
$\vdots$	$\vdots$
9	187,000
10	842,000

Wasserman (2004) expresses it this way: Having 842,000 observations in a ten-dimensional problem is like having four observations in a one-dimensional problem. Another way to dramatize this is to imagine a large number of dimensions; 20,000 is common for many fields such as microarray analysis. Suppose you had to use the NW estimator to estimate $\hat{f}(\boldsymbol{x})$ when $p = 20,000$ and data collection was not rapid. Then, humans could well have evolved into a different species rendering the analysis meaningless, before the NW estimator got close to the true function.

2.3.6 Kernel and Bandwidth Selection

There are still several choices to be made in forming a kernel estimator: the kernel itself and the exact choice of $h = h_n$. The first of these choices is easy because differences in Ks don't matter very much. The choice of h is much more delicate, as will be borne out in Section 2.5.

2.3.6.1 Optimizing over K

Observe that the expression for the minimal AMISE in (2.3.15) depends on the two kernel constants $\mu_2(K)$ and $S(K)$ through

$$V(K)B(K) = [S(K)]^2\mu_2(K) = \left[\int K^2(t)dt\right]^2 \left[\int t^2K(t)dt\right].$$

The obvious question is how to minimize over K. One of the major problems in seeking an optimum is that the problems of finding an optimal kernel K^* and an optimal bandwidth h are coupled. These must be uncoupled before comparing two candidate kernels. The question becomes: What are the conditions under which two kernels can use the same bandwidth (i.e., the same amount of smoothing) and still be compared to see which one has the smaller MISE?

The concept of canonical kernels developed by Marron and Nolan (1988) provides a framework for comparing kernels. For the purposes of the sketch below, note that the standardization $V(K) = B(K)$ makes it possible to optimize MISE as a function of K. So, the original goal of minimizing $V(K)B(K)$ becomes

$$\text{minimize} \quad \int K^2(t)dt$$

subject to

$$\text{(i)} \int K(t)dt = 1, \quad \text{(ii)} \quad K(t) = K(-t), \quad \text{and} \quad \text{(iii)} \quad \mu_2(K) = 1.$$

Using Lagrange multipliers on the constraints, it is enough to minimize

$$\int K^2(t)dt + \lambda_1 \left[\int K(t)dt - 1 \right] + \lambda_2 \left[\int t^2 K(t)dt - 1 \right].$$

Letting $\triangle K$ denote a small variation from the minimizer of interest gives

$$2 \int K(t) \triangle K(t)dt + \lambda_1 \left[\int \triangle K(t)dt \right] + \lambda_2 \left[\int t^2 \triangle K(t)dt \right] = 0,$$

which leads to

$$2K(t) + \lambda_1 + \lambda_2 t^2 = 0.$$

It can be verified that the Epanechnikov kernel, defined by

$$K(t) = \frac{3}{4}(1 - t^2) \qquad \text{for} \qquad -1 \le t \le 1,$$

satisfies the conditions and constraints above, and is therefore the optimum.

Although the Epanechnikov kernel emerges as the optimum under (2.32), there are other senses of optimality that result in other kernels; see Eubank (1988). Nevertheless, it is interesting to find out how suboptimal commonly used kernels are relative to Epanechnikov kernels. Hardle (1990) addresses this question by computing the efficiency of suboptimal kernels, with respect to the Epanechnikov kernel K^*, based on $V(K)B(K)$. The natural ratio to compute is

$$D(K^*, K) = \left[\frac{V(K)B(K)}{V(K^*)B(K^*)} \right]^{\frac{1}{2}},$$

and some values for it for certain kernels are provided in the table below.

Kernel name	Expression	Range	$V(K)B(K)$	$D(K^*,K)$
Epanechnikov	$K(v) = (3/4)(1 - v^2)$	$-1 \le v \le 1$	$9/125$	1
Biweight	$K(v) = \frac{15}{16}(1 - v^2)^2$	$-1 \le v \le 1$	$25/343$	1.0061
Triangle	$K(v) = (1 - \|v\|)$	$-1 \le v \le 1$	$2/27$	1.0143
Gaussian	$K(v) = e^{-v^2/2}/\sqrt{2\pi}$	$-\infty < v < \infty$	$1/4\pi$	1.0513

The table above makes it clear that if minimizing the MISE by examining $V(K)B(K)$ is the criterion for choosing a kernel, then nonoptimal kernels are not much worse than the Epanechnikov kernel. Indeed, in most applications there will be other sources of error, the bandwidth for instance, that contribute more error than the choice of kernel.

2.3.6.2 Empirical Aspects of Bandwidth Selection

The accuracy of kernel smoothers is governed mainly by the bandwidth h. So, write $d_\bullet(h)$ in place of $d_\bullet(\hat{f}, f)$. This givesd_I for the ISE, d_A for the ASE, and d_M for the MSE. The first result, used to help make selection of h more data driven, is the surprising observation that the ASE and ISE are the same as the MSE in a limiting sense. Formally, let H_n be a set of plausible values of h defined in terms of the dimension p of the covariate $\boldsymbol{X}$, and the sample size n. For the theorem below, H_n is the interval $H_n = [n^{\delta-1/d}, n^{-\delta}]$, with $0 < \delta < 1/(2d)$. In fact, one can put positive constants into the expressions defining the endpoints while retaining the essential content of the theorem; see Hardle (1990), Chapters 4 and 5 and also Eubank (1988), Chapter 4 and Marron and Härdle (1986).

Theorem: Assume that the unknown density $p(x)$ of the covariate X and the kernel function K are Hölder continuous and that $p(x)$ is positive on the support of the $w(x)$. If there are constants C_k, $k = 1, \ldots, \infty$ so that $\mathbb{E}(Y^k | X = x) \le C_k < \infty$, then for kernel estimators

$$\sup_{h \in H_n} \frac{|d_A(h) - d_M(h)|}{d_M(h)} \to 0 \quad \text{a.s.,}$$

$$\sup_{h \in H_n} \frac{|d_I(h) - d_M(h)|}{d_M(h)} \to 0 \quad \text{a.s.}\square$$

This theorem gives insight about optimal bandwidth, but identifying a specific choice for h remains.

The importance of choosing h correctly has motivated so many contributions that it would be inappropriate to survey them extensively here. It is enough to note that none seem to be comprehensively satisfactory. Thus, in this subsection, it will be enough to look at one common method based on CV because a useful method must be given even if it's not generally the best. It may be that a systematically ideal choice based on p, the data, K, and the other inputs to the method (including the true unknown function) just does not exist apart from local bandwidth concepts discussed briefly in Section 2.4.1 and indicated in Silverman's theorem in Section 3.3.2.

Clearly, the bandwidth should minimize an error criterion over a set of plausible values of h. For instance, consider selecting the bandwidth that achieves the minimum of $d_M(h) = \mathsf{MISE}(\hat{f}_h)$ over H_n; i.e., let

$$\hat{h} = \arg \min_{h \in H_n} d_M(h).$$

In this expression, it is pragmatically understood that H_n just represents an interval to be searched and that H_n shrinks to zero. Note that $d_I(h)$ and $d_M(h)$ cannot be computed but that $d_A(h)$ can be computed because it is the empirical approximation to MISE.

The theorem above assures $d_A(h)$ is enough because, for $\delta > 0$,

$$\frac{d_A(h)}{d_M(h)} \xrightarrow{\text{a.s.}} 1$$

uniformly for $h \in H_n$. Therefore, the minimizer of $d_A(h)$ is asymptotically the same as the minimizer of $d_M(h)$, the desired criterion. By writing

$$d_A(h) = \frac{1}{n}\sum_{i=1}^{n} w(X_i)\hat{f}_h^2(X_i) + \frac{1}{n}\sum_{i=1}^{n} w(X_i)f^2(X_i) - 2C(h),$$

where $C(h) = \frac{1}{n}\sum_{i=1}^{n} w(X_i)\hat{f}_h(X_i)f(X_i)$, it is easy to see that the middle term does not depend on h and so does not affect the minimization. Dropping it leaves

$$\hat{h} = \arg\min_{h \in H_n} d_A(h) \approx \arg\min_{h \in H_n} \left(\frac{1}{n}\sum_{i=1}^{n} w(X_i)\hat{f}_h^2(X_i) - 2\hat{C}(h) \right). \tag{2.3.17}$$

Note that, to get the approximation, $C(h)$ is replaced by $\hat{C}(h)$, in which Y_i is used in place of $f(X_i)$. That is,

$$\hat{C}(h) = \frac{1}{n}\sum_{i=1}^{n} w(X_i)\hat{f}_h(X_i)Y_i.$$

On the right-hand side of (2.3.17), complete the square by adding and subtracting $\frac{1}{n}\sum_{i=1}^{n} w(X_i)Y_i^2$, which does not depend on h. The resulting objective function is

$$\pi(h) = \frac{1}{n}\sum_{i=1}^{n} w(X_i)(Y_i - \hat{f}_h(X_i))^2.$$

This leads to defining

$$\hat{h}_\pi = \arg\min_{h \in H_n} \pi(h). \tag{2.3.18}$$

It is important to note that $\hat{h}_\pi$ and $\hat{h}$ are different since they are based on slightly different objective functions when n is finite, even though they are asymptotically equivalent; i.e., $\hat{h} \approx \hat{h}_\pi$ in a limiting sense. After all, the objective function for $\hat{h}_\pi$ is derived to approximate $\text{bias}^2 + \text{variance}$ while $\hat{h}$ is the "pure" bandwidth, effectively requiring knowledge of the unknown f. It is easy to imagine optimizing other objective functions that represent different aspects of bias and variance.

Despite the apparent reasonableness of (2.3.18), the bandwidth $\hat{h}_\pi$ is not quite good enough; it is a biased estimate of $\arg\min d_A(h)$. Indeed, using Y_i in the construction of $\hat{f}_h(X_i)$ means that $|Y_i - \hat{f}_h(X_i)|$ will systematically be smaller than $|Y_i - f(X_i)|$; i.e., $\pi(h)$ will typically underestimate $d_A(h)$. So, one more step is required.

This is where CV comes in. It permits removal of the Y_i from the estimate used to predict Y_i. That is, the bias can be removed by using the optimal bandwidth

$$\hat{h}_{CV} = \arg \min_{h \in H_n} CV(h),$$

where

$$CV(h) = \sum_{i=1}^{n} w(X_i)(Y_i - \hat{f}_h^{(-i)}(X_i))^2$$

and $\hat{f}_h^{(-i)}$ is the estimator of f obtained without the ith observation. Intuitively, each term in the sum forming $CV(h)$ is the prediction of a response not used in forming the prediction. This seems fundamental to making $CV(h)$ less prone to bias than $\pi(h)$.

There is a template theorem available from several sources (Hardle 1990 is one) that ensures the CV-generated bandwidth works asymptotically as well as bandwidths selected by using $d_A(h)$ directly.

Theorem: Let H_n be a reasonable interval, such as $[n^{\delta - 1/d}, n^{-\delta}]$. Suppose f, K, $p(x)$, and the moments of ε are well behaved. Then, the bandwidth estimate $\hat{h}_{CV}$ is asymptotically optimal in the sense that

$$\frac{d_A(\hat{h}_{CV})}{\inf_{h \in H_n} d_A(h)} \xrightarrow{\text{a.s.}} 1 \qquad \text{for} \quad n \to \infty.$$

Although $\hat{h}_{CV}$ is now well defined and optimal in certain cases, CV is computationally onerous. The need to find n estimates of $\hat{f}_h(\cdot)$ becomes prohibitive even for moderately large sample sizes. Fortunately, the burden becomes manageable by rewriting the expression for $CV(h)$ as

$$CV(h) = \sum_{i=1}^{n} v_i (Y_i - \hat{f}_h(X_i))^2,$$

where

$$v_i = \left[1 - \frac{K(0)}{\sum_{j=1}^{n} K\left(\frac{x_i - x_j}{h}\right)} \right]^{-2}.$$

In this form, the estimate $\hat{f}_h$ is found only once; the rest of the task consists of searching for the minimum of $CV(h)$ over H_n.

To conclude this subsection, the role of the weight function $w(\cdot)$ in ASE, see (2.3.1) or (2.3.17), bears comment. Recall that outliers and extreme values are different. Outliers are anomalous measurements of Y and extreme values are values of X far from the bulk of X measurements. Extreme values, valid or not, are often overinfluential, and sometimes it is desirable to moderate their influence. Choice of w is one way to do this. That is, if necessary in a particular application, one can choose w to stabilize $\pi(h)$ to prevent it from being dominated by an extreme point, or outlier. The stability is in terms of how sensitive the choice of h is to small deviations in the data. Roughly, one can choose w to be smaller for those values of X_i that are far from a measure of location of the Xs, such as $\bar{X}$, provided the values of X are clustered around their central value, say $\bar{X}$. When a Y_i is "far" from where it "should" be the problem is more acute and specialized, requiring techniques more advanced than those here.

2.3.7 Linear Smoothers

It was observed in (2.1.1) that a linear form was particularly desirable. In this section it is seen that the NW estimator (and PC estimator) are both linear because the smoothing they do gives a weighted local average $\hat{f}$ to estimate the underlying function f. The $W_j(x)$s for $j = 1, \ldots, n$ are a sequence of weights whose size and form near x are controlled by h from the (rescaled) kernel K_h. By their local nature, kernel smoothers have weights $W_j(x)$ that are large when x is close to X_j and that are smaller as x moves away from X_j.

It can be checked that the NW kernel smoother admits a linear representation with

$$W_j(x) = \frac{K_h(x - X_j)}{\sum_{i=1}^{n} K_h(x - X_i)}.$$

However, the linearity is only for fixed h; once h becomes dependent on any of the inputs to the smoother, such as the data, the linearity is lost.

Let $\mathbf{W} = (\mathbf{W_{ij}})$ with $W_{ij} = W_j(x_i)$. Then, (2.1.1) can be expressed in matrix form as

$$\hat{\mathbf{f}} = \mathbf{W}\boldsymbol{y}, \tag{2.3.19}$$

where

$$\hat{\mathbf{f}} = \begin{bmatrix} \hat{f}(x_1) \\ \hat{f}(x_2) \\ \vdots \\ \hat{f}(x_n) \end{bmatrix}, \mathbf{W} = \begin{bmatrix} W_1(x_1) & W_2(x_1) & \cdots & W_n(x_1) \\ W_1(x_2) & W_2(x_2) & \cdots & W_n(x_2) \\ \vdots & \vdots & \vdots & \vdots \\ W_1(x_n) & W_2(x_n) & \cdots & W_n(x_n) \end{bmatrix}, \text{ and } \quad \boldsymbol{y} = \begin{bmatrix} y_1 \\ y_2 \\ \vdots \\ y_n \end{bmatrix}.$$

It will be seen that spline smoothers, like LOESS and kernel smoothers, are linear in the sense of (2.1.1).

One immediate advantage of the linear representation is that important aspects of $\hat{\mathbf{f}}$ can be expressed in interpretable forms. For instance,

$$\text{Var}(\hat{\mathbf{f}}) = \text{Var}(\mathbf{W}\boldsymbol{y}) = \mathbf{W}\text{Var}(\boldsymbol{y})\mathbf{W}^{\mathsf{T}}.$$

In the case of IID noise with variance σ^2, this reduces to

$$\text{Var}(\hat{\mathbf{f}}) = \sigma^2 \mathbf{W}\mathbf{W}^{\mathsf{T}}, \tag{2.3.20}$$

generalizing the familiar form from linear regression.

2.4 Nearest Neighbors

Consider data of the form $(Y_i, \boldsymbol{X}_i)$ for $i = 1, ..., n$, in which the $\boldsymbol{X}$s and the Ys can be continuous or categorical. The hope in nearest-neighbor methods is that when covariates are close in distance, their ys should be similar. The idea is actually more natural to classification contexts, but generalizes readily to regression problems. So, the discussion here goes back and forth between classification and regression. In its regression form, nearest neighbors is a Classical method in the same spirit as kernel regression.

In first or 1-nearest neighbors classification, the strategy for binary classification is to assign a category, say 0 or 1, to Y_{new} based on which old set of covariates $\boldsymbol{x}_i$ is closest to $\boldsymbol{x}_{new}$ the covariates of Y_{new}. Thus, the 1-nearest-neighbor rule for classification of Y_{new} based on $\boldsymbol{x}_{new} = (x_{1,new}, ..., x_{p,new})$ looks for the set of covariates $x_{closest}$ that has already occurred that is closest to $\boldsymbol{x}_{new}$ and assigns its y-value. Thus, $\hat{y}_{new} = y_{closest}$. More generally, one looks at the k closest $\boldsymbol{x}_i$s to $\boldsymbol{x}_{new}$ to define the k nearest neighbor rule, k-NN. Thus, find the k sets of covariates closest to $\boldsymbol{x}_{new}$ that have already occurred, and assign the majority vote of their y values to be $\hat{y}_{new}$. More formally, if $\boldsymbol{x}_{i_1}, ..., \boldsymbol{x}_{i_k}$ are the k closest sets of covariates to $\boldsymbol{x}_{new}$, in some measure of distance on $\mathbb{R}^p$, then $(1/k)\sum_{j=1}^{k} y_{i_j} \geq (1/2)$ implies setting $\hat{y}_{new} = 1$.

The same procedure can be used when Y is continuous. This is called k-NN regression and the k nearest (in x) y-values are averaged.

The main free quantities are the value of k and the choice of distance. It is easiest to think of k as a sort of smoothing parameter. A small value of k means using few data points and accepting a high variance in predictions. A high value of k means using many data points and hence lower variance at the cost of high bias from including a lot of data that may be too far away to be relevant. In principle, there is an optimal value of k that achieves the best trade-off.

Good values of k are often found by cross-validation. As in other settings, divide the sample into, say, ℓ subsets (randomly drawn, disjoint subsamples). For a fixed value of k, apply the k-NN procedure to make predictions on the ℓth subset (i.e., use the $\ell - 1$ subsets as the data for the predictions) and evaluate the error. The sum of squared errors is the most typical choice for regressions; for classification, the most typical choice is the accuracy; i.e., the percentage of correctly classified cases. This process is then successively applied to each of the remaining $\ell - 1$ subsets. At the end, there are ℓ errors; these are averaged to get a measure of how well the model predicts future outcomes. Doing this for each k in a range permits one to choose the k with the smallest error or highest accuracy.

Aside from squared error on the covariates, typical distances used to choose the nearest neighbors include the sum of the absolute values of the entries in $\boldsymbol{x} - \boldsymbol{x}'$ or their maximum (sometimes called the city block or Manhattan distance).

Given k and the distance measure, the k-NN prediction for regression is

$$\hat{Y}_{new}(\boldsymbol{x}_{new}) = \frac{1}{k} \sum_{i \in K(\boldsymbol{x}_{new})} Y_i,$$

where $K(x_{new})$ is the set of covariate vectors in the sample closest to x_{new}.

In the general classification setting, suppose there are K classes labeled 1,..., K and $\boldsymbol{x}_{new}$ is given. For $j = 0, 1$, let $C_j(x_{new})$ be the data points x_i among the k values of $\boldsymbol{x}$ closest to $\boldsymbol{x}_{new}$ that have $y_i = j$. Then the k nearest-neighbor assignment is the class j having the largest number of the k data points' $\boldsymbol{x}_i$s in it,

$$\hat{Y}_{new}(\boldsymbol{x}_{new}) = \text{argmax}\{j \mid \#(\hat{C}_j(\boldsymbol{x}_{new}))\}.$$

To avoid having to break ties, k is often chosen to be odd.

An extension of k-NN regression, or classification, is distance weighting. The idea is that the $\boldsymbol{x}$ closest to an $\boldsymbol{x}_{new}$ should be most like it, so its vote should be weighted higher than the second closest $\boldsymbol{x}_i$ or the 3rd closest $\boldsymbol{x}_i$, and so on to the kth closest. Indeed, once the form of the weights is chosen, k can increase so that all the data points are weighted giving a sort of ensemble method over data points. In effect, a version of this is done in linear smoothing, see (2.1.1).

Let $\boldsymbol{x}_{(i),new}$ be the ith closest data point to $\boldsymbol{x}_{new}$. Given a distance d, weights w_i for $i = 1, ..., k$ can be specified by setting

$$w_i(\boldsymbol{x}_{new}, \boldsymbol{x}_{(i),new}) = \frac{e^{d(\boldsymbol{x}_{new}, \boldsymbol{x}_{(i),new})}}{\sum_{i=1}^{k} e^{d(\boldsymbol{x}_{new}, \boldsymbol{x}_{(i),new})}}.$$

Now, $\sum_{i=1}^{k} w_i(\boldsymbol{x}_{new}, \boldsymbol{x}_{(i),new}) = 1$. For classification, one takes the class with the maximum weight among the k nearest-neighbors. For regression, the predictor is

$$\hat{y}_{new} = \sum_{i=1}^{k} w_i(\boldsymbol{x}_{new}, \boldsymbol{x}_{(i),new}) y_{\boldsymbol{x}_{(i),new}},$$

in which $y_{\boldsymbol{x}_{(i),new}}$ is the value of the ith closest data point to $\boldsymbol{x}_{new}$. In either case, it is conventional to neglect weights and associate an assessment of variance to the predicted value from

$$\widehat{\text{Var}}(y_{new}) = \frac{1}{k-1} \sum_{i=1}^{k} (y - y_{\boldsymbol{x}_{(i),new}})^2.$$

2.4.0.1 Some Basic Properties

Like kernel methods, nearest neighbors does not really do any data summarization: There is no meaningful model of the relationship between the explanatory variables and the dependent variable. It is a purely nonparametric technique so k-NNs is not really interpretable and so not as useful for structure discovery or visualization as later methods.

Second, another problem with k-NNs is that it is sensitive to predictors that are useless when their values are close to those in an $\boldsymbol{x}_{new}$. This is the opposite of many New Wave methods (like recursive partitioning and neural networks), which often exclude irrelevant explanatory variables easily. To work well, k-NNs needs good variable selection or a distance measure that downweights less important explanatory variables.

Third, on the other hand, k-NNs has the robustness properties one expects. In particular, when data that can safely be regarded as nonrepresentative are not present, the removal of random points or outliers does not affect predictions very much. By contrast, logistic regression and recursive partitioning can change substantially if a few, even only one, data point is changed.

Overall, k-NNs is useful when a reliable distance can be specified easily but little modeling information is available. It suffers the Curse on account of being little more than a sophisticated look-up table.

Perhaps the biggest plus of k-NNs is its theoretical foundation, which is extensive since k-NNs was one of the earliest classification techniques devised.

One theoretical point from the classification perspective is decision-theoretic and will be pursued in Chapter 5 in more detail. Let $L_{j,k}$ be the loss from assigning the jth class when the kth class is correct. The expected loss or misclassification risk is $R_j = \sum_{k=1}^{M} L_{jk} P(k|x)$, where $W(k|x)$ is the posterior probability of class k given x. Now, the best choice is

$$j^{\text{opt}} = \arg \min_{i \le j \le M} R_j(\boldsymbol{x}),$$

which reduces to the modal class of the posterior distribution on the classes when the cost of misclassification is the same for all classes. In this case, the risk is the misclassification rate and one gets the usual Bayes decision rules in which one can use estimates $\hat{W}(j|\boldsymbol{x})$. It is a theorem that the classification error in $k = 1$ nearest-neighbor classification is bounded by twice the Bayes error; see Duda et al. (2000). Indeed, when both k and n go to infinity, the k nearest-neighbor error converges to the Bayesian error.

An even more important theoretical point is the asymptotics for nearest-neighbor methods. Let $f_j(\boldsymbol{x}) = \mathbb{E}(Y_i|\boldsymbol{x})$; in classification this reduces to $f_(\boldsymbol{x}) = \mathbb{P}(j|\boldsymbol{x}) = \mathbb{P}(Y_j = 1|\boldsymbol{x})$. Clearly, the "target" functions f_j satisfy

$$f_j(\boldsymbol{x}) = \arg \min_{f} \mathbb{E}((Y_j - f(\boldsymbol{x}))^2|\boldsymbol{x}),$$

where $0 \le f_j(\boldsymbol{x}) \le 1$ and $\sum_j f_j(\boldsymbol{x}) = 1$. One way to estimate these functions is by using the NW estimator

$$\hat{f}_j(\boldsymbol{x}) = \frac{\sum_{i=1}^{n} Y_i K_h(\boldsymbol{x} - \boldsymbol{X}_i)}{\sum_{i=1}^{n} K_h(\boldsymbol{x} - \boldsymbol{X}_i)},$$

where $h > 0$ is a smoothing parameter and the kernel K is absolutely integrable.

An alternative to the NW estimator based on k-NN concepts is the following. Suppose the f_js are well behaved enough that they can be locally approximated on a small neighborhood $R(\boldsymbol{x})$ around $\boldsymbol{x}$ by an average of evaluations. Then,

$$f_j(\boldsymbol{x}) \approx \frac{1}{\#(\tilde{R}(\boldsymbol{x}))} \sum_{x' \in \tilde{R}(\boldsymbol{x})} f_j(\boldsymbol{x}'),$$

where $\tilde{R}(\boldsymbol{x})$ is a uniformly spread out collection of points in a region $R(\boldsymbol{x})$ around $\boldsymbol{x}$. Then, approximating again with the data gives an estimate. Let

$$\hat{f}_j(\boldsymbol{x}) = \frac{1}{\#(\hat{R}(\boldsymbol{x}))} \sum_{x_i \in \hat{R}(\boldsymbol{x})} y(\boldsymbol{x}_i),$$

in which $\hat{R}(\boldsymbol{x}) = \{\boldsymbol{x}_i | \boldsymbol{x}_i \in R(\boldsymbol{x})\}$. In the classification case, this reduces to

$$\hat{f}_j(\boldsymbol{x}) = \frac{1}{\#(\hat{R}(\boldsymbol{x}))} \#(\{\boldsymbol{x}_i \in R(\boldsymbol{x}) | y_i = j\}).$$

Note that $\hat{R}(\boldsymbol{x})$ contains all the points close to $\boldsymbol{x}$. There may be more than k of them if $R(\boldsymbol{x})$ is fixed and n increases.

Heuristically, to ensure that $\hat{f}_j$ converges to f_j on the feature space, it is enough to ensure that these local approximations converge to their central value $f_j(\boldsymbol{x})$ for each $\boldsymbol{x}$. This means that as more and more data points accumulate near $\boldsymbol{x}$, it is important to be more and more selective about which points to include in $\hat{R}(\boldsymbol{x})$ when taking the local average. That is, as n gets large, $\#(\hat{R}(\boldsymbol{x}))$ must get large and, most importantly, the sum must be limited to the k closest points $\boldsymbol{x}_i$ to $\boldsymbol{x}$. The closest points get closer and closer to $\boldsymbol{x}$ as n increases. This is better than including all the points that are close (i.e., within a fixed distance of $\boldsymbol{x}$) because it means that the k closest points are a smaller and smaller fraction of the n data points available.

However, it turns out that fixing k will not give convergence: k must be allowed to increase, albeit slowly, with n so that the points near $\boldsymbol{x}$ not only get closer to $\boldsymbol{x}$ but become more numerous as well. Thus, to ensure $f_j(\boldsymbol{x}) = \lim_{n\to\infty} \hat{f}_j(\boldsymbol{x})$, both $n \to \infty$ and $k \to \infty$ are necessary. An extra condition that arises is that the points in the sum for a given $\boldsymbol{x}$ mustn't spread out too much. Otherwise, the approximation can be harmed if it becomes nonlocal. So, a ratio such as $k/n \to \infty$ also needs to be controlled. Finally, it can be seen that if the $\hat{f}_j$s converge to their target f_js the limiting form of the nearest-neighbor classifier or regression function achieves the minimum risk of the Bayes rule.

Exactly this sort of result is established in Devroye et al. (1994). Let $(\boldsymbol{X}_1, Y_1)$,..., $(\boldsymbol{X}_n, Y_n)$ be independent observations of the $p \times 1$ random variable $(\boldsymbol{X}, Y)$, and let μ be the probability measure of $\boldsymbol{X}$. The Y_is are the responses, and the $\boldsymbol{X}_i$s are the feature vectors. The best classifier, or best regression function, under squared error loss is $f(\boldsymbol{x}) = \mathbb{E}(Y|\boldsymbol{X} = \boldsymbol{x})$. Write the k-NN estimate of $f(\boldsymbol{x})$ as

$$\hat{f}_n(\boldsymbol{x}) = \sum_{i=1}^{n} W_{ni}(\boldsymbol{x}; \boldsymbol{X}_1, \ldots, \boldsymbol{X}_n) Y_i;$$

it is a linear smoother in which $W_{ni}(\boldsymbol{x}; \boldsymbol{X}^n)$ is $1/k$ when $\boldsymbol{X}_i$ is one of the kth nearest neighbors of $\boldsymbol{x}$ among the $\boldsymbol{X}_1, \ldots, \boldsymbol{X}_n$ and zero otherwise. Clearly, $\sum_i W_{ni} = 1$.

The desired result follows if

$$J_n = \int |f(\boldsymbol{x}) - \hat{f}_n(\boldsymbol{x})| d\mu(\boldsymbol{x}) \to 0$$

without knowing μ.

Theorem (Devroye et al., 1994): Let Y be bounded, $|Y| \leq M$. Then,

$$\lim_{n\to\infty} k = \infty \quad \text{and} \quad \lim_{n\to\infty} \frac{k}{n} = 0,$$

taken together, imply that $\forall \varepsilon > 0$, $\exists N_0$, so that $n \geq N_0$ ensures

$$P(J_n \geq \varepsilon) \leq e^{-n\frac{\varepsilon^2}{8M^2c^2}},$$

where c is a constant related to the minimal number of cones centered at the origin of angle $\pi/6$ required to cover the feature space. A converse also holds in the sense that the conclusion is equivalent to $J_n \to 0$ in probability or with probability 1. □

The proof of such a general result is necessarily elaborate and is omitted.

It is worth noting that nearest neighbors suffers the Curse as dramatically as the other techniques of this chapter. Indeed, as p increases, the best value for k, k^{opt} goes to zero because of the curious fact that, as p increases, the distance between points becomes nearly constant; see Hall et al. (2008). This means that knowing any number of neighbors actually provides no help; in binary classification terms, it means that the classifier does no better than random guessing because noise swamps the signal. In fact, any time the convergence rates decrease with increasing dimension, the Curse follows. The implication of this is that variable selection methods are essential.

2.5 Applications of Kernel Regression

To conclude the present treatment of regression, it is worthwhile to see how the methods work in practice. Two computed examples are presented next. The first uses simulated data to help understand the method. The second uses real data to see how the method helps us understand a physical phenomenon. An important source for data for testing machine learning procedures is `http://www.ics.uci.edu/~mlearn/MLRepository.html`.

2.5.1 A Simulated Example

Before comparing kernel regression and LOESS computationally, note that user-friendly computer packages, contributed by researchers and practitioners from around the world, are readily available. In the statistical context, most of the packages are written in R, although a decent percentage are in Matlab and pure C. Here, the examples are based on the R language unless indicated otherwise. For those readers who are still new to the R environment, a good starting point is to visit `http://www.r-project.org/`, download both the R package and the manual.

For starters, the function for implementing LOESS is a built-in R function (no download needed). Simply type help(loess) from the R prompt to see how to input arguments for obtaining LOESS function estimation. The function loess can also be used to compute the NW kernel regression estimate, even though there exists yet another function written exclusively for the NW regression, namely ksmooth. Again, use help(ksmooth) to see how to use it. Among other functions and packages, there is the R package locfit provided by Catherine Loader, which can be obtained from the website `http://www.locfit.info/` or directly installed from within R itself. locfit allows the computation of the majority of statistics discussed here in nonparametric kernel regression: estimation of the bandwidth, construction of the cross validation plot, and flexibility of kernel choice, to name just a few. Use library(locfit) to load the package and help(locfit) to see the way arguments are passed to it.

Another package of interest is lokern, which has both a global bandwidth form through its function glkerns and a local bandwidth form through lokerns. Recall that by default the NW uses a single (global) bandwidth h for all the neighborhoods. However, for some functions, a global bandwidth cannot work well, making it necessary to define different local bandwidths for each neighborhood. The idea is that the bandwidth h is treated as a function $h(x)$ and a second level of nonparametric function estimation is used to estimate $h(x)$. Usually, it is a simple estimator such as a bin smoother which assigns a value of h to each bin of x-values, but in lokerns it is another kernel estimate. Deciding when to use a local bandwidth procedure and when a global bandwidth is enough is beyond the present scope; it seems to depend mostly on how rapidly the function increases and decreases. The package lokern uses polynomial kernels, defaulting to two plus the number of derivatives, but concentrates mostly on bandwidth selection. The output automatically uses the optimal h.

A reasonable first example of kernel regression computation is with simulated data. Let the function underlying the observations be

$$f(x) = \frac{\sin\left(\frac{\pi}{2}x\right)}{1+2x^2(\text{sign}(x)+1)} \quad \text{with} \quad x \in [-\pi, \pi].$$

Suppose $n = 100$ equally spaced training points are generated from $[-\pi, \pi]$ and denoted x, and the corresponding response values denoted y are formed as $y_i = f(\mathbf{x}_i) + \varepsilon_i$, where the independent noise terms ε_i follow a zero-mean Gaussian distribution with standard deviation 0.2. Since the signal-to-noise ratio is pretty high, good estimation is expected. To see what happens, use the package lokern since it returns an optimal bandwidth along with the estimates of variances and fits. So, call glkerns(x, y) to get the fit using a global bandwidth and lokerns(x, y) to get the fit with local bandwidths:

```
glkfit <- glkerns(x, y) # fit with global kernel
lokfit <- lokerns(x, y) # fit with local kernel
```

The variance of the noise term in this case is estimated by the function glkerns to be $\sigma^2 \approx 0.044$, which is gratifyingly close to the true value 0.04. The global bandwidth returned by glkerns is $\hat{h} \approx 0.456$, but the fit under global bandwidth does not seem to differ substantially from the fit under local bandwidth, see the top panel in Fig. 2.9. It is important to note that the curves are formed by joining the estimated function values at each sampled data point by straight lines. This provides an illusion of continuity in the estimate that is not really justified: It only makes sense to estimate function values at points where there really data unless some more sophisticated routine is used with concomitantly stronger assumptions.

To explain the 2 LOESS plots in Fig. 2.9 and why a kernel package like glkerns would generate them it is important to look at the structure of equations (2.2.7), (2.2.9), and (2.2.10). In effect, LOESS is a kernel method. The tri-cube weight (2.2.9) is a de facto compact support kernel, and d_q plays the same role as h in terms of representing how concentrated each component in the estimator is. Moreover, locally optimizing in (2.2.9) gives a weighted least squares estimator of the same form as the NW estimator. (Indeed, recall that $\hat{\beta}_1 = \sum(x_i - \bar{x})(y_i - \bar{y}) / \sum(x_i - \bar{x})^2$ is a one-dimensional least squares minimization. This is a covariance divided by a variance and for data centered at their means is of the same form as the NW estimator or the local form of LOESS.)

Given this, it makes sense that the fraction of data included in a bin could be regarded as an analog to the bandwidth h. Indeed, if q is the number of data points to be included in a bin, then the fraction is $\alpha = q/n$ and is an analog to the bandwidth. Typically $.25 \leq \alpha \leq .5$. So, using the value $\hat{h} = 0.4561029$ from the earlier fit (as if it were relevant) and degree 2 (the maximum loess accepts) for the polynomial, the function loess produced the middle fit in Fig. 2.9. As before, the curve looks continuous only because the points at which estimates are made are joined by straight lines.

Confidence bands can be found from the normal-based expressions in Section 2.2.2. These are pointwise confidence intervals, not really a confidence region in function space. Again, the confidence interval for each function value $\hat{f}(x_i)$ is found; the upper endpoints are joined with straight lines, as are the lower endpoints. The LOESS fit along with these 2 SD confidence bands is the bottom panel in Fig. 2.9. (Confidence bands for the NW estimator will be seen in the next subsection.)

2.5.2 Ethanol Data

To see how kernel methods fare in practice, consider a well-known data set gathered in the context of the study of exhaust emissions, called ethanol.

In the ethanol data set plotted in Fig. 2.10, the variable NOx represents the exhaust emissions from a single-cylinder engine and is the response variable. The two predictor variables are E (the engine's equivalence ratio) and C (compression ratio). From Fig. 2.10, it can be seen that there are five levels for C and that the shapes of the points in each of the five planes are roughly similar, suggesting a one dimensional analysis will suffice. Thus, collapsing all the points into the NOx and E planes gives the scatterplot

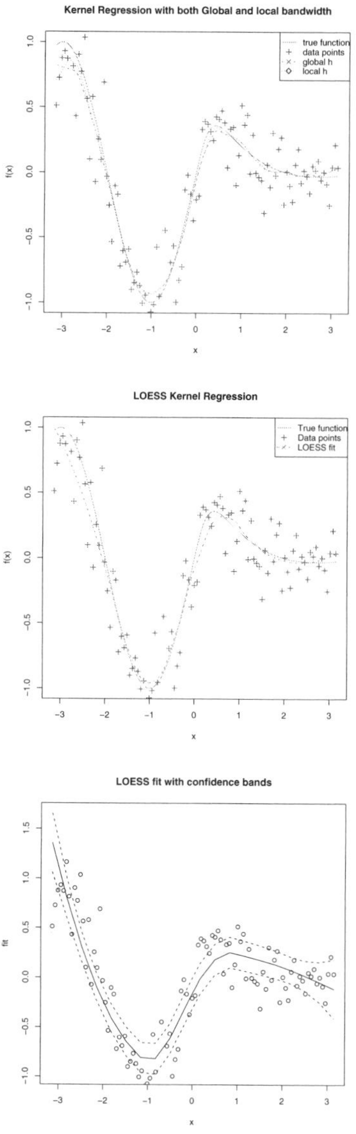

Fig. 2.9 Kernel regression estimation. The top plot shows the true function with two NW estimates, under global and local bandwidth selection. It is seen that they are pretty close. The middle plot shows the true function with a LOESS fit, which is, to the eye, only a little worse. The bottom plot shows what a practitioner would get with LOESS: the fitted curve and the 2SD confidence bands.

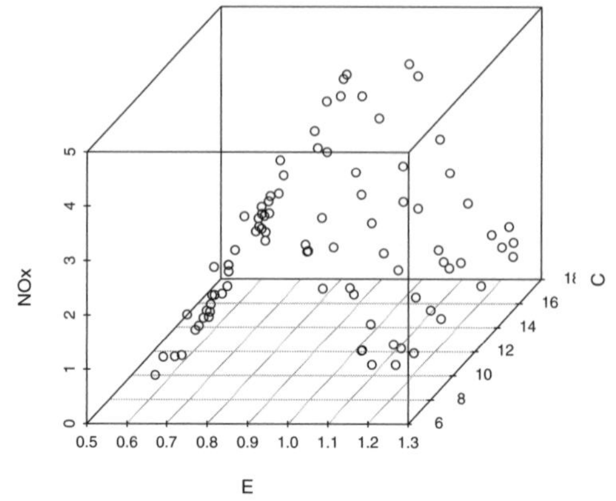

Fig. 2.10 3D scatterplot of the ethanol data set. It is easy to see that the data points for fixed values of C appear to form a ridge.

of data to be smoothed, as seen on the left-hand panel of Fig. 2.11. Moreover, from the right-hand panel of Fig. 2.11, the variability in C seems to be independent of the value of C, suggesting it is not very important and can be neglected. It will therefore be enough to focus on the relationship between NOx and E.

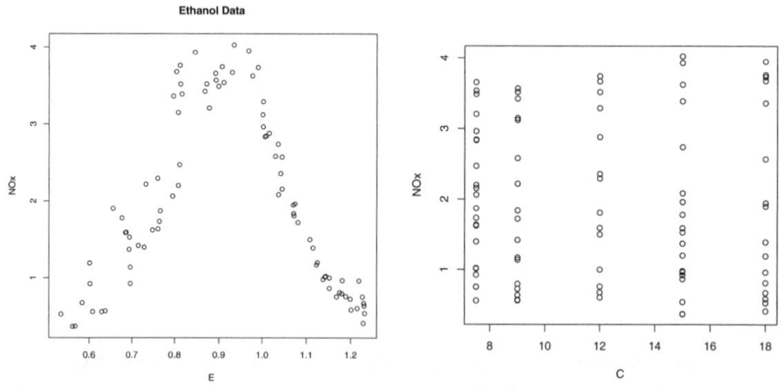

Fig. 2.11 Ethanol data scatter for the two predictor variables; the left-hand panel shows the data used in the numerical work, and the right-hand panel suggests that C is not important.

It now makes sense to model the response NOx as a univariate function of a single predictor variable E. For the NW estimator, it is enough to find the optimal bandwidth and find the sum of kernels at each point. First, it turns out that using GCV to search the interval $[0.1, 0.7]$ for values of h gives an optimal bandwidth h_{opt} of around 0.4

as can be seen from the left-hand panel is Fig. 2.12. The left hand panel in Fig. Fig. 2.12 shows the NW curve with h_{opt}. Clearly, the curve tracks the data, but arguably not very well: In the middle, it underestimates, while at the two ends it has a tendency to overestimate. Note that although the NW curve is potentially defined for all points, not just those at which data were collected, in fact the functional form of the NW curve is used only at the data points. As in the simulated case, the values at the data points are joined by straight lines. The reason is that on intermediate points the NW estimator would be making a point prediction whose variance could not be assessed without more sophisticated techniques.

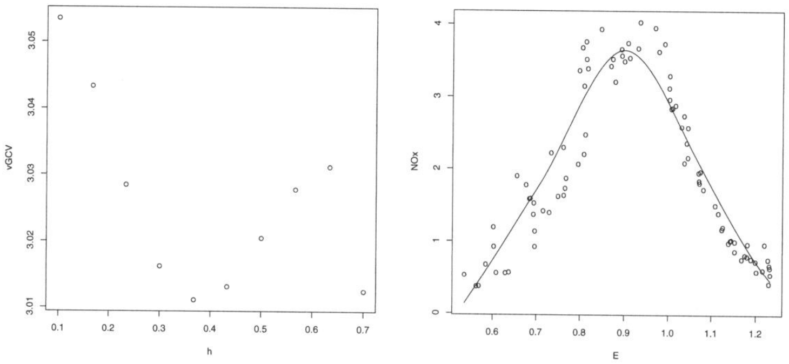

Fig. 2.12 NW best fit for NOx as a function of E, along with GCV plot suggesting the best h.

Figure 2.13 shows typical results from non-optimal values of h, for which the NW estimator generally produces worse fits. Indeed, for small h, the fit is erratic, as expected from theory, and the error tends to inflate the estimate of the variance of the noise term. At the other end, large h, the fit is too smooth and hence has a high bias.

Another aspect that might be of concern is the effect of the kernel. However, a polynomial kernel of order two is used by default, and the difference in fit from one kernel to another is small. It usually contributes less to the overall variability than the errors contributed from other sources, such as variability in the response or the variability implicit in h. Although not shown, the choice of the kernel here does not affect the fit substantially.

Finally, it is helpful to have some assessment of the overall variability of the fitted curve. There are two senses in which this can be done. First, formulating the NW estimator as a linear smoother as in Section 2.3.7 makes it easy to write down an expression for its variance at the sampled x_is. Indeed, from (2.3.20) it is enough to estimate σ because $\mathbf{W}$ is fully specified from its definition once h has been chosen. Under a normality assumption and the pretense that bias doesn't matter, it is possible to find confidence limits using the variance from (2.3.20) in which σ^2 is estimated as usual by the residual sum of squares over the degrees of freedom (recall that the degrees of freedom in this nonparametric context is estimated using the trace of some function of the smoothing matrix). Doing this at the x_is, joining the upper and lower confidence

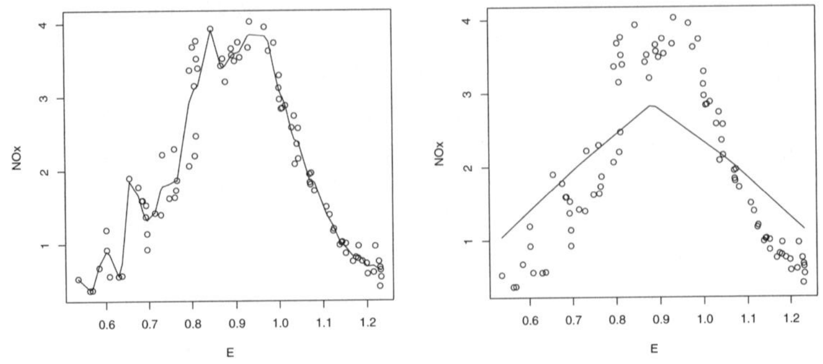

Fig. 2.13 Bad examples: the two extreme NW fits of the ethanol data. The left-hand panel shows a fit that has low bias but high variance. The right-hand panel shows a fit that has low variance but high bias.

limits with straight lines, and choosing a normal threshold of, say, 1.96 would give curves qualitatively like those in Fig. 2.14a plotted on either side of the estimated curve. In fact, the left-hand panel in Fig. 2.14 was generated by the lokern package, which does not use a normal approximation but rather a sophisticated approach to obtain a threshold more exact than 1.96. See Sun and Loader (1994) for more details on this advanced topic. The signal-to-noise ratio for the ethanol data is pretty high, so it is unclear how much improvement these techniques can give.

In greater generality, it would be desirable to use equations (2.3.11), (2.3.12), and (2.3.13) over all real values of x. However, while it is easy enough to get point predictions for xs not in the sample, and in principle a bootstrap technique might be used to estimate $B(x)$ and $V(x)$, it is extremely difficult to get variances for the predictions. In practice, (2.3.12) and (2.3.13) are ignored in favor of (2.3.20), and the normality of (2.3.11) is only invoked at the sampled x_is.

A second way to assess overall variability is to find MSE bands for the fitted curve. Instead of a confidence interpretation, these admit a prediction interpretation. Examining the MSE by looking at pointwise bias and variance separately in a function estimation context is deferred to the end of Chapter 3. Here, it is enough to look at the MSE as a whole by using the variance estimate already obtained at the sampled x_is and adding an estimate of bias to it. Clearly, the estimate of the bias requires a resampling technique such as the bootstrap for the bias. The procedure used here is to find the base NW curve using all the data, draw n independent bootstrap samples from the data to form n bootstrap estimates of the curve, and then take the average of their n distances from the base curve. The right-hand panel in Figure 2.14 shows the resulting bootstrap estimates of the curves $\hat{\mathbf{f}} \pm 2\widehat{MSE}$ plotted around the fitted base curve. As ever, the illusion of continuity arises from joining the values at the sampled x_is with straight lines. The bands are seen to be rough, or excessively wiggly, most likely due to the variance of the bootstrap estimates. The MSE bands are much wider than the confidence bands and the two have different interpretations (confidence versus prediction) but the widths of the bands vary similarly in the two cases.

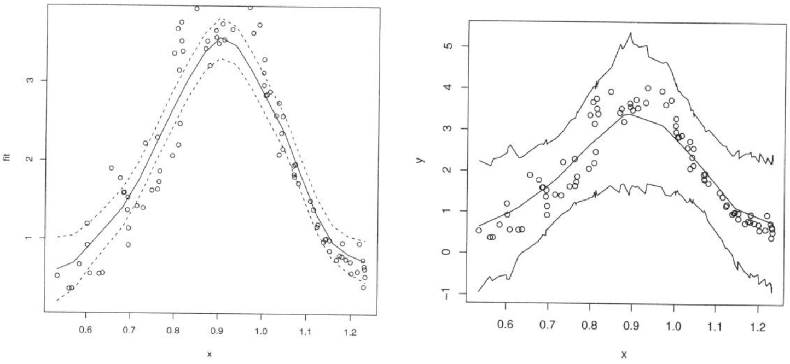

Fig. 2.14 On the left-hand panel, confidence bands for the NW estimator for the Ethanol data plotted with the fitted curve. On the right-hand panel, raw MSE bands for the NW estimator for the Ethanol data plotted with the fitted curve.

2.6 Exercises

Exercise 2.1. Let Y be a response variable with $\mathbb{E}(Y) = \mu$ and $\mathbb{E}(Y^2) < \infty$, and let X be an explanatory variable. The model is

$$Y = f(X) + \varepsilon,$$

where the noise term ε has a distribution such that the signal-to-noise ratio allows the signal to be recovered.

1. Deduce that, for a fixed x,
$$f(x) = \mathbb{E}(Y|X = x)$$
minimizes the quadratic risk $\mathbb{E}[(Y - f(X))^2]$.
2. For a parametric model such as simple linear regression, the empirical counterpart of the quadratic risk $\mathbb{E}[(Y - f(X))^2]$ is the least squares deviation
$$\hat{R}_2(\beta_0, \beta_1) = \sum_{i=1}^{n} (y_i - (\beta_0 + \beta_1 x))^2,$$
from which the estimates of β_0 and β_1 are obtained to form the desired approximating function. Least squares estimates are, however, sensitive to outliers.

a. Show why least squares estimates are sensitive to outliers.

b. One of the measures commonly used to circumvent the outlier problem is to change the loss function from the squared error loss $\ell(u,v) = (u-v)^2$ to the absolute deviation loss $\ell(u,v) = |u-v|$, so that the empirical risk is

$$\hat{R}_1(\beta_0, \beta_1) = \sum_{i=1}^{n} |y_i - (\beta_0 + \beta_1 x)|.$$

c. Why is $\hat{R}_1$ less sensitive to outliers than $\hat{R}_2$?

d. Why is $\hat{R}_1$ not used as much as its desirable properties suggest it should be?

e. Suggest a way to estimate the desired parameters when $\hat{R}_1$ is used. Can you make any statement about their asymptotic properties such as consistency and asymptotic normality?

Exercise 2.2. Consider using the polynomial

$$g_p(x) = \beta_0 + \beta_1 x + \beta_2 x^2 + \cdots + \beta_{p-1} x^{p-1} + \beta_p x^p$$

to approximate a function underlying a random sample $\{(x_i, y_i), i = 1, 2, \cdots, n\}$ under squared error loss.

1. Show that the coefficients $\beta_0, \beta_1, \cdots, \beta_p$ that minimize the squared error loss are given by the solution to the set of linear equations

$$\sum_{l=0}^{p} S_{k+l} \beta_l = T_k \tag{2.6.1}$$

for $k = 0, 1, \cdots, p$, where

$$S_k = \sum_{i=1}^{n} x_i^k, \qquad T_k = \sum_{i=1}^{n} x_i^k y_i.$$

2. Consider the "homogeneous" equations corresponding to the "nonhomogeneous" system (2.6.1); i.e.,

$$\sum_{l=0}^{p} S_{k+l} \beta_l = 0. \tag{2.6.2}$$

Write (2.6.2) in matrix form and verify the determinant is nonzero.

3. Assume that all the x_is are uniformly distributed in the interval $[0,1]$.

a. Show that

$$S_k \approx \frac{p}{k+1}.$$

b. Deduce that the homogeneous system is now

$$\mathbf{B}\beta = 0, \qquad \text{with} \quad B_{jk} = \frac{1}{j+k+1}.$$

It can be shown that the determinant of $\mathbf{B}$ for a polynomial of degree p,

$$\det(\mathbf{B}) = \left| \frac{1}{j+k+1} \right| \qquad j,k = 0,1,2,\cdots,p,$$

is

$$H_p = \frac{p!(p+1)!\cdots(2p-1)!}{[1!2!3!\cdots(p-1)!]^3}.$$

Tabulate the values of H_p for $p = 1,2,3,\cdots,9$, and comment on what happens to the value of the determinant as p increases. (The ambitious student can try to derive H_p, as well.)

c. What does this mean for polynomial curve fitting?

Exercise 2.3. Consider a function f defined on the interval $[-1,+1]$. If f is interpolated on a set of knots x_is in $[-1,+1]$ by a polynomial $g_k(x)$ of degree $\leq k$, then as k increases, the interpolant oscillates ever more at the endpoints -1 and 1 for a large class of functions.

1. Show that the interpolation error tends to infinity at the endpoints as the degree k increases; i.e.,

$$\lim_{k\to\infty} \left(\max_{-1\leq x\leq 1} \left| f(x) - g_k(x) \right| \right) = \infty.$$

2. What is the intuitive justification for such a limitation? In general, can you identify where on the domain this bad behavior typically occurs and for which functions it occurs?
3. How can you fix this?

Exercise 2.4. Consider the Runge function defined on the interval $[-1,+1]$. Using either Matlab or R (or your favorite package), do the following:

1. Generate equally spaced points in the interval along with their Runge function values.
2. Estimate the coefficients of the interpolating polynomial for various degrees (start from 1 and go to something really large such as 8, 9, 10, or even 11).
3. Tabulate your results, indicating the coefficient in the column and the degree of the polynomial in the row, so that the cell contains the estimated value of the coefficient for that degree of polynomial. (Since the number of coefficients increases with the degree of the polynomial, your table should be triangular).
4. Comment on what you notice in light of the theoretical assertion made regarding the limitations of the polynomial in curve fitting seen in exercise (2.3).

Exercise 2.5. Generate n values x_i for $i = 1,\ldots,n$ by setting $x_i = i/n$. Then, let $Y_1,\ldots,Y_n$ be IID draws from a $Bernoulli(p)$ for some $p \in (0,1)$. Thus, treated as a data set, the collection (x_i, y_i) for $i = 1, \ldots, n$ is just noise. Now, instead of defining a model, consider two predictive schemes. The first is Scheme A which is the 1-nearest-neighbor method. The second is Scheme B which will be the trivial predictor always predicting zero.

1. What are the expected mean square training error and the expected mean square leave-one-out CV error for Scheme A?
2. What are the expected mean square training error and the expected mean square leave-one-out CV error for Scheme B?

Exercise 2.6. Consider a data set x_i for $i = 1, \ldots, n$ where the x_is are unidimensional and distinct. Suppose n is even – take $n = 100$ for definiteness – and that exactly half the x_is are positive and exactly half are negative. To explore the properties of leave-one-out CV and nearest-neighbors, try the following.

1. Can you specify the x_is so that the leave-one-out CV error for 1-nearest-neighbors is 0 % ? Explain.
2. Can you specify the x_is so that the leave-one-out CV error for 1-nearest-neighbors is 100 % but for 3-nearest-neighbors in 0 % ? Explain.

Exercise 2.7. It is a theorem that as the amount of training data increases, the error rate of the 1-nearest-neighbor classifier is bounded by twice the Bayes optimal error rate. Here one proof of the theorem is broken down into steps so you can prove it for binary classification of real inputs. (An alternative proof can be found at `http://www.ismll.uni-hildesheim.de/lehre/ml-08w/skript/nearest.pdf`.) Let x_i for $i = 1, \ldots, n$ be the training samples with corresponding class labels $y_i = 0, 1$. Think of x_i as a point in p-dimensional Euclidean space. Let $p_y(x) = p(x|Y = y)$ be the true conditional density for points in class y. Assume $0 < p_y(x) < 1$ and $\theta = \mathbb{P}(Y = 1) \in (0, 1)$.

1. Write the true probability $q(x) = \mathbb{P}(Y = 1|X = x)$ that a data point x belongs to class 1. Express $q(x)$ in terms of $p_0(x)$, $p_1(x)$, and θ.
2. Upon receipt of a datum x, the Bayes optimal classification assign the class
$$\arg\max_y P(Y = y|X = x)$$
to maximize the probability of correct classification. Under $q(x)$, what is the probability that x will be misclassified using the Bayes optimal classifier?
3. Recall that the 1-nearest-neighbor classifier assigns x to the class of its closest training point x'. Given x and its x', what is the probability, terms of $q(x)$ and $q(x')$, that x will be misclassified?
4. Suppose that in the limit of large n, the number of training examples in both classes goes to infinity in such as way as to fill out the space densely. This means that the nearest neighbor $x' = x'(n)$ of x satisfies $q(x') \to q(x)$ as n increases. Using this substitution in item 3, express the asymptotic error for the 1-nearest-neighbor classifier at x in terms of the limiting value $q(x)$.

5. Now show that the asymptotic error obtained in item 4 is bounded by twice the Bayes optimal error obtained in item 2.
6. Why doesn't the asymptotic bound hold for finite n as well? What goes wrong?

Exercise 2.8. Let $\{(x_i, Y_i), i = 1, \cdots, n\}$ be a data set and let $w_i(x) = K((x_i - x)/h)$ be a weight function. One way to justify the NW estimator is by the following optimization. Consider finding the $c = \hat{f}_n(x)$ that minimizes the weighted sum of squared errors

$$\sum_{i=1}^{n} w_i(x)(Y_i - c)^2.$$

1. Show that the minimum is achieved for

$$c = \hat{f}_n(x) = \frac{\sum_{i=1}^{n} w_i(x) Y_i}{\sum_{i=1}^{n} w_i(x)}.$$

   This means that the NW estimator is the local constant kernel estimator.
2. Consider trying to improve the local constant kernel estimator to a local polynomial kernel estimator. To do this, write the Taylor expansion on a neighborhood of x,

$$g_x(u; \mathbf{c}) = c_0 + c_1(x - u) + \frac{c_2}{2!}(x - u)^2 + \cdots + \frac{c_p}{p!}(x - u)^p.$$

   In this form, the goal is to find the $\hat{\mathbf{c}} = (\hat{c}_0, \hat{c}_1, \cdots, \hat{c}_n)^\top$ that minimizes a sum of squares modified by a local weight w_i,

$$\sum_{i=1}^{n} w_i(x)(Y_i - g_x(x_i; \mathbf{c}))^2.$$

   a. Why does $\hat{\mathbf{c}}$ now depend on x?
   b. Show that $\hat{f}_n(x) = g_x(x; \hat{\mathbf{c}}) = \hat{c}_0(x)$. Why is this different from the local constant kernel regressor?
   c. To establish linearity, derive the expression for $\hat{f}_n(x)$ in matrix form; i.e., define an appropriate $L_n(x)$ so that

$$\hat{f}_n(x) = L_n(x)\mathbf{y}.$$

   d. Deduce the expression of $\hat{f}_n(x)$ for the local linear kernel regression estimator for $p = 1$.
   e. Suggest how to estimate the derivative $\hat{f}_n'(x)$ of $\hat{f}_n(x)$.

Exercise 2.9 (Statistical comparison of local constant and local linear estimators). Let $Y_i = f(X_i) + \varepsilon_i$, where ε_i is mean zero with variance σ^2, and let K be a kernel function. The variance for both the local constant and the local linear kernel regression estimators of a function $f(x)$ based on a design with density function $p(x)$ is

$$\frac{\sigma^2(x)}{nh_n p(x)} \int t^2 K(t)dt + O_P\left(\frac{1}{nh_n}\right).$$

1. Explain why the expression of the variance is the same for both estimators.
2. Show that the bias for the local linear estimator is given by

$$\frac{1}{2}h_n^2 f''(x) \int t^2 K(t)dt + O_P(h_n^2).$$

3. The bias for the NW estimator is

$$h_n^2\left(\frac{1}{2}f''(x) + \frac{f'(x)p'(x)}{p(x)}\right) \int t^2 K(t)dt + O_P(h_n^2),$$

as can be inferred from (2.3.12), or derived from (2.3.9). (It is a good exercise to do this derivation.)

4. Based on the expressions of the bias for both estimators, what are the strengths and the weaknesses of the two kernel methods?

Exercise 2.10. Let $X_1, X_2, \cdots, X_n$ be drawn IID from a distribution with density p and let $\hat{p}_n$ be the kernel density estimator using the boxcar kernel:

$$K(x) = \begin{cases} 1 & -\frac{1}{2} < x < \frac{1}{2} \\ 0 & \text{otherwise.} \end{cases}$$

1. Show that

$$\mathbb{E}(\hat{p}_n(x)) = \frac{1}{h}\int_{x+h/2}^{x-h/2} p(y)dy$$

and

$$\mathrm{Var}(\hat{p}_n(x)) = \frac{1}{nh^2}\left[\int_{x+h/2}^{x-h/2} p(y)dy - \left(\int_{x+h/2}^{x-h/2} p(y)dy\right)^2\right].$$

2. Show that if $h \longrightarrow 0$ and $nh \longrightarrow \infty$ as $n \longrightarrow \infty$, then

$$\hat{p}_n(x) \xrightarrow{P} p(x).$$

Exercise 2.11. Consider the following kernels encountered earlier in kernel smoothing: (1) the uniform kernel $K(u) = 1/2$ for $-1 < u < 1$, (2) the triangular kernel $K(u) = 1 - |u|$ for $-1 < u < 1$, the quartic kernel $K(u) = (15/16)(1-u^2)^2$ for $-1 < u < 1$, the Epanechnikov kernel $K(u) = (3/4)(1-u^2)$ for $-1 < u < 1$, and the Gaussian kernel $K(u) = (2\pi)^{-\frac{1}{2}}\exp(-0.5u^2)$ for $-\infty < u < \infty$.

1. To contrast the optimal bandwidths for for local constant regression evaluated at a point x under AMSE optimality for different kernels, show that the AMSE-optimal bandwidth using the Epanechnikov kernel is $(10/3)^{1/5}$ times the AMSE-optimal bandwidth for the uniform kernel.

2. Again, fix x but now compare minimal AMSE's rather than bandwidth. Show that the minimal AMSE at x using the Epanechnikov kernel is 94.3% of the minimal AMSE at x using the uniform kernel.
3. Redo the first two items, but use the normal kernel in place of the uniform kernel. Which kernel gives the most efficient estimator? How important is this?
4. Let f be a regression function and suppose $f''(x)$ is overestimated by a factor of 1.5. If local constant regression were used to estimate $f(x)$, what effect would this have on the estimate of the optimal bandwidth? To be specific, suppose $f''(x) = \sigma = 1$. Now, using the Epanechnikov kernel, find the AMSE at x using the optimal bandwidth and the estimated bandwidth.

Exercise 2.12. Let $Y = f(X) + \varepsilon$, where ε is mean zero with variance σ^2. Let the estimator $\hat{f}$ of f be linear. That is, given a collection of outcomes $\{(x_1, y_1), \cdots, (x_n, y_n)\}$ the vector $\hat{\boldsymbol{f}} = (\hat{f}(x_1), \cdots, \hat{f}(x_n))$ can be written as $\hat{\boldsymbol{f}} = S\mathbf{y}$, where S is a smoothing matrix. Define $\nu_1 = \text{trace}(S)$ and $\nu_2 = \text{trace}(S^\top S)$ and let

$$s^2 = \frac{1}{n - 2\nu_1 + \nu_2} \sum_{i=1}^{n} (y_i - \hat{f}(x_i))^2.$$

1. Show that

$$\mathbb{E}\left[\sum_{i=1}^{n} (y_i - \hat{f}(x_i))^2\right] = \sigma^2(n - 2\nu_1 + \nu_2) + \boldsymbol{f}^\top (I - S)^\top (I - S)\boldsymbol{f}.$$

2. Comment on the properties of s^2 as an estimator of σ^2.

Exercise 2.13. The goal of this exercise is to perform a computational comparison of the bias and the variance associated with NW smoothing, and local polynomial smoothing. To do this, compute empirical bias and empirical variance based on $m = 1000$ data sets $\mathscr{D}_1, \cdots, \mathscr{D}_m$, each of size $n = 199$, simulated from the model $Y_i = f(x_i) + \varepsilon_i$ with

$$f(x) = -x + \sqrt{2}\sin(\frac{1}{10}\pi^2/x^2), \qquad x \in [0,3], \tag{2.6.3}$$

where $\varepsilon_1, \varepsilon_2, \cdots, \varepsilon_n$ are IID $N(0, .2^2)$.

1. Consider the (deterministic) fixed design with equidistant points in $[0,3]$.

   a. For each data set $\mathscr{D}_j$, compute the NW and the local polynomial estimates at every point in $\mathscr{D}_j$.

   b. At each point x_i, compute the empirical bias $\widehat{\text{Bias}\{f(x_i)\}}$ and the empirical variance $\widehat{\text{Var}\{f(x_i)\}}$, where

$$\widehat{\text{Bias}\{f(x_i)\}} = \frac{1}{m}\sum_{j=1}^{m} \hat{f}^{(j)}(x_i) - f(x_i),$$

and

$$\widehat{\mathrm{Var}\{f(x_i)\}} = \frac{1}{m-1} \sum_{j=1}^{m} (\hat{f}^{(j)}(x_i) - f(x_i))^2.$$

c. Plot these quantities against x_i for each estimator. (Plotting is straightforward with the command matplot along with apply to get the means and the variances if the values needed are stored in a matrix.)

d. Provide a thorough analysis of what the plots suggest.

Hint: Here is some R code that may help.

```
## Generate n=101 equidistant points in [-1,1]
m <- 1000
n <- 101
x <- seq(-1,1, length=n)

# Initialize the matrix of fitted values
fvnw <- fvlp <- fvss <- matrix(0, nrow = n, ncol = m)

# Fit the data and store the fitted values
for (j in 1:m){
## Simulate  y-values
y <- f(x) + rnorm(length(x))

## Get the estimates and store them
fvnw[,j] <- ksmooth(x, y, kernel = "normal",
bandwidth=0.2, x.points = x)$y
fvlp[,j] <- predict(loess(...), newdata = x)
fvss[,j] <- predict(smooth.spline(...), x=x)$y}
```

Exercise 2.14. Repeat Exercise 2.13, this time with a design that has nonequidistant points. The following R commands can be used to generate the design points:

```
set.seed(79)
x <- sort(c(0.5, -1 + rbeta(50,2,2), rbeta(50,2,2)))
```

Use span = 0.3365281 for loess and spar = 0.7162681 for smooth.spline to get the same degrees of freedom.

Exercise 2.15. Consider the Mexican hat function

$$f(x) = (1-x^2)\exp(-0.5x^2), \qquad x \in [-2\pi, 2\pi].$$

This function is known to pose a variety of estimation challenges. Construct a simulation study like the one described in Exercise 2.13 to explore the difficulties inherent in the study of this function. Consider both the statistical challenges and the computational ones.

Exercise 2.16. The definition of linearity for smoothers, (2.1.1), was that the vector $\hat{\mathbf{y}}$ of fitted values can be written as $\hat{\mathbf{y}} = \mathbf{S}\mathbf{y}$, where $\mathbf{S}$ is the smoothing matrix depending on the x_is and the smoothing technique. In this definition, it is usually assumed that the outcomes $(x_1, y_1), \cdots, (x_n, y_n)$ were drawn independently from the model $Y = f(x) + \varepsilon$ where ε is mean zero with $\mathrm{Var}(\varepsilon_i) = \sigma^2$.

1. Show that
$$\sum_{i=1}^{n} \mathrm{Cov}(\hat{y}_i, y_i) = \mathrm{trace}(\mathbf{S})\sigma^2.$$

2. Recall that in linear regression $\mathbf{Y} = \mathbf{X}\beta + \varepsilon$ and the vector $\hat{\mathbf{y}}$ of fitted values comes from the projection matrix $\mathbf{H}$; i.e., $\hat{\mathbf{y}} = \mathbf{X}(\mathbf{X}^\top \mathbf{X})^{-1}\mathbf{X}\mathbf{y} = \mathbf{H}\mathbf{y}$.

    a. Compute $\mathrm{trace}(\mathbf{H})$.
    b. Show that the same identity holds for $\mathbf{H}$ in place of $\mathbf{S}$; i.e., $\sum_{i=1}^{n} \mathrm{Cov}(\hat{y}_i, y_i) = \mathrm{trace}(\mathbf{H})\sigma^2$.
    c. What does this suggest about the degrees of freedom of linear smoothers?

Exercise 2.17 (Computationally efficient cross validation for linear smoothers). Consider independent outcomes $(x_1, y_1), \cdots, (x_n, y_n)$ drawn from the model $Y = f(x) + \varepsilon$ with $\mathrm{Var}(\varepsilon) = \sigma^2$. As in Exercise 2.16, or (2.1.1), if a smoother is linear, it can be represented by a matrix $\mathbf{S}$ so that the vector $\hat{\mathbf{y}}$ of fitted values is $\hat{\mathbf{y}} = \mathbf{S}\mathbf{y}$. The smoothing matrix $\mathbf{S}$ often depends delicately on the smoothing parameter h (as well as on the x_is and the smoothing procedure itself). Often h is estimated by CV. That is, h is chosen to minimize the leave-one-out objective function

$$CV(h) = \sum_{i=1}^{n} \left(\{y_i - \hat{f}_{n-1}^{(-i)}(x_i)\}^2 \right). \tag{2.6.4}$$

Here, $\hat{f}_{n-1}^{(-i)}$ is the estimator of f based on the deleted data; i.e., after leaving out one of the x_i's.

1. Show that, for linear smoothers,
$$CV(h) = \sum_{i=1}^{n} \left(\frac{y_i - \hat{f}_n(x_i)}{1 - S_{ii}(h)} \right)^2,$$
where $S_{ii}(h)$ is the ith diagonal element of S.
2. Show that, for the traditional linear model,
$$CV = \sum_{i=1}^{n} \left(\frac{y_i - x_i^\top \hat{\beta}}{1 - h_{ii}} \right)^2,$$
where h_{ii} is the ith diagonal element of H.
3. In what senses can you argue that the expressions in items 1 and 2 improve on (2.6.4) as an objective function?

Exercise 2.18. Recall that linearity for a smoother is defined in (2.1.1) and that a moving-average smoother is defined in Section 2.1. Suppose a data set of the form $(\mathbf{x}_1, y_1), \cdots, (\mathbf{x}_n, y_n))$ is available.

1. Verify that a moving-average smoother is linear by finding L_n^{MA} so that (2.1.1) is satisfied.
2. Write the kernel regression smoother in the same notation, identifying its matrix L_n^{KR}.
3. Can L_n^{MA} or L_n^{MA} fail to be symmetric?
4. Can L_n^{MA} or L_n^{MA} have eigenvalues strictly greater than 1?

Hint: To address items 3 and 4, it may help to generate a small data set and see what the smoothing matrices look like. Note that this exercise can be done for other smoothers as well, such as bin smoothing.

Exercise 2.19 (This exercise is aimed at showing that using CV or GCV to choose the bandwidth h overfits; i.e., the h is too small). Let f be the tooth function

$$f(x) = x + \frac{9}{4\sqrt{2\pi}} \exp\left[-4^2(2x-1)^2\right], \qquad x \in [0,1].$$

Generate independent outcomes $(x_i, y_i), i = 1, \cdots, n$ from the model $Y = f(x) + \varepsilon$ ε is $N(0, \sigma^2)$ for a reasonable choice of σ, say $\sigma = .2$. Use the NW estimator to get an estimate $\hat{f}_n^h$. The R command is ksmooth().

1. Plot the simulated data and the fitted curve from the NW estimator. Explain how to perform the computations to find the overfit caused by CV and GCV.
2. For contrast, try the same procedure using the Akaike criterion in place of CV and GCV; i.e.,

   $$\mathrm{AIC}_c(h) = n \log \hat{\sigma}^2(h) + n \frac{1 + \mathrm{trace}(S_h)/n}{1 - \{\mathrm{trace}(S_h) + 2\}/n},$$

   where

   $$\hat{\sigma}^2(h) = n^{-1} \sum_{i=1}^{n} (y_i - \hat{f}(x_i))^2.$$

   Use the AIC on the same data to fit a new NW curve and compare your results to those from item 1.

Chapter 3
Spline Smoothing

In the kernel methods of Chapter 2, the estimator of a function is defined first and then a measure of precision is invoked to assess how close the estimator is to the true function. In this chapter, this is reversed. The starting point is a precision criterion, and the spline smoother is the result of minimizing it. Before presenting and developing the machinery of smoothing splines, it is worthwhile to introduce interpolating splines. This parallels the discussion of Early smoothers and the considerations leading to local smoothers, thereby giving insight into the formulation of smoothing splines. Like the kernel-based methods of the last chapter, splines suffer from the Curse of Dimensionality. Nevertheless, there is a parallel theory for multivariate splines. Eubank (1988), Chapter 6.2.3 touches on it with some references. Such Laplacian smoothing splines are neglected here, as are partial splines, which generalize splines to include an extra nonparametric component.

3.1 Interpolating Splines

First, a spline is a piecewise polynomial function. More formally, let $a = x_1 < x_2 < \cdots < x_n = b$ be ordered design points, and partition the interval $\mathscr{X} = [a,b]$ into subintervals $[x_i, x_{i+1})$, $i = 1, 2, \cdots, n-1$. When each piece of a spline is a polynomial of degree d, the corresponding spline is said to be a spline of order d. Thus, generically, a spline of order d is of the form

$$s(x) = \begin{cases} s_1(x), & x \in [x_1, x_2); \\ s_2(x), & x \in [x_2, x_3); \\ \vdots & \vdots \\ s_{n-1}(x), & x \in [x_{n-1}, x_n], \end{cases}$$

where each $s_i(x)$ is a polynomial of degree d.

Splines have long been used by numerical analysts as a technique for constructing interpolants of functions underlying data because they avoid Runge's phenomenon

B. Clarke et al., *Principles and Theory for Data Mining and Machine Learning*, Springer Series in Statistics, DOI 10.1007/978-0-387-98135-2_3, © Springer Science+Business Media, LLC 2009

arising with high-degree polynomials. However, in general, the spline of degree d that interpolates a data set is not uniquely defined. Consequently, additional degrees of freedom must be specified to get uniqueness. Unlike splines, global polynomial interpolation often yields a unique interpolant.

Bin smoothers and running line smoothers can be regarded as special cases of low-order splines. For instance, let $(x_1,y_1),(x_2,y_2),\cdots,(x_n,y_n)$ be a set of observations. The simplest spline is of zero order and corresponds to a piecewise constant function defined by

$$s(x) = y_i, \qquad x_i \le x < x_{i+1}.$$

Zero-order splines are step functions, essentially bin smoothers, and not continuous.

Setting $d = 1$ gives linear interpolating splines, somewhat like running line smoothers. A linear interpolating spline s_i is given by

$$s_i(x) = y_i + (y_{i+1} - y_i)\left[\frac{x - x_i}{x_{i+1} - x_i}\right], \qquad x_i \le x < x_{i+1}.$$

This means data points are graphically connected by straight lines. Now,

$$s_i(x_i) = y_i + (y_{i+1} - y_i)\left[\frac{x_i - x_i}{x_{i+1} - x_i}\right] = y_i$$

and

$$s_{i-1}(x_i) = y_{i-1} + (y_i - y_{i-1})\left[\frac{x_i - x_{i-1}}{x_i - x_{i-1}}\right] = y_i.$$

Hence,

$$s_i(x_i) = s_{i+1}(x_i), \qquad i = 1,\cdots,n-1.$$

So, first-order splines are continuous at each data point, unlike zero-order splines. However, the first derivative of a linear spline is a zero-order spline and so not continuous. The loss of continuity is one rung higher on the ladder of differentiability. (Technically, the low-order polynomials used in LOESS are also splines, although they do not have any continuity requirements at the design points.)

Linear splines can be quite good. Figure 3.1 shows the result of linear spline interpolation on the function $f(x) = (\sin x)/x$ for $x \in [-10,10]$. For comparison, the corresponding graph for noisy data is also plotted. It is the second graph that reveals spline performance for statistical function estimation with noisy responses. The next level of splines is quadratic. However, cubic splines have very appealing properties for statistical function estimation. So, the focus is usually on them.

Let $a = x_1 < x_2 < \cdots < x_n = b$ be ordered design points. Given observations (x_i,y_i) for $i = 1,\ldots,n$, the cubic interpolating spline function corresponding to this design is $s \in C^2$, a twice continuously differentiable function with each polynomial of degree three, $s_i(x)$, defined on the subinterval $[x_i,x_{i+1})$, $i = 1,\cdots,n-1$. In other words, the function $s(x)$ is a cubic interpolating spline on the interval $\mathscr{X} = [a,b]$ if

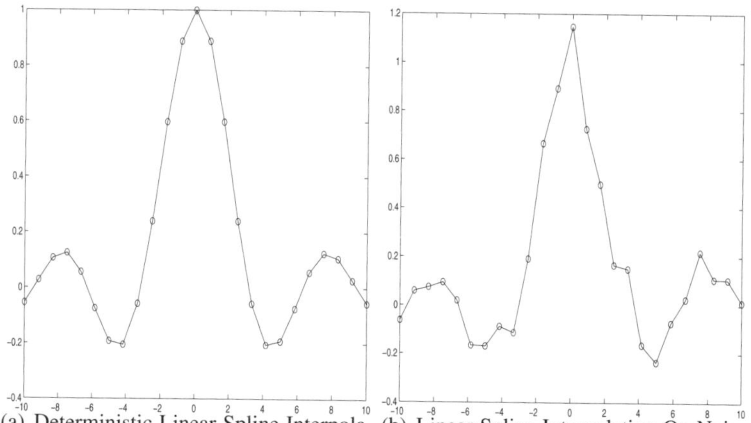

(a) Deterministic Linear Spline Interpolation

(b) Linear Spline Interpolation On Noisy data

Fig. 3.1 The plot on the left shows the linear spline interpolant for a nonnoisy (deterministic) response case. The plot on the right shows the linear spline interpolant applied to data with noisy response variables. The curve on the right is more "wiggly" or "rough", due to noise.

- s matches the data (i.e., $s(x_i) = y_i$, $i = 1, \cdots, n$);
- s is continuous (i.e., $s_{i-1}(x_i) = s_i(x_i)$, $i = 2, \cdots, n-1$);
- s is twice continuously differentiable, in particular
  - ☐ $s'_{i-1}(x_i) = s'_i(x_i)$, $i = 2, \cdots, n-1$ (first derivative continuity);
  - ☐ $s''_{i-1}(x_i) = s''_i(x_i)$, $i = 2, \cdots, n-1$ (second derivative continuity);
- each piece $s_i(x)$ of s is a cubic polynomial in each subinterval $[x_i, x_{i+1})$:

$$s_i(x) = \beta_{3i}(x - x_i)^3 + \beta_{2i}(x - x_i)^2 + \beta_{1i}(x - x_i) + \beta_{0i} \quad x \in [x_i, x_{i+1}].$$

The dimension of spaces of splines can be evaluated. Observe that each cubic polynomial piece $s_i(x)$ of $s(x)$ requires four conditions for its construction apart from the specification of the design points x_i, $i = 1, ..., n$. As a result, a total of $4(n-1)$ conditions are needed to determine the $n-1$ cubic polynomial pieces of $s(x)$. The interpolation requirement provides n of these; the continuity constraints for the 0th, 1st, and 2nd derivatives at the x_is for $i = 2, ..., n-1$ each provide $n-2$ more. These four kinds of requirements for a cubic spline interpolant provide a total of

$$n + n - 2 + n - 2 + n - 2 = 4n - 6 = 4(n-1) - 2$$

constraints. On the other hand, it is easy to see that there are $4(n-1)$ real numbers that must be specified, apart from the design points x_i, for the $n-1$ pieces: Each of $n-1$ cubics has four coefficients. So, to specify a spline function uniquely, two more conditions are required. Since the behavior of the spline function at its endpoints is, so far, relatively free, pinning down the behavior of s'' at x_1 and x_n finishes the specification.

It turns out that conditions on the second derivative at the endpoints work well, so

$$s''(x_1) = s''(x_n) = 0$$

can be imposed, and the resulting functions are called natural cubic splines.

The obvious question is how much the cubic spline interpolant improves on the linear spline interpolant. The improvement can be exemplified by using the same function, $f(x) = (\sin x)/x$ for $x \in [-10, 10]$, as before. In Fig. 3.2, it is obvious that, in the absence of noise in the response, the cubic interpolant recovers the underlying function nearly perfectly. However, the subfigure on the right shows that noise complicates things. Despite the smoothness of the cubic spline interpolant pieces themselves, the overall spline is still very rough or wiggly. This is due to the noise, and the spline fails to capture the more slowly varying trend characteristic of the true function. From a statistical perspective, the smoothness of the estimate alone is therefore not satisfactory. The roughness from the rapid fluctuations due to the error must be reduced to recover the underlying function from noisy data.

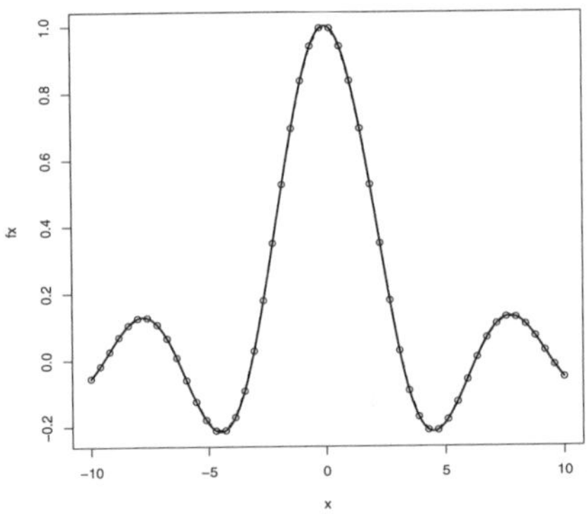

(a) Deterministic Cubic Spline Interpolation

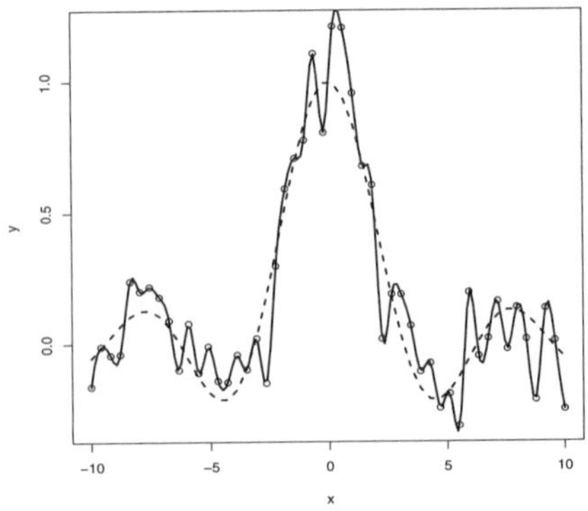

(b) Cubic Spline Interpolation On Noisy data

Fig. 3.2 In both the plots, the dashed curve is the $(\sin x)/x$ function and the solid curve represents the cubic interpolating spline. The plot on the left shows the cubic spline for the nonnoisy (deterministic) case; the plot on the right shows it when the response is noisy.

A natural way to measure roughness, incorporate it into the smoothing procedure, and thereby get smoothing that is more representative of the underlying function is to penalize it. This can be done readily using norms on derivatives. Let $f \in C^2$ be a twice continuously differentiable function defined on $\mathscr{X} = [a, b] \subset \mathbb{R}$. A measure of roughness is given by the total curvature penalty of a function f,

$$J(f) = \int_a^b (f''(x))^2 dx.$$

For estimation purposes, $J(f)$ is a good measure of the oscillation of the function for which interpolation is being carried out. After all, the second derivative forces constant and linear dependence of f on x to zero, but higher-order fluctuations remain. Of course, a natural question is why we use this roughness penalty rather than something else. Wouldn't the square of any sum of derivatives be a suitable penalty? In fact, many other penalty terms can be used to derive estimates with moderate fluctuations. Green and Silverman (1994) discuss the choice of roughness penalty in more detail.

The way a roughness measure such as J gets used in regularized optimizations such as splines is through penalized least squares. Observe that when there is no restriction on the form of f, the traditional least squares criterion automatically interpolates (passes through all the points). Due to response noise and other sources of variation, this leads to unacceptably high variability. Restricting the class of functions – not by imposing a form as in the parametric context, but by specifying properties such as smoothness that these functions should have – is a natural way to resolve the problem since it typically permits optimization over a nonparametric set of alternatives.

For smoothing splines, the function space is determined by penalizing the sum of squared errors by the total curvature (roughness) for $f \in \mathscr{C}^2$. That is, the cubic spline smoother is (implicitly) defined as

$$\hat{f} = \arg \min_{f \in \mathscr{H}^2(\mathscr{X})} E_\lambda(f) \tag{3.1.1}$$

where

$$E_\lambda(f) = \frac{1}{n} \sum_{i=1}^{n} (Y_i - f(X_i))^2 + \lambda \int_a^b (f''(t))^2 dt \tag{3.1.2}$$

and $\mathscr{H}^2[a,b]$ is the Sobolev space defined by

$$\mathscr{H} \equiv \mathscr{H}^2[a,b] = \Big\{ f : [a,b] \to \mathbb{R} : f \text{ and } f' \quad \text{are absolutely continuous} \\ \text{and } \int_a^b [f''(t)]^2 dt < \infty \Big\}.$$

In this formulation, given a λ, one minimizes (3.1.2). It will be seen that the result is a cubic spline with coefficients that best fit the data; this is a generalization of the linear regression. It's as if the terms in a regression function are basis elements of a spline space and their coefficients correspond to the basis expansion of the unknown f. The benefit of splines is that they track local behavior well, like LOESS, but have continuity properties while avoiding the poor behavior of global polynomials.

The class of regularized risk problems exemplified by (3.1.2) generalizes to other measures of risk (i.e., not just squared error) and to other penalty functions (i.e., not just regularizers expressible in terms of an inner product). Many instances of these – LASSO, Bridge, CART, GLMs, SVMs and so on – will be encountered in later chapters. In all these cases the optimization is merely an extra constraint to enable estimation of the coefficients in the regression function.

The smoothing parameter λ helps achieve the balance between two antithetic goals: (a) goodness of fit of $\hat{f}$ to the observations $(X_1,Y_1),\cdots,(X_n,Y_n)$ and (b) smoothness of $\hat{f}$ in the sense that smoothness captures the trends in the data that vary more slowly than the error terms. Intuitively, λ controls the trade-off between bias expressed through the empirical risk $E_{\text{emp}}(f) = \frac{1}{n}\sum_{i=1}^{n}(Y_i - f(X_i))^2$ and variance expressed by $J(f)$, the roughness of the function. Figure 3.3 shows how λ controls this trade-off.

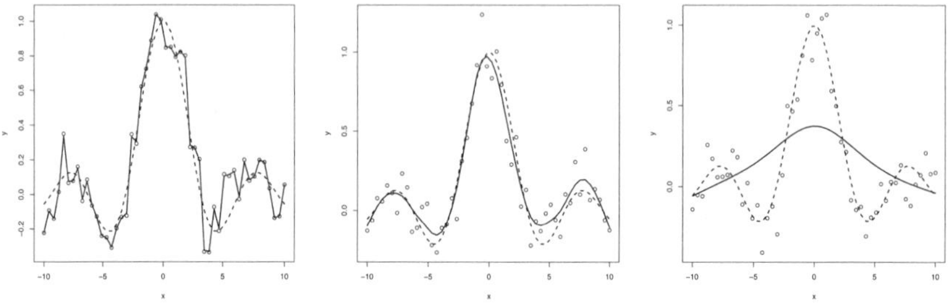

(a) Less Smoothing (b) Optimal smoothing (c) Too much smoothing

Fig. 3.3 Panel (a) shows that with less smoothing (small λ) the fitted curve is rougher, while more smoothing in panel (b) makes the fitted curve track the shape of the true function well, and too much smoothing (large λ) in panel (c) makes the fitted curve too close to a straight line. CV was used to find the optimal value $\lambda = 0.545$ used in panel (b); $\lambda = 0.1$ for panel (a) and $\lambda = 0.9$ for panel (c).

Another interpretation of (3.1.2) is as a regularized empirical risk, so that λ summarizes the trade-off between residual error and local variation. For small values of λ, the criterion $E_\lambda(f)$ is dominated by the residual sum of squares $E_{\text{emp}}(f)$ and the corresponding curve in Fig. 3.3(a) tends to interpolate the data. For large values of λ, the criterion $E_\lambda(f)$ is dominated by $J(f)$ and the corresponding curve in Fig. 3.3(c) displays very little curvature (roughness) since $J(f)$ is forced to be small. In the limiting case of λ close to infinity, the curve in Fig. 3.3(c) would simply be a straight line because $J(f)$ would be forced to zero. It is reasonable therefore to expect an intermediate value of λ to be most desirable as in Fig. 3.3(b).

Note that λ in (3.1.2) for splines plays the same role as h for kernel methods. Indeed, in both cases there is a concept of risk – MISE for kernel methods, (3.1.2) for splines – in which a parameter controls the tradeoff between bias and variability. The qualitative behavior of these risk functionals is similar and is seen in Fig. 3.4.

It turns out that minimizing $E_\lambda(f)$ in (3.1.2) is a particular instance of a much larger class of problems. Some of these occur naturally in the context of Reproducing Kernel Hilbert spaces (RKHSs), Hilbert spaces equipped with a 2-argument function that reproduces elements of the space in terms of the inner product. Indeed, it will be seen that RKHSs are the natural setting for spline methods in general. The optimality of cubic spline smoothers transfers readily to the RKHS framework, in which the generalization of (3.1.2) is to minimize

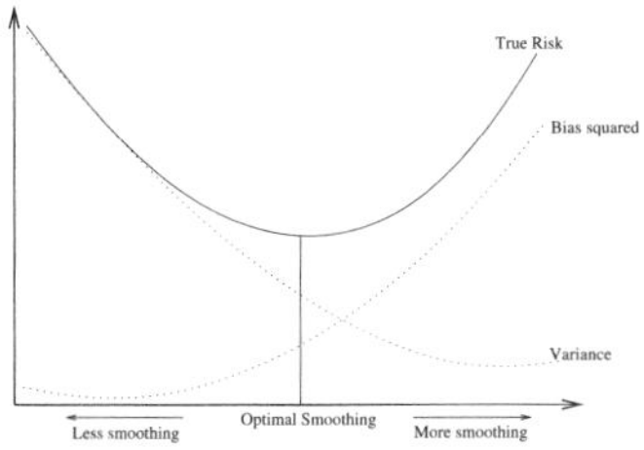

Fig. 3.4 Qualitative behavior of the dependence of bias versus variance on a trade-off parameter such as λ or h. For small values, the variability is too high; for large values the bias gets large.

$$E_\lambda(f) = \frac{1}{n}\sum_{i=1}^{n}(Y_i - F_i(f))^2 + \lambda \int_a^b [(Lf)(t)]^2 dt, \tag{3.1.3}$$

where the F_is are continuous linear functionals and L is a linear differential operator of order $d \geq 1$, say

$$(Lf)(t) = f^{(d)}(t) + \sum_{j=0}^{d-1} w_j(t) f^{(j)}(t), \tag{3.1.4}$$

in which the weight functions $w_j(\cdot)$ are real-valued and continuous. With this generalization, the minimization in (3.1.3) is over f in the Sobolev space,

$$\mathscr{H} \equiv \mathscr{H}^d(\mathscr{X}) = \Big\{ f : \mathscr{X} \to \mathbb{R} : f^{(j)}, j = 0, 1, \cdots, d-1 \text{ are absolutely continuous, and } \int_{\mathscr{X}} [f^{(d)}(t)]^2 dt < \infty \Big\}.$$

Expression (3.1.2) corresponds to $F_i(f) = f(x_i)$ and $Lf = f''$ in (3.1.3). In various applications, different choices of F_i and L are required to capture the functional dependencies underlying the data. This will be taken up briefly later. Indeed, it will be argued that without prior information to help restrict the search space, the whole smoothing problem becomes ill-posed in Hadamard's sense. For now, however, the focus is on the traditional way smoothing splines are computed in practice.

3.2 Natural Cubic Splines

Let s be a natural cubic spline with knots at $x_1 < \cdots < x_n$. The representation of s from Green and Silverman (1994) in terms of second derivatives provides insights into how to compute a natural cubic spline (NCS) in practice. To see this, first write the vector of function values of s at its knots as $\boldsymbol{s} = (s_1, s_2, \cdots, s_n)^\mathsf{T}$, where $s_i = s(x_i)$. The dual use of s as a spline and as a vector will not be confusing because the context will make clear which is meant. Next, let $\boldsymbol{\gamma} = (\gamma_1, ..., \gamma_n)$ be the vector of second derivatives of s at its knots. That is, $\gamma_i = s''(x_i)$ for $i = 1, \cdots, n$. For NCSs, $s''(x_1) = s''(x_n) = 0$, so $\gamma_1 = \gamma_n = 0$. Taken together, the vectors $\boldsymbol{s}$ and $\boldsymbol{\gamma}$ specify s completely, however, the values of s_i and γ_i must be restricted to ensure there is an NCS s corresponding to them. Of course, in many applications, setting the second derivatives to zero at the endpoints is not "natural" at all and so is not used. However, when all of these derivative constraints can be used, the extra structure is striking.

For instance, these restrictions can be expressed in terms of two "band" matrices. First, let $\delta_i = x_{i+1} - x_i$ for $i = 1, \cdots, n-1$ be the distance between successive design points. Denote the first band matrix by $A = (a_{i,j})$, where $i = 1, ...n$ and $j = 2, ..., n-1$, so that A is $n \times (n-2)$. The entries in A are defined to be

$$a_{j-1,j} = \delta_{j-1}^{-1}, \quad a_{jj} = -\delta_{j-1}^{-1} - \delta_j^{-1}, \quad \text{and} \quad a_{j+1,j} = \delta_j^{-1},$$

for $j = 2, \cdots, n-1$, and otherwise $a_{ij} = 0$ for $|i-j| \geq 2$. Clearly, A is a band matrix in the sense that there is a band of nonzero values along the main diagonal. That is, A looks like

$$A = \begin{bmatrix} a_{12} & 0 & \cdots & \cdots & \cdots & \cdots & 0 \\ a_{22} & a_{23} & 0 & \cdots & \cdots & \cdots & 0 \\ a_{32} & a_{33} & a_{34} & 0 & \cdots & \cdots & 0 \\ \vdots & \vdots & \vdots & \vdots & \vdots & \vdots & \vdots \\ 0 & \cdots & \cdots & 0 & a_{n-2,n-3} & a_{n-2,n-2} & a_{n-2,n-1} \\ 0 & \cdots & \cdots & \cdots & 0 & a_{n-1,n-2} & a_{n-1,n-1} \\ 0 & \cdots & \cdots & \cdots & \cdots & 0 & a_{n,n-1} \end{bmatrix}.$$

To forestall frustration, it is important to note that A is not indexed in the usual way: A starts with a_{12} in the upper left corner.

The second band matrix, B, is $(n-2) \times (n-2)$, symmetric, and has entries $b_{i,j}$ for $i, j = 2, \cdots, n-1$ defined by

$$b_{ii} = \frac{1}{3}(\delta_{i-1} + \delta_i) \quad i = 2, \cdots, n-1,$$

$$b_{i,i+1} = b_{i+1,i} = \frac{1}{6}\delta_i \quad i = 2, \cdots, n-2,$$

and $b_{ij} = 0$ for $|i-j| \geq 2$. Note that the elements are defined by differences in various x_is. Like A, B is not indexed in the usual way: B starts with b_{22}. It turns out that B is strictly positive definite, so B^{-1} exists. Now, define

$$K = AB^{-1}A^{\mathsf{T}}.$$

The main result, due to Green and Silverman (1994), that gives the "compatibility" conditions a spline must satisfy is the following.

Theorem (Green and Silverman, 1994): The vectors $\boldsymbol{s}$ and $\boldsymbol{\gamma}$ specify a natural cubic spline s if and only if

$$A^{\mathsf{T}}\boldsymbol{s} = B\boldsymbol{\gamma} \tag{3.2.1}$$

is satisfied. If (3.2.1) is satisfied, then the roughness penalty $J(s)$ will satisfy

$$\int_a^b [s''(x)]^2 dx = \boldsymbol{\gamma}^{\mathsf{T}} B \boldsymbol{\gamma} = \mathbf{s}^{\mathsf{T}} K \mathbf{s}.$$

This theorem dramatizes the fact that splines, natural cubic or otherwise, are not the same as all local polynomials on the partition of an interval $[a,b]$. Indeed, the collection of all local polynomials of degree d or less will have dimension $(d+1)(n-1)$ for $n-1$ intervals defined by n partition points. The dimension of splines is much lower because forcing continuity up to, say, second-order derivatives severely restricts the class of functions. It is important to note that splines of degree d have a leading term in x of degree d with a nonzero coefficient. The preceding theorem represents all those constraints on the collection of local polynomials. It can be seen that the constraints represented by (3.2.1) lead to the uniqueness of the NCS interpolant, as stated in the next theorem.

Theorem (Uniqueness of NCSs): Suppose $n \geq 2$ and $x_1 < \cdots < x_n$. Given any set of responses $y_1, \cdots, y_n$, there is a unique natural cubic spline interpolant s with knots at the x_is satisfying

$$s(x_i) = y_i, \qquad i = 1, \cdots, n. \square$$

Given these results, the NCSs exist and are uniquely characterized. The most important result is that they are optimal in a roughness sense.

Proposition: Among all interpolating, twice differentiable functions, NCSs minimize the total curvature of the function. In other words, if $s(x)$ is an NCS and $z(x)$ denotes any other function in $\mathscr{C}^2$, then

$$\int_a^b (s''(t)^2 dt \leq \int_a^b z''(t)^2 dt, \quad \text{where} \quad s''(a) = s''(b) = 0.$$

Proof: The proof amounts to a verification that for any other twice continuously differentiable interpolating function $z(t)$,

$$\int_a^b (s''(t))^2 dt - \int_a^b (z''(t))^2 dt \leq 0.$$

Consider the following expansion:

$$\begin{aligned}\int_a^b (s''(t) - z''(t))^2 dt &= \int_a^b (s''(t))^2 dt + \int_a^b (z''(t))^2 dt - 2\int_a^b s''(t) z''(t) dt \\ &= \int_a^b (z''(t))^2 dt - \int_a^b (s''(t))^2 dt - 2\int_a^b s''(t)(z''(t) - s''(t)) dt.\end{aligned}$$

Since the left-hand side is nonnegative, it is enough to show that the last term on the right-hand side is zero.

Consider $h(t) = z(t) - s(t)$. Using integration by parts and the fact that $s'''(t)$ is piecewise constant (i.e., $s'''(t) = \delta_i$) for $t_i \leq t < t_{i+1}$ one can examine

$$
\begin{aligned}
Q &= \int_a^b s''(t)h''(t)dt = \int_a^b s''(t)(z''(t) - s''(t))dt \\
&= s''(t)h'(t)\Big|_a^b - \int_a^b s'''h'(t)(t)dt \\
&= s''(b)h'(b) - s''(a)h'(a) - \sum_i \int_{t_i}^{t_{i+1}} s'''(t)h'(t)dt \\
&= s''(b)h'(b) - s''(a)h'(a) - \sum_{i=1}^{n-1} \delta_i[h(t_{i+1}) - h(t_i)].
\end{aligned}
$$

The endpoint conditions are $s''(a) = s''(b) = 0$, and continuity of $s(t)$ and $z(t)$ implies $h(t_{i+1}) = h(t_i)$ for all i. Thus, all the terms in Q are 0. □

3.3 Smoothing Splines for Regression

Up to this point, the focus has been on interpolation. However, it is clear that interpolation is not satisfactory from a statistical perspective because noise in the data causes rapid local fluctuations in the interpolant. So, turning to smoothing splines for regression, it is time to prove that the optimal solution to the penalized least squares objective function in (3.1.1) is a cubic smoothing spline. This can be shown using vectors $\boldsymbol{\gamma}$ and $\boldsymbol{s}$ and matrices A, B, and K.

Recall that the objective function is

$$
E_\lambda(f) = \frac{1}{n}\sum_{i=1}^{n}(Y_i - f(x_i))^2 + \lambda \int_a^b [f''(t)]^2 dt,
$$

where the (fixed) knots $x_1 < \cdots < x_n$ have random responses $Y_1, \cdots, Y_n$. The penalized least squares error $E_\lambda(f)$ can be rewritten as

$$
\begin{aligned}
E_\lambda(f) &= (\mathbf{Y} - \boldsymbol{f})^\mathsf{T}(\mathbf{Y} - \mathbf{f}) + \lambda \boldsymbol{f}^\mathsf{T} K \boldsymbol{f} \\
&= \boldsymbol{f}^\mathsf{T}(I + \lambda K)\boldsymbol{f} - 2\mathbf{Y}^\mathsf{T}\mathbf{f} + \mathbf{Y}^\mathsf{T}\mathbf{Y}.
\end{aligned} \tag{3.3.1}
$$

Because $I + \lambda K$ is strictly positive definite, (3.3.1) has a unique minimum over f corresponding to

$$
\boldsymbol{f} = (I + \lambda K)^{-1}\mathbf{Y}. \tag{3.3.2}
$$

It is easy to show that this is indeed a minimum simply by noticing that

$$
\frac{\partial^2 E_\lambda(f)}{\partial \boldsymbol{f}^\mathsf{T} \partial \mathbf{f}} = 2(I + \lambda K),
$$

which is positive definite. More formally, the optimality of the NCS smoother can now be stated.

Theorem (Optimality of NCSs for smoothing): Suppose $n \geq 3$, and consider the knots $x_1 < \cdots < x_n \in [a,b]$ with corresponding random response values $Y_1, \cdots, Y_n$. Let λ be a strictly positive smoothing parameter, and let $\hat{f}$ be the natural cubic spline with knots at the points $x_1 < \cdots < x_n$ for which

$$\boldsymbol{f} = (I + \lambda K)^{-1}\mathbf{Y}.$$

Then, $\hat{f}$ is the minimizer of $E_\lambda(f)$ over the class $\mathscr{C}^2[a,b]$,

$$\forall f \in \mathscr{C}^2[a,b], \quad E_\lambda(\hat{f}) \leq E_\lambda(f),$$

with equality only if f and $\hat{f}$ are identical.

Proof: Omitted. □

Despite this theorem, the expressions that allowed derivation of the existence, uniqueness, and optimality of NCS-based smoothers are not easy to deal with computationally. Since a cubic spline is just a piecewise function whose pieces are polynomials of degree three, one way to find a cubic smoothing spline for regression would start by expressing $s(x)$ as

$$s(x) = \beta_0 + \beta_1 x + \beta_2 x^2 + \beta_3 x^3 + \sum_{i=2}^{n-1} \theta_i (x - x_i)_+^3,$$

where

$$u_+ = \begin{cases} u, & \text{if } u > 0; \\ 0, & \text{otherwise.} \end{cases}$$

Justification for this representation will become clear from the connection between smoothing splines and RKHSs. Also, by plugging this expression for $s(x)$ into the expression for the penalized least squares, it becomes clear that the notation and the computation usually become unwieldy. Indeed, the matrices derived from this representation are often ill-conditioned, causing the construction of the smoother to be unstable. As a consequence, using regression splines to construct cubic spline smoothers is usually avoided in practice. An alternative representation is provided by basis B-splines; their appeal is their simplicity of interpretation and the fact that the ensuing matrices are banded, typically leading to more computationally stable estimation procedures.

3.3.0.1 Cubic Spline Smoothers Through Basis B-Splines

Unlike the two representations mentioned earlier, basis B-splines turn out to provide a great advantage of both simplicity and computational stability. The ith B-spline basis function of degree j is usually defined by the Cox-de Boor recursion formula

$$B_{i,0}(x) = \begin{cases} 1, & x_i \le x < x_{i+1}; \\ 0, & \text{otherwise}, \end{cases}$$

$$B_{i,j}(x) = \left(\frac{x - x_i}{x_{i+j-1} - x_i} \right) B_{i,j-1}(x) + \left(\frac{x_{i+j} - x}{x_{i+j} - x_i} \right) B_{i+1,j-1}(x). \quad (3.3.3)$$

As written, this recursion is for natural splines; i.e., $s''(x_0) = s''(x_n) = 0$ for the cubic case. However, it extends to include free values at the boundary; see Zhou et al. (1998).

Clearly, for given j, $B_{i,j}(x)$ is computed from two B-spline basis functions of degree $j-1$, each of which is computed from two B-spline basis functions of degree $j-2$, and so on. Therefore, $B_{i,j}(x)$ is constructed recursively starting from degree 0 and giving step functions on the subintervals of $[a,b]$. For $j = 1$, it can be verified that the splines are "triangles" on adjacent subintervals and zero elsewhere. The index i ranges from 1 to $n-1$ and the index j ranges from 0 to d, but from the second term in (3.3.3) it is seen that $i + j \le n$. Thus, for fixed j, there are $n - j$ linearly independent elements that span the space of splines of degree j. The elements $B_{i,d}$ define the dth order (natural) B-spline basis because all the splines at iteration j are polynomials of degree j. To find the dth order B-spline basis, one ends up finding dn functions along the way in the recursion but it is only the last iteration for d that forms the basis elements. When the knots are equidistant, B-splines are uniform; otherwise they are nonuniform. With uniform (natural) B-splines, the whole machinery is considerably simplified.

For convenience, the doubly indexed B-spline basis elements $B_{i,d}$ can be reindexed to $B_i(x)$, in which the d is understood, for $i = 1, ..., n-d$. Since the B-splines of a given order d form a basis for the collection of splines of that order, smoothing is achieved by expressing the cubic spline as a linear combination of basis elements. There are $n-d$ basis elements; however, this was determined by counting knots with multiplicity one (i.e., assuming they are distinct). More commonly, the left- and right-hand knots are counted with multiplicity d, adding $2d$ more spline functions. Note that B_{ij} is nonzero on (x_i, x_{i+j+1}), so the effect is to include in the basis all the higher-order polynomial splines as long as there is a region on which they are strictly positive. Doing this gives the B-spline basis expansion

$$\hat{f}(x) = \sum_{i=1}^{n+d} \beta_i B_i(x) \quad (3.3.4)$$

for some collection $\beta = (\beta_1, ..., \beta_{n+d})$ of coefficients. This can be compactly expressed by defining the matrix $\mathbf{U}$ with elements $u_{ji} = B_i(x_j)$, and the vector $\hat{\boldsymbol{f}}$ with elements $\hat{\boldsymbol{f}} = (\hat{f}(x_1)\hat{f}(x_2), \cdots, \hat{f}(x_n))^\mathsf{T}$, so that

$$\hat{\boldsymbol{f}} = \mathbf{U}\boldsymbol{\beta}. \quad (3.3.5)$$

Now, the original problem of minimizing (3.3.1) or (3.1.2) can be reformulated as

$$\hat{\boldsymbol{\beta}} = \arg \underset{\boldsymbol{\beta}}{\text{argmin}}\, (\boldsymbol{y} - \boldsymbol{U}\boldsymbol{\beta})^\mathsf{T}(\boldsymbol{y} - \boldsymbol{U}\boldsymbol{\beta}) + \lambda \boldsymbol{\beta}^\mathsf{T} \boldsymbol{V} \boldsymbol{\beta}, \quad (3.3.6)$$

where the elements v_{ij} of the matrix $\boldsymbol{V}$ are

$$v_{ij} = \int_a^b B_i''(x)B_j''(x)dx,$$

in which $a = x_0$ and $b = x_n$.

From simple matrix algebra, it is easy to see that the minimization in (3.3.6) can be solved from

$$\frac{\partial}{\partial\beta}\{(\mathbf{y}-\mathbf{U}\beta)^{\mathsf{T}}(\mathbf{y}-\mathbf{U}\beta)+\lambda\beta^{\mathsf{T}}\mathbf{V}\beta\} = -2\mathbf{U}^{\mathsf{T}}(\mathbf{y}-\mathbf{U}\beta)+2\lambda\mathbf{V}\beta = \mathbf{0},$$

which leads to

$$\hat{\beta} = (\mathbf{U}^{\mathsf{T}}\mathbf{U}+\lambda\mathbf{V})^{-1}\mathbf{U}^{\mathsf{T}}\mathbf{y}. \tag{3.3.7}$$

Incidentally, using (3.3.7) in (3.3.5) shows that $\hat{\boldsymbol{f}}$ from cubic splines is linear in the sense of (2.1.1). That is, for fixed λ, there is a weight matrix $\mathbf{W}^{\lambda}$ such that

$$\hat{\boldsymbol{f}} = \mathbf{W}^{\lambda}\mathbf{y}, \tag{3.3.8}$$

which can be rewritten as

$$\hat{f}(x_i) = \sum_{j=1}^{n} W_j^{(\lambda)}(x_i)Y_j, \tag{3.3.9}$$

where $(W_1^{\lambda}(x_i), W_2^{\lambda}(x_i), \cdots, W_n^{\lambda}(x_i))^{\mathsf{T}}$ is the ith row of a matrix $\mathbf{W}^{\lambda}$ and

$$\mathbf{W}^{\lambda} = \mathbf{U}(\mathbf{U}^{\mathsf{T}}\mathbf{U}+\lambda\mathbf{V})^{-1}\mathbf{U}^{\mathsf{T}}.$$

3.3.1 Model Selection for Spline Smoothing

As with the bandwidth h in kernel methods, the smoothing parameter λ plays a central role and there is a small industry devoted to developing techniques to estimate it. Here, for completeness, one simple method is described as a parallel to the method for obtaining $\hat{h}$ in the last section.

Again, the method is CV. Given λ, the expression is

$$\mathrm{CV}(\lambda) = \frac{1}{n}\sum_{i=1}^{n}(Y_i - \hat{f}_{\lambda}^{(-i)}(X_i))^2,$$

where $\hat{f}_{\lambda}^{(-i)}(X_i)$ is the spline smoothing estimator for the given λ without using observation i. Unfortunately, CV is computationally very intensive and hence sometimes impractical, even for moderate sample sizes. The GCV criterion mentioned in Chapter

1 is more computationally feasible. By using the linear form of the spline smoother from (3.3.8), one can verify that the GCV criterion finds the value of λ that minimizes

$$\mathsf{GCV}(\lambda) = \left[\frac{n}{\text{trace}\left(\mathbf{I} - \mathbf{W}^{(\lambda)}\right)} \right]^2 \mathsf{MASE}(\lambda),$$

where the mean average square error is

$$\mathsf{MASE}(\lambda) = \frac{1}{n} \sum_{i=1}^{n} (Y_i - \hat{f}_\lambda(X_i))^2,$$

intended to approximate $\mathsf{MASE} = E(\mathsf{ASE}(h)|x_1, x_2, ..., x_n)$. The $\mathsf{ASE}(h)$ is an average, but still has stochasticity in it; the MASE is the mean value of ASE. See Eubank (1988) and Wahba (1990) for more sophisticated techniques for choosing λ and the order of the spline basis.

3.3.2 Spline Smoothing Meets Kernel Smoothing

A local approach to bandwidth selection only mentioned in the Ethanol example at the end of the last chapter was writing $h = h(x)$ and then estimating the bandwidth function itself nonparametrically. This would be akin to saying that no fixed bandwidth was a good choice because the degree of smoothing was a local property. In some regions, bias might be a bigger problem, while in others variance might be dominant, corresponding to too much or too little smoothing relative to a single fixed bandwidth. Essentially, this enlarges the minimization from a search over constants h to functions $h(x)$, giving a stronger minimum and improving the function estimation. For an in depth review on kernel smoothing with variable kernels, see Hardle (1990), Fan and Gijbels (1996), and Green and Silverman (1994), among others.

A surprising insight, due to Silverman (1984) (see also Eubank (1988), Chapter 6 for discussion) is that, in the enlarged minimization permitting $h = h(x)$, spline smoothing and kernel smoothing are much the same. That is, the linear transformation of $Y_1, ..., Y_n$ done by spline smoothing in (3.3.8) corresponds to appropriately weighting the individual kernel's contributions in the NW estimator (i.e., choosing the linear transformation in (2.3.19) the right way).

To state an informal version of Silverman (1984)'s result, let $p(x)$ be the density generating the design points $X_1, ..., X_n$, and choose the kernel

$$K_{\mathsf{S}}(u) = \frac{1}{2} e^{-|u|/\sqrt{2}} \sin\left(\frac{|u|}{\sqrt{2}} + \frac{\pi}{4} \right). \tag{3.3.10}$$

Consider $\{X_1, X_2, \cdots, X_n\} \subseteq [a, b]$ and a point x_i away from the boundary.

Theorem (Silverman, 1984): Suppose $\lambda \to 0$ for $n \to \infty$, and choose

$$h(x) = \left(\frac{\lambda}{np(x)} \right)^{1/4}.$$

Then, with $W_j^{(\lambda)}(x_i)$ defined as in equation (3.3.8),

$$W_j^{(\lambda)}(x_i) \approx \frac{1}{p(X_j)} \frac{1}{h(X_j)} K_{\mathsf{S}} \left(\frac{x_i - X_j}{h(X_j)} \right) \tag{3.3.11}$$

when $n \to \infty$. □

Silverman's theorem states that, asymptotically, spline smoothers yield NW estimates with a variable bandwidth $h(x)$ that depends on both the global smoothing parameter λ and the local density $p(x)$ around x. The rate of $1/4$ used in h for $\lambda/(np(x))$ is necessary for the equivalence in (3.3.11); the issue may be that λ should not go to zero too fast or else there will be too little smoothing. The kernel in (3.3.10) arises from a Fourier transform of a damping function; see Eubank (1988), Chapter 6.3.1. The kernel does not have compact support and has a different shape from the optimal Epanechnikov kernel. The limitation that x_i be away from the boundary can be overcome but is more technical than needed here.

Since Silverman (1984), other authors have started from spline smoothing, reformulating it as kernel smoothing with variable bandwidth. For instance, Huang (2001), building on work of Terrell (1990), shows how kernel smoothing can be obtained under the roughness-penalty framework. Indeed, the de facto equivalence of splines and kernels for large classes of functions invites speculation that there could be a more general (and richer) theory for statistical curve estimation.

3.4 Asymptotic Bias, Variance, and MISE for Spline Smoothers

In principle, (3.3.8) could be used to adapt the asymptotics for kernel methods to the present case of splines. However, explicit, direct rates of convergence for the bias-variance tradeoff and MISE achieved by splines are available from Zhou et al. (1998). This subsection is largely based on their work; formally only the deterministic design case will be stated, but the random design case leads to the same asymptotic forms.

Recall the natural B-spline basis described in (3.3.3) for a fixed degree d, indexed to consist of $n+d$ elements $B_i(\cdot)$, corresponding to n subintervals and degree d splines, in (3.3.4) leading to the matrix $\mathbf{U}$ in (3.3.5). Following Zhou et al. (1998), there is no need to restrict to natural splines, so the free values at the endpoints mean the spline basis will have more elements. Set $0 = x_0 < \ldots < x_n = 1$ for $[a,b] = [0,1]$, so the natural spline basis of order d has $d+n$ elements because the multiplicity used in (3.3.3) at zero and one is d.

Next, set $\delta_i = x_i - x_{i-1}$ and let $\hat{F}_n(\cdot)$ be the empirical distribution, assumed to get close to a distribution F in Kolmogorov-Smirnov distance, where F has strictly positive density f. Now, for a fixed design, the asymptotic bias and variance can be stated.

Theorem (Zhou et al., 1998): Suppose that

$$\max_{1 \le i \le n} |\delta_i - \delta_{i-1}| = o\left(\frac{1}{n}\right) \quad \text{and} \quad \frac{\delta}{\min\limits_{1 \le i \le n} \delta_i} \le M$$

for some $M > 0$, where $\delta = \max\limits_{1 \le i \le n} \delta_i$. Also assume the design points $\{x_i\}_{i=1}^n$ are deterministic but

$$\sup_{x \in [0,1]} |\hat{F}_n(x) - F(x)| = o\left(\frac{1}{k}\right).$$

Let $\hat{f}(x)$ be the spline regression estimator for $f \in \mathscr{C}^d[0,1]$ based on the fixed design x_i for $i = 1, ..., n$. Then, for any $x \in (x_{i-1}, x_i]$,

$$\mathbb{E}[\hat{f}(x)] - f(x) = b(x) + o(\delta^d)$$

and

$$\text{Var}[\hat{f}(x)] = \frac{\sigma^2}{n} \mathbf{U}^\mathsf{T}(x) G^{-1}(f) \mathbf{U}(x) + o\left(\frac{1}{n\delta}\right),$$

where

$$b(x) = -\frac{f^{(d)}(x)\delta_i^d}{d!} C_d\left(\frac{x - x_i}{\delta_i}\right) \quad \text{and} \quad G(f) = \int_0^1 \mathbf{U}(x)\mathbf{U}^\mathsf{T}(x) f(x) dx. \square$$

In the above, $C_d(\cdot)$ is the dth Bernoulli polynomial defined recursively by

$$C_0(x) = 1 \quad \text{and} \quad C_i(x) = \int_0^x i C_{i-1}(z) dz + b_i,$$

where $b_i = i \int_0^1 \int_0^x C_{i-1}(z) dz dx$ is the ith Bernoulli number. □

As noted by Zhou et al. (1998), the assumptions of the theorem and the expressions for both the asymptotic bias and the asymptotic variance reveal that n controls the trade-off between the bias and the variance associated with $\hat{f}(x)$. But observe that the theorem assumes the knots are chosen to be the design points. In fact, one can choose the knots to be t_i for $i = 1, ..., k$ and let the x_is be design points $i = 1, ..., n$, and then ensure convergence in the argument x. In this more general version, Zhou et al. (1998) show that when $k = Cn^{1/(2d+1)}$ for some constant $C > 0$, the bias and variance have asymptotic rates

$$\max_{x \in [0,1]} |b(x)| = O(n^{-d/(2d+1)})$$

and

$$\max_{x \in [0,1]} \text{Var}[\hat{f}(x)] = O(n^{-2d/(2d+1)}).$$

As a result,

$$\hat{f}(x) - f(x) = O_P(n^{-d/(2d+1)}) \quad \text{uniformly for } x \in [0,1].$$

The consequence of this (see Zhou et al. (1998)) is that, for any probability μ,

$$\mathsf{IMSE}(\hat{f}) = \sup_{f \in C(d,\delta)} \int_0^1 \mathbb{E}\left[\hat{f}(x) - f(x)\right]^2 d\mu(x) = O(n^{-2d/(2d+1)}),$$

where $C(d,\delta) = \{f \in \mathscr{C}^d[0,1] : |f^{(d)}| \leq \delta\}$ for some $\delta > 0$. This agrees with the best rates one can get from kernels and in fact is optimal. Nonparametric optimality will be discussed in Chapter 4.

Turning at last to asymptotic normality, also for the fixed-design case, Zhou et al. (1998) establish the following.

Theorem (Zhou et al., 1998): Suppose that the assumptions of the theorem hold. Assume in addition that the error terms $\{\varepsilon_i\}_{i=1}^n$ are independently and identically distributed with mean 0 and common variance σ^2 and that the number of knots is $k \geq Cn^{1/(2d+1)}$ for some constant $C > 0$. Then, for any fixed $x \in [0,1]$,

$$\frac{\hat{f}(x) - (f(x) + b(x))}{\sqrt{\mathrm{Var}[\hat{f}(x)]}} \xrightarrow{d} N(0,1). \quad \square$$

Since $b(x)$ tends to zero uniformly in x at the same rate as $\sqrt{\mathrm{Var}[\hat{f}(x)]}$, it is possible that the scaled bias does not go to zero asymptotically. However, Zhou et al. (1998), Theorem 4.2, shows that the scaled bias only distorts the confidence region by modifying the critical value.

Zhou et al. (1998) also provide results analogous to the above for the case where the design is stochastic. Basically, for the random design, $\mathbb{E}[\hat{f}(x)]$ is replaced by $\mathbb{E}[\hat{f}(x)|x_1,\cdots,x_n]$, while $\mathrm{Var}[\hat{f}(x)]$ is replaced by $\mathrm{Var}[\hat{f}(x)|x_1,\cdots,x_n]$. The key is that the asymptotic quantities now depend on the distribution $F(x)$ and the distribution of the knots, unlike the deterministic design case.

3.4.1 Ethanol Data Example – Continued

To illustrate the performance of smoothing splines, consider again the ethanol data set from Section 2.4.2 and model E as a function of NOx. In fitting a spline model, the main choices that must be made are (a) the degree of the local polynomial, (b) the choice of the knots, and (c) the value of the smoothing parameter λ. These are primarily interesting in how they affect the bias–variance trade-off in estimation and prediction.

As a pragmatic point, the degree is usually chosen to be 2 or 3, sometimes 4. Choosing 0 or 1 is often too rough, while choosing degrees higher than 4 defeats the purpose of local fits, which is in part to use few degrees of freedom. If higher degrees seem necessary, possibly there are not enough x_is. The plots in this subsection use cubic splines; in Section 2.6, local polynomials of degree 2 and 3 were used. (Recall that the NW estimator can be regarded as zeroth order, and a running line smoother would be first order.)

Next, in statistics, the knots are often chosen to be the values of x_i. The Zhou et al. (1998) theorem permits a general selection of knots, both random and fixed. This includes equispaced knots, which are regarded as optimal in numerical analysis, where the knots are essentially never regarded as randomly sampled. Note that knots represent points of local change of the function. In view of this, it is unclear how effective conventional statistical practice is even though it is motivated by insisting on having variances that are only available at the sampled xs. Common statistical practice is followed here for the ethanol data, but see Sun and Loader (1994).

The selection of the smoothing parameter λ parallels bandwidth selection in kernel regression. In the spline context, λ can be regarded as a sort of reverse degrees of freedom. Indeed, recall that if the smoothing parameter λ is zero, the optimal function interpolates the data. This can be regarded as modeling the data with an infinite degree polynomial. At the other extreme, when λ equals infinity, the data are modeled as a straight line that has 2 degrees of freedom. These two extreme cases are shown in Fig. 3.5. As expected, when λ is too large, the bias is obviously large, and when λ is too small, the variability gets large and the fit looks ever more like an interpolation. Heuristically, values of λ between 0 and infinity correspond to intermediate values of degrees of freedom. One formalization of the relationship between λ and a concept of degrees of freedom is given in Exercise 2.16.

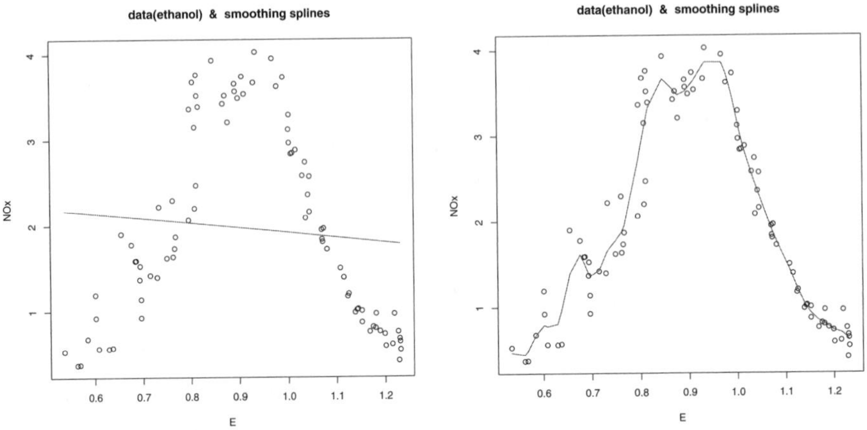

Fig. 3.5 Cubic spline fits for the ethanol data for very small and very large degrees of freedom. Note that, as λ gets larger, it corresponds to fewer degrees of freedom, in which case the optimal curve is in the nullspace of the penalty term regarded as a functional.

In practice, it is often reasonable to use GCV to find an optimal value for λ. For the ethanol data set, the GCV curve for λ (regarded as a degrees of freedom) and the fit for the optimal λ are given in Fig. 3.6. The graph does not show a unique minimum; this sometimes occurs. In general, it is not clear when to expect a unique minimum or when either the bias side or the variance side will not rise as expected. In these indeterminate cases, usually one chooses the value on the horizontal axis that looks like it would be the lowest if the other side were to rise. Sometimes this is called the knee or elbow of the curve. Here, 8 is a reasonable choice.

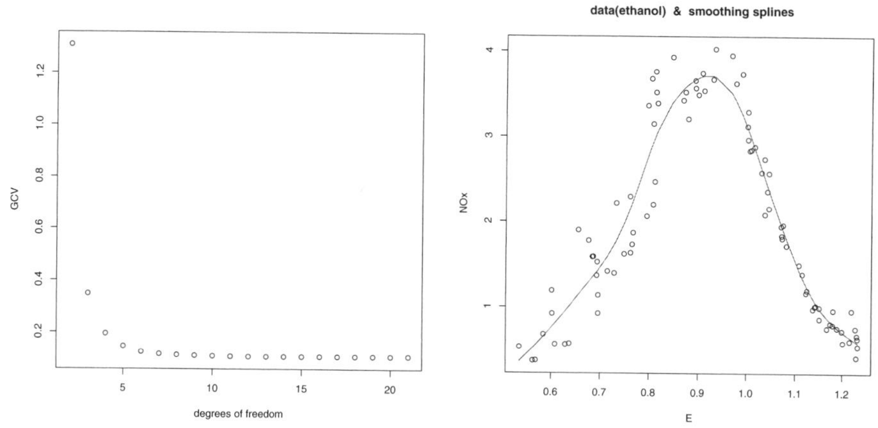

Fig. 3.6 GCV plot for cubic spline smoothing for the ethanol data. The value of λ corresponding to 8 degrees of freedom was chosen.

From (3.3.8), it is seen that spline smoothers are linear. As with kernels, the variance of $\hat{f}$ can be derived; expression (2.3.20) continues to hold but for $\mathbf{W}^{\lambda}$ as in (3.3.9). Since σ^2 can again be estimated by the residual sum of squares, $\mathrm{Var}(\hat{f})$ can be estimated at the points x_i. Recall that the variance is unspecified for values x outside the sampled x_is, and the curves in Fig. 3.7 consequently only appear continuous because the endpoints of confidence intervals at the x_is have been joined by straight lines. Since the bias at the sampled x_is can be taken as zero asymptotically, the work of Zhou et al. (1998) implies that normal-based confidence intervals at the x_is centered at $\hat{f}(x_i)$ using the estimate $\widehat{\mathrm{Var}}(\hat{f})$ just described can be used with a normal cutoff of, say, 1.96. As seen in kernel regression for these data, the signal-to-noise ratio is pretty high, so inference is pretty good.

The package gss by Chong Gu is a general framework for implementing smoothing splines. The package allows the choice not only of different degrees for the local polynomial but also different bases. The procedure ssanova is called from the package gss; here it is used to construct a cubic spline fit to the ethanol data. The procedure predict can then be used to obtain predictions for the training data along with the standard error estimates. Finally, although Gu calls confidence bands generated from the standard error estimates Bayesian, they are constructed using normal cutoffs that can be equally well regarded as frequentist. In the call to ssanova, the arguments permitted for method allow a choice of the technique for estimating the smoothing parameter. (One can use “v” for GCV, “u” for Mallows; C_p, or “m” for REML.) The results for the ethanol data are given in Fig. 3.7.

The actual R code was:

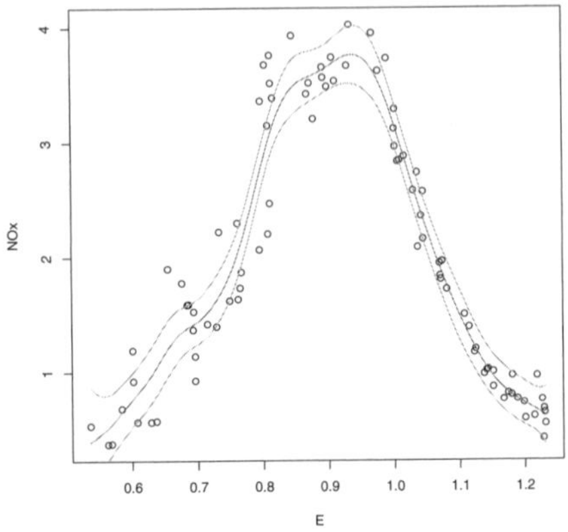

Fig. 3.7 Fit of the ethanol data with confidence bands, using the gss package.

```
cubic.fit <- ssanova(y~x, type = "cubic", method="m")
new <- data.frame(x=seq(min(x),max(x),len=length(x)))
## Obtain estimates and standard errors on a grid
est <- predict(cubic.fit,new,se=TRUE)
## Plot the fit and the Bayesian confidence intervals
plot(x,y,col=1, xlab="E", ylab="NOx");
lines(new$x,est$fit,col=2)
lines(new$x,est$fit+1.96*est$se,col=3)
lines(new$x,est$fit-1.96*est$se,col=3)
```

Estimates of the MSE and the corresponding raw MSE curves can also be obtained for the fitted spline model as was done in Section 2.4.2 for the NW estimator, but this is left as an exercise.

3.5 Splines Redux: Hilbert Space Formulation

The main goal in reformulating smoothing splines in RKHSs is to have a framework that allows for a more elegant and general way to solve the type of optimization problem that implicitly defines a smoothing spline. In addition, Wahba (2005) gives reasons why the RKHS framework is appealing in general. The underlying motivations for spline smoothing include:

- The RKHS framework provides methods for model tuning that readily allow the optimization of the bias–variance trade-off.
- Models based on RKHSs are the foundation of penalized likelihood estimation, and regularization methods more generally. They can handle a wide variety of

distributions – Gaussian, general exponential families – and summarize a variety of inference criteria for estimation, prediction, and robustness.

- In the RKHS framework, constraints such as positivity, convexity, linear inequality constraints, and so forth can be easily incorporated into the models.
- Estimates obtained via the RKHS framework have a dual interpretation as Bayes estimates (up to a point).

It is important to remember that the objective functions dealt with throughout this section are functionals (that is, operators acting on functions), as opposed to objective functions with arguments that are real variables as in the parametric paradigm. To emphasize this, rewrite the penalized least squares equation (3.1.2) as

$$E_\lambda(f) = L(f) + \lambda J(f). \tag{3.5.1}$$

In typical settings, $L(f)$ will involve function evaluations; the continuity of evaluation functionals is central to RKHSs.

Splines were initially defined as local polynomials with constraints to ensure continuous derivatives, apart from (3.1.1) and its generalization (3.1.3), (3.1.4). This second approach meant that splines also solve a minimization problem in which the knots are the x_is in a regression problem. This last property permits generalization of splines and reveals the sense in which they are a shrinkage method. Indeed, Fig. 3.5 shows that, as λ increases, the curve shifts closer to a straight line. In general this is called shrinkage because the solution "shrinks" closer to the functions in the kernel (or nullspace) of the penalty term. For the cubic spline case, the kernel of the penalty term in (3.1.1) consists of linear functions. Choosing other penalty terms and shrinking to their kernels may give functions different from local polynomials with smoothness constraints.

The formulas (3.1.2), (3.1.3), and (3.1.4) are themselves special cases of a regularized error functional on a Hilbert space. First, the total curvature $J(f)$ can be replaced by the integrated square of a higher-order derivative of f; this yields polynomial splines. Replacing derivatives by general differential operators and reformulating the minimization so it occurs in a Hilbert space of functions (also called splines) generalizes beyond polynomial splines. In the regression spline case, it will be seen that the optimization remains an extra constraint on the collection of functions, so a unique minimizer can be identified and its coefficients estimated. Here, the Hilbert space formulation, based on reproducing kernels, will be presented. Note that now the word kernel has a different meaning from the last section. Here, a kernel is a function of two variables that acts something like the inverse of the matrix defining an inner product. Heckman (1997) provides a concise and deft overview of the material developed here.

To keep focused on problems that can be solved, it is important to have a notion of ill-posed. Hadamard's definition is one choice, see Canu et al. (2005).

The problem of interpolating, or more generally of obtaining a linear smoother, comes down to the following. Let $\mathscr{X}$ and $\mathscr{Y}$ be two sets, and let $\mathscr{F}$ be the collection of functions from $\mathscr{X}$ to $\mathscr{Y}$. Given a sample $S_n = \{(x_i, y_i) | x_i \in \mathscr{X}, y_i \in \mathscr{Y}, i = 1, \cdots, n\}$, let Q be a linear operator from $\mathscr{F}$ to $\mathscr{Y}$ with domain D_Q so that

$$\forall f \in D_Q \subseteq \mathscr{F};\; Qf \triangleq (q_1(f), q_2(f), \cdots, q_n(f)). \tag{3.5.2}$$

In (3.5.2), the q_is are evaluations of f at the x_is so that $q_i(f) = f(x_i)$ and the q_is are continuous linear functionals from $D_Q \subseteq \mathscr{F}$ to $\mathscr{Y}$. Denoting $\mathbf{y} = (y_1, y_2, \cdots, y_n)$, the interpolation f, if it exists, of the n points of S_n is the solution to

$$Qf = \mathbf{y}. \tag{3.5.3}$$

Hadamard's definition of an ill-posed problem can now be stated in terms of (3.5.2) and (3.5.3). Let $(\mathscr{F}, \mathscr{Y}^n)$ be a pair of metric spaces equipped respectively with the metrics $d_{\mathscr{F}}$ and $d_{\mathscr{Y}}$. The problem $Qf = \mathbf{y}$ is said to be ill-posed on $(\mathscr{F}, \mathscr{Y}^n)$ in Hadamard's sense if it does not satisfy one of the following three conditions:

1. Existence: There exists a solution to the problem; the rank of the image of the operator Q equals the rank of $\mathscr{Y}^n$.
2. Uniqueness: The solution is unique; i.e.,

$$\forall f_1, f_2 \in \mathscr{F}, \quad Qf_1 = Qf_2 \implies f_1 = f_2.$$

3. Stability: The solution is stable in the sense that the inverse Q^{-1} of Q exists and is continuous on $\mathscr{Y}^n$.

Intuitively, the third condition on stability means that a small perturbation of the data leads to only a small change in the solution. Explicitly, this is

$$\forall \varepsilon > 0,\ \exists \delta_\varepsilon, \qquad d_{\mathscr{Y}}(y_\varepsilon, y) < \delta_\varepsilon \implies d_{\mathscr{F}}(f_\varepsilon, f) < \varepsilon,$$

where $f_\varepsilon = Q^{-1} y_\varepsilon$ and $f = Q^{-1} y$.

The reformulation of the penalized least squares problem in the RKHS framework will help show that cubic smoothing splines are the solution to least squares objective functions penalized by the integrated squared derivative as in (3.1.2). The exposition rests on some foundational definitions and notations that can be skipped if they are already familiar.

3.5.1 Reproducing Kernels

Henceforth, $\mathscr{H}$ is a linear space equipped with an inner product, a positive definite bilinear form, denoted by $\langle \cdot, \cdot \rangle$. A common choice is, for $f, g \in \mathscr{H}$, defined by

$$\langle f, g \rangle = \int_{\mathscr{X}} f(x) g(x) dx.$$

Once $\mathscr{H}$ has an inner product $\langle \cdot, \cdot \rangle$, it is an inner product space and hence has a norm defined by $\forall f \in \mathscr{H}$,

$$\|f\| = \sqrt{\langle f, f \rangle}.$$

The norm provides a metric on $\mathscr{H}$ that can be used to measure the distance between two elements $f, g \in \mathscr{H}$:$D(f,g) = \|f - g\|$. Now, in an inner product space such as $\mathscr{H}$, a sequence $\{f_n\}$ of functions converges uniformly to f^* if

$$\forall \varepsilon > 0,\ \exists N,\ \text{such that}\ \|f_n - f^*\| < \varepsilon, \quad \forall n \geq N.$$

The notation used is $\lim_{n\to\infty} \|f_n - f^*\| = 0$ or $\lim_{n\to\infty} f_n = f^*$ or simply $f_n \to f^*$.

The definitions of Section 2.3.1 for Cauchy sequence, completeness, and Banach and Hilbert space carry over. As a generality, Banach spaces are interesting in and of themselves, for their topology for instance. By contrast, because of the inner product, it is typically the (linear) operators on a Hilbert space that are interesting not the space itself.

Recall that an operator is a function on a space; a functional is a real-valued function of a function-valued argument (on a space). A linear operator, or more typically a linear functional, L, defined on a linear space $\mathscr{H}$, is a functional that satisfies two properties: $\forall f, g \in \mathscr{H}$ and $\forall \alpha \in \mathbb{R}$,

- $L(f+g) = Lf + Lg$.
- $L(\alpha f) = \alpha Lf$.

In addition, on a Hilbert space, an operator is continuous if and only if it is bounded, which means, in the linear case, it has a finite norm. Thus, a linear functional L on $\mathscr{H}$ is continuous if and only if, for a given sequence $\{f_n\}$ of functions in $\mathscr{H}$, we have

$$\lim_{n\to\infty} f_n = f^* \Longrightarrow \lim_{n\to\infty} Lf_n = Lf^*,$$

and for linear functionals this happens exactly when $||L||_{\mathscr{H}}$ is finite. In general, a closed linear subspace of $\mathscr{H}$ is itself a Hilbert space. Also, the distance between an element $f \in \mathscr{H}$ and a closed linear subspace $\mathscr{G} \subset \mathscr{H}$ is

$$D(f, \mathscr{G}) = \inf_{g\in\mathscr{G}} \|f - g\|.$$

An important implication of requiring evaluation functionals to be continuous is the Riesz representation theorem from, say, Royden (1968). This deep theorem guarantees that every linear operator can be obtained by regarding one entry in the inner product as the argument of a function while the other argument defines the operator.

Theorem (Riesz Representation): Let L be a continuous linear functional on a Hilbert space $\mathscr{H}$. Then, there exists a unique $g_L \in \mathscr{H}$ such that

$$\forall f \in \mathscr{H} \quad Lf = \langle g_L, f \rangle. \quad \square$$

While the Reisz representation theorem holds for any Hilbert space, within the collection of Hilbert spaces, there is a subclass of particular relevance to splines. These Hilbert spaces have what is called a reproducing kernel. This is a function like g_L in the theorem, but here denoted as $K(\boldsymbol{x})$, that has a reproducing property. The idea is that the kernel makes the inner product act like an identify function; i.e., for any f in the Hilbert space it "reproduces" f. Naturally, these spaces are called reproducing kernel Hilbert spaces (RKHS).

More formally, let $\mathscr{X}$ be an arbitrary function domain (index set), and let $\mathscr{S}$ be a Hilbert space of real-valued functions on $\mathscr{X}$. Denote the functional on $\mathscr{S}$ that evaluates a function f at a point $x \in \mathscr{X}$ by $[\mathbf{x}]$. An RKHS is a Hilbert space of functions in which all the evaluation functionals are continuous. That is, the space $\mathscr{H} \subset \mathscr{S}$ is an RKHS if and only if any linear evaluation functional defined by

$$\begin{aligned} [\boldsymbol{x}](\cdot) : \mathscr{H} &\longrightarrow \mathbb{R} \\ f &\longmapsto [\boldsymbol{x}]f = f(\boldsymbol{x}) \end{aligned}$$

is continuous for any $\boldsymbol{x} \in \mathscr{X}$. When this holds, the Riesz representation theorem gives that, for any $\mathbf{x} \in \mathscr{X}$, there exists an element $K_{\boldsymbol{x}}(\cdot) \in \mathscr{H}$, which is the representer of the evaluation functional $[\boldsymbol{x}](\cdot)$ such that

$$\langle K_{\boldsymbol{x}}, f \rangle = f(\boldsymbol{x}), \quad \forall f \in \mathscr{H}.$$

The symmetric function $K(\boldsymbol{x},\boldsymbol{y}) = K_{\boldsymbol{x}}(\boldsymbol{x}) = \langle K_{\boldsymbol{x}}, K_{\boldsymbol{x}} \rangle$ is called the reproducing kernel of the space $\mathscr{H}$. It is easy to check that $K(\cdot,\cdot)$ has the reproducing property

$$\forall \boldsymbol{x} \quad \langle K(\boldsymbol{x},\cdot), f(\cdot) \rangle = f(\mathbf{x}). \tag{3.5.4}$$

Thus, in principle, one can start with a Hilbert space and obtain a reproducing kernel.

One can also do the reverse and generate an RKHS from a kernel function K. Let $\mathscr{X}$ be a domain, or index set. An RKHS on $\mathscr{X}$ is a Hilbert space of real-valued functions that is generated by a bivariate symmetric, positive definite function $K(\cdot,\cdot)$, known as the kernel, provided K has the reproducing property from (3.5.4),

$$\langle K(\boldsymbol{x},\cdot), f(\cdot) \rangle = f(\boldsymbol{x}),$$

again for all $\boldsymbol{x}$. Requiring positive definiteness of $K(\cdot,\cdot)$ is essential for constructing RKHSs. In fact, the Aronszajn theorem, stated below, only gives a one-to-one correspondence between positive definite functions and reproducing kernel Hilbert spaces.

Consider the domain $\mathscr{X}$. Recall that a bivariate symmetric function $K(\cdot,\cdot)$ defined on $\mathscr{X} \times \mathscr{X}$ is positive definite (PD) if, for every n and $\boldsymbol{x}_1, \cdots, \boldsymbol{x}_n \in \mathscr{X}$ and every $a_1, \cdots, a_n$,

$$\sum_{i=1}^{n} \sum_{j=1}^{n} a_i a_j K(\boldsymbol{x}_i, \boldsymbol{x}_j) \geq 0.$$

For a reproducing kernel K, $K(\boldsymbol{x},\boldsymbol{y}) = K_{\boldsymbol{x}}(\boldsymbol{y})$, from which

$$\sum_{i=1}^{n} \sum_{j=1}^{n} a_i a_j K(\boldsymbol{x}_i, \boldsymbol{x}_j) = \left\| \sum_{i=1}^{n} a_i K_{\boldsymbol{x}_i} \right\|^2 \geq 0.$$

As a result, reproducing kernels are PD.

This definition for PD functions is very general, applying to a large variety of choices of $\mathscr{X}$. For a discrete set $\mathscr{X} = \{1,2,\cdots,N\}$, for instance, K reduces to an $N \times N$ matrix. More typically, however, $\mathscr{X} = [0,1]$, or $\mathscr{X} = [a,b]$ for real a,b. Sometimes, $\mathscr{X} \subseteq \mathbb{R}^p$ for high-dimensional observations in Euclidean space. Wahba (2005) provides a variety of choices of $\mathscr{X}$ from a simple discrete set to a collection of graphs, a collection of gene microarray chips, or even a Riemannian manifold.

Wahba (2005) also emphasizes the importance of PD functions in machine learning techniques by stating the following important fact: Since a PD function, or kernel, defined on $\mathscr{X} \times \mathscr{X}$ defines a metric on a class of functions defined on $\mathscr{X}$ possessing in inner product, the PD function provides a way to find solutions to optimization, clustering, and classification problems.

The key features defining an RKHS are summarized in Aronszajn (1950), stated here in the following theorem.

Theorem (Aronszajn, 1950): For every PD function K on $\mathscr{X} \times \mathscr{X}$, there is a unique RKHS $\mathscr{H}_K$ of real-valued functions on $\mathscr{X}$ having $K(\cdot,\cdot)$ as its reproducing kernel. Let $\langle\cdot,\cdot\rangle$ be the inner product associated with $\mathscr{H}_K$, and define $K_{\boldsymbol{x}}(\cdot) = K(\boldsymbol{x},\cdot)$. Then, for every $f \in \mathscr{H}_K$ and every $\boldsymbol{x} \in \mathscr{X}$,

$$\langle K_{\boldsymbol{x}}, f\rangle = f(\boldsymbol{x}).$$

Conversely, for every reproducing kernel Hilbert space $\mathscr{S}$ of functions on $\mathscr{X}$, there corresponds a unique reproducing kernel $K(\cdot,\cdot)$, which is positive definite. □

A detailed proof of this theorem can be found in Gu (2002). Note that this theorem means that to construct an RKHS all one requires is the reproducing kernel.

3.5.2 Constructing an RKHS

Choose a positive definite function $K(\cdot,\cdot)$ as a kernel on $\mathscr{X} \times \mathscr{X}$. Fix $\boldsymbol{x} \in \mathscr{X}$, and define the function $K_{\boldsymbol{x}}$ on $\mathscr{X}$ by

$$K_{\boldsymbol{x}}(\cdot) = K(\boldsymbol{x},\cdot).$$

From K, one can construct a unique function space $\mathscr{H}_K$:

- First, for $\forall \boldsymbol{x} \in \mathscr{X}$, put $K_{\boldsymbol{x}}(\cdot) \in \mathscr{H}_K$.
- Second, for all finite m and $\{a_i\}_{i=1}^m$, and fixed $\{\boldsymbol{x}_i\}_{i=1}^m \in \mathscr{X}$, include the function

$$f(\cdot) = \sum_{i=1}^{m} a_i K_{\boldsymbol{x}_i}(\cdot) \in \mathscr{H}_K. \tag{3.5.5}$$

- Third, define the inner product in $\mathscr{H}_K$ by

$$\langle K_{\boldsymbol{x}}, K_{\boldsymbol{x}}\rangle = K(\boldsymbol{x},\boldsymbol{y})$$

so that

$$\left\langle f,g\right\rangle = \left\langle \sum_{i=1}^{m} a_i K_{\mathbf{x}_i}, \sum_{j=1}^{m} b_j K_{\mathbf{y}_j} \right\rangle = \sum_{i=1}^{m}\sum_{j=1}^{m} a_i b_j K(\mathbf{x}_i, \mathbf{y}_j).$$

Clearly, this procedure gives a linear space. The effectiveness of this procedure for generating a Hilbert space with a reproducing kernel is assured by the following two results. First, the reproducing property holds.

Proposition: Let $\mathscr{H}_K$ be the function space constructed from the positive definite kernel K. Then, for $f \in \mathscr{H}_K$,

$$\langle K_{\mathbf{x}}, f\rangle = f(\mathbf{x}),$$

and K is a reproducing kernel.

Proof: With $f \in \mathscr{H}_K$ defined as in the second step,

$$f(\mathbf{x}) = \sum_{i=1}^{m} a_i K_{\mathbf{x}_i}(\mathbf{x})$$

for each $\mathbf{x} \in \mathscr{X}$. Using the inner product in $\mathscr{H}_K$ from Step 3 gives

$$\begin{aligned}\langle K_{\mathbf{x}}, f\rangle &= \left\langle K_{\mathbf{x}}, \sum_{i=1}^{m} a_i K_{\mathbf{x}_i} \right\rangle = \sum_{i=1}^{m} a_i \langle K_{\mathbf{x}_i}, K_{\mathbf{x}}\rangle = \sum_{i=1}^{m} a_i K(\mathbf{x}_i, \mathbf{x}) \\ &= \sum_{i=1}^{m} a_i K_{\mathbf{x}_i}(\mathbf{x}) = f(\mathbf{x}).\square\end{aligned}$$

Second, given that $\mathscr{H}_K$ is a linear space with a reproducing kernel, it remains to verify completeness. That is, it remains to show that every Cauchy sequence of functions in $\mathscr{H}_K$ converges to a limit in $\mathscr{H}_K$. So, recall the Cauchy-Schwartz inequality linking the inner product and norm: $\langle \mathbf{x}, \mathbf{y}\rangle \leq \|\mathbf{x}\|\|\mathbf{y}\|$.

Proposition: Let $\{f_n\}$ be a sequence of functions in $\mathscr{H}_K$. Then strong convergence implies pointwise convergence. That is, if $\{f_n\}$ is a Cauchy sequence, so that

$$\|f_n - f_m\| \longrightarrow 0 \quad \text{as} \quad n,m \to \infty,$$

then, for every $\mathbf{x} \in \mathscr{X}$,

$$|f_n(\mathbf{x}) - f_m(\mathbf{x})| \longrightarrow 0.$$

Proof: Let $f_n, f_m \in \mathscr{H}_K$. Then, for every $\mathbf{x} \in \mathscr{X}$,

$$|f_n(\mathbf{x}) - f_m(\mathbf{x})| = |\langle K_{\mathbf{x}}, f_n - f_m\rangle| \leq \|K_{\mathbf{x}}\|\|f_n - f_m\|.\square$$

Now it is reasonable to include all the limits of Cauchy sequences of functions in $\mathscr{H}_K$ and call the overall space $\mathscr{H}$. Part of what gives RKHSs their structure is that when K is square integrable it leads to a spectral representation for the kernel in terms of its eigenfunctions, which are orthonormal and form a basis, and its eigenvalues. This important structure undergirds almost all kernel methods.

Theorem (Mercer-Hilbert-Schmidt): Let $\mathscr{X}$ be an index set, and let $K(\cdot,\cdot)$ be a positive definite function on $\mathscr{X} \times \mathscr{X}$ satisfying

$$\int_{\mathscr{X}} \int_{\mathscr{X}} K^2(\boldsymbol{x},\boldsymbol{x})d\boldsymbol{x}d\boldsymbol{x} = C < \infty.$$

Then there exists an orthonormal set of eigenfunctions of K as an integral operator, $\{\phi_i\}_{i=0}^{\infty}$ on $\mathscr{X}$; i.e.,

$$\int_{\mathscr{X}} \int_{\mathscr{X}} K(\boldsymbol{x},\boldsymbol{y})\phi_i(x)d\boldsymbol{x}d\boldsymbol{x} = \lambda_i \phi_i(y)$$

and

$$\int_{\mathscr{X}} \phi_i(\boldsymbol{x})\phi_j(\boldsymbol{x})d\boldsymbol{x} = \begin{cases} 1, \ i = j; \\ 0, \ otherwise, \end{cases}$$

with all $\lambda_i \geq 0$, such that $\sum_{i=0}^{\infty} \lambda_i^2 = C$ and

$$K(\boldsymbol{x},\boldsymbol{x}) = \sum_{i=0}^{\infty} \lambda_i \phi_i(\boldsymbol{x})\phi_i(\boldsymbol{x}). \tag{3.5.6}$$

With the kernel K defined by (3.5.6), the inner product $\langle \cdot,\cdot \rangle$ has the representation

$$\langle f,g \rangle = \sum_{i=0}^{\infty} \frac{(f,\phi_i)(g,\phi_i)}{\lambda_i},$$

where $(h_1,h_2) = \int_{\mathscr{X}} h_1(\mathbf{x})h_2(\mathbf{x})d\mathbf{x}$ for every h_1,h_2 defined on $\mathscr{X}$. □

In practice, there is little need to find the eigenvalues λ_i and the eigenfunctions ϕ_i of K. For most RKHS-based solutions, it will suffice to know the positive definite kernel K. This will be the case with support vector machines in Chapter 5, for instance. Wahba (2005) observes that the eigenfunction decomposition above is a generalization of the finite dimensional case in which the the inner product is defined by a PD matrix.

Still, the key question remains: Given an objective functional like (3.5.1), how does the RKHS help express a solution? The answer is that a recurring strategy for solving problems in the RKHS framework is to decompose the RKHS of interest into tensor sums, taking advantage of the relative ease of construction of subspaces of the RKHS. Gu (2002) provides the following theorem. It is the key result that guides tensor sum constructions. Note that here the subscript K is retained on some Hilbert spaces since more than one kernel function will be used.

Theorem: If the reproducing kernel K of a function space $\mathscr{H}_K$ on domain $\mathscr{X}$ can be decomposed into $K = K_0 + K_1$, where K_0 and K_1 are both positive definite, $K_0(\boldsymbol{x},\cdot)$ and $K_1(\boldsymbol{x},\cdot)$ in $\mathscr{H}_K$, for all $\boldsymbol{x}$ in $\mathscr{X}$, and also $\langle K_0(\boldsymbol{x},\cdot), K_1(\boldsymbol{y},\cdot)\rangle = 0, \ \forall \boldsymbol{x},\boldsymbol{y} \in \mathscr{X}$, then the spaces $\mathscr{H}_0$ and $\mathscr{H}_1$ corresponding to K_0 and K_1 form a tensor sum decomposition of $\mathscr{H}$.

Conversely, if K_0 and K_1 are both positive definite and their corresponding RKHSs $\mathscr{H}_0$ and $\mathscr{H}_1$ are such that $\mathscr{H}_0 \cap \mathscr{H}_1 = \{0\}$, then the space $\mathscr{H} = \mathscr{H}_0 \oplus \mathscr{H}_1$ has $K = K_0 + K_1$ as its reproducing kernel. □

The general procedure for using this theorem in smoothing splines has three steps; see Pearce and Wand (2005). Given a positive definite kernel K, the RKHS $\mathscr{H}_K$ may be constructed by:

1. Determining the eigenfunction decomposition of K: This means finding a sequence of eigenvalues $\{\lambda_i\}_{i=0}^{\infty}$ and a sequence of eigenfunctions $\{\phi_i\}_{i=0}^{\infty}$ so that, for every $\boldsymbol{x}, \boldsymbol{y} \in \mathscr{X}$,

$$K(\boldsymbol{x}, \boldsymbol{y}) = \sum_{i=0}^{\infty} \lambda_i \phi_i(\boldsymbol{x}) \phi_i(\boldsymbol{y}). \tag{3.5.7}$$

   For bounded K, the Mercer theorem ensures this.

2. With the eigenfunctions $\{\phi_i\}_{i=0}^{\infty}$, define the function space $\mathscr{H}_K$ to be

$$\mathscr{H}_K = \left\{ f : f = \sum_{i=0}^{\infty} a_i \phi_i \right\}. \tag{3.5.8}$$

3. Equip $\mathscr{H}_K$ with an inner product $\langle \cdot, \cdot \rangle_{\mathscr{H}_K}$,

$$\langle f, g \rangle_{\mathscr{H}_K} = \left\langle \sum_{i=0}^{\infty} a_i \phi_i, \sum_{i=0}^{\infty} b_i \phi_i \right\rangle_{\mathscr{H}_K} = \sum_{i=0}^{\infty} a_i b_i / \lambda_i. \tag{3.5.9}$$

   With the decomposition of K (3.5.7), the form of the functions in $\mathscr{H}_K$ as in (3.5.8), and the inner product (3.5.9), the reproducing property of K,

$$\langle K(\boldsymbol{x}, \cdot), K(\mathbf{y}, \cdot) \rangle_{\mathscr{H}_K} = K(\boldsymbol{x}, \boldsymbol{x}),$$

   follows and the norm of f in $\mathscr{H}_K$ is $\|f\|^2_{\mathscr{H}_K} = \sum_{i=0}^{\infty} a_i^2 / \lambda_i$.

The key question in a given application will be: What positive definite kernel should be used to construct the RKHS? In general, this is a hard question because it is de facto equivalent to choosing the basis for a function space and hence much like model selection. However, to hint at a procedure, consider a quick example. Suppose the (unknown) true function underlying a data set is the "tooth" function

$$f(x) = x + \frac{9}{4\sqrt{2\pi}} \exp\left[-4^2(2x-1)^2\right] \qquad x \in \mathscr{X} = [0, 1].$$

Given a data set $\{(x_i, y_i), i = 1, \cdots, n\}$, the task is to find a positive definite kernel K and construct the corresponding RKHS, say $\mathscr{H}_K$, so that the estimate

$$\hat{f}_\lambda = \arg\min_{f \in \mathscr{H}_K} \left\{ L(f) + \lambda \|f\|^2_{\mathscr{H}_K} \right\} \tag{3.5.10}$$

can be found. Following the Pearce and Wand (2005) steps, choose prespecified knots $t_1 < \cdots < t_m$ so the intervals for the spline are of the form $[t_j, t_{j+1})$. First, try choosing a kernel that can generate an RKHS solution to (3.5.10). There are many possible choices; however, the simpler, the better. Mercer's theorem suggests looking at a sum of products of simple functions, and the simplest are linear. Thus, try

$$K(x,y) = 1 + xy + \sum_{k=1}^{m} (x - t_k)_+ (y - t_k)_+. \tag{3.5.11}$$

One can verify that the eigenfunctions associated with K are

$$\phi_0(x) = 1,\ \phi_1(x) = x,\ \phi_{k+1}(x) = (x - t_k)_+ \qquad k = 1, \ldots m-1, \tag{3.5.12}$$

and the associated RKHS is

$$\mathcal{H}_K = \left\{ f : f(x) = \beta_0 + \beta_1 x + \sum_{k=1}^{m} \alpha_k (x - t_k)_+ \right\} \tag{3.5.13}$$

with corresponding inner product

$$\langle f, g \rangle_{\mathcal{H}_K} = \left\langle \beta_0 + \beta_1 x + \sum_{k=1}^{m} \alpha_k'(x - t_k)_+, \beta_0' + \beta_1' x + \sum_{k=1}^{m} \alpha_k'(x - t_k)_+ \right\rangle_{\mathcal{H}_K}$$

$$= \beta_0 \beta_0' + \beta_1 \beta_1' + \sum_{k=1}^{m} \alpha_k \alpha_k'.$$

From this definition of the inner product, it can be seen that

$$\|f\|^2_{\mathcal{H}_K} = \|\beta\|^2 + \|\alpha\|^2.$$

Since the norm for this RKHS is a sum of two finite-dimensional Euclidean norms, the tensor sum theorem implies that the penalized spline RKHS is isomorphic to $\mathbb{R}^{m+2}$. This is a particularly simple Hilbert space, as one would normally expect an RKHS to be infinite-dimensional.

The decomposition of an RKHS into a direct sum of two subspaces reflects the structure of the optimality criterion: There is one subspace for each of the two terms. The smoothing parameter λ controls the trade-off between the two terms; i.e., between goodness of fit and amount of roughness. Large values of λ put a lot of weight on the roughness penalty, forcing it to be small; i.e., forcing the minimum to be very smooth, in particular close to the nullspace of the differential operator in the sense of the discussion after (3.5.1). For the tooth function, Fig. 3.8(a) shows the fit when $J(f) = 0$; this corresponds to the estimated function of x being a straight line. Small values of λ put little weight on the smoothness and (relatively) a lot of weight on goodness of fit, forcing the error term to be small. This forces good fit at the cost of more roughness, see Fig. 3.8(b); the estimated function is mostly in the orthogonal complement of the null-space of J.

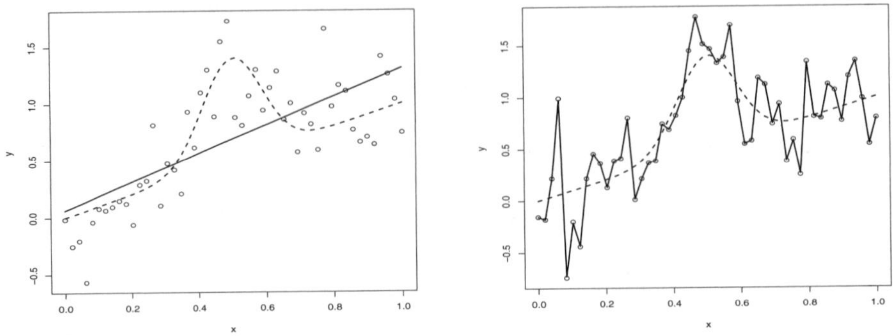

(a) Null Space of Roughness penalty

(b) Orthogonal complement of Nullspace

Fig. 3.8 Like Fig. 3.5, the two graphs show how large and small values of λ affect spline estimates of the tooth function. If $\mathscr{H} = \mathscr{H}_0 \oplus \mathscr{H}_1$, where $\mathscr{H}_0$ is the null-space of J with orthogonal complement $\mathscr{H}_1$ and f is the true unknown function, then $f = f_{0,\lambda} + f_{1,\lambda}$ for $f_{i,\lambda} \in \mathscr{H}_i$. If $\hat{f} = \hat{f}_{0,\hat{\lambda}} + \hat{f}_{1,\hat{\lambda}}$, then (a) is the case where $\hat{\lambda}$ makes $\|\hat{f}_{1,\hat{\lambda}}\|$ very small and (b) is the case where $\hat{\lambda}$ makes $\|\hat{f}_{2,\hat{\lambda}}\|$ very small, so the part of $\hat{f}$ in the other subspace dominates, respectively.

3.5.3 Direct Sum Construction for Splines

To develop an intuition for the use of RKHS methods, it is worthwhile to examine the properties of the construction of the penalized sum of squares optimization and the structure of the Hilbert space. Recall that the subspace of unpenalized functions in $\mathscr{H}_K$ consists of linear functions; they form the null-space of the total curvature $J(f)$:

$$\mathscr{H}_0 = \{f : f(x) = \beta_0 + \beta_1 x, \quad \forall x \in \mathscr{X}\} = \left\{f : J(f) = \int_0^1 [f''(x)]^2 dx = 0\right\}.$$

Its orthogonal complement,

$$\mathscr{H}_1 = \mathscr{H}_0^{\perp} = \left\{f : f(x) = \sum_{k=1}^{m} \alpha_k (x - t_k)_+ \right\},$$

is generated by the spline basis. It is easy to see that the kernel K in (3.5.11) is $K = K_0 + K_1$, where

$$K_0(x,y) = 1 + xy \quad \text{and} \quad K_1(x,y) = \sum_{k=1}^{m} (x - t_k)_+ (y - t_k)_+.$$

Consequently, any $f \in \mathscr{H}$ can be written $f = f_0 + f_1$ where $f_0 \in \mathscr{H}_0$ and $f_1 \in \mathscr{H}_1$, so that $\mathscr{H} = \mathscr{H}_0 \oplus \mathscr{H}_1$.

All the functions given by (3.5.13) as solutions to (3.5.10) are penalized by an amount governed by λ. For a variety of technical and computational reasons, it is often desired in practice that some of the functions in $\mathscr{H}_K$ be unpenalized. Such functions are the elements of the null-space $\mathscr{H}_0$ and would ideally be minimizers of the unconstrained $L(f)$. Therefore, only the projections of $f \in \mathscr{H}$ onto the orthogonal complement of $\mathscr{H}_0$, which in this case is $\mathscr{H}_1 = \mathscr{H}_0^{\perp}$, should be penalized. If P_1 denotes the linear operator corresponding to the projection onto $\mathscr{H}_1$, $\mathscr{H}_0$ is the null space of P_1. With respect to the null space $\mathscr{H}_0$, the solution can be restated as

$$\hat{f}_\lambda = \arg\min_{f \in \mathscr{H}_K} \left\{ L(f) + \lambda \|P_1 f\|^2_{\mathscr{H}_K} \right\}.$$

For the simple example of the last subsection,

$$P_1 f = P_1 \left(\beta_0 + \beta_1 x + \sum_{k=1}^{m} \alpha_k (x - t_k)_+ \right) = \sum_{k=1}^{m} \alpha_k (x - t_k)_+.$$

As a result, $\|P_1 f\|^2_{\mathscr{H}_K} = \|\alpha\|^2$. The solution $\hat{f}_\lambda$ now reduces to

$$\hat{f}_\lambda(x) = \mathbf{X}_x \hat{\beta} + \mathbf{Z}_x \hat{\alpha},$$

where $\mathbf{X}_x$ and $\mathbf{Z}_x$ are evaluations of the basis elements of $\mathscr{H}_0$ and $\mathscr{H}_1$ at x with coefficients obtained from the data string $\mathbf{y}$ by

$$(\hat{\beta}, \hat{\alpha}) = \arg\min_{\beta,\alpha} \left\{ \|\mathbf{y} - [\mathbf{X}\beta + \mathbf{Z}\alpha]\|^2 + \lambda \|\alpha\|^2 \right\},$$

in which $\mathbf{X}$ and $\mathbf{Z}$ are the design matrices with elements given by evaluations of the basis functions for $\mathscr{H}_0$ and $\mathscr{H}_1$, namely $\{1, x\}$ and the functions $(x - t_k)_+$ for $k = 1, \ldots, m$, at the design points x_i for $i = 1, \ldots, n$.

3.5.3.1 A Generalization

In the particular case of the squared error loss, $L(f)$ can be written as

$$L(f) = \frac{1}{n} \sum_{i=1}^{n} (y_i - f(\boldsymbol{x}_i))^2 = \frac{1}{n} \sum_{i=1}^{n} (y_i - \langle \eta_i, f \rangle)^2,$$

where η_i is the representer of the evaluation functional for the point $\mathbf{x}_i \in \mathscr{X}$,

$$\langle \eta_i, f \rangle = f(\boldsymbol{x}_i).$$

Using the evaluation functional and an arbitrary loss function $\mathscr{L}$, one can find

$$\hat{f}_\lambda(x) = \arg\min_{f \in \mathscr{H}_K} \left\{ \frac{1}{n} \sum_{i=1}^{n} \mathscr{L}(y_i, \langle \eta_i, f \rangle) + \lambda \|P_1 f\|^2_{\mathscr{H}_K} \right\}. \tag{3.5.14}$$

Observe that this general formulation allows the use of various projection operators and various evaluation functionals. Also, because of its direct connection to the null-space $\mathscr{H}_0$, this formulation allows a step-by-step construction of $\mathscr{H}$, starting from the solution of the differential equation defining the null-space $\mathscr{H}_0$ and adding its orthogonal complement to form $\mathscr{H}_K$ as a direct sum of $\mathscr{H}_0$ and $\mathscr{H}_0^{\perp}$. Finally, (3.5.14) indicates how general RKHS methods are since many regularizations can be formulated expressed in the same form. (This may be very hard to implement pragmatically, but the conceptual unity is satisfying.)

3.5.3.2 A More Detailed Example of RKHS Construction

For the sake of intuition, it is worthwhile seeing how the foregoing case for the curvature penalty extends to more general penalties that integrate squares of higher-order derivatives. To motivate this by local function approximation, let $f \in \mathscr{C}^m[0,1]$ and write the Taylor series expansion in the neighborhood of zero using the integral form of the remainder. This gives

$$f(x) = \sum_{j=0}^{m-1} \frac{x^j}{j!} f^{(j)}(0) + \int_0^x \frac{(x-u)^{m-1}}{(m-1)!} f^{(m)}(u)du.$$

Observe that the remainder term finds an integral for $0 \leq t \leq x$ and for $x \in [0,1]$, $x - t \geq 0$. So, the remainder term is unchanged if the domain becomes $[0,1]$, provided the positive part is used in the integral. Thus,

$$f(x) = \sum_{j=0}^{m-1} \frac{x^j}{j!} f^{(j)}(0) + \int_0^1 \frac{(x-u)_+^{m-1}}{(m-1)!} f^{(m)}(u)du.$$

Now, with some hindsight, one can seek a reasonable two-term inner product. It is seen that this generalizes (3.5.10) – (3.5.13). The form of the Taylor series expansion suggests

$$\langle f, g \rangle = \sum_{j=0}^{m-1} f^{(j)}(0) g^{(j)}(0) + \int_0^1 f^{(m)}(x) g^{(m)}(x)dx.$$

Now try defining

$$K_x(y) = \sum_{j=0}^{m-1} \frac{x^j}{j!}\frac{y^j}{j!} + \int_0^1 \frac{(x-u)_+^{m-1}}{(m-1)!}\frac{(y-u)_+^{m-1}}{(m-1)!}du.$$

□ It is seen that the first reproducing kernel,

$$K_0(x,y) = \sum_{j=0}^{m-1} \frac{x^j}{j!}\frac{y^j}{j!},$$

provides the key ingredient for constructing the subspace

$$\mathscr{H}_0 = \left\{ f : f^{(m)} = 0 \right\}$$

with inner product

$$\langle f, g \rangle_0 = \sum_{j=0}^{m-1} f^{(j)}(0) g^{(j)}(0).$$

☐ A second reproducing kernel,

$$K_1(x,y) = \int_0^1 \frac{(x-u)_+^{m-1}}{(j-1)!} \frac{(y-u)_+^{m-1}}{(j-1)!} du,$$

is seen to generate the orthogonal complement of $\mathscr{H}_0$, namely

$$\mathscr{H}_1 = \left\{ f : f^{(j)}(0) = 0,\ j = 0, 1, \cdots, m-1, \text{ and} \int_0^1 [f^{(m)}(x)]^2 dx < \infty \right\},$$

with the inner product derived from the boundary conditions:

$$\langle f, g \rangle_1 = \int_0^1 f^{(m)}(x) g^{(m)}(x) dx.$$

Notice that this example generalizes cubic splines – the case $m = 2$. In general, this construction gives polynomial splines, which means that the null-space of the operator is a collection of polynomials corresponding to $D^m f = 0$. As seen in Gu (2002) Chapter 4.3, many other differential operators L have been studied, leading to trigonometric splines, hyperbolic splines, exponential splines, splines on the circle, and so forth. The case of higher dimensions is of particular importance; see Wahba (1990), Chapter 2. Indeed, it seems possible to generalize further to the use of kernels that are not in general PD; see Canu et al. (2005) in this regard.

3.5.4 Explicit Forms

So far, the results presented have been specific examples or general properties. In this subsection, it is important to state two results that provide explicit descriptions, at least in principle, for general optimizations like (3.5.14). The first result rests on differential equations, the second on Hilbert space optimization.

3.5.4.1 Using Differential Equation Tools to Construct RKHSs

To identify reproducing kernels for various contexts, two definitions are needed.

First, we define the Wronskian matrix. Let $f_1, f_2, \cdots, f_d$ be d functions that are $d-1$ times continuously differentiable. The Wronskian matrix $\mathbf{W}$ is constructed by putting the functions in the first row, the first derivative of each function directly under it in the second row, and so on through to the $(d-1)$th derivative. So, the Wronskian matrix $\mathbf{W}$ associated with $f_1, f_2, \cdots, f_d$ is

$$\mathbf{W}(f_1, f_2, \cdots, f_d) = \begin{bmatrix} f_1 & f_2 & \cdots & f_d \\ f_1' & f_2' & \cdots & f_d' \\ \vdots & \vdots & \ddots & \vdots \\ f_1^{(d-1)} & f_2^{(d-1)} & \cdots & f_d^{(d-1)} \end{bmatrix}.$$

For $x \in \mathscr{X}$, the Wronskian matrix is $\mathbf{W}(x) = [\mathsf{w}_{ij}(x)]$, where

$$\mathsf{w}_{ij}(x) = f_i^{(j-1)}(x) \qquad i, j = 1, 2, \cdots, d.$$

The determinant of a Wronskian matrix is simply called the Wronskian and plays an important role in differential equations.

Second are Green's functions. These arise in solving inhomogeneous ordinary differential equations (ODEs)with boundary conditions. Consider a differential operator L such as appears in the integrand of penalty terms. If a function h is given and one tries to solve $L(x)f(x) = h(x)$, subject to boundary conditions on, say, an interval $[0, \ell]$, then there is a unique solution of the form

$$u(x) = L^{-1}h(x) = \int_0^\ell h(s)G(x,s)ds,$$

where G is the Green's function associated with L. Green's functions satisfy

$$L(x)G(x,s) = \delta(x-s),$$

where δ is the Dirac's δ operator at 0. It is easy to see that if G exists, then $L(x)u(x) = h(x)$.

Green's functions have several useful properties. First, they exist in great generality. Second, they express solutions for inhomogeneous equations where the coefficients of the differential operators may be functions of x. Third, they are particularly well suited to settings where the difficulty in solving the equation arises from satisfying the boundary conditions. As suggested by the Dirac operator, certain derivatives of Green's functions have discontinuities; it turns out that the jumps can be expressed in terms of the coefficient functions of the highest-order derivative in L. Thus, in spline problems, where part of the difficulty is to ensure smoothness constraints, Green's functions are a natural way to express solutions. This is not the place to digress on such problems, so the reader is referred to any of numerous books that treat this class of ODEs.

To return to spline optimization, let $\mathscr{X} = [a,b]$, and define the inner product

$$\langle f, g \rangle = \sum_{j=0}^{m-1} f^{(j)}(a)g^{(j)}(a) + \int_a^b (Lf)(t)(Lg)(t)dt. \tag{3.5.15}$$

Now, the decomposition theorem for identifying RKHSs can be stated; see Gu (2002).

Theorem: Let $\{u_1, u_2, \cdots, u_m\}$ be a basis for the null-space of L (i.e., all f with $Lf \equiv 0$), and let $\mathbf{W}(t)$ be the associated Wronskian matrix. Then, under the inner product (3.5.15), $\mathscr{H}$ is an RKHS with reproducing kernel

$$K(s,t) = K_0(s,t) + K_1(s,t),$$

where

$$K_0(s,t) = \sum_{i=1}^{d} \sum_{j=1}^{m} C_{ij} u_i(s) u_j(t) \quad \text{with } C_{ij} = \left[\{\mathbf{W}(a)\mathbf{W}^{\mathsf{T}}(a)\}^{-1} \right]_{ij}$$

and

$$K_1(s,t) = \int_{u=a}^{u=b} G(s,u)G(t,u)du,$$

with $G(\cdot,\cdot)$ being the Green's function associated with the differential operator L. Furthermore, $\mathscr{H}$ can be partitioned into the direct sum of two subspaces,

$$\mathscr{H} = \mathscr{H}_0 \oplus \mathscr{H}_1,$$

where

$$\mathscr{H}_0 = \{f \in \mathscr{H} : (Lf)(t) = 0 \text{ almost everywhere on } \mathscr{X}\}$$

and

$$\mathscr{H}_1 = \left\{ f \in \mathscr{H} : f^{(j)}(t) = 0, \qquad j = 0, 1, \cdots, m-1 \right\},$$

so that $\mathscr{H}_0$ has reproducing kernel K_0 and $\mathscr{H}_1$ has reproducing kernel K_1. □

3.5.4.2 Form of Solutions

Surprisingly, there is a closed-form expression for solving spline optimization problems in some cases. In particular, consider the special case of (3.5.14) for squared error loss,

$$\hat{f}_\lambda(x) = \arg\min_{f \in \mathscr{H}_K} \left\{ \frac{1}{n} \sum_{i=1}^{n} (y_i - \langle \eta_i, f \rangle)^2 + \lambda \|P_1 f\|^2_{\mathscr{H}_K} \right\}, \tag{3.5.16}$$

where η_i is the representer of the evaluation functional at $\mathbf{x}_i$ and P_1 is the orthogonal projection of a given Hilbert space of functions $\mathscr{H}$ onto $\mathscr{H}_1$ with the orthogonal complement $\mathscr{H}_0$ given by the null-space of the operator L defining the penalty term. Wahba (1990) establishes the following theorem.

Theorem: Let $u_1, \ldots, u_m$ be a basis for $\mathscr{H}_0$, and suppose the $n \times m$ matrix

$$T = T_{n \times m} = (\eta_i(u_\nu))_{i=1,\ldots,n;\nu=1,\ldots m}$$

has full column rank. Then, for fixed λ, the minimum of (3.5.16) is given by

$$f_\lambda = \sum_{\nu=1}^{m} \delta_\nu u_\nu + \sum_{i=1}^{n} c_i \xi_i,$$

where $\xi_i = P_1 \eta_i$, and the coefficients are defined as follows. Let $\Sigma = \Sigma_{n\times n} = (\langle \xi_i, \xi_j \rangle)$ and $M = \Sigma + n\lambda I_{n\times n}$. Then

$$\delta' = (\delta_1, ..., \delta_m)' = (T'M^{-1}T)^{-1}T'M^{-1}\mathbf{y},$$

$$\mathbf{c}' = (c_1, ..., c_n)' = M^{-1}(I_{n\times n} - T(T'M^{-1}T)^{-1}T'M^{-1})\mathbf{y}. \quad \square$$

As a final point, although statisticians typically choose the x_is themselves as the knots, the knots can be, in principle, estimated, chosen, or inferred by other, possibly Bayesian, techniques.

3.5.5 Nonparametrics in Data Mining and Machine Learning

So far, focus has been on the one-dimensional case for kernel smoothing and spline smoothing. It was seen that the two smoothers essentially coincided in the unidimensional case. In practice, for a variety of choices of tuning parameter and signal-to-noise ratio in the data, numerical results confirm that kernel smoothing and spline smoothing have equivalent performance. The estimated curve from one method looks like it was printed on top of the curve from the other method, on graphs at any reasonable scale. More generally, this is not surprising because the two techniques are fundamentally instances of the same smoothing paradigm, local polynomial fitting, see Silverman (1984), Huang (2001).

However, the unidimensional case is comparatively unimportant in DMML since an important feature of many DMML problems is their (highly) multivariate settings.

Spline smoothing and kernel smoothing differ in the way they scale up to higher dimensions. Essentially, kernel smoothing does not scale up in any meaningful way; it suffers the Curse. The reasoning for this is intuitive: In multivariate settings, kernel smoothers use product kernels, one for each coordinate, to build up a smoother that fills out the input space. As mentioned in Section 2.3.5, this causes the technique to require enormous amounts of data for good inference, hence the Curse. By contrast, the spline smoothing formulation leads to RKHS techniques that can be scaled up, as seen below. For this reason, RKHS techniques provide a flexible framework for solving a large class of approximation, estimation, and optimization problems.

To see how RKHS techniques evade the Curse, consider a brief description of thin plate splines. In this case, the kernel function evaluations are based on norms of the entire input vectors, with no need to address each coordinate separately. It is as if the norm of a vector is being treated as a summary statistic for the whole vector of measurements.

Let the predictor variable be a p-dimensional vector $\boldsymbol{x} = (x_1, x_2, \cdots, x_p)^{\mathsf{T}}$. Suppose that the objective functional remains

$$E_\lambda(f) = \frac{1}{n}\sum_{i=1}^{n}(y_i - f(\boldsymbol{x}_i))^2 + \lambda J_m(f),$$

where the penalty function is

$$J_m(f) = \int_{\mathscr{X}} \sum_{|\alpha|=m} \frac{m!}{\alpha_1! \cdots \alpha_p!} \left(\frac{\partial^m f}{\partial x_1^{\alpha_1} \cdots \partial x_p^{\alpha_p}} \right)^2 d\boldsymbol{x},$$

where $|\alpha| = \sum_i \alpha_i$, $2m > p$, and the smoothing parameter $\lambda > 0$ controls the balance between fit and smoothness.

In this context, one of the most commonly used members of the thin-plate spline smoother family corresponds to the case $p = 2$ and $m = 2$, in which the penalty term is

$$J(f) = \int_{\mathscr{X} \subset \mathbb{R}^2} \left(\frac{\partial^2 f}{\partial x_1^2} \right)^2 + 2\left(\frac{\partial^2 f}{\partial x_1 \partial x_2} \right)^2 + \left(\frac{\partial^2 f}{\partial x_2^2} \right)^2 d\boldsymbol{x}.$$

This roughness penalty, along with the constraint that the second derivatives at the end-points be zero, provides an immediate generalization of natural cubic splines. It turns out that the minimizing function under this penalty is of the form

$$f(\boldsymbol{x}) = \sum_{i=1}^{n} \alpha_i K^{\mathsf{tps}}(\boldsymbol{x}_i, \boldsymbol{x}) + \beta^{\mathsf{T}}\boldsymbol{x} + \gamma, \tag{3.5.17}$$

in which

$$K^{\mathsf{tps}}(\boldsymbol{x}_i, \boldsymbol{x}) = \|\boldsymbol{x} - \boldsymbol{x}_i\|^2 \log \|\boldsymbol{x} - \boldsymbol{x}_i\|^2$$

is the thin-plate spline kernel. A series of interesting kernels will be discussed in Sections 5.4.7 and 5.4.8 in the context of support vector machines. The point for now is to note that although many applications of thin-plate splines are for two-dimensional design points, (3.5.17) extends naturally to higher dimensions.

Indeed, (3.5.17) is just one instance of a more general solution to an RKHS-based approach to function approximation. The general solution is called the representer theorem. The earliest statement of this important theorem is in Kimeldorf and Wahba (1971), but see also Scholkopf and Smola (2002), Wahba (1998), Wahba (2005), and Smola and Scholkopf (1998), among others.

Representer Theorem: Let $\Omega : [0, \infty) \to \mathbb{R}$ be a strictly monotonic increasing function, $\mathscr{X}$ be a set, and $c : (\mathscr{X} \times \mathbb{R}^2)^n \to \mathbb{R} \cup \{\infty\}$ be an arbitrary loss function. Then each minimizer $f \in \mathscr{H}$ of the regularized risk functional

$$c((\boldsymbol{x}_1, y_1, f(\boldsymbol{x}_1)), \cdots, (\boldsymbol{x}_n, y_n, f(\boldsymbol{x}_n))) + \Omega(\|f\|_{\mathscr{H}})$$

admits a representation of the form

$$\hat{f}(\boldsymbol{x}) = \sum_{i=1}^{n} \alpha_i K(\boldsymbol{x}_i, \boldsymbol{x}). \square$$

The central point of (3.5.17) and the conclusion of the representer theorem is that the reproducing kernel K is defined on $\mathscr{X} \times \mathscr{X}$, where $\mathscr{X}$ is a p-dimensional domain for arbitrary $p \geq 1$. That is, the form of the solution in an RKHS is insensitive to the dimension of the domain of the underlying function f being approximated. There are n terms in the minimum where n is independent of p. The Curse has been avoided, arguably by using of the norm.

An even more general form of the theorem is given in the semiparametric settings by Scholkopf and Smola (2002) and Wahba (1998).

Semiparametric Representer Theorem: Suppose that in addition to the assumptions of the representer theorem, the set of real-valued functions $\{\phi_j\}_{j=1}^m : \mathscr{X} \to \mathbb{R}$ has the property that the $n \times m$ matrix $(\phi_j(\boldsymbol{x}_i))_{ij}$ has rank m. Let $\tilde{f} = f + h$, with $f \in \mathscr{H}$ and $h \in \text{span}\{\phi_j\}$. Then, minimizing the regularized risk

$$c((\boldsymbol{x}_1, y_1, \tilde{f}(\boldsymbol{x}_1)), \cdots, (\boldsymbol{x}_n, y_n, \tilde{f}(\boldsymbol{x}_n))) + \Omega(\|f\|_{\boldsymbol{H}})$$

over $\tilde{f}$ results in a representation of the form

$$\tilde{f}(\boldsymbol{x}) = \sum_{i=1}^{n} \alpha_i K(\boldsymbol{x}_i, \boldsymbol{x}) + \sum_{j=1}^{m} \beta_j \phi_j(\boldsymbol{x})$$

with $\beta_j \in \mathbb{R}$ for all $j = 1, ..., m$. $\square$

The form of the solution in these representer theorems is ubiquitous in DMML and highlights the importance of the kernel function. It is not just that the kernel defines the space of functions but that the kernel defines the span of the solutions within the space it defines. In essence, the kernel gives the terms in which a linear model is expressed. It is therefore reasonable to regard a kernel as a continuously parametrized collection of basis functions $K_{\boldsymbol{z}}(\cdot)$, where the parameter $\boldsymbol{z}$ has the same dimension as the design points. Taken together, this means that the variability due to choice of kernel is akin to model uncertainty in conventional statistical contexts because perturbations of the model lead to alternative estimates and predictions just as varying a kernel does. This is reinforced by the fact that the RKHS formulation of spline smoothers has a Bayesian interpretation (see Exercises 6.7 and 6.8). Also, the penalty term in (3.5.10) or (3.5.16) can be regarded as the logarithm of a prior, see Chapter 10.4.4

3.6 Simulated Comparisons

It is actually quite easy to compare the techniques discussed so far, namely LOESS, NW, and spline smoothing using R, because most of the routines are already built into it. The function $\text{sinc}(\boldsymbol{x})$, defined by

$$f(\boldsymbol{x}) = \frac{\sin(\boldsymbol{x})}{\boldsymbol{x}} \qquad \boldsymbol{x} \in [-10, +10],$$

is a good test case because it has been used by various authors in the machine learning literature and has several reasonable hills and valleys a good technique should match. The main way techniques are compared is through MSE or bias and variance.

To be specific, the true model is $Y_i = f(\boldsymbol{x}_i) + \varepsilon_i$, in which the noise terms ε_i are IID normal with constant variance $\sigma^2 = 0.2^2$. The function f is, of course, not known to the technique; only the data are. Here, the xs are assumed to be from a fixed design with equally spaced design points. The sinc function is graphed in Fig. 3.9 with a scatterplot of data points generated from it.

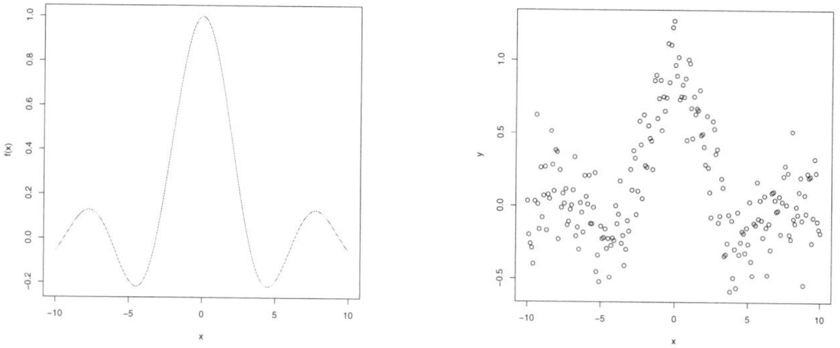

Fig. 3.9 The left-hand panel shows a graph of the sinc function. The right-hand panel shows the data generated from it using error variance $.02^2$.

To analyze the data to estimate f, consider a sequence of three estimators depending on the order of the polynomials that form them. First is the NW estimate which can be regarded as a local constant polynomial estimate; i.e., a LOESS estimate with the local polynomial being a constant. Second is a degree 2 LOESS, or local polynomial, fit. Third is a spline smoother based on natural cubic splines with all the smoothness conditions imposed to guarantee the existence of a solution. These three estimators have smoothing parameter inputs h, α, and λ. The first and third were chosen as before by GCV. The bandwidth was found to be $h_{opt} = 3.5$, a largish value but valid for the whole interval $[-10, 10]$. The fraction of data used in bins for the local polynomial plot was $\alpha = .25$. The actual GCV plot to find λ, or the degrees of freedom it represents, is given in Fig. 3.10 and $\lambda_{opt} = 11$.

The fitted curves generated by the three estimators are given in Fig. 3.11. By eye, it appears that the NW estimate is the worst and the spline estimator is the best. That is, as the degree increases, the fitted curve becomes smoother and matches the function more closely. In fact, a more detailed examination reveals that the choice among the three is not so obvious.

Instead of drawing confidence or MSE bands as in Section 3.4 or Section 2.5, observe that the same techniques as used to find the MSE give its constituent pieces, the bias and variance. In particular, the biases for the three estimators can be estimated by bootstrapping as before. The variances can be estimated by either (i) using the linearity

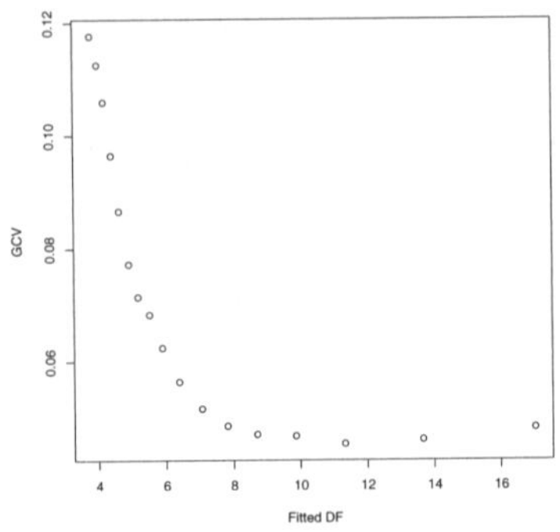

Fig. 3.10 Plot of the generalized CV error. It decreases to about 11 after which it starts to increase.

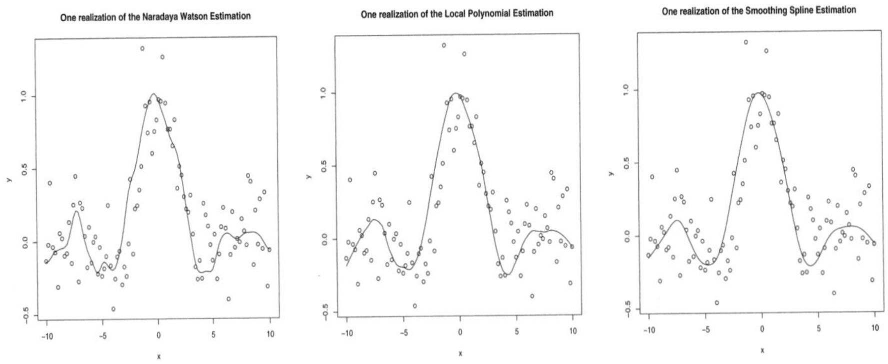

Fig. 3.11 Fit of the data for the three estimators, NW, local polynomial of degree 2, and cubic splines.

of the smoothers (see (2.3.20) and (3.3.8)) and normal approximation since n is large or (ii) using the lokern and gss packages, the first of which invokes a more sophisticated procedure to find variances. Since n is large, these will be the same; lokern and gss were used here for convenience. Figure 3.12 gives the results of estimating both the squared bias and the variance for the sinc(x) based on $m = 1000$ samples, each of size $n = 100$. As in earlier cases, the values are only valid at the sampled x_is but they are joined by straight lines.

The curves in Fig. 3.12 qualitatively confirm intuition and theory. First, the natural cubic spline smoother has the highest squared bias. This is plausible because NCSs explicitly impose many conditions to obtain a unique function estimate. All those conditions lead to strong restrictions on the function space, which translates into a high bias, bigger than for the other two estimators. The flip side is that strong restrictions tend to reduce the dimension of the space of estimators and consequently give the smallest variance. One can argue, a little flippantly, that this means splines may estimate the wrong value really well.

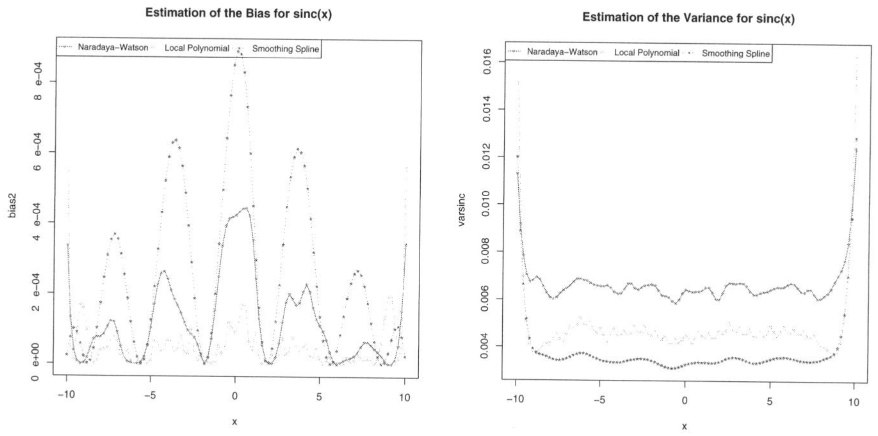

Fig. 3.12 On the left, the squared biases are plotted. The highest is for the spline smoother, next is NW, and the smallest is for local polynomials. On the right, variances are plotted. The highest is for NW, next is local polynomials, and the smallest is for smoothing splines.

Second, the NW estimator's only restriction on the function space is through the choice of the kernel function. Otherwise the span of the NW estimators is unrestricted. (Note that, in the theory of Chapter 2, the Lipschitz-like assumptions are applied only to the unknown function to get rates of convergence, not to the estimation process itself.) Unsurprisingly, therefore, the NW estimator has a smaller squared bias than the spline smoother. However, the cost of the large span of the class of estimators is that the NW estimator has the highest variance.

Third, the smallest squared bias is from the local polynomial estimator of degree 2. This is due to the fact that there are few restrictions on the piecewise degree 2 polynomials so they have much higher degrees of freedom than the NCS estimator and can therefore track the unknown function quite well if there are enough design points. The cost is that their variance is higher than that of NCS, though less than that for NW. They would seem to achieve a better variance–bias trade-off overall than NCS once degrees of freedom are taken into account. Note that there are two variance–bias trade-offs operating. The first is within a class of estimators – that is how the h or the λ was chosen; in principle, α could have been chosen by a similar optimization strategy. The second is across the spaces from which the estimators are drawn. The NCS space is arguably too small, and the space over which the NW estimator varies is arguably too large. This might leave the space of local polynomials of degree 2 as preferred.

3.6.1 What Happens with Dependent Noise Models?

In the last subsection, the noise was IID normal. In reality, noise terms are usually just approximately normal and often have a nontrivial dependence structure that is just treated as independent. Just to see what happens, consider the sinc function example but use an AR(1) noise term so the true model becomes

$$Y = f(X_i) + \varepsilon_i \text{ and } \varepsilon_i = \varphi\varepsilon_{i-1} + \eta_{i-1},$$

in which φ is a constant, ε_i is the time-dependent error in the function, and η_i is the independent increment to the error process, often taken, as here, to be $N(0,\sigma^2)$. For definiteness, set $\sigma = .2$ as before and choose some value between -1 and 1, say $\varphi = .72$ as a test case. In generating the data, the xs were taken from left to right, thereby giving the serial dependence of the AR(1) error.

Once the data were generated, the NW, local polynomial, and NCS estimators from the last subsection were found and graphed over the scatterplot of the data as in Fig. 3.13, using optimal GCV values $h_{opt} = .95$ and $\lambda_{opt} = .6$; α remained .25. The effect of dependence in the error term is seen in the plots because there tend to be runs of data points all deviating in the same direction from the function values. The small values of Y around $x = -5$ and the large ones that form a cluster at 0 are instances of this. One can argue that such departures from independence will be detected from scatterplots, hypothesis tests, or residual analysis and so modeled. While this is true in the present slightly extreme case $\varphi = .72$, it is unclear, in general, whether a case like $\varphi = .35$ would be similar. The extra variability from the dependence structure could be swamped by all the other sources of variability and so be undetectable graphically.

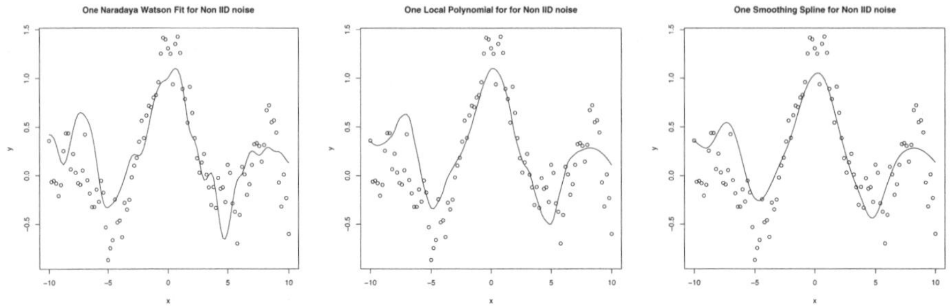

Fig. 3.13 Fits for an AR(1) noise model. Note that the fits weakly track runs in the data but shrink to the correct curve and are otherwise similar to the IID case.

As before, it is instructive to look at curves for the bias and variance of the three estimators. They are given in Fig. 3.14. They indicate the same variance–bias decomposition as before; however, the scales on the horizontal axes in both graphs have expanded. Both the bias and the variance are elevated. This is not a surprise since the extra variability in the error tends to make quantities harder to estimate (larger variance), and unmodeled terms like the η_is add to the bias. The curves also appear choppier than before, with sharper peaks and valleys. These qualitative features would be more pronounced if φ were larger or other terms were included in the noise model.

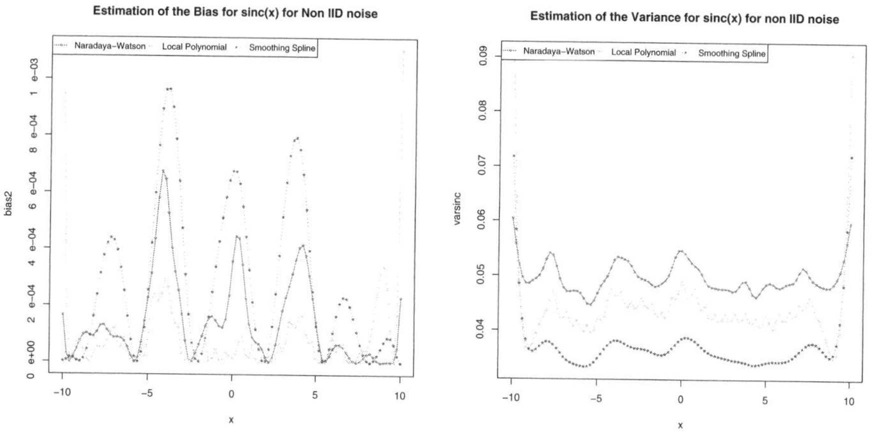

Fig. 3.14 Squared bias and variance for an AR(1) noise model for splines, NW, and local polynomials. The ordering is the same as before but the curves are choppier.

3.6.2 Higher Dimensions and the Curse of Dimensionality

How readily do kernel estimators, spline smoothers, and local polynomial fitting extend to multivariate settings? The short answer for kernel estimators is not well at all; they fare poorly in three or more dimensions and are mediocre in two dimensions. Spline smoothers and local polynomials scale up to higher dimensions somewhat better than kernel estimators (which does not say much) but at the cost of mathematical complexity and some loss of tractability. Overall, they are often adequate but not particularly compelling. This is why more recent techniques, like those in the next chapter, are essential. To illustrate the limitation of the techniques so far, consider doing kernel regression with two predictor variables rather than one. Let $\boldsymbol{x}$ be a two-dimensional vector with coordinates x_1 and x_2, and consider a 2D version of $\mathsf{sinc}(\boldsymbol{x})$:

$$f(\boldsymbol{x}) = \frac{\sin ||\boldsymbol{x}||}{||\boldsymbol{x}||}.$$

This is graphed in Fig. 3.15. Clearly f is rotationally symmetric and undulating; not an easy function to estimate but only moderately difficult.

Assuming a true model $Y = f(\boldsymbol{x}) + \varepsilon$ with a normal mean-zero error having variance the same as before (i.e. $\sigma^2 = 0.2^2$) consider using a two-dimensional version of the NW kernel regression estimator. First, the kernel has to be redefined for the two-dimensional case. In one dimension, the summands were of the form $K((x - x_i)/h)$, K being one of the kernels identified earlier. The natural extension to two dimensions is the product $K((x - x_i)/h)K((y - y_i)/h)$, and this is the most commonly used form. Already this is a restriction on which functions can be estimated well: They have to be parsimoniously expressible as sums of rectangular neighborhoods so that neighborhood distances would be compatible with the unknown function. Better would be to

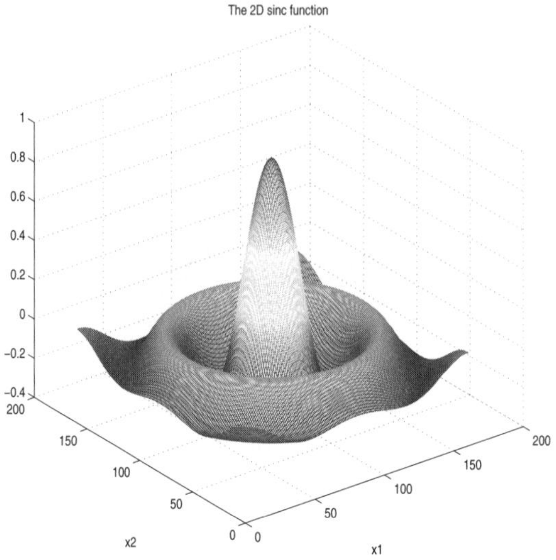

Fig. 3.15 3D plot of the true function, a two dimensional version of sinc.

use the vectors $\boldsymbol{x}$ and $\boldsymbol{x}_i$ and compute distances more generally using forms like

$$K(\boldsymbol{x},\boldsymbol{x}_i) = g(\|\boldsymbol{x}-\boldsymbol{x}_i\|,h).$$

However, the details are beyond the goals of this chapter.

The basic point is that kernel regression estimators perform poorly and splines, naively used, only perform a little better in two or more dimensions. Both just need too much data to learn the function underlying the observations. This is suggested by Fig. 3.16. The middle and right fit are given by NW and smoothing splines using $n = 100$ observations on an evenly spaced grid. For the two-dimensional sinc($\boldsymbol{x}$) function, both fits are already poor. Even increasing the sample size from 100 to 1000 does not improve the fit greatly. This is another demonstration of the Curse of Dimensionality.

For the sake of completeness, it's worth noting that, for the ethanol data, the two-dimensional NW estimator does quite well. This is shown is Fig. 3.17. The plot on the left shows the response NOx modeled with both predictors E and C. The surface looks like an arch. This is not surprising because, as the scatterplot of the full data set showed, the response is unaffected by C, effectively making the problem unidimensional.

A plot for splines like that in Fig. 3.17 for kernels would look very similar, possibly a little smoother. The improvement would be much like the improvement of splines over NW in Fig. 3.16 but less since the NW fit is already pretty good. This kind of comparison will often hold because, as seen in Silverman's theorem, splines and kernels are somewhat equivalent (for a specific kernel – but the exact form of the kernel is not very important) in the sense that variable-bandwidth kernel regression is equivalent to spline fitting given $p(x)$. Since the choice of kernel makes little difference, if the optimal constant-bandwidth kernel regression is really good, variable-bandwidth kernel regression cannot do much better and so neither can spline smoothing. Good

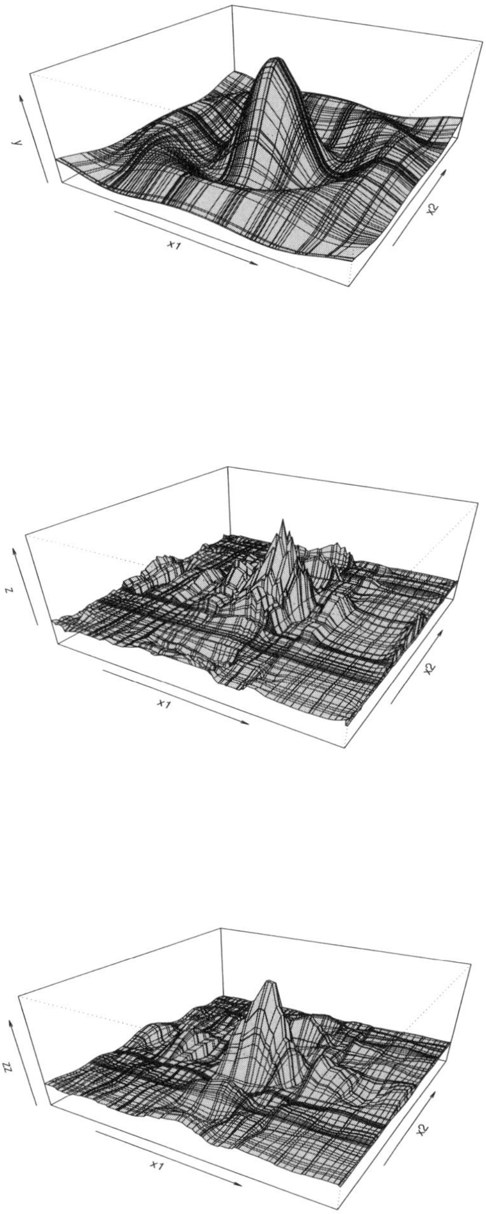

Fig. 3.16 The top panel shows the two-dimensional sinc function. It is included to help assess the quality of the fit. The middle panel shows the fit obtained using the locfit package, which implements a two dimensional NW estimator on the sample of size $n = 100$. The locfit smoothing parameter is $h = 0.45$. The fit obtained using Gu's smoothing spline ANOVA (ssanova) function within his gss package is shown in the bottom panel. The ssanova function internally finds the optimal value of the smoothing parameter. The spline fit is a little better (i.e., smoother) than the NW fit, as can be seen from the more rounded peak at 0.

constant-bandwidth kernel estimators correspond to the optimal variable-bandwidth reducing as $h(x) \approx h_{opt}$ where h_{opt} is the optimal constant bandwidth. Thus, the λ for optimal spline fitting would be determined by $p(x)$. If the xs were chosen roughly uniformly, then $p(x) \approx C$, for some constant C so the best λ would be determined by h_{opt}, suggesting the three fits (constant-bandwidth NW, variable-bandwidth NW, and spline) should be roughly equivalent. Even so, splines will typically be a little better than kernels because they involve an extra optimization in approximating the projection of the function in an RKHS as well as choosing the λ optimally.

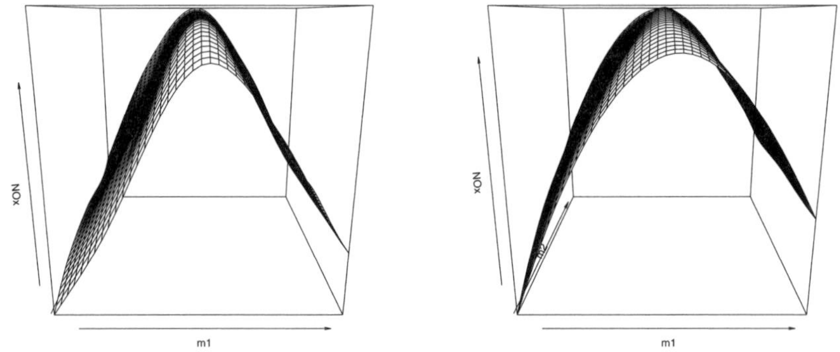

Fig. 3.17 Fits obtained with the locfit package. On the left, the bandwidth is $h = .45$, and on the right $h = .75$ is the optimal value. The fit on the right is closer to a plane and is smoother because the weight of the kernel is spread over a larger neighborhood. When h is small, the weight of the kernel is more concentrated, giving a rougher appearance.

The difference in performance between smoothing spline techniques and kernel regression more generally arises because, given a reproducing kernel, the regularization structure of splines is not harmed as much by the dimension of the input space as kernel regression is. The basic idea, heuristically and concisely, is the following. Given a spline optimization of the form $L(f) + \lambda \|f\|_{\mathscr{H}_K}$, the reproducing kernel defines a collection of eigenfunctions by Mercer's theorem. Some of the eigenfunctions span the null-space of the penalty, while the others form its orthogonal complement. Minima can therefore be expressed in terms of the eigenfunctions, somewhat independent of the dimension of $\boldsymbol{x}$. Also, the kernel defining the RKHS is itself a sum of products of the eigenfunctions.

Even better, the representer theorem gives a representation of the minimum in terms of evaluations of the kernel function. The representation is equivalent to an expression in terms of the eigenfunction basis, but the basis elements are chosen by the data, as if one of the (continuous) indices of the kernel actually indexed a basis for the RKHS. Thus, the terms in the representer theorem representation can be expressed in terms of the Mercer Theorem expression for the kernel (i.e., in terms of the eigenfunctions), giving a solution dependent on the sample size n, not the dimension of the data p.

Of course, having as many terms as data points poses a problem for most estimation techniques, but the representer theorem reduction is a great start; further efforts to achieve parsimony can improve inference.

3.7 Notes

3.7.1 Sobolev Spaces: Definition

Let $\mathscr{X} = [a,b] \subset \mathbb{R}$ be a domain on the real line. Recall that the space $L^2(a,b)$ of square-integrable functions on $\mathscr{X}$ is the space

$$L^2(a,b) = \left\{ f : [a,b] \to \mathbb{R}\ s.t.\ \|f\|_{L^2}^2 = \int_a^b (f(x))^2 dx < \infty \right\},$$

where the norm is defined through the inner product

$$\langle f, g \rangle_{L^2} = \int_a^b f(x)g(x)dx,$$

so that $\|f\|_{L^2}^2 = \langle f, f \rangle_{L^2}$. Sobolev spaces are Banach spaces where the norm involves derivatives or at least something other than just function values. The simplest of Sobolev spaces is $H^1(a,b)$, which is the space of functions defined by

$$H^1(a,b) = \{ f \mid f \in L^2(a,b),\ f' \in L^2(a,b).\}$$

The Sobolev space $H^1(a,b)$ is endowed with the inner product

$$\langle f, g \rangle = \int_a^b (f(x)g(x) + f'(x)g'(x))dx,$$

from which the norm for $H^1(a,b)$ is given by

$$\|f\|_{H^1}^2 = \|f\|_{L^2}^2 + \|f'\|_{L^2}^2.$$

Higher-order Sobolev spaces can be defined as

$$H^k(a,b) = \left\{ f : [a,b] \to \mathbb{R} \mid f, f', \ldots, f^{(k)} \in L^2(a,b) \right\}$$

with corresponding inner product

$$\langle f, g \rangle = \int_a^b \sum_{j=0}^k f^{(j)}(x) g^{(j)}(x)dx,$$

from which the norm is obtained as

$$||f||_{H^k}^2 = \sum_{j=0}^{k} ||f^{(j)}||_{L^2}^2 = ||f^{(j)}||_{H^{k-1}}^2 + ||f^{(k)}||_{L^2}^2 = ||f'||_{H^{k-1}}^2 + ||f||_{L^2}^2.$$

Note: A more general notation for Sobolev spaces is

$$W^{k,p}(a,b) = \left\{ f : [a,b] \to \mathbb{R} \mid f, f', \ldots, f^{(k)} \in L^p(a,b) \right\},$$

where the norm is defined by

$$||f||_k^p = \sum_{j=0}^{k} ||f^{(j)}||_{L^p}^p = ||f^{(j)}||_{H^{k-1}}^p + ||f^{(k)}||_{L^p}^p = ||f'||_{H^{k-1}}^p + ||f||_{L^p}^p.$$

Some examples of simple Sobolev spaces are:

1. $W^{1,1}(0,1)$, the space of absolutely continuous functions on $[0,1]$.
2. $W^{1,\infty}(a,b)$, the space of Lipschitz functions on $[a,b]$.

This definition of Sobolev spaces is based on the fact that the functions involved are one-dimensional. In fact, to guarantee that $H^1(a,b)$ is well defined, one may require that f be absolutely continuous,

$$f(y) = f(x) + \int_x^y f'(z)dz.$$

However, for higher-dimensional domains, $\mathscr{X} \subset \mathbb{R}^p$, absolute continuity may be hard to achieve. So, other approaches are used to define Sobolev spaces.

3.8 Exercises

Exercise 3.1. Use either Matlab or R to explore the difference in magnitude of the interpolation error. The goal is to compare global polynomial interpolation with a piecewise polynomial interpolation technique such as splines. The point here is to show computationally that the interpolation error grows exponentially with n for global polynomials but is a small constant for piecewise polynomial interpolants.

1. Generate $n = 16$ points as before with the Runge function.
2. Use the modified norm based on the Lagrange basis to compute both $||T_{poly}||_\ell$ for the polynomial interpolant and $||T_{spline}||_\ell$ for the spline interpolant. Compare and comment.
3. Repeat Steps 1 and 2 for $n = 25, 36, 64$ and tabulate the norms for polynomial and spline side by side. What pattern emerges?

Exercise 3.2. Consider the smoothing splines objective function

$$E_\lambda(f) = \frac{1}{n}\sum_{i=1}^{n}(Y_i - f(x_i))^2 + \lambda \int_a^b [f''(t)]^2 dt,$$

where $Y_i = f(x_i) + \varepsilon_i$ for $i = 1, \ldots, n$, f is an unknown function, the knots $x_1 < \cdots < x_n$ are given and the error term is mean-zero with variance σ^2.

1. If f has two derivatives, use integration by parts to show that

$$\int_a^b [f''(t)]^2 dt = \boldsymbol{f}^\top K \boldsymbol{f},$$

   where $\boldsymbol{f} = (f(x_1), f(x_2), \cdots, f(x_n))^\top$ and $\mathbf{K} = A^\top B^{-1} A$ as defined in Section 3.2
2. Now, show that the regularized risk $E_\lambda(f)$ can be written as

$$E_\lambda(f) = (\mathbf{Y} - \boldsymbol{f})^\top (\mathbf{Y} - \mathbf{f}) + \lambda \boldsymbol{f}^\top K \boldsymbol{f}.$$

3. Deduce that the minimizer of $E_\lambda(f)$ is

$$\hat{\boldsymbol{f}} = (I + \lambda K)^{-1}\mathbf{Y}.$$

Exercise 3.3. Consider the function $f(\boldsymbol{x}) = e^{\boldsymbol{x}}$ on the interval $[0, 1]$.

1. Find the least squares approximation of $f(\boldsymbol{x})$ on $[0, 1]$ among all polynomials of degree at most 2; i.e., find the numerical values

$$\hat{\boldsymbol{c}} = \arg \min_{\boldsymbol{c} \in \mathbb{R}} \int_0^1 [e^{\boldsymbol{x}} - (c_0 + c_1 \boldsymbol{x} + c_2 \boldsymbol{x}^2)]^2 d\boldsymbol{x}.$$

   where $\boldsymbol{c} = (c_0, c_1, c_2)^\top$ and $\hat{\boldsymbol{c}} = (\hat{c}_0, \hat{c}_1, \hat{c}_2)^\top$.
2. Graph $e^{\boldsymbol{x}}$, the first three terms of its power series expansion, and $\hat{c}_0 + \hat{c}_1 \boldsymbol{x} + \hat{c}_2 \boldsymbol{x}^2$. How do they compare?

Exercise 3.4 (Hilbert spaces and norms). Consider the space $\mathscr{H} = \mathscr{C}([0,1])$ of continuous real-valued functions $\mathscr{X} = [0, 1]$, and equip it with the inner product

$$\langle f, g \rangle = \int_0^1 f(x) g(x) dx. \tag{3.8.1}$$

Consider the sequence $\{f_n\}$ with

$$f_n(x) = \begin{cases} (2x)^{n/2} & 0 \le x \le 1/2 \\ 1 - (2(1-x))^{n/2} & 1/2 \le x \le 1. \end{cases}$$

1. Show that $\{f_n\}$ is a Cauchy sequence under (3.8.1).
2. Show that $\{f_n\}$ is convergent pointwise, but not to a continuous limit.
3. Deduce that the inner product space induced by (3.8.1) is not complete.

Exercise 3.5 (Reproducing kernel Hilbert spaces). Consider the space $\mathscr{H}$ of functions on $[0,1]$ with square integrable second derivatives. Equip $\mathscr{H}$ with the inner product

$$\langle f,g\rangle = f(0)g(0) + \int_0^1 f'(x)g'(x)dx \tag{3.8.2}$$

and the kernel

$$K(s,t) = 1 + \min(s,t).$$

Write $K_t(\cdot) = K(\cdot,t)$ and $K_s(\cdot) = K(s,\cdot)$, so that $K_t(s) = K(s,t)$.

1. Show that $K(\cdot,\cdot)$ is a reproducing kernel; i.e., that

$$\langle f,K_t\rangle = f(t), \quad \forall t \in [0,1].$$

2. Fix $2n$ constants $0 \le t_1 \le t_2 \le \cdots \le t_n \le 1$ and $c_1, c_2, \cdots, c_n$ and assume that $K_{t_i}(s)$ is piecewise linear in s for each t_i. Define the linear combination

$$\bar{f} = \sum_{i=1}^n c_i K_{t_i},$$

and verify that $\bar{f}(s)$ is also piecewise linear in s.

3. Let $\bar{\mathscr{H}}$ be the space spanned by the K_{t_i}s and let P be an orthogonal projection of $f \in \mathscr{H}$ onto it. Now,

$$\mathscr{H} = \bar{\mathscr{H}} \oplus \bar{\mathscr{H}}^{\perp}.$$

Write $Pf = \bar{f}_{\mathscr{H}}$ for the orthogonal projection of an element $f \in \mathscr{H}$ onto $\bar{\mathscr{H}}$.

a. Show that $\langle f - \bar{f}_{\mathscr{H}}, K_{t_i}\rangle = 0, \quad i = 1,2,\cdots,n.$

b. Consider the system of equations

$$\sum_{j=1}^n \langle K_{t_j}, K_{t_i}\rangle c_j^* = \langle f, K_{t_i}\rangle = f(t_i), \quad i = 1,2,\cdots,n. \tag{3.8.3}$$

Show that the coefficients of $\bar{f}_{\mathscr{H}}$ denoted by $c_1^*, c_2^*, \cdots, c_n^*$ can be found by solving (3.8.3).

c. Show that

$$\bar{f}_{\mathscr{H}}(t_i) = \sum_{j=1}^n c_j^* \langle K_{t_j}, K_{t_i}\rangle = f(t_i).$$

d. Deduce that $\bar{f}_{\mathscr{H}}$ interpolates f at the points $0 \le t_1 \le t_2 \le \cdots \le t_n \le 1$.

e. Show that $\bar{f}_{\bar{\mathscr{H}}}$ minimizes

$$\|f - \bar{f}\|_{\mathscr{H}} = \left\{ (f(0) - \bar{f}(0))^2 + \int_0^1 (f'(x) - \bar{f}'(x))^2 dx \right\}^{1/2}.$$

Exercise 3.6 (Discrete finite-dimensional RKHSs). Hilbert spaces can be finite dimensional and this exercise shows one way to construct them using a kernel. The idea is to use restrict the entries of the kernel to be elements of a finite index set say $\mathscr{I} = \{1,2,\cdots,n\}$. Then, let $\Sigma = (\sigma_{ij}), i, j = 1,2,\cdots,n$ be an $n \times n$ strictly positive definite matrix and set $K(i,j) = \sigma_{ij}$. Next consider two vectors of length n, $\boldsymbol{f}$ and $\boldsymbol{g}$, and define the inner product

$$\langle \boldsymbol{f}, \boldsymbol{g} \rangle = \boldsymbol{f}^\top \Sigma^{-1} \boldsymbol{g}.$$

Let $\boldsymbol{c}_i = (\sigma_{1i}, \sigma_{2i}, \cdots, \sigma_{ni})^\top = K(\cdot, i)$ denote the ith column of the matrix Σ.

1. Show K satisfies

$$\langle \boldsymbol{c}_i, \boldsymbol{c}_j \rangle = \sigma_{ij} = \langle K(\cdot,i), K(\cdot,i) \rangle = K(i,j).$$

2. Show that K has the reproducing property

$$\langle \boldsymbol{c}_i, \boldsymbol{f} \rangle = \langle K(\cdot,i), \boldsymbol{f} \rangle = f_i.$$

3. Parallel to the Mercer theorem, let $[\Phi_1, \Phi_2, \cdots, \Phi_n]$ be a set of orthogonal vectors and $\lambda_1, \cdots, \lambda_n$ be a collection of positive real numbers such that the values of Σ are given by

$$\sigma_{ij} = K(i,j) = \sum_{k=1}^{n} \lambda_k \Phi_k(i) \Phi_k(j).$$

Derive a new expression of the inner product $\langle \boldsymbol{f}, \boldsymbol{g} \rangle = \boldsymbol{f}^\top \Sigma^{-1} \boldsymbol{g}$ in terms of $\{\Phi_k\}$ and $\{\lambda_k\}$.

Exercise 3.7 (Comparing splines to NW and local polynomials).

1. Refer to Exercises 2.13, 2.14, 2.15. Redo them using spline smoothing in place of the NW estimator to generate plots of how biases and variances for spline smoothers, for equidistant and non-equidistant design points, as functions of x compare to those for NW and local polynomials.
2. Refer to Exercise 2.19. Redo it for spline smoothing to see how CV, GCV, and AIC compare for finding an estimate $\hat{f}_n^\lambda$. The R command for this is `smooth.spline()`.

Exercise 3.8 (RKHS for penalized linear regression splines). Consider the function space

$$\mathscr{H}_K = \left\{ f : f(x) = \beta_0 + \beta_1 x + \sum_{k=1}^{K} u_k (x - \kappa_k), \forall x \in \mathscr{X} \right\},$$

equipped with the inner product

$$\langle f, g \rangle_{\mathscr{H}_K} = \left\langle \beta_0 + \beta_1 x + \sum_{k=1}^{K} u_k (x - \kappa_k), \tilde{\beta}_0 + \tilde{\beta}_1 x + \sum_{k=1}^{K} \tilde{u}_k (x - \kappa_k) \right\rangle = \boldsymbol{\beta}^\top \tilde{\boldsymbol{\beta}} + \boldsymbol{u}^\top \tilde{\boldsymbol{u}},$$

for any two functions f and g in $\mathscr{H}_K$, where $\mathscr{X} = [a,b]$. Now define the kernel

$$K(s,t) = 1 + st + \sum_{k=1}^{K} (s - \kappa_k)_+ (t - \kappa_k)_+$$

for $s,t \in [a,b]$ and assume you have an IID sample $\mathscr{D} = \{(\boldsymbol{x}_i, y_i), i = 1, \cdots, n\}$. The penalized least squares empirical risk functional is

$$R_\lambda(\boldsymbol{f}) = \frac{1}{n} \sum_{i=1}^{n} (y_i - f(\boldsymbol{x}_i))^2 + \lambda \|f\|^2_{\mathscr{H}_K}.$$

Do the following:

1. Write down the expression of $\|f\|_{\mathscr{H}_K}$ in terms of the inner product.
2. Let $\mathscr{H}_0$ be the subspace of all those functions that are not penalized. Then, $\mathscr{H}_K$ as $\mathscr{H}_K = \mathscr{H}_0 + \mathscr{H}_1$, where $\mathscr{H}_1 = \mathscr{H}_0^\perp$ is the orthogonal complement of $\mathscr{H}_0$. Let P_1 be the orthogonal projection operator onto $\mathscr{H}_1$. What is $\|P_1 f\|^2$ for a given function $f \in \mathscr{H}_K$?
3. Using item 2, rewrite the expression of the empirical risk functional $R_\lambda(\boldsymbol{f})$.
4. Using your expression from item 3, derive an expression for the estimators of the unknown quantities.
5. How would the process change if you used quadratic splines in $\mathscr{H}_K$ rather than linear splines?

Exercise 3.9. Suppose you have the artificial data set

$$\mathscr{D} = \{(-2,-1), (-1,1), (0,2), (1,4), (2,-2)\}$$

drawn from the model model

$$y = \beta_0 + \beta_1 x + \beta_2 (x-1)_+ + \varepsilon. \tag{3.8.4}$$

Assume that $\varepsilon \sim N(0,1)$ and that the observations are independent.

1. Do the following:
   a. Use a boxcar kernel and a bandwidth of 2.0 to fit a local constant regression curve to $\mathscr{D}$. (See Exercise 2.8.)
   b. Evaluate the estimator at $-1, 0, 1$.
2. Now compare this to a least squares approach:
   a. Estimate the parameters in the model.
   b. Find a 95% confidence interval for β_2. Does this analysis suggest the knot at $x = 1$ is worth including?
3. Assume x ranges over $[-2, 2]$.

a. Obtain the formulas for the three linear B-splines with a knot at $x = 1$ and sketch their graphs.

b. Using the derived B-splines as your basis functions, find the least squares fit to the artificial data set. How does the estimate of σ^2 obtained using linear B-splines compare with the estimate of σ from item 2?

4. Now, let's look at quadratic B-splines.

a. Assume that $B_{0,1}(x) \equiv 0$ and find the formulas for the four quadratic B-splines with a knot at $x = 1$ and sketch their graphs.

b. Using the quadratic B-splines as your basis, find the least squares fit to the data set $\mathscr{D}$. How does the estimate of σ^2 obtained via quadratic B-splines compare with the one obtained with the model of equation (3.8.4)?

5. Use the R functions matplot() and bs() to check the graphs you sketched earlier.
6. Compare the fitted values from the local constant regression with those from the linear and quadratic B-splines.

Exercise 3.10. Let

$$H^2(0,1) = \{f : [0,1] \to \mathbb{R} \mid f, f', f'' \in L^2(0,1)\}.$$

be the second-order Sobolev space on $[0,1]$. Define the inner product

$$\langle f, g \rangle = \alpha f(0)g(0) + \beta f'(0)g'(0) + \int_0^1 f''(t)g''(t)dt$$

and let $K(\cdot,\cdot)$ be the reproducing kernel for $H^2(0,1)$. Consider the integral operator as a linear functional $L(f) = \int_0^1 f(t)dt$.

1. Verify $L(f)$ is continuous.
2. Find the representer of $L(f)$ using the inner product.
3. Find the optimal weights w_i and the optimal design points $t_i \in (0,1)$ such that $\sum_{i=1}^n w_i f(t_i)$ is the best approximation of $L(f)$.

Exercise 3.11 (Dimension of a spline space). Fix an interval (a,b) and let $z_1, z_2, \cdots, z_n \in (a,b)$. Denote the space of polynomial splines of order r with simple knots at the z_is by $S_r(z_1, z_2, \cdots, z_k)$ Here you can prove that $\dim(S_r(z_1, z_2, \cdots, z_k)) = r + n$.

1. First consider the $r+n$ functions $1, \boldsymbol{x}, \boldsymbol{x}^2, \cdots, \boldsymbol{x}^{r-1}, (\boldsymbol{x} - z_1)_+^{r-1}, \cdots, (\boldsymbol{x} - z_n)_+^{r-1}$. Verify they are elements of $S_r(z_1, z_2, \cdots, z_n)$.
2. Verify the $r+n$ functions are linearly dependent.
3. Let $s \in S_r(z_1, z_2, \cdots, z_k)$ and let p_i be the polynomial, of degree at most $r-1$, that coincides with s on the interval (z_i, z_{i+1}).

a. Prove that there exist constants c_i such that

$$p_{i+1}(\boldsymbol{x}) - p_i(\boldsymbol{x}) = c_i(\boldsymbol{x} - z_i)^{r-1}.$$

b. From this conclude that s is a linear combination of the $r+n$ functions.

Exercise 3.12. Let

$$H_m(0,1) = \left\{ f : [0,1] \to \mathbb{R} \mid f(t) = \sqrt{2} \sum_{\nu=1}^{\infty} a_\nu \cos(2\pi\nu t) + \sqrt{2} \sum_{\nu=1}^{\infty} b_\nu \sin(2\pi\nu t) \right.$$
$$\left. \text{with} \sum_{\nu=1}^{\infty} (a_\nu^2 + b_\nu^2)(2\pi\nu)^{2m} < \infty \right\}.$$

That is, $H_m(0,1)$ is the space of Fourier expansions with coefficients decreasing at a rate determined by m.

1. Show that the mth derivatives of functions $f \in H_m(0,1)$ satisfy

$$\int_0^1 \left(f^{(m)}(u) \right)^2 du = \sum_{\nu=1}^{\infty} (a_\nu^2 + b_\nu^2)(2\pi\nu)^{2m}.$$

2. Let $K(s,t)$ denote the reproducing kernel for $H_m(0,1)$. Show that

$$K(s,t) = \sum_{\nu=1}^{\infty} \frac{2}{(2\pi\nu)^{2m}} \cos[2\pi\nu(s-t)].$$

3. Show that

$$\int_0^1 f^{(k)}(u)du = 0 \qquad \text{for } k = 0, 1, 2, \cdots, m.$$

4. Define the extension of $H_m(0,1)$, $W_m(0,1) = \{1\} \oplus H_m(0,1)$, with norm

$$\|f\|^2 = \left[\int_0^1 f(u)du \right]^2 + \int_0^1 \left(f^{(m)}(u) \right)^2 du$$

on $W_m(0,1)$ and verify that

$$R(s,t) = 1 + K(s,t)$$

is its reproducing kernel.

Chapter 4
New Wave Nonparametrics

By the late 1980s, Classical nonparametrics was established as "classical". Concurrently, however, the beginnings of a different stream of nonparametric thinking were already under way. Indeed, its origins go back to the 1970s if not earlier. The focus here is not on large spaces of functions but on classes of functions intended to be tractable representations for intermediate tranches. The models retain much of the flexibility of Classical methods but are much more interpretable; not as interpretable as many subject matter specialists might want but possessing much more structure than the methods of Chapters 2 and 3. In practice, computer-intensive procedures pervade these more recent techniques. This permits iterative fitting algorithms, cross-validation for model selection, bootstrapping for pointwise confidence bands on the estimated functions as seen earlier, and much more besides.

New Wave nonparametrics focuses on the intermediate tranche. This is where the dimension is increasing without bound and the techniques rely explicitly on approximating an unknown function to an adequate accuracy. That is, bias is admitted but controlled; one chooses, in effect, how far a sequence of approximations should be taken even though in principle they could be taken to limits that realize a whole infinite-dimensional space. In some cases, the search is over directions, as in projection pursuit, and amounts to a sequential approximation because there is no bound on the number of parameters introduced.

The Alternative regression methods, to be seen in Chapter 6, differ from these in that multiple models are usually considered, perhaps implicitly. Also, the usual goal is unabashedly predictive rather than model identification, although some of the Alternative methods do that, too.

This chapter discusses several of the famous New Wave nonparametric regression techniques, including additive models, generalized additive models, projection pursuit, neural nets, recursive partitioning, multivariate adaptive regression splines, sliced inverse regression, alternating conditional variances, and additivity and variance stabilization.

B. Clarke et al., *Principles and Theory for Data Mining and Machine Learning*, Springer Series in Statistics, DOI 10.1007/978-0-387-98135-2_4, © Springer Science+Business Media, LLC 2009

4.1 Additive Models

The main problem with multiple linear regression is that the estimate is always flat. However, the class of all possible smooth models is too large to fit, and the Curse makes such fits inadequate in high dimensions. The class of additive models is one useful compromise. Essentially, the additive assumption reduces the size of the function space in which the regression is done. Rather than general p-variate functions of the form $f(x_1,...,x_p)$, one uses a sum of p univariate functions, $f_1(x_1)+...+f_p(x_p)$. Of course, when p itself is large, the Curse remains.

At their root, additive models in DMML are a generalization of additive models in ANOVA. Recall the standard ANOVA model

$$Y_{k,j,i} = \alpha_k + \beta_j + \varepsilon_{k,j,i},$$

in which $k = 1,...,K$ and $j = 1,...,J$ are the levels for factors A and B, and $i = 1,...,n$ are the samples. In essence, α_k is a function $\alpha = \alpha(k) = \alpha_k$, where k is an explanatory variable taking K values. Likewise, β_k is a function $\beta = \beta(j) = \beta_j$ is a function of J, an explanatory variable taking J values. Thus, $Y = Y_{k,j}$ is a function of two discrete explanatory variables assumed to decompose into a sum of two univariate functions, one for each variable. If k and j are taken as continuous variables, α and β are taken as functions of them, and p such functions are permitted in the representation of Y, the result is the class of additive models used in DMML. Another way to see this is to replace the X_ks in a p-dimensional linear repression model with general smooth functions $f_k(X_k)$ for $k = 1,...,p$.

More formally, the additive model for a response is

$$Y = \beta_0 + \sum_{k=1}^{p} f_k(x_k) + \varepsilon, \tag{4.1.1}$$

where the f_k are unknown smooth functions fit from the data. Thus, additive models preserve additivity but lose linearity in the parameters. Often, one writes $\mathbb{E}(Y|\boldsymbol{X})$ in place of Y and drops the error term. The basic assumptions are as before, except that $\mathbb{E}[f_k(X_k)] = 0$ and $\mathbb{E}(\varepsilon|\boldsymbol{X}) = 0$ are required to prevent nonidentifiability; e.g., confounding means with β_0. Additive models are biased unless Y really is the sum of the terms on the right-hand side, which is not common. The greatest benefit from using additive models occurs when Y is reasonably well approximated by the right-hand side so that the bias is small and the reduction in variance from the representation in (4.1.1) is substantial.

Observe that the parameters in the additive model are $\{f_k\}$, β_0, and σ^2. First, recall that in the linear model it is the parameters that enter linearly and estimating a parameter costs one degree of freedom. Fitting smooth functions costs more, depending on what kind of univariate smoother is used, and, for smoothers, linearity is not in the parameters (they do not exist) but rather in the sense of (2.1.1). Second, given $Y_1,...,Y_n$, the location β will be estimated by $\hat{\beta} = (1/n)\sum_{i-1}^{n} Y_i$. So, without loss of generality, take $\beta_0 = 0$, if necessary, by replacing the Y_is with $Y_i - \bar{Y}$. Third, since $\text{Var}(\varepsilon|\boldsymbol{X}) = \sigma^2$ is typically assumed, it will usually be enough to use the residuals to estimate σ.

Clearly, the central issue in fitting a model such as (4.1.1) is estimating the f_ks. Depending on how one does this, one can force the selected f_ks to lie in a specific family such as linear, locally polynomial, or monotone. One can also enlarge the collection of additive models by including functions of the X_ks. For instance, X_1X_2 can be included as a $p+1$ explanatory variable. Indeed, one can force the inclusion of certain prechosen higher-dimensional smooths such as $f(X_1, X_2)$, or $f(X_1X_2)$ if such an interaction is desired. Thus, any function of the form $f(a_1(\boldsymbol{X}), \ldots, a_u(\boldsymbol{X}))$ can be included for fixed u and known a_is. However, the larger u is, the less benefit the additive structure gives. The key benefit of additive models is that, in (4.1.1), transformation of each explanatory variable is done automatically by the marginal smoothing procedure. The main technique for fitting additive models is called the backfitting algorithm. It permits the use of an arbitrary smoother (e.g., spline, LOESS, kernel) to estimate the $\{f_k\}$s.

4.1.1 The Backfitting Algorithm

The backfitting algorithm is a central idea that recurs in a variety of guises. So, it's worthwhile to provide an overview before turning to the technicalities.

Overview: Suppose the additive model is exactly correct. Then, for all $k = 1, \ldots, p$,

$$\mathbb{E}\left[Y - \sum_{k \neq j} f_k(X_k) \,|\, x_j\right] = f_j(x_j). \tag{4.1.2}$$

The backfitting algorithm solves these p equations for the f_ks iteratively. At each stage, it replaces the conditional expectation of the delete-j residuals on the left-hand side with a univariate smooth.

To see this, it helps to use vectorized notation for the smooth functions. Let $\boldsymbol{Y}$ be the vector of responses and let $\boldsymbol{X}$ be the $n \times p$ matrix of explanatory values with columns $\boldsymbol{X}_{k,\cdot}$ representing the n outcomes of the kth explanatory variable. Then define $\boldsymbol{f}_k = (f_k(X_{k,1}), \ldots, f_k(X_{k,n}))$ to be the vector of the n values f_k takes on the outcomes $X_{k,i}$. To represent the actual estimation of the univariate smooths, define $L(\boldsymbol{Z}|\boldsymbol{C}_{k,\cdot})$ to be the smooth from the scatterplot of $\boldsymbol{Z}$ against the values of the kth explanatory variable. Note that in this notation $\boldsymbol{Z} \in \mathbb{R}^n$ typically is an iterate of $\boldsymbol{Y} - \sum_{k \neq j} f_k$ and $L(\cdot|\boldsymbol{C}_{k,\cdot})$ is the linear smooth as an operator on $\boldsymbol{Z}$. The conditioning indicated by $\boldsymbol{C}_{k,\cdot}$ typically represents $\boldsymbol{X}_{k,\cdot}$, which is used to form the univariate smooth f_k being fit.

The backfitting procedure, the Gauss-Seidel algorithm in the additive model context, is attributed to Buja et al. (1989). The term backfitting arises because the procedure iteratively omits one of the occurrences of one of the summands when it occurs, replacing it with an improved value. One version of their procedure is as follows:

Given $Y_1, \ldots, Y_n$ and n outcomes of the p-dimensional explanatory variable $\boldsymbol{X}$:

- ☐ Initialize: Set $\hat{\beta}_0 = \bar{Y}$ and set the f_k functions to be something reasonable (e.g., a linear regression). Set the $\boldsymbol{f}_k$ vectors to match.
- ☐ Cycle: For $k = 1, \ldots, p$, set

$$f_k = L\left(\boldsymbol{Y} - \hat{\beta}_0 - \sum_{j \neq k} \boldsymbol{f}_j(\boldsymbol{X}_{\cdot,j}\right)$$

  and update the $\boldsymbol{f}_k$s to match.
- ☐ Iterate: Repeat the cycle step until the change in f_k between iterations is sufficiently small.

One may use different smoothers $L(\cdot|C_{\cdot,k})$ for different variables or bivariate smoothers for predesignated pairs of explanatory variables.

Technical Description: Following Hastie and Tibshirani (1990), consider the optimization

$$\min_{f: f(\boldsymbol{x}) = \sum_{k=1}^{p} f_k(x_k)} \mathbb{E}(Y - f(\boldsymbol{X}))^2,$$

meaning one is searching for the best additive predictor for the overall minimum $\mathbb{E}(Y|\boldsymbol{X})$. It can be verified that there is a unique minimum within the additive class of fs and that this minimum satisfies p equations of the form

$$\mathbb{E}((Y - f(\boldsymbol{X}))|X_k) = 0.$$

These equations are equivalent to (4.1.2).

The convention is to write these k equations in matrix form as

$$\begin{pmatrix} \mathbb{E}(\cdot|\mathscr{F}) & \mathbb{E}(\cdot|X_1) & \ldots & \mathbb{E}(\cdot|X_1) \\ \mathbb{E}(\cdot|X_2) & \mathbb{E}(\cdot|\mathscr{F}) & \ldots & \mathbb{E}(\cdot|X_2) \\ \cdot & \cdot & \cdot & \cdot \\ \mathbb{E}(\cdot|X_p) & \ldots & \mathbb{E}(\cdot|X_p) & \mathbb{E}(\cdot|\mathscr{F}) \end{pmatrix} \begin{pmatrix} f_1(X_1) \\ f_2(X_2) \\ \cdot \\ f_p(X_p) \end{pmatrix} = \begin{pmatrix} \mathbb{E}(Y|X_1) \\ \mathbb{E}(Y|X_2) \\ \cdot \\ \mathbb{E}(Y|X_p) \end{pmatrix}, \quad (4.1.3)$$

in which $\mathbb{E}(\cdot|\mathscr{F})$ is the conditional expectation with respect to the overall σ-field and so acts like the identity operator. For brevity, write (4.1.3) as

$$\boldsymbol{Pf} = \boldsymbol{QY}, \quad (4.1.4)$$

in which $\boldsymbol{Q}$ is the linear transformation with the p projection operators $\mathbb{E}(\cdot|X_k)$ on its main diagonal and zeros elsewhere.

To use (4.1.4), one needs expressions for the $\mathbb{E}(\cdot|X_k)$s that can be used for regressing Y on X_k. So, let S_k be a collection of linear transformations like $L_n(\cdot)$ in (2.3) that give a linear smooth for Y as a function of X_k by acting on $\mathbf{y}$. That is, let L_k be an $n \times n$ matrix from a linear smoothing technique for the univariate regression of Y on X_k so that the data-driven quantity $S_k\mathbf{y}$ is an estimate of the theoretical quantity $(\mathbb{E}(Y_1|X_{k,1}), \ldots, \mathbb{E}(Y_n|X_{k,n}))$. Now,

$$L_k \boldsymbol{y} \approx \begin{pmatrix} \mathbb{E}(Y_1|X_{k,1}) \\ \dots \\ \mathbb{E}(Y_n|K_{k,n}) \end{pmatrix}.$$

Substituting L_k for $\mathbb{E}(\cdot|X_k)$ for $k = 1, ..., p$ in (4.1.4) gives the data form

$$\begin{pmatrix} I_n & L_1 & \dots & L_1 \\ L_2 & I & \dots & L_2 \\ \cdot & \cdot & \cdot & \cdot \\ L_p & \dots & L_p & I_n \end{pmatrix} \begin{pmatrix} \hat{f}_1(X_1) \\ \hat{f}_2(X_2) \\ \cdot \\ \hat{f}_p(X_p) \end{pmatrix} = \begin{pmatrix} L_1 & 0 & \dots & 0 \\ 0 & L_2 & \dots & 0 \\ \cdot & \cdot & \cdot & \cdot \\ 0 & \dots & 0 & L_p \end{pmatrix} \begin{pmatrix} \boldsymbol{Y} \\ \boldsymbol{Y} \\ \cdot \\ \boldsymbol{Y} \end{pmatrix}, \tag{4.1.5}$$

in which each entry in the $p \times p$ matrices indicated is itself an $n \times n$ matrix, giving an overall dimension of $np \times np$. The entries in the two vectors are similarly defined but are $np \times p$. On the right-hand side of (4.1.5), the application of the L_ks is written explicitly unlike in (4.1.3) where the $\mathbb{E}(\cdot|X_k)$s are not written in a separate matrix from the $\boldsymbol{Y}$s. For brevity, write

$$\hat{\boldsymbol{P}}\hat{\boldsymbol{f}} = \hat{\boldsymbol{Q}}\boldsymbol{Y} \tag{4.1.6}$$

parallel to (4.1.4). Also, note that $\mathbb{E}(\cdot|X_k)$ takes the expectation over all directions $X_{k'}$ for $k \neq k'$, but L_k ignores all values of the $X_{k'}$s. Ignoring data is not the same as taking an expectation over its distribution; however, if all the data are random, then ignoring some of the variables can be much like using the marginal for the rest.

In principle, the linear system (4.1.6) can be solved for the $(\hat{f}_k(X_{k,1}), ..., \hat{f}_k(X_{k,n}))$s for $k = 1, ..., p$. However, when p and n get large, this becomes difficult. What's more, $\hat{\boldsymbol{P}}$ is often difficult to work with. So, instead of direct solutions such as Gaussian elimination, iterative methods are used. Often these are superior for sparse matrices. One such technique is called the Gauss-Seidel algorithm, see Hastie and Tibshirani (1990), Chapter 5.2. This structure ensures that the backfitting algorithm converges for smoothers that correspond to a symmetric smoothing matrix with all eigenvalues in $(0, 1)$. This includes smoothing splines and the Nadaraya-Watson estimator, but not LOESS or local polynomial regression for degrees larger than 0. Empirically, however, it is usually observed that the eigenvalues of most kernel smoothers are in (0,1). (Counterexamples are possible, but hard.)

Implementation of the procedure from the last subsection is as follows. Given starting values $\hat{f}_k^{\,0}$, the iterates for $m = 1, 2, ...$ are

$$\hat{f}_k^{\,m} = L_k \left(\boldsymbol{Y} - \sum_{k' \neq k} \hat{f}_{k'}^{\,m-1} \right),$$

and one iterates until a Cauchy criterion is satisfied (i.e., the distance between successive iterates is satisfactorily small). This formalizes the univariate regression of $\boldsymbol{Y}$ on $\boldsymbol{X_k}$ using the partial residuals $\boldsymbol{Y} - \sum_{k' \neq k} \hat{f}_{k'}^{\,m-1}$ instead of $\boldsymbol{Y}$ for each m.

Clearly, this depends on the formulation of the Cauchy criterion and the order in which the univariate regressions are done. However, there are general assumptions guaranteeing the convergence of backfitting estimators. One is due to Opsomer (2000).

Rearrange (4.1.6) to give $\hat{\boldsymbol{f}} = \hat{\boldsymbol{P}}^{-1}\hat{\boldsymbol{Q}}\boldsymbol{Y}$, and set $W_k = E_k \hat{\boldsymbol{P}}^{-1}\hat{\boldsymbol{Q}}$, where E_k is a partitioned $n \times np$ matrix with the $n \times n$ identity matrix as the kth block on its main diagonal. Now, each component estimator is of the form $\hat{f}_k = W_k\boldsymbol{Y}$, and the estimator $\hat{\boldsymbol{F}}$ for $f(x_1,...,x_p) = f_1(x_1) + ... + f_p(x_p)$ can be written as

$$\hat{\boldsymbol{f}} = \sum_{k=1}^{p} \hat{f}_k = (W_1 + ... + W_p)\boldsymbol{Y} \equiv \boldsymbol{W}_Q\boldsymbol{Y},$$

in which $\boldsymbol{W}_Q$ is the additive smoother matrix for the additive model. Let $\boldsymbol{W}_Q^{-k}$ be the additive smoother matrix for the additive model with the kth variable deleted; i.e., for the model $Y = f_1(x_1) + ... + f_{k-1}(x_{k-1}) + f_{k+1}(x_{k+1}) + ... + f_p(x_p)$. It is important to realize that $\boldsymbol{W}_Q^{-k}$ is not the sum $\sum_{k' \neq k} \boldsymbol{W}_{k'}$.

Corollary 4.3 of Buja et al. (1989) showed that $||\boldsymbol{L}_1\boldsymbol{L}_2|| < 1$ is a sufficient condition for the convergence of the backfitting algorithm in a bivariate additive model, and this condition can only be satisfied when the univariate smoother matrices $\boldsymbol{L}_k$ are centered in the sense that they are replaced by $\tilde{\boldsymbol{L}}_k = (\boldsymbol{I} - \boldsymbol{1}\boldsymbol{1}'/n)\boldsymbol{L}_k$. Opsomer (2000), Lemma 2.1 established a p-dimensional generalization of Buja et al. (1989). Specifically, the backfitting algorithm with smoothers $\boldsymbol{L}_1,...,\boldsymbol{L}_p$ converges to a unique solution if

$$||\boldsymbol{L}_k\boldsymbol{W}_Q^{-k}|| < 1 \tag{4.1.7}$$

for $k = 1,...,p$ and any matrix norm $|| \cdot ||$, in which case the additive smoother with respect to the kth covariate is

$$\boldsymbol{W}_k = \boldsymbol{I} - (\boldsymbol{I} - \boldsymbol{L}_k\boldsymbol{W}_Q^{-k})^{-1}(1 - \boldsymbol{L}_k). \tag{4.1.8}$$

In the simple bivariate case, $p = 2$, $\boldsymbol{W}_Q^{-1} = L_2$, and $\boldsymbol{W}_Q^{-2} = \boldsymbol{L}_1$, however, the importance of additivity is far greater in higher dimensions. Indeed, from (4.1.8) it is seen that a pth order additive model has smoother matrices $\boldsymbol{W}_k$ that are expressed in terms of the smoother matrices from the corresponding $(p-1)$-order additive model, $\boldsymbol{W}_Q^{-k}$, and the univariate smoother $\boldsymbol{L}_k$. Therefore, the recursion can be built up from $p = 2$ cases to $p = 3$ and so forth.

For the case of local polynomial regression smoothers, Opsomer (2000) also establishes that the hypotheses of his result hold – (4.1.7) in particular – after centering, in the absence of concurvity (which will be discussed at the end of this section). Indeed, asymptotically valid expressions for the conditional bias and variance of the kth component can be obtained. That is, it can be shown that

$$\mathbb{E}(f_k(\hat{X}_{k,i}) - f_k(X_{k,i})|\boldsymbol{X}) = C + \mathcal{O}_p\left(\frac{1}{\sqrt{n}}\right) + o_p(h_k^{r+1}), \tag{4.1.9}$$

where h_k is the bandwidth for the kth univariate smoother, $r = r_k$ are odd numbers giving the degrees of the polynomials for each k, the bias is evaluated at an observation point $X_{k,i}$, and C is a leading term involving derivatives of f_k and other quantities in the estimator. It can be shown that the $\mathcal{O}_p(1/\sqrt{n})$ term is smaller than the leading term $C = \mathcal{O}_p(\sum_{k=1}^{p} h_k^{r+1})$, which dominates $o_p(h_k^{r+1})$. Now, the recursivity implicit in C motivates a corollary,

$$\mathbb{E}(f_k(\hat{X}_{k,i}) - f_k(X_{k,i})|\boldsymbol{X}) = \mathcal{O}_p\left(\sum_{k=1}^{p} h_k^{r+1}\right), \tag{4.1.10}$$

showing how the bandwidths add to bound the dimension effects. Moreover, the conditional variance is

$$\mathrm{Var}(\hat{f}_k(X_{k,i})|\boldsymbol{X}) = \sigma^2 \frac{R(K)}{nh_k} \frac{1}{f_k(X_{k,i})} + o_p\left(\frac{1}{nh_k}\right), \tag{4.1.11}$$

in which $R(K)$ is the integral of the square of the kernel K defined by the local polynomials. Convergence requires $n \longrightarrow \infty$, $h_k \longrightarrow 0$, and $(nh_k/\log n) \longrightarrow \infty$ for both variance and bias to go to zero. Expressions (4.1.7) – (4.1.11) can be combined to give an evaluation of the MSE; however, the dependence among the X_ks makes this difficult outside certain independence assumptions. Nevertheless, an optimal rate can be found by balancing the variance and squared bias.

A projection-based approach to backfitting is presented in Mammen et al. (1999), and is known to achieve the oracle efficiency bound. That is, projection-based methods give an expression for the bias of a single f_k separate from the biases of the $f_{k'}$s with $h' \neq k$, the other additive components, while the backfitting estimator only has this property if the X_ks are independent; see Opsomer (2000), Corollary 3.2. A general discussion of this approach is found in Mammen et al. (2001). Overall, they find that many smoothing methods (kernels, local polynomials, smoothing splines, constrained smoothing, monotone smoothing, and additive models) can be viewed as a projection of the data with respect to appropriate norms. The benefit of this approach is that it unifies several seemingly disparate methods, giving a template for the convergence of backfitting; see Mammen et al. (1999) for a version of the convergence of the backfitting algorithm for a Nadaraya-Watson type estimator. It is important to realize that, in addition to the varieties of smoother one can use in fitting the components of an additive model, there are numerous variants of the backfitting procedure. Indeed, one needn't use the same smoother in all components, and even if one did, Mammen and Park (2005) give three ways to do bandwidth selection in additive models. Moreover, there are several backfitting type procedures that can be derived from projection optimizations; see Mammen et al. (1999).

4.1.2 Concurvity and Inference

As a generality, if the backfitting algorithm converges, the solution is unique unless there is concurvity, in which case the solution depends on the initial conditions. Roughly, concurvity occurs when the $\{\boldsymbol{X}_i\}$ values lie on a smooth lower-dimensional manifold in $\mathbb{R}^p$. In the present context, a manifold is smooth if the smoother used in backfitting can interpolate all the $\{\boldsymbol{x}_i\}$ perfectly. The picture to keep in mind is a range of smooth deformations of a very twisted curve snaking through a high-dimensional space, with the attendant sparsity of data. Qualitatively, this is analogous to the nonuniqueness of regression solutions when the $\boldsymbol{X}$ matrix is not full rank.

The concurvity space of (4.1.4) is defined to be the set of additive functions $f(\boldsymbol{x}) = \sum_{j=1}^{p} f_j(x_j)$ such that $\boldsymbol{Pf} = \boldsymbol{0}$. That is,

$$f_j(x_j) + \mathbb{E}\left[\sum_{k \neq j} f_k(x_k) \,|\, x_j\right] = 0.$$

When the explanatory values do not lie exactly on a smooth lower-dimensional submanifold but tend to fall near one, then there are the same instability problems that arise in multiple linear regression when the data are nearly collinear. In terms of (4.1.6), nonuniqueness occurs when there is a vector $\boldsymbol{a}$ with $\hat{\boldsymbol{P}}\boldsymbol{a} = 0$, for then any solution $\hat{\boldsymbol{f}}$ has an associated vector space of solutions $\hat{\boldsymbol{f}} + \lambda \boldsymbol{a}$ for $\lambda \in \mathbb{R}$.

It is evident that concurvity in practice occurs when certain relationships among the smoother matrices S_k hold. So, suppose the eigenvalues of all the S_ks are in $[0,1]$, and let $\mathcal{N}_k$ be the vector space of eigenvectors of S_k with eigenvalue 1; these are the vectors that are unchanged by the smoother matrix S_k. (This is nothing like a nulls-pace; it is like an identity space.) Then $\hat{\boldsymbol{P}}\boldsymbol{a} = 0$ can occur, for instance, when $\sum_{k=1}^{p} \boldsymbol{a}_k = 0$ for some choices $\boldsymbol{a}_k \in \mathcal{N}_k$. There are versions of the backfitting algorithm that reduce the effect of this linear dependence among vectors; see Hastie and Tibshirani (1990). However, in the presence of concurvity, backfitting tends to break down.

In principle, one way to reduce concurvity is to eliminate some of the components f_k in the model. That is, once a model has been fit by the backfitting algorithm for a fixed p, one wants to test whether setting the component to zero would lead to a model with substantially less explanatory power.

Testing whether a term f_k in an additive model is worth keeping has been studied by Fan and Jiang (2005), where a generalized likelihood ratio test has been developed along with results similar to Wilks' theorem. To a large extent, this is a straightforward extension of the usual goodness-of-fit statistic based on normality. If the likelihood for the alternative hypothesis (i.e., the full model) is much larger than the likelihood for the reduced model (i.e., with one of the components set to zero), then one is led to reject the null. Thus, one needs critical values, techniques for nuisance statistics, and some applicable asymptotics. Unfortunately, generalized likelihood ratio (GLR) tests cannot be used directly unless a distribution for ε is specified. So, one wants a test that is robust to the distribution of ε and valid for a wide range of smoothers that might be used in backfitting (provided they converge).

Consider the hypothesis test

$$\mathscr{H}_0 : f_p \equiv 0 \quad \text{vs.} \quad \mathscr{H}_1 : f_p \not\equiv 0$$

for whether the pth variable has a contribution to the model for Y. Since the distribution of ε is unknown, the likelihood function cannot be written down. However, if ε were $N(0, \sigma^2)$, the log likelihood function would be

$$-\frac{n}{2}\log(2\pi\sigma^2) - \frac{1}{2\sigma^2}\sum_{i=1}^{n}\left(Y_i - \beta_0 - \sum_{k=1}^{p} f_k(X_{k,i})\right),$$

in which $\hat{\beta}_0$ and $\hat{f}_k$ based on K and bandwidths h_k can be substituted. Doing so, setting

$$RSS_1 = \sum_{i=1}^{n}\left(Y_i - \hat{\beta}_0 - \sum_{k=1}^{p}\hat{f}_k(X_{k,i})\right)^2,$$

and maximizing the result over σ gives the normal-based likelihood under the alternative:

$$-\frac{n}{2}\log\frac{2\pi}{n} - \frac{n}{2}\log RSS_1 - \frac{n}{2}.$$

Set $\ell(\mathscr{H}_1) = -(n/2)\log RSS_1$. Similarly, under the null, the same procedure gives a log likelihood for use under $\mathscr{H}_0$. Let

$$RSS_0 = \sum_{i=1}^{n}\left(Y_i - \hat{\beta}_0 - \sum_{k=1}^{p-1}\hat{f}_k(X_{k,i})\right)^2,$$

in which $\hat{f}_k$ remains the estimator of f_k under $\mathscr{H}_0$, using the same bandwidths and backfitting algorithm. Now set $\ell(\mathscr{H}_0) = -(n/2)\log RSS_0$. Fan and Jiang (2005) define the GLR statistic

$$\lambda_n(\mathscr{H}_0) = \ell(\mathscr{H}_1) - \ell(\mathscr{H}_0) = \frac{n}{2}\log\frac{RSS_0}{RSS_1} \approx \frac{n}{2}\frac{RSS_0 - RSS_1}{RSS_1},$$

rejecting when λ_n is too large.

Although this uses normality, Fan and Jiang (2005) identify general hypotheses under which λ_n has a Wilks' theorem under the condition that local polynomial smoothing (of order p_k for $\boldsymbol{X}_k$) with a kernel function K is used as the univariate smoother in the backfitting algorithm. Their hypotheses are reasonable, including (i) K is bounded, with bounded support, and Lipschitz-continuous, (ii) the individual X_ks and the pairs $(X_k, X_{k'})$ have Lipschitz continuous densities, bounded support, and are bounded away from zero, (iii) $p+1$ derivatives of the f_k exist, the moment $E\varepsilon^4 < \infty$, and (iv) $nh_k/\log n \to \infty$ for all k.

Their main theorem is the following.

Theorem (Fan and Jiang, 2005): Under the four conditions, the asymptotic behavior of $\lambda_n(\mathscr{H}_0)$ for testing $\mathscr{H}_l$ against $\mathscr{H}_1$ is given by

$$P\left(\frac{(\lambda_n(\mathscr{H}_0) - \mu_n - d_{1,n})}{\sigma_n} < t|\mathscr{X}\right) \to \Phi(t),$$

where

$$d_{1,n} = \mathcal{O}_p\left(1 + \sum_{k=1}^{p} nh_k^{2p_k+1} + \sum_{k=1}^{p} \sqrt{n}h_k^{p_k+1}\right),$$

in which p_k is the dimension of the local polynomial for the kth variable, $\mathcal{X}$ represents the design points assumed to be chosen so the backfitting algorithm converges, and Φ is the standard normal probability. Also, if $nh_k^{2p_k+1} \to 0$ for $k = 1, ..., p$, then, conditional on $\boldsymbol{X}$, a Wilks' theorem holds:

$$r_K \lambda_n(\mathcal{H}_0) \sim \chi^2_{r_K \mu_n}. \quad \square$$

The expressions for μ_n, σ_n, and r_K are complicated and are given in terms of various evaluations of K. Fan and Jiang (2005), Section 3, give some further simplifications and extensions. They also characterize the rates at which this testing procedure can detect alternatives; these rates indicate how fast the minimum norm of f_i may decrease when testing is at a fixed level α and power at least β.

The fact that normal-based tests continue to perform well suggests that many regression diagnostics also generalize to additive models. In particular, ideas from weighted regression generalize to handle heteroscedasticity. Also, as a pragmatic point, one can use the bootstrap to set pointwise confidence bands on the f_k.

4.1.3 Nonparametric Optimality

In a pair of papers, Stone (1982), Stone (1985) established results that characterize the optimal behavior of nonparametric function estimators and their use in additive models. These results apply not just to estimating a function directly but also to estimating its derivatives; these have been seen when optimizing the bias–variance terms in the MSE over bandwidths in kernel estimation in Section 2.3 for instance. Accordingly, one wants optimal rates of decrease for the norms of the difference between functions and their estimators, even when the function is a derivative. Stone (1982), Theorem 1 shows that if $f(\boldsymbol{x})$ is r times differentiable in $x_1,..., x_p$, with mth derivative denoted $f^{(m)}$ for $m = 0, 1, ..., r$, and has estimators denoted $\widehat{f^{(m)}}$, then

$$||\widehat{f^{(m)}} - f^{(m)}||_q = \mathcal{O}\left(\frac{1}{n^{(r-m)/(2r+p)}}\right), \qquad (4.1.12)$$

where $||\cdot||_q$ is the Lebesgue q norm for some $q \in \mathbb{R}^+$. If $q = \infty$, then the optimal rate is $\mathcal{O}((\log n)/n)^{(r-m)/(2r+p)}$. Clearly, $m = 0$ is the case one wants for estimation of f. IID data from continuous or discrete distribution families are usually included in Stone's conditions, and the optimal rates are achieved by kernel, spline, and nearest-neighbor methods.

Expression (4.1.12) extends to additive models because each component in an additive model can be estimated at an optimal rate so that the overall p-variate estimator has the same rate as the individual univariate estimators. In particular, Stone (1985) shows, under relatively mild conditions, that estimators of the individual components are optimal. That is, when the fitting is done using local polynomials such as splines on $[0,1]$ partitioned into $N_n \approx n^{1/(2r+1)}$ equal intervals, the error is

$$E(||\hat{f}_k^{(m)} - f_k^{(m)}||_2^2|\boldsymbol{X}) = \mathcal{O}_p\left(\frac{1}{n^{2(r-m)/(2r+p)}}\right),$$

for $k = 1, ..., p$, with the anticipated convergence of the constant,

$$E((\bar{Y}\mu)^2|\boldsymbol{X}) = \mathcal{O}_p\left(\frac{1}{n}\right) = \mathcal{O}_p\left(\frac{1}{n^{2r/(2r+p)}}\right).$$

(Stone's norm actually has an extra weight function in it; here this is set to one.) Furthermore, when $m = 0$, Stone (1985) obtains

$$E(||\hat{f} - f||^2|\boldsymbol{X}) = \mathcal{O}_p\left(\frac{1}{n^{2r/(2r+p)}}\right),$$

indicating that the errors merely add. Of course, if p is large, the constant in the $\mathcal{O}$ term will increase, and if one takes limits as p increases, the asymptotic behavior is undetermined. Nevertheless, the point remains that the errors in estimating the terms in additive models are additive.

4.2 Generalized Additive Models

It is clear how the additive model (4.1.1) extends the linear model (2.0.1)). Another way to extend the linear model is by introducing a "link function" in which a function of the conditional mean is modeled by the regression function. These are called generalized linear models (GLMs). The link function generalization can be applied to additive models as well, giving generalized additive models GAMs.

To see this, first recall the definition of GLMs. Concisely, GLMs are formed by replacing Y in (2.0.1) with $g(\mathbb{E}(Y|\boldsymbol{X}))$ to give

$$g(\mathbb{E}(Y|\boldsymbol{X})) = \beta_0 + \beta_1 X_1 + \ldots + \beta_p X_p. \tag{4.2.1}$$

The g is called the link function because if one writes $\mathbb{E}(Y|\boldsymbol{X}) = \mu$ and $g(\mu) = \boldsymbol{X}\beta$, then g is the "link" between the conditional mean of Y given $\boldsymbol{X}$ and a representation in terms of the explanatory variables. As before, $(Y|\boldsymbol{X}) = \mu(\boldsymbol{X}) + \varepsilon$, in which the IID error terms have constant variance σ^2, independent of $x_1, \ldots, x_p$, and usually have normal distributions. Taking $\mathbb{E}$ of both sides and applying g gives the general expression $g(\mathbb{E}(Y|\boldsymbol{X})) = \mu(\boldsymbol{X})$, which in GLMs is represented as in (4.2.1). Alternatively, one can assume g^{-1} exists, apply it to the right side of (4.2.1), and recover an expression for $\mathbb{E}(Y|\boldsymbol{X})$.

This procedure is mathematically valid but conceptually a little awkward because the transformation g applies only to the mean. That is, the variability in $(Y|\boldsymbol{X})$ as a random variable does not necessarily transform by g the same way its mean does. Indeed, $g((Y|\boldsymbol{X})) = g(\mathbb{E}(Y|\boldsymbol{X}) + \varepsilon) \neq g(\mathbb{E}(Y|\boldsymbol{X}))$. Thus, using a regression model on $g((Y|\boldsymbol{X}))$ is not the same as using a regression model on $g(\mathbb{E}(Y|\boldsymbol{X}))$ because the error terms have different meanings. See McCullagh and Nelder (1989) for more details on residual analysis, joint modeling of means, and dispersions.

In practice, GLMs have proved most effective when the response variable Y comes from an exponential family such as the binomial or Poisson. For instance, if Y is Bernoulli, then $\mathbb{E}[Y\,|\boldsymbol{X} = \boldsymbol{x}] = p(\boldsymbol{x}) = \mathbb{P}[Y = 1\,|\,\boldsymbol{x}]$. Then it is natural to set

$$g(p(\boldsymbol{x})) = \text{logit}(p(\boldsymbol{x})) = \ln \frac{p(\boldsymbol{x})}{1 - p(\boldsymbol{x})},$$

which yields logistic regression. McCullagh and Nelder (1989) develop the techniques for this and several other cases, thereby establishing the usefulness of the common GLM structure.

Additive models can be generalized in the same spirit by expressing the link function as an additive, rather than linear, function of $\boldsymbol{x}$. For $\mu(\boldsymbol{X}) = \mathbb{E}(Y|\boldsymbol{X})$, set

$$g(\boldsymbol{\mu}) = \beta_0 + \sum_{j=1}^{p} f_k(x_k), \tag{4.2.2}$$

so the left-hand side of (4.2.2) is a transformation of the conditional mean response variable but everything else is the same as for the additive model.

Domain knowledge is usually required to choose the link function g. For example, the additive version of logistic regression is

$$\text{logit}(p(\boldsymbol{x})) = \beta_0 + \sum_{k=1}^{p} f_k(x_k),$$

and this would often be used when the responses are binary and the probability of a specific response is believed to depend smoothly on the explanatory variables. If $g(\mu) = \log \mu$, then one gets the log-additive model often used for Poisson (count) data. The gamma and negative-binomial sampling models also have natural link functions. Note that all four of these cases are exponential families that lead to generalized linear models which can be extended to GAMs.

Entertainingly, one can choose the functions f_k to depend on two explanatory variables or different numbers of them, or to include a linear model term; a simple case is $g(\mu) = \beta_0 + x_1\beta_1 + \sum_{k=2}^{p} f_k(x_k)$. Factor levels can also be incorporated as indicator functions.

However GAMs arise, the central task is estimating the f_ks. Often they are estimated by the Classical, flexible smoothing methods discussed, namely LOESS, splines, and kernels, even nearest neighbors. To see how this can be done, denote the data by $(\boldsymbol{x}_i, y_i)$ for $i = 1, \ldots, n$, where the y_is are response outcomes and the $\boldsymbol{x}_i$s are explanatory variables, $\boldsymbol{x}_i = (x_{1,i}, \ldots, x_{p,i})$. Now suppose the link function is just the identity (i.e., the goal is to fit $Y = f(\boldsymbol{X}) + \varepsilon$). As seen earlier, finding $\hat{f}$ that minimized $\sum_i (y_i - f(\boldsymbol{x}_i))^2$ would yield an interpolant that was quite rough. However, the roughness of $\hat{f}$ could be reduced either by narrowing the class of functions used in the optimization or by changing the objective function by adding a penalty term.

In the GAM context, write $f(\boldsymbol{x}) = f_1(x_1) + \cdots + f_p(x_p)$ and consider a linear smoother like cubic splines for each f_k. These are piecewise cubic polynomials that arise from optimizing

$$\sum_{i=1}^{n} \left(y_i - \sum_{k=1}^{p} f_k(\boldsymbol{x}_{k,j}) \right)^2 + \sum_{k=1}^{p} \lambda_k \int f_k''(t)^2 dt$$

for fixed λ. The knots of the spline occur at the observed values of each X_k. So, a version of the backfitting algorithm described in the previous section can be applied to obtain $\hat{f}_k$s. The core idea is to use a cubic spline smoother on residuals $y_i - \sum_{k \neq k'} f_k(x_{k,i})$ for each variable $x_{k'}$ in turn, continuing the process until the estimated $\hat{f}_k$s stabilize. As noted, this is a version of the Gauss-Seidel algorithm for solving matrix equations. When the link function is not the identity, the procedure is applied to $g(\boldsymbol{\mu})$ rather than Y, as will be seen below for an additive logistic model.

More formally, Hastie and Tibshirani (1996) note that this procedure is equivalent to solving a set of estimating equations. Let S_k be the (linear) smoothing spline operator for smoothing the kth variable, for instance. Then, the backfitting equations can be written as

$$f_k(x_k) = S_k(y - f_1(x_1) - \ldots - \xi_k - \ldots - f_p(x_p)) \tag{4.2.3}$$

for $k = 1, ..., p$. The collection (4.2.3) is a set of estimating equations, and when S_k is linear, the ξ_k can be found. In other words, the n vectors $(y_i, f_1(x_{1,i}), ..., f_p(x_{p,i}))$ can be found by solving (4.2.3). Other choices for S_k are possible, giving different answers.

To see explicitly what happens when the link function is not the identity consider the case of additive logistic regression, as in Hastie and Tibshirani (1990), Chapter 6. The heart of the matter is writing

$$\log \frac{p(y_i|x_{1,i}, \ldots, x_{p,i})}{1 - p(y_i|x_{1,i}, \ldots, x_{p,i})} = \beta_0 + f_1(x_{1,i}) + \ldots + f_p(x_{p,i})$$

for $i = 1, ..., n$. Using some technique such as linear logistic regression, one can get starting values for β_0 and the f_ks, say $\beta_{0,init}$ and $f_{k,init}$, for each i. This gives $\eta_{i,init} = \beta_{0,init} + \sum_{k=1}^{p} f_{k,init}(x_{k,i})$. Now, one cycle of the backfitting algorithm, with Newton-Raphson, forms adjusted dependent variables, replacing the y_is for $i = 1, \ldots, n$ with

$$z_{1,i} = \eta_{i,init} + \frac{(y_i - p_{i,init})}{p_{i,init}(1 - p_{i,init})},$$

in which the $p_{i,init}$s come from the initial linear logistic model. The weights $w_{i,1} = p_{i,init}(1 - p_{i,init})$ then lead to a new vector $\boldsymbol{\eta}_1 = \boldsymbol{A}_w \boldsymbol{z}_1$, where $\boldsymbol{\eta}_1 = (\eta_{1,1}, \ldots, \eta_{1,n})^{\mathsf{T}}$, $\boldsymbol{z} = (z_{1,1}, ..., z_{1,n})^{\mathsf{T}}$, and $\boldsymbol{A}_w$ is the operator that uses the weights. The procedure is repeated to get the vector $\boldsymbol{\eta}_2$, and so forth, until convergence is observed. Analogous procedures can be developed for other link functions.

4.3 Projection Pursuit Regression

Another extension of the additive model is projection pursuit regression (PPR). The idea is to use linear transformation of the space of explanatory variables so that the mean response can be represented in terms of univariate functions of projections of the explanatory variables. That is, after recoordinatization, one can fit an additive model. The axes in the recoordinatized space are expressed in terms of vectors $\boldsymbol{\beta}_k$. So, the $f_k(\boldsymbol{x})$s in (4.1.2) are replaced by $f_k(\boldsymbol{x}\boldsymbol{\beta})$s. The PPR model is

$$Y = f(\boldsymbol{x}) + \varepsilon = \beta_0 + \sum_{k=1}^{r} f_k(\boldsymbol{x}'\boldsymbol{\beta}_k) + \varepsilon, \tag{4.3.1}$$

in which r replaces p because the number of terms need not be equal to the number of variables. In (4.3.1), one seeks linear combinations of the explanatory variables that give good additive model fits when r is small; the linear combinations that one pursues are projections of the data. There are a variety of techniques for finding r, the $\boldsymbol{\beta}_k$s, and the smooths f_k. This was popularized by Friedman and Stuetzle (1981), but the approach originates with Kruskal (1969).

The original motivation for PPR was to automate the selection of low-dimensional projections of a high-dimensional data cloud, similar to the local dimension discussed in Chapter 1. It is seen that picking out a linear combination is equivalent to choosing a one-dimensional projection of $\boldsymbol{X}$. For example, take $r = 1$, $\boldsymbol{\beta}^{\mathsf{T}} = (1,1)$, and $\boldsymbol{x} \in \mathbb{R}^2$. Write an arbitrary vector $\boldsymbol{x} \in \mathbb{R}^2$ as $\boldsymbol{x} = \boldsymbol{x}_1 + (\boldsymbol{x}_1)^{\perp}$, where $\boldsymbol{x}_1$ is the component of $\boldsymbol{x}$ in the direction of $(1,1)^{\mathsf{T}}$ and $(\boldsymbol{x}_1)^{\perp}$ is its component in the orthogonal complement of $(1,1)^{\mathsf{T}}$. Then, $\boldsymbol{x}\boldsymbol{\beta} = \boldsymbol{x}_1$ because it projects $\boldsymbol{x}_1^{\perp}$ to zero. Since $\boldsymbol{x}_1$ is in the direction of $\boldsymbol{\beta}$, it is of the form $\lambda\boldsymbol{\beta}$ for some $\lambda \neq 0$ giving the space $\mathscr{S}$, as shown in Fig. 4.1. If $r = 1$, then the fitted PPR surface is constant along lines orthogonal to $\mathscr{S}$. If f_1 were the sine function, then the surface would look like a sheet of corrugated aluminum, oriented so that the ridges were perpendicular to $\mathscr{S}$. When $r \geq 2$, the surface is hard to visualize, especially since the $\boldsymbol{\beta}_1, \ldots, \boldsymbol{\beta}_r$ need not be mutually orthogonal. Consequently, one expects PPR to outperform other methods when the primary trend in the data lies in different directions than the natural axes.

The PPR procedure seeks the $\{\hat{f}_j\}$ and $\{\hat{\boldsymbol{\beta}}_j\}$ that minimize

$$\sum_{i=1}^{n} \left[Y_i - \sum_{k=1}^{r} \hat{f}_k(\boldsymbol{x}_i^{\mathsf{T}} \hat{\boldsymbol{\beta}}_k) \right]^2. \tag{4.3.2}$$

Backfitting can be used to estimate the f_ks for fixed β_ks, which are then updated by a Gauss-Newton search. The steps iterate until convergence is observed. Separate from

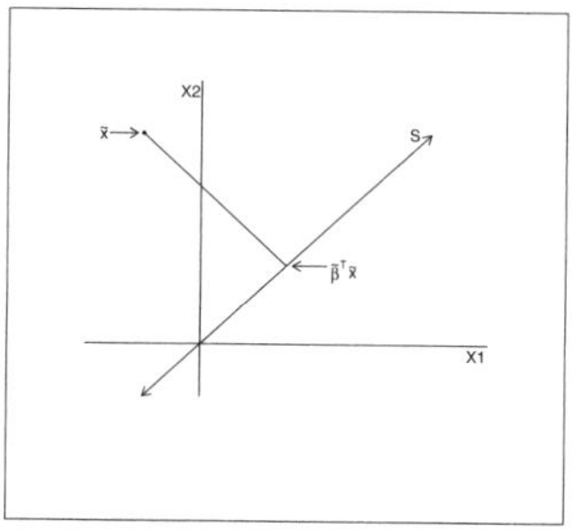

Fig. 4.1 This graph shows an arbitrary $\boldsymbol{x} \in \mathbb{R}^2$ and its projection under $\boldsymbol{\beta} = (1,1)^\top$ onto the line $x_1 = x_2$. The space S is the axis that a function $f(\boldsymbol{x}\boldsymbol{\beta})$ uses to track the response.

this iterative procedure, extra terms are added (by a univariate search on r) until a fitness criterion is satisfied.

The basic PPR procedure, modified from Friedman and Stuetzle (1981), is as follows:

Given $Y_1, ..., Y_n$ and n outcomes of the p-dimensional explanatory variable $\boldsymbol{X}$:

- ☐ Initialize: Start with $\hat{\phi}_r(\boldsymbol{X}) = \sum_{k=1}^r \hat{f}_k(\boldsymbol{X}^\top \hat{\boldsymbol{\beta}}_k)$ as the fitted model for some $r \geq 0$ with $\hat{f}_k$s and $\hat{\boldsymbol{\beta}}_k$s specified. Form the current residuals $e_i = y_i - \hat{\phi}_r(\boldsymbol{x}_i)$ for $i = 1, ...n$. (When $r = 0$, set $e_i = y_i$.)
- ☐ Check fit: Evaluate a goodness of fit measure on ϕ_{r+1} to see if it is worth adding another term. For instance, let S_{r+1} give a smooth representation for the residuals (i.e., univariate nonparametric regression of the residuals on the $\boldsymbol{x}_i^\top \hat{\beta}$s), and set

$$F(\boldsymbol{\beta}_{r+1}) = 1 - \sum_{i=1}^n (e_i - S_{r+1}(\boldsymbol{x}_i^\top \boldsymbol{\beta}_{r+1}))^2 / \sum_{i=1}^n e_i^2.$$

  Then, let $\boldsymbol{\beta}_{r+1} = \arg\max_\beta F(\boldsymbol{\beta})$. (Solving this optimization problem may necessitate backfitting; i.e., cycling through $k = 1, ..., r$ to solve

$$\hat{f}_k(\boldsymbol{x}^\top \hat{\boldsymbol{\beta}}_k) = S\left(\boldsymbol{Y} - \sum_{k' \neq k} \hat{f}_{k'}(\boldsymbol{x}^\top \hat{\boldsymbol{\beta}}_{k'}) \mid \hat{\boldsymbol{\beta}}_k\right)$$

  iteratively.)
- ☐ If $\max F(\boldsymbol{\beta}_{r+1})$ is small enough, the extra term adds little, so stop and do not include it. Otherwise, add an extra term $f_{r+1}(\boldsymbol{x}^\top \beta_{r+1})$, formed by minimizing in (4.3.2). That is, iterate over Gauss-Newton to find an optimal $\boldsymbol{\beta}_{r+1}$ and backfitting to identify the new term. Once convergence is achieved, return to the initialization step.

Hall (1989) used a kernel-based smoother S in his version of the PPR procedure. This enabled him to establish formulas for the variance and bias of kernel-based projection pursuit estimators; see Hall (1989), Section 4. He obtains rates of convergence identical to unidimensional problems, analogous to the way additive models have unidimensional rates. It seems that the major source of error is in the bias of the estimate of the $\boldsymbol{\beta}_k$s.

One can also think of (4.3.1) as a sort of expansion of f in terms that summarize local, lower-dimensional behavior. Then the successive approximation by residuals amounts to finding the next term in the expansion to minimize the error in (4.3.2). Following Huber (1985), Section 9, observe that any square-integrable function f can be approximated in an L^2 sense as in (4.3.1): For appropriate choices of β_k and f_k, one can ensure

$$\int \left(f(\boldsymbol{x}) - \sum_{k=1}^{r} f_k(\boldsymbol{x}^\top \boldsymbol{\beta}_k) \right)^2 d\mathbb{P} \to 0,$$

for various choices of probability $\mathbb{P}$ governing $\boldsymbol{X}$, by any of a wide variety of series expansions (e.g., Fourier). If we already have $\boldsymbol{\beta}_1, \ldots, \boldsymbol{\beta}_r$ and $f_1, \ldots, f_r$, then each iteration of the procedure seeks a minimum of

$$\int e_{r+1}(\boldsymbol{x})^2 d\mathbb{P} = \int \left(f(\boldsymbol{x}) - \sum_{k=1}^{r+1} f_k(\boldsymbol{x}^\top \boldsymbol{\beta}_k) \right)^2 d\mathbb{P}$$

over $\boldsymbol{\beta}_r$ and f_{r+1}. For fixed $\boldsymbol{\beta}_k$s, the minimum under squared error is achieved by $f_{r+1}^{opt}(z) = \mathbb{E}(f(\boldsymbol{X}) - \Sigma_{k=1}^{r+1} f_k(\boldsymbol{X}\boldsymbol{\beta}_k) | \boldsymbol{X}\boldsymbol{\beta}_{r+1} = z)$, where the $\mathbb{P}$ is the distribution of $\boldsymbol{X}$ used in the conditional expectation. Also, the residual norm

$$\mathbb{E}(e_{r+1}(\boldsymbol{x}) - f_{r+1}^{opt}(\boldsymbol{x}^\top \boldsymbol{\beta}_{r+1}))^2 = \mathbb{E}(e_{r+1}^2(\boldsymbol{x})) - \mathbb{E} f_{r+1}^{opt}(\boldsymbol{X})^2$$

is decreased the most by choosing $\boldsymbol{\beta}_{r+1}$ to maximize the marginal norm $\mathbb{E} f_{r+1}^{opt}(\boldsymbol{X})^2$; if $||\boldsymbol{\beta}_{r+1}||$ is a unit vector, then a maximum must exist. This shows that the norm of the term added at the $r+1$ stage, $f_{r+1}^{opt}(\boldsymbol{x})$, goes to zero, although that does not immediately imply that e_{r+1} goes to zero. However, it would be strange if it didn't, and sufficient conditions are given by Jones (1987).

An extra wrinkle with PPR is that representations such as (4.3.1) are not unique. In fact, Huber (1985) notes that when $p = 2$, the function $f(\boldsymbol{x}) = x_1 x_2 = (1/4ab)[(ax_1 + bx_2)^2 - (ax_1 - bx_2)^2$ for any a, b. Thus, f has infinitely many projection pursuit additive models. Despite this, there is a uniqueness result that establishes that the difference between two representations for the same function is a polynomial. In addition, there are functions that cannot be represented as a sum like (4.3.1), such as $f(\boldsymbol{x}) = e^{x_1 x_2}$.

As with the backfitting algorithm, there are numerous variants of PPR. One that is particularly important is due to Chen (1991), who used a polynomial spline smoother for the S in the generic procedure above. This variant of PPR is more complicated than others but permits characterization of the rates of convergence of $\hat{\phi}_r(\cdot)$ to $f(\cdot)$ and verification that optimal rates are achieved.

To present the main result, some careful definitions are needed. First, the r projection vectors $\boldsymbol{\beta}_k$ are assumed to lie in a set $A_r = \{\boldsymbol{\beta}_1, \ldots, \boldsymbol{\beta}_r\} \subset S^{p-1}$, where S^{p-1} is the unit sphere in p dimensions. Also, the angle between any $\boldsymbol{\beta}_k$ and the hyperplane generated by the $\boldsymbol{\beta}_{k'}$s for $k' \neq k$ is assumed bounded below by a constant. Essentially, this ensures the terms in $\hat{\phi}_r$ will not proliferate excessively and lead to redundancy in the regression function. Also, the domain of the $\boldsymbol{X}$s must be restricted to $B_p(0,1)$, the unit ball in p dimensions. The space of permitted approximands is now all sums of polynomial splines of degree q on $[-1,1]$, with equispaced knots at a distance of $2/N$, denoted

$$S(A_r) = S(A_r, q, N) = \left\{ s_{A_r}(\boldsymbol{x}) = \mu + \sum_{k=1}^{r} s_k(\boldsymbol{x}^{\mathsf{T}} \boldsymbol{\beta}_k) \right\}.$$

As a vector space, $S(A_r)$ has finite dimension $rN + r(q-1) + 1$.

For a set $U \supset B_p(0,1)$ and a fixed A_r, let $U(A_r)$ be the p-dimensional preimage of $B_r(0,1)$ in U under the projections in A_r. That is, set

$$U(A_r) = \left\{ \boldsymbol{x} \mid \sum_{\boldsymbol{\beta}_k \in A_r} (\boldsymbol{x}^{\mathsf{T}} \boldsymbol{\beta}_k)^2 \leq 1, \boldsymbol{x} \in U \right\}.$$

Now, the estimator for f is $\hat{\phi}_r \in S(A_r)$, given by

$$\hat{\phi}(\boldsymbol{x}) = \hat{\phi}_{r,n,A_r}(\boldsymbol{x}) = \hat{\mu}_{A_r} + \sum_{k=1}^{r} \hat{\psi}_{k,A_r}(\boldsymbol{x}^{\mathsf{T}} \boldsymbol{\beta}_k), \tag{4.3.3}$$

which achieves

$$\arg \min_{\hat{\phi} \in S(A_r)} \sum_{i=1}^{n} (y_i - \hat{\phi}(\boldsymbol{x}_i))^2 1_{U(A_r)}(\boldsymbol{x}_i),$$

a special case of (4.3.2). Chen (1991) shows that (4.3.3) has a unique solution; let it be defined by the (linear) smoother $S_{n,A_r} = S_{n,A_r,q,N}$.

Finally, Chen (1991) uses the following procedure:

- ☐ For given r and A_r, find $\hat{\phi}_{r,A_r}$.
- ☐ Given $\hat{\phi}_r$, find the residual sum of squares

$$RSS_n(A_r, q, N) = \sum_{i=1}^{n} (y_i - \hat{\phi}_{r,n,A_r}(\boldsymbol{x}_i))^2 1_{B_p(0,1)}(\boldsymbol{x}_i)$$

and the corrected form of it,

$$FPE_n(A_r, q, N) = \frac{n_{B_p(0,1)} + \mathrm{tr}(S_{n,A_r} I_{B_p(0,1)})}{n_{B_p(0,1)} - \mathrm{tr}(S_{n,A_r} I_{B_p(0,1)})} \times \frac{RSS_n(A_r, q, N)}{n_{B_p(0,1)}},$$

where $n_{B_p(0,1)} = \#\{i \mid \boldsymbol{x}_i \in B_p(0,1)\}$ and $I_{B_p(0,1)}$ is the $n \times n$ diagonal matrix with ith diagonal equal to 1 when $\boldsymbol{x}_i \in B_p(0,1)$ and zero otherwise.

□ The estimate $\hat{\phi}_r$ is taken as any of the $\hat{\phi}_{r,A_r}$s with $r \leq p$ that achieve the minimum of $FPE_n(A_r, q, N)$ over the A_rs satisfying the minimum angle requirement.

For this procedure, Chen (1991) establishes optimal convergence. One of the issues is that the models in PPR can have large rs and so be very flexible. This means one must distinguish carefully between finding structure in the data that is true and incorrectly finding structure that has arisen purely by chance. Two of the four major conditions for Chen's theorem help to avoid spurious findings. They are (i) the density of $\boldsymbol{X}$ is bounded away from 0 and infinity on its support and (ii) $\inf_{\boldsymbol{x}} \mathrm{Var}(Y|\boldsymbol{X}=\boldsymbol{x}) > 0$ and, for a large enough τ, $\sup_{\boldsymbol{x}} E(|Y - \phi(\boldsymbol{x})|^\tau | \boldsymbol{X} = \boldsymbol{x})$ is bounded. Condition (i) ensures the smoother is nontrivial, and condition (ii) ensures $y_i - \phi(\boldsymbol{x}_i)$ is not too small.

Theorem (Chen, 1991): In addition to (i) and (ii), assume (iii) that the true function ϕ_0 is of the form (4.3.1) and can be written as

$$\phi_0(\boldsymbol{x}) = \mu_0 + \sum_{k=1}^{r_0} \psi_k(\boldsymbol{x}\boldsymbol{\beta}_k),$$

for a collection of r_0 bounded, q times differentiable, Lipshitz-continuous functions ψ_k of order ℓ, and (iv) that A_{K_0} in ϕ satisfies the minimum angle bound.

Let $N \approx n^{1/(2p+1)}$, where $\nu = q + \ell$. Then

$$\lim_{n\to\infty} \sup_{\substack{f \text{ satisfying (4.3.1)} \\ \text{with } \Sigma f_k \text{ as in (iii)}}} P_f\left(\frac{1}{n}\sum_{i=1}^{n}(\hat{\phi}(\boldsymbol{x}_i) - \phi_0(\boldsymbol{x}_i))^2 1_{B_p(0,1)} \geq \frac{c}{n^{2\nu/(2\nu+1)}}\right) = 0. \quad \square$$

Thus, as long as the true function is representable as a sum of functions of univariate projections, the PPR fit is a consistent estimator of smooth surfaces. Informally, the PPR estimator converges to a unique solution under essentially the same conditions as for the additive model; i.e., the values of the explanatory variables do not lie in the concurvity space of the smoother, and the functions in f are not too small or functions of nearby projections. Incidentally, Chen's theorem also establishes consistency and optimality for purely additive models if the β_k's are chosen to have 1 in the kth place and zeros elsewhere. Note that Chen's method uses splines, obviating backfitting.

Zhao and Atkeson (1991) have a theorem that is conceptually foundational: PPR escapes the Curse on spaces that do not admit a finite-dimensional parametrization if the function space is suitably restricted otherwise; see also Zhao and Atkeson (1994). In their theorem, the restrictions include square differentiability of the target function (which is mild) as well as two other sorts of properties whose restrictiveness is more difficult to assess. One is that the true f can be represented as an integral of ridge functions,

$$f(\boldsymbol{x}) = \int_{\Omega_d} g(\boldsymbol{x}\beta) w(\beta) d\beta,$$

where Ω_d is the unit sphere in d dimensions, equipped with a weight function w on it, and g is some fixed function. The other assumptions are smoothness requirements. The Zhao-Atkeson theorem seems to go beyond its predecessor Barron's theorem (see

Section 4) in that it may reduce to the case for single hidden-layer neural networks if the f_ks are known to be sigmoids. Moreover, PPR regression includes much beyond single hidden-layer neural networks. (General neural networks do not reduce to the PPR case in any obvious way, either.) The proof of Barron's theorem is explained below and has some features in common with the more complicated proof of the Zhao-Atkeson theorem which is not presented here.

Overall, PPR works best when some explanatory variables are commensurate; e.g., in predicting life span, similar biometric measurements might be bundled into one linear combination, and income-related measurements might form another. Also, heuristically, PPR avoids the Curse in a limited sense: Because it is intended for settings in which most of the regions of a high-dimensional space are empty, it focuses on low-dimensional linear projections. This means that the force of the Curse kicks in when the structures are highly nonlinear; i.e., typically only at relatively refined levels of approximation. As can be surmised from the procedure, PPR can be computationally demanding. However, using principal components as the explanatory functions in a regression, see Chapter 9, is a special case of PPR that can be easily implemented, as it is based on the eigenvectors of the estimated covariance matrix.

4.4 Neural Networks

Originally, mathematical neural networks were intended to model the learning and pattern recognition done by physiological neurons. This is described in Hebb (1949), who modelled a synapse as a link from the output of one node to the input of another node, with a weighting related to the correlation in their activity. Thus, groups of neurons could be linked and thinking modeled as the activation of such assemblies. Rosenblatt (1958) continued the Hebb model, focusing on how the links from neuron to neuron could be developed; in particular he proposed the basic mathematical model still used for (artificial) neural networks (NNs). His basic unit was called the perceptron (now called a "node"), which upon receipt of a signal would either respond or not, depending on whether a function exceeded a threshold. Although this model has not been considered applicable by neurophysiologists since the 1970s, the mathematical structure remains of great importance in classification and regression contexts.

The class of NNs is very large and somewhat complicated because it involves network architecture as well as estimation of parameters once the network is fixed. The simplest NN has the form

$$Y = \beta_0 + \sum_{k=1}^{r} \gamma_k \psi(\boldsymbol{x}^{\mathsf{T}} \boldsymbol{\beta}_k + \nu_k) + \varepsilon, \tag{4.4.1}$$

where $\boldsymbol{x}$ is a p-dimensional vector of explanatory variables, the β_ks are projection vectors as in PPR and ν_k shifts the argument of the sigmoid function ψ to locate the projected vectors in the right place. The typical choice for the sigmoid function is

$$\psi(\boldsymbol{x}) = \psi_{\nu,\beta}(\boldsymbol{x}) = \frac{1}{1+\exp(\nu+\boldsymbol{x}^\top\boldsymbol{\beta})}, \tag{4.4.2}$$

which is shown in Fig. 4.2. More generally, any nondecreasing $\mathbb{R}$-valued function ψ satisfying $\psi(t) \to 1,0$ as $t \to \pm\infty$ will do.

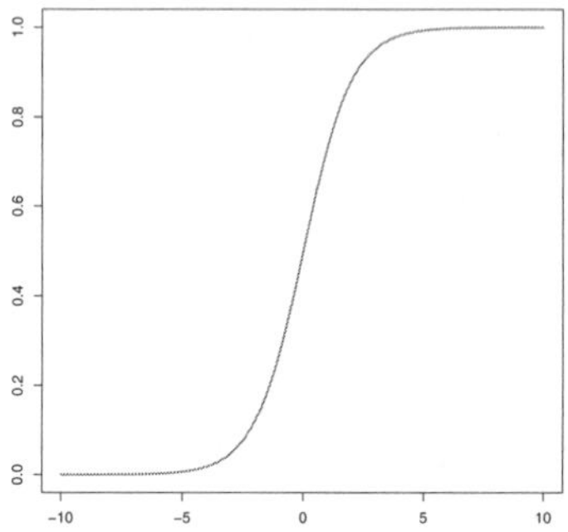

Fig. 4.2 The logistic sigmoid from (4.4.2) is one choice for the node function.

The network associated with (4.4.1) is shown in Fig. 4.3. It has a single "hidden layer" with r nodes, each corresponding to a term $\psi_k(\cdot)$. The layer is called hidden because only the linear combination of their outputs is actually seen as the sum Σ. The network is feedforward in the sense that the outputs from the hidden layer do not affect the earlier inputs. The p-dimensional input $\boldsymbol{X}$ is written as having been partitioned into N subvectors $\underline{X}_j$ so $\boldsymbol{X} = (\underline{X}_1, \ldots, \underline{X}_N)$. The $\underline{X}_j$s represent blocks of data that can be treated differently if desired. Proponents of NNs argue one of the main strengths of NNs is their ability to accommodate multitype data.

Like GAMs and PPR, single hidden layer NNs are variants on basic additive models. GAMs, for instance, merely adds a link function. Note that single hidden layer NNs are a special case of PPR – regard the node functions as in (4.4.2) as fixed versions of the univariate f_ks so that only the projection vectors need to be identified. The general relationship between multi-layer NNs and PPR is less clear: In NNs the functions are fixed and there are more parameters while in PPR the functions must be estimated but there are fewer parameters. So, it is difficult to compare the spaces they span.

Overall, however, feedforward NNs, like PPR, are best regarded as a rich class for nonlinear regression. Indeed, NN structures go far beyond additive models by iterating the composition of nodes. For instance, a two hidden layer neural network with s nodes in the second hidden layer extends (4.4.1) by treating the r outputs from the first hidden layer as inputs to another layer of nodes. That is, set $\psi_k = \psi(\boldsymbol{x}\boldsymbol{\beta}_{1,k} + \nu_{1,k})$ for $k = 1, \ldots, r = r_1$ and form

$$\phi_j = \psi((\psi_1, \ldots, \psi_r) \cdot \boldsymbol{\beta}_{2,j} + \nu_{2,j}) \tag{4.4.3}$$

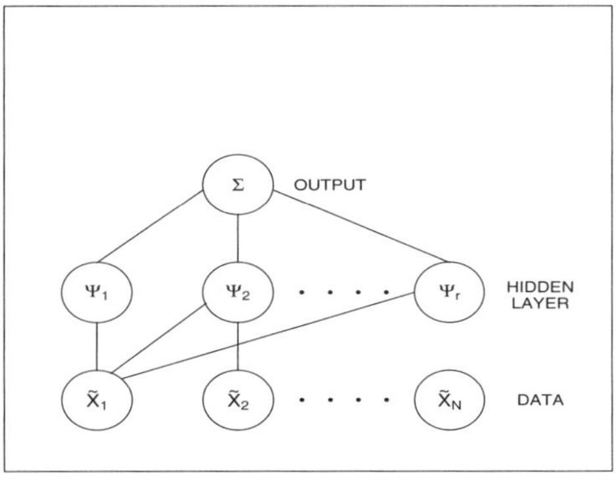

Fig. 4.3 Data of N types is fed into the first (and here only) hidden layer. Each node ψ_k in the hidden layer is the composition of the sigmoid ψ with an affine function of $\boldsymbol{X}$ defined by $\boldsymbol{\beta}_k$ and ν_k. The outputs from the hidden layer are combined linearly to give the overall output Σ.

for $j = 1, ..., r_2$, where the $\boldsymbol{\beta}_{2,j}$s are r_1-dimensional and the $\nu_{2,j}$s are real. Then the overall output from the second layer is

$$\Sigma = \beta_0' + \sum_{j=1}^{r_2} \gamma_{2,j}\phi_j.$$

That is, a single hidden layer NN structure is fed into the second hidden layer, which is combined as before. Evidently, this can be repeated to form many layers, and the structure becomes even more complicated if the output of one layer need not be fed forward to the next layer but can skip one or more layers. A class of NNs that includes more than one hidden layer is a reasonable class of models for nonlinear regression but does not obviously contain PPR or additive models, nor do they necessarily contain NNs with more than one hidden layer.

Consequently, when implementing an NN model, one often must choose the network architecture first before estimating the coefficients within each layer. Architecture selection, r in the case of (4.4.1), can be done by cross-validation or by simulated annealing (both discussed briefly in Chapter 1). Even so, it is often important to limit the number of nonzero parameters; this can be done by regularization (also sometimes called shrinkage methods), which amounts to penalizing the squared error, usually by λ times a squared error penalty on the parameters.

Observe that multilayer neural nets reduce to linear models if ψ is linear. Indeed, if the input vector to the first layer is X, then the output of a first-layer node is $Y = W_1X$ for weights W_1. If this is fed into a linear second-layer node, then the output is $Z = W_2W_1X$, again a linear function of the inputs. Repeating this under squared error makes the overall SSE of fitting the NN, $\hat{E}$, reduce to the usual SSE in linear models, although the linearity only holds if the parameters are redefined to absorb the γ_k's into the β_ks. Another choice for ψ is thresholding, so that the output of a node will be, say, 0 or 1, depending on the value of its argument, say $\sum w_i x_i$. This is often not very good because thresholding is relatively inflexible. As but one instance, thresholds are unable to express an "exclusive or" function using linear functions of its inputs. For instance, if we want a large value when either $x_1 = 1$ or $x_2 = 1$ but not when $x_1 = x_2 = 0, 1$, a simple threshold will often assign 1 when $x_1 = x_2 = 1$ or $x_1 = x_2 = 0$.

Since the point of NNs is to make use of a rich class of functions, there are many theorems about their ability to approximate functions; one will be stated in Subsection 4.4.3 . Some experts take the view that single hidden layer neural networks are a large enough class because large enough sums of them can approximate essentially any function. Others argue that neural network performance can be significantly improved in many applications by careful selection within larger classes of NN models. In other words, the decrease in bias from using a deeper net is predictively better than combining many nodes in a single layer. It is an open question how the variability from model selection (either the number of nodes or the configuration of the multilayer net) affects the prediction, but some authors have examined this question by using simulated annealing repeatedly to assess the impact of model selection. A good overview of these issues can be found in Bullinaria (2004). Here, it will be enough to outline the main computational procedure for estimation, called backpropagation, discuss some straightforward aspects of inference and approximation, and see that, surprisingly, NNs evade the Curse, at least in a formal sense.

4.4.1 Backpropagation and Inference

If the number of nodes, r in (4.4.1) or (r,s) in (4.4.3) for instance, is assumed big enough that the degree of approximation is adequate (i.e., model misspecification can be neglected), then the only task is to estimate the parameters. If one must estimate all the parameters, the main technique is called backpropagation; it is an example of a gradient descent. "Backprop", sometimes called a general delta rule, is an iterative fitting technique based on improving an initial estimate by shifting it in a direction in the parameter space along which the empirical error decreases.

Start with a measure of performance such as the sum of squared errors

$$\hat{E}(\boldsymbol{W}) = \sum_{i=1}^{n} (y_i - \mathrm{Net}(x_i, \boldsymbol{W}),)^2 \tag{4.4.4}$$

in which Net is the function described by the NN and W generically indicates the parameters in the NN. In (4.4.1), W is β_0, the $\boldsymbol{\beta}_k$s, the ν_ks, and the γ_ks. For (4.4.3), W includes β_o', the $\boldsymbol{\beta}_{2,j}$s, the $\nu_{2,j}$s, and the $\gamma_{2,j}$s as well.

In general, estimators $\hat{w}_n$ satisfying $(1/n)\sum_{i=1}^{n} m(Z_i, \hat{w}_n) \to 0$ are consistent for a solution to $\mathbb{E}\, m(Z_i, w) = 0$, where $Z_i = (Y_i, \boldsymbol{X}_i)$ and m represents a general optimality criterion. If m is chosen to represent squared error as in (4.4.4), then it is possible to find $\hat{w} = \arg\min_{\boldsymbol{W}} \hat{E}(\boldsymbol{W})$ that is consistent for the minimizing $w_{opt} = \arg\min_{\boldsymbol{W}} \mathbb{E}(Y - \text{Net}(X, \boldsymbol{W}))^2$; see White (1981). Moreover, such $\hat{w}$s are typically asymptotically normal. Indeed,

$$\sqrt{n}(\hat{w} - w) \to N(0, A^{-1}BA^{-1})$$

in distribution, where $A = \mathbb{E}\, \nabla^2 \hat{E}(\boldsymbol{w}_{opt})$, and $B = \text{Var}(\sqrt{n}\nabla \hat{E}(\boldsymbol{w}_{opt}))$. Consistent estimators are given in White (1989) as $\hat{A} = \nabla^2 \hat{E}(\hat{\boldsymbol{w}})$ and

$$\hat{B} = (1/n) \sum_{i=1}^{n} \nabla \text{Net}(x_i, \hat{\boldsymbol{w}})' \nabla \text{Net}(x_i, \hat{\boldsymbol{w}})(y_i - \text{Net}(x_i, \hat{w}))^2.$$

Despite the seeming simplicity of this procedure, it can be computationally demanding because the error surface as a function of $\boldsymbol{W}$ is very complicated, with many local maxima and minima in which a solution can get trapped. Even so, a method based on Taylor expansions (i.e., gradient descent) or backpropagation is often used. Start with a guess $\boldsymbol{w}_0$ for the value of $\boldsymbol{w}_{opt}$. Choose a sequence of "learning rates" $\eta = \eta_k > 0$ and consider the recursion

$$\boldsymbol{w}_i = \boldsymbol{w}_{i-1} + \eta (\nabla \textbf{Net})^{\mathsf{T}}_{i-1}(y_i - \textbf{Net}_{i-1}), \tag{4.4.5}$$

in which $\textbf{Net}_{i-1} = \text{Net}(x_i, \boldsymbol{w}_{i-1})$, and $\nabla \textbf{Net}_{i-1}$ is the Jacobian matrix from $\textbf{Net}_{i-1}$. The factor η is a learning rate in the sense that it specifies how much $\boldsymbol{w}_i$ changes as a consequence of the Newton updates; in some cases $\eta = \eta_i$. Note that this version of backprop implicitly does as many iterations as there are data points, and the formal theorems assume n increases indefinitely.

A more explicit form for this local gradient descent can be derived for the case where the nodes in the network are fully connected from layer to layer in a grid. Thus, each node $j = 1, ..., r_\ell$ at layer ℓ, $\ell = 1, \ldots, L$ receives inputs from all $r_{\ell-1}$ nodes at the earlier layer but from no other nodes. If the βs are suppressed by incorporating them into a coefficient on a constant variable at each layer, then the generic node function is

$$\psi_{\ell,j} = \psi\left(\sum_{u=1}^{r_\ell - 1} \psi_{\ell-1,u} w_{\ell,j,u} \right) \tag{4.4.6}$$

for $j = 1, ..., r_\ell$ and weights $\boldsymbol{W} = (w_{\ell,j,u} | \ell, j, u)$. Now, if there are L layers, the error can be written in terms of the Lth layer as

$$\hat{E}(\boldsymbol{W}) = \sum_{i=1}^{n} \left(y_i - \sum_{u=1}^{r_L} \psi_{L,u} w_{L,u} \right)^2 \tag{4.4.7}$$

since $r_L = 1$ means there is only one node function after the Lth layer to match the response Y. Clearly, (4.4.6) can be substituted into (4.4.7) to effect the composition of functions. So, for instance, the outputs from the second layer of a two-layer network that get linearly combined to fit the Y_is are

$$\psi_{2,j}(\cdot) = \psi\left(\sum_{u=1}^{r_1} \psi_{1,u} w_{1,j,u}\right) = \psi\left(\sum_{u=1}^{r_1} \psi\left(\sum_{v=1}^{N} \underline{X}_v w_{0,u,v}\right) w_{1,j,u}\right) \tag{4.4.8}$$

for $j = 1, ..., r_2$. In this form, it is easy to see how to take the partial derivatives of $\psi_{2,j}$, and hence $E(\boldsymbol{W})$, with respect to any of the weights $w_{r_\ell,j,u}$.

It is seen that there are two sorts of backprop algorithms. The first, from (4.4.5), assumes an infinite sequence of variables and uses their average properties to see that they converge to a minimum. If the minimum is unique, then the process is convergent to a useful estimate. More typically, the roughness of the error surface for a fixed sample as a function of the parameter w is so high that there are numerous minima. So, the theoretical backprop cycles through all the minima like a mixture over the limit points. The usefulness of this construction is the formal obtention of consistency and asymptotic normality.

The second sort of backprop is pure gradient descent. This rests on Newton-Raphson, and the iteration is over the location of w_i as determined by derivatives of the empirical error, not cycling over the individual data points (which would formally limit the procedure to n iterations unless repetitions were allowed). This kind of procedure takes derivatives explicitly in (4.4.7) for use in (4.4.5) for the Newton step so as to adjust the weights $\boldsymbol{W}$ to reduce $\hat{E}(\boldsymbol{W})$. It has a tendency to converge to a unique limit (which is good) but does not deal with the possibility that the limit is purely local, not global (which is bad). In practice, a variety of starting values w_0 can be used to search over limits in an effort to ensure a global minimum has been found.

The central idea in backprop is to make a change in one or more of the weights so that $\hat{E}(W)$ is reduced. A large change in a weight $w_{\ell,j,u}$ only makes sense if (i) there is a big discrepancy between the actual output and desired output of a node, (ii) the discrepancy depends on the weight $w_{\ell,j,u}$, and (iii) the change in the weights leads to a correspondingly large change in $\hat{E}$. In stating these conditions, model uncertainty means that the true regression function is arbitrary and need not be any neural network close to the current estimate. Since $w_{\ell,j,i}$ is the weight on a connection between a node in layer ℓ and a node in layer $\ell - 1$, a change in the input to a node results in a change in its output that depends on the slope of the function, its sigmoid in particular. The steeper the sigmoid, the faster the learning but the harder it is to get stable values.

Alternatively, and more typically, a penalty function (usually the sum of squared parameters) is added to $\hat{E}(\boldsymbol{W})$ so that the solution is smoothed, as with splines. This decreases the chance that a solution will get trapped in a local minimum. Because the landscape (i.e., the error as a function of the architecture and parameter values) of NNs is so rough, this is a major problem for NNs. Indeed, there may be changes in architecture and parameter values that are well within any reasonable confidence regions that give a substantially better fit and prediction. The penalty term ensures the objective function is more bowl shaped, for instance in the squared error case, so that the bottom of the bowl is likely to represent a better model.

To obtain formal statements for the consistency and asymptotic normality of backprop estimators from (4.4.5), at least for single hidden layer NNs, a small detour into recursive m-estimators is helpful; results for backprop will be special cases. To state this, let

$Z_i = (Y_i, \boldsymbol{X}_i)$ be a sequence of IID $1+p$-dimensional random vectors with Euclidean norm $\|Z_i\| \leq \Delta < \infty$ and let $m : \mathbb{R}^{1+p} \times \mathbb{R}^\ell \to \mathbb{R}^\ell$ be smoothly differentiable with mean $M(w) = \mathbb{E}m(Z_1, w) < \infty$ for $w \in \mathbb{R}^\ell$. Given η_i and an initial w_0, the recursive m-estimator at the ith step is

$$\tilde{w}_i = \tilde{w}_{i-1} + \eta_i m(Z_i, \tilde{w}_{i-1}).$$

The next result is from White (1989), who credits Huber (1967) and Ljung (1977).

Theorem (White, 1989): Suppose $< \eta_j >$ is a divergent sequence so that

$$\sum_{i=1}^{\infty} \eta_i = \infty, \quad \sup_i \left(\frac{1}{\eta_i} - \frac{1}{\eta_{i-1}} \right) < \infty, \quad \text{and} \quad \sum_{i=1}^{\infty} \eta_i^r < \infty,$$

for some $r > 1$, and suppose that there is a smooth $Q : \mathbb{R}^\ell \to \mathbb{R}$ so that, $\forall w \in \mathbb{R}^\ell$, the inequality $(\nabla Q(w))M(w) \leq 0$ holds.

Then, either $\tilde{w}_i \to \Omega \equiv \{w : (\nabla Q(w))M(w) = 0\}$ as $n \to \infty$, in the sense that $\inf \|w - \tilde{w}_i\| \to 0$ for $w \in \{w : \nabla Q(w)M(w) = 0\}$, or $\tilde{w}_i \to \infty$, with probability 1. □

Extensions to this theorem ensure $M(w^*) = 0$ for limit points w^* of $\tilde{w}_i$ and that $\tilde{w}_i$ tends to a local minimum of $Q(w)$. Moreover, White (1989) establishes the results for the case of multidimensional outputs.

Now, the consistency of backprop in single hidden layer NNs is guaranteed by the following. Recall that the empirical error is $\hat{E}(w)$, and $\nabla \text{Net}(\boldsymbol{x}, w)$ is the $1 \times \ell$ Jacobian matrix of Net wrt $\boldsymbol{w}$. Set $E(w) = \mathbb{E}L_i(w)$, where $L_i(w) = (y_i - \text{Net}(\boldsymbol{x}_i, w))^2$, so that $\nabla L_i(w) = -2\nabla \text{Net}(y_i - \text{Net}(\boldsymbol{x}_i, w))$. For notational convenience, set

$$\nabla L_i^* = \nabla L_i(w^*), \quad \widetilde{\text{Net}}_i = \widetilde{\text{Net}}(\boldsymbol{x}_i, \tilde{w}_{i-1}), \quad \text{and} \quad \widetilde{\nabla \text{Net}}_i = \nabla(\widetilde{\text{Net}}(\boldsymbol{x}_i, \tilde{w}_{i-1})),$$

so that (4.4.5) becomes

$$\tilde{w}_i = \tilde{w}_{i-1} + \eta_i (\widetilde{\nabla \text{Net}}_{i-1}^{\mathsf{T}})(y_i - \widetilde{\text{Net}}_{i-1}). \tag{4.4.9}$$

Corollary (White, 1989): Assume (i) $Z_i = (Y_i, \boldsymbol{X}_i)$ are IID $1+p$ dimensional random vectors, (ii) the output $\text{Net}(\boldsymbol{x}, w)$ from the single hidden layer NN to fit Y is of the form (4.4.8) with $r_2 = 1$, or, equivalently, ψ on a linear combination of outputs from (4.4.3), and (iii) the sequence $< \eta_i >$ satisfies the conditions of the theorem.

Then, the backprop procedure in (4.4.9) has iterates $\tilde{w}_i$ that converge to

$$\Omega^* = \{w : \mathbb{E}\, \nabla L_n(w) = 0\}$$

with probability 1 or diverge to ∞ with probability 1. Moreover, if $E(w)$ has isolated stationary points with $J^* = \mathbb{E}((\nabla L_i^*)^{\mathsf{T}} \nabla L_i^*)$ positive definite for $w^* \in \Omega^*$, then $\tilde{w}_i$ converges to a local minimum of $E(w)$ with probability 1 or to ∞ with probability 1. □

Note that White's result is for a single output formed from the nodes in a single hidden layer NN. The reason is that White's result extends to multidimensional Y. This suggests the whole framework can be extended to multiple hidden layers. However, such an extension would be difficult involving theory from nonlinear least squares estimation. White's formulation also permits the analogous conclusion using different sigmoids at different nodes or no outer sigmoid. Indeed, the setting of Barron (1993) discussed in the next subsection does not use such a final sigmoid and demonstrates how to estimate the NN with risk $\mathscr{O}(1/n)$. Of course, it is easy to see that for any sigmoid of a linear combination of first-layer outputs, there will be a linear combination of a (possibly larger) collection of first-layer nodes that can approximate it to any desired accuracy.

Given the consistency for backprop from the proposition, a result for asymptotic normality can be stated. The proof involves techniques from Gaussian processes that are not central to the development here, for which reason they are omitted.

Theorem (White, 1989): Strengthen assumption (ii) of the corollary to ensure that the derivatives of the output function Net and its constituents exist and are bounded. In place of (iii), assume $\eta_i = \delta/n$ for some $\delta > 0$. Then assume that $\tilde{w}_i \to w^*$ a.s. for a stationary point w^* of $E(w)$ and that J^* is PD. If $\delta > 1/2\lambda^*$, where λ^* is the smallest eigenvalue of $\nabla^2 E(w^*)$, then

$$\sqrt{n}(\tilde{w}_i - w^*) \to N(0, PHP^{-1})$$

in distribution, where P is the orthogonal matrix such that $P\Lambda P^{-1} = \nabla^2 E(w^*)$, in which Λ is the diagonal matrix containing the eigenvalues $\lambda_1, ..., \lambda_\ell$ of $\nabla^2 E(w^*)$ in decreasing order and H is the $\ell \times \ell$ matrix with elements $h_{i,j} = \delta^2(\delta\lambda_i + \delta\lambda_j - 1)^{-1} K^*_{i,j}$ for $i, j = 1, ..., \ell$, where the matrix $K^* = [K_{i,j}] = P^{-1} J^* P$. □

It is seen that this result can be difficult to apply. Indeed, such results are necessarily difficult because although a given true function f has a unique representation in terms of a limit of single-layer feedforward neural nets, the parameters defining the approximation at each step may be quite different. In practice, it's as if several different parameter vectors give the same functional form but cannot be distinguished. Even when this can be avoided, as a generality, $PHP^{-1} - A^{-1}BA^{-1}$ is positive semidefinite; see White (1989), Section 5. Thus, in contrast to the nonlinear least squares estimator, backprop is not efficient because $A^{-1}BA^{-1}$ is the best possible. There are ways to improve backprop, but they are beyond the present scope. Despite this, backprop as a technique is often used because it does readily give point estimates and bootstrapping can be used to indicate precision. In all of these methods, it is unclear in general how quickly the asymptotics backprop, nonlinear least squares, risk-based methods, and so forth become dominant.

Germane to the problem of identifying NNs is the asymptotic behavior of least squares estimators in nonlinear regression settings, namely consistency and asymptotic normality. Standard results of this sort, for general nonlinear models, can be found in Gallant (1987). One problem is that asymptotic normality in NNs can occur regardless of whether the model whose parameters are being estimated is true. Consequently, White (1981), Section 4 develops goodness of fit analogous to the χ^2 test but for general models, based on squared residuals.

There is also a well-developed Bayesian theory of NNs; it rests on putting priors on the number of node functions and their architecture while also assigning priors to the parameters in each NN model. As is typical in this kind of Bayes context, the computational implementation to find the posterior is a major challenge. See Lee (2004) for a good treatment.

4.4.2 Barron's Result and the Curse

In 1991, Andrew Barron startled statistics by showing that neural networks can evade the Curse of Dimensionality. Because NN can be related to other classes of models, analogous results are expected for other settings. Indeed, Zhao and Atkeson (1991), Zhao and Atkeson (1994) give such a result for PPR. Here, an intuitive sketch for Barron's theorem is given; the full result is stated at the end of this section and proved in the Notes at the end of this chapter.

Recall that for function estimates $\hat{f}(\boldsymbol{x})$ of a true function $f(\boldsymbol{x})$, one typical measure of distance is the

$$\text{MISE}[\hat{f}] = \mathbb{E}_F \left[\int [\hat{f}(\boldsymbol{x}) - f(\boldsymbol{x})]^2 \, d\boldsymbol{x} \right],$$

where the expectation is taken with respect to the randomness in the data $\{(Y_i, \boldsymbol{X}_i)\}$.

Before Barron (1991), it had been thought that the Curse implied that, for any regression procedure, the MISE had to grow faster than linearly in p, the dimension of the data. Barron showed that neural networks could achieve an MISE of order $\mathscr{O}(r^{-1}) + \mathscr{O}(rp/n) \ln n$, where r is the number of hidden nodes.

Barron's theorem is a tour de force. It applies to the class of functions $f \in \Gamma_c$ on $\mathbb{R}^p$ whose Fourier transforms $\tilde{g}(\omega)$ satisfy

$$\int |\omega| \tilde{g}(\omega) \, d\omega \leq c,$$

where the integral is in the complex domain and $|\cdot|$ denotes the complex modulus. The importance of the class Γ_c is that it is thick, meaning that it cannot be parametrized by a finite-dimensional parameter. However, as it is defined in terms of Fourier transformations, Barron's set is not the full nonparametric function space. It may be best regarded as an unusually flexible version of a parametric model. Indeed, it excludes important functions such as hyperflats. However, it contains open sets in the topology.

With much oversimplification, the strategy in Barron's proof is:

- Show that, for all $f \in \Gamma_c$, there exists a neural net approximation $\hat{f}^*$ such that $\|f - \hat{f}^*\|^2 \leq c^*/n$.
- Show that the MISE in estimating any of the approximations is bounded.
- Combine these results to obtain a bound on the MISE of a neural net estimate $\hat{f}$ for an arbitrary $f \in \Gamma_c$.

Barron's theorem ensures that these NNs can approximate any element in a large space with n terms to order $O(1/n)$ independently of the dimension p of $\boldsymbol{x}$.

Since PPR is a generalization of single-layer feedforward NNs, observe that if a function f on $\mathbb{R}^p$ admits a representation in both function spaces, we have that

$$f(x) = \sum_{k=1}^{r} f_k(\boldsymbol{x}'\theta_k) = \sum_{k=1}^{r} \gamma_k \psi(\beta_k x + \nu_k).$$

PPR permits r distinct functions, absorbing the coefficients γ_k and the locations ν_k into them, while NNs restrict the generality of the representation. The function space one would associate with PPR is clearly larger. Since the Zhao and Atkeson (1991) approach relies on smoothness classes, it is unclear how much larger their function space is than the collection of all NNs. The import of their result is that, like NNs, one can obtain rates of convergence of an L_2 error that goes to zero independently of p when the number of terms in the PPR sum is n and the rate is $\mathcal{O}(1/n)$.

Indeed, one can use Barron's theorem on each term in the PPR sum of f to get rm terms (m nodes for each NN that approximates an f_k) that are needed to approximate f to order $O(1/r)$, provided the functions f_k are in Γ_c, the space of functions used in Barron's Theorem. This will give a generalization to a class of PPRs, but is a weaker statement with a narrower domain of application.

4.4.3 Approximation Properties

Neural nets are an exceedingly rich class of models that can be very unstable because very different NNs may fit the same set of data equally well. In other words, small differences in the data, or estimation procedure can lead to very different networks with large differences in performance. This arises because they are so flexible: The number of nodes, the architecture, the large number of parameters, and the sigmoid function can all be chosen from very broad classes. This instability is one reason NNs are so hard to interpret and why regularization is so important. In Chapter 7, the computed examples will show that even with regularization NNs give fairly irregular curves.

NNs can also be used in a classification context. Binary classification can formally be regarded as a special case of function estimation in which the function to be estimated takes only values ± 1 so the task is to identify the set on which one of the values is assumed. Thus, NNs can be used to find the decision boundary. For simplicity, assume the Y_i's are binary, taking values -1, $+1$, and the model is of the usual form $Y = f(\boldsymbol{x}) + \varepsilon$. (Technically, this model is incorrect. However, it is useful.) A NN classifier should provide an estimate $\hat{f}$, from a class of NNs, that tries to express Y as a step function of the features with regions for -1 and $+1$.

To see that NNs can do classification as well as regression, it is enough to show an approximation theorem: Essentially any continuous function can be approximated by an NN of sufficiently large and complicated structure in any "reasonable" measure of distance in any function space. One way to state this more formally, for single hidden layer NNs, is the following.

Theorem: For any sigmoid function ψ, any continuous function $f(\boldsymbol{x})$ for $\boldsymbol{x} \in [0,1]^p$, and any $\varepsilon > 0$, there is an integer r and real constants γ_k, $\beta_{k,j}$, and ν_k for $j = 1,...,p$ and $k = 1,...,r$ such that

$$\left| f(x_1,...,x_p) - \sum_{k=1}^{r} \gamma_k \psi\left(\sum_{j=1}^{p} \beta_{k,j} \boldsymbol{x}_j - \nu_k \right) \right| < \varepsilon. \quad \square$$

Note that this theorem only requires single hidden layer networks, a relatively small subclass of all NNs.

It is obvious that when the true function f assumes values ± 1 and $\{f(\boldsymbol{x}) = 1\}$ has a smooth boundary that NNs can give a continuous approximation ψ to f that is within any preassigned ε of f away from the boundary. Then, one can replace the continuous approximation ψ with

$$\psi_{class}(\boldsymbol{x}) = \begin{cases} 1 & \text{if } \psi(\boldsymbol{x}) \geq 0, \\ -1 & \text{if } \psi(\boldsymbol{x}) < 0. \end{cases}$$

Alternatively, one can choose a sequence of continuous functions ψ_ℓ converging to the discontinuous function f as $\ell \to \infty$. Thus, step functions with nice boundaries can be approximated in a limiting sense.

Even when cross-validation or simulated annealing can be used to fix a network architecture and gradient descent minimizing $E(W) = \sum_{i=1}^{n}(y_i - \text{Net}(x_i, W))^2$ can be used to give the weights in an NN, the results may not be satisfactory. One way to investigate this is to examine the old nemeses of bias and variance. Recall that from standard decision theory the optimal theoretical minimizer under squared error is the conditional mean. So, if possible, $g(x) = \mathbb{E}(Y|x)$ is a natural choice to be estimated by finding the weights (and architecture) achieving the minimum of $(\mathbb{E}(Y|x) - \text{Net}(x,W))^2$ on average over random data D. That is, write

$$\begin{aligned} \mathbb{E}(\mathbb{E}(Y|x) - \text{Net}(x,W,D))^2 &= (\mathbb{E}(Y|x) - \mathbb{E}\text{Net}(x,W,D))^2 \\ &\quad + \mathbb{E}(\mathbb{E}\text{Net}(x,W,D) - \text{Net}(x,W,D))^2. \end{aligned}$$

The first term is a bias, or approximation error, and the second is the variance of the approximation $\text{Net}(x,W,D)$ over possible data sets D. Thus, the optimal net, as ever, is a trade-off between variance and bias. The two extreme cases would be a constant network that has zero variance but terrible bias and a network so large that for D with sample size n it could fit every data point perfectly giving zero bias but terrible variance. In practice, one deals with these by using networks that are neither too big nor too small and tries to ensure that the error is small over the range of NNs that can be estimated adequately. As will be seen in the computations of Chapter 7, regularization, or penalty terms, must be used on NN models to help stabilize variability.

A standard criticism of NNs, apart from instability, is that they are hard to interpret physically. However, there are partial answers to this. For instance, once an NN has been obtained, one may want to eliminate irrelevant explanatory variables and re-estimate the NN – with weights alone or weights and architecture together. Then, reversing the procedure, one can use the NN to partition the data by looking at the outputs from the last hidden layer. If these clusters made sense, then the model would be partially validated. This procedure can be repeated on any layer in the NN as a check on how the data are grouped.

Another strategy for developing interpretations for NNs is given in Feraud and Clerot (2002). Their definitions help formalize stability and plausibility arguments and can be used in both classification and regression settings even though they are more natural for classification. First, one can investigate the optimality of the NN by looking at the second derivative of the error. For brevity, write the NN function as $f(\boldsymbol{x})$. Then, for the kth value of input i,

$$f_{k,i}(a) = f(x_1^k, ..., x_i^k + a, ..., x_p^k)$$

represents the stability of the class or value assigned as a varies. An extension of this called the causal importance is

$$CI(a|\boldsymbol{x}_i, f) = \int_{x_i} f_i(x_i + a) p_i(x_i) dx_i,$$

where p_i is the marginal distribution of the ith explanatory variable and f_i is $f_{k,i}$ for an arbitrary location (not necessarily a data point). Once irrelevant or correlated variables have been removed, the "saliency" of each input is

$$S(\boldsymbol{x}_i|f) = \int_a \left| \int_{x_i} p_i(x_i)[f_i(x_i + a) - f_i(x_i)] dx_i \right| da,$$

or, to take into account the input values,

$$S(x_i|f) = \int_a \left| \int_{x_i} p_i(x_i) p(a|x_i)[f_i(x_i + a) - f_i(x_i)] dx_i \right| da.$$

While these definitions do not themselves give a general interpretation for parameters or architectures, they do assess key properties, akin to specificity and sensitivity, that help characterize the robustness of an NN under local perturbations.

4.4.4 Barron's Theorem: Formal Statement

Recall that Barron's theorem is a formal demonstration that NNs escape the Curse of Dimensionality at least for some function classes that have an interior and are not parametrizable by any finite number of parameters. To present Barron's result, let $f_r(\boldsymbol{x})$ be the right-hand side of (4.4.1), the generic expression for a single hidden layer feed-forward NN, suppressing the weights W for concision. The Curse will be evaded if the approximation error, as a function of the number of terms r, is of order $\mathcal{O}(1/r)$, for, if so, the number of parameters to be estimated is $(p+2)r+1$, linear in r, which can be estimated with error $\mathcal{O}(r)$. Other regression techniques have an approximation error rate of $(1/r)^{(2/p)}$ (i.e., slower and dependent on p) or need exponentially many parameters, see Barron (1993).

The generic result is that, for a large class of functions f on $\mathbb{R}^p$, for each r an f_r can be found so that

$$||f - f_r|| \leq C_f / r,$$

where

$$C_f = \int_{\mathbb{R}^p} |w| |\tilde{f}(\omega)| d\omega,$$

in which $\tilde{f}(\omega)$ is the Fourier transform of f,

$$f(\boldsymbol{x}) = \int_{\mathbb{R}^p} e^{i\omega \cdot \boldsymbol{x}} \tilde{f}(\omega) d\omega.$$

The large class is defined via Fourier transforms as follows. For a function f on $\mathbb{R}^p$, write its magnitude distribution as $F(d\omega)$ and its phase at frequency ω as $\theta(\omega)$. Then, the Fourier distribution of f is a unique complex-valued measure $\tilde{F}(d\omega) = e^{i\theta(\omega)} F(d\omega)$. Now,

$$f(\boldsymbol{x}) = \int e^{i\omega \cdot \boldsymbol{x}} \tilde{F}(d\omega) = f(0) + \int (e^{i\omega \cdot \boldsymbol{x}} - 1) \tilde{F}(d\omega).$$

The second expression holds more generally than the first, a distinction ignored here. Now, let $B \subset \mathbb{R}^p$ be a bounded set containing $\boldsymbol{x} = 0$ and let

$$\Gamma_B = \left\{ f : B \to R \mid \forall \boldsymbol{x} \in B, f(\boldsymbol{x}) = f(0) + \int (e^{i\omega \cdot \boldsymbol{x}} - 1) \tilde{F}(d\omega) \right\}.$$

It is implicit, as part of the definition of Γ_B, that $\tilde{F}$ has a magnitude distribution F for which $\int |\omega| F(d\omega)$ is finite. The set to be actually used in the theorem is a restriction of Γ_B denoted $\Gamma_{B,C}$: Letting $C > 0$ be a bound on B for use with $|\omega|_B = \sup_{x \in B} |\omega \cdot x|$, set

$$\Gamma_{B,C} = \left\{ f \in \Gamma_B : \exists \tilde{F}, \text{ representing } f \text{ on } B, \text{ so that } \int |\omega|_B F(d\omega) \leq C \right\}.$$

The unexpected result in this context is that the error of approximating $f \in \Gamma_{B,C}$ by sums of r sigmoids is C_f / r, independent of p.

Theorem (Barron, 1993): For any $f \in \Gamma_{B,C}$, any sigmoidal function ϕ, any probability measure μ, and any $r \geq 1$, there exists a linear combination of r sigmoidal functions $f_r(\boldsymbol{x})$ so that

$$\int_B (f(\boldsymbol{x}) - f_r(\boldsymbol{x}))^2 \mu(d\boldsymbol{x}) \leq \frac{(2C)^2}{r}.$$

The coefficients in f_r can be assumed to satisfy $\sum_{j=1}^{r} |c_j| \leq 2C$ and $c_0 = f(0)$.

Barron's proof has five major steps, some necessitating extra definitions. These steps are outlined in the Notes, some in detail where it's needed and some cursorily where the statements are more intuitive.

4.5 Recursive Partitioning Regression

The partitioning meant here is of the domain of the regression function. That is, the space of $\boldsymbol{x}$s is partitioned into subsets on each of which a local regression function can be specified. The initial partition is coarse, often just two subsets, but gets refined with each iteration because sets in the partition are split, usually into two subsets but sometimes more. Often the local regression function on a partition element is just a constant – the mean of the Ys arising from $\boldsymbol{x}$s in a set from a given partition. In these cases, the regression function is like a bin smoother that defines bins by the splits. The benefit of the recursive selection of elements in a final partition is that it is adaptive and so can capture functions whose surfaces represent interaction structure.

Formally, recursive partitioning fits a model of the form

$$Y = \sum_{j=1}^{r} \beta_j I(\boldsymbol{x} \in R_j) + \varepsilon, \tag{4.5.1}$$

where the regions R_j form a partition of the space of explanatory variables and the coefficients β_j are estimated from the data. Model (4.5.1) is another extension of the generalized additive model; the smooths are just average values on regions that must be estimated from the data. As written, (4.5.1) uses a constant function on each R_j; below, a generalization of this is used in which the β_js are replaced by arbitrary regression functions, one for each R_j. This permits different models to be used for different regions of the explanatory variables. Recursive partitioning models are usually called trees because the sequence of partitions can be represented as a tree under containment.

The most famous recursive partitioning method is CART, an acronym for classification and regression trees, see Breiman et al. (1984). As the name implies, the tree structure of the recursive partitioning method applies to both regression and classification. However, the techniques used with trees for regression and for classification are quite different owing to the nature of the problem. By contrast, NNs can be used for both regression and classification as well, but the estimation techniques are more similar. This section focuses on regression primarily; in Chapter 5, the classification perspective is primary.

Tree structure helps makes regression results interpretable. For instance, suppose $p = 2$ and a recursive partitioning procedure has fit a model of the form (4.5.1), generating an $r = 3$ element partition: $\{x_2 > 1\}$, $\{x_2 \leq 1, x_1 > 2\}$, and $\{x_2 \leq 1, x_1 \leq 2\}$, with $(\hat{\beta}_1, \hat{\beta}_2, \hat{\beta}_3) = (7, 5, 3)$. Then, the partition itself can be visualized as three regions in the plane or as a decision tree structure; see Fig. 4.4. Clearly, the tree indicates how the partition was found; the top node is called the root and the termini are called the leaves. Coupled with a rule for estimating the function in each partition element, this is equivalent to the model of the form (4.5.1), or to the diagram on the right (with the values on each partition), when the rule is to estimate the function by its average on the partition element. (More complicated function estimation procedures can be used on the partition elements. Often this is not feasible because there are not enough data to choose a good partition and estimate a complicated regression function on it.)

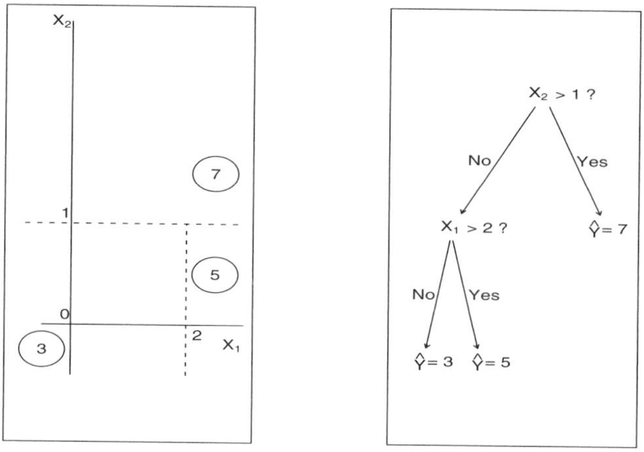

Fig. 4.4 On the left is the partition in the plane, with boundaries parallel to the axes. The decision tree on the right indicates the equivalent sequence of splits a recursive partitioning procedure would find.

One of the benefits of regression trees is that the influence of some variables can be localized to some regions of the domain space and not matter on others. That is, the values of the regression function on different regions can be so different that it is efficient to allow function values on one region to be unaffected by values on other regions. This may be useful when a response really does require different explanations on different regions. NNs can also localize the influence of variables to a particular region, at least somewhat, by using nodes in the last hidden layer. However, the localization in NNs uses a sigmoid of a linear combination, whereas the localization in trees is on regions that roughly reflect an interaction between variables.

Regression trees are a very different generalization from linear regression than additive models are in at least four ways. First, they encapsulate region-dependent interactions, usually by using a decision threshold. This is qualitatively different from the usual product term popular in multiple regression, which is not localized. Second, the procedure fits regression functions that are discontinuous at the boundaries of the regions; this is a drawback if Y is believed to be smooth. However, from a predictive standpoint, this is rarely a problem. Third, like any method, there are some functions that are difficult for models such as (4.5.1) to approximate and estimate. In the case of recursive partitioning, functions that cut across the decision boundaries of the R_js are a problem. For instance, with boundaries parallel to the axes (as in Fig. 4.4) it is difficult to approximate functions that are linear, or additive in a small number of variables. As an example, if $p = 1$ and the true f is a straight line, the best approximation will look like a staircase. It is hard to decide when a complex recursive partitioning model is close to a model that would be simple if another set of regressors was chosen. However, the reverse holds as well: Some functions that are easily described by regions (e.g., have localized interactions) will be more accessible to recursive partitioning than to other additive models. Finally, from an empirical standpoint, recursive partitioning methods are better adapted to high dimensions than many additive methods are because the sparsity of the data, naturally leads to rougher models. Indeed, recursive partition methods are often not competitive in low dimensions.

In practice, trees are generally less wiggly than NNs (even with regularization), probably because of the pruning. However, trees and NNs have similar expressive power, seeing as how they can both approximate any reasonable function in a limiting sense. Consequently, in many cases one expects that the stabilities of estimators from trees and NNs will be roughly comparable.

Because of its flexibility, recursive partitioning is virtually a template method, admitting numerous variations at each stage of fitting. Arguably, there are three typical core stages. First, the data are used to generate a maximal tree. Second, a collection of subtrees of the maximal tree must be chosen. Finally, a particular member of that collection must be chosen. Ideally, because these three stages are disjoint, they should be done with disjoint, equal, randomly chosen subsets of the data. Of course, in practice this is not always reasonable. Nevertheless, the presentation here will assume this for ease of exposition, and the sample sizes will be denoted n_1, n_2, and n_3 as needed, where $n = n_1 + n_2 + n_3$. Fitting procedures that combine two of the three stages are often used.

4.5.1 Growing Trees

Suppose n_1 of the n data points are to be used to grow a tree, T_{max}. Any tree-growing procedure must have a way to:

1. Select splits at intermediate nodes.
2. Declare a node to be terminal.
3. Estimate Y for the set at a terminal node.

The first two parts are often related: If one has a criterion by which to select splits, it can also be used to stop selecting splits. The third part is usually the most straightforward: Use the sample average of the data at each terminal node. In more complicated settings, linear regression is often used. In principle, any function estimation procedure can be applied to any of the terminal nodes, subject to having enough data. In addition to linear regression, the other methods in this chapter are obvious candidates. So, the focus in this subsection will be on selecting splits and stopping criteria.

The goal of splitting is to partition the training sample into increasingly homogeneous groups, thereby inducing a partition on the space of explanatory variables. Homogeneity refers to the adequacy of the model at the terminal node for describing the cases at the terminus, essentially in terms of an error criterion. Splitting usually stops when a satisfactorily high degree of homogeneity is achieved at the terminal nodes and hence in the corresponding regions of the $\boldsymbol{x}$-space.

The usual approach is to select splits of the predictor variables used to predict values of the response variable. One can search for splits in many ways; often the methods come down to some kind of cluster analysis to find a good split (choose a variable and cluster its values by some technique such as those described in chapter 8) or to a predictive criterion (propose a split on the basis of some of the data and evaluate how well it fits the other data). In general, the split at a node is intended to find the greatest improvement in predictive accuracy; these algorithms are greedy. However, this is evaluated within a sample, initially by some kind of node impurity measure – an assessment of the relative homogeneity or fit of the data to the model at the node. If one ends in a node for which all the values are well described by the same model, the impurity is effectively zero, the homogeneity is maximal, the terminal model fits, and the within-sample "prediction" or fit is perfect. The problem is that out-of-sample prediction can be terrible.

To minimize bad out-of-sample prediction, three standard techniques are used. First is a splitting rule that stops before overfitting is too serious. A simple rule is minimum n: Disallow any further splits when the number of data points at a node is at or below a threshold. This ensures a minimum number of data points are available for estimating the coefficients in the regression at the node. A variant on this is to stop when a specified fraction of ill-fitting points at a node has been achieved. Second is to impose a condition that one stops splitting when the data at a node are just similar enough, but not over-similar. This is clearer in the classification context: One can have numerous small sets of outcomes $\boldsymbol{x}_i$ giving the same class, as permitted by the Gini coefficient, to be discussed in Chapter 5. In regression, the corresponding condition would be insisting a variance such as $\sum_{i \in v}(y_i - \bar{y}(v))^2$, in which $\bar{y}$ is the average of the y_i-values at node v, be small but not too small. Third, and probably best, is to grow a large tree and then prune it back in some way. This is discussed in detail in the next subsection.

Different partitioning algorithms use different methods for assessing improvement in homogeneity and for stopping splitting. Because they represent different design criteria, they grow different sorts of trees. Four popular techniques are: (i) defined boundaries, (ii) squared error, (iii) Gini, and (iv) twoing. Twoing, like Gini, is more appropriate for classification and so is deferred to Chapter 5; the first two are discussed here. Hybrid methods can switch homogeneity criteria as they move down the decision tree; some methods maximize the combined improvement in homogeneity on both sides of a split, while other methods choose the split that achieves maximum homogeneity on one side or the other (see Buja and Lee (2001)). It should be noted that in small data sets these four techniques may not lead to particularly different tree structures after pruning. In larger data sets, however, the resulting trees can be substantially different because they have more splits, and later splits depend more and more on the class of trees each technique favors.

Boundaries can be defined in many ways, the simplest being parallel to the axes of the $\boldsymbol{x}$-space. The simplest splitting rule is to choose a variable, order its values, and split at or near the median. Alternatively, one can cluster the values of that variable and split between clusters. Generically, there are three obvious kinds of splits based on $\boldsymbol{x}$:

1. Is $x_i \le t$? (univariate split).
2. Is $\sum_{i=1}^{p} w_i x_i \le t$? (linear combination split).
3. Is $x_i \in U$? (categorical split, used if x_i is a categorical variable).

One can do a univariate search over values of t, more complicated searches over $\{w_i\}$, or search over subsets U of the category values. In all cases, the search seeks the split that separates the cases in the training sample into two groups with maximum increase in overall homogeneity.

More statistically, one can use the within-region variance (i.e., a squared error criterion). This is quite popular, and is standard in many implementations. The idea is to let $g \in L^2$, possibly of the form (4.5.1), have the associated error

$$\hat{E}(g) = \frac{1}{n_1} \sum_{i=1}^{n_1} (Y_i - g(\boldsymbol{X}_i))^2.$$

Minima are found by solving

$$\hat{g} = \arg \min_{g^* \in \mathscr{T}} \hat{E}(g^*),$$

where g^* varies over the set of piecewise constant (say) estimators of g defined on the leaves of a tree T assumed to be in the class $\mathscr{T}$ of trees. When $\mathbb{E}$ is used in place of $\hat{E}$, it gives the population value of the error; i.e., the expectation over the error and $\boldsymbol{X}$.

To see how this works, let $\mathscr{S}$ be a collection of sets; the simplest choice is to set $\mathscr{S}$ to be the collection of halfspaces with boundaries parallel to the axes of $\boldsymbol{x}$. This defines the possible splits and the class $\mathscr{T} = \mathscr{T}_{\mathscr{S}}$. One strategy is to grow a binary tree out to a defined maximal size, for instance one data point per terminal node, starting with the whole $\boldsymbol{X}$-space as the root. Following Gey and Nedelec (2005), let $split$ vary over $\mathscr{S}$. Since $\mathscr{S}$ consists of halfspaces with boundaries parallel to the axes, the optimal first split from the root is

$$\widehat{split} = \arg \min_{split \in \mathscr{S}} \left[E(g_{split}) + E(g_{split^c}) \right], \tag{4.5.2}$$

in which g_{split} is an estimator of g using constant functions on the sets defined by $split$. In (4.5.2), it is assumed that the $\boldsymbol{x}$s (as data or as instances of the random variable $\boldsymbol{X}$) are also partitioned according to $split$. This gives $\hat{g}_{split} = g_{\widehat{split}} = a 1_{\widehat{split}} + b 1_{\widehat{split}^c}$ as the minimum least squares estimator of g using constant functions at the daughter nodes; it is easy to see that under the squared error criterion the constants are the means over $split$ and $split^c$.

Once the root has been split once, each of the resulting nodes, n_L and n_R, can be split analogously: New sets $\mathscr{S}_L$ and $\mathscr{S}_R$ can be defined to replace $\mathscr{S}$. From them one obtains $\widehat{split}_L$ and $\widehat{split}_R$ with resulting $\hat{g}_{split,L}$, $\hat{g}_{split,R}$. Splitting continues in this fashion until a tree is produced with the desired homogeneity at each terminal node. Often, one grows a maximal tree; i.e., one that has a single data point at each leaf.

There is some debate as to how much these splitting criteria matter because in some contexts, large differences haven't been noticed very often. This is partially explained by noting that when data sets are small and highly accurate, trees can be generated easily and the particular splitting rule does not matter much. However, in many data mining problems with large inchoate data sets, obtaining a good answer is genuinely difficult and there is evidence that splitting rules matter a lot because of the kind of trees they favor finding. In this view, choosing a splitting rule is like choosing a prior – it favors some trees and disfavors others, though this can be overwhelmed by the data. Related to this is the fact that with small sample sizes (relative to p or other measures of complexity) trees exhibit great variability: Two trees may be predictively similar but mathematically quite different.

4.5.2 *Pruning and Selection*

When generating trees, it is usually optimal to grow a larger tree than is justifiable and then prune it back. The main reason this works well is that stop splitting rules do not look far enough forward. That is, stop splitting rules tend to underfit, meaning that even if a rule stops at a split for which the next candidate splits give little improvement, it may be that splitting them one layer further will give a large improvement in accuracy. Here, it is supposed that n_2 data points are used to develop a set of subtrees of a large, possibly maximal tree and the last n_3 points are used to choose among the subtrees. Clearly, one could use all $n_2 + n_3$ data points to prune down to a single tree rather than dividing the generation of candidate subtrees from selecting among them.

One way to generate a sequence of trees is to apply minimum cost-complexity pruning. In this process, one creates a nested sequence of subtrees (indexed by α in the cost-complexity function below) of an initial large tree by weakest link cutting. That is, given a large tree T generated from a technique in the last subsection, one prunes off all the nodes that arise from a fixed nonterminal node. The cost-complexity criterion chooses the nonterminal node to be ever closer to the root as the trade-off between error and complexity shifts more and more weight to the complexity penalty. If two nodes are approximately equal in terms of the cost-complexity values, one prefers to prune off the larger number of nodes. The result is a sequence of ever smaller subtrees.

Formally, the cost-complexity measure for a tree is

$$C(T;\alpha) = E(\hat{g}) + \alpha|T|, \tag{4.5.3}$$

where the number of terminal nodes in the tree T defined by $\hat{g}$, denoted $|T|$, summarizes the complexity of the tree. The weight α determines the relative importance of fit

$E(\hat{g})$ and the complexity. The goal is to find trees that achieve small values of $C(T,\alpha)$. Large αs penalize large trees heavily, making small trees optimal. Small values of α permit large trees susceptible to overfitting. Thus, in the limit of large α, the one node tree consisting of just the root is optimal; in the limit of small α, the large tree T itself is optimal. Note that pruning means that regions R_j are being joined so they have a common node function.

By starting with $\alpha = 0$ and letting α increase, one generates a sequence of subtrees by weakest link cutting. That is, consider the sequence of trees generated by

$$T_\alpha = \arg\min_{T' \subset T} C(T', \alpha)$$

as α ranges from 0 to infinity. It can be shown that $T_{\alpha_1} \subset T_{\alpha_2}$ when $\alpha_1 \geq \alpha_2$ and that the sequence itself is nested; see Breiman et al. (1984). In this way, one generates a nested sequence of subtrees T_{α_j} corresponding to functions $\hat{g}_j$ for $j = 1, ..., J$.

Given such a sequence from the middle n_2 data points, the last n_3 data points can be used to select an element $\hat{g}_{\alpha_{j*}}$ of the sequence, for instance by cross-validation. Hopefully, by choosing a subtree, one effectively chooses α to be a compromise value indexing a tree with the right complexity and good fit. Here, right means minimal predictive error in future cases. Most commonly, squared error loss is used in this procedure, but any measure of goodness of fit can be used in principle.

It is worth noting that the criterion (4.5.3) has a common form. Recall that the optimality criterion defining spline smoothing in Chapter 3 has the same form: a sum of two terms, one being an assessment of fit and the other an assessment of wiggliness. Indeed, it was commented that some of the instability of NNs could be smoothed out if the squared error fit was moderated by a complexity penalty on the parameters. All three of these cases are instances of complexity regularization because the penalty term regularizes the overall estimation by penalizing some aspect of solutions that can be broadly interpreted as a complexity.

4.5.3 Regression

The tree-based regression function resulting from the method presented so far is effectively of the form

$$\hat{g} = \hat{g}_{\alpha_{j*}} = \arg\min_{j=1,\ldots,J} E(\hat{g}_{\alpha_j}). \tag{4.5.4}$$

In this expression, n_1 data points were used to find the maximal tree T_{max} by a splitting rule. Then, n_2 data points were used with (4.5.3) to generate the sequence of trees T_{α_j} that represent $\hat{g}_{\alpha_j}$. The trees T_{α_j} are subtrees of T_{max}, and the α_js are the values of α that order the subtrees of T_{max} under (4.5.3). The sum of squared errors, or other criterion for fit, in (4.5.4) is formed using the last n_3 data points.

If the error term in (4.5.1) is IID $N(0,\sigma^2)$ and the $\boldsymbol{X}$s are drawn from a distribution μ, Gey and Nedelec (2005) have established an important property of $\hat{g}$: Its conditional expected L^2 distance from g_{true} is less than the smallest conditional expected distance from g_{true} for any of the trees in the sequence T_{α_j} plus a $\mathscr{O}(1/n)$ term. This is a sort of Oracle inequality because it ensures that even if the optimal subtree of T_{max} were known, the estimate $\hat{g}$ would not perform much worse than for approximating g_{true}.

To derive a weak, informal version of the Gey and Nedelec (2005) result, suppose $g_{true} \in L^2(\mu)$ and let μ_{n_3} be the empirical distribution formed from $\boldsymbol{X}_1,...,\boldsymbol{X}_{n_3}$. Let $||\cdot||_{n_3}$ be the norm from $L^2(\mu_{n_3})$ (i.e., the norm with respect to the empirical distribution), and recall that for any $g \in L^2(\mu)$, the sum of squared errors is $\hat{E}_3 = \hat{E}_{n_3}(g) = (1/n_3)\sum_{i=1}^{n_3}(Y_i - g(\boldsymbol{X}))^2$. Two easy identities are

$$\mathbb{E}(\hat{E}_3(g_{true})|\boldsymbol{X}^{n_3} = \boldsymbol{x}^{n_3}) = \sigma^2,$$

where $\boldsymbol{X}^n = (\boldsymbol{X}_1,...,\boldsymbol{X}_n)$ and $\boldsymbol{x}^n = (\boldsymbol{x}_,...,\boldsymbol{x}_n)$, and, for any $g \in L^2(\mu)$,

$$||g_{true} - g||_{n_3}^2 = \mathbb{E}(E_{n_3}(g) - E_{n_3}(g_{true})|\boldsymbol{X}_1,...,\boldsymbol{X}_{n_3}).$$

Now, adding and subtracting an arbitrary $\hat{g}_{\alpha_j}$ gives

$$\begin{aligned} ||g_{true} - \hat{g}||_{n_3}^2 &= ||g_{true} - \hat{g}_{\alpha_j}||_{n_3}^2 + \mathbb{E}\left(\left[\hat{E}_3(\hat{g}) - \hat{E}_3(\hat{g}_{\alpha_j})\right] |\boldsymbol{X}^n = \boldsymbol{x}^n\right) \pm \hat{E}_3(\hat{g}) \pm \hat{E}_3(\hat{g}_{\alpha_j}) \\ &= ||g_{true} - \hat{g}_{\alpha_j}||_{n_3}^2 + (\hat{E}_3(\hat{g}) - \hat{E}_3(\hat{g}_{\alpha_j})) \\ &+ (\hat{E}_3(\hat{g}_{\alpha_j}) - \mathbb{E}(\hat{E}_3(\hat{g}_{\alpha_j})|\boldsymbol{x}^n)) \\ &- (\hat{E}_3(\hat{g}) - \mathbb{E}(\hat{E}_3(\hat{g})|\boldsymbol{x}^n)). \end{aligned} \tag{4.5.5}$$

The Gey and Nedelec (2005) approach is to recognize that the second term on the right is negative and can be dropped, giving an upper bound of the form

$$||g_{true} - \hat{g}_{\alpha_j}||_{n_3}^2 + ||\hat{g}_{\alpha_j} - \hat{g}||_{n_3}^2 \left(\frac{\bar{E}_3(\hat{g}_{\alpha_j} - \bar{E}_3(\hat{g})}{||\hat{g}_{\alpha_j} - \hat{g}||_{n_3}^2} \right),$$

in which $\bar{E}_3$ represents the centered forms of $\hat{E}_3$ given in the last two terms of (4.5.5). For this last expression, one can give bounds on the second term as in Gey and Nedelec (2005) so that after taking an expectation conditional on the first $n_1 + n_2$ data points an infimum over j gives a uniform bound on the second term of order $\mathscr{O}(1/n)$. So, $||g_{true} - \hat{g}||_{n_3}^2$ is bounded by an infimum over j of $||g_{true} - \hat{g}_{\alpha_j}||_{n_3}^2$ plus $\mathscr{O}(1/n)$. The actual proof relies on a substantial collection of other reasoning from empirical process theory that can be found in the references in Gey and Nedelec (2005).

4.5.4 Bayesian Additive Regression Trees: BART

Bayesian versions of recursive partitioning have also been developed. Often they use a Bayes testing approach to decide whether a split at a node is worthwhile. That is, a parametric prior is put on the number of nodes and on any of the parameters in the regression function, while a nonparametric prior (typically a Dirichlet process prior determined by the empirical distribution) is used on the splits themselves. The worth of including a split is then decided by a Bayes factor from the appropriate tests. Bayesian nonparametrics is discussed more fully in Chapter 6.

Here, it is worth describing a variant on CART from a Bayesian standpoint that is due to Chipman et al. (1996) and Chipman et al. (2005). It uses a sum of small trees, often called stumps, say $g_j(\boldsymbol{x})$, to model a response:

$$Y = f(\boldsymbol{x}) + \varepsilon \approx g_1(\boldsymbol{x}) + \ldots + g_k(\boldsymbol{x}) + \hat{\varepsilon}.$$

The stumps, g_k, can be regarded as small, biased models in their own right, so that an estimate $\hat{f}$ is an ensemble method; such methods will be discussed in more detail in Chapter 6 as well. The BART procedure treats the g_js as terms in the larger model $\hat{f}$ rather than models in their own right. Thus, conceptually, BART is a single model with tree terms, not a model average in which each term explains part of f. Thus, predictions from several BART models could be averaged to give better overall predictions.

BART is a trade-off between using individual trees as in this section and combining full tree models as developed in Chapter 5 in random forests. In this middle ground, it is important that the individual trees be weak learners. If one of them becomes bigger, and hence more able to explain f, it can thereby dominate, paradoxically degrading performance. The paradox is resolved by realizing that if one of the trees is good enough that it explains too much, the stumps lose their descriptive power because they were weak learners from the outset. Computations suggest that there is improvement in performance as the (fixed) k is permitted to increase; see Breiman's theorem in Chapter 5. Of course, ensuring good behavior in BART depends delicately on prior specification, both within individual trees and across the collection of trees.

Overall, BART is a Bayesian intermediate tranche technique fitting a parameter-rich model, using extensive prior information, to ensure balance among complexity, bias, and variability. Computationally BART is implemented by a backfitting MCMC algorithm rather than by a gradient descent approach.

4.6 MARS

Multivariate adaptive regression splines (MARS) is a hybrid template method with a conceptually singular position in statistics. It was invented by Friedman (1991) – with much discussion, still ongoing. As a hybrid, MARS combines recursive partitioning regression and additive models, although taxonomists would mention splines too. The core idea is to express the regression function as a sum of functions (a la additive models), each of which has its support on a region (a la CART). Within a region, the regression function reduces to a product of simple functions that are initially constant but can be chosen as splines. The points defining the boundary of a region, like knots in splines, are obtained from the data.

The basic building block of the MARS regression function is a univariate function, identically zero up to a knot, after which it rises linearly. That is, the root element is $(x-t)^+$, where the $+$ indicates the positive part, the knot defining the support is t, and the shape is as in Fig. 4.5. Each term in the regression function is formed from these

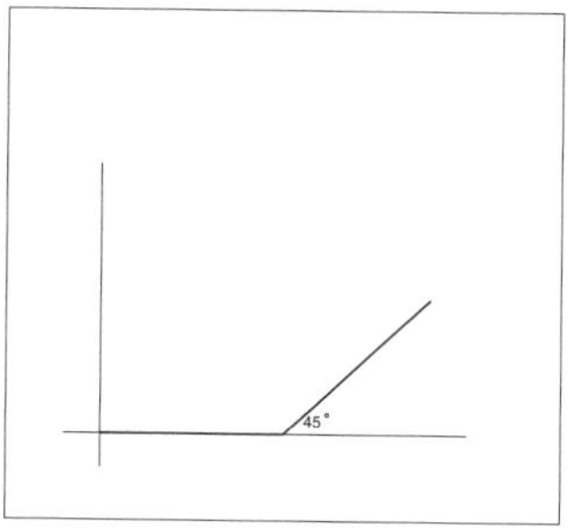

Fig. 4.5 The basic function $(x-t)^+$ from which MARS is built, t is at on the horizontal axis at its intersection with the 45°. The indicator for $\mathbb{R}^+$ can be applied to this to give positive values only on certain regions.

root elements by summing products of them. If the products are composed with indicator functions for disjoint regions R_j before summing, the result is a recursive partitioning model with spline-type node functions (having disjoint support) and a splitting rule based on the lack of fit of the whole model rather than just the individual nodes. If the indicator functions are not included, the model is arguably a more general MARS. Thus, there are (at least) two flavors of MARS – recursive and general.

Formally, MARS models, recursive or general, can be described as follows. Let $I_0(x)$ be the indicator for $x \in \mathbb{R}^+$, and consider $I_0(s_k(x_k - t_k))$, which is 1 when $s_k(x_k - t_k) \geq 0$. A product of such functions over $j = 1, ...p$ has the form $\prod_{k=1}^{p} I_0(s_k(x_k - t_k))$ and is positive in the region $[t_1, \infty) \times ... \times [t_p, \infty)$ when all the s_ks are one. Thus, it is a constant on a set with edges parallel to the axes of $\boldsymbol{x}$. To be more general, note that the product need not be over all $k = 1, ..., p$. Let $\mathscr{U}$ be a subset of js. The product over $k \in \mathscr{U}$ is only nontrivial for some k; for $k \notin \mathscr{U}$, the indicator function of the kth element does not appear.

Now, the MARS model is to write

$$Y = \sum_{j=1}^{r} \beta_j I(\boldsymbol{x} \in R_j) B_j(\boldsymbol{x}) + \varepsilon, \tag{4.6.1}$$

where typically

$$B_j(\boldsymbol{x}) = \prod_{k \in \mathscr{U}_j} I_0(s_{kj}(x_k - t_{kj})) \quad \text{or} \quad B_j(\boldsymbol{x}) = \prod_{k \in \mathscr{U}_j} [s_{kj}(x_k - t_{kj})]^+ \qquad (4.6.2)$$

for $s_{kj} = \pm 1$ and $\mathscr{U}_j$ is the subset of the explanatory variables appearing in the jth term. The first case in (4.6.2) is sometimes called the recursive partitioning version, and the second is sometimes called the forward stepwise version. The indicators for the regions R_j may be included or not. Thus, B_j is a product of functions on regions determined by the knots $\{t_{kj}\}$ for $k \in \mathscr{U}_j$ and the R_js. Clearly, the regression function is not continuous because of the indicators. However, omitting the indicator functions or the positive parts in (4.6.2) (i.e., just adding products of the hockey stick shaped functions as in Fig. 4.5) does give a continuous regression function. More generally, one can use spline basis functions of the form $[s_{kj}(x_k - t_{kj})^+]^q$ as factors in the second form in (4.6.2).

Fitting a MARS model is where the conceptual singularity begins. First, the core method is to start with a maximal number of terms, say $r = M_{max}$, for the model (4.6.1), an initial model comprised of one term $B_1(\boldsymbol{x}) = 1$, and a lack-of-fit criterion set to a large value, say ∞. GCV is one choice, but many may be considered. Friedman (1991), Section 3 suggests several; Barron and Xiao (1991) suggest minimum description length or the Bayes information criterion. The value M_{max} is often chosen to be 2 or 3 times larger than the anticipated correct number of terms; this is important because the MARS procedure merely searches a class of functions for one that has a good fit. The larger the class, the better the fit – but generalization error remains to be properly examined. In (4.6.2), r gives the number of regions for the spline-type basis functions. Within the large class, the MARS procedure is to construct new terms until there are too many as measured by the lack-of-fit criterion, here GCV. Here is one version of the MARS template procedure for the recursive partitioning case.

Let $B_1 = 1$. Search over k for $k = 2, 3 \ldots M_{max}$ terms with $GCV(initial) = \infty$.

- ☐ For each of the terms in the model, look at the n outcomes for variable $j = 1, \ldots, p$, where $x_{j,v}$ is assumed not to be in the term already.
- ☐ For each j and each $m = 1, \ldots, k-1$, examine the function

$$\sum_{u=1}^{k-1} \beta_u B_u(\boldsymbol{x}) + \beta_k B_m(\boldsymbol{x}) I_0((x_j - t)) + \beta_{k+1} B_m(\boldsymbol{x}) I_0(-(x_j - t)) \qquad (4.6.3)$$

  to find the value t among the n outcomes of the jth variable that gives the smallest lack of fit.
- ☐ Now, the optimal values of t, j, and m, say t^*, j^*, and m^*, can be used to elaborate one of the terms in the model into two new terms as in the last step. The term to be elaborated is chosen to minimize the lack-of-fit criterion.
- ☐ The new model is of the same form as the old model (4.6.1), with the optimally chosen term indexed by m^* replaced by two terms: $B_{m^*}(\boldsymbol{x}) I_0((x_{j^*} - t^*))$ and $B_{m^*}(\boldsymbol{x}) I_0(-(x_{j^*} - t^*))$.

☐ This procedure continues until the whole collection of models is searched or some other error criterion is met.

It is easy to see how to use this procedure for the forward stepwise version; just replace $I_0((x_j - t))$ and $I_0(-(x_j - t))$ with $(x_j - t)^+$ and $-(x_j - t)^+$ in step two. There is also a backwards elimination version of the procedure above that may be used to prune a model. Note that a term with $\mathscr{U}$ factors in it can only emerge from this procedure at the $\#(\mathscr{U})$th iteration or later.

Several aspects bear comment. First, like recursive partitioning, MARS is order dependent. That is, changing the order in which new splits (different js for instance) are included may change the final function found. Second, MARS is not really an estimation procedure but an approximation procedure. That is, it is the scope of the search and the goodness-of-fit criterion that determine the adequacy of the fit. Changing the class can, in principle, change the fit a lot; there is little objective validity in this sense. Third, whether or not the indicator function is included greatly changes the character of the approximation. If the indicator is included, then MARS is a particular case of recursive partitioning. MARS can also be framed as an approximation using a "tensor product" basis, so no explanatory variable appears twice in any term B_k, though this could be relaxed. In fact, MARS is a template for a collection of methods that includes recursive partitioning and much more besides.

It is seen that additive effects are captured by splitting the B_ks on several variables, and nonlinear effects are captured by allowing splits of B_ks on the same variable more than once with knots at different values of the x_js (or more general locations). If the indicator functions are included, then the products of indicators in the two terms of (4.6.3) combine to give a single indicator function. However, if the indicator functions are omitted, the procedure permits the terms to have overlapping regions. What happens is that, when a region is split on a variable, one can retain the function on the combined region while adding the two functions generated, one for each side of the split. As with trees, the strategy is to overfit and then prune back, typically by backwards elimination under the same lack-of-fit criterion as used to generate the MARS model in the first place. The pruning is not included in the algorithm above, but again a variety of methods parallel to trees or to conventional linear regression can be applied, e.g., backwards elimination or cost-complexity.

Comments on MARS have suggested problems and improvements. For instance, Barron and Xiao (1991) observe that in spline methods there can be nonrobustness when knots are too close together because small changes in $\boldsymbol{x}$ can lead to large changes in Y. They propose a roughness penalty on (4.6.3) to help smooth out such oversensitivity. By contrast, Gu and Wahba (1991) suggest that the class of splines used will not perform well with rotationally invariant problems and that stepwise procedures like MARS will often be confused by concurvity or nonparsimony. Thus, confidence intervals may have to be weakened to coverage bands or predictive intervals based on bootstrapping. Note that in this class of models, as with trees, the basic strategy is to keep reducing the bias by improving the approximation until the class is searched and then making sure that the variance is not too large by pruning. That is, there is always likely to be nonzero bias, but it should not contribute too much to an MSE relative to the overall variance.

MARS is not really interpretable, but Friedman (1991) observes an ANOVA-esque decomposition in which the terms dependent on a fixed number of variables are gathered into cumulative expressions,

$$\hat{f}(\boldsymbol{x}) = \beta_0 + \sum_{j \in \mathscr{J}} f_j(x_j) + \sum_{(j,k) \in \mathscr{K}} f_{jk}(x_j, x_k) + \ldots,$$

where β_0 is the coefficient of the $B_1 \equiv 1$ basis function, the first sum is over those basis functions that involve a single explanatory variable, the second sum is over those basis functions that involve exactly two explanatory variables, and so forth. These terms can be thought of as the grand mean, the main effects, the two-way interactions, and so forth. Thus, in MARS, there are several sorts of variability: the number of terms, the regions, and the estimates of parameters β_k on regions.

The intertwining of model variability with parameter variability highlights a key feature of MARS: It is purely a procedure for generating approximations, not really a statistical model. The model (4.6.1) has an error term in it, but this is a backformation from Friedman's original class of approximations. There is no genuine statistical model and hence no associated distribution for inference of any sort. One cannot really argue that overfit or underfit exists without further criteria, or that model identification has been successful, much less parameter estimation or prediction. Realistically, the best one can do to quantify reliability for predictive purposes is to use the bootstrap to get something that could be called a predictive distribution once a minimum GCV model of some reasonable size had been found.

This is the conceptual singularity evinced from MARS: Is it acceptable, statistically, to obtain estimates by using a method solely defined by a procedure when so little formal inference can be done? One can argue that this is acceptable under some circumstances. For instance, one set of reasonable conditions might be: (i) there is an algorithmic approximation method which is not close to any feasible, genuinely statistical procedure, (ii) empirical evaluations such as GCV and bootstrapping provide enough predictive guidance for meaningful implications to account for model and parameter uncertainty, and (iii) the procedure itself has been systematically examined, via simulations for instance, to find settings where it performs well or poorly (i.e., individual data sets, however numerous, are inadequate because they do not generalize). On the other hand, one could argue that any numerical approximation procedure amenable to GCV, bootstrapping, predictive analyses, and related computational techniques is valid even though not hitherto seen as part of the traditional statistical framework. Overall, this is part of the charm of DMML.

4.7 Sliced Inverse Regression

Sliced inverse regression (SIR) invented by Li (1991), is a way to combine inverse regressions on disjoint subsets of the range of Y to identify optimal directions of the explanatory variables. The basic model is to write

$$Y = f(\boldsymbol{X}\beta_1, ..., \boldsymbol{X}\beta_r, \varepsilon), \tag{4.7.1}$$

in which $r < p$, $f : \mathbb{R}^{k+1} \to \mathbb{R}$ is an unknown function, and the error term ε has conditional mean zero, $\mathbb{E}(\varepsilon|\boldsymbol{X}) = 0$. Regarding ε as an argument of f contrasts sharply with the earlier models because there is no longer a disjoint signal, and the sum of squares due to error, $\hat{E}$, is no longer the relevant quantity. Moreover, the loss of additivity, in contrast to PPR, means that the span of the β_js only defines a subspace of dimension r. That is, expression (4.7.1) models Y as a function of r linear functions of the $\boldsymbol{X}$s so that the span of the $\boldsymbol{X}\beta_i$s is an effective-dimension reduction space (see Chapter 9) of dimension r on which f is supported or at least well approximated. It is the subspace that expresses the dimension reduction, not the specific directions β_j.

Recall that inverse regression is literally an effort to invert a regression line. Instead of seeking an estimate for the unidimensional function $\mathbb{E}(Y|\boldsymbol{X})$ as a function of p variables in the context of $Y = \boldsymbol{X}\beta + \varepsilon$, one seeks the p dimensional function $\mathbb{E}(\boldsymbol{X}|Y)$ as a function of Y. Locally inverting this on S intervals of Y, called slices, and tying them together gives an approximation to $\mathbb{E}(Y|\boldsymbol{X})$, whence SIR. To do this, it is assumed for the rest of this section that $\boldsymbol{X}$ is drawn from a nondegenerate, elliptically symmetric distribution. This includes the often assumed normal family but much more besides. While this is a restriction, it is often not severe. Moreover, it is clearly specified and, in principle, can be checked by scatterplots of the X_{ij}s.

Even though the goal is to estimate $\mathbb{E}(Y|X = x)$, it may be helpful to examine $\mathbb{E}(X|Y = y)$ because finding p functions of a single real variable is less complicated than finding a single real-valued function of p real variables. Then, given p univariate regressions of the components of $\boldsymbol{X}$ on Y, implementing (4.7.1) requires knowing r, the β_is, and s. In this section, estimation of f is ignored; techniques such as those presented earlier must be used to estimate it.

Following Duan and Li (1991), set $\xi_1(y) = \mathbb{E}(\boldsymbol{X}|y)$ in the context of the model

$$Y = g(\alpha + \boldsymbol{x}^{\mathsf{T}}\beta, \varepsilon), \tag{4.7.2}$$

where $\varepsilon|\boldsymbol{X} \sim F(\varepsilon)$ is independent of $\boldsymbol{X}$; g is sometimes called a link function. The function ξ_1 is the step that helps minimize the Curse of Dimensionality. Note that this is essentially (4.7.1) for $r = 1$; models like (4.7.2) will be combined on the slices of Y. To use the elliptical symmetry, Li (1991) introduces a condition on the conditional expectations of linear combinations of coordinates of $\boldsymbol{X}$ in terms of the β_js. This is

$$\forall b \in \mathbb{R}^p \;\; \mathbb{E}(\boldsymbol{X}^{\mathsf{T}}b|\boldsymbol{X}^{\mathsf{T}}\beta_1 = \boldsymbol{x}^{\mathsf{T}}\beta_1, ..., \boldsymbol{X}^{\mathsf{T}}\beta_p = \boldsymbol{x}^{\mathsf{T}}\beta_p) = c_0 + \sum_{j=1}^{p} c_j \boldsymbol{x}^{\mathsf{T}}\beta_j \tag{4.7.3}$$

for some sequence $c_0,...,c_p$. Cook and Weisberg (1991) observe that elliptical symmetry is equivalent to (4.7.3). That is, under a model of the form (4.7.1), $\boldsymbol{X}$ is elliptically symmetric if and only if (4.7.3) holds.

The main result that makes SIR feasible is the following. Its importance is that it shows that the centered regression line in the univariate case varies over the space spanned by the vectors $\mathrm{Cov}(\boldsymbol{X})\beta_j$, for $j=1,\ldots,p$.

Theorem (Duan and Li, 1991): In model (4.7.2), the inverse regression function $\xi_1(y)$ satisfies

$$\xi_1(y) - \mathbb{E}\boldsymbol{X} = \mathrm{Cov}(\boldsymbol{X})\beta \frac{\mathbb{E}((\boldsymbol{X}-\mathbb{E}(\boldsymbol{X}))^\top \beta | y)}{\beta^\top \mathrm{Cov}(\boldsymbol{X})\beta}. \tag{4.7.4}$$

Proof: Following Li (1991), suppose $\mathbb{E}(\boldsymbol{X})=0$, and let $\boldsymbol{b}$ be an element of the orthogonal complement of the span of $\mathrm{Cov}(\boldsymbol{X})\beta_j$ for $j=1,\ldots,p$. In the context of model (4.7.1),

$$\begin{aligned}\boldsymbol{b}^\top \mathbb{E}(\boldsymbol{X}|Y=y) &= \mathbb{E}[\mathbb{E}(\boldsymbol{b}^\top \boldsymbol{X}|\boldsymbol{X}^\top \beta_j, j=1,...,p, Y=y)|y] \\ &= \mathbb{E}[\mathbb{E}(\boldsymbol{b}^\top \boldsymbol{X}|\boldsymbol{X}^\top \beta_j, j=1,...,p)|y]. \end{aligned} \tag{4.7.5}$$

So, to show that the centered regression line varies over the space spanned by the $\mathrm{Cov}(\boldsymbol{X})\beta_j$s, it is enough to show that the inner conditional expectation $\mathbb{E}(\boldsymbol{b}^\top \boldsymbol{X}|\boldsymbol{X}^\top \beta_k, k=1,...,p)$ is zero, or equivalently its square is; i.e., $\mathbb{E}(\mathbb{E}(\boldsymbol{b}^\top \boldsymbol{X}|\boldsymbol{X}^\top \beta_k, k=1,...,p)^2)=0$. This follows from using conditioning and the elliptical symmetry. Indeed, it is seen that the square is

$$\begin{aligned}\mathbb{E}[\mathbb{E}(\boldsymbol{b}^\top \boldsymbol{X}|\beta_j \boldsymbol{x}, j=1,\ldots,p)\boldsymbol{x}^\top \boldsymbol{b}] &= \mathbb{E}\left[\left(c_0 + \sum_{j=1}^{p} c_j \beta_j \boldsymbol{X}\right)\boldsymbol{x}^\top b^\top\right] \\ &= \sum_{j=1}^{p} c_j \beta_j \mathrm{Cov}(\boldsymbol{X})\boldsymbol{b}^\top = 0.\end{aligned}$$

As noted in Duan and Li (1991), the elliptical symmetry can also be used to obtain

$$\mathbb{E}(\boldsymbol{X}|\boldsymbol{X}^\top \beta) - \mathbb{E}\boldsymbol{X} = \frac{\mathrm{Cov}(\boldsymbol{X})\beta\beta^\top(\boldsymbol{X}-\mathbb{E}\boldsymbol{X})}{\beta^\top \mathrm{Cov}(\boldsymbol{X})\beta}.$$

So, taking the conditional expectation gives $\xi_1(y) = \mathbb{E}(\mathbb{E}(\boldsymbol{X}|\boldsymbol{X}'\beta)|y)$. $\square$

From the standpoint of parameter estimation rather than ξ_1, this theorem also gives that $\beta \propto \mathrm{Cov}(\boldsymbol{X})^{-1}(\xi_1(y) - \mathbb{E}\boldsymbol{X})$ and identifies the constant of proportionality as the fraction in (4.7.4), which is seen to be a real number dependent on β but not y.

Now consider the standardized variable $\boldsymbol{Z} = \mathrm{Cov}(\boldsymbol{X})^{-1/2}(\boldsymbol{X}-\mathbb{E}(\boldsymbol{X}))$, and form the p-dimensional inverse regression function $\xi(y) = \xi_p(y) = \mathbb{E}(\boldsymbol{Z}|Y=y)$. The theorem continues to hold since $\boldsymbol{Z}$ is a linear transformation of $\boldsymbol{X}-\mathbb{E}(\boldsymbol{X})$. Thus, for each y, $\xi(y)$ is a point in the span $\mathscr{S}$ of $\{\mathrm{Cov}(\boldsymbol{X})^{1/2}\beta_1,...,Cov(\boldsymbol{X})^{1/2}\beta_r\}$. If b is orthogonal to $\mathscr{S}$, then $\xi(y)'b=0$; using (4.7.1) to express ξ in terms of $\mathscr{S}$ gives

$$\text{Cov}(\xi(Y))b = \mathbb{E}(\xi(y)\xi(y)^{\mathsf{T}})b = 0.$$

This means that $\text{Cov}(\mathbb{E}(\boldsymbol{Z}|y))$, is degenerate in every direction orthogonal to $\mathscr{S}$.

Given these results, the overall SIR strategy for data $(Y_i, \boldsymbol{X}_i)$ for $i = 1, ..., n$, is to rewrite (4.7.1) as

$$Y = f(\boldsymbol{Z}\eta_1, ..., \boldsymbol{Z}\eta_r, \varepsilon) \tag{4.7.6}$$

since the exact representation of the r-dimensional effective dimension-reduction space is not important. Then, partitioning the range of Y into H slices, one can form an estimate of the inverse regression curve on each slice. The pooled estimator of the $p \times p$ $\text{Cov}(\xi(y))$ matrix, based on the H slices of the range of Y, has a principal components decomposition. So, finding the r largest eigenvalues (out of p) of an estimate $\widehat{\text{Cov}(\xi(y))}$ and transforming the corresponding r eigenvectors of the standardized variable in terms of the η_js gives estimates of the $\hat{\beta}_j$s. More formally, the SIR procedure is the following.

Estimate $\Sigma = \text{Cov}(\boldsymbol{X})$ by the sample covariance matrix $\hat{\Sigma}$, define the standardized data $z_i = \hat{\Sigma}^{-1/2}(\boldsymbol{x}_i - \bar{\boldsymbol{x}})$, and partition the range of Y into S slices, H_s for $s = 1, ..., S$. Let n_s be the number of observations y_i in slice H_s, so that $n_s = \sum_{i=1}^n I_{H_s}(y_i)$.

- ☐ Find the mean of the z_i on each slice:

$$\bar{z}_s = \frac{1}{n_s} \sum_{i=1}^{n} z_i I_{H_s}(y_i).$$

  This mean will serve as a crude (constant) estimate for the inverse regression curve $\mathbb{E}(Z|Y)$.

- ☐ Estimate $\text{Cov}(\xi(y))$:

$$\widehat{\text{Cov}(\xi(y))} = \frac{1}{n} \sum_{u=1}^{S} n_u \bar{z}_u \bar{z}_u'.$$

- ☐ Obtain the eigenvalues $\hat{\lambda}_i$ and the eigenvectors $\hat{\eta}_i$ of $\widehat{\text{Cov}(\xi(y))}$. (This is the principal components analysis for the $\bar{z}_i$s.)
- ☐ Transform the eigenvectors $\hat{\eta}_i$ corresponding to the r largest eigenvalues $\hat{\lambda}_i$ by applying $\hat{\Sigma}^{-1/2}$. Thus, obtain $\hat{\beta}_j = \hat{\Sigma}^{-1/2}\hat{\eta}_{(p-j+1)}$ for use in (4.7.1).

Once the estimates $\hat{\beta}_j$ are obtained, it is desirable to do inference on them. The usual consistency, asymptotic normality and identification variance theorems have been established; see Li (1991) and Duan and Li (1991). The extra bit that's interesting here is that under the admittedly strong assumption of a normal distribution for $\boldsymbol{X}$ one can identify asymptotic sampling distributions for the eigenvalues; parallel results for the eigenvectors exist but are more complicated. Although the eigenvectors and eigenvalues are well defined and identifiable, they are only important because of their span. Any other set of β_js with the same span would do as well.

Asymptotic rates for the convergence of the β_js are easy to identify. Indeed, Li (1991) uses the following reasoning. The central limit theorem gives that the $\bar{z}_s$s converge to the $\mathbb{E}(\bar{z}_s)$s at rate $\mathcal{O}(1/\sqrt{n})$. Consequently, the estimate $\widehat{\mathrm{Cov}(\xi(y))}$ of $\mathrm{Cov}(\xi(y)) = \sum_{s=1}^{S} \pi_s \mathbb{E}(\bar{z}_s)\mathbb{E}(\bar{z}_s')$, where $\pi_s = \lim(n_s/n)$ converges at a $\mathcal{O}(1/\sqrt{n})$ rate. Thus, the eigenvectors of $\widehat{\mathrm{Cov}(\xi(y))}$, the $\hat{\eta}_i$s, converge to the corresponding eigenvectors of $\mathrm{Cov}(\xi(y))$ at a $\mathcal{O}(1/\sqrt{n})$ rate. By using the theorem, one obtains that the standardized $\boldsymbol{x}_i$s (i.e., the z_is) give an inverse regression curve $\mathbb{E}(\mathbf{Z}|y)$ that is contained in the span of $\eta_1,\ldots,\eta_r$. So, since $\mathbb{E}(\mathbf{Z}) = \mathbb{E}(\mathbb{E}(\mathbf{Z}|y)|y \in S_s)$, the largest r eigenvectors of $\mathrm{Cov}(\xi(y))$ are in the space generated by the standardized vectors $\eta_1,\ldots,\eta_r$. Since $\hat{\Sigma}$ converges to Σ, $\hat{\Sigma}^{-1/2}$ converges to $\Sigma^{-1/2}$, so the corresponding $\hat{\beta}_j = \hat{\Sigma}^{-1/2}\hat{\eta}_j$ also converge at rate $\mathcal{O}(1/\sqrt{n})$.

Although the rate determination is seen to be straightforward, identifying the constant in the $\mathcal{O}(1/\sqrt{n})$ rate is not easy. It is done by Duan and Li (1991), Section 4, who establish $\sqrt{n}(\hat{\beta} - \beta) \to N(0, \boldsymbol{V})$, as $n \to \infty$, where $\beta = (\beta_1, \ldots, \beta_r)$ and $\boldsymbol{V}$ is a matrix that depends delicately on the choice of slices.

Like f, r remains to be estimated. One approach is to construct models using several different rs, searching for the model with the smallest cross-validation error. Alternatively, the usual criteria for a principal components analysis can be invoked; e.g., the knee in the error curve. Further topics related to SIR and sufficient dimension reduction more generally are taken up in Chapter 9.

4.8 ACE and AVAS

So far, the methods presented have focused exclusively on representing Y using a class of functions of $\boldsymbol{X}$. However, this is only one side of the story. The other side is that one can transform Y as well – or instead of – $\boldsymbol{X}$. In an additive model context gives

$$g(Y) = \sum_{j=1}^{p} f_j(X_j) + \varepsilon \tag{4.8.1}$$

as a more general model class. Mathematically, it is as reasonable to transform Y as $\boldsymbol{X}$; however, many resist it on the grounds that it is Y that one measures, not $g(Y)$, and that introducing g, especially in addition to the f_js, leads to such a large increase in instability that little can be said reliably. Nevertheless, (4.8.1) is a generalization of GAMs that avoids having to choose a link function.

Study of this model class, summarized in Hastie and Tibshirani (1990), is incomplete, and the two methods briefly discussed here alternating conditional expectations (ACE) and additivity and variance stabilization (AVAS) are variants of each other. Thus, both can be regarded as instances of a single template method. They are interesting here not so much for their current usefulness as for their potential. Both involve techniques based on alternating the way one takes conditional expectations, which may be an important idea for dealing with transformations on Y. It is easy to imagine further variants that may yield better results than those obtained so far. If this interesting class became tractable, the benefits would be large and pervasive.

The ACE algorithm is symmetric in its treatment of the conditional expectations of g and the f_js. It originates in Breiman and Friedman (1985) and seeks $f_1, \ldots, f_p$ and g to maximize the correlation between

$$g(Y) \text{ and } \sum_{j=1}^{p} f_j(X_j),$$

a generalization of canonical correlation. This is equivalent to minimizing

$$\mathbb{E}\left[\left(g(Y) - \sum_{j=1}^{p} f_j(X_j)\right)^2\right] / \mathbb{E}[g^2(Y)],$$

where the expectation is taken over $(Y_i, \boldsymbol{X}_i)$. Thus, ACE minimizes a variant on the mean squared error; one can readily imagine other variants.

One version of the ACE algorithm (see Hastie and Tibshirani (1990)) can be summarized as follows.

Start with $g(y_i) = (y_i - \bar{y})/s_y$ and $f_j(x_j)$ as the linear regression of Y on X_j.

- □ Find $f(\boldsymbol{X}) = \sum_{j-1}^{p} f_j(X_j) = \mathbb{E}(g(Y)|\boldsymbol{X})$ as an additive model, possibly by the backfitting algorithm. This gives a new $g(y)$ in terms of new functions $f_1(x_1), \ldots, f_p(x_p)$.
- □ Use smoothing to estimate

$$\tilde{g}(y) = \mathbb{E}\left[\sum_{j=1}^{p} f_j(x_j) \,|\, Y_i = y_i\right],$$

and standardize a new $g(y)$ as

$$g(y) = \tilde{g}(y)/\sqrt{\text{Var}[\tilde{g}(y)]}.$$

(This standardization ensures that the trivial solution $g \equiv 0$ does not arise.)

- □ Alternate: Repeat the last two steps until $\mathbb{E}[(g(Y) - \sum_{j=1}^{p} f_j(X_j))^2]$ is satisfactorily small.

As shown in Breiman and Friedman (1985), Section 5 and Appendix 3, there are unique optimal transformations g and f_j, and the ACE algorithm converges weakly to them. The proof rests on the recognition of conditional expectation as a projection operator on suitably defined Hilbert spaces of functions and using its eigenfunctions. There are settings, however, where this is not enough.

Unfortunately, from the standpoint of nonparametric regression, ACE has several undesirable features: (i) For the one-dimensional case, $g(Y) = f(X) + \varepsilon$, ACE generally will not find g and f but rather $u \circ g$ and $u \circ f$ for some function u. This is a sort of nonidentifiability. (ii) The solution is sensitive to the marginal distributions of the explanatory variables and therefore is often nonrobust against outliers in the data. (iii) ACE treats the explanatory and response variables in the same way, reflecting correlation, whereas, arguably, regression should be asymmetric. (iv) ACE (and AVAS below) only minimize correlation, so when correlation is not very high the model only captures part of the relationship between the dependent and explanatory variables.

AVAS is a modification of ACE that addresses item (iii) by purposefully breaking the symmetry between g and $\sum f_j$, thereby removing some of the undesirable features of ACE; see Tibshirani (1988). The central difference between ACE and AVAS is that instead of using the standardization indicated in Step (2) immediately, a variance-stabilizing transformation is applied first: If a family of distributions for Z has mean μ and variance $V(\mu)$, then the asymptotic variance-stabilizing transformation for Z is

$$h(t) = \int_0^t V(s)^{-1/2}\, ds,$$

as can be verified by a delta method argument. In AVAS, one finds $g(y)$ given $E(g(Y)|\boldsymbol{X} = \boldsymbol{x})$ as before but then standardizes $h \circ g(Y)$ (if necessary) rather than $g(Y)$ for the next iteration.

As a generality, model selection in an ACE or AVAS context is difficult when cross-validation cannot be readily applied. Wong and Murphy (2004) develop some techniques and are able to give spline estimates of g in (4.8.1). However, model uncertainty with ACE and AVAS can be high because they often do not perform well (i) when the explanatory variables are correlated, (ii) because they can give different models depending on the order of inclusion of variables (permitting functions f_j of univariate functions of the components of $\boldsymbol{X}$ makes this problem very difficult), and (iii) because they can be sensitive to omitted variables or spuriously included variables.

Similar to MARS, ACE and AVAS are, at their root, procedures for finding good approximations in contexts where inference in any conventional sense may be intractable. Work going beyond the short description here includes de Leeuw (1988), who has a more general approach for which both ACE and AVAS are special cases. Despite all their limitations, it is clear that the template of which ACE and AVAS are instances has enormous potential.

4.9 Notes

4.9.1 Proof of Barron's Theorem

First, for a given sigmoidal function ψ, let

$$G_\psi = \{\, \gamma\psi(a \cdot \boldsymbol{x} + b) : |\gamma| \le 2C,\; a \in \mathbb{R}^p,\; b \in \mathbb{R} \,\}$$

be the collection of bounded multiples of the sigmoid composed with an affine function. Apart from the constant term, Barron's theorem bounds the error of approximating $f(x) - f(0)$ by convex combinations of functions in G_ψ for functions $f \in \Gamma_{B,C}$. To begin the function approximation step of the proof, let G be a bounded set in a Hilbert space (i.e., $\forall g \in G, ||g|| \leq b$) and let $\overline{\text{chull}}(G)$ be the closed convex hull of G.

Step 1: For any $\bar{f} \in \overline{\text{chull}}(G)$, any $n \geq 1$, and every $c' > b^2 - ||\bar{f}||^2$, $\exists f_n$, a convex combination of n points in chull(G), such that

$$||\bar{f} - f_n||^2 \leq \frac{c'}{n}.$$

Proof: Let $n \geq 1$ and $\delta > 0$, and $f^* \in \text{chull}(G)$ such that $||\bar{f} - f^*|| \leq \delta/n$. Now,

$$f^*(x) = \sum_{k=1}^{m} \gamma_k \, g_k^*(x)$$

for some m, with $\gamma_k \geq 0$ and $\sum_k \gamma_k = 1$ for a set of g_k^*s in G.

Let G now be a random object taking values in the finite set $\{g_1^*, ..., g_m^*\}$ with probabilities given by $P(G = g_k^*) = \gamma_k$. Let $g_1, ..., g_r$ be r independent outcomes of G with sample average $\bar{g}_r$. Clearly, as functions, $E\bar{G}_r(\boldsymbol{x}) = f^*(\boldsymbol{x})$. By the usual rules for variance,

$$E||\bar{G}_r - f^*||^2 = \frac{1}{r}E||G - f^*||^2 = \frac{1}{r}E||G||^2 - ||f^*||^2 \leq \frac{1}{r}(b^2 - ||f^*||^2).$$

That is, as a random variable, $\bar{g}_r$ approximates f^* within $1/r$ in expectation. This is possible only if there is an outcome of the random variable that achieves the bound. So, there must be a fixed $g_1, ..., g_r$ for which $||\bar{g}_r - f^*||^2 \leq (1/r)(b^2 - ||f^*||^2)$. Using this, choosing δ small enough, and applying the triangle inequality to $||\bar{f} - \bar{g}_r||^2$ gives the result. □

To proceed, three sets of functions must be defined; Steps 2, 3, and 4 will give some of their containment relationships. The first set is

$$G_{cos} = \left\{ \frac{\gamma}{|\omega|_B} [\cos(\omega \cdot x + b) - \cos(b)] : \omega \neq 0, |\gamma| \leq C, b \in \mathbb{R} \right\};$$

it is seen that G_{cos} depends on B and C. It will be seen that, heuristically, G_{cos} is the smallest of the sets used in the proof. For convenience, set $\hat{f} = f(x) - f(0)$ to make use of the Fourier representation. The next step shows that $\Gamma_{B,C}$ is almost in G_{cos}.

Step 2: For $f \in \Gamma_{B,C}$, $\hat{f} \in \overline{\text{chull}}(G_{cos})$, the closure of the convex hull of G_{cos}.

Proof: Let $\boldsymbol{x} \in B$. For real-valued functions,

$$\begin{aligned} f(\boldsymbol{x}) - f(0) &= \mathbb{R}e\left[\int (e^{i\omega\cdot\boldsymbol{x}} - 1)\tilde{F}(d\omega)\right] \\ &= \int \cos(\omega\cdot\boldsymbol{x} + \theta(\omega)) - \cos(\theta(\omega))F(d\omega) \\ &= \int_\Omega \frac{C_{f,B}}{|\omega|_B}\cos(\omega\cdot\boldsymbol{x} + \theta(\omega)) - \cos(\theta(\omega))\Lambda(d\omega) \\ &= \int_\Omega g(\boldsymbol{x},\omega)\Lambda(d\omega), \end{aligned}$$

in which $C_{f,B} = \int |\omega|_B F(d\omega) \le C$ is a variant on C_f and is assumed bounded. The new probability measure is $\Lambda(d\omega) = |\omega|_B F(d\omega)/C_{f,B}$. (The restriction to $\Omega = \mathbb{R}^p - 0$ is needed to make $|\omega|_B$ well defined.)

It is seen that $|g(\boldsymbol{x},\omega)| \le C|\omega\cdot\boldsymbol{x}|/|\omega|_B \le C$ for $\boldsymbol{x} \in B$ and nonzero ω. Thus, the functions $g(\boldsymbol{x},\omega)$, as functions of $\boldsymbol{x}$, are in G_{cos}. So, any function $\hat{f}$, as an integral of gs, is an infinite convex combination of functions in G_{cos} and hence is in $\bar{\text{chull}}(G_{cos})$.

Formally, to see this last statement, if F has a continuous density on $\mathbb{R}^p$, one can verify that Riemann sums in terms of the gs converge to $\hat{f}$. More generally, the claim follows from an L_2 law of large numbers omitted here. □

The two remaining sets needed for the proof are special cases of G_ψ when ψ is a step function. Setting $\text{step}(z) = 1_{\{z\ge 0\}}$, let

$$G_{step} = \{\ \gamma\text{step}(\alpha\cdot\boldsymbol{x} - t) :\ |\gamma| \le 2C,\ |\alpha|_B = 1,\ |t| \le 1\ \}.$$

(The role of the $2C$ will be apparent shortly.) Also, consider restricting t to the continuity points of the distribution of $z = \alpha\cdot\boldsymbol{x}$ induced by the measure μ on $\mathbb{R}^p$. This is a dense set in $\mathbb{R}$. Let T_α be its intersection with the closed interval $[-1,1]$ and set

$$G^{\mu}_{step} = \{\gamma\ \text{step}(\alpha\cdot x - t) : |\gamma| \le 2C, |\alpha|_B = 1, t \in T_\alpha\} \subset G_{step}.$$

Next, the goal is to show that functions in G_{cos} are in $\bar{\text{chull}}(G_\psi)$ for any ψ. Consider first the special case that $\psi = $ step.

Step 3: (i) $G_{cos} \subset \bar{\text{chull}}(G_{step})$.

(ii) $G_{cos} \subset \bar{\text{chull}}(G^{\mu}_{step})$.

(iii) $G^{\mu}_{step} \subset \bar{G}_\psi$, the closure of G_ψ in $L_2(\mu, B)$.

Proof: Begin with (i). Each function in G_{cos} is a composition of a sinusoidal function $g(z)$ with a linear function $z = \alpha\cdot\boldsymbol{x}$, where $\alpha = \omega/|\omega|_B$ for some $\omega \ne 0$. When $\boldsymbol{x} \in B$, $z = \alpha\cdot\boldsymbol{x}$ is in $[-1,1]$, so it's enough to restrict attention to that interval. Since g is uniformly continuous on $[-1,1]$, it is uniformly well approximated by piecewise constant functions on any sequence of partitions of $[-1,1]$ with maximum interval length tending to zero. Each piecewise constant function is a linear combination of step functions.

As a representative case, consider the restriction of a function $g(z)$ to $[0,1]$ and fix a partition $0 = t_0 < t_1 < \ldots < t_k = 1$. The function

$$g_{k,+}(z) = \sum_{i=1}^{k-1} (g(t_i) - g(t_{i-1})) 1_{\{z \geq t_i\}}$$

is a piecewise constant interpolation of g at the t_is (for $i \leq k-1$). It is also a linear combination of step functions. Since $g' \leq C$ on $[0,1]$,

$$\sum_{i=1}^{k-1} |g(t_i) - g(t_{i-1})| \leq C.$$

Similarly, define

$$g_{k,-}(z) = \sum_{i=1}^{k-1} (g(t_i) - g(t_{i-1})) 1_{\{z \leq -t_i\}}.$$

Now, $g_k(z) = g_{k,+}(z) + g_{k,-}(z)$ is a piecewise constant function on $[-1,1]$ uniformly close to g for a fine enough partition and the sum of the absolute values of the coefficients is bounded by $2C$. Hence, (i) follows.

For (ii), the T_α is dense in $[-1,1]$ so an extra limit over choices of t is enough to get the result.

For (iii), observe that the limit of any sequence of sigmoidal functions $\psi(|a|(\alpha \cdot \boldsymbol{x} - t))$ as $|a| \to \infty$ is step$(\alpha \cdot \boldsymbol{x} - t)$ unless $\alpha \cdot \boldsymbol{x} - t = 0$, which has μ measure zero. By the DCT, the limit holds in $L_2(\mu, B)$ so (iii) follows. □

Now, without proof, the main statements to be used with Step 1 to prove the theorem are given in the following. Full details can be found in Barron (1993).

Step 4: With closures taken in $L_2(\mu, B)$,

$$\Gamma^0_{B,C} \subset \bar{\text{chull}}(G_{\cos}) \subset \bar{\text{chull}}(G^\mu_{\text{step}}) \subset \bar{\text{chull}}(G_\psi),$$

where $\Gamma^0_{B,C}$ is the set of functions in $\Gamma_{B,C}$ with $f(0) = 0$.

Proof: These containments follow from Steps 2, 3(ii), and 3(iii). □

Step 5: The conclusion of Barron's theorem holds.

Proof: To see that the constant in the theorem can be taken to be $(2C)^2$, for functions $f \in \Gamma_{B,C}$ note that the approximation bound is trivially true if $\bar{f} = 0$, for then $f(\boldsymbol{x})$ is just a constant a.e. on B.

So, suppose $||f|| > 0$, and consider two cases: (i) The sigmoid is bounded by 1 or (ii) it is not.

In case (i), the functions in G_ψ are bounded by $b = 2C$, as seen for the step functions. Thus, for any $c' > (2C)^2 - ||\bar{f}||^2$, Step 1 holds. The conclusion is the theorem: There is a convex combination of functions in G_ψ for which the squared norm (in $L_2(\mu, B)$) is bounded by c'/r.

For case (ii), Step 1 and $\Gamma^0_{B,C} \subset \bar{\text{chull}}(G^\mu_{step})$ from Step 4 give that there is a convex combination of functions in G^μ_{step} for which the squared $L_2(\mu, B)$ norm of the error of approximation is bounded by $(1/r)[(2C)^2 - ||\hat{f}||^2/2]/n$. Then, by Step 3(iii), with a suitable choice of scale of the sigmoidal function, one can replace the step functions by sufficiently accurate approximations in terms of ψ that the resulting convex combination of r functions in G_ψ yields a square $L_2(\mu,B)$ norm bounded by $(2C)^2/r$, completing the proof. □

4.10 Exercises

Exercise 4.1. Consider independent outcomes $(x_1,y_1),\cdots,(x_n,y_n)$ drawn from the model $Y=f(x,\theta)+\varepsilon$, where ε is $N(0,\sigma^2)$ and θ is a vector of real parameters with true value θ_T. Write $\mathbf{y}=(y_1,...,y_n)'$ and $f(\theta)=(f(x_1,\theta),...,f(x_n,\theta))'$ and consider the function $S_n(\theta)=(1/n)\|\mathbf{y}-f(\theta)\|^2$

1. Derive an expression for $S_n(\theta,\theta_T)=\mathbb{E}S_n(\theta)$. (Remember $\mathbb{E}$ is taken under θ_T.)
2. Let
$$S^*(\theta,\theta_T)=\sigma^2+\int (f(x,\theta_T)-f(x,\theta))^2dx.$$
Argue that $S_n(\theta)$ and $S_n(\theta,\theta_T)$ converge pointwise to $S^*(\theta,\theta_T)$.

Exercise 4.2. In the context of the last exercise, let $\hat\theta=\arg\min S_n(\theta)$ It is known that under appropriate regularity conditions $\hat\theta\to\theta_T$ in distribution. Let $S^2=SSE(\hat\theta)/(n-p)$, where $p=\dim(\theta)$ and $SSE(\theta)=\|\mathbf{y}-f(\hat\theta)\|^2$.

1. Let $\partial f/\partial\theta|_{\theta_T}$ be the matrix with typical row $(\partial/\partial\theta)f(x_i,\theta_T)$ and assume
$$\hat\theta=\theta_T+\left(\frac{\partial}{\partial\theta}f^{\mathrm{T}}\frac{\partial}{\partial\theta}f\right)^{-1}\left(\frac{\partial}{\partial\theta}f\right)^{\mathrm{T}}\varepsilon+0_P\left(\frac{1}{\sqrt{n}}\right).$$
Show that
$$\hat\theta\sim N_p\left(\theta_T,\sigma^2\left(\frac{\partial}{\partial\theta}f^{\mathrm{T}}\frac{\partial}{\partial\theta}f\right)^{-1}\right)$$
approximately.
2. Parallel to the usual theory, argue that
$$\frac{(n-p)S^2}{\sigma^2}\sim\chi^2_{n-p}$$
asymptotically.
3. Consider testing $\mathscr{H}_0: h(\theta)=0$ vs. $\mathscr{H}_1: h(\theta)\neq 0$ for some well behaved function h representing a constraint on the p-dimensional parameter space where q is the number of restrictions imposed on θ by h.

   Let $\hat\theta$ estimate θ under the full model so that $SSE(\text{full})=SSE(\hat\theta)$ as before, and let
$$\tilde\theta=\arg\min SSE(\theta)\ \text{ subject to }\ h(\theta)=0$$

with $SSE(\text{reduced})$ denoting the minimum. Argue that

$$\frac{(SSE(\text{reduced}) - SSE(\text{full}))/q}{SSE(\text{full})/(n-p)}$$

is asymptotically an F statistic.

Exercise 4.3. [Sequential regression.] Let η be a learning rate parameter, $\beta_i \in \mathbb{R}^p$ be a sequence of n vectors, and suppose a sequence of data $(\boldsymbol{x}_1, y_1), \cdots, (\boldsymbol{x}_n, y_n))$ is revealed one at a time. Consider the following procedure:

- ☐ Initialize: Set $\beta_1 = 0$.
- ☐ Get an outcome of the explanatory variable: $\boldsymbol{x}_1$.
- ☐ Predict the dependent variable: $\hat{y}_1 = \beta_1 \cdot \boldsymbol{x}_1$.
- ☐ Receive the correct answer: y_1.
- ☐ Update: Set $\beta_2 = \beta_1 - \eta(\hat{y}_1 - y_1)\boldsymbol{x}_1$.
- ☐ Repeat with $\boldsymbol{x}_2$, $\boldsymbol{x}_3$ and so on.

Now suppose that the linear model is correct; i.e., there is some $\beta \in \mathbb{R}^p$ and $M > 0$ such that, for all i, $y_i = \beta \cdot \boldsymbol{x}_i$ and $\|\boldsymbol{x}_i\|_2 \leq M$ for all i on the domain of the explanatory variables. Show that the cumulative squared error from the above procedure can be bounded by

$$\sum_{i=1}^{n} (y_i - \hat{y}_i)^2 \leq \|\beta\|_2^2.$$

Hint: First show the identity

$$\frac{1}{2}\|\beta - \beta_i\|_2^2 - \frac{1}{2}\|\beta - \beta_{i+1}\|_2^2 = \left(\eta - \frac{\eta}{2}M^2\right)(y_i - \hat{y}_i)^2.$$

Then, optimize over η and sum over i.

Exercise 4.4. Think of Net as the function from a neural network, and consider the sigmoidal function

$$f(Net) = a\tanh(bNet) = a\left[\frac{1 - e^{-bNet}}{1 + e^{-bNet}}\right] = \frac{2a}{1 + e^{-bNet}} - a,$$

where $a, b \geq 0$.

1. Verify that $f'(Net)$ can be written in terms of $f(Net)$ itself.
2. Find $f(Net)$, $f'(Net)$, and $f''(Net)$ when Net tends to $-\infty$, 0, and ∞.
3. What are the extrema of $f''(Net)$?

Exercise 4.5. Consider a neural net with a single output neuron, defined as $y = Net(\boldsymbol{x}, \boldsymbol{w})$, where $\boldsymbol{x}$ and $\boldsymbol{w}$ are in $\mathbb{R}^p$. Suppose a data set of the form $(\boldsymbol{x}_1, y_1), \cdots, (\boldsymbol{x}_n, y_n))$ is available and the average squared error on the data set is

$$\hat{E} = \frac{1}{2n}\sum_{i=1}^{n}(y_i - \hat{y}_i)^2,$$

where y_i is the outcome for $\boldsymbol{x}_i$ and $\hat{y}_i$ is the "fitted" value produced by *Net* for $\boldsymbol{x}_k$ using the eight vector $\boldsymbol{w}$. Show that the Hessian matrix $\mathbf{H}$ evaluated at a local minimum of the error surface defined by $\hat{E}$ may be approximated by

$$\mathbf{H} \simeq \frac{1}{n}\sum_{k=1}^{n}\mathbf{g}_k\mathbf{g}_k^\top,$$

where

$$\mathbf{g}_k = \frac{\partial Net(\boldsymbol{w}, \boldsymbol{x}_k)}{\partial \boldsymbol{w}}.$$

Exercise 4.6. Suppose a data set of the form $(\boldsymbol{x}_1, y_1), \cdots, (\boldsymbol{x}_n, y_n))$ is available and that the jth components of the $\boldsymbol{x}_i$s are written simply as $x_1, \cdots, x_n$. Let $c \in \mathbb{R}$ be a candidate split point for the x_is. Then, the best left prediction, p_L, is the average of the observations with $x_i < c$ and the best right prediction, p_R, is the average of the observations with $x_i > c$. The issue is how to define "best".

1. Suppose first that best is in the sense of squared error loss. So find the value of c which minimizes

$$L(p_L, p_R, c) = \sum_{i=1}^{n}\mathbf{1}_{x:x_j<c}(y_i - p_L)^2 + \mathbf{1}_{x:x_j>c}(y_i - p_R)^2.$$

   How can you use this to generate splits at a node in a tree model?
2. Redo item 1, but with absolute error loss. How can you use this to generate splits at a node in a tree model?

Exercise 4.7 (Variance–bias trade-off and regularization). The point of this exercise is to see the effect of the number of nodes in the hidden layer of a single hidden layer NN on the overall performance of the NN. Under squared error loss, you should see the variance–bias trade-off. This should lead you to suggest seeking a trade-off through regularization.

So, suppose the true function f defined on $[-10, 10]$ is

$$f(\mathbf{x}) = \frac{\sin x}{x}.$$

Using this f, generate a sample of $n = 200$ data points $(\boldsymbol{x}_i, y_i)$ for $i = 1, \cdots, 200$ from

$$Y = f(\mathbf{x}) + \varepsilon,$$

where the εs are IID $N(0, \sigma^2)$. with $\sigma = 0.2$, and the x_is equally spaced in $[-10, 10]$. The error to be used in this exercise is

$$E(\boldsymbol{w}) = \sum_{i=1}^{n}(y_i - Net(\boldsymbol{x}_i, \boldsymbol{w}))^2,$$

where $Net(\mathbf{x}_i, \mathbf{w})$ is the output produced by the NN for $\mathbf{x}_i$.

1. Construct the scatterplot of the data, and superimpose the true function on it in a different color.
2. Randomly split the data set into two portions, one with $n_{\text{train}} = 120$ data points to be used as the training set and the rest with $n_{\text{test}} = 80$ points to be used as the test set. (Other choices for n_{train} and n_{test} are possible.)
3. Download and install the R Neural Network Package **nnet** or the MATLAB Neural Network package **NETLAB**. Let r_1 denote the number of nodes in the hidden layer. Use $r_1 = 1, 2, 4, 6, 8, 16, 20$ and, for the sigmoid, use the function tanh. Do the following for each value of r_1:

   a. Estimate the weights of the NN, storing them and the fitted values.
   b. Compute the SSE; i.e., the training error.
   c. Using the optimal weights from item 1, compute the predicted values and find the SSE for the test set (i.e., the test error).

4. Plot both the training error and the test error as a function of r_1 and interpret what you see.

Exercise 4.8 (Extension of Exercise 4.7). In this exercise, use the regularized squared error loss

$$E_{\text{reg}}(\mathbf{w}) = \sum_{i=1}^{n} (y_i - Net(\mathbf{x}_i, \mathbf{w}))^2 + \lambda \sum_{j=1}^{n_1} \|\mathbf{w}_j\|^2,$$

where $Net(\mathbf{x}_i, \mathbf{w})$ is the NN output for $\mathbf{x}_i$. The point is to explore the effect of the regularization parameter λ and the number of nodes in the hidden layer on both the training error and the test error. Use the same data generation procedure as in Exercise 4.7 and either the R package **nnet** or the MATLAB package **NETLAB**.

1. For $r_1 \in \{4, 8, 16\}$, repeat the following for $\lambda = \frac{2}{10} i, i = 1, \cdots, 10$:

   a. Estimate the weights of the NN, storing them and the fitted values.
   b. Compute the SSE; i.e., the training error.
   c. Using the optimal weights from item 1, compute the predicted values and find the SSE for the test set (i.e., the test error).

2. Plot the training error and the test error as a function of λ.
3. For each r_1, compute the mean and variance of the test error as a function of λ. For each λ, plot them as a function of r_1.
4. Are the results what you expected? Explain.
5. What effect does increasing or decreasing σ^2 have on the mean and variance of the test error?

This process can be repeated for a multilayer neural network; i.e., one for which $\ell \geq 2$. If the network is rectangular (the same number of nodes in each hidden layer), the results should be similar. For irregular networks, the results should be more diffuse, but the curious reader is invited to seek a similar trade-off.

Exercise 4.9 (Ridge function representations). Functions whose argument is a one dimensional projection, $f(\mathbf{x}\beta)$, are called ridge functions because the projection divides the p-dimensional space into positive and negative parts. Ridge functions are the terms in PPR.

1. For an arbitrary function g write out the three term decomposition for estimating it by a technique such as PPR or GAMs. Note that, given the class of functions to be used to approximate g, there is a best approximand g^* that gives the optimal approximation error. After this, the usual bias–variance decomposition can be used on g^*.
2. To see that the decomposition in item 1 is too simple, let $p = 2$ and consider $g(x_1, x_2) = x_1 x_2$. Find two different representations of g^* as a sum of two ridge functions. Find a third. How many are there?
3. Find a function that is not a sum of finitely many ridge functions.
4. GAMs suffers the same problem as PPR. So, fix a link function and find a function that cannot be represented as a link function on an additive model.
5. To explore the effect of bias as n increases, use a normal error in $Y = f(x) + \varepsilon$ and generate data from

$$Y = f(\mathbf{x}) = \frac{\cos\left[\frac{1}{p}\sum_{j=1}^{p}\cos r_i(x_i)\right]}{1+\cos\left[\frac{1}{p}\sum_{j=1}^{p}\cos s_i(x_i)\right]},$$

   using various choices of r_i, s_i, and p; start with $p = 3$, $r_i(x) = s_i(x) = x$. Use equally spaced design points for convenience. Do trees or neural nets – which can give zero bias in the limit of large data – perform better, worse, or about the same as the biased techniques like GAMs or PPR in small sample sizes? Does the same problem occur for SIR and MARS? Explain. (Don't forget about variable selection.)

Exercise 4.10 (Projection pursuit regression). This is a generic comparison of several methods. Choose a complicated function f and generate responses of the form $Y = f(X_1, ..., X_p) + \varepsilon$ for some $p \geq 3$, and generate data from it. (The Friedman function in Chapter 7 is one instance.) You can generate an error term from a $N(0, \sigma_0^2)$ for various choices of σ or use other unimodal distributions. A few ways to choose the design points are the following: Choose each $X_i \sim N(0, B^2/16)$ for some fixed $B > 0$, choose $X_i \sim Unif[0, B]$ and choose $X_i \sim Unif[-B/2, B/2]$.

To do two-dimensional PPR, an excellent package is called XGobi and available from the StatLib archive. A version of Xgobi for the freeware R statistical software package is at CRAN. The software associated with Friedman (1987) is also available from the StatLib archive.

1. Fit a linear model to the data. Does residual analysis suggest any linear model captures the key feature of your function well?
2. Fit an additive nonparametric model, and compare it with the linear one. Any individual smoother will do.
3. Perform PPR using various smoothing methods and different numbers of terms (one, two, or three). Choose the most reasonable PPR. What are the resulting explanatory variables? Compare the results with those obtained in (a) and (b).
4. Now redo the problem using neural nets (with regularization). What are the differences?
5. Hardy spirits may try the ACE, AVAS, or MARS procedures. Do the transformations from ACE or AVAS hint at any parametric model?

Exercise 4.11 (Another comparison).

Parallel to the previous exercise, consider the (one-dimensional) Doppler function

$$g(x) = \sqrt{(x*(1-x))} * \sin((2*\pi*(1-e))/(x+e)),$$

where $e = 0.05$. Generate a data set by adding $N(0,\sigma^2)$ noise to $g(x)$ at uniformly chosen design points $x_i = i/500$ for $i = 1,...,500$.

1. Plot the original Doppler signal and the generated noisy one.
2. Use the techniques of Chapters 2 and 3 to estimate g, namely the Nadaraya-Watson estimator, nearest neighbors, and a cubic or other spline estimator. Find the MSE for these two techniques and generate residual plots.
3. Use a single hidden layer neural network (with and without a penalty term) to generate an approximation to g. What happens if you allow extra hidden layers?
4. Now, use a recursive partitioning (tree) based approach. The Gini index is standard (see Chapter 5) but other selection techniques are possible; use standard cost-complexity pruning. Again, plot g and the tree-based approximation and give an MSE. Compare the residual analyses of these techniques.

Exercise 4.12. In this problem the goal is to establish conditions under which backfitting converges and to identify the limit. So, refer back to (4.1.5).

1. Verify that (4.1.5) with $p = 2$ is equivalent to

$$\boldsymbol{f}_1 = \boldsymbol{L}_1(\boldsymbol{y} - \boldsymbol{f}_2) \quad \text{and} \quad \boldsymbol{f}_2 = \boldsymbol{L}_2(\boldsymbol{y} - \boldsymbol{f}_1),$$

   where L_1 and L_2 are the smoother matrices for two linear procedures as in the definition (2.1.1).
2. Use the expressions in item 1 to follow the backfitting procedure. That is, start with $f_{2,init}$ and use the first equation to get f_1^1. Use f_1^1 in the second equation to get f_2^1 for use in the first equation again. Write the iterates as $\boldsymbol{f}_1^j$ and $\boldsymbol{f}_2^j$. Verify that

$$\boldsymbol{f}_1^j = \boldsymbol{L}_1(\boldsymbol{y} - \boldsymbol{f}_2^{j-1}) \quad \text{and} \quad \boldsymbol{f}_2^j = \boldsymbol{L}_2(\boldsymbol{y} - \boldsymbol{f}_1^j).$$

3. The goal is to show that if $||\boldsymbol{L}_1\boldsymbol{L}_2|| < 1$, then $\boldsymbol{f}_1^j \to \boldsymbol{f}_1$ and $\boldsymbol{f}_2^j \to \boldsymbol{f}_2$. So, start by using induction on j to derive

$$\boldsymbol{f}_1^J = \boldsymbol{y} - \sum_{j=0}^{J-1} (\boldsymbol{L}_1\boldsymbol{L}_2)^j (I_p - \boldsymbol{L}_1)\boldsymbol{y} - (\boldsymbol{L}_1\boldsymbol{L}_2)^{J-1}\boldsymbol{L}_1\boldsymbol{f}_{2,init},$$

$$\boldsymbol{f}_2^J = \boldsymbol{L}_2 \sum_{j=0}^{J-1} (L_1L_2)^j (I_p - L_1)\boldsymbol{y} + L_2(L_1L_2)^{J-1}\boldsymbol{f}_{2,init},$$

for given J.

4. Choose the usual norm on matrices; i.e., for a $p \times p$ matrix $\mathbf{M}$, let $||\mathbf{M}|| = \sup_{||\mathbf{v}||=1} ||\mathbf{M}\mathbf{v}\}$, where the norm on $\mathbf{v}$ is Euclidean. Argue that the expressions in item 3 converge when $||\boldsymbol{L}_1\boldsymbol{L}_2|| < 1$.

5. Now, given $||\boldsymbol{L}_1\boldsymbol{L}_2|| < 1$, solve the system of equations in item 4 to obtain the limiting expressions

$$\begin{aligned}\boldsymbol{f}_{1,\infty} &= (I_p - (I_p - \boldsymbol{L}_1\boldsymbol{L}_2)^{-1}(I_p - \boldsymbol{L}_1))\boldsymbol{y} \\ \boldsymbol{f}_{2,\infty} &= \boldsymbol{L}_2(I_p - \boldsymbol{L}_1\boldsymbol{L}_2)^{-1}(I_p - \boldsymbol{L}_1)\boldsymbol{y} \\ &= (I_p - (I_p - \boldsymbol{L}_2\boldsymbol{L}_1)^{-1}(I_p - \boldsymbol{L}_2))\boldsymbol{y}.\end{aligned}$$

6. Argue that the fitted values are $\hat{y} = \boldsymbol{f}_{1,\infty} + \boldsymbol{f}_{2,\infty}$.

Chapter 5
Supervised Learning: Partition Methods

Basically, supervised learning is what statisticians do almost all the time. The "supervision" refers to the fact that the Y_is are available, in contrast to unsupervised learning, the topic of Chapter 8, where Y_is are assumed unavailable. The term "learning" is used in a heuristic sense to mean any inferential procedure that can, in principle, be tested for validity. Having measurements on Y available means that model identification, decision making, prediction, and many other goals of analysis can all be validated.

In practice, supervised learning more typically refers to a topic that is less familiar to statisticians but is the focus of this chapter: classification. This is like a regression in which the dependent variable assumes one of finitely many values, representing classes. The task remains to identify a regression function but now it is called a classifier because the goal is mostly to treat a new feature vector as an input and take its value as the class label. Implicit in this is the primacy of prediction because classifiers are evaluated almost exclusively on future data (or holdout sets). Roughly every regression technique gives a classifier if it is applied for discrete responses, and every classification procedure corresponds to a regression problem, although the Y may only take two values.

The point of classification in general is to slot objects in a population – patients, cars, images, etc. – into one of two or more categories based on a set of features measured on each object. For patients, this might include sex, age, income, weight, blood pressure, and so forth. The categories are known and, in general, not ordered. However, it is easy to imagine classifying patients into low-, medium-, and high-risk groups, for instance.

Like regression, the typically data consist of n outcomes Y_i with corresponding covariates or explanatory variables $\boldsymbol{X}_i$ of length p. In the supervised (and unsupervised) learning context, explanatory variables are often called features and the data are often called a training set. The Machine Learning Repository at the University of California-Irvine has well-documented training sets often used in DMML for testing new methods. The first goal is to use the data to choose which components of $\boldsymbol{X}$, or possibly functions of them, are most important for determining which category Y_i came from; this topic will be addressed in Chapter 10. The second goal is to use this information to use the data to find a function of the explanatory variables that will identify the class for a given $\boldsymbol{x}$. This analog to the regression function is often called the classification

B. Clarke et al., *Principles and Theory for Data Mining and Machine Learning*, Springer Series in Statistics, DOI 10.1007/978-0-387-98135-2_5, © Springer Science+Business Media, LLC 2009

rule, $\hat{f}(\cdot)$. It is equivalent to specifying the decision boundary in feature space that best separates the classes. Strictly speaking, only the second of these two goals is classification, but variable selection, is often a necessary step for constructing a classifier.

In a training set, each data point is a vector of length $p+1$ consisting of Y and the p covariates that gave rise to it. It is assumed that there is an ideal classifier f that $\hat{f}$ estimates, just as in regression there is a true function that the fitted curve estimates. Thus, for a new observation Y_{new}, one looks at the covariate values $\boldsymbol{X}_{new}$ and predicts that Y_{new} will be in the class $\hat{f}(\boldsymbol{X}_{new})$. The basic way a classifier such as $\hat{f}$ is assessed is through its probability of misclassification, $P(\hat{f}(\boldsymbol{X}_{new}) \neq Y_{new})$. The probability P may be conditional on the training set, i.e., neglect the randomness in estimating f or P may be as written and include the variability in $\hat{f}$. Note that the existence of an ideal classifier does not in general figure in the evaluation of an estimated classification rule; this is unlike regression, where rates are on the norm between an estimate and a true function. Another consequence of the discreteness of Y is that the loss function in classification is usually discrete as well; for instance, 0-1 loss is typical, as opposed to a continuous loss like squared or absolute error.

Classification procedures themselves can be classified into several classes. One is called partitioning methods. These methods partition the feature space into disjoint, usually exhaustive, regions. The hope is that the responses for feature vectors from the same region should be similar to each other. Often these regions are found recursively, and in the simplest cases they are rectangles. For instance, one can start with a variable x_1 and partition first by $x_1 < c$ versus $x_1 \geq c$. Then, within each rectangle, the partition can be repeated on another variable.

Classification can also be done nonrecursively. In some cases, a discriminant function, say $d_k(\boldsymbol{x}_{new})$, is used to assess how representative $\boldsymbol{x}_{new}$ is of each class $k = 1, \ldots, K$, assigning class $k_{opt} = \arg\max_k \; d_k(x_{new})$ to $\boldsymbol{x}_{new}$. Discriminant functions can be linear or not, in which case they are called flexible. Even when they are not linear, discriminant functions are sometimes only a monotone transformation away from being linear. For instance, it will be seen that logistic regression can be used to obtain a classifier. In the two-class setting, let

$$\mathbb{P}(Y=1|X=x) = \frac{e^{\beta_0+\beta' x}}{1+e^{\beta_0+\beta' x}}$$

and

$$\mathbb{P}(Y=2|X=x) = \frac{1}{1+e^{\beta_0+\beta' x}}$$

be discriminant functions for the two classes. Then, the monotone transformation given by the logit, $p \to \log(1/(1-p))$, gives

$$\log \frac{\mathbb{P}(Y=1|X=x)}{\mathbb{P}(Y=2|X=x)} = \beta_0 + \beta' x,$$

and it is seen that the line $\beta_0 + \beta' x = 0$ is the boundary between class 1 and class 2.

Other partitioning methods include trees, neural networks, and SVMs. Indeed, SVMs are one of the most important: They have linear and nonlinear forms, can be used with separable or nonseparable data, and, to a great extent, evade the Curse because they make use of kernel methods. Separable data means that the classes, or more exactly the data, really are disjoint; nonseparable, which is more typical, means that the regions for each class have boundaries that are much more general and the data appear to overlap.

Nonpartitioning classification procedures include nearest neighbor methods, discussed in Chapter 2, and the relevance vector machine, to be discussed in Chapter 6. There are not as many nonpartitioning methods; they tend to identify representative points and predict based on proximity to them. In fact, nonpartitioning methods can be expressed so they give a partition. However, the partition is so variable that it is not really helpful to think of the method in those terms.

All of the methods treated here are crisp. That is, the complication that Y may only be probabilistically determined by $\boldsymbol{X}$ is not considered. In other word, the setting that, for a given feature vector $\boldsymbol{x}$, it may be that both $P(Y(\boldsymbol{x}) = 0) > 0$ and $P(Y(\boldsymbol{x}) = 1) > 0$ is ruled out. For instance, one can have high blood pressure but not be at risk for a heart attack. Blood pressure increases when someone is in pain. So, a heart-attack classifier assuming Y is essentially determined by $\boldsymbol{x}$ may improperly identify some one with back pain as actually having a heart attack on the basis of elevated blood pressure when a more sophisticated technique would only report a probability (less than one) of heart attack. Non-crisp methods are a level of complexity beyond the present scope.

This section presents three basic partitioning methods in order of increasing complexity. The first and third, discriminant analysis and SVMs, are non-recursive. The second, based on trees, is recursive. At the end, a short discussion of neural nets (also nonrecursive) in a classification context is given.

5.1 Multiclass Learning

The simplest classification problems separate a population into two classes labeled 1 and 2. These binary classification problems almost always generalize to multiclass classification problems. Although binary classification is the paradigm case and convenient to examine first, it is no harder to state some of the formalities for the general K class setting.

The task is to find a decision function to discriminate among data from K different classes, where $K \geq 2$. The training set consists of samples $(\boldsymbol{x}_i, y_i)$ for $i = 1, \ldots, n$, where $\boldsymbol{x}_i \in \mathbb{R}^p$ are the feature vectors and $y_i \in \{1, \ldots, K\}$ is the class label for the ith sample. The main task is to learn a decision rule,

$$f(\boldsymbol{x}) : \mathbb{R}^p \to \{1, \ldots, K\},$$

used to separate the K classes and predict the class label for a new input $\boldsymbol{x} = \boldsymbol{x}_{new}$.

A trained multiclassifier is generally associated with a K-function vector

$$\boldsymbol{d}(\boldsymbol{x}) = (d_1(\boldsymbol{x}), \ldots, d_K(\boldsymbol{x})),$$

where $f_k(\boldsymbol{x})$ represents the strength of evidence that $\boldsymbol{x}$ belongs to class k. The induced classifier ϕ from $\boldsymbol{f}$ is defined as

$$f(\boldsymbol{x}) = \arg \max_{k=1,\ldots,K} d_k(\boldsymbol{x}). \tag{5.1.1}$$

The decision boundary between classes k and l is the set

$$\{\boldsymbol{x} \in \mathbb{R}^p : d_k(\boldsymbol{x}) = d_l(\boldsymbol{x})\}, \quad \forall k \neq l.$$

If the decision boundaries are linear in $\boldsymbol{x}$, the problem is linearly separable. Figure 5.1, cf. Liu and Shen (2006), shows a simple three-class classification problem, where the training set can be perfectly separated by linear functions. Relatively few classification

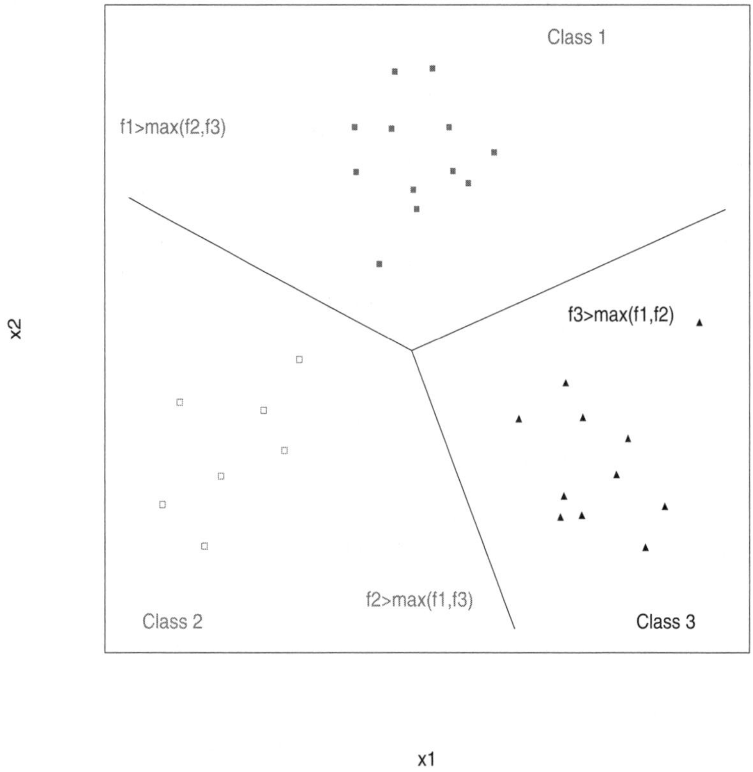

Fig. 5.1 An example of a three-class problem with linear boundaries.

problems are linearly separable. However, the concept is useful, and often it is possible to transform the feature space so that the boundaries in the transformed space are linear.

If K is not too large, one way to simplify multiclass problems is to transform them into a series of binary problems. Two popular choices are the one-versus-rest approach (also known as one-versus-all or OVA) and the pairwise comparison approach (also known as all-pairs, all-versus-all, or AVA). The OVA method trains K binary classifiers. Each

$\hat{d}_k(\boldsymbol{x})$ is trained to separate class k from the rest. These K binary classifiers are then combined to give a final classification

$$\hat{f}(\boldsymbol{x}) = \arg \max_{k=1,\ldots,K} \hat{d}_k(\boldsymbol{x}). \tag{5.1.2}$$

The OVA method is easy to implement and hence is a popular choice in practice. However, it may give poor classification performance in the absence of a dominating class when none of the $p_k(\boldsymbol{x})$s is greater than $1/2$; see Lee et al. (2004). Another disadvantage of the OVA method is that the resulting binary problems are often very unbalanced; see Fung and Mangasarian (2004). By contrast, the pairwise classifications in AVA train $K(K-1)/2$ binary classifiers, each separating a pair of classes. The final class predicted for an $\boldsymbol{x}$ is decided by a voting scheme among all the classifiers. One criticism of these methods is that a large number of training tasks may be involved, especially for AVA, when K is not small.

In the rest of this chapter, the goal will be to develop the general K case. However, discriminant methods and SVMs are most naturally presented in the binary $K = 2$, case and this will be done below. Details on extensions to general K will be discussed where possible. It is only for tree-based classifiers, and NNs that the general K case is no harder to present than the binary case.

5.2 Discriminant Analysis

As noted earlier, the idea behind discriminant analysis is that K functions $d_k(\cdot)$ are derived so that $d_k(\boldsymbol{x}_{new})$ can be used to assess how representative $\boldsymbol{x}_{new}$ is of each class $k = 1,\ldots,K$, as in (5.1.1) or (5.4.23). The natural classifier then assigns class

$$k_{opt} = \arg\max_k \; d_k(x_{new})$$

to $\boldsymbol{x}_{new}$. In practice, one does not go too far wrong to think of d_k as representing something like a distance between the sample mean of the kth class and a new value of $\boldsymbol{X}$, remembering that projections are closely related to distances.

The paradigm case for discriminant analysis is that $K = 2$ and the two classes correspond to values of two explanatory variables that concentrate on two parallel, elliptically shaped regions in the plane, one for each class. In this case, the two classes are linearly separable because there is a linear decision boundary, a line in the plane such that essentially all the class 1 cases are on one side and all class 2 cases are on the other. This corresponds to having linear discriminant functions. More generally, linear separability means that the classes can be separated by a linear combination of features. For instance, when $p = 3$, a linear decision boundary would be a plane and when $p = 4$ a linear decision boundary would be a hyperplane. Squares of the explanatory variables can be included in the d_ks to give quadratic discriminant functions with quadratic boundaries, although the term quadratic separability is not in common parlance. In general, decision boundaries obtained from discriminant functions d_k can be any partition of the feature space.

Discriminant functions fall roughly into three conceptual classes – distance-based, Bayes, and probability-based even though the techniques in the classes have some overlap. (For instance, the Bayes classifier turns out to be distance-based in many cases.) Distance-based classifiers were the earliest and led to the linear discriminant functions pioneered by Fisher in the 1930s, now often just called collectively Fisher's linear discriminant analysis (LDA). Bayesian classification came later and has a decision-theoretic foundation; it also leads to linear discriminants but of a slightly different form. In the third class, probability based, the Bayes rule is estimated. First, conditional probabilities $p(\boldsymbol{x}) = \mathbb{P}(Y = k|\boldsymbol{X} = \boldsymbol{x})$ are estimated, often using a model, and then sample points $\boldsymbol{x}$ are classified according to the highest probability $\hat{p}_k(\boldsymbol{x})$. It can be argued that most standard statistical approaches for multiclass discrimination are probability-based because the Bayes rule is optimal (even if hard to implement) and any good classifier should approximate it satisfactorily.

5.2.1 Distance-Based Discriminant Analysis

To start thinking about distance-based classification, start with $K = 2$, so $Y = 1, 2$, and consider looking for a linear function that predicts the class for a new value $\boldsymbol{X} = \boldsymbol{x}$. The simplest case is that $\mathbb{P}(\boldsymbol{X}|Y = 0)$ and $\mathbb{P}(\boldsymbol{X}|Y = 1)$ are normally distributed with means μ_0 and μ_1 and the same covariance matrix Σ, assumed to be of full rank. Given data, it makes intuitive sense to find $\bar{\boldsymbol{x}}_1$ and $\bar{\boldsymbol{x}}_2$, the means of the observations with values $y = 1$ and $y = 2$, and assign $\boldsymbol{x}$ to the class i with a smaller value of $\|\bar{\boldsymbol{x}}_i - \boldsymbol{x}\|$, where the norm uses the inner product on $\mathbb{R}^p$ defined from Σ. It will be seen that this is equivalent (apart from estimating parameters) to using the optimal classifier that assigns whichever of class 1 or 2 has a higher value of $\mathbb{P}(Y = y|\boldsymbol{X} = \boldsymbol{x}) = \Sigma^{-1}(\mu_1 - \mu_2) \cdot \boldsymbol{x}$. That is, distance-based classification often converts to a projection, and projections used for classification often have a distance-based interpretation. Even better, both perspectives can lead to linear discriminants. Such procedures typically extend from the binary case to the multiclass setting and have a Bayesian interpretation if the proportions of the classes are regarded as a prior.

Noting that variances are average squared distances, Fisher ignored conditional probabilities and derived a linear discriminant using a criterion based on the difference between two projections of $\boldsymbol{x}$ scaled by the variances of the classes. This did not require normality or equal variance matrices. To see this, suppose there are two classes with means μ_0 and μ_1 and variance matrices Σ_1 and Σ_2, and consider a linear combination $L = \boldsymbol{w} \cdot \boldsymbol{x}$. Then, $\mathbb{E}(L|Y = y) = \boldsymbol{w}\mu_y$ and $\mathrm{Var}(L|Y = y) = \boldsymbol{w}\Sigma_y\boldsymbol{w}$. Let the separation between the two classes be defined by

$$S = \frac{(\boldsymbol{w} \cdot \mu_2 - \boldsymbol{w} \cdot \mu_1)^2}{\boldsymbol{w}\Sigma_1\boldsymbol{w} + \boldsymbol{w}\Sigma_2\boldsymbol{w}},$$

a sort of signal-to-noise ratio. It can be verified that S is maximized when $\mathbf{w} = (\Sigma_1 + \Sigma_2)^{-1}(\mu_2 - \mu_1)$, giving a projection form of Fisher's LDA, and directly generalizing beyond the normal case which required $\Sigma_1 = \Sigma_2$.

Again visualize two long, thin, parallel, elliptically shaped regions in a plane, each ellipse representing the values of an explanatory variable for a class. If the class information is ignored, the scatterplot of the full data set may look like a single thicker ellipse of the same length or, better, like two thin ellipses of the same length without much overlap. That is, the full data set can be treated as a single ellipse in which the major axis parallel to the two thin ellipses represents the larger of two variances and the minor axis, crossing the two small ellipses, is the smaller of two variances.

Imagine separating the two classes by a decision boundary. If the data points were projected onto the major axis and then a one-dimensional classification were done, the result would be terrible because the points from the two classes would be intermingled. The decision boundary would be transverse to the ellipses. However, if the data points were projected onto the minor axis, the result would be quite good; the decision boundary would be a line between the two ellipses, and the separation between the two classes would be relatively clean. As noted above, the basic task of linear discriminants is to find good decision boundaries, and this can be done by finding useful projections. There are many ways this can be done. Two are as follows. First, a relevant objective function can be maximized to give a useful projection for classification in general; this is another form of Fisher's LDA. Second, one can seek an optimal linear transformation corresponding to the optimal directions for classification generalizing Σ^{-1} in Fisher's LDA. Both of these will be developed below.

5.2.1.1 Direct Maximization

Instead of maximizing the separation S between two classes, one can use ANOVA reasoning. Recall that the main way the different cells in an ANOVA are contrasted is by comparing the between-cell variation to the within-cell variation. The analogous treatment for classification can be developed defining between-scatter and within-scatter matrices. In terms of the data, these are S_B and S_W, defined by

$$S_B = \sum_{k=1}^{K} n_k(\bar{\mathbf{x}}_k - \bar{\mathbf{x}})(\bar{\mathbf{x}}_k - \bar{\mathbf{x}})^{\mathsf{T}} \text{ and } S_W = \sum_{k=1}^{K}\sum_{i=1}^{n_k}(\mathbf{x}_i - \bar{\mathbf{x}}_k)(\mathbf{x}_i - \bar{\mathbf{x}}_k)^{\mathsf{T}}, \quad (5.2.1)$$

in which n_k is the number of samples from class k, $\sum_k n_k = n$, $\bar{\mathbf{x}}_k$ is the average of the samples in class k, and $\bar{\mathbf{x}}$ is the average over all the samples. It can be checked that

$$S_B + S_W = S_T = \sum_{i=1}^{n}(\mathbf{x}_i - \bar{\mathbf{x}})(\mathbf{x}_i - \bar{\mathbf{x}})^{\mathsf{T}},$$

where S_T is the total scatter.

Now, the analog of the ratio of variances in ANOVA is

$$J(\boldsymbol{w}) = \frac{\boldsymbol{w}^\mathsf{T} S_B \boldsymbol{w}}{\boldsymbol{w}^\mathsf{T} S_W \boldsymbol{w}}.$$

Maximizing $J(\boldsymbol{w})$ is equivalent to maximizing the total scatter $\boldsymbol{w}^\mathsf{T} S_T \boldsymbol{w}$ while minimizing the within-scatter $\boldsymbol{w}^\mathsf{T} S_W \boldsymbol{w}$, parallel to the ANOVA sum of squares decomposition.

Since $J(\boldsymbol{w}) = J(\alpha \boldsymbol{w})$ is homogeneous for any α and the denominator is a squared norm, it makes sense to maximize over the unit sphere in p dimensions in the norm defined by using S_W as an inner product. Thus, it is enough to solve

$$\begin{aligned} \text{Find} \quad & \min_{\boldsymbol{w}} -\frac{1}{2} \boldsymbol{w}^\mathsf{T} S_B \boldsymbol{w} \\ \text{subject to} \quad & \boldsymbol{w}^\mathsf{T} S_W \boldsymbol{w} = 1. \end{aligned}$$

This is a familiar quadratic optimization problem solvable by Lagrange multipliers; see Johnson and Wichern (1998) and Welling (2005). The Lagrange multiplier λ can be found to satisfy $S_W^{-1} S_B \boldsymbol{w} = \lambda \boldsymbol{w}$. If $S_W^{-1} S_B$ were symmetric, this would be an eigenvalue problem. However, $S_W^{-1} S_B$ is not always symmetric. To get around this, one can use a transformation based on the fact that S_B is positive symmetric and therefore has a square root that can be represented as $S_B^{1/2} = U \Lambda^{1/2} U$, where U diagonalizes S_B to Λ, i.e., $S_B = U \Lambda U$. Writing $\boldsymbol{v} = S_B^{1/2} \boldsymbol{w}$ converts the Lagrange multiplier condition to

$$S_B^{-1/2} S_W^{-1} S_B^{-1/2} \boldsymbol{v} = \lambda \boldsymbol{v}.$$

So, the possible solutions for λ are seen to be the eigenvalues of $S_B^{-1/2} S_W^{-1} S_B^{-1/2}$ with eigenvectors $\boldsymbol{v}_k$ given corresponding $\boldsymbol{w}$s as $\boldsymbol{w}_k = S_B^{-1/2} \boldsymbol{v}_k$. Putting this back into the numerator of $J(\boldsymbol{w})$ gives that

$$J(\boldsymbol{w}) = \frac{\boldsymbol{w}^\mathsf{T} S_B \boldsymbol{w}}{\boldsymbol{w}^\mathsf{T} S_W \boldsymbol{w}} \leq \lambda_p \frac{\boldsymbol{w}_p' S_W \boldsymbol{w}_p}{\boldsymbol{w}_p^\mathsf{T} S_W \boldsymbol{w}_p} = \lambda_p$$

when λ_p is the largest eigenvalue.

This means that the projection defined by $\boldsymbol{w}_p$, the eigenvector corresponding to the largest eigenvalue, is the optimal projection for maximizing the between-class scatter relative to the within-class scatter, i.e., the discriminant function is

$$\hat{d}_k(\boldsymbol{x}) = \hat{\boldsymbol{w}}_p(\boldsymbol{x} - \bar{\boldsymbol{x}}_k) \tag{5.2.2}$$

because of the centering in the scatter matrices. Note that this remains a projection onto a unidimensional space and so may not be particularly good if the classes are not properly aligned so their minor axes are roughly parallel. This deficiency can be partially fixed by a projection onto a space of dimension higher than one; see below.

5.2.1.2 Noise in LDA

It is important at this point to look briefly at the effect of noise on distance-based classification. The effect of noise is most easily analyzed using a form of Fisher's LDA that will be derived in the next subsection. It is Bayesian and obtains expressions for conditional probabilities of the form $\mathbb{P}(Y = k|\boldsymbol{X} = \boldsymbol{x})$. The discriminant function it gives reduces to assigning $\boldsymbol{x}$ to the class k achieving the maximum of

$$d_k(\boldsymbol{x}) = d_M(\boldsymbol{x}, \bar{x}_k) = [(x - \bar{x}_k)^{\mathsf{T}} \Sigma^{-1} (x - \bar{x}_k)]^{1/2} \tag{5.2.3}$$

when the variance matrices of the K classes are the same. That is, $\boldsymbol{x}$ is assigned to the class whose sample mean is closest to the observation in Mahalanobis distance, a norm derived from an inner product defined by the inverse variance matrix. It can be seen that this is equivalent, apart from estimating parameters, to the normal LDA rule for two classes given at the start of this section and is of a form similar to the other linear discriminants. If Σ is unknown, then the usual estimate can be used and, in practice, it is helpful to use robust estimators for both μ_k and Σ.

To analyze the effect of noise in linear discriminant analysis, assume a fixed sample size of n and that the common variance matrix is diagonal, $\Sigma = \sigma^2 I$. Write the estimates of the means as

$$\hat{\mu}_1 = \mu_1 + \frac{\sigma}{\sqrt{n}} \boldsymbol{V}_1 \quad \text{and} \quad \hat{\mu}_2 = \mu_2 + \frac{\sigma}{\sqrt{n}} \boldsymbol{V}_2$$

for unknown vectors $\boldsymbol{V}_1$ and $\boldsymbol{V}_2$. Suppose that the new observation to classify is $\boldsymbol{X} = \mu_0 + \sigma \boldsymbol{V}$. Here, $\boldsymbol{V}_1$, $\boldsymbol{V}_2$, and $\boldsymbol{V}$ can be treated as $N_p(0, I)$ random errors.

Now, the Mahalanobis form of Fisher's LDA assigns class 1 if $d_M(\boldsymbol{X}, \hat{\mu}_1) < d_M(\boldsymbol{X}, \hat{\mu}_2)$, and this is equivalent to $(\boldsymbol{X} - \hat{\mu}_1)^{\mathsf{T}} (\boldsymbol{X} - \hat{\mu}_1) < (\boldsymbol{X} - \hat{\mu}_2)^{\mathsf{T}} (\boldsymbol{X} - \hat{\mu}_2)$. Writing $\boldsymbol{X}$, $\hat{\mu}_1$, and $\hat{\mu}_2$ in terms of $\boldsymbol{V}$, $\boldsymbol{V}_1$, and $\boldsymbol{V}_2$ shows this is equivalent to

$$\begin{aligned} \left((\mu_1 - \mu_2) + \sigma \boldsymbol{V} - \frac{\sigma}{\sqrt{n}} \boldsymbol{V}_2 \right)^{\mathsf{T}} & \left((\mu_1 - \mu_2) + \sigma \boldsymbol{V} - \frac{\sigma}{\sqrt{n}} \boldsymbol{V}_2 \right) \\ & > \left(\sigma \boldsymbol{V} - \frac{\sigma}{\sqrt{n}} \boldsymbol{V}_2 \right)^{\mathsf{T}} \left(\sigma \boldsymbol{V} - \frac{\sigma}{\sqrt{n}} \boldsymbol{V}_1 \right). \end{aligned} \tag{5.2.4}$$

As $n \to \infty$, this criterion converges to the rule that assigns class 1 when $2\sigma \boldsymbol{V} \cdot (\mu_1 - \mu_2) + \|\mu_1 - \mu_2\|^2 > 0$. So, when $p = 1$, the asymptotic probability of misclassification is seen to be $P[\boldsymbol{V} > |\mu_1 - \mu_2|/2\sigma]$ and the error rate depends on the signal-to-noise ratio $|\mu_1 - \mu_2|/\sigma$. Analogous expressions for larger p can also be derived but are left as an exercise.

Without using asymptotics, one can use (5.2.4) to show that Fisher's LDA specifies a plane in p dimensions that partitions the feature space. Raudys and Young (2004), Sec. 3 give formulas for the probability of misclassification.

Note that fully three forms of Fisher's LDA have now been seen. The original version based on separation led to the projection form using $\boldsymbol{w} = (\Sigma_0 + \Sigma_1)^{-1}(\mu_1 - \mu_0)$; the ratio of scatter matrices led to another projection form

$$\hat{d}_k(\boldsymbol{x}) = \hat{\boldsymbol{w}}_p(\boldsymbol{x} - \bar{\boldsymbol{x}}_k).$$

Third, the Mahalanobis form just seen uses a distance explicitly and is derived from a Bayesian approach. In fact, the Bayes approach more generally leads to yet a fourth form for Fisher's LDA. All of these are called Fisher's LDA, although, in practice, it is the Bayes version that is most commonly meant.

5.2.1.3 Optimal Linear Transformations

Maximizing the between-scatter matrix relative to the within-scatter matrix can be done in many ways; there are many functions that are reasonable besides taking the matrices as inner products. In fact, taking inner products may not reflect the spread of a matrix at all: The largest single eigenvalue says little about the $(p-1)$-dimensional subspace corresponding to the eigenvectors of the other eigenvalues. The dispersion of a matrix may therefore be better represented by some function of its entries that treats the eigenvalues symmetrically, for instance the trace. Moreover, it may be desirable to be able to choose the dimension of the projection.

To set this up, follow Ye (2007) and consider an $\ell \times p$ real matrix G and for any $\boldsymbol{x}_i$ write $G\boldsymbol{x}_i = \boldsymbol{x}_i^L$. It would be nice to maximize $\text{trace}(S_B)/\text{trace}(S_W)$ or perhaps $\det(S_B)/\det(S_W)$, but this seems hard. So, recall the matrix form of the conservation of variation equation, $S_T = S_B + S_W$, and transform it by G. Set

$$S_W^L = GS_WG^\top, \quad S_B^L = GS_BG^\top, \quad \text{and} \quad S_T^L = GS_TG^\top.$$

Writing G as column vectors $G = (\boldsymbol{g}_1, \ldots, \boldsymbol{g}_\ell)$, it is seen that, for $\boldsymbol{x} \in \mathbb{R}^p$, $G\boldsymbol{x} = (\boldsymbol{g}_1 \cdot \boldsymbol{x}, \ldots, \boldsymbol{g}_\ell \cdot \boldsymbol{x}) \in \mathbb{R}^\ell$.

Now, it would be nice to find G that maximizes $\text{trace}(S_B^L)$ and minimizes $\text{trace}(S_W^L)$ at the same time. So, it is reasonable to look at

$$\max_G \ \text{trace}(S_W^L)^{-1} S_B^L \ \text{and} \ \min_G \ \text{trace}(S_B^L)^{-1} S_W^L.$$

As noted in Ye et al. (2004), these optimizations, like the last direct maximization, are equivalent to a generalized eigenvalue problem. Indeed, for the maximization, set $S_B\boldsymbol{x} = \lambda S_W\boldsymbol{x}$ for $\lambda \neq 0$ assuming S_W is nonsingular and use an eigendecomposition of $S_W^{-1}S_B$. (If S_B is nonsingular, use an eigendecomposition of $S_B^{-1}S_W$.) Since $\text{rank}(S_B) \leq K-1$, there are at most $K-1$ nonzero eigenvalues. The minimization is similar. Hopefully, G will preserve the class structure of the data in $\mathbb{R}^p$ while projecting it to $\mathbb{R}^\ell$.

These two optimizations are also equivalent to finding G that maximizes $\text{trace}(S_B^L)$ and minimizes $\text{trace}(S_T^L)$. A standard argument (see Fukunaga (1990)) gives that the optimal G^{LDA} satisfies

$$G^{LDA} = \arg\max_G \ \text{trace}(S_B^L(S_T^L)^{-1})$$

and consists of the ℓth largest eigenvectors of $S_B(S_T)^{-1}$ (with nonzero eigenvalues), assuming S_T is nonsingular. Clearly, these three optimizations generalize the optimization in the last subsection in which G was just the eigenvector corresponding to the largest eigenvalue in (5.2.2).

It is straightforward to verify (see Ye (2007)) that when there are $K = 2$ classes Fisher's LDA and linear regression of the $\boldsymbol{x}$s on the class labels are equivalent. Indeed, suppose the labels are ± 1 and the data points are centered; i.e., $\boldsymbol{x}_i$ is replaced by $\boldsymbol{x}_i - \bar{\boldsymbol{x}}$ and y_i is replaced by $y_i - \bar{y}$, and the linear model $f(\boldsymbol{x}) - \boldsymbol{x}^\top \boldsymbol{\beta}$ is fit for $\boldsymbol{\beta} \in \mathbb{R}^p$. Then, under squared error loss, $\hat{\boldsymbol{\beta}} = (X^\top X)^{-1} X^\top y^n$, where X is the design matrix and $y^n = (y_1, \ldots, y_n)^\top$ is the vector of classes. Using $X^\top X = nS_T$ and $Xy^n = (2n_1n_2/n)(\bar{\boldsymbol{x}}_1 - \bar{\boldsymbol{x}}_2)$, it can be verified that

$$\hat{\boldsymbol{\beta}} = \frac{2n_1n_2}{n^2} S_T^{-1}(\bar{\boldsymbol{x}}_1 - \bar{\boldsymbol{x}}_2)$$

when S_T is nonsingular, so that the optimal G is $G_F = S_T^{-1}(\bar{\boldsymbol{x}}_1 - \bar{\boldsymbol{x}}_2)$ for Fisher's LDA, which is the solution to G^{LDA} in this case. If S_T is singular, a generalized inverse may be used in place of S_T^{-1}. Ye (2007) verifies that this equivalence between linear regression and LDA holds quite generally.

5.2.2 Bayes Rules

Abstractly, the Bayes classifier rests on a conditional density $p(Y|X_1, ..., X_p)$ for a dependent class variable $Y = 1, ..., K$, given explanatory variables X_1 through X_p. Bayes' theorem gives

$$p(Y|X_1, ..., X_p) = \frac{p(Y)p(X_1, \ldots, X_p|Y)}{p(X_1, \ldots, X_p)}, \tag{5.2.5}$$

but interest only focuses on the numerator because the denominator is a constant because the $X_j = x_j$ must be given, independent of the value of $Y = y$. Although $p(y)$ is not known, it represents the actual proportion of class y in the population and so plays the role of the prior. Given an outcome $(x_1, \ldots, x_p)$, the Bayes classifier is the mode of (5.2.5),

$$\arg\max_y \ p(y)p(x_1, \ldots, x_p|Y = y).$$

The mode of a posterior is the Bayes optimal under certain loss functions, as will be seen shortly.

An interesting special case of this is called the Idiot's Bayes procedure. The idea is to ignore any dependence structure among the X_js so the X_js are like independent data. Thus, the numerator in (5.2.5) is written as

$$p(Y)p(X_1|Y)p(X_2|Y,X_1)p(X_3|Y,X_1,X_2)\dots p(X_p|Y,X_1,\dots,X_{p-1}) = \prod_{j=1}^{p} p(X_j|Y).$$

So, the conditional distribution for Y becomes

$$p(Y|X_1,\dots,X_p) = \frac{1}{K(X_1,\dots,X_p)} p(Y) \prod_{i=1}^{p} p(X_i|Y),$$

where K is a normalizing constant depending on X_1^p. Again, the natural classifier is the mode, which is now

$$\arg\max_y \; p(y)\Pi_{j=1}^p p(x_j|\, y)$$

if the $p(\cdot|c)$s are known; otherwise they are usually parametrized by, say, α as in $p_\alpha(x|y)$, which must be estimated as in the next subsection.

The rest of this subsection examines the Bayes classifier for normally distributed classes, deriving the Bayes version of Fisher's LDA. Then, a more general decision-theoretic framework is given to motivate the use of the mode of the posterior. Finally, Fisher's LDA is examined once more to identify the decision boundaries.

5.2.2.1 Bayes Classification in the Normal Case

The basic idea of the Bayes approach is to assign a prior to the classes, say $W(\cdot)$, so that the posterior probability for Y assuming the value of the kth class upon receipt of $\boldsymbol{X} = \boldsymbol{x}$ is

$$W(Y = k|\boldsymbol{X} = \boldsymbol{x}) = \frac{W(k) f_k(\boldsymbol{x})}{W(1) f_1(\boldsymbol{x}) + \ldots + W(K) f_K(\boldsymbol{x})}, \tag{5.2.6}$$

where the f_ks are the class densities. The modal class is therefore

$$\arg\max_k W(Y = k|\boldsymbol{X} = \boldsymbol{x}) = \arg\max_k f_k(\boldsymbol{x}) W(k).$$

If the $f_k(\cdot)$s are normal with common variance (i.e., $N(\mu_k, \Sigma)$ densities) then the Bayes classification rule is

$$\begin{aligned} f(\boldsymbol{x}) &= \arg\max_k W(Y = k|\boldsymbol{X} = \boldsymbol{x}) \\ &= \arg\max W(k) f_k(\boldsymbol{x}) \\ &= \arg\max_k \left(-\frac{1}{2} (\boldsymbol{x} - \mu_k)^{\mathsf{T}} \Sigma^{-1} (\boldsymbol{x} - \mu_k) + \ln W(k) \right) \\ &= \arg\max_k \left(\boldsymbol{x}^{\mathsf{T}} \Sigma^{-1} \mu_k - \frac{1}{2} \mu_k^{\mathsf{T}} \Sigma^{-1} \mu_k + \ln W(k) \right) \end{aligned} \tag{5.2.7}$$

because the common variance term drops out. It makes sense to define $d_k(\boldsymbol{x})$ to be the parenthetical quantity in (5.2.7) so that the Bayes classification rule is $f(\boldsymbol{x}) =$

$\arg\max_k d_k(\boldsymbol{x})$. This means that the decision boundary between two classes k and ℓ is $\{\boldsymbol{x} | d_k(\boldsymbol{x}) = d_\ell(\boldsymbol{x})\}$.

The Bayesian approach extends to the setting in which the variance matrices depend on the classes. This is sometimes called quadratic discrimination analysis and results in the quadratic discriminant function

$$d_k(\boldsymbol{x}) = -\frac{1}{2}\ln\det(\Sigma) - \frac{1}{2}(\boldsymbol{x}-\mu_k)'\Sigma^{-1}(\boldsymbol{x}-\mu_k) + \ln W(k),$$

so that the classification rule is, again, to assign class $\arg\max_k d_k(\boldsymbol{x})$ to $\boldsymbol{x}$. However, decision boundaries of the form $\{d_k(\boldsymbol{x}) = d_m(\boldsymbol{x})\}$ become more complicated to express and there are many more parameters from the variance matrices that must be estimated.

For contrast, recall that the first discriminant given at the start of this section used a projection based on $\boldsymbol{w} = \Sigma^{-1}(\mu_2 - \mu_1)$. Here, it is worth noting that, when $K = 2$, setting $\boldsymbol{v} = (v_1, ..., v_p) = \Sigma^{-1}(\mu_2 - \mu_1)$ and

$$v_0 = \ln(W(1)/W(2)) + (1/2)(\mu_2 - \mu_1)^{\mathsf{T}}\Sigma^{-1}(\mu_1 + \mu_2)$$

means the Bayes classification rule reduces to assigning $\boldsymbol{x}$ to class 1 when

$$v_0 + \boldsymbol{v}\cdot\boldsymbol{x} > 0.$$

5.2.2.2 Decision-Theoretic Justification

In classification problems, the data $(y_i, \boldsymbol{x}_i), i = 1, \cdots, n$, are assumed to be independently and identically drawn from a joint distribution $\mathbb{P}_{\mathbf{X},Y}(\boldsymbol{x}, y)$. Deterministic designs common to regression do not occur, and $\mathbb{P}$ summarizes the error structure instead of a perturbation ε. It is $\mathbb{P}$ that defines the conditional probabilities of Y given X as

$$p_k(\boldsymbol{x}) = \mathbb{P}(Y = k | \boldsymbol{X} = \boldsymbol{x}), \quad k = 1, \ldots, K.$$

Earlier, the mode of a posterior was used to give a classification. Here, it will be seen that the mode is the Bayes action under a specific loss structure.

Let C be the $K \times K$ cost matrix associated with classification, i.e., an entry $C(k,l)$ represents the cost of classifying a data point from class k to class l. In general, $C(k,k) = 0$ for $k = 1, \ldots, K$ since correct classifications should not be penalized. For any classifier f, the risk is the expected cost of misclassification,

$$\begin{aligned}\mathbb{E}_{\mathbf{X},Y}(C(Y, f(\boldsymbol{X}))) &= \mathbb{E}_{\boldsymbol{X}}\left(\sum_{k=1}^{K} C(k, f(\boldsymbol{x}))\Pr(Y = k|\boldsymbol{X} = \boldsymbol{x})\right)\\ &= \mathbb{E}_{\boldsymbol{X}}\left(\sum_{k=1}^{K} C(k, f(\boldsymbol{x}))p_k(\boldsymbol{x})\right). \qquad (5.2.8)\end{aligned}$$

The Bayes rule, which minimizes the risk functional, is given by

$$f_B(\boldsymbol{x}) = \arg\min_{k=1,\ldots,K}\left(\sum_{l=1}^{K} C(k,l)p_k(\boldsymbol{x})\right). \tag{5.2.9}$$

In the special case of equal misclassification costs, the risk in (5.2.8) is equivalent to the expected misclassification rate

$$\mathbb{E}[f(\boldsymbol{X}) \neq Y] = \mathbb{E}_{\boldsymbol{X}}\left(\sum_{k=1}^{K} I(f(\boldsymbol{x}) \neq k)\mathrm{P}(Y = k|\boldsymbol{X} = \boldsymbol{x})\right), \tag{5.2.10}$$

and the Bayes rule becomes

$$f_B(\boldsymbol{x}) = \arg\min_{k=1,\ldots,K}[1 - p_k(\boldsymbol{x})] = \arg\max_{k=1,\ldots,K} p_k(\boldsymbol{x}), \tag{5.2.11}$$

which assigns $\boldsymbol{x}$ to the most likely class. Thus, the modal class is the optimal Bayes rule and would be computable if the underlying conditional distribution $\mathbb{P}(Y|\boldsymbol{X})$ were known. Many classification methods have been proposed to estimate, or approximate, the Bayes rule directly or indirectly. Based on their learning schemes, existing classification methods fall into two main categories: probability-based, which are discussed in the next subsection, and margin-based, of which SVMs taken up in Section 5.4 are the most popular.

5.2.2.3 LDA Redux

Recall (5.2.6), which gives the conditional probability of a class $Y = y$ given $\boldsymbol{X}$. Sometimes explicitly, but often implicitly, LDA assumes each class density is a p-dimensional multivariate Gaussian $\mathrm{MVN}(\boldsymbol{\mu}_k, \Sigma_k)$. Often, equal variances are assumed, $\Sigma_k = \Sigma$ for all k. Therefore,

$$f_k(\boldsymbol{x}) = (2\pi)^{-p/2}|\Sigma|^{-1/2}\exp\left\{-\frac{1}{2}(\boldsymbol{x}-\boldsymbol{\mu}_k)^T\Sigma^{-1}(\boldsymbol{x}-\boldsymbol{\mu}_k)\right\}.$$

The log ratio of a sample belonging to class k and belonging to class l is

$$\log\frac{\mathbb{P}(Y = k|\boldsymbol{X} = \boldsymbol{x})}{\mathbb{P}(Y = l|\boldsymbol{X} = \boldsymbol{x})} = \log\frac{\pi_k}{\pi_l} - \frac{1}{2}(\boldsymbol{\mu}_k+\boldsymbol{\mu}_l)^T\Sigma^{-1}(\mu_k-\boldsymbol{\mu}_l)+\boldsymbol{x}^T\Sigma^{-1}(\boldsymbol{\mu}_k-\boldsymbol{\mu}_l).$$

For each class k, the associated discriminant function is defined as

$$d_k(\boldsymbol{x}) = \boldsymbol{x}^T\Sigma^{-1}\mu_k - \frac{1}{2}\boldsymbol{\mu}_k^T\Sigma^{-1}\boldsymbol{\mu}_k + \log\pi_k.$$

The decision rule is given by

$$\hat{Y}(\boldsymbol{x}) = \arg\max_{k=1,\ldots,K} d_k(\boldsymbol{x}).$$

Note that we can rewrite $d_k(\boldsymbol{x})$ as

$$d_k(\boldsymbol{x}) = -\frac{1}{2}(\boldsymbol{x}-\boldsymbol{\mu}_k)^T\Sigma^{-1}(\boldsymbol{x}-\boldsymbol{\mu}_k) + \frac{1}{2}\boldsymbol{x}^T\Sigma^{-1}\boldsymbol{x} + \log\pi_k.$$

If prior probabilities are the same, the LDA classifies $\boldsymbol{x}$ to the class with centroid closest to $\boldsymbol{x}$ using the squared Mahalanobis distance based on the within-class covariance matrix. The decision boundary of the LDA is linear in $\boldsymbol{x}$ due to the linear form of d_k,

$$d_k(\boldsymbol{x}) = \boldsymbol{w}_k^T\boldsymbol{x} + b_k, \quad \text{where } \boldsymbol{w}_k = \Sigma^{-1}\boldsymbol{\mu}_k, \quad k = 1, \cdots, K.$$

The boundary function between class k and class l is then

$$(\boldsymbol{w}_k - \boldsymbol{w}_l)^T\boldsymbol{x} + (b_k - b_v) = \tilde{\boldsymbol{w}}^T\boldsymbol{x} + \tilde{b} = 0,$$

where $\tilde{\boldsymbol{w}} = \Sigma^{-1}(\boldsymbol{\mu}_k - \boldsymbol{\mu}_l)$. The normal vector $\tilde{\boldsymbol{w}}$ is generally not in the direction of $\boldsymbol{\mu}_k - \boldsymbol{\mu}_l$, and it attempts to minimize the overlap for Gaussian data. In practice, the parameters are estimated from the training data

$$\hat{\pi}_k = n_k/n, \quad \hat{\mu}_k = \sum_{Y_i=k} \boldsymbol{x}_i/n_k,$$

where n_k is the number of observations in class k. The common covariance is often estimated by the pooled within-class sample variances

$$\hat{\Sigma} = \sum_{k=1}^{K}\sum_{Y_i=k}(\boldsymbol{x}_i - \hat{\mu}_k)(\boldsymbol{x}_i - \hat{\mu}_k)^T/(n-K).$$

However, in the presence of outliers, these estimates can be unstable, so more robust estimators are often preferred.

5.2.3 Probability-Based Discriminant Analysis

In probability-based discriminant functions, the optimal Bayes rule is estimated or approximated directly. This can be done by estimating the probabilities or densities that appear in (5.2.6). For instance, if $K = 2$, $\mathbb{P}(Y = y|\boldsymbol{x})$ can be obtained by estimating the densities $f_1(x)$ and $f_2(x)$ by $\hat{f}_1$ and $\hat{f}_2$ and the proportions of the two classes by the sample proportions $\hat{\pi}_1$ and $\hat{\pi}_2$. The basic idea is to form

$$\hat{\mathbb{P}}(Y = k|x) = \frac{\hat{\pi}_k\hat{f}_k(x)}{\hat{\pi}_1\hat{f}_1(x) + \hat{\pi}_2\hat{f}_2(x)}$$

and choose the value of k that maximizes it. Alternatively, a model for the conditional probability can be proposed and the Bayes rule estimated by it. Both approaches bear some discussion.

5.2.3.1 Multiple Logistic Regression Models

Given IID observations $(\boldsymbol{x}_i, y_i), i = 1, \ldots, n$, the counts for the K classes can be treated as multinomial with probabilities $\{p_1(\boldsymbol{x}), \ldots, p_K(\boldsymbol{x})\}$. The multiple logistic regression model (MLR) is merely one successful way to obtain a classifier by using a model based estimate of the Bayes rule. The MLR simultaneously proposes a specific form for the log odds for all $\binom{K}{2}$ pairs of classes, so the log odds can be estimated.

By regarding the last class as a baseline, the MLR model assumes the linear form for the following $K-1$ logits:

$$\log \frac{p_k(\boldsymbol{x})}{p_K(\boldsymbol{x})} = \eta_k(\boldsymbol{x}), \quad k = 1, \ldots, K-1. \tag{5.2.12}$$

With the model (5.2.12), the conditional probabilities $p_k(\boldsymbol{x})$ become

$$p_k(\boldsymbol{x}) = \frac{\exp\{\eta_k(\boldsymbol{x})\}}{1 + \sum_{k=1}^{K-1} \exp\{\eta_k(\boldsymbol{x})\}}, \quad k = 1, \ldots, K. \tag{5.2.13}$$

In particular, the linear MLR model assumes the linear forms for the logits

$$\log \frac{p_k(\boldsymbol{x})}{p_K(\boldsymbol{x})} = \alpha_k + \boldsymbol{\beta}_k^{\mathrm{T}} \boldsymbol{x}, \quad k = 1, \ldots, K-1, \tag{5.2.14}$$

with $\alpha_K = 0, \boldsymbol{\beta}_K = \boldsymbol{0}$.

Maximum likelihood estimation can be used to estimate $\boldsymbol{\theta} = \left(\alpha_1, \boldsymbol{\beta}_1^{\mathrm{T}}, \ldots, \alpha_{K-1}, \boldsymbol{\beta}_{K-1}^{\mathrm{T}}\right)$. For the ith sample, let the vector $\boldsymbol{z}_i = (z_{i1}, \ldots, z_{iK})^{\mathrm{T}}$ represent its membership characteristic; i.e., $z_{ik} = 1$ if $y_i = k$ and $z_{ik} = 0$ otherwise. Note that $\sum_{k=1}^{K} z_{ik} = 1$. The log likelihood function is then given by

$$\begin{aligned}
l(\boldsymbol{\theta}) &= \log \prod_{i=1}^{n} \left[\prod_{k=1}^{K} p_k(\boldsymbol{x}_i)^{z_{ik}} \right] \\
&= \sum_{i=1}^{n} \left\{ \sum_{k=1}^{K-1} z_{ik} (\alpha_k + \boldsymbol{\beta}_k^{\mathrm{T}} \boldsymbol{x}_i) - \log \left[1 + \sum_{k=1}^{K} \exp(\alpha_k + \boldsymbol{\beta}_k^{\mathrm{T}} \boldsymbol{x}_i) \right] \right\} \\
&= \sum_{k=1}^{K-1} \alpha_k \left(\sum_{i=1}^{n} z_{ik} \right) + \sum_{k=1}^{K-1} \sum_{j=1}^{d} \beta_{jk} \left(\sum_{i=1}^{n} x_{ij} z_{ik} \right) - \sum_{i=1}^{n} \log \left[1 + \sum_{k=1}^{K-1} \exp(\alpha_k + \boldsymbol{\beta}_k^{\mathrm{T}} \boldsymbol{x}_i) \right].
\end{aligned}$$

The negative log likelihood function is convex, so the Newton-Raphson algorithm is often used to optimize the function. Denote the maximum likelihood estimates by $\widehat{\boldsymbol{\theta}}$ and $\hat{p}_k(\boldsymbol{x})$, $k = 1, \ldots, K$. The MLR decision rule classifies data with an input $\boldsymbol{x}$ to its most probable class, i.e., $f(\boldsymbol{x}) = \arg\max_{k=1,\ldots,K} \hat{p}_k(\boldsymbol{x})$.

5.2.3.2 Variants on the Linear MLR

Consider the linear version of the MLR, but, following Dreiseitl and Ohno-Machado (2002), write it as

$$\mathbb{P}(Y = 2|\boldsymbol{x}, \boldsymbol{\beta}) = \frac{1}{1 + e^{-\boldsymbol{\beta}\cdot x}} \tag{5.2.15}$$

for the binary case, where $\mathbb{P}(Y = 1|\boldsymbol{x}, \boldsymbol{\beta}) = 1 - P(Y = 2|\boldsymbol{x}, \boldsymbol{\beta})$. It is seen that the hyperplane $\boldsymbol{\beta} \cdot \boldsymbol{x} = 0$ is the decision boundary between class 1 and class 2 and corresponds to the case where the two classes are equally likely, $\mathbb{P}(Y = 2|\boldsymbol{x}, \boldsymbol{\beta}) = \mathbb{P}(Y = 1|\boldsymbol{x}, \boldsymbol{\beta})$. The linear function is written $\beta \cdot \boldsymbol{x}$, but this can be generalized to nonlinear functions in the covariates.

In particular, it is seen that the logistic model (5.2.15) is a one node neural net if $1/(1 + e^{-t})$ is taken as the sigmoid. So, the whole model can be generalized by replacing $\boldsymbol{\beta} \cdot \boldsymbol{x}$ to give

$$\mathbb{P}(Y = 2|\boldsymbol{x}, \boldsymbol{\beta}) = \text{out}_r = \frac{1}{1 + e^{-\beta \cdot \text{out}_{r-1}(\boldsymbol{\beta}_{r-1}, \boldsymbol{x}) + \beta_0}},$$

where $\text{out}_{r-1}(\boldsymbol{\beta}_{r-1}, \boldsymbol{x})$ is the vector of outcomes of the nodes at the $r - 1$ hidden layer, with all parameters at that layer denoted by $\boldsymbol{\beta}_{r-1}$.

Typically, one uses a maximum likelihood approach to estimate β. That is, one uses

$$\hat{\boldsymbol{\beta}} = \arg\max_{\boldsymbol{\beta}} \Pi_{i=1}^{n} P(y_i|x_i, \boldsymbol{\beta}).$$

Alternatively, Bayesian estimates can be developed. Given a prior on β, the posterior density is

$$p(\boldsymbol{\beta}|x_i, y_i : i = 1, \ldots n) = \frac{p(\boldsymbol{\beta})\Pi_{i=1}^{n} p(y_i|x_i, \boldsymbol{\beta})}{\int_{\boldsymbol{\beta}} p(\boldsymbol{\beta})\Pi_{i=1}^{n} p(y_i|x_i, \boldsymbol{\beta}) d\boldsymbol{\beta}}.$$

This becomes sharply peaked at its mode $\hat{\boldsymbol{\beta}}_{pm}$, which will generally be close to $\hat{\boldsymbol{\beta}}_{ML}$; it's as if $\hat{\boldsymbol{\beta}}_{pm}$ is the same as $\hat{\boldsymbol{\beta}}_{ML}$ but using an extra finite number of data points corresponding to the information in the prior. Clearly, one can imagine using gradient descent and other techniques from NNs as well.

Since Idiots Bayes, logistic regression, and neural nets can be regarded as successive generalizations, some comments on model evaluation and discriminant analysis more generally may be helpful.

The two main criteria for model evaluation are discrimination and calibration. Discrimination asks how well the classes are separated; in the $K = 2$ case, common measures of discrimination include the familiar concepts of sensitivity and specificity (i.e., the proportion of class 1 and class 2 cases correctly identified). This is often summarized graphically by plotting the true positive rate against the false positive rate, the "sensitivity versus 1 minus specificity". The graph is called a receiver operating characteristic (ROC) curve. An ideal classifier rises from $(0, 0)$ straight up to one and then is constant to $(1,1)$. This never happens in practice, so classifiers with a large area under the ROC curve are preferred.

Calibration asks how accurately the model probability $p(\cdot|\boldsymbol{x},\boldsymbol{\beta})$ estimates the correct probability $p(y|\boldsymbol{x})$. This is harder to automate because the true probability function is never known. So, one may be led instead to compare two different estimates of the probability, a sort of robustness check. Another is to divide the sample into smaller subsets, estimate $\boldsymbol{\beta}$s on each, and then calculate the sum of the predictions and the sum of the outcomes for each subset (i.e., for each $\boldsymbol{\beta}$). If these are close, then the modeling may be good.

Overall, discriminant analysis is fundamental to the traditional ways classification was done. This is partially because it is relatively easy to understand and it is computationally relatively easy: It mostly devolves to matrix calculations. Linear discriminants are related to principal component analysis because the rows of the data matrix, and the class mean vectors, can be regarded as points in a p-dimensional space. The strategy then, whether based on maximum separation or Bayes methods, determines discriminating axes in this space. This accounts for the multiplicity of roughly equivalent forms for Fisher's LDA. Mathematically, the problem becomes finding the eigenvectors of a symmetric, real matrix because the eigenvalues represent the ability of the eigenvectors to discriminate along the eigendirections. One of the most important improvements to conventional use of LDA is the use of robust estimates for the μ_k and Σs since these define the discriminants.

However, when p is large, the sample sizes required to estimate the parameters well can be prohibitive. Consequently, over time, discriminant analysis has been modernized to deal with more complex settings where model selection and uncertainty have been central to finding good choices for $P(\boldsymbol{x}|Y=k)$. Indeed, even when the number of explanatory variables p is not particularly large, it may be important to include more than univariate functions of the components of $\boldsymbol{x}$. Then, merely including the second-order terms x_i^2 and x_ix_j increases the dimension of the parameter space (Σ and the μ_ks) beyond most reasonable data sets. The problem becomes worse if general univariate functions of the components of $\boldsymbol{x}$ are permitted, let alone general functions of two or more components. In these contexts, traditional dimension-reduction techniques such as principal components and factor analysis have been used to reduce the number of explanatory variables; see Chapter 9.

Recently, numerous variants on discriminant analysis have been proposed. Some authors have combined discriminant analysis with boosting to get better predictive performance or with regularization methods to do difficult model selection. Others have used kernel methods to recast $J(\boldsymbol{w})$ so as to get more general decision boundaries with a linear form in the nonlinearly transformed feature space.

The problem is even more difficult when the nonindependence of the explanatory variables is considered. In fact, it is unclear whether modeling the dependence is worthwhile when p gets large. Bayes methods are often less sensitive to such dependencies than frequentist methods, and Idiot's Bayes is a colloquial term indicating the neglect of correlation between explanatory variables in a Bayes framework. In a counterintuitive paper, Bickel and Levina (2004) demonstrate that Idiot's Bayes may not be so idiotic in a classification context: Idiot's Bayes often outperforms "intelligent" linear discriminant analysis that models the dependencies when $p >> n$.

5.3 Tree-Based Classifiers

In Chapter 4, tree structures were used for regression, although some of the terminology (e.g., homogeneity of nodes) was appropriate for classification as well. This section focuses on the classification story. Regression and classification differ in their formalities but have many common features. For instance, both network architecture (the recursive partitioning) and parameter estimation are needed to specify the tree.

5.3.1 Splitting Rules

Again, a model of the form

$$Y = \sum_{j=1}^{r} \beta_j I(\boldsymbol{x} \in R_j) \tag{5.3.1}$$

is to be found, but Y is categorical taking values from 1 to K and the node functions are constant; in principle, one can set $\beta_k = k$ for $k = 1, \ldots, K$ for appropriate regions R_j, so the node functions identify the classes. The error in modeling represents the mismatch between the estimated and correct regions on which a class is identified. Assuming there is one fixed optimality criterion for evaluating classification tree performance (e.g., zero-one loss or some other cost of misclassification) the main issue is choosing the partition of the range of the $\boldsymbol{X}$s (i.e., selecting the splits). Two methods for selecting the splits were discussed in Chapter 4: clustering on a variable so as to separate the clusters and finding a split to minimize a sum of squared errors. This was in the regression context. Here, three further methods are given. The first two, hypothesis testing and finding optimal directions, could be used in regression as well; the third replaces the squared error criterion for finding optimal splits for regression with other criteria appropriate for classification.

First, to select splits, consider the situation at a node, possibly the root, but more generally any terminal node in a growing tree. At the node, one can choose which variable to split on by hypothesis testing, for instance. Suppose the jth explanatory variable has values $X_{j,i}$ for $i = 1, \ldots, n$. One can do a test of dependence, for instance the χ^2, between the Y_is and the $X_{j,i}$s for each j to find the variable with the lowest p-value. Using the most dependent variable, one can cluster its values (using the techniques of Chapter 8) and then choose a split to partition the clusters. There are techniques from Bayesian clustering (see Chapter 8) that test whether a candidate partition of the data at a node is optimal. Clearly, there are many tests and clustering procedures.

Second, an alternative to testing to find a single variable to split on is to find splits based on linear combinations of the X_js. This is a more elaborate procedure than finding a single variable since there are many more choices for directions than for variables. One way to find a direction comes down to looking for eigenvectors of the design matrix $\underline{X}$. For square matrices, this is called principal components, which will be discussed in Chapter 9 in detail as a dimension-reduction technique. Here, because design matrices are usually not square, a slightly different version of eigenvector decomposition called a singular value decomposition (SVD) must be used. An SVD helps find the internal structure of $\underline{X}$ as an operator on a linear space.

Consider the subdesign matrix, also denoted $\underline{X}$ for convenience, formed from the overall design matrix by including only those rows that correspond to the data points assigned to a given node. These are the vectors of explanatory variables that might be split at the next iteration of tree formation. Essentially, an SVD represents a rectangular matrix, such as the $n \times p$ matrix $\underline{X}$, as a product of three matrices, the middle one of which has nonzero elements only on its main diagonal. The SVD theorem states that for any matrix $\underline{X}$ there exist matrices $\underline{U}$ and $\underline{V}$ such that $\underline{U}^{\top}\underline{U} = Id_{n\times n}$ and $\underline{V}^{\top}\underline{V} = \underline{I}_{p\times p}$ so that

$$\underline{X} = \underline{U}_{n\times n}\underline{S}_{n\times p}\underline{V}^{\top}_{p\times p}, \tag{5.3.2}$$

in which $\underline{S}_{n\times p}$ is an $n \times p$ diagonal and $\underline{U}$ is unitary. Clearly, this is a generalization to rectangular matrices of the familiar eigenvalue decomposition of square matrices.

Looking at (5.3.2), $\underline{S}$ contains the eigenvalues, now square rooted and called singular values, while $\underline{U}$ and $\underline{V}$ contain the analogs of eigenvectors. However, there are right eigenvectors and left eigenvectors, now called singular vectors. The columns of $\underline{U}$ are the left singular vectors, say $\boldsymbol{u}_k$ for $k = 1, \ldots, n$, that comprise an orthonormal basis and the rows of $\underline{U}$ are the right singular vectors, say $\boldsymbol{v}_k$, also comprising an orthonormal basis for $k = 1, \ldots, p$. The diagonal values s_k for $k = 1, \ldots, \min p, n$ of $\underline{S}$ are assumed to be in decreasing order. Let $\underline{S}_\ell$ be an $n \times p$ submatrix formed from using only the first ℓ singular values in $\underline{S}$. An important property of the SVD is that

$$\underline{X}_\ell = \underline{U}_{n\times n}\underline{S}_\ell\underline{V}^{\top}_{p\times p} = \sum_{k=1}^{\ell} \boldsymbol{u}_k s_k \boldsymbol{v}_k^{\top} \tag{5.3.3}$$

is the best rank $\ell \le \min(p,n)$ approximation to $\underline{X}$ in Frobenius norm (basically, squared error applied to the elements of a matrix as if it were a vector).

In principle, it is not hard to find the SVD for $\boldsymbol{X}$. Write

$$\underline{X}^{\top}\underline{X} = \underline{V}\underline{S}^2\underline{V}^{\top}$$

and then find the eigenvalues for $\underline{S}$ and the corresponding eigenvectors for $\underline{V}$ by the usual diagonalization procedure. Now, $\underline{U} = \underline{X}\underline{V}\underline{S}^{-1}$. If there are r nonzero s_ks, then the remaining $n - r$ columns of $\underline{V}$ are ignored in the last matrix multiplication. Choices for those $n - r$ singular vectors in $\underline{V}$ (or $\underline{V}$) may be found by Gram-Schmidt (or any other method for filling out the dimensions). It is worth noting that although this method can be used, it is often numerically unstable when any of the singular values are near zero. In practice, more sophisticated procedures are generally used.

Given an SVD for $\underline{\boldsymbol{X}}$, one can find the principal components as if calculated from an empirical covariance matrix. If the columns of $\underline{\boldsymbol{V}}$ are centered, then $\underline{\boldsymbol{X}}^{\mathsf{T}}\underline{\boldsymbol{X}} = \Sigma_i s_i^2 \boldsymbol{v}\boldsymbol{v}^{\mathsf{T}}$ is seen to be proportional to the empirical covariance matrix. This means that the diagonalization of $\underline{\boldsymbol{X}}^{\mathsf{T}}\underline{\boldsymbol{X}}$ that gives $\underline{\boldsymbol{V}}$ also gives the principal components, with variances s_i^2. A similar argument holds if the rows of $\underline{\boldsymbol{X}}$ are centered. In this case, $\underline{\boldsymbol{X}}^{\mathsf{T}}\underline{\boldsymbol{X}} = \sum_i s_i^2 \boldsymbol{u}_i \boldsymbol{u}_i^{\mathsf{T}}$ and the left singular vectors are the principal components with variances s_i^2. In statistics, it is usually the left principal components that are used unless specified otherwise. (The other side can be related to factor analysis; see Chapter 9.)

Now, the PCs generated from an SVD can be used to choose splits at a node. For instance, the first PC, the vector u_1, corresponding to the element of $\underline{\boldsymbol{S}}_{n\times p}$ with the largest absolute value, can be used to give a projection of the data points $\boldsymbol{x}_i \cdot \boldsymbol{u}_1$ along the direction $\boldsymbol{u}_1$. Clustering these values for the $\boldsymbol{x}_i$s at the node gives a way to find a threshold for splitting at a node. Recalling that in the case of two thin, parallel ellipses, it was the direction with smallest variance (the last principal component) that gave a good split, not the direction with the smallest variance, it is clear that more complicated node-splitting rules than just using the first PC may be necessary for a good fit.

A third possibility is the regression style search of all possible splits using an error criterion. This was discussed in Chapter 4, where $E(g)$ or, more empirically, $\hat{E}(g^*)$ was minimized. Instead of squared error, suppose the $\boldsymbol{X}_i$s consist entirely of categorical variables, and consider what happens at a node. The jth predictor variable X_j may assume, say, ℓ_{node} levels among the data points at the node. So, in principle, there are $2^{\ell(node)} - 1$ ways to divide the data points into two sets using X_j alone. If each X_j has ℓ possible values, then the only general upper bound for the splits at a node is trivial, $(2^{\ell})^p - 1$, from taking the product over all p variables. This is very large so simplifications, such as choosing specific X_js, are necessary.

One way to simplify is to use indices of impurity. Such indices can be used to choose splits as well as to decide when to stop splitting. The Gini index is specifically for discrete variables and is a measure of inequality. Given a data set $\mathscr{D}$ of size n, say, with data points ranging over K classes, let n_k be the number of data points in class k for $k = 1, \ldots, n$ so that $n_1 + \ldots + n_K = n$. Then $\hat{p}_k = n_k/n$ is the relative frequency of class k in the data. The Gini index for $\mathscr{D}$ is

$$Gini(\mathscr{D}) = 1 - \sum_{k=1}^{K} \hat{p}_k^2.$$

Clearly, $Gini \leq 1$ with equality if and only if $K \to \infty$ and all the $\hat{p}_k$s tend to zero. Also, $Gini \geq 0$ with equality if all $\hat{p}_k$s are zero except for one, which assumes the value one. (The only if part fails; it is enough for the $\hat{p}_i$s to lie on the unit sphere.)

To obtain a split of the points at a node, write

$$Gini_{split}(\mathscr{D}) = \frac{N_1}{n} Gini(\mathscr{D}_1) + \frac{N_2}{n} Gini(\mathscr{G}_2),$$

where $\mathscr{D}_1 \cup \mathscr{D}_2 = \mathscr{D}$, with cardinalities $N_1 = \#(\mathscr{D}_1)$ and $N_2 = \#(\mathscr{D}_2)$. The optimal split of the data in $\mathscr{D}$ is the one that minimizes $Gini_{split}$. This can be done when $\mathscr{D}$ is defined as the set of p-dimensional vectors $\boldsymbol{x}_i$ or for the individual explanatory variables that are components of the $\boldsymbol{x}_i$s.

Essentially, minimizing the Gini index to find splits is a way to try to pull off distinct classes sequentially in order of size. Suppose there is a collection of homogeneous classes of distinct sizes perfectly represented by the data at the root node. If Gini were used, and worked perfectly, the first split would set a threshold to reduce the Gini index by finding the split that optimally reduced the inequality. This would set a threshold to pull off the largest class by itself. So, the first split would separate it from the other classes. Since the first class would be homogeneous, Gini would not split it further. However, Gini would split the other data points since they comprised the reduced set of homogeneous distinct classes. So, the second split would set a threshold to reduce the Gini index of the reduced set by finding the split that again optimally reduced inequality. Thus, the second largest homogeneous class would be pulled off. The procedure would then repeat until the classes were all pulled off, one at each iteration. If successful, the final tree would ideally look like a long skinny branch with single leaves coming off it along the way, one for each class.

There is a version of the Gini index appropriate for continuous data; it too is a measure of inequality. The idea is to choose a Lorenz curve (i.e., a curve of the form $f(x) = \int_0^x x dF(x)/\mu$) as the density for some positive random variable X on [0,1]. Then, the Gini index of X is

$$Gini = 2\int (x - f(x))dx = \frac{1}{2\mathbb{E}(X)} \int_0^\infty \int_0^\infty |x-y| f(x) f(y) dx dy.$$

Empirically, if $\mu = \mathbb{E}(X)$, the Gini index can be estimated by

$$\widehat{Gini} = \frac{1}{2n^2 \hat{\mu}} \sum_{i=1}^n \sum_{j=1}^n |x_i - x_j|;$$

it ranges from zero, when all xs are the same, to a maximum of one.

The entropy is another notion of impurity. It is

$$H(\mathscr{D}) = \sum_{k=1}^K \hat{p}_k \log \frac{1}{\hat{p}_k}.$$

The entropy has an interpretation in terms of information gain. It often gives results similar to Gini.

The idea behind twoing is the opposite of Gini. In idealized form, twoing first splits the points at the root node into two groups, attempting to find groups that each represent 50 percent of the data. This contrasts with Gini, trying to pull off a single class. Twoing then searches for a split to partition each of the two subgroups, again into two groups, each containing now 25 percent of the data. That is, Twoing seeks equal-sized leaves from a node, under some splitting rule. Although twoing tries to ensure the leaves from a split are equal, this can be difficult, especially at terminal nodes from the same parent. As with Gini, most real-world classification problems will not allow impurity measures to give such a clean result.

Different splitting rules may result in the same tree, but usually they predispose the procedure to a class of trees – twoing favors equal binary splits giving bushy, shortish trees, whereas Gini favors trees that often have long, straggly branches.

Because of their high variability, it is important to evaluate trees comparatively. The two basic techniques for generating alternatives within a class of trees and selecting among them have already been discussed in Chapter 4: pruning back on the basis of some cost-complexity criterion and using some version of cross-validation to choose a tree in the class with the least average error. It is important to realize that there can be bias in classification as a consequence of variable selection as well as variability in the trees produced; see Strobl (2004) and the references therein. A survey of these methodologies from the information theory standpoint can be found in Safavian and Landgrebe (1991). They also contrast tree structures with neural net classification.

5.3.2 Logic Trees

A different tree-based tack for classification, called logic trees, is due to Ruczinski et al. (2003). The idea is to recognize that the classification trees presented so far are really "decision" trees, based on regression and incorporating variability in the usual statistical way, but that there are alternatives that are more overtly rule-based. Logic trees can be regarded as a sort of generalized linear models approach where the terms in the model are Boolean functions.

The basic structure is as follows. Let $\boldsymbol{X} = (X_1, \ldots, X_p)$ be a sequence of 0-1 predictors and Y be a class indicator. The logic tree model is to write

$$g(E(Y)) = \beta_0 + \sum_{k=1}^{K} \beta_k L_k,$$

where the β_ks are coefficients, g is the function that makes the linear model "generalized", and L_k is a Boolean function of the covariates X_j. For instance, if X_k^c is the opposite of X_k, then one might have $L(\boldsymbol{X}) = (X_1 \vee X_2) \wedge X_3^c$. If $p = 3$, the value $\boldsymbol{X} = (0,1,0)$ gives $L(0,1,0) = 1$ because $0 \vee 1$ gives 1 and $1 \wedge 0^c$ gives 1. The task is to estimate the L_ks and the β_ks; the link function g is chosen by the user.

Note that the linear combination of logic functions is a tree not in the recursive partitioning sense but in a Boolean function sense. However, functions of the form of the right-hand side span the space of all real-valued functions when the X_js are categorical, as does recursive partitioning when it is applied to categorical variables. Thus, logic trees and recursive partitioning trees are merely different representations for the same function space. The difference is in which functions are conveniently expressed in each form. In both recursive partitioning and logic trees an "and" function such as $(X_1 = 1) \vee (X_2 = 1) \vee (X_3 = 1) \vee (X_4 = 1)$ when the X_js are binary is parsimoniously represented as a single long branch or as a single term, respectively. In both recursive partitioning and logic trees, the components $X_j = 1$ can be adjoined (or removed) sequentially. However, a function such as $((X_1 = 1) \vee (X_2 = 1)) \wedge ((X_3 = 1) \vee (X_4 = 1))$ is easy to find as one term in a logic tree since it is not hard to build sequentially from its individual components, whereas in recursive partitioning the cases $X_1 = 0$ and $X_1 = 1$ would correspond to two branches that would have similar structures and be hard for recursive partitioning to find.

Building logic trees consists of adding or removing Boolean functions of individual variables sequentially, and model selection can be done on the basis of MSE in simulated annealing; see Ruczinski et al. (2003). This can be extended to continuous predictors as well.

5.3.3 Random Forests

So far, attention has focused on obtaining a specific tree for classification. However, this is a model and hence subject to model misspecification. Thus, choosing a single tree and not giving some assessment of its MSE or other measure of variability that includes variability over model selection gives a falsely precise notion of how good a classifier is. Although it is unclear how to assign a standard error or bias to a tree, it is worthwhile to look into techniques that take averages over trees because the average will typically reduce the variability from what one would have with a single tree. In practice, this means that averaging trees will often give better classifiers.

Random forests are a generalization of recursive partitioning that combines a collection of trees called an ensemble. However, random forests is best seen as a bootstrapped version of a classification tree generating procedure. It was invented by Breiman (2001) and substantially developed by Breiman and Cutler (2004). A random forest is a collection of identically distributed trees whose predicted classes are obtained by a variant on majority vote. Another way to say this is that random forests are a bagged version of a tree classifier – improved by two clever tricks. The term bagging, or bootstrap aggregation, will be discussed in detail in the next chapter; roughly one uses the bootstrap to generate the members of an ensemble, which are then aggregated in some way. As discussed in Breiman and Cutler (2004), random forests may also be used for regression, but their advantages are less clear.

Operationally, one starts with a data set. From that, one draws multiple bootstrap samples, constructing a classification rule for each, for instance by recursive partitioning. The random forest consists of the trees formed from the bootstrap samples. However, no pruning of the trees is done. The tree is just grown until each terminal node contains only members of a single class. Usually, about 100 trees are generated, each from an independent bootstrap sample.

To classify a new observation, one uses each of the trees in the forest. If a plurality of the trees agree on a given classification, then that is the predicted category of the observation. Note that there is a built-in estimate of the uncertainty in each new classification: The distribution of the votes indicates whether nearly all trees agree or whether there are many dissenters.

There are two clever "tricks" that make random forests particularly effective. The first is an "out of bag" error estimate; the second is random feature selection.

The first trick is a technique to get an estimate of the misclassification error. Indeed, the estimate is often claimed to be unbiased when the training set is a random sample from the population. The idea is to estimate the expected predictive accuracy by running each observation through all of the trees that were formed without using it. It will be seen below that, for each data point, approximately one-third of the bootstrap samples will not contain it, so approximately two-thirds of the trees generated can be used. If k be the class that gets the most votes for the ith data point. The proportion of times that k differs from the class of the initial data point is the out-of-bag error estimate.

Before going on to the second trick, it's worth looking closely at the bootstrapping procedure in more detail since random forests relies on it.

5.3.3.1 Occupancy

For the bootstrap, start with a sample $\boldsymbol{x}_1,\ldots,\boldsymbol{x}_n$ and then sample from it with replacement until a new sample of size n, say $y_1,\ldots,y_n$, has been generated. It is possible that some of the y_is will be repeated and that some of the $\boldsymbol{x}_i$s will not be among the y_is. Since each bootstrap sample is drawn with replacement, about a third of the original sample is not chosen. This is important because the effectiveness of bootstrapping depends on the overlap of the resampling. If one has a fixed sample of size n and intends to take n bootstrap samples of size n from it, then, asymptotically, $1/e$ of the original sample is not chosen. This is seen by the elementary argument that $P(\boldsymbol{x}_1 \text{ not chosen}) = (1-1/n)^n \to 1/e$.

More generally, this result follows from a central limit theorem because selecting a bootstrap sample can be modeled by a ball-and-urn scheme. Let n urns represent the n original data points, the $\boldsymbol{x}_i$s. Imagine randomly dropping balls into the urns one at a time and independently. If n balls are dropped, then the number of balls in the ith urn represents how many times $\boldsymbol{x}_i$ occurs in the first bootstrap sample.

This is seen to be a multinomial problem: Given that n is fixed and n balls are dropped at random, define X_k to be the (random) count of how many balls are in urn k, $k = 1,\ldots,n$. Now,

$$\mathbb{P}(X_1 = j_1,\ldots,X_n = j_n) = \frac{n!}{j_1!\ldots j_n!} p_1^{j_1}\ldots.p_n^{j_n},$$

in which $\sum j_k = n$. If all $p_j = 1/n$, then

$$\mathbb{P}(X_1 = j_1,\ldots,X_n = j_n) = \frac{n!}{j_1!\ldots j_n!}\frac{1}{n^n}.$$

The study of the properties of these urns is called the occupancy problem; see Barbour et al. (1992). If n increases, a classical central limit theorem for the occupancy problem has been established by Rahmann and Rivals (2000).

Theorem (Rahman and Rivals, 2000): Let N_k and M_k be sequences of natural numbers such that $N_k \to \infty$ and $M_k \to \infty$ with $(N_k/M_k) \to \lambda$ as $k \to \infty$. Let W_k be the sequence of random variables denoting the number of empty urns after N_k balls have been dropped independently into M_k urns. Then, as $k \to \infty$,

$$\mathbb{E}(W_k/M_k) \to e^{-\lambda},$$

$$\text{Var}(W_k/\sqrt{M_k}) \to (e^{\lambda} - 1 - \lambda)e^{-2\lambda},$$

and we have that

$$\frac{W_k - M_k e^{-\lambda}}{\sqrt{M_k(e^{\lambda} - 1 - \lambda)e^{-2\lambda}}} \to N(0,1)$$

in distribution.

Proof: See Johnson and Kotz (1977); Harris (1966) gives a superb reference list and observes that the proof goes back to Geiringer (1937). □

This result shows, in principle, how bootstrapping on discretized data tends to operate as the discretization, represented by M_k, gets finer and the number of samples of size N_k increases.

One of the important implications of the out-of-bag prediction error analysis is that random forests do not overfit. Breiman's theorem below shows that the average misclassification rate decreases to a fixed value. As the number of trees N increases, the estimated predictive error converges to a value bounded above by the ratio of the average correlation between trees to a function of the "strength" of a set of classifiers. The correlation decreases as the size of the training sample increases and the strength increases as the signal in the data grows.

5.3.3.2 Random Feature Selection

The second clever trick is random feature selection. Recall that, at each step in growing a tree, classic recursive partitioning examines all p variables to determine the best split. By contrast, random forests (usually) picks $\sqrt{p}$ of the variables at random, taking the best split among them. This extra level of selection makes the different trees in the forest less similar by allowing highly correlated variables to play nearly equivalent roles (otherwise, the slightly more predictive variable would always be chosen). Random feature selection decorrelates the trees, lowering the prediction error.

As part of feature selection, random forests can be used to estimate the relative importance of, say, the jth explanatory variable. To do this, recall that a random forest runs each observation through all the trees for which the observation is out-of-bag and counts the number of votes for the correct class each observation gets. This vote count can be compared to the corresponding vote count after randomly permuting the values of variable j among the samples. So, permute the values of X_j and run them down the corresponding trees, again counting the number of correct votes. The difference in the number of correct votes under the two procedures is the raw importance for variable j. Usually this is standardized so the scores sum to one.

Rather than looking at features, one can look at observations and create a local measure of importance specific to each observation in the training sample. For observation i, run it down all the trees for which it is out-of-bag. Repeat the process with randomly permuted values of variable j, and look at the difference in the number of correct votes. This is a measure of how important variable j is for classifying cases in the neighborhood of observation i.

5.3.3.3 Breiman's Theorems

So far, the discussion of random forests has been informal, describing heuristics that can be calculated to evaluate performance. So it is important to present several of the formal results that ensure random forests will work well. These are primarily due to Breiman (2001).

First note that a good classifier h has a high value of $\mathbb{P}(h(\boldsymbol{X}) = Y)$, where the probability $\mathbb{P}$ is in the joint probability of $\boldsymbol{X} \times Y$. It is assumed that Y ranges over $\{1, ..., K\}$, and, as ever, the vector $\boldsymbol{X} = \boldsymbol{x}$ represents explanatory variables and Y is the response. In practice, usually an IID sample $(\boldsymbol{x}_1, y_1), ..., (\boldsymbol{x}_n, y_n)$ is available from which to learn h. Estimation of h can be done as described in the first part of this section, or as in the other sections of this chapter, so the construction of finitely many distinct classifiers h_i can be assumed. Random forests as a procedure takes a set of h_is as an input and produces an improved classifier from them.

Recall that effectively each h_j is equivalent to a partition of the range of the $\boldsymbol{x}$s based on the values of Y. Although random forests can be applied to many types of classifiers, here it will only be used for tree-based methods. Thus, it will be enough to consider only those partitions of the feature space that correspond to trees. To state this formally, define a function $h : X \to \{1, \ldots, K\}$ to be tree-structured if and only if the partition of the domain it induces is described by a finite sequence of inequalities involving individual x_js in $\boldsymbol{x} = (x_1, ..., x_p)$. Now, a classifier $h(\boldsymbol{x}, \Theta)$ is tree-structured if and only if for each outcome $\Theta = \theta$, $h(\boldsymbol{x}, \theta)$ is tree-structured as a function of $\boldsymbol{x}$. A random forest is a classifier derived from a collection of tree-structured classifiers, $\{h_j(\boldsymbol{x}) = h(\boldsymbol{x}, \theta_j) | j = 1, ..., J\}$, where $\theta_1, ..., \theta_J$ are IID outcomes of Θ.

For a collection of tree-structured classifiers h_j, consider the random forest classifier formed by taking a vote of the individual tree-structured classifiers comprising it. The average number of correct classifications is the proportion of classifiers identifying the right class and is

$$AV(Y) = \frac{1}{J} \sum_{j=1}^{J} 1_{\{h_j(\boldsymbol{X}) = Y\}},$$

in which 1_A is the indicator function for a set A. The average number of misclassifications of type k, for $k \neq Y$, is the proportion of h_js misclassifying Y as k,

$$AV(k) = \frac{1}{J} \sum_{j=1}^{J} 1_{\{h_j(\boldsymbol{X})=k\}}.$$

For good classification, $AV(Y)$ should be large relative to all the $AV(j)$s with $j \neq Y$, for all $(\boldsymbol{X}, Y)$. The worst case occurs for the value of k achieving $\max_{k \neq Y} AV(k)$. Since it is enough for $AV(Y)$ to be large relative to the worst case, let

$$CA = CA(\boldsymbol{X}, Y) = AV(Y) - \max_{j \neq Y} AV(j)$$

be the classification accuracy. CA represents how many more of the classifiers get the right class than get the wrong class; a good classifier would give large CA. If CA were negative and $K = 2$, it would indicate that interchanging the predicted classes would give a better classifier. On average, the behavior of CA is described by the probability of error

$$PE = \mathbb{P}(CA(\boldsymbol{X}, Y) < 0),$$

in which $\boldsymbol{X}, Y$ are both treated as random variables. It is seen that PE is the probability that the correct classification by the pooled classifiers is given less often than the most likely of the wrong classifications. Clearly, it is desirable for PE to be small.

Suppose the J classifiers are tree-structured with $h_j(\boldsymbol{x}) = h(\boldsymbol{x}, \theta_j)$ and the random forest generated from them is formed by majority vote. In this notation, θ summarizes the extra variability used in constructing the tree, for instance the dropping of n balls into K urns in the occupancy theorem scenario. Breiman's random forest theorem states that as J increases (i.e., more and more trees are aggregated) PE, which depends on J, the size of the forest, converges to a limiting value. This limiting value, $P(CA)$, is derived from CA by replacing indicator functions with expectations. Formally, we have the following theorem.

Theorem (Breiman, 2001): As the number of tree-structured classifiers in the random forest increases, PE converges a.s. to $P(CA)$. That is, as $j \to \infty$

$$PE \to \mathbb{P}\left[P_\Theta(h(X, \Theta) = Y) - \max_{k \neq Y} P_\Theta(h(X, \Theta) = k) < 0\right],$$

in which P_Θ is the probability for Θ.

Remark: This result is asymptotic in the number of trees in the forest, not the sample size. Thus, it ensures that, as more trees are added, the random forest does not overfit. Entertainingly, no assumption has been made that any of the trees in the forest actually are good classifiers, only that they are randomly generated. This will include some good classifiers, but many poor ones, too.

Proof: Fix a value k. As $J \to \infty$, it is intuitive that

$$\frac{1}{J}\sum_{j=1}^{J} 1_{\{h(\boldsymbol{X},\theta_j)=k\}} \to P_\Theta(h(\boldsymbol{X},\Theta)=k). \tag{5.3.4}$$

However, it is not enough just to use an LLN to take expectations because the average in (5.3.4) is over functions $h(\cdot,\Theta)$, not individual random variables.

To verify (5.3.4), first observe that the J functions are tree-structured. So, together they induce a minimal partition of, say, L elements $S_1,\ldots,S_L$ of the feature space. That is, each $h(\cdot,\theta_j)$ corresponds to a partition of the feature space into K sets, each a finite union of hyperrectangles, representing the points on which the classifier assumes the K values in S. These J partitions can be represented as subpartitions of a larger but finite partition, and it is the smallest of these, of size L, that is most useful.

Fix $k \in \{1,\ldots,K\}$ and define a function $\phi(\theta)=\phi_k(\theta)$ by setting

$$\phi(\theta)=\ell \Leftrightarrow \{\boldsymbol{x}|h(\boldsymbol{x},\theta)=k\}=S_\ell \Leftrightarrow h_\theta^{-1}(k)=S_\ell, \tag{5.3.5}$$

i.e., for a given k, $\phi(\theta)$ gives the index of the partition element for which $h(\cdot,\theta)=k$. (In fact, the partition element may be a finite union of partition elements; this abuse of notation should not be confusing.) Set $\phi(\theta)=0$ if $h(\theta,\cdot)$ is never 1. To use (5.3.5), let N_ℓ count the number of times $\phi(\theta_k)=\ell$ on the sequence $\theta_1,\ldots,\theta_K$. That is,

$$N_\ell=\sum_{j=1}^{J} 1_{\{\phi(\theta_j)=\ell\}}. \tag{5.3.6}$$

The usual LLN applies to (5.3.6) to give

$$\frac{N_\ell}{J} \to P_\Theta(\phi(\Theta)=\ell).$$

Now, the strategy of proof is to convert the sum over J outcomes of the tree-structured classifier in (5.3.4) to a sum over the partition elements since the LLN can be applied to all L of them individually. Thus, for each k,

$$\begin{aligned}
\frac{1}{J}\sum_{j=1}^{J} 1_{\{h(\boldsymbol{X},\theta_j)=k\}} &= \frac{1}{J}\sum_{j=1}^{J}\sum_{\ell=1}^{L} 1_{\{h(\boldsymbol{X},\theta_j)=k\}} 1_{\{\phi(\theta_j)=\ell\}} \\
&= \frac{1}{J}\sum_{j=1}^{J}\sum_{\ell=1}^{L} 1_{\{\boldsymbol{X}\in S_\ell,\phi(\theta_j)=\ell\}} \\
&= \frac{1}{J}\sum_{\ell=1}^{L}\left(\sum_{j=1}^{J} 1_{\{\phi(\theta_j)=\ell\}}\right) 1_{\{\boldsymbol{X}\in S_\ell\}} = \sum_{\ell=1}^{L}\frac{N_\ell}{J} 1_{\{\boldsymbol{X}\in S_\ell\}} \\
&= \sum_{\ell=1}^{L} P_\Theta(\phi(\Theta)=\ell)1_{\{\boldsymbol{X}\in S_\ell\}} + \sum_{\ell=1}^{L}\left(\frac{N_\ell}{J} - P_\Theta(\phi(\theta)=\ell)\right) \\
&= \sum_{\ell=1}^{L} P_\Theta(\phi(\Theta)=\ell)P_\Theta(h(\Theta,\boldsymbol{X})=k|\phi(\Theta)=\ell)+o_{P_\Theta}(1) \\
&= P_\Theta(h(\Theta,\boldsymbol{X})=k)+o_{P_\Theta}(1).
\end{aligned} \tag{5.3.7}$$

Clearly, (5.3.7) is another way to write (5.3.4). Now, a series of Slutsky arguments gives the result: The limiting value in the theorem is

$$\sum_{k=1}^{K} \mathbb{P}(P_\Theta(h(\boldsymbol{X},\Theta) = k) - \max_{u:u\neq k} P_\Theta(h(\boldsymbol{X},\Theta) = u) < 0 | Y = k)\mathbb{P}(Y = k).$$

So, as long as the quantities conditional on k converge to their limits as above as J increases, $AV(Y)$ and the $AV(k)$s converge as in the theorem. □

Breiman's theorem has several implications. First, it shows that the reason there is no overfit is that outcomes of θ effectively search the model space for scenarios at random. So, probability cannot pile up at any one model on account of an apparent ability to explain the data. That is, a model that explains the data perfectly may be found by chance by θ, but only as one of many models that are found. A model that was good as a consequence of overfit would be washed out by averaging over other models that more fairly represented the explanatory power of the data. Second, the proof leads one to suspect that, in general, averaging can be used to avoid overfit. The avoidance of overfit may be a general property of many ensemble methods discussed in the next chapter. Indeed, in practice, ensemble methods based on randomness do not appear to lead to overfit unless used improperly (e.g., the selection of splits for the tree is unduly narrow). Third, the theorem assumes endless outcomes θ_j at random. With samples of finite size n, typically one does not take more than n bootstrap samples. If one were to take "too many" bootstrap samples, the result would be like taking a limit in the empirical distribution; this has good properties, as seen in Chapter 1, but would give a false sense of convergence because the empirical distribution function itself is only an approximation. In the present context, it would typically take many more than n bootstrap samples to fill out the empirical distribution over trees.

Another important theorem from Breiman (2001) bounds PE. To state this result, two definitions are required. First is the strength s of a classifier, and second is the standardized correlation $\bar{\rho}$. Following the structure of Breiman (2001), let the limiting random variable from CA be

$$LCA(\boldsymbol{X},Y) = P_\Theta(h(\boldsymbol{X},\Theta) = Y) - \max_{k\neq Y} P_\Theta(h(\boldsymbol{X},\Theta) = k).$$

Take the expectation of this, over Θ, to get the "strength" of the set of classifiers:

$$s = \mathbb{E}(LCA(X,Y)).$$

This summarizes the strength of the entire set of classifiers because LCA is the performance measure for a given θ and s is the average performance over classifiers generated by the randomness in Θ. It is a measure of how good the collection of tree-structured classifiers is.

The second definition is more involved. It encapsulates the correlation between the performances of two randomly generated classifiers. The tricky part is the need to standardize the expected correlation (over θ) by the variability in θ. To set up this correlation, observe that another expression for LCA comes from identifying the best wrong classifier. Let

$$\hat{k}(X,Y) = \arg\max_{k \neq Y} P_\Theta(h(\boldsymbol{X},\Theta) = k),$$

so that *LCA* can be written as

$$\begin{aligned} LCA &= P_\Theta(h(\boldsymbol{X},\Theta) = Y) - P_\Theta(h(\boldsymbol{X},\Theta) = \hat{k}) \\ &= E_\Theta\left[1_{\{h(\boldsymbol{X},\Theta)=Y\}} - 1_{\{h(\boldsymbol{X},\Theta)=\hat{j}\}}\right], \end{aligned}$$

in which the subscript Θ indicates that the probability measure involved only looks at the randomness in Θ, not $\boldsymbol{X}$ or Y. Nevertheless, the expression in brackets is a function of Θ, $\boldsymbol{X}$, and Y, indicating how much better the best classifier does given that both were generated from θ. Let this be the empirical classification accuracy,

$$ECA(\Theta,\boldsymbol{X},Y) = 1_{\{h(\boldsymbol{X},\Theta)=Y\}} - 1_{\{h(\boldsymbol{X},\Theta)=\hat{j}\}}.$$

It is seen that $\mathbb{E}(ECA) = LCA$.

At its root, it is the *ECA* that characterizes the correlation between the classifiers corresponding to independent copies of Θ. When the correlation is nonzero, it arises from the $\boldsymbol{X}$ and Y in the two indicator functions in *ECA*. Indeed, it is important to distinguish conceptually between what is random and what is not, (i.e., expectations over Θ, E_Θ versus expectations over $(\boldsymbol{X},Y)$ where $\mathbb{E}$ does not have a subscript).

Suppose independent copies of Θ, say Θ and Θ', have been generated, the expectation of their product factors giving

$$LCA^2 = E_{\Theta,\Theta'}\left[ECA(\Theta,\boldsymbol{X},Y)ECA(\Theta',\boldsymbol{X},Y)\right],$$

in which $\boldsymbol{X}$ and Y remain random. The variance over $\boldsymbol{X}$ and Y is

$$\begin{aligned} \mathrm{Var}(LCA) &= E_{\Theta,\Theta'}\mathrm{Cov}(ECA(\Theta,\boldsymbol{X},Y), ECA(\Theta',\boldsymbol{X},Y)) \\ &= E_{\Theta,\Theta'}\rho(\Theta,\Theta')SD(\Theta)SD(\Theta'), \end{aligned}$$

in which $\rho(\Theta,\Theta')$ is the correlation between $ECA(\Theta,\boldsymbol{X},Y)$ and $ECA(\Theta',\boldsymbol{X},Y)$ holding Θ and Θ' fixed, and *SD* is the standard deviation of *ECA*, again holding Θ or Θ' fixed. Finally, the standardized correlation is

$$\bar{\rho} = \frac{E_{\Theta,\Theta'}\rho(\Theta,\Theta')SD(\Theta)SD(\Theta')}{E_{\Theta,\Theta'}SD(\Theta)SD(\Theta')},$$

in which the quantities inside the expectation are random only in Θ and Θ'. It is seen that $\bar{\rho}$ is a measure of how correlated two randomly chosen trees are on average, standardized by their variability. The result is the following.

Theorem (Breiman, 2001): The generalization error can be bounded:

$$PE \leq \frac{\bar{\rho}(1-s^2)}{s^2}. \tag{5.3.8}$$

Proof: Chebyshev's inequality can be applied to *PE*:

$$PE = \mathbb{P}(CA < 0) \leq \mathbb{P}([CA - E(LCA)]^2 > E(LCA)^2) \leq \frac{\text{Var}(LCA)}{s^2}.$$

Next, the numerator on the right-hand side is

$$\text{Var}(LCA) = \bar{\rho}[E_\Theta SD(\Theta)]^2 \leq \bar{\rho} E_\Theta(\text{Var}(\Theta)) \leq \bar{\rho} E_\Theta(\mathbb{E}ECA(\Theta, X, Y))^2 - s^2$$

by Jensen's inequality (twice) and the definition of *ECA* and *LCA*. Putting these together gives the theorem. The right-hand side is bounded by $\bar{\rho}(1 - s^2)$ since ECA^2 is bounded by 1. The ratio in (5.3.8) follows. □

The qualitative behavior of the ratio in (5.3.8) is intuitively reasonable. When the strength increases or decreases, the bound on *PE* tightens or loosens. When the correlation is small or large, then repeated sampling gives more or less information and, again, the bound tightens or loosens. A strong set of classifiers with little correlation gives a strong bound.

Overall, random forests of trees tend to give good classifiers, competitive with SVMs, the topic of the next section. Unlike SVMs they often work relatively well with missing data, too. Random forests can give decent results even when as much as 80% of the data are missing at random.

5.4 Support Vector Machines

The core idea behind support vector machines (SVMs) as developed by Vapnik (1998), among others, is to define a boundary between two classes by maximal separation of the closest points. It turns out that, in addition to nice theoretical properties, SVMs give exceptionally good performance on classification tasks and avoid the Curse.

5.4.1 Margins and Distances

The underlying intuition of SVMs can be seen in the basic problem of building a classifier for a linearly separable, two-class problem, in the plane. This case will help introduce the key ideas underlying the SVM machinery.

It is intuitive that a good separating hyperplane is one that is far from the data; this will be formalized in the concept of a large margin classifier, where the term margin refers to the width of the blank strip separating two data clouds. Figure 5.2 shows a near trivial case where two classes are perfectly separated. Clearly, even for this data set, there is an infinity of hyperplanes that can separate the observations. The problem of which boundary to choose for classifying future observations remains and is dramatized by Fig. 5.3 which shows four superficially reasonable choices for how to choose boundaries to separate the two classes in Fig. 5.2. The boundary in Fig. 5.3d most accurately represents the margin and so would be preferred.

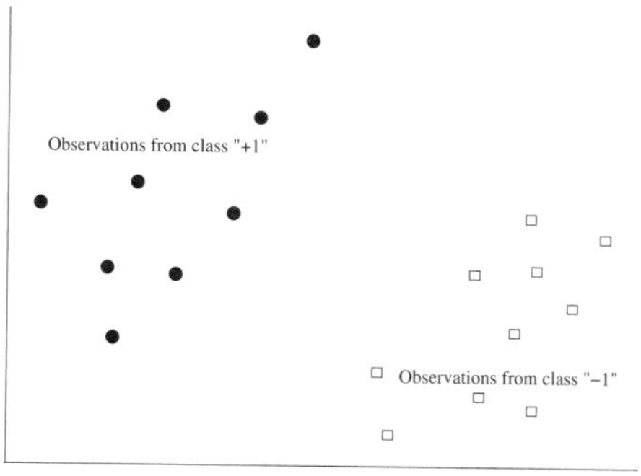

Fig. 5.2 The solid circles and the empty squares indicate the $\boldsymbol{x}$-values for two classes labeled 1 and -1. It is seen that the two classes are linearly separable.

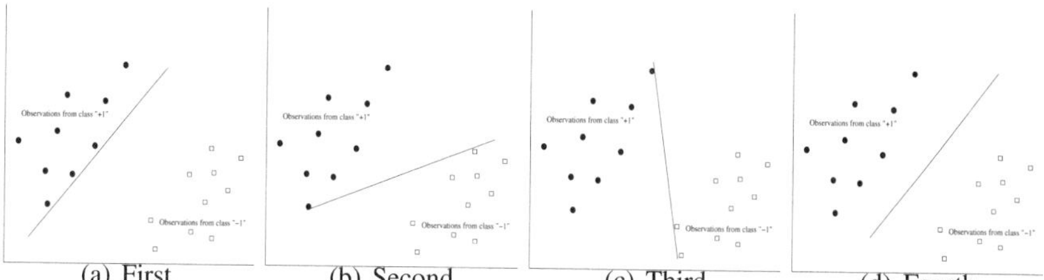

Fig. 5.3 These panels show four ways to separate the two classes. In (a), the boundary tries to draw a path around one class. In (b), and (c), the boundary is strongly affected by points that may not be representative of the data cloud. In (d), the boundary runs down the middle of the margin between the data clouds.

More formally, the margin is defined as the shortest perpendicular distance between the hyperplane and the observations. In effect, this is a minimax distance, as will be seen in the next subsection, because it maximizes over planes that satisfy a minimal distance criterion. Figure 5.4 shows the ideal case and gives the terminology. The dashed lines in Fig. 5.5 indicate the two hyperplanes closest to, or in this case on, the data points, realizing the smallest perpendicular distance. The margin is the distance between the two "outer" hyperplanes and is equidistant from them. The solid line indicates the large margin classifier or optimal separating hyperplane.

To formalize the central concept of margin, there are 4 tasks: (i) compute the distance between a point and the separating hyperplane, (ii) determine the minimum of such a distance from a given set of observations, (iii) compute the distance between two parallel hyperplanes, and (iv) verify that the two outer hyperplanes are equidistant from the separating hyperplane. These tasks are geometrically intuitive but necessitate some definitions.

Let $\boldsymbol{w} = (w_1, w_2, \cdots, w_p)^\mathsf{T} \in \mathbb{R}^p$ be a vector of coefficients and $b \in \mathbb{R}$ be a constant. Write the linear function $h : \mathbb{R}^p \to \mathbb{R}$ as

$$h(\boldsymbol{x}) = \boldsymbol{w}^\mathsf{T}\boldsymbol{x} + b.$$

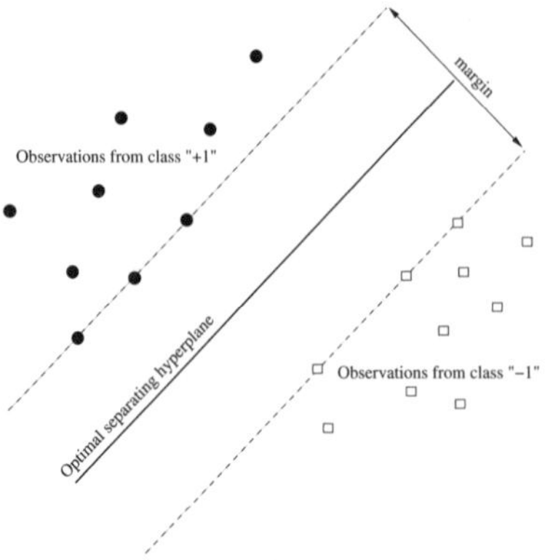

Fig. 5.4 The margin is the minimal perpendicular distance between the points in the two data clouds. The optimal separating hyperplane, the solid line, runs down the middle of the margin; it can be shown to be equidistant from the closest points on each of its sides. The dashed lines run through these points, indicating hyperplanes that are closest to, or indeed right on, certain of the data points.

Now, for a given constant $c \in \mathbb{R}$,

$$H_c(\boldsymbol{w}, b) = \left\{\boldsymbol{x} : \quad h(\boldsymbol{x}) = c\right\}$$

is a $(p-1)$-dimensional hyperplane. If $c = 0$, $H_{c=0}(\boldsymbol{w}, b)$ is denoted $H(\boldsymbol{w}, b)$.

A vector is a direction vector of a hyperplane if it is parallel to that hyperplane and is a normal vector of a hyperplane if it is perpendicular to all possible direction vectors of that hyperplane. Clearly, the vector $\boldsymbol{w}$ is a normal vector to $H_c(\boldsymbol{w}, b)$ for any c. Indeed, $\forall \boldsymbol{x}_i, \boldsymbol{x}_j \in H_c(\boldsymbol{w}, b)$, $\boldsymbol{w}^{\mathsf{T}}\boldsymbol{x}_i + b = c = \boldsymbol{w}'\boldsymbol{x}_j + b$, so $\boldsymbol{w}^{\mathsf{T}}(\boldsymbol{x}_i - \boldsymbol{x}_j) = 0$.

The formulation of the SVM classifier requires the expression of the perpendicular distance between a point in $\mathbb{R}^p$ and a hyperplane. This is indicated in Fig. 5.5 and stated in the following theorem.

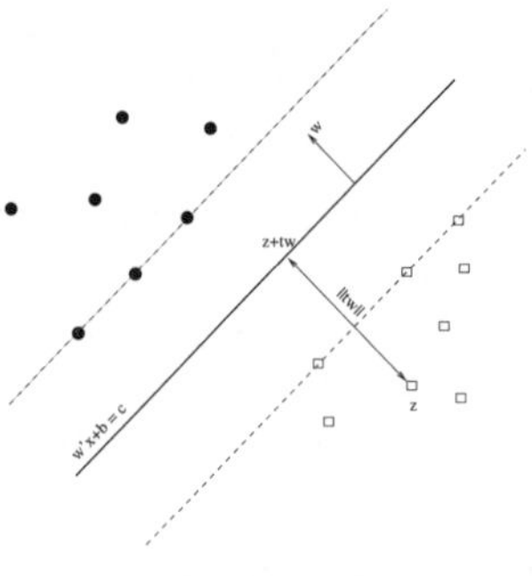

Fig. 5.5 The point $\boldsymbol{z}$ is t units away from the hyperplane indicated by the solid line in direction $\boldsymbol{w}$.

Theorem: The perpendicular distance $d(\boldsymbol{z}, H_c(\boldsymbol{w}, b))$ between the hyperplane $H_c(\boldsymbol{w}, b)$ and a point $\boldsymbol{z} \in \mathbb{R}^p$ is given by

$$d(z, H_c(\boldsymbol{w}, b)) = \frac{|\boldsymbol{w}^\mathsf{T} z + b - c|}{\|\boldsymbol{w}\|}.$$

Proof: Let $t \in \mathbb{R}$. Since $d(z, H_c(\boldsymbol{w}, b))$ is a perpendicular distance, it has to be traveled in the direction of the vector $\boldsymbol{w}$, which is normal to the hyperplane $H_c(\boldsymbol{w}, b)$. As a result, the vector $z^* = z + t\boldsymbol{w}$ in Fig. 5.4 must lie on $H_c(\boldsymbol{w}, b)$, giving $\boldsymbol{w}^\mathsf{T} z^* + b = c$. This is equivalent to $\boldsymbol{w}^\mathsf{T}(z + t\boldsymbol{w}) + b = c \Longleftrightarrow \boldsymbol{w}^\mathsf{T} z + t\boldsymbol{w}^\mathsf{T}\boldsymbol{w} + b = c$. Therefore

$$t = \frac{c - b - \boldsymbol{w}^\mathsf{T} z}{\boldsymbol{w}^\mathsf{T}\boldsymbol{w}} = \frac{c - b - \boldsymbol{w}^\mathsf{T} z}{\|\boldsymbol{w}\|^2}.$$

This expression for t gives

$$\|t\boldsymbol{w}\|^2 = |t|^2 \|\boldsymbol{w}\|^2 = \frac{|c - b - \boldsymbol{w}^\mathsf{T} z|^2}{\|\boldsymbol{w}\|^2},$$

which in turn gives

$$d(z, H_c(\boldsymbol{w}, b)) = \|t\boldsymbol{w}\| = \frac{|\boldsymbol{w}^\mathsf{T} z + b - c|}{\|\boldsymbol{w}\|}. \quad \square$$

Theorem: The perpendicular distance between two parallel hyperplanes $H_c(\boldsymbol{w}, b)$ and $H_{c'}(\boldsymbol{w}, b)$ is given by

$$d(H_c(\boldsymbol{w}, b), H_{c'}(\boldsymbol{w}, b)) = \frac{|c - c'|}{\|\boldsymbol{w}\|}. \tag{5.4.1}$$

Proof: This distance is found by choosing a point in one of the two hyperplanes and then using the previous theorem to compute the distance between that point and the other hyperplane. Without loss of generality, choose a point from the hyperplane $H_c(\boldsymbol{w}, b)$. Suppose it has the form $z = ((c - b)/w_1, 0, \cdots, 0)$. Then, the distance between the hyperplane $\boldsymbol{w}^\mathsf{T}\boldsymbol{x} + b = c'$ and z is

$$d(H_c(\boldsymbol{w}, b), H_{c'}(\boldsymbol{w}, b)) = \frac{|c' - b - \boldsymbol{w}^\mathsf{T} z|}{\|\boldsymbol{w}\|} = \frac{\left|c' - b - w_1\left(\frac{c-b}{w_1}\right)\right|}{\|\boldsymbol{w}\|} = \frac{|c - c'|}{\|\boldsymbol{w}\|}. \quad \square$$

5.4.2 Binary Classification and Risk

Now the generic problem of binary classification starts with an input domain $\mathscr{X} \subseteq \mathbb{R}^p$ to represent the values of the explanatory variables and a class domain $\mathscr{Y} = \{-1, +1\}$ to act as labels. Note that for mathematical convenience below, the values of Y are now ± 1 rather than $Y = 1, 2$. Let $\mathscr{D} = \{(\boldsymbol{x}_1, y_1), (\boldsymbol{x}_2, y_2), \cdots, (\boldsymbol{x}_n, y_n)\}$ be a data set. Assuming linear separability, the goal in binary classification is to use the data to estimate the p dimensional vector $\boldsymbol{w} = (w_1, w_2, \cdots, w_p)^\mathsf{T}$ and the constant b, so that the hyperplane

$$H(\boldsymbol{w},b) = \left\{\boldsymbol{x} : \boldsymbol{w}^\mathsf{T}\boldsymbol{x} + b = 0\right\}$$

separates the observations $\boldsymbol{x}_i$ into their class labels -1 and $+1$. More formally, the goal is to find a function $f : \mathscr{X} \longrightarrow \{-1,+1\}$ so that

$$\mathsf{class}(\boldsymbol{x}_i) = f(\boldsymbol{x}_i) = \text{sign}(\boldsymbol{w}^\mathsf{T}\boldsymbol{x}_i + b) = \begin{cases} +1, & \boldsymbol{w}^\mathsf{T}\boldsymbol{x}_i + b > 0, \\ -1, & \boldsymbol{w}^\mathsf{T}\boldsymbol{x}_i + b \leq 0, \end{cases}$$

for $\boldsymbol{x}_i \in \mathscr{D}$.

Clearly, all the separating hyperplanes of Fig. 5.2 achieve such a separation. However, it is apparent that although the separating hyperplanes in Figs. 5.2a-c separate the data in hand, they may be more likely to misclassify future data than the hyperplane in Fig. 5.2d is under the assumption of linear separability. That is, these three classifiers may not generalize as well as the one in Fig. 5.2d does. One can argue that the key property making Fig. 5.2d more desirable is that its separating hyperplane is further from the data than the separating hyperplanes in the others. In general, it is intuitive that by making the boundaries between the two classes as far apart as possible, little changes in the data are less likely to make the classifier change value. Thus, stability (or robustness) is a desirable consequence of the large margin concept on which SVMs are based.

Overall, an SVM classifier rests on a five-step concept: (i) Identify the class of functions from which the decision boundary is to be chosen (so far, this has been linear functions of $\boldsymbol{x}$); (ii) formally define the margin, which is the minimal distance between a candidate decision boundary and the points in each class; (iii) choose the decision boundary from the class in (i) (so far, this has been a hyperplane); (iv) evaluate the performance of the chosen decision boundary on the training set (this is usually the empirical risk); and (v) evaluate the anticipated classification performance on new data points (this is the generalization error).

It will be seen that accomplishing these tasks for SVMs is somewhat lengthy, as it requires a logical and conceptual development. See also Burges (1998) for a good, slightly different overview.

To begin, the quantities in (i), (iv) and (5) should be identified. So, consider the function class $\mathscr{F}$ of functions built from hyperplanes in $\mathbb{R}^p$,

$$\mathscr{F} = \{f : \mathbb{R}^p \longrightarrow \{-1,+1\}, \text{s.t. } \forall \boldsymbol{x} \in \mathscr{X},\ f(\boldsymbol{x}) = \text{sign}(h(x)),\ h \text{ is hyperplane in } \mathbb{R}^p\}.$$

As an assessment of error, 0-1 loss is often used: If the true class of $\boldsymbol{x}$ is y and the classifier delivers $f(\boldsymbol{x})$, the loss incurred, or the misclassification error, is

$$\ell(y, f(\boldsymbol{x})) = \mathbf{I}(f(\boldsymbol{x}) \neq y), \tag{5.4.2}$$

which is 0 if the classifier is correct and 1 otherwise, hence the name. As a point of terminology, in machine learning parlance, the classifier f is often called a hypothesis, and the class $\mathscr{F}$ from which f is taken is called the hypothesis class. Expression (5.4.2) leads to the generalization error, also called the prediction error, or theoretical risk. Let

$\mathscr{P}$ be a population with two variables, $X \in \mathscr{X} \subseteq \mathbb{R}^p$ and $Y \in \mathscr{Y} = \{-1,+1\}$, and a (usually unknown) probability $\mathbb{P}(X,Y)$ on $\mathscr{X} \times \mathscr{Y}$. Let f be a classifier (hypothesis) defined on $\mathscr{X}$, with $\ell(\cdot,\cdot)$ as in (5.4.2). The generalization error of f is

$$R(f) = \mathbb{E}[\ell(Y, f(\boldsymbol{X}))] = \int_{\mathscr{X}\times\mathscr{Y}} \ell(y, f(\boldsymbol{x}))d\mathbb{P}(\boldsymbol{x},y). \tag{5.4.3}$$

This is the probability that f will misclassify a randomly selected member of $\mathscr{X}$.

The ideal would be to find the best f, namely

$$f^* = \arg\inf_f R(f). \tag{5.4.4}$$

Unfortunately, finding f^* presupposes the ability to derive closed-form expressions for $R(f)$ and a way to search the space of functions $\{-1,+1\}^{card(\mathscr{X})}$. Clearly, the goal of (5.4.4) is an ill-posed problem in Hadamard's sense, as defined in Chapter 3. So, it is helpful to use some prior knowledge to restrict the class of possible hypotheses. Thus, the minimization is done over a class say $\mathscr{F}$; i.e., one seeks

$$f^\circ = \arg\inf_{f\in\mathscr{F}} R(f).$$

The risk under 0-1 loss, $R(f)$, is not really available because only a small fraction of the population is in the random sample. However, since $R(f)$ is also the probability of misclassification, it can be estimated by the empirical risk, or training error, $\hat{R}_n(f)$, computed from the size n sample. For a hypothesis space $\mathscr{H}$ of classifiers, a data set $\mathscr{D} = \{(\boldsymbol{x}_1,y_1),(\boldsymbol{x}_2,y_2),\cdots,(\boldsymbol{x}_n,y_n)\}$ and $f \in \mathscr{H}$,

$$\hat{R}_n(f) = \frac{1}{n}\sum_{i=1}^{n} \mathbf{I}(f(\boldsymbol{x}_i) \neq y_i). \tag{5.4.5}$$

This functional, $\hat{R}_n(f)$, is the empirical fraction of points misclassified by f. Minimizing (5.4.5) means finding the classifier $\hat{f}$ that minimizes the empirical risk functional,

$$\hat{f} = \arg\min_{f\in\mathscr{F}} \hat{R}_n(f). \tag{5.4.6}$$

Observe that (5.4.5) and (5.4.6) can be directly generalized to any loss function even though they have only been defined for 0-1 loss.

Superficially, (5.4.6) looks reasonable. However, because $\hat{f}$ is chosen to make $\hat{R}_n(\hat{f})$ as small as possible, $\hat{R}_n(\hat{f})$ tends to underestimate $R(\widehat{f_{true}})$, the true error rate. This follows because the optimal classifier from (5.4.6) is likely to achieve a zero empirical risk at the cost of overfitting from merely memorizing the sample. Quantifying the degree of underestimation requires a trade-off between ensuring (i) that the true risk $R(\hat{f})$ associated with $\hat{f}$ is as close as possible to $R(f^\circ) = \min_{f\in\mathscr{F}} R(f)$, the lowest risk achieved in $\mathscr{F}$, and (ii) that $\hat{R}_n(\hat{f})$ is as small as it can be.

More formally, the empirical risk $\hat{R}_n(f)$ should converge to the true risk $R(f)$. So, for preassigned δ and ε and uniformly over $f \in \mathscr{F}$,

$$\mathbb{P}\left(|\hat{R}_n(f) - R(f)| < \varepsilon\right) \geq 1 - \delta, \quad \forall f \in \mathscr{F}. \tag{5.4.7}$$

Essentially, (5.4.7) is a generalized variance–bias trade-off: The approximation error – the event in (5.4.7) – is a sort of bias, and $\hat{R}(\cdot)$ itself is comprised of both a variability term and a bias term. In other words, what is being sought is not a zero empirical risk but the smallest empirical risk that also corresponds to a small approximation error so that the combination of the two has a low prediction error.

If one assumes that the class $\mathscr{F}$ contains the best function f^*, then there is no approximation error from using f°. However, in practice, little is known a priori about f^*, so no assumptions can be made about its form or properties. Inference in this context is called agnostic learning. In agnostic learning, it is usually necessary to quantify uncertainty about f as an estimator of f^* using bounds.

5.4.3 Prediction Bounds for Function Classes

First, consider both the pointwise and the uniform convergences of the empirical risk to the true risk in the case of a finite function class. Without further comment, this subsection uses the Hoeffding inequality; it is stated and proved in the Notes at the end of the chapter.

Let $\mathscr{F} = \{f_1, f_2, \cdots, f_m\}$ be a finite list of classifiers. By the law of large numbers, the empirical risk $\hat{R}_n(f)$ converges pointwise for $f \in \mathscr{F}$ almost surely to the true risk $R(f)$ for any fixed classifier $f \in \mathscr{F}$. As usual, one usually wants a stronger mode of convergence than pointwise. Thus, uniform convergence for $f \in \mathscr{F}$ is defined as follows. If the space $\mathscr{F}$ of classifiers is finite, then $\forall \varepsilon > 0$,

$$\lim_{n \to \infty} \mathbb{P}\left(\max_{f \in \mathscr{F}} \left|\hat{R}_n(f) - R(f)\right| < \varepsilon\right) = 1.$$

In fact, since it is mostly upper bounds that will matter for controlling risk, one-sided uniform convergence is enough. This amounts to removing the absolute value bars and requiring

$$\lim_{n \to \infty} \mathbb{P}\left(\max_{f \in \mathscr{F}} \{\hat{R}_n(f) - R(f)\} < \varepsilon\right) = 1.$$

Now, a small empirical risk is likely to correspond to a small true risk.

However, for practical purposes, asymptotic verifications are not usually enough; it is important to quantify how fast the empirical risk converges to the true risk. Risk bounds provide insights into the difference between empirical and true risk. For finite sets of classifiers, the uniform risk bound is $\forall \varepsilon > 0$,

$$\mathbb{P}\left(\max_{f \in \mathscr{F}} \left|\hat{R}_n(f) - R(f)\right| > \varepsilon\right) \leq 2|\mathscr{F}|e^{-2n\varepsilon^2}. \tag{5.4.8}$$

Expression (5.4.8) leads to a confidence interval for $R(\hat{f})$. Let $\hat{h}$ minimize the empirical risk minimizer and let $1-\eta$ be the desired confidence level. Set

$$\varepsilon = \sqrt{\frac{1}{2n}\log\left(\frac{2|\mathscr{F}|}{\eta}\right)}.$$

Then,

$$\mathbb{P}\left(|\hat{R}_n(\hat{f}) - R(\hat{f})| > \varepsilon\right) \leq \mathbb{P}\left(\max_{f\in\mathscr{F}}|\hat{R}_n(f) - R(f)| > \varepsilon\right) \leq 2|\mathscr{F}|e^{-2n\varepsilon^2} = \eta,$$

which is equivalent, after inversion, to

$$\mathbb{P}\left(|\hat{R}_n(\hat{f}) - R(\hat{f})| < \varepsilon\right) \geq 1 - 2|\mathscr{F}|e^{-2n\varepsilon^2} = 1-\eta,$$

which gives the traditional form of $(1-\eta)$ confidence intervals. That is, $\hat{R}_n(\hat{f}) \pm \varepsilon$ is a $1-\eta$ confidence interval for $R(\hat{f})$.

Now, the most common one-sided uniform risk bound can be obtained. For finite $\mathscr{F}$ and $\forall n > 0$ and $\forall \eta > 0$,

$$\mathbb{P}\left[R(f) - \hat{R}_n(f) < \sqrt{\frac{1}{2n}\log\left(\frac{|\mathscr{F}|}{\eta}\right)}\right] \geq 1-\eta \qquad \forall f \in \mathscr{F}$$

which means that, with probability at least $1-\eta$,

$$R(f) < \hat{R}_n(f) + \sqrt{\frac{\log|\mathscr{F}| + \log(1/\eta)}{2n}}$$

for all $f \in \mathscr{F}$ and all $\eta > 0$.

Although this last bound is not valid when $\mathscr{F}$ is not finite, the basic result generalizes if the Vapnik-Chervonenkis dimension of a collection of functions is used in place of the cardinality, as in (5.4.9) below. The VC dimension is treated in the Notes at the end of this chapter. For the present, the following will suffice. Say that a set of points U in $\mathbb{R}^p$ is "shattered" by a set of functions $\mathscr{F}$ if and only if every partition of U into two subsets can be represented in terms of a function in $\mathscr{F}$. Here, represented means that for each subset $V \subset U$, there is a function $f \in \mathscr{F}$ such that the points in V are on one side of the surface generated by f. There are $2^{\#(U)}$ subsets, so any set of functions $\mathscr{F}$ that can shatter a set U with cardinality at least 3 must have at least eight elements. Any set $\mathscr{F}$ of eight or fewer functions cannot shatter more than three points. Now, the VC dimension of $\mathscr{F}$ is the cardinality of the largest set of points that can be shattered by $\mathscr{F}$ and written $\zeta = \text{VCdim}(\mathscr{F})$; see Vapnik and Chervonenkis (1971).

Theorem (Vapnik and Chervonenkis, 1971): Let $\mathscr{F}$ be a class of functions with VC dimension ζ with risks as in (5.4.3) and (5.4.5). Then, with uniformly high probability, $\forall f \in \mathscr{F}$ the theoretical risk is bounded by the empirical risk plus a term of order $\zeta/\sqrt{n}$. That is, for given $\eta > 0$ and $\forall f \in \mathscr{F}$,

$$R(f) \leq \hat{R}_n(f) + \sqrt{\frac{\zeta\left(\log\frac{2n}{\zeta}+1\right) - \log\frac{\eta}{4}}{n}} \tag{5.4.9}$$

with probability of at least $1-\eta$. □

The right-hand side of (5.4.9) is the VC bound. One of its greatest appeals is that even though it holds for arbitrary $\mathscr{F}$s, it preserves the same form as for finite $\mathscr{F}$s. One heuristic interpretation that justifies this is the following. Since ζ is a natural number, there are approximately ζ equal-sized subsets, of size $2n/\zeta$, among $2n$ points. If one chooses ζ of these subsets at random, with replacement, there are $(2n/\zeta)^\zeta$ vectors of length ζ. Roughly, if $\zeta = \text{VCdim}(\mathscr{F})$, then the functions in $\mathscr{F}$ can shatter point sets of size ζ. Replacing the ζ points in such a set with the vectors of length ζ means that the number L of possible labeling configurations obtainable from $\mathscr{F}$ with $\text{VCdim}(\mathscr{F}) = \zeta$ over $2n$ points satisfies

$$L \leq \left[\frac{en}{\zeta}\right]^\zeta. \tag{5.4.10}$$

The VC bound is seen to be obtained by replacing $\log|\mathscr{F}|$ with L in the expression of the risk bound for finite-dimensional $\mathscr{F}$. Thus, the generalization ability of a learning machine depends on both the empirical risk and the complexity of the class of functions used, and the bounds are distribution-free. This means that, in general, the predictive performance of a class of machines can be improved by finding the best trade-off between small VC dimension and minimization of the empirical risk, which is a measure of fit. In a sense, the VC bound is acting like the number of parameters since both are assessments of the complexity of $\mathscr{F}$.

A key result in SVM classification is that, within the class of separating hyperplanes, those hyperplanes based on larger margins achieve a better predictive performance. This can be seen intuitively by noting that the larger the margin, the less the corresponding classifier is sensitive to small perturbations in the data. The size of the margin is a measure of how much one can move the separating hyperplane without misclassifying future points. In addition, large margin classifiers are also associated with a lower VC dimension. Recall that the VC dimension is a sort of generalization of the real dimension obtained by looking at the largest number of points that can be partitioned in all possible ways into two subsets obtainable from a class $\mathscr{F}$ of classifiers. Since these partitions correspond to binary labelings of the data and SVMs amount to choosing a partition of the data, the class of separating hyperplanes based on large margins corresponds to fewer labelings than the class of all separating hyperplanes.

Note that the classification by SVMs depends on the points closest to each other in the two classes. This means SVMs are driven only by the vectors on the margin. These are called support vectors since, as will be seen, their coefficients in the expression for the classifier are positive. Since the number of support vectors is much less than n, the number of labelings needed is enormously reduced. The relationship between the VC dimension and number of labelings, as in (5.4.10), suggests that lower numbers of labelings correspond to smaller VC dimensions. Therefore, with their reduced number of possible labelings, large margin classifiers correspond to lower VC bounds, which in turn reflect a lower VC dimension. This intuition is formalized in the following theorems; see Vapnik (1995) for detailed proofs.

Proposition (Vapnik, 1995): Let $\mathscr{X} \subseteq \mathbb{R}^p$ be the domain of a set of functions implementing binary SVM classification. Let $\mathscr{F}$ denote the space of all separating hyperplanes and $\zeta = \text{VCdim}(\mathscr{F})$. Assume that all training sample points are contained in a ball of radius at most γ; i.e.,

$$\boldsymbol{x}_i \in \text{ ball of radius } \gamma, \qquad \forall \boldsymbol{x}_i \in \mathscr{D}.$$

Let M be the margin of the SVM under consideration. Then,

$$\zeta \leq \min(\gamma^2/M^2, p) + 1. \quad \square$$

This result shows that large margin classifiers reduce the VC bound, which translates into a lower prediction error or better generalization on new data points. In fact, SVM classifiers are also motivated by their link to cross-validation.

Theorem (Vapnik, 1995): Let E_n be the classification error rate for an SVM classifier, estimated by leave-one-out cross-validation using a sample of size n. Let s be the number of support vectors used in the SVM procedure. Then,

$$E_n \leq \frac{s}{n}. \quad \square$$

That is, the fewer support vectors needed, the less likely SVMs are to misclassify new examples.

5.4.4 Constructing SVM Classifiers

Given the last proposition, it is desirable to find a classifier that maximizes the margin $M = 1/\|\boldsymbol{w}\|$, which corresponds to minimizing $\|\boldsymbol{w}\|/2$. In Fig. 5.6, cases (a) and (b) are examples of separating hyperplanes with margins that are so narrow as to cause instability of the decision boundary and hence poor generalization. Case (c) indicates the best choice of hyperplane from the standpoint of good generalization or equivalently, large margin and VC dimension.

More formally, given $\mathscr{D} = \{(\boldsymbol{x}_1, y_1), \cdots, (\boldsymbol{x}_n, y_n)\}$, with $\boldsymbol{x}_i \in \mathbb{R}^p$ and $y_i \in \{-1, +1\}$, the margin maximization principle applied to linear classifiers is:

Find the function $h(\boldsymbol{x}) = \boldsymbol{w}^\mathsf{T}\boldsymbol{x} + b$ that achieves

$$\max_{\boldsymbol{w}, b} \left[\min_{y_i=+1} d(\boldsymbol{x}_i, H(\boldsymbol{w}, b)) + \min_{y_i=-1} d(\boldsymbol{x}_i, H(\boldsymbol{w}, b)) \right]$$

subject to $y_i(\boldsymbol{w}^\mathsf{T}\boldsymbol{x}_i + b) \geq 1, \quad \forall i = 1, \cdots, n.$

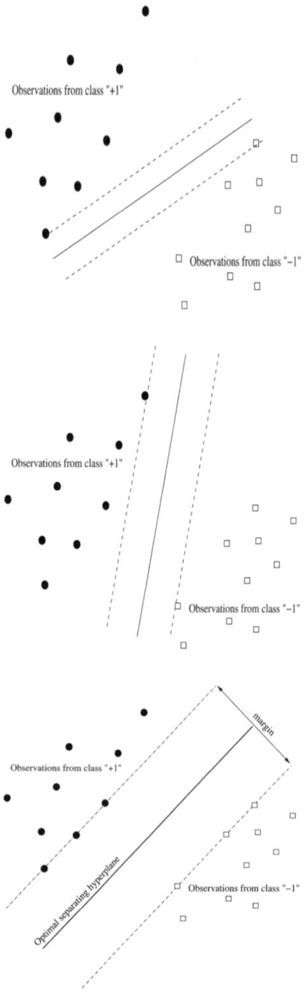

Fig. 5.6 The top panel has a smaller margin than the middle panel does, which is smaller than the (visually) optimal separating hyperplane in the bottom panel.

Using the fact that the distance from a point $\boldsymbol{x}_i$ to hyperplane $H(\boldsymbol{w}, b)$ is given by

$$d(\boldsymbol{x}_i, H(\boldsymbol{w}, b)) = \frac{|\boldsymbol{w}^\mathsf{T}\boldsymbol{x}_i + b|}{\|\boldsymbol{w}\|},$$

the formulation above translates more succinctly into

Find the function $h(\boldsymbol{x}) = \boldsymbol{w}^\mathsf{T}\boldsymbol{x} + b$ that achieves

$$\max_{\boldsymbol{w},b}\left[\min_{y_i=+1}\left\{\frac{|\boldsymbol{w}^\mathsf{T}\boldsymbol{x}_i+b|}{\|\boldsymbol{w}\|}\right\} + \min_{y_i=-1}\left\{\frac{|\boldsymbol{w}^\mathsf{T}\boldsymbol{x}_i+b|}{\|\boldsymbol{w}\|}\right\}\right]$$

subject to $y_i(\boldsymbol{w}^\mathsf{T}\boldsymbol{x}_i + b) \geq 1, \quad \forall i = 1, \cdots, n.$

As written, this optimization problem can be solved, but not uniquely; there are an infinite number of solutions. This can be seen by observing that if $\boldsymbol{w}$ and b are solutions, then for any constant $\kappa \geq 1$, $\kappa\boldsymbol{w}$ and κb also give a solution. Indeed, when $\|\kappa\boldsymbol{w}\| \geq \|\boldsymbol{w}\|$, the boundaries of the classes move closer to the hyperplane $H(\boldsymbol{w}, b)$, contrary to the goal of large margins.

It would be nice to have a unique solution that makes $\boldsymbol{w}$ and b as small as possible. To do this, one seeks the smallest $\kappa \in (0,1]$ such that $\kappa\boldsymbol{w}$ and κb remain solutions to the constrained optimization problem. Since small κs push the boundaries of the classes away from the separating hyperplane, the constraint $y_i(\boldsymbol{w}^\mathsf{T}\boldsymbol{x}_i + b) \geq 1$ must be enforced. So, solutions are characterized by the fact that at least one pair $(\boldsymbol{x}_i, y_i)$ exists such that $y_i(\boldsymbol{w}^\mathsf{T}\boldsymbol{x}_i + b) = 1$ holds for each of the two classes. This means that it is enough for the points to be correctly classified; how far within its class a point is does not matter.

Now, since $y_i = \pm 1$, an optimal solution to the minimization of the distance is characterized by $\boldsymbol{w}^\mathsf{T}\boldsymbol{x}_i^* + b = +1$ for some points $\boldsymbol{x}_i^*$ with $y_i^* = +1$ and $\boldsymbol{w}^\mathsf{T}\boldsymbol{x}_i^* + b = -1$ for some other points $\boldsymbol{x}_i^*$ with $y_i^* = -1$. In this context, for given $\boldsymbol{w}$ and b, the canonical hyperplane for points of class $+1$ is defined by

$$H_{+1}(\boldsymbol{w}, b) = \{\boldsymbol{x} \in \mathscr{X} : \boldsymbol{w}^\mathsf{T}\boldsymbol{x} + b = +1, \text{ for } y = +1\},$$

and the canonical hyperplane for points of class -1 is defined by

$$H_{-1}(\boldsymbol{w}, b) = \{\boldsymbol{x} \in \mathscr{X} : \boldsymbol{w}^\mathsf{T}\boldsymbol{x} + b = -1, \text{ for } y = -1\}.$$

Using the results above, the expression for the margin is

$$\begin{aligned} M &= \min_{y_i=+1}\left\{\frac{|\boldsymbol{w}^\mathsf{T}\boldsymbol{x}_i+b|}{\|\boldsymbol{w}\|}\right\} + \min_{y_i=-1}\left\{\frac{|\boldsymbol{w}^\mathsf{T}\boldsymbol{x}_i+b|}{\|\boldsymbol{w}\|}\right\} \\ &= \frac{1}{\|\boldsymbol{w}\|} + \frac{1}{\|\boldsymbol{w}\|} = \frac{2}{\|\boldsymbol{w}\|}. \end{aligned}$$

Alternatively, the margin could have been calculated directly using the distance (5.4.1) between two canonical hyperplanes. This is

$$M = d(H_{+1}(\boldsymbol{w}, b), H_{-1}(\boldsymbol{w}, b)) = \frac{|1-(-1)|}{\|\boldsymbol{w}\|} = \frac{2}{\|\boldsymbol{w}\|}.$$

Finding the maximum perpendicular distance that satisfies the SVM constraint implies finding a hyperplane $H(\mathbf{w},b)$ such that the minimal distance from it to points of class $+1$ equals the minimal distance from it to points of class -1.

Now, the optimization problem can be reformulated as

Find the function $h(\mathbf{x}) = \mathbf{w}^{\mathsf{T}}\mathbf{x} + b$ that achieves

$$\max_{\mathbf{w},b} \frac{2}{\|\mathbf{w}\|}$$

subject to $y_i(\mathbf{w}^{\mathsf{T}}\mathbf{x}_i + b) \geq 1, \quad \forall i = 1, \cdots, n.$

Clearly, maximizing $M = 2/\|\mathbf{w}\|$ is equivalent to minimizing $\|\mathbf{w}\|/2$, which in turn is equivalent to minimizing $(1/2)\|\mathbf{w}\|^2$. As a result, support vector classification under the linearly separable assumption becomes the solution to

Find the function $h(\mathbf{x}) = \mathbf{w}^{\mathsf{T}}\mathbf{x} + b$ that achieves

$$\min_{\mathbf{w},b} \frac{1}{2}\|\mathbf{w}\|^2$$

subject to $y_i h(\mathbf{x}_i) \geq 1, \quad \forall i = 1, \cdots, n.$

5.4.4.1 Constrained Optimization Problems

The last formulation of SVM classification is as a nonlinear constrained convex optimization. To understand the solutions, some key definitions of convex optimization under inequality constraints must be mastered. Consider a domain $\mathscr{X}$ and a nonlinear constrained optimization problem of "primal" form,

$$\begin{aligned} &\text{minimize } f(\mathbf{w}) \\ &\text{subject to } h_i(\mathbf{w}) \geq 0, i = 1, \cdots, n. \end{aligned}$$

In a more general setting, there are both equality and inequality constraints on the objective function $f(\mathbf{w})$. Here, it will be enough to focus on the inequality constraints, which are often more difficult. The set of is for which the inequality constraints become equalities is particularly important.

Definition an inequality constraint $h_i(\mathbf{w}) \geq 0$ to be active (or effective) at a point $\mathbf{w}^*$ if $h_i(\mathbf{w}^*) = 0$. Then, given $\mathbf{w}^*$, the set of constraints that are active at $\mathbf{w}^*$ is

$$\mathscr{A}(\mathbf{w}^*) = \left\{ i : h_i(\mathbf{w}^*) = 0 \right\}.$$

In nonlinear constrained optimization, the Karush-Kuhn-Tucker conditions (also known as the Kuhn-Tucker or the KKT conditions) are necessary and a useful way to construct a solution.

Theorem (KKT necessary conditions for local minima): Let $\mathscr{X} \subseteq \mathbb{R}^p$ and $f : \mathscr{X} \to \mathbb{R}$, with continuously differentiable constraint functions $h_i : \mathscr{X} \to \mathbb{R}$, $i = 1, 2, \cdots, n$. Suppose $\boldsymbol{w}^*$ is a local minimum of f on the set

$$\mathscr{S} = \mathscr{X} \cap \{\boldsymbol{w} \in \mathbb{R}^p \mid h_i(\boldsymbol{w}) \geq 0,\ i = 1, \cdots, n\}$$

and that $\{\nabla h_i(\boldsymbol{w}^*) \mid i \in \mathscr{A}(\boldsymbol{w}^*)\}$, the derivatives of the active constraints at $\boldsymbol{w}^*$ are a set of independent vectors. Then $\exists\ \alpha_1^*, \alpha_2^*, \cdots, \alpha_n^* \in \mathbb{R}$ so that

$$\alpha_i^* \geq 0, \qquad i = 1, 2, \cdots, n,$$

$$\nabla f(\boldsymbol{w}^*) - \sum_{i=1}^{n} \alpha_i^* \nabla h_i(\boldsymbol{w}^*) = 0,$$

$$\alpha_i^* h_i(\boldsymbol{w}^*) = 0, \qquad i = 1, 2, \cdots, n. \quad \square$$

The n conditions $\alpha_i^* h_i(\boldsymbol{w}^*) = 0$ are called the complementary slackness conditions. By the KKT local minimum conditions, it can be shown that

$$\alpha_i^* > 0 \text{ whenever } h_i(\boldsymbol{w}^*) = 0$$

and

$$\alpha_i^* = 0 \text{ whenever } h_i(\boldsymbol{w}^*) > 0,$$

i.e., α_i^* and $h_i(\boldsymbol{w}^*)$ cannot both be zero. The α_i^*s are called the KKT multipliers, in the same spirit as the Lagrange multipliers. Clearly, $\alpha_i^* > 0$ only on the set of active constraints, as will be seen later with SVMs. The support vectors will be defined as those points $\boldsymbol{x}_i$ that have nonzero coefficients.

5.4.4.2 From Primal to Dual Space Formulation

In optimization parlance, the initial problem of this subsection is the optimization formulation in "primal" space, usually just called the primal problem. The primal problem is often transformed into an unconstrained one by way of Lagrange multipliers, and the result is called the dual problem. The Lagrangian corresponding to the primal space formulation is

$$E_P(\boldsymbol{w}, \boldsymbol{\alpha}) = f(\boldsymbol{w}) + \sum_{i=1}^{n} \alpha_i h_i(\boldsymbol{w}).$$

Reformulating the primal problem into dual space makes certain aspects of the problem easier to manipulate and also makes interpretations more intuitive. Basically, the intuition is the following. Since the solution of the primal problem is expressed in terms of $\boldsymbol{\alpha}^\top = (\alpha_1, \cdots, \alpha_n)$, plugging such a solution back into the Lagrangian yields a new objective function where the roles are reversed; i.e., $\boldsymbol{\alpha}$ becomes the objective variable. More specifically, the Lagrangian of the dual problem is

$$E_D(\boldsymbol{\alpha}) = \inf_{\boldsymbol{w} \in \mathscr{X}} E_P(\boldsymbol{w}, \boldsymbol{\alpha}),$$

and, since the KKT conditions give that $\alpha_i^* \leq 0$ at the local minimum $\boldsymbol{w}^*$, the dual problem can be formulated as

$$\begin{array}{ll} \text{Minimize} & E_D(\boldsymbol{\alpha}) \\ \text{subject to} & \boldsymbol{\alpha} \geq \mathbf{zero}. \end{array}$$

One of the immediate benefits of the dual formulation is that the constraints are simplified and generally fewer in number. Also, if the primal objective function is quadratic, then the dual objective will be quadratic. Finally, by a result called the duality theorem (not discussed here), the solution to the dual problem coincides with the solution to the primal problem.

5.4.4.3 SVMs as a Constrained Optimization

Given the SVM problem statement and the mini-review on constrained optimization, SVM classification in primal space can be written as

$$\begin{array}{l} \text{Find the function } h(\boldsymbol{x}) = \boldsymbol{w}^{\mathsf{T}}\boldsymbol{x} + b \text{ that} \\ \text{minimizes } \dfrac{1}{2}\|\boldsymbol{w}\|^2 \\ \text{subject to } y_i(\boldsymbol{w}^{\mathsf{T}}\boldsymbol{x}_i + b) \geq 1, \quad i = 1, \cdots, n. \end{array}$$

The Lagrangian objective function for "unconstrained" optimization is

$$E_P(\boldsymbol{w}, b, \boldsymbol{\alpha}) = \frac{1}{2}\|\boldsymbol{w}\|^2 - \sum_{i=1}^{n} \alpha_i [y_i(\boldsymbol{w}^{\mathsf{T}}\boldsymbol{x}_i + b) - 1],$$

where $\alpha_i \in \mathbb{R}$, for all $i = 1, 2, \cdots, n$, are the Lagrange multipliers.

To solve the problem mathematically, start by computing the partial derivatives and solve the corresponding equations. Clearly,

$$\frac{\partial}{\partial \boldsymbol{w}} E_P(\boldsymbol{w}, b, \boldsymbol{\alpha}) = \boldsymbol{w} - \sum_{i=1}^{n} \alpha_i y_i \boldsymbol{x}_i \quad \text{and} \quad \frac{\partial}{\partial b} E_P(\boldsymbol{w}, b, \boldsymbol{\alpha}) = -\sum_{i=1}^{n} \alpha_i y_i.$$

Solving $\nabla E_P(\boldsymbol{w}, b, \boldsymbol{\alpha}) = 0$ for both $\boldsymbol{w}$ and b, a local minimum must satisfy

$$\boldsymbol{w} = \sum_{i=1}^{n} \alpha_i y_i \boldsymbol{x}_i \quad \text{with} \quad \sum_{i=1}^{n} \alpha_i y_i = 0.$$

Based on this local minimum, the KKT conditions imply that there exists an $\boldsymbol{\alpha}^*$ such that $\alpha_i^* = 0$ for all $\boldsymbol{x}_i$ satisfying $y_i(\boldsymbol{w}^\mathsf{T}\boldsymbol{x}_i + b) > 1$. So, as noted after the KKT Theorem, for all $i \in \{1, 2, \cdots, n\}$, it follows that

$$\alpha_i^* = 0 \quad \text{when} \quad y_i(\boldsymbol{w}^\mathsf{T}\boldsymbol{x}_i + b) > 1$$

and

$$\alpha_i^* > 0 \quad \text{when} \quad y_i(\boldsymbol{w}^\mathsf{T}\boldsymbol{x}_i + b) = 1.$$

The vectors $\boldsymbol{x}_i$ for which $\alpha_i > 0$ (i.e., the solution has strictly positive weight) are the support of the solution and hence called the support vectors.

This definition is reasonable because support vectors belong to the hyperplanes forming the boundary of each class, namely $H_{+1} = \{\boldsymbol{x} : \boldsymbol{w}^\mathsf{T}\boldsymbol{x} + b = +1\}$ or $H_{-1} = \{\boldsymbol{x} : \boldsymbol{w}^\mathsf{T}\boldsymbol{x} + b = -1\}$, thereby providing the definition of the margin. This is depicted in Fig. 5.7, where the support vectors lie on either of the two hyperplanes parallel to the optimal separating hyperplane.

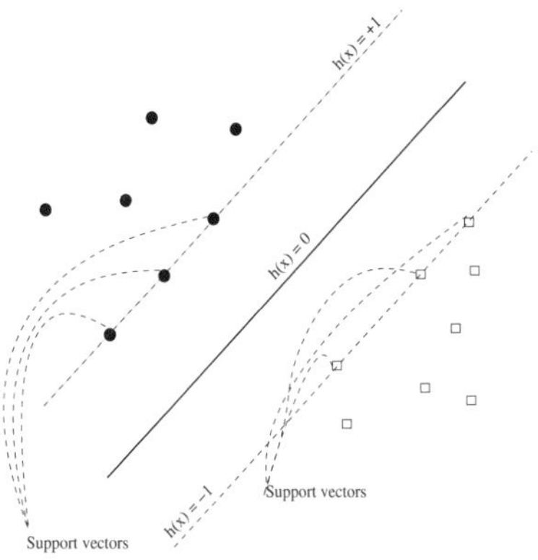

Fig. 5.7 Each of the two hyperplanes has three of the data points on it. Midway between them is the actual separating hyperplane. All other data points are outside the margin.

Figure 5.7 shows the desirable case in which most of the α_is are zero, leaving only very few $\alpha_i > 0$. When this is the case, the solution (i.e., the separating hyperplane) is a sparse representation of the function of interest (i.e., the optimal boundary). But note that it is dual space sparsity that is meant, and this is different from the traditional sparsity, or parsimony, based on the primal space formulation in terms of the inputs directly. The next subsection will clarify this.

5.4.4.4 Dual Space Formulation of SVM

The dual space formulation for the SVM problem is easily derived by plugging

$$w = \sum_{i=1}^{n} \alpha_i y_i \boldsymbol{x}_i$$

into the original objective function. It is easy to see that $E_P(\boldsymbol{w}, b, \boldsymbol{\alpha})$ becomes

$$\begin{aligned} E_P(\boldsymbol{w}, b, \boldsymbol{\alpha}) &= \frac{1}{2}\boldsymbol{w}^\mathsf{T}\boldsymbol{w} - \sum_{i=1}^{n} \alpha_i [y_i(\boldsymbol{w}^\mathsf{T}\boldsymbol{x}_i + b) - 1] \\ &= \frac{1}{2}\sum_{i=1}^{n}\sum_{j=1}^{n} \alpha_i \alpha_j y_i y_j \boldsymbol{x}_i^\mathsf{T}\boldsymbol{x}_j \\ &- \sum_{i=1}^{n}\sum_{j=1}^{n} \alpha_i \alpha_j y_i y_j \boldsymbol{x}_i^\mathsf{T}\boldsymbol{x}_j - b\sum_{i=1}^{n} \alpha_i y_i + \sum_{i=1}^{n} \alpha_i. \end{aligned}$$

Since the new objective function has neither $\boldsymbol{w}$ nor b, denote it $E_D(\boldsymbol{\alpha})$. Now, the dual space formulation of linear SVM classification is

Maximize

$$E_D(\boldsymbol{\alpha}) = \sum_{i=1}^{n} \alpha_i - \frac{1}{2}\sum_{i=1}^{n}\sum_{j=1}^{n} \alpha_i \alpha_j y_i y_j \boldsymbol{x}_i^\mathsf{T}\boldsymbol{x}_j$$

subject to

$$\sum_{i=1}^{n} \alpha_i y_i = 0 \quad \text{and } \alpha_i \geq 0, \quad i = 1, \cdots, n.$$

This last formulation is particularly good for finding solutions because it devolves to a quadratic programming problem for which there is a large established literature of effective techniques. In fact, defining the $n \times n$ matrix $\mathbf{Q} = (Q_{ij})$, where

$$Q_{ij} = y_j y_i \boldsymbol{x}_j^\mathsf{T}\boldsymbol{x}_i,$$

and the n-dimensional vector $\mathbf{c} = (-1, -1, \cdots, -1)^\mathsf{T}$, training a linear SVM classifier boils down to finding

$$\hat{\boldsymbol{\alpha}} = \arg\max_{\boldsymbol{\alpha}} \left\{ -\mathbf{c}^\mathsf{T}\boldsymbol{\alpha} - (1/2)\boldsymbol{\alpha}^\mathsf{T}\mathbf{Q}\boldsymbol{\alpha} \right\}.$$

That is, all the abstract manipulations undergirding linear SVM classification problems can be summarized in a recognizable quadratic minimization problem:

Minimize

$$E_D(\boldsymbol{\alpha}) = \frac{1}{2}\boldsymbol{\alpha}^\mathsf{T}\boldsymbol{Q}\boldsymbol{\alpha} + \mathbf{c}^\mathsf{T}\boldsymbol{\alpha}$$

subject to

$$\sum_{i=1}^{n} \alpha_i y_i = 0 \quad \text{and } \alpha_i \geq 0, \quad i = 1, \cdots, n.$$

The matrix $\mathbf{Q}$ is guaranteed to be positive semidefinite, so traditional quadratic programming algorithms will suffice.

To finish this development, note that from the determination of the α_is, the vector $\boldsymbol{w}$ can be deduced so all that remains is the determination of the constant b. Since $y_i(\boldsymbol{w}^\mathsf{T}\boldsymbol{x}_i + b) = 1$ for support vectors, write

$$\hat{b} = -\frac{1}{2}\left(\min_{y_i=+1}\left\{\hat{\boldsymbol{w}}^\mathsf{T}\boldsymbol{x}_i\right\} + \max_{y_i=-1}\left\{\hat{\boldsymbol{w}}^\mathsf{T}\boldsymbol{x}_i\right\}\right). \tag{5.4.11}$$

A simpler way to find $\hat{b}$ is to observe that the KKT conditions give

$$\alpha_i(y_i(\boldsymbol{w}^\mathsf{T}\boldsymbol{x}_i + b) - 1) = 0, \quad \forall i = 1, \cdots, n.$$

So, for support vectors $\alpha_i \neq 0$, it is seen that

$$\hat{b} = y_i - \hat{\boldsymbol{w}}^\mathsf{T}\boldsymbol{x}_i.$$

Equivalently, this gives $\hat{b} = -1 - \min_{y_i=-1}\{\hat{\boldsymbol{w}}^\mathsf{T}\boldsymbol{x}_i\}$ and $\hat{b} = 1 - \min_{y_i=+1}\{\hat{\boldsymbol{w}}^\mathsf{T}\boldsymbol{x}_i\}$, again giving (5.4.11). Finally, the SVM linear classifier can be written as

$$f(\boldsymbol{x}) = \operatorname{sign}\left(\sum_{i=1}^{n} \hat{\alpha}_i y_i \boldsymbol{x}_i^\mathsf{T}\boldsymbol{x} + \hat{b}\right).$$

To emphasize the sparsity gained by SVM, one could eliminate zero terms and write

$$f(\boldsymbol{x}) = \operatorname{sign}\left(\sum_{j=1}^{|\boldsymbol{s}|} \hat{\alpha}_{s_j} y_{s_j} \boldsymbol{x}_{s_j}^\mathsf{T}\boldsymbol{x} + \hat{b}\right), \tag{5.4.12}$$

where $s_j \in \{1, 2, \cdots, n\}$, $\boldsymbol{s}^\mathsf{T} = (s_1, s_2, \cdots, s_{|\boldsymbol{s}|})$, and $|\boldsymbol{s}| \ll n$.

5.4.5 SVM Classification for Nonlinearly Separable Populations

So far, the data have been assumed linearly separable. However, truly interesting real-world problems are not that simple. Figure 5.8 indicates a more typical case where two types of points, dark circles and empty boxes, are scattered on the plane and a (visually optimal) decision boundary is indicated. The points labeled A are on the boundary (support vectors), but other points, labeled B, are in the margin. SVM has misclassified points labeled F. This is a setting in which the data are not linearly separable. Even so, one might want to use a linear SVM and somehow correct it for nonseparability.

For problems like those in Fig. 5.8, there is no solution to the quadratic programming formulation given above; the optimization will never find the optimal separating hyperplane because the margin constraints in the linearly separable case are too "hard".

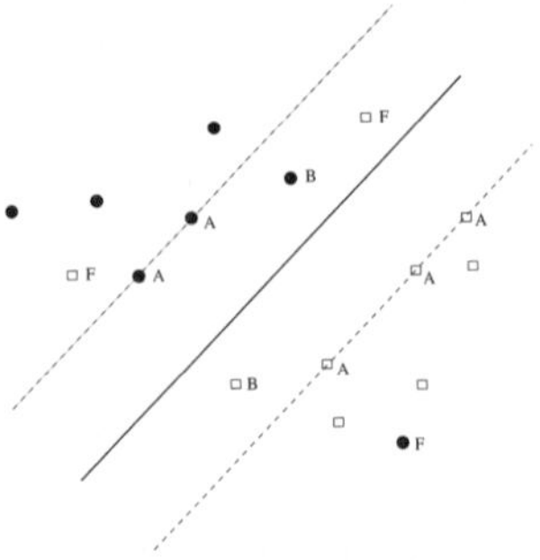

Fig. 5.8 Nonlinearly separable data. Dark circles and empty boxes indicate two types of points. Points labeled A or B are classified correctly, on the boundary or in the margin. The misclassified points labeled F can be on either side of the separating hyperplane.

They will not allow points of type B, for instance, to be correctly classified. It's worse for points labeled F, which are on the wrong side of the margin.

To circumvent this limitation, one can try to "soften" the margins by introducing what are called "slack" variables in the constraints. This gives a new optimization problem, similar to the first, to be solved for an optimal separating hyperplane. In other words, when the hard constraints $y_i(\mathbf{w}^\top \mathbf{x}_i + b) - 1 \geq 0$ cannot be satisfied for all $i = 1, 2, \cdots, n$, replace them with the soft constraint

$$y_i(\mathbf{w}^\top \mathbf{x}_i + b) - 1 + \xi_i \geq 0, \quad \xi_i \geq 0,$$

in which new ξ_is are the slack variables. With the $\xi_i \geq 0$, the new form of the classification rule is: For $i = 1, 2, \cdots, n$,

$$\mathbf{w}^\top \mathbf{x}_i + b \geq +1 - \xi_i \quad \text{for} \quad y_i = +1$$

and

$$\mathbf{w}^\top \mathbf{x}_i + b \leq -1 + \xi_i \quad \text{for} \quad y_i = -1.$$

Now, the indicator for an error is a value $\xi_i > 1$ so

$$\text{number of errors} = \sum_{i=1}^{n} \mathbf{I}(\xi_i > 1). \tag{5.4.13}$$

Next, the optimization problem resulting from including slack variables in the constraints must be identified. It is tempting to use the same objective function as before apart from noting the difference in the constraints. Unfortunately, this doesn't work because it would ignore the error defined in (5.4.13), resulting typically in the trivial solution $\mathbf{w} = \mathbf{zero}$. To fix the problem, at the cost of introducing more complexity, one can add a penalty term to the objective function to account for the errors made. It is natural to consider

$$E_P(\mathbf{w}, \xi) = \frac{1}{2}\|\mathbf{w}\|^2 + C\sum_{i=1}^{n} \mathbf{I}(\xi_i > 1) \tag{5.4.14}$$

for some $C > 0$ as a possible penalized objective function. Unfortunately, (5.4.14) is hard to optimize because it is nonconvex.

Something more needs to be done. Traditionally, the problem is simplified by dropping the indicator function and using the upper bound ξ_i in place of $\mathbf{I}(\xi_i > 1)$. The new primal problem that can be solved is

$$\begin{aligned}
&\text{Find the function } h(\boldsymbol{x}) = \boldsymbol{w}^{\mathsf{T}}\boldsymbol{x} + b \text{ and } \xi \text{ that} \\
&\text{minimizes } E_P(\boldsymbol{w}, \xi) = \frac{1}{2}\|\boldsymbol{w}\|^2 + C\sum_{i=1}^{n}\xi_i \\
&\text{subject to } y_i(\boldsymbol{w}^{\mathsf{T}}\boldsymbol{x}_i + b) \geq 1 - \xi_i \text{ and } \xi_i \geq 0 \qquad i = 1, \cdots, n.
\end{aligned}$$

As with trees and RKHS methods, among others, there is a trade-off between complexity and error tolerance controlled by C. Large values of C penalize the error term, whereas small values of C penalize the complexity.

Having written down a primal problem that summarizes the situation, the next question is what the dual problem is. Interestingly, the dual problem turns out to be essentially the same as before. Unlike the primal problem, it is enough to record the difference in the constraint formulation. More precisely, the dual problem is

$$\begin{aligned}
&\text{Maximize} \\
&\qquad E_D(\boldsymbol{\alpha}) = \sum_{i=1}^{n}\alpha_i - \frac{1}{2}\sum_{i=1}^{n}\sum_{j=1}^{n}\alpha_i\alpha_j y_i y_j \boldsymbol{x}_i^{\mathsf{T}}\boldsymbol{x}_j \\
&\text{subject to} \\
&\qquad \sum_{i=1}^{n}\alpha_i y_i = 0 \quad \text{and } 0 \leq \alpha_i \leq C, \quad i = 1, \cdots, n.
\end{aligned}$$

With this new definition of the constraints, the complete description of the KKT conditions is not as clean as in the separable case. Parallel to the last subsection, it can be verified (with some work) that the KKT conditions are equivalent to

$$\sum_{i=1}^{n}\alpha_i y_i = 0 \quad \text{and} \quad (C - \alpha_i)\xi_i = 0 \quad \text{and} \quad \alpha_i(y_i(\boldsymbol{w}^{\mathsf{T}}\boldsymbol{x}_i + b) - 1 + \xi_i) = 0.$$

Vapnik (1998) shows that the KKT conditions in the nonlinearly separable case reduce to the following three conditions:

$$\begin{aligned}
\alpha_i = 0 \quad &\Rightarrow \quad y_i(\boldsymbol{w}^{\mathsf{T}}\boldsymbol{x}_i + b) \geq 1 \quad \text{and} \quad \xi_i = 0, \\
0 < \alpha_i < C \quad &\Rightarrow \quad y_i(\boldsymbol{w}^{\mathsf{T}}\boldsymbol{x}_i + b) = 1 \quad \text{and} \quad \xi_i = 0, \\
\alpha_i = C \quad &\Rightarrow \quad y_i(\boldsymbol{w}^{\mathsf{T}}\boldsymbol{x}_i + b) \leq 1 \quad \text{and} \quad \xi_i \geq 0.
\end{aligned}$$

From this, it is seen that there are two types of support vectors in the nonlinearly separable case:

- **Margin support vectors:** These correspond to those points lying on one of the hyperplanes H_{+1} or H_{-1} parallel to the "optimal" separating hyperplane. These are controlled by the second of the three KKT conditions above and correspond to points of type A in Fig. 5.8.
- **Nonmargin support vectors:** The condition of the third equation contains the case where $0 \leq \xi_i \leq 1$ and $\alpha_i = C$. Points satisfying these conditions are correctly classified and correspond to points of type B in Fig. 5.8.

Points within the margin but not correctly classified are not support vectors, but are errors, and likewise for any points outside the margin. Indeed, the third equation implies that points satisfying $\alpha_i = C$ and $\xi_i > 1$ are misclassified and correspond to errors. In Fig. 5.8, these are points of type F.

Using all the details above, the SVM classifier for the nonlinearly separable case has the same form as in (5.4.12), namely

$$f(\boldsymbol{x}) = \text{sign}\left(\sum_{j=1}^{|\boldsymbol{s}|} \hat{\alpha}_{s_j} y_{s_j} \boldsymbol{x}_{s_j}^{\mathsf{T}} \boldsymbol{x} + \hat{b} \right),$$

where $s_j \in \{1, 2, \cdots, n\}$, $\boldsymbol{s}^{\mathsf{T}} = (s_1, s_2, \cdots, s_{|\boldsymbol{s}|})$, and $|\boldsymbol{s}| \ll n$. However, it is important to note that the clear geometric interpretation of support vectors is now lost because of the use of slack variables.

By permitting errors, slack variables represent a compromise between linear solutions that are too restrictive and the use of nonlinear function classes, which, although rich, can be difficult. This is a sort of complexity–bias trade-off: The use of the ξ_is reduces the complexity that would arise from using a nonlinear class of functions but of course is more complicated than the original linear problem. However, even as it reduces bias from the linear case, it can allow more bias than the nonlinear problem would have.

5.4.6 SVMs in the General Nonlinear Case

The intuitive idea in SVM classification for nonlinear problems lies in replacing the Euclidean inner product $\boldsymbol{x}_j^{\mathsf{T}} \boldsymbol{x}$ in

$$h(\boldsymbol{x}) = \text{sign}\left(\sum_{j=1}^{n} \alpha_j y_j \boldsymbol{x}_j^{\mathsf{T}} \boldsymbol{x} + b \right), \tag{5.4.15}$$

the expression for the linear SVM classifier, with $\boldsymbol{\Phi}(\boldsymbol{x}_j)^{\mathsf{T}} \boldsymbol{\Phi}(\boldsymbol{x})$, to give

$$h(\boldsymbol{x}) = \text{sign}\left(\sum_{j=1}^{n} \alpha_j y_j \boldsymbol{\Phi}(\boldsymbol{x}_j)^{\mathsf{T}} \boldsymbol{\Phi}(\boldsymbol{x}) + b \right). \tag{5.4.16}$$

The Euclidean inner product in (5.4.15) is computed in the input space of the primal problem, and the generalization (5.4.16) uses a transformation $\boldsymbol{\Phi}$ that converts an input vector $\boldsymbol{x}$ into a point in a higher-dimensional feature space. Using $\boldsymbol{\Phi}$ allows the inclusion of more features in the vectors making them easier to separate with hyperplanes.

Figure 5.9 is inspired by an example from Scholkopf and Smola (2002). It provides a visual for what a transformation like the $\boldsymbol{\Phi}$ helps achieve. In Fig. 5.9, a suitable $\boldsymbol{\Phi}$ is as follows. Let $\boldsymbol{x}^{\mathsf{T}} = (x_1, x_2)$, and consider feature vectors $\boldsymbol{z}^{\mathsf{T}} = (z_1, z_2, z_3)$ in the feature space, Euclidean $\mathbb{R}^3$. Define $\boldsymbol{\Phi} : \mathbb{R}^2 \to \mathbb{R}^3$ by

$$\boldsymbol{\Phi}(\boldsymbol{x}) = \boldsymbol{\Phi}(x_1, x_2) = (x_1^2, \sqrt{2}x_1 x_2, x_2^2) = \boldsymbol{z}^{\mathsf{T}}.$$

With this $\boldsymbol{\Phi}$, a difficult nonlinear classification problem in 2D is converted to a standard linear classification task in 3D. In general, $\boldsymbol{\Phi} : \mathscr{X} \longrightarrow \mathscr{F}$ transforms an input space $\mathscr{X}$ to a feature space $\mathscr{F}$ of much higher dimension, so that inclusion of more features makes the data in $\mathscr{F}$ linearly separable.

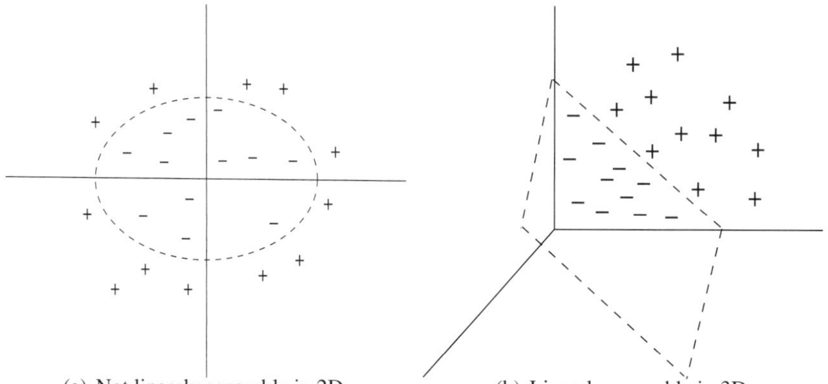

(a) Not linearly separable in 2D (b) Linearly separable in 3D

Fig. 5.9 Panel (a) shows original data in the plane. They cannot be separated linearly. However, a transformation may be used so that a higher-dimensional representation of the pluses and minuses becomes linearly separable.

The core problem in implementing this strategy is to know which, if any, transformation $\boldsymbol{\Phi}$ will make the data separable in feature space. Clearly, if the transformation does not linearize the task, the effort is wasted. The central development to follow will be a systematic way to determine the right transformation to "linearize" a given nonlinear classification task.

5.4.6.1 Linearization by Kernels

It is evident that, to have a linear solution to the classification problem, the image of $\boldsymbol{\Phi}$ must be of higher dimension than its inputs. Otherwise, the transformation is just the continuous image of $\mathbb{R}^p$ and unlikely to be any more linearly separable than its inputs. On the other hand, if Φ constructs a feature vector of much higher dimension than the input vector, the Curse of Dimensionality may become a problem. This can be avoided in principle, as will be seen.

In fact, these concerns are largely bypassed by the kernel trick. In the context of Fig. 5.9, the kernel trick can be developed as follows. Let $\boldsymbol{x}^\top = (x_1, x_2)$ and $\mathbf{y}^\top = (y_1, y_2)$ be two vectors in input space $\mathscr{X} = \mathbb{R}^2$, and consider the transformation to 3D used earlier. Let $\Phi(\boldsymbol{x})$ and $\Phi(\mathbf{y})$ be two feature vectors generated by $\boldsymbol{x}$ and $\mathbf{y}$. Now, look at the inner product $\Phi(\boldsymbol{x})^\top \Phi(\mathbf{y})$ in feature space. It is

$$\begin{aligned}\Phi(\boldsymbol{x})^\top \Phi(\mathbf{y}) &= (x_1^2, \sqrt{2}x_1x_2, x_2^2)(y_1^2, \sqrt{2}y_1y_2, y_2^2)^\top \\ &= (x_1y_1 + x_2y_2)^2 = (\boldsymbol{x}^\top \mathbf{y})^2 = K(\boldsymbol{x}, \mathbf{y}).\end{aligned} \tag{5.4.17}$$

Equation (5.4.17) shows how an inner product based on Φ converts to a function of the two inputs. Since choosing an inner product and computing with it in feature space can quickly become computationally infeasible, it would be nice to choose a function K, again called a kernel, so as to summarize the geometry of feature space vectors and ignore Φ entirely.

Now the kernel trick can be stated: Suppose a function $K(\cdot,\cdot) : \mathscr{X} \times \mathscr{X} \to \mathbb{R}$ operating on input space can be found so that the feature space inner products are computed directly through K as in (5.4.17). Then, explicit use of Φ has been avoided and yet results as if Φ were used can be delivered. This direct computation of feature space inner products without actually explicitly manipulating the feature space vectors themselves is known as the kernel trick.

Assuming that a kernel function K can be found so that $K(\boldsymbol{x}_j, \boldsymbol{x}) = \Phi(\boldsymbol{x}_j)^\top \Phi(\boldsymbol{x})$, the classifier of (5.4.16) can be written as

$$h(\boldsymbol{x}) = \text{sign}\left(\sum_{j=1}^{n} \alpha_j y_j K(\boldsymbol{x}_j, \boldsymbol{x}) + b\right). \tag{5.4.18}$$

Equation (5.4.18) is a solution of the optimization problem

Maximize

$$E_D(\boldsymbol{\alpha}) = \sum_{i=1}^{n} \alpha_i - \frac{1}{2}\sum_{i=1}^{n}\sum_{j=1}^{n} \alpha_i\alpha_j y_i y_j K(\boldsymbol{x}_i, \boldsymbol{x}_j)$$

subject to

$$\sum_{i=1}^{n} \alpha_i y_i = 0 \quad \text{and } 0 \le \alpha_i \le C, \quad i = 1, \cdots, n.$$

These expressions are the same as before except that $K(\boldsymbol{x}_i, \boldsymbol{x}_j) = \Phi(\boldsymbol{x}_i)^\top \Phi(\boldsymbol{x}_j)$ replaces the inner product $\boldsymbol{x}_i^\top \boldsymbol{x}_j$ in (5.4.15).

By the kernel trick, the optimization problem for the nonlinear case has the same form as in the linear case. This allows use of the quadratic programming machinery for the nonlinear case. To see this more explicitly, one last reformulation of the generic optimization problem in quadratic programming form is

Minimize

$$E_D(\boldsymbol{\alpha}) = \frac{1}{2}\boldsymbol{\alpha}^\mathsf{T}\mathbf{K}\boldsymbol{\alpha} + \mathbf{c}^\mathsf{T}\boldsymbol{\alpha}$$

subject to

$$\sum_{i=1}^{n} \alpha_i y_i = 0 \quad \text{and } 0 \le \alpha_i \le C, \quad i = 1, \cdots, n,$$

where $\mathbf{K} = (K_{ij})$ with $K_{ij} = y_i y_j K(\boldsymbol{x}_i, \boldsymbol{x}_j)$ is called the Gram matrix. The only drawback in this formulation is that the matrix $\mathbf{K}$ is not guaranteed to be positive semidefinite. This means the problem might not have a solution. Nevertheless, when it does, this is a convenient form for the problem.

Since bivariate functions K do not necessarily yield positive semidefinite matrices $\mathbf{K}$, the question becomes how to select a kernel function K that is positive definite and represents an underlying feature space transformation $\boldsymbol{\Phi}$ that makes the data linearly separable. It turns out that if K corresponds to an inner product in some feature space $\mathscr{F}$, then the matrix $\mathbf{K}$ is guaranteed to be positive definite. It remains to determine the conditions under which a bivariate function K corresponds to an inner product $K(\boldsymbol{x}, \boldsymbol{y}) = \boldsymbol{\Phi}(\boldsymbol{x})^\mathsf{T}\boldsymbol{\Phi}(\boldsymbol{x})$ for some $\boldsymbol{\Phi} : \mathscr{X} \to \mathscr{F}$. The answer is given by Mercer's conditions, discussed next.

5.4.6.2 Mercer's conditions and Mercer's kernels

For the sake of completeness, it is worthwhile restating the Mercer-Hilbert-Schmidt results as they arise in this slightly different context. The reader is referred to Chapter 3 for the earlier version; as there, proofs are omitted. The core Mercer theorem is the following.

Theorem (Mercer conditions): Let $\mathscr{X}$ be a function domain, and consider a bivariate symmetric continuous real-valued function K defined on $\mathscr{X} \times \mathscr{X}$. Then K is said to fulfill Mercer's conditions if, for all real-valued functions on $\mathscr{X}$,

$$\int g(\boldsymbol{x})^2 d\boldsymbol{x} < \infty \Longrightarrow \int K(\boldsymbol{x}, \boldsymbol{y}) g(\boldsymbol{x}) g(\boldsymbol{y}) d\boldsymbol{x} d\boldsymbol{y} \ge 0. \quad \square$$

This theorem asserts that K is well behaved provided it gives all square-integrable functions finite inner products. As seen in Chapter 2, the link between Mercer kernels and basis expansions must be made explicitly.

Theorem: Let $\mathscr{X}$ be a function domain, and consider a bivariate symmetric continuous real-valued function K defined on $\mathscr{X} \times \mathscr{X}$. Now, let $\mathscr{F}$ be a feature space. Then there exists a transformation $\boldsymbol{\Phi} : \mathscr{X} \to \mathscr{F}$ such that

$$K(\boldsymbol{x}, \boldsymbol{y}) = \boldsymbol{\Phi}(\boldsymbol{x})^\mathsf{T}\boldsymbol{\Phi}(\boldsymbol{y})$$

if and only if K fulfills Mercer's conditions. □

Taken together, these theorems mean that the kernel under consideration really has to be positive definite. Recall that the discussion in Chapter 3 on Mercer's theorem led to using an eigenfunction decomposition of any positive definite bivariate function K to gain insight into the corresponding reproducing kernel. Here, by Mercer's theorem, we can write the decomposition

$$K(\boldsymbol{x},\boldsymbol{y}) = \sum_{i=1}^{\infty} \lambda_j \psi_i(\boldsymbol{x}) \psi_i(\boldsymbol{y})$$

with

$$\int K(\boldsymbol{x},\boldsymbol{y}) \psi_i(\boldsymbol{y}) d\boldsymbol{y} = \lambda_i \psi_i(\boldsymbol{x}).$$

Then, by defining $\phi_i(\boldsymbol{x}) = \sqrt{\lambda_i}\psi_i(\boldsymbol{x})$, it follows that

$$K(\boldsymbol{x},\boldsymbol{y}) = \Phi(\boldsymbol{x})^{\mathsf{T}} \Phi(\boldsymbol{y}).$$

For a given bivariate function K, verifying the conditions above might not be easy. In practice, there exist many functions that have been shown to be valid kernels, and fortunately many of them deliver good performance on real-world data. A short annotated list is compiled at the end of this subsection.

5.4.6.3 SVMs, RKHSs and the Representer Theorem

For completeness, it's worth seeing that the SVM classifier fits the regularized approximation framework discussed in Chapter 3.

Consider the formulation of the SVM classification:

$$\begin{aligned}
&\text{Find the function } h(\boldsymbol{x}) = \mathbf{w}^{\mathsf{T}}\boldsymbol{x} + b \text{ and } \xi \text{ that} \\
&\text{minimizes } E_P(\mathbf{w}, \xi) = \frac{1}{2}\|\mathbf{w}\|^2 + C\sum_{i=1}^{n} \xi_i \\
&\text{subject to } y_i h(\boldsymbol{x}_i) \geq 1 - \xi_i \text{ and } \xi_i \geq 0 \qquad i = 1, \cdots, n.
\end{aligned}$$

Now consider the following regularized optimization:

$$\begin{aligned}
&\underset{\mathbf{w},b}{\text{Minimize}} \left\{ \sum_{i=1}^{n} [1 - y_i h(\boldsymbol{x}_i)]_+ + \lambda \|\mathbf{w}\|^2 \right\} \\
&\text{subject to } h(\boldsymbol{x}) = \mathbf{w}^{\mathsf{T}}\boldsymbol{x} + b.
\end{aligned}$$

Sometimes the product $y_i h(\boldsymbol{x}_i)$ is called the margin and $[1 - y_i h(\boldsymbol{x}_i)]_+$ is called the hinge loss; this is another sense in which SVM is a maximum margin technique.

Theorem: These two optimization problems are equivalent when $\lambda = 1/2C$.

Proof: First, the constraints can be rewritten as $[1 - y_i h(\boldsymbol{x}_i)] \leq \xi_i$ with $\xi_i \geq 0$. Clearly, if $y_i h(\boldsymbol{x}_i) > 1$, then the constraint is satisfied, since ξ must be positive. However, when $y_i h(\boldsymbol{x}_i) < 1$ instead, the corresponding positive quantity $[1 - y_i h(\boldsymbol{x}_i)]$ is compared with another positive quantity, ξ_i. Therefore, the bulk of the constraint lies in cases corresponding to $[1 - y_i h(\boldsymbol{x}_i)] > 0$, so that it is enough to minimize the positive part of $[1 - y_i h(\boldsymbol{x}_i)]$, denoted by $[1 - y_i h(\boldsymbol{x}_i)]_+$. For a given ξ_i, seek the function $h(\boldsymbol{x})$ such that $[1 - y_i h(\boldsymbol{x}_i)]_+ \leq \xi_i$, which, by ignoring ξ_i, boils down to making $[1 - y_i h(\boldsymbol{x}_i)]_+$ as small as possible. Finally, dividing $E_P(\boldsymbol{w}, \xi)$ by C and taking $\lambda = 1/2C$, the desired result follows (see Fig. 5.10).□

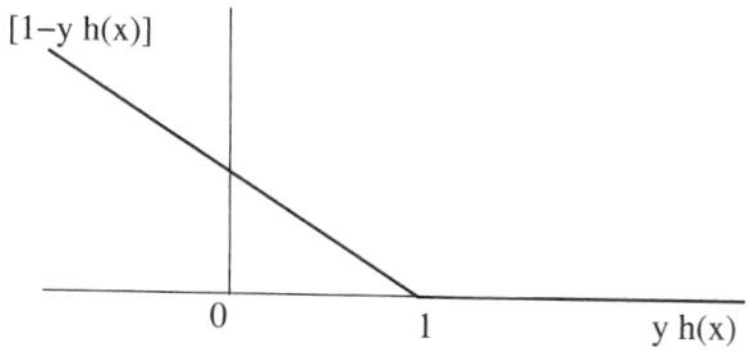

Fig. 5.10 This graph shows the hinge loss function $[1 - y_i h(\boldsymbol{x}_i)]_+$. The theorem states that the SVM formulation is equivalent to a decision problem using the hinge loss.

In fact, this machinery fits the RKHS paradigm from Chapter 3. For instance, the theorem casts SVM in the classical framework of regularization theory, where the more general form

$$\underset{\boldsymbol{w},b}{\text{minimize}} \left\{ \sum_{i=1}^{n} \ell(y_i, h(\boldsymbol{x}_i)) + \lambda \|h\|^2_{\mathscr{H}_K} \right\} \tag{5.4.19}$$

was defined. In (10.3.3), $\ell(\cdot,\cdot)$ is the loss function and $\|\cdot\|^2_{\mathscr{H}_K}$ is the penalty defined in the RKHS used to represent the function h. For the SVM classifier in the nonlinear decision boundary case, reintroduce the feature space transformation Φ, and then the regularized optimization formulation becomes

$$\text{Minimize}_{\ \boldsymbol{w},b} \left\{ \sum_{i=1}^{n} [1 - y_i h(\boldsymbol{x}_i)]_+ + \lambda \|\boldsymbol{w}\|^2 \right\}$$

$$\text{subject to } \ h(\boldsymbol{x}) = \boldsymbol{w}^\top \Phi(\boldsymbol{x}) + b, \tag{5.4.20}$$

where the norm $\|\boldsymbol{w}\|^2$ is now computed in the feature space that constitutes the image of Φ. The transformation Φ can be derived from an appropriately chosen Mercer kernel K guaranteeing that $K(\boldsymbol{x}_i, \boldsymbol{x}_j) = \Phi(\boldsymbol{x}_i)^\top \Phi(\boldsymbol{x}_j)$. So, considering results on RKHSs from Chapter 3, $\|\boldsymbol{w}\|^2$ is the norm of the function h in the RKHS corresponding to the kernel that induces Φ.

From all this, equation (5.4.20) can be written as

$$\text{Minimize}_{\,\mathbf{w},b} \left\{ \sum_{i=1}^{n} \ell(y_i, h(\boldsymbol{x}_i)) + \lambda \|h\|^2_{\mathcal{H}_K} \right\}$$

$$\text{subject to } h(\boldsymbol{x}) = \mathbf{w}^{\mathsf{T}} \Phi(\boldsymbol{x}) + b, \tag{5.4.21}$$

where $\ell(y_i, h(\boldsymbol{x}_i)) = [1 - y_i h(\boldsymbol{x}_i)]_+$ is the hinge loss function shown earlier. The formulation (5.4.21) contains all the ingredients of the RKHS framework and is essentially an instance of (10.3.3). As a result, the representer theorem applies, so that the solution to (5.4.21) is of the form

$$h(\boldsymbol{x}) = \sum_{i=1}^{n} \alpha_i K(\boldsymbol{x}_i, \boldsymbol{x}) + b$$

as determined earlier.

5.4.7 Some Kernels Used in SVM Classification

To conclude the formal treatment, it is worth listing several of the variety of kernels used most regularly. The simplest kernel choice is linear. The linear kernel corresponds to the identity transformation as defined by the Euclidean inner product

$$K(\boldsymbol{x}_i, \boldsymbol{x}_j) = \langle \boldsymbol{x}_i, \boldsymbol{x}_j \rangle.$$

This is the one underlying the SVM classifier for linearly separable data.

Slightly more elaborate is the polynomial kernel defined in its homogeneous form,

$$K(\boldsymbol{x}_i, \boldsymbol{x}_j) = (\langle \boldsymbol{x}_i, \boldsymbol{x}_j \rangle)^d.$$

This was seen in the 3D example for $d = 2$; see (5.4.17). The nonhomogeneous version of the polynomial kernel is defined by

$$K(\boldsymbol{x}_i, \boldsymbol{x}_j) = (\langle \boldsymbol{x}_i, \boldsymbol{x}_j \rangle + c)^d.$$

The greatest advantage of the polynomial family of kernels lies in the fact that they are direct generalizations of the well-known Euclidean norm and therefore intuitively interpretable. Indeed, it is straightforward, though tedious, to obtain representations for the ψ_is that correspond to these kernels. (For $p = 2$, say, let $\boldsymbol{x} = (x_1, x_2)$ and $\boldsymbol{x}^* = (x_1^*, x_2^*)$ and start with $c = 1$ and $d = 3$. Derive a polynomial expression for $K(\boldsymbol{x}, \boldsymbol{x}^*)$, and recognize the ψ_is as basis elements.)

The Laplace radial basis function (RBF) kernel is

$$K(\boldsymbol{x}_i, \boldsymbol{x}_j) = \exp\left(-\frac{1}{2\sigma} ||\boldsymbol{x}_i - \boldsymbol{x}_j|| \right).$$

As the cusp at 0 suggests, this kernel might be more appropriate than others in applications where sharp nondifferentiable changes in the function of interest are anticipated. Using the Laplace RBF kernel for very smooth functions understandably gives very poor results, lacking sparsity and having a large prediction error. This occurs with relevance vector machines (RVMs) as well. (Roughly, RVMs are a Bayesian version of SVMs based on recognizing the prior as a penalty term in a regularization framework; see Chapter 6.) This is consistent with regarding kernel selection as similar to selecting a model list.

Arguably, the most widely used kernel is the Gaussian RBF kernel defined by

$$K(\boldsymbol{x}_i, \boldsymbol{x}_j) = \exp\left(-\frac{1}{2\sigma^2}||\boldsymbol{x}_i - \boldsymbol{x}_j||^2\right).$$

The parametrization of such kernels by σ creates a large, flexible class of models. The class of kernels is large enough that one can be reasonably sure of capturing the underlying function behind a wide variety of data sets, provided σ is well tuned, usually by cross-validation.

Sigmoid kernels are used in feedforward neural network contexts, as studied in Chapter 4. One sigmoid is defined by

$$K(\boldsymbol{x}_i, \boldsymbol{x}_j) = \tanh(\kappa \boldsymbol{x}_i^{\mathsf{T}} \boldsymbol{x}_j + \gamma).$$

In contexts where it is important to be able to add steps in a smooth way (e.g., handwritten digit recognition), this kernel is often used.

To finish, two other kernels that arise are the Cauchy kernel

$$K(\boldsymbol{x}_i, \boldsymbol{x}_j) = \frac{1}{\pi}\frac{1}{1 + ||\boldsymbol{x}_i - \boldsymbol{x}_j||^2},$$

which is a further variant on the Laplace or Gaussian RBF kernels (to give more spread among the basis functions), and the thin-plate spline kernel

$$K(\boldsymbol{x}_i, \boldsymbol{x}_j) = ||\boldsymbol{x}_i - \boldsymbol{x}_j|| \log ||\boldsymbol{x}_i - \boldsymbol{x}_j||,$$

implicitly encountered in Chapter 3.

5.4.8 Kernel Choice, SVMs and Model Selection

The kernel plays the key role in the construction of SVM classifiers. Steinwart (2001) provides a theoretical discussion on the central role of the kernel in the generalization abilities of SVMs and related techniques. Genton (2001) discussed the construction of kernels with details on aspects of the geometry of the domain, particularly from a statistical perspective. More generally, the webpage `http://www.kernel-machines.org/jmlr.htm` has many useful references.

In overall terms, partially because of the use of a kernel, SVMs typically evade the Curse. However, the selection of the kernel itself is a major issue, and the computing required to implement an SVM solution (which can be used in certain regression contexts, too) can be enormous.

On the other hand, conceptually, SVMs are elegant and can be regarded as a deterministic method with probabilistic properties as characterized by the VC-dimension. Indeed, prediction on a string of data such as characterized by Shtarkov's theorem and follow-on techniques from the work of Cesa-Bianchi, Lugosi, Haussler, and others is similar in flavor: Predictions and inferences are done conditionally on the string of data received but characterized in the aggregate by probabilistic quantities. In principle, it is possible to choose a kernel in an adaptive way (i.e., data-driven at each time step), but this approach has not been investigated.

At its root, choosing a kernel is essentially the same as choosing an inner product, which is much like choosing a basis. Usually, one wants a basis representation to be parsimonious in the sense that the functions most important to represent can be represented with relatively few terms, so that for fixed bias tolerance, the variance from estimating the coefficients will be small. Thus, in a regression or classification context, selecting the kernel is like selecting a whole model space or a defined model class to search; fixing a K can be likened to a specific model selection problem in the traditional statistical sense. In other words, different Ks correspond to different model space coordinatizations, not to individual models within a space.

5.4.9 Support Vector Regression

The key difference between support vector classification and support vector regression lies in the noise model and loss function; the paradigm of maximization of a margin remains the same. Vapnik calls the loss function used for support vector regression ε-insensitive loss, defined as follows: Let $\varepsilon > 0$ and set

$$\ell(u) \equiv |u|_\varepsilon \equiv \begin{cases} 0, & |u| < \varepsilon \\ |u| - \varepsilon, & \text{otherwise.} \end{cases}$$

It is seen in Fig. 5.11 that this loss function assigns loss zero to any error smaller than ε, whence the name. This means that any function closer than ε to the data is a good candidate. Pontil et al. (1998) observe that the ε-insensitive loss function also provides some robustness against outliers.

Using ε-insensitive loss for regression amounts to treating the regression function as a decision boundary as sought in classification. This is valid because the ε-insensitive loss corresponds to a margin optimization interpretation. That is, support vector regression estimates the true function by constructing a tube around it. The tube defines a margin outside of which the deviation is treated as noise.

Given a data set $\{(\boldsymbol{x}_i, y_i), i = 1, \cdots, n\}$, support vector regression is formulated in the same way as the optimization underlying support vector classification, i.e., one seeks

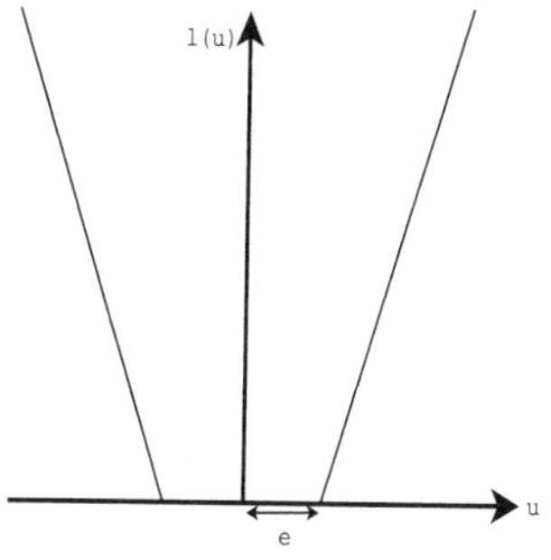

Fig. 5.11 The ε-insensitive loss function.

the f achieving

$$\min \tfrac{1}{2}\|\mathbf{w}\|^2 \text{ subject to}$$
$$\ell(y_i - f(\mathbf{x}_i)) < \varepsilon,\ \forall i = 1,\cdots,n,$$

where f is of the form

$$f(\mathbf{x}) = \mathbf{w}\mathbf{x} + b.$$

An equivalent formulation in a single objective function consists of finding the function $f(\mathbf{x}) = \mathbf{w}\mathbf{x} + b$ that minimizes

$$R_{\mathsf{emp}}(f) = \frac{1}{n}\sum_{i=1}^{n} \ell(y_i - f(\mathbf{x}_i)) + \frac{\lambda}{2}\|\mathbf{w}\|^2.$$

When the constraints are violated (i.e., some observations fall within the margin), then, just like in classification, slack variables are used. The regression problem is then

$$\begin{aligned} &\min \frac{1}{2}\|\mathbf{w}\|^2 + C\sum_{i=1}^{n}(\xi_i + \xi_i^*) \\ &\text{subject to} \\ &\qquad y_i - f(\mathbf{x}_i) < \varepsilon + \xi_i, \text{ and} \\ &\qquad f(\mathbf{x}_i) - y_i < \varepsilon + \xi_i^* \\ &\forall i = 1,\cdots,n,\ \xi_i, \xi_i^* \geq 0, C > 0. \end{aligned}$$

In the formulation above, C controls the trade-off between the flatness of $f(\mathbf{x})$ and the amount up to which deviations larger than the margin ε are tolerated. From a computational standpoint, the estimator, just as in support vector classification, is obtained by solving the dual optimization problem rather than the primal one. As with support vector classification, this is tackled by forming the primal Lagrangian function,

$$L_P = \frac{1}{2}\|\boldsymbol{w}\|^2 + C\sum_{i=1}^{n}(\xi_i + \xi_i^*)$$
$$- \sum_{i=1}^{n} \alpha_i(\varepsilon + \xi_i - y_i + \boldsymbol{w}^{\mathsf{T}}\boldsymbol{x}_i + b)$$
$$- \sum_{i=1}^{n} \alpha_i^*(\varepsilon + \xi_i^* + y_i - \boldsymbol{w}^{\mathsf{T}}\boldsymbol{x}_i - b)$$
$$- \sum_{i=1}^{n} (\beta_i\xi_i + \beta_i^*\xi_i^*),$$

where $\alpha_i, \alpha_i^*, \beta_i, \beta_i^* \geq 0$ are the Lagrange multipliers. Classical optimization of L_P proceeds by setting derivatives equal to zero,

$$\frac{\partial L_P}{\partial \mathbf{w}} = 0, \quad \frac{\partial L_P}{\partial b} = 0, \quad \frac{\partial L_P}{\partial \xi} = 0, \quad \frac{\partial L_P}{\partial \xi^*} = 0,$$

and using the resulting equations to convert L_P into the dual problem. The constraint $\partial L_P / \partial \mathbf{w} = 0$ gives

$$\boldsymbol{w}^* = \sum_{i=1}^{n}(\alpha_i - \alpha_i^*)\boldsymbol{x}_i,$$

so the dual problem is to maximize

$$L_D(\langle\alpha_i\rangle, \langle\alpha_i^*\rangle) = \frac{1}{2}\sum_{i,j=1}^{n}(\alpha_i - \alpha_i^*)(\alpha_j - \alpha_j^*)\mathbf{x}_i^{\mathsf{T}}\mathbf{x}_j - \varepsilon\sum_{i=1}^{n}(\alpha_i - \alpha_i^*),$$

where $0 \leq \alpha_i \leq C$ and $0 \leq \alpha_i^* \leq C$. Note that the intercept b does not appear in L_D, so maximizing L_D only gives the values α_i and α_i^*. However, given these and using $\boldsymbol{w}^*$ in $f(\boldsymbol{x}) = \mathbf{w}^{\mathsf{T}}\boldsymbol{x} + b$ results in the desired estimator

$$f(\boldsymbol{x}) = \sum_{i=1}^{n}(\alpha_i - \alpha_i^*)\boldsymbol{x}_i^{\mathsf{T}}\boldsymbol{x} + b.$$

The correct value of b^* can be found by using the Karush-Kuhn-Tucker conditions. In fact, these conditions only specify b^* in terms of a support vector. Consequently, common practice is to average over the b^*s obtained this way. To a statistician, estimating b directly from the data, possibly by a method of moments argument, may make as much sense.

Clearly, this whole procedure can be generalized by replacing $\mathbf{w}^{\mathsf{T}}\mathbf{x}$ with $\mathbf{w}^{\mathsf{T}}\boldsymbol{\Phi}(\mathbf{x})$ and setting $K(\mathbf{x}, \mathbf{x}') = \boldsymbol{\Phi}(\mathbf{x})'\boldsymbol{\Phi}(\mathbf{x})$. Then, an analogous analysis leads to

$$f(\boldsymbol{x}) = \sum_{i=1}^{n}(\alpha_i - \alpha_i^*)K(\boldsymbol{x}_i, \boldsymbol{x}) + b.$$

More details on the derivation of support vector regression and its implementation can be found in Smola and Scholkopf (2003).

5.4.10 Multiclass Support Vector Machines

As noted in Section 5.1, there are two ways to extend binary classification to multiclass classification with $K \geq 3$. If K is not too large, the AVA case of training $K(K-1)/2$ binary classifiers can be implemented. However, here it is assumed that K is too large for this to be effective, so an OVA method is developed.

The geometric idea of margin – perpendicular distance between the points closest to a decision boundary – does not have an obvious natural generalization to three or more classes. However, the other notion of margin, $yf(\boldsymbol{x})$, which is also a measure of similarity between y and f, does generalize, in a way, to multiclass problems. Following Liu and Shen (2006), note that for an arbitrary sample point $(\boldsymbol{x}, y)$, a correct decision vector $\boldsymbol{f}(\boldsymbol{x})$ should encourage a large value for $f_y(\boldsymbol{x})$ and small values for $f_k(\boldsymbol{x}), k \neq y$. Therefore, it is the vector of relative difference, $f_y(\boldsymbol{x}) - f_k(\boldsymbol{x}), k \neq y$, that characterizes a multicategory classifier. So, define the $(K-1)$-dimensional $\boldsymbol{g}$-vector

$$\boldsymbol{g}(\boldsymbol{f}(\boldsymbol{x}), y) = (f_y(\boldsymbol{x}) - f_1(\boldsymbol{x}), \ldots, f_y(\boldsymbol{x}) - f_{y-1}(\boldsymbol{x}), f_y(\boldsymbol{x}) - f_{y+1}(\boldsymbol{x}), \ldots, f_y(\boldsymbol{x}) - f_K(\boldsymbol{x})). \tag{5.4.22}$$

It will be seen that the use of $\boldsymbol{g}$ simplifies the representation of generalized hinge loss for multiclass classification problems.

Several multiclass SVMs, MSVMs, have been proposed. Similar to binary SVMs (see (10.3.3)) these MSVMs can be formulated in terms of RKHSs. Let $\boldsymbol{f}(\boldsymbol{x}) \in \prod_{k=1}^{K}(\{1\} + \mathcal{H}_K)$ be the product space of K reproducing kernel Hilbert spaces $\mathcal{H}_K$. In other words, each component $f_k(\boldsymbol{x})$ can be expressed as $b_k + h_k(\boldsymbol{x})$, where $b_k \in R$ and $h_k \in \mathcal{H}_K$. Then the MSVM can be defined as the solution to the regularization problem

$$\frac{1}{n}\sum_{i=1}^{n} l(y_i, \boldsymbol{f}(\boldsymbol{x}_i)) + \lambda \sum_{k=1}^{K} ||h_k||^2_{\mathcal{H}_K}, \tag{5.4.23}$$

where $l(\cdot,\cdot)$ is the loss function. The basic idea behind the multiclass SVM is, for any point $(\boldsymbol{x}, y)$, to pay a penalty based on the relative values given by $f_k(\boldsymbol{x})$s.

In Weston and Watkins (1999), a penalty is paid if

$$f_y(\boldsymbol{x}) < f_k(\boldsymbol{x}) + 2, \quad \forall k \neq y.$$

Therefore, if $f_y(\boldsymbol{x}) < 1$, there is no penalty provided $f_k(\boldsymbol{x})$ is sufficiently small for $k \neq y$. Similarly, if $f_k(\boldsymbol{x}) > 1$ for $k \neq y$, there is no penalty if $f_y(\boldsymbol{x})$ is sufficiently larger. Therefore, the loss function can be represented as

$$\sum_{i=1}^{n} l(y_i, \boldsymbol{f}(\boldsymbol{x}_i)) = \sum_{i=1}^{n}\sum_{k \neq y_i} [2 - \{f_{y_i}(\boldsymbol{x}_i) - f_k(\boldsymbol{x}_i)\}]_+ . \tag{5.4.24}$$

In Lee et al. (2004), a different loss function,

$$\sum_{i=1}^{n} l(y_i, \boldsymbol{f}(\boldsymbol{x}_i)) = \sum_{i=1}^{n}\sum_{k \neq y_i} [f_k(\boldsymbol{x}_i) + 1)]_+, \tag{5.4.25}$$

is used, and the objective function is minimized subject to a sum-to-zero constraint,

$$\sum_{k=1}^{K} f_k(\boldsymbol{x}) = 0.$$

If the generalized margin $\boldsymbol{g}_i = \boldsymbol{g}(\boldsymbol{f}(\boldsymbol{x}_i), y_i)$ defined in (5.4.22) is used, then (5.4.25) becomes

$$\sum_{i=1}^{n} l(y_i, \boldsymbol{f}(\boldsymbol{x}_i)) = \sum_{i=1}^{n} V(\boldsymbol{g}_i),$$

where $V(u) = \sum_{k=1}^{K-1}[(\sum_{j=1}^{K-1} u_j)/K - u_k + 1]_+$.

The following result establishes the connection between the MSVM classifier and the Bayes rule.

Proposition (Lee et al., 2004): Let $\mathbf{f}(\mathbf{x}) = (f_1(\mathbf{x}), \ldots, f_k(\mathbf{x}))$ be the minimizer of $\mathbb{E}[L(Y, \mathbf{f}(\mathbf{x}))]$ defined in (5.4.25) under the sum-to-zero constraint. Then

$$\arg\max_{l=1,\cdots,k} f_l(\mathbf{x}) = f_B(\mathbf{x}).$$

5.5 Neural Networks

Recall from Chapter 4 that a single hidden layer feedforward neural network (NN) model is of the form

$$Y = \beta_0 + \sum_{u=1}^{r} \gamma_k \psi(\boldsymbol{x}^{\mathsf{T}} \beta_u + \nu_u) + \varepsilon, \tag{5.5.1}$$

where ψ is a sigmoidal function and each term is a node, or neuron, and ε is an error term. When $r = 1$ and ψ is a threshold, the simple model is often called a perceptron. The β_js are weights, and the ν_js are sometimes called biases. More complicated neural net models permit extra layers by treating the r outputs from (5.5.1) as inputs to another layer.

One extension of NNs from regression to classification is based on using categorical variables and regarding the likelihood as multinomial rather than normal. However, this rests on a multivariate generalization of regression networks because K class classification problems must be transformed to a regression problem for a collection of K indicator functions.

First, in a multivariate response regression problem, regard an output as $\boldsymbol{Y} = (Y_1, ..., Y_K)$, where each Y_j is an indicator for class j taking values zero and one. Then, for each outcome $Y_{j,i}$ for $i = 1, ..., n$ of Y_j, there is an NN model of the form

$$Y_j = \beta_{0,j} + \sum_{u=1}^{r} \gamma_{u,j} \psi(\boldsymbol{x}^{\mathsf{T}} \beta_u) + \varepsilon_j, \tag{5.5.2}$$

in which, for simplicity, all the sigmoids are the same, the νs are absorbed into the βs by taking a constant as an explanatory variable, and it is seen that the βs do not depend on j. This means that the indicator functions for all K classes will be exhibited as linear combinations of the same r sigmoids that play the role of a basis. Explicitly, a logit sigmoid gives

$$\psi(\boldsymbol{x}^{\mathsf{T}}\beta_u) = \frac{1}{1+e^{-\nu_u - \sum_{h=1}^{p} \beta_{u,h} x_h}}. \tag{5.5.3}$$

Next, in the classification problem, let Z be the response variable assuming values in $\{1,...,K\}$. Following Lee (2000), represent the categorical variable $\boldsymbol{Z}$ as a vector $\boldsymbol{Y} = (Y_1,...,Y_K)$ of length K, where Y_j is the indicator for $Z = j$; i.e., $Y_j = 1$ when $Z = j$ and $Z_j = 0$ otherwise. Now, the response Y_is have regression functions as in (5.5.2). Suppose the Z_is are independent, and write

$$f(Z^n|\boldsymbol{p}) = \Pi_{i=1}^{n} f(Z_i|p_1,...,p_K),$$

in which $p_j = P(Z = j) = P(Y_j = 1)$ and

$$f(Z_i|p_1,...,p_K) \propto p_1^{y_{1,i}} ... p_K^{y_{K,i}}.$$

The $\hat{p}_j$s are estimated from the regression model by finding

$$\hat{W}_{i,k} = \beta_{0,k} + \sum_{u=1}^{r} \beta_{u,k} \psi_u(\boldsymbol{x}_i^{\mathsf{T}}\beta_u)$$

and setting

$$\hat{p}_k = \frac{e^{\hat{W}_k}}{\sum_{h=1}^{p} e^{\hat{W}_h}} \tag{5.5.4}$$

using (5.5.3). Note that the $\hat{W}_k$s are the continuous outputs of the regression model in (5.5.2), which are transformed to the probability scale of the p_ks in (5.5.4). In practice, one of the W_ks must be set to zero, say W_K, for identifiability. It is seen that there are r nodes, rk real parameters from the γs (K for each node), and $r(p+1)+K$ parameters from the βs ($p+1$ for each node and K offsets).

Despite the logical appeal of using a multinomial model, many practitioners use a normal type model, even for classification. This is valid partially because they intend to derive a discriminant function from using K networks, say $\widehat{Net}_k(\boldsymbol{x})$, of the form (5.5.2). Then $\arg\max_k \widehat{Net}_k(\boldsymbol{x}_{new})$ can be taken as a discriminant function to assign a class to Y_{new}. A related point is that the estimation procedures in NNs, as in basic linear regression, rest on optimizations that are independent of the error term. In fact, using techniques like bootstrapping and cross-validation, some inferences can be made about NNs that are also independent of the error term. The error term really only figures in when it is important to get estimates of parameters.

5.6 Notes

Here Hoeffding's inequality is presented, followed by some details on VC dimension.

5.6.1 *Hoeffding's Inequality*

For completeness, a statement and proof of Hoeffding's inequality are provided.

Lemma: If Z is a random variable with $\mathbb{E}[Z]=0$ and $a \le Z \le b$, then

$$\mathbb{E}[e^{sZ}] \le e^{\frac{s^2(b-a)^2}{8}}.$$

Proof: By the convexity of the exponential function, for $a \le Z \le b$,

$$e^{sZ} \le \frac{Z-a}{b-a}e^{sb} + \frac{b-Z}{b-a}e^{sa}.$$

Now,

$$\begin{aligned}
\mathbb{E}[e^{sZ}] &\le \mathbb{E}\left[\frac{Z-a}{b-a}e^{sb}\right] + \mathbb{E}\left[\frac{b-Z}{b-a}\right]e^{sa} \\
&= \frac{b}{b-a}e^{sa} - \frac{a}{b-a}e^{sb} \quad \text{since } \mathbb{E}[Z]=0 \\
&= (1-t+te^{s(b-a)})e^{-ts(b-a)}, \quad \text{where } t = \frac{-a}{b-a}.
\end{aligned}$$

Let $u = s(b-a)$ and $\phi(u) = -tu + \log(1-t+te^u)$. Then,

$$\mathbb{E}[e^{sZ}] \le e^{\phi(u)}.$$

It is easy to see that $\phi(0)=0$, with the Taylor series expansion of $\phi(u)$ given by

$$\phi(u) = \phi(0) + u\phi'(0) + \frac{u^2}{2}\phi''(v), \quad \text{where } v \in [0,u].$$

It is easy to check that $\phi'(0)=0$ since $\phi'(u) = -t + \frac{te^u}{1-t+te^u}$. Also,

$$\phi''(u) = \frac{te^u}{1-t+te^u} - \frac{te^u}{(1-t+te^u)^2} = \frac{te^u}{1-t+te^u}\left[1 - \frac{te^u}{1-t+te^u}\right],$$

which can be written as $\phi''(u) = \pi(1-\pi)$, where $\pi = (te^u)/(1-t+te^u)$. The maximizer of $\phi''(u)$ is $\pi^* = 1/2$. As a result,

$$\phi''(u) \le \frac{1}{4}, \quad \text{so that } \phi''(u) \le \frac{u^2}{8} = \frac{s^2(b-a)^2}{8}.$$

Therefore,

$$\mathbb{E}[e^{sZ}] \le e^{\frac{s^2(b-a)^2}{8}}. \quad \square$$

Theorem (Hoeffding's inequality): Let $Y_1, Y_2, \cdots, Y_n$ be bounded independent random variables such that $a_i \le Y_i \le b_i$ with probability 1. Let $S_n = \sum_{i=1}^n Y_i$. Then, for any $t > 0$,

$$\mathbb{P}(|S_n - \mathbb{E}[S_n]| \ge t) \le 2e^{\frac{-2t^2}{\sum_{i=1}^n (b_i - a_i)^2}}.$$

Proof: The upper bound of the lemma above is applied directly to derive Hoeffding's inequality. Now,

$$\begin{aligned}
\mathbb{P}(S_n - \mathbb{E}[S_n] \ge t) &\le e^{-st} \prod_{i=1}^n E\left[e^{s(L_i - \mathbb{E}[L_i])}\right] \\
&\le e^{-st} \prod_{i=1}^n e^{\frac{s^2(b_i-a_i)^2}{8}} \\
&\le e^{-st} e^{s^2 \sum_{i=1}^n \frac{(b_i-a_i)^2}{8}} \\
&= e^{\frac{-2t^2}{\sum_{i=1}^n (b_i-a_i)^2}},
\end{aligned}$$

where s is replaced by $s = \frac{4t}{\sum_{i=1}^n (b_i - a_i)^2}$. $\square$

5.6.2 VC Dimension

The point of the VC dimension is to assign a notion of dimensionality to collections of functions that do not necessarily have a linear structure. It often reduces to the usual real notion of independence – but not always. The issue is that just as dimension in real vector spaces represents the portion of a space a set of vectors can express, VC dimension for sets of functions rests on what geometric properties the functions can express in terms of classification. In DMML, the VC dimension helps set bounds on the performance capability of procedures.

There are no less than three ways to approach defining the VC dimension. The most accessible is geometric, based on the idea of shattering a set of points.

Since the VC dimension h of a class of functions $\mathscr{F}$ depends on how they separate points, start by considering the two-class discrimination problem with a family $\mathscr{F}$ indexed by θ, say $f(x, \theta) \in \{-1, 1\}$. Given a set of n points, there are 2^n subsets that can be regarded as arising from labeling the n points in all 2^n possible ways with $0, 1$. Now, fix any one such labeling and suppose there is a θ such that $f(x_i, \theta)$ assigns 1 when x_i has the label 1 and -1 when x_i has the label 0. This means that $f(\cdot, \theta)$ is a member of $\mathscr{F}$ that correctly assigns the labels. If for each of the 2^n labelings there is a member of $\mathscr{F}$ that can correctly assign those labels to the n points, then the set of points is "shattered" by $\mathscr{F}$. The VC dimension for $\mathscr{F}$ is the maximum number of points that can be shattered by the elements of $\mathscr{F}$ – a criterion that is clearly relevant to classification.

Now, the VC dimension of a set of indicator functions $I_\alpha(z)$, generated by, say, $\mathscr{F}$, where $\alpha \in \Lambda$ indexes the domains on which $I = 1$, is the largest number h of points that can be separated into two different classes in all 2^n possible ways using that set of indicator functions. If there are n distinct points $z_1, ..., z_n$ (in any configuration) in a fixed space that can be separated in all 2^n possible ways, then the VC dimension h is at least n. That is, it is enough to shatter one set of n vectors to show the dimension is at least n.

If, for every value of n, there is a set of n vectors that can be shattered by the $I(z,\alpha)$s, then $\mathscr{F}$ has VC dimension infinity. So, to find the VC dimension of a collection of functions on a real space, one can test each $n = 1,2,3...$ to find the first value of n for which there is a labeling that cannot be replicated by the functions. It is important to note that the definition of shattering is phrased in terms of all possible labelings of n vectors as represented by the support of the indicator functions, which is in some fixed space. So, the VC dimension is for the set $\mathscr{F}$ whose elements define the supports of the indicator functions, not the space itself.

In a sense, the VC dimension is not of $\mathscr{F}$ itself so much as the level sets defined by $\mathscr{F}$ since they generate the indicator functions. Indeed, the definition of VC dimension for a general set of functions $\mathscr{F}$, not necessarily indicator functions, is obtained from the indicator functions from the level sets of $\mathscr{F} = \{f_\alpha(\cdot) : \alpha \in \Lambda\}$. Let $f_\alpha \in \mathscr{F}$ be a real-valued function. Then the set of functions

$$I_{\{z: f_\alpha(z) - \beta \geq 0\}}(z) \text{ for } \alpha \in \Lambda, \beta \in \left(\inf_{z,\alpha} f_\alpha(z), \sup_{z,\alpha} f_\alpha(z)\right)$$

is the complete set of indicators for $\mathscr{F}$. The VC dimension of $\mathscr{F}$ is then the maximal number h of vectors $z_1,...,z_h$ that can be shattered by the complete set of indicators of $\mathscr{F}$, for which the earlier definition applies.

To get a sense for how the definition of shattering leads to a concept of dimension, it's worth seeing that the VC dimension often reduces to simple expressions that are minor modifications of the conventional dimension in real spaces.

An amusing first observation is that the collection of indicator functions on $\mathbb{R}$ with support $(-\infty, a]$ for $a \in \mathbb{R}$ has VC dimension 2 because it cannot pick out the larger of two points x_1 and x_2. However, the collection of indicator functions on $\mathbb{R}$ with support $(a,b]$ for $a, b \in \mathbb{R}$ has VC dimension 3 because it cannot pick out the largest and smallest of three points, x_1, x_2, and x_3. The natural extensions of these sets in $\mathbb{R}^n$ have VC dimension $d+1$ and $2d+1$.

Now, consider planes through the origin in $\mathbb{R}^n$. That is, let $\mathscr{F}$ be the set of functions of the form $f_\theta(x) = \theta \cdot x = \sum_{i=1}^n \theta_i x_i$ for $\theta = (\theta_1, ..., \theta_n)$ and $\boldsymbol{x} = (x_1, ..., x_n)$. The task is to determine the highest number of points that can be shattered by $\mathscr{F}$. It will be seen that the answer depends on the range of $\boldsymbol{x}$.

First, suppose that $\boldsymbol{x}$ varies over all of $\mathbb{R}^n$. Then the VC dimension of $\mathscr{F}$ is $n+1$. To see this, recall that the shattering definition requires thinking in terms of partitioning point sets by indicator functions. So, associate to any f_θ the indicator function $I_{\{x:f_\theta(x)>0\}}(x)$, which is 1 when $f_\theta > 0$ and zero otherwise. This is the same as saying the points on one side of the hyperplane $f_\theta(x) \geq 0$ are coded 1 and the others 0. (A minus sign gives the reverse, 0 and 1.) Now ask: How many points in R^n must accumulate before they can no longer be partitioned in all possible ways? More formally, if there are k points, how large must k be before the number of ways the points can be partitioned by indicator functions $I_{\{x:f_\theta(x)>0\}}(x)$ falls below 2^k?

One way to proceed is to start with $n = 2$ and test values of k. So, consider $k = 1$ point in $\mathbb{R}^2$. There are two ways to label the point, 0 and 1, and the two cases are symmetric. The class of indicator functions obtained from $\mathscr{F}$ is $I_{\{x:f_\theta(x)>0\}}(x)$. Given any labeling of the point by 0 or 1, any $f \in \mathscr{F}$ gives one labeling and $-f$ gives the other. So, the VC dimension is at least 1.

Next, consider two points in $\mathbb{R}^2$: There are four ways to label the two points with 0 and 1. Suppose the two points do not lie on a line through the origin unless one is the origin. It is easy to find one line through the origin so that both points are on the side of it that gets 1 or that gets zero. As long as the two points are not on a line through the origin (and are distinct from the origin), there will be a line through the origin so that one of the points is on the side of the line that gets 1 and the other will be on the side of the line that gets 0. So, there are infinitely many pairs of points that can be shattered. Picking one means the VC dimension is at least 2.

Now, consider three points in $\mathbb{R}^2$. To get VC dimension at least three, it is enough to find three points that can be shattered. If none of the points is the origin, typically they cannot be shattered. However, if one of the points is the origin and the other two are not collinear with the origin, then the three points can be shattered by the indicator functions. So, the VC dimension is at least 3.

In fact, in this case, the VC-dimension cannot be 4 or higher. There is no configuration of four points, even if one is at the origin, that can be shattered by planes through the origin. If $n = 3$, then the same kind of argument produces four points that can be shattered (one is at the origin) and four is the maximal number of points that can be shattered. Higher values of n are also covered by this kind of argument and establish that $\text{VCdim}(\mathscr{F}) = n+1$.

The real dimension of the class of indicator functions $I_{\{x:f_\theta(x)>0\}}(x)$ for $f \in \mathscr{F}$ is, however, n, not $n+1$. The discrepancy is cleared up by looking closely at the role of the origin. It is uniquely easy to separate from any other point because it is always on the boundary of the support of an indicator function. As a consequence, the linear independence of the components x_i in $\boldsymbol{x}$ is indeterminate when the values of the x_is are zero. Thus, if the origin is removed from $\mathbb{R}^n$ so the new domain of the functions in $\mathscr{F}$ is $\mathbb{R}^n \setminus \{\mathbf{0}\}$, the VC dimension becomes n.

For the two-class linear discrimination problem in $\mathbb{R}^p$, the VC dimension is $p+1$. Thus, the bound on the risk gets large as p increases. But, if $\mathscr{F}$ is a finite-dimensional vector space of measurable functions, it has a VC dimension bounded by $\dim \mathscr{F} + 2$.

Moreover, if ϕ is a monotonic function, its set of translates $\{\phi(x-a) : a \in \mathbb{R}\}$ has VC dimension 2.

One expects that as the elements of $\mathscr{F}$ get more flexible, then the VC dimension should increase. But the situation is complex. Consider the following example credited to E. Levin and J. Denker. Let $\mathscr{F} = \{f(x,\theta)\}$, where

$$f(x,\theta) = \begin{cases} 1 & \text{if } \sin(\theta x) > 0, \\ -1 & \text{if } \sin(\theta x) \leq 0. \end{cases}$$

Select the points $\{x_i = 10^{-i}\}$ for $i = 1, \ldots, n$, and let $y_i \in \{0,1\}$ be the label of x_i. Then one can show that, for any choice of labels,

$$\theta = \pi \left[1 + \sum_{i=1}^{n} \frac{1}{2}(1-y_i)10^i \right]$$

gives the correct classification. For instance, consider $y_1 = -1$ and $y_i = 1$ for $i \neq 1$. Then, $\sin(x_1\theta) = \sin(\pi(1+10^{-1})) < 0$, correctly leading to -1 for x_1 and, for $i \neq 1$, $\sin(x_i\theta) = \sin(\pi(1/10^i + 1/10^{i+1})) > 0$ correctly leading to 1. The other labelings for the x_is in terms of the y_is arise similarly. Thus, a sufficiently flexible one-parameter family can have infinite VC dimension.

5.7 Exercises

Exercise 5.1 (Two normal variables, same mean). Suppose $\mathbf{X} \in \mathbb{R}^p$ and is drawn from one of two populations $j = 0$ and $j = 1$ having (conditional) density $p(\boldsymbol{x}|j)$ given by either $N(0,\Sigma_0)$ or $N(0,\Sigma_1)$, where the variance matrices are diagonal; i.e., $\Sigma_j = \text{diag}(\sigma_{j,1}^2, \cdot, \sigma_{j,p}^2)$ for $j = 0,1$, and distinct. Show that there exists a weight vector $\mathbf{w} \in \mathbb{R}^p$ and a scalar α such that $\Pr(j = 1|\boldsymbol{x})$ can be written in the form

$$\mathbb{P}(j = 1|\boldsymbol{x}) = \frac{1}{1+\exp(-\boldsymbol{w}^\top \boldsymbol{x} + \alpha)}.$$

Exercise 5.2. Let $H : g(\boldsymbol{x}) = \boldsymbol{w}^\top \boldsymbol{x} + w_0 = 0$ be a hyperplane in $\mathbb{R}^p$ with normal vector $\boldsymbol{w} \in \mathbb{R}$ and let $\boldsymbol{x}' \in \mathbb{R}^p$ be a point.

1. Show that the perpendicular distance $d(H,\boldsymbol{x}')$ from H to the point $\boldsymbol{x}'$ is $|g(\boldsymbol{x}')|/\|\boldsymbol{w}\|$ and this can be found by minimizing $\|\boldsymbol{x}-\boldsymbol{x}'\|^2$ subject to $g(\boldsymbol{x}) = 0$. That is, show that
$$d(H,\boldsymbol{x}') = \frac{|g(\boldsymbol{x}')|}{\|\boldsymbol{w}\|} = \min_{g(\boldsymbol{x})=0} \left(\|\boldsymbol{x}-\boldsymbol{x}'\|^2\right).$$
2. Show that the projection of an arbitrary point $\boldsymbol{x}'$ onto H is
$$\boldsymbol{x}' - \frac{g(\boldsymbol{x}')}{\|\boldsymbol{w}\|^2}\boldsymbol{w}.$$

Exercise 5.3 (Quadratic discriminant function). Consider the generalization of the linear discriminant function in Exercise 2 given by the quadratic discriminant function

$$g(\boldsymbol{x}) = w_0 + \sum_{j=1}^{p} w_i x_i + \sum_{j=1}^{p}\sum_{k=1}^{p} w_{ij} x_i x_j = w_0 + \boldsymbol{w}^\top \boldsymbol{x} + \boldsymbol{x}\mathbf{W}\boldsymbol{x},$$

where $\boldsymbol{w} \in \mathbb{R}^p$ and $\mathbf{W} = (w_{ij})$ is a symmetric nonsingular matrix. Show that the decision boundary defined by this discriminant function can be described in terms of the matrix

$$\mathbf{M} = \frac{\mathbf{W}}{\mathbf{w}^\top \mathbf{W}^{-1}\mathbf{w} - 4w_0}$$

in terms of two cases:

1. If $\mathbf{M}$ is positive definite, then the decision boundary is a p-dimensional ellipsoid.
2. If $\mathbf{M}$ has both positive and negative eigenvalues, then the decision boundary is a hyperboloid, also in p dimensions.

   Note that item 1 gives a p-dimensional sphere when all the axes of the ellipsoid are the same length.
3. Suppose $\mathbf{w} = (5, 2, -3)^\top$ and

$$\mathbf{W} = \begin{bmatrix} 1 & 2 & 0 \\ 2 & 5 & 1 \\ 0 & 1 & -3 \end{bmatrix}.$$

   What does the decision boundary look like?
4. Suppose $\mathbf{w} = (2, -1, 3)^\top$ and

$$\mathbf{W} = \begin{bmatrix} 1 & 2 & 3 \\ 2 & 0 & 4 \\ 3 & 4 & -5 \end{bmatrix}.$$

   What does this decision boundary look like?

Exercise 5.4 (Single node NNs can reduce to linear discriminants). Consider a "network" of only a single output neuron, i.e., there are no hidden layers. Suppose the network has weight vector $\boldsymbol{w} \in \mathbb{R}^p$, the input $\boldsymbol{x}$ has p entries and the sigmoid in the output neuron is

$$\phi(u) = \frac{1}{1 + \exp(-u)}.$$

Thus, the network function is

$$f(\boldsymbol{x}) = \phi\left(\sum_{k=0}^{p} w_k x_k\right)$$

and has, say, threshold w_0.

1. Show that the single output neuron implements a decision boundary defined by a hyperplane in $\mathbb{R}^p$. That is, show that f is a linear discriminant function with boundary of the form

$$\sum_{j=0}^{p} w_j x_j = 0.$$

2. Illustrate your answer to item 1 for $p = 2$.

Exercise 5.5 (Continuation of Exercise 5.4).

1. Redo Exercise 5.4 item 1, but replace the sigmoid with a Heaviside function; i.e., use

$$f(\boldsymbol{x}) = H\left(\sum_{j=0}^{p} w_j x_j\right)$$

where $H(u) = 1$ if $u > 0$, $H(u) = -1$ if $u < 0$, and $H(u) = 0$ if $u = 0$.

2. How can you make this classifier able to handle nonlinearly separable $\boldsymbol{x}$s?

Exercise 5.6 (Gradient descent to find weights in a NN). Consider a data set $\{(\boldsymbol{x}_i, y_i),\ i = 1, \cdots, n\}$ where $\boldsymbol{x}_i$ is an input vector and $y_i \in \{0,1\}$ is a binary label indicating the class of $\boldsymbol{x}_i$. Suppose that given a fixed weight vector $\boldsymbol{w}$, the output of the NN is $f(\boldsymbol{x}) = f(\boldsymbol{x}, \boldsymbol{w})$. To choose $\boldsymbol{x}$, define the binomial error function

$$E(\boldsymbol{w}) = -\sum_{i=1}^{n} [y_i \ln f(\boldsymbol{x}_i, \boldsymbol{w}) + (1 - y_i) \ln(1 - f(\boldsymbol{x}_i, \boldsymbol{w}))].$$

1. Verify that $\boldsymbol{g} = \partial E / \partial \boldsymbol{w}$ has entries given by

$$g_j = \frac{\partial E}{\partial w_j} = \sum_{i=1}^{n} -(y_i - f(\boldsymbol{x}_i, \boldsymbol{w})) x_{ij},$$

for $j = 1, ..., p$.

2. Why does the derivative in item 1 suggest gradient descent is possible for estimating $\boldsymbol{w}$? How would you do it?

Hint: Observe that for each j, $g_j = \sum_i^n -(y_i - f(\boldsymbol{x}_i; \boldsymbol{w})) x_{ij}$.

Exercise 5.7 (Examples of kernels).

1. To see how the concept of a kernel specializes to the case of discrete arguments, let S be the set of strings of length at most ten, drawn from a finite alphabet $\mathscr{A}$; write $s \in S$ as $s = a_1, ..., a_{10}$ where each $a_j \in \mathscr{A}$. Now, let $K : S \times S \to \mathbb{Z}$ be defined for $s_1, s_2 \in S$ by $K(s_1, s_2)$ is the number of substrings s_1 and s_2 have in common, where the strings needn't be consecutive.

Prove that K is a kernel. To do this, find a pair $(\mathscr{H}, \boldsymbol{\Phi})$ with $\boldsymbol{\Phi} : S \to \mathscr{H}$ so that $K(s, s') = \langle \boldsymbol{\Phi}(s), \boldsymbol{\Phi}(s') \rangle$ for every $s' \in S$.

2. Here is another discrete example. Let $x, x' \in \{1, 2, ..., 100\}$ and set

$$K(x, x') = \min(x, x').$$

Show that K is a kernel.

3. In the continuous case, let d be a positive integer, and $c \in \mathbb{R}^+$. Let

$$K(\boldsymbol{x}_1, \boldsymbol{x}_2) = (\boldsymbol{x}_1^\top \boldsymbol{x}_2 + c)^d.$$

Show that K is a kernel.

Hint: Try induction on d.

4. Show that if K_1 and K_2 are kernels then so is $K_1 + K_2$.

5. Show that if K is a kernel with feature map Φ, then the normalized form of K,

$$\tilde{K}_1 = \frac{K_1(x, z)}{\sqrt{K_1(x, x)K_1(z, z)}},$$

is a kernel for $\tilde{\Phi} = \Phi(x)/\|\Phi(x)\|$.

6. To see that not every function of two arguments is a kernel, define

$$K(x, s') = e^{\|x - x'\|^2}$$

for $x, x' \in \mathbb{R}^n$. Prove that this K is *not* a kernel.

Hint: Find a counterexample. For instance, suppose K is kernel and find a contradiction to Mercer's theorem or some other mathematical property of kernels.

Exercise 5.8. Let N be a node in a tree based classifier and let $r(N)$ be the proportion of training points at N with label 0 rather than 1. Let *psi* be a concave function with $\psi_{max} = \psi(1/2)$ and $\psi(0) = \psi(1) = 0$. Write $i(N) = \psi(r(N))$ to mean the impurity of node N under ψ.

Verify the following for such impurities.

1. Show that i is concave in the sense that, if N is split into nodes N_1 and N_2, then

$$i(N) \geq r(N_1)i(N_1) + r(N_2)i(N_2). \tag{5.7.1}$$

2. Consider a specific choice for ψ, namely the misclassification impurity in which $\psi(r) = (1 - \max(r, 1 - r))$. Note that this ψ is triangle shaped and hence concave, but not strictly so. Suppose you are in the unfortunate setting where a node has, say, 70 training points, 60 from class 0 and 10 from class 1 and there is no way to split the class to get a daughter node which has a majority of class 1 points in it. (This can happen in one dimension if half the class 0 points are on each side of the class 1 points.) Show that in such cases equality holds in (5.7.1), i.e.,

$$i(N) = r(N_1)i(N_1) + r(N_2)i(N_2).$$

3. Now, let i be the Gini index and suppose there is a way to split the points in N into N_1 and N_2 so that all 10 class 1 points are in N_1 along with 20 class 0 points, and the remaining 40 class 0 points are in N_2. Write Gini in terms of r and an appropriate ψ. Show that in this case,

$$i(N) > r(N_1)i(N_1) + r(N_2)i(N_2).$$

4. What does this tell you about the effect of using the Gini index as an impurity versus the misclassification impurity?

Exercise 5.9 (Toy data for a tree classifiers). Consider a classification data set in the plane, say $(y_i, \boldsymbol{x}_i)$ for $i = 1,...,9$, with $y_i = 0,1$ and $\boldsymbol{x}_i \in \mathbb{R}^2$. Suppose the first point is class 1 at the origin, i.e., $(y_1, \boldsymbol{x}_1) = (0,(0,0))$, and the other eight points are of the form $(1,(R\sin(2\pi i/8), R\cos(2\pi i/8)))$ for $i = 1,...,8$, i.e., class 2 points, equally spaced on the circle of radius R centered at the origin.

Suppose you must find a classifier for this data using only linear discriminant functions; i.e., a decision tree where the node functions assign class 1 when $f(x_1,x_2) = \text{sign}(a_1x_1 + a_2x_2 + b) > 0$ for real constants a_1, a_2 and b.

1. What is the smallest tree that can classify the data correctly?
2. Choose one of the impurities in the previous exercise (Gini or triangle) and calculate the decrease in impurity at each node of your tree from item 1.
3. If the eight points were located differently on the circle of radius R, could you find a smaller tree than in item 1? If the eight points were located differently on the circle of radius R, would a larger tree be necessary?

Exercise 5.10 (Ridge regression error). Another perspective on regularization in linearly separable classification comes from the ridge penalty in linear regression. Consider a sample $(\mathbf{x}_1, y_1),..., (\mathbf{x}_n, y_n)$ in which $\mathbf{x_i} \in \mathbb{R}^p$ and $y_i \in \{-1,1\}$ and define the span of the design points to be the vector space

$$V = \left\{ \sum_{i=1}^{n} \alpha_i \mathbf{x}_i | \alpha_i \in \mathbb{R} \right\}.\}$$

Let $C > 0$ and define the regularized risk

$$E(\mathbf{w}) = \sum_{i=1}^{n} (\mathbf{w} \cdot \mathbf{x}_i - y_i)^2 + C\|\mathbf{w}\|_2^2. \tag{5.7.2}$$

1. Show that the minimizer $\hat{\mathbf{w}} = \arg\ \min_{\mathbf{w}} E(\mathbf{w})$ is an element of V.

   Hint: Although it is worthwhile to rewrite (5.7.2) in matrix notation and solve for the projection matrices, there is a more conceptual proof. Let $\mathbf{w} \in V$ and let $\mathbf{v}$ be a vector in $\mathbb{R}^p$ that is orthogonal to all the $\mathbf{x}_i$s. Show that adding any such $\mathbf{v}$ to $\mathbf{w}$ will always increase $E(\mathbf{w})$.

2. Show that the argument in item 1 is not unique to (5.7.2) by giving another loss function and penalty for which it can be used to identify a minimum.

Exercise 5.11 (Pruning trees by hypothesis tests). Once a tree has been grown, it is good practice to prune it a bit. Cost-complexity is one method, but there are others. Hypothesis testing can be used to check for dependence between the "Y" and the covariate used to split a node. The null hypothesis would be that the data on the two sides of the split point were independent of Y. If this hypothesis cannot be rejected, the node can be eliminated. Although not powerful, the chi-square test of independence is the simplest well-known test to use. Suppose there are two covariates X_1 and X_2, each taking one of two values, say T and F, in a binary classification problem with $Y = \pm 1$. For splitting on X_1 to classify Y, imagine the 2×2 table of values (Y, X_1). Then, the chi-square statistic is

$$\chi_s^2 = \sum_{j=1,k=1}^{2,2} \frac{(O_{j,k} - E_{j,k})}{E_{j,k}},$$

where $O_{j,k}$ is the number of observations in cell (i,k) and $E_{j,k} = np(j)p(k)$ is the expected number of observations in cell (j,k) under independence; i.e., $p(j)$ is the marginal probability of $Y = j$ and $p(k)$ is the marginal probability of $X_1 = k$. Under the null, $\chi_s^2 \sim \chi_1^2$ and the null is rejected for large values of χ_s^2. The same reasoning holds for (Y, X_2) by symmetry. (All of this generalizes to random variables assuming any finite number of values.)

Now, consider the data in the table for an unknown target function $f : (X_1 1, X_2 2) \to Y$. Each 4-tuple indicates the value of Y observed, the two values (X_1, X_2) that gave it, and how many times that triple was observed.

Y	X_1	X_2	count	Y	X_1	X_2	count
+1	T	T	5	-1	T	T	1
+1	T	F	4	-1	T	F	2
+1	F	T	3	-1	F	T	3
+1	F	F	2	-1	F	F	5

1. Generate a classification tree and then prune it by cost-complexity and by using a χ^2 test of independence.
2. Now, examine the splits in the cost-complexity generated tree. Use the chi-square approach to see if the splits are statistically significant; i.e., if you do a chi-square test of independence between Y and the covariate split at a node, with level $\alpha = 0.10$, do you find that dependence?
3. The sample entropy of a discrete random variable Z is $\hat{H}(Z) = \sum \hat{P} \ln(1/\hat{P})$, where the $\hat{P}$s are the empirical probabilities for Z. What is the sample entropy for Y using the data in the table?
4. What is the sample entropy for X_1 and X_2, (Y, X_1), (Y, X_2), and (Y, X_1, X_2)?
5. Sometimes the Shannon information $I(Y;X) = H(Y) - H(Y|X)$ is called the information gain, in which the conditional entropy is $H(Y|X) = \sum_x \sum_y P(Y = y|X = x) \ln(1/P(Y = y|X = x))$.
6. What is the information gain after each split in the trees in item 1?
7. What is the information gain $I(Y;X_1)$ for this sample?

8. What is the information gain $I(Y;X_2)$ for this sample?

Exercise 5.12. Consider the two quadratic functions $f(x) = x^2$ and $g(x) = (x-2)^2$ and suppose you want to find

$$\text{minimize } f(x) \text{ subject to } g(x) \leq 1.$$

1. Solve the problem by Lagrange multipliers.
2. Write down the Lagrangian $L(x,\lambda)$ and solve the problem using the KKT conditions from Section 5.4.4.1.
3. Give a closed form expression for the dual problem.
4. Plot the function $y = L(x,\lambda)$ in $\mathbb{R}^3$ as a function of x and λ. On the surface, find the profile of x; i.e., identify $y = \max_\lambda L(x,\lambda)$, and the profile of λ; i.e., identify $y = \min_x L(x,\lambda)$. At what point do the intersect?
5. Suppose the constraint $g(x) \leq 1$ is replaced by the constraint $g(x) \leq 1$. If $a \neq 1$, do the results change?

Exercise 5.13. In the general nonlinearly separable case, support vector machine classification was presented using a fixed but arbitrary kernel. However support vector regression was only presented for the kernel corresponding to the inner product: $K(\mathbf{x},\mathbf{y}) = \mathbf{x}\cdot\mathbf{y}$. Using the support vector machine classification derivation, extend the support vector machine regression derivation to general kernels.

Exercise 5.14 (LOOCV error for SVMs). Recall (5.4.12), the final expression for a linear SVM on a linearly separable data set $((y_1,\boldsymbol{x}_1),\ldots,(y_n,\boldsymbol{x}_n))$. Note that s is the number of support vectors. Although CV is usually used to compare models, the fact that CV can be regarded as an estimator for the predicted error makes it reasonable to use CV to evaluate a single model, such as that in (5.4.12).

1. Show that the leave-one-out CV error for the linear SVM classifier in (5.4.12) is bounded by s/n for linearly separable data.

   Hint: In the leave-one-out CV error, note that each $\boldsymbol{x}_i$ is either a support vector or is a nonsupport vector. So, there are two cases to examine when leaving out one data point.

2. Suppose the data have been made linearly separable by embedding them in a high-dimensional feature space using a transformation $\boldsymbol{\Phi}$ from a general Mercer kernel. Does the bound in item 1 continue to hold? Explain.

   Hint: The $\boldsymbol{\Phi}$ is not unique.

Chapter 6
Alternative Nonparametrics

Having seen Early, Classical and New Wave nonparametrics, along with partitioning-based classification methods, it is time to examine the most recently emerging class of techniques, here called Alternative methods in parallel with contemporary music. The common feature all these methods have is that they are more abstract. Indeed, the four topics covered here are abstract in different ways. Model-averaging methods usually defy interpretability. Bayesian nonparametrics requires practitioners to think carefully about the space of functions being assumed in order to assign a prior. The relevance vector machine (RVM) a competitor to support vector machines, tries to obtain sparsity by using asymptotic normality; again the interpretability is mostly lost. Hidden Markov models pre-suppose an unseen space to which all the estimates are referred. The ways in which these methods are abstract vary, but it is hard to dispute that the degree of abstraction they require exceeds that of the earlier methods.

As a generality, Alternative techniques are evaluated mostly by predictive criteria only secondarily by goodness of fit. It is hard to overemphasize the role of prediction for these methods since they are, to a greater or lesser extent, black box techniques that defy physical modeling even as they give exceptionally good performance. This is so largely because interpretability is lost. It is as if there is a trade-off: As interpretability is allowed to deteriorate, predictive performance may improve and conversely. This may occur because, outside of simple settings, the interpretability of a model is not reliable: The model is only an approximation, of uncertain adequacy, to the real problem. Of course, if a true model can be convincingly identified and the sample size is large enough, the trade-off is resolved. However, in most complex inference settings, this is just not feasible. Consequently, the techniques here typically give better performance than more interpretable methods, especially in complex inference settings.

Alternative techniques have been developed as much for classification as regression. In the classification context, most of the techniques are nonpartitioning. This is typical for modelaveraging techniques and is the case for the RVM. Recall that nonpartitioning techniques are not based on trying to partition the feature space into regions. This is a slightly vague classification because nearestneighbor methods are non-partitioning but lead naturally to a partition of the feature space, as the RVM does. The point, though, is that nonpartitioning techniques are not generated by directly evaluating partitions.

B. Clarke et al., *Principles and Theory for Data Mining and Machine Learning*, Springer Series in Statistics, DOI 10.1007/978-0-387-98135-2_6, © Springer Science+Business Media, LLC 2009

The biggest class of Alternative techniques is ensemble methods. The idea is to back off from choosing a specific model and rest content with averaging the predictions from several, perhaps many, models. It will be seen that ensemble methods often improve both classification and regression. The main techniques are Bayes model averaging (BMA) bagging (bootstrap aggregation), stacking, and boosting. All of these extend a set of individual classifiers, or regression functions, by embedding them in a larger collection formed by some kind of averaging. In fact, random forests, seen in the previous chapter, is an ensemble method. It is a bagged tree classifier using majority vote. Like random forests, ensemble methods typically combine like objects; e.g., combining trees with other trees or neural nets with other neural nets rather than trees with neural nets. However, there is no prohibition on combining like with unlike; such combinations may well improve performance.

Bayesian nonparametrics has only become really competitive with the other nonparametric methods since the 1990s, with the advent of high-speed computing. BMA was the first of the Bayesian nonparametric methods to become popular because it was implementable and satisfied an obvious squared error optimality. As noted above, it is an ensemble method. In fact, all Bayes methods are ensemble based because the posterior assigns mass over a collection of models. Aside from computing the posterior, the central issue in Bayesian nonparametrics is the specification of the prior, partially because its support must be clearly specified. One of the main benefits of the Bayesian approach is that the containment property of Bayes methods (everything is done in the context of a single probability space) means the posterior fully accounts for model variability (but not bias).

A third Alternative method is the RVM. These are not as well studied as SVMs, but they do have regression and classification versions. Both of these are presented. RVMs rest on a very different intuition than SVMs and so RVMs often give more sparsity than SVMs. In complex inference problems, this may lead to better performance. Much work remains to be done to understand when RVMs work well and why; their derivation remains heuristic and their performance largely unquantified. However, they are a very promising technique.

As a final point, a brief discussion of Hidden Markov Models (HMMs) is provided for the sake of expressing the intuition. These are not as generally applicable, at present, as the other three techniques, but HMMs have important domains of application. Moreover, it is not hard to imagine that HMMs, with proper development, may lead to a broad class of methods that can be used more generally.

6.1 Ensemble Methods

Ensemble methods have already arisen several times. The idea of an ensemble method is that a large model class is fixed and the predictions from carefully chosen elements of it are pooled to give a better overall prediction. As noted, Breiman's random forests is an ensemble method based on bootstrapping, with trees being the ensemble. In other words, random forests is a "bagged" (i.e., bootstrap aggregated) version of trees. It will be seen that bagging is a very general idea: One can bag NNs, SVMs, or any other model class.

Another way ensemble methods arise is from model selection principles (MSPs). Indeed, MSPs are equivalent to a class of ensemble methods. Given a collection of models, every MSP assigns a worth to each model on a list. If the predictions from the models are averaged using a normalized version of the worths, the result is an ensemble method. Conversely, any ensemble method that corresponds to a collection of weights on the models implicitly defines an MSP since one can choose the model with the maximum of those weights. In this way, Bayesian model selection and CV or GCV generate BMA and stacking.

A third way ensemble methods arise is from reoptimizing sequentially and averaging the solutions. This is done by boosting, which uses a class of "weak learners" as its ensemble. A weak learner is a poor model that still captures some important aspect of the data. So, it is plausible that pooling over the right collection of weak learners will create a strong learner; i.e., a good inference technique. Boosting is for classification; a variant of it has been developed for regression but is less successful than many other techniques and is not presented here.

Overall, there are two central premises undergirding ensemble methods. First is the fact that pooling models represents a richer model class than simply choosing one of them. Therefore the weighted sum of predictions from a collection of models may give improved performance over individual predictions because linear combinations of models should give a lower bias than any individual model, even the best, can. Clemen (1989) documents this principle in detail. The cost, however, may be in terms of variance, in that more parameters must be estimated in a model average than in the selection of a single model.

The second central premise of ensemble methods is that the evaluation of performance is predictive rather than model-based. The predictive sum of squares is one measure of error that is predictive and not model-based (i.e., it is not affected by changing the model); there are others. Predictive evaluations can be compared across model classes since they are functions only of the predictions and observations. Risk, by contrast, is model-based and confounds the method of constructing a predictor with the method for evaluating the predictor, thereby violating the Prequential principle see Dawid (1984).

Combining predictions from several models to obtain one overall prediction is not the same as combining several models to get one model. The models in the ensemble whose predictions are being combined remain distinct; this makes sense because they often rest on meaningfully different assumptions that cannot easily be reconciled, and they have different parameters with different estimates. This means submodels of a fixed supermodel can be used to form an ensemble that improves the supermodel.

Indeed, ensemble-based methods only improve on model selection methods when the models in the ensemble give different predictions. An elementary version of this can be seen by noting that if three uncorrelated classifiers with the same error rates $p < 1/2$ are combined by majority voting, then, in the binary case, the combined classifier will have a lower error rate than any of the individual classifiers. While this example is an ideal case, it is often representative of ensemble methods in situations involving high complexity and uncertainty.

Aside from evaluating predictive performance, ensemble methods can be evaluated by use of Oracle inequalities. The idea is to compare the performance of a given method to the theoretically best performance of any such method. If a given method is not too much worse, in a risk sense, than the best method that would be used by an all-knowing Oracle, the method satisfies an Oracle inequality. Two such inequalities will be seen after several ensemble methods have been presented.

6.1.1 Bayes Model Averaging

The key operational feature of Bayesian statistics is the treatment of the estimand as a random variable. Best developed for the finite-dimensional parametric setting, the essence is to compare individual models with the average of models (over their parameter values) via the posterior distribution, after the data have been collected. The better a summary of the data the model is, the higher the relative weight assigned to the model. In all cases, the support of the prior defines the effective model class to which the method applies. Intuitively, as the support of the prior increases, parametric Bayesian methods get closer to nonparametrics.

Beyond the finite-dimensional parametric case, Bayes methods often fall into one of two categories: Bayes model averaging (BMA) treated in this section, and general Bayesian nonparametrics, treated in the next section. In BMA, one typically uses a discrete prior on models and continuous priors on the (finitely many) parameters within models. In practice, BMA usually uses finitely many models, making the whole procedure finite-dimensional. However, the number of parameters can be enormous. One can also imagine that such a BMA is merely the truncated form of countably many models, and indeed BMAs formed from countably infinite sums of trees, NNs, or basis expansions or other model classes can be used. Even in the finite case, if the support of the discrete prior includes a wide enough range of models, the BMA often acts like a purely nonparametric method.

In general Bayesian nonparametrics, a prior distribution is assigned to a class of models so large that it cannot be parametrized by any finite number of parameters. In these cases, usually there is no density for the prior with respect to Lebesgue measure. While the general case is, so far, intractable mathematically, there are many special cases in which one can identify and use the posterior. Both BMA and Bayesian nonparametrics are flexible ensemble strategies with many desirable properties.

The central idea of BMA can be succinctly expressed as follows. Suppose a finite list $\mathscr{E}$ of finite-dimensional parametric models, such as linear regression models involving different selections of variables, $f_j(\boldsymbol{x}) = f_j(\boldsymbol{x}|\theta_j)$ is to be "averaged". Equip each $\theta_j \in \mathbb{R}^{p_j}$ with a prior density $w(\theta_j|M_j)$, where M_j indicates the jth model f_j from the ensemble $\mathscr{E}$, and let $w(M_j)$ be the prior on $\mathscr{E}$. Let $\mathscr{S} \subset \mathscr{E}$ be a set of models. Given data $D = \{(\boldsymbol{X}_i, Y_i) : i = 1, ..., n\}$, the posterior probability for $\mathscr{S}$ is

$$W(\mathscr{S}|D) = \sum_{M_j \in \mathscr{E}} \int w(M_j, \theta_j|D) I_{\{f_j \in \mathscr{S}\}}(\theta_j) d\theta_j$$

$$= \sum_{M_j \in \mathscr{E}} \int w(M_j|D) w(\theta_j|D, M_j) I_{\{f_j \in \mathscr{S}\}}(\theta_j) d\theta_j. \quad (6.1.1)$$

The expression for $W(\mathscr{S}|D)$ in (6.1.1) permits evaluation of the posterior probability of different model choices so in principle one can do hypothesis tests on sets of models or individual models. Using (6.1.1) when $\mathscr{S}$ is a single point permits formation of the weighted average

$$\hat{Y}_B(\boldsymbol{x}) = \sum_{M_j \in \mathscr{E}} W(M_j|D) f_j(\boldsymbol{x}|\mathbb{E}(\Theta_j|D)) \quad (6.1.2)$$

to predict the new value of Y at $\boldsymbol{x}$. Note that the more plausible model M_j is, the higher its posterior probability will be and thus the more weight it will get. Likewise, the more plausible the value θ_j is in M_j, the more weight the posterior $w_j(\cdot|x^n)$ assigns near the true value of θ_j and the closer the estimate of the parameter in (6.1.2), $\mathbb{E}(\Theta_j|D)$, is to the true θ_j as well. It is seen that the BMA is the posterior mean over the models in $\mathscr{E}$, which is Bayes risk optimal under squared error loss. For this reason, the posterior mean $\mathbb{E}(\Theta_j|D)$ is used; however, other estimates for θ_j may also be reasonable. One can readily imagine forming weighted averages using coefficients other than the posterior probabilities of models as well.

Theoretically, Madigan and Raftery (1984) showed that BMA (6.1.1) provides better predictions under a log scoring rule than using any one model in the average, possibly because it includes model uncertainty. It should be noted that, depending on the criteria and setting, non-Bayes averages can be predictively better than Bayes averages when the prior does not assign mass near the true model; in some cases, non-Bayes optima actually converge to the Bayesian solution; see Shtarkov (1987), Wong and Clarke (2004), and Clarke (2007).

Despite extensive use, theoretical issues of great importance for BMA remain unresolved. One is prior selection. The first level of selection is often partially accomplished by using objective priors of one sort or another. On the continuous parameters in a BMA, the θ_js, uniform, normal, or Jeffreys priors are often used. The more difficult level of prior selection is on the models in the discrete set $\mathscr{E}$. A recent contribution focussing on Zellner's g-prior is in Liang et al. (2008), who also give a good review of the literature. Zellner's g-prior is discussed in Chapter 10.

Another issue of theoretical importance is the choice of $\mathscr{E}$. This is conceptually disjoint from prior selection, but the two are clearly related. The problem is that if $\mathscr{E}$ has too many elements close to the true model, then the posterior probability may assign all of them very small probabilities so that none of them contribute very much to the overall average. This phenomenon, first identified by Ed George, is called dilution and can occur easily when $\mathscr{E}$ is defined so that BMA searches basis expansions. In these cases, often $\mathscr{E}$ is just the 2^p set defined by including or omitting each of the basis elements. Thus, as the approximation by basis elements improves, the error shrinks and the probability associated with further terms is split among many good models.

For the present, it is important to adapt existing intuition about posterior distributions to the BMA setting. First, when the true model is on the list of models being averaged, its weight in the BMA increases to 1 if $\mathscr{E}$ and the priors are fixed and $n \to \infty$. If the true model is not in $\mathscr{E}$ but $\mathscr{E}$ and the priors are fixed, then the weight on the model closest to the true model in relative entropy goes to 1. This is the usual posterior consistency (see Berk (1966)) and holds because any discrete prior gives consistency provided the true model is identifiable.

The problem gets more complicated when n is fixed and $\mathscr{E}$ changes. Suppose $\mathscr{E}$ is chosen so large that model list selection can be reduced to prior selection. That is, the support of the prior W, $\mathscr{W}$ in $\mathscr{E}$, is large enough to define a good collection of models to be averaged. Moreover, suppose the priors within models are systematically chosen, perhaps by some objective criterion, and so can be ignored. Then, if the models in $\mathscr{W}$ have no cluster points, are not too dispersed over the space $\mathscr{F}$ in which f is assumed to lie, and for the given n are readily distinguishable, the usual posterior consistency reasoning holds.

However, suppose $\mathscr{W}$ is permitted to increase and that $\mathscr{E}$ is replaced by $\mathscr{E}_m$ which increases as $m = m(n)$ increases, thereby including more and more functions that are close to f but still distinguishable. Then the posterior probability of each element in $\mathscr{E}_m$ – even the true model – can go to zero; this is called vague convergence to zero because the distribution converges to zero pointwise on each model even though each $\mathscr{E}_m$ has probability one overall. This, too, is a sort of dilution because the probability is split among ever more points that are good approximations to f given the sample size. This problem only becomes worse as the dimension p increases because there are more models that can be close to the true model.

A partial resolution of this problem comes from Occam's window approaches; see Madigan and Raftery (1984). The idea is to restrict the average to include only those models with a high enough posterior probability since those are the weights in the BMA. This violates the usual data independence of priors Bayesians often impose since the support of the prior depends on the sample. However, this may be necessary to overcome the dilution effect. Moreover, there is some reason to believe that the data independence of the prior derived from Freedman and Purves (1969) may not always be reasonable to apply; see Wasserman (2000), Clarke (2007).

6.1.2 Bagging

Bagging is a contraction of bootstrap aggregation, a strategy to improve the predictive accuracy of a model as seen in random forests. Given a sample, fit a model $\hat{f}(\boldsymbol{x})$ called the base and then consider predicting the response for a new value of the explanatory vector $\boldsymbol{x}_{new}$. A bagged predictor for $\boldsymbol{x}_{new}$ is found by drawing B bootstrap samples from the training data; i.e., draw B samples of size n from the n data points with replacement. Each sample of size n is used to fit a model $\hat{f}_i(\boldsymbol{x})$ so that the the bagged prediction is

$$\hat{f}_{bag}(\boldsymbol{x}_{new}) = \frac{1}{B}\sum_{i=1}^{B}\hat{f}_i(\boldsymbol{x}_{new}). \tag{6.1.3}$$

Note that, unlike BMA, the terms in the average are equally weighted (by $1/B$). Thus, this method relies on the resampling to ensure that the models in the average are representative of the data.

A good procedure is one that searches a large enough class of models that it can, in principle, get a model that has small misclassification error (i.e., in zero-one loss) or a small error in some other sense. However, even if a procedure does this, it may be unstable. Unstable procedures are those that have, for instance, a high variability in their model selection. Neural nets, trees, and subset selection in linear regression are unstable. Nearest-neighbor methods, by contrast, are stable. As a generality, bagging can improve a good but unstable procedure so that it is close to optimal.

To see why stability is more the issue than bias is, recall Breiman (1994)'s original argument using squared error. Let $\hat{\phi}(x,D)$ be a predictor for Y when $\boldsymbol{X} = \boldsymbol{x}$, where the data are $D = \{(y_1,\boldsymbol{x}_1),\ldots,(y_n,\boldsymbol{x}_n)\}$. The population-averaged predictor is

$$\phi_A(\boldsymbol{x},\mathbb{P}) = \mathbb{E}\hat{\phi}(\boldsymbol{x},D),$$

the expectation over the joint probability $\mathbb{P}$ for $(\boldsymbol{X},Y)$. In squared error loss, the average prediction error for $\hat{\phi}(\boldsymbol{x},D)$ is

$$APE(\hat{\phi}) = \mathbb{E}_D\mathbb{E}_{Y,\boldsymbol{X}}(Y - \phi(\boldsymbol{X},D))^2$$

over the probability space for $n+1$ outcomes, and the error for the population-averaged predictor is

$$APE(\phi_A) = \mathbb{E}_{Y,\boldsymbol{X}}(Y - \phi_A(\boldsymbol{X},\mathbb{P}))^2.$$

Jensen's inequality on x^2 gives

$$APE(\hat{\phi}) = \mathbb{E}Y^2 - 2\mathbb{E}Y\phi_A + \mathbb{E}_{Y,\boldsymbol{X}}\mathbb{E}_D\hat{\phi}^2(\boldsymbol{X},Data) \geq \mathbb{E}(Y-\phi_A)^2 = APE(\phi_A).$$

The difference between the two sides is the improvement due to aggregation.

So, $APE(\hat{\phi}) - APE(\phi_A)$ should be small when a procedure is stable. After all, a good, stable procedure $\hat{\phi}$ should vary around an optimal predictor ϕ_{opt} so that $\phi_A \approx \hat{\phi} \approx \phi_{opt}$. On the other hand, $APE(\hat{\phi}) - APE(\phi_A)$ should be large when $\hat{\phi}$ is unstable because then aggregation will stabilize it, necessarily close to a good procedure because the procedure itself was good; i.e., it had high variance not high bias. This suggests aggregating will help more with instability than bias, but is unlikely to be harmful. At root, bagging improves $\hat{\phi}$ to ϕ_B, a computational approximation to the unknown ϕ_A.

Using misclassification error rather than squared error gives slightly different results. To set up Breiman (1994)'s reasoning, consider multiclass classification and let $\phi(\boldsymbol{x},D)$ predict a class label $k \in \{1,\ldots,K\}$. For fixed data D, the probability of correct classification is

$$r(D) = \mathbb{P}(Y = \phi(\boldsymbol{X}, D)|D) = \sum_{k=1}^{K} \mathbb{P}(\phi(\boldsymbol{X}, D) = k|Y = k, D)\mathbb{P}(Y = j).$$

Letting $\mathbb{Q}(k|\boldsymbol{x}) = \mathbb{P}_D(\phi(\boldsymbol{x}, D) = k)$, the probability of correct classification averaged over D is

$$r = \sum_{k=1}^{K} \mathbb{E}(\mathbb{Q}(k|\boldsymbol{X})|Y = k)\mathbb{P}(Y = k) = \sum_{k=1}^{K} \int \mathbb{Q}(k|\boldsymbol{x})\mathbb{P}(k|\boldsymbol{x})\mathbb{P}_X(d\boldsymbol{x}),$$

where $\mathbb{P}_X$ is the marginal for $\boldsymbol{X}$. The Bayes optimal classifier is the modal class, so if $\mathbb{Q}$ were correct, ϕ_{orig} would be $\phi_{orig}(\boldsymbol{x}) = \arg\ \max_k\ \mathbb{Q}(k|\boldsymbol{x})$, in which case

$$r_{orig} = \sum_{k=1}^{K} \int I_{\phi_{opt}(\boldsymbol{x})=k}\mathbb{P}(k|\boldsymbol{x})\mathbb{P}_X(d\boldsymbol{x}).$$

So, let $C = \{\boldsymbol{x}| \arg\ \max_k\ \mathbb{P}(k|\boldsymbol{x}) = \arg\ \max_k\ \mathbb{Q}(k|\boldsymbol{x})\}$ be the set, hopefully large, where the original classifier matches the optimal classifier. It can be seen that the size of C helps analyze how well the original classifier performs. Breiman (1994)'s argument is the following: For $\boldsymbol{x} \in C$ we have the identity

$$\sum_{k=1}^{K} I_{\arg\ \max_j\ \mathbb{Q}(j|\boldsymbol{x})=k}\mathbb{P}(k|\boldsymbol{x}) = \max_j\ \mathbb{P}(j|\boldsymbol{x}).$$

So the domain in r_{orig} can be partitioned into C and C^c to give

$$r_{orig} = \int_{\boldsymbol{x}\in C} \max_j\ \mathbb{P}(j|\boldsymbol{x})\mathbb{P}_X(d\boldsymbol{x}) + \int_{\boldsymbol{x}\in C^c} \sum_{k=1}^{K} I_{\phi_A(\boldsymbol{x})=k}\mathbb{P}(k|\boldsymbol{x})\mathbb{P}_X(d\boldsymbol{x}).$$

Since $\mathbb{P}$ is correct, the best classification rate is achieved by

$$Q^*(\boldsymbol{x}) = \arg\ \max_j\ \mathbb{P}(j|\boldsymbol{x}),$$

and has rate

$$r^* = \int \max_j\ \mathbb{P}(j|\boldsymbol{x})\mathbb{P}_X(d\boldsymbol{x}).$$

Observe that if $\boldsymbol{x} \in C$, it is possible that sum

$$\sum_{k=1}^{K} \mathbb{Q}(k|\boldsymbol{x})\mathbb{P}(k|\boldsymbol{x}) < \max_j\ \mathbb{P}(j|\boldsymbol{x}).$$

So, even when C is large (i.e., $\mathbb{P}_X(C) \approx 1$) the original predictor can be suboptimal. However, ϕ_{opt} may be nearly optimal. Taken together, this means that aggregating can improve good predictors into nearly optimal ones but, unlike in the squared error case, weak predictors can be transformed into worse ones. In other words, bagging unstable classifiers usually improves them; bagging stable classifiers often worsens them. This is the reverse of prediction under squared error. The discussion of bagging in Sutton

(2005), Section 5.2 emphasizes that, in classification, bagging is most helpful when the bias of the procedures being bootstrapped is small.

There have been numerous papers investigating various aspects of bagging. Friedman and Hall (2000) and Buja and Stuetzle (2000a) Buja and Stuetzle (2000b) all give arguments to the effect that, for smooth estimators, bagging reduces higher order variation. Specifically, if one uses a decomposition into linear and higher-order terms, bagging affects the variability of higher-order terms. If one uses U-statistics, then it is the second-order contributions of variance, bias, and MSE that bagging affects.

Buhlman and Yu (2002) tackle the problem of bagging indicator functions. Their work applies to recursive partitioning, for instance, which is known to be somewhat unstable. The basic insight is that hard decision rules (i.e., deciding unambiguously which class to say Y_{new} belongs to) create instability and that bagging smooths hard decision rules so as to give smaller variability and MSE. It also smooths soft decision rules (i.e., decision rules that merely output a probability that Y_{new} is in a given class), but as they are already relatively smooth, the improvement is small.

The template of the Buhlman-Yu argument is the following: Consider a predictor of the form

$$\hat{\theta}(x) = 1_{\hat{d}_n \le x},$$

where $x \in \mathbf{R}$ and the threshold $\hat{d}_n$ is asymptotically well behaved. That is, (i) there is a value d_0 and a sequence $< b_n >$ such that $(\hat{d}_n - d_0)(b_n/\sigma_\infty)$ is asymptotically standard normal, where σ_∞ is the asymptotic variance, and (ii) the bootstrapped version of $\hat{d}_n$, say $\hat{d}_n^*$, is asymptotically normal in the sense that

$$\sup_{v \in \mathbf{R}} |\mathbb{P}^*(b_n(\hat{d}_n^* - \hat{d}_n) \le v) - \Phi(v/\sigma_\infty)| = o_P(1),$$

in which $\mathbb{P}^*$ is the probability from the bootstrapping; i.e., the distribution functions converge. Denote the bootstrapped version of $\hat{\theta}_n$ by $\hat{\theta}_{n,B}$; in essence, it is the expectation of $\hat{\theta}_n$ in the empirical distribution from the bootstrapping procedure, which the choice of a specific number of bootstrap samples approximates. Then, Buhlman and Yu (2002) show the following.

Theorem (Buhlman and Yu, 2002): For $x = x_n(c) = d_0 + c\sigma_\infty/b_n$:

(i) The pure predictor has a step function limit,

$$\hat{\theta}_n(x_n(c)) \to 1_{Z \le c},$$

and (ii) The bagged predictor has a normal limit,

$$\hat{\theta}_{n,B}(x_n(c)) \to \Phi(c - Z),$$

where Z is $N(0,1)$.

Proof: Both statements follow from straightforward manipulations with limiting normal forms. □

An interesting feature of bagging is that it reuses the data to get out more of the information in them but at the same time introduces a new level of variability, that of the model. It is a curious phenomenon that sometimes appearing to use more variation as in the resampling actually gives better inference. Likely this is because the extra level of variability permits a large enough reduction in bias that the overall MSE decreases. This phenomenon occurs with ratio estimators in survey sampling, for instance. Ratio estimators often outperform direct estimators even though the numerator and denominator are both random. Similarly, it is easy to show that two $\sqrt{n}$ consistent estimators are closer to each other than either is to θ_T: $\sqrt{n}(\hat{\theta} - \tilde{\theta}) \to 0$ even though both $\sqrt{n}(\hat{\theta} - \theta_T)$ and $\sqrt{n}(\tilde{\theta} - \theta_T)$ are asymptotically normal. Paradoxically, more variability can be an improvement.

6.1.3 Stacking

Stacking is an adaptation of cross-validation to model averaging because the models in the stacking average are weighted by coefficients derived from CV. Thus, as in BMA, the coefficient of a model is sensitive to how well the model fits the response. However, the BMA coefficients represent model plausibility, a concept related to, but different from, fit. In contrast to bagging, stacking puts weights of varying sizes on models rather than pooling over repeated evaluations of the same model class. Although not really correct, it is convenient to regard stacking as a version of BMA where the estimated weights correspond to priors that downweight complex or otherwise ill-fitting models. That is to say, informally, stacking weights are smaller for models that have high empirical bias or high complexity.

Here's the basic criterion: Suppose there is a list of K distinct models $f_1,...,f_K$ in which each model has one or more real-valued parameters that must be estimated. When plug-in estimators for the parameters in f_k are used, write $\hat{f}_k(\boldsymbol{x}) = f_k(\boldsymbol{x}|\hat{\theta}_k)$ for the model used to get predictions. The task is to find empirical weights $\hat{w}_k$ for the $\hat{f}_k$s from the training data and then form the stacking prediction at a point $\boldsymbol{x}$,

$$\hat{f}_{stack}(\boldsymbol{x}) = \sum_{k=1}^{K} \hat{w}_k \hat{f}_k(\boldsymbol{x}).$$

The $\hat{w}_k$s are obtained as follows. Let $f_k^{(-i)}(\boldsymbol{x})$ be the prediction at $\boldsymbol{x}$ using model k, as estimated from training data with the ith observation removed. Then the estimated weight vector $\hat{w} = (\hat{w}_1, ..., \hat{w}_K)$ solves

$$\hat{w} = \arg \min_{w} \sum_{i=1}^{n} \left[y_i - \sum_{k=1}^{K} w_k \hat{f}_k^{(-i)}(\boldsymbol{x}_i) \right]^2 . \tag{6.1.4}$$

This puts low weight on models that have poor leave-one-out accuracy in the training sample (but beware of the twin problem).

Stacking was invented by Wolpert (1992) and studied by Breiman (1996), among others. Clearly, (6.1.4) can be seen as an instance from a template so that, rather than linearly combining the models with the above weights, one could use a different model class. For instance, one could find the coefficients for a single hidden layer neural net with the $\hat{f}_k$s as inputs or use a different measure of distance.

Several aspects of (6.1.4) deserve discussion.

First, the optimization over w is an exercise in quadratic optimization of varying difficulty. In parallel with BMA, one can impose the constraint that the w_ks are positive and sum to one. Alternatively, one can get different solutions by supposing only that they are positive or only sum to one. They can also be unconstrained.

Second, this procedure, like BMA, assumes one has chosen a list of models to weight. Computationally, BMA permits one to use a larger model list more readily than stacking does. In either case, however, the selection of the procedures to combine, the f_ks, is a level of variability both methods neglect. In effect, these methods are conditional on the selection of a suitable list. The difference in performance between one model list and another can be substantial. For instance, if one list consists of extremely complicated models and another consists of extremely simple models, one expects the predictions from the first to be highly variable and predictions from the second to be more biased even if they both achieve the same MSE. Breiman (1996) argues that one should choose the f_ks to be as far apart as possible given the nature of the problem.

Third, the deficiencies of leave-one-out cross-validation are well known. Often one uses a fifths approach: Leave out 1/5 of the data chosen at random in place of leaving out one data point, and then cycle through the fifths in turn. Whether fifths or other fractions are better probably depends in part on the placement of the actual model f_T relative to the candidate models f_k.

Fourth, the stacking optimization can be applied very generally. It can give averages of densities (see Wolpert and Smyth (2004)) and averages of averages (a convex combination of, for instance, a BMA, a bagged predictor, a boosted predictor, and a stacking average itself). It can be applied to regression problems as well as classifiers or indeed to any collection of predictors. The stacked elements need not be from a common class of models: One can stack trees, linear models, different nonparametric predictors, and neural networks in the same weighted average if desired. Stacking can also be combined with bagging (see Wolpert and Macready (1996)): Either stack the models arising from the bootstrapping or bootstrap a stacking procedure.

Fifth, stacking and Bayes are often similar when the model list chosen is good – at least one member of the list is not too far from f_T. If the model list is perfect (i.e., there is an i such that $f_i = f_T$) BMA usually does better because BMA converges quickly (posterior weights converge exponentially fast) and consistently. However, as model list misspecification increases, stacking often does better in predictive performance relative to BMA. This is partially because stacking coefficients do not depend on the likelihood as posterior weights do; see Clarke (2004) for details.

This highlights an important interpretational difference between BMA and stacking. In BMA, the prior weights represent degrees of belief in the model. Strictly speaking, this means that if one a priori believes a proposed model is incorrect but useful, the conventional Bayes interpretation necessitates a prior weight of zero. Thus, the two methods are only comparable when the f_ks are a priori possibly true as models. More general comparisons between the two methods go beyond the orthodox Bayes paradigm. If one regards the f_ks not as potentially "true" but rather as actions that might be chosen in a decision theory problem aimed at predicting the next outcome, then the prior weights are no longer priors, but merely an enlargement of the space of actions to include convex combinations such as BMA of the basic actions f_k. In this context, cross-validation is Bayesianly or decision-theoretically acceptable because it is merely a technique to estimate the coefficients in the mixture. So, BMA and stacking remain comparable, but in a different framework.

It is easy to see that stacking is clearly better than BMA because stacking does not require any interpretation involving a prior. Stacking can be seen as an approximation technique to find an expansion for f_T treating the f_is as elements in a frame. The flexibility makes stacking easier to apply.

It is equally easy to see that stacking is clearly worse than BMA because BMA permits an interpretation involving a prior. The Bayes or decision theory framework forces clear-minded construction and relevance of techniques to problems. The constraints make BMA easier to apply.

6.1.4 Boosting

Classification rules can be weak; that is, they may only do slightly better than random guessing at predicting the true classes. Boosting is a technique that was invented to improve certain weak classification rules by iteratively optimizing them on the set of data used to obtain them in the first place. The iterative optimization uses (implicitly) an exponential loss function and a sequence of data-driven weights that increase the cost of misclassifications, thereby making successive iterates of the classifier more sensitive. Essentially, one applies iterates of the procedure primarily to the data in the training sample that were misclassified, thereby producing a new rule. The iterates form an ensemble of rules generated from a base classifier so that ensemble voting by a weighted sum over the ensemble usually gives better predictive accuracy.

Boosting originated in Schapire (1990) and has subsequently seen rapid development. To review this, a good place to start is with the Adaboost algorithm in Freund and Schapire (1999); its derivation and properties will be discussed shortly.

Begin with data $(\boldsymbol{x}_1, y_1),...,(\boldsymbol{x}_n, y_n)$, in which, as ever, $\boldsymbol{x}_i \in \mathbf{R}^p$ and $y_i = 1, -1$. Choose an integer T to represent the number of iterations to be performed in seeking an improvement in a given initial (weak) classifier $h_0(\boldsymbol{x})$. At each iteration, a distribution in which to evaluate the misclassification error of h_t is required. Write this as $D_t = (D_t(1),...,D_t(n))$, a sequence of $T+1$ vectors, each of length n, initialized with $D(0) = (1/n,...,1/n)$.

☐ Starting with h_0 at $t = 0$, define iterates h_t for $t = 1,...,T$ as follows. Write the stage t misclassification error as

$$\varepsilon_t = P_{D_t}(h_t(\boldsymbol{X}_i) \neq Y_i) = \sum_{i:h_t(\boldsymbol{x}_i)\neq y_i} D_t(i). \quad (6.1.5)$$

This is the probability, under D_t, that the classifier h_t misclassifies an $\boldsymbol{x}_i$.

☐ Set

$$\alpha_t = \frac{1}{2}\log\frac{1-\varepsilon_t}{\varepsilon_t},$$

and update D_t to D_{t+1} by

$$D_{t+1}(i) = \frac{D_t(i)e^{-\alpha_t y_i h_t(\boldsymbol{x}_i)}}{C_t}, \quad (6.1.6)$$

in which C_t is a normalization factor to ensure D_{t+1} is a probability vector (of length n). In (6.1.6), the exponential factor upweights the cost of misclassifications and downweights the cost of correct classifications.

☐ Set $h^*_{t+1}(x)$ to be

$$h^*_{t+1}(x) = \arg\min_{g\in G}\sum_{i=1}^{n} D_t(i)1_{\{y_i\neq g(x_i)\}},$$

and, with each iteration, add h^*_{t+1} to a growing sum, which will be the output.

☐ The updated weighted-majority-vote classifier is

$$h_{t+1}(x) = \text{sign}\left(\sum_{s=0}^{t+1}\alpha_t h^*_s(x)\right), \quad (6.1.7)$$

and the final classifier in this sequence, h_T, is the boosted version of h_0.

Several aspects of this algorithm bear comment. First, note that it is the distribution D_t that redefines the optimization problem at each iteration. The distribution D_t is where the exponential reweighting appears; it depends on the n pairs and α_t. In fact, D_{t+1} is really in two parts: When $h_t(x_i) = y_i$, the factor is small, $e^{-\alpha_t}$, and when $h_t(x_i) \neq y_i$, the factor is large, e^{α_t}. In this way, the weights of the misclassification errors of the weak learner are boosted relative to the classification successes, and the optimal classifier at the next stage is more likely to correct them. Note that D_t depends on h^*_t, the term in the sum, not the whole partial sum at time t. The α_t, a function of the empirical error ε_t, is the weight assigned to h^*_t. It is justified by a separate argument, to be seen shortly.

The basic boosting algorithm can be regarded as a version of fitting an additive logistic regression model via Newton-like updates for minimizing the functional

$$J(F) = \mathbb{E}(e^{-YF(X)}.$$

These surprising results were demonstrated in Friedman et al. (2000) through an intricately simple series of arguments. The first step, aside from recognizing the relevance of J, is the following.

Proposition (Friedman et al., 2000): J is minimized by

$$F(x) = \frac{1}{2}\log\frac{\mathbb{P}(y=1|x)}{\mathbb{P}(Y=-1|x)}.$$

Hence,

$$\mathbb{P}(y=1|x) = \frac{e^{F(x)}}{e^{F(x)}+e^{-F(x)}} \quad \text{and} \quad \mathbb{P}(y=-1|x) = \frac{e^{-F(x)}}{e^{F(x)}+e^{-F(x)}}.$$

Proof: This follows from formally differentiating

$$\mathbb{E}(e^{-yF(x)}|x) = \mathbb{P}(Y=1|x)e^{-F(x)} + \mathbb{P}(Y=-1|x)e^{F(x)}$$

with respect to F and setting the derivative to zero. □

The usual logistic function does not have the factor $(1/2)$ but

$$\mathbb{P}(Y=1|x) = \frac{e^{2F(x)}}{1+e^{2F(x)}},$$

so minimizing J and modeling with a logit are related by a factor of 2.

Now the central result from Friedman et al. (2000) is as follows; a version of this is in Wyner (2003), and a related optimization is in Zhou et al. (2005).

Theorem (Friedman et al., 2000): The boosting algorithm fits an additive logistic regression model by using adaptive Newton updates for minimizing $J(F)$.

Proof: Suppose $F(\boldsymbol{x})$ is available and the task is to improve it by choosing a c and an f and forming $F(\boldsymbol{x})+cf(\boldsymbol{x})$. For fixed c and $\boldsymbol{x}$, write a second-order expansion about $f(\boldsymbol{x}) = 0$ as

$$\begin{aligned} J(F(x)+cf(x)) &= \mathbb{E}(e^{-y(F(x)+cf(x))}) \\ &\approx \mathbb{E}(e^{-yF(x)}(1-ycf(x)+c^2y^2f(x)^2/2)) \\ &= \mathbb{E}(e^{-yF(x)}(1-ycf(x)+c^2/2)); \end{aligned} \tag{6.1.8}$$

the last inequality follows by setting $|y| = |f| = 1$. (The contradiction between $|f| = 1$ and $f(\boldsymbol{x}) = 0$ is resolved by noting that in (6.1.8) the role of f is purely as a dummy variable for a Taylor expansion on a set with high probability.)

Now, setting $W(\boldsymbol{x},y) = e^{yF(\boldsymbol{x})}$ and incorporating it into the density with respect to which $\mathbb{E}$ is defined gives a new expectation operator, $\mathbb{E}_W$. Let $\mathbb{E}_W(\cdot|\boldsymbol{x})$ be the conditional expectation given $\boldsymbol{x}$ from $\mathbb{E}_W$ so that for any $g(\boldsymbol{x},y)$, the conditional expectation can be written as

$$\mathbb{E}_W(g(\boldsymbol{X},Y)|\boldsymbol{X}=\boldsymbol{x}) = \frac{\mathbb{E}(W(\boldsymbol{X},Y)g(\boldsymbol{X},Y)|\boldsymbol{X}=\boldsymbol{x})}{\mathbb{E}(W(\boldsymbol{X},Y)|\boldsymbol{X}=\boldsymbol{x})}.$$

Thus, the posterior risk minimizing action over $f(\boldsymbol{x}) \in \{1,-1\}$, pointwise in $\boldsymbol{x}$, is

$$\hat{f} = \arg\min_f \; c\,\mathbb{E}_W((1+c^2/2)/c - Yf(\boldsymbol{X})|\boldsymbol{x}). \tag{6.1.9}$$

Intuitively, when $c > 0$, the minimum is achieved when $Yf(\boldsymbol{X})$ is large. So, although pointwise in x, (6.1.9) is equivalent to maximizing

$$\mathbb{E}_W(Yf(\boldsymbol{X}) = -\mathbb{E}_W(Y - f(\boldsymbol{X}))^2/2 + 1,$$

on average, using $f^2(\boldsymbol{X}) = Y^2 = 1$. It can be seen that the solution is

$$f(\boldsymbol{x}) = \begin{cases} 1 & \text{if } \mathbb{E}_W(Y|\boldsymbol{x}) = \mathbb{P}_W(Y = 1|\boldsymbol{x}) - \mathbb{P}_W(Y = -1|\boldsymbol{x}) > 0 \\ -1 & \text{else.} \end{cases}$$

Thus, minimizing a quadratic approximation to the criterion gives a weighted least squares choice for $f(x) \in \{1, -1\}$. (This defines the Newton step.)

Next, it is necessary to determine c. Given $f(x) \in \{1, -1\}$, $J(F + cf)$ can be minimized to find c,

$$\hat{c} = \arg \min_c \mathbb{E}_W e^{-cyf(x)} = \frac{1}{2} \log \frac{1-\varepsilon}{\varepsilon},$$

in which ε is the misclassification probability now under W (i.e., $\varepsilon = \mathbb{E}_W 1_{\{y \neq \hat{f}(x)\}}$) rather than under D_t as in the procedure.

Combining these pieces, it is seen that $F(x)$ is updated to $F(x) + (1/2)\log[(1 - \varepsilon)/\varepsilon]\hat{f}(x)$ and that the updating term $\hat{c}\hat{f}(x)$ updates $W_{old}(x,y) = e^{-yF(x)}$ to

$$W_{new}(x,y) = W_{old}(x,y)e^{-\hat{c}\hat{f}(x)y}.$$

Equivalently, since $y\hat{f}(x) = 2 1_{\{y \neq \hat{f}(x)\}} - 1$, the updated W_{old} can be written

$$W_{new}(x,y) = W_{old}(x,y)e^{\log[(1-\varepsilon)/\varepsilon] 1_{\{y \neq \hat{f}(x)\}}}.$$

These function updates and weight updates are the same as given in the boosting algorithm when $\hat{c} = \alpha_m$, $D_t = W_{old}$, and $D_{t+1} = W_{new}$. □

It is clear that boosting is not the same as SVMs. However, there is a sense in which boosting can be regarded as a maximum margin procedure. Let

$$M_\alpha(\boldsymbol{x}, y) = \frac{y \sum_{t=1}^T \alpha_t h_t(\boldsymbol{x})}{\sum_{t=1}^T \alpha_t}. \tag{6.1.10}$$

Freund and Schapire (1999) observe that M_α is in $[-1, 1]$ and is positive if and only if h_T correctly classifies $(\boldsymbol{x}, y)$. The function M_α can be regarded as the margin of the classifier since its distance from zero represents the strength with which the sign is believed to be a good classifier. Clearly, a good classifier has a large margin.

In parallel with the vector $\alpha = (\alpha_1, ..., \alpha_T)$, write $h(x) = (h_1(x), ..., h_T(x))$. Now, maximizing the minimum margin means seeking the h that achieves

$$\max_\alpha \min_{i=1,\ldots,n} M(\boldsymbol{x}_i, y_i) = \max_\alpha \min_{i=1,\ldots,n} \frac{(\alpha \cdot h(\boldsymbol{x}_i))y_i}{||\alpha||_1 ||h(x_i)||_\infty} \tag{6.1.11}$$

since $||\alpha||_1 = \sum_{t=1}^T |\alpha_t|$ and, when $h_t \in \{1, -1\}$, $||h(\boldsymbol{x})||_\infty = \max_t |h_t(\boldsymbol{x})| = 1$.

By contrast, SVMs rely on squared error. The goal of SVMs is to maximize a minimal margin of the same form as (6.1.11) using

$$||\alpha||_2 = \sqrt{\sum_{t=1}^{T} \alpha_t^2} \text{ and } ||h(\boldsymbol{x})||_2 = \sqrt{\sum_{t=1}^{T} h_t(\boldsymbol{x})^2} \tag{6.1.12}$$

in place of the L_1 and L_∞ norms in the denominator of (6.1.11). Note that in both cases, (6.1.11) and (6.1.12), the norms in the denominator of the optimality criteria are dual because L^p spaces are dual to L^q spaces in the sense that they define each other's weak topology. Moreover, in both cases, the quantity being optimized is bounded by one because of the Cauchy-Schwartz inequality. Indeed, it would be natural to ask how the solution to an optimization like

$$\max_{\alpha} \min_{i=1,\ldots,n} \frac{(\alpha \cdot h(\boldsymbol{x}_i))y_i}{||\alpha||_p ||h(\boldsymbol{x}_i)||_q}$$

would perform.

To finish this short exposition on boosting, two results on training and generalization error, from Schapire et al. (1998), are important to state. First, let

$$\phi_\theta(z) = \begin{cases} 1 & \text{if } z \leq 0, \\ 1 - z/\theta & \text{if } 0 < z \leq \theta, \\ 0 & \text{if } z \geq \theta, \end{cases}$$

for $\theta \in [0, 1/2]$. It is seen that ϕ_θ is continuous for $\theta \neq 0$ and that, as θ shrinks to zero, the range of z that gives $\phi_\theta \neq 0, 1$ also shrinks to the right.

Next, call the function

$$\rho(f) = yf(\boldsymbol{x})$$

the margin of f as a classifier. Then, for any function f taking values in $[-1, 1]$, its empirical margin error is

$$\hat{L}^\theta(f) = (1/n) \sum_{i=1}^{n} \phi_\theta(y_i f(x_i)),$$

in which taking $\theta = 0$ gives the usual misclassification error and $\theta_1 \leq \theta_2$ implies $\hat{L}^{\theta_1}(f) \leq \hat{L}^{\theta_2}(f)$. The empirical margin error for zero-one loss is

$$\tilde{L}^\theta(f) = (1/n) \sum_{i=1}^{n} 1_{y_i f(x_i) \leq \theta},$$

and, since $\phi_\theta(yf(x)) \leq 1_{yf(x) \leq \theta}$, it follows that $\hat{L}^\theta(f) \leq \tilde{L}^\theta(f)$. So, to bound $\hat{L}^\theta$, it is enough to bound $\tilde{L}^\theta$.

Let $\varepsilon(h, D)$ be the empirical misclassification error of h under D as before,

$$\varepsilon_t(h_t, D_t) = \sum_{i=1}^{n} D_t(i) 1_{y_i \neq h_t(x_i)},$$

with the same normalization constant C_t. The training error can be bounded as in the following.

Theorem (Schapire et al., 1998): Assume that, at each iteration t in the boosting algorithm, the empirical error satisfies $\varepsilon(h_t, D_t) \leq (1/2)(1-\gamma_t)$. Then the empirical margin error for h_T satisfies

$$\hat{L}^{\theta}(f_T) \leq \Pi_{t=1}^{T}(1-\gamma_t)^{(1-\theta)/2}(1+\gamma_t)^{(1+\theta)/2},$$

where f_T is the final output from the boosting algorithm.

Proof: As given in Meir and Ratsch (2003), there are two steps to the proof. The first step obtains a bound; the second step uses it to get the result. Recall $h_t \in \{1, -1\}$.

Step 1: Start by showing

$$\hat{L}^{\theta}(f_T) \leq \exp\left(\theta \sum_{t=1}^{T} \alpha_t\right)\left(\Pi_{t=1}^{T} C_t\right) \tag{6.1.13}$$

for any sequence of α_ts.

It can be verified that $f_T = \sum_{t=1}^{T} \alpha_t h_t / (\sum \alpha_t)$, so the definition of f_T gives that

$$y f_T(x) \leq \theta \Rightarrow \exp\left(-y \sum_{t=1}^{T} \alpha_t h_t(x) + \theta \sum_{t=1}^{T} \alpha_t\right) \geq 1,$$

which implies

$$1_{Y f_T(X) \leq \theta} \leq \exp\left(-y \sum_{t=1}^{T} \alpha_t h_t(x) + \theta \sum_{t=1}^{T} \alpha_t\right). \tag{6.1.14}$$

Separate from this, the recursive definition of $D_t(i)$ can be applied to itself T times:

$$D_{T+1}(i) = \frac{D_T(i) e^{-\alpha_T y_i h_T(x_i)}}{C_T} = \ldots = \frac{e^{-\sum_{t=1}^{T} \alpha_t y_i h_t(x_i)}}{n \Pi_{t=1}^{T} C_t}. \tag{6.1.15}$$

Now, using (6.1.14) and (6.1.15),

$$\tilde{L}^\theta(f) = \frac{1}{n}\sum_{i=1}^n 1_{y_i f_t(x_i)\le\theta} \le \frac{1}{n}\sum_{i=1}^n \left[\exp\left(-y_i\sum_{t=1}^T \alpha_t h_t(x_i) + \theta\sum_{t=1}^T \alpha_t\right)\right]$$
$$= \frac{1}{n}\exp\left(\sum_{t=1}^T \alpha_t\right)\sum_{i=1}^n \exp\left(-y_i\sum_{t=1}^T \alpha_t h_t(x_i)\right)$$
$$= \exp\left(\sum_{t=1}^T \alpha_t\right)(\Pi_{t=1}^T C_t)\sum_{i=1}^n D_{T+1}(i)$$
$$= \exp\left(\sum_{t=1}^T \alpha_t\right)(\Pi_{t=1}^T C_t), \quad (6.1.16)$$

which gives (6.1.13).

Step 2: Now the theorem can be proved. By definition, the normalizing constant is

$$C_t = \sum_{i=1}^n D_t(i)e^{-y_i\alpha_t h_t(x_i)}$$
$$= e^{-\alpha_t}\sum_{i:y_i=h_t(x_i)} D_t(i) + e^{\alpha_t}\sum_{i:y_i\ne h_t(x_i)} D_t(i)$$
$$= (1-\varepsilon_t)e^{-\alpha_t} + \varepsilon_t e^{\alpha_t}. \quad (6.1.17)$$

As before, set $\alpha_t = (1/2)\log((1-\varepsilon_t)/\varepsilon_t)$ to see that

$$C_t = 2\sqrt{\varepsilon_t(1-\varepsilon_t)}.$$

Using this in (6.1.13) gives

$$\tilde{L}^\theta(f) \le \Pi_{t=1}^T \sqrt{4\varepsilon_t^{1-\theta}(1-\varepsilon_t)^{1+\theta}},$$

which, combined with $\varepsilon_t = (1/2)(1-\gamma_t)$ and $\hat{L}^\theta(f) \le \tilde{L}^\theta(f)$, gives the theorem. □

To see the usefulness of this result, set $\theta = 0$ and note the training error bound

$$\hat{L}(f_T) \le e^{-\sum_{t=1}^T \gamma_t^2/2}.$$

So, it is seen that $\sum_{t=1}^T \gamma_t^2 \to \infty$ ensures $\hat{L}(f_T) \to 0$. In fact, if $\gamma_t \ge \gamma_0 > 0$ then for $\theta \le \gamma_0/2$ it follows that $\hat{L}^\theta(f_T) \to 0$.

Next consider generalization error. For the present purposes, it is enough to state a result. Recall that the concept of VC dimension, VCdim($\mathscr{F}$), is a property of a collection $\mathscr{F}$ of functions giving the maximal number of points that can be separated by elements in $\mathscr{F}$ in all possible ways. A special case, for classification loss, of a more general theorem for a large class of loss functions is the following.

Theorem (Vapnik and Chervonenkis, 1971; quoted from Meir and Ratsch, 2003): Let $\mathscr{F}$ be a class of functions on a set χ taking values in $\{-1,1\}$. Let $\mathbb{P}$ be a probability on $\chi \times \{-1,1\}$, and suppose the n data points (x_i, y_i), $i = 1, \ldots, n$ are IID $\mathbb{P}$ and give empirical classification error $\hat{\mathbb{P}}(Y \ne f(X))$. Then there is a constant C such that $\forall n$, with probability at least $1-\delta$, all sets of data of size n, and $\forall f \in \mathscr{F}$,

$$\mathbb{P}(Y \neq f(X)) \leq \hat{\mathbb{P}}(Y \neq f(X)) + C\sqrt{\frac{\text{VCdim}(\mathscr{F}) + \log(1/\delta)}{n}}. \quad \square$$

Finally, we give some background to boosting. Boosting was originally intended for weak learners, and the paradigm weak learner was a stump – a single-node tree classifier that just partitioned the feature space. A stump can have a small error rate if it corresponds to a good model split, but usually it does not, so improving it in one way or another is often a good idea. Curiously, comparisons between stumps and boosted stumps have not revealed typical situations in which one can be certain how boosting will affect the base learner.

That being admitted, there are some typical regularities exhibited by boosting. Like bagging, boosting tends to improve good but unstable classifiers by reducing their variances. This contrasts with stacking and BMA, which often give improvement primarily by overcoming model misspecification to reduce bias. (There are cases where boosting seems to give improvements primarily in bias (see Schapire et al. (1998), but it is not clear that this is typical).

Friedman et al. (2000), as already seen, attribute the improved performance from boosting to its effective search using forward stagewise additive modeling. The role of T in this is problematic: Even as T increases, there is little evidence of overfitting, which should occur if stagewise modeling is the reason for the improvement. There is evidence that boosting can overfit, but its resistance to overfitting is reminiscent of random forests, so there may be a result like Breiman's theorem waiting to be proved. An issue that does not appear to have been studied is that the summands in the boosted classifier are dependent.

There is some evidence that neither bagging nor boosting helps much when the classifier is already fairly good – stable with a low misclassification error. This may be so because the classifier is already nearly optimal, as in some LDA cases.

There is even some evidence that boosting, like bagging, can make a classifier worse. This is more typical when the sample size is too small: There is so much variability due to lack of data that no averaging method can help much. Stronger model assumptions may be needed.

There are comparisons of boosting and bagging: however, in generality, these two methods are intended for different scenarios and don't lend themselves readily to comparisons. For instance, stumps are a weak classifier, often stable, but with high bias. In this case, the benefits of boosting might be limited since the class is narrow, but bagging might perform rather well. Larger trees would be amenable to variance reduction but would have less bias and so might be more amenable to boosting. In general, it is unclear when to expect improvements or why they occur.

Finally, the methods here are diverse and invite freewheeling applications. One could bag a boosted classifier or boost a bagged classifier. One is tempted to stack classifiers of diverse forms, say trees, nets, SVMs, and nearest neighbors and then boost the bagged version, or take a BMA of stacked NNs and SVMs and then boost the result.

The orgy of possibilities can be quite exhausting. Overall, it seems that, to get improvement from ensemble methods, one needs to choose carefully which regression or classification techniques to employ and how to employ them. This amounts to an extra layer of variability to be analyzed.

6.1.5 Other Averaging Methods

To complete the overview of averaging strategies that have been developed, it is worth listing a few of the other ensemble methods not dealt with here and providing a brief description of the overall class of ensemble predictors.

Juditsky and Nemirovskiii (2000) developed what they call functional aggregation. Choose $f_1, \ldots, f_K$ models and find the linear combination of f_ks by α_ks achieving

$$\min_{\alpha_j} \int \left(f(\boldsymbol{x}) - \sum_{k=1}^{K} \alpha_j f_k(\boldsymbol{x}) \right)^2 d\mu(\boldsymbol{x}),$$

in which α_j ranges over a set in the L^1 unit ball. This is a variation of what stacking tries to do. The main task is to estimate the α_ks and control the error. More generally, one can take $f_k(\boldsymbol{x}) = f_k(\boldsymbol{x}|\theta)$, where the θ indexes a parameterized collection of functions such as polynomial regression, NNs or trees. In addition, distinct nonparametric estimators can be combined. Unless K is large, this may be unrealistic as p increases.

Lee et al. (1996) have an optimality criterion like Juditsky and Nemirovskiii (2000) but derived from information theory. They call their technique agnostic learning and their setting is more general: They only assume a joint probability for $(\boldsymbol{X}, Y)$ and seek to approximate the probabilistic relationship between $\boldsymbol{X}$ and Y within a large class of functions. This technique is intended for NNs and they establish a bound on performance. It is primarily in the technique of proof that they introduce averages of models. Of course, one can regard NN as a model average as well: The input functions to the terminal node are combined with weights and a sigmoid. An information-theoretic way to combine models is called data fusion by Luo and Tsitsiklis (1994). Functions from different sources in a communications network are combined to get one message.

Jones (1992), Jones (2000) used greedy approximation, with a view to PPR and NNs, to approximate a function in an L^2 space by a function in a subspace of L^2 by evaluating it at a linear combination of the variables. The best linear combination at each iteration of the fit produces a residual to which the procedure is applied again. The resulting sum of functions evaluated at linear combinations of explanatory variables converges to the true function in L^2 norm at a rate involving n.

Apart from the plethora of ensemble methods primarily for regression, Dietterich (1999), Section 2, reviews ensemble methods in the context of classification. Given an ensemble and a data set, there are basically five distinct ways to vary the inputs to generate ensemble-based predictors. In a sense, this provides a template for generating models to combine. First, one can reuse the data, possibly just subsets of it, to generate more predictors. This works well with unstable but good predictors such as trees or NNs. Stable predictors such as linear regression or nearest neighbors do not seem to benefit from this very much. Bagging and boosting are the key techniques for this.

Second, one can manipulate the explanatory variables (or functions) for inclusion in a predictor. So, for instance, one can form neural nets from various subsets of the input variables using all the values of the chosen inputs. This may work well when the input explanatory variables duplicate each other's information. On the other hand, a third technique is the opposite of this: One can partition the available outputs differently. For instance, in a classification context, one can merge classes and ask for good future classification on the reduced set of classes.

Fourth, one can at various stages introduce randomness. For instance, many neural network or tree based methods do a random selection of starting points or a random search that can be modified.

Finally, and possibly most interesting, there are various weighting schemes one can use for the regression functions or classifiers generated from the ensemble. That is, the way one combines the predictors one has found can be varied. The Bayes approach chooses these weights using the posterior, and stacking uses a cross-validation criterion; many others are possible. In this context, Friedman and Popescu (2005) use an optimality criterion to combine "base learners", a generic term for either regression functions or classifiers assumed to be oversimple. Given a set of base learners $f_k(\boldsymbol{x})$, for $k = 1, \ldots, K$, and a loss function ℓ, they propose a LASSO type penalty for regularized regression. That is, they suggest weights

$$\{\alpha_k \mid k = 1, .., K\} = \arg\min_{\{\alpha_k\}} \left[\sum_{i=1}^{n} \ell \left(y_i, \alpha_0 + \sum_{k=1}^{K} \alpha_k f_k(\boldsymbol{x}_i) \right) + \lambda \sum_{k=1}^{K} |\alpha_j| \right],$$

where λ is a trade-off between fit and complexity permitting both selection and shrinkage at the same time. Obviously, other penalty terms can be used.

The kind of combination is intriguing because it's as if the list of base learners is a sort of data in its own right: The f_ks are treated as having been drawn from a population of possible models. So, there is uncertainty in the selection of the set of base learners to be combined as well as in the selection of the base learner from the ensemble. In a model selection problem, this is like including the uncertainty associated with the selection of the list of models from which one is going to select in addition to the selection of an individual model conditional on the list. Adaptive schemes that reselect the list of models from which one will form a predictor amount to using the data to choose the model list as well as select from it. From this perspective, it is not surprising that ensemble methods usually beat out selection methods: Ensemble methods include uncertainty relative to the ensemble as a surrogate for broader model uncertainty.

6.1.6 Oracle Inequalities

From a predictive standpoint, one of the reasons to use an averaging method is that the averaging tends to increase the range of predictors that can be constructed. This means that the effect of averaging is to search a larger collection of predictors than using any one of the models in the average would permit. The effectiveness of this search remains to be evaluated. After all, while it is one thing to use a model average and assign an SE to it by bootstrapping or assign a cumulative risk to a sequence of its predictions, it is another thing entirely to ask if the enlarged collection of predictors actually succeeds in finding a better predictor than just using a model on its own.

Aside from BMA, which optimizes a squared error criterion, it is difficult to demonstrate that a model average is optimal in any predictive sense. Indeed, even for BMA, it is not in general clear that the averaging will be effective: Dilution may occur, and BMA can be nonrobust in the sense of underperforming when the true model is not close enough to the support of the prior.

One way to verify that a class of inference procedures is optimal is to establish an Oracle inequality; as the term oracle suggests, these are most important in predictive contexts. The basic idea of an Oracle inequality is to compare the risk of a given procedure to the risk of an ideal procedure that permits the same inference but uses extra information that would not be available in practice – except to an all-knowing Oracle. Since Oracle inequalities are important in many contexts beyond model averaging, it is worth discussing them in general first.

The simplest case for an Oracle inequality is parametric inference. Suppose the goal is to estimate an unknown θ using n observations and the class of estimators available is of the form $\{\hat{\theta}(t) | \, t \in \mathscr{T}\}$, where $\mathscr{T}$ is some set. Then, within $\mathscr{T}$ there may be an optimal value t_{opt} such that

$$R(t_{opt}, n, \theta) = \min_{t \in \mathscr{T}} \mathbb{E}_\theta \|\hat{\theta}_t - \theta\|^2,$$

where R is the risk from the squared error loss. The value t_{opt} is unknown to us, but an Oracle would know it. So, the Oracle's risk is

$$R(\text{Oracle}, n, \theta) = R(t_{opt}, n, \theta), \tag{6.1.18}$$

and the question is how close a procedure that must estimate t_{opt} can get to the oracle risk $R(\text{Oracle}, n, \theta)$. The paradigm case is that $\hat{\theta}$ is a smoother of some sort, say spline or kernel, and t is the smoothing parameter h or λ. Another case occurs in wavelet thresholding: Any number of terms in a wavelet expansion might be included; however, an Oracle would know which ones were important and which ones to ignore. The thresholding mimics the knowledge of an Oracle in choosing which terms to include. Note that t may be continuous or not; usually, in model averaging settings $\mathscr{T}$, is finite.

An Oracle inequality is a bound on the risk of a procedure that estimates t in terms of a factor times (6.1.18), with a bit of slippage. A ideal form of an Oracle inequality for an estimator $\hat{\theta}$ relative to the class $\theta(t)$ of estimators is a statement such as

$$\mathbb{E}_\theta \|\hat{\theta} - \theta\|^2 \leq K_n \left[R(\text{Oracle}, n, \theta) + \frac{1}{n} \right]. \tag{6.1.19}$$

That is, up to a bounded coefficient K_n, the estimator $\hat{\theta}$ is behaving as if it were within $1/n$ slippage of the risk of an Oracle. The term $1/n$ comes from the fact that the variance typically decreases at rate $\mathscr{O}(1/n)$. Variants on (6.1.19) for different loss functions can be readily defined.

To return to the function estimation setting, often the form of (6.1.19) is not achievable if θ is replaced by f. However, a form similar to (6.1.19) can often be established by using the differences in risks rather than the risks themselves. That is, the bound is on how much extra risk a procedure incurs over a procedure that uses extra information and so has the least possible risk. Parallel to (6.1.19), the generic form of this kind of Oracle inequality is

$$\chi_{S_n}\left(R(\hat{f}) - R(f_{opt})\right) \leq \chi_{S_n} K_n \left(R(f_{true}) - R(f_{opt}) + W_n\right),$$

where S_n is the set where the empirical process defined from the empirical risk converges at a suitable rate, $\hat{f}$ estimates f_{true} or more exactly f_{opt}, the element of the space closest to f_{true}, usually by minimizing some form of regularized empirical risk, and W_n is a term representing some aspect of estimation error as opposed to approximation error) (see Van de Geer (2007)), which might involve knowing the ideal value of a tuning parameter, for instance.

Oracle inequalities have been established for model averages in both the regression and classification settings. One important result for each case is presented below; there are many others. The coefficients used to form the averages in the theorems are not obviously recognizable as any of the model averages discussed so far. However, these model averages are implementable, may be close to one of the model averages discussed, and are theoretically well motivated. At a minimum, they suggest that model averages are typically going to give good results.

6.1.6.1 Adaptive Regression and Model Averaging

Yang's Oracle inequality constructs a collection of model averages and then averages them again to produce a final model average that can be analyzed. This means that it is actually an average of averages that satisfies the Oracle inequality. It is the outer averaging that is "unwrapped" so that whichever of the inner averages is optimal ends up getting the most weight and providing the upper bound.

The setting for Yang's Oracle inequality is the usual signal-plus-noise model, $Y = f(\boldsymbol{X}) + \sigma(\boldsymbol{x})\varepsilon$, in which the distribution of the error term ε has a mean zero density h. Let $\mathscr{E} = \{\delta_j\}$ be an ensemble of regression procedures for use in the model, in which, using data $Z_i = \{(\boldsymbol{X}, Y_k) : k = 1, ..., i\}$, δ_j gives an estimator $\hat{f}_{j,i}$ of $f(\boldsymbol{x})$. That is, δ_j gives a collection of estimators depending on the input sequence of the data. The index set for j may be finite or countably infinite. Now, the risk of δ_j for estimating f from i data points is

$$R(\hat{f}_{j,i}, i, f) = R(\delta_j, i, f) = \mathbb{E}\|f - \hat{f}_{j,i}\|^2$$

under squared error loss. If $i = n$, this simplifies to $R(\delta_j, n, f) = \mathbb{E}\|f - \hat{f}_j\|^2$, in which $\hat{f}_{j,n} = \hat{f}_j$.

The idea behind Yang's procedure is to assign higher weights to those elements of the ensemble that have residuals closer to zero. This is done by evaluating h at the residuals because h is typically unimodal with a strong mode at zero: Residuals with a smaller absolute value contribute more to the mass assigned to the model that generated them than residuals with larger absolute values. Moreover, it will be seen that the average depends on the order of the data points, with later data points (higher index values) having a higher influence on the average than earlier data points (lower index values).

To specify the average, let $N = N_n$ with $1 \le N_n \le n$. It is easiest to think of n as even and $N = n/2 + 1$ since then the lower bounds in the products giving the weights start at $n/2$. Let the initial weights for the procedures δ_j be $W_{j,n-N+1} = \pi_j$, where the π_js sum to 1. Now consider products over subsets of the data ranging from $n - N + 1$ up to $i - 1$ for each i between $n - N + 2$ and n. These give the weights

$$W_{i,j} = \frac{\pi_j \Pi_{\ell=n-N+1}^{i-1} h(y_{\ell+1} - \hat{f}_{j,\ell}(\boldsymbol{x}_{\ell+1})/\hat{\sigma}_{j,\ell}(\boldsymbol{x}_{\ell+1}))}{\sum_{j=1}^{\infty} \pi_j \Pi_{\ell=n-N+1}^{i-1} h(y_{\ell+1} - \hat{f}_{j,\ell}(\boldsymbol{x}_{\ell+1})/\hat{\sigma}_{j,\ell}(\boldsymbol{x}_{\ell+1}))}. \tag{6.1.20}$$

In (6.1.20), the $\hat{\sigma}_{j,\ell}$ are the estimates of σ based on the indicated data points. It is seen that $\sum_j W_{j,i} = 1$ for each $i = n - N + 1, \ldots, n$. Now, the inner averages, over j, are formed from the $\hat{f}_{j,i}$s for the procedures δ_j for fixed is. These are

$$\tilde{f}_i = \sum_j W_{j,i} \hat{f}_{j,i}(\boldsymbol{x}). \tag{6.1.21}$$

These are aggregated again, this time over sample sizes, to give the outer average

$$\bar{f}_n(\boldsymbol{x}) = \frac{1}{N} \sum_{i=n-N+1}^{n} \tilde{f}_i(\boldsymbol{x}) \tag{6.1.22}$$

as the final model average.

Clearly, (6.1.22) depends on the order of the data, and data points earlier in the sequence are used more than those later in the sequence. Under the IID assumption, the estimator can be symmetrized by taking the conditional expectation given the data but ignoring order. In applications, the order can be permuted randomly several times and a further average taken to approximate this.

The main hypotheses of the next theorem control the terms in the model and the procedures δ_j as follows. (i) Suppose $|f| \le A < \infty$, the variance function satisfies $0 < \underline{\sigma} \le \sigma(\boldsymbol{x}) \le \overline{\sigma} < \infty$, and the estimators from each δ_j also satisfy these bounds, and (ii) for each pair $s_0 \in (0,1)$ and $T > 0$, there is a constant $B = B(s_0, T)$ such that the error density satisfies

$$\int h(u) \log \frac{h(u)}{(1/s)h((u-t)/s)} du < B((1-s)^2 + t^2), \tag{6.1.23}$$

for $s \in (s_0, 1/s_0)$ and $t \in (-T, T)$. These assumptions permit the usual special cases: (B) is satisfied by the normal, Student's t with degrees of freedom at least 3, and the double exponential, among other errors. Also, the values of the constants in (A) are not actually needed to use the procedure; their existence is enough.

Theorem (Yang, 2001): Use $\mathscr{E}$ to construct $\bar{f} = \bar{f}_n$ as in (6.1.22), and suppose that (i) and (ii) are satisfied. Then

$$R(\bar{f},\ n\ , f) \tag{6.1.24}$$
$$\leq C_1 \inf_j \left(\frac{1}{N} \log \frac{1}{\pi_j} + \frac{C_2}{N} \sum_{\ell=n-N+1}^{n} \left(\mathbb{E}\|\sigma^2 - \hat{\sigma}_{j,\ell}\|^2 + \mathbb{E}\|f - \hat{f}_{j,\ell}\|^2 \right) \right)$$

where C_1 depends on A and $\overline{\sigma}$ and C_2 depends on A, $\overline{\sigma}/\underline{\sigma}$ and h. The inequality (6.1.24) also holds for the average risks:

$$\frac{1}{N} \sum_{i=n-N+1}^{n} \mathbb{E}\|f - \tilde{f}_i\|^2 \leq C_1 \inf_j \left(\frac{1}{N} \log \frac{1}{\pi_j} \right. \tag{6.1.25}$$
$$\left. + \frac{C_2}{N} \sum_{\ell=n-N+1}^{n} \left(\mathbb{E}\|\sigma^2 - \hat{\sigma}_{j,\ell}\|^2 + \mathbb{E}\|f - \hat{f}_{j,\ell}\|^2 \right) \right).$$

Proof: See Subsection 6.5.1 of the Notes at the end of the chapter. □

6.1.6.2 Model Averaging for Classification

In binary classification, the pairs $(\boldsymbol{X}, Y)$ have joint distribution $\mathbb{P}$ and are assumed to take values in a set $\mathscr{X} \times \{-1, 1\}$. The marginal for $\boldsymbol{X}$ is also denoted $\mathbb{P} = \mathbb{P}_X$ as long as no confusion will result. Under squared error loss, the conditional probability $\eta(\boldsymbol{x}) = E(1_{Y=1}|\boldsymbol{X} = \boldsymbol{x})$ gives a classifier f that is zero or one according to whether $\boldsymbol{x}$ is believed to be from class 1 or class 2. The misclassification rate of f is $R(f) = \mathbb{P}(Y \neq f(\boldsymbol{X}))$, suggestively written as a risk (which it is under 0-1 loss). It is well known that the Bayes rule is

$$\min_f R(f) = R(f^*) \equiv R^*, \ \text{ i.e., } \ f^* = \arg\min_f R(f)$$

where f varies over all measurable functions and $f^*(\boldsymbol{x}) = \text{sign}(2\eta(\boldsymbol{x}) - 1)$.

Given n IID data points $D = \{(\boldsymbol{X}_i, Y_i)_{i=1,\ldots,n}\}$, let $\hat{f}(\boldsymbol{x}) = \hat{f}_n(\boldsymbol{x})$ estimate the Bayes rule classifier. Without loss of generality, assume $\hat{f}$ only takes values ± 1. Then, the generalization error of $\hat{f}$ is $\mathbb{E}(R(\hat{f}))$, where $R(\hat{f}) = \mathbb{P}(Y \neq \hat{f}(\boldsymbol{X})|D_n)$. The excess risk of $\hat{f}$ is the amount by which its risk exceeds the minimal Bayes risk. That is, the excess risk of $\hat{f}$ is $\mathbb{E}(R(\hat{f}) - R^*)$.

The setting for model averaging in classification supposes K classifiers are available, say $\mathscr{F} = \{f_1, \ldots, f_K\}$. The task is to find an $\hat{f}$ that mimics the best f_k in $\mathscr{F}$ in terms of having excess risk bounded by the smallest excess risk over the f_ks. Here, a theorem of LeCué (2006) will be shown. It gives an Oracle inequality in a hinge risk sense for a classifier obtained by averaging the elements of $\mathscr{F}$.

Since Oracle inequalities rest on a concept of risk, it is no surprise that constructing a classifier based on risk makes it easier to prove them. The two losses commonly used in classification are the zero-one loss, sometimes just called the misclassification loss, and the hinge loss $\phi(x) = \max(0, 1-x)$ seen in the context of support vector machines. Among the many risk-based ways to construct a classifier, empirical risk minimization using zero-one loss gives a classifier by minimizing $R_n(f) = (1/n)\sum_{i=1}^n 1_{y_i f(\boldsymbol{x}_i)}$ as a way to minimize the population risk R. This kind of procedure has numerous good theoretical properties. By contrast, the (population) hinge risk is, say, $A(f) = \mathbb{E}\max(0, 1 - Yf(\boldsymbol{x}))$ for any f. The optimal hinge risk is $A^* = \inf_f A(f)$, and the corresponding Bayes rule f^* achieves A^*.

The link between empirical risk minimization under 0-1 loss and the hinge risk is

$$R(f) - R^* \leq A(f) - A^*, \tag{6.1.26}$$

a fact credited to Zhang (2004), where $R(f)$ is understood to be the misclassification error of sign(f). Consequently, if minimizing hinge loss is easier than using the misclassification error directly, it may be enough to provide satisfactory bounds on the right-hand side of (6.1.26).

Accordingly, LeCué (2006) establishes an Oracle inequality under hinge loss for the average of a finite collection of classifiers under a low-noise assumption. This is a property of the joint distribution $\mathbb{P}$ of $(\boldsymbol{X}, Y)$ and depends on a value $\kappa \in [1, \infty)$. Specifically, $\mathbb{P}$ satisfies the low-noise assumption $MA(\kappa)$ if and only if there is a $K > 0$ so that, for all fs taking values ± 1,

$$E(|f(\boldsymbol{X}) - f^*(\boldsymbol{X})|) \leq C[R(f) - R^*]^{1/\kappa}. \tag{6.1.27}$$

The meaning of (6.1.27), also called a margin condition, stems from the following reasoning. Suppose f is an arbitrary function from which a classifier is derived by taking the sign of its value. Then,

$$\begin{aligned} R(f) - R^* &= \mathbb{P}(Y \neq f(\boldsymbol{X})) - \mathbb{P}(Y \neq \text{sign}(2\eta(\boldsymbol{X}) - 1)) \\ &\leq \mathbb{P}_{\boldsymbol{X}}(f(\boldsymbol{x}) \neq \text{sign}(\eta(\boldsymbol{x}) - 1/2)). \end{aligned} \tag{6.1.28}$$

Equality holds in (6.1.28) if η is identically 0 or 1. The left-hand side of (6.1.28) is the bracketed part of the right-hand side of (6.1.27), and the right hand side of (6.1.28) is the left hand side of (6.1.27). So, the $MA(\kappa)$ condition is an effort to reverse the inequality in (6.1.28). If $\kappa = \infty$, then the assumption is vacuous and for $\kappa = 1$ it holds if and only if $|2\eta(\boldsymbol{x}) - 1| \geq 1/C$. In other words, (6.1.27) means that the probability that f gives the wrong sign relative to f^* is bounded by a power of a difference of probabilities that characterizes how f differs from f^*.

To state Lecue's theorem for model-averaged classifiers – often called aggregating classifiers – let $\mathscr{F} = \{f_1, \ldots, f_K\}$ be a set of K classifiers, and consider trying to mimic the best of them in terms of excess risk under the low-noise assumption. First, a convex combination must be formed. So, following LeCué (2006), let

$$\tilde{f} = \sum_{k=1}^{K} w_k^n f_k, \quad \text{where } w_k^n = \frac{e^{\sum_{i=1}^n Y_i f_k(\mathbf{X}_i)}}{\sum_{k=1}^K e^{\sum_{i=1}^n Y_i f_k(\mathbf{X}_i)}}. \tag{6.1.29}$$

Since the f_ks take values ± 1, the exponential weights can be written as

$$w_k^n = \frac{e^{-nA_n(f_k)}}{\sum_{k=1}^K e^{-nA_n(f_k)}} \quad \text{where } A_n(f) = \frac{1}{n}\sum_{i=1}^n \max(0, 1 - Y_i f(\mathbf{X}_i)). \tag{6.1.30}$$

Clearly, A_n is the empirical hinge risk. Indeed, it can be verified that $A_n(f_k) = 2R_n(f_k)$ for $k = 1, \ldots, K$, so the weights w_k^n can be written in terms of the 0-1 loss-based risk.

A weak form of the Oracle inequality is the following.

Proposition (Lecue, 2006): Let $K \geq 2$, and suppose the f_ks are any $\mathbb{R}$-valued functions. Given n, the aggregated classifier $\tilde{f}$ defined in (6.1.29) and (6.1.30) satisfies

$$A_n(\tilde{f}) \leq \min_{k=1,\ldots,K} A_n(f_k) + \frac{\log M}{n}. \tag{6.1.31}$$

Proof: Since hinge loss is convex, $A_n(\tilde{f}) \leq \sum_{k=1}^K w_k^n A_n(f_k)$. Let

$$\hat{k} = \arg\min_{k=1,\ldots,K} A_n(f_k).$$

So, for all k,

$$A_n(f_k) = A_n(f_{\hat{k}}) + \frac{1}{n}[\log w_{\hat{k}}^n - \log w_k^n]$$

from the definition of the exponential weights. Averaging over the w_k^ns (for fixed n) gives (6.1.31). □

Now the proper Oracle inequality can be given.

Theorem (Lecue, 2006): Suppose $\mathscr{F}$ is a set of K classifiers with closed convex hull $\mathscr{C}$ and that $\mathbb{P}$ satisfies the $MA(\kappa)$ condition (6.1.27) for some $\kappa \geq 1$. Then, for any $a > 0$, the aggregate $\tilde{f}$ from (6.1.29) and (6.1.30) satisfies

$$E(A(\tilde{f}_n) - A^*) \leq (1+a)\min_{f \in \mathscr{C}}[A(f) - A^*] + C\left(\frac{\log M}{n}\right)^{\kappa/(2\kappa-1)}, \tag{6.1.32}$$

where $C = C(a) > 0$ is a constant.

Proof: See Subsection 6.5.2 of the Notes at the end of the chapter. □

6.2 Bayes Nonparametrics

Recall that the parametric Bayesian has a p-dimensional parameter space Ω and a prior density w on it. The model-averaging Bayesian generalizes the parametric Bayesian by using a class of models indexed by, say, $j \in J$, a prior w_j within each model on its parameter space Ω_j, and a prior across models (i.e., on J) to tie the structure together. The overall parameter space is (j, Ω_j) for $j \in J$. In turn, the pure nonparametric Bayesian generalizes the BMA Bayesian by working in the logical endpoint of the Bayesian setting. That is, the nonparametric Bayesian starts with a set $\mathscr{X}$ and considers $\mathscr{M}(\mathscr{X})$, the collection of all probability measures on $\mathscr{X}$, fixing some σ-field for $\mathscr{X}$, usually the Borel. Thus, $\mathscr{M}$, the collection of all reasonable probabilities on $\mathscr{X}$, is the set on which a prior must be assigned. If the explanatory variables are assumed random, then $\mathscr{M}(\mathscr{X})$ contains all the models discussed so far; however, $\mathscr{X} = \mathbb{R}$ is the most studied case because other bigger sets of probabilities remain intractable outside special cases. That is, $\mathscr{M}(\mathbb{R})$ is the collection of all probabilities on $\mathbb{R}$ and is the most studied.

The key Bayesian step is to define a prior on $\mathscr{M}(\mathscr{X})$. In fact, usually only a prior probability can be specified since densities do not generally exist. Starting with results that ensure distributions on $\mathscr{M}(\mathscr{X})$ exist and can be characterized by their marginals at finitely many points, Ghosh and Ramamoorthi (2003) provide a sophisticated treatment dealing with the formalities in detail. Doob (1953) has important background. Here, the technicalities are omitted for the sake of focusing on the main quantities.

There are roughly three main classes of priors that have received substantial study. The first two are the Dirichlet process prior and Polya tree priors. These are for $\mathscr{M}(\mathbb{R})$. The third, Gaussian processes, includes covariates by assigning probabilities to regression function values. In all three cases, the most important expressions are the ones that generate predictions for a new data point.

6.2.1 *Dirichlet Process Priors*

Recall that the Dirichlet distribution, D, has a k-dimensional parameter $(\alpha_1, \ldots, \alpha_k) \in \mathbb{R}^k$ in which $\alpha_j \geq 0$ and support $S_k = \{(p_1, \ldots, p_k) \mid 0 \leq p_j \leq 1, \Sigma_j p_j = 1\}$; i.e., the set of k dimensional probability vectors. The density of a Dirichlet distribution, denoted $D(\alpha_1, \ldots, \alpha_k)$, is

$$w(p_1, \ldots, p_k) = \frac{\Gamma(\Sigma_{j=1}^k \alpha_i)}{\Pi_{j=1}^k \Gamma(\alpha_j)} p_1^{\alpha_1 - 1} \ldots p_{k-1}^{\alpha_{k-1} - 1} \left(1 - \sum_{j=1}^{k-1} p_i\right)^{\alpha_k - 1}.$$

The idea behind the Dirichlet process prior is to assign Dirichlet probabilities to partition elements. Note that the "random variable" the Dirichlet distribution describes is the probability assigned to a set in a partition, not the set itself.

To use the Dirichlet distribution to define a prior distribution on the set of distributions $\mathscr{M}(\mathscr{X})$, start with a base probability, say α on $\mathscr{X}$, and an arbitrary partition $\mathscr{B} = \{B_1, ..., B_k\}$ of $\mathscr{X}$ of finitely many, say k, elements. Since α assigns probabilities $\alpha(B_j)$ to the B_js in the partition, these can be taken as the values of the parameter. So, set $(\alpha_1, ..., \alpha_k) = (\alpha(B_1), ..., \alpha(B_k))$ to obtain, for $\mathscr{B}$, a distribution on the values of the probabilities of the B_js. That is, the Dirichlet process prior assigns Dirichlet probabilities to the probability vector for $\mathscr{B}$ by using

$$(P(B_1), ..., P(B_k)) \sim D(\alpha(B_1), ..., \alpha(B_k)),$$

in which the P of a set is the random quantity. The Dirichlet process, DP, is a stochastic process in the sense that it assigns probabilities to the distribution function derived probabilities $F(t_1)$, $F(t_2) - F(t_1)$, ...,$F(t_n) - F(t_{n-1})$ for any partition.

The Dirichlet process has some nice features – consistency, conjugacy, and a sort of asymptotic normality of the posterior, for instance. Here, consistency means that the posterior concentrates on the true distribution. In addition, the measure α is the mean of the Dirichlet in the sense that $\mathbb{E}(P(A)) = \alpha(A)$ for any set A. The variance is much like a binomial: $\text{Var}(P(A)) = \alpha(A)(1-\alpha(A))/2$. Sometimes an extra factor γ, called a concentration, is used, so the Dirichlet is $D(\gamma\alpha(B_1), ..., \gamma\alpha(B_k))$. If so, then the variance changes to $\text{Var}(P(A)) = \alpha(A)(1-\alpha(A))/(1+\gamma)$, but the mean is unchanged. Here, $\gamma = 1$ for simplicity.

To see the conjugacy, let α be a distribution on Ω and let $P \sim DP(\alpha)$. Consider drawing IID samples according to P denoted θ_1,..., θ_n from Ω. To find the posterior for P given the θ_is, let B_1,..., B_k be a partition of Ω and let $n_j = \#(\{i \mid \theta_i \in B_j\})$ for $j = 1, ..., k$. Since the Dirichlet and the multinomial are conjugate,

$$(P(B_1), ..., P(B_k))|\theta_1, ..., \theta_n \sim D(\alpha(B_1)+n_1, ..., \alpha(B_k)+n_k).$$

Because the partition was arbitrary, the posterior for P must be Dirichlet, too. It can be verified that the posterior Dirichlet has concentration $n+1$ and base distribution $(\alpha + \sum_{i=1}^n \delta_{\theta_i})/(n+1)$, where δ_{θ_i} is a unit mass at θ_i so that $n_j = \sum_{i=1}^n \delta_{\theta_i}(B_j)$. That is,

$$P \mid \theta_1, ..., \theta_n \sim DP\left(n+1, \frac{1}{n+1}\alpha + \frac{n}{n+1}\frac{\sum_{i=1}^n \delta_{\theta_i}}{n}\right).$$

It is seen that the location is a convex combination of the base distribution and the empirical distribution. As n increases, the mass of the empirical increases, and if a concentration γ is used, the weight on the base increases if γ increases.

As noted, it is the predictive structure that is most important. So, it is important to derive an expression for $(\Theta_{n+1} \mid \theta_1, ..., \theta_n)$. To find this, consider drawing $P \sim DP(\alpha)$ and that $\theta_i \sim P$ IID for $i = 1, ..., n$. Since

$$\Theta_{n+1} \mid P, \theta_1, ..., \theta_n \sim P$$

for any set A,

$$\mathbb{P}(\Theta \in A \mid \theta_1, ..., \theta_n) = \mathbb{E}(P(A) \mid \theta_1, ..., \theta_n) = \frac{1}{n+1}\left(\alpha(A) + \sum_{i=1}^{n} \delta_{\theta_i}(A)\right).$$

Now, marginalizing out α gives

$$\Theta_{n+1} \mid \theta_1, ..., \theta_n \sim \left(\alpha + \sum_{i=1}^{n} \delta_{\theta_i}\right).$$

That is, the base distribution of the posterior given θ_1, ..., θ_n is the predictive.

An important property of the *DP* prior is that it concentrates on discrete probabilities. This may be appropriate in some applications. However, more typically it means the support of the *DP* prior is too small. For instance, if we have$\alpha \neq \alpha'$, then $D(\alpha)$ and $D(\alpha')$ are mutually singular and so give mutually singular posteriors. Thus, small differences in the base probability can give big differences in the posteriors. Since the support of the posterior is the same as the support of the prior, this means that the collection of posteriors $\{W_\alpha(\cdot|\boldsymbol{X}^n)\}$ as α varies is too small. Indeed, if α is continuously deformed to α', $W_\alpha(\cdot|\boldsymbol{X}^n)$ does not continuously deform to $W_\alpha(\cdot|\boldsymbol{X}^n)$. As a function of α, $W_\alpha(\cdot|\boldsymbol{X}^n)$ is discontinuous at each point!

Even so, Dirichlets are popular for Bayesian estimation of densities and in mixture models. Indeed, there are two other constructions for the *DP*, called stick-breaking and the Chinese restaurant process. Although beyond the present interest, these interpretations suggest the usefulness of *DP*s in many clustering contexts as well.

6.2.2 Polya Tree Priors

Polya trees are another technique for assigning prior probabilities to $\mathcal{M}(\mathcal{X})$. They are a variation on Dirichlet process priors in two senses. First, Polya trees use the $Beta(\alpha_1, \alpha_2)$ density, for $\alpha_1, \alpha_2 \geq 0$ which is a special case of the Dirichlet for $p = 2$. Second, Polya trees involve a sequence of partitions, each a refinement of its predecessor. This means that the partitions form a tree under union: If a given partition is a node, then each way to split a set in the partition into two subsets defines a more refined partition that is a leaf from the node. Formalizing this is primarily getting accustomed to the notation.

The Polya tree construction is as follows. Let E_j denote all finite strings of 0s and 1s of length j, and let $E^* = \cup_{j=1}^{\infty} E_j$ be all finite binary strings. For each k, let $\tau_k = \{B_\varepsilon | \varepsilon \in E_k\}$ be a partition of $\mathbb{R}$. This means τ_1 has two elements since the only elements of E_1 are $\varepsilon = 0, 1$. Likewise, τ_2 has four elements since E_2 has $\varepsilon = (\varepsilon_1, \varepsilon_2)$, in which each $\varepsilon_i = 0, 1$ and similarly for τ_3. The intuition is that the first element of ε is ε_1, taking values zero or one, indicating $\mathbb{R}^-$ and $\mathbb{R}^+$. The second element of ε is ε_2, again taking values zero or one. If $\varepsilon_1 = 0$ indicates $\mathbb{R}^-$, then ε_2 indicates one of two subintervals of $\mathbb{R}^-$ that must be chosen. The successive elements of ε indicate further binary divisions of one of the intervals at the previous stage. It is seen that the partitions

τ_k in the sequence must be compatible in the sense that, for any $\varepsilon \in E^*$, there must be a j such that $\varepsilon \in E_j$ and there must be a set $B_\varepsilon \in \tau_j$ whose refinements $B_{(\varepsilon,0)}$ and $B_{(\varepsilon,1)}$ form a partition of B_ε within $(\varepsilon,0),(\varepsilon,1) \in E_{j+1}$.

Now suppose a countable sequence of partitions τ_k is fixed. A distribution must be assigned to the probabilities of the sets. This rests on setting up an equivalence between refining partitions and conditioning. Let $\alpha = \{\alpha_\varepsilon \in \mathbb{R}^+ | \varepsilon \in E^*\}$ be a net of real numbers. (A net generalizes a sequence. Here, the α_εs form a directed system under containment on the partitions indexed by ε.) The Polya tree prior distribution PT_α on $\mathscr{M}(\mathscr{X})$ assigns probabilities to the partition elements in each τ_k by using independent *Beta* distributions. That is, the probabilities drawn from $PT(\alpha)$ satisfy

$$\text{i)}\forall k \;\; \forall \varepsilon \in \cup_{j=1}^{k-1} E_j : P(B_{\varepsilon,0}|B_\varepsilon) \quad \text{are independent}$$

in which the values indicated by P are the random variables satisfying

$$\text{ii)}\forall k \;\; \forall \varepsilon \in \cup_{j=1}^{k-1} E_j : P(B_{\varepsilon,0}|B_\varepsilon) \sim Beta(\alpha_{\varepsilon,0}, \alpha_{\varepsilon,1}).$$

As with the Dirichlet process, (i) and (ii) specify the distributions for the probabilities of intervals. That is, for fixed $\varepsilon \in E^*$, the distribution of the probability of the set B_ε is now specified. In other words, if $B_\varepsilon = (s,t]$, then the probability $F(t) - F(s)$ is specified by the appropriate sequence of conditional densities. If the partitions are chosen so that each point $t \in \mathbb{R}$ is the limit of a sequence of endpoints of intervals from the τ_ks, then the distributions that never assign zero probability to sets of positive measure get probability one.

Like the *DP* prior, Polya tree priors also give consistency in that the posteriors they give concentrate at the true distribution if it is in their support. More specifically, Polya trees can, in general, be constructed to concentrate arbitrarily closely about any given distribution and can be constructed so they assign positive mass to every relative entropy neighborhood of every finite entropy distribution that has a density. These two results fail for the *DP* priors but give more general consistency properties for *PT* priors.

Also like the *DP* priors, Polya tree priors are conjugate. It can be verified that if $\theta_1,...,\theta_n \sim P$ are IID and $P \sim PT(\alpha)$, then

$$P|\theta_1,...,\theta_n \sim PT(\alpha(\theta_1,...,\theta_n)), \quad \text{where} \quad \forall\varepsilon: \quad \alpha_\varepsilon(\theta_1,...,\theta_n) = \alpha_\varepsilon + \sum_{i=1}^n \delta_{\theta_i}(B_\varepsilon).$$

Of great importance is the prediction formula for Polya trees. It is more complicated than for the *DP* priors, but no harder to derive. It is expressed in terms of the partition elements. Let $n_\varepsilon = \#(\{\theta_i \in B_\varepsilon\})$. Then,

$$P(\theta_{n+1} \in B_{\varepsilon_1,...,\varepsilon_k}) = \frac{\alpha_{\varepsilon_1} + \sum_{i=1}^n \delta_{\theta_i}(B_{\varepsilon_1})}{\alpha_0 + \alpha_1 + n} \times \frac{\alpha_{\varepsilon_1,\varepsilon_2} + \sum_{i=1}^n \delta_{\theta_i}(B_{\varepsilon_1,\varepsilon_2})}{\alpha_{\varepsilon_1,0} + \alpha_{\varepsilon_1,1} + n_{\varepsilon_1}}$$
$$\times \ldots \times \frac{\alpha_{\varepsilon_1,...,\varepsilon_k} + \sum_{i=1}^n \delta_{\theta_i}(B_{\varepsilon_1,...,\varepsilon_k})}{\alpha_{\varepsilon_1,...,\varepsilon_{k-1},0} + \alpha_{\varepsilon_1,...,\varepsilon_{k-1},1} + n\varepsilon_1,...,\varepsilon_{k-1}}.$$

Polya tree priors have other nice features. In particular, they have a much larger support than the *DP* priors. It is a theorem that if λ is a continuous measure, then there is a *PT* prior that has support equal to the collection of probabilities that are absolutely continuous with respect to λ. Indeed, the support of $PT(\alpha)$ is $\mathscr{M}(\mathscr{X})$ if and only if $\alpha_\varepsilon > 0$ for all $\varepsilon \in E^*$. Thus, the support of a *PT* prior corresponds to an intuitively reasonable class of probabilities.

One of the main uses of *PT* priors is on the error term in regression problems. However, this and more elaborate mixtures of Polya trees are beyond the present scope.

6.2.3 Gaussian Process Priors

Gaussian processes (GPs) do not assign probabilities to sets in $\mathscr{M}(\mathscr{X})$, the probability measures on $\mathscr{X}$, but to the values of regression functions. Thus *GP* priors are more general than *DP* or *PT* priors. Roughly, the function values are treated as the outcomes of a GP so that finite selections of the function values (at finitely many specified design points) have a normal density with mean given by the true function. *GP* priors are surprisingly general and are closely related to other methods such as splines, as seen in Chapter 3.

6.2.3.1 Gaussian Processes

To start, a stochastic process $\langle Y_x \rangle|_{x\in I}$ is Gaussian if and only if the joint distribution of every finite subset of Y_{x_i}s, $i = 1,...,n$, is multivariate normal. Commonly, the index set I is an interval, x is the explanatory variable, and there is a mean function $\mu(x)$ and a symmetric covariance function $r(x,x')$. The variance matrix for any collection of Y_{x_i}s has entries $r(x_i,x_j)$ and is assumed to be positive definite. Thus, formally, given any n and any values $x_1,...,x_n$, a Gaussian process satisfies

$$(Y(x_1),...,Y(x_n))^t \sim N((\mu(x_1),...,\mu(x_n))^t, [(r(x_i,x_j)]_{i,j=1,...,n}).$$

A theorem, see Doob (1953), ensures $\mathbb{E}Y_x = \mu(x)$ and $r(x,x') = \mathbb{E}Y_xY_{x'} - \mu(x)\mu(x')$. (The main step in the proof is setting up an application of Kolmogorov's extension theorem to ensure that for given μ and r the finite-dimensional marginals coalesce into a single consistent process.) Together the μ and r play the same role as α does for the Dirichlet or *PT* distributions.

Although they have fixed marginal distributions, GPs (surprisingly) approximate a very wide class of general stochastic processes up to second moments. Indeed, let U_x be a stochastic process with finite second moments for $x \in I$. Then there is a Gaussian process $\langle Y_x \rangle|_{x\in I}$ defined on some (possibly different) measure space so that $\mathbb{E}Y_x = 0$ and $\mathbb{E}Y_xY_{x'} = \mathbb{E}U_xU_{x'}$. That is, if concern is only with the first and second moments, there is no loss of generality in assuming the process is Gaussian. Moreover, if continuous, or even differentiable, sample paths are desired, then conditions to ensure $Y_x(\omega)$ and $Y_{x'}(\omega)$ are close when x and x' must be imposed. In practice, these come down to choosing a covariance function to make the correlations among the values get larger as the x_is get closer to each other. One common form is $r(x,x') = \rho(|Y_x - Y_{x'}|)$ for some function ρ; a popular choice within this class is $r(x,x') = \sigma^2 e^{(x-x')^2/sv^2}$. Clearly, σ^2 is the maximum covariance, and it can be seen that as x and x' get closer the values on a sample path become perfectly correlated. High correlation makes the unknown sample path (i.e., function) smoother, and low correlation means that neighboring points do not influence each other, permitting rougher sample paths (i.e., rougher functions).

One of the motivations for using GPs is that they generalize least squares approximations. In particular, if $(Y, X_1, ..., X_p)$ has a multivariate normal distribution with mean 0, then the difference $Y - \sum_{j=1}^p a_j X_j$ has expectation zero and is uncorrelated with the X_js when the $a_j = \mathbb{E}(X_j Y)$. Thus, the conditional $(Y|X_1, ..., X_p)$ is normal and

$$\mathbb{E}(Y|X_1,...,X_p) = \sum_{j=1}^p a_j X_j;$$

see the elliptical assumption used in SIR. Since $Y - \sum_{j=1}^p a_j X_j$ is independent of any square-integrable function of $X_1, ..., X_p$, the sum of squares decomposition

$$\mathbb{E}\left|Y - f(X_1,...,X_p)\right|^2 = \mathbb{E}\left|Y - \sum_{j=1}^p a_j X_j\right|^2 + \mathbb{E}\left|\sum_{j=1}^p a_j X_j - f(X_1,...,X_p)\right|^2$$

is minimized over f by choosing $f(x_1, ..., x_p) = \sum_{j=1}^p a_j x_j$. The optimal linear predictor will emerge automatically from the normal structure.

Having defined GPs, it is not hard to see how they can be used as priors on a function space. The first step is often to assume the mean function is zero so that it is primarily the covariance function that relates one function value at a design point to another function value at another design point. Now there are two cases, the simpler noise-free data setting and the more complicated (and useful) noisy data setting.

The noise-free case is standard normal theory, but with new notation. Choose $x_1,...,$ x_n and consider unknown values $f_i = f(x_i)$. Let $\boldsymbol{f} = (f_1, ..., f_n)^\mathsf{T}$, $\boldsymbol{x} = (x_1, ..., x_n)$, and write $K(\boldsymbol{x},\boldsymbol{x})$ to mean the $n \times n$ matrix with values $r(x_i, x_j)$. Now let x_{new} be a new design point, and consider estimating $f_{new} = f(x_{new})$. Since the covariance function will shortly be related to a kernel function, write $K(\boldsymbol{x}, x_{new}) = (r(x_1, x_{new}), \ldots, r(x_n, x_{new}))$. Now, the noise-free model is $Y(x) = f(x)$ with a probability structure on the values of f. Thus, the model is

$$\begin{pmatrix} \boldsymbol{f} \\ f_{new} \end{pmatrix} \sim N\left(\begin{pmatrix} 0 \\ 0 \end{pmatrix}, \begin{pmatrix} K(\boldsymbol{x},\boldsymbol{x}) & ,K(\boldsymbol{x},x_{new}) \\ K(x_{new},\boldsymbol{x}) & ,K(x_{new},x_{new}) \end{pmatrix}\right). \tag{6.2.1}$$

Since x_{new} is a single design point, it is easy to use conventional normal theory to derive a predictive distribution for a new value:

$$f_{new} \mid \boldsymbol{x}, \boldsymbol{f}, x_{new} \tag{6.2.2}$$
$$\sim \ N(K(x_{new},x)K^{-1}(\boldsymbol{x},\boldsymbol{x})\boldsymbol{f},\ K(x_{new},x_{new}) - K(x_{new},\boldsymbol{x})K^{-1}(\boldsymbol{x},\boldsymbol{x})K(\boldsymbol{x},x_{new})).$$

Note that (6.2.1) is noise free in the sense that the function values f_i are assumed to be directly observed without error. In fact, they are usually only observed with error, so a more realistic model is $Y = f(x) + \varepsilon$, in which $\text{Var}(\varepsilon) = \sigma^2$. Now, it is easy to derive that

$$\text{Cov}(Y_{x_i}, Y_{x_j}) = K(x_i, x_j) + \sigma^2 \delta_{i,j}, \tag{6.2.3}$$

where $\delta_{i,j} = 0, 1$ according to whether $i = j$ or $i \neq j$. Letting $\boldsymbol{Y} = (Y_1, \ldots, Y_n)^\top = (Y_{x_1}, \ldots, Y_{x_n})$ with realized values $\mathbf{y} = (y_1, \ldots, y_n)^\top$, (6.2.3) is $\text{Cov}(\boldsymbol{Y}, \boldsymbol{Y}) = K(\boldsymbol{x}, \boldsymbol{x}) + \sigma^2 I_n$ in more compact notation. Now, (6.2.1) is replaced by

$$\begin{pmatrix} \mathbf{y} \\ f_{new} \end{pmatrix} \sim N\left(\begin{pmatrix} 0 \\ 0 \end{pmatrix}, \begin{pmatrix} K(\boldsymbol{x},\boldsymbol{x}) + \sigma^2 I_n, K(\boldsymbol{x}, x_{new}) \\ K(x_{new}, \boldsymbol{x}) \qquad ,K(x_{new}, x_{new}) \end{pmatrix}\right). \tag{6.2.4}$$

The usual normal theory manipulations can be applied to (6.2.4) to give

$$f_{new} \mid \boldsymbol{x}, \mathbf{y}, x_{new} \sim N(\mathbb{E}(f_{new} \mid \boldsymbol{x}, \mathbf{y}, x_{new}), \text{Var}(f_{new})),$$

parallel to (6.2.2). Explicit expressions for the mean and variance in (6.2.5) are

$$\begin{aligned} \mathbb{E}(f_{new} \mid \boldsymbol{x}, \mathbf{y}, x_{new}) &= K(x_{new}, \boldsymbol{x})(K(\boldsymbol{x},\boldsymbol{x}) + \sigma^2 I_n)^{-1}\mathbf{y}, \\ \text{Var}(f_{new}) &= K(x_{new}, x_{new}) - (K(\boldsymbol{x},\boldsymbol{x}) + \sigma^2 I_n)^{-1} K(\boldsymbol{x}, x_{new}). \end{aligned}$$

It is seen that the data $\mathbf{y}$ only affect the mean function, which can be written

$$\mathbb{E}(f_{new} \mid \boldsymbol{x}, \mathbf{y}, x_{new}) = \sum_{i=1}^{n} \alpha_i K(x_i, x_{new}), \tag{6.2.5}$$

in which the α_is come from the vector $\boldsymbol{\alpha} = (\alpha_1, \ldots \alpha_n)^\top = (K(\boldsymbol{x},\boldsymbol{x}) + \sigma^2 I_n)^{-1}\mathbf{y}$. It is seen that (6.2.5) is of the same form as the solution to SVMs, or as given by the representer theorem.

6.2.3.2 GPs and Splines

It has been noted that the results of *GP* priors are similar or identical to those of using least squares regression, SVMs or the representer theorem. To extend this parallel, this section presents four important links between GPs and spline methods.

First, the covariance function $r(\boldsymbol{x}, \boldsymbol{x}')$ of a GP can be identified with the RK from an RKHS as suggested by the notation of the last subsection. Consequently, every mean-zero GP defines and is defined by an RKHS. This relationship is tighter than it first appears. Consider a reproducing kernel K. The Mercer-Hilbert-Schmidt theorem guarantees there is an orthonormal sequence $\langle \psi_j \rangle$ of functions with corresponding λ_js decreasing to zero so that

$$\int K(\boldsymbol{x},\boldsymbol{x}')\psi_j(\boldsymbol{x}')d\boldsymbol{x}' = \lambda_j \psi_j(\boldsymbol{x}); \tag{6.2.6}$$

see Wahba (1990) and Seeger (2004). This means that the ψ_js are eigenfunctions of the operator induced by K. Then, not only can one write the kernel as

$$K(\boldsymbol{x},\boldsymbol{x}') = \sum_j \lambda_j \psi_j(\boldsymbol{x})\psi_j(\boldsymbol{x}') \tag{6.2.7}$$

but the zero-mean GP, say $Y(\boldsymbol{x})$ with covariance $r(\boldsymbol{x},\boldsymbol{x}') = k(x,x')$, can be written as

$$Y(\boldsymbol{x}) = \sum_j Y_j \psi_j(\boldsymbol{x}),$$

in which the Y_js are independent mean-zero Gaussian variables with $\mathbb{E}(Y_j^2) = \lambda_j$. In this representation, the Y_js are like Fourier coefficients: $Y_j = \int Y(\boldsymbol{x})\psi_j(\boldsymbol{x})d\boldsymbol{x}$. It turns out that one can (with some work) pursue this line of reasoning to construct an RKHS that is spanned by the sample paths of the GP $Y(\boldsymbol{x})$. This means GPs and RKHSs are somewhat equivalent.

Second, separate from this, one can relate GPs to the penalty term in the key optimization to obtain smoothing splines. In fact, roughness penalties correspond to priors in general, and the GP process prior is merely a special case. Consider the functional

$$G(f(\boldsymbol{x})) = G(f(\boldsymbol{x}),\beta,\lambda) = -\frac{\beta}{2}\sum_{i=1}^{n}(y_i - f(\boldsymbol{x}_i))^2 - \frac{\lambda}{2}\int [f^{(m)}(\boldsymbol{x})]^2 d\boldsymbol{x}. \tag{6.2.8}$$

The first term is the log likelihood for normal noise, and the second term is a roughness penalty. Now, link spline optimizations to GPs by treating the roughness penalty as the log of a prior density, specifically a *GP* prior on the regression function. Then, the two terms on the right in (8.3.50) sum to the log of the joint probability for the regression function and the data. The benefit of this Bayesian approach is that smoothing splines are seen to be posterior modes: Up to a normalizing constant depending on the data, G is the log of the posterior density. So, maximizing G, which leads to splines, also gives the mode of the posterior. In addition, when the prior and noise term are both derived from normals, the spline solution is the posterior mean, as will be seen shortly.

To see these points explicitly, replace the p-dimensional $\boldsymbol{x}$ by the unidimensional x. Then, try to backform a Gaussian prior from the penalty in (8.3.50) by setting

$$-\frac{1}{2}\boldsymbol{f}(\boldsymbol{x})\boldsymbol{\Lambda}(\boldsymbol{x})\boldsymbol{f}(\boldsymbol{x}) = \log \Pi(f(x)) \approx -\frac{1}{2}\int [f^{(m)}(x)]^2 dx, \tag{6.2.9}$$

in which $\boldsymbol{f}(\boldsymbol{x}) = (f(x_1),...,f(x_n))^{\mathsf{T}}$, and

$$\boldsymbol{\Lambda}(\boldsymbol{x}) = (f^{(m)}(x_1),...,f^{(m)}(x_n))^T(f^{(m)}(x_1),...,f^{(m)}(x_n)),$$

the matrix with entries given by products of evaluations of the dth derivative of f at the x_i and x_j. The approximation in (8.3.50) arises because the penalty term in splines is an integral, but when doing the optimization numerically, one must discretize. It can be imagined that if the x_is are chosen poorly, then the approximation can be poor. This is especially the case with repeated x_is for instance, which can lead to Λ not being of full rank. Despite these caveats, it is apparent from (8.3.50) that the usual spline penalty is roughly equivalent to using the mean-zero *GP* prior $\Pi(f)$ with covariance function Λ. That is, the non-Bayes analog of a *GP* prior is the squared error spline penalty.

As an observation, recall that in Chapter 3 it was seen that roughness penalties correspond to inner products and thence to Hilbert spaces of functions equipped with those inner products. Also, as just noted, roughness penalties and priors are closely related. So, there is an implied relationship between inner products and priors. To date, this seems largely unexplored.

In the case where $\boldsymbol{x}$ is p-dimensional, the situation is similar but messier. To obtain a form for Π, the covariance function of the GP has to be specified, often in terms of a norm on the $\boldsymbol{x}$s. This can be done quite loosely, however: The spline penalty is an integral and the GP has a huge number of sample paths. So, one does not expect to match these two in any general sense; it is enough to ensure that the particular points on the sample path of the GP are close to the value of the penalty for the data accumulated. That is, the finite-dimensional approximation must be good, but the approximation need not match throughout the domain of the integral in the penalty or on all the entire sample paths. So, one choice for the covariance function is $r(\boldsymbol{x}_i, \boldsymbol{x}_j) = e^{-||x_i - x_j||^2}$; there are many others. It is usually enough that $r(\boldsymbol{x}_i, \boldsymbol{x}_j)$ be small when $\boldsymbol{x}_i - \boldsymbol{x}_j$ are far apart. Thus, one can get a discretized form of a spline roughness penalty and recognize a form for Λ as the covariance matrix from a GP that makes the two of them close. This matching is important and leads to many applications as well as more theory; see Genton (2001) for merely one example. However, this becomes quite specialized, and the present goal was only to argue the link between GPs and certain spline penalties.

Third and more abstractly, Seeger (2004), Section 6.1 credits Kimmeldorf and Wahba (1971) for relating GPs to the Hilbert space structure of spline penalties. Let K be a positive semidefinite spline kernel corresponding to a differential operator with an m-dimensional null-space. Let $\mathscr{H} = \mathscr{H}_1 \bigoplus \mathscr{H}_2$, where $\mathscr{H}_1$ is the Hilbert space spanned by the orthonormal basis $g_1, ..., g_m$ of the nulls-pace and $\mathscr{H}_2$ is the Hilbert space associated with K. Consider the model

$$Y_i = F(\boldsymbol{x}_i) + \varepsilon_i = \sum_{j=1}^{m} \beta_j g_j(\boldsymbol{x}_i) + \sqrt{b} U(\boldsymbol{x}_i) + \varepsilon_i, \tag{6.2.10}$$

where $U(\boldsymbol{x})$ is a mean-zero GP with covariance function $r = K$, the εs are independent $N(0, \sigma^2)$, and the β_js are independent $N(0, a)$. Next, let f_λ be the minimum of the regularized risk

$$\frac{1}{n} \sum_{i=1}^{n} (y_i - f(\boldsymbol{x}_i))^2 + \lambda ||\mathscr{P}_2 f||_2^2 \tag{6.2.11}$$

in $\mathscr{H}$, where $\mathscr{P}_2$ is the projection of $\mathscr{H}$ onto $\mathscr{H}_2$. Expression (6.2.11) is clearly a general form of the usual optimization done to find splines. Kimmeldorf and Wahba (1971) show that f_λ is in the span of $\mathscr{H}_1 \cup \{K(\cdot, \boldsymbol{x}_i) | i = 1, ..., n\}$, a result nearly equivalent to the representer theorem (in Chapter 5). If F in (6.2.10) is denoted F_a, they also show that

$$\forall \boldsymbol{x} \quad \lim_{a \to \infty} \mathbb{E}(F_a(\boldsymbol{x}) | y_1, ..., y_n) = f_\lambda(\boldsymbol{x}),$$

and $\lambda = \sigma^2/nb$.

Fourth and finally, after all this, one can see that spline estimators are posterior means under GP priors if the Hilbert spaces are chosen properly. This will be much like the example in Chapter 3 of constructing the Hilbert space since attention is limited to the case of polynomial splines for unidimensional x using the integral of the squared m-derivative as a roughness penalty. That is, the focus is on polynomial smoothing splines that minimize

$$\frac{1}{n}\sum_{i=1}^{n}(Y_i - f(x_i))^2 + \lambda \int_0^1 (f^{(m)})^2 d\boldsymbol{x} \tag{6.2.12}$$

in the space $\mathscr{C}^{(m)}[0,1]$. As seen in Chapter 3, an RKHS and an RK can be motivated by Taylor expansions.

For $f \in \mathscr{C}^{(m)}[0,1]$, derivatives from the right give the Taylor expansion at 0,

$$f(x) = \sum_{j=1}^{m-1} \frac{x^j}{j!} f^{(j)}(0) + \frac{1}{(m-1)!}\int_0^1 (x-t)^{m-1} f^{(m)}(t)dt, \tag{6.2.13}$$

in which the last term is a transformation of the usual integral form of the remainder. The two terms in (6.2.13) can be regarded as "projections" of f, one onto the space $\mathscr{H}_0$ of monomials of degree less than or equal to $m-1$ and the other onto the orthogonal complement $\mathscr{H}_1$ of $\mathscr{H}_0$ in $\mathscr{C}^{(m)}[0,1]$. The problem is that the orthogonal complement needs an inner product to be defined. So, the task is to identify two RKHSs, $\mathscr{H}_0$ and $\mathscr{H}_1$, one for each projection, and an RK on each, and then to verify that the sum of their inner products is an inner product on $\mathscr{C}^{(m)}[0,1]$.

As in Chapter 3, for $f_0, g_0 \in \mathscr{H}_0$, the first term in (6.2.13) suggests trying

$$\langle f_0, g_0 \rangle_0 = \sum_{j=1}^{m-1} f_0^{(j)}(0) g_0^{(j)}(0)$$

as an inner product, and for $f_1, g_1 \in \mathscr{H}_1$ the second term suggests

$$\langle f_1, g_1 \rangle_1 = \int_0^1 f_1^{(m)}(x) g_1^{(m)}(x) dx,$$

so an inner product on $\mathscr{C}^{(m)}[0,1] = \mathscr{H}_0 \bigoplus \mathscr{H}_1$ for $f = f_0 + f_1$ and $g = g_0 + g_1$ is

$$\langle f, g \rangle = \langle f_0, g_0 \rangle_0 + \langle f_1, g_1 \rangle_1, \tag{6.2.14}$$

where the subscript on a function indicates its projection onto $\mathscr{H}_i$.

To get RKHSs, RKs must be assigned to each inner product $\langle \cdot, \cdot \rangle_0$ and $\langle \cdot, \cdot \rangle_1$. It can be shown (see Gu (2002), Chapter 2) that

$$R_0(x,y) = \sum_{j=1}^{m-1} \frac{x^j y^j}{j!j!} \quad \text{and} \quad R_1(x,y) = \int_0^1 \frac{(x-t)^{m-1}(y-t)^{m-1}}{(m-1)!(m-1)!} dt \qquad (6.2.15)$$

are RKs on $\mathscr{H}_0$ and $\mathscr{H}_1$ under $\langle \cdot, \cdot \rangle_0$ and $\langle \cdot, \cdot \rangle_1$, respectively. So, $R(x,y) = R_0(x,y) + R_1(x,y)$ is an RK on $\mathscr{C}^{(m)}[0,1]$ under $\langle \cdot, \cdot \rangle$.

Now in the model $Y = f(x) + \varepsilon$, where ε is $N(0, \sigma^2)$, suppose $f(x) = f_0(x) + f_1(x)$, where f_i varies over $\mathscr{H}_i$ and each f_i is equipped with a *GP* prior having mean zero and covariance functions derived from the RKs. That is, set

$$\mathbb{E}(f_0(x)f_0(y)) = \tau^2 R_0(x,y) \quad \text{and} \quad \mathbb{E}(f_1(x)f_1(y)) = bR_1(x,y). \qquad (6.2.16)$$

Then, one can derive an expression for the posterior mean $\mathbb{E}(f(x)|Y)$. In fact, Gu (2002), Chapter 2, proves that a spline from (6.2.12) has the same form as $\mathbb{E}(f(x)|Y)$, where f_0 has a finite-dimensional normal distribution on $\mathscr{H}_0$ and f_1 has a *GP* prior with mean zero and covariance function $bR_1(x,y)$.

This argument generalizes to penalty terms with arbitrary differential operators showing that smoothing splines remain Bayesian estimates for more general Hilbert spaces and reproducing kernels (see Gu (2002), Chapter 2). It is less clear how the argument generalizes to p-dimensional $\boldsymbol{x}$s; for instance, the integral form of the remainder in (6.2.13) does not appear to hold for $p \geq 2$.

6.3 The Relevance Vector Machine

The relevance vector machine (RVM) introduced in Tipping (2001), was motivated by the search for a sparse functional representation of the prediction mechanism in a Bayesian context. Clearly, for representations that are weighted sums of individual learners, the function evaluations for the learners can be computationally burdensome if there are many learners or if the learners are hard to compute. Thus, for the sake of computational effectiveness as well as the desire to keep predictors simple, it is important to trim away as many individual learners as possible, provided any increase in bias is small.

The question becomes whether a sparse solution, in the sense of few learners or other simplicity criteria, can also provide accurate predictions. It turns out that RVM achieves both relatively well in many cases. In fact, one of RVMs advantages is that, unlike using regularization to achieve sparsity, which can be computationally demanding, RVM achieves sparsity by manipulating a Gaussian prior over the weights in the expansion. Taking advantage of the normality of its main expressions, the RVM framework simplifies and therefore speeds computations while maintaining sparsity.

6.3.1 RVM Regression: Formal Description

To present the RVM, let D denote the data $\{(\mathbf{x}_i, y_i), i = 1, ..., n\}$, with $\boldsymbol{x}_i \in \mathbb{R}^p$ and $y_i \in \mathbb{R}$. Using the representer theorem, kernel regression assumes that there exists a kernel function $K(\cdot,\cdot)$ such that, for each new design point $\boldsymbol{x}$, the response Y is a random variable that can be expressed as a weighted sum of the form

$$Y = \sum_{j=1}^{n} w_j K(\boldsymbol{x}, \boldsymbol{x}_j) + \varepsilon, \tag{6.3.1}$$

in which, without loss of generality, the intercept w_0 has been set to zero. For notational convenience, let $\boldsymbol{w} = (w_1, w_2, \cdots, w_n)^\mathsf{T}$ and define the n-dimensional vector $\mathbf{h}(\mathbf{x}) = (K(\mathbf{x}, \mathbf{x}_1), K(\mathbf{x}, \mathbf{x}_2), ..., K(\boldsymbol{x}, \boldsymbol{x}_n))^\mathsf{T}$. Now, (6.3.1) can be rewritten as $Y = \eta(\boldsymbol{w}, \boldsymbol{x}) + \varepsilon$, where

$$\eta(\boldsymbol{w}, \boldsymbol{x}) = \boldsymbol{w}^\mathsf{T} \mathbf{h}(\boldsymbol{x}). \tag{6.3.2}$$

The representation in (6.3.1) is in data space and therefore is not a model specification in the strict classical sense. The vector $\boldsymbol{w}$ is therefore not a vector of parameters in the classical sense, but rather a vector of weights indicating the contribution of each term to the expansion. In classical dimension reduction, the dimension of the input space is reduced with consequent reductions in the dimension of the parameter space. Here, in RVM (and SVM) the number of data points used in the expansion is reduced which therefore reduces the dimension of $\boldsymbol{w}$.

Achieving a sparse representation for RVM regression requires finding a vector $\boldsymbol{w}^* = (w_1^*, w_2^*, \cdots, w_k^*)$ of dimension $k << n$ and the corresponding function

$$\mathbf{h}^*(\boldsymbol{x}) = (K(\boldsymbol{x}, \boldsymbol{x}_1^*), K(\boldsymbol{x}, \boldsymbol{x}_2^*), ..., K(\boldsymbol{x}, \boldsymbol{x}_k^*))',$$

so that

$$\eta^*(\boldsymbol{x}, \boldsymbol{w}) = \boldsymbol{w}^{*\mathsf{T}} \mathbf{h}^*(\boldsymbol{x}). \tag{6.3.3}$$

Equation (6.3.3) corresponds to the sparse design and is based only on k well-chosen support points $\boldsymbol{x}_1^*, \boldsymbol{x}_2^*, \cdots, \boldsymbol{x}_k^*$, called relevant vectors. Relevant vectors in RVMs are analogous to support vectors in SVM.

There are a variety of ways to derive (6.3.3) from (6.3.2), each corresponding to a way to choose the relevant vectors.

In principle, a regularization approach is possible. Given a p-dimensional predictor variable $\boldsymbol{x}^\top = (x_1, x_2, \cdots, x_p)$, a response variable Y, and the traditional linear model $\mathbb{E}(Y|\mathbf{X} = \mathbf{x}) = \alpha + \sum_j \beta_j x_j$ with IID normal noise and squared error loss, the LASSO estimate $(\hat{\alpha}, \hat{\beta})$ of the parameters α and $\beta^\top = (\beta_1, \beta_2, \cdots, \beta_p)$ is

$$(\hat{\alpha}, \hat{\beta}) = \arg\min \left\{ \sum_{i=1}^{n} \left(y_i - \alpha - \sum_j \beta_j x_{ij} \right)^2 \right\} \quad \text{subject to } \sum_j |\beta_j| \leq \lambda, \tag{6.3.4}$$

where λ must be chosen by an auxiliary technique. In the present context, the regression function is $\eta(\mathbf{w},\mathbf{x})$ and the L_1 penalty would be used to reduce it to $\eta^*(\boldsymbol{x},\boldsymbol{w})$, thereby selecting which terms go into the expansion.

A simpler approach is truncation using a posterior threshold for a hyperparameter in a Bayesian context. This approach seeks to circumvent the computational difficulties by using a Gaussian prior on the weights $\boldsymbol{w}$ to induce closed-form expressions for almost all the important estimation and prediction equations. To specify the RVM regression in greater detail, write the matrix form of equation (6.3.1) as

$$\mathbf{Y} = \mathbf{H}\boldsymbol{w} + \boldsymbol{\varepsilon}, \tag{6.3.5}$$

where $\boldsymbol{y} = (y_1, y_2, \cdots, y_n)^\mathsf{T}$, $\boldsymbol{w} = (w_1, w_2, \cdots, w_n)^\mathsf{T}$, $\boldsymbol{\varepsilon} = (\varepsilon_1, \varepsilon_2, \cdots, \varepsilon_n)^\mathsf{T}$, and

$$\mathbf{H} = \begin{bmatrix} K(\boldsymbol{x}_1,\boldsymbol{x}_1) & K(\boldsymbol{x}_1,\boldsymbol{x}_2) & \cdots & K(\boldsymbol{x}_1,\boldsymbol{x}_n) \\ K(\boldsymbol{x}_2,\boldsymbol{x}_1) & K(\boldsymbol{x}_2,\boldsymbol{x}_2) & \cdots & K(\boldsymbol{x}_2,\boldsymbol{x}_n) \\ \vdots & \vdots & \vdots & \vdots \\ K(\boldsymbol{x}_n,\boldsymbol{x}_1) & K(\boldsymbol{x}_n,\boldsymbol{x}_2) & \cdots & K(\boldsymbol{x}_n,\boldsymbol{x}_n) \end{bmatrix}. \tag{6.3.6}$$

Also, suppose the error term in (6.3.5) is $\varepsilon_i \overset{IID}{\sim} N(0,\sigma^2)$ and the likelihood function for an IID sample D is normal; i.e.,

$$p(\mathbf{y}|\mathbf{H},\boldsymbol{w},\sigma^2) = N(\mathbf{y}|\mathbf{H}\boldsymbol{w},\sigma^2 I_n). \tag{6.3.7}$$

Now, the parameter vector $\theta = (\boldsymbol{w},\sigma^2)$ is $(n+1)$-dimensional, and there are n data points for estimating it. This underdetermination is a problem because it leads to non-unique solutions. This problem disappears (technically) if the variance σ^2 is known, but this is unrealistic in practice.

An alternative solution used in Tipping (2001) uses independent zero-mean normal priors for the coefficients in w_i. Thus, set

$$p(w_i|\alpha_i) = N(w_i|0,\alpha_i^{-1}), \tag{6.3.8}$$

so that $p(\boldsymbol{w}|\alpha) = N_n(\mathbf{0},\mathbf{D})$, in which $\mathbf{D} = \mathrm{Diag}(\alpha_1^{-1},\ldots,\alpha_n^{-1})$ and α denotes the vector $(\alpha_1,\alpha_2,\ldots,\alpha_n)^\mathsf{T}$. Although a Gaussian prior will not usually give sparsity, it turns out that using a Gamma hyperprior for each α_i yields a Student-t marginal prior for w_i when α_i is integrated out, and this leads to sparsity of a sort. Even though each individual Student t for w_i is no candidate for sparsity, their product $p(\boldsymbol{w}) = \prod_{i=1}^n p(w_i)$ has a surface that induces a sparsity pressure on the space of $\boldsymbol{w}$. More specifically, with

$$p(\alpha_i|a,b) = \mathrm{Gamma}(\alpha_i|a,b), \tag{6.3.9}$$

the marginal prior density for w_i is

$$p(w_i) = \int p(w_i|\alpha_i)p(\alpha_i)d\alpha_i = \frac{b^a\Gamma(a+\frac{1}{2})}{(2\pi)^{\frac{1}{2}}\Gamma(a)}(b+w_i^2/2)^{-(a+(1/2))}. \tag{6.3.10}$$

With such a marginal prior for each w_i, the prior for the vector $\boldsymbol{w}$ is a product of independent Student-t distributions over the w_is whose density surprisingly exhibits sparsity. Tipping (2001) uses a two-dimensional case $\boldsymbol{w} = (w_1, w_2)^{\mathsf{T}}$ to show that with such a prior the probability is concentrated at the origin and along the spines where one of the weights is zero. Figure 6.1 shows the marginal density $p(\boldsymbol{w})$ for the two-dimensional case $\boldsymbol{w} = (w_1, w_2)^{\mathsf{T}}$. The concentration of the prior on the ridges over the coordinate axes induces a pressure toward sparsity. Products of univariate priors with this property are sometimes called porcupine priors.

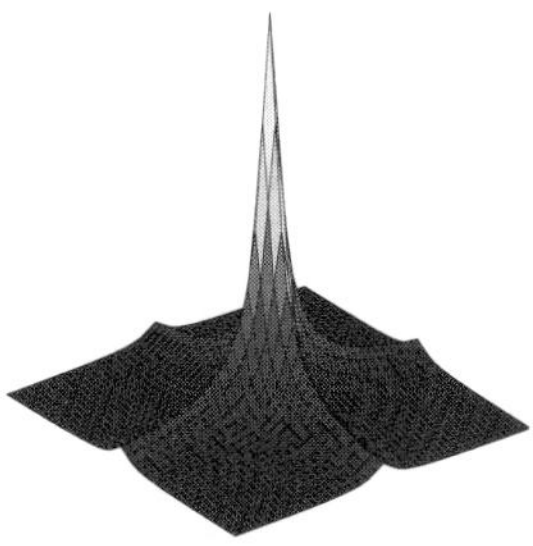

Fig. 6.1 The two-dimensional marginal prior for $\boldsymbol{w} = (w_1, w_2)^{\mathsf{T}}$ for $a = 1$ and $b = 1$.

Now, relevant vectors can be defined. First, suppose a weight w_i has corresponding variance α_i^{-1} tending to zero. Then, by the prior specification of (6.3.8), the distribution of w_i is sharply peaked at zero, and the corresponding vector $\boldsymbol{x}_i$ is irrelevant. All the vectors for which the variance α_i^{-1} does not tend to zero are relevant vectors.

In practice, the RVM is easily determined by truncation: Choose a large threshold for α_i, and set α_i^{-1} to zero if α_i is greater than the threshold. This means that the relevant vectors are only those for which the data do not permit the distribution of α_i to be too concentrated at zero. Even though this notion of relevance and the geometry of $p(\boldsymbol{w})$ suggests the sparsity of the prior, it remains to be seen how the prior combines with the likelihood to give a sparse posterior density.

For simplicity, assume that σ^2 is known. Therefore, from a Bayesian perspective, the posterior is

$$p(\mathbf{w}, \alpha|\mathbf{y}) \propto p(\mathbf{y}|\mathbf{H}, \mathbf{w}, \sigma^2) p(\mathbf{w}|\alpha) p(\alpha|a, b)$$

where the likelihood under normal noise with variance σ^2 is

$$p(\mathbf{y}|\mathbf{H}, \mathbf{w}, \sigma^2) = (2\pi\sigma^2)^{-\frac{n}{2}} \exp\left\{-\frac{1}{2\sigma^2}\|\mathbf{y} - \mathbf{H}\mathbf{w}\|^2\right\},$$

with

$$p(\mathbf{w}|\alpha) = \prod_{i=1}^{n} p(w_i|\alpha_i) \qquad \text{and} \qquad p(\alpha|a, b) = \prod_{i=1}^{n} p(\alpha_i|a, b).$$

The marginal posterior $p(\mathbf{w}|\mathbf{y})$ is obtained from the joint posterior $p(\mathbf{w}, \alpha|\mathbf{y})$ by integration and specification of the hyperparameters a and b. Ideally, this is done to ensure that $p(\boldsymbol{w}|\mathbf{y})$ has the same qualitative form as in $p(\boldsymbol{w})$ in Fig. 6.1. That is, the effect of the prior specification is to favor values of α along the axes or at the origin. Notice that even with σ^2 assumed known, the marginal posteriors,

$$p(\boldsymbol{w}|\mathbf{y}) = \int p(\boldsymbol{w}, \alpha|\mathbf{y}) d\alpha = \frac{p(\mathbf{y}|\boldsymbol{w})p(\boldsymbol{w})}{p(\mathbf{y})}$$

and

$$p(\alpha|\mathbf{y}) = \int p(\boldsymbol{w}, \alpha|\mathbf{y}) d\boldsymbol{w} = \frac{p(\mathbf{y}|\alpha)p(\alpha)}{p(\mathbf{y})},$$

cannot be computed in closed form. However, empirical Bayes approximation techniques can be used to obtain estimates of $\boldsymbol{w}$ and α. Alternatively, Markov chain Monte Carlo techniques can also be used to explore the joint posterior $p(\boldsymbol{w}, \alpha|\mathbf{y})$.

A quick derivation gives that the conditional posterior density for $\boldsymbol{w}$ is

$$p(\boldsymbol{w}|\alpha, \sigma^2, \mathbf{y}) = N(\boldsymbol{w}; \mu, \mathbf{V}),$$

where

$$\mathbf{V} = (\mathbf{H}^\top \sigma^2 I_n \mathbf{H} + \mathbf{A})^{-1} \text{ and } \mu = \mathbf{V}\mathbf{H}^\top \sigma^2 . I_n \mathbf{y}.$$

A more elaborate yet still straightforward derivation shows that the marginal likelihood $p(\mathbf{y}|\alpha, \sigma^2)$ is given by

$$p(\mathbf{y}|\alpha, \sigma^2) = N(\mathbf{y}; \mathbf{0}, \sigma^2 I_n + \mathbf{H}\mathbf{A}^{-1}\mathbf{H}^\top),$$

where $\mathbf{A} = \text{Diag}(\alpha_1, \alpha_2, \ldots, \alpha_n) = \mathbf{D}^{-1}$.

The two most important quantities, namely α and σ^2, are estimated by finding the values that maximize $\ln p(\mathbf{y}|\alpha, \sigma^2)$. As shown in Tipping (2001), it turns out that finding

$$(\hat{\alpha}, \hat{\sigma}^2) = \arg \max_{\alpha, \sigma^2} \ln p(\mathbf{y}|\alpha, \sigma^2)$$

reduces to a two-step iterative procedure: Initialize α and σ^2, and use them to obtain the posterior covariance matrix $\mathbf{V}$ and the posterior mean μ. Then, let μ_i be the ith component of μ and $\gamma_i = 1 - \alpha_i V_{ii}$. The iteration proceeds by setting

$$\alpha_i^{(\text{new})} = \frac{\gamma_i}{\mu_i^2}, \tag{6.3.11}$$

finding

$$(\sigma^2)^{(\text{new})} = \frac{\|\mathbf{y} - \mathbf{H}\mu\|^2}{n - \sum_{i=1}^n \gamma_i}, \tag{6.3.12}$$

and then recalculating α and σ^2 until convergence is reached. Using prior specifications (6.3.8) and (6.3.9), RVM typically gives a regression function that is more sparse than SVM does and therefore often gives better predictive performance.

Statistically, there is some characterization of the relevance vectors. Fokoue and Goel (2006) suggest the relevant points yielded by RVM regression (or classification) coincide with the support points found through the D-optimality criterion. This suggests that the optimal relevance vectors will often be relatively far apart. This is consistent with Breiman's observation that with stacking one gets better prediction if the regression functions being "stacked" are relatively dissimilar. Indeed, as the kernel achieves a closer approximation to the function class in which the true function lives, the relevant vectors seem to converge to the support points of the design.

6.3.2 RVM Classification

RVMs lead to classifiers in two ways. The first, which is not what is usually meant by RVM classification, just uses RVM regression directly in a logistic classifier. The second, which is what is commonly meant by RVM classification, couples a different sort of logistic regression approach with Laplace's method to achieve sparsity.

The first RVM-type classifier just uses the optimal model from the representer theorem as in RVM regression to fit a logistic model to the probabilities

$$\mathbb{P}(Y_i = 1|\boldsymbol{x}_i) = \frac{1}{1+e^{-h(\boldsymbol{x}_i)}} = g(h(\boldsymbol{x}_i)),$$

where $g(z) = 1/(1+e^{-z})$ and $h(\boldsymbol{x}_i)$ is defined via the kernel K as before; i.e.,

$$h(\boldsymbol{x}_i) = w_0 + \sum_{j=1}^{n} w_j K(\boldsymbol{x}_i, \boldsymbol{x}_j).$$

This is the logistic regression classifier of Chapter 5, but with a different regression function, and can be logically developed as follows: As in the RVM regression, one can assign priors to the w_js by setting $p(w_j|\alpha_j) = N(0, 1/\alpha_j)$ in which the αs are hyperparameters. A Gamma prior on the α_js will give a Student's t distribution marginally for the w_js as before. So, the same sort of sparsity can be obtained. In principle, imposing a threshold on the α_is by another procedure, empirical Bayes (essentially a maximization with an ML type estimate), MCMC, or a variant on the earlier iterative procedure is also possible. In general, defaulting to a product of normals as degrees of freedom in the ts increase tends to give sparsity. Indeed, most products of independent priors will lead to sparsity because they give porcupine priors.

The use of porcupine priors forces the distributions of w_j to concentrate at zero, and w_js that are close enough to zero can be taken as zero, reducing the number of terms in $h(\boldsymbol{x})$. The classifier becomes

$$\hat{Y}(\boldsymbol{x}) = \begin{cases} 1 & P(Y=1|\boldsymbol{X}=\boldsymbol{x}) > 1/2 \\ 0 & P(Y=1|\boldsymbol{X}=\boldsymbol{x}) \le 1/2. \end{cases}$$

The $\boldsymbol{x}_j$s for which $\alpha_i \neq \infty$ or, more generally, with $w_j \neq 0$, are the relevant vectors.

To see the reason a second type of RVM classifier might be useful, note that this first classifier ignores the fact that the regression function only needs to assume the values 0 and 1. The actual RVM classifier makes use of this restriction – and much more besides. Indeed, the framework uses not just the representer theorem (which already gives some sparsity) but also independence priors with hyperparameters (that are not tied together by having been drawn from the same distribution) to achieve sparsity.

For binary classification, the actual RVM classifier still starts by assuming a logistic regression model: For $(\boldsymbol{x}_i, y_i)$ with $i = 1, \ldots, n$, write

$$p(\mathbf{y}|\mathbf{w}) = \prod_{i=1}^{n} [g(h(\boldsymbol{x}_i))]^{y_i} (1 - g(h(\boldsymbol{x}_i)))^{1-y_i}. \tag{6.3.13}$$

Clearly, the likelihood in (6.3.13) is not normal, so the posterior for the w_js will not be normal, and getting the sparsity by way of forcing some α_js large enough to get concentration in the distribution of some w_js will require a few extra steps. To be specific, suppose $N(0, 1/\alpha_i)$ priors continue to be used for the w_is and that σ is known; more generally, a prior can be assigned to σ, too. Now, for n data points, $y^n = (y_1, ..., y_n)$, and n parameters $w^n = (w_1, ..., w_n)$, the conditional density given the n hyperparameters $\alpha = \alpha^n = (\alpha_1, ..., \alpha_n)$ is

$$p(y^n|\alpha^n) = \int p(y^n|w^n) p(w^n|\alpha^n) dw^n. \tag{6.3.14}$$

Tipping (2001) observes that (6.3.14) can be approximated, under some conditions, by a constant independent of α^n. The technique for seeing this is the Laplace approximation. Let $N(w^{n*})$ be a small neighborhood around the mode of $p(w^n|y^n)$, and approximate $p(y^n|\alpha^n)$ by

$$\int_{N(w^{n*})} \exp\left[-n\left(\frac{1}{n}\ln\frac{p(y^n|w^{n*})p(w^{n*}|\alpha^n)/p(y^n|\alpha^n)}{p(y^n|w^n)p(w^n|\alpha^n)/p(y^n|\alpha^n)}\right)\right] dw^n \tag{6.3.15}$$

times $p(y^n|w^{n*})p(w^{n*}|\alpha^n)$. The exponent in (6.3.15) is a log ratio of posteriors because Bayes' rule gives that $p(w^n|y^n) = p(y^n|w^n)p(w^n|\alpha)/p(y^n|\alpha)$. As a function of w^n, this is minimized at $w^{n*} = w^{n*}(y^n)$, so the exponent is maximized by $w^n = w^{n*}$.

This is important because the Laplace approximation rests on the fact that the biggest contribution to the integral is on small neighborhoods around the maximum of the integrand, at least as n increases in some uniform sense in the Y^ns. It is seen that the maximum of the integrand is at the mode of the posterior, w^{n*}. So, a standard second-order Taylor expansion of $p(y^n|w^n)p(w^n|\alpha^n)$ at w^{n*} has a vanishing first term (since it is evaluated at w^{n*}, for which the first derivative is zero), giving the term $(w^n - w^{n*})' J (w^n - w^{n*})$ in the exponent, in which J is the matrix of second partial derivatives of $\ln p(y^n|w^n)p(w^n|\alpha^n)$ with respect to w^n. This means that the approximation to (6.3.15) is independent of the value of α^n, at least when the reduction to the neighborhood $N(w^{n*})$ is valid and higher-order terms in the expansion are ignored.

The convergence of this Laplace approximation has not been established and cannot hold in general because the number of parameters increases linearly with the number of data points. However, the RVM classification strategy is to assume the Laplace approximation works, thereby getting a normal approximation to $p(w^n|y^n,\alpha^n)$ that can then be treated as a function of the hyperparameters α_j only. When this is valid, the approximation can be optimized over the α_js. Thus, in principle, large α_js can be identified and treated as infinite, resulting in distributions for their respective w_js being concentrated at 0, giving sparsity.

To see this optimization, begin with the Laplace approximation. The log density in the exponent is

$$\begin{aligned}\ln p(w^n|y^n,\alpha^n) &= \ln \frac{p(y^n|w^n)p(w^n|\alpha^n)}{p(y^n|\alpha^n)} \qquad (6.3.16)\\ &= \sum_{i=1}^{n} y_i \ln g(h(\boldsymbol{x}_i)) + (1-y_i)\ln g(h(\boldsymbol{x}_i)) - \frac{1}{2}(w^n)'Aw^n + \text{Error},\end{aligned}$$

in which $A = \text{Diag}(\alpha_1,\alpha_2,\ldots,\alpha_n)$ and Error is a list of terms that do not depend on the α_is; it is here that the approximation of the denominator in (6.3.16) by the Laplace argument on (6.3.15) is used. (Note that the approximation depends on w^n and J, which is neglected in the next step; this is the standard argument and, although not formally justified in general, appears to give a technique that works well in many cases.)

Taking Error as zero and differentiating with respect to the w_is (remember h depends on w) and setting the derivatives equal to zero gives the maximum (provided the second derivative is negative). A quick derivation gives that $g' = g(1-g)$, so simplifying gives

$$\nabla \ln p(w^n|y^n,\alpha^n) = \Phi'(y^n - (gh)^n) - Aw^n, \qquad (6.3.17)$$

in which $\Phi = (K(\boldsymbol{x}_i,\boldsymbol{x}_j))$. It is seen that the jth component of the left side is the derivative with respect to w_j and the jth component of the right side is

$$(K(\boldsymbol{x}_j,\boldsymbol{x}_1),\ldots,K(\boldsymbol{x}_j,\boldsymbol{x}_n))\cdot(y^n - (g(h(\boldsymbol{x}_1)),\ldots,g(h(\boldsymbol{x}_n))) - \alpha_j w_j.$$

Solving gives $(w^n)* = A^{-1}\Phi'(y^n - (gh)^n)$.

The second-order derivatives of (6.3.16) with respect to the w_is give the variance matrix for the approximating normal from Laplace's method (after inverting and putting in a minus sign). The second-order derivatives can be obtained by differentiating $\nabla \ln p(w^n|y^n,\alpha^n)$ in (6.3.17). The y^n term differentiates to 0, and the w^n drops out since it is linear. Recognizing that differentiating the $(gh)^n$ term gives another Φ and a diagonal matrix $B = \text{Diag}((gh)(\boldsymbol{x}_1)(1-(gh)(\boldsymbol{x}_1)),\ldots,(gh)(\boldsymbol{x}_n)(1-(gh)(\boldsymbol{x}_n)))$, the second-order term in the Taylor expansion is of the form

$$\nabla\nabla \ln p(w^n|y^n,\alpha^n) = -(\Phi' B \Phi + A).$$

So, the approximating normal for $p(w^n|y^n,\alpha^n)$ from Laplace's method is

$$N((w^n)^*,(\Phi' B\Phi + A)^{-1}),$$

in which the only parameter is α^n and the y^n has dropped out; the $\mathbf{x}_i$s remain, but they are chosen by the experimental setup. Putting these steps together, Laplace's method gives an approximation for $p(w^n|y^n,\alpha^n) \approx p(w^n|\alpha^n)$ in (6.3.16), which can also be used in a Laplace's approximation to

$$p(y^n|\alpha^n) = \int p(y^n|w^n)p(w^n|\alpha^n)dw^n,$$

in place of the conditional density of w^n; again this gives a normal approximation. Also using the Laplace method principle – evaluating at $(w^n)^*$ – gives

$$p(y^n|\alpha^n) \approx p(y^n|(w^n)^*)p((w^n)^*|\alpha)(2\pi)^{n/2}\det(\Phi' B\Phi + A)^{-1}.$$

The right-hand side can be optimized over α^n by a maximum likelihood technique. There are several ways to do this; one of the most popular is differentiating with respect to each α_i, setting the derivative to zero, and solving for α_i in terms of w_i^* for each i. If σ is not assumed known, a variant on the two-step iterative procedure due to Tipping (2001), described in (6.3.11) and (6.3.12), leads to a solution for the α_is and σ together.

6.4 Hidden Markov Models – Sequential Classification

To conclude this chapter on Alternative methods, a brief look at a class of models based on functions of a Markov chain is worthwhile. These are qualitatively different from the earlier methods because hidden Markov models posit an incompletely seen level from which observations are extracted. This is important in some application areas such as speech recognition.

Unlike the earlier methods that supposed a single explanatory vector $\boldsymbol{X}$ could be used to classify Y, suppose now a sequence of observations $\boldsymbol{x}_i$ and the task is to choose a sequence of classes y_i when the classes are hidden from us. Moreover, the relationship between the $\boldsymbol{X}_i$s and Y_is is indirect: Y is a state in principle never seen, and there is one observation $\boldsymbol{x}_i$ given $Y_i = y_i$ from $p(\boldsymbol{x}_i|y_i)$ and the states $y_1,...,y_n$ are themselves evolving over time with a dependence structure.

The simplest model in this class is called a hidden Markov model. The states y_i evolve according to a discrete time, discrete space Markov chain, and the $\boldsymbol{x}_i$s arise as functions of them. It's as if state y_i has a distribution associated to it, say $p(\cdot|y_i)$, and when y_i occurs, an outcome from $p(\cdot|y_i)$ is generated. The conditional distributions $p(\cdot|y_i)$, where $y_i \in \{1,...,K\}$ for K classes, or states as they are now called, are not necessarily constrained in any way (lthough often it is convenient to assume they are distinct and that the observed xs assume one of $M < \infty$ values). Sometimes this is represented compactly by defining a Markov chain on the sequence Y_i and taking only $X_i = f(Y_i)$ as observed. In this case, it's as if the conditional distribution for X_i is degenerate at $f(y_i)$. This is not abuse of notation because the distribution is conditional on y_i. However, without the conditioning, there is still randomness not generally represented by a deterministic function f.

An example may help. Suppose you are a psychiatrist with a practice near a major university. Among your practice are three patients: a graduate student, a junior faculty member, and a secretary all from the same department. Exactly one of these three sees you each month, reporting feelings of anxiety and persecution; because of confidentiality constraints, they cannot tell you who is primarily responsible for their desperation. Each of these people corresponds to a value an observed X_i might assume, so $M = 3$.

After some years, you realize there is an inner sanctum of senior professors in the department who take turns oppressing their underlings. In a notably random process, each month one senior professor takes it upon him or herself to immiserate a graduate student, a junior colleague, or the secretary. Thus, the number of hidden states is the number of senior professors in the inner cabal. If there are four senior professors, then $K = 4$. Given a senior professor in month i, say y_i, the victim x_i is chosen with probability $p(x_i|y_i)$ determined by the whims of the senior professor. The Markov structure would correspond to the way the senior cabal take turns; none of them would want to do all the dirty work or get into a predictable pattern lest the anxiety of the lower orders diminish.

The task of the psychiatrist given the sequence of patient visits would be to figure out K, $p(x|y)$, the transition matrix for the sequence of y_is, and maybe the sequence y_1,..., y_n from the sequence x_1,..., x_n.

To return to the formal mathematical structure, let Y_1,...,Y_t be a K state stationary Markov chain with states a_1,...,a_K, initial distribution $\pi(\cdot)$, and a $K \times K$ transition matrix T with entries

$$\tau_{i,j} = P(Y_2 = a_j | Y_1 = a_i). \tag{6.4.1}$$

(This definition extends to other time steps by stationarity.) Now let X_i be another sequence of random variables, each assuming one of M values, b_1,...,b_M. The Xs depend on the Ys by the way their probabilities are defined. So, let B be a $K \times L$ matrix with entries

$$b_{l,m} = P(X_1 = b_m | Y_1 = a_l). \tag{6.4.2}$$

Thus, the future chain value and the current observation both depend on Y_1. By using the Markov property (6.4.1) and the distribution property (6.4.2), given $Y_t = y_t$ there is a distribution on the b_ms from which is drawn the outcome $\boldsymbol{x}_t$ seen at time t. The model is fully specified probabilistically: If the a_is, b_is, T, B, M, K, and π are known, then the probability of any sequence of states or observations can be found.

However, when there are several candidate models, the more important question is which of the candidates matches the sequence of observations better. One way to infer this is to use the model that makes the observed sequence most likely. In practice, however, while the a_is and b_is, and hence the K and M, can often be surmised from modeling, both T and B usually need to be estimated from training data and π chosen in some sensible way (or the Markov chain must evolve long enough that the initial distribution matters very little).

The sequential classification problem is often called the decoding problem. In this case, the goal is not to find a model that maximizes the probability of observed outcomes but to infer the sequence of states that gave rise to the sequence of outcomes. This, too, assumes the a_is, b_is, T, B, and π are known. So, given that a sequence of outcomes $X_1 = x_{i_1},...,X_n = x_{i_n}$ is available, the task is to infer the corresponding states $Y_1 = a_{j_1},...,Y_n = a_{j_n}$ that gave rise to them. This is usually done by the Viterbi algorithm.

Naively, one could start with time $t = 1$ and ask what a_{i_1} is most likely given that x_{i_1} was seen. One would seek

$$j_1 = \arg \max_i \mathbb{P}(Y_1 = a_i | X_1 = x_{i_1}),$$

and then repeat the optimization for time $t = 2$ and so on. This will give an answer, but often that answer will be bad because the observations are poor or it will correspond to transitions for which $t_{i,j} = 0$. The Viterbi algorithm resolves these problems by imagining a $K \times t$ lattice of states over time $1,...,t$. States in neighboring columns are related by the Markov transitions, and the task is to find the route from column 1 to column t that has the maximum probability given the observations. The core idea is to correct the naive algorithm by including the $t_{i,j}$ and $b_{l,m}$ in the optimization.

There are several other problems commonly addressed with this structure. One is to estimate any or all of K, M, T, or B given the Xs and Ys. Sometimes called the learning problem, this is usually done by a maximum likelihood approach. One way uses the Baum-Welch algorithm, sometimes called the forward-backward algorithm; the forward algorithm is used to solve the evaluation problem of finding the probability that a sequence of observations was generated in a particular model. This can also be done through a gradient descent approach. Another problem is, given the Xs, the Ys, and two HMMs, which one fits the data better? While some ways to address this problem are available, it is not clear that this problem has been completely resolved. See Cappe et al. (2005) for a relatively recent treatment of the area.

6.5 Notes

6.5.1 Proof of Yang's Oracle Inequality

This is a detailed sketch; for full details, see Yang (2001). Write

$$p_{f,\sigma}(\boldsymbol{x},y) = \frac{1}{\sigma(\boldsymbol{x})} h\left(\frac{y - f(\boldsymbol{x})}{\sigma(\boldsymbol{x})}\right)$$

for the joint density of $(\boldsymbol{X},Y)$ under f and σ^2. For ease of notation, let superscripts denote ranges of values as required. Thus, $z_i^j = \{(\boldsymbol{x}_\ell, y_\ell) | \ell = i,\ldots,j\}$ and, for the case $i = 1$, write $z_1^j = z^j$.

Now, for each $i = n - N + 1,...,n$, let

$$q_i(\boldsymbol{x}, y; z^{n-N+2}) = \frac{\sum_{j\geq 1} \pi_j \left(\Pi_{\ell=n-N+1}^{i-1} p_{\hat{f}_{j,\ell},\hat{\sigma}_{j,\ell}}(\boldsymbol{x}_{\ell+1}, y_{\ell+1}) \right) p_{\hat{f}_{j,i},\hat{\sigma}_{j,i}}(\boldsymbol{x}, y)}{\sum_{j\geq 1} \pi_j \left(\Pi_{\ell=n-N+1}^{i-1} p_{\hat{f}_{j,\ell},\hat{\sigma}_{j,\ell}}(\boldsymbol{x}_{\ell+1}, y_{\ell+1}) \right)},$$

the weighted average of the $p_{\hat{f}_{j,i},\hat{\sigma}_{j,i}}(\boldsymbol{x}, y)$s over the procedures indexed by j.

The error distribution has mean zero given $\boldsymbol{x}$, but the distributions from $q_i(\boldsymbol{x}, y; Z^{n-N+2})$ have mean $\sum_j W_{j,i}\hat{f}_{j,i}(\boldsymbol{x})$ (in Y). Taking the average over the $q_i(\boldsymbol{x}, y; z^{n-N+2})$s gives

$$\hat{g}_n(y|\boldsymbol{x}) = \frac{1}{N} \sum_{i=n-N+1}^{n} q_i(\boldsymbol{x}, y, Z^i), \tag{6.5.1}$$

which is seen to be a convex combination of densities in y of the form $h((y-b)/a)$ and scales that depend on the data but not the underlying distribution $\mathbb{P}_X$. In fact, $\hat{g}_n$ is an estimator for the conditional density of $(Y|\boldsymbol{X} = \boldsymbol{x})$ and satisfies $\mathbb{E}\hat{g}_n(Y|\boldsymbol{X} = \boldsymbol{x}) = \bar{f}_n(\boldsymbol{x})$.

Now consider a predictive setting in which there are n pairs of data $(Y_i, \boldsymbol{X}_i)$ for $i = 1, \ldots, n$ and the goal is to predict $(Y, \boldsymbol{X}) = (Y_{n+1}, \boldsymbol{X}_{n+1})$. Denoting $i_0 = n - N + 1$, simple manipulations give

$$\sum_{\ell=i_0}^{n} \mathbb{E}\, D\,(p_{f,\sigma} \| q_\ell) \tag{6.5.2}$$

$$= \int \Pi_{\ell=i_0}^{n} p_{f,\sigma}(\boldsymbol{x}_{\ell+1}, y_{\ell+1}) \log \left(\frac{\Pi_{\ell=i_o}^{n} p_{f,\sigma}(\boldsymbol{x}_{\ell+1}, y_{\ell+1})}{g^{(n)}(z_{i_0+1}^{n+1})} \right) d(\mu \times \mathbb{P}_X)_{i_0+1}^{n+1},$$

in which D is the relative entropy and $D(f\|g) = \int f \log(f/g) d\mu$ for densities f and g with respect to μ. Also, in (6.5.2), $g^{(n)}$ is

$$g^{(n)}(z_{i_0+1}^{n+1}) = \sum_{j\geq 1} \pi_j g_j(z_{i_0+1}^{n+1}) \quad \text{and} \quad g_j(z_{i_0+1}^{n+1}) = \Pi_{\ell=i_0}^{n} p_{\hat{f}_{j,i},\hat{\sigma}_{j,i}}(\boldsymbol{x}, y).$$

Note that $p_{\hat{f}_{j,i},\hat{\sigma}_{j,i}}(\boldsymbol{x})$ is a density estimator that could be denoted $\hat{p}_{j,i}(\boldsymbol{x}, y; x^\ell)$ because the ℓ data points get used to form the estimator that is evaluated for $(\boldsymbol{x}, y)$.

Since $g^{(n)}$ is a convex combination and log is an increasing function, an upper bound results from looking only at the jth term in $g^{(n)}$. The integral in (6.5.2) is bounded by $\log(1/\pi_j)$ plus

$$\int \Pi_{\ell=i_0}^{n} p_{f,\sigma}(\boldsymbol{x}_{\ell+1}, y_{\ell+1}) \log \left(\frac{\Pi_{\ell=i_o}^{n} p_{f,\sigma}(\boldsymbol{x}_{\ell+1}, y_{\ell+1})}{g_j(z_{i_0+1}^{n+1})} \right) d(\mu \times \mathbb{P}_X)_{i_0+1}^{n+1}. \tag{6.5.3}$$

Manipulations similar to that giving (6.5.2) for the convex combination of g_js give

$$\sum_{\ell=i_0}^{n} \mathbb{E}D(p_{f,\sigma}\|\hat{p}_{j,\ell})$$

$$= \int \Pi_{\ell=i_0}^{n} p_{f,\sigma}(\boldsymbol{x}_{\ell+1},y_{\ell+1}) \log\left(\frac{\Pi_{\ell=i_o}^{n} p_{f,\sigma}(\boldsymbol{x}_{\ell+1},y_{\ell+1})}{g_j(z_{i_0+1}^{n+1})}\right) d(\mu \times P_X)_{i_0+1}^{n+1}$$

for the individual g_js. These individual summands $D(p_{f,\sigma}\|\hat{p}_{j,\ell})$ can be written as

$$\int\left(\int \frac{1}{\sigma(\boldsymbol{x})} h\left(\frac{y-f(\boldsymbol{x})}{\sigma(\boldsymbol{x})}\right) \log \frac{1/(\sigma(\boldsymbol{x}))h((y-f(\boldsymbol{x})/\sigma(\boldsymbol{x})))}{1/(\hat{\sigma}_{j,\ell}(\boldsymbol{x}))h((y-\hat{f}_{j,\ell}(\boldsymbol{x})/\sigma_{j,\ell}(\boldsymbol{x})))} \mu(dy)\right) d(\mu \times \mathbb{P})$$

$$= \int\int h(y) \log \frac{h(y)}{(\sigma(\boldsymbol{x})/\hat{\sigma}_{j,\ell}(\boldsymbol{x}))h((\sigma(\boldsymbol{x})/\hat{\sigma}_{j,\ell}(\boldsymbol{x}))y+(f(\boldsymbol{x})-\hat{f}_{j,\ell}(\boldsymbol{x}))/\hat{\sigma_{j,\ell}}(\boldsymbol{x}))} d(\mu \times \mathbb{P})$$

$$\leq B\left(\int\left(\frac{\hat{\sigma}_{j,\ell}(\boldsymbol{x})}{\sigma(\boldsymbol{x})}-1\right)^2 \mathbb{P}(dx) + \int\left(\frac{f(\boldsymbol{x})-\hat{f}_{j,\ell}(\boldsymbol{x})}{\sigma(\boldsymbol{x})}\right)^2 \mathbb{P}(dx)\right)$$

$$\leq \frac{B}{\underline{\sigma}^2}\left(\int (\sigma(\boldsymbol{x})-\hat{\sigma}_{j,\ell}(\boldsymbol{x}))^2 \mathbb{P}(d\boldsymbol{x}) + \int (f(\boldsymbol{x})-\hat{f}_{j,\ell}(\boldsymbol{x}))^2 \mathbb{P}(d\boldsymbol{x})\right).$$

In this sequence of expressions, the first is definition, the second follows from a linear transformation, the third follows from assumption (ii), and the fourth from (i) on the variances.

Taken together, this gives

$$\sum_{i=i_0}^{n} \mathbb{E}D(p_{f,\sigma}\|q_\ell) \leq \log\frac{1}{\pi_j} + \frac{B}{\underline{\sigma}^2}\sum_{\ell=i_0}^{n}\left(\mathbb{E}\|\sigma^2-\hat{\sigma}_{j,\ell}^2\|^2 + \mathbb{E}\|f-\hat{f}_{j,\ell}\|^2\right). \quad (6.5.4)$$

The bound that is most central to the desired result is the fact that the convexity of the relative entropy in its second argument gives

$$\mathbb{E}D(p_{f,\sigma}\|\hat{g}_n) \leq \frac{1}{N}\sum_{\ell=i_0}^{n} \mathbb{E}D(p_{f,\sigma}\|q_\ell).$$

Using this, (6.5.4), and minimizing over j gives that $\mathbb{E}D(p_{f,\sigma}\|\hat{g}_n)$ is bounded above by

$$\inf_j \left[\frac{1}{N}\log\frac{1}{\pi_j} + \frac{B}{N\underline{\sigma}^2}\sum_{\ell=i_0}^{n}\mathbb{E}\|\sigma^2-\hat{\sigma}_{j,\ell}^2\|^2 + \frac{B}{N\underline{\sigma}^2}\sum_{\ell=i_0}^{n}\mathbb{E}\|f-\hat{f}_{j,\ell}\|^2\right]. \quad (6.5.5)$$

The strategy now is to show that the left side of (6.5.4) upper bounds the squared error of interest; that way the right-hand side (which is the desired upper bound) will follow. There are two steps: First get a lower bound for the left side of (6.5.4) in terms of the Hellinger distance and then verify that the Hellinger distance upper bounds the risk.

Now, since the squared Hellinger distance d_H^2 is bounded by the relative entropy D, (6.5.5) gives that $Ed_H^2(p_{f,\sigma}, \hat{g}_n)$ also has the right-hand side of (6.5.5) as an upper bound. This is the first step.

For the second step, note that, for each $\boldsymbol{x}$,

$$\bar{f}_n(\boldsymbol{x}) = \mathbb{E}_{\hat{g}_n} Y = \int y \hat{g}_n(y|\boldsymbol{x}) \mu(dy) \text{ estimates } f(\boldsymbol{x}) = \int y p_{f(\boldsymbol{x}),\sigma(\boldsymbol{x})}(\boldsymbol{x}, y) \mu(dy).$$

So, $(f(\boldsymbol{x}) - \mathbb{E}_{\hat{g}_n} Y)^2$ equals the square of

$$\int y \left[\sqrt{p_{f(\boldsymbol{x}),\sigma(\boldsymbol{x})}(\boldsymbol{x}, y)} + \sqrt{\hat{g}_n(y|\boldsymbol{x})}\right] \left[\sqrt{p_{f(\boldsymbol{x}),\sigma(\boldsymbol{x})}(\boldsymbol{x}, y)} - \sqrt{\hat{g}_n(y|\boldsymbol{x})}\right] \mu(dy). \quad (6.5.6)$$

Using Cauchy-Schwarz and recognizing the appearance of $d_H^2(p_{f(\boldsymbol{x}),\sigma(\boldsymbol{x})}(\boldsymbol{x}, \cdot), \hat{g}_n(\cdot, \boldsymbol{x}))$ gives that (6.5.6) is bounded above by

$$2\left(f^2(\boldsymbol{x}) + \sigma^2(\boldsymbol{x}) + \int y^2 \hat{g}_n(y|\boldsymbol{x}) \mu(dy)\right) \times d_H^2(p_{f(\boldsymbol{x}),\sigma(\boldsymbol{x})}(\boldsymbol{x}, \cdot), \hat{g}_n(\cdot, \boldsymbol{x})), \quad (6.5.7)$$

in which the constant factor is bounded by $2(A^2 + \overline{\sigma}^2)$. Integrating (6.5.7) over $\boldsymbol{x}$ gives

$$\int (f(\boldsymbol{x}) - \mathbb{E}_{\hat{g}_n} Y)^2 \mathbb{P}_X(d\boldsymbol{x}) \leq 2(A^2 + \bar{\sigma}^2) \int d_H^2(p_{f(\boldsymbol{x}),\sigma(\boldsymbol{x})}(\boldsymbol{x}, \cdot), \hat{g}_n(\cdot, \boldsymbol{x})) \mathbb{P}_X(d\boldsymbol{x}).$$

Thus, finally,

$$\begin{aligned}
\mathbb{E} \int (f(\boldsymbol{x}) - \bar{f}_n(\boldsymbol{x}))^2 \mathbb{P}_X(d\boldsymbol{x}) &\leq 2(A^2 + \bar{\sigma}^2) \mathbb{E} d_H^2(p_{f,\sigma}, \hat{g}_n) \\
&\leq 2(A^2 + \bar{\sigma}^2) \inf_j \left[\frac{1}{N} \log \frac{1}{\pi_j} + \frac{B}{N\underline{\sigma}^2} \sum_{\ell=i_0}^{n} \mathbb{E}\|\sigma^2 - \hat{\sigma}_{j,\ell}^2\|^2 \right. \\
&\quad \left. + \frac{B}{N\underline{\sigma}^2} \sum_{\ell=i_0}^{n} \mathbb{E}\|f - \hat{f}_{j,\ell}\|^2 \right],
\end{aligned}$$

establishing the theorem. □

6.5.2 *Proof of Lecue's Oracle Inequality*

Let $a > 0$. By the Proposition in Section 1.6.2, for any $f \in \mathscr{F}$,

$$\begin{aligned}
A(\tilde{f}_n) - A^* &= (1+a)(A_n(\tilde{f}_n) - A_n(f^*)) + A(\tilde{f}_n) - A^* - (1+a)(A_n(\tilde{f}_n) - A_n(f^*)) \\
&\leq (1+a)(A_n(f) - A_n(f^*)) + (1+a)\frac{\log M}{n} \\
&\quad + A(\tilde{f}_n) - A^* - (1+a)(A_n(\tilde{f}_n) - A_n(f^*)). \quad (6.5.8)
\end{aligned}$$

Taking expectations on both sides of (6.5.8) gives

$$\mathbb{E}[A(\tilde{f}_n)-A^*] \le (1+a)\min_{f\in\mathscr{F}}(A(f)-A^*)+(1+a)\frac{\log M}{n} + \mathbb{E}\left(A(\tilde{f}_n)-A^*-(1+a)(A_n(\tilde{f}_n)-A_n(f^*))\right). \quad (6.5.9)$$

The remainder of the proof is to control the last expectation in (6.5.9).

Observe that the linearity of the hinge loss on $[-1,1]$ gives

$$A_n(\tilde{f}_n)-A^* - (1+a)(A_n(\tilde{f}_n)-A_n(f^*)) \le \max_{f\in\mathscr{F}}[A_n(f)-A^*-(1+a)(A_n(f)-A_n(f^*))]. \quad (6.5.10)$$

Also, recall Bernstein's inequality that if $|X_j|\le M$, then

$$\mathbb{P}\left(\sum_{i=1}^n Z_i > t\right) \le e^{-\frac{t^2/2}{\Sigma EZ_i^2+Mt/3}} \quad (6.5.11)$$

for any $t>0$.

Now,

$$\begin{aligned}
&\mathbb{P}\ (A_n(\tilde{f}_n)-A^*-(1+a)(A_n(\tilde{f}_n)-A_n(f^*))\ge\delta)\\
&\le \sum_{f\in\mathscr{F}}\mathbb{P}(A_n(f)-A^*-(1+a)(A_n(f)-A_n(f^*))\ge\delta)\\
&\le \sum_{f\in\mathscr{F}}\mathbb{P}\left(A_n(f)-A^*-(A_n(f)-A_n(f^*))\ge\frac{\delta+a(A(f)-A^*)}{1+a}\right)\\
&\le \sum_{f\in\mathscr{F}}\exp\left(-\frac{n(\delta+a(A(f)-A^*)^2)}{2(1+a)^2(A(f)-A^*)^{1/\kappa}+(2/3)(1+a)(\delta+a(A(f)-A^*))}\right),
\end{aligned} \quad (6.5.12)$$

in which (6.5.10), some manipulations, and (6.5.11) were used. Note that in the application of Bernstein's inequality, $Z_i=\mathbb{E}\max(0,1-Yf(X_i))-\mathbb{E}\max(0,1-Yf^*(X_i))-(\max(0,1-Yf(X_i))-\max(0,1-Yf^*(X_i)))$, and the identity $A_n=2R_n$ is used. Reducing the difference of two maxima to the left-hand side of (6.1.27) by recognizing the occurrences of probabilities where f and f^* are different gives (6.5.12).

The main part of the upper bound in (6.5.12) admits the bound for all $f\in\mathscr{F}$,

$$\frac{(\delta+a(A(f)-A^*)^2)}{2(1+a)^2(A(f)-A^*)^{1/\kappa}+(2/3)(1+a)(\delta+a(A(f)-A^*))}\ge C\delta^{2-1/\kappa}, \quad (6.5.13)$$

where $C=C(a)$, not necessarily the same from occurrence to occurrence.

Taken together, this gives a bound for use in (6.5.9):

$$\mathbb{P}(A_n(\tilde{f}_n)-A^*-(1+a)(A_n(\tilde{f}_n)-A_n(f^*))\ge\delta)\le Ke^{-nC\delta^{2-1/\kappa}}. \quad (6.5.14)$$

To finish the proof, recall that integration by parts gives

$$\int_a^{\infty} e^{bt^{\alpha}} dt \leq \frac{e^{-ba^{\alpha}}}{\alpha b a^{\alpha} - 1}.$$

So, for any $u > 0$, the inequality $EZ = \int_0^{\infty} P(Z \geq t)dt$ can be applied to the positive and negative parts of the random variable in (6.5.14) to give

$$\mathbb{E}(A_n(\tilde{f}_n) - A^* - (1+a)(A_n(\tilde{f}_n) - A_n(f^*))) \leq 2u + M \frac{e^{-nCu^{2-1/\kappa}}}{nCu^{1-1/\kappa}}. \quad (6.5.15)$$

Now, some final finagling gives the theorem. Let $X = Me^{-X}$ have the solution $X(M)$. Then $\log(M/2) \leq X(M) \leq \log M$, and it is enough to choose u such that $X(M) = nCu^{2-1/\kappa}$. □

6.6 Exercises

Exercise 6.1. Consider a data set $\mathscr{D} = \{(y_i, \boldsymbol{x}_i), i = 1, ..., n\}$ and suppose $\phi_n(\boldsymbol{x})$ is a predictor formed from $\mathscr{D}$. Fix a distribution P on $\mathscr{D}$ and resample independently from $\mathscr{D}$. For each bootstrap sample there will be a $\hat{\phi}_n(\boldsymbol{x})$. Aggregate the $\hat{\phi}_n$s by taking an average, say, and call the result $\phi(\boldsymbol{x}, \mathscr{D})$. For any predictor ϕ write

$$E(\phi) = \mathbb{E}_P(Y - \phi(\boldsymbol{x}))^2$$

for its pointwise (in $\boldsymbol{x}$) error. Thus,

$$E(\phi(\boldsymbol{x}, \mathscr{D}) = \mathbb{E}_P(Y - \phi(\boldsymbol{x}, \mathscr{D}))^2$$

is the pointwise error of $\phi(\boldsymbol{x}, \mathscr{D}$ and

$$E_(\phi_A) = \mathbb{E}_P(Y - \phi_A(\boldsymbol{x}, \mathscr{D})))^2,$$

where

$$\phi_A(\boldsymbol{x}, \mathscr{D}) = E_{\mathscr{D}} \phi(\boldsymbol{x}, \mathscr{D})$$

is the average bootstrap predictor.

Show that $E(\phi_A) \leq E(\phi(\boldsymbol{x}, \mathscr{D})$.

Exercise 6.2 (Stacking terms). Consider a data set $\mathscr{D} = \{(y_i, \boldsymbol{x}_i), i = 1, ..., n\}$ generated independently from a model $Y = f(\boldsymbol{x}) + \varepsilon$ where ε is mean-zero with variance σ^2. Suppose f is modeled as a linear combination of terms $B_j(\boldsymbol{x})$ assumed to be uncorrelated. That is, let

$$\sum_{j=1}^{J} a_j B_j(\boldsymbol{x})$$

be the regression function. One estimate of f is

$$\hat{f}(\boldsymbol{x}) = \sum_{j=1}^{J} \hat{\beta}_j B_j(\boldsymbol{x}),$$

in which the $\hat{a}_j$s are estimated by least squares; i.e.,

$$\{\hat{\beta}_1, \ldots, \hat{\beta}_J\} = \arg \min_{\{\beta_1,\ldots,\beta_J\}} \sum_{i=1}^{n} \left(y_i - \sum_{j=1}^{J} \beta_j B_j(\boldsymbol{x}) \right)^2 .$$

1. Show that the bias is
$$E\left(\hat{f}(\boldsymbol{x}) - f(\boldsymbol{x})\right)^2 = \frac{J\sigma^2}{n}.$$
2. What does this result say about "stacking" the functions B_j? Can you generalize this to stacking other functions, sums of B_js for instance?

Exercise 6.3 (Boosting by hand). Consider a toy data set in the plane with four data points that are not linearly separable. One of the simplest consists of the square data $Y = 1$ when $\boldsymbol{x} = (1,0), (-1,0)$ and $Y = -1$ when $\boldsymbol{x} = (0,-1), (0,1)$. Note that these are the midpoints of the sides of a square centered at the origin and parallel to the axes.

1. Choose a simple weak classifier such as a decision stump, i.e., a tree with exactly two leaves, representing a single split. Improve your decision stump as a classifier on the square data by using $T = 4$ iterations of the boosting procedure. For each $t = 1,2,3,4$, compute ε_t, α_t, C_t, and $D_t(i)$ for $i = 1,2,3,4$. For each time step, draw your weak classifier.
2. What is the training error of boosting?
3. Why does boosting do better than a single decision stump for this data.

Exercise 6.4 (Relative entropy and updating in boosting). Recall that the boosting weight updates are

$$D_{t+1}(i) = \frac{D_t(i) e^{-\alpha_t y_i h_t(\mathbf{x}_i)}}{C_t},$$

where

$$C_t = \sum_{i=1}^{n} D_t(i) e^{-\alpha_t y_i h_t(\mathbf{x}_i)}$$

is the normalization. It is seen that D_{t+1} and is a "tilted" distribution, i.e., multiplied by an exponential and normalized. This form suggests that one way that D_{t+1} could have been derived is to minimize a relative entropy subject to constraints. The relative entropy between two successive boosting weights is

$$RE(D_{t+1} \| D_t) = \sum_{i=1}^{n} D_{t+1}(i) \ln \frac{D_{t+1}(i)}{D_t(i)}$$

and represents the redundancy of basing a code on D_t when D_{t+1} is the true distribution. Show that given D_t, D_{t+1} achieves a solution to

$$\begin{array}{ll} \text{minimize} & RE(D_{t+1}\|D_t) \\ \text{subject to} & \forall\, i\; D_{t+1}(i) \geq 0,\;\; \sum_{i=1}^{n} D_{t+1}(i) = 1, \\ & \sum_{i=1}^{n} D_{t+1}(i) y_i h_t(x_i) = 0. \end{array} \tag{6.6.1}$$

(Notice that the last constraint is like insisting on a sort of orthogonality between h_t and D_{t+1} implying that a weak learner should avoid choosing an iterate too close to h_t.)

Exercise 6.5 (Training error analysis under boosting). Let h_t be a weak classifier at step t with weight α_t, denote the final classifier by

$$H(\boldsymbol{x}) = \operatorname{sign}(f(\boldsymbol{x})), \quad \text{where } f(\boldsymbol{x}) = \sum_{t=1}^{T} \alpha_t h_t(\boldsymbol{x}),$$

and recall the training error of h_t is $\varepsilon_t = \sum_{i=1}^{n} D_t(i) \mathbf{1}_{h_t(\boldsymbol{x}_i) \neq y_i}$. (See Section 6.1.4 or the previous Exercise for more notation.)

The point of this exercise is to develop a bound on the training error.

1. Show that the final classifier has training error that can be bounded by an exponential loss; i.e.,
$$\sum_{i=1}^{n} \mathbf{1}_{H(\boldsymbol{x}_i) \neq y_i} \leq \sum_{i=1}^{n} e^{-f(\boldsymbol{x}_i) y_i},$$
where y_i is the true class of $\boldsymbol{x}_i$.
2. Show that
$$\frac{1}{n} \sum_{i=1}^{n} e^{-f(\boldsymbol{x}_i) y_i} = \Pi_{t=1}^{T} C_t.$$
(Remember $e^{\sum_i g_i} = \Pi_i e^{g_i}$, $D_1(i) = 1/n$, and $\sum_i D_{t+1}(i) = 1$.)
3. Item 2 suggests that one way to make the training error small is minimize its bound. That is, make sure C_t is as small as possible at each step t. Indeed, items 1 and 2 together give that $\varepsilon_{training} \leq \Pi_{t=1}^{T} C_t$. Show that C_t can be written as
$$C_t = (1 - \varepsilon_t) e^{-\alpha_t} + \varepsilon_t e^{\alpha_t}.$$
(**Hint**: Consider the sums over correctly and incorrectly classified examples separately.)
4. Now show that, C_t is minimized by
$$\alpha_t = \frac{1}{2} \ln\left(\frac{1 - \varepsilon_t}{\varepsilon_t}\right).$$

5. For this value of α_t, show that $C_t = 2\sqrt{\varepsilon_t(1-\varepsilon_t)}$.
6. Finally, let $\gamma_t = 1/2 - \varepsilon_t$. Show

$$\varepsilon_{training} \leq \Pi_t C_t \leq e^{-2\sum_t \gamma_t^2}.$$

(Note the importance of this conclusion: If $\gamma_t \geq \gamma$ for some $\gamma > 0$, then the error decreases exponentially.)

7. Show that $\varepsilon_t \leq .5$ and that when $\varepsilon_t = .5$ the training error need not go to zero. (**Hint**: Note that, in this case, the $D_t(i)$s are constant as functions of t.)

Exercise 6.6. Kernel methods are flexible, in part because when a decision boundary search is lifted form a low-dimensional space to a higher-dimensional space, many more decision boundaries are possible. Whichever one is chosen must still be compressed back down into the original lower-dimensional data space. RVMs and SVMs are qualitatively different processes for doing this lifting and compressing. Here is one way to compare their properties, computationally.

1. First, let $K(\boldsymbol{x},\boldsymbol{y}) = (\boldsymbol{x}\cdot\boldsymbol{y}+1)^2$ be the quadratic kernel. Verify that when $\boldsymbol{x} = (x_1,x_2)$ the effect of this K is to lift $\boldsymbol{x}$ into a five dimensional space by the feature map

$$\Phi(\boldsymbol{x}) = (x_1^2 x_2^2, \sqrt{2}x_1x_2, \sqrt{2}x_1, \sqrt{2}x_2, 1).$$

2. What is the corresponding Φ if $K(\boldsymbol{x},\boldsymbol{y}) = (\boldsymbol{x}\cdot\boldsymbol{y}+1)^3$?
3. Generate a binary classification data set with two explanatory variables that cannot be linearly separated but can be separated by an SVM and an RVM. Start by using a polynomial kernel of degree 2. Where are the support points and relevant points situated in a scatterplot of the data? (Usually, support vectors are on the margin between classes whereas relevant vectors are in high density regions of the scatterplot as if they were a typical member of a class.)
4. What does this tell you about the stability of RVM and SVM solutions? For instance, if a few points were changed, how much would the classifier be affected? Is there a sensible way to associate a measure of stability to a decision boundary?

Exercise 6.7 (Regularization and Gaussian processes). Consider a data set $\mathscr{D} = \{(\boldsymbol{x}_i,y_i) \in \mathbb{R}^{p+1}, i = 1,2,\cdots,n\}$ generated independently from the model

$$Y = f(\boldsymbol{x}) + \varepsilon,$$

with $\varepsilon \overset{\text{IID}}{\sim} N(0,\sigma^2)$ but that no function class for f is available. To make the problem feasible, write

$$\boldsymbol{y} = \boldsymbol{f} + \boldsymbol{\varepsilon},$$

where $\boldsymbol{f} = (f(\boldsymbol{x}_1), f(\boldsymbol{x}_2),\cdots,f(\boldsymbol{x}_n))^\top = (f_1,f_2,\cdots,f_n)^\top$, $\boldsymbol{y} = (y_1,y_2,\cdots,y_n)^\top$, and $\boldsymbol{\varepsilon} = (\varepsilon_1,\varepsilon_2,\cdots,\varepsilon_n)^\top$. Also, assume $\boldsymbol{f}$ and $\boldsymbol{\varepsilon}$ are independent and that $\boldsymbol{f} \sim N(\boldsymbol{0}, b\boldsymbol{\Sigma})$ for some strictly positive definite matrix $\boldsymbol{\Sigma}$.

1. Let $\lambda = \sigma^2/b$. Show that

$$\mathbb{E}(\boldsymbol{f}|\mathbf{y}) = \boldsymbol{\Sigma}(\boldsymbol{\Sigma}+\lambda\mathbf{I})^{-1}\mathbf{y}. \tag{6.6.2}$$

Hint: Note that $(\boldsymbol{f}^\top \mathbf{y}^\top)^\top$ is normal with mean $\mathbf{0}$ and covariance matrix that can be written as $\begin{pmatrix} A & C \\ C^\top & B \end{pmatrix}$, where $A = \text{Var}(\boldsymbol{f})$, $B = \text{Var}(\mathbf{y})$, and $C = \text{Cov}(\boldsymbol{f},\mathbf{y})$. standard results giving the conditional distributions from multivariate normal random variables give that $\mathbb{E}[\boldsymbol{f}|\mathbf{y}] = CB^{-1}\mathbf{y}$.

2. Deduce that the posterior mean $\hat{\boldsymbol{f}}_n = \mathbb{E}(\boldsymbol{f}|\mathbf{y})$ can be represented as a linear smoother.
3. Consider classical ridge regression; i.e., find $\boldsymbol{f}$ to minimize the regularized risk

$$R_\lambda(\boldsymbol{f}) = \|\mathbf{y}-\boldsymbol{f}\|^2 + \lambda \boldsymbol{f}^\top \boldsymbol{\Sigma}^{-1}\boldsymbol{f}.$$

a. Show that $R_\lambda(\boldsymbol{f})$ is minimized by

$$\boldsymbol{f}_\lambda = (\mathbf{I}+\lambda\boldsymbol{\Sigma}^{-1})^{-1}\mathbf{y}.$$

b. Show that $\boldsymbol{f}_\lambda$ can be written in the form of (6.6.2).

c. Show that the posterior mean of $\boldsymbol{f}$ given $\mathbf{y}$ is the ridge regression estimate of $\boldsymbol{f}$ when the penalty is $\boldsymbol{f}^\top \boldsymbol{\Sigma}^{-1}\boldsymbol{f}$.

Exercise 6.8. Using the setting and results of Exercise 6.7, do the following.

1. Verify that the risk functional $R_\lambda(\boldsymbol{f})$ in item 3 of 6.7 can be written as

$$R_\lambda(\boldsymbol{f}) = \frac{1}{n}\sum_{i=1}^{n}(y_i - f(\boldsymbol{x}_i))^2 + \lambda \|\boldsymbol{f}\|^2_{\mathscr{H}_K},$$

where $\mathscr{H}_K$ is a reproducing Kernel Hilbert space and $\|f\|^2_{\mathscr{H}_K}$ is the squared norm of f in $\mathscr{H}_K$ for some kernel function $K(\cdot,\cdot)$.

2. Identify a Gaussian process prior for which $\|f\|^2_{\mathscr{H}_K}$ reduces to $\boldsymbol{f}^\top\boldsymbol{\Sigma}^{-1}\boldsymbol{f}$.
3. How does the kernel K appear in the representation of the estimator $\hat{f} = \mathbb{E}[\boldsymbol{f}|\mathbf{y}]$?

Chapter 7
Computational Comparisons

Up to this point, a great variety of methods for regression and classification have been presented. Recall that, for regression there were the Early methods such as bin smoothers and running line smoothers, Classical methods such as kernel, spline, and nearest-neighbor methods, New Wave methods such as additive models, projection pursuit, neural networks, trees, and MARS, and Alternative methods such as ensembles, relevance vector machines, and Bayesian nonparametrics. In the classification setting, apart from treating a classification problem as a regression problem with the function assuming only values zero and one, the main techniques seen here are linear discriminant analysis, tree-based classifiers, support vector machines, and relevance vector machines. All of these methods are in addition to various versions of linear models assumed to be familiar.

The obvious question is when to use each method. After all, though some examples of these techniques have been presented, a more thorough comparison remains to be done. A priori, it is clear that no method will always be the best; the "No Free Lunch" section at the end of this chapter formalizes this. However, it is reasonable to argue that each method will have a set of functions, a type of data, and a range of sample sizes for which it is optimal – a sort of catchment region for each procedure. Ideally, one could partition a space of regression problems into catchment regions, depending on which methods were under consideration, and determine which catchment region seemed most appropriate for each method. This ideal solution would amount to a selection principle for nonparametric methods. Unfortunately, it is unclear how to do this, not least because the catchment regions are unknown.

There are three ways to characterize the catchment regions for methods. (i) One can, in principle, prove theorems that partition the class of problems into catchment regions on which one method is better than the others under one or another criterion. (ii) One can, systematically, do simulation experiments, using a variety of data types, models, sample sizes, and criteria in which the methods are compared. (iii) The methods can be used on a wide collection of real data of diverse types, and the performance of methods can be assessed. Again, the conclusions would depend on the criterion of comparison. Cross-validation would give different results from predictive mean integrated squared error, for instance. The use of real data can highlight limitations of methods and

B. Clarke et al., *Principles and Theory for Data Mining and Machine Learning*, Springer Series in Statistics, DOI 10.1007/978-0-387-98135-2_7, © Springer Science+Business Media, LLC 2009

provide assessments of their robustness. An extra point often ignored is that real data are often far more complex than simulated data. Indeed, methods that work well for simulated data may not give reasonable results for highly complex data.

The first approach promises to be ideal, if it can be done, since it will give the greatest certainty. The cost is that it may be difficult to use with real data. On the other hand, the last two approaches may not elucidate the features of a data set that make it amenable to a particular method. The benefit of using data, simulated or real, is that it may facilitate deciding which method is best for a new data set. This is possible only if it can be argued that the new data have enough in common with the old data that the methods should perform comparably. Obviously, this is difficult to do convincingly; it would amount to knowing what features of the method were important and that the new data had them, which leads back to the theoretical approach that may be infeasible.

Here, the focus will be on simulated data since the point is to understand the method itself; this is easier when there really is a true distribution that generated the data. Real data can be vastly more difficult, and a theoretical treatment seems distant.

The next section uses various techniques presented to address some simple classification tasks; the second section does the same, but for regression tasks. In most cases, the R code is given so the energetic reader can redo the computations easily – and test out variations on them. The third section completes the discussion started at the end of Section 7.2 by reporting on one particular study that compares fully ten regression methods. After having seen all this, it will be important to summarize what the overall picture seems to say – and what it does not.

7.1 Computational Results: Classification

The effectiveness of techniques depends on how well they accomplish a large range of tasks. Here, only two tasks will be examined. The first uses Fisher's iris data – a standard data set used extensively in the literature as a benchmark for classification techniques. The second task uses a more interesting test case of two-dimensional simulated data just called Ripley's data.

7.1.1 Comparison on Fisher's Iris Data

As a reminder, Fisher's iris data are made up of 150 observations belonging to three different classes, with each class supposed to have 50 observations. Each sample has four attributes relating to petal and sepal widths and lengths; one class is linearly separable from the other two but those two are not linearly separable from each other. Let's compare the performances of three of the main techniques: recursive partitioning, neural nets, and SVMs. (Nearest neighbors, for instance, is not included here because the focus of this text is on methods that scale up to high dimensions relatively easily.) It is seen that these three techniques represent completely different function classes – step functions, compositions of sigmoids with linear functions, and margins from optimization criteria. Accordingly, one may be led to conjecture that their performances would be quite different. Curiously, this intuition is contradicted; it is not clear why.

First, we present the results. The contributed R package rpart gives a version of recursive partitioning. Here is the R code:

```
library(rpart)
sub <- c(sample(1:50, 25), sample(51:100, 25),
sample(101:150, 25))
fit <- rpart(Species ~ ., data=iris, subset=sub)
table(predict(fit, iris[-sub,], type="class"),
iris[-sub, "Species"])
```

The data are randomly split into a training set with 75 observations and a test set with 75 observations. The application of rpart to the iris data produces the confusion matrix in Table 7.1 from its predictions. The confusion matrix gives the actual and predicted classifications done by a classification system so that classifiers can be evaluated. In the 2×2 case, there are a true and false positive rate and a true and false negative rate. More generally, the proportion correct is called the accuracy. For Fisher's iris data, recursive partitioning has an accuracy of 72/75, or 96%.

	setosa	versicolor	virginica
setosa	25	0	0
versicolor	0	25	3
virginica	0	0	22

Table 7.1 Confusion matrix for rpart and Fisher's iris data. There are exactly three misclassifications, out of 75, of virginica as versicolor.

The contributed R package nnet gives (see Table 7.2) a version of recursive partitioning. Since neural nets are rough (i.e., the surface formed by a squared error fit criterion as a function of the architecture and coefficients in a large class of neural nets has many peaks and valleys, often with very complicated topography) it is rare to fit neural nets as is. Typically, they are regularized. That is, a penalty term with a decay coefficient is added. In nnet, a squared error fit criterion is regularized by λ times a squared error penalty on the coefficients in the net. (Note that this is a variance–bias trade-off, conceptually separate from a random search as in simulated annealing, for instance, which can be used with neural networks to avoid finding a solution that gets trapped in a local optimum far from the global optimum.)

As with rpart, nnet trains a neural net procedure on 75 randomly selected observations. For a single hidden layer, the default value $\lambda = 5 \times e^{-4}$ gives 19 nodes in the hidden layer. Alternatively, the number of nodes in the hidden layer could have been chosen by cross-validation or by any other procedure that achieves a good trade-off between the bias of too few nodes and the excess variance from too many nodes. Now, the confusion matrix produced by the test points is:

	setosa	versicolor	virginica
setosa	21	0	0
versicolor	0	25	0
virginica	4	0	25

Table 7.2 Confusion matrix for `nnet` and Fisher's iris data. There are exactly four misclassifications, out of 75, of setosa as virginica.

For Fisher's iris data, neural nets have an accuracy of 71/75, or 94.67%.

The contributed R package `e1071` gives (see Table 7.3) an implementation of support vector machines. Again, the data are split into a training set of 75 randomly selected observations, with the remaining 75 set aside for testing. The kernel used for the SVM in this case is Gaussian with a bandwidth of 0.25. On the 75 training points, SVM selects 36 support vectors and produces the following confusion matrix:

	setosa	versicolor	virginica
setosa	25	0	0
versicolor	0	24	0
virginica	0	1	25

Table 7.3 Confusion matrix for `e1071` and Fisher's iris data. There is exactly one misclassification, out of 75, of versicolor as virginica.

Note that, in general, SVMs will not fit a training set perfectly. The optimization finds the best margin but is not usually flexible enough to match all test cases perfectly. This can be seen as a liability or as an effort to summarize the variation in the data efficiently, if imperfectly.

The confusion matrix on the remaining 75 test points is given in Table 7.4.

	setosa	versicolor	virginica
setosa	24	0	0
versicolor	1	24	2
virginica	0	1	23

Table 7.4 Confusion matrix for `e1071` and Fisher's iris data. There are exactly four misclassifications, out of 75, of setosa as versicolor, of versicolor as virginica and two of virginica as versicolor.

For Fisher's iris data, SVMs have an accuracy of 71/75, or 94.67%.

The curious fact is that all three methods perform well, with almost the same accuracy. This can have several interpretations. First, it is possible that the three methods really are comparable on this data set. Indeed, it is possible that the classification task is actually rather easy, so that routine tuning of the three methods can readily perform well. Second, it is possible that one of the methods is much better or much worse than the other two and this is being masked by the relatively coarse tuning and application here. If the best (although unknown) method were applied carefully (e.g., with a well-chosen tuning parameter or kernel or other variant on the defaults used here), then it would be revealed as better than at least one of the two others. Third, it is possible that many standard methods have been applied to many standard data sets such as Fisher's iris data, and most packages will therefore do rather well since they have been developed on benchmark data sets. Without careful work, it would be impossible to discriminate among these possibilities.

Ensemble methods, which can scale up quite effectively, are not examined computationally here because they are often complex, as they are not individual functions. Ensemble-based procedures are typically better for prediction than individual regressors or classifiers. This is partially because they are the result of a more involved optimization and partially because ensemble-based procedures take model uncertainty into account. In fact, R computing packages for boosting, random forests, and BART (among other techniques) are readily available to complete the comparison of procedures this section only initiates for the reader.

7.1.2 Comparison on Ripley's Data

As a second comparison of some of the classification methods here, consider the well-known 2D Ripley data set. This consists of two classes, where the data for each class were generated by a mixture of two normal distributions. The training set has 250 observations, with 125 from each of the two classes. The test set has 1000 observations, with 500 from each class. The training set is contained in the file synth.tr. The test set is in the file synth.te. Both are easy to find on the web. Each of the 350 rows of data has three entries: a point in the plane and a code of 0 or 1 for the class. The two classes are strongly overlapping, so the performance of any classifier is limited. That is, Ripley's data are not linearly separable in the space in which they are represented as can be seen in Fig. 7.1.

Again, the performance of recursive partitioning, NNs, and SVMs can be compared. A fourth classifier, RVMs, is also included because it seems to be extremely effective for achieving sparsity. Note that the use of real data is generally a poor initial evaluation of the effectiveness of a technique. It is only on simulated data, or other data where the correct answer is known, that one can determine whether a technique has come close to uncovering the correct answer. Attempting to use a real but ill-understood phenomenon to understand a new and not understood technique can all too easily be a case of obscurum per obscurius – explaining the obscure by the more obscure. It is only after a technique is at least provisionally understood that it makes sense to use the technique to say something about real-world data generators – or to let real-world generators that are well understood say something new about the technique.

7.1.2.1 Recursive Partitioning

Again, using the R package rpart and defaulting to its standard inputs will give a sense of how recursive partitioning works. For instance, the splitting rule is Gini, and ten-fold cross-validation is used to find an optimal tree. Thus, within each cross-validation run, minimal cost-complexity pruning is used to find the optimal tree for that run. Note that even though a specific tree generated from one-tenth of the data is chosen from all the cross-validation candidates, all of the data are used in providing the collection of trees being compared as well as in the validation. The R code is:

```
library(rpart)
data(synth.tr)
data(synth.te)
t.tr <- class.ind( c(rep("0", 125), rep("1", 125)) )
t.te <- class.ind( c(rep("0", 500), rep("1", 500)) )
rpart.rip.fit <- rpart(t.tr ~ xs + ys, data=synth.tr)
test.cl <- function(true, pred)
true <- max.col(true)
cres <- max.col(pred)
table(true, cres)
test.cl(t.tr, predict(rpart.rip.fit, synth.tr[,-3]))
test.cl(t.te, predict(rpart.rip.fit, synth.te[,-3]))
```

The confusion matrices are in Table 7.5. The training error from the sample of 250 is seen to be $22/250 = 8.8\%$ from the left-hand table. The test error from the sample of 1000 is seen to be $100/1000 = 10\%$. As expected, the training error (or fitted error) that results from applying the optimal tree to the data that generated it is a little better than the test error (or generalization error, or out-of-sample error) that results from applying the optimal tree to new data not used in forming it.

	class 1	class 2
class 1	120	5
class 2	17	107

	class 1	class 2
class 1	472	28
class 2	72	428

Table 7.5 Confusion matrices for rpart and Ripley's 2D data. The training set is used on the right, the test set on the left.

An important point is that trees are a very rich class and therefore have a tendency to be unstable in the sense that small changes in the procedure or data that generated a tree can lead to sharply different trees. Another way to say this is that if there were a standard error one could associate to a tree, it would be larger than for many other model classes that are smaller or in which estimation is more robust.

7.1.2.2 Neural Networks

As with the iris data, using the R package nnet and defaulting to its standard inputs will give a sense for how neural nets perform. The nnet package iteratively minimizes the squared error criterion, possibly including a penalty term, using a technique similar to, but more sophisticated than, standard backpropagation. The optimization is called the Broyden-Fletcher-Goldfarb-Shanno (BFGS) method after its creators. It is regarded as one of the best techniques for solving nonlinear optimization problems (in the absence of constraints).

Like backpropagation, the BFGS method is derived from Newton's method. Both are "hill-climbing" procedures that seek the point where the gradient of a function is 0. Recall that Newton's method locally approximates the function by a quadratic on a region around the optimum, using the first and second derivatives to find the stationary point. The BFGS method uses a quasi-Newton approach. The Hessian matrix of second derivatives of the objective function does not need to be computed; instead it is updated by using gradient vectors. The details of the procedure are beyond the present scope; they are numerical analysis.

In practice, nnet only applies to single hidden layer networks. It defaults to the logistic function as a sigmoid and to 100 iterations of the procedure for finding a solution. Another default is that the factor λ on a squared error penalty term is set to zero. However, as with the iris data, the standard value $\lambda = e^{-4}$ is used here. There seems to be some insensitivity to the choice of λ in Ripley's data, but there are also many cases where there is high sensitivity, too, in which case λ should be chosen more carefully, possibly by CV or by any one of numerous other search techniques that have been developed. It can be argued that nonzero values of λ give solutions that are prior driven since the squared error penalty term may be seen as a normal prior on a normal likelihood with its exponent given by the squared error loss term. However, the penalty term helps ensure that the solution gives a good fit on a rough surface, which is more important unless the value of λ used is unreasonably large.

The R code is:

```
library(MASS)
library(nnet)
data(synth.tr)
data(synth.te) t.tr <- class.ind( c(rep("0", 125),
rep("1",125)) )
rpart.rip.fit <- rpart(t.tr ~ xs + ys,
data=synth.tr)
test.cl <- function(true, pred)
true <- max.col(true)
cres <- max.col(pred)
table(true, cres)
nnet.rip.fit <- nnet(t.tr ~ xs + ys, size=16,
decay=5e-4, data=synth.tr)
test.cl(t.tr, predict(nnet.rip.fit, synth.tr[,-3]))
```

```
t.te <- class.ind( c(rep("0", 500), rep("1", 500)) )
test.cl(t.te, predict(nnet.rip.fit, synth.te[,-3]))
```

To use nnet effectively, it is standard to run the program on the same data for a variety of choices of the number of nodes in the hidden layer. Here, the range from 10 to 20 was used – with some hindsight from a series of test runs. For each number of nodes, a fitted error can be found or CV can be used. The result is that 16 hidden nodes gives the best trade-off between high bias when the number of hidden nodes is too small and high variance when the number of hidden nodes is too large. The result is the typical V-shape associated with bias–variance trade-off. With nnet, this error is calculated internally.

Of some concern in the present is that if no initial value for the parameter vector is given, then nnet chooses an initialization at random. Thus, in general, running the same routine on the same data without changing any component in the function call generates different solutions. This is not surprising since neural nets suffer from multiple local optima because they are very rough and strongly nonlinear. It is plausible that the differences between solutions result from the underlying minimizer getting trapped in local minima. This instability can be reduced by pooling the predictions from multiple runs, possibly by majority vote, to get improved prediction; essentially this makes NNs into an ensemble method.

In Tables 7.6 and 7.7, are two runs of the nnet routine on Ripley's 2D classification data set. Each row corresponds to one run; on the left is the confusion matrix based on the original 250 observations from the sample, and on the right is the confusion matrix generated from predicting the class memberships of the 1000 out-of-sample observations. It is seen that the confusion matrices from the two runs are meaningfully different.

	class 1	class 2
class 1	115	10
class 2	12	113

	class 1	class 2
class 1	460	40
class 2	76	424

Table 7.6 Confusion matrices for one run of nnet on Ripley's 2D data. The training set is used on the left, the test set on the right.

	class 1	class 2
class 1	119	6
class 2	13	112

	class 1	class 2
class 1	460	40
class 2	88	412

Table 7.7 Confusion matrices for a second run of nnet on Ripley's 2D data. The training set is used on the left, the test set on the right.

On average, therefore, the training error is $(1/2)(22+19)/250 = 8.2\%$, while the test error remains around $(1/2)(116+128)/1000 = 12.2\%$. Recall that the training error produced by recursive partitioning was 8.8%, just a little higher than for NNs. The test error of trees, however, was much smaller at 10%, compared with 12.2% for neural networks. This is surprising, as one would expect more overfitting from trees than from neural nets (since trees are far more flexible for identifying regions) and therefore a better generalization error from neural nets than from trees.

Note that, for Ripley's data, both neural nets and trees give slightly worse performances than they did for the iris data. This may be because Ripley's classification problem is harder even though its dimension is lower. The nonseparability of classes limits the effectiveness of any technique.

7.1.2.3 SVMs

As with the Iris data, using the R package e1071 and defaulting to its standard inputs will give a sense for how SVMs perform. Recall that the central mathematical idea in SVM optimization is the conversion of the geometric problem to a quadratic optimization problem, possibly with a kernel. Many quadratic optimizers exist; any optimization package that solves linearly constrained convex quadratic problems can be used. The svm command uses a procedure based on an interior point method for nonlinear programming (see Vanderbei and Shanno (1999)) a concise summary is in `homepages.cwi.nl/~sem/Lunteren/vanderbei1.ps`.

The svm command is also good for regression using the ε-insensitive loss (see below), sometimes called C-classification to emphasize the cost C appearing in the optimization problem. In svm, the default value is $C = 1$ in the Lagrangian. In addition, the default bandwidth in the RBF kernel is $1/p$, where p is the dimension of $\boldsymbol{x}$. For Ripley's data, this is $1/2$; for the iris data, it was $1/4$. The default error criterion for the optimization is .001. Finally, the λ is chosen internally as a tuning parameter. It could be done using a CV technique but in fact is done by a routine called sigest within the kernlab package. sigest uses a technique based on looking at quantiles of the norms of differences between the explanatory variables. In practice, any quantile between the 10th and 90th is believed to work satisfactorily.

Here, svm was used on Ripley's data with radial basis function kernels (other kernels are possible). The test error was 9%, a little better than for trees or neural nets, probably because of the sparsity. The decision boundary for SVMs, along with the data set, can be seen in the top panel in Fig. 7.1. The two classes are denoted by pluses (top part of the panel) and "x"s (mostly in the lower part); when they are on a black line, it means that the data point was a support vector for its class. The decision boundary is the lighter streak running from .8 up to the top of the frame and then turning down to give the lighter boundary meeting the bottom f the frame near .45. Although unusual in two dimensions, this is not surprising because the point of kernel methods is to get linearity in the transformed space. Once this is transformed back to the original space, the appearance can be highly nonlinear.

There are 103 support vectors; this is a large number relative to the 250 vectors in the training set, so the SVM solution is not sparse at all. In terms of test error, SVMs do a little better than neural nets and trees. Note that some of the support vectors are quite far from the two-dimensional boundary. In the transformed space, this distance may be

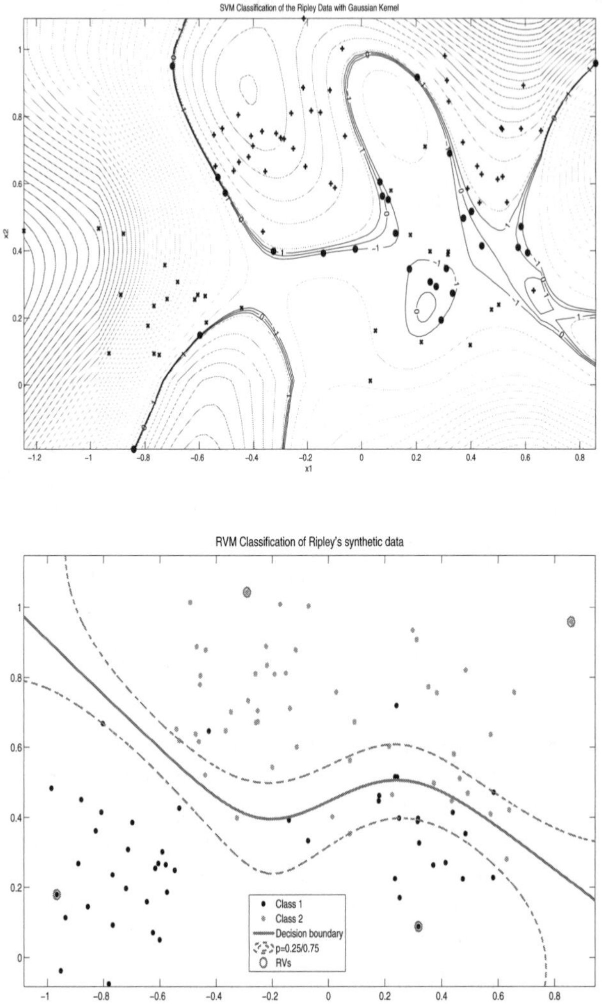

Fig. 7.1 Top: SVM on Ripley's data. The points in the two classes are indicated by plus and x. Solid dots indicate support points and it is seen that the decision boundary passes through many of them. Note that in contrast to the SVM classification in the lower panel, the curves are much more undulating and complicated. This is so because they are projected down into two dimensions from a much higher dimensional space. (Although svm is the easiest package to use, the top panel was generated from another package which gives more esthetic black-and-white graphics.) Bottom: RVM fit on Ripley's data. The points on the lower portion are mostly class 1 and the points on the upper portion are mostly class 2. Relevance vectors are indicated with small circles. The curves above and below the decision boundary indicate where the probability of membership in a class was between .25 and .75; each point is in its class of higher probability.

much smaller because nonlinear, nonseparable SVMs often do a good job of isolating regions. This occurs because a large, diverse collection of functions are permitted to define a decision boundary.

7.1.2.4 RVMs

A Matlab implementation of SparseBayes V1.0 is available from Tipping's webpage

`http://research.microsoft.com/mlp/RVM/default.htm`

and using its standard inputs will give a sense of how RVMs perform. Despite their differences, it is reasonable to compare SVMs and RVMs because both generate solutions that depend on selecting data points to give sparsity. The key difference is that SVMs are a maximum marginal distance criterion over a collection of functions that represent decision boundaries, while RVMs (for classification) come from Bayesian logistic regression and a series of normal approximations, so the prior forces sparsity.

The implementation of RVMs used Laplace's method for asymptotic normality. In particular, the differentiation with respect to the α_is to maximize the likelihood for y^n with the w_is integrated out is called a Type II maximum likelihood since it pertains to the hyperparameters rather than to the parameters themselves. The hyperparameters are thresholded as in the two-step iterative procedure.

In Fig. 7.1, a radial basis function kernel is used on the 2D Ripley data. The two classes are denoted by light and dark dots. There are four relevant vectors, indicated by circles around the dots. The decision boundary is the middle curve. The upper and lower curves result from connecting points that have high enough probabilities of being in their respective classes. That is, the upper margin connects the points that have probability at least .75 of being class 2. This is similar to, but different from, saying one-quarter of the data are above the upper curve and one-quarter below the lower curve, leaving 50% between them.

Like SVMs, RVMs give a closed-form expression for the decision boundary in terms of some data points and the kernel but in terms of the four relevant vectors rather than the 103 support vectors. Thus, as is typical, RVMs give more sparsity than SVMs and in this example RVMs also give a smaller test error, 8.6% versus 9% for SVMs. Also, as is seen in this example, the relevant vectors tend to be close to where typical points from classes would be in contrast to SVMs, where the support vectors track, indeed define, the boundary. Indeed, it is tempting to identify the four relevant vectors with the means of the four normals used to generate Ripley's data.

The good performance of RVM's for the Ripley data, better than the other three methods, is probably typical for this class of examples, but of course will not necessarily extend to others. Recall that RVM is a succession of approximations from the way it handles the logistic model to the Laplace approximation that it uses to turn non-normal conditional likelihoods into approximate normal quantities for analytic and computational tractability. Thus, the structure of Ripley's data, mixtures of normals, may make the normal approximation perform better than the originally hypothesized logistic model. Indeed, if the data to be classified comprise groups that are approximately normal or mixtures of groups that are approximately normal, then the normal approximation may make the model fit the data better. This would lead RVMs to have better performance than would otherwise be the case.

7.2 Computational Results: Regression

In this section, the performance of six of the methods presented in Chapters 4, 5, and 6 are assessed on two functions. The first, called Vapnik's sinc function is

$$y_V = \frac{\sin x}{x} \quad \text{on} \quad [-5\pi, 5\pi].$$

Its appearance can be seen in any of the graphs in the next section. Despite its mathematical form, it is a smooth real-valued function of a single real variable. It is a "nice" function in that it doesn't have discontinuities, corners, or even high curvature. It is symmetric with one global maximum at 0 and smaller local maxima and minima of decreasing amplitude as x moves away from 0.

The second function, called Friedman's function is

$$y_F = 10\sin(\pi x_1 x_2) + 20(x_3 - 5)^2 + 10x_4 + 5x_5 \quad \text{on} \quad [0,1]^5.$$

It is a real-valued function of five variables. When used as a test case, often extra variables are simulated so that the regression task involves finding the correct variables as well as finding the correct functional form for the correct variables. Being a surface in five dimensions, it is difficult to describe the shape of this function. However, it is a smooth function, and its intersections with the coordinate planes are harmless, as can be seen in figures from the subsection on GAMs.

The six regression methods to be applied to these functions are trees, neural nets, SVMs, RVMs, GAMs and Gaussian processes. This covers all the major classes, apart from spline based methods, examples of which were seen at the end of Chapter 3. Note that the comparisons given here, like the earlier ones for classification, are based on defaults. Thus, the conclusions might change if a more sophisticated application of the methods were used. Moreover, the two functions considered here are both relatively nice and simple. They do not represent functions typical of applications, which will often be more complicated. As a consequence, methods intended for much more difficult functions may not be well suited to Y_V and Y_F. Indeed, there is evidence that complexity concepts are germane to regression: More complex methods (e.g., trees) may perform better on refractory functions than simpler methods do, while simpler methods (e.g., Gaussian processes) may perform better on simpler problems than more complicated methods do.

7.2.1 *Vapnik's* sinc *Function*

To get data representative of the Vapnik function, choose 1000 equally spaced points x in the interval $[-5\pi, 5\pi]$. For each of these points, generate a Y_V value from

$$Y_V = \frac{\sin x}{x} + \varepsilon,$$

where ε is $N(0, 0.2^2)$. Next, randomly draw $n = 100$ points from the 1000 values of x. These x_is with their corresponding $y_{V,i}$s, for $i = 1, ..., 100$, form the data to be used for training a model. The remaining 900 points are set aside for estimation of the test error. The R-code for defining the function is:

```
f <- function(x)
r <- sqrt(apply($x^2$,1,sum))
r <- x
z <- sin(r)/r
z[is.na(z)] <- 1
return(z),
```

and the R code for generating the data is:

```
m <- 1000
n <- 100
pop <-  seq(-5*pi,5*pi, length=m)
train <- sample(m,n)
x <-  sort(pop[train])
w <-  sort(pop[-train])
noise <- rnorm(length(x), mean=0, sd = 0.2)
y <- f(x) + noise
v <-  f(w) + rnorm(length(w), mean=0, sd = 0.2)
```

The six techniques are evaluated on the basis of training error and test error, as well as the root mean square error for these. Also, the estimated functions are plotted along with the data and the actual function so the fit can be seen. For trees and neural nets the R packages rpart and nnet continue to be used. For generalized additive models, the package gamair is used. For the remaining three regression techniques, SVM, RVM, and Gaussian process priors for regression (GPs) the contributed R package kernlab is used. It provides ready-to-use R functions for many supervised (and unsupervised) learning techniques.

As with classification, the use of real data to determine how well a regression technique performs is generally not possible unless a fair amount is already known about the data. It may be possible to compare techniques with each other in a predictive sense on a particular data set, but the conclusions will not necessarily generalize. Here, the task is to understand the basics of the techniques, so the examples involve simulated data.

7.2.1.1 Regression Trees on sinc

For regression trees, the rpart default is to ANOVA splitting, meaning that the impurity criterion is the reduction in the sum of squares as a consequence of a binary split at a node. Other splitting criteria are possible but are often used for more specialized purposes, like survival trees. The cost-complexity function is also based on squared error, and tenfold cross-validation is used, as in the classification setting. The R code is the following:

```
sinct <- data.frame(w,v)
sinc  <- data.frame(x,y)
rpart.sinc <- rpart(y~x,data=sinc)
rpart.sinc.fit <- predict(rpart.sinc,sinc)
rpart.sinc.tst <- predict(rpart.sinc, sinct)
sqrt(mean((v-rpart.sinc.tst$)^2$))
plot(x, y)
lines(x, f(x), col = 2)
lines(x, rpart.sinc.fit, col = 4)
```

The fitted or training error is $\sum_{i=1}^{100}(y_i - \hat{y}_i)^2/100 = .0586$; the corresponding expression for the test error on the holdout data is much larger at .2285. Thus, the $RMSE = .478$. A better evaluation would be to compute the test error for a large number of independent data sets and take the average. Each data set would generate a graph as in Fig. 7.2, which shows the plot of the fitted curve, the data, and sinc. (In fact, one could take the average of such curves to give predictions.) It is common for fitted errors to be smaller than generalization errors because of the variance bias tradeoff: Usually, the variance for the fitted and predicted values will be much the same, but the bias for the predicted will be much larger than for the fitted. In most of the examples given here, this will typically be observed. The only way to avoid this would be to tolerate more bias in the fitted values, but there is no guarantee there will be a corresponding reduction in the bias of the predictions.

Looking closely at the figure, it is seen that there are seven leaves – one for each value of the step function; the value of the fitted function at a point is roughly the average of the data points in the bin it represents. That is, the value about .8 on $[-2,2]$ is roughly the average of the six data points above .8 and the six or seven data points below .8. The fitted function tracks the five biggest modes of the sinc function but flatlines beyond ± 10 and so misses the finer structure. Of course, trees are discontinuous and sinc is smooth, so even though there are limiting senses in which trees can approximate any function, the approximation can be poor for any reasonable number of leaves if the leaves divide one region finely but leave other regions summarized by too few leaves. In addition, the class of trees is very large, and multiple trees can provide roughly equally good fits to a given data set.

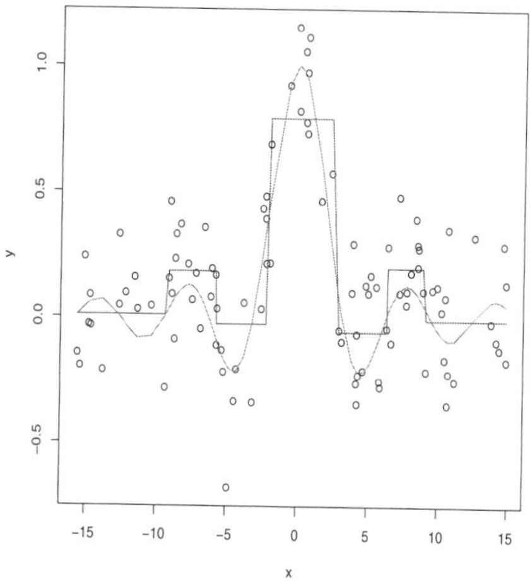

Fig. 7.2 Recursive partitioning model fit on the sinc function.

7.2.1.2 NNs on sinc

As with classification, the R package nnet permits selection of a NN model on the basis of the number of nodes in a single hidden layer (the size) and the magnitude of the weight or decay on the penalty term used for regularization. While NNs have impressive theoretical properties, in practice they can be exquisitely sensitive to the data and optimization routine. This makes them somewhat unstable unless the analyst is very experienced. One consequence of this is that the test error is not as reliable an indicator for fit as one might want: The entire curve can be biased up but have small test error or can have some points where fit is extremely good interspersed with points where it is poor, giving an extremely rough approximation. Indeed, often "connecting the dots" from estimating at the design points gives a curve that is quite different in appearance from the true model. It is usually necessary to examine a residual plot and look at the estimated function to see if it looks reasonable.

To explore the neural net fits, consider using two, three, four or five nodes and three values of λ, say 5, 2.5, and 1.25. The R code to generate the fits is:

```
library(nnet)
nn.model <- nnet(t(x), t(y), size=3, linout = FALSE,
entropy = FALSE, softmax = FALSE, censored = FALSE,
skip = FALSE, rang = 0.7, decay = 0.05, maxit = 100,
Hess = FALSE, trace = TRUE, MaxNWts = 1000,
abstol = 1.0e-4, reltol = 1.0e-8)
nn.pred <- predict(nn.model,t(x))
plot(x, y) lines(x, f(x), col = 2) lines(t(x), nn.pred,
col = 4)
```

The actual coefficients can also be obtained, but usually these are not very informative due to lack of interpretability. The test and training errors can be compiled into tables for the 12 cases. The test errors are

	2	3	4	5
5	.1291	.1671	.1223	.1467
2.5	.1107	.1819	.1904	.0990
1.25	.1634	.1337	.1021	.0948

and the training errors are

	2	3	4	5
5	.0515	.4664	.0401	.0784
2.5	.0308	.0247	.0175	.1028
1.25	.1634	.0092	.0048	.1944

Taken together, these tables suggest the best fit has three or four nodes, but the predictively best nets have five nodes (or more). In fact, NNs with five nodes have a fairly large number of parameters relative to the sample size $n = 100$. (Using six nodes leads to bizarre fits, sometimes even flat lines.) So, for the sake of sparsity, it's better to look further at three or four nodes; predictively, it is well known that more parameters will often give better predictions because of better fit, but only if there are enough data to estimate them. In addition, the fit seems to be better with smaller values of λ, say less than 1.25. The small value for $\lambda = 2.5$ and two nodes is a little isolated from the other small values and may be an anomaly.

Before examining the new cases, this first collection of errors suggests, it's good to look at the actual fits of the 12 neural nets generated (see Figs. 7.3–7.6). It can be seen that two or five nodes give poor fit: Neural nets with two nodes seem unable to match the central mode of sinc well and five nodes leads to biased-looking curves.

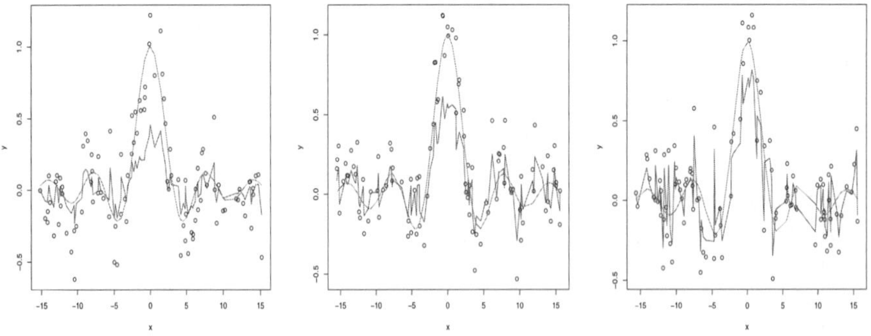

Fig. 7.3 Neural network fits for the sinc function for $h = 2$ nodes in the single hidden layer and decay $\lambda = 5, 2.5, 1.25$. These curves have a hard time fitting the central mode.

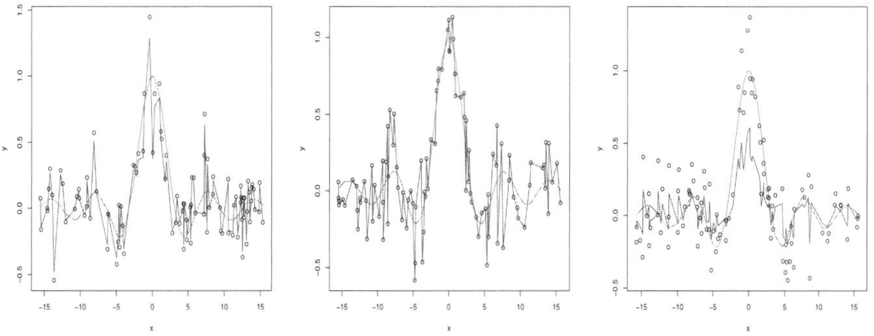

Fig. 7.4 Neural network fits for the sinc function for $h = 3$ nodes in the single hidden layer and decay $\lambda = 5, 2.5, 1.25$. These curves fit the central mode but are quite rough in the tails.

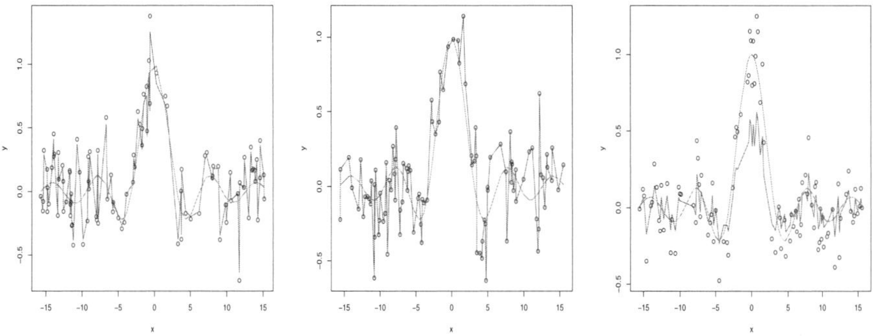

Fig. 7.5 Neural network fits for the sinc function for $h = 4$ nodes in the single hidden layer and decay $\lambda = 5, 2.5, 1.25$. These curves fit the central mode but are even rougher in the tails.

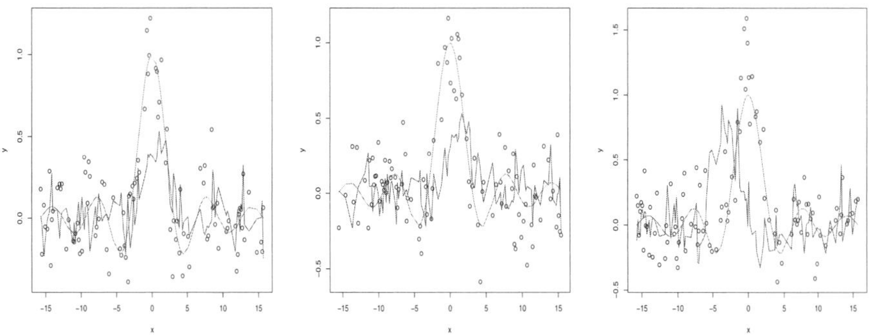

Fig. 7.6 Neural network fits for the sinc function for $h = 5$ nodes in the single hidden layer and decay $\lambda = 5, 2.5, 1.25$. These curves are so unstable that they have high variability and high bias; they do not capture any features of sinc accurately.

Given this, consider four further plots for three or four nodes and values .5 and 5×10^{-4} for the decay. The corresponding table for the test errors is

	3	4
.5	.2951	.2499
.0005	.2054	.2511

and for the training errors is

	3	4
.5	.0012	.0010
.0005	3.29×10^{-9}	1.49×10^{-9}

The corresponding fits are seen in Figs. 7.7 and 7.8.

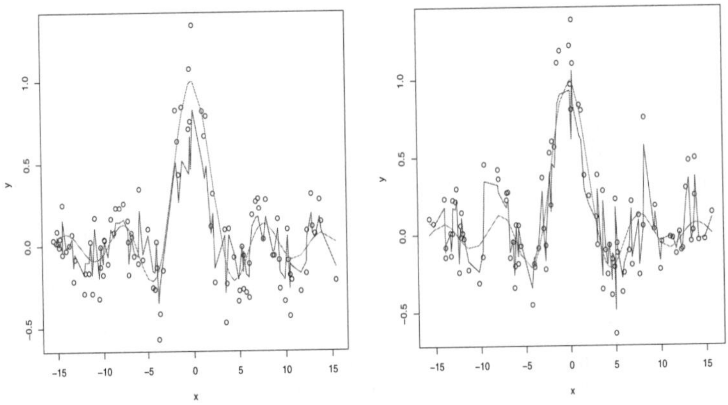

Fig. 7.7 Neural network fits for the sinc function for $h = 3$ nodes in the single hidden layer and decay $\lambda = .5, 10 \times 10^{-4}$.

These plots roughly suggest that the better the neural net fits the central mode, the more unstable it appears to the left and right of the main mode. There is a continual tendency for the fitted curve to be systematically below the sinc curve around the main mode. Overall, it looks as though four nodes with .5 decay provides as good a fit as will be achieved, although three nodes with a larger decay also looks reasonable.

7.2.1.3 SVMs on sinc

Recall that SVMs were developed in Chapter 5. kernlab is an R package authored by Alexandros Karatzoglou, Alex Smola, and Kurt Hornik to implement some of the

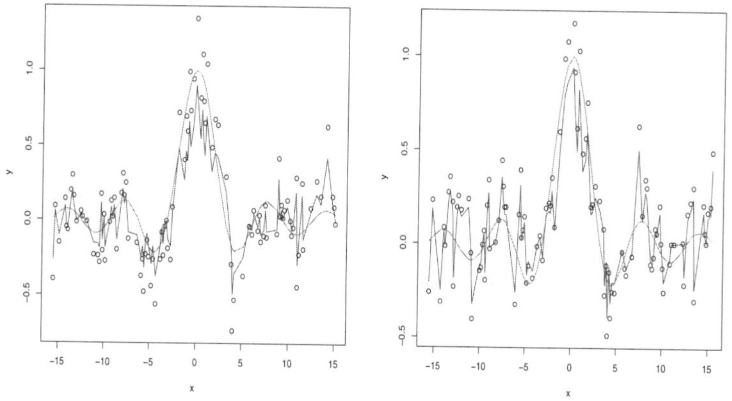

Fig. 7.8 Neural network fits for the sinc function for $h = 4$ nodes in the single hidden layer and decay $\lambda = .5, 10 \times 10^{-4}$.

most popular kernel methods used in statistical machine learning. The version of kernlab used here was published in the *Journal of Statistical Software*, (Vol. 11, Issue 9, 2004). Of relevance here, kernlab covers the RVM (only regression), the SVM (regression and classification), and Gaussian processes. kernlab also covers a variety of other kernel-based techniques, such as kernel PCA and kernel canonical correlation, as well as providing some quadratic programming routines and Cholesky decompositions. Like all R packages, it can be downloaded straight from the R project site, and readily used by invoking library(kernlab). A good way to start is to download Karatzoglou et al. (2004) or get the manual from within R itself. Typing help(ksvm) from the R prompt (command line) provides details on how kernlab implements SVMs including kernel choice, estimating tuning parameters, the default value of some constants, and so forth. This help also lists references to the main research articles used in implementing the package.

For the SVM estimation, the function ksvm is used. In the present case, the R code for using a radial basis function kernel is:

```
library(kernlab)
ksvm.model $\leftarrow$ ksvm(x, y, scaled = TRUE,
kernel ="rbfdot", kpar = "automatic", type="eps-svr",
epsilon = 0.1, cross = 3, fit = TRUE, tol = 0.001)
ksvm.fit <-  predict(ksvm.model, x)
ksvm.tst <- predict(ksvm.model, w)
x11()
plot(x,y, lty = 0, main = "SVM estimation")
lines(x, f(x), lty = 1)
lines(x,ksvm.fit, lty = 2)
legend("topright", c("data", "true", "SVM fit"),
```

```
lty = c(0, 1, 2), pch = c(0, 1, 2), merge = TRUE)
[this computes training and test errors]
ksvm.tr.err <- sqrt(mean((y-ksvm.fit$)^2$))
ksvm.ts.err <- sqrt(mean((v-ksvm.tst$)^2$))
```

The earlier discussion on the defaults in svm for classification continues to apply here. In addition, the default for ε-insensitive loss regression in SVM is $\varepsilon = .1$ and for the cost is $C = 1$.

Now, the estimation yields the fit shown on the right in Fig. 7.9. On the left, the same commands are used, but for the Laplace kernel. The training and testing errors are shown in the table below; as in other cases, they are square rooted so the error is the RMSE. The output from these computations gives that there are 93 support vectors with the Laplace kernel and a few less, 89, with the RBF kernel.

	RBF	Laplace
Train	.1930	.1021
Test	.2293	.2601

The errors shown in the table are typical in that the test error on the holdout set is larger than the fitted error for each kernel. Note that the RBF kernel has a larger fitted error than the Laplace but a smaller test error than the Laplace. This suggests that the cusp in the Laplace kernel helps pile mass on the data points but that this strategy does not generalize as well as letting the mass spread out a little more the way the RBF kernel may do. As can be seen in the figure, the Laplace kernel gives a choppier appearance, which fits well in regions of high curvature but not so well in regions of low curvature. Otherwise, the fitted curves are similar because both Laplace and RBF are unimodal.

Kernels that are polynomial in the inner product may not do as well as kernels that are more general unimodal functions of the inner product unless the components of the function are conveniently expressed in terms of inner products; e.g., polynomials in $\langle \mathbf{x}_n, \mathbf{x}_{new} \rangle$. Thus, RBF or other unimodal kernels may summarize the components of general functions better. Indeed, as expected, spline and polynomial kernels with the sinc function data yielded poor results: The fitted curve was essentially a straight line with a slight positive slope, nothing like the sinc function.

Although not shown here, the support vectors tend to track the curve; in fact, SVM regression was designed for good generalization error. An important fact about SVMs in general is Cover's theorem (see Cover (1965)) which states that any data set becomes arbitrarily separable as the dimension of the space in which it is embedded grows. That is, one can always find a space with high enough dimension to separate points linearly. The limitation is that it may be hard to find Φ and it may have poor generalization error. In classification, it is clear that the support points track the boundary; indeed define the boundary, however, in SVM regression, it is hard to state an analog of this. It is not clear how the support points situate themselves relative to the curve in general.

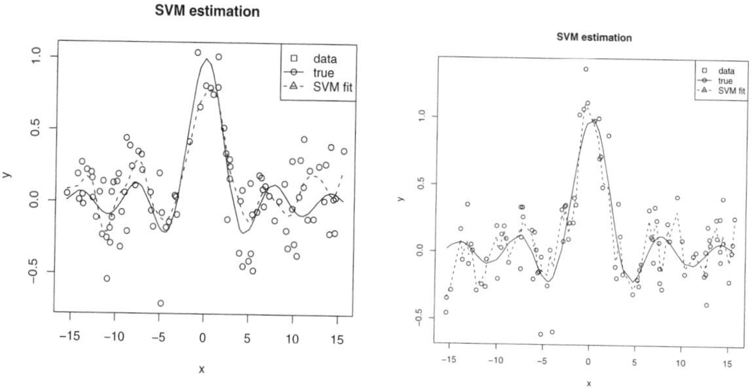

Fig. 7.9 SVM fit on data drawn from a noisy sinc function. The RBF kernel was used to generate the left panel and the Laplace kernel was used on the right.

7.2.1.4 RVMs on sinc

Recall, RVMs were developed in Chapter 6. To assess the performance of the RVM method on sinc, the same package, kernlab, is used, this time through its function rvm. The R-code is:

```
library(kernlab)
rvm.model <- rvm(x, y, type = "regression",
kernel = "rbfdot", kpar = "automatic", alpha = 5, var
= 0.1, iterations = 100, verbosity = 0, tol =
.Machine\$double.eps, minmaxdiff = 1e-3, cross = 0,
fit =TRUE)
rvm.fit <- predict(rvm.model, x)
rvm.tst <- predict(rvm.model, w)
x11() <- plot(x,y, lty = 0, main = "RVM estimation")
lines(x, f(x), lty = 1)
lines(x,rvm.fit, lty = 2)
legend("topright", c("data", "true", "RVM fit"), lty = c(0, 1, 2),
pch = c(0, 1, 2), merge = TRUE)
[compute the training and test errors]
rvm.tr.err <- sqrt(mean((y-rvm.fit$)^2$))
rvm.ts.err <- sqrt(mean((v-rvm.tst$)^2$))
```

Here, kpar indicates the parameters (if any) in the kernel; the defaults for the RBF are used above. Also, alpha indicates the initial vector of regression coefficients, var indicates the initial noise variance, and iterations, verbosity, tol, and minmaxiff indicate properties of the numerical convergence. (cross = 0 means that CV is not used.)

The fits provided by RVM for the RBF, Laplace, and spline kernels are given in Fig. 7.10. The corresponding training (estimation) and test (out-of-sample) errors for these cases are in the table below.

	RBF	Laplace	Spline
Train	.2108	.1703	.200
Test	.2098	.2100	.3500

Surprisingly, the test error for the RBF is smaller than its training error. This is likely just an anomaly since the two are very close. The Laplace and RBF kernels are roughly comparable, but the spline kernel has an unusually large test error, indicating that while it can fit the data it doesn't readily permit generalization, possibly because a function like $x\log x$ is increasing for $x > 1$ and so is much like using a linear or polynomial kernel, whereas RBF and Laplace kernels are unimodal

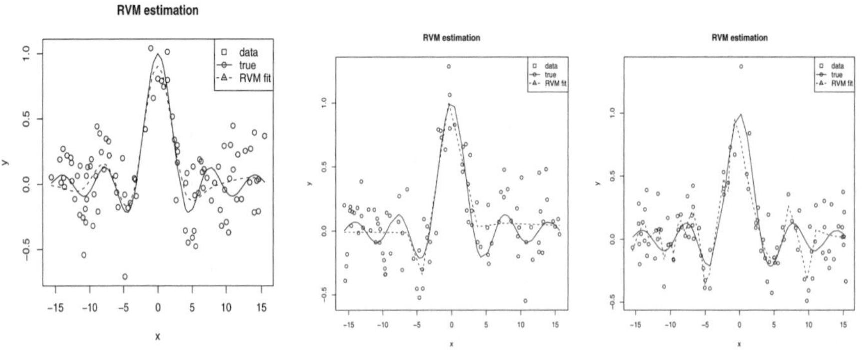

Fig. 7.10 Relevance vector machine fits on data drawn from a noisy sinc function using, from left to right, the RBF, Laplace, and spline kernels.

The computations also show that there are 32 relevant vectors for the spline kernel, seven for the RBF, and eight for the Laplace. This is much lower than the number of support vectors for SVMs, but RVMs were designed mainly for sparsity, as opposed to generalization error. In classification contexts, the relevant points often appear to be prototypical points so that data points often cluster around them. It is not clear what the analogous interpretation for RVM regression is.

Note that the regression functions used for predictions from both the SVM and RVM regression procedures are of the form

$$f(\boldsymbol{x}) = \sum_{i=1}^{n} \alpha_i K(\boldsymbol{x}_i, \boldsymbol{x}),$$

where most of the α_is are zero. The terms with $\alpha_i \neq 0$ correspond to the $\boldsymbol{x}_i$s that are support vectors or relevant vectors. That is, the prediction of a new value is based on how far $\boldsymbol{x}_{new}$ is from certain carefully chosen data points already obtained. Typically, it is the support or relevant vectors farthest from $\boldsymbol{x}_{new}$ that contribute most to the function value at $\boldsymbol{x}_{new}$ since $K(\boldsymbol{x}_i, \boldsymbol{x})$ is largest for them. This is counterintuitive since most nonparametric methods try to approximate function values at new design points by the function values at the design points closest to them. However, this seems to be a general property of kernel methods emerging from the representer theorem (see Zhu and Hastie (2001)) where kernel logistic regression is developed and a regularized classification scheme based on logistic regression is presented, leading to the "import vector machine".

7.2.1.5 Gaussian Processes on sinc

Recall that GPs were discussed in Chapter 6 under Bayesian nonparametrics. Because of the recent interest in GPs as a technique for regression, it is interesting to see how they compare with the other frequentist techniques here. Within the kernlab package, the function gausspr can be used to obtain a Gaussian process fit for the sinc function.

GPs are specified by a mean and a covariance function. The mean is a function of $\boldsymbol{x}$ (which is often the zero function), and the covariance is a function $r(\boldsymbol{x}, \boldsymbol{x}')$, which expresses the expected covariance between the value of the function f at the points $\boldsymbol{x}$ and $\boldsymbol{x}'$. The actual function $f(\boldsymbol{x})$ in any data-modeling problem is assumed to be a single sample from this Gaussian distribution. Various rs can be chosen; here they are taken to be kernels so that $r(\boldsymbol{x}, \boldsymbol{x}') = K(\boldsymbol{x}, \boldsymbol{x}')$ and the RBF kernel is the default.

In the model $Y = f(\boldsymbol{x}_i) + \varepsilon_i$, where the ε_is are assumed $N(0, \sigma^2)$, the posterior distribution is

$$p(y^n | f(\boldsymbol{x}_1), \ldots, f(\boldsymbol{x}_n)) = [\Pi_{i=1}^n p(y_i - f(\boldsymbol{x}_i))] \times \frac{e^{(1/2)\mathbf{f}K^{-1}\mathbf{f}}}{\sqrt{(2\pi)^n \det(K)}},$$

where $K = (K(\boldsymbol{x}_i, \boldsymbol{x}_j))_{i,j=1,\ldots,n}$, $\mathbf{f} = (f(\boldsymbol{x}_i), \ldots, f(\boldsymbol{x}_n))$, and $p(\cdot)$ indicates the density of ε_i. Writing $\mathbf{f} = K\alpha$ and taking the log gives

$$\ln p(\alpha | y^n) = -\frac{1}{2\sigma^2} \|y^n - K\alpha\|^2 - \frac{1}{2}\alpha' K \alpha + \text{const},$$

so maximizing the log over α leads to the posterior mode $\hat{\alpha} = (K + \sigma^2)^{-1} y^n$ and therefore leads to fitted values and predictions from

$$y(\boldsymbol{x}_{new}) = (K(\boldsymbol{x}_{new}, \boldsymbol{x}_1), \ldots, K(\boldsymbol{x}_{new}, \boldsymbol{x}_n))' \hat{\alpha}.$$

Essentially, this is like taking n observations and treating the $n+1$ as a random variable, getting its conditional distribution given all the other variables, and taking the conditional mean. Thus, the posterior given the first n observations becomes the prior for predicting the $n+1$.

The R code for implementing this is:

```
library{kernlab}
```

```
gpr.sinc $\leftarrow$ gausspr(x, y)
gpr.sinc.fit <- predict(gpr.sinc, x)
gpr.sinc.tst <- predict(gpr.sinc, w)
x11()
plot(x,y,lty = 0,main ="Gaussian process estimation")
lines(x, f(x),    lty = 1)
lines(x,gpr.sinc.fit, lty = 4, col=4)
legend("topright", c("data", "true", "GP fit"),
lty = c(0, 1, 2), pch = c(0, 1, 2), merge = TRUE)
[Compute the training and test error] \newline
(gpr.tr.err <- sqrt(mean((y-gpr.sinc.fit$)^2$)))
(gpr.ts.err <- sqrt(mean((v-gpr.sinc.tst$)^2$)))
```

The fit is shown in Fig. 7.11; the RMS training (estimation) and test (out-of-sample) errors are .1967 and .2164 for the left panel and .17 (train) and .207 (test) for the right panel, respectively. Doing this twice for a new set of identically generated data gives an indication of how stable the results would be. In this case, the similarity is very high in terms of RMSE, but the fit on the right is clearly better in that it matches the secondary modes. These comparisons are even more important for higher-dimensional problems, which are harder to visualize, as will be seen with the Friedman function. In the present case, the RBF kernel is particularly well suited to the normal noise in the data and another kernel is not likely to perform as well.

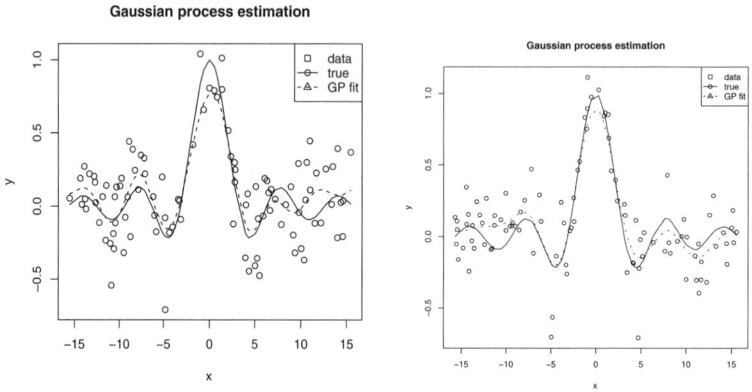

Fig. 7.11 Gaussian process prior fit on data drawn from a noisy sinc function. The left panel has predictive error .2164 and the right has .207.

7.2.1.6 Generalized Additive Models on sinc

Unlike some other methods, the generalized additive model fit is smooth right away. The R package used here, gamair, provides a simple routine for fitting GAMs with a variety of choices for the random component (Gaussian, binomial, Poisson, etc.). Next, we present a fit obtained for the sinc function, under Gaussian noise.

The gam command fits the specified GAM to the data. Each term is estimated using regression splines; in the penalized case, the smoothing parameters are selected by GCV or by an unbiased risk estimator akin to the AIC. This uses a combination of Newton's method in multiple dimensions and steepest descent to adjust a set of relative smoothing parameters iteratively. For univariate terms in the GAM, cubic regression splines are used. (In the two-dimensional case, thin-plate splines are used.) Below, the choice of Gaussian means that the identity link function is called; others are possible. The notation $y \sim s(x)$ means that the target function is approximated by a single smooth based on x that gives the predicted values. The R code is:

```
library(gamair)
library(mgcv)
[This is the library that actually has the GAM code]
gam.sinc.model  <- gam(y~s(x), family=gaussian)
oldx <- data.frame(x)
newx <- data.frame(w)
gam.sinc.fit <- predict.gam(gam.sinc.model,oldx)
[This provides the fit]
gam.sinc.tst <- predict.gam(gam.sinc.model,newx)
[This provides the test]
(tr.err <- mean((y-gam.sinc.fit$)^2$))
(ts.err <- mean((v-gam.sinc.tst$)^2$))
plot(gam.sinc.model,pages=1,residuals=TRUE,
all.terms=TRUE, shade=TRUE, shade.col=2)
```

For the sinc function data, the training error is .0481 and the test error is .1093, which is relatively good compared with the other methods. However, the one-dimensional case is not a severe test: A GAM devolves to a spline smoother which is good in simple settings. The output here is graphed in Fig. 7.12; the shaded portion indicates a 95% confidence interval for each of the function values at the design points, connected so as to look like a region rather than a collection of intervals at the x_is. The curve itself runs along the center of the shaded region.

7.2.2 *Friedman's Function*

The Friedman #1 function has been used extensively to assess the performance of regression techniques. All the predictors (explanatory variables) x_i are drawn from a

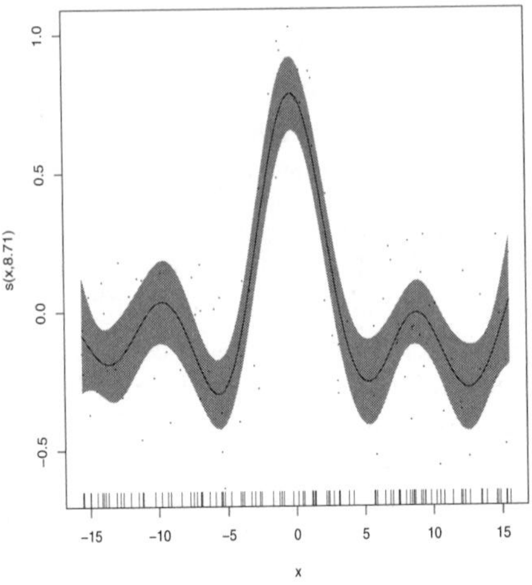

Fig. 7.12 Generalized additive model fit on the sinc function.

uniform distribution in $[0,1]$, so the input space is the unit hypercube in five dimensions. The noise term is a mean-zero normal with $SD = 1$. All the training errors here are based on a single run for each method on a sample of size $n = 200$. The test errors are empirical averages based on $r = 100$ runs each on a new sample of size 200 drawn from the data generator used in training.

7.2.2.1 Test and Training Errors

The six methods used for the sinc function were applied using the same defaults and packages to the Friedman function. First, the training and test errors are recorded in the table below. The standard deviation of the test errors is in brackets on the last row of the table.

	Trees	NN	GP	SVM	RVM	GAM
Training Error	2.4229	0.8314	1.5325	1.5026	1.1023	1.3558
Test Error	6.6751	1.3384	6.6124	6.7931	7.0259	6.7073
	(0.2741)	(0.0678)	(0.2775)	(0.2804)	(0.2871)	(0.3042)

In the results given in this table, the RBF kernel was used for both SVM and RVM, with automatic determination of the bandwidth parameter. In this case, the SVM method retained 149 support vectors while the RVM method only kept 33 relevant vectors a substantial difference. For neural networks, the optimal model seemed to result from using nine nodes in the single hidden layer with a decay of 5×10^{-4}. For trees, the package rpart selected a model with 16 leaves and, as before, the tree model assigned the mean to the points in the leaf rather than fitting a more complicated model. In this case, the tree model did quite well: It was quite sparse and, given the splitting rule as before, had few parameters. Knowing the support often counts more than knowing the functional form, especially in smaller sample sizes or higher dimensions.

Recall that, as seen for the results for the sinc function, each of the methods has a tuning parameter that must be selected. These selections are done internally to the packages to ensure a trade-off between variance and bias that is well motivated, even if its performance is not always ideal or comparable from method to method. NNs, GAMs, RVM and SVM all have tuning parameters representing the strength of a penalty term; GAMs because of the penalty on the splines. Trees have tuning parameters related to the cost-complexity function. In GPs, the kernel itself tunes the posterior because it defines the dependence structure.

It is difficult to assess how well any of the methods fit the data. Indeed, with $p \geq 3$ explanatory variables, only 2D slices intersecting the hypersurface in $p+1$ dimensions can be easily visualized to give an assessment of fit; the problem is that there are so many of them that it is hard to search the whole p-dimensional domain effectively.

Nevertheless, in this example, the computations show that NNs do best among the six methods in a predictive sense. This is reasonable given the richness and smoothness of the model class. In particular, NNs escape the Curse of Dimensionality, at least in a limited sense, and NNs can readily approximate any continuous function, so the theoretical properties of NNs match what one wants for a good approximation to the Friedman #1 function. The counterintuitive part in their good performance is that they are so rough, as seen earlier with the sinc function. It may be that the roughness is harder to detect as the dimension increases because distances between points tend to get bigger. That is, increasing the dimension may give the impression of smoothing out some roughness because of the extra "space" that the errors do not detect because there are no data there. In other words, the assessments of error may not be very sensitive in the regions where the roughness and consequent poor performance may be lurking. Of course, these statements are necessarily tentative because choosing a different target function may give very different results and, however good a method is, a poor implementation of it may have misleadingly poor performance. It is well known that many methods' performances are exquisitely sensitive to the choice of tuning parameters, and it is clear that some methods will be better for some function classes than other methods are.

7.2.2.2 GAMs

The performance of GAMs on the Friedman function is noteworthy because it did better than three of the methods that use much larger function spaces. The function space that GAMs use to approximate the Friedman function is actually quite limited: Unlike the other methods, no GAM function approximation can converge to the Friedman function since it does not have an additive form. The bias will always be bounded away from zero and may be large away from the coordinate axes. Thus it is of particular interest to examine it a bit further. In particular, because of the additive structure, it is natural to compare the GAM function approximation with the Friedman function on the coordinate axes.

Here is the R code and some output:

```
Family: gaussian Link function: identity

Formula: y ~ s(x1) + s(x2) + s(x3) + s(x4) + s(x5)
Parametric coefficients:
             Estimate Std. Err.  t-value  Pr(>|t|)
(Intercept)14.6756  0.1133     129.5    <2e-16    ***
Approximate significance of smooth terms:
        edf   Est.rank    F         p-value
s(x1)  2.989       6     41.43     <2e-16 ***
s(x2)  2.788       6     66.31     <2e-16 ***
s(x3)  3.302       7     25.68     <2e-16 ***
s(x4)  1.000       1     579.23    <2e-16 ***
s(x5)  4.098       9     16.39     <2e-16 ***
Signif. codes:  0 *** 0.001 ** 0.01 * 0.05 . 0.1   1
R-sq.(adj) =  0.902   Deviance explained = 90.9%
GCV score = 2.7776   Scale est. = 2.5669    n = 200
```

It is seen that all five terms are worth including in the model. This can be visually confirmed by looking at planes through the axes of the six-dimensional space in which the Friedman function represents a five-dimensional hypersurface. These five planes are depicted in Fig. 7.13. It is seen that the smooths track the function values rather well in these planes, suggesting that bigger discrepancies between the function and its estimate (if they exist) may be far from the coordinate axes. Searching these regions becomes increasingly difficult as the dimension increases. Note that the data points have also been projected into the coordinate planes, and the apparent spread is much larger than the bands around the estimate of the function, consistent with the possibility that regions of poor fit are far from the coordinate planes.

For Friedman's function, GAMs could be improved by doing some model selection, possibly leading to the inclusion of terms that are based on univariate functions of the input x_is, such as x_1x_2, rather than just using the raw x_is alone. A residual analysis could be used to detect the bias that would suggest missing terms in a GAM. Another way to think of this is that GAMs use a smaller function space than trees, GPs or SVMs and this reduces the variance so much that the increase in bias does not increase the overall error above that of trees, GPs or SVMs.

7.2.3 Conclusions

Estimating a whole function is already a difficult problem because it's as if each function value $f(\boldsymbol{x})$ is a parameter so there are as many parameters as there are $\boldsymbol{x}$s. The

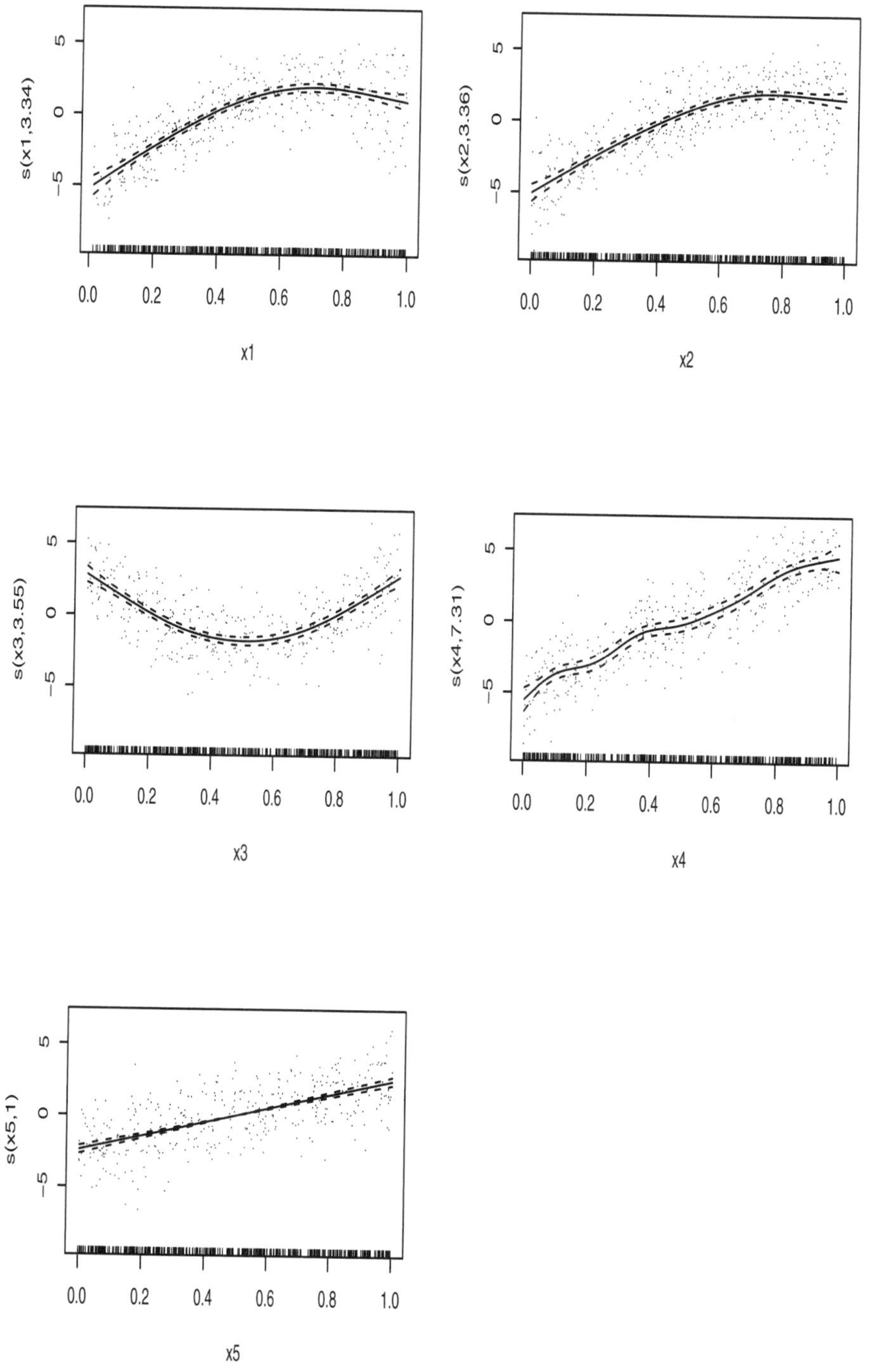

Fig. 7.13 Fit of each term in the GAM estimate for the Friedman function in its coordinate plane. It is seen that the smooths track the target function reasonably closely.

problem is easier when a function is smooth because nearby values can be used to approximate each other so that pooling data locally helps give decent estimates. Even in the one-dimensional case, however, there are difficulties: A function like sinc has multiple modes of decreasing size, so the ability of any method to resolve them given a fixed n is limited.

The six methods used here have test errors for sinc as follows: .2285 (trees), .2054 (neural nets), .2293 (SVM regression), .2098 (RVM regression), .207 (GPs), and .1093 (GAMs). Clearly GAMs are best in this sense, but this is not in general representative of its performance because the GAM package used here defaults to a spline smoother in one dimension. All this shows is that, for some simple problems, splines may do remarkably well. Aside from GAMs, the techniques are all broadly comparable from a predictive standpoint.

However, as can be seen from the figures, some obviously give a better fit than others. GPs gave better fits than RVMs which flatlined away from the dominant mode (or got choppy), or SVMs which got a little unstable away from the dominant mode. The neural net fits looked the most different from the true function on account of their high fluctuations. While trees caught all the modes, they gave dubious fits because of their discontinuities; better than neural nets but arguably worse than the others, except for the fact that they caught all the gross structure (including the secondary modes) that some methods missed. For functions such as sinc, this suggests that GPs may be the best; arguably the effect of a prior provides the right degree of smoothing and sparsity for fit and prediction.

For the Friedman function, the situation is murkier. The increase in dimension changes a lot because of the Curse of Dimensionality. The three big features that practitioners confront as dimension increases are (i) that there are many more directions away from each point, (ii) the number of points in a grid increases exponentially in the dimensions (a grid in two dimensions with k points per side has k^2 points, but in ten dimensions this is k^{10}), and (iii) the data are patchy in the sense that there is so much more space that it is as if there are patches of space with data that can be modeled and vast areas where there are no data so nothing can be said. The Friedman function as used here has five input dimensions, so although the Curse comes into play, it is still relatively mild compared with other settings.

In this sense, NNs may be doing well because their apparent roughness smooths out as dimension increases because there is so much more space in which they can wiggle and gyrate. The extra sparsity of RVM over SVM (fewer relevant vectors than support vectors) may be introducing bias, and the other methods seem to be roughly comparable. Using the optimal number of leaves or support vectors does about as well as the additive approximation of GAMs or the smoothing in GPs.

An important common feature of all the methods that escape the Curse to a greater or lesser extent is that they all lead to a small set of representatives that can encapsulate something like a patch of data in the high-dimensional space. NNs do this with nodes and trees do this with leaves. Support vectors and relevant vectors are analogous to nodes or leaves because they, too, summarize the data into representatives of the solution. GPs rest on the use of a kernel, like SVMs or RVMs, and although they do not obviously have a small set of representatives, the support vectors or relevant vectors probably provide the most important terms in the *GP* regression function. Finally, GAMs by their definition reduce the complexity of a multivariate function to a collection of univariate functions. Unless the dimension is very high (say in the hundreds or thousands), this, too, de facto provides a small set of representatives.

The methods here also typically invoke some kind of tuning parameter to control the bias–variance trade-off. Reducing to representatives clearly achieves some sparsity and helps to reduce the variance, thereby giving better prediction. The bias, however, is seen mostly in the fit, and good fit suggests that the representatives a method uses to express a solution are good enough that other sources of error are greater. Of course, as dimension increases, the bias–variance relationship changes character as the variance term becomes more and more important to control in regions where predictions are made from little data. This is partially because fixed sample sizes become relatively smaller as dimension increases. In these cases, more sophisticated techniques are required than just New Wave regression. These techniques often involve regularization, and will be seen in Chapter 10; this was already seen with penalty functions in splines, the penalty on neural nets, and the cost-complexity pruning in trees, and was implicit in the optimization that led to SVMs or RVMs.

More generally, in passing from a binary classification problem to a univariate sinc regression problem, to the five variable Friedman function regression problem, some patterns emerge. First, kernel-based techniques such as SVM and RVM are often superior to other techniques because they are almost insensitive to passage from one dimension to higher dimensions. Actually, GPs can be regarded as a kernel-based technique because kernels are covariance functions. Not surprisingly, GPs with typically occurring kernels tend to be much smoother than NNs which are usually the roughest; although NNs are continuous, unlike trees, they are usually the wiggliest.

Trees are clearly good for classification but in general not so good for function approximation in regression because they are not smooth. However, as seen above, they often predict well even though they are somewhat unstable due to their high variance (in the presence of low, asymptotically zero, bias). This is why random forests are important as a smoothing technique to stabilize individual trees.

GAMs, of course, are a little clumsy because choosing the link function is somewhat arbitrary and the dimension of the problem represents something like a complexity parameter (since extra univariate functions can be included to improve the model) rather than a modeling feature. Moreover, GAMs span a much reduced class of functions, so nonadditive functions cannot be approximated arbitrarily well; i.e., there is unavoidable bias. Of course, this may not be too bad for some predictive purposes.

The effect of dimensions greater than three is always a problem because they become increasingly hard to visualize. The root difficulty is how to choose the right sense of distance as the dimension increases. That is, problems in lower versus higher dimension are essentially the same qualitatively but bigger quantitatively in higher dimensions. This is so because there's more space to explain with a consistent notion of distance in high dimensions so that diverse regions are contrasted fairly.

Like choosing a good kernel, choosing a distance is much at one with model selection. Since we always project down from higher to lower dimensions, the task is to find the right measurement of similarity between points in the input or feature space on which the projection is performed. The right measure of distance affects regression because a whole function is being broken down into local functions on patches. This means that the distance properties between the function and its approximation, and the comparisons of the discrepancy on one region to the discrepancy on another region, must be fair – not an easy task. Even bandwidth selection is difficult in general because it is the way the distance measures are scaled to get the local measure of distance right on each patch; the curvature of the function affects the way the local distance should be measured. The same type of problem occurs in classification with many values because each local patch in regression is like a class in a multiclass classification. In the binary case, the problem is simpler, but getting the sense of distance right between the regions on which the two classes occur is the root of the challenge.

7.2.3.1 RVM versus SVM?

It has been suggested that GAMs performed best for sinc and neural nets performed best for Friedman's function. The inference from all these comparisons is a counterfactual: Unless you know the function you are approximating, you don't really know which technique to use, but of course you are using the technique because you don't know the true function. In truth, it is not quite so bad because skilled practitioners develop a sixth sense for when one technique should be preferred over another. However, it's still pretty bad because that sort of sixth sense is hard to develop and even harder to formalize or explain. For instance, some engineers argue that neural nets are best when combining multitype data. This cannot be true for all multitype data sets but may well have validity for some commonly occurring large classes of multitype data. Likewise, some people argue that. when sparsity is of greatest importance, RVMs are best because they typically give the sparsest solutions. However, as with all Bayesian procedures, there is little control of the bias, so for complicated functions it is easy to imagine that the sparsity will not help and some other technique will work better.

For a more direct comparison, consider RVM and SVM. In the same way as trees and NNs are similar in that they are local, RVM and SVM are similar in that they are quadratic optimizations based on a kernel that summarizes a function space. (GPs are representative of Bayesian techniques; GAMs are representative of additive methods that include projection pursuit.) Also, it is helpful to compare SVM and RVM because they are relatively not as well understood as most of the other methods here.

Tipping (2001) provides part of such a comparison. He argues that RVM typically performs better than SVM for regression, while SVM tends to outperform RVM for classification. This should not come as a big surprise because the SVM was derived from the need to construct accurate classification rules; the adaptation of the SVM paradigm to regression only came later. In fact, the ε-insensitive loss is already an approximation. On the other hand, the RVM starts with a penalized squared error loss, making RVM natural for regression. However, when it comes to classification, the RVM is clearly a succession of semijustified approximations.

This leads to the No Free Lunch Principle: Methods perform well if the conditions of their derivation are met, with no method capable of covering all possible conditions. More strongly put, each good method has a domain on which it may be best, and different methods have different domains so the task is to characterize those domains and then figure out which domain a given problem's solution is likely to be in. This of course, is extraordinarily difficult in its own right.

In the present chapter, RVM performed better than SVM for the sinc function and for Ripley's data, while SVM did better with Friedman's function. Despite this outperformance of SVM by RVM in Ripley's toy classification problem, the relative superiority of SVM over RVM in classification is evidenced by the following real life-data sets:

- The Titanic survivors data has $p = 3$ attributes measured on $n = 150$ people, with the response being binary (survived, did not survive). In this data set, SVM yields a test error of 22.1%, while RVM produced a test error of 23%.
- The Wisconsin Breast Cancer data set is now a benchmark data set in machine learning, with $p = 9$ attributes measured on $n = 400$ people. The test error for SVM in this case is 26.9%, which is lower than the 29.9% test error yielded by RVM.
- The USPS handwritten digit data set has now circulated to almost every machine learning researcher's computer. With $p = 256$ attributes and $n = 7291$ observations, this is clearly a relatively large dataset, with the added difficulty that there are ten classes rather than two as earlier. For these data, SVM produces a test error of 4.4%, while RVM comes up with a 5.1% test error.

It is important to note that no method wins all the time. Each method seems to be best sometimes, and the trick is to know when. There seems to be an art to practical machine learning.

7.3 Systematic Simulation Study

In an effort to characterize which methods can be expected to work well on different domains in a function space, systematic simulations are invaluable. There are many such studies comparing classification methods with each other and comparing regression methods with each other.

One that is more extensive than the others is in Banks et al. (2003). These authors compare ten regression methods. Eight are LOESS, additive models (AM), projection pursuit regression (PPR), neural networks (NN), ACE, AVAS, recursive partitioning regression (RPR), and MARS. The other two are multiple linear regression (MLR) and stepwise linear regression (SLR). The ten methods are applied to five target functions, each with three choices of dimension ($p = 2, 6, 12$), at three sample sizes ($n = 2^p k$ for $k = 4, 10, 25$), and three (normal) noise levels ($\sigma = .02, .1, .5$). In addition, for each level of p, three cases in which some of the variables entered with coefficient zero were used. This tested the tendency of the procedures to generate spurious structure. The fractions of zero-coefficient variables were all, half, and none. Only five functions were used: a linear function, a multivariate normal density with correlations all 0 or all .8, a mixture of two correlation-zero multivariate normal densities, and the product of the variables. In all cases, only the permitted variables appeared in the target function.

It is evident that even though this study is small compared with the general problem of function estimation in many dimensions, it necessitates a huge amount of computing – apart from having access to suitable implementations of the ten methods. The error criterion was MISE, as defined in Chapter 2, and the simulation procedure was fairly standard. (Generate the random sample of $\boldsymbol{x}$s uniformly in the cube, generate normal errors, choose a function to evaluate to generate the $Y_i = f(\boldsymbol{x}_i) + \varepsilon_i$s, use each of the ten methods on the $\boldsymbol{x}_i, y_i$)s, and estimate the MISE also by simulation, repeating the steps enough times to get reliable results.)

The results of the Banks et al. (2003) computations are an enormous number of tables listing each case – the setting (choice of true function, σ, n, p, and so forth), the regression method, and the MISE. It is important to remember that MISE values are simulated, so that slightly different values may be equivalent.

An ideal summarization would be to treat the analysis as a regression problem with $Y = MISE$ and explanatory variables corresponding to the method, target function, dimension, sample size, proportion of irrelevant variables, and noise. However, metastatistics (i.e., treating the results of statistical analyses of statistical methods as the data for another statistical analysis) is a form of self-referentiality that is too cutely post-modern. Moreover, it can be quite difficult and may not yield the kind of heuristics practitioners want. Herewith follows a comparative discussion that may be more helpful in guiding applications.

First, it is seen that the range of functions for which the methods were tested is likely the most oversimplified. This is unavoidable. Second, Banks et al. (2003) suggest the noise level was not as important as the levels of other factors for the ranges they considered. Other authors have found high sensitivity to noise levels, sometimes dependent on the class of functions used or the true function assumed.

Third, the nonparametric regression techniques segregate into 6 classes, each worth several comments. The following is a precis of Banks et al. (2003).

(1) MLR, SLR, and AM form one class, and they seem to perform roughly similarly. They are never terrible, but rarely the best. It is easy to find functions for which SLR and MLR are better than AM and the reverse because AM permits general functions of the x_is. SLR is usually somewhat better when there are spurious variables, but less so as the number of spurious variables increases and the nonlinearities become more pronounced. This class of methods did poorly with the product function, possibly because of the curvature or possibly because a simple nonlinear transformation (log) turns the product into a sum that is easily modeled.

(2) Because their algorithms are so similar, one expects ACE and AVAS to be similar. Broadly, they are. Both methods were in the midrange of performance as the number of variables and fraction of spurious variables was varied. The more interesting instances are where they differ, which depended strongly on the function taken as true. ACE is decisively better than AVAS for the product function, and AVAS is better for the constant function. Both ACE and AVAS are the best methods for the product function

(as expected – the log transformation produces a linear relationship) but among the worst for the constant function and the mixture of Gaussians.

(3) MARS, surprisingly, did not do well in higher dimensions, especially when all variables were used and especially for the linear function. It had a tendency to overdetect variables and structure and therefore make more errors. For lower dimensions, MARS did adequately over the different functions.

(4) RPR was typically poor in low dimensions but sometimes extremely good in high dimensions, especially when all variables were used. The ability of RPR to do variable selection (i.e., avoid spurious variables) was poor. Unsurprisingly, RPR was poor for the linear function: This is typical for functions in which there is little relationship between any two variables.

(5) PPR and NN are theoretically similar methods, but PPR outperformed NN in all cases except the correlated Gaussian. (This may stem from the CASCOR implementation of neural nets.) PPR was often among the best when the target function was the Gaussian, correlated Gaussian, or mixture of Gaussians, but among the worst with the product and constant functions. PPRs variable selection was also generally good. By contrast, NN was generally poor, except for the correlated Gaussian when $p = 2,6$ and all variables were used and when $p = 6$ and half the variables were used. Clearly, a response such as a correlated Gaussian that can be approximated well by a small number of sigmoidal functions with orientations determined by the data is where NNs should work well. In practice, NN methods may be successful because the response function has a sigmoidal shape in many applications.

(6) LOESS did well in low dimensions with the Gaussian, correlated Gaussian, and mixture of Gaussians. It was not as successful with the other target functions, especially the constant function. Often, it was not bad in higher dimensions, though its relative performance tended to deteriorate.

Fourth, a separate approach to comparison looks at how each method fares with respect to other factors.

(1) In terms of behavior with increasing dimensions, LOESS deteriorated in relative performance but also in the sense that computational time increases rapidly with p. In the CASCOR implementation of NNs used, computing demands were also high because of the cross-validated selection of the hidden nodes. (Alternative NN methods fix these a priori, making fewer computational demands, but this is equivalent to imposing strong, though complex, assumptions.) LOESS, NN, and sometimes AVAS proved infeasible in high dimensions. With the computing system and implementations they used, these authors reported that, typically, fitting a single high-dimensional data set with either LOESS or NN took more than two hours. AVAS was faster, but the combination of high dimension and large sample size also required substantial time.

(2) In terms of functions, or the classes they may be taken to represent, the results displayed few regularities. For the linear function, MLR and SLR were consistently strong, as expected. For the constant function, MARS was good when $p = 2$, SLR was good when $p = 6$, and RPR was good when $p = 12$. Few other regularities could be discerned from the available results.

(3) In terms of variable selection, two strategies were used by the ten methods: global variable selection, as practiced by SLR, ACE, AVAS, and PPR, and local variable reduction, as practiced by LOESS, MARS, and RPR. Generally, local selection does better in high dimensions, but performance depends on the target function.

Obviously, these conclusions are tentative and incomplete – a wide variety of methods and implementations of them have not been included. The remarks here can only serve as guidelines in a few settings.

7.4 No Free Lunch

In a series of papers starting in the early 1990s, David Wolpert alarmed (and possibly annoyed) the computational learning community. His results are summarized in Wolpert (2001). The central point is that the class of models for which a given learning technique is good is limited: No one technique will work well for all problems. More formally, a technique that is good when the true model is f_1 may be poor compared with another technique when the true model is f_2. For instance, for a very complex but low-dimensional model, nearest neighbors might give the best solution in an expected squared error sense when data are limited whereas SVMs might be better at expressing the behavior of simpler models with moderate sample sizes in another sense of goodness of fit, especially when there are few outliers or inliers (since the SVM boundary is determined by only a few points).

To game theorists or statisticians, this kind of statement is a de facto folk theorem. Indeed, in most statistical investigations, it is essential to compare several techniques because it is not clear a priori how well each will do. For instance, the No Free Lunch Principle leads to the conjecture that mixing over diverse predictive strategies, when the strategies have roughly disjoint domains of optimality, and updating the weights in response to an error criterion is asymptotically as good as knowing which domain contains the true function. The collection of models on which a model selection principle (MSP) performs better than others is its catchment area; the idea is to identify the catchment area by weighting the prediction each MSP gives by how well it performs. Thus, the nonexistence of a universally good method is a justification for ensemble methods. Naturally, advocates of specific approaches find this vexatious and counterargue that any ensemble method is just another method susceptible to the same problem: The ensemble method will have its catchment area, perhaps larger than for a specific technique but limited nonetheless.

Here, the reasoning behind the No Free Lunch Theorems is presented not because they are controversial (they are not), nor because the proofs are compelling (the mathematics is abstract and needs development). Rather, the use of an extension to the Bayes paradigm is intriguing and with development promises more insight. Also, it provides a view for statisticians into the structure that computational learners take as natural because the phrasing of the No Free Lunch Principle is in terms of conventional statistical settings. Moreover, the dramatization of the No Free Lunch Principle in a decision context is rather nice.

Essentially, the extended Bayes formulation includes an extra model, and model class, different from the models in the model class assumed to have the true model in it. The extra model, or indeed class of models, represents the formation of an estimate in the ignorance of a true model f and its model class $\mathscr{F}$. This is very realistic.

Start with a quadruple denoted $Q = (d, f, C, g)$. The data are d, a collection of pairs $(x_1, y_1),...,(x_n, y_n)$. The true model is f, a relation between X and Y. The cost function, C, is an evaluation of the loss of using the estimate g from a set $\mathscr{G}$ for f in a set $\mathscr{F}$ when d is available. To give a Bayesian structure, assume priors on $\mathscr{F}$ are assigned, generically denoted Q_i. Also represent the estimate g as arising from a model, but write only $\rho(g|d)$, a posterior for g, in which the model $\rho(d|g)$, the prior $\nu(g)$, and perforce the mixture of them are unspecified. The model ρ varies over a class Ω, which is also not specified further. The relationship between the class of "learners" Ω and the class of possible true models $\mathscr{F}$ is the central feature of the generalization.

The cost C is

$$C(f, g, d) = \sum_x \pi(x)(1 - \delta(f(x), g(x))),$$

in which g is formed from data d by using of some ρ and ν, $f \in \mathscr{F}$, and π is a weighting on the space of explanatory variables indicating how important it is that f and g agree at a given point; here, δ is the Kronecker delta.

Now, there is a probability space for the random variables F, G, and D that have realizations g, f, and d. It is seen that, as a random variable, $C = C(F, G, D)$. Formally, it must be imposed that the densities satisfy $p(g|f, d) = p(g|d)$; i.e., the formation of the estimate g of f does not depend on f. This is the familiar requirement that the estimator must not depend on the estimand. The complicating issue is that g actually has a distribution inherited from d, possibly different from one assumed to be in Ω. That is, the true distribution of G, separate from the class of learners available, is not defined yet. The resolution seems to be that g actually has the correct posterior $\rho_c(g|d)$ and that d has some distribution f from $\mathscr{F}$ that is not otherwise known. Marginally, it's as if G actually has the mixture distribution $\int \rho_c(g, d) f(d(x, y)) Q_i(df)$ even though only $\rho(g|d)$ is available.

Assume that all the quantities are discrete for convenience. Then,

$$E(C|d) = \sum_f \left[\sum_g (C(g, f, d)\rho(g|d)) \right] P(f|d);$$

see Wolpert (2001). This is a sort of posterior risk, averaged over distributions in $\mathscr{F}$ and Ω. It can also be seen as a weighted L^2 inner product between $P(g|d)$ and $P(f|d)$. Note that the conditional $P(f|d)$ is like the correct posterior distribution, and the model implicit in $P(d|f)$ is only known to be in $\mathscr{F}$. The conditional $\rho(g|d)$ is the learner's posterior using Ω.

Write

$$E_1 = \int C(f, g, d) w_1(g|d) dg \quad E_2 = \int C(f, g, d) w_2(g|d) dg,$$

in which w_1 and w_2 are two posteriors for g, i.e., "learners" for f. That is, w_1 and w_2 represent two different inference methods for forming estimates g. In the parametric Bayes context, this is like having two different models, priors, and data sets, corresponding to two different experiments, for estimating the same unknown parameter, which may not follow either of the models used in the two posteriors.

Clearly, if f were known, one would prefer the experiment that gives $\min(E_1, E_2)$. However, f varies over $\mathscr{F}$. So, it is reasonable to assume there exist distributions Q_1 and Q_2 on $\mathscr{F}$ such that

$$\int C(f,g,d)w_1(g|d)dgQ_1(f)df = \int C(f,g,d)w_2(g|d)dgQ_2(f)df.$$

That is, whichever of E_1 and E_2 is smaller will depend on which $f \in F$ is true. All one needs to do is choose Q_1 to put high mass on fs with large $C(f,g)$ and Q_2 to put high mass on fs with small $C(f,g)$, or the reverse. More dramatically, if one experimenter learns f by method w_1, then there will be true models for which the risk performance obtained is worse than an experimenter who uses w_2 and conversely.

Essentially, Wolpert (2001) is enlarging the problem to include model uncertainty or misspecification in the estimand f as well as uncertainty in the model used to form the estimator by distinguishing between the two models, their classes, and the assessments of variability on them. Indeed, the two model classes may be quite different: The model class for learning, Ω, might be purposefully simpler than the actually valid model class $\mathscr{F}$. On this level, the generalization is truly compelling.

7.5 Exercises

Exercise 7.1. Consider the blocky function of Donoho and Johnstone given in Matlab code as follows:

```
pos = [ .1 .13 .15 .23 .25 .40 .44 .65 .76 .78 .81 ];
hgt = [ 4  -5   3   -4  5 -4.2  2.1  4.3  -3.1  2.1
-4.2 ];
sig = zeros(size (t) );
for j=1:length(pos) ;
sig = sig +  (1 + sign(t - pos(j)) ) . * (hgt(j)/2);
end f= sig + 10
```

1. Use your favorite recursive partitioning package (rpart is one possibility) to generate a bagged regression tree estimate for blocky.
2. Use your favorite NN package (nnet is one possibility) to generate a bagged NN estimate for blocky. You might want to include selection of the number of nodes by using CV.

3. Now, form a third estimate of blocky by stacking the bagged tree and bagged NN estimators.
4. Which of the three estimates does best for blocky? Can you argue why this method is best based on generic properties of blocky?
5. Suppose you were to estimate a tree and a NN for blocky, then stack them and then repeat this, effectively, bagging the stacked tree and NN. Would this be different that the result of item 3 in which the stacking was done after the bagging?
6. Would you speculate that further layers of aggregation (e.g., bagging the result of item 2) would be much help? As you increase the levels of averaging, what effects would you expect?

Exercise 7.2. Pick a collection of functions to play the role of Vapnik's sinc function. Two possible choices are the Mexican hat wavelet

$$Y_M = \frac{1}{\sqrt{2\pi\sigma^3}}\left(1 - \frac{t^2}{\sigma^2}\right)e^{\frac{-t^2}{2\sigma^2}} + \varepsilon$$

on $[-5,5]$, which is the normalized second derivative of a normal density, and the tooth function

$$Y_T = x + (9/(2\pi 4^2)^{1/2})e^{-4^2(2x-1)^2} + \varepsilon$$

on [0,1]. Using uniformly spaced points (to start) generate data with a $N(0,\sigma^2)$ noise term for $\sigma = .2$, or nearby values.

1. Use a curvilinear regression model to fit either of the functions and obtain the usual residual plots to evaluate a reasonable polynomial regression function. You can try the same procedure with other basis functions.
2. Next, generate the SVM regression fit for your function, finding the testing and training errors and the appropriate residual plots. (Choose a radial basis function kernel or a polynomial kernel.)
3. Do the same for RVM regression.
4. Now, implement some of the methods used in this chapter (trees, neural nets, Gassian processes, and so forth.)
5. If you have done variable selection before, do standard variable selection techniques such as Mallows' C_p, AIC or BIC perform well if used in the procedures of items 1-4? (Variable selection methods will be presented in Chapter 10.)
6. What statements, if any, can you make about the comparative performance of methods (with and without variable selection) for these two functions? Is there any way to examine the residual plot from a fit using one model class to surmise that another model class should be chosen?

Exercise 7.3 (Comparing BMA and stacking).

In this exercise the point is to find conditions under which stacking linear models or Bayes averaging over linear models gives better predictive performance.

1. Compare stacking to Bayes Model averaging assuming fixed σ, with standard normal priors on all parameters in a linear regression model with the X_ks generated by a normal as well. Verify that when the true model is on the model list, Bayes usually gives better predictions; i.e., has smaller predictive error.
2. Continue in the setting of item 1, but now suppose the true model is not on the model list. Verify that when the true model is sufficiently different from the models in the list that stacking generally gives better predictions.
3. Verify that the prior in a Bayesian technique can be regarded as penalty term so that item 1 may be reasonable. How do you explain item 2?
4. Change the error term in items 1 and 2 to a Cauchy. Verify that now the Bayes model average typically outperforms the stacking average in a prediction sense, whether or not the true function is on the model list.

Note that this procedure can be done for any model class – trees, neural nets, GAMs and so forth – not just linear regression models. However, it is not known (yet) whether the conclusions continue to hold.

Exercise 7.4. Here is a way to combine classification techniques on different domains of a feature space.

1. Consider building a composite classifier. Start with a tree-based classifier, but instead of using Gini or some other measure of impurity, to decide whether or not to split, use a kernel based classifier (such as SVM) that gives an explicit decision boundary. Then, treat each side of the decision boundary as a daughter node from the root. Do this again on each node until the resulting daughter nodes are pure, i.e., all from one class. Now, repeat this several times with a variety of kernels to generate a collection of trees. Use a majority vote on the trees to classify new points. Compare this with a classical tree-based classifier and with either SVM for classification or RVM for classification on simulated data; e.g., Ripley's data.
2. By simulating classification data, from a mixture distribution for instance, determine how much bagging, boosting, or stacking the result help.
3. Repeat this procedure, but use a neural net classifier to define the decision boundary at each node. What if you cycled through the various classification procedures as the tree deepened? (For instance, use SVM at the first split, neural nets at the second, some other technique at the third, and so forth.)

Chapter 8
Unsupervised Learning: Clustering

In contrast to supervised learning, unsupervised learning fits a model to observations assuming there is no dependent random variable, output, or response. That is, a set of input observations is gathered and treated as a set of random variables and analyzed as is. None of the observations is treated differently from the others. An informal way to say this is that there is no Y. For this reason, sometimes classification data that includes the Y as the class is called labeled data but clustering data is called unlabeled. Then, it's as if the task of clustering is to surmise what variable Y should have been measured (but wasn't). Another way to think of this is to assume that there are n independent data vectors $(X_1, \ldots, X_p, Y)$ but that all the Y_is are missing, and in fact someone has even hidden the definition of Y.

A question that must be addressed immediately is: Is there any information in the $\boldsymbol{X}_i$s that is useful in the absence of Y? The answer is: Yes. Lots. In particular, if n is large enough, the distribution of $\boldsymbol{X}$ can be estimated well. In this case, the Y might correspond to classes identified with modes of the marginal distribution. The modes may just be the structure of a single complicated distribution for $\boldsymbol{X}$ or they might indicate that the distribution of $\boldsymbol{X}$ is the mixture of many distributions, each representing a class and giving a mode. In this case, the task of clustering is to distinguish which component of the mixture a given $\boldsymbol{x}_i$ represents. At a minimum, the information in the $\boldsymbol{X}_i$s indicates where the density of any reasonable Y must be relatively high.

An extreme case of this is to imagine a collection of $\boldsymbol{x}_i$s concentrated on three circles in the plane centered at the origin with radii one, two, and three. It would be natural to assume three clusters, one for each circle. However, if more $\boldsymbol{x}_i$s accumulated and all of them were either within the circle of radius 1 or between the two circles of radii 2 and 3, a very different structure on Y would be inferred – classes 2 and 3 would merge, leaving two clusters.

Having used the term clustering, it is important to give it a definition, however imprecise. Briefly, clustering is the collection of procedures used to describe methods for grouping unlabeled data, the $\boldsymbol{X}_i$s, into subsets that are believed to reflect the underlying structure of the data generator. The techniques for clustering are many and diverse, partially because clustering cuts across so many domains of application and partially because clustering is a sort of preprocessing often needed before applying any

B. Clarke et al., *Principles and Theory for Data Mining and Machine Learning*, Springer Series in Statistics, DOI 10.1007/978-0-387-98135-2_8, © Springer Science+Business Media, LLC 2009

model-based inferential technique. Consequently, only a fraction of techniques can be presented here.

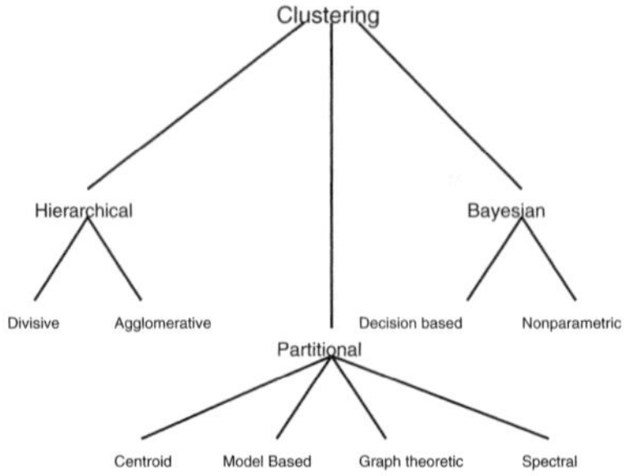

Fig. 8.1 The collection of clustering techniques can itself be clustered into three groups – hierarchical, partitional, and Bayesian – and then subdivided. Note that the leaves are not exclusively defined. For instance, there are techniques that are primarily spectral (i.e., based on the spectrum of a matrix of distances, and hence noted here as partitional) but yield a hierarchical clustering. However, most spectral techniques developed so far are partitional rather than hierarchical.

Figure 8.1 shows a taxonomy of clustering techniques. The three main classes of clustering techniques are hierarchical, partitional, and Bayesian. It is true that Bayesian techniques are often either hierarchical or partitional; however, their character is often so different from what is typically understood by those terms that it is better to put them in a separate class. Moreover, techniques do not in general fit into one leaf of the tree uniquely, as will be seen below. The tree in Fig. 8.1 just indicates one reasonable and typical structure for the techniques. The leaves indicate the topics to be presented.

The idea of a hierarchical technique is that it generates a nested sequence of clusterings. So, usually one must choose a threshold value indicating how far along in the procedure to go to find the best clustering in the sequence. The nesting can be decreasing or increasing. If it's decreasing, usually it's started by treating each data point as a singleton cluster. These procedures are called agglomerative since clusters are merged. If it's increasing, usually it's started by treating the whole data set as one big cluster that is decomposed into smaller ones. These procedures are called divisive since clusters are split. It will be seen that a graph-theoretic technique called minimal spanning trees yields a nested sequence of clusters and can be regarded as hierarchical, even though it will be treated here as a graph-theoretic technique.

Hierarchical methods also typically require a dissimilarity, which is a measure of distance on individual data points that lifts to a measure of distance on groups of data points. This does not particularly distinguish hierarchical clustering techniques because partitional clustering techniques often require a distance that functions like a dissimilarity. However, the way a dissimilarity is used is somewhat more consistent and central across hierarchical techniques as a class.

By contrast, nonhierarchical methods such as partitional methods usually require that the number of clusters K and an initial clustering be specified as an input to the procedure, which then tries to improve the initial assignment of data points. The initial clustering takes the place of a dissimilarity and threshold value in hierarchical procedures. It is true that hierarchical methods require an initial clustering; however, it is trivial: n clusters or one cluster. Algorithmically, the difference between hierarchical and partitional methods is that with hierarchical algorithms clusters are found using previously established clusterings, while partitioning methods try to determine all clusters optimally in one shot.

Bayesian clustering techniques are different from the first two classes because they try to generate a posterior distribution over the collection of all partitions of the data, with or without specifying K; the mode of this posterior is the optimal clustering. Bayesian techniques are closer to hierarchical techniques than partitional techniques because often there is an ordering on the partitions. All of these techniques require the specification of a prior and usually come down to some kind of hypothesis test interpretation.

Two problems that can arise in any technique are clumping and dissection. Clumping is the case where a single object fits into two or more clusters that are therefore allowed to overlap. An instance of this is document retrieval: The same word can have two different meanings, so a text cannot be readily fit into exactly one cluster. Overlapping clusters will not be examined here, but awareness of the problem is important. Likewise, the concept of dissection will not be studied here, apart from noting that it is the case where there is a single population that does not contain meaningful clusters but the goal is still to cluster the data for some other purpose. For instance, there may be no meaningful way to find clusters in a homogeneous city, but a post office may still partition the city into administrative regions so letter carriers can be more efficient.

Other terms often associated with clustering are important to note. A clustering procedure is monothetic if it looks at the components of the $\boldsymbol{x}_i$s one at a time to generate an overall clustering. Polythetic techniques, which are more common, consider the entire vector $\boldsymbol{x}_i$. In the regression context, trees are monothetic because splits can be on a single variable, whereas projection pursuit is polythetic because it uses all the components of $\boldsymbol{x}_i$ at the same time. A clustering is hard if it unambiguously puts each $\boldsymbol{x}_i$ in a single cluster. Fuzzy clusterings allow partial membership: A point may belong 70% to one cluster and 30% to another. Fuzzy clustering comes up naturally in evaluating the stability of a clustering.

The single most common clustering technique used is probably K-means, which is a centroid based technique invented by MacQueen (1967). It rests on assigning points to whichever centroid is closest to them, recomputing the centroid, and repeating the procedure, hoping the clustering converges. This uses a dissimilarity and so can be regarded as hierarchical; however, K-means also starts with an initial assignment of the clustering, which it refines. One can argue that the dissimilarity is only used as a distance on the $\boldsymbol{x}_i$s, so the initial clustering is more important to the procedure.

Hierarchical agglomerative clustering developed in the biological sciences and as a class is the most widely used. Mixture model methods, which originate in work by Wolfe (1963), Wolfe (1965), Wolfe (1967), Wolfe (1970) have recently enjoyed more attention (see Fraley and Raftery (2002)). The key procedure here is called the E-M algorithm, of which there are many variants. In general, methods based on mixture models make the strongest assumptions about the data being a random sample from a population; these are also the methods that best support inference unless the posterior from a Bayes procedure is used for inference beyond just choosing a clustering. In addition, there are numerous other clustering procedures that have been developed but do not really fit well into any of the classes of Fig. 8.1. These include information-theoretic clustering, among others, but these are not discussed here.

It is important to remember that unsupervised learning is often just one step in a problem. For instance, one could cluster data into groups, do some kind of variable selection within each group, and then use Bayesian methods to generate conditional probabilities for one of the random variables given the others. Note that the last step actually becomes supervised because one of the Xs, or a function of them, has been found to be a good choice for Y. As another instance, bioinformaticians often cluster their data using some of the explanatory variables, then do principal components analysis on the explanatory variables for each group, then select the top two or three principal components to include in a model for the response, and finally use shrinkage methods on the coefficients in the principal components to eliminate the smaller terms for the sake of sparsity. The first steps are seen to be unsupervised; they set up the actual analysis, which is supervised because it uses the response.

To fix notation, consider a collection of data $\boldsymbol{x}_i$, for $i = 1,...,n$, that can be treated as IID outcomes of a variable $\boldsymbol{X}$. The goal is to find a class of sets $\mathscr{C} = \{C_1,\ldots,C_K\}$ such that each x_i is in one set C_k and the union of the C_ks is the whole sample $x^n = (\boldsymbol{x}_1,...,\boldsymbol{x}_n)$. The key interpretative point is that the elements within a C_k are much more similar to each other than to any element from a different $C_{k'}$. The natural comparison, if possible, is between a known clustering, which is essentially classification data, and whatever clustering a clustering procedure outputs. The difference is an assessment of how little information the algorithm loses relative to perfect information. Even though this is not formalized, it does represent a standard for comparison that could be readily formalized. It is an analog of the concept of goodness of fit.

Here, centroid-based methods are presented first because they are based on K-means. Then, hierarchical methods are presented because they are the next most popular. After that, there are a plethora of other partitional methods, followed by Bayesian techniques and a discussion of cluster validity.

8.1 Centroid-Based Clustering

The centroid of a cluster can be thought of as the pure type the cluster represents, whether that object actually exists or is just a mathematical construct. It may correspond to a particular data point in the cluster as in K-medoids or to a point in the convex hull of the cluster, such as the cluster mean. Given an initial clustering, centroid-based methods find the centroids of the clusters, reassign the data points to new clusters defined by proximity to the centroids, and then repeat the procedure. The similarity between two clusters is the similarity between their respective centroids. Thus, the key choices to be made, aside from an initial clustering, are the distance to be used, the centroids to be used, and how many iterations of the procedure to use.

Most of these procedures are variants on K-means clustering, which is worth presenting in detail since it readily captures the geometric intuition. Then the use of other measures of location or distances becomes straightforward.

8.1.1 K-Means Clustering

Intuitively, the K-means clustering approach is the following. The analyst picks the number of clusters K and makes initial guesses about the cluster centers. The procedure starts at those K centers, and each center absorbs nearby points, based on distance; often the distances are found using a covariance matrix to define a norm. Then, based on the absorbed cases, new cluster centers, usually the mean, are found. The procedure is then repeated: The new centers are allowed to absorb nearby points based on a norm, new centers are found, and so on.

Usually this is done using the Mahalanobis distance from each point $\boldsymbol{x}_i$ to the current K centers,

$$d(\boldsymbol{x}_i, \bar{\boldsymbol{x}}_k) = [(\boldsymbol{x}_i - \bar{\boldsymbol{x}}_k)' S^{-1} (\boldsymbol{x}_i - \bar{\boldsymbol{x}}_k)]^{1/2},$$

where $\bar{\boldsymbol{x}}_k$ is the center of the current cluster k and S is the within-cluster covariance matrix defined by $S = (s_{j,\ell})$ for $j, k = 1, ..., p$ as

$$s_{j,\ell} = \frac{1}{n-1} \sum_{i=1}^{n} (x_{i,j} - \bar{x}_j)(x_{i,\ell} - \bar{x}_\ell).$$

Smart computer scientists can do K-means clustering (or approximately this) very quickly, sometimes involving the pooled within-cluster covariance matrix $W = (w_{j,\ell})$, where, for $j, \ell = 1, ..., p$,

$$w_{j,\ell} = \frac{1}{K} \sum_{k=1}^{K} \sum_{i=1}^{n_k} d_{i,k} (x_{i,j} - \bar{x}_{k,j})(x_{i,\ell} - \bar{x}_{k,\ell}),$$

in which n_k is the number of points in cluster k, $\bar{x}_{k,\ell}$ is the mean of the ℓth variable in cluster k, and $d_{i,k} = 1$ if observation i is in cluster k and 0 otherwise. In practice, K-means clustering is often the only clustering procedure that is computationally feasible for large p, small n data sets.

The K-means procedure works because the cluster centers change from iteration to iteration, hopefully converging to the correct cluster centers, if K is correct. If the centers move too much, or for too many iterations, the clusters may be unstable. In fact, there is no guarantee that a unique solution for K-means clustering exists: If K is correct and two starting points fall within the same true cluster, then it can be hard to discover new clusters since the two points will often get closer together and only $K-1$ clusters will be found. This means that it is very important to know K, even though it can be hard to determine. Unfortunately, specifying some values seems to be necessary for all clustering procedures. To use K-means, one can do a univariate search over k:

Try many values of k, and pick the one at the knee of some lack-of-fit curve; e.g., the ratio of the average within-cluster to between-cluster sum of squares.

To explain the K-means procedure more formally and see how it is similar (or not) to other procedures, define what's called a dissimilarity measure. In essence, this is a generalization of a metric: Any metric is a dissimilarity; not all dissimilarities are metrics. The idea is that a dissimilarity expresses how different two points are: A dissimilarity measure should be high when two $\boldsymbol{x}_i$s differ in many entries and be low otherwise. Likewise, a similarity measure should be low when many entries are equal or close.

One way to construct a dissimilarity is componentwise. Let d_j be a dissimilarity acting on the jth coordinate of $\boldsymbol{x}$ for $j = 1,...,p$. Now, for $\boldsymbol{x}$ and $\boldsymbol{x}'$, the vector $(d_1(x_1,x'_1),...,d_p(x'_p,x'_p))$ is the coordinatewise difference between them. This lifts to a dissimilarity d on the entire vectors by defining $d(\boldsymbol{x}',\boldsymbol{x}) = \sum_{j=1}^p w_j d_j(x'_j,x_j)$. Now, $\boldsymbol{x}_1,..., \boldsymbol{x}_n$ leads to the $n \times n$ matrix of the dissimilarities $D = (d(\boldsymbol{x}_i,\boldsymbol{x}_j))$ for $i, j = 1,...,n$. Sometimes the dissimilarity matrix D is called a proximity matrix. In general, the d_js are zero when their entries are the same, so the main diagonal of D is all zeros. The weights w_i need not be identical but are often set to 1 for convenience. There are dissimilarities d on $\boldsymbol{x}_i$s that are neither metrics nor derived from coordinatewise distances. For instance, a dissimilarity can be derived from measures of association between $\boldsymbol{x}$ and $\boldsymbol{x}'$ such as $d(\boldsymbol{x},\boldsymbol{x}') = \text{Corr}(\boldsymbol{x},\boldsymbol{x}')$.

Thus, for K-means, the eponymous case uses squared error and calculates the sample mean within clusters. So, for $x = (x_1,...,x_p)$ and $\boldsymbol{x}' = (x'_1,...,x'_p)$, set all the d_js to be squared error so that $d(\boldsymbol{x}',\boldsymbol{x})$ is the Euclidean distance $||\cdot||$. Write m_k for $k = 1,...,K$ for the (vector) means of the unknown classes C_k under cluster assignment $\mathscr{C}$. Now define the membership function C, which assigns data points to clusters, so $C(i) = k \Leftrightarrow \boldsymbol{x}_i \in C_k$ under the clustering $\mathscr{C}$. The ideal K-means solution is $\mathscr{C}_{K-means} = (C_1^*,...,C_{K_{opt}}^*)$, satisfying

$$\mathscr{C}_{K-means} = \arg\left(\min_K \min_{m_1,...,m_K} \sum_{k=1}^K \sum_{i:C(i)=k} ||\boldsymbol{x}_i - m_k||^2\right), \tag{8.1.1}$$

where $C(i)$ is the membership function of the clustering with centers $m_1,...,m_K$ under the Euclidean distance.

The optimization over the number of clusters is done, in principle, for each possible K to find the minimizing value. However, in practice, one finds the value only for several values of K; the one with the smallest error is used as "true". (This can be done with the R function kmeans(); often one selects the initial values of the cluster means m_i at random from the closed convex hull of the data.) It may take several tries to get the minimal distance because one can get caught in a local minimum, so it is prudent to use the minimum of (say) ten guesses as the global minimum.

To see where (8.1.1) came from, recall that under Euclidean distance the mean achieves the minimum error. In the one-dimensional case, this is

$$\bar{x} = \arg\min_{\mu} \sum_{i=1}^{p} (x_i - \mu)^2.$$

Now, recall the ANOVA decomposition of the total variability into a sum of the between-cell variability and within-cell variability and apply it to clusters. By analogy with the ANOVA case, define the total variability of a clustering $\mathscr{C}$ with membership function C under a dissimilarity d as

$$\begin{aligned}
T &= \sum_{i=1}^{n}\sum_{i'=1}^{n} d(\boldsymbol{x}_i, \boldsymbol{x}_{i'}) \\
&= \sum_{k_1,k_2=1}^{K} \sum_{C(i)=k_1} \sum_{C(i')=k_2} d(\boldsymbol{x}_i, \boldsymbol{x}_{i'}) \\
&= \sum_{C(i)\neq C(i')} d(\boldsymbol{x}_i, \boldsymbol{x}_{i'}) + \sum_{C(i)=C(i')} d(\boldsymbol{x}_i, \boldsymbol{x}_{i'}) \\
&= \sum_{k=1}^{K} \sum_{C(i)=k} \sum_{C(i')\neq k} d(\boldsymbol{x}_i, \boldsymbol{x}_{i'}) + \sum_{k=1}^{K} \sum_{C(i)=k} \sum_{C(i')=k} d(\boldsymbol{x}_i, \boldsymbol{x}_{i'}) \\
&= B(\mathscr{C}) + W(\mathscr{C}).
\end{aligned} \tag{8.1.2}$$

The last expressions in (8.1.2) are the between-cell and within-cell variabilities of $\mathscr{C}$. So, for fixed T (independent of the clustering), minimizing $W(\mathscr{C})$ should lead to the clustering, for a given K, that is tightest about its cluster means. Equivalently, maximizing $B(\mathscr{C})$ should lead to the clustering, for a given K, that has cluster centers as far apart as possible.

Hastie et al. (2001) observe that the within-cluster variability for a given K and set of centers m_k is

$$\begin{aligned}
W(C) &= \sum_{k=1}^{K} \sum_{C(i)=k} \sum_{C(i')=k} ||\boldsymbol{x}_i - \boldsymbol{x}_{i'}||^2 \\
&= 2\sum_{k=1}^{K} \sum_{C(i)=k} ||\boldsymbol{x}_i - \bar{\boldsymbol{x}}_k||^2
\end{aligned} \tag{8.1.3}$$

by adding and subtracting the cluster means in each cluster.

So, setting $m_k = \arg\ \min_m \sum_{C(i)=k} ||\boldsymbol{x}_i - m|| = \bar{\boldsymbol{x}}_k$ in (8.1.1) will find the clustering with the smallest value of $W(\mathscr{C})$. Putting this clustering into the objective function in (8.1.1) and reminimizing will give a new clustering, giving a new set of m_ks and so on. This is sometimes called an iterative descent algorithm since the objective function $W(\mathscr{C})$ is typically decreasing.

In general, this is most appropriate for normal data, and one would prefer to use the Mahalanobis distance because that way the shapes of the neighborhoods will better reflect the mode structure of the true underlying density.

8.1.2 Variants

There are numerous variations on the K-means theme, usually based on changing the dissimilarity or centering. The common feature is calculating a set of summary statistics for locations and maybe a variance matrix. Two of these are given here.

8.1.2.1 K-medians

In this case, the squared error is replaced by the absolute error, and the mean is replaced with a median-like object arising from the optimization (because there is some ambiguity about how to define a median in two or more dimensions). The defining feature is that the cluster centers in this case minimize the overall L^1 distance to each point. The covariance matrix is often ignored because it doesn't scale absolute error neighborhoods the way it scales squared error neighborhoods.

The objective function is

$$\sum_{k=1}^{K} \sum_{C(i)=k} |\boldsymbol{x}_i - m_k|. \tag{8.1.4}$$

As before, K is chosen and the vectors m_k are found iteratively from an initial clustering. There is some evidence that this procedure is more resistant to outliers or strong non-normality than regular K-means. However, like all centroid-based methods, it works best when the clusters are convex.

8.1.2.2 K-medoids

The K-medoid algorithm, like the K-means, tries to minimize a squared error criterion but rather than taking a mean as the cluster center chooses one of the data points. The medoid of a cluster is the element of the cluster whose average dissimilarity to all the objects in the cluster is minimal. It is the prototypical data point of the cluster because it is the most centrally located.

Again, K and an initial clustering must be given, and the basic procedure is similar to that of K-means or K-medians but iteratively replaces each of the medoids by one of the nonmedoid points to search for a smaller error. That is, after selection of the K medoid points, each data point is associated with its closest medoid. Then a nonmedoid data point is selected at random and used to replace one of the medoids to see if the error is reduced. If so, the two points are interchanged. If not, another nonmedoid point is tested. This is repeated until there is no change in the medoid. Obviously, this does not scale up well to large data sets; however, it is more robust to noise and outliers than K-means.

8.1.2.3 Ward's Minimum Variance Method

Actually, Ward's minimum variance method is a form of hierarchical agglomerative clustering, but it is based on the same analysis of the clustering problem as K-means. Motivated by (8.1.3), consider the two sorts of error sums of squares for an initial clustering $\mathscr{C}$ given by

$$ESS(\mathscr{C},k) = \sum_{C(i)=k} \|\boldsymbol{x}_i - \bar{\boldsymbol{x}}_k\|^2 \text{ and } ESS(\mathscr{C}) = \sum_{k=1}^{K} \sum_{C(i)=k} \|\boldsymbol{x}_i - \bar{\boldsymbol{x}}_k\|^2. \quad (8.1.5)$$

There are $\binom{K}{2}$ possible pairs of clusters. For each pair (k_1,k_2), find $ESS(\mathscr{C}(k_1,k_2)$, where $\mathscr{C}(k_1,k_2)$ is the clustering derived from $\mathscr{C}$ by merging the k_1th and k_2th clusters. Clearly, $ESS(\mathscr{C}(k_1,k_2)) > ESS(\mathscr{C})$ for each (k_1,k_2), so find the pair (k_1,k_2) giving the smallest increase over $ESS(\mathscr{C})$. Then repeat the process for the clustering $\mathscr{C}(k_1,k_2)$ to merge two more clusters and so forth. An analogous procedure could be used for medians and L^1, or medoids.

This procedure tends to join small clusters first and is biased toward producing clusters with roughly the same shape and roughly the same size. This doesn't scale up to large n or large p very well, is sensitive to outliers, and most importantly is rigid in that once a merger is made, it is locked in and never reconsidered. On the other hand, usually Ward's method gives clusters that appear well defined, which may explain its popularity. Of course, using $\|\boldsymbol{x}_i - \bar{\boldsymbol{x}}_k\|^4$, or even $\|\boldsymbol{x}_i - \bar{\boldsymbol{x}}_k\|^6$ would seem to do better – but this is fallacious. The better approach is to use $\sqrt{ESS}$ so that one is not misled by the nonlinearity of the square. A separate point is that Ward's method does not necessarily find the clustering that minimizes ESS, unlike K-means; it is limited to merging initial clusters, not seeking entirely new ones, so if the initial clusters are suboptimal, Ward's method will typically exacerbate this.

8.2 Hierarchical Clustering

As noted before, hierarchical clustering techniques give a nested sequence of clusterings. So, the process of merging or dividing clusters can be represented as a tree, usually called a dendrogram in the clustering context, where the branching is based on inclusion. The process of generating a dendrogram is usually based on evaluations of distances between points or sets of points at each stage of the clustering procedure. So, hierarchical clustering procedures tend to be rigid in the sense that once a merge or split has been done it cannot be undone. Of course, this is not necessary; in principle one could write algorithms that include a depth-first search to carry out a procedure without locking in earlier choices.

Hierarchical clustering procedures are either agglomerative or divisive. Agglomerative clustering means that, starting with all the data points as singleton sets, clusters are merged under a dissimilarity criterion until the entire data set is a trivial cluster.

Divisive clustering means the reverse: The starting point takes the entire data set as a cluster and successively splits it until at the end the trivial clustering of n singleton sets results. In both cases, a threshold for how much merging or splitting is to be done must be chosen. There are a great variety of ways to choose the merges and splits and while a split may divide a cluster into more than two subsets and a merge may unite more than two clusters, it is conventional to limit all merges to two input clusters and all splits to two output clusters.

8.2.1 Agglomerative Hierarchical Clustering

At root, hierarchical agglomerative clustering merges sets according to fixed rules. The basic template for hierarchical agglomerative clustering is:

Start with a sample $\boldsymbol{x}_i$, $i = 1, ..., n$, regarded as n singleton clusters, and a dissimilarity d defined for all pairs of disjoint, nonvoid subsets of the sample. If desired, a fixed number of clusters to form can be chosen. Otherwise, the procedure can be allowed to iterate $n-1$ times.

- ☐ At the first step, join the two singletons $\boldsymbol{x}_i$ and $\boldsymbol{x}_j$ that have the minimum dissimilarity $d(\boldsymbol{x}_i, \boldsymbol{x}_j)$ among all possible pairs of data points.
- ☐ At the second step, there are $n-1$ clusters $C_{1,1}, ..., C_{1,n-1}$. Find

$$(j^*, j'^*) = \arg\min_{j \neq j'} \; d(C_{1,j}, C_{1,j'}), \tag{8.2.1}$$

  and merge clusters C_{1,j^*} and C_{1,j'^*}. Now there are $n-2$ clusters, say $C_{2,1}, ..., C_{2,n-2}$.
- ☐ Continue until the n points have been agglomerated into the desired number of clusters or into a single cluster of size n.

Alternatively, choose an error criterion. Find the errors for clusterings with different numbers of clusters. Choose the largest number of clusters for which the drop in error from permitting one more cluster is small.

The two main inputs to this template are the stopping rule and the dissimilarity d, which is taken up in the next subsection.

The stopping rule basically is a way to decide where to draw a horizontal line in the dendrogram. Even if the procedure runs $n-1$ iterations, the choice of when to stop merging must be made to specify a particular clustering. The last step in the template is called finding the knee (or elbow). This is an informal way to choose the number of clusters when it is not preassigned. Consider a graph of the number of clusters K versus an error criterion for clustering; the most common error criterion is the ESS in (8.1.5), regarded as a function of K.

In practice, as K increases, the less information there is per extra cluster because there is a finite amount of information in the data. Intuitively, when K is small, there is a high amount of information per cluster because the information in the data is being summarized by a small number of cluster centers. As K increases, however, the information in the data accurately captured by using extra clusters decreases. This means there is a K beyond which permitting one more cluster reduces the error by such a small amount that the extra cluster is not worth including. In these cases, a graph of K versus the error criterion usually has a dent, or a corner, indicating a sudden dropoff of information per extra cluster, indicating a point of diminishing returns. This is called the knee or elbow of the curve and is usually found just by looking at the graph. Even so, there are ambiguous cases, and it can be hard to decide when to stop agglomerating and declare the clustering.

The result of a hierarchical agglomerative cluster analysis is often displayed as a tree and is called a dendrogram; see Fig. 8.2. The lengths of the edges in the binary tree shows the order in which cases or sets were joined together and the magnitude of the distance between them. If three clusters were to be found, the natural place to draw the horizontal line would be just above where $\{\boldsymbol{x}_2\}$ is merged with $\{\boldsymbol{x}_5, \boldsymbol{x}_3\}$, so that the three clusters are $\{\boldsymbol{x}_1\}$, $\{\boldsymbol{x}_4\}$, and $\{\boldsymbol{x}_2, \boldsymbol{x}_3, \boldsymbol{x}_5\}$.

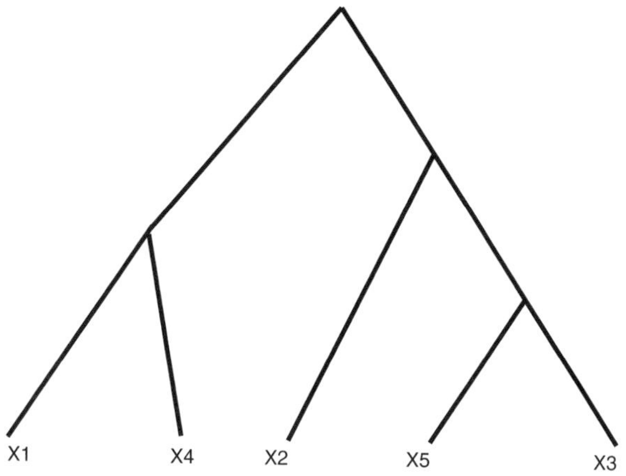

Fig. 8.2 In this hierarchical agglomerative clustering procedure on five points, $\boldsymbol{x}_5$ and $\boldsymbol{x}_3$ are joined first and then $\boldsymbol{x}_1$ and $\boldsymbol{x}_4$ are joined. Then $\boldsymbol{x}_2$ is merged with the first pair, and finally the two sets of two and three elements are merged. Depending on where one draws a horizontal line, one can get one, two, three, four or five clusters.

8.2.1.1 Choice of d

Note that in (8.2.1) the optimization depends on d. In fact, the choice of d determines the set of clusterings from which a final clustering can be chosen. Different choices of d favor different classes of clusterings much like different choices of stop-splitting rules favor qualitatively different classes of trees in recursive partitioning. So, it is important to list the most frequently used dissimilarities along with some of their properties.

Let $\mathscr{A}$ and $\mathscr{B}$ be two disjoint subsets of $\{\boldsymbol{x}_1,...,\boldsymbol{x}_n\}$. To use the agglomerative clustering template from the last subsection, it is enough to define $d(\mathscr{A},\mathscr{B})$ in general for various choices of d. Since the goal is to merge sets of points successively, the d are often called linkages because they control the lengths of line segments joining, or linking, points $\boldsymbol{x}_i$ to each other. Four of the most common dissimilarities are:

☐ Nearest-neighbor or single linkage: A metric d on $\mathbb{R}^p$ is minimized to define a dissimilarity on sets, also denoted d, given by

$$d(\mathscr{A},\mathscr{B}) = \min_{a\in\mathscr{A},b\in\mathscr{B}} d(a,b).$$

☐ Complete linkage: Here a metric d is maximized to define a dissimilarity on sets given by

$$d(\mathscr{A},\mathscr{B}) = \max_{a\in\mathscr{A},b\in\mathscr{B}} d(a,b).$$

☐ Centroid linkage: The idea is to merge at iteration $k+1$ those clusters whose centroids at stage k were closest. That is,

$$d(\mathscr{A},\mathscr{B}) = d(\bar{a},\bar{b})$$

where $\bar{a}$ is a centroid for $\mathscr{A}$ and $\bar{b}$ is a centroid for $\mathscr{B}$.

☐ Average linkage: In this case, the metric distances between points in $\mathscr{A}$ and $\mathscr{B}$ are averaged:

$$d(\mathscr{A},\mathscr{B}) = \frac{1}{\#(\mathscr{A})\#(\mathscr{B})} \sum_{i\in\mathscr{A},j\in\mathscr{B}} d(\boldsymbol{x}_i,\boldsymbol{x}_j).$$

(This is the reverse of centroid linkage, where the distance between the averages is used.)

Single-linkage clustering is essentially a nearest-neighbor criterion: Two points, or clusters, are linked if they are or have each others closest data point; only one link is required to join them. The extension of the single link connection between points to sets ensures there will always be a path with shortest lengths connecting all the points in a cluster. Thus, single-linkage clustering also admits a graph-theoretic interpretation. While this has many good properties, it has some deficiencies: Two clusters of any shape that are well separated apart from a few points on a line between them will be joined if the points between them are close enough to each other. This is called the chaining problem, and it is very serious in many data sets. The tendency toward such barbell-shaped regions can be quite strong because single-linkage clustering tends to form a few big clusters early in the hierarchy. Worse, in the presence of outliers, it can join two sets of points when most of the points are distant. Odd-shaped regions do occur naturally, so it's unclear in general whether such properties are features or bugs. Indeed, if one anticipates regions that are separated, irregular or not, single-linkage clustering can be a good way to find them. Often single-linkage clustering runs in $\mathscr{O}(n^2)$ time (range $\mathscr{O}(n\log n) - \mathscr{O}(n^5)$) for fixed p.

Complete linkage means that all points in the two clusters are joined by lines of length less than the upper bound on the maximum. This means that all the linkages between points are within the distance at which the cluster was formed. Compared with single linkage, complete linkage tends to form many smaller, tighter clusters. So, complete-linkage dendrograms often have more structure to them and can therefore be more informative than single-linkage dendrograms. Also, because of the maximum, complete linkage is sensitive to outliers of the clusters themselves. Arguably, like support vector machines, complete linkage is often driven by a small set of points that may be outliers or inliers and not particularly representative of the data set. This may also lead to non-robustness of the clustering in that small changes in the data can result in big changes to the clustering. Often complete-linkage clustering runs in $\mathscr{O}(n^3)$ for fixed p, possibly less for sparse similarity matrices.

Centroid linkage calculates the distance between means of clusters. This average can be a good trade-off between the extremes of single and complete linkage. The limitation is that the distances at which clusters are formed are averages and do not correspond to any actual links between data points. A downside is that this linkage is not monotonic: If d_j is the dissimilarity level for a given iteration j, it may be higher or lower than d_{j+1}; a reversal of the ordering is called an inversion. In contrast, single linkage and complete linkage are monotonic in that the dissimilarity is always increasing. So, the clusters from centroid linkage can be more difficult to interpret.

Average linkage is another compromise between single and complete linkage that is more demanding computationally because of the summation. The chaining problem is far less common for this method as compared with single linkage, and the effect of outliers is reduced by the averaging. So, often the result is more, looser clusters than in single linkage but fewer, tighter clusters than in complete linkage. For these reasons, average linkage is fairly popular in applications; AGNES, or agglomerative nesting, is one commonly used package. Average linkage often runs in $\mathscr{O}(n^2)$ for fixed p.

Many other dissimilarities have been used. Ding and He (2002) propose setting

$$d(\mathscr{A},\mathscr{B}) = d(\mathscr{A},\mathscr{B})/d(\mathscr{A},\mathscr{A})d(\mathscr{B},\mathscr{B})$$

for a dissimilarity d and suggested a good choice for d based on graph-theoretic considerations. Note that when the self-similarities $d(\mathscr{A},\mathscr{A})$ or $d(\mathscr{B},\mathscr{B})$ are small, the dissimilarity between $\mathscr{A}$ and $\mathscr{B}$ is large. So the tendency will be to merge loose clusters together when the dissimilarity between them is not too large. A related case is Ward's method, which essentially looks at the ratio of between-set and within-set sums of squares. There are also choices of d based on medoids, or representing the cluster by the median, rather then centroids, often giving more robustness. Usually, whatever choice of d is used, it reduces to the underlying metric when $\mathscr{A}$ and $\mathscr{B}$ are singleton sets, as in the four instances above.

Loosely, Ward's method, K-means, and centroid linkage tend to give homogeneous clusters around central values, means or medians, so the clusters can be readily summarized. To a greater or lesser extent, these methods refine a given clustering rather than generating a clustering from the start.

In addition to AGNES, PAM (partitioning around medoids) and CLARA (clustering large applications) are popular packages for clustering. In PAM the collection of all pairwise distances between data points or sets of data point objects is stored using $\mathcal{O}(n^2)$ memory space. So, PAM doesn't scale up to large n well. CLARA avoids this problem by only finding the dissimilarity matrix for subsets of the data to reduce computation time and storage requirements. Roughly, CLARA uses two steps. A subsample of the data is drawn and divided into K clusters, as in PAM, and the two steps are to choose successive medoids to find the smallest possible average distance between the data points in the sample and their most similar representative data points and then to try to decrease the average distance by replacing representative data points. Then each data point not in the sample is assigned to the nearest medoid. However, PAM and CLARA are similar in that they try to find K central values for their clusters.

Aside from these heuristics, the various clustering algorithms have been compared extensively, theoretically and empirically. Fisher and Ness (1971) compare two versions of K-means and single, complete, and centroid linkages under a variety of geometric criteria. The most important ones are convexity of the clusters, the ability to reconstruct the dissimilarity matrix from the clusters, whether the similarity between two elements of the same cluster is higher than between elements of different clusters, and several versions of the Sibson-Jardine stability principles (see Sibson and Jardine (1971)) discussed below.

Roughly, in terms of the desirable properties they examined, single linkage is best, complete linkage is second best, variants on K-means are third, and a method based on centroids did worst. However, as noted before, because hierarchical clustering requires several passes through the data, efficiency of the algorithm may limit how well it scales up, and it is K-means that seems to scale up best to complex problems. This is roughly consistent with the observations on the methods above and in the exercises at the end of this chapter, as well as with the Sibson-Jardine theory discussed below. However, while average linkage does not have such nice theoretical properties as yet, it is regarded by many practitioners as giving better results.

Separate from how clustering procedures work and what type of clusters they favor, there are a large variety of ways to evaluate clusterings. Obviously, the within- and between-cluster sums of squares used in K-means or Ward's method can be used more generally. More pragmatically, given a clustering, one can insist on giving it an interpretation and then ask whether the points in the cluster are consistent with that interpretation. For example, in document clustering, one might have clusters that seem to correspond to different subject matter. In principle, a human can read the documents and determine whether each document actually belongs to its cluster and so compute the proportion of documents correctly assigned. Essentially, this turns a clustering problem into a classification problem.

More generally, the entries $d(\boldsymbol{x}_i, \boldsymbol{x}_j)$ of the dissimilarity matrix D can be used to define criteria a good clustering should satisfy. Representative of this class of optimality criteria is the average distance of points in a cluster $C_i = \{\boldsymbol{x}_{i_1}, \ldots, \boldsymbol{x}_{i_u}\}$, given by

$$\bar{d}_i = \frac{1}{u(u-1)} \sum_{j=1}^{u-1} \sum_{h=j+1}^{u} d(\boldsymbol{x}_{i_h}, \boldsymbol{x}_{i_j})$$

which can be summed for K clusters to give the total average distance

$$T(C_1,...,C_K) = \frac{1}{\#(C_1)+...+\#(C_K)} \sum_{i=1}^{K} \#(C_i)\bar{d}_i$$

much at one with the expressions obtained for K-means clustering. Obviously, many different functions $d(\boldsymbol{x}_i, \boldsymbol{x}_j)$ can be used.

As has been noted, different clustering procedures perform well under various criteria. In addition, different clustering procedures give different clusterings for the same data and on different data sets a given clustering procedure may perform quite differently.

As a practical matter, when n or p is large, the fastest hierarchical agglomerative clustering is usually single linkage, which creates a minimal spanning tree (although K-means is faster, in general, than most hierarchical methods). Indeed, to get a useful clustering from a single-linkage procedure, sometimes it is enough just to remove the longest edges, as shown in Fig. 8.3. This procedure may give results similar to K-means when K is chosen correctly and the clusters are roughly convex. In fact, this simple technique will be adequate for separated clusters, but this is a very strong assumption. One of the more difficult tests for a clustering procedure is where the clusters are formed by concentric annuli with even integer inner radii. In this case, one can choose two points, one from each of two clusters that are closer to each other than either is to most of the points in their respective clusters. Losing convexity of regions and messing up the within-cluster distances challenges most clustering procedures that scale up well. (Nearest-neighbor methods can sometimes be adjusted to handle this, but they scale up poorly.)

When p is large and most of the measurements are noise, it is very difficult to discover true cluster structure because the signal that distinguishes one cluster from another gets lost in the chance variation among the spurious variables. This can lead to multiple cluster structures over different runs that seem equally plausible initially. Sometimes this is due to the instability of the procedure – a method based on means will be less stable than one based on the median. However, more typically the $\boldsymbol{x}_i$s will show strong clustering with respect to one subset of variables and comparably strong but different clustering with respect to a different subset of the variables. Recall that polythetic procedures use all p entries in the $\boldsymbol{x}_i$s as opposed to monothetic strategies in which the entries in the $\boldsymbol{x}_i$s are used one at a time. In monothetic strategies, a defined clustering may be found before all the entries have been examined so that not all the coordinates of the data points necessarily affect the procedure. Since monothetic clusterings depend on which entries are examined first, monothetic clustering strategies are usually more susceptible to spurious variables than polythetic ones are.

To dramatize this, consider a data visualization technique for multivariate data $\boldsymbol{x}_1,...,\boldsymbol{x}_n$ from Inselberg (1985) called a parallel coordinate plot. Imagine p real lines as in Fig. 8.4, labeled by the entries in $\boldsymbol{x}_i$. Then each $\boldsymbol{x}_i = (x_{1,i},...,x_{p,i})$ can be represented as a piecewise linear curve intersecting the p real lines so that in the kth real line $x_{k,i}$ is plotted. This gives n piecewise linear curves to represent the n data points. For six data points, this is shown in Fig. 8.4. Using the first three coordinates, A, B, C and D, E, F form two clusters. However, using x_6 and x_7, C, E, B and F, A, D form two

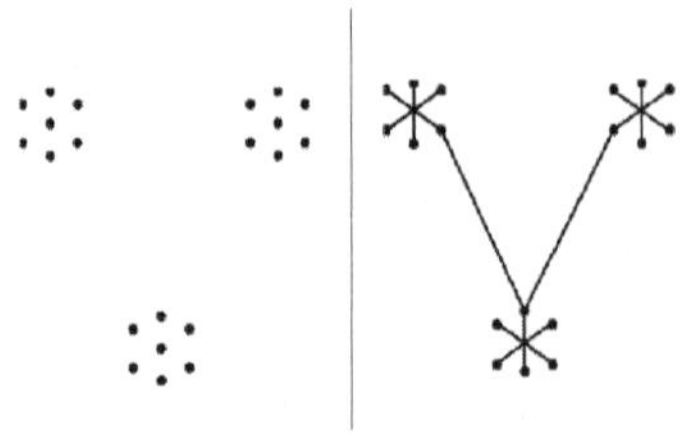

Fig. 8.3 In this hypothetical case, the clusters indicated in the left panel were obtained by deleting the long lines connecting them in the right panel.

different clusters. While this kind of visualization technique helps sort out different monothetic clusterings for a small number of dimensions, it usually does not provide enough summarization for dimensions much above, say, $p = 20$.

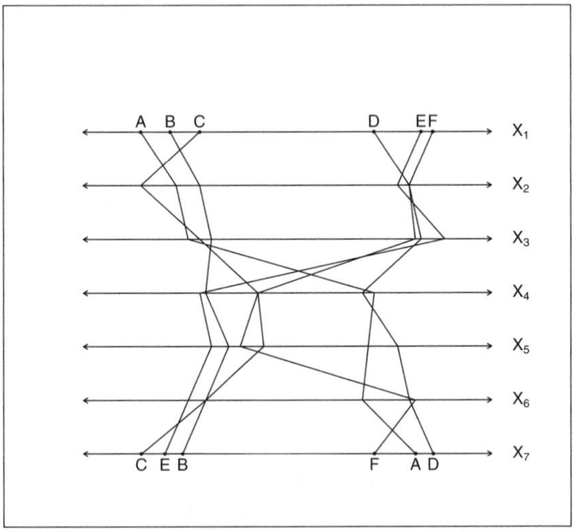

Fig. 8.4 In this diagram the coordinates of the points A–F are indicated by the intersections of the jagged lines with the axes labeled by the variables. It is seen that clustering on X_1, X_2, or X_3 gives different results from clustering on X_6 or X_7.

Friedman and Meulman (2004) discuss the problem of using only a few of the entries of the $\boldsymbol{x}_i$s to produce a clustering; i.e., of doing variable selection on $\boldsymbol{x}$. Variable selection will be treated in Chapter 10, but it is worth stating their criterion here. To state their optimality criterion, recall the membership function $C(i) = k \Leftrightarrow \boldsymbol{x}_i \in C_k$ for a collection of C_ks in the clustering $\mathscr{C}$. Using this, the clustering goal can be stated as finding the choice of $C(\cdot)$, say, $C^*(\cdot)$ that minimizes how bad a clustering is. Intuitively, a clustering is good if the cases in the clusters are more similar to each other than they are to cases in other cluster, and poor if not. If the criterion is denoted $Q(C)$, then one might seek, for instance,

$$C^* = \arg\min_{\phi} Q(C) = \arg\min_{C} \sum_{k=1}^{K} \frac{W_k}{n_k^2} \sum_{C(i)=k} \sum_{C(j)=k} d(\boldsymbol{x}_i, \boldsymbol{x}_j), \tag{8.2.2}$$

where $n_k = \#(C_k)$ and W_k for $k = 1, \ldots, K$ are weights assigned to the clusters. Obviously, there are other choices for Q in (8.2.2). Now, define

$$d(\boldsymbol{x}_i, \boldsymbol{x}_\ell) = \sum_{j=1}^{p} w_j d_{i,\ell,j} = \sum_{j=1}^{p} w_j d_j(\boldsymbol{x}_i, \boldsymbol{x}_\ell),$$

where d_j is the distance measure on the jth entry in the $\boldsymbol{x}_i$s and the weights w_j are positive and sum to 1. The distances can also be normalized, as in Friedman and Meulman (2004), by replacing d_j by

$$\frac{d_j(\boldsymbol{x}_i, \boldsymbol{x}_\ell)}{\frac{1}{n^2} \sum_{i=1}^{n} \sum_{\ell=1}^{n} d_k(\boldsymbol{x}_i, \boldsymbol{x}_\ell)}.$$

If all $w_j = 1/p$ in (8.2.3), then all explanatory variables have the same influence on (8.2.2) and so on C^*. However, the weights need not be uniform, for instance if there is prior information that some variables are more important than others. The equal-weights clustering criterion is

$$Q(C) = \sum_{k=1}^{K} W_k \left(\frac{1}{p} \sum_{j=1}^{p} \left(\frac{1}{n_k^2} \sum_{C(i)=k} \sum_{C(\ell)=k} d_{i,\ell,j} \right) \right). \tag{8.2.3}$$

A natural choice for $d_{i,\ell,j}$ is $(x_{ij} - x_{\ell j})^2 / s_j^2$.

Since $Q(C)$ in (8.2.3) depends also on the w_ks, one can rewrite it as $Q(C, \boldsymbol{w})$ and minimize jointly over C and $\boldsymbol{w}$. An optimum $\boldsymbol{w}^*$ will put small weight on explanatory variables that do not affect the clustering and large values on those that do. If this problem is extended so that different clusters are permitted to have different weights on the entries of the $\boldsymbol{x}_i$s assigned to them and a penalty is imposed for putting weight on too few explanatory variables, then it is likely that all the explanatory variables will be used on the clustering unless there is strong evidence that some do not matter (and so get a zero coefficient). Using a negative entropy penalty (that is minimized when the weights are all the same), scaled by a tuning parameter λ, Friedman and Meulman (2004) show, in a series of examples, that optimal hierarchical agglomerative clustering using (8.2.3) tends to identify clusters whose data points are comparatively close based on different subsets of the explanatory variables.

8.2.2 Divisive Hierarchical Clustering

Divisive hierarchical clustering treats the data initially as being one group that gets split successively using a distance measure, in principle until each subset consists of a single element. Clearly, divisive hierarchical clustering is the reverse of agglomerative clustering. In fact, every technique for agglomerative clustering ultimately unites all the data points into one cluster, so doing it backwards gives a technique for divisive clustering. Conversely, every technique for divisive clustering can be done backwards to give an agglomerative technique. However, the different starting points for the two sorts of hierarchical clustering make different techniques seem more natural even though they are mathematically equivalent.

At its root, hierarchical divisive clustering splits sets according to fixed rules. The basic template for hierarchical divisive clustering is:

> Start with a sample $\boldsymbol{x}_i$, $i = 1, \ldots, n$ regarded as a single cluster of n data points and a dissimilarity d defined for all pairs of points in the sample. Fix a threshold t for deciding whether or not to split a cluster.
>
> - ☐ First, determine $d(\boldsymbol{x}_i, \boldsymbol{x}_j)$, the distance between all pairs of data points, and choose the pair with the largest distance d_{max} between them.
> - ☐ Compare d_{max} to t. If $d_{max} > t$, then divide the single cluster in two by using the selected pair as the first elements in two new clusters. The remaining $n-2$ data points are put into one of the two new clusters: $\boldsymbol{x}_\ell$ is added to the new cluster containing $\boldsymbol{x}_i$ if $d(\boldsymbol{x}_i, \boldsymbol{x}_\ell) < d(\boldsymbol{x}_j, \boldsymbol{x}_\ell)$; otherwise $\boldsymbol{x}_\ell$ is added to the new cluster containing $\boldsymbol{x}_i$.
> - ☐ At the second stage, the values of $d(\boldsymbol{x}_i, \boldsymbol{x}_j)$ are found for $\boldsymbol{x}_i$, $\boldsymbol{x}_j$ within one of the two new clusters to find the pair in the cluster with the largest distance d_{max} between them. If $d_{max} < t$, the division of the cluster stops and the other cluster is considered. Then the procedure repeats on the clusters generated from this iteration.
>
> The procedure can be run up to $n-1$ times, which gives n singleton clusters, in which case a level in the dendrogram must be chosen to specify a clustering, or the procedure can be run until $d_{max} > ts$ for all the existing clusters.

Divisive methods are often more computationally demanding than agglomerative methods because decisions must be made about dividing a cluster in two in all possible ways. In fact, the template shows there are two basic problems that must be resolved to implement a divisive procedure. First, a cluster to split must be chosen and then the actual split must be found.

The first problem has three standard answers. (i) Split every cluster until a complete binary dendrogram is obtained. (ii) Always split the largest cluster. (iii) If $\boldsymbol{w}$ is the center of the cluster, choose the cluster with the largest variability around $\boldsymbol{w}$; i.e., the j with the highest value of $\sum_{i\in\mathscr{C}_j} \|\boldsymbol{x}_i - \boldsymbol{w}\|^2$. The first two are simple: The first ignores whether the clusters are meaningful and lets the practitioner choose where along the dendrogram to stop. Always splitting the largest cluster tends to generate dendrograms where all leaves are about the same size. The third may be the best because it is sensitive to the scatter of points in clusters. It will tend to produce clusterings that are tighter than the others. Other cluster selection rules can be readily formulated; the specific rule chosen tends to act like extra information, favoring one class of clusterings over others; see Savaresi et al. (2002). Ding and He (2002) present examples suggesting that splitting the loosest cluster (i.e., the cluster with the smallest average similarity or largest average dissimilarity) gives good performance.

As with agglomerative clustering, there are many procedures that implement the template. Here, it is enough to describe two packages and two procedures, one based on K-means and the other based on a matrix decomposition. To an extent, other popular methods are variants of them.

The two packages are MONA (monothetic analysis) and DIANA (divisive analysis). MONA is a divisive monothetic procedure that applies to binary variables only. Each split is based on one variable, hopefully well chosen. Given the variable, a starting cluster is divided into two clusters, one having data points with value 1 for that variable and the other having value, say, 0 for that variable. Each cluster is divided until all the data points in the same cluster are identical. Choosing the variable on which to split is the key feature: It is the variable with the maximal total association to the other variables within the cluster to be split. To define this measure of dependency, define the association between two binary variables X_i and X_j as

$$A(i,j) = P(X_i = 1, X_j = 1)P(X_i = 0, X_j = 0) - P(X_i = 1, X_j = 0)P(X_i = 0, X_j = 1).$$

The total association of a variable X_i is then

$$TA(i) = \sum_{j\neq i} A(i,j).$$

Treating the components X_i for $i = 1, \ldots, p$ of $\boldsymbol{x}$ in this way, MONA estimates the $TA(i)$s and splits on x_i, where $i = \arg\max_i TA(i)$. This procedure does not work with missing values, so missing values in a data point are filled in by using a separate procedure also based on total association.

DIANA is an implementation of the divisive template for the dissimilarity $\bar{D}$,

$$\bar{D}(\boldsymbol{x}_i, C_k) = \begin{cases} \frac{1}{\#(C_k)-1} \sum_{\boldsymbol{x}_j \in C_k, j\neq i} d(\boldsymbol{x}_i, \boldsymbol{x}_j) & \text{if } \boldsymbol{x}_i \in C_k; \\ \frac{1}{\#(C_k)} \sum_{\boldsymbol{x}_j \in C_k} d(\boldsymbol{x}_i, \boldsymbol{x}_j) & \text{if } \boldsymbol{x}_i \notin C_k. \end{cases}$$

Clearly, $\bar{D}$ is an average dissimilarity based on d. The full data set is treated as a single cluster and split until each cluster contains only a single object. At each iteration, the cluster with the largest diameter is selected; the diameter of a cluster is the largest dissimilarity between any two of its objects. For this cluster, DIANA first finds the data point that has the largest $\bar{D}$ distance to the other objects of the selected cluster. This data point starts another cluster. Data points are then assigned to this new cluster if they are closer to it than to the old cluster they were in. The result is a split of the selected cluster into two new clusters.

8.2.2.1 Divisive *K*-means

Recall that the basic K-means procedure is to select K points as initial centers, assign the data points to their closest center, recompute the centers, and then repeat until the centers stop changing too much. Savaresi et al. (2002) proposes a divisive form of this, called bisecting K-means, that can be summarized as follows. Note that the suggestive notation $||\cdot||$ is used in place of the generic dissimilarity d.

> Start with a sample $\boldsymbol{x}_i$ for $i = 1,...,n$ regarded as a single cluster of n data points and a norm $||\cdot||$. Initialize by selecting a point $\boldsymbol{c}_0 \in \mathbb{R}^p$ and finding some location $\boldsymbol{w}$ of the sample, such as a mean. Define $\boldsymbol{c}_1 = \boldsymbol{w} - (\boldsymbol{c}_0 - \boldsymbol{w})$.
>
> ☐ Divide the sample into subsets S_0 and S_1 by
>
> $$\forall i: \text{if } \ ||\boldsymbol{x}_i - \boldsymbol{c}_0|| \leq ||\boldsymbol{x}_i - \boldsymbol{c}_1||, \ \text{ then put } \ \boldsymbol{x}_i \in S_0,$$
>
> and
>
> $$\forall i: \text{if } \ ||\boldsymbol{x}_i - \boldsymbol{c}_0|| > ||\boldsymbol{x}_i - \boldsymbol{c}_1||, \ \text{ then put } \ \boldsymbol{x}_i \in S_1.$$
>
> ☐ Find the centers $\boldsymbol{w}_0$ and $\boldsymbol{w}_1$ of S_0 and S_1, respectively.
>
> ☐ If $\boldsymbol{w}_0 = \boldsymbol{c}_0$ and $\boldsymbol{w}_1 = \boldsymbol{c}_1$, stop. Otherwise, take $\boldsymbol{c}_0$ as $\boldsymbol{w}_0$ and $\boldsymbol{c}_1 = \boldsymbol{w}_1$ and repeat, systematically searching through the clusters to minimize the *SSE* from the bisection in terms of $||\cdot||$.
>
> The last step can be modified so the procedure stops when $||\boldsymbol{w}_0 - \boldsymbol{c}_0||$ and $||\boldsymbol{w}_1 - \boldsymbol{c}_1||$ are less than a threshold.

Comparing this with the K-means algorithm, one sees that rather than proposing an initial set of K-means, the data set is bisected to find the means that can be used to define clusters. The division is done in the first step, and each $\boldsymbol{x}_i$ is put in the cluster whose center is closest to it. The analogy is even closer if the procedure stops when a pre-assigned number of clusters has been found. The time complexity of this bisecting version is linear in n (i.e., $\mathcal{O}(n)$), making it more efficient than regular K-means clustering: There is no need to compare every point to every cluster center since to bisect a cluster one just looks at the points in the cluster and their distances to two centers.

8.2.2.2 Principal Direction Divisive Partitioning

The PDDP method was invented by Boley (1998) and was developed in the context of document retrieval, where the data points $\boldsymbol{x}_i = (x_{i,1}, ..., x_{i,p})$ represent the relative frequency of p words, suitably scaled. For ease of exposition, let $\boldsymbol{M} = [\boldsymbol{x}_1, ..., \boldsymbol{x}_n]$ be the $p \times n$ matrix formed by concatenating the data points as column vectors. This can be centered by subtracting the mean coordinatewise to give $\boldsymbol{M}_c = \boldsymbol{M} - \bar{\boldsymbol{x}}\mathbf{1}'$, where $\bar{\boldsymbol{x}} = (\bar{x}_1, ..., \bar{x}_p)'$. Although $\boldsymbol{M}_c$ is not a square matrix in general, the strategy is to find the eigenvector corresponding to its largest eigenvalue and use this to identify a direction for a good split. The premise is that the direction of greatest stretching is the right axis on which to split to find clusters. For nonsquare matrices, the analog of an eigenspace decomposition is called a singular value decomposition (SVD). Essentially, it generalizes the usual diagonalization procedure by identifying the subspace (of dimension less than or equal to $\min(p,n)$) on which there really is a diagonal form. The remaining dimensions are treated as a null space for the matrix as a linear operator.

Consider a real $p \times n$ matrix $\boldsymbol{A}$ with $p > n$. It is possible to define the eigenvectors of $\boldsymbol{A}$, but there will be at most n of them, even though they are elements of a p-dimensional space. Let the matrix of eigenvectors of $\boldsymbol{A}$ be $\boldsymbol{P}$. Since $\boldsymbol{P}$ is not square, it is singular so there is no basis transformation in which $\boldsymbol{A}$ is diagonal with eigenvalues on its main diagonal. Nevertheless, $\boldsymbol{A}$ can be written as

$$\boldsymbol{A} = \boldsymbol{U}\boldsymbol{D}\boldsymbol{V}', \tag{8.2.4}$$

in which $\boldsymbol{U}$ is $p \times n$, $\boldsymbol{D}$ is $n \times n$ diagonal, and $\boldsymbol{V}$ is an $n \times n$ square matrix. Both $\boldsymbol{U}$ and V can be chosen to have orthogonal columns, so

$$\boldsymbol{U}'\boldsymbol{U} = \boldsymbol{V}'\boldsymbol{V} = Id.$$

The elements of $\boldsymbol{D}$ are called singular values; at most n of them are nonzero. The proof uses the Gram-Schmidt orthogonalization technique to identify submatrices corresponding to the spaces of $\boldsymbol{x}$s and $\boldsymbol{y}$s on which the equation $\boldsymbol{A}\boldsymbol{x} = \lambda \boldsymbol{y}$ can be solved. The relationship between $\boldsymbol{P}$ and $\boldsymbol{A}$, $\boldsymbol{U}$, or $\boldsymbol{V}$ is beyond the present scope. (In fact, $\boldsymbol{V}$ and $\boldsymbol{P}$ do not appear in the PDDP procedure.) However, it is clear that, without loss of generality, the diagonal entries of $\boldsymbol{D}$ can be assumed to decrease. Assume that the first element is largest and the last element is nonnegative.

Given these formalisms, Savaresi et al. (2002) present PDDP as follows.

> Start with a sample $\boldsymbol{x}_i$ for $i = 1, ..., n$ regarded as a single cluster of n data points. Find the mean $\bar{\boldsymbol{x}}$ and $\boldsymbol{M}_c = \boldsymbol{M} - \bar{\boldsymbol{x}}\mathbf{1}'$.
>
> ☐ Let $\boldsymbol{A} = \boldsymbol{M}_c$, and find the $\boldsymbol{U}$, $\boldsymbol{V}$ in the SVD of $\boldsymbol{M}_c$ as in (8.2.4).
>
> ☐ Take the first column of $\boldsymbol{U}$, say $u = U_{\cdot,1}$, and partition the data points $\boldsymbol{x}_1$, ...,$\boldsymbol{x}_n$ from $\boldsymbol{M}$ into two subsets $\boldsymbol{M}_0$ and $\boldsymbol{M}_1$ using the inequalities
>
> $$\forall i: \text{if } \ u'(\boldsymbol{x}_i - \boldsymbol{w}) \leq 0, \ \text{ then put } \ \boldsymbol{x}_i \in \boldsymbol{M}_0,$$

and

$$\forall i : \text{if } u'(x_i - w) > 0, \text{ then put } x_i \in M_1.$$

- ☐ Iterate this procedure with M_0 and M_1 in place of M: Center M_0 and M_1, find the SVDs of the centered matrices, and then use the first columns of the new U matrices on each cluster.

In fact, this is close to the spectral methods discussed below since it uses the spectrum of the data matrix, though not the spectrum of a dissimilarity matrix. Moreover, it is enough to find the first singular value and column of U, rather than the whole SVD, to effect this split.

The bisecting k-means and PDDP algorithms are somewhat similar. Both split M through the centroid $\bar{x}$, but the first does so using a plane perpendicular to the line joining the centroids w_0 and w_1 of S_0 and S_1, while the second uses a plane perpendicular to the direction of the centered data M_c that has the largest variance. Consequently, the actual splits the two procedures give can often be close; the differences emerge more from the choice of clusters to split. The main contribution of Savaresi et al. (2002) is a technique for taking the shape of clusters into account. Their procedure does so by "looking ahead" – if there is little scatter to the points relative to the distance between centroids, then there may be two separable clusters, while if there is a lot of scatter, then there may be one large cluster. This criterion also indicates which clusters are most important to split.

In fact, Boley (1998) actually developed PDDP using principal components, a technique for decomposing covariance matrices. However, the derivation for principal components uses an SVD and in fact is nearly equivalent to an SVD.

8.2.3 Theory for Hierarchical Clustering

First, it is important to note that clustering methods are generally just procedures and the notion of fit derives from dissimilarities, not probabilities. Probabilistic properties, when invoked, are mostly for evaluating some aspect of performance of the clustering procedure rather than generating clusters.

Second, clustering data effectively means regarding them as having come from a single source, even if that source is a mixture of components. The single source is a density on X so clustering corresponds to searching for the local maxima of a density. In general, where to draw the boundary between different modes is arbitrary unless extra information compels a particular choice. Given the variety of possible densities for a p-dimensional random variable, it is likely unreasonable to expect to find an optimal clustering procedure for completely general settings. Despite this caveat, there is reason to hope a comprehensive theory will emerge over time. The widespread importance of clustering means the search for good theory to motivate and develop clustering procedures, and interpret results from them, is worthwhile.

Third, the range of current theory runs from optimists who remain hopeful a comprehensive theory can be developed (see Luxburg and Ben-David (2005)) to pessimists, such as Kleinberg (2003), who have theorems showing that no clustering procedure can satisfy an innocuous list of desirable properties.

An intermediate position that continues to motivate investigation is taken in Sibson and Jardine (1971) and is intended only for hierarchical clustering procedures. They introduce a collection of desiderata that a good clustering procedure should satisfy. They regard a clustering procedure C as acting on the dissimilarity matrix for a data set using a dissimilarity d and giving an equivalence relation or hierarchy on the data points. This can be thought of as an "ultrametric" (see Janowitz (2002)) a level of study deeper than needed here.

Several of the Sibson and Jardine (1971) conditions are worth stating. First, they want the clustering to be independent of the labeling of the data points. Second, they want a clustering to be well-defined, meaning that any fixed dissimilarity matrix will always give the same clustering under the procedure. A third condition is a fitting-together condition: Removing only a few data points should change the clustering only a little. One way to formalize this is to insist that if the data points on a branch are removed, then clustering the points on the branch and on the data points in the complement separately should give the same tree structure as in the original clustering. Fourth, the dendrogram should be independent of scale; i.e., multiplying the dissimilarity matrix should not change the clusters it generates. Fifth, the clusters should be stable under growth in that as more data accumulate the clusters shouldn't change much. There are several other criteria, but the most controversial of all of them is the continuity axiom: The clustering function C should be continuous as a function of the dissimilarities. Using these properties, Sibson and Jardine (1971) provide a (unique) characterization of single-linkage clustering. The problem is that single-linkage clustering suffers the chaining problem and doesn't scale up particularly well to higher dimensions.

To elaborate on the debate, let d be a distance on a set S identified with the set $\{1,...,n\}$, and define a clustering function to be any function C that takes a distance function d on S to a partition Γ of S. That is, $C(d) = \Gamma$. Kleinberg (2003) defines three properties a good C should satisfy:

- Scale Invariance: For any d and $\alpha > 0$, $C(d) = C(\alpha d)$.
- Richness: Range(C) equals the set of all partitions of S.
- Consistency: For two distances d_1 and d_2, if $C(d_1) = \Gamma$ and d_2 is a γ transformation of d_1 (i.e, d_2 satisfies (i) $\forall i, j \in S$ belonging to the same cluster $d_2(i,j) \leq d_1(i,j)$ and (ii) $\forall i, j \in S$ belonging to different clusters $d_2(i,j) \geq d_1(i,j)$), then $C(d_2) = \Gamma$ as well.

Intuitively, scale invariance means that the clustering function should be insensitive to changes in the unit of measurement. Richness means that, given any partition Γ, $C^{-1}(\Gamma)$ is a well-defined, nonvoid set. The consistency condition encapsulates the idea that shrinking the distance between points in the same cluster or expanding the distance between points in different clusters should not affect the clustering itself. Using these,

Kleinberg (2003) establishes an impossibility result: There is no clustering procedure that satisfies all three of these properties.

To prove the theorem, the concept of an antichain must be defined. First, Γ' is a refinement of Γ if each set in Γ' is a subset of a set in Γ. The binary relation of refinement between partitions is a partial ordering, so the set of partitions of S under the refinement relation is a partially ordered set. Now, a collection of partitions is an antichain if it does not contain two distinct partitions such that one is a refinement of the other; that is, an antichain is a sequence of partitions whose elements when compared under refinement are not comparable.

Theorem (Kleinberg, 2003): If a clustering function C satisfies scale invariance and consistency, then Range(C) is an antichain.

Since the collection of all partitions is not an antichain, Kleinberg observes the following immediate consequence.

Corollary: There is no clustering function C that satisfies scale invariance, richness, and consistency.

Proof of the Theorem: To set up the proof, two definitions are required. First, given a partition Γ, a distance d is said to (a,b)-conform if there is an a and b such that for any i, j in the same cluster $d(i,j) \leq a$ and for i, j in different clusters $d(i,j) \geq b$. Second, given a clustering function C, $(a,b) \in \mathbb{R}^2$ is Γ-forcing if every d that (a,b)-conforms to Γ satisfies $C(d) = \Gamma$.

Now, let C be a clustering function on S that satisfies consistency.

Step 1: For any partition $\Gamma \in$ Range(C), $\exists a, b \in \mathbb{R}^+$ with $a < b$ so that (a,b) is Γ-forcing.

Let $\Gamma \in$ Range(C). Then $\exists d$ with $C(d) = \Gamma$. Let

$$a^* = \min_{i,j \text{ in the same cluster of } \Gamma} d(i,j)$$

and

$$b^* = \max_{i,j \text{ in different clusters of } \Gamma} d(i,j).$$

Next, let $(a,b) \in \mathbb{R}^2$ with $a < b$ so that $[a^*, b^*] \subset (a,b)$. Now, any distance d^* that (a,b)-conforms to Γ must be a Γ transformation of d, so, by Consistency, $C(d^*) = \Gamma$ and (a,b) is Γ-forcing.

Suppose now that C also satisfies scale invariance and that there are distinct partitions $\Gamma_0, \Gamma_1 \in \text{Range}(C)$ so that Γ_0 is a refinement of Γ_1.

Step 2: This last supposition leads to a contradiction.

Let (a_u, b_u) be a Γ_u-forcing pair with $a_u < b_u$ for $u = 1, 2$. Step 1 ensures that these requirements can be satisfied.

Let $a_2 \leq a_1$ and let $\varepsilon \in (0, a_0 a_2 / b_0)$. Now, there is a distance function d such that (1) $d(i,j) \leq \varepsilon$ when i, j belong to the same cluster of Γ_0; (2) $a_2 \leq d(i,j) \leq a_1$ when pairs

i, j belong to the same cluster of Γ_1 but not to the same cluster of Γ_0; and (3) $d(i, j) \geq b_1$ when i, j are not in the same cluster of Γ_1.

The contradiction arises as follows. The distance d (a_1, b_1)-conforms to Γ_1 so $C(d) = \Gamma_1$. So, set $\alpha = b_0/a_2$ and $d' = \alpha d$. Scale invariance implies $C(d') = C(d) = \Gamma_1$.

However, for points i, j in the same cluster of Γ_0,

$$d'(i, j) \geq \frac{a_2 b_0}{b_0} \geq b_0,$$

meaning that d' (a_0, b_0)-conforms to Γ_0, and therefore $C(d') = \Gamma_0$ and so $\Gamma_0 = \Gamma_1$, a contradiction. □

Kleinberg's theorem holds if the distances are metrics, and Kleinberg (2003) verifies that any two of the three properties can be satisfied.

Despite Kleinberg's theorem, other approaches to the development of a theory remain promising. The central issue may be not to work with the distance to which a clustering corresponds, but to define a distance between clusterings. Luxburg and Ben-David (2005) introduce two ideas for how to do this. The first is an extension operator: If $\mathscr{C}_1$ is a clustering of S_1 and $\mathscr{C}_2$ is a clustering of S_2, where S_1 and S_2 are subsets of $\mathscr{X}$, then both $\mathscr{C}_u$s can be extended to a clustering on the whole space and the extended clusterings can be compared. For instance, one can take the K means found for a clustering and generate a clustering for $\mathscr{X}$ by assigning any point in $\mathscr{X}$ to its closest center. Now, the two partitions of $\mathscr{X}$ from the $\mathscr{C}_u$s are comparable and any way to assign extra data points to clusters defines an extension.

The second idea of Luxburg and Ben-David (2005) is to define a quality measure, say $q(\mathscr{C})$, for a clustering $\mathscr{C}$ and use $d(\mathscr{C}_1, \mathscr{C}_2) = |q(\mathscr{C}_1) - q(\mathscr{C}_2)|$ as a distance measure. Two similar clusterings should have qs that are close together, but it is possible for quite different clusterings to have similar qs as well. Steinbach et al. (2000), Section 4, suggest several possibilities; one is $q(\mathscr{C}) = \sum_{k=1}^{K} (1/\#(C_k)^2) \sum_{\boldsymbol{x}_i, \boldsymbol{x}_j \in C_k} \cos(\boldsymbol{x}_i, \boldsymbol{x}_j)$, where the cosine is of the angle between the vectors in its argument.

Given a distance measure on clustering and its extensions, it is straightforward to define convergence of clusterings. Since they depend on the initial data $\boldsymbol{x}_1, \ldots, \boldsymbol{x}_n$, the usual modes of convergence (distribution, probability, etc.) apply. One consequence of this is that if the distance measure is bounded, its expectation

$$A(\mathscr{A}_1, \mathscr{A}_2, P, n) = \mathrm{Ex}_P(d(\mathscr{C}_1(S_{1,n}), \mathscr{C}_2(S_{2,n})))$$

exists, where, for $u = 1, 2$, $\mathscr{A}_u$ is a clustering procedure giving clusterings $\mathscr{C}_u$ for independent samples of n points, $S_{u,n}$. Under reasonable conditions, it is possible to establish probabilistic bounds of the form

$$P(|d(\mathscr{C}_1(S_{1,n}), \mathscr{C}_2(S_{2,n})) - A(\mathscr{A}_1, \mathscr{A}_2, P, n)| > t) \leq e^{-\alpha_n t^2}$$

using Hoeffding's inequality, for instance. It remains to be seen how well this strategy generates and evaluates clustering procedures.

8.3 Partitional Clustering

In contrast to hierarchical techniques, which give a nested sequence of clusterings, usually based on a dissimilarity, partitional clustering procedures start with an initial clustering (a partition of the data) and refine it iteratively, typically until the final clustering satisfies an optimality criterion. In general, the clusterings in the sequence are not nested, so no dendrogram is possible. On the other hand, this means that partitional clusterings do not lock in bad choices; in principle, a bad choice for forming a cluster in one iteration can be undone at a later iteration. In a trivial sense, any hierarchical clustering can be regarded as partitional by ignoring the nesting of the clusters at each iteration. However, aside from the nesting, hierarchical methods are usually based on a dissimilarity, while partitional procedures are usually based on an objective function. So, if a hierarchical method has a fixed stopping rule and its iterations try to satisfy an optimality principle, the sense need not be so trivial. Since it is hard even to think of a general partitional method as hierarchical, it is easy to see that partitional methods are the larger class.

While objective functions are not necessarily part of the definition of partitional methods, they occur commonly enough to justify discussion. So, consider a function $F = F(K_0, \mathscr{C}(K_0))$, where K_0 is an initial number of clusters and $\mathscr{C}_0(K_0)$ is an initial clustering of the $\boldsymbol{x}_i$s. Given F, a natural procedure for partitional clustering is to rearrange points to maximize F:

Start with a sample $\boldsymbol{x}_i$ for $i = 1,...,n$ and an integer K_0. Let $\mathscr{C}_0(K_0)$ be written $C_{0,1},...,C_{0,K_0}$.

☐ Find $F(K_0, \mathscr{C}_0(K_0))$, and set

$$F_0 = F(K_0, C_{0,1}, ..., C_{0,K_0}).$$

☐ Rearrange the points assigned to the $C_{0,k}$s, possibly rechoosing K_0 to obtain a new clustering, $\mathscr{C}_1(K_1)$, with K_1 clusters, $(C_{1,1}, ..., C_{1,K_1})$.

☐ Find the new value of the objective function

$$F_1 = F(K_1, C_{1,1}, ..., C_{1,K_1}).$$

☐ Continue iterating, searching through values of K and clusterings $C_{j,1},...,C_{j,K_j}$ until F_j is small enough and the final choice of K_{fin} and $C_{fin,1},...,C_{fin,K_{fin}}$ obtained.

The initial clustering $\mathscr{C}(K_0)$ may be random or defined by K_0 statistics such as cluster means.

The big gap in this template is the choice of F. There are many possibilities. Indeed, most partitional clustering techniques have a specific F they optimize. For instance,

(8.1.1) is the optimality criterion that K-means optimizes and (8.3.1) below is another objective function from the perspective of graph theoretic clustering.

Several important classes of objective functions are identified by Zhao and Karypis (2002). One is the class of internal criteria. These objective functions express the desire that the clustering should maximize the homogeneity of the points in each cluster rather than the between cluster spread; e.g., try to make $W(\mathscr{C})$ small rather than trying to make $B(\mathscr{C})$ large. External criteria are the reverse: They focus on how clusters are different from each other rather than how similar the elements of a cluster are to each other. This would correspond to maximizing $B(\mathscr{C})$ rather than trying to make $W(\mathscr{C})$ small. An evaluation of a wide variety of clustering procedures, including more objective functions than noted below, can be found in Jain et al. (2004). Zhao and Karypis (2005) note some examples:

- ☐ One internal criterion is motivated by K-means which focuses on maximizing the similarity between cluster centers and cluster elements. It can be written

$$IC_1 = \sum_{k=1}^{K} n_k \sum_{\boldsymbol{x}_i \in C_k} \frac{\boldsymbol{x}_i' \bar{\boldsymbol{x}}_k}{\|\boldsymbol{x}_i\| \|\bar{\boldsymbol{x}}_k\|}. \tag{8.3.1}$$

  (Recall that $\cos(\boldsymbol{x}_i, \boldsymbol{x}_j) = \boldsymbol{x}_i'\boldsymbol{x}_j / \|\boldsymbol{x}_i\| \|\boldsymbol{x}_j\|$, the angle between the vectors.)

- ☐ Another internal criterion is to maximize the sum of normalized pairwise similarities between the elements in each cluster. This effectively separates each cluster from the entire data set as much as possible (i.e., makes the average squared distances between the cluster centers as large as possible). This can be expressed as maximizing

$$IC_2 = \sum_{k=1}^{K} n_k \left(\frac{1}{n_k^2} \sum_{\boldsymbol{x}_i, \boldsymbol{x}_j \in C_k} \frac{\boldsymbol{x}_i' \boldsymbol{x}_j}{\|\boldsymbol{x}_i\| \|\boldsymbol{x}_i\|} \right) = \sum_{k=1}^{K} \frac{\|n_k \bar{\boldsymbol{u}}_k\|^2}{n_k},$$

  where $\boldsymbol{u}_k$ is the average of the normalized elements of cluster k.

- ☐ An external criterion might focus on minimizing the angle between cluster centers and the overall mean of the data set, for instance in

$$EC = \sum_{k=1}^{K} N_k \frac{\bar{\boldsymbol{x}}_k' \bar{\boldsymbol{x}}}{\|\bar{\boldsymbol{x}}_k\| \|\bar{\boldsymbol{x}}\|}.$$

Numerous other objective functions have been proposed; IC_1/EC and IC_2/EC are among them as efforts to combine internal and external criteria.

Once an objective function is chosen, there remain many ways to search the possible clusterings of a data set to find one that optimizes it. Obviously, one could exhaustively search all possible clusterings $(C_1, ..., C_K)$ for each $K = 1, ..., n$. This would be computationally prohibitive, so effective search strategies for finding optima, possibly only local, are needed. The simplest, used in K-means, is to refine an initial clustering by finding the new centers and then reassigning each data point to its closest center.

This can be repeated but usually makes most of its gains in the first few iterations. Sometimes this is called iterated assign to nearest.

More flexibly, one can start with an initial clustering and then use split or join rules to divide a cluster into two smaller ones or to condense two clusters into a single one; see Cutting et al. (1992). The basic idea of splitting is to find a cluster that scores poorly on a homogeneity criterion and split it using some procedure (e.g., K-means) using $K = 2$. To join two clusters that are not usefully distinguished by the difference between their centers, one can look for data points that are close, possible only on some of the p entries in $\boldsymbol{x}_i$.

A greedy approach can also be employed; it will always give a local minimum but can be restarted with different initial clusters to help find a global minimum. Given an initial clustering, visit the n points $\boldsymbol{x}_i$ in random order. For each $\boldsymbol{x}_i$, find the change in F from moving $\boldsymbol{x}_i$ from its cluster to another cluster by testing each alternative cluster. Then, put $\boldsymbol{x}_i$ into the cluster giving the highest improvement. If there is no move that increases F, then $\boldsymbol{x}_i$ remains in its cluster. If no $\boldsymbol{x}_i$ can be moved, the iterations stop. One can improve this by recomputing the cluster centers each time an $\boldsymbol{x}_i$ is moved.

Next, three different classes of partitional clustering procedures will be explained. These complement the first partitional procedure, K-means, that was explained at the beginning on account of its popularity. The three classes are model-based, graph-theoretic, and spectral.

8.3.1 Model-Based Clustering

The idea is to propose an overall model for the observed data by assigning distributions to subsets of the data. The most typical case is to regard the clusters as representing local modes of the true distribution, which is then modeled as a mixture of components, one for each mode. While a mixture model might be true, for clustering purposes it is enough to assume that the true model can be readily approximated by a mixture.

A less typical case, but possibly more general, is to recall the cherry-picking technique of House and Banks (2004), see Chapter 1, Section 2.2, which is an effort to decompose a collection of points into subsets with lower dimensional structure. Given n points $\boldsymbol{x}_i$, one can choose $q+2 < n$ points and use q of them to define a q-dimensional subspace of the p-dimensional space. The other two points are needed to estimate a variance and to test goodness of fit. Once a set of points has been found that defines a useful subspace, any other points that are not too far from the hyperplane can be added to the $q+2$ points and declared part of the same cluster. Then the process can be repeated on the remaining $n-q-2$ points.

This kind of strategy may find clusters that are lower-dimensional and dispersed. The human eye tends to be good at picking out the shapes of clusters that are tight or relatively well filled. The eye does not do so well with dispersed clusters, even when they have just as much structure. For instance, a small-ish component of the form $N(0,M)$ for a large M could look like outliers rather than a cluster. However, model-based cherry picking might find this kind of cluster because goodness of fit is a global criterion, whereas proximity is a local condition.

Regardless of the clustering procedure, identifying a partition and unambiguously assigning each $\boldsymbol{x}_i$ to a partition element often leads to points that sit near a boundary. This may be unsatisfactory because the choice of boundary includes unquantified variability: A different run of the experiment would have given different data and thus, probably, different boundaries. There is a robustness problem in the boundaries as well. Tweaking the procedure used to get the clustering procedure will typically give slightly different clusters. Imagine replacing single-linkage clustering with a sort of double-linkage clustering (the average of the two closest points is within a specified distance), for instance. A variant on this problem is that it is easy to generate small samples from many unimodal distributions, such as an exponential, that lead to (weakly) bimodal histograms. So, it is important to assess how accurately the clusters represent the source as well as assessing whether the clusters are otherwise reasonable. These questions can be answered in part by asking how stable a clustering is.

Addressing these questions may be where model-based procedures are most useful. To see why, note that one way to represent the uncertainty in whether a point is in a cluster is to back off from determining absolute membership of each $\boldsymbol{x}_i$ in a cluster and ask only for a degree of membership. That is, one can use soft rather than hard assignment: Rather than saying $\boldsymbol{x}_i$ is or is not in cluster C_k, the statement is of the form that $\boldsymbol{x}_i$ is in C_k with probability $\pi_{i,k}$, where $\sum_k \pi_{i,k} = 1$. One can recover a hard clustering, if desired, by saying $\boldsymbol{x}_i$ is in the cluster C_k with the highest probability; i.e., choose $k = \arg\max_k \pi_{i,k}$. Since the clusters themselves can be regarded as random, too, this procedure is usually done conditional on a fixed clustering.

The idea of soft membership can be formalized in a mixture model. Suppose f is the density of $\boldsymbol{X}$, and write

$$f(\boldsymbol{x}) = \sum_{k=1}^{K} \pi_k p_k(\boldsymbol{x} \,|\, \boldsymbol{\theta}_k), \tag{8.3.2}$$

where the π_k are positive and sum to one and the p_ks are density functions in a family indexed by $\boldsymbol{\theta}$. The p_ks are called components and the π_ks are weights indicating what fraction of the data arise from each component. In fact, most densities can be represented, to any desired level of approximation, as a finite sum of densities drawn from a large enough class, in most distance measures. So, mixture models are more general than they may appear at first.

Models like (8.3.2) are familiar from the Bayesian formulation. Consider the discrete Bayesian model in which the parameter values are θ_k for $k = 1, \ldots, K$ with prior probabilities $\pi(\theta_k)$. When θ_k is true, the model for the data is $p_k(\boldsymbol{x}|\theta_k)$. So, the posterior is the ratio of the joint density for $(\Theta, \boldsymbol{X})$ to the marginal for the data,

$$m(\boldsymbol{x}) = \sum_{k=1}^{K} \pi(\theta_k) p_k(\boldsymbol{x}|\theta_k). \tag{8.3.3}$$

Writing $\pi(\theta_k) = \pi_k$ turns (8.3.3) into (8.3.2). In other contexts where the values of θ_k can be regarded as missing data, they are often called latent variables.

Implementing a model such as (8.3.2) necessitates choosing K and the densities p_k and then estimating the weights π_k and the parameters θ_k.

Effectively, choosing K amounts to choosing the number of clusters. So, methods of clustering already discussed can be used to give an indication of the range of K. In addition, as with K-means, several values of K can be chosen in (8.3.2) and the resulting model evaluated to see how well it fits the data. The issue is how to evaluate the fit, especially when p is large relative to n. The Bayes information criterion is one way to choose K; it amounts to regarding $f(\boldsymbol{x})$ as $f(\boldsymbol{x}^n|\pi_1,...,\pi_K,\theta_1,...,\theta_K)$ (i.e., a parametric family with $2K$ parameters) and then maximizing

$$BIC(K) = -2\log f(\boldsymbol{x}|\hat{\pi}_1,...,\hat{\pi}_K,\hat{\theta}_1,...,\hat{\theta}_K) + K'\log n, \tag{8.3.4}$$

where K' is the number of real-valued parameters; see Chapter 10, Section 2.2, on variable selection, or Fraley and Raftery (2002). It is well known that this expression gives the mode of the posterior on a class of models indexed by K. Sarle (1996) has a thorough discussion and reference list for choosing the number of clusters. When the g_ks are normal, Goldberger and Roweis (2004) propose a technique that uses a value of K known to be too large so as to collapse (8.3.2) down to the correct K; their theorem ensures the K found by their procedure is a local minimum of an objective function.

Given that K has been chosen, the choices of p_ks are varied. One problem that must be resolved is that the resulting $f(\boldsymbol{x})$ is not identifiable. For instance, with normal g_k, consider two instances of (8.3.2). It is seen that both

$$\text{Mixture Density 1}: \{\pi_1 = .5, p_1 = N(0,1);\ \pi_2 = .5, p_2 = N(0,1)\} \tag{8.3.5}$$

and

$$\text{Mixture Density 2}: \{\pi_1 = 1, p_1 = N(0,1/2);\ \pi_2 = 0, p_2 = N(-10,7)\} \tag{8.3.6}$$

have the same mean and variance and so describe exactly the same data generator. In practice, identifiability problems only arise on sets of measure zero in the model space, usually corresponding to boundaries. Lindsay (1995) gives a convexity argument for solving the identifiability problem, but in practice typical mixture modeling ignores the identifiability question, using a computational approach (the EM algorithm to be discussed shortly) that generally gives decent results.

Although the p_ks can correspond to any distribution, discrete or continuous, usually they are chosen for convenience, typically being of a familiar parametric form (Gamma, logistic, log normal, t, Poisson, Weibull, multinomial, or normal). However, it is obvious that many different choices for the set of K densities p_k are possible and that different choices would give different mixture models. For instance, if the p_ks are normal but the clusters are not elliptical, or even convex, then clusterings generated under a normal assumption are not likely to be representative of the data. The choice of models p_k implicitly makes relatively strong assumptions about the clusters to be expected; in higher dimensions, this can be a major problem.

In practice, it is often assumed that the differences arising from different choices of the p_ks are small for commonly occurring cases. This is used to justify leaving the variability due to selection of the p_ks unquantified. As a pragmatic point, most authors just fix a K and choose the p_ks to be normal. Thus, $\theta_k = (\boldsymbol{\mu}_k, \boldsymbol{\Sigma}_k)$, and for p-dimensional data, f has $K(p + p(p-1)/2)$ real-valued parameters in the p_ks plus $K-1$ independent weights.

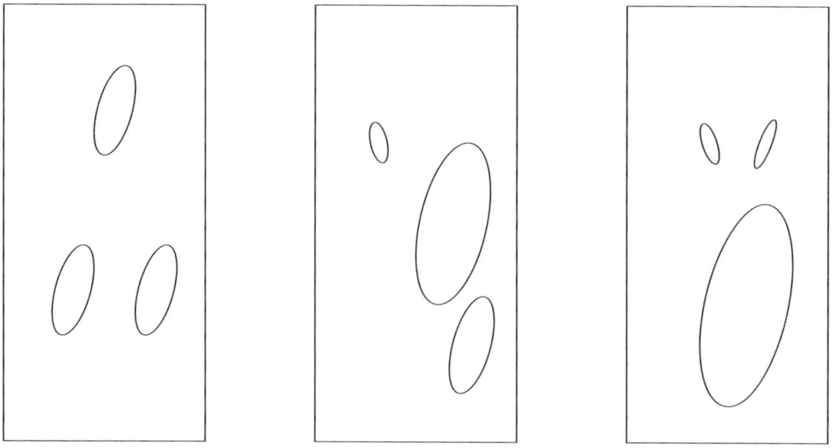

Fig. 8.5 These figures show contour plots of densities in two dimensions. Each has three local modes; the location of the modes and regions around the modes differ. In the first, the modes of the three peaked regions form an equilateral triangle and they have the same indicated spread. In the second, the modes are unrelated and the regions have spreads that are small, medium, and large. In the third, the modes are unrelated and the spreads are small, small, and large.

Figure 8.5 shows three contour plots from two-dimensional densities each corresponding to three clusters. If $K = 3$ and the components in (8.3.2) are normal (i.e., $p_k(\boldsymbol{x}|\theta_k)) = \phi(\boldsymbol{x}|\boldsymbol{\mu}_k, \boldsymbol{\Sigma}_k)$), then the centers of the ellipses estimate the $\boldsymbol{\mu}_k$s and the relative sizes of the ellipses indicate the spread of the normals corresponding to properties of the $\boldsymbol{\Sigma}_k$s, tighter ellipses corresponding to smaller variances of the entries of $\boldsymbol{x}$. Clearly, if the level sets of a density were not convex, for instance a kidney or annular shape, then a normal mixture might not approximate the true density well – unless the normals were allowed to be so close that their modes formed a ridge mapping out the nonconvex shape.

Note that Fig. 8.5 also shows how obtaining (8.3.2) leads to a clustering. The number of clusters is K, often chosen by BIC. The soft membership function for a typical data point $\boldsymbol{x}_i$ is the vector

$$\frac{1}{\pi_1 p_1(\boldsymbol{x}_i|\theta_1) + \ldots + \pi_K p_K(\boldsymbol{x}_i|\theta_K)} (\pi_1 p_1(\boldsymbol{x}_i|\theta_1), \ldots, \pi_K p_K(\boldsymbol{x}_i|\theta_K)).$$

This can be converted to a hard membership by assigning each $\boldsymbol{x}_i$ to the cluster for which its soft membership is highest. Thus, if the model (8.3.2) is fit, the clustering is determined. Indeed, it can be seen that the result of the maximization step in the EM algorithm below effectively gives the soft clustering entries.

Fixing a K and taking the p_ks to be normals with unknown means and variance matrices, the task remaining is to estimate the parameters. In a single normal distribution, the MLEs for $\boldsymbol{\mu}$ and $\boldsymbol{\Sigma}$ are well known and well behaved. However, when two or more normal densities are mixed, the unboundedness of the density as a function of the variance becomes a problem. Consider a standard $N(0,1)$ and an $N(\mu_2, \sigma_2^2)$, with $\pi_1 = \pi_2 = 1/2$. Given a data point x, one can set $\mu_2 = x$. Now, by choosing σ_2 to be ever smaller, the likelihood can be made arbitrarily large, making estimation by an MLE impossible. The next subsection presents a computational technique called the EM algorithm. Under some conditions, it produces estimates that converge to their true values at the usual rates.

8.3.1.1 The EM Algorithm: Two Normal Components

To fix ideas, suppose that a random variable with $p = 1$ has been measured repeatedly, leading to the smoothed histogram depicted in Fig. 8.6. Since two modes appear, it is reasonable to fit a $K = 2$ component mixture model. In the absence of other information, since the neighborhood of the modes looks symmetric, it's probably acceptable to choose the $p_k(x) = \phi(\mu_k, \sigma_k^2)(x)$ to be normal. Setting $\boldsymbol{\theta}_k = (\mu_k, \sigma_k^2)$ in (8.3.3) means that an observation has a density of the form

$$f(x) = \pi\phi(x \mid \mu_1, \sigma_1^2) + (1-\pi)\phi_2(x \mid \mu_2, \sigma_2^2). \tag{8.3.7}$$

Like (8.3.3), (8.3.7) is a mixture; it assumes that x is drawn from $\phi(x \mid \mu_1, \sigma_1^2)$ with probability π and from $\phi_2(x \mid \mu_2, \sigma_2^2)$ with probability $1-\pi$ but that the observer does not know which of the two densities produced a given observation. It's as if the data set is of the form $((Y_1, X_1), \ldots, Y_n, X_n))$, where $Y_i = 1, 2$ indicates the component sampled, but Y_i is missing so only the marginal density for the data can be used. The component index behaves like a parameter in a Bayesian treatment; the task here is to estimate its π as well as the parameters in the components.

Consider the standard approach to finding the MLE. A product of n factors of the form (8.3.7) becomes a sum of logs so the log likelihood is

$$\ell(\boldsymbol{\theta}, \boldsymbol{x}) = \sum_{i=1}^{n} \ln[\pi\phi_1(x_i \mid \mu_1, \sigma_1^2) + (1-\pi)\phi_2(x_i \mid \mu_2, \sigma_2^2)], \tag{8.3.8}$$

in which the parameter is five-dimensional, $\boldsymbol{\theta} = (\pi, \mu_1, \mu_2, \sigma_1^2, \sigma_2^2)'$. Maximizing (8.3.8) is hard because the argument of the log has more than one term.

Although useful expressions for the MLE do not exist, a computational approach called the EM algorithm is usually successful. The EM algorithm alternates between an

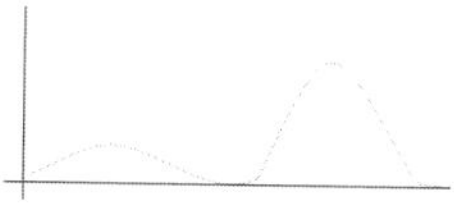

Fig. 8.6 This figure depicts the smoothed histogram from a one-dimensional density with two modes or clusters.

expectation step and a maximization step to converge to an estimator of $\boldsymbol{\theta}$ with many of the same properties as the MLE.

The idea behind the EM algorithm is to reconstruct the unobserved Y_is that would have indicated which component in (8.3.7) generated the observation. Since the correct Y_is were unobserved, write the reconstructed versions as latent variables denoted Z_i, $i = 1, ..., n$. That is, the Z_is indicate which component produced each X_i:

$$Z_i = \begin{cases} 0 & \text{if } x_i \sim \phi_2 \\ 1 & \text{if } x_i \sim \phi_1, \end{cases} \tag{8.3.9}$$

where $\mathbb{P}[Z_i = 1] = \pi$ and for brevity write $Z^n = (Z_1, ..., Z_n)$ and $X^n = (X_1, ..., X_n)$. Using (8.3.9), (8.3.7) can be written as

$$\begin{aligned} X_{1,i} &\sim N(\mu_1, \sigma_1^2), \\ X_{2,i} &\sim N(\mu_2, \sigma_2^2), \\ X_i &= Z_i X_{1,i} + (1 - Z_i) X_{2,i}. \end{aligned}$$

Clearly, if Z^n were known, it would be easy to get MLEs for (μ_j, σ_j^2) separately for $j = 1, 2$ as well as estimate π. However, Z^n is unknown and so can't be put back except in a probabilistic sense. This amounts to taking a mean, but it's subtle because π is unknown, and even if it were known, different X_is could still come from different components.

Overall, the strategy for estimating $\boldsymbol{\theta}$ can be introduced as the following:

Let $\{\boldsymbol{x}_i \mid i = 1, ..., n\}$ be drawn from (8.3.7), in which $\boldsymbol{\theta} = (\pi, \mu_1, \mu_2, \sigma_1^2, \sigma_2^2)$ is unknown.

- ☐ Start the procedure by choosing $\boldsymbol{\theta}_0$, the initial values for $\hat{\pi}$, $\hat{\mu}_1$, $\hat{\mu}_2$, $\hat{\sigma}_1^2$, and $\hat{\sigma}_2^2$. Often $\hat{\pi} = .5$ is reasonable, the $\hat{\mu}_j$s can be taken as well-separated sample values, and the $\hat{\sigma}_j^2$s are both set to the sample variance of the full data set.
- ☐ The Expectation step: For each i, estimate the conditional probability that X_i came from component one given the data and $\boldsymbol{\theta}$, i.e., estimate

$$\zeta_i = \mathbb{E}(Z_i|\boldsymbol{\theta}, x^n) = \mathbb{P}(Z_i = 1|\boldsymbol{\theta}, x^n).$$

These can be taken as posterior probabilities that x_i came from component one, suggesting

$$\hat{\zeta}_i = \frac{\hat{\pi}\phi_1(x_i \mid \hat{\mu}_1, \hat{\sigma}_1^2)}{\hat{\pi}\phi_1(x_i \mid \hat{\mu}_1, \hat{\sigma}_1^2) + (1-\hat{\pi})\phi_2(x_i \mid \hat{\mu}_2, \hat{\sigma}_2^2)}. \tag{8.3.10}$$

☐ The Maximization step: Update the estimate $\hat{\boldsymbol{\theta}}_0$ to $\hat{\boldsymbol{\theta}}_1$ by finding

$$\hat{\boldsymbol{\theta}}_1 = \arg\max_{\theta} \mathbb{E}_{Z^n|X^n, \hat{\boldsymbol{\theta}}_0} \log p(X^n, Z^n|\boldsymbol{\theta}). \tag{8.3.11}$$

☐ Use (8.3.10) to weight the x_is. This gives $\hat{\pi}_1 = \sum_{i=1}^n \hat{\zeta}_i/n$. For (8.3.7), the expressions are:

$\hat{\mu_1} = \sum_{i=1}^n \hat{\zeta}_i x_i/n\hat{\pi}_1$ $\qquad \hat{\mu_2} = \sum_{i=1}^n (1-\hat{\zeta}_i)x_i/n(1-\hat{\pi}_1)$
$\hat{\sigma}_1^2 = \sum_{i=1}^n \hat{\zeta}_i(x_i - \hat{\mu}_1)^2/n\hat{\pi}_1$ $\qquad \hat{\sigma}_2^2 = \sum_{i=1}^n (1-\hat{\zeta}_i)(x_i - \hat{\mu}_2)^2/n(1-\hat{\pi}_1).$

☐ Cycle through the steps until a convergence criterion is satisfied.

Note that the central feature of this procedure is that the proportion of outcomes from component one, π, is estimated by an average of the ζ_is that represents the probability that an X_i came from component one. Then, these weights are just carried through to give weights for inclusion in the means and variances of the components. Exercise 8.10 asks you to derive (8.3.10) for the case of two normal distributions and verify that the four expressions in Step 4 follow from (8.3.11).

It can already be surmised that the EM algorithm produces a sequence of values $\boldsymbol{\theta}_j$ that tends to converge to a local, possibly not global, maximum. In fact, both the E and the M steps tend to increase the objective function, giving a high convergence rate. Thus, even though the EM algorithm does not solve the identifiability problem, it does find a local maximum.

8.3.1.2 EM Algorithm: Derivation of the General Case

What is now called the EM algorithm was originally proposed by Hartley (1958). It was studied by Baum and Petrie (1966), and developed by Baum and various collaborators in the early 1970s, for which reason it is sometimes called the Baum-Welch algorithm. (Curiously, the role of Welch is unclear.) The EM algorithm was introduced to statistics by Dempster et al. (1977). In the following subsections, the EM algorithm will be derived more generally and some of its properties given. In fact, the EM algorithm is considerably more flexible than the two-component case indicates. Dempster et al. (1977) show how the EM algorithm can be applied to a wide variety of other models, including imputation in missing, truncated, or censored data problems, variance components, iteratively reweighted least squares, and factor analysis. Also,

Bilmes (1998) shows how the EM algorithm applies to estimating the parameters in a hidden Markov Model, as noted in Chapter 6.

The treatment here covers many of the basic facts about the procedure for the context of clustering but does not develop its broad applicability. Here, it will be enough to observe that any modeling strategy for which the unknown parameters can be related to incomplete data will tend to fit the EM framework.

To introduce the general EM algorithm, consider a density $p(x|\theta)$ and denote the log likelihood by $\ell(\theta) = \log p(x^n|\theta)$. Implicitly take x real; vector-valued xs follow the same derivation. If $\theta = \theta_0$ is an initial value for θ, and θ_k is the value after k iterations, then, at the $k+1$ iteration, one seeks θ_{k+1} with $\ell(\theta_k) \geq \ell(\theta_{k+1})$.

Now suppose that for each $\boldsymbol{X}_i$ there is a discrete random variable Z_i that is unknown. In the previous section, the Z_is were 0-1 random variables indicating the component. The marginal for $\boldsymbol{X}$ is now

$$\ell(\theta) = p(x^n|\theta) = \sum_{z^n \in \mathscr{Z}^n} p(\boldsymbol{x}, z^n|\theta) = \sum_{z^n \in \mathscr{Z}^n} p(x^n|z^n, \theta) p(z^n|\theta),$$

reminiscent of (8.3.7). Following Borman (2004), the problem of maximizing $\ell(\theta)$ is seen to be equivalent to maximizing

$$\begin{aligned} \ell(\theta) - \ell(\theta_k) &= \log \sum_{z^n \in \mathscr{Z}^n} p(x^n|z^n, \theta) p(z^n|\theta) - \log p(x^n|\theta_k) \\ &= \log \sum_{z^n \in \mathscr{Z}^n} p(x^n|z^n, \theta) \frac{p(x^n|z^n, \theta) p(z^n|\theta)}{p(z^n|x^n, \theta_k)} - \log p(x^n|\theta_k). \end{aligned} \tag{8.3.12}$$

It is tempting to get a lower bound using Jensen's inequality in expression (8.3.12) to bring the log inside the expectation over $\boldsymbol{z}$ represented by the sum.

Giving in to temptation leads to

$$\begin{aligned} \ell(\theta) - \ell(\theta_k) &\geq \sum_{z^n \in \mathscr{Z}^n} p(z^n|x^n, \theta_k) \log \frac{p(\boldsymbol{x}|z^n, \theta) p(z^n|\theta)}{p(z^n|x^n, \theta_k) p(x^n|\theta_k)} \\ &\equiv \Delta(\theta, \theta_k). \end{aligned} \tag{8.3.13}$$

So it is reasonable to set

$$\ell(\theta|\theta_k) \equiv \ell(\theta_k) + \Delta(\theta, \theta_k). \tag{8.3.14}$$

Now, (8.3.13) and (8.3.14) give

$$\ell(\theta) \geq \ell(\theta|\theta_k). \tag{8.3.15}$$

Strictly, $\ell(\theta|\theta_k)$ is not a likelihood and $\Delta(\theta, \theta_k)$ is not a distance. However, the abuse of notation is suggestive. Indeed, since (8.3.15) is just a desired property of the $k+1$ stage value of θ, the hope is that the lower bound from Jensen's inequality is tight enough that using it as an objective function for the next iteration will move θ_k to some

θ_{k+1} closer to a local maximum of $\ell(\theta)$. That is, hopefully it is enough to optimize the lower bound $\ell(\theta|\theta_k)$ in (8.3.15).

To see that it is reasonable to assume (8.3.15) is tight, observe that

$$\begin{aligned} \ell(\theta_k|\theta_k) &= \ell(\theta_k) + \Delta(\theta_k, \theta_k) \\ &= \ell(\theta_k) + \sum_{z^n \in \mathscr{Z}^n} p(z^n|x^n, \theta_k) \log \frac{p(x^n|z^n, \theta_k) p(z^n|\theta_k)}{p(z^n|x^n, \theta_k) p(x^n|\theta_k)}, \end{aligned} \tag{8.3.16}$$

and rearranging in (8.3.16) gives $\ell(\theta_k|\theta_k) = \ell(\theta_k)$; i.e., the actual log likelihood equals the contrived log likelihood at θ_k. So, if there is a local maximum of $\ell(\theta)$ near θ_k, perhaps $\ell(\theta|\theta_k)$ will track the increase of $\ell(\theta)$ well enough that obtaining a θ_{k+1} from $\ell(\theta|\theta_k)$ will give a larger value of $\ell(\theta)$ than $\ell(\theta_k)$.

Supposing it is reasonable to use $\ell(\theta|\theta_k)$ as an objective function, then the EM criterion can be derived as follows. Write the maximization problem as:

$$\theta_{k+1} = \arg\max_{\theta} \ell(\theta|\theta_k).$$

Recalling the definition of $\ell(\theta|\theta_k)$ and Δ in (8.3.13) and (8.3.14), it is seen that the terms involving θ_k in the denominator of (8.3.13) do not depend on θ. So,

$$\theta_{k+1} = \arg\max_{\theta} \sum_{z^n \in \mathscr{Z}^n} p(z^n|x^n, \theta_k) \log p(x^n|z^n, \theta) p(z^n|\theta).$$

Writing the conditional densities in the log as ratios of joint densities and canceling gives a conditional expectation. In fact,

$$\begin{aligned} \theta_{k+1} &= \arg\max_{\theta} \sum_{z^n \in \mathscr{Z}^n} p(z^n|x^n, \theta_k) \log p(x^n, z^n|\theta) \\ &= \arg\max_{\theta} \left(E_{Z^n|x^n, \theta_k} \log p(x^n, Z^n|\theta) \right) \end{aligned} \tag{8.3.17}$$

in which the expectation is over Z^n and x^n, θ_k are held constant.

The EM algorithm can now be succinctly stated as an E step in which the conditional expectation in (8.3.17) is found and an M step in which the conditional expectation is maximized, leading to a new E step. The difference between the EM algorithm and regular MLEs is that (8.3.17) takes into account the unobserved Z^n.

8.3.1.3 EM algorithm: *K* components.

To see how the EM algorithm can be applied more generally, consider a general mixture family with K components for a vector $\boldsymbol{x}$,

$$p(\boldsymbol{x}|\theta) = \sum_{k=1}^{K} \pi_k p_k(\boldsymbol{x}|\theta_k), \tag{8.3.18}$$

and write $\theta = (\pi_1, ..., \pi_K, \theta_1, ..., \theta_K)$, in which $\sum \pi_i = 1$, $\pi_i \geq 0$, and each p_i is a density function parametrized by θ_i. The log likelihood is

$$\ell(\theta|\boldsymbol{x}^n) = \sum_{i=1}^{n} \log\left(\sum_{k=1}^{K} \pi_k p_k(\boldsymbol{x}_i|\theta_k)\right).$$

Augmenting each $\boldsymbol{x}_i$ by a z_i to indicate which component produced it gives

$$\ell(\theta|\boldsymbol{x}^n, z^n) = \sum_{k=1}^{n} \log p(\boldsymbol{x}_i|z_i,\theta)p(z_i|\theta) = \sum_{k=1}^{n} \log(\pi_{z_i} p_{z_i}(\boldsymbol{x}_i|\theta_{z_i})) \qquad (8.3.19)$$

as the log likelihood, if the z_is were known.

Since in fact the z_is are unknown, the first task is to get an expression for their distribution. Let $\theta_0 = (\pi_{1,0}, ..., \pi_{K,0}, \theta_{1,0}, ..., \theta_{K,0})$ be an initial guess for use in (8.3.19). This means that $p_k(\boldsymbol{x}_i|\theta_{k,0})$ is known for each i and k. Since $\pi_k = P(Z_i = k)$ for each i and the mixing probabilities constitute a prior on the components, Bayes' rule gives

$$p(z_i|\boldsymbol{x}_i, \theta_0) = \frac{\pi_{z_i,0} p_{z_i}(\boldsymbol{x}_i|\theta_{z_i,0})}{p(\boldsymbol{x}_i|\theta_0)}, \qquad (8.3.20)$$

in which the denominator of (8.3.20) is of the form (8.3.18). In addition,

$$p(\boldsymbol{z}^n|\boldsymbol{x}^n, \theta) = \Pi_{i=1}^{n} p(z_i|\boldsymbol{x}_i, \theta). \qquad (8.3.21)$$

Using (8.3.19) and (8.3.21) in (8.3.17) gives

$$\begin{aligned} G(\theta,\theta_0) &= \sum_{\boldsymbol{z}^n \in \mathscr{Z}^n} p(\boldsymbol{z}^n|\boldsymbol{x}^n,\theta_0) \log p(\boldsymbol{x}^n, z^n|\theta) \\ &= \sum_{\boldsymbol{z}^n \in \mathscr{Z}^n} \sum_{i=1}^{n} \log(\pi_{z_i} p_{z_i}(\boldsymbol{x}_i|\theta_{z_i})) \Pi_{j=1}^{n} p(z_j|\boldsymbol{x}_j,\theta_0) \end{aligned}$$

as the objective function.

Following the reasoning laid out lucidly in Bilmes (1998), one can express the sum over $\boldsymbol{z}^n$ as n sums over $z_i \in \mathscr{Z}$. Introducing an extra summation over $u = 1, ..., K$ with a Kronecker δ-function δ_{u,z_i} and some algebra gives that

$$G(\theta,\theta_0) = \sum_{u=1}^{K} \sum_{i=1}^{n} \log(\pi_u p_u(\boldsymbol{x}_i|\theta_u)) \sum_{z_1=1}^{K} \cdots \sum_{z_n=1}^{K} \delta_{u,z_i} \Pi_{j=1}^{n} p(z_j|\boldsymbol{x}_j,\theta_0). \quad (8.3.22)$$

Fortunately, this can be simplified. Fix a value of u in $1, ..., K$. Then the inner sequence of summations in (8.3.22) is

$$\sum_{z_1=1}^{K} \cdots \sum_{z_n=1}^{K} \delta_{u,z_i} \, \Pi_{j=1}^{n} p(z_j|\boldsymbol{x}_j,\theta_0)$$

$$= \left(\sum_{z_1=1}^{K} \cdots \sum_{z_{i-1}=1}^{K} \sum_{z_{i+1}=1}^{K} \cdots \sum_{z_n=1}^{K} \Pi_{j=1,j\neq i}^{n} p(z_j|\boldsymbol{x}_j,\theta_0) \right) p_u(\boldsymbol{x}_i,\theta_0)$$

$$= \Pi_{j=1,j\neq i}^{n} \left(\sum_{z_j=1}^{K} p(z_j|\boldsymbol{x}_j,\theta_0) \right) p(Z_i = u|\boldsymbol{x}_i,\theta_0) \tag{8.3.23}$$

$$= p(Z_i = u|\boldsymbol{x}_i,\theta_0), \tag{8.3.24}$$

since the summation in brackets in (8.3.23) is one.

Using (8.3.24) in (8.3.22) gives

$$G(\theta,\theta_0) = \sum_{u=1}^{K}\sum_{i=1}^{n} \log(\pi_u p_u(\boldsymbol{x}_i|\theta_u)) p(Z = u|\boldsymbol{x}_i,\theta_0)$$

$$= \sum_{u=1}^{K}\sum_{i=1}^{n} \log(\pi_u) p(Z = u|\boldsymbol{x}_i,\theta_0)$$

$$+ \sum_{u=1}^{K}\sum_{i=1}^{n} \log(p_u(\boldsymbol{x}_i|\theta_u)) p(Z = u|\boldsymbol{x}_i,\theta_0), \tag{8.3.25}$$

in which one term is free of π_us and the other term is free of θ_us.

Maximizing the first term in (8.3.25) gives an expression for the π_us. To see this, recall the constraint $\sum_u \pi_u = 1$, let λ be the Lagrange multiplier, and try to solve

$$\frac{\partial}{\partial \pi_u} \left(\sum_{u=1}^{K}\sum_{i=1}^{n} \log(\pi_u) p(Z = u|\boldsymbol{x}_i,\theta_0) + \lambda \left(\sum_{u=1}^{K} \pi_u - 1 \right) \right) = 0. \tag{8.3.26}$$

Expression (8.3.26) simplifies to

$$\frac{1}{\pi_u} \sum_{i=1}^{n} p(u|\boldsymbol{x}_i,\theta_0) + \lambda = 0.$$

Rearranging and summing over u gives $\lambda = -n$, so that

$$\hat{\pi}_u = \frac{1}{n} \sum_{i=1}^{n} p(Z = u|\boldsymbol{x}_i,\theta_0), \tag{8.3.27}$$

generalizing the $K = 2$ case, see (8.3.10). A separate argument, not given here, shows that the $\hat{\pi}_u$s give a maximum.

Up to this point, the treatment has been fully general, but the second term in (8.3.25) is hard to maximize without extra assumptions about the analytic form of the components. The normal is the easiest case, and the result was stated for $K = 2$. More generally, the p-dimensional normal with mean μ and variance Σ has density

$$p_u(\boldsymbol{x}|\mu_u, \Sigma_u) = \frac{1}{(2\pi)^{p/2}|\Sigma_u|^{1/2}} e^{-(1/2)(\boldsymbol{x}-\mu_u)'\Sigma_u^{-1}(\boldsymbol{x}-\mu_u)},$$

which can be used as $p_u(\boldsymbol{x}_i|\theta_u)$ so that closed form expressions can be derived to give the maximum of the second term in (8.3.25).

In fact, it is straightforward to derive the updating equations for the normal case using (8.3.25), thereby generalizing (8.3.10). Take the log of the p-dimensional normal, substitute it into the second term of (8.3.25), and take derivatives to solve for the maximum. Using $\theta = (\mu, \Sigma)$, the first step is to write

$$\Sigma_{u=1}^{K} \sum_{i=1}^{n} \log(p_u(\boldsymbol{x}_i|\mu_u, \Sigma_u))p(Z = u|\boldsymbol{x}_i, \theta_0) \tag{8.3.28}$$
$$= \sum_{u=1}^{K}\sum_{i=1}^{n} \left(-\frac{1}{2}\log(|\Sigma_u|) - \frac{1}{2}(\boldsymbol{x}_i - \mu_u)\Sigma_u^{-1}(\boldsymbol{x}_i - \mu_u)\right) p(Z = u|\boldsymbol{x}_i, \theta_0).$$

The derivative of (8.3.28) with respect to μ_u is

$$\sum_{i=1}^{n} \Sigma_u^{-1}(\boldsymbol{x}_i - \mu_u)p(Z = u|\boldsymbol{x}_i, \theta_0) = 0,$$

which can be readily rearranged to give the solution for the μ_us, namely

$$\mu_u = \frac{\Sigma_{i=1}^{n} \boldsymbol{x}_i p(Z = u|\boldsymbol{x}_i, \theta_0)}{\Sigma_{i=1}^{n} p(Z = u|\boldsymbol{x}_i, \theta_0)}, \tag{8.3.29}$$

which can be verified to be a maximum, generalizing (8.3.10)

Maximizing over the variance matrix is harder, and the derivatives need to be defined properly; this is recalled in the Notes at the end of this chapter. (This was necessary for (8.3.28) too, but vector derivatives are more familiar.) Again, Bilmes (1998) gives a thorough treatment which is followed closely here.

So, to solve for Σ_u, write (8.3.28) as

$$\sum_{u=1}^{K} \left(-\frac{1}{2}\log(|\Sigma_u|) \sum_{i=1}^{n} p(Z = u|\boldsymbol{x}_i, \theta_0) - \frac{1}{2}\sum_{i=1}^{n} p(Z = u|\boldsymbol{x}_i, \theta_0)\mathrm{tr}(\Sigma_u^{-1}\boldsymbol{M}_{u,i}), \right), \tag{8.3.30}$$

where $\boldsymbol{M}_{u,i} = (\boldsymbol{x}_i - \mu_u)(\boldsymbol{x}_i - \mu_u)'$, and the matrices were rearranged under the trace. Fixing a value of u and differentiating (8.3.30) with respect to Σ_u^{-1} leaves only

$$\frac{1}{2}\sum_{i=1}^{n} p(Z=u|\boldsymbol{x}_i,\theta_0)(\ 2\ \Sigma_u - \text{diag}(\Sigma_u))$$

$$- \frac{1}{2}\sum_{i=1}^{n} p(Z=u|\boldsymbol{x}_i,\theta_0)(2\boldsymbol{M}_{u,i} - \text{diag}(\boldsymbol{M}_{u,i}))$$

$$= \frac{1}{2}\sum_{i=1}^{n} p(Z=u|\boldsymbol{x}_i,\theta_0)(2\boldsymbol{Q}_{u,i} - \text{diag}(\boldsymbol{Q}))$$

$$= 2\boldsymbol{U} - \text{diag}(\boldsymbol{U}), \tag{8.3.31}$$

where $\boldsymbol{Q}_{u,i} = \Sigma_u - \boldsymbol{M}_{u,i}$ and $\boldsymbol{U} = (1/2)\sum_i p(Z=u|\boldsymbol{x}_i,\theta_0)\boldsymbol{Q}_{u,i}$.

Setting the derivative (8.3.31) equal to zero gives $\boldsymbol{U} = 0$ so that

$$\sum_{i=1}^{n} p(Z=u|\boldsymbol{x}_i,\theta_0)(\Sigma_u - M_{u,i}) = 0.$$

Finally,

$$\Sigma_u = \frac{\sum_{i=1}^{n} p(Z=u|\boldsymbol{x}_i,\theta_0)\boldsymbol{M}_{u,i}}{\sum_{i=1}^{n} p(Z=u|\boldsymbol{x}_i,\theta_0)}$$

$$= \frac{\sum_{i=1}^{n} p(Z=u|\boldsymbol{x}_i,\theta_0)(\boldsymbol{x}_i-\mu_u)(\boldsymbol{x}_i-\mu_u)'}{\sum_{i=1}^{n} p(Z=u|\boldsymbol{x}_i,\theta_0)}. \tag{8.3.32}$$

Note that (8.3.32), like (8.3.29), is a generalization of (8.3.10).

So, to update in the EM algorithm in this case, form the $\hat{\pi}_u$s from (8.3.27) and the μ_us from (8.3.29) and then use the new μ_us in (8.3.32) to get the updated estimate for the Σ_us. These can be used in the next expectation step to update θ_0 and so on until convergence is observed.

8.3.1.4 EM Algorithm: Exponential Family Case

For completeness, it is important to state the form of the EM algorithm for exponential families like

$$p(\boldsymbol{y}|\theta) = a(\theta)b(\boldsymbol{y})e^{\eta(\theta)'\boldsymbol{t}(\boldsymbol{y})},$$

where $\boldsymbol{t}$ is a vector of m sufficient statistics, $\eta(\theta) = (\eta_1(\theta),...,\eta_m(\theta))$, b is a real-valued function of $\boldsymbol{y}$, and $a(\theta)$ is a normalizing constant. In general, $\theta = (\theta_1,...,\theta_d)$. Now, regard $\boldsymbol{y}$ as the complete data of the form $(\boldsymbol{x},\boldsymbol{z})$ in which $\boldsymbol{x}$ represents the actual data collected and $\boldsymbol{z}$, as before, represents the extra, hidden information, for instance the knowledge of which component in the mixture gave a given data point. Thus, the incomplete but available data are $\boldsymbol{x} = \boldsymbol{x}(\boldsymbol{y})$, a function of the complete data $\boldsymbol{y}$. Formally, the density for $\boldsymbol{x}$ is

$$p(\boldsymbol{x}|\theta) = \int_{\{\boldsymbol{y}:\boldsymbol{x}(\boldsymbol{y})=\boldsymbol{x}\}} p(\boldsymbol{y}|\theta)d\boldsymbol{y}.$$

For exponential families, the objective function in (8.3.17), also called the E-step, can be written as

$$\sum_{i=1}^{n} E_{Z|\boldsymbol{x},\theta_k}\left[\log b(\boldsymbol{x}_i,\boldsymbol{Z}_i)+\eta(\theta)'\boldsymbol{t}^k(\boldsymbol{x}_i,\boldsymbol{Z}_i)\right]+\log a(\theta).$$

The first term does not depend on θ, so it does not affect the maximization in (8.3.17). The other two terms become

$$\sum_{i=1}^{n} \eta(\theta)'\boldsymbol{t}^k(\boldsymbol{x}_i,\theta_k)+n\log a(\theta), \tag{8.3.33}$$

where $\boldsymbol{t}^k = E(\boldsymbol{t}(\boldsymbol{X},\boldsymbol{Z})|\boldsymbol{x},\theta_k)$. For an exponential family in its natural parametrization, (8.3.33) becomes

$$\frac{1}{n}\sum_{i=1}^{n}\boldsymbol{t}^k(\boldsymbol{x}_i,\theta_k)=\frac{\partial \log a(\theta)}{\partial\theta}.$$

So, the EM algorithm reduces to an E step, find the function $\boldsymbol{t}^k = E(\boldsymbol{t}(\boldsymbol{X},\boldsymbol{Z})|\boldsymbol{x},\theta_k)$, and an M step, use the $\boldsymbol{t}^k$s in expression (8.3.33), optimize over θ by differentiating, and repeat until convergence.

8.3.1.5 EM Algorithm: Properties

First, it is important to see that the EM algorithm does increase the likelihood, hopefully to a maximum. With some abuse of notation, let the conditional density of $\boldsymbol{Y}$ given $\boldsymbol{X}=\boldsymbol{x}$ be

$$p(\boldsymbol{y}|\boldsymbol{x},\theta)=\frac{p(\boldsymbol{y}|\theta)}{p(\boldsymbol{x}|\theta)}.$$

(The abuse is that the relationship between the two measure spaces, for $\boldsymbol{y}$ and for $(\boldsymbol{x},\boldsymbol{z})$, has not been stated and that the sample size n has been absorbed into the vector notation, in both cases for simplicity.)

Now, the complete data log likelihood is

$$\ell_c(\theta)=\log p(\boldsymbol{y}|\theta)=\ell(\theta)+\log p(\boldsymbol{x}|\theta).$$

Doing the E step from (8.3.17) means taking the conditional expectation over the conditional for the missing data $\boldsymbol{Z}$; i.e., with respect to $\boldsymbol{Z}|\boldsymbol{x},\theta_k$. Then, denoting the objective function by $EM(\theta,\theta_k)$, in which the dependence on $\boldsymbol{x}$ is suppressed,

$$\begin{aligned}EM(\theta,\theta_k) &= E_{Z|\boldsymbol{x},\theta_k}\log p(\boldsymbol{x},\boldsymbol{Z}|\theta)\\ &= \ell(\theta)-E_{Z|\boldsymbol{x},\theta_k}\log p(\boldsymbol{Y}|\boldsymbol{x},\theta)\equiv \ell(\theta)+H(\theta,\theta_k),\end{aligned}$$

where H is almost a conditional entropy.

The difference in log likelihoods for two EM iterations is $\ell(\theta_{k+1})-\ell(\theta_k)$, which equals

$$[EM(\theta_{k+1},\theta_k) - EM(\theta_k,\theta_k)] + [H(\theta_k,\theta_k) - H(\theta_{k+1},\theta_k)] \geq 0. \qquad (8.3.34)$$

This follows by (i) Jensen's inequality on the log in the relative entropy to see that the second bracketed term on the right in (5.37) is nonnegative and (ii) the EM procedure, which always seeks a θ_{k+1} that never decreases the value of $EM(\theta_{k+1},\theta_k)$. Thus, exponentiating the result from (8.3.34) gives that the likelihoods are increasing over iterations; i.e.,

$$\ell(\theta_{k+1}) \geq \ell(\theta_k). \qquad (8.3.35)$$

Second, the EM function itself defines stationary points. Let the EM function M be defined by an iteration of the EM algorithm as

$$\theta_{k+1} = M(\theta_k).$$

Then, extending (8.3.35), it can be shown (see McLachlan and Krishnan (1997), Section 3.5) that

$$\log \ell(M(\theta)) \geq \log \ell(\theta)$$

with equality if and only if

$$EM(M(\theta),\theta) = EM(\theta,\theta) \text{ and } p(\mathbf{y}|\mathbf{x},M(\theta)) = p(\mathbf{y}|\mathbf{x},\theta).$$

Little and Rubin (2002) have results that ensure the sequence θ_k will converge to a solution of the likelihood equation.

Third, using a convergence theorem from optimization theory, Wu (1983) corrects Theorems 2 and 3 in Dempster et al. (1977) and establishes a series of results that characterize the performance of the EM iterates. Let $L(\theta) = \log p(\mathbf{x}|\theta)$, the log likelihood for the incomplete available data. Then, Wu (1983)) can be summarized as follows.

Theorem (Wu, 1983): Suppose $EM(\theta,\theta')$ is continuous in θ and θ'. Then:

(i) All limit points of any sequence θ_k of an EM algorithm are stationary points of L and $L(\theta_k)$ converges monotonically to $L^* = L(\theta^*)$ for some stationary point θ^*.

(ii) If, in addition,

$$\sup_{\theta'} EM(\theta',\theta) > EM(\theta,\theta),$$

then all the limit points of any sequence θ_k are local maxima of $L(\theta)$, and $L(\theta_k)$ converges monotonically to $L^* = L^*(\theta^*)$ for some local maximum θ^*.

(III) If $L(\theta)$ is unimodal, with a unique stationary point θ^*, and the first derivatives of $EM(\theta',\theta)$ with respect to the entries of θ' are continuous in θ and θ', then any EM sequence θ_k converges to the unique maximum θ^* of $L(\theta)$.

Proof: Omitted, see Wu (1983). □.

Finally, the rate of convergence of the EM algorithm is an issue. Dempster et al. (1977) argue that the rate of convergence over iterations in the EM algorithm is linear and depends on the proportion of information in the complete data. If a large portion of the data are missing then convergence slows.

Meng (1994) observes that if θ_k converges to θ^* and $M(\cdot) = (M_1(\cdot), ..., M_d(\cdot))$ is differentiable, then Taylor expanding $\theta_{k+1} = M(\theta_k)$ at $\theta_k = \theta^*$ gives

$$\theta_{k+1} - \theta^* = J(\theta^*)(\theta_k - \theta^*),$$

where J is the $d \times d$ Jacobian matrix of M, i.e., with (i, j)th entry $\partial M_i / \partial \theta_j$. So, the EM algorithm is a linear iteration with rate matrix $J(\theta^*)$. Since the rate is determined by the largest eigenvalue, in principle an explicit rate can be identified; see Meng (1994) for some details.

Arguably, the major drawbacks of EM procedures in general include slow convergence (sometimes), choosing initial values θ_0, and sometimes the E or M steps are intractable. Meng and Rubin (1991) have some methods for assigning covariances; see Ng et al. (2004) for a general discussion.

8.3.2 Graph-Theoretic Clustering

Graph-theoretic clustering comes in two forms. In one, a graph (usually directed, sometimes weighted) is given. An instance of this form is given by reference lists. Each paper on a reference list refers to other papers, which in turn refer to earlier papers, and often there is an overlap between the references in one paper and the references in one of the papers it cites. Abstractly, this is a directed, unweighted graph. The task is to partition it into subgraphs that correspond to meaningful clusters such as topics. The other form of graph-theoretic clustering uses a similarity d on a collection of data points $\boldsymbol{x}_i$ $i = 1, ..., n$. One can form a fully connected, undirected graph from the points in the sample, with weights given by a similarity measure; sometimes this is called a similarity graph. Again, the task is to partition the graph into subgraphs that correspond to meaningful clusters. Here, the focus will be on this second formulation of the graph-theoretic clustering problem on the grounds that many of the techniques apply to the first, albeit imperfectly.

Consider a graph $\mathscr{G} = (\mathscr{V}, \mathscr{E})$ in which $\mathscr{V}$ is the set of vertices of $\mathscr{G}$ and $\mathscr{E}$ is the set of edges of $\mathscr{G}$. To describe the techniques, some basic definitions from graph theory are needed. The degree of a vertex in an undirected graph is the number of edges connected to it. If the graph is directed, there are two degrees, the inward and the outward degree for each vertex. The inward degree is the number of edges arriving at the vertex; the outward degree is the number of edges leaving from the vertex. A tree, graph-theoretically, is a graph in which any two vertices are connected by exactly one path. A clique in an undirected graph is a set of vertices which is fully connected; i.e., the graph restricted to those vertices is complete. The cut of two subsets of $\mathscr{V}$ is a general concept of adding up the weights of the edges between points in the two sets.

Now, given these definitions, two classes of techniques that have proved important are those based on minimal spanning trees (MST) and those based on local connectivity properties of graphs, for instance the degrees of vertices. Since many of these are recent developments in a large and fast-moving field, only a small fraction of the existing techniques can be surveyed here.

8.3.2.1 Tree-Based Clustering

The central concept in many tree-based clustering procedures is the minimal spanning tree. Given a connected, undirected graph, a spanning tree is a subgraph, which is a tree connecting all the vertices; such trees are not unqiue. If the graph is weighted, a minimum spanning tree is a spanning tree with the smallest possible sum of edge weights; this is unique provided all the edge weights are distinct.

To see why an MST might be useful, consider the similarity graph generated by n data points $\boldsymbol{x}_i$. If an MST is found, one can immediately see that forming a clustering of K clusters by deleting the $K-1$ edges with the largest weights results in a divisive procedure giving the point sets represented by the connected components of the K subgraphs as the clusters. Of course, one need not choose K at the outset of this procedure; it can emerge if a threshold is set for how large the weights must be before an edge is deleted. The problem is that there may be relationships among the points that are not well represented within the MST, and the edges with the largest weights need not in general correspond to points that belong in different clusters. Single-linkage agglomerative clustering also gives subgraphs of the MST and has analogous problems, including chaining. (By contrast, complete-linkage agglomerative clustering corresponds to maximal complete subgraphs.)

Another way that MSTs can be seen to arise comes from noting that the levels sets of a density estimator from a data set are nested. That is, write

$$L(\lambda, p) = \{\boldsymbol{x} \mid p(\boldsymbol{x}) > \lambda\} \tag{8.3.36}$$

for a density p and a parameter $\lambda > 0$. If $p = \hat{p}$ estimates the density of $\boldsymbol{X}$, the highest density clusters correspond to maximally connected subsets of $L(\lambda, p)$. If two high density clusters A and B are chosen, then either one contains the other ($A \subset B$ or $B \subset A$) or their intersection is void, $A \cap B = \phi$. Stuetzle (2003) recursively defines the resulting hierarchical structure and calls it a cluster tree: Each node N of the tree represents a subset $D(N)$ of the support of p, $D(N) \subset L(0, p)$, and is defined by a value $\lambda = \lambda(N)$. Thus, the root node is $L(0, p)$ and $\lambda(root) = 0$. The descendants of the root node are determined by the lowest value of λ, say λ_d, for which $L(\lambda, p) \cap L(0, p)$ becomes disconnected. Repeating the procedure on the components generates a sequence of λ_ds and splits the support of p until there is no λ_d giving two connected components. This means a leaf of the tree has been found. Thus the cluster tree represents the modal structure of the unknown density.

Stuetzle (2003) shows that the cluster tree of a nearest-neighbor density estimate can be found from the MST of the data, and it is isomorphic to the dendrogram from single-linkage clustering. Indeed, write the nearest-neighbor density estimator as

$$\hat{p}(\boldsymbol{x}) = \frac{1}{nVd(\boldsymbol{x}, \mathscr{D})^p}, \tag{8.3.37}$$

where V is the volume of the unit ball in $\mathbb{R}^p$ and $d(\boldsymbol{x}, \mathscr{D}) = \min_i d(\boldsymbol{x}, \boldsymbol{x}_i)$. Now, fix λ and set

$$r(\lambda) = \left(\frac{1}{nV\lambda}\right)^{1/p} \tag{8.3.38}$$

so that

$$\hat{p}(\boldsymbol{x}) > \lambda \Leftrightarrow d(\boldsymbol{x}, \mathscr{D}) = \min_i d(\boldsymbol{x}, \boldsymbol{x}_i) < r(\lambda). \tag{8.3.39}$$

This means that the set $L(\lambda, \hat{p})$ is a union of open balls of radius $r(\lambda)$ centered at the $\boldsymbol{x}_i$; i.e.,

$$L(\lambda, \hat{p}) = \cup_i B(\boldsymbol{x}_i, r(\lambda)). \tag{8.3.40}$$

Now, Stuetzle (2003) shows the following, but credits it to Hartigan (1985).

Proposition: The sets $L_k = \cup_{i \in C_k} B(\boldsymbol{x}_i, r(\lambda))$ are the connected components of $L(\lambda, \hat{p})$.

Proof: First note that each L_k is connected because the maximum weight on an edge in the subgraph corresponding to the kth cluster is below $2r(\lambda)$, so the kth subgraph of the MST is a subset of L_i.

To see that $L_k \cap L_j$ is void unless $k = j$, consider a proof by contradiction. Suppose T is an MST and that there are observations $\boldsymbol{x}^*$ and $\boldsymbol{x}^{**}$ in C_k and C_j, respectively, such that $d(\boldsymbol{x}^*, \boldsymbol{x}^{**}) < 2r(\lambda)$. Then, an edge of length at least $2r(\lambda)$ in the path connecting C_k and C_j could be replaced by a series of edges containing an edge connecting $\boldsymbol{x}^*$ and $\boldsymbol{x}^{**}$ giving a spanning tree of smaller total weight, a contradiction. □

Given the importance of the MST, it is important to describe how to find it for a data set. There are two common techniques; one is called Kruskal's algorithm the other is called Prim's algorithm. Kruskal's procedure basically removes edges one at a time based on how large their weight is. Formally, start with a set of trees $\mathscr{T}$ and a set of edges $\mathscr{S}$. Initialize these by $\mathscr{T} = \mathscr{V}$ and $\mathscr{S} = \mathscr{E}$, so the procedure starts with n trivial trees of a single vertex and the set of edges of the graph. Choose the edge in $\mathscr{S}$ with the smallest weight: If it connects two of the trees in $\mathscr{T}$ then include it and merge the trees. Otherwise, discard the edge and continue until $\mathscr{S}$ is empty.

Prim's algorithm is similarly easy: Start with any $\boldsymbol{x} = \boldsymbol{x}_i$. Choose an edge from $\mathscr{E}$ with minimal weight joining $\boldsymbol{x}$ to some other vertex, $\boldsymbol{x}'$. The result is $\boldsymbol{x}$, $\boldsymbol{x}'$ and the edge connecting them. Repeat the procedure choosing an $\boldsymbol{x}''$ that can be joined to either $\boldsymbol{x}$ or $\boldsymbol{x}'$ by an edge of the next smallest weight. Continue in this way until all the vertices are connected. Proof that these two algorithms give minimal spanning trees is in the notes at the end of this chapter.

8.3.2.2 Degrees of Vertices

In many real-world cases, only local connectivity properties of vertices are available or are most important. In this setting, the main property of a vertex is the list of vertices linked to it by one edge. So, these clustering procedures tend to focus on the degree of vertices as a main way to form clusters.

Perhaps the most basic of these is the k-degree algorithm, see Jennings et al. (2000). Fix two parameters k and D_{max}, where k is a guess as to the highest degree a vertex might have and D_{max} is a bound on the cluster radius, the number of edges away from a given vertex one is willing to travel. The procedure starts by looking at vertices with degree at least k and simultaneously constructs a cluster around each. The high-degree vertices will often compete for vertices that are midway between them in terms of number of edges. Starting with a high-degree vertex, add vertices layer by layer stepping out one edge at each layer. The growth of the layer stops when either the maximum radius D_{max} is reached or the number of vertices added is too few. Formally, this can be done by setting $\ell(u)$ to be the number of vertices in the uth layer and continuing as long as at least k vertices are added at each layer; i.e., there are at least $k\ell(u)$vertices at the $u+1$ layer. Then the same procedure can be repeated using $k-1$ in place of k to continue the search for highly connected subgraphs. Jennings et al. (2000) report that this procedure tends to give relatively few clusters of variable size and radius, arguably revealing the structure of the real clusters.

An alternative to this layered procedure is proposed by Hartuv and Shamir (2000). Let $d = d(\mathcal{G})$ be the minimum number of edges that need to be removed to disconnect $\mathcal{G}$. Thus, if $\mathcal{G}$ is an undirected graph with all edgeweights 1, then a minimum cut $\mathcal{S}$ consists of exactly d edges; the higher the degree of the vertices, the more edges must be removed to disconnect it. Define a graph $\mathcal{G}$ to be highly connected if $d(\mathcal{G}) \geq n/2$; a highly connected subgraph $\mathcal{H}$ has $d(\mathcal{H}) \geq \#(\mathcal{V}_{\mathcal{H}})/2$, where $\mathcal{V}_{\mathcal{H}}$ is the set of vertices of $\mathcal{H}$. The higher the degree of the vertices, the easier it is for a graph to be highly connected. Indeed, the diameter of every highly connected graph is at most 2: When all the edges of a minimum degree vertex are removed, the graph is disconnected so $d(\mathcal{G}) \leq \min_v \deg(v)$. If $\mathcal{G}$ is highly connected then $\#(\mathcal{V})/2 \leq d(\mathcal{G})$. Taken together, these inequalities mean that every vertex is adjacent to at least half the vertices of $\mathcal{G}$. Every two vertices are distance 2 apart since have a common neighbor.

Now, if highly connected subgraphs are regarded as clusters, it is enough to search for highly connected subgraphs in order of size from largest to smallest. The largest subgraphs will correspond to the minimum-sized cut. So, the Hartuv and Shamir (2000) procedure starts by looking at $\mathcal{G}$. If $\mathcal{G}$ is highly connected, then the procedure stops. Otherwise, it starts with the minimum cuts and searches the resulting bipartite partitions of $\mathcal{G}$ for highly connected subgraphs. Then the minimum cut is augmented by one and the search repeated. The cut size continues to be augmented until all the highly connected subgraphs are found. To avoid triviality, singleton vertices are not considered clusters. It can be proved that if the smaller of $\mathcal{H}$ and $\mathcal{H}^c$ contains ℓ vertices, then $\#(\mathcal{S}) \leq \ell$, where $\mathcal{S}$ is a minimum cut, with equality only if $\min(\mathcal{H}, \mathcal{H}^c)$ is a clique. Although this procedure uses more than just local connectivity properties, degrees of vertices are at the heart of it and the procedure can be speeded up substantially by preprocessing to remove low-degree vertices so that the higher-degree vertices will give the desired clusters.

An even more sophisticated procedure effectively regards the neighborhoods of a vertex probabilistically. Aksoy and Haralick (1999) define the neighborhood $N(v)$ of a vertex v to be the vertices linked to it by one edge. Then a conditional count can be defined by

$$D(u|v) = \#(\{w \in \mathscr{V} | (w,u) \in \mathscr{E} \text{ and } (x,w) \in \mathscr{E}\}).$$

This leads to the notion of a region $Z(v,\ell)$ around a vertex v for a fixed ℓ:

$$Z(v,\ell) = \{u|D(u|v) \geq \ell\}.$$

Such regions are said to be dense and a dense region candidate is of the form $Z(v) = Z(v,J)$ around v where $J = \max\{\ell|\#(Z(u,\ell)) \geq \ell\}$. Clearly, if M is a clique of size L (i.e., a fully connected subgraph of L vertices), then $(u,v) \in M$ implies $D(u,v) \geq L$. So, $M \subset Z(v,\ell)$ and $\ell \leq L \leq \#(Z(u,\ell)$. Another conditional probability idea is called the association of a vertex v to a subset B of $\mathscr{V}$. Let

$$A(v|B) = \frac{\#(\{N(v) \cap B\}}{\#(B)},$$

so that $0 \leq A(v|B) \leq 1$. Finally, the compactness of a set $B \subset \mathscr{V}$ is the normalized the association

$$C(B) = \frac{1}{\#(B)} \sum_{v \in B} A(v|B).$$

Again, $0 \leq C(B) \leq 1$.

Aksoy and Haralick (1999) now give two algorithms one for finding dense regions and one for aggregating dense regions into clusters that may not be disjoint. To determine a dense region B around a vertex v, their procedure is as follows.

Start with a sample $\boldsymbol{x}_i$ for $i = 1, ..., n$ regarded as a point set in $\mathbb{R}^p$ giving $\mathscr{V}$. $\mathscr{G}$ may be fully connected or only have a subset of the edges connecting the $\boldsymbol{x}_i$s. Fix a vertex v.

- ☐ Compute $D(u|v)$ for each $u \in \mathscr{V}$.
- ☐ Find a dense region candidate $Z(v|\ell')$, where

$$\ell' = \max\{\ell|\#\{u|D(u|v) \geq \ell\} \geq \ell\}.$$

- ☐ Remove the vertices from the candidate set if their association to v is too low.
- ☐ Ensure the remaining vertices have high enough compactness (average association) and give a large enough set.

When $C(B) \geq 1$ and $B = \{w \in Z(v)|A(w|Z(v)) \geq 1\}$ for v, the regions found are cliques of the graph.

The Aksoy and Haralick (1999) procedure for turning dense regions into clusters is to start with dense regions of the graph B_1 and B_2. If

$$\min\left(\frac{\#(B_1 \cap B_2)}{\#(B_1)}, \frac{\#(B_1 \cap B_2)}{\#(B_2)}\right) \geq T$$

where T is a threshold chosen in advance, then merge B_1 and B_2. Iterate over all the Bs produced from their first algorithm to get the clustering. Depending on the choice of T, it is possible that clusters overlap.

8.3.3 Spectral Clustering

Spectral techniques are based on graph theory as well, however, they focus on the spectrum and eigenvectors of the graph Laplacian rather than the usual connectivity properties of a graph. The graph Laplacian is an $n \times n$ matrix of the form $\boldsymbol{L} = \boldsymbol{D} - \boldsymbol{W}$, where $\boldsymbol{W}$ is the matrix of edgeweights $w(i, j)$ and $\boldsymbol{D}$ is the adjacency matrix, a generalization of the concept of the degrees of the vertices to take the weighting into account. Specifically, let $\Delta_i = \sum_j w(i, j)$ be defined as the degree of vertex i so that $w(i, j) = 0$ means that i and j are not connected. If all the edgeweights are one, this reduces to the earlier definition of the degree of a vertex. Now, $\boldsymbol{D} = \text{diag}(\Delta_1, ..., \Delta_n)$. In some cases, the graph Laplacian can be derived to represent a specific optimality principle based on the cut of a graph. More generally, $\boldsymbol{D}$ reflects the weighting and connectivity of the graph so that $w(i, j)$ is related to $d(\boldsymbol{x}_i, \boldsymbol{x}_j)$ but operations on it such as finding its eigenvalues are not as readily interpretable in terms of the structure of a graph. When the edgeweights are based on the distances between the $\boldsymbol{x}$s, spectral methods are akin to kernel methods, which are also distance based in that the kernel provides a representation for an inner product so as to evade the Curse of Dimensionality.

In practice, given data $\boldsymbol{x}_i$, there are three types of graphs whose Laplacians are studied. The obvious example is a fully connected graph in which $w(i, j) = d(\boldsymbol{x}_j, \boldsymbol{x}_j)$. There are many choices for the similarity d; one is Gaussian, $d(\boldsymbol{x}_i, \boldsymbol{x}_j) = \exp[-\|\boldsymbol{x}_i - \boldsymbol{x}_j\|^2/2\sigma^2]$, in which large σs make the data points farther apart. For a given set of vertices $\boldsymbol{V}$, the other extreme is to form an edge set $\boldsymbol{E}$ by only joining $\boldsymbol{x}_i$s for which $d(\boldsymbol{x}_i, \boldsymbol{x}_j) \leq \varepsilon$ for some preassigned $\varepsilon > 0$. This includes the local connectivity of the points, but in achieving sparsity it may fail to model more distant relations. Between fully connected and ε-neighborhood graphs are k-nearest-neighbor graphs. The idea is to include an edge from vertex $\boldsymbol{x}_i$ to $\boldsymbol{x}_j$ only when $\boldsymbol{x}_j$ is one of the k-nearest neighbors of $\boldsymbol{x}_i$. The result is a directed graph because the nearest-neighbor relationship is not symmetric. However, the graph can be made directed by ignoring the directions or by only including edges for which both directions are present.

Spectral techniques have been developing rapidly because, when p is large, reducing to $\boldsymbol{L}$ converts a large p, small n problem into a manageable $n \times n$ problem. Of the numerous methods available, only a few can be discussed.

8.3.3.1 Minimizing Cuts

In general, a graph-theoretic cut is a partition of the vertices of the graph into two sets. For a graph $\mathscr{G} = (\mathscr{V}, \mathscr{E})$, one writes $\mathscr{V} = \mathscr{S} \cup \mathscr{T}$, a disjoint union. An edge e with vertices $s \in \mathscr{S}$ and $t \in \mathscr{T}$ is a cut edge and "crosses the cut". In a weighted graph, the size of a cut is the sum of the weights of the edges crossing the cut; that is,

$$\mathsf{Cut}(\mathscr{S}, \mathscr{T}) = \sum_{s \in \mathscr{S}, t \in \mathscr{V}} w(s,t). \tag{8.3.41}$$

Note that if a point in $\mathscr{S}$ and a point in $\mathscr{T}$ are connected by a path of length two or more, this does not give a term in Cut; the point of controlling Cut is to control the paths of length one between the two sets. More generally, the requirement that $\mathscr{V} = \mathscr{S} \cup \mathscr{T}$ can be dropped in (8.3.41) and the definitions do not change substantially.

Cuts can be used to define optimality criteria a clustering procedure should satisfy. In these cases, it is easiest to think of the edgeweights as dissimilarities such as the inverse of the distance between two vertices. So, usually, cuts in one way or another are minimized on the grounds that very dissimilar points should be in different clusters. Often, this kind of strategy will give clusters that are not unreasonable, or at least identify highly connected components of a graph corresponding to points that might reasonably belong to the same cluster. Wu and Leahy (1993) probably had the earliest criterion of this form. For a given K, they partitioned a graph into K subgraphs so that the maximum cut between any two subgraphs is minimized. While effective in many cases, this had the drawback that it favored leaving small sets of isolated vertices. This arises because the cut is a sum of edgeweights and fewer vertices in a subgraph tends to give a smaller sum.

Shi and Malik (2000) improved on Wu and Leahy (1993) by changing the optimality criterion to a normalized cut,

$$\mathsf{Ncut}(\mathscr{S}, \mathscr{T}) = \frac{\mathsf{Cut}(\mathscr{S}, \mathscr{T})}{\mathsf{Assoc}(\mathscr{S}, \mathscr{V})} + \frac{\mathsf{Cut}(\mathscr{S}, \mathscr{T})}{\mathsf{Assoc}(\mathscr{T}, \mathscr{V})}, \tag{8.3.42}$$

in which the association Assoc is defined by

$$\mathsf{Assoc}(\mathscr{T}, \mathscr{V}) = \sum_{u \in \mathscr{T}, v \in \mathscr{V}} w(u,v)$$

and is the edge sum from the vertices in $\mathscr{T}$ to the vertices in the whole graph. In this optimality criterion, a cut that gives a small set of isolated points will have a large Ncut since the cut will be a large fraction of the edge sum. Similarly, the association can be normalized to

$$\mathsf{Nassoc}(\mathscr{S}, \mathscr{T}) = \frac{\mathsf{Assoc}(\mathscr{S}, \mathscr{S})}{\mathsf{Assoc}(\mathscr{S}, \mathscr{V})} + \frac{\mathsf{Assoc}(\mathscr{T}, \mathscr{T})}{\mathsf{Assoc}(\mathscr{T}, \mathscr{V})}.$$

It is not hard to see that the Nassoc summarizes how tightly vertices in a set are connected to each other and that

$$\mathsf{Ncut}(\mathscr{S},\mathscr{T}) = 2 - \mathsf{Nassoc}(\mathscr{S},\mathscr{T}).$$

Shi and Malik (2000) establish that minimizing the Ncut even for $K = 2$ is NP complete but convert the problem into a generalized eigenvalue problem that is more amenable to solution. Suppose $\mathscr{S} \cup \mathscr{T} = \mathscr{V}$, $\mathscr{S} \cup \mathscr{T} = \phi$ and search over partitions of the graph into two subgraphs. Since $\mathscr{S} = \mathscr{T}^c$ in $\mathscr{V}$, Shi and Malik (2000) use a sequence of clever steps to show

$$\min_{\mathscr{S}} \mathsf{Ncut}(\mathscr{S},\mathscr{T}) = \min_{\boldsymbol{\tau}} \frac{\boldsymbol{\tau}'(\boldsymbol{D}-\boldsymbol{W})\boldsymbol{\tau}}{\boldsymbol{\tau}'\boldsymbol{D}\boldsymbol{\tau}}, \tag{8.3.43}$$

where $\boldsymbol{\tau} = (\tau_1,...,\tau_n) = (\mathbf{1}+\boldsymbol{v}) - b(\mathbf{1}-\boldsymbol{v})$ and $\boldsymbol{v} = (v_1,...,v_n)$ with $v_i = 1$ when $\boldsymbol{x}_i \in \mathscr{S}$ and -1 otherwise. The constant b is

$$b = \frac{\sum_{v_i>0}\sum_j w(i,j)}{\sum_{v_i<0}\sum_j w(i,j)}.$$

Thus, in (8.3.43), $\tau_i \in \{1, -b\}$ and the constraint $\boldsymbol{\tau}'\boldsymbol{D}\mathbf{1} = 0$ must be imposed. If the τ_is are permitted to assume real values, then (8.3.43) reduces to solving the generalized eigenvalue system

$$(\boldsymbol{D}-\boldsymbol{W})\boldsymbol{\tau} = \lambda \boldsymbol{D}\boldsymbol{\tau}. \tag{8.3.44}$$

Clearly, (8.3.44) is a criterion based on finding the spectrum of $\boldsymbol{L}$.

Shi and Malik (2000) verify that solutions to (8.3.44) automatically satisfy $\boldsymbol{\tau}'\boldsymbol{D}\mathbf{1} = 0$. Indeed, (8.3.44) is

$$\boldsymbol{D}^{-1/2}(\boldsymbol{D}-\boldsymbol{W})\boldsymbol{D}^{-1/2}\boldsymbol{z} = \lambda \boldsymbol{z}, \tag{8.3.45}$$

where $\boldsymbol{z} = \boldsymbol{D}^{-1/2}\boldsymbol{\tau}$, so $\boldsymbol{z}_0 = \boldsymbol{D}^{1/2}\mathbf{1}$ is an eigenvector of (8.3.45) with eigenvalue 0. Since $\boldsymbol{L}$ is positive semidefinite, zero is the smallest eigenvalue of $\boldsymbol{D}^{-1/2}(\boldsymbol{D}-\boldsymbol{W})\boldsymbol{D}^{-1/2}$, and therefore the first eigenvector corresponding to a nonzero eigenvalue is orthogonal to $\boldsymbol{z}_0$. It is a theorem that, for any matrix $\boldsymbol{A}$ with real eigenvalues, the minimum of $\boldsymbol{x}'\boldsymbol{Z}\boldsymbol{x}$ over unit length vectors $\boldsymbol{x}$ orthogonal to the first j eigenvectors of $\boldsymbol{A}$ is achieved by the eigenvector of the next smallest eigenvalue. Thus, the eigenvector $\boldsymbol{z}_1$ from the second smallest eigenvalue is the real solution to the minimum normalized cut problem

$$\boldsymbol{z}_1 = \arg \min_{z:\, z \perp z_0} \frac{\boldsymbol{z}'\boldsymbol{D}^{-1/2}(\boldsymbol{D}-\boldsymbol{W})\boldsymbol{D}^{-1/2}\boldsymbol{z}}{\boldsymbol{z}'\boldsymbol{z}},$$

and so

$$\boldsymbol{\tau} = \arg \min_{\tau:\, \tau D\mathbf{1}=0} \frac{\boldsymbol{\tau}'\boldsymbol{D}^{-1/2}(\boldsymbol{D}-\boldsymbol{W})\boldsymbol{D}^{-1/2}\boldsymbol{\tau}}{\boldsymbol{\tau}'\boldsymbol{\tau}}. \tag{8.3.46}$$

This is converted back to the discrete case where $\tau_i \in \{1, -b\}$ by choosing a grid of values and finding the gridpoint that gives the minimal Ncut.

A clustering into two classes now comes from the $n \times 2$ matrix $[z_0, z_1]$. If $\boldsymbol{r}_j$ is the vector from the j-row, $j = 1,..,n$ then cluster the $\boldsymbol{r}_i$s into two clusters, C_1' and C_2', using, say, the K-means algorithm. Setting $C_k = \{\boldsymbol{x}_j | \boldsymbol{r}_j \in C_k'\}$ gives the clustering of the original data points. If the first ℓ eigenvectors are found so that an $n \times \ell$ matrix is found, then K-means is used on vectors of length ℓ to cluster the $\boldsymbol{x}_i$s into K clusters.

The Shi and Malik (2000) procedure can be extended to graph partitions of more than two elements in other ways. One is a divisive procedure that permits some subjectivity. Simply examine the bipartite partition for stability and to make sure the Ncut is below a prespecified value. If either of these desiderata fail, one of the partition elements is split again using the same technique. An alternative is to generate a K class partition for some large K' and then merge clusters so as to minimize the K-way normalized cut criterion

$$\mathsf{Ncut}_K = \frac{\mathsf{Cut}(\mathscr{S}_1, \mathscr{V} - \mathscr{S}_1)}{\mathsf{Assoc}(\mathscr{S}_1, \mathscr{V})} + \ldots + \frac{\mathsf{Cut}(\mathscr{S}_K, \mathscr{V} - \mathscr{S}_K)}{\mathsf{Assoc}(\mathscr{S}_K, \mathscr{V})}.$$

In addition to (8.3.42), another variant on Wu and Leahy (1993) comes from Ding and He (2002). Instead of normalizing a cut by an association, observe that an ideal bipartite partition of a graph into two sets $\mathscr{S}$ and $\mathscr{T}$ should have small $\mathsf{Cut}(\mathscr{S}, \mathscr{T})$, but large values for the edgeweights within $\mathscr{S}$ and $\mathscr{T}$. So, another cut-based criterion would be to minimize

$$\mathsf{Mcut}(\mathscr{S}, \mathscr{T}) = \frac{\mathsf{Cut}(\mathscr{S}, \mathscr{T})}{\mathsf{Cut}(\mathscr{S})} + \frac{\mathsf{Cut}(\mathscr{S}, \mathscr{T})}{\mathsf{Cut}(\mathscr{T})}. \qquad (8.3.47)$$

(Here, $\mathsf{Cut}(\mathscr{T}) = \mathsf{Cut}(\mathscr{T}, \mathscr{T})$.) It turns out that approximately minimizing (8.3.47) can also be converted to a generalized eigenvalue problem in terms of $\boldsymbol{L}$. This criterion will tend to generate clusters with similar sizes by extending it to K clusters as in (8.3.47).

Another cut-based criterion that has been studied results from changing the denominators in (8.3.47). The ratio cut criterion is to minimize

$$\mathsf{Rcut}(\mathscr{S}, \mathscr{T}) = \frac{\mathsf{Cut}(\mathscr{S}, \mathscr{T})}{\#(\mathscr{S})} + \frac{\mathsf{Cut}(\mathscr{S}, \mathscr{T})}{\#(\mathscr{T})}.$$

Again, this is amenable to a generalized eigenvalue formulation in terms of $\boldsymbol{L}$. Optimization of these cut-based criteria has been improved by using concepts of linkage across the cut. The idea is to search the vertices adjacent to the cut to see if any of them have a higher linkage to the other cluster than the one they are assigned to.

Finally, an even more elaborate cut-based clustering procedure is given by Flake et al. (2004). The idea is to form minimum-cut trees from a graph $\mathscr{G}$. The minimum-cut tree has the same vertices as $\mathscr{G}$ but the edges are derived from cuts. Specifically, the minimum cut between any two vertices u, v in $\mathscr{G}$ can be found by looking at the path in the min-cut tree connecting them. The edge of minimum capacity connecting u to v in the min-cut tree corresponds to the minimum cut between u and v in $\mathscr{G}$. Removing the edge of minimum capacity in the min-cut tree gives two sets of vertices corresponding to the two clusters. This procedure can be elaborated to give a hierarchical procedure.

However, finding the min-cut tree may not be easy. So, Flake et al. (2004) have several heuristics to speed the search for min-cut trees.

8.3.3.2 Some General Properties

Having seen that the graph Laplacian arises in procedures, it is worth seeing some of its general properties. First, note that $\mathrm{Cut}(\mathscr{S},\mathscr{T})$ can be represented in terms of $\boldsymbol{L}$. With $\boldsymbol{v}$ as before,

$$\mathrm{Cut}(\mathscr{S},\mathscr{T}) = \sum_{v_i>0, v_j<0} -w(i,j)v_i v_j = (\mathbf{1}+\boldsymbol{v})'(\boldsymbol{D}-\boldsymbol{W})(\mathbf{1}+\boldsymbol{v}). \quad (8.3.48)$$

In addition,

$$\mathrm{Assoc}(\mathscr{A},\mathscr{V}) = \sum_{v_i>0} \Delta_i = (b/(1+b))\mathbf{1}'\boldsymbol{D}\mathbf{1}. \quad (8.3.49)$$

Taken together, these suggest how (8.3.42) gets converted to (8.3.43). Moreover, (8.3.48) and (8.3.49) suggest that in general cuts can be represented in terms of graph Laplacians, which therefore represent aspects of how graphs can be partitioned. A proposition from Luxburg (2007) helps summarize this.

Proposition: The graph Laplacian $\boldsymbol{L}$ satisfies:

(i) $\forall \boldsymbol{v} \in \mathbb{R}^n$,

$$\boldsymbol{v}'\boldsymbol{L}\boldsymbol{v} = \frac{1}{2}\sum_{i,j=1}^{n} w(i,j)(v_i - v_j)^2.$$

(ii) $\boldsymbol{L}$ is symmetric and positive definite; $\boldsymbol{L}$ has n eigenvalues

$$0 = \lambda_1 \le \lambda_2 \le \ldots \le \lambda_n,$$

in which the eigenvector of 0 is $\mathbf{1}$.

Proof: Only the first part requires proof. By definition of $\boldsymbol{D}$ in terms of the Δ_is,

$$\begin{aligned}\boldsymbol{v}'\boldsymbol{L}\boldsymbol{v} &= \sum_{i=1}^{n}\Delta_i v_i^2 - \sum_{i,j=1}^{n} v_i v_j w(i,j)\\ &= \sum_{i=1}^{n}\Delta_i v_i^2 - 2\sum_{i,j=1}^{n} v_i v_j w(i,j) + \sum_{j=1}^{n}\Delta_j v_j = \frac{1}{2}\sum_{i,j=1}^{n} w(i,j)(v_i - v_j)^2. \quad \square\end{aligned}$$

The graph Laplacian also summarizes the number of connected components.

Proposition: Let $\mathscr{G}$ be an undirected graph with nonnegative weights. Then, the multiplicity of the eigenvalue 0, say m, gives the number of connected components $\mathscr{A}_1,\ldots,\mathscr{A}_m$ of $\mathscr{G}$ and the eigenspace is the span of the vectors $\mathbf{1}_{\mathscr{A}_1},\ldots,\mathbf{1}_{\mathscr{A}_m}$, the indicator vectors for the vertices of the components.

Proof: Suppose $\boldsymbol{v}$ is an eigenvector of 0; i.e.,

$$0 = \boldsymbol{v}\boldsymbol{L}\boldsymbol{v} = \sum_{i,j=1}^{n} w(i,j)(v_i - v_j)^2.$$

Then, for $w(i,j) \geq 0$, the terms in the sum must be zero. So, if vertices i and j are connected and $w(i,j) > 0$, $v_i = v_j$. Consequently, any vector $\boldsymbol{v}$ must be constant on connected components. □

Two variants on $\boldsymbol{L}$ are called normalized graph Laplacians. They are

$$\boldsymbol{L}_{sym} = D^{-1/2}\boldsymbol{L}D^{-1/2} = \boldsymbol{I} - D^{-1/2}\boldsymbol{W}D^{-1/2}$$

and

$$\boldsymbol{L}_{rw} = \boldsymbol{D}^{-1}\boldsymbol{L} = \boldsymbol{I} - \boldsymbol{D}^{-1}\boldsymbol{W},$$

called the symmetric and random walk forms. Both are positive, semi-definite and have n real valued eigenvalues, the smallest being 0. In parallel to the last proposition, Luxburg (2007) shows the following.

Proposition: (i) For any $\boldsymbol{v} \in \mathbb{R}^n$,

$$\boldsymbol{v}'\boldsymbol{L}\boldsymbol{v} = \frac{1}{2}\sum_{i,j=1}^{n} w(i,j)\left(\frac{v_i}{\sqrt{d_i}} - \frac{v_j}{\sqrt{d_i}}\right)^2.$$

(ii) The following are equivalent: (I) λ is an eigenvalue of L_{rw} with eigenvector $\boldsymbol{v}$; (II) λ is an eigenvalue of L_{sym} with eigenvector $\boldsymbol{w} = \boldsymbol{D}^{1/2}\boldsymbol{v}$; and (III) λ and $\boldsymbol{v}$ solve the generalized eigenvalue problem $\boldsymbol{L}\boldsymbol{v} = \lambda\boldsymbol{D}\boldsymbol{v}$. □

A special case of this is when $\lambda_1 = 0$, so that $\mathbf{1}$ is the eigenvector of L_{rw} and $\boldsymbol{D}^{1/2}\mathbf{1}$ is the eigenvector of $\boldsymbol{L}_{sym}$.

With these results stated, it is important to observe that there are numerous variants on the Shi and Malik (2000) procedure. For instance, in place of finding the first ℓ eigenvectors of the generalized eigenvalue problems $\boldsymbol{L}\boldsymbol{v} = \lambda\boldsymbol{D}\boldsymbol{v}$, one can use the first ℓ eigenvalues of $\boldsymbol{L}$ by themselves; see Wu and Leahy (1993). Alternatively, one can find the first ℓ eigenvectors of $\boldsymbol{L}_{sym}$, say $z_1,...,z_\ell$, and convert to a matrix $\boldsymbol{U}$ in which $u_{i,j} = (z_{i,j}/\sum_j z_{i,j}^2)^{1/2}$. Then set $\boldsymbol{r}_i$ to be the ith row of $\boldsymbol{U}$, see Ng et al. (2002).

Typically, the eigenvectors of normalized graph Laplacians converge while the eigenvectors of unnormalized graph Laplacians often do not, see Luxburg et al. (2008). Verma and Meila (2003) compare several spectral clustering methods. The methods presented here appear to perform well when the correct number of clusters is moderate to large, say $K \geq 6$. A different sort of spectral clustering algorithm (normalized multiway cuts, based on finding the cut values between elements of a partition of a graph into K disjoint pieces rather than two (see Meila and Xu (2002)) appears to do well when $K \leq 5$.

8.4 Bayesian Clustering

Fundamentally, the goal of Bayesian clustering is to obtain a posterior distribution over partitions of the data set $\mathscr{D}$, denoted by $\mathscr{C} = \mathscr{C}(K) = \{C_1, \ldots, C_K\}$, with or without specifying K, so that the modal partition can be identified. Several methods have been proposed for how to do this. Usually they come down to specifying a hierarchical model mimicking the partial order on the class of partitions so that the procedure is also hierarchical, usually agglomerative. At its root, Bayesian clustering is a much more probabilistic approach than the earlier clustering techniques. The first procedure below is an attempt to use a straightforward application of probability modeling to extract optimal partitions. More generally, Bayesian nonparametrics extend the probability modeling and lead to more elaborate hierarchical models. The second is simpler in that the probability model is used only to decide merges on the basis of Bayes factors. One benefit of pure Bayesian techniques is their built-in treatment of the variability of the clustering.

8.4.1 Probabilistic Clustering

Possibly the first effort to do Bayesian clustering was the hierarchical technique due to Makato and Tokunaga (1995). Starting with the data $\mathscr{D} = \{\boldsymbol{x}_1, \ldots, \boldsymbol{x}_n\}$ as n clusters of size 1, the idea is to merge clusters when the probability of the merged cluster $P(C_k \cup C_j)$ is greater than the probability of the individual clusters $P(C_k)P(C_j)$. Thus, the clusters themselves are treated as random variables, an idea that is developed in later work. It's an open question how well the Makato and Tokunaga (1995) method performs; their major contribution may have been encapsulating the clustering problem within the Bayesian framework.

8.4.1.1 Makato and Tokunaga Procedure

Write $\mathscr{C}_\ell$ to mean the clustering $\{C_1, \ldots., C_{K_\ell}\}$ at the ℓth stage of merging, $\ell = 0, \ldots, n-2$. Each C_k is a cluster with elements drawn from the $\boldsymbol{x}_i$s; at $\ell = 0$, each is a singleton set. Interpreting $P(\mathscr{C}|\mathscr{D})$ to mean the probability that a data set $\mathscr{D}$ is clustered into a clustering $\mathscr{C}$, the task is to find the clustering $\mathscr{C}_{opt}$ with maximal conditional probability. This optimization is hard to do in full generality, so a search over clusterings is used starting with the trivial clustering of n clusters and merging.

The general step of merging or not merging two clusters in passing from step ℓ to step $\ell + 1$ is as follows. Suppose the $\boldsymbol{x}_i$s have been partitioned into a clustering $\mathscr{C} = \mathscr{C}_\ell$ with clusters $C_1, \ldots, C_K = C_{K_\ell}$, generically denoted C. The posterior probability of a clustering $\mathscr{C}$ is

$$\begin{aligned}
P(\mathscr{C}|\mathscr{D}) &= \prod_{C\in\mathscr{C}} P(C|\mathscr{D}) = \prod_{C\in\mathscr{C}} \frac{P(\mathscr{D}|C)P(C)}{P(\mathscr{D})} \\
&= \left(\prod_{C\in\mathscr{C}} \frac{P(C)}{P(\mathscr{D})}\right) \prod_{C\in\mathscr{C}} \left[\prod_{\boldsymbol{x}\in C} P(\boldsymbol{x}|C)\right] \qquad (8.4.1)\\
&= \frac{PC(\mathscr{C})}{P(\mathscr{D})^{\#(\mathscr{C})}} \prod_{C\in\mathscr{C}} SC(C),
\end{aligned}$$

in which $SC(C)$ is the quantity in square brackets in (8.4.1), while $PC(\mathscr{C})$ is the product of $P(C)$s over $C \in \mathscr{C}$ and the clusters and data points have been treated as independent.

To derive a merge rule, observe that if $\mathscr{C}_{\ell+1} = \mathscr{C}_\ell \setminus \{C_k \cup C_j\} \cup \{C_k, C_j\}$, then the ratio of the posterior probabilities is

$$\frac{P(\mathscr{C}_{\ell+1}|\mathscr{D})}{P(\mathscr{C}_\ell|\mathscr{D})} = \frac{PC(\mathscr{C}_{\ell+1})}{PC(\mathscr{C}_\ell)} \frac{SC(\{C_k, C_j\})}{SC(C_k)SC(C_j)}. \qquad (8.4.2)$$

The ratio of PCs functions like a prior on clusters; the information in the data is mostly only in the ratio of SCs. So, it makes sense to merge clusters C_k and C_j in $\mathscr{C}_\ell$ at the next, $\ell+1$, stage when they achieve

$$\max_{C_k, C_j \in \mathscr{C}_\ell} \frac{SC(\{C_k, C_j\})}{SC(C_k)SC(C_j)}.$$

Choosing this merge as optimal at each step also means that from merge step to merge step only the new merge factor $SC(\{C_k, C_j\})$ must be found since the other factors are already available.

Implementing this search for clusters to merge requires some interpretation. First, note that $P(\mathscr{D})$ drops out of the maximization so only $P(C)$ and $P(\boldsymbol{x}|C)$ need to be given. Treating $P(C)$ as a prior, it may be reasonable to set $P(C) \propto 1/(A)^{\#(C)}$ so that clusters have probabilities that respond to their sizes. Other choices may be reasonable, too. Specifying $P(\boldsymbol{x}|C)$ usually requires more work. For instance, if there is a random variable T such that C and $\boldsymbol{x}$ are conditionally independent given T, then it may be reasonable to set

$$P(\boldsymbol{x}|C) = P(\boldsymbol{x}) \sum_t \frac{P(T=t|\boldsymbol{x})P(T=t|C)}{P(T=t)}.$$

In the text classification context, T corresponds to choosing a term at random from a set of terms. In this case, $P(t|\boldsymbol{x})$ can be estimated by the relative frequency of a term in a document $\boldsymbol{x}$, $P(t|C)$ can be estimated by the relative frequency of t in C, and $P(T=t)$ is the relative frequency of the term in the whole data set. So, it is seen that the Makato and Tokunaga technique is an incomplete template: Extra information specific to the subject matter must be added to have a useful implementation.

8.4.1.2 General Nonparametric Clustering

The last method extends to more general assignments of probabilities so as to get posterior probabilities on partitions. This subsection draws heavily from Quintana (2006), who provides a clear summary.

For simplicity, regard $\mathscr{D}$ as $\mathscr{S} = \{1,...,n\}$ so $\mathscr{C} = \{C_1,...,C_K\}$ is a collection of sets of integers. The membership function is still $C(i) = k$ to mean integer i is in cluster k and without loss it may be assumed that the C_is are in ascending order; i.e., the smallest elements in the C_ks are increasing with k. Now, defining a probability model on $\mathscr{P}(\mathscr{S})$, the collection of all partitions of $\mathscr{S}$, can be done by factorization. Let $\rho \in \mathscr{P}(\mathscr{S})$ such that ρ is a specific clustering of $\mathscr{S}$. Then a probability on $\mathscr{P}(\mathscr{S})$ is given by specifying the conditional probabilities from the membership function

$$P(C(1),...,C(n)) = \prod_{i=2}^{n} p(C(i)|C(i-1),...,C(1)). \tag{8.4.3}$$

Because of the ascending order, $P(C(1) = 1) = 1$. For convenience, a no void clusters restriction is assumed: If $C(i) = k > 1$ for some i, then there exist at least $k-1$ values $i_1,...,i_{k-1} \in S$ such that $C(i_j) = k$ for $j = 1,...,k-1$. There are many ways to do this; one involves the Dirichlet process.

Suppose a probability vector F is drawn from $\mathscr{D}(\alpha)$. Then, sample $X_1,...,X_n$ drawn IID from F can be used to induce a clustering on S as follows. Since the X_is are discrete, if n is bigger than the number of clusters to be created, there will be ties among the X_is. So, define an equivalence relation on S by

$$i \sim j \Leftrightarrow X_i = X_j.$$

The equivalence classes give a partition of S by looking at the multiplicity of the X_is. Formally, $C(1) = 1$, so set $X_1^* = X_1$. The value X_1^* represents the location of the first cluster, or the first canonical representative of an equivalence class to be specified. If $X_2 = X_1$, let $C(2) = C(1)$, but if $X_2 \neq X_1$, create a new cluster by setting $C(2) = C(1)+1$ and set $X_2^* = X_2$. Thus, X_2^* is the first element of the new cluster to be formed. It is the second canonical value to emerge from the multiplicity in $X_1,...,X_n$. Proceeding in this way, the general step is to put the next X_i into a previously occurring cluster if $X_i = X_j$ for some $j < i$ and to use X_i to start a new cluster otherwise.

Continuing this procedure defines the conditional distributions in (8.4.3). It can be verified that

$$P(C(i+1) = k|C(i),...,C(1)) = \begin{cases} \frac{n_{k,i}}{1+i} & \text{if } 1 \leq k \leq m_i \\ \frac{1}{1+k} & \text{if } k = m_i + 1, \end{cases}$$

where m_i is the number of clusters formed up to the point where $i+1$ is put into a cluster. This is a version of the Chinese restaurant process in which customers, the X_is, enter one at a time and either sit at a table X_j^* where other people are already sitting or start a new table X_i^* where people arriving later can sit.

Another way to specify conditional distributions in (8.4.3) is by using product partition models, PPMs; see Quintana (2006). These are based on factoring a model conditional on $\mathscr{C}$, i.e., writing

$$p(\boldsymbol{x}_1,...,\boldsymbol{x}_n|\mathscr{C}) = \Pi_{k=1}^K p_{C_k}(\boldsymbol{x}_k), \tag{8.4.4}$$

where $\boldsymbol{x}_k = \{\boldsymbol{x}_i : i \in C_k\}$, the subvector of $\boldsymbol{x}$s with indices in C_k, with probability mass function p_{C_k} dependent only on C_k. Then, treating $\mathscr{C}$ as a parameter, assign it a prior by writing

$$P(\mathscr{C}) \propto \Pi_{k=1}^K \chi(C_k), \tag{8.4.5}$$

in which $\chi(C_k)$ is the "cohesion" of the set C_k. Since there are finitely many partitions of S, for a fixed choice of χ the normalizing constant makes P into a probability. Now, the posterior probability from (8.4.4) and (8.4.5) is

$$p(\mathscr{C}|\boldsymbol{x}_1,...,\boldsymbol{x}_n) = \frac{\Pi_{k=1}^K p_{C_k}(\boldsymbol{x}_k)\Pi_{k=1}^K \chi(C_k)}{\sum_{\mathscr{C}} \Pi_{k=1}^K p_{C_k}(\boldsymbol{x}_k)\Pi_{k=1}^K \chi(C_k)}.$$

Finally, a hierarchical Bayes model can be assigned to partitions. For fixed K, write

$$\begin{aligned} \boldsymbol{X}_1,...,\boldsymbol{X}_n|\theta_1^*,...,\theta_k^*,\mathscr{C}(K),\psi &\sim p(\boldsymbol{X}_i|\theta_i,\psi), \quad \text{where } \theta_i = \theta_{C(i)}^* \\ \theta_1^*,...,\theta_K^*|\mathscr{C},\psi &\sim p_0(\cdot \mid \psi), \\ \mathscr{C} &\sim P_1, \end{aligned}$$

in which $p_0(\theta_1^*,...,\theta_K^*|\psi)$ is a density chosen by the analyst and P_1 is a distribution specified by its conditionals as in (8.4.3).

8.4.2 Hypothesis Testing

The simplest Bayesian approach to clustering may be the Bayesian hierarchical clustering BHC of Heller and Ghahramani (2005). This procedure has the same structure as any other agglomerative clustering procedure, but the merge rule is based on marginal probabilities; essentially the merge rule finds the probability that the data points in a potential merge come from the same component in a mixture. This is done by finding the relevant Bayes factors, which are parallel to the left side of (8.4.2). This is different from the Makato and Tokunaga (1995) method because no probability is assigned to partitions; it is Bayesian only in the sense that merges are determined by Bayes factors. Of course, this procedure is sensitive to the choice of models to form the components and to the priors. The normal is the most tractable; however, many methods work well for normal components, so the complexity of BHC may be most worthwhile when normal components do not approximate all the clusters well.

To state the procedure, let $\mathscr{D}$ denote the n data points and start with n singleton sets, each containing one data point. Write $\mathscr{D}_T$ to mean the data in the leaves of a tree T, where T is understood to represent the sequence of merges of $\mathscr{D}_T$ into a single root node. Thus, each singleton set is a trivial tree structure with one node. The idea of BHC is to use Bayesian hypothesis testing, starting with the trivial trees, and decide at each stage whether or not to merge two sets of data points, or clusters, $\mathscr{D}_{T_1}$ and $\mathscr{D}_{T_2}$, by deciding whether or not to merge the two trees T_1 and T_2 into a single larger tree. Doing this gives a sequence of trees that grow stepwise into a dendrogram for all of $\mathscr{D}$.

Fix a collection of models $p(\cdot \mid \theta)$, with a prior $w(\theta|\psi)$, where the ψ represents any hyperparameters. To decide whether or not to merge given subtrees T_i and T_j into a bigger subtree $T_{i,j}$, the two hypotheses must be specified. The first is $\mathscr{H}_1$: All the data in $T_{i,j}$ were generated IID from a single model $p(\cdot \mid \theta)$, with θ unknown with prior $w(\theta|\psi)$. This means that $\mathscr{H}_1$ specifies the model

$$p(D_{T_{i,j}}|\mathscr{H}_1) = \int p(T_{i,j}|\theta)w(\theta|\psi)d\theta = \int \Pi_{\boldsymbol{x}\in T_{i,j}} p(\boldsymbol{x}|\theta)w(\theta|\psi)d\theta. \quad (8.4.6)$$

Expression (8.4.6) evaluates how well the points in $T_{i,j}$ fit into one cluster.

The alternative $\mathscr{H}_2$ is that the T_i and T_j represent different clusters so the subtrees should not be merged. In this case, $\mathscr{H}_2$ includes the assumption that the data points in the trees are independent and is

$$p(D_{T_{i,j}}|\mathscr{H}_2) = p(D_{T_i} \cup D_{T_j}|T_i, T_j) = p(D_{T_i}|T_i)p(D_{T_i}|T_i), \quad (8.4.7)$$

where $p(D_{T_i}|T_i)$ is defined recursively from the initial choice of models $p(\boldsymbol{x}|\theta)$ for individual data points.

In practice, this step can be the hardest to implement: The natural model to use for the whole data set is of the form

$$p(\boldsymbol{x}_i|\phi) = \sum_{k=1}^{K} p(\boldsymbol{x}_i|\theta_k)p(\boldsymbol{x}_i \in C_k|\boldsymbol{p}),$$

where θ_k is the value for cluster k, and k and $\boldsymbol{p}$ are the parameters of multinomial distribution with $p(\boldsymbol{x}_i \in C_k) = p_k$. If $p(\phi|\alpha)$ is the prior, then the mixture can be denoted $p(\mathscr{D}|\alpha) = p(\boldsymbol{x}_1, ..., \boldsymbol{x}_n|\alpha)$. For subsets $\mathscr{D}_T$ of $\mathscr{D}$, the restriction of this model to the available data points can be used, giving $p(\mathscr{D}_T|\alpha)$. When testing a merge step, the mixture over data in a set of the form $\mathscr{D}_T$ must be done over all partitions of the data that are consistent with the tree structure; i.e., that represent the possible permutations of the data that result in the same clusters. This is denoted by $p(\mathscr{D}_T|T)$, where the conditioning indicates the restriction of $p(\mathscr{D})$ to the T-consistent permutations. Heller and Ghahramani (2005) evaluate this factor for the Dirichlet processes.

To do the test, choose $\pi = P(\mathscr{H}_1)$. Then the marginal probability for $T_{i,j}$ is

$$p(D_{T_{i,j}}|T_{i,j}) = \pi(D_{T_{i,j}}|\mathscr{H}_1) + (1-\pi)p(D_{T_i}|T_i)p(D_{T_i}|T_i) \quad (8.4.8)$$

from (8.4.6) and (8.4.7). So, the posterior probability that the two trees (i.e., the clusters they represent) should be merged is

$$\frac{\pi(D_{T_{i,j}}|\mathscr{H}_1)}{\pi(D_{T_{i,j}}|\mathscr{H}_1)+(1-\pi)p(D_{T_i}|T_i)p(D_{T_i}|T_i)}.$$

Note that these tests are defined recursively; the first term in (8.4.8) is for the merged clusters and the second one is over the other clusterings that are consistent with the emerging tree structure. Thus, the joint probabilities are built up successively.

As written, different versions of this procedure can give different results depending on the order in which subtrees are considered. In principle, this can be overcome as in the Makato and Tokunaga (1995) procedure by seeking the highest posterior probability merges and doing them.

Note that this is a procedure that is not the same as doing a Bayesian search over a model class comprised of trees. Indeed, the procedure neglects model uncertainty in the trees since there is no distribution over the collection of trees. In addition, it can be computationally complex to implement. On the other hand, this procedure gives a predictive distribution $m(\boldsymbol{x}_{new}|\mathscr{C})$. Indeed, Bayes testing is the Bayes action under zero-one loss so this procedure has an optimality property of sorts. The delicate use of tree-consistent permutations can even be regarded as a kind of surrogate for model uncertainty.

8.5 Computed Examples

Having described a wide variety of clustering methods, it is worth seeing what results they give in some familiar cases. Recall Ripley's two mixtures of two normal sources of data in the plane, and Fisher's iris data, which have four explanatory variables and three classes. These were described and used in Chapter 7. Both of these are supervised in the sense that the classes are known; however, in this section, the class indicators will be ignored and the remaining data treated as an unsupervised problem. This provides a useful technique for evaluating the performance of clustering methods because the absolute performance of a procedure (relative to the truth) can be found as well as the relative performance of one procedure over another.

Eight clustering methods will be compared here. Four are hierarchical: single linkage, complete linkage, average linkage, and DIANA. The first three are agglomerative and the last is divisive. Then, two methods based on centroids will be compared, K-means and Ward's method. Ward's method is actually agglomerative, but it is natural to compare it with K-means since the two are based on assigning points to whichever cluster mean is closest to them. The last two methods are the mixture of components model, for which the EM algorithm is appropriate, and the third class is spectral, an implementation of the Ng et al. (2002) procedure.

Unfortunately, graph-theoretic and Bayesian techniques are not exemplified here. Indeed, graph-theoretic techniques are only covered implicitly in the sense that the minimal spanning tree procedure is equivalent to single linkage. For this method, a fully connected graph of the data was used, although sparser graphs would naturally be considered in data sets with high n. More general graph partitioning techniques have been implemented by the Karypis lab. However, the Karypis software, METIS, CLUTO, and PAFI, doesn't interact well with R, although it is otherwise usable with practice. In addition, spectral clustering procedures, such as specc, used below, often boil down to graph-partitioning criteria. Bayes methods, however, remain difficult even to implement outside of special cases; e.g., Heller and Ghahramani (2005) focus on Dirichlet processes. No software seems to be available for the Makato and Tokunaga (1995) method.

As in the computed examples of Chapter 7, numerous extra features must be specified to implement these eight methods. In particular, the distance or dissimilarity is often the single most important feature. So, for comparison purposes, wherever reasonably possible, two distances were used to define dissimilarities, Euclidean and absolute error (also called the city block, L^1, or Manhattan metric). Actually, a variant on K-means can be done with the absolute error but is not shown here.

Before proceeding to the results, it is important to comment that clustering in general is not a very well-defined goal. It is easiest to think of clusters as local modes of the overall density; however, since all points must be put in clusters, some will be far from modes. Regarding some points as outliers is also problematic. Usually, they are valid points but inappropriate for cluster formation or they are on a boundary between two clusters and so contribute to instability of the clustering. This makes preanalysis data cleaning somewhat subjective. Alternatively, one can think of clustering as unsupervised classification. This is correct, except when there aren't any classes. Even when there are classes, this is little help because classification optimality and techniques require a response Y. It may be best to regard clustering like data summarization and visualization; i.e., as a search technique to uncover structure.

Clustering is a relatively big problem since the number of clusterings one is searching through is the power set of a (often high-dimensional) real space. Consequently, clustering is relatively informal, and there are many heuristics or "sanity checks" that can be used, depending on the data. For instance, one can form a histogram of the dissimilarities $d(\boldsymbol{x}_i, \boldsymbol{x}_j)$ for $j \neq j$ between the points and look for modes. The modes represent common distances; a global mode at a small distance might correspond to the distances between points within the same cluster. Later, smaller modes might indicate distances between points in different clusters. The number of modes minus one might represent the number of different distances between cluster centers; the height of these modes might indicate how large the clusters are. In addition, clusterings on different subsets of the variables may be generated and compared for consistency.

Partial information, if available, can be used to discriminate among different clustering procedures and in the choice of K, the distance, initial cluster centers, and so forth. If partial information is not available, then the clustering procedure can be varied (e.g., choose different Ks, starting centers, weights, or which edges to include) to give several clusterings and justify a sort of consensus clustering. In general, at the end of the procedure, subject matter information must be invoked to see if any of the clusterings are reasonable. One can even hypothesize properties of the data generator and see if they are borne out from a clustering. For instance, if a collection of variables that are conjectured to indicate some general trait do not appear together in a cluster, it calls into question how reasonable it is to associate them. For example, in glaucoma, if high intraocular pressure did not appear in the afflicted group, it might suggest thatreducing intraocular pressure would not be helpful.

Informally, Koepke (2008b) suggests using K-means with a purposefully large K and then combining clusters agglomeratively to find irregular shapes. This kind of idea is built into hierarchical methods since they split or merge sets but would expand the range of clusterings partitional methods would be able to find. The main criterion would be to ensure that the agglomeration combines the large number of small clusters into representative bigger ones.

The Ripley data example is presented first because it is two-dimensional and therefore more easily visualized. Recall that there are two clusters. The iris data are known to have three. So, for brevity, where K must be specified (K-means, mixture models, and spectral clustering), only the results for values $K = 2,3,4$ are shown.

8.5.1 Ripley's Data

Recall that Ripley's data are analyzed from a classification perspective in Chapter 7, where several plots of them can be found. Both clusters are mixtures of two normals; roughly, there is a top cluster and a bottom cluster. What makes this synthetic data set difficult is that over many regions it is impossible to cluster (or classify) perfectly because at many points there is a nonzero probability of each class occurring. Since they are so commonly used, and easy to access, hierarchical methods are presented first, followed by centroid-based, mixture model, and spectral procedures. It will be seen that complete linkage, Ward's method, K-means, and spectral procedures with polynomial or Laplace kernels did well, while mixtures were passable and other methods were poor.

8.5.1.1 Ripley, Hierarchical

The hierarchical methods discussed here, and others, can be found in the hclust contributed R package. The basic form of the commands to generate dendrograms for these cases is

```
data(synth.tr)
xx<-synth.tr[,-3]
hc <- hclust(dist(xx, method="euclidean"),
method = "single")
postscript("hSingleEuclidRipley.eps")
plot(hc,  labels = FALSE, hang = -1,
main = "Hierarchical Clustering",
```

```
sub = "Single Linkage on Ripley's data using the
Euclidean distance")
dev.off()
```

in which the data set synth.tr is clustered using the Euclidean distance and single linkage. The dendrogram plotted is in Fig. 8.7. Making the obvious substitutions for the absolute error (called the Manhattan distance in the command) and other forms of linkage (average and complete) generates the rest of the panels in Fig. 8.7.

The top row of Fig. 8.7 shows that the chaining problem associated with single linkage is severe for this data set. Single points are pulled off one or a few at a time by the nearest-neighbor criterion, making it difficult to find natural break points to identify clusters. Note that aside from a few large decreases in dissimilarity at the start of the procedure, the subsequent decreases in dissimilarity are quite small on the vertical axis, indicating that only a little similarity is gained at each split. This means that single linkage clustering is not giving good results; since it is equivalent to finding a minimal spanning tree, the simplest graph-theoretic technique is also ineffective.

The middle row of Fig. 8.7 shows the dendrograms for average linkage. It is seen that there is a large decrease in dissimilarity with the initial split, from around .9 to a little over .8. In both dendrograms, the next split on the left branch gives a small cluster at the extreme left. These likely represent near-outliers because if they are ignored one can discern four clusters – two from the first split on the right branch and two more from the second split on the left branch. Past this point, the decreases in dissimilarity become small. Taken together, this suggests four clusters are reasonable for this data set. This makes sense because both classes in the Ripley data set consist of two normal components. Since all four normal components have weights that are not close to zero, the clustering procedure detects them, as well as the two classes that emerge from the initial split.

The third row of Fig. 8.7 shows the dendrograms for complete linkage. This is at the other extreme from single linkage in that it considers all neighbors rather than nearest neighbors. As with average linkage, four clusters are detected, but they are detected a bit more cleanly because the small collection of outliers at the extreme left in average linkage does not appear in complete linkage. The points are within one of the four clusters found at the second split. With hindsight, therefore, complete linkage appears to do best in this case.

If the Euclidean distance is replaced by absolute error, Fig. 8.7 does not change much. So, these results are not shown. Neither distance gives splits substantially better reduction in dissimilarity at any level; neither distance finds or fails to find any meaningful substructure. The effect of the change in distance is probably comparable in size to other sources of error.

Recall that DIANA is divisive, so the dendrograms are formed from the bottom up rather than the top down as with agglomerative procedures. In Fig. 8.8, it is seen that both distances give a large reduction in dissimilarity, of about the same amount, at the last merge (first split if the procedure is done in reverse). This probably corresponds to the two classes. The second to last merge (second split) gives four well-defined

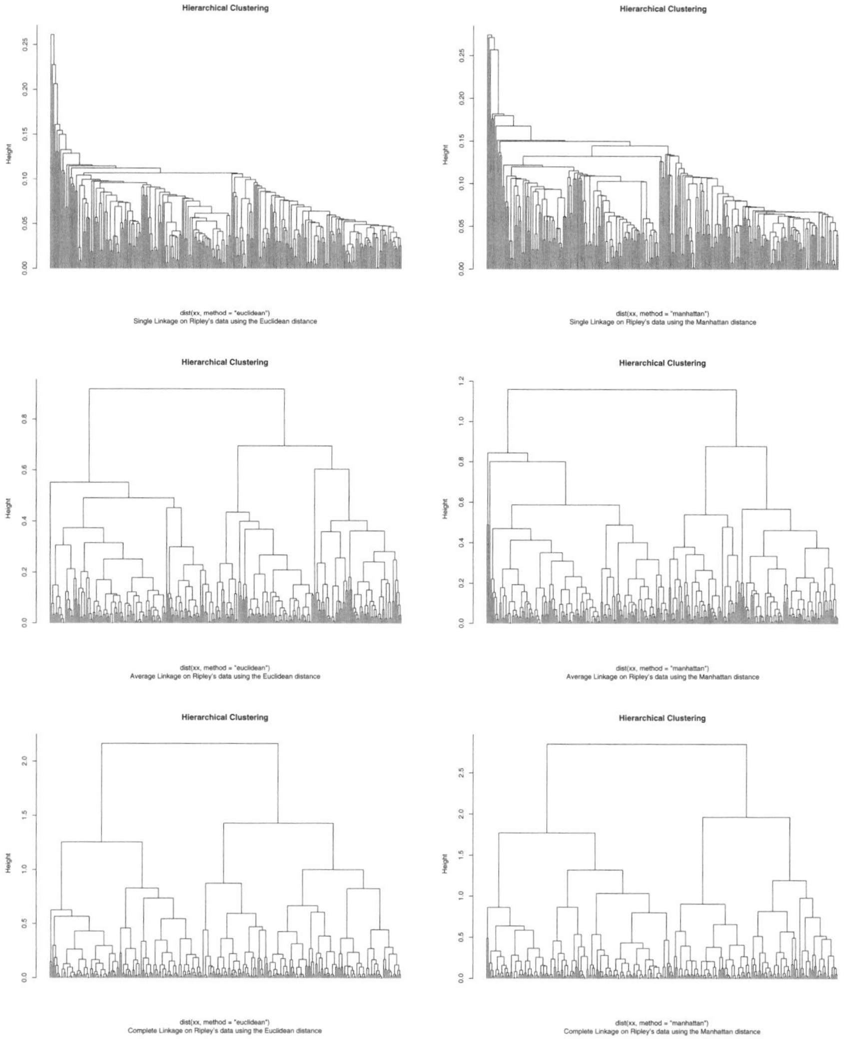

Fig. 8.7 Hierarchical agglomerative procedures on the Ripley data. The left column uses the Euclidean distance and the right column uses absolute error. The rows correspond to single linkage, average linkage, and complete linkage.

groups in each case, although the dissimilarity is a bit lower for absolute error than for Euclidean distance. Looking one step further down on the absolute case, it is seen that the decrease in dissimilarity is roughly comparable for all the merges, whereas it is more variable for the Euclidean case. That is, the appearance of the absolute error graph is more regular in terms of reducing dissimilarity. However, the four clusters are more similar in size for the Euclidean case than for absolute error. In the absence of knowledge of the four-component structure of the data, one would be led to use absolute error over Euclidean – if the dissimilarity were believed appropriate. However, the Euclidean version is slightly better at finding the correct clusters.

Comparing the complete-linkage hierarchical dendrograms with DIANA with absolute error, it is seen that DIANA seems to identify too many clusters because the decrease in dissimilarity is still relatively large compared with the decreases close to the tips of the branches. Complete linkage tends to give four well-defined clusters and so, again with hindsight, might be preferred.

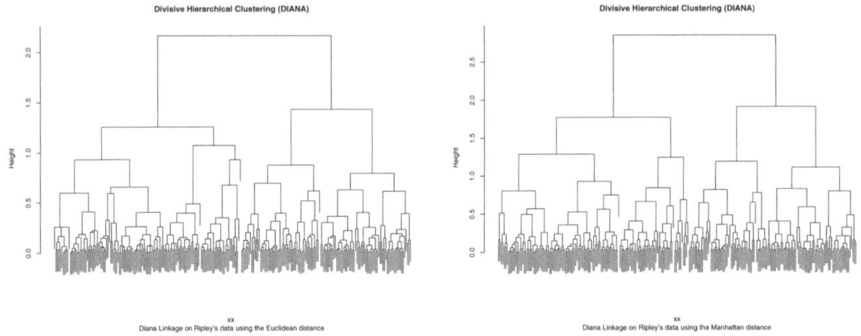

Fig. 8.8 Hierarchical division; DIANA on Ripley. On the left, the Euclidean distance is used; on the right, absolute error.

8.5.1.2 Ripley, Centroid Based

In its top row, Fig. 8.9 shows Ward's method on Ripley for the Euclidean and absolute error distances. Although Ward is a hierarchical agglomerative procedure, it agglomerates clusters defined by their means and sums of squares to keep the overall sum of squares as small as possible. Thus, Ward is methodologically more comparable to K-means than to other hierarchical procedures. The command for Ward is similar to the commands for the other hierarchical procedures in hclust noted above. Also in Fig. 8.9, three panels show K-means clusters for $K = 2, 3, 4$. The data roughly form two clouds, one on the left (three components) and one on the right (one component). However, there are enough points between the two clouds that one can imagine a single cloud, like a backwards S rotated counterclockwise by 90°. Note that, in the $K = 4$ panel, the larger cluster on the left has centroids (stars) that roughly line up to suggest a single kidney-shaped cluster. The last panel in Fig. 8.9 shows the result from using median absolute error linkage; median Euclidean linkage is similar. Both give slightly bizarre results for Ripley, although one could be led to identify two or four clusters. However, this can only be said with hindsight.

It is worth looking at panels 1 through 5 in more detail.

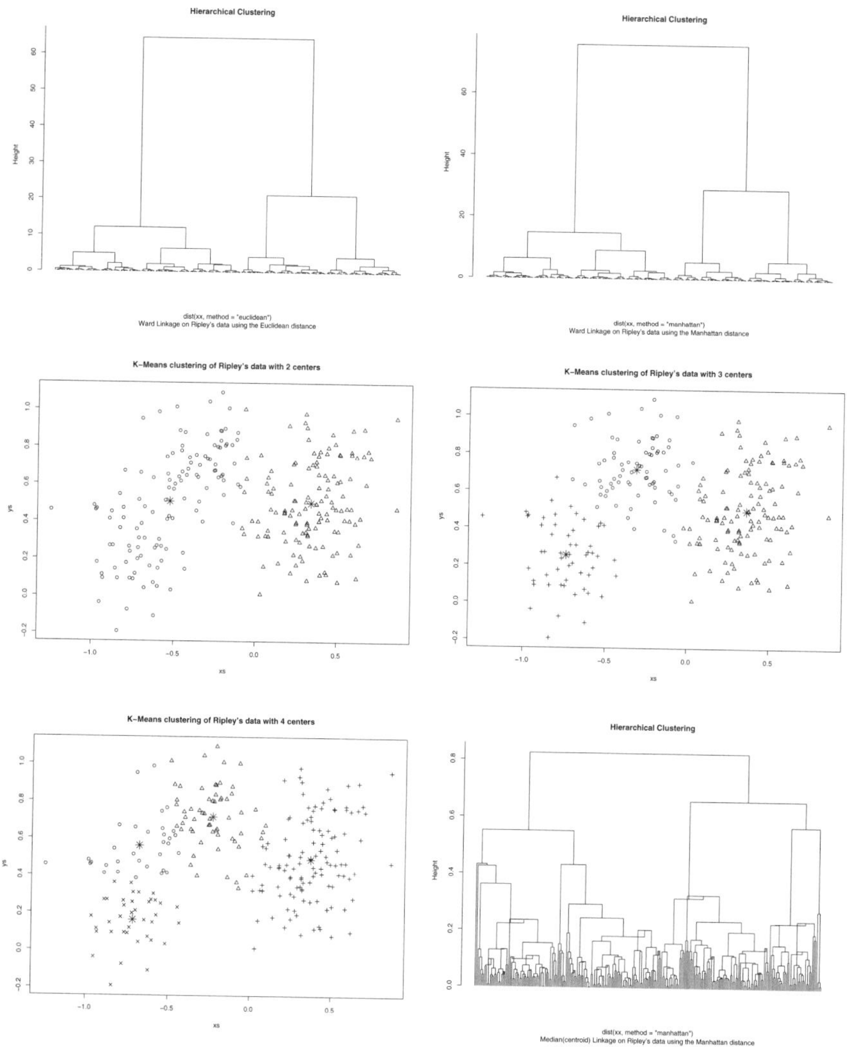

Fig. 8.9 The top row is Ward's method on Ripley data, Euclidean distance on the left, absolute error on the right. The next three panels are for K-means on Ripley with $K = 2, 3, 4$; different geometric shapes indicate the cluster to which each data point belongs, and stars indicate the location of the centers. The lower right corner show median linkage on Ripley with absolute error.

It is seen that Ward's method actually picks out 4 clusters fairly unambiguously, as it is well adapted to finding compact spherical clusters. The initial drop in squared or absolute error is quite large and, after the second split indicated in the dendrograms, there is very little dissimilarity to be explained. (Indeed, after the third split, the dissimilarity is effectively zero.) In both distances, Ward appears to outperform the earlier complete linkage and the other hierarchical methods. It must be remembered that Ward's method rests on having an initial clustering to agglomerate. The hclust documentation does not indicate how this initial clustering is found by the software, but clearly it can have a major impact if done poorly. In practice, one could just use K-means with large K and random starts. Or, one could use a leader algorithm in which an initial data point is chosen at random and later centers are chosen to be at least some fixed distance δ from the centers found. Then the points are assigned to their closest center.

The K-means procedure in R uses the Hartigan-Wong algorithm starting with K random centers. The command for $K = 2$ is:

```
data(synth.tr)
xx<-synth.tr[,-3]
kmRipley2 <- kmeans(xx,2)
kmRipley2$centers
kmRipley2$size
kmRipley2$withinss
```

As seen in Fig. 8.9, the K-means procedure on Ripley with $K = 2$ gives left- and right-hand clusters (circles and triangles) with the two centers, noted by stars, roughly in the middle of the clusters, as confirmed by the tables below. With $K = 3$, the data cloud is treated as a sort of arc and partitioned into three sectors, indicated by plus, circle, and triangle from left to right. The centers are again roughly in the middle of each cluster. When $K = 4$, the pattern is similar. There are four sectors, indicated by xs (circles, triangles, and plus signs, from left to right). The centers are in the middle of their clusters. Essentially, as K increases, the data get partitioned into smaller and smaller convex regions, sometimes called Voronoi sets because each point is assigned to the cluster center closest to it. On the face of it, therefore, there is no way to choose which K is best or which clusters are more meaningful.

In some cases, practitioners repeat the K-means procedure several times (with different starting values, usually random) to search for better clusterings. Even so, there are data sets for which the pattern seen in the Ripley data is quite strong: The data cloud gets partitioned into convex regions with representative centers so that the optimal number of clusters is hard to determine.

In these cases, it can be worthwhile to examine the output more carefully. The table below summarizes the numerical results for $K = 2,3,4$. For each K, the results of K-means on the Ripley data with the Euclidean distance and K cluster centers are in the first two columns, the size of the clusters is in column three and the sum of squares within the clusters is in the last column.

For $K = 2$, the total sum of squares is $SST(\mathscr{C}(2)) = 119 * 17.2 + 131 * 11.8 = 3593$; for $K = 3$, the total sum of squares is $SST(\mathscr{C}(3)) = 71 * 3.9 + 124 * 10.2 + 55 * 3.1 = 1712$; and for $K = 4$, the total is $SST(\mathscr{C}(4)) = 32 * 1.8 + 60 * 2.9 + 120 * 9.7 + 38 * 1.3 = 1445$. Larger Ks will give smaller total sums of squares, at the risk of overfitting by using too many clusters. In these cases, often one looks for the knee in the curve $(K, SST(\mathscr{C}(K)))$ as a function of K. In the present case, the knee appears at $K = 4$, correctly identifying the most refined number of clusters despite $K = 4$ being correct. However, the appearance of the panel for $K = 2$ in Fig.8.9 looks most esthetic – and is correct. This illustrates a problem with Voronoi type methods: The convexity of the clusters makes all clusterings seem not unreasonable a priori. Indeed, finding too many clusters may correspond to finding structure within the clusters rather than differences between clusters.

	x	y	size	$SS(W)$
C_1	$-.53$	.51	119	17.2
C_2	.34	.50	131	11.8
C_1	$-.32$	.71	71	3.9
C_2	.36	.49	124	10.2
C_3	$-.74$	.26	55	3.1
C_1	$-.68$	.56	32	1.8
C_2	$-.23$	.71	60	2.9
C_3	.37	.49	120	9.7
C_4	$-.72$	.16	38	1.3

There are many creative ways to justify a choice of K. In some cases, a normal model is invoked and K can be chosen by an AIC or BIC type of criterion. For instance, one can find the K that minimizes

$$SST(\mathscr{C}(K)) + \lambda pK \log n,$$

in which λ is a decay factor and pK is the number of entries in the K centers in p dimensions. However, there will always be anomalies. Consider a data set of three separated clusters in the plane. If two clusters are small and close together while the third is large but very far from the other two, it can be optimal to use one center to represent the two small clusters and two centers to represent the big one!

8.5.1.3 Ripley, Model-Based

The main model-based method is fitting a mixture of components by using the EM algorithm. This has been implemented for multivariate normal mixtures in the contributed R package mclust. The command for the Ripley data is

```
yy<-synth.tr[,-3]
RipleyMixCluster <- Mclust(yy)
plot(RipleyMixCluster)
```

Because the procedure is model-based, the BIC (see (8.3.4)) can be used to choose the correct number of clusters K; see Chapter 10, Section 2.2 for more details on the BIC. The default range of K is one to nine, and as seen on the horizontal axis in Fig. 8.10, the BIC value is on the vertical axis. Even though the mixture consists only of normals, the covariance can be parametrized in many different ways to represent the shapes of the clusters; this affects the penalty term in the BIC. The legend at the lower right of the figure codes for ten of these parametrizations. For instance, EII means that all the variance matrices in the normals are of the form λI_p; i.e., they are diagonal and all diagonal elements are the same. VII is a little more general; the covariance matrices are of the form $\lambda_k I_p$ (i.e., they are diagonal and each component is allowed a different diagonal value). EEI and VEI indicate variance matrices of the form λA and $\lambda_k A$; i.e., the same general covariance matrix for all components and component dependent factors on general covariance matrices. The remaining six cases continue to increase the generality of the class of covariance matrices permitted.

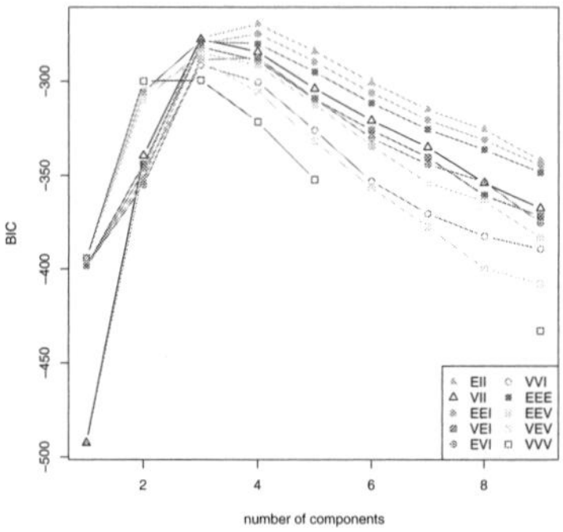

Fig. 8.10 BIC values for ten mixture of normal components models fit by the EM algorithm for the Ripley data.

It is seen in Fig. 8.10 that the most general mixture model VVV (indicated by empty squares) chooses $K = 2$ or $K = 3$, while the most restricted mixture model (indicated by solid triangles) chooses $K = 4$. Of the remaining eight parametrizations, two choose $K = 4$, five choose $K = 3$, and one is indeterminate. In general, the richer the model class, the fewer components will be needed, although the BIC penalty reduces the extent to which this is true. Unlike the earlier techniques, which reveal two or four clusters, consistent with the structure of the data generator (two mixtures of two components), the smoothing effect of the BIC tends to give three as an intermediate value, roughly corresponding to the $K = 3$ case in K-means. This is wrong for the Ripley data set, but not terribly so, and is probably quite good for many real data sets. Graphs of this can be given but are omitted for brevity.

8.5.1.4 Ripley, Spectral

The contributed R package specc in kernlab is an implementation of the Ng et al. (2002) variant on the Shi and Malik (2000) procedure. In these cases, the dissimilarity $d(\boldsymbol{x}, \boldsymbol{x}')$ is usually a kernel function. For the radial basis function kernel with $K = 2$, the command is:

```
data(synth.tr)
```

```
xx<-synth.tr[,-3]
xx<- as.matrix(xx)
spRbfRipley2 <- specc(xx, kernel = "rbfdot", centers=2)
centers(spRbfRipley2)
size(spRbfRipley2)
withinss(spRbfRipley2)
```

The command for polynomial and Laplace kernels is similar. Embedding the data into a nonlinear space by using a kernel corrects one of the drawbacks of other procedures, such as K-means , that cannot capture clusters that are not linearly separable in their input space. For the Ripley data, it is unclear a priori how much this will help because the nonseparability of the clusters by linear functions is a property of the data generator itself. Kernels generally have a parameter that must be estimated so they are correctly scaled. Built into the specc command above is the estimation of the variance parameter; this can be done in several ways. The default was used here; it assumes the dimensions of the data are all on the same scale. The two other kernels used here, polynomial and Laplace, also have parameters to be estimated, and the commands for them are similar. The results are given in Fig. 8.11. The documentation does not seem to make it clear what the default number of eigenvectors m is to give the columns of the matrix to which K-means is applied or how m is chosen.

In looking at the left column, it is seen that, as with K-means, the clustering partitions the data cloud roughly into Voronoi regions. For $K = 2$ there are a left and right clusters, triangles and circles, as in K-means with $K = 2$. For $K = 3$, there are three sectors – circles, triangles, and pluses – again as in K-means with $K = 3$. For $K = 4$, however, the right cluster is partitioned into three clusters; the circles indicate the largest, the triangles the next largest, and the multiplication signs the smallest cluster – quite different from K-means with $K = 4$. The polynomial kernel is similar to the RBF kernel for $K = 2,3$ but unlike the RBF kernel gives clusters that continue to be roughly the same as the K-means even for $K = 4$. (The Laplace kernel also gives results much the same as for the polynomial kernel, i.e., it, too, matches the K-means results.) Taken together, it seems that squaring and exponentiating the squared error distance as in the RBF kernel moves it too far away from the K-means optimality. Merely exponentiating the L^1 distance as in the Laplace kernel, or just taking a polynomial power distorts the data less, giving more representative clusters. This is reasonable because the correct clusters in the data themselves overlap. Forcing a linear separation with the RBF kernel actually decreases the adequacy of a clustering. Thus, the Laplace or polynomial kernels seem to do better than the RBF.

As with K-means, the selection of K can be difficult with spectral clustering. For comparison with the output of the K-means procedure, the $SS(W)$s for the three values of K are compiled into a table for the polynomial kernel case. The corresponding table for the Laplace kernel is essentially identical. (The RBF kernel is not shown because it has already been seen to give poor results for $K = 4$.) Consistent with Fig. 8.11, the table shows $C_2(2) = C_3(2)$, so that increasing K from 2 to 3 splits $C_2(1)$. Likewise, $C_3(1) \approx C_4(4)$, again indicating a split.

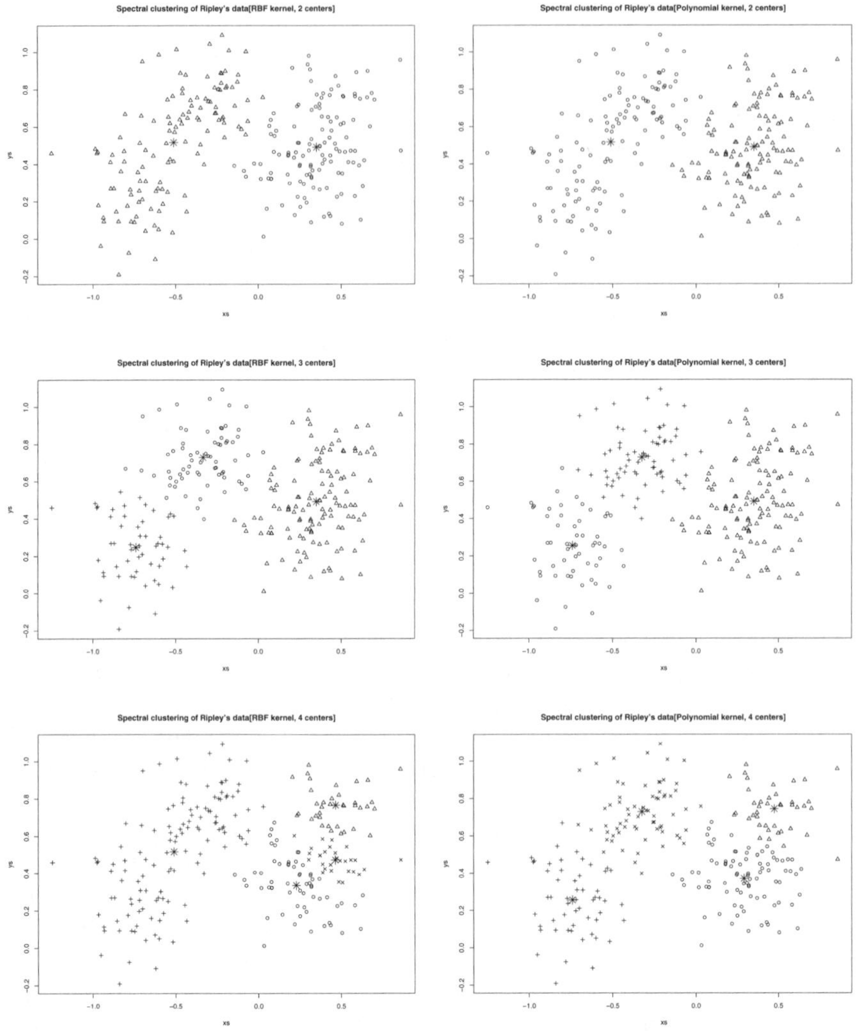

Fig. 8.11 Spectral clustering on the Ripley data. The left column uses the radial basis function and the right column uses the polynomial kernel. The top row uses $K = 2$, the middle row uses $K = 3$, and the bottom row uses $K = 4$.

It is left as an exercise to see that $K = 4$ does better than $K = 2, 3$ but that the total sum of squares does not decrease as rapidly as it did for K-means. However, spectral clustering does not optimize K-means any more than K-means tries to embed the data into a subspace of the eigenvectors of the dissimilarity matrix (although it must be noted that this spectral clustering uses K-means on the matrix of eigenvectors with a random start). Nevertheless, if further Ks were computed, a scree plot of $(K, TSS(\mathscr{C}(K)))$ could be found and the knee value of K chosen. This procedure or BIC would likely work as well for choosing K in general but correspond to different, additional, assumptions. Nevertheless, the two criteria, spectral and squared errors are really quite different, and it is difficult to decide a priori which will be better in a specific context.

Poly	x	y	size	$SS(W)$
C_1	-.51	.52	123	146
C_2	.35	.49	127	13
C_1	-.74	.25	55	58
C_2	.35	.49	127	13
C_3	-.32	.73	68	75
C_1	.29	.37	87	5.3
C_2	-.33	.73	69	78
C_3	.47	.75	40	5
C_4	-.74	.25	54	55

8.5.2 *Iris Data*

Recall that Fisher's iris data were analyzed from a classification perspective in Chapter 7. In this section, they are reanalyzed from an unsupervised perspective, so $K = 3$ can be taken as correct. Since the procedures and software have already been explained in the last subsection and the commands for the iris data are similar to those for the Ripley data, they are omitted. This section will just look at the analysis of the data. Plots are not given because the data is four dimensional.

Average linkage hierarchical, Ward's method, and K-means worked well here, DIANA and model-based methods were passable, and the other methods did poorly.

8.5.2.1 Iris, Hierarchical

In Fig. 8.12, six dendrograms are given, as in the last subsection. It is seen immediately that single linkage suffers the chaining problem for both absolute error and Euclidean distance. Complete linkage does better; however, it is easy to "overcluster". By the time the dissimilarity is decreased to a satisfying level, too many clusters have been found. If you stop at dissimilarity 4 on the left (Euclidean), the results are good, but the natural temptation is to see six clusters by using the dissimilarity level of two. Likewise, any dissimilarity level between 5 and 8 on the right (absolute error) gives good results, but the temptation is to choose a level below 5, again giving the wrong number of clusters.

It can be seen, with hindsight, that average linkage does better than single or complete. At any persuasive dissimilarity level, say around 2 on the right or 4 on the left, three clusters are found. They look fairly stable because decreasing the dissimilarity in small

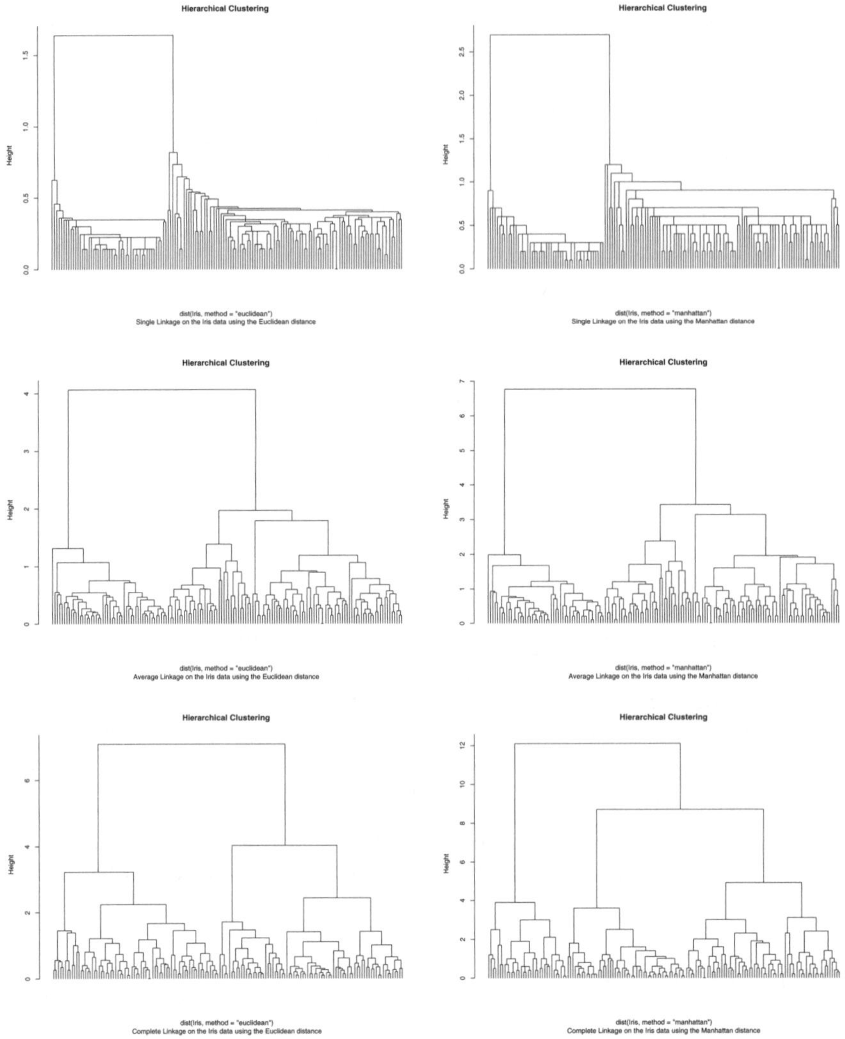

Fig. 8.12 Hierarchical agglomerative procedures on the iris data. The left column uses the Euclidean distance and the right column uses absolute error. The rows correspond to single linkage, average linkage, and complete linkage.

increments only pulls off small numbers of data points that look like outliers more than collections of points that seem like a unit.

Figure 8.13 gives the dendrograms for DIANA. The results are passable, better than for complete linkage but worse than for average linkage. Indeed, both dendrograms reveal three clusters as reasonable (dissimilarity between 3 and 5 on the left or 5 and 10 on the right) as well as tempting one to use a lower dissimilarity and overcluster to get seven clusters (at levels 2 or 3, respectively). DIANA tends to give more esthetic dendrograms which lures one to overcluster. One can argue that in such cases one should informally prune back the clustering much as trees must be pruned back sometimes by cost-complexity.

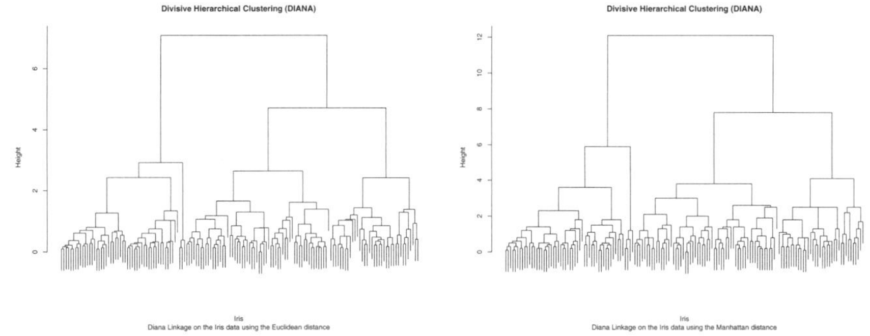

Fig. 8.13 Hierarchical division; DIANA on iris data. On the left, the Euclidean distance is used; on the right, absolute error.

8.5.2.2 Iris, Centroid-Based

Like the comparable section for Ripley data, the two methods used here were Ward and K-means.

In Fig. 8.14, the dendrograms for Ward's method are given. Recall that Ward is conceptually similar to K-means, even though it is hierarchical. It is seen that Ward's method does well: Only two or three clusters are at all reasonable. This is consistent with the data because two of the clusters overlap substantially. Whether or not one would choose 3 as the correct number of clusters depends on how much one resists the desire to overfit – i.e., insist on an unjustifiably low dissimilarity. Here, that temptation actually leads to the correct answer, unlike complete linkage or DIANA.

Like Ward, K-means clustering does well. However, since the data is four dimensional, it is difficult to generate scatterplots. One could look at projections into three dimensions, but here it will be enough to examine the sums of squares. These are easily given, as in the table below. It is seen that $C_1(2) \approx C_1(3)$, so in passing from $K = 2$ to $K = 3$, roughly $C_2(2)$ is split. Likewise, $C_2(3) \approx C_4(4)$ and $C_3(3) \approx C_1(4)$, indicating that, roughly, $C_1(3)$ was split into $C_2(4)$ and $C_3(4)$.

If you knew to choose $K = 3$, the result would likely be a reasonable clustering, given that two of the clusters are hard to distinguish. Indeed, $C_2(2)$ and $C_2(3)$ probably have many points (10 to 14) that could be put into either cluster, so cluster stability or validation might be difficult. Since it can be verified that the knee of the curve of total sum of squares (i.e., of $(K, SST(\mathscr{C}(K)))$) is at $K = 3$, K-means is giving a reasonable result, just like Ward.

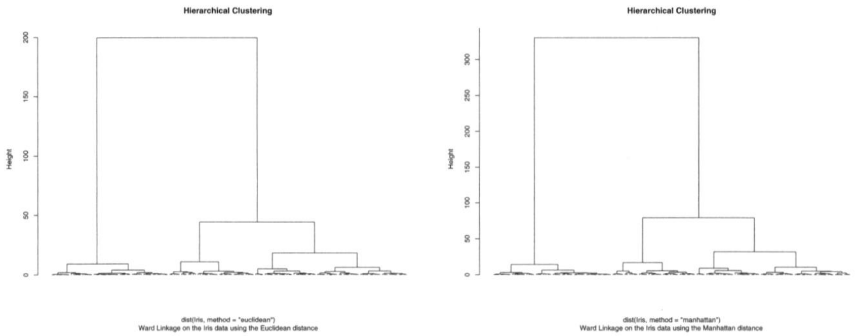

Fig. 8.14 Ward's method on iris data; Euclidean distance is on the left, absolute error on the right. There are no plots for K-means for iris because it's four-dimensional.

	x_1	x_2	x_3	x_4	size	$SS(W)$
C_1	5.0	3.4	1.6	.3	53	28.6
C_2	6.3	2.9	5.0	1.7	97	123.8
C_1	5.0	3.4	1.5	.2	50	15.1
C_2	6.8	3.1	5.7	2.1	38	23.9
C_3	5.9	2.7	4.4	1.4	62	39.8
C_1	5.9	2.7	4.4	1.4	62	40.0
C_2	5.4	3.8	1.5	0.3	17	2.6
C_3	4.8	3.2	1.4	0.2	33	5.4
C_4	6.8	3.1	5.7	2.1	38	23.9

8.5.2.3 Iris, Model-Based

Figure 8.15 shows the results for mclust on the iris data. As before, there are ten BIC curves as a function of K. This time, they segregate into two classes, of four (top group) and six (bottom group). The top group has well defined maxima and the bottom group has knees. The difference between the two groups is that all the techniques in the top group use richer model classes; i.e., they are more flexible. With more flexible models, it is more typical to see a maximum – to the left, the curve is low, indicating bias and to the right it is low, indicating excess variance. With narrower model classes, curves with a knee are more typical: The idea is that, past a certain point, the gains are small relative to the extra complexity introduced.

The top group has modes at 2 (two curves), 3 (one curve), and 4 (one curve). All the curves in the bottom group have knees at 3. With hindsight, this suggests the top group of richer classes is more unstable, possibly from overfitting the data. The lower group appears to give the right K uniformly. So, the larger group of versions of this method works rather well, except for the fact that two of the clusters must be unstable since they are not easily separated. As a group, the results are not as good as centroid methods or average linkage. But, if the narrower classes are used to avoid overfit (i.e., one requires a well defined-knee) then this is a good way to choose K.

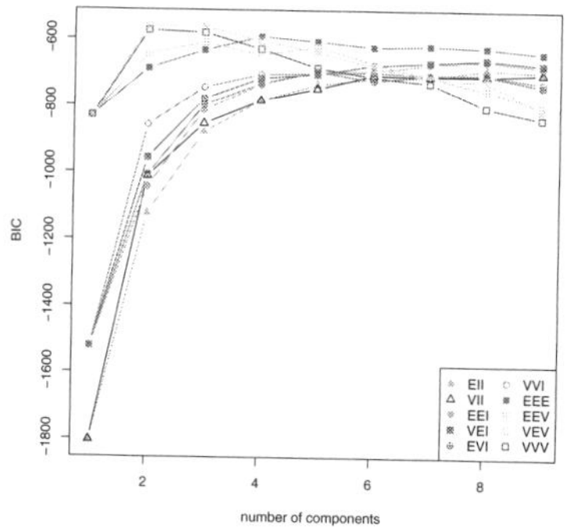

Fig. 8.15 BIC values for 10 mixture of normal components models fit by the EM algorithm for the Iris data.

8.5.2.4 Iris, Spectral

With spectral clustering, K must be chosen, and since $p = 4$, it is hard to generate scatter plots without reducing the dimension. Results were found for $K = 2, 3, 4$ using the RBF, Laplace, and polynomial kernels.

The RBF kernel, as before, did poorly. The table below for $K = 2$ indicates the problem. The two classes that are hard to separate are grouped together as $C_2(2)$. If K increases, this clustering changes very little: For $K = 4$, the clusters have 4, 50, 2, and 94 members. That is, increasing K does not split the clusters but just forces the method to slice off a few points from the larger cluster that do not fit as well with fewer clusters. In effect, these points are treated as outliers more than the start of new clusters.

RBF	x_1	x_2	x_3	x_4	size	$SS(W)$
C_1	5.0	3.4	1.5	0.2	50	1313.2
C_2	6.2	2.9	4.9	1.7	100	2656.1

For $K = 2, 3$, both Laplace and polynomial kernels give the same results as the RBF kernel: They have approximately the same table as the RBF for $K = 2$ (above), and for $K = 3$ they do not separate clusters 2 and 3.

When $K = 4$, the Laplace and polynomial kernels do better than the RBF kernel. (Recall that Laplace and polynomial kernels did better than RBF for the Ripley data.) They were very similar; the table below is for the Laplace kernel. It is seen that merging $C_3(4)$ and $C_4(4)$ would give a better result; however, this is not any of the clusters when $K = 3$.

Overall, spectral methods in this example give reasonable results when $K = 2,4$ but when $K = 3$, the correct value, the results are poor.

Lapl	x_1	x_2	x_3	x_4	size	$SS(W)$
C_1	5.0	3.4	1.5	.2	50	1313.2
C_2	6.6	2.9	5.4	2.0	55	1489.1
C_3	5.9	2.8	4.4	1.4	6	141.5
C_4	5.9	2.8	4.2	1.3	39	900.0

8.6 Cluster Validation

Having seen a plethora of clustering techniques, there are two natural questions, aside from implementation. The first is which clustering procedure to choose; the second is how to evaluate a clustering a procedure produces. The first question does not readily admit a serviceable answer: If someone knew enough about the phenomenon to choose unambiguously which clustering method to use, then one would de facto know what the clusters were likely to look like. Many methods work well for relatively compact, relatively separated, and elliptical clusters in relatively small dimensions. Outside that convenient setting, the situation is murkier and mostly unresolved. In practice, one chooses whichever clustering technique is not contraindicated by some feature of the data or phenomenon and tunes it to get as reasonable an answer as it will give by choosing it inputs – K, models, or distances.

Clearly, this is not entirely satisfactory, so the second question of evaluating clusterings becomes central. There are two senses in which clusters can be evaluated: The first is in terms of the subject matter discipline for which the clustering is intended; the second is the statistical analysis of stability of the clustering. Of these, the first is usually more important – when it can be done, which is not often. After all, if a given clustering accurately encapsulates a physically reasonable model structure, it may not matter much what its stability properties are. The natural questions are: Is the K a reasonable number for the clusters physically? Does each cluster represent a definable, recognizable subgroup of the population? Does the distance from cluster to cluster match some aspect of the similarity between the units the clusters summarize? Are the cluster boundaries close to boundaries between subgroups of the whole population? In any specific application, there will often be even more criteria for how a clustering should match the physical problem. (In text mining, infantry should be clustered with soldier, not infant. In genomics, the genes in a cluster should serve related purposes.)

Regardless of problem-specific criteria, a clustering can also be evaluated statistically. These techniques generally come down to measuring how different one clustering is from another. This is a sort of robustness or sensitivity analysis; it is the clustering analog of model uncertainty and deserves to be called clustering uncertainty.

At the heart of this is the choice of dissimilarity. As noted at the end of Chapter 7, in high dimensions assigning distances between points is one of the main features determining inferences. The question comes down to what makes two points far apart:

many small differences in the entries of a vector all contributing to the overall distance as with Euclidean distance or a few really large differences in entries of a vector, regardless of the other coordinates. Where one is situated on this spectrum substantially determines the metric topology of the solution. Another way to phrase this is to ask, for each similarity (or dissimilarity) what the shape of its unit ball is. This is well known for the L^p distances but is unknown in general.

In addition to the usual notion of distance between points, there are extra criteria one might wish to impose so that lifting a dissimilarity on individual points to a dissimilarity on clusters is more reasonable. Two conditions often imposed are that $d(C \cup \{\mathbf{y}\}, C') \geq d(C, C')$ if $\mathbf{y}$ is not in C, C' and that $d(C \cup \{\mathbf{y}\}, C' \cup \{\mathbf{y}\}) \leq d(C, C')$. These criteria emphasize that dissimilarity should decrease if one cluster is changed and should decrease if both are changed the same way. These conditions are a sort of replacement for the triangle inequality, which typically fails for dissimilarities. After all, there is no need for clusters C and C' to be close just because there is a third cluster C'' that is metrically close to each of them.

Apart from the effect of choice of dissimilarity, it is important to examine the stability of clustering procedures because K is unknown. Even when K is known, in high dimensions it is exceedingly easy to obtain spurious clusterings. Indeed, many procedures will generate clusters even if they do not exist. For instance, for any K, K-means will give a clustering of K clusters. Any hierarchical method will give a complete nested sequence of subsets. Thus, guarding against spurious clusterings is essential. This problem is accentuated in high dimensions, where it is easy to find clusterings that are persuasive – even stable – but spurious in that under further data collection they disappear. This probably arises because data are sparse in high dimensions.

Regardless of the manifold pitfalls, there are three basic techniques for evaluating cluster stability, called external, internal, and relative.

In external cluster validity assessments a clustering is given as a reference and the data-driven clustering is compared with it using some d. For instance, if $\mathscr{C} = \{C_1, ..., C_K\}$ is the clustering obtained from data and an alternative is $\mathscr{P} = \{P_1, ..., P_{K'}\}$, then one can measure the distance between $\mathscr{C}$ and $\mathscr{P}$ as follows. Recall that there are $C(n,2)$ distinct pairs of points that can be chosen from the $\boldsymbol{x}_i$s. For each pair $(\boldsymbol{x}_i, \boldsymbol{x}_j)$, there are four cases:

- Both $\boldsymbol{x}_i$ and $\boldsymbol{x}_j$ are in the same cluster C_k for some k in $\mathscr{C}$ and in the same cluster $P_{k'}$ for some k' in $\mathscr{P}$. Call the number of pairs a.
- Both $\boldsymbol{x}_i$ and $\boldsymbol{x}_j$ are in the same cluster C_k for some k in $\mathscr{C}$ but are in different clusters $P_{k'}$ and $P_{k''}$ for some k', k'' in $\mathscr{P}$. Call the number of pairs b.
- The $\boldsymbol{x}_i$ and $\boldsymbol{x}_j$ are in different clusters C_k and $C_{k'}$ for some k, k' in $\mathscr{C}$ but are in the same cluster $P_{k''}$ for some k'' in $\mathscr{P}$. Call the number of pairs c.
- The $\boldsymbol{x}_i$ and $\boldsymbol{x}_j$ are in different clusters C_k and $C_{k'}$ for some k, k' in $\mathscr{C}$ and are in different clusters $P_{k''}$ and $P_{k'''}$ for some k'', k''' in $\mathscr{P}$. Call the number of pairs d.

Clearly, $a+b+c+d = C(n,2)$. Commonly occurring indices for gauging clustering similarity include the Rand index, $R = (a+d)/C(n,2)$, and the Jaccard index, $J = a/(a+b+c)$, but there are many others.

Internal criteria measures are an effort to test how well the clustering reflects the internal structure of the data set. Let $d(C_j, C_k) = d_{j,k}$ be the dissimilarity between two clusters and s_k be a measure of dispersion for cluster C_k. Then a similarity measure $R_{j,k}$ between C_j and C_k should satisfy two conditions (i) $R_{j,k} \geq 0$, $R_{j,k} = R_{k,j}$, $s_j = s_k = 0$ implies $R_{j,k} = 0$, and (ii)

$$\left[s_j > s_k \text{ and } d_{i,j} = d_{i,k}\right] \text{ or } \left[s_j = s_k \text{ and } d_{i,j} < d_{i,k}\right] \Rightarrow R_{i,j} > R_{i,k}.$$

One choice for the similarity between clusters is $R_{j,k} = (s_j + s_k)/d_{j,k}$. Often, these are combined into a one-dimensional index by setting $R_j = \max_{k|j \neq j} R_{j,k}$ and using the Davies-Bouldin Index, $DB = (1/C(n,2)) \sum_j R_j$.

The silhouette index is another internal criterion defined for a point $\boldsymbol{x}_i$ as

$$s(i) = \frac{b(i) - a(i)}{\max(a(i), b(i))},$$

in which $a(i) = (1/\#(C)) \sum_{\boldsymbol{x} \in C} d(\boldsymbol{x}_i, \boldsymbol{x})$ is the average dissimilarity between $\boldsymbol{x}_i$ and the data points $\boldsymbol{x}$ in the cluster C that contains $\boldsymbol{x}_i$ and $b(i) = \min_{C_k \neq C} d(\boldsymbol{x}_i, C_k)$. Essentially, this is a scaled difference between the dissimilarities. Often this index works well for compact separated clusters.

Relative criteria are an effort to choose among a collection of clusterings one that is best in the sense that it satisfies an objective function. For instance, a clustering procedure can be run several times under a collection of different conditions in an effort to find the best clustering for the data set. The goal is that the clusterings should be relatively similar to each other, perhaps in an internal or external sense.

As a generality, there are many ways to check the internal consistency of a clustering. Two heuristics are: (i) Given a clustering, one can sample from it and treat the cluster indices 1 through K as classes so the data are supervised. The accuracy of a classifier is then a measure of the cluster stability. (ii) Given a clustering, choose a collection of, say m, clusters from it. Rerun the clustering procedure to make sure that m remains the best number of clusters for the data. A special case of this is choosing a single cluster and making sure the data in it are best described by one cluster.

Recently, a Bayesian approach to cluster validation has been given in Koepke (2008a). The formulation rests on a notion of soft membership because the amount of a given data point apportioned to a cluster varies over $(0,1)$. Being Bayesian, one conditions on the data rather than perturbing them and the prior controls the distribution of a factor determining how hard it is to shift a point from one cluster to another. There is evidence that this technique gives better performance for evaluating the representativity of a base clustering than other cluster validation techniques. Although the technique as presented in Koepke (2008a) is internal, it can be readily reformulated to be external or relative.

Finally, it must be remembered that the point of clustering is to find the modes of a complex distribution, but that this introduces a dependence structure probabilistically. Thus, one can search for clusters or evaluate their reasonableness by searching for events that commonly occur together. This is the general goal of using association rules, which is a way to search for events that have high joint or conditional probability. Interest is not on the form of the rule per se, since the rules themselves are instances of traditional nonparametric statistical quantities, but rather in how to use them to search for interesting events. Since the main concepts are familiar, an example may help explain the terminology.

Consider a collection of n patients, each with a health history. The health history is a long vector, of length say p, having one entry for each conceivable question. It is well known that the answers to some health questions will be probabilistically dependent on the answers to others in the population. A statistician searching for associations between two questions might ask if a patient has taken drug A or drug B, or both, in the past. A health economist might want to know whether A and B are consistently prescribed together.

Interpret the statement "If A, then B" to mean if a patient has taken drug A then he or she has also taken drug B, and call such an implication a "rule". The first "if" part is called the antecedent; the second "then" part is called the consequent. The two, A and B, can often be interchanged as well.

The "support" for this rule is the number of patients who have taken both A and B. This is the frequency of the intersection of the two events $A = \{\text{have taken drug}A\}$ and $B = \{\text{have taken drug}B\}$. This is $nP(A \cap B)$. The "confidence" is the ratio of the number of patients who have taken both A and B to the number of patients who have taken A. This is the conditional probability $P(B|A)$. The "expected confidence" is the proportion of patients who have taken drug B. This is the marginal probability of the consequent, $P(B)$. The "lift" is the ratio of confidence to expected confidence. That is,

$$Lift = \frac{P(B|A)}{P(B)} = \frac{P(A \cap B)}{P(A)P(B)},$$

a familiar measure of association. If A and B are independent, $Lift$ is one. Otherwise, $Lift$ can be any positive number. Quantities like $|Lift - 1|$ occur in the ϕ-mixing condition in probability theory.

Practical data miners have numerous sophisticated ways to search over events to identify those that commonly occur together or not under a variety of dependency criteria. To date, practical data miners do not seem to have used other measures of association such as the Pearson correlation, the Spearman rank correlation, or Kendall's τ, whose theory under conditioning, is also well established; see Daniel (1990). However, the combination of clustering techniques and cluster validation promises great improvements in the near future.

8.7 Notes

8.7.1 Derivatives of Functions of a Matrix:

Consider a real-valued function F with a matrix argument $\boldsymbol{M}$. The derivative of F with respect to $\boldsymbol{M}$ is taken coordinatewise,

$$\frac{\partial F}{\partial \boldsymbol{M}} = \left[\frac{\partial F(\boldsymbol{M})}{\partial m_{i,j}}\right],$$

where $\boldsymbol{M} = (m_{i,j})_{i,j=1}^{p}$. Thus the derivative of the matrix $\boldsymbol{M}$ is the matrix with all p^2 entries equal to 1. A vector is a $p \times 1$ matrix, so the definition applies to real-valued functions of vector arguments as well. For instance, the derivative of a squared norm is

$$\frac{\partial \boldsymbol{x}'\boldsymbol{M}\boldsymbol{x}}{\partial \boldsymbol{x}} = (\boldsymbol{M} + \boldsymbol{M}')\boldsymbol{x},$$

which reduces to the unidimensional case when $p = 1$.

Other useful facts can be readily derived. Let $\text{Cof}(\boldsymbol{M})_{i,j}$ be the (i,j)th cofactor (signed minor) of $\boldsymbol{M}$. The derivative of a determinant is

$$\frac{\partial |\boldsymbol{M}|}{\partial m_{i,j}} = \begin{cases} \text{Cof}(\boldsymbol{M})_{i,j} & \text{if } i = j, \\ 2\text{Cof}(\boldsymbol{M})_{i,j} & \text{if } i \neq j, \end{cases}$$

and the derivative of the log of a determinant is

$$\frac{\partial \log |\boldsymbol{M}|}{\partial m_{i,j}} = \begin{cases} \frac{\text{Cof}(\boldsymbol{M})_{i,j}}{|\boldsymbol{M}|} & \text{if } i = j \\ \frac{2\text{Cof}(\boldsymbol{M})_{i,j}}{|\boldsymbol{M}|} & \text{if } i \neq j \end{cases}$$
$$= 2\boldsymbol{M}^{-1} - \text{diag}(\boldsymbol{M}^{-1}).$$

(This uses the definition of the inverse matrix as the transpose of the matrix of cofactors over the determinant.) Finally, the trace, of a matrix $\boldsymbol{M}$ satisfies

$$\frac{\partial \text{trace}(\boldsymbol{T}\boldsymbol{M})}{\partial \boldsymbol{T}} = \boldsymbol{M} + \boldsymbol{M}' - \text{diag}(M).$$

8.7.2 Kruskal's Algorithm: Proof

Let T_n be the output of Kruskal's algorithm. It is enough to show that T_n is a single tree that is the MST for the points. It is easy to see that the output cannot have any cycles (or else the last edge added would not have been between two trees) and so is a spanning tree. The minimality follows by an argument by contradiction: If T_n is not minimal, let T' be an MST. Let e be the first edge in $\mathscr{S}$ that is in T_n but not T'. Then $T' \cup e$ has a cycle. The cycle must contain another edge $e' \in \mathscr{S}$, which is only considered by the procedure after e. Now, $T'' = T' \cup e \setminus e'$ is a spanning tree with weight less than or equal to T'. If equality holds, repeat this procedure for the next edge in T_n not in T'. At the end, either another minimal spanning tree has been constructed or there is a contradiction to the minimality of T'.

8.7.3 Prim's Algorithm: Proof

Now let T_n be the output of Prim's algorithm. This T_n is a minimal spanning tree as well: Clearly T_n is connected and is a tree because the edge and vertex added are connected as well. To see the minimality, let T' be an MST. If $T_n \neq T'$, let e be the first edge added to make T_n that is not in T', and let V' be the vertices added up to the step where e is added. Then, one vertex of e is in V' and the other is in V'^C and there must be a path in T' joining them. On this path there must be an edge e' joining a vertex in V' to one that is outside V'. At the step where e was added in the algoritm, e' could have been added if its weight had been less than that of e. However, e was added. So, the weight of e must have been bounded by the weight of e'. Now, if $T'' = Y' \cup e \setminus e'$, T'' is connected, has the same number of edges as T', and has total weight bounded by the weight of T'. So, T'' is also a minimal spanning tree containing e and all the edges added in Prim's algorithm before V'. Repeating this gives a spanning tree that is identical to T_n with weight less than or equal to the weight of T'. Thus, T_n is an MST.

8.8 Exercises

Exercise 8.1 (Cosine similarity). Define the cosine similarity between two p-dimensional vectors $\boldsymbol{a}$ and $\boldsymbol{b}$ as

$$c(\boldsymbol{a}, \boldsymbol{b}) = \frac{\boldsymbol{a}^T \boldsymbol{b}}{\|\boldsymbol{a}\| \|\boldsymbol{b}\|}.$$

If the entries in $\boldsymbol{a}$ and $\boldsymbol{b}$ only take values 1 and -1, give an interpretation for c. Find an expression for $\|\boldsymbol{a} - \boldsymbol{b}\|_2^2$ in terms of $c(\boldsymbol{a}, \boldsymbol{b})$.

Exercise 8.2 (Sequential K-means). Because of its tractability in high-dimensional problems, K-means has received a great deal of study. Here is a sequential modification of the K-means procedure; it updates the means one data point at a time rather than all at once.

- ☐ Start with initial guesses for the means, $m_1(0), ..., m_K(0)$, and let $c_k = \#C_k$ for $k = 1, ..., K$, initially set to zero.
- ☐ For $i = 0, ..., n-1$ get the next data point, $\boldsymbol{x}_{i+1}$.

☐ If $\boldsymbol{x}_{j+1}$ is closest to m_k, increase c_k to c_k+1 and update $m_k(i)$ to $m(i+1) = m_k(i) + (1/c_k)(\boldsymbol{x}_{i+1} - m_k(i))$.

☐ Repeat until the data points are all used.

This procedure might be appropriate if data is gathered over time and clustered as they are received.

Verify that the final cluster centers $m_1(n),\ldots, m_K(n)$ give the batch K-means clustering, i.e., each $m_k(n)$s is the average of the data points $\boldsymbol{x}_i$ that were closest to it.

Exercise 8.3 (Computations with a dissimilarity matrix). Consider seven points in a real space of dimension p, and suppose a (symmetric) dissimilarity d has been used to generate the following dissimilarity matrix, in which the (i, j) entry is $d(\boldsymbol{x},\boldsymbol{x}_j)$:

	P1	P2	P3	P4	P5	P6	P7
P1	0	.1	.4	.55	.65	.7	1.5
P2		0	.6	.5	.95	.8	2
P3			0	.45	.85	.7	1.75
P4				0	.75	.1	.2
P5					0	.3	.1
P6						0	.2
P7							0

Using your favorite package, do the following:

1. Generate the dendrograms corresponding to several different linkage criteria such as single, complete, and average linkage.
2. Re-cluster the data using a divisive procedure.

Exercise 8.4 (Clustering and density). Suppose you have a data set consisting of $n = 3r$ points in the plane. One-third of the points are evenly distributed on the unit circle centered at the origin, one-third are evenly distributed on a circle of radius 10 centered at the point $(11, 11)$ and one-third of the points are evenly distributed on a circle of radius 20 centered at $(22, 22)$.

1. If you use K-means to cluster this data set, how would you expect the cluster means to be distributed? The same number in all three sets? More in the densest set? More in the least dense set? Explain what happens as n increases.
2. What would the K-means procedure give if the data were distributed even over the combined area of all three sets so that the smallest set had the fewest points?
3. Suppose the circles were replaced with non-intersecting ellipses that were relatively long and thin. What kind of answer would you expect from K means in if the data were as in item 1 or item 2?

Exercise 8.5 (Spurious clusterings). Generate a sample of size $n = 10$ with $p = 100$; use a $N(0,4)$ or a $Uniform[-4,4]$.

1. Using any clustering method you like, cluster the vectors by looking only at the ten first coordinate values. How many clusters are found in the first coordinate? In the second, third and so forth? What is the average number of clusters per coordinate?
2. Now, take the coordinates two at a time, or three at a time and repeat the procedure. Does the number of clusters decrease, increase, or remain constant with the number of coordinates used?
3. Now find a set of data with a smallish n and a largish p. Can you correct for the number of spurious clusters you would find in it by comparing the results of clustering it to clustering a data set of the same size and dimension but having data generated from studentizing in the coordinates? (That is, generate n data points for each entry using the $N(\bar{x}_j, s_j^2)$ where $\bar{x}_j$ and s_j are the mean and standard deviation from the actual data values in the jth coordinates, $j = 1, ..., p$.)

Exercise 8.6 (Graph Laplacian and cuts). Consider a graph with cu vertices made by fully connecting c cliques each of size of size u but leaving the cliques disconnected. Suppose also that the edgeweights within each clique are 1.

1. What is the weight matrix $\boldsymbol{W}$ and the (unnormalized) graph Laplacian?
2. Find the smallest c eigenvalues of the Laplacian and their eigenvectors. What does the result mean in terms of clustering?
3. Verify that the multiplicity of 0 for the graph Laplacian equals the number of connected components. What is the eigenvector correspond to eigenvalue 0?
4. Now, connect the cliques with $c-1$ edges to get a graph with a beads-on-a-string appearance and suppose the edgeweights between any two graphs are the same value, w. What is the graph Laplacian now?
5. For$c = 2$, find the smallest two eigenvalues and their eigenvectors as a function of w. What happens when $w = 1$? What does this mean in terms of clustering?
6. Verify that the second smallest eigenvector provides a good indication for the min-cut for a connected graph.

Exercise 8.7 (Best-merge persistence). Define a condition called best-merge persistent as follows. Suppose that at some stage of an agglomerative hierarchical clustering the clusters are $\{C_1, \ldots, C_K\}$ and consider C_j for some $1 \leq j \leq K$. Suppose that the best cluster to merge with C_j is C_k, $k \neq j$ and suppose that in fact C_j is merged with C_ℓ to form $C_{j,\ell} = C_j \cup C_\ell$. Then, a clustering is best-merge persistent if, for all j, k, and ℓ, the best cluster to merge with $C_{j,\ell}$ remains C_k since it was the best merge for C_j. In other words, the best cluster to merge with the merged cluster is one of the best-merge clusters of the components of the merged cluster.

Show that single-linkage clustering is best-merge persistent, but that centroid, complete, and average linkage clusterings are not.

Exercise 8.8 (Nearest-neighbors and single linkage). Single-linkage clustering roughly corresponds to a nearest-neighbors criterion, and this can be formalized by the use of a minimal spanning tree for the data. Construct a fully connected graph from n data points and weight each edge by the distance between its vertices. The minimal spanning tree is a connected tree, with $n-1$ edges, that goes through all the points and has a minimal sum of weights. Let the edgeweights of the minimal spanning tree be $\Delta_1 > \ldots \Delta_{n-1}$. Show that the merges from a single-linkage clustering procedure generate a minimal spanning tree and that any minimal spanning tree can be produced by a single-linkage clustering procedure applied to the data.

Exercise 8.9 (Bisecting K-means versus K-means). Another way to describe bisecting K-means is as follows. Start with the whole data set as a cluster and find the best way to bisect it using K-means. Find the best bisection; i.e., the one with the smallest SSE. Then, redo this procedure on each of the two clusters, making the SSE as small as possible, until K clusters are found.

Give an example of a data set which is most naturally partitioned into three clusters for which K-means would likely find the correct clusters, but bisecting K-means would not. Can you give an example of the reverse; i.e., bisecting K-means would find the correct clusters but K-means would not?

Exercise 8.10 (EM for two normals). Reread Sections 8.3.1.1 and 8.3.1.2 where the EM algorithm for two normal components and the general algorithm are stated. Derive the form of the EM algorithm for the case $p = 1$, $K = 2$, and the two components have the same variance; i.e., $\sigma_1 = \sigma_2$. Now suppose the available data are is p dimensional. Rederive the EM algorithm for $K = 2$ when the variance matrices are diagonal, but not necessarily equal. (This is the case VII in Fraley and Raftery (2002).)

Hint: Some steps can be found at the Wikipedia entry under "Expectation-maximization algorithm".

Exercise 8.11. The EM algorithm is a way to make up for missing data in a probabilistic way in general. Indeed, the missing data do not have to be different from the available data. This means that the EM algorithm can be used in a predictive sense as well, although really it's more like imputation since the new data point is assumed to be from the same population.

To see this, imagine that a ten patients undergo the same operation and each either lives or dies afterward. You are told that seven of the ten lived and you model the outcomes as independent *Bernoulli*(π). An 11th patient is undergoing the same operation and during the operation the patients family asks you what the chance for survival is.

Your task is to describe an EM algorithm to use the data, including the missing value, to estimate p. Write out the complete data likelihood, the conditional likelihood, the function $G(\theta, \theta_0)$ in (8.3.17), the E step and the M step. Mimic the derivation in Section 8.3.1.3 to derive an explicit form for the estimator of π.

Implement your procedure. Does including the 11th data point in the EM algorithm give an answer different from estimating π by its MLE using only the ten data points? Explain why or why not.

Exercise 8.12. Consider 100 points on a 10×10 evenly distributed on the unit square to form a regular grid, and 100 points randomly chosen from the uniform distribution on the unit square.

1. If you used a K-means procedure to cluster both data sets, would you always get $K = 1$? Explain.
2. Which of the two data sets is likely to have a smaller SSE?
3. Can you formalize a sense in which, on average, the clusterings from of random points will be the same as for the grid? (You may need to let n increase.)
4. What does this tell you about the importance of ruling out spurious clusters, especially in high dimensions?

Exercise 8.13. The zero-inflated Poisson is a distribution that modifies the usual Poisson(λ) to permit an extra component representing zeros, above what would be expected from the Poisson alone, for any λ. Its probability mass function is the mixture of a Poisson and a point mass at zero,

$$p(X_i = x_i|\lambda, \pi) = (1-\pi)\frac{e^{-\lambda}\lambda^{x_i}}{x_i!} + \pi I_{x_i=0},$$

where each x_i is in $0, 1, 2, \ldots$. Derive an EM algorithm for estimating π and λ. Specify the complete data likelihood by identifying Z_i and give the conditional likelihood, the E step, and the M step.

(Zero inflation also occurs for the geometric and binomial. Can you derive an EM algorithm for these cases?)

Exercise 8.14. Consider the Rand, Jaccard, and silhouette indices from Section 8.6.

1. For a toy data set consisting of points at the integers $1, 2, \ldots, K$, find the value of the silhouette index for each point.
2. Do the same, but now assume that $K = r^2$ and that the data points form a regular $r \times r$ grid.
3. Now consider 100 data points that form a regular 10×10 grid (use (i, j) for $i, j = 1, 10$) and choose 100 data points independently generated from the uniform distribution on $[.5, 10.5] \times [.5, 10.5]$. Cluster both sets of data using the same number of clusters. What statements (if any) can you make using the Rand or Jaccard indices?

Exercise 8.15 (Spectral clustering in a simple setting). Consider a data set evenly spread over two disjoint rectangles in the plane, say $[0, 1] \times [2, 4]$ and $[4, 5] \times [1, 2]$. Clustering for this data would usually be trivial; here it is simple enough that it will help us understand some aspects of spectral clustering.

1. Spectral clustering requires a dissimilarity matrix, $\boldsymbol{W}$. So, consider a zero-one dissimilarity based on Euclidean distance

$$w_{i,j} = w(j, i) = \begin{cases} 1 & \text{if } \|\boldsymbol{x}_i - \boldsymbol{x}_j\|_2 \le t \\ 0 & \text{else} \end{cases}. \tag{8.8.1}$$

What values of the threshold t are most helpful?

2. Let D be the dissimilarity matrix formed using the weights $w_{i,j}$. Let n be anything reasonable and assume the data are $Unif(R)$ where R is the union of the two rectangles. Find the first K eigenvalues of D, where K is the number of clusters desired.
3. For the problem with two uniform clusters in the plane set K=2. Is there a value of t for which the eigenvectors corresponding to the first two eigenvalues can be found readily? Find them – or explain why it's hard.
4. Let U be the matrix with K columns given by the first K normalized eigenvectors of D. Cluster the rows of this matrix by K-means (or any other clustering procedure). For $K = 2$, what are the centroids of the two clusters?
5. Given the two centroids from item 3, try a distance based approach. Say that a point $\boldsymbol{x}_i$ is in cluster j if and only if the ith row of U is in cluster j. What are the final two clusters given by this procedure?
6. What clusters you would expect (intuitively) from this kind of data?
7. Does the result of item 5 match item 6? Explain. Does the result generalize to data drawn from a uniform with three disjoint regions?
8. Redo the problem but for data in a pair of concentric rings. To be definite, suppose the two rings are centered at 2.5 and that the first ring has inner and outer diameter given by 2 and 3, respectively, and that the second has uses 4 and 5. In this case, use a radially symmetric dissimilarity such as

$$w_{i,j} = e^{\|\boldsymbol{x}_i - \boldsymbol{x}_j\|_2/2t^2}.$$

Exercise 8.16 (Graph clustering). The CHAMELEON algorithm by Karypis, Han, and Kumar is a graph-based clustering template with three steps. (i) Given a data set, a sparse graph is first constructed for it. Usually, this is the K-nearest-neighbors graph formed, say $\mathscr{G}$, formed by including a directed edge from each vertex to its nearest neighbors. (ii) A graph partitioning procedure is then applied to $\mathscr{G}$ to get an initial clustering into a large number of small subgraphs. The graph partitioning in CHAMELEON uses a min-cut bisection approach to split $\mathscr{G}$ into disjoint, roughly equal subgraphs $\mathscr{G}_1,...,\mathscr{G}_K$ with minimum edgeweights between the subgraphs. (Often this can only be done approximately; e.g., by trying to ensure $\max_k \#(\mathscr{G}_k) \leq (1+\varepsilon)\#(\mathscr{G})/K$.) (iii) An agglomerative hierarchical clustering procedure is then used to merge the small subgraphs into meaningful clusters by connectivity and closeness measures; i.e., degree of vertices and edgeweights. CHAMELEON can be downloaded from `http://glaros.dtc.umn.edu/gkhome/`; note also the programs METIS, gCLUTO, and CLUTO.

By using the software on test cases, explain what the effect of varying K would be for various sets of artificial test data; e.g., normal data in the plane with tight clusters and bulls-eye data (uniform on concentric rings around zero), also in the plane.

Exercise 8.17. Recall that $\mathscr{S} = \{1,...,n\}$. Write $(S_1(j),...,S_{k_j}(j))$ to mean the partition of the first j elements of $\mathscr{S}$ as determined by the cluster memberships of $s_1,...,s_j$. Using (8.4.5), show that the conditional probabilities as in (8.4.3) for product partition models are of the form

$$p(s_{j+1} = i | s_j, ..., s_1) \propto \begin{cases} \frac{\chi(S_i(j) \cup \{i\})}{\chi(S_i(j))} & \text{if } 1 \leq i \leq k_j, \\ \chi(\{i\}) & \text{if } i = k_j + 1, \end{cases}$$

see Quintana (2006) for more details.

Exercise 8.18. The GGobi software is an open-source visualization program intended for high-dimensional data. The webpage `http://www.ggobi.org/book/` has a collection of interesting, relatively well-understood data sets to which clustering techniques can be applied and evaluated in part by sophisticated visualization techniques. Any of these can be studied using the methods of this chapter. So, apply the various techniques from Section 8.5 (or other sections) to see how they perform in these cases.

Three data sets of particular note, in increasing order of difficulty, are Flea Beetles, Olive Oils, and PRIM7.

1. Flea Beetles is an easy six-dimensional data set. It has three elliptical clusters that are well separated. Ignoring the class variables, centroid-based clustering procedures such as Ward and K-means can be applied to find the answer. Other methods, such as average or complete linkage, do poorly.
2. Olive Oils is a medium-difficulty eight dimensional data set. Stripping away the class variables for region and area, it has nine actual clusters. There is some separation between the clusters. Some clusters get detected by many clustering methods and some don't; most clustering methods pick up four or five of the classes accurately but miss the others. The clusters are mostly spherical (with different variances), but one is crescent shaped.
3. PRIM7 is a hard seven dimensional data set from physics. It has no known classes, but there are seven clusters and some outliers. What makes this so hard is that the clusters are low-dimensional structures in the seven-dimensional space. One cluster is roughly an isosceles triangle, a two-dimensional structure situated in seven dimensions. Two one dimensional structures stick out of each vertex of the triangle; whence seven clusters. These six unidimensional clusters stick out at odd angles so as to require all seven dimensions to contain the data. Essentially no clustering method finds this structure, so it's a good test case. Low-dimensional structures are typically found more by visualization methods than by clustering, although variable selection techniques often help, too.

Chapter 9
Learning in High Dimensions

A typical data set can be represented as a collection of n vectors $\boldsymbol{x} = (x_1, ..., x_p)$ each of length p. They are usually modeled as IID outcomes of a single random variable $\boldsymbol{X} = (X_1, ..., X_p)$. Classical data sets had small values of p and small to medium values of n, with $p < n$. Currently emerging data sets are much more complicated and diverse: The sample size may be so large that a mean cannot be calculated in real time. The dimension p may be so large that no realistic sample size will ever be obtained. The $\boldsymbol{X}$ may summarize a waveform, a graph with many edges and vertices, an image, or a document. Often data sets are multitype, meaning they combine qualitatively different classes of data. In all these cases, and many others, the complexity of the data – to say nothing of the model – is so great that inference becomes effectively impossible. This means that in one guise or another, dimension reduction – literally reducing the number of random variables under consideration – becomes essential.

There are two senses in which dimension can be reduced. First, conditional on having chosen functions of the explanatory variables, one can choose those that are most important. This includes long-established model selection techniques such as AIC, BIC, cross-validation, Mallows C_p, forward and backward elimination, and so forth. As well, there are more recent techniques such as the LASSO and regularization more generally. These cases, covered in Chapter 10, do selection over a set of variables believed to be good for modeling purposes, often just the outcomes themselves.

However, one needn't be restricted to using the explanatory variables as is. One can choose functions of the data to use in a model instead and search the class of models they define. This second sort of dimension reduction might be called feature extraction since a collection of features (i.e., functions of the explanatory variables, say $\mathscr{F}$) must be defined before a model can be proposed. Thus, beyond variable selection, feature extraction includes identifying which functions of the data are most important.

This is done in techniques such as SVMs and RVMs where the model (such as it is) is found by using a kernel evaluated at a well-selected subset of the data points. In effect, a kernel function $K(\boldsymbol{x}, \boldsymbol{x}')$ is a continuously parametrized collection of functions $K_{\boldsymbol{x}'}(\boldsymbol{x})$ of $\boldsymbol{x}$ as, say, $\boldsymbol{x}'$ varies. For each fixed $\boldsymbol{x}'$, $K_{\boldsymbol{x}'}(\boldsymbol{x})$ is like an element of a basis that might be used to express a solution. Thus, SVMs and RVMs perform dimension reduction in the sense of feature extraction because they search $\mathscr{F} = \{K_{\boldsymbol{x}'}(\boldsymbol{x}) | \boldsymbol{x} \in \mathbb{R}^p\}$ to return a

B. Clarke et al., *Principles and Theory for Data Mining and Machine Learning*, Springer Series in Statistics, DOI 10.1007/978-0-387-98135-2_9, © Springer Science+Business Media, LLC 2009

linear combination of a few elements. Once the SVM or RVM model is chosen, one can seek a secondary sort of dimension reduction, variable selection, by removing the explanatory variables from $\boldsymbol{x}$ that do not contribute enough to the features.

It is the second notion of dimension reduction as feature extraction that is the focus of this chapter. The notion of dimension reduction meant by variable selection is taken up in the next chapter. It is as reasonable to treat these two senses of dimension reduction in this order as it is to use the reverse order. However, in practice an order must be chosen, and it is more typical to delay variable selection until a model class is identified. This is partially because feature extraction may be unsupervised, whereas, as will be seen in Chapter 10, variable selection is almost always supervised. That is, it is standard to extract information from the covariates by themselves before bringing in the response. Hence, the goal here is to condense the information in the X_is into functions that have the information most relevant to modeling, regardless of the response. Then the information in the Ys can be used for variable selection on the extracted features.

The concept of dimension reduction can be interpreted in terms of a variance–bias trade-off. Variable selection is an effort to throw out variables that contribute too much variance to be worth including for the amount of bias they eliminate. Feature extraction is an effort to find a collection of functions so that variable selection over the new collection of features will give an even lower MSE because fewer features will be required to express a model for the response.

Dimension reduction techniques in the feature extraction sense are presented here using three classes: The biggest conceptual division is between linear and nonlinear, but linear techniques are divided further into those that are based on second-order statistics and those that use higher-order statistics.

The two most common second-order procedures are principal components and factor analysis. These are covered in Sections 9.1 and 9.2 and can be applied quite generally. However, when a distribution is not characterized by its first two moments, or the optimal dimension reduction is not via linear functions of the data, the results of these techniques may not be appropriate. Thus, it is worthwhile developing techniques that are not limited to second-moment criteria or even linearity. Sections 9.3 and 9.4 present linear methods that include dependence on higher moments such as projection pursuit and independent components analysis. The first is a variant of projection pursuit regression seen in Chapter 4, but here it is unsupervised. The second minimizes the dependence among the components in the representation of the data. This technique is easiest to apply in the linear case but is more general.

Nonlinear techniques for dimension reduction are more difficult, less is known about them, and they are not as commonly used – yet. However, nonlinear methods may summarize complicated data sets better or more efficiently than linear methods do. Many of these techniques arise from generalizing concepts already familiar from other contexts, for instance nonlinear principal components and independent component analysis. Separately, curves and surfaces have also been proposed for feature extraction, although they remain somewhat undeveloped.

For completeness, two supervised dimension reduction techniques are presented. The first is called partial least squares and the second is called sufficient dimension reduction. These are intended for regression settings and are presented on the grounds that it is natural to compare unsupervised dimension reduction with supervised dimension reduction to evaluate the benefit of including the extra information in the Y.

Because they are more complex, nonlinear and supervised dimension reductions are only covered discursively in Sections 9.5-9.8; the interested reader is referred to the growing literature on these topics.

Finally, it must be remembered that some data sets are so complicated, perhaps because of low sample size and high dimension, that formal statistical techniques may be effectively powerless without a lot of (unavailable) modeling information. In these cases, the best that can be done is to search for suggestive patterns. Chapter 8, on clustering, included some of these ideas; however, the field of data visualization has even more insight to offer for the most complicated settings. In addition to the usual graphical presentations of data, techniques that yield visual insight into the structure of the data include multidimensional scaling, self-organizing maps, and a variety of computer intensive approaches such as brushing (coloring parts of the data) and spinning (looking at projections of the data into lower-dimensional subspaces). Sections 9.9 and 9.10 are a treatment of these ideas just to provide an overview.

9.1 Principal Components

Principal components, PCs, is one of a collection of techniques (including canonical correlation and factor analysis) that extend linear regression by trying to define underlying factors that explain a response. PCs were quite early, already in use by the 1930s, see Hotelling (1933). At its root, the idea behind PCs is to find a rotation of the original coordinate system in which to express the p-variate $\boldsymbol{X}_i$s so that each coordinate expresses as much of the variability in the $\boldsymbol{X}$ as a linear combination of the p entries can. More formally, let $\boldsymbol{U} = (U_1, \ldots, U_p)$ be a random vector and write $\boldsymbol{U} = \boldsymbol{A}\boldsymbol{X}$ with $\mathrm{Var}(\boldsymbol{X}) = \Sigma$. The goal of PCs is to find $\boldsymbol{A} = (\boldsymbol{a}_1, .., \boldsymbol{a}_p)^\top$ in which each $\boldsymbol{a}_j = (a_{j,1}, \ldots, a_{j,p})$ such that

$$\forall j = 1, \ldots, p \;\; U_j = \boldsymbol{a}_j^\top \boldsymbol{X} = \sum_{k=1}^{p} a_{j,k} X_k$$

and $\mathrm{Var}(U_j) = \boldsymbol{a}_j^\top \Sigma \boldsymbol{a}_j$ is as high as possible subject to being uncorrelated with the other $U_j = \boldsymbol{a}_j^\top \boldsymbol{X}$s; i.e., $\mathrm{Cov}(U_j, U_k) = \boldsymbol{a}_j^\top \Sigma \boldsymbol{a}_k = 0$ for $j \neq k$. A consequence of this will be that the $\mathrm{Var}(U_j)$s are decreasing. It will be seen that PCs are eigenvectors of the covariance matrix, ranked in order of the size of their eigenvalues, but that normality of $\boldsymbol{X}$ does not need to be assumed.

The reason to search for linear combinations with high variance is that they are the ones that affect the response the most. When p is so large that some variables must be omitted, the variables least worth including are those with the smallest variability, i.e., the ones whose deviations affect the response the least. If most of the variation comes from the first few PCs then it is enough to use them because the other linear combinations vary so little from subject to subject that they can safely be ignored.

9.1.1 Main Theorem

The PCs can be peeled off one at a time from Σ by a sequence of optimizations. Start by finding $U_1 = \boldsymbol{a}_1^\top \boldsymbol{X}$, where

$$\boldsymbol{a}_1 = \arg\max_{\|\boldsymbol{a}\|=1} \operatorname{Var}(\boldsymbol{a}^\top \boldsymbol{X}). \tag{9.1.1}$$

The $\boldsymbol{a}_1$ from (9.1.1) is the direction in the $\boldsymbol{X}$-space along which the variability is maximized; i.e., the eigenvector of Σ with maximal eigenvalue. To find U_2, or equivalently $\boldsymbol{a}_2$, set

$$\boldsymbol{a}_2 = \arg \max_{\|\boldsymbol{a}\|=1, \operatorname{Cov}(\boldsymbol{a}_1^\top \boldsymbol{X}, \boldsymbol{a}^\top \boldsymbol{X})=0} \operatorname{Var}(\boldsymbol{a}^\top \boldsymbol{X}). \tag{9.1.2}$$

Then, $U_2 = \boldsymbol{a}_2^\top \boldsymbol{X}$. It will be seen that $\boldsymbol{a}_1 \perp \boldsymbol{a}_2$ when $\lambda_1 > \lambda_2$ and can be made orthogonal otherwise. Later PCs are defined analogously; $\boldsymbol{a}^\top \boldsymbol{X}$ is assumed uncorrelated with all the previous $\boldsymbol{a}_j\boldsymbol{X}$s (i.e., $\boldsymbol{a}$ is in the orthogonal complement of the eigenspaces of the earlier eigenvectors).

Provided Σ is positive definite, it has a full set of p real eigenvalues $\lambda_1 \geq \ldots \geq \lambda_p > 0$. Then, it turns out that the correct $\boldsymbol{A}$ has columns given by the eigenvectors $\boldsymbol{e}_1$, ..., $\boldsymbol{e}_p$ of Σ and the variances of the PCs are the eigenvalues. This can be formalized in the following theorem.

Theorem: Let $\operatorname{Cov}(\boldsymbol{X}) = \Sigma$ have eigenvectors $\boldsymbol{e}_1, \ldots, \boldsymbol{e}_p$ with corresponding eigenvalues $\lambda_1 \geq \ldots \geq \lambda_p > 0$. Then:

(i) The jth PC is $U_j = \boldsymbol{e}_j^\top \boldsymbol{X} = e_{j,1}X_1 + \ldots + e_{j,p}X_p$ for $j = 1, \ldots, p$.

(ii) The variances of the U_j are $\operatorname{Var}(U_j) = \boldsymbol{e}_j^\top \Sigma \boldsymbol{e}_j = \lambda_j$.

(iii) The covariances between the PCs are $\operatorname{Cov}(U_j, U_k) = \boldsymbol{e}_j^\top \Sigma \boldsymbol{e}_k = 0$.

The proof of this theorem is so important that it is worth looking at twice. The first version is based on Lagrange multipliers; an informal version is presented here. The second proof is cleaner but rests on an auxiliary inequality.

9.1.1.1 Proof: Lagrange Multipliers

To begin, center $\boldsymbol{X}$ so that $\mathbb{E}\boldsymbol{X} = 0$. Let $\boldsymbol{a} = (a_1, \ldots, a_p)$ have norm 1 so

$$\operatorname{Var}(\boldsymbol{a}^\top X) = \mathbb{E}(\boldsymbol{a}^\top X)^2 = \boldsymbol{a}^\top \Sigma \boldsymbol{a}.$$

Following Anderson (1984), Chapter 11, and Mizuta (2004), the task is to find $\boldsymbol{a}$ such that the normalized form of $\boldsymbol{a}^{\top}\boldsymbol{X}$ has maximal variance; i.e., to solve

$$\arg\max_{\|\boldsymbol{a}\|=1} \mathrm{Var}(\boldsymbol{a}^{\top}\boldsymbol{X}).$$

So, consider the equivalent Lagrange multiplier problem and maximize

$$\phi_1(\boldsymbol{a}) = \boldsymbol{a}^{\top}\Sigma\boldsymbol{a} - \lambda(\boldsymbol{a}^{\top}\boldsymbol{a} - 1),$$

where λ (without a subscript) is the Lagrange multiplier. Setting the vector of partial derivatives equal to zero gives

$$\nabla_{\boldsymbol{a}}\phi_1 = \frac{\partial \phi}{\partial \boldsymbol{a}} = 2\Sigma\boldsymbol{a} - 2\lambda\boldsymbol{a} = 0 \Rightarrow (\Sigma - \lambda \boldsymbol{I}_{p\times p})\boldsymbol{a} = 0. \tag{9.1.3}$$

Expression (9.1.3) is satisfied for λ that are eigenvalues with corresponding eigenvectors $\boldsymbol{a}$. So it is enough to solve $\det(\Sigma - \lambda) = 0$ for λ. Since this equation is a p-dimensional polynomial, there are p solutions and, for $\boldsymbol{a}$s of norm 1, the first derivative condition gives

$$\boldsymbol{a}^{\top}\Sigma\boldsymbol{a} = \lambda\boldsymbol{a}^{\top}\boldsymbol{a} = \lambda.$$

So, the λ_js are the variances of $\boldsymbol{a}_j^{\top}\boldsymbol{X}$s, where the $\boldsymbol{a}_j$s are recognized to be the eigenvectors $\boldsymbol{e}_j$. That is, if $\boldsymbol{a}_1$ satisfies $(\Sigma - \lambda_1)\boldsymbol{a}_1 = 0$, set the first entry of $\boldsymbol{U}$ to be $U_1 = \boldsymbol{e}_1\boldsymbol{X}$ and therefore the first column of $\boldsymbol{A}$ to be $\boldsymbol{e}_1$. Now, $\Sigma\boldsymbol{e}_1 = \lambda_1\boldsymbol{e}_1$.

As in Anderson (1984), repeating the procedure and ensuring that U_2 is uncorrelated with U_1, the extra condition on the next $\boldsymbol{a}$ is

$$0 = \mathbb{E}\boldsymbol{a}^{\top}\boldsymbol{X}U_1 = \mathbb{E}\boldsymbol{a}^{\top}\boldsymbol{X}\boldsymbol{X}^{\top}\boldsymbol{a}_1 = a^{\top}\Sigma\boldsymbol{a}_1 = \lambda_1\boldsymbol{a}^{\top}\boldsymbol{a}_1. \tag{9.1.4}$$

This leads to a Lagrange multiplier problem with two constraints, one for norm 1 with λ and the other from (9.1.2) with ν_1 for orthogonality. Thus

$$\phi_2 = \boldsymbol{a}^{\top}\Sigma\boldsymbol{a} - \lambda(\boldsymbol{a}^{\top}\boldsymbol{a} - 1) - \nu_1\boldsymbol{a}^{\top}\Sigma\boldsymbol{a}_1, \tag{9.1.5}$$

where λ and ν_1 are the Lagrange multipliers, is the expression to be maximized. The first derivative condition from (9.1.5), with (9.1.4), implies that the optimal $\boldsymbol{a}$ must be an eigenvector for λ_2. So, $U_2 = \boldsymbol{a}_2^{\top}\boldsymbol{X}$ and $\boldsymbol{a}_2 = \boldsymbol{e}_2$ is the second column of $\boldsymbol{A}$. This procedure continues until $\boldsymbol{A}$ is the matrix of p eigenvectors.

Thus, there is an orthogonal matrix $\boldsymbol{A}$ such that $\boldsymbol{U} = \boldsymbol{A}\boldsymbol{X}$ so that the covariance matrix of $\boldsymbol{U}$ is $\mathbb{E}\boldsymbol{U}\boldsymbol{U}^{\top} = \mathrm{diag}(\lambda_1, \ldots, \lambda_p)$, the eigenvalues of Σ in decreasing order. Each U_j has maximum variance among normalized linear combinations of the $\boldsymbol{X}$s uncorrelated with the $\boldsymbol{U}_1, \ldots, \boldsymbol{U}_{j-1}$.

9.1.1.2 Proof: Quadratic Maximization

A standard result on maximizing quadratic forms on the unit sphere is that the kth eigenvalues of a matrix $\boldsymbol{B}$ give the maxima of $\boldsymbol{x}^{\top}\boldsymbol{B}\boldsymbol{x}/\|\boldsymbol{x}\|^2$ over $\boldsymbol{x} \perp \boldsymbol{e}_1, ..., \boldsymbol{e}_{k-1}$, which are achieved by the eigenvectors. Using this, Johnson and Wichern (1998) set $\boldsymbol{B} = \Sigma$, giving

$$\max_{a \neq 0} \frac{\boldsymbol{a}\Sigma\boldsymbol{a}}{\boldsymbol{a}^{\top}\boldsymbol{a}} = \lambda_1 \text{ and } \arg\max_{a \neq 0} \frac{\boldsymbol{a}\Sigma\boldsymbol{a}}{\boldsymbol{a}^{\top}\boldsymbol{a}} = \alpha\boldsymbol{e}_1,$$

where $\alpha \neq 0$. Thus, for $\alpha = 1$, $U_1 = \boldsymbol{e}_1\boldsymbol{X}$ and $\text{Var}(U_1) = \text{Var}(\boldsymbol{e}_1\boldsymbol{X}) = \boldsymbol{e}_1\Sigma\boldsymbol{e}_1$.

For $j = 2, ..., p$, the inequality gives

$$\max_{a \perp \boldsymbol{e}_1, ..., \boldsymbol{e}_{j-1}} \frac{\boldsymbol{a}\Sigma\boldsymbol{a}}{\boldsymbol{a}^{\top}\boldsymbol{a}} = \lambda_j \text{ and } \arg\max_{a \perp \boldsymbol{e}_1, ..., \boldsymbol{e}_{j-1}} \frac{\boldsymbol{a}\Sigma\boldsymbol{a}}{\boldsymbol{a}^{\top}\boldsymbol{a}} = \alpha\boldsymbol{e}_j.$$

So, one can set $\boldsymbol{a} = \boldsymbol{e}_j$, giving $U_j = \boldsymbol{e}_j^{\top}\boldsymbol{X}$, $\boldsymbol{e}_{j+1}^{\top}\boldsymbol{e}_k = 0$ for $k \leq j$, and $\text{Var}(U_j) = \boldsymbol{e}_j^{\top}\Sigma\boldsymbol{e}_j = \lambda_j$. Finally, if $\boldsymbol{e}_j \perp \boldsymbol{e}_k$ for $j \neq k$, then $\text{Cov}(U_j, U_k) = \boldsymbol{e}_j\Sigma\boldsymbol{e}_k 0$. If the λ_is are distinct, then the $\boldsymbol{e}_j$s are orthogonal. Otherwise, an orthogonal basis can be chosen for eigenspaces of dimension two or greater.

9.1.2 Key Properties

Having obtained a convenient form for PCs, it is important to see how they characterize variability in $\boldsymbol{X}$. Clause (ii) in the previous theorem leads to regarding $\lambda_j/\sum\lambda_j$ as the proportion of variation of $\boldsymbol{X}$ explained by U_j. Separately, the relationship between the U_js and the X_ks may be looser or tighter; this, too, is mostly given by the entries of the eigenvectors of Σ. Given this, the normal case will help to visualize what using PCs means geometrically.

9.1.2.1 Characterizations of Variability

It is not just that PCs re-express the explanatory $\boldsymbol{X}$ so that the biggest contributions to variance can be identified. The PCs and their structure permit sparsity in many cases because one can, for instance, regress on relatively few of the PCs or summarize data by using the PCs with, say, variances above a prechosen threshold. In principle, one can even use shrinkage arguments to get sparse representations for the PCs themselves. To do this, two important properties of PC in terms of how they characterize variability in $\boldsymbol{X}$ are helpful.

Theorem: Suppose $\text{Cov}(\boldsymbol{X}) = \Sigma$ and Σ has p eigenvalues $\lambda_1 \geq \ldots \geq \lambda_p$ with corresponding eigenvectors $\boldsymbol{e}_j = (e_{j,1}, ..., e_{j,p})^{\top}$.

(i) The sum of the variances of the U_js is

$$\sum_{j=1}^{p} \text{Var}(U_j) = \lambda_1 + \cdots + \lambda_p = \sum_{j=1}^{p} \text{Var}(X_j) = \sum_{j=1}^{p} \sigma_{jj}.$$

(ii) The correlation between U_j and X_k is

$$\rho(U_j, X_k) = \frac{e_{jk}\sqrt{\lambda_j}}{\sqrt{\sigma_{kk}}}.$$

Proof: For (i), write $\Sigma = \boldsymbol{P}\Lambda\boldsymbol{P}$, where Λ is the diagonal matrix of eigenvalues λ_j and $\boldsymbol{P} = (\boldsymbol{e}_1, \ldots, \boldsymbol{e}_p)$. Now, $\boldsymbol{P}\boldsymbol{P}^\top = I_{p\times p}$, so rearrange in trace$(\Sigma)$ under the trace.

For (ii), let $\boldsymbol{a}_k$ be the vector of zeros with 1 in the kth location so $X_k = \boldsymbol{a}_k^\top \boldsymbol{X}$. Use

$$\text{Cov}(X_k, U_j) = \text{Cov}(\boldsymbol{a}^\top \boldsymbol{X}, \boldsymbol{e}_j \boldsymbol{X}) = \boldsymbol{a}_k^\top \Sigma \boldsymbol{e}_j = \lambda_j e_{jk}$$

in the definition of correlation. □

Note that (ii) invites one to set the e_{jk}s to zero when they are small in absolute value, in addition to using only the first few PCs. This is a way to get more sparsity, albeit at the cost of orthogonality. Alternatively, one can look for which entries of $\boldsymbol{X}$ get the highest weight in the first few normalized PCs; these X_js might have more information in them relative to a response. See also Chipman and Gu (2001), who explore simplifications of PCs.

9.1.2.2 Normal Case Interpretation

If $\boldsymbol{X} \sim N_p(\boldsymbol{\mu}, \Sigma)$, then the density is constant on ellipses of the form

$$(\boldsymbol{x} - \boldsymbol{\mu})^\top \Sigma^{-1} (\boldsymbol{x} - \boldsymbol{\mu}) = C^2.$$

Ellipses of this form have axes along $\pm C\sqrt{\lambda_j}\boldsymbol{e}_j$, determined by the eigenvalues and eigenvectors of Σ. Set $\boldsymbol{\mu} = 0$, and use the spectral representation of Σ as $\Sigma = \sum_{j=1}^{p} \lambda_j \boldsymbol{e}_j \boldsymbol{e}_j^\top$ to give

$$C^2 = \boldsymbol{x}^\top \Sigma^{-1} \boldsymbol{x} = \sum_{j=1}^{p} \frac{1}{\lambda_j} (\boldsymbol{e}_j^\top \boldsymbol{x})^2.$$

It is seen that $\boldsymbol{e}_j^\top \boldsymbol{x}$ is the component of $\boldsymbol{x}$ in the direction of $\boldsymbol{e}_j$, the PCs. Writing $u_j = \boldsymbol{e}_j^\top \boldsymbol{x}$ gives

$$C^2 = \frac{1}{\lambda_1} u_1^2 + \cdots + \frac{1}{\lambda_p} u_p^2,$$

an ellipse in standard form in the $(\boldsymbol{e}_1, \ldots, \boldsymbol{e}_p)$ coordinate system.

This means that the PCs $U_j = \boldsymbol{e}_j \boldsymbol{X}$ lie in the directions of the axes of a constant-density ellipse. That is, any geometric point on the jth axis of the ellipse has coordinates in the $\boldsymbol{x}$-frame proportional to $\boldsymbol{e}_j$ and in the $\boldsymbol{u}$-frame (of the PCs) has coordinates proportional to $\boldsymbol{a}_j$.

9.1.3 Extensions

Given the form and interpretation of PCs, it is important to see how to use them correctly. First, it is often better to use the correlation matrix $\boldsymbol{\rho}$ than to use Σ. Also, since neither $\boldsymbol{\rho}$ nor Σ are usually available, it is important to be able to obtain PCs from data. These two variations are amenable to the same procedure as before. Second, the point of PCs is to use only the first few, say K. There are several ways to choose K, but none are entirely satisfactory in all cases.

An extension in a different direction, not covered here, is called canonical correlation analysis (CCA). The essence of the method is to partition $\boldsymbol{X}$ into two vectors with p_1 and p_2 entries, $p = p_1 + p_2$, $\boldsymbol{X} = (\boldsymbol{X}_1, \boldsymbol{X}_2)^{\mathsf{T}}$. Then, search for linear combinations L_1 and L_2 of the entries of $\boldsymbol{X}_1$ and $\boldsymbol{X}_2$ so that $\mathrm{Corr}(L_1, L_1^*)$ is maximized. As with PCs, this can be done successively, giving $\min(p_1, p_2)$ pairs $(L_1, L_1^*), (L_2, L_2^*), \ldots$ having unit variance, so that, at the rth stage, L_r and L_r^* are uncorrelated with the vectors from the $r-1$ and earlier stages. Like PCs, these pairs of vectors emerge from solving an eigenvector problem on $\mathrm{Var}(\boldsymbol{X})$. Finding maximally correlated linear combinations among explanatory variables will not be a suitable way to reduce dimension if the goal is to explain $\boldsymbol{Y}$ unless there is some way to remove a linear combination that was highly correlated with another that was already in a model. However, CCA may serve other dimension reduction purposes.

9.1.3.1 Correlation PCs and Empirical PCs

Standardizing variances, in addition to enforcing the mean-zero criterion, is important when the data have dependence structures. So, instead of decomposing Σ, consider applying the method for deriving PCs to the correlation matrix for $\boldsymbol{X}$, say $\boldsymbol{\rho}$. Write $Z_j = (X_j - \mu_j)/\sqrt{\sigma_{jj}}$ so that $\boldsymbol{Z} = (Z_1, ..., Z_p) = \boldsymbol{V}^{-1/2}(\boldsymbol{X} - \boldsymbol{\mu})$, where $\boldsymbol{V} = \mathrm{diag}(\sqrt{\sigma_{11}}, \ldots, \sqrt{\sigma_{pp}})$. Now, $\mathbb{E}\boldsymbol{Z} = 0$ and $\mathrm{Cov}(\boldsymbol{Z}) = \boldsymbol{V}^{-1/2}\Sigma\boldsymbol{V}^{-1/2} = \boldsymbol{\rho}$, the correlation matrix of $\boldsymbol{X}$. This standardization ensures that all the Z_js are on the same scale. Otherwise, X_js with larger scales will dominate an analysis.

The PCs found from $\boldsymbol{Z}$ are not, in general, numerically the same as those found from $\boldsymbol{X}$. Nevertheless, their forms and properties are the same and follow by proofs that are only slight modifications from before.

Theorem: Let $(\lambda_1, \boldsymbol{e}_1)$,..., $(\lambda_p, \boldsymbol{e}_p)$ be the eigenvalue, eigenvector pairs from $\boldsymbol{\rho}$, with $\lambda_1 \geq \ldots \geq \lambda_p \geq 0$. Then,

(i) The PCs of $\boldsymbol{\rho}$ are $U_j = \boldsymbol{e}_j^\top \boldsymbol{Z} = \boldsymbol{e}_j^\top \boldsymbol{V}^{-1/2}(\boldsymbol{X} - \boldsymbol{\mu})$.

(ii) Variances are preserved in the sense that $\sum_j \mathrm{Var}(U_j) = \sum_j \mathrm{Var}(Z_j) = p$.

(iii) Correlations between U_j and Z_k are expressed in terms of the eigenvalues and eigenvectors of $\boldsymbol{\rho}$, $\rho(U_j, Z_k) = e_{j,k}\sqrt{\lambda_j}$. □

Similarly, empirical forms of the PCs can be given. These result from using the PC procedure on estimates of Σ or $\boldsymbol{\rho}$. For instance, consider $\hat{\Sigma} = (1/n)\sum_{i=1}^n \boldsymbol{x}_i \boldsymbol{x}_i^\top$ or $\hat{\boldsymbol{\rho}} = \hat{\boldsymbol{V}}^{1/2}\hat{\Sigma}\hat{\boldsymbol{V}}^{1/2}$, although any other estimate of Σ or $\hat{\boldsymbol{\rho}}$ could be used as well. The empirical PCs for $\boldsymbol{X}$ are $U_j = \hat{\boldsymbol{e}}_1 \boldsymbol{X}$, where $\hat{\boldsymbol{e}}_j$ is the eigenvector corresponding to the jth largest eigenvalue $\hat{\lambda}_j$ of $\hat{\Sigma}$ or $\hat{\boldsymbol{\rho}}$. The other properties of the PCs remain the same in both cases, apart from using estimates in place of population values.

It is worth noting that the estimates of the eigenvalues λ_j and eigenvectors $\boldsymbol{e}_j$ are both asymptotically normal, centered at the true values with variances

$$2\lambda_j^2 \quad \text{and} \quad \lambda_j \sum_{k=1, k\neq j}^{p} [\lambda_k/(\lambda_k - \lambda_j)^2]\boldsymbol{a}_k \boldsymbol{a}_k^\top,$$

respectively, where $\boldsymbol{a}_k$ is a vector of zeros with one in the kth entry.

9.1.3.2 How Many PCs?

The point of PCs is dimension reduction. So, let K be the number of PCs to be retained in a model, $1 \leq K \leq p$. Obviously, the first K PCs corresponding to the largest eigenvalues should be chosen, but the value of K is undetermined. If $K = p$, there is no reduction; the variables have just been transformed by $\boldsymbol{A}$. If $K = 1$, then a one-dimensional model, perhaps involving all the components of $\boldsymbol{X}$, is the result. Unidimensionality may be desirable since it gives a linear scale; however, in most cases, $K \geq 2$ to capture all the useful variability in $\boldsymbol{X}$.

The three common ways to choose K are as follows. First, one can fix a proportion α of the variation to be explained by the PCs and let K be the smallest number of PCs required to achieve that. Thus, one chooses K just large enough that

$$\frac{\hat{\lambda}_1 + \ldots + \hat{\lambda}_K}{\sum_{j=1}^p \hat{\lambda}_j} \geq \alpha.$$

This is somewhat arbitrary and rests on the notion that it is reasonable to interpret α as the proportion of information retained.

Second, perhaps most commonly, one can produce a scree plot. This is a graph of $\hat{\lambda}_j$ as a function of j. If the points are connected, then one can look for the knee in the curve, the point at which adding another PC adds relatively little explanatory power. This is hard to formalize but often works well in practice.

Third, and less common, is to invoke a physical interpretation. This is problematic because one uses PCs because a physical interpretation is unavailable and the PCs themselves are a mathematical construct that does not have a standard physical interpretation. Nevertheless, one can look at the $\hat{\lambda}_j$s and recall that they represent the width of an ellipse of constant density, at least in the normal case. Thus, if a $\hat{\lambda}_j$ is small, the ellipse is narrow in the jth dimension and so the jth dimension may be neglected for physical reasons if the most important contributing components in $\hat{\boldsymbol{e}}_j$ are known to be unimportant at that scale. By the same logic, if $\hat{\lambda}_1 \approx \hat{\lambda}_2$, it would be unreasonable to drop $\hat{\lambda}_2$ without dropping $\hat{\lambda}_1$ unless there were reason to believe that the components in $\hat{\boldsymbol{e}}_2$ that had the greatest contribution were not helpful enough – for instance, there might be many of them contributing a little each and so represent little more than noise.

9.2 Factor Analysis

Heuristically, the idea of factor analysis (FA) is to partition $\boldsymbol{X}$ into m strings of components $(X_1,...,X_{p_1}),...,(X_{p_{K-1}+1},...,X_p)$ with $p = p_K$ with the property that the correlations within each string are high and the correlations between components from different strings are low. When this can be done, it may be reasonable to summarize each string by a single construct or factor.

Thus, FA is a generalization of PCs in which, rather than seeking a full-rank linear transformation with second-moment properties, one allows non-full-rank linear transformations. That is, instead of finding $\boldsymbol{U} = \boldsymbol{A}\boldsymbol{X}$ and dropping some of the components of $\boldsymbol{U}$, consider modeling $\boldsymbol{X}$ as

$$\boldsymbol{X} - \boldsymbol{\mu} = \Lambda \boldsymbol{f} + \boldsymbol{T}, \tag{9.2.1}$$

where $E\boldsymbol{X} = \boldsymbol{\mu}$. More explicitly, for the $j = 1,...,p$ entries of $\boldsymbol{X}$, (9.2.1) is

$$X_j = \sum_{k=1}^{K} \lambda_{j,k} f_l + T_j + \mu_j. \tag{9.2.2}$$

In (9.2.1) or (9.2.2), Λ is a fixed $p \times K$ real matrix; its entries are called "loadings". As noted in Fodor (2002), the loadings $\lambda_{j,k}^2$ indicate how much X_j is affected by f_k; if several X_js have high values of $\lambda_{j,k}$ for a given factor f_k, then one may surmise that those X_js emanate from the same unobservable quantity and are therefore redundant. The $K \times 1$ vector $\boldsymbol{f}$ can be nonrandom but is assumed random here. It represents the common factors, the intrinsic traits that underlie an observed $\boldsymbol{x}$. The random $p \times 1$ vector $\boldsymbol{T}$ represents the specific factors that underlie the particular experiment performed. Both $\boldsymbol{T}$ and $\boldsymbol{f}$ are unobservable. The goal is to explain the outcomes of $\boldsymbol{X}$ using fewer variables, the K unobserved factors in $\boldsymbol{f}$, with $K << p$.

To make (9.2.2) tractable, the random quantities $\boldsymbol{f}$ and $\boldsymbol{T}$ are standardized so

$$\mathbb{E}\boldsymbol{f} = 0, \mathbb{E}\boldsymbol{T} = 0, \tag{9.2.3}$$

and the second moments are

$$\text{Cov}(\boldsymbol{f},\boldsymbol{T}) = 0, \text{Cov}(T_j,T_{j'}) = 0 \text{ for } j \neq j', \text{Cov}(\boldsymbol{f}) = Id_{K\times K}. \tag{9.2.4}$$

Usually one sets $\text{Var}(\boldsymbol{T}) = \text{diag}(\psi_1,...,\psi_K) = \boldsymbol{\psi}$. Expression (9.2.1) also leads to second-moment properties of $\boldsymbol{X}$ such as

$$\text{Cov}(\boldsymbol{X},\boldsymbol{f}) = \Lambda, \quad \text{Cov}(X_j,X_\ell) = \lambda_{j,1}\lambda_{\ell,1} + \ldots + \lambda_{j,K}\lambda_{\ell,K} \quad \text{and} \quad \text{Cov}(X_j,F_\ell) = \lambda_{j,\ell}.$$

To see how (9.2.1) and (9.2.2), with (9.2.3) and (9.2.4), lead to dimension reduction, observe that (9.2.1) gives

$$\Sigma = \Lambda\Lambda^{\mathsf{T}} + \boldsymbol{\psi}. \tag{9.2.5}$$

That is, the p variables correspond to K variables, and the $p(p-1)/2$ entries in the variance matrix are reduced to $K(K-1)/2+K$ entries. When $K < p$, the savings can be substantial. In many cases, a K-factor model can provide a better explanation of a data set than a full covariance model using $\text{Var}(\Sigma)$.

When (9.2.5) is valid, it follows that, for $j = 1,...p$,

$$X_j = \sum_{k=1}^{K} \lambda_{j,k} f_k + u_j$$

leading to

$$\sigma_{j,j} = \sum_{k=1}^{K} \lambda_{j,k}^2 + \psi_j = h_j^2 + \psi_j. \tag{9.2.6}$$

The h_j^2 in (9.2.6) is called the communality; it represents the part of the variance of X_j that comes from the underlying factors. The second term, ψ_i, is the specific variance, contributed from T_j, summarizing deviations from the variability that the common factors, the f_ks, can express.

As noted in Hardle and Simar (2003), FA reduces to PCs when $\boldsymbol{T} = 0$ and the last $p-K$ eigenvalues of $\Sigma = \text{Var}(\boldsymbol{X})$ are zero. In this case, write $\Sigma = \boldsymbol{\Gamma D \Gamma}^{\mathsf{T}}$ with the last eigenvalues in $\boldsymbol{D}$ zero, $d_{K+1},...,d_p = 0$, so that the upper left $K \times K$ block of $\boldsymbol{D}$ is $\boldsymbol{D}_1 = \text{diag}(d_1,...,d_K)$ and only the upper left $K \times K$ block $\boldsymbol{\Gamma}_1$ of $\boldsymbol{\Gamma}$ matters.

Since PCs come from writing $\boldsymbol{U} = \boldsymbol{\Gamma}^{\mathsf{T}}(\boldsymbol{X}-\boldsymbol{\mu})$, it follows that $\boldsymbol{X}-\boldsymbol{\mu} = \boldsymbol{\Gamma U} = \boldsymbol{\Gamma}_1\boldsymbol{U}_1 + \boldsymbol{\Gamma}_2\boldsymbol{U}_2$, where $\boldsymbol{U}_1$ and $\boldsymbol{U}_2$ are the first K and last $p-K$ components of $\boldsymbol{U}$ and $\boldsymbol{\Gamma}_2$ is the lower right block of $\boldsymbol{\Gamma}$. Thus, $\boldsymbol{U}_2$ is trivial: It has mean and variance 0. So, $\boldsymbol{X}-\boldsymbol{\mu} = \boldsymbol{\Gamma}_1\boldsymbol{U}_1$, which is

$$\boldsymbol{X} = \boldsymbol{\Gamma}_1\boldsymbol{D}_1^{1/2}\boldsymbol{D}_1^{-1/2}\boldsymbol{U}_1 + \boldsymbol{\mu}_1.$$

Setting $\Lambda = \boldsymbol{\Gamma}_1\boldsymbol{D}_1^{1/2}$ and $\boldsymbol{f} = \boldsymbol{D}_2^{-1/2}\boldsymbol{U}_1$ gives the reduced form of (9.2.1)–(9.2.4).

It is seen that there are three sources of ambiguity in FA models: the choices of Λ, K, and $\boldsymbol{f}$. Taken together, they represent an enormous subjectivity that can be readily abused to give ridiculous results.

To deal with these three ambiguities in turn, observe first that Λ is only determined up to an orthogonal transformation. So, regard (9.2.1)–(9.2.4) as a components-of-variance model with unidentifiable components. That is, let $K \geq 2$ and let $\boldsymbol{V}$ be any $K \times K$ orthogonal matrix, $\boldsymbol{V}\boldsymbol{V}^{\mathsf{T}} = \boldsymbol{V}^{\mathsf{T}}\boldsymbol{V} = I_{K\times K}$. The model (9.2.1) can be written as

$$\boldsymbol{X} - \boldsymbol{\mu} = (\Lambda \boldsymbol{V})(\boldsymbol{V}^{\mathsf{T}}\boldsymbol{f}) + \boldsymbol{T} = \Lambda^* \boldsymbol{f}^* + \boldsymbol{T} \tag{9.2.7}$$

since $E(\boldsymbol{f}^*) = 0$ and $\mathrm{Cov}(\boldsymbol{f}^*) = \boldsymbol{V}^{\mathsf{T}}\mathrm{Cov}(\boldsymbol{f})\boldsymbol{V} = I_{K\times K}$. This means that outcomes of $\boldsymbol{X}$ cannot be used to distinguish Λ from Λ^*; i.e., the model is not identifiable. In fact, $\Sigma = \Lambda\Lambda + \boldsymbol{\psi} = \Lambda^*\Lambda^* + \boldsymbol{\psi}$. So, the communalities given by the elements of $\Lambda\Lambda = \Lambda^*\Lambda^*$ are also unchanged by $\boldsymbol{V}$. Thus, extra conditions must be imposed to get unique estimates of Λ and $\boldsymbol{\psi}$. In some cases, Λ can be purposefully rotated by $\boldsymbol{V}$ to make the results interpretable. More generally, estimating Λ and $\boldsymbol{\psi}$ is essential because they permit estimation of the factor scores f_j in $\boldsymbol{f}$.

9.2.1 *Finding Λ and ψ*

As with PCs, the estimation procedures rest on $\hat{\Sigma}$. Let $\bar{\boldsymbol{x}}$ denote the sample mean from $\boldsymbol{x}_1$, ..., $\boldsymbol{x}_n$, and denote the sample covariance matrix by $\hat{\Sigma}$ and the sample correlation matrix by $\hat{R}$. Setting $\hat{\sigma}_{j,j} = s_{j,j}$, the usual estimate of the standard deviation for an X_j, write

$$\hat{\Sigma} = \hat{\Lambda}\hat{\Lambda}^{\mathsf{T}} + \hat{\psi} \text{ and } \hat{\sigma}_{j,j} = \sum_{k=1}^{K} \hat{\lambda}_{j,k}^2 + \hat{\boldsymbol{\psi}}_{j,j}. \tag{9.2.8}$$

An analogous expression can be written for $\hat{R}$. The problem is to identify estimators $\hat{\Lambda}$ and $\hat{\boldsymbol{\psi}}$. Two standard ways are ML factors and principal factors.

9.2.1.1 Maximum Likelihood

This method uses the distributional assumptions. When all the (f_i, T_j)s are bivariate normal, the $\boldsymbol{X}$s are normal too. The likelihood is

$$L(\boldsymbol{\mu}, \Sigma) = \frac{|\Sigma|^{-n/2}}{(2\pi)^{-np/2}} e^{-(1/2)\mathrm{trace}[\Sigma^{-1}\sum_{i=1}^{n}(\boldsymbol{x}_i - \bar{\boldsymbol{x}})(\boldsymbol{x}_i - \bar{\boldsymbol{x}})^{\mathsf{T}} + n(\bar{\boldsymbol{x}} - \mu)(\bar{\boldsymbol{x}} - \mu)]}. \tag{9.2.9}$$

It is tempting to fix K, substitute using $\Sigma = \Lambda\Lambda^{\mathsf{T}} + \boldsymbol{\psi}$, and maximize over Λ, $\boldsymbol{\psi}$. However, this will not give a unique solution because of the rotation problem.

One way around this is to impose the extra constraint $\Lambda^{\top}\boldsymbol{\psi}\Lambda = \Delta$, where Δ is diagonal. It is a fact that under this constraint (9.2.9) can be maximized to find the MLEs $\hat{\Lambda}$, $\hat{\boldsymbol{\psi}}$, and $\hat{\boldsymbol{\mu}} = \bar{\boldsymbol{x}}$, usually by a numerical method. The MLEs for the communalities are

$$\hat{h}_j^2 = \sum_{k=1}^{K} \hat{\lambda}_{j,k},$$

and the proportion of total variance explained by the jth factor is

$$(\hat{\lambda}_{1,j}^2 + \ldots + \hat{\lambda}_{p,j}^2) / \sum_{j=1}^{p} s_{jj}.$$

The same procedure can be applied to the correlation matrix $\boldsymbol{\rho}$ for $\boldsymbol{X}$. As before, write $\boldsymbol{Z} = \boldsymbol{V}^{-1/2}(\boldsymbol{X} - \boldsymbol{\mu})$, where $\boldsymbol{V} = \text{diag}(\sqrt{\sigma_{11}}, \ldots, \sqrt{\sigma_{pp}})$. Now, $\boldsymbol{\rho} = \text{Cov}(\boldsymbol{Z}) = \boldsymbol{V}^{-1/2}\Sigma\boldsymbol{V}^{-1/2} = (\boldsymbol{V}^{-1/2}\Lambda)(\boldsymbol{V}^{-1/2}\Lambda)^{\top} + \boldsymbol{V}^{-1/2}\boldsymbol{\psi}\boldsymbol{V}^{-1/2}$, like (9.2.8) for Σ. However, the loading matrix is $\Lambda_2 = \boldsymbol{V}^{-1/2}\Lambda$, not Λ, and the specific variance matrix is $\boldsymbol{\psi}_2 = \boldsymbol{V}^{-1/2}\boldsymbol{\psi}\boldsymbol{V}^{-1/2}$.

9.2.1.2 Principal Factors

Principal factors is an alternative technique for finding $\hat{\boldsymbol{\Lambda}}$ and $\hat{\boldsymbol{\psi}}$ given K. As the name suggests, it is related to PCs. In fact, when $\boldsymbol{\psi} = 0$ or $K = p$, the eigenvector decomposition in PCs gives the FA representation. Indeed, the basic idea is to start with the spectral decomposition of Σ and write

$$\begin{aligned}
\Sigma &= \sum_{j=1}^{p} \lambda_j \boldsymbol{e}_j \boldsymbol{e}_j^{\top} \approx \sum_{j=1}^{K} \lambda_j \boldsymbol{e}_j \boldsymbol{e}_j^{\top} \\
&= (\sqrt{\lambda_1}\boldsymbol{e}_1, \ldots, \sqrt{\lambda_1}\boldsymbol{e}_K) \times (\sqrt{\lambda_1}\boldsymbol{e}_1^{\top}, \ldots, \sqrt{\lambda_1}\boldsymbol{e}_K)^{\top} \\
&= \Lambda\Lambda,
\end{aligned} \tag{9.2.10}$$

in which the jth column of Λ is $\sqrt{\lambda_1}\boldsymbol{e}_j$. That is, the jth factor loading comes from the jth PC and is exact if $K = p$, in which case $\boldsymbol{\psi} = 0$. Of course, this is usually done on $\hat{\Sigma} = (1/n)\sum_{i=1}^{n} \boldsymbol{x}_i \boldsymbol{x}_i^{\top}$ using $\hat{\boldsymbol{e}}_j$ and $\hat{\lambda}_j$ to give $\hat{\Lambda}$ for a given K. In this case, $\hat{\boldsymbol{\psi}}$ is usually not zero and it is typical to set $\hat{\psi}_j = s_{jj} - \sum_{k=1}^{K} \hat{\lambda}_{j,k}^2$ so $\text{diag}(\hat{\Sigma}) = \text{diag}(\hat{\Lambda}\hat{\Lambda} + \hat{\boldsymbol{\psi}})$. The communalities are then $\hat{h}_j^2 = \sum_{k=1}^{K} \hat{\lambda}_{jk}^2$.

The same procedure can be applied to the correlation matrix $\boldsymbol{\rho}$. Explicitly, if $(\alpha_j, \boldsymbol{\gamma}_j)$ are the eigenvalue, eigenvector pairs for $\hat{\boldsymbol{R}} = \hat{\boldsymbol{\rho}} = \hat{\boldsymbol{V}}^{-1/2}\hat{\Sigma}\hat{\boldsymbol{V}}^{-1/2}$, then (9.2.10) becomes

$$\hat{\boldsymbol{R}} - \hat{\boldsymbol{\psi}} = \sum_{j=1}^{K} \alpha_j \boldsymbol{\gamma}_j \boldsymbol{\gamma}_j^{\top} \tag{9.2.11}$$

for fixed K and approximates $\boldsymbol{\rho} - \boldsymbol{\psi}$. As before, $\hat{\boldsymbol{\psi}}$ is usually chosen so the diagonal entries match. Again, the decomposition for Σ and $\boldsymbol{\rho}$ is usually different, and often it is preferable to use the correlation for scale standardization of the X_js.

9.2.2 *Finding K*

As with PCs, the degree of dimension reduction depends on how small K can be without losing too much information. There are several methods for ensuring K is reasonable. The first is really only for principal factors; the second is primarily for ML factors.

9.2.2.1 Bound on the Approximation

When the FA model is found using PCs, the natural way to evaluate how well it fits is to look at how good the approximation of Σ is. It can be shown that

$$\|\hat{\Sigma} - \hat{\Lambda}\hat{\Lambda} - \hat{\boldsymbol{\psi}}\| \leq \hat{\lambda}_{K+1}^2 + \ldots + \hat{\lambda}_p^2, \tag{9.2.12}$$

in which the norm on the right is the sum of squares of the entries of the matrix. So as K increases, the bound tightens.

However, the goal is small K, meaning that the contributions from a small number of factors f_j to the sample variance should be large enough that the other factors can be neglected. The contribution to $s_{jj} = s_j^2$ from the first factor f_1 is $\hat{\lambda}_{j1}^2$. So, the contribution of f_1 to the total sample variance trace$(\hat{\Sigma}) = s_{11} + \ldots + s_{pp}$ is

$$\sum_{j=1}^{p} \hat{\lambda}_{j1}^2 = (\sqrt{\hat{\lambda}_1}\hat{\boldsymbol{e}}_1)^\mathsf{T}(\sqrt{\hat{\lambda}_1}\hat{\boldsymbol{e}}_1) = \hat{\lambda}_1,$$

the $(1,1)$ entry of $\hat{\Lambda}\hat{\Lambda}^\mathsf{T}$, where $\hat{\Lambda} = (\sqrt{\hat{\lambda}_1}\hat{\boldsymbol{e}}_1), \ldots, (\sqrt{\hat{\lambda}_p}\hat{\boldsymbol{e}}_p)$, and this holds for 2, 3,..., p. The consequence is that the proportion of the total sample variance attributable to the jth factor is $\hat{\lambda}_j / \sum_{j=1}^{p} s_{jj}$ and $\hat{\lambda}_j / p$ when factor analysis is applied to $\hat{\Sigma}$ or $\hat{\boldsymbol{\rho}}$, respectively. Since the eigenvalues are decreasing, the techniques for choosing K from the section on choosing the correct number of PCs continue to apply.

9.2.2.2 Large Samples

When n is large and normal distributions can be assumed, the value of K can be inferred from a likelihood ratio test of $\mathscr{H}_0$: the model with K factors is correct, versus $\mathscr{H}_1$: the full model with Σ is correct when $K \leq p-1$. To set this up, recall that in the context of (9.2.1)–(9.2.4), when $\boldsymbol{T}$ and $\boldsymbol{f}$ are multivariate normals, there is a well-defined likelihood that can be maximized. In fact, writing $\boldsymbol{X} \sim N_p(\boldsymbol{\mu}, \Sigma)$ as in (9.2.1) and (9.2.9) and assuming data $\boldsymbol{x}_1, ..., \boldsymbol{x}_n$, denoted $\mathscr{D}$, the log likelihood is

$$\begin{aligned}
\ell(\boldsymbol{\mu}, \boldsymbol{\Sigma}|\mathscr{D}) &= -\frac{n}{2}\log|2\pi\Sigma| - \frac{1}{2}\sum_{i=1}^{n}(\boldsymbol{x}_i - \boldsymbol{\mu})^{\mathsf{T}}\Sigma^{-1}(\boldsymbol{x}_i - \boldsymbol{\mu}) \\
&= -\frac{n}{2}\log|2\pi\Sigma| - \frac{n}{2}\text{trace}(\Sigma^{-1}\hat{\Sigma}) - \frac{n}{2}(\bar{\boldsymbol{x}} - \boldsymbol{\mu})^{\mathsf{T}}\Sigma^{-1}(\bar{\boldsymbol{x}} - \boldsymbol{\mu}) \\
&= -\frac{n}{2}\left(\log|2\pi\Sigma| + \text{trace}(\Sigma^{-1}\hat{\Sigma})\right), \qquad (9.2.13)
\end{aligned}$$

where $\boldsymbol{\mu}$ was replaced by $\hat{\boldsymbol{\mu}} = \bar{\boldsymbol{x}}$ in the last expression. When $K = p$, (9.2.13) can be maximized to give the usual MLE $\hat{\Sigma}$ for Σ and $\bar{\boldsymbol{x}}$ for $\boldsymbol{\mu}$. When $K \leq p-1$, the maximization subject to the diagonal constraint ($\Lambda^{\mathsf{T}}\psi\Lambda = \Delta$) can be done.

Assuming the ML estimates $\hat{\boldsymbol{\Lambda}}$ and $\hat{\boldsymbol{\psi}}$ for $\boldsymbol{\Lambda}$ and $\boldsymbol{\psi}$ have been found, one can follow the usual recipe. Two instances of (9.2.13), under the null and the alternative, give that the test statistic is of the form

$$LR = -2\log\left(\frac{\text{ML under } \mathscr{H}_0}{\text{ML under } \mathscr{H}_1}\right) = n\log\left[\frac{|\hat{\boldsymbol{\Lambda}}\hat{\boldsymbol{\Lambda}}^{\mathsf{T}} + \hat{\boldsymbol{\psi}}|}{|\hat{\boldsymbol{\Sigma}}|}\right]. \qquad (9.2.14)$$

Wilks' theorem applies and gives that LR has a χ^2 distribution on $(1/2)[(p-K)^2 - p - K]$ degrees of freedom, as $n \to \infty$. (Bartlett's correction can also be used to speed the convergence of (9.2.14) to its limiting chi-square.)

9.2.3 Estimating Factor Scores

The choice of K gives the degree of dimension reduction from p, but it remains to convert the p-dimensional data points $\boldsymbol{x}_i$ to new points $\hat{\boldsymbol{f}}_i$, called factor scores, in K dimensions. Note that the choice of K, Λ, and $\boldsymbol{\psi}$ is determined by all the $\boldsymbol{x}_i$s, so adding another data point may change the whole model. In this sense, factor scores are more of an imputation method than a dimension reduction method: If the FA model is not updated as data accumulate, then the factor scores for new data points are forced to look like they came from the same distribution as the earlier data points, even when this is not reasonable. In other words, this dimension reduction technique is very sensitive to any change in the population to which it is applied.

The basic problem is that there are n known values, the $\boldsymbol{x}_i$s, but $2n$ unknown values, the $\boldsymbol{\varepsilon}_i$s and the $\boldsymbol{f}_i$s. Nevertheless, there are two common ways to convert $\boldsymbol{x}_i$s to factor scores. The first is weighted least squares and the second is based on regression. Both methods start with fixed values for Λ, $\boldsymbol{\mu}$, and $\boldsymbol{\psi}$ assumed known; usually these are estimates. Also, both methods treat $\boldsymbol{T}$ as if it were an error term, $\boldsymbol{\varepsilon}$. So, write the model as $\boldsymbol{X}_i - \boldsymbol{\mu} = \Lambda\boldsymbol{f}_i + \boldsymbol{\varepsilon}_i$ for $i = 1, \ldots, n$ and consider determining $\boldsymbol{f}_i$ for $\boldsymbol{x}_i$.

The least squares strategy for overcoming the indeterminacy is as follows. Observe that the sum of squares due to error is

$$SSE = \sum_{j=1}^{p} \varepsilon_j^2/\psi_j = \boldsymbol{\varepsilon}\boldsymbol{\psi}^{-1}\boldsymbol{\varepsilon} = (\boldsymbol{X} - \boldsymbol{\mu} - \Lambda \boldsymbol{f})^{\mathsf{T}} \boldsymbol{\psi}^{-1} (\boldsymbol{X} - \boldsymbol{\mu} - \Lambda \boldsymbol{f}).$$

Minimizing it gives $\boldsymbol{f}_{min} = (\Lambda^{\mathsf{T}} \boldsymbol{\psi}^{-1} \Lambda)^{-1} \Lambda^{\mathsf{T}} \boldsymbol{\psi}^{-1} (\boldsymbol{x} - \boldsymbol{\mu})$. So, for $i = 1, ..., n$, when Λ and $\boldsymbol{\psi}$ are estimated by the ML method (and $\hat{\Lambda}\hat{\boldsymbol{\psi}}^{-1}\Lambda = \hat{\Delta}$), it is natural to set

$$\hat{\boldsymbol{f}}_i = (\hat{\Lambda}^{\mathsf{T}} \hat{\boldsymbol{\psi}}^{-1} \hat{\Lambda})^{-1} \hat{\Lambda}^{\mathsf{T}} \hat{\boldsymbol{\psi}}^{-1} (\boldsymbol{x}_i - \bar{\boldsymbol{x}}).$$

This procedure on the correlation matrix with $z_i = \hat{\boldsymbol{V}}^{-1/2}(\boldsymbol{x}_i - \bar{\boldsymbol{x}})$ and $\hat{\boldsymbol{\rho}} = \hat{\Lambda}_2 \hat{\Lambda}_2^{\mathsf{T}} + \hat{\boldsymbol{\psi}}_2$ gives $\hat{f}_i = (\hat{\Lambda}_2^{\mathsf{T}} \hat{\boldsymbol{\psi}}^{-1} \hat{\Lambda}_2)^{-1} \hat{\Lambda}_2^{\mathsf{T}} \hat{\boldsymbol{\psi}}^{-1} z_i$.

When the principal factors are used, the $\hat{\boldsymbol{\psi}}$ drops out: The results are

$$\hat{\boldsymbol{f}}_i = (\hat{\Lambda}^{\mathsf{T}} \hat{\Lambda})^{-1} \hat{\Lambda}^{\mathsf{T}} (\boldsymbol{x}_i - \bar{\boldsymbol{x}})$$

for Σ, which can be recognized as the first K scaled PCs evaluated at $\boldsymbol{x}_i$. For the correlation matrix, $\hat{f}_i = (\hat{\Lambda}_2^{\mathsf{T}} \hat{\Lambda}_2)^{-1} \hat{\Lambda}_2^{\mathsf{T}} z_i$ for the correlation.

The regression method for overcoming the indeterminacy is as follows. Since $(\boldsymbol{X} - \boldsymbol{\mu}) = \Lambda \boldsymbol{f} + \boldsymbol{\varepsilon} \sim N(0, \Sigma = \Lambda\Lambda^{\mathsf{T}} + \boldsymbol{\psi})$, the joint distribution for $(\boldsymbol{X} - \boldsymbol{\mu}, \boldsymbol{f})$ is a 2×2 block matrix of size $p+K$, with Σ and I_K on the main diagonal and Λ or Λ^{T} otherwise. Now,

$$(\boldsymbol{f}|\boldsymbol{X} = \boldsymbol{x}) \sim N(E(\boldsymbol{f}|\boldsymbol{x}), \mathrm{Cov}(\boldsymbol{f}|\boldsymbol{x})) = N(\Lambda^{\mathsf{T}} \Sigma^{-1} (\boldsymbol{x} - \boldsymbol{\mu}), I - \Lambda^{\mathsf{T}} \Sigma^{-1} \Lambda).$$

Regarding the mean as a regression, it is reasonable to set $\hat{\boldsymbol{f}}_i = \hat{\Lambda}^{\mathsf{T}} \hat{\Sigma}^{-1} (\boldsymbol{x}_i - \bar{\boldsymbol{x}})$, in which the estimate $\hat{\Sigma}$ is either via ML or principal factors. For the correlation matrix, $\hat{f}_i = \hat{\Lambda}_2^{\mathsf{T}} \hat{\rho}^{-1} z_i$.

9.3 Projection Pursuit

High-dimensional data often have important structure obscured by high variability. The hope in projection pursuit (PP) is that projecting the high-dimensional data down to lower dimensions will retain only the most useful information, enhancing the informativity of the data because so much of the variability is removed.

Consider the following somewhat extreme example. Suppose $\boldsymbol{X}$ is p-dimensional, in which X_1 is $N(0, 100)$, X_2 is $N(Z/2, 1/16)$ with $Z = \pm 1$ with probability 1/2 on each value, and the last $p-2$ X_js are $N(0, 1)$ and all coordinates are independent. Given n independent outcomes $\boldsymbol{x}_i$, $i = 1, ..., n$, projecting onto the first two coordinates will give a data cloud that looks like two horizontally narrow, vertically stretched elliptical data clouds centered at $(-1/2, 0)$ and $(1/2, 0)$.

If the task is to find the structure in the data, without making any modeling assumptions, PCs will fail: The first PC will detect X_1, the next $p-2$ PCs will detect the last $p-2$ X_js, and only at the end will the really interesting X_2 be found. Thus, the first PC with maximal variance would parallel the two clusters, so that projecting the clusters onto a vertical line would mix them rather than reveal them. Only at the pth PC would one project the data onto the horizontal axis and see the structure. This means that the strategy of using PCs with large eigenvalues performs poorly. Similarly, factor analysis performs poorly: The mean is $\boldsymbol{\mu} = 0$ and $\boldsymbol{T} = 0$ because $\boldsymbol{X}$ is normal. So, principal factors fail just as PCs do, and the ML version, while somewhat different, usually gives roughly comparable results, therefore performing poorly, too. The problem with PCs and FA is that second-moment characterizations are just not enough for many data sets because the data are too far from normality in the interesting directions.

The search for interesting directions onto which high-dimensional data can be projected is more intuitive than it sounds. Consider a direction $\beta \in \mathbb{R}^p$. Then $\beta^{\mathsf{T}}\boldsymbol{x}_i$, $i = 1,\ldots,n$ gives the portion of each $\boldsymbol{x}_i$ in the direction of β. For many directions, $\beta^{\mathsf{T}}\boldsymbol{x}_i$ will look roughly normal, as noted in Chapter 1. So, those βs are not very informative. However, if a criterion can be formulated to express the idea of nonnormality, then it can be optimized to give good βs. Then, the new data set $\beta^{\mathsf{T}}\boldsymbol{x}_i$ for $i = 1,\ldots n$ can be analyzed and it is unlikely to be normal (i.e., uninteresting).

This procedure can be formalized. The extra information required to choose a good β is called a projection index, say $I : L_2(\mathbb{R}) \to \mathbb{R}$, so the central task is to find

$$\hat{\boldsymbol{\beta}} = \arg \max_{\boldsymbol{\beta}:\|\boldsymbol{\beta}\|=1} I(\boldsymbol{X}^{\mathsf{T}}\boldsymbol{\beta}). \tag{9.3.1}$$

The natural extension to K dimensions is to seek

$$\hat{\boldsymbol{A}} = \arg \max_{\boldsymbol{A}:\boldsymbol{a}_j\cdot\boldsymbol{a}_k=\delta_{jk}} I(\boldsymbol{A}\boldsymbol{X}), \tag{9.3.2}$$

where $\boldsymbol{A}$ is a $K \times p$ matrix whose rows $\boldsymbol{a}_j$ are orthogonal and $I : \mathbb{R}^K \to \mathbb{R}$. Expressions like (9.3.2) are usually optimized sequentially, and one row of $\boldsymbol{A}$ is identified at each of K iterations.

The template for finding (9.3.2) is:

Start with a sample $\boldsymbol{x}_i$, $i = 1,\ldots,n$, and index I, an initial $\boldsymbol{\beta}_0$.

☐ Sphere the data: For $i = 1,\ldots,n$, transform the outcomes to $z_i = \hat{\boldsymbol{\Sigma}}^{-1/2}(\boldsymbol{x}_i - \bar{\boldsymbol{x}})$, where $\bar{\boldsymbol{x}}$ is the sample mean and $\hat{\boldsymbol{\Sigma}}^{-1/2}$ is the sample covariance matrix.

☐ Form the criterion

$$\hat{I}(z^n) = \sum_{i=1}^{n} I(z_i^{\mathsf{T}}\boldsymbol{\beta}), \tag{9.3.3}$$

estimating any quantities required for evaluation of $\hat{I}$.

☐ Optimize (9.3.3) numerically (if necessary), for instance by a gradient descent method, starting with $\boldsymbol{\beta}_0$.

☐ Find K acceptable $\hat{\boldsymbol{\beta}}_k$s and then analyze the K-dimensional projected data, $\boldsymbol{\beta}_k \boldsymbol{x}_i$, for each $k = 1, ..., K$ and $i = 1, ..., n$.

Thus the usefulness of PP for extracting nonnormal features of a data set depends on the choice of I. For instance, if $I(\boldsymbol{X}^\top \boldsymbol{\beta}) = \text{Var}(\boldsymbol{X}^\top \boldsymbol{\beta})$, then the maximum in (9.3.1) is the first eigenvalue of the variance matrix and $\boldsymbol{\beta}$ from (9.3.1) is the first PC. (The formula (9.3.2) would give successive PCs.) Thus, PCs are an instance of PP in which the projection obtained from variance maximization does not reveal the important structures for some data sets.

Three general classes for I have been identified by Huber (1985) based on invariances of I. They are:

- Class I : location scale equivariance: $I(a\boldsymbol{X} + b) = aI(\boldsymbol{X}) + b$,
- Class II : location invariance, scale equivariance: $I(a\boldsymbol{X} + b) = |a| I(\boldsymbol{X})$,
- Class III : affine invariance: $I(a\boldsymbol{X} + b) = I(\boldsymbol{X})$,

where $a, b \in \mathbb{R}$. It is seen that PCs are a Class II approach. Class III is probably the most important because it expresses invariance to affine transformations of the data space, thereby enabling the discovery of structure not captured by correlation. Usually, I is required to satisfy $I(X_1 + X_2) \leq \max\{I(X_1), I(X_2)\}$). Huber (1985) Section 5, shows that a sufficient condition for I to be Class III is that it have the form $I(\boldsymbol{X}) = h(S_1(\boldsymbol{X})/S_2(\boldsymbol{X}))$, where h is increasing and the S_i are Class II functionals with $S_1^2(X_1 + X_2) \leq S_1^2(X_1) + S_1^2(X_2)$ and $S_2^2(X_1 + X_2) \geq S_2^2(X_1) + S_2^2(X_2)$. Thus, many Class III indices can be found.

Commonly used projection indices depend on more than just the first two moments and usually are constructed to increase as the underlying data deviate from normality so that the optimal βs will detect nonnormal features rather than the usual uninteresting normal projections. For instance, standardized absolute cumulants sometimes find interesting directions. In this case,

$$I_{sac}(\boldsymbol{X}) = \frac{|\kappa_m(\boldsymbol{X})|}{\kappa_2(\boldsymbol{X})^{m/2}}$$

for some $m \geq 3$. (Recall that, for the univariate case, the mth order cumulant of X is $\kappa_m = (1/i^m) d/dt^m \log \mathbb{E}(e^{ixt})$.) The first two cumulants of the normal are zero, so this will not detect normality. A variation on this is

$$I_{cum}(\boldsymbol{X}) = \kappa_3^2 + \frac{\kappa_4^2}{4},$$

which may be regarded as an approximation to the entropy projection index, $I_{ent}(\boldsymbol{X}) = \int p(\boldsymbol{x}) \ln p(\boldsymbol{x}) d\boldsymbol{x}$, in some cases; see Jones and Sibson (1987). Also, the Fisher information has been proposed as a good projection index; see Huber (1985).

Many other indices have been proposed. Some are based on distances between normal densities and orthogonal series density estimators based on Hermite polynomials. However, dependence on higher moments often emphasizes nonnormality in the tails and can be oversensitive to outliers. Others are chosen to express some geometric aspect of the data.

9.4 Independent Components Analysis

Independent component analysis (ICA) starts with a model somewhat like (9.2.1) but imposes a criterion that ensures the analog of the factors will be independent. Formally, the general ICA model is

$$\boldsymbol{X}_i = \mathscr{T}(\boldsymbol{S}_i, \boldsymbol{\varepsilon}_i) \text{ for } i = 1, \ldots, n, \tag{9.4.1}$$

(see Eriksson (2004)), in which the $\boldsymbol{X}_i$s are observed, $\boldsymbol{S}_i$ is the original source signal, $\boldsymbol{\varepsilon}_i$ is a noise term, and $\mathscr{T}$ is the system operator. The idea is that a signal $\boldsymbol{S}$ with K independent components is generated but, when it is observed as $\boldsymbol{X}$, $\dim(\boldsymbol{X}) = p$ through the data collection procedure $\mathscr{T}$. That is, $\mathscr{T}$ mixes all the components of $\boldsymbol{S}$ together. Sometimes (9.4.1) is called a blind source separation system model because there is little or no information about how the components of the signal have been combined to form $\boldsymbol{X}$. For instance, no training set of pairs $(\boldsymbol{s}_i, \boldsymbol{x}_i)$ is available to help identify $\mathscr{T}$. The task of the analyst is to use the n measurements of the mixed components to uncover the original signal $\boldsymbol{S}$ or identify the system operator $\mathscr{T}$, possibly both.

Expression (9.4.1) models the multiple-input/multiple-output setting. Imagine K people at a meeting all talking at the same time, their voices picked up by p microphones. Suppose no one listens to anyone else, so their voices are independent. The waveforms from the microphones represent p different superpositions (i.e., mixtures) of the K waveforms from the voices. Although a single human ear can disentangle the K voices, mechanically separating the K waveforms representing the voices may require the outputs of many, p, microphones. But, this will take work because separating the voices rests on the mixing process, which is unknown.

9.4.1 Main Definitions

If the system is memoryless (i.e., the measurements for time i are independent of those for any other time i'), the system is called instantaneous; if the noise term in (9.4.1) cannot be ignored, the system is noisy; otherwise it is noiseless. If the signals have no time structure, it is reasonable to regard them as IID outcomes of a K-dimensional random vector, $\boldsymbol{S}$, so that $\boldsymbol{X}$, also IID, is the mixture from $\boldsymbol{S}$ via $\mathscr{T}$. Thus, the instantaneous, time-independent, noiseless ICA model is

$$X_i = \mathscr{T}(S_i) \text{ for } i = 1, ..., n, \tag{9.4.2}$$

where $\boldsymbol{S}$ and $\mathscr{T}$ may be unknown but the $\boldsymbol{X}$s are available. When $\mathscr{T}$ is assumed to be a matrix, ICA is a linear method. The simplification to linearity is not necessary, but permitting nonlinear $\mathscr{T}$s introduces substantial difficulties; see Eriksson (2004) for a partial treatment. Thus, the focus is usually on the special case of (9.4.2)

$$\boldsymbol{X} = \boldsymbol{T}\boldsymbol{S}, \tag{9.4.3}$$

where $\boldsymbol{T}$ is a $p \times K$ real matrix.

By definition, a linear ICA representation of $\boldsymbol{X}$ is any pair $(\boldsymbol{S}, \boldsymbol{T})$ such that (i) (9.4.3) holds and (ii) the components of $\boldsymbol{S} = (S_1, ..., S_K)$ are as independent of each other as possible, in the sense of minimizing some measure $\Delta(S_1, ..., S_K)$ of dependence. Clearly, (9.4.3) is similar in spirit to (9.2.1), but the focus here is on the nonnormal case. Moreover, the optimization in clause (ii) imposes a stringent condition. Independence is a much stronger condition than, for instance, noncorrelation, which only depends on the first two moments. When the components of $\boldsymbol{S}$ are independent, or nearly so, the reduction from $\boldsymbol{X}$ to $\boldsymbol{S}$ is efficient since independent random variables typically have more information than the same variables would if they were dependent. Note that, like FA and PCA, there is some ambiguity in K. If p is given, often $p = K$ is chosen. Sometimes one is forced to a model with $p < K$. However, the focus here will be on $p \geq K$ since this corresponds to dimension reduction; deciding how small to make K is a separate problem not treated here, except to note that $K = p$ is the default value.

Representations $(\boldsymbol{S}, \boldsymbol{T})$ are not usually unique. Indeed, from simply writing $\boldsymbol{X} = \boldsymbol{T}\boldsymbol{S} = (\boldsymbol{T}\boldsymbol{M})(\boldsymbol{M}^{-1})\boldsymbol{S}$ and setting $\boldsymbol{M} = \boldsymbol{\Lambda}\boldsymbol{P}$, where $\boldsymbol{\Lambda}$ is a full-rank diagonal matrix and $\boldsymbol{P}$ is a permutation, it is seen that the representation is unchanged. That is, the variances of the S_ks cannot be specified since any scalar multiple of an S_k can be cancelled by using its inverse in the kth column of $\boldsymbol{T}$. Also, the order of the elements in $\boldsymbol{S}$ cannot be determined. Without loss of generality, it is therefore conventional to "sphere" the data (i.e., replace $\boldsymbol{x}_i$ with $\hat{\Sigma}^{-1/2}(\boldsymbol{x}_i - \bar{\boldsymbol{x}})$ so that $\mathbb{E}(S_k) = 0$ and $\mathbb{E}(S_k^2) = 1$, forcing $\boldsymbol{\Lambda} = Id_K$), but the permutation ambiguity remains, unlike with PCs for instance. If two or more of the S_ks are normal, then choosing $\boldsymbol{M}$ to be orthogonal on those dimensions means the representation is unchanged, the same as in FA.

To reduce these ambiguities, representations $(\boldsymbol{S}, \boldsymbol{T})$ for $\boldsymbol{X}$ are usually assumed to be reduced; i.e., no two columns in $\boldsymbol{T}$ are collinear. If a representation is not reduced, then it is hard to obtain any kind of uniqueness because if one column is a multiple of another, then $\bar{\boldsymbol{x}}$ has a representation with only $K - 1$ of the S_ks by combining the two columns with arbitrary coefficients. Even worse, as noted by Eriksson (2004), if one of the source random variables has a divisible distribution, then $\boldsymbol{X}$ would have representations for many $K' > K$. (A divisible distribution is one whose random variable can be written as a sum of other distributions of the same form. Distributions such as the normal, Cauchy, Poisson, and Gamma are infinitely divisible and so give representations for any $K' > K$.) In addition, different Δs give different ICA representations unless actual independence is achieved.

Comon (1994), Corollary 13 established the first rigorous result ensuring model (9.4.3) can be implemented. Eriksson and Koivunen (2003) and Eriksson and Koivunen (2004) refined those results, defining a hierarchy of criteria for representations.

In the parlance of signal processing, the model (9.4.3) is

- ☐ identifiable if, for any two reduced representations of (9.4.3), $(\boldsymbol{T}_1, \boldsymbol{S}_1)$ and $(\boldsymbol{T}_2, \boldsymbol{S}_2)$ for $\boldsymbol{X}$, every column of $\boldsymbol{T}_1$ is a multiple of a column in $\boldsymbol{T}_2$ and the reverse;
- ☐ unique if, in addition to identifiability, the sources $\boldsymbol{S}_1$ and $\boldsymbol{S}_2$ have the same distribution, apart from location and scale;
- ☐ separable if, for every matrix $\boldsymbol{W}$ for which $\boldsymbol{WX}$ has independent components, there is a diagonal matrix $\boldsymbol{\Lambda}$ with all diagonal entries nonzero and a permutation matrix $\boldsymbol{P}$ such that

$$\boldsymbol{\Lambda PS} = \boldsymbol{WX}.$$

Identifiability is a property of the system operator; without identifiability there is no single well-defined mixing matrix. Here, uniqueness means that both $\boldsymbol{T}$ and $\boldsymbol{S}$ are unambiguously specified. Separability essentially removes all but permutation indeterminacy and ensures $\boldsymbol{W}$ can be regarded as virtually a pseudoinverse of $\boldsymbol{T}$. Trivially, uniqueness implies identifiability; it will be seen that separability implies uniqueness. However, the implications are not reversible. Eriksson and Koivunen (2003) gives examples of nonidentifiability, identifiability without uniqueness or separability, and uniqueness without separability.

9.4.2 Key Results

Eriksson and Koivunen (2003), Eriksson and Koivunen (2004) establish a series of results that give (1) sufficient conditions for identifiability, (2) conditions equivalent to separability, and (3) that separability implies uniqueness. To start the reasoning, recall a result from Kagan et al. (1973) stemming from the characterization of the normal as invariant under affine transformations.

Theorem (Kagan et al., 1973): Let $(\boldsymbol{T}_1, \boldsymbol{S}_1)$ and $(\boldsymbol{T}_2, \boldsymbol{S}_2)$ be two representations of $\boldsymbol{X}$, where $\boldsymbol{T}_1$ is $p \times K_1$ and $\boldsymbol{T}_2$ is $p \times K_2$ with $\dim(S_1) = K_1$ and $\dim(S_2) = K_2$. If the ith column of $\boldsymbol{T}_1$ is not a multiple of any column in $\boldsymbol{T}_2$, then S_i is normal.

Proof: Omitted. ☐

Now (1) can be stated and proved.

Theorem (Eriksson and Koivunen 2004): Part (i): If $\boldsymbol{X}$ has no X_i that is normally distributed, then model (9.4.3) is identifiable among all representations $(\boldsymbol{T}, \boldsymbol{S})$.

Part (ii): If $\boldsymbol{T}$ is of full column rank and at most one entry in $\boldsymbol{X}$ is normal, then again model (9.4.3) is identifiable among all representations $(\boldsymbol{T}, \boldsymbol{S})$.

Proof (sketch): Part (i) If there are no normal random variables, then the last theorem implies each column of $\boldsymbol{T}$ must be linearly dependent on some column in any other representation, so the model is identifiable.

Part (ii) By the theorem of Kagan et al., the columns corresponding to nonnormal random variables in $(\boldsymbol{T}_1, \boldsymbol{S}_1)$ and $(\boldsymbol{T}_2, \boldsymbol{S}_2)$ are identifiable. By the assumption on the column ranks and the fact that $\text{rank}(\boldsymbol{T}_1) = \text{rank}(\boldsymbol{T}_2) = K$, there are either $K-1$ or K columns in both representations. Since representations are assumed reduced, every column in one representation is a multiple of a column in the other. So, both have the same number, 0 or 1, of normal entries. If both have 0, this is Part (i).

If both have 1, then without loss of generality the first $K-1$ columns can be regarded as the same. Now, let $\boldsymbol{T}_1^- = (\boldsymbol{T}_1^\mathsf{T}\boldsymbol{T}_1)^{-1}\boldsymbol{T}_1^\mathsf{T}$ be the pseudoinverse of $\boldsymbol{T}_1$. If the two representations are simultaneously valid, then

$$\boldsymbol{S}_1 = \boldsymbol{T}_1^-\boldsymbol{X} = \boldsymbol{T}_1^-\boldsymbol{T}_2\boldsymbol{S}_2 = [I_{K-1,0}\boldsymbol{S}_{2,K-1}, \boldsymbol{T}_1^-\boldsymbol{t}_{2,K}S_{2,K}],$$

where the right-hand side is a matrix defined by its columns: $\boldsymbol{S}_{2,K-1}$ and $\boldsymbol{S}_{2,K}$ are the first $K-1$ entries and the Kth entry of $\boldsymbol{S}_2$, $I_{K-1,0}$ is the $K-1 \times K-1$ identity with K zeros attached as a Kth row at the bottom, and $\boldsymbol{t}_{2,K}$ is the last (Kth) column of $\boldsymbol{T}_2$, which is nonzero because the sources are nondegenerate.

An argument with characteristic functions based on the partition $\boldsymbol{T}_1 = [\boldsymbol{T}_{1,K-1}, \boldsymbol{t}_{1,K}]$ in which $\boldsymbol{T}_{1,K-1}$ is the first $K-1$ columns and $\boldsymbol{t}_{1,K}$ is the last column of $\boldsymbol{T}_1$ gives that $\boldsymbol{t}_{1,K}$ is a multiple of $\boldsymbol{t}_{2,K}$; see Eriksson and Koivunen (2004), p. 603, and Lemma A.1 in Eriksson and Koivunen (2003). □

Using (1), (2) can be stated and proved in the following.

Theorem (Eriksson and Koivunen, 2004): Model (9.4.3) is separable if and only if $\boldsymbol{T}$ is of full column rank and at most one source variable is normal.

Proof: Suppose (9.4.3) is separable and that there are two normal variables in $\boldsymbol{S}$. By way of contradiction, to see that at most one variable is normal, recall that multiplying any bivariate normal random vector by an orthogonal matrix and then by a diagonal matrix preserves normality. So, $\exists \boldsymbol{W}$ of full column rank K so that $\boldsymbol{WT}$ is not of the form $\boldsymbol{\Lambda P}$ and $\boldsymbol{WX}$ has K independent components. Thus, the model is not separable, contradicting the hypothesis. To see that $\boldsymbol{T}$ is of full column rank, observe that $\text{rank}(\boldsymbol{T}) \geq \text{rank}(\boldsymbol{WT}) = \text{rank}(\boldsymbol{\Lambda P}) = K$ and it is seen that the $K \times p$ matrix $\boldsymbol{T}$ is of full rank K.

For the other direction, let $\boldsymbol{T}$ be of full column rank for (9.4.3) and assume that at most one source is normal.

Suppose another matrix of full column rank separates (9.4.3); that is, there is a $\boldsymbol{W}$ such that $\boldsymbol{WX} = \boldsymbol{R}$ has independent components. Then $\boldsymbol{X} = \boldsymbol{TS} = \boldsymbol{W}^-\boldsymbol{R}$. By Part II of the last theorem, $\boldsymbol{W}^-$ must be of the form $\boldsymbol{T}$ times diagonal times permutation; i.e., $\boldsymbol{W}^- = \boldsymbol{T\Lambda P}$. Now,

$$\boldsymbol{S} = \boldsymbol{T}^-\boldsymbol{X} = \boldsymbol{T}^-\boldsymbol{W}^-\boldsymbol{R} = \boldsymbol{T}^-\boldsymbol{T\Lambda PR} = \boldsymbol{\Lambda P}. \ \square$$

Finally, (3) can be extracted from (2) as a corollary.

Corollary (Eriksson and Koivunen, 2004): If (9.4.3) is separable, then it is unique.

Proof (sketch): The identifiability follows from the theorem and Part II of the theorem before. If that proof is written out in detail, it will be seen that the distribution of $\boldsymbol{S}$ is unique up to an affine transformation. □

Equipped with these results, it is important to define which class of measures of dependence can be used as objective functions. Comon (1994) restricts to contrasts. A contrast is any measure of dependence $\Delta(S_1, ..., S_K)$ that is (i) invariant under permutation of the S_ks, (ii) invariant under diagonal invertible transformations (that is, $\Delta(\boldsymbol{S}) = \Delta(\boldsymbol{AX})$ for any matrix $\boldsymbol{A}$) and (iii) $\Delta(\boldsymbol{AX}) \leq \Delta(\boldsymbol{S})$ for $\boldsymbol{A}$ invertible. Examples of contrasts include entropy-based measures (e.g., the Shannon mutual information) or Csiszar ϕ-divergences. Indeed, Hyvarinen (1999), motivated by Jones and Sibson (1987), suggests the negentropy $J(\boldsymbol{X}) = H(\boldsymbol{X}_G) - H(\boldsymbol{X})$, where H is the usual entropy and $\boldsymbol{X}_G$ is a normal variable with the same variance matrix as $\boldsymbol{X}$ (both have mean zero). These, however, can be hard to implement computationally. So, Hyvarinen (1999) uses maximum entropy methods to obtain approximations of the form

$$J(\boldsymbol{X}) \approx \sum_{j=1}^{p} k_j [\mathbb{E}(G_j(\boldsymbol{X})) - \mathbb{E}(G_j(\boldsymbol{v}))]^2, \qquad (9.4.4)$$

where $k_j > 0$, v is $N(0,1)$, and G_j are nonquadratic functions depending on the optimization. Choices $G(x) = (1/a)\ln\cosh(ax)$ and $G(x) = -\exp(x^2/2)$ have been found useful.

9.4.3 Computational Approach

The intuition for identifying the $\boldsymbol{S}$ for a representation of $\boldsymbol{X}$ using $\boldsymbol{T}$ in a linear ICA is that "nonnormal is independent". To see this, suppose that all the S_j are identical. Then, to find S_1, consider $y = \boldsymbol{w}^{\top}\boldsymbol{x}$, where $\boldsymbol{w} \in \mathbb{R}^p$ is to be determined. If $\boldsymbol{w}$ were a row in $\boldsymbol{T}^{-1}$, then y would actually equal one of the S_js. In an effort to find $\boldsymbol{w}$, follow Hyvarinen and Oja (2000) in setting $\boldsymbol{z} = \boldsymbol{T}^{\top}\boldsymbol{w}$ so that $y = \boldsymbol{w}^{\top}\boldsymbol{TS} = \boldsymbol{z}^{\top}$. Using the principle that a sum of variables is more normal than any of the variables in the sum, the fact that y is a linear combination of the S_js with weights given by $\boldsymbol{z}$ means that y is least normal when it equals one of the S_js, in which case only one entry in $\boldsymbol{z}$ is nonzero. So, it makes sense to choose $\boldsymbol{w}$ to maximize the nonnormality of y, for instance by using the negentropy $J(\boldsymbol{X})$ in (9.4.4), hoping to find a $\boldsymbol{w}$ for which $\boldsymbol{w}^{\top}\boldsymbol{x} = \boldsymbol{z}^{\top}\boldsymbol{S}$ and gives one of the S_js. There are $2K$ local maxima, one for each of $\pm S_j$. To help identify them, one can insist that the estimates of the S_js be uncorrelated with the previous ones found.

In practice, one needs an algorithm for maximizing contrast functions such as those of the form (9.4.4). One popular approach is called FastICA; it is a fixed-point iteration scheme to find a unit vector $\boldsymbol{w}$ such that the projection $\boldsymbol{w}^{\top}\boldsymbol{x}$ is maximally nonnormal as measured by $J(\boldsymbol{w}^{\top}\boldsymbol{X})$. Assume that the data have been sphered; i.e., relocated to mean zero by subtracting the sample mean and transformed to have identity covariance by multiplying by $\hat{\Sigma}^{-1/2}$. Letting g be the derivative of G, the basic form of FastICA to find a single component is the following.

Start with a sphered sample x_i, $i = 1, ..., n$, and an index I, an initial w_0.

☐ Find $w_1^+ = \mathbb{E}(Xg(w^\top X)) - \mathbb{E}(g'(w^\top X)w)$.

☐ Replace w_0 with $w_1 = w_1^+/\|w_1^+\|$.

☐ Similarly, find w_2, w_3 until a convergence criterion is met.

The basic justification (see Hyvarinen and Oja (2000)) is that the maxima of the approximation of the negentropy of $w^\top x$ are obtained at certain optima of $\mathbb{E}(G(w^\top x))$. The Kuhn-Tucker conditions give that the optima of $\mathbb{E}(G(wX))$ subject to $\mathbb{E}(w^\top X)^2 = \|w\|^2 = 1$ occur when w satisfies

$$F(w) = \mathbb{E}(Xg(w^\top x)) - \beta w = 0,$$

where β is a Lagrange multiplier. Applying Newton's method, the Jacobian matrix $JF(w)$ of the left-hand side is

$$JF(w) = \mathbb{E}(XX^\top g'(w^\top X)) - \beta I_p.$$

Since this matrix must be inverted, it helps to approximate the first term by

$$\mathbb{E}(XX^\top g'(w^\top X)) \approx \mathbb{E}(XX^\top)\mathbb{E}(g'(w^\top X)) = \mathbb{E}(g'(w^\top X))I_p,$$

which is diagonal and easily inverted. Now, the Newton iteration is

$$w^+ = w - \frac{\mathbb{E}(Xg(w^\top x)) - \beta w}{\mathbb{E}(g'(wX)) - \beta},$$

which can be further simplified by multiplying by $\beta - E(g'(wX))$ on both sides. Some algebraic simplification gives the FastICA iteration above. Of course, in practice, sample-based estimates would be used for the population quantities. Also, to find successive S_js, a Gram-Schmidt-based decorrelation procedure would be imposed; see Hyvarinen and Oja (2000) for details.

9.5 Nonlinear PCs and ICA

Aside from being more difficult, nonlinear versions of linear ideas exhibit a wider variety. Thus, there are several definitions of nonlinear PCs and many nonlinear ICA models. A few of these are mentioned here to indicate the range of possibilities. The hope is that well-chosen nonlinear functions will do a better job of encapsulating the information in a collection of random variables than linear functions of them do.

9.5.1 Nonlinear PCs

Recall that linear PCs are based on finding eigenvalues and eigenvectors of $\mathrm{Var}(\boldsymbol{X}) = \Sigma$. There are at least two ways to generalize PCs to reflect the behavior of nonlinear functions. The most direct is due to Mizuta (1983), Mizuta (2004). The idea is to choose real-valued functions f_k of $\boldsymbol{x}$ for $k = 1, ..., p$ and do the usual PC analysis on the covariance matrix of $\boldsymbol{f}(\boldsymbol{x}) = (f_1(\boldsymbol{x}), ..., f_p(\boldsymbol{x}))$. So, in place of Σ, consider $\Sigma^* = \mathrm{Cov}(\boldsymbol{f}, \boldsymbol{f})$ and suppose Σ^* has eigenvalue, eigenvector pairs $(\lambda_1^*, \boldsymbol{e}_1^*)$, ..., $(\lambda_p^*, \boldsymbol{e}_p^*)$ with $\lambda_1^* \geq \ldots \geq \lambda_p^*$. Then, the nonlinear PC vector is $\boldsymbol{U} = (U_1, ..., U_p)$, in which the jth nonlinear principal component is

$$U_j = \sum_{k=1}^{p} \boldsymbol{e}_{j,k}^* f_k(\boldsymbol{x}) = \boldsymbol{e}_j^{*\mathsf{T}} \boldsymbol{f}(\boldsymbol{x}).$$

Clearly, if $f_k(\boldsymbol{x}) = x_k$, this reduces to PCs. Quadratic PCs, QPCs, are obtained by choosing f to have components obtained from the linear and second order terms in the x_js. For $p = 2$, this is five-dimensional, with $f_1(\boldsymbol{x}) = x_1$, $f_2(\boldsymbol{x}) = x_2$, $f_3(\boldsymbol{x}) = x_1^2$, $f_4(\boldsymbol{x}) = x_1 x_2$, and $f_5(\boldsymbol{x}) = x_2^2$.

Usually, the f_ks are assumed (i) linearly independent and (ii) continuous. Moreover (iii), for any orthogonal matrix $\boldsymbol{T}$, it is assumed that there is a matrix $\boldsymbol{W}$ such that $\boldsymbol{f}(\boldsymbol{T}\boldsymbol{x}) = \boldsymbol{W}\boldsymbol{f}(\boldsymbol{x})$; this is useful for understanding how orthogonal coordinate transformations affect the results. It can be shown (see Mizuta (1983), Mizuta (2004)), that orthogonal transformations of the $\boldsymbol{X}$ coordinates do not affect the new, $\boldsymbol{U}$ coordinates, and this is typically the case under the three standard conditions.

A second way to define nonlinear PCs is due to Karhunen et al. (1998) and introduces the nonlinearity via the objective function. Let $\boldsymbol{f} : \mathbb{R}^p \to \mathbb{R}^p$ be a nonlinear function, $\boldsymbol{W}$ a $p \times p$ real weight matrix, and $\|\cdot\|$ a norm such as L_2. Set

$$J(\boldsymbol{W}) = E\|\boldsymbol{X} - \boldsymbol{W}\boldsymbol{f}(\boldsymbol{W}^{\mathsf{T}}\boldsymbol{X})\|. \tag{9.5.1}$$

Now, J can be optimized, by gradient descent for instance, to find $\boldsymbol{W}$. (With learning rate c, one update is $\Delta\boldsymbol{W} = c(\boldsymbol{x} - \boldsymbol{W}\boldsymbol{f}(\boldsymbol{W}^{\mathsf{T}}\boldsymbol{x}))\boldsymbol{f}(\boldsymbol{x}^{\mathsf{T}}\boldsymbol{W})$.) The goal is to find $\boldsymbol{U} = \boldsymbol{W}\boldsymbol{X}$, where $\boldsymbol{W}$ depends on $\boldsymbol{f}$. Sensible $\boldsymbol{f}$s are usually odd functions with domain and range $\mathbb{R}$. In practice, (9.5.1) is usually optimized using estimates in place of population values, and the data are usually sphered.

A variant on this (see Fodor (2002)) is to use the usual sequential optimization in the PC procedure on the variance of $f(\boldsymbol{w}^{\mathsf{T}}\boldsymbol{x})$. So, the first PC is

$$\boldsymbol{w}_1 = \arg\max_{\|\boldsymbol{w}\|=1} \mathrm{Var}(f(\boldsymbol{w}^{\mathsf{T}}\boldsymbol{x})).$$

The procedure can be repeated to find $\boldsymbol{w}_2$ among vectors orthogonal to $\boldsymbol{w}_1$. It is unclear how this method works. A further variant would be to insist that subsequent PCs be defined as in (9.1.2) but with the orthogonality between $f(\boldsymbol{w}_1^{\mathsf{T}}\boldsymbol{x})$ and the next $f(\boldsymbol{w}^{\mathsf{T}}\boldsymbol{x})$. More recently, de Leeuw (2005) has given a different perspective on the general problem of nonlinear PCs.

9.5.2 Nonlinear ICA

Nonlinear ICA is more complicated than nonlinear PCs and amounts to treating (9.4.2) assuming only the memoryless property and noiselessness. Eriksson (2004) describes a nonlinear mixture model

$$X_j = f_j\left(\sum_{k=1}^{K} a_{jk}S_k\right) = f_j(\boldsymbol{AS}),$$

in which each X_i is a linear model distorted by f_j. In this case, a separating structure is a pair $(\boldsymbol{G},\boldsymbol{W})$, where $\boldsymbol{G}$ is also a componentwise nonlinear function such that $\boldsymbol{WG}(\boldsymbol{X})$ has K mutually independent components. The model is separable if and only if, for any separating structure $(\boldsymbol{G},\boldsymbol{W})$, the random vector $\boldsymbol{WG}(\boldsymbol{X})$ is a permuted scaled version of $\boldsymbol{S}$. This notion of separability is satisfied when (i) $K = p$, (ii) $\boldsymbol{A}$ is invertible and has at least two nonzero entries in each row and column, (iii) the f_js are differentiable and invertible so that $G_k \circ f_k$ has nonzero derivative, and (iv) the density of each S_js is zero at least at one point; see Taleb et al. (1999) for a formal proof.

Eriksson (2004) established that if the class of system operators of which $\mathscr{T}$ is a member is restricted to being a group, then a separable ICA model results. The idea is to write $\mathscr{T}$ as a function $\boldsymbol{f}(\boldsymbol{S}) = (f_1(\boldsymbol{S}),...,f_p(\boldsymbol{S}))$, where each f_j is a group action on $\boldsymbol{S}$. Sometimes these are called addition theorem models because the group operation is built up from continuous, strictly increasing functions tied together by several steps involving composition and inversion.

More general nonlinear ICA models often do not admit unique solutions. Indeed, if the space of the nonlinear mixing functions f_j is not limited, multiple solutions will always exist; see Karhunen (2001), who discusses the problem more generally.

There are other nonlinear dimension reduction techniques such as that in Globerson and Tishby (2003), which uses Shannon information; latent semantic indexing which uses a singular value decomposition; and a variety of maximum entropy methods along with techniques such as multidimensional scaling and self-organizing maps, to be discussed later in this chapter, as visualization techniques. All of these methods are currently under development, including efforts to kernelize them to help escape the Curse of Dimensionality; see Scholkopf et al. (1998).

9.6 Geometric Summarization

The goal here is to reduce the dimension of the space of explanatory variables by finding lower-dimensional structures, usually submanifolds, within it. If the realized values of the explanatory variables concentrate on those submanifolds, then the space of explanatory variables can be recoordinatized with fewer dimensions. The paradigm example would be collinearity in linear regression. For instance, if $p = 2$ and $X_1 = \alpha X_2 + \eta$, where $\mathrm{Var}(\eta)$ is small, then it makes sense to model a response Y

as $Y = \beta_0 + \beta_1 X_1 + \varepsilon$ instead of $Y = \beta_0 + \beta_1 X_1 + \beta_2 X_2 + \varepsilon$. In effect, X_1 parametrizes a one-dimensional submanifold of (X_1, X_2) space on which the explanatory variables concentrate. More generally, the problem is finding the lower-dimensional submanifold and verifying that the dimension reduction provides a good summarization.

One way to frame this problem is in terms of algebraic shapes. An algebraic shape is the set of zeros of a set of polynomials. Specifically, consider a function $\boldsymbol{f} : \mathbb{R}^p \to \mathbb{R}^K$, where $\boldsymbol{f} = (f_1, ..., f_K)$ and each f_k is a polynomial in $\boldsymbol{x} \in \mathbb{R}^p$. A (p,K) algebraic curve or surface is the zero set of the f_ks. That is,

$$Z(\boldsymbol{f}) = \{\boldsymbol{x} \in \mathbb{R}^p | \ f_1(\boldsymbol{x}) = ... = f_K(\boldsymbol{x}) = 0\}$$

is a (p,K) algebraic shape. If $p - K = 1$, it defines a curve in $\mathbb{R}^p$; if $p - K = 2$, it defines a surface in $\mathbb{R}^p$. In general, K constraints on p variables gives a solution set of dimension $p - K$. Clearly, most curves and surfaces in any dimension can be well approximated by an algebraic shape since polynomials can be used to approximate any function. When the f_ks are parametrized as well (i.e., each f_k is of the form $f_k(\boldsymbol{x}|\theta_k)$), then $\theta = (\theta_1, \ldots, \theta_K)$ parametrizes a collection of submanifolds and the result is a parametrized class of (p,K) algebraic shapes. Below, some of the geometric properties of these shapes are discussed. Then, some of their statistical properties are given. It will be seen that these areas admit substantial further development because techniques for evaluating the adequacy of the summarization are not yet available.

9.6.1 Measuring Distances to an Algebraic Shape

Consider using a parametrized class of (p,K) algebraic curves or surfaces to summarize a data set $\boldsymbol{x}_1, ..., \boldsymbol{x}_n$. The adequacy of summarization depends on the distance of the data points to the curves or surfaces; i.e., the distances from the $\boldsymbol{x}_i$s as elements of $\mathbb{R}^p$ to any shape $Z(\boldsymbol{f}_\theta)$ in the class. An empirical optimality criterion for fit is, for fixed θ,

$$L(Z(\boldsymbol{f}), \boldsymbol{x}_1, ..., \boldsymbol{x}_n) = \sum_{i=1}^{n} \text{dist}(\boldsymbol{x}_i, Z(\boldsymbol{f})) = \sum_{i=1}^{n} \inf_{z \in Z(\boldsymbol{f})} ||\boldsymbol{x}_i - \boldsymbol{z}||, \tag{9.6.1}$$

where $Z(\boldsymbol{f}) = Z(\boldsymbol{f}_\theta)$. A second minimization would be required to find the θ indexing the manifold giving the empirically optimal data summarization.

Unfortunately, it can be hard to identify the distance from a given point $\boldsymbol{x}_i \notin Z(f)$ to the closest point $\boldsymbol{x}$ in $Z(\boldsymbol{f})$ in closed-form expressions. For instance, Mizuta (2004) Section 6.3.2.3 gives a direct technique based on partial differential equations; this is hard to use because such equations are not easy to solve and the parameters they introduce must be estimated. Fortunately, several other approximations have been proposed. The crudest is the algebraic distance

$$\text{dist}(\boldsymbol{x}, Z(f)) \approx \sum_{k=1}^{K} f_k(x)^2,$$

which makes sense because it gives zero if $\boldsymbol{x} \in Z(\boldsymbol{f})$. The Taubin approximation (see Taubin (1988), and Taubin et al. (1994)) rests on writing $f(\boldsymbol{x}) - f(\hat{\boldsymbol{x}}) = (\boldsymbol{x} - \hat{\boldsymbol{x}})\nabla f(\boldsymbol{x})$, where $\boldsymbol{x}$ is a given point and $\hat{\boldsymbol{x}}$ is the point in $Z(\boldsymbol{f})$ closest to it. Since $f(\hat{\boldsymbol{x}}) = 0$, rearranging gives

$$\text{dist}(\boldsymbol{x}, Z(\boldsymbol{f})) \approx \frac{\sum_{k=1}^{K} f_k(\boldsymbol{x})^2}{\|\nabla \boldsymbol{f}(\boldsymbol{x})\|^2}, \tag{9.6.2}$$

which can be used in (9.6.1). Since (9.6.2) gives anomalous results when $\nabla \boldsymbol{f}(\boldsymbol{x})$ is near zero, a further refinement is found in Taubin et al. (1994), exp. 4.

Minimizing (9.6.1) for various classes of $Z(\boldsymbol{f})$ is not stochastic: It finds a minimum irrespective of any distributional structure, just like least squares fitting of a hyperplane in linear regression. Indeed, minimizing in (9.6.1) is a sort of generalization of least squares fitting. To see this, note that, in principle, minimization can be done directly on the distance or approximate distance in (9.6.1) or (9.6.2). In this case, the class of $Z(\boldsymbol{f})$ comes from the zero sets of a collection of polynomials $f_k(S\boldsymbol{x}|\theta_k)$ in which θ_k is multidimensional and represents the coefficients of monomials $x_1^{d_1}...x_p^{d_p}$ appearing in f_k, $\boldsymbol{\theta} = (\theta_1, ..., \theta_K)$. Now, the content of (9.6.1) can be expressed in a more familiar form. Let $\hat{\boldsymbol{x}}_i$ be the point in $Z(\boldsymbol{f})$ closest to $\boldsymbol{x}_i$ for $i = 1, ..., n$. Then the $\hat{\boldsymbol{x}}_i$s depend on $\boldsymbol{\theta}$ and the natural squared error is

$$R = \sum_{i=1}^{n} (\boldsymbol{x}_i - \hat{\boldsymbol{x}}_i)^{\mathsf{T}} (\boldsymbol{x}_i - \hat{\boldsymbol{x}}_i). \tag{9.6.3}$$

Clearly, R depends on $\boldsymbol{\theta}$, the data, and $Z(\boldsymbol{f})$.

Like the linear regression case, the partial derivatives of R with respect to the components of $\boldsymbol{\theta}$ can be found, and used to give a set of equations that can be solved; see Mizuta (2004). This is called the Levenberg-Marquardt method. See Faber and Fisher (2001a) and Faber and Fisher (2001b) for a detailed description and comparative examples. Taubin et al. (1994) proposed a representation of the space of polynomials that enables fitting to be done more readily. More recently, Helzer et al. (2004) refined the Taubin approximate approach to get better robustness with noisy data.

One limitation of this approach is that the estimates obtained by minimizing (9.6.3), or expressions like it, do not have a sampling distribution. So, inferences from an algebraic approach do not have confidence or credibility associated with them. Nevertheless, the results can be suggestive and effective. The next section reports on several efforts to stochasticize algebraic fitting.

9.6.2 Principal Curves and Surfaces

Principal curves and surfaces were invented by Hastie (1984); they represent a curve or surface as the expectation of a conditional population. This is one way to include an error structure in (9.6.1).

Consider curves first. A curve in p dimensions is the image of a function $\boldsymbol{f}: \mathbb{R} \to \mathbb{R}^p$ of the form $\boldsymbol{f} = (f_1(s), ..., f_p(s))$; it is conventional to use s to mean the arclength parametrization of the curve (i.e., s is the length of travel from $f(0)$ to $\boldsymbol{f}(s)$ along the curve and, for all s, $\|\boldsymbol{f}'(s)\| = 1$). Now, suppose a p-dimensional random variable $\boldsymbol{X}$ with distribution q is mean zero and has finite second moments. The projection of a value $\boldsymbol{X} = \boldsymbol{x}$ onto the curve $\boldsymbol{f}$ can be written as

$$s_f(\boldsymbol{x}) = \sup\{s : \|\boldsymbol{x} - \boldsymbol{f}(\boldsymbol{s})\| = \inf_u \|\boldsymbol{x} - \boldsymbol{f}(u)\|\}; \qquad (9.6.4)$$

i.e., $\boldsymbol{x}$ is associated with the value of s for the point $\boldsymbol{f}(s)$ that is closest to $\boldsymbol{x}$, and ties are broken by choosing the largest. In well-behaved cases, a hyperplane of points $\boldsymbol{x}$ gets projected to the same s_f.

It is now reasonable to define the curve $\boldsymbol{f}$ as a principal curve of q if it can be represented as the conditional expectation of $\boldsymbol{X}$ over such $\boldsymbol{x}$s. Formally, $\boldsymbol{f}$ is a principal curve of q if and only if

$$\boldsymbol{f}(s) = E_q(\boldsymbol{X} | s_f(\boldsymbol{x}) = s). \qquad (9.6.5)$$

This says that, for each s, averaging over the points in the $(p-1)$-dimensional manifold $s_f(\boldsymbol{x}) = s$ defined by s gives the point on the curve at distance s. Empirically, one would be tempted to average over $\boldsymbol{x}_i$s in the ε-envelope at s, $\{|s_f(\boldsymbol{x}) - s| < \varepsilon\}$, as an approximation to (9.6.5), with ε chosen to ensure a sufficient number of $\boldsymbol{x}_i$s were included; this would give a piecewise approximation to the curve defined by $\boldsymbol{f}$.

Hastie and Stuetzle (1989) note that, for any spherically symmetric distribution, any straight line through its mean vector is a principal curve; e.g., in a standard two-dimensional normal, any line through the origin is a principal curve. More generally, for ellipsoidal distributions, the principal curves include the principal components. The inclusion may be proper because for spherically symmetric distributions a circle of radius $\mathbb{E}\|X\|$ is also a principal curve. Furthermore, if a version of (9.6.5) is applied to surfaces, these properties generalize. A further complication is that Carreira-Perpinan (1997) observes that if a straight line satisfies (9.6.5), then it is a principal component, but that in a model of the form $\boldsymbol{X} = \boldsymbol{g}(s) + \varepsilon$, where $g : \mathbb{R} \to \mathbb{R}^p$ and ε is spherically symmetric, $\boldsymbol{f}$ seems not to be a principal curve. The bias, though, may be small and decrease to zero if $\text{Var}(\varepsilon)/\kappa(f) \to 0$, where κ is the radius of curvature of $\boldsymbol{f}$. (The radius of curvature of $\boldsymbol{f}(s)$ is the radius of a circle that would be tangent to the curve at $\boldsymbol{f}(s)$; it gets large as $\boldsymbol{f}$ gets closer to a straight line, which has infinite radius of curvature.)

Hastie and Stuetzle (1989) propose a simple algorithm for finding principal curves. The idea is to start with the first PC, use it to define the projection in (9.6.4), and then use this projection on the right-hand side of (9.6.5) to calculate a new curve onto which points can be projected, cycling until convergence is observed. Applying this "expectation-projection" algorithm data is not hard but involves numerous subjective choices. To state the procedure, represent the principal curve by lines determined from n points $(s_i, \boldsymbol{f}_i)$. Mizuta (2004) gives the following template:

Start with $\boldsymbol{x}_i$ for $i = 1, ..., n$.

- ☐ Initialize by setting $\boldsymbol{f}^0(s) = \bar{\boldsymbol{x}} + \boldsymbol{u}s$, where $\boldsymbol{u}$ is the first PC.
- ☐ Expectation Step to update $\boldsymbol{f}(s)$: For $j = 1, 2, ...,$ smooth over the entries of $\boldsymbol{X}$ separately given s. Call the result $\boldsymbol{f}^j(s)$. That is, for each s, set

$$\boldsymbol{f}^j(s) = E(\boldsymbol{X} | s_{f^{j-1}}(\boldsymbol{x}) = s).$$

- ☐ Projection Step to update s as a function of $\boldsymbol{x}$: For each $\boldsymbol{x}_i$, search for the point on the curve $\boldsymbol{f}^j(s)$ closest to it and assign it the corresponding s. That is, for each $\boldsymbol{x}$, set

$$s^j(\boldsymbol{x}) = s_{f^j}(\boldsymbol{x})$$

and transform the left-hand side to get the new arclength parametrized curve.
- ☐ If a convergence criterion is met, stop. Otherwise repeat from the expectation step.

One natural criterion for the convergence is to set

$$D(h, \boldsymbol{f}^j) = \mathbb{E}_{s^j}\left(\mathbb{E}\|\boldsymbol{X} - f(s^j(\boldsymbol{X}))\|^2 | \, s^j(\boldsymbol{x})\right),$$

where $h = h_j$ is the distribution of the $\boldsymbol{X}$ satisfying $s_{f^{j-1}}(\boldsymbol{x}) = s$ and require $(D(h, f^{j-1}) - D(h, f^j))/D(h, f^{j-1})$ to be suitably small.

In practice, in this E step, one calculates the expectation with respect to h and uses $f^j(s)$ as an estimate for it. Likewise, in the P step, one projects the data to the curve $f^j(s)$ and assigns $s^j(\boldsymbol{x})$.

This procedure automatically fits a line. It can be generalized, more or less directly, to fit a 2D surface or a hypersurface. Choosing which geometric shape to use as a data summary is a difficult matter for model selection; there is no obvious metric to permit comparisons across dimensions.

A variety of conceptual problems remain with principal curves and surfaces. However, Delicado (2001) has carefully reformulated the definition while retaining its intuition and has a general result ensuring that principal curves exist. This treatment ensures that the theoretical problems from conditioning on sets of measure zero implicit in (9.6.5) are resolved. Moreover, Delicado (2001) proposes algorithms that seem to produce curves appropriately. Principal surfaces remain more difficult, but Chang and Ghosh (2001) have a computational method called probabilistic principal surfaces based on applying (9.6.4) at specific points, and LeBlanc and Tibshirani (1994) have a treatment of surfaces using MARS, effectively replacing the p univariate smoothings in the E step with a single multivariate smoother. Amato et al. (2004) extend probabilistic principal surfaces to an ensemble setting using bagging and stacking in a classification, or prediction, context. A more complete listing of references can be found at `http://www.iro.umontreal.ca/~kegl/research/pcurves/references/`.

9.7 Supervised Dimension Reduction: Partial Least Squares

One of the obvious criticisms of unsupervised dimension reduction is that it explains $\boldsymbol{X}$ but there is no reason to be sure that the result explains a response Y. In addition, since supervised techniques use the extra information in Y, it is reasonable to believe they will usually do better than unsupervised techniques. Therefore, supervised techniques may define a standard for assessing how well an unsupervised technique performs. For both of these reasons, among others, it is worthwhile to look at unsupervised dimension reduction as well.

One of the most important supervised dimension reduction techniques is called partial least squares (PLS). In fact, PLS is a collection of techniques with two common properties: (i) PLS maximizes correlation between Y and $\boldsymbol{X}$ (rather than maximizing the variance of $\boldsymbol{X}$ only, for instance) and (ii) PLS can be interpreted, somewhat like FA, in terms of seeking underlying factors of $\boldsymbol{X}$ that are also underlying factors of Y. Some PLS techniques are defined by procedures; others have a theoretical justification.

There are two versions of PLS presented here. The first presents PLS as a variant of PCs but using correlation in place of variance. The second identifies the model and gives a procedure explicitly. In this context, some general properties of PLS can be stated easily.

9.7.1 Simple PLS

Given data $(Y_i, \boldsymbol{X}_i)$ for $i = 1, ..., n$, let

$$\widehat{\boldsymbol{\Sigma}} = \frac{1}{n} \sum_{i=1}^{n} (\boldsymbol{x}_i - \bar{\boldsymbol{x}})(\boldsymbol{x}_i - \bar{\boldsymbol{x}})^{\mathsf{T}}$$

and

$$\hat{s}_{x,y} = \widehat{\mathrm{Cov}}(Y, \boldsymbol{X}) = \frac{1}{n} \sum_{i=1}^{n} (y_i - \bar{y})(\boldsymbol{x}_i - \bar{\boldsymbol{x}}).$$

Let $d \leq p$ be the number of linear combinations of $\boldsymbol{X}$ believed necessary to model the response, and denote them $\boldsymbol{t}_j$, $j = 1, ..., d$. To find the $\boldsymbol{t}_j$s, write them in terms of a matrix $\boldsymbol{W}$ so that $\boldsymbol{t}_i = \boldsymbol{W}\boldsymbol{x}_i$, where $\boldsymbol{W} = \boldsymbol{W}_d$ is a $d \times p$ weight matrix, $\boldsymbol{W} = [\boldsymbol{w}_1, ..., \boldsymbol{w}_d]^{\mathsf{T}}$. Now, the simple PLS factors for the $\boldsymbol{x}_i$s are given by the $\boldsymbol{w}_j$s, where

$$\boldsymbol{w}_j = \arg\max \left\{ \mathrm{Cov}(Y, \boldsymbol{w}^{\mathsf{T}}) | \; \|\boldsymbol{w}\| = 1, \forall k = 1, ..., j-1 : \boldsymbol{w}^{\mathsf{T}} \widehat{\boldsymbol{\Sigma}} \boldsymbol{w}_k = 0 \right\}. \quad (9.7.1)$$

Note that the first PLS maximizes the correlation between Y and $\boldsymbol{X}$, the second one maximizes the correlation subject to being orthogonal to the first, and so on. That is, the PLS dimensions are linear combinations of the X_js whose vectors of coefficients have norm 1 and are orthogonal to the earlier PLS linear combinations.

Once found, Y can be regressed on the $\boldsymbol{t}_j$s; see Hinkle and Rayens (1994). Typically, the resulting variances of the regression coefficients are smaller than under least squares. However, they can be biased. Even so, this often leads to a lower MSE overall. One standard algorithm for finding the PLS coefficients is in Rosipal and Trejo (2001); Wegelin (2000) provides another variant in the factor analytic setting. Hastie et al. (2001) provides a slightly different formulation of the procedure and a different statement of the optimality PLS vectors satisfy.

9.7.2 PLS Procedures

The most popular way to find PLS dimensions is Wold's nonlinear iterative partial least squares (NIPALS) procedure. It rests on a model similar to FA in which both the explanatory variables $\boldsymbol{X}$ and a q-dimensional response, $\boldsymbol{Y} = (Y_1, \ldots, Y_q)$, are represented in terms of underlying related factors. The relationship is seen in the second step of the procedure below.

9.7.2.1 The PLS Model and NIPALS

At its root, the idea is to use one FA on $\boldsymbol{X}$ and another FA on $\boldsymbol{Y}$ but tie them together by the procedure used to identify the factors. So, following Rosipal and Kramer (2006), write the joint ℓ factor model for $\boldsymbol{X}$ and $\boldsymbol{Y}$ as

$$\begin{aligned} \boldsymbol{X} &= \boldsymbol{T}\boldsymbol{P} + \boldsymbol{E} \\ \boldsymbol{Y} &= \boldsymbol{U}\boldsymbol{Q} + \boldsymbol{F}, \end{aligned} \tag{9.7.2}$$

but regard $\boldsymbol{X} = (\boldsymbol{X}_1, \ldots, \boldsymbol{X}_n)^{\mathsf{T}}$ as an $n \times p$ data matrix and $\boldsymbol{Y} = (\boldsymbol{Y}_1, \ldots, \boldsymbol{Y}_n)^{\mathsf{T}}$ as the corresponding $n \times q$ response matrix; both $\boldsymbol{X}$ and $\boldsymbol{Y}$ are assumed to be relocated to have mean zero. In (9.7.2), $\boldsymbol{T}$ and $\boldsymbol{U}$ are both $n \times \ell$ matrices of the ℓ factors. The $\boldsymbol{P}$, $p \times \ell$, and $\boldsymbol{Q}$, $q \times \ell$ are the factor loadings in the two FA models and the $\boldsymbol{E}$ and $\boldsymbol{F}$ are the $n \times p$, $n \times q$ residual matrices.

The goal is to find ℓ linear combinations from $\boldsymbol{X}$ and ℓ linear combinations of $\boldsymbol{Y}$ to use as new dimensions. Write the first of these as $\boldsymbol{t} = \boldsymbol{X}\boldsymbol{w}$ and $\boldsymbol{u} = \boldsymbol{Y}\boldsymbol{c}$; these will give the columns of $\boldsymbol{T}$ and $\boldsymbol{U}$. Parallel to (9.7.1), the NIPALS procedure is to find $\boldsymbol{w}$ and $\boldsymbol{c}$ such that

$$\mathrm{Cov}^2(\boldsymbol{t}, \boldsymbol{u}) = \mathrm{Cov}^2(\boldsymbol{X}\boldsymbol{w}, \boldsymbol{Y}\boldsymbol{c}) = \max_{\|\boldsymbol{r}\| = \|\boldsymbol{s}\| = 1} \mathrm{Cov}^2(\boldsymbol{X}\boldsymbol{r}, \boldsymbol{Y}\boldsymbol{s}). \tag{9.7.3}$$

NIPALS is an iterative procedure that produces a solution to (9.7.3) in two parts. The first part is called deflation; it sets up the optimization problem that will yield a single factor. The second part solves the optimization problem to find the single factor. Thus, the two parts must be used ℓ times to get ℓ factors.

The second part of the NIPALS procedure is much easier than the first and so is better to explain first. It can be expressed in two steps, one that produces an iterate of $\boldsymbol{t}$ and one that produces an iterate of $\boldsymbol{u}$. Now, the procedure for the second part of NIPALS (i.e., finding a single PLS factor) is the following.

Given a mean zero $n \times p$ data matrix $\boldsymbol{X}$ and a mean zero $n \times q$ response matrix $\boldsymbol{Y}$, choose an initial $\boldsymbol{u}$ to start the procedure.

- ☐ Set $\boldsymbol{w}^* = \boldsymbol{X}^\top \boldsymbol{u}/(\|\boldsymbol{X}^\top \boldsymbol{u}\|)$, $\boldsymbol{w} = \boldsymbol{w}^*/\|\boldsymbol{w}^*\|$, and use this to find $\boldsymbol{t} = \boldsymbol{X}\boldsymbol{w}$.
- ☐ Using $\boldsymbol{t}$, do the same procedure to find $\boldsymbol{u}$: Let $\boldsymbol{c}^* = \boldsymbol{Y}^\top \boldsymbol{t}/(\boldsymbol{t}^\top \boldsymbol{t})$, $\boldsymbol{c} = \boldsymbol{c}^*/\|\boldsymbol{c}^*\|$, and use this to find $\boldsymbol{u} = \boldsymbol{Y}\boldsymbol{c}$.
- ☐ Cycle by replacing the $\boldsymbol{u}$ in the first step with $\boldsymbol{u} = \boldsymbol{Y}\boldsymbol{c}$ from the last step.

Note that it is in the second step that the underlying factors of $\boldsymbol{X}$ are related to the underlying factors of $\boldsymbol{Y}$ by the expression for $\boldsymbol{c}^*$ and that, if $q = 1$, then $\boldsymbol{U} = \boldsymbol{Y}$ and the procedure converges after one iteration.

The first part of the PLS procedure is harder; it gives the sequence of maximization problems to be solved. The first of these maximization problems has already been identified: It is (9.7.3). Later iterates, to find successive PLS linear combinations, are more difficult and can be found in several ways.

- The simplest way to iterate over the minimization problems is to compute the loading vectors at the kth iteration and use them to form a rank one approximation to $\boldsymbol{X}$ and $\boldsymbol{Y}$ based on $\boldsymbol{t}$ and $\boldsymbol{u}$. Then $\boldsymbol{X}$ and $\boldsymbol{Y}$ are "deflated" by the approximation and the result used to find the next single PLS factor. That is, write the column of $\boldsymbol{P}$ and $\boldsymbol{Q}$ in (9.7.2) as

$$\boldsymbol{p} = \frac{\boldsymbol{X}^\top \boldsymbol{t}}{\boldsymbol{t}^\top \boldsymbol{t}} \quad \text{and} \quad \boldsymbol{q} = \frac{\boldsymbol{Y}^\top \boldsymbol{u}}{\boldsymbol{u}^\top \boldsymbol{u}}$$

and replace $\boldsymbol{X}$ and $\boldsymbol{Y}$ in (9.7.3) with the deflated forms

$$\boldsymbol{X} - \boldsymbol{t}\boldsymbol{p}^\top \quad \text{and} \quad \boldsymbol{Y} - \boldsymbol{u}\boldsymbol{q}^\top. \tag{9.7.4}$$

Then find the next single factor using the same procedure as used to find the last. In this way, hopefully, $\boldsymbol{X}$ and $\boldsymbol{Y}$ are accurately approximated by ℓ dimensions. This is called PLS Mode A (see Rosipal and Kramer (2006)) and was the original formulation.

- An alternative way to deflate $\boldsymbol{X}$ and $\boldsymbol{Y}$ that is more true to the model is to assume

$$\boldsymbol{U} = \boldsymbol{T}\boldsymbol{D} + \boldsymbol{H}$$

at the kth stage, where $\boldsymbol{D}$ is $k \times k$ diagonal and $\boldsymbol{H}$ is the residual matrix. Using (9.7.5) assumes the $\boldsymbol{t}_m$ for $m = 1, \ldots, k$ are good predictors of $\boldsymbol{Y}$, see Rosipal and Kramer (2006). Now, the score vectors, the $\boldsymbol{t}_m$s, can be used to deflate $\boldsymbol{Y}$ as well as $\boldsymbol{X}$. Thus, at each iteration,

$$X = X - tp^{\mathsf{T}} \text{ and } Y = Y - \frac{tt^{\mathsf{T}}}{t^{\mathsf{T}}t} = Y - tc.$$

This guarantees the mutual orthogonality of the t_ms and is called PLS1 (unidimensional Y) or PLS2 (multidimensional Y).

- It can be shown that the first eigenvector of the eigenvalue problem $X^{\mathsf{T}}YY^{\mathsf{T}}Xw = \lambda w$ is w. Computing all the eigenvalues of $X^{\mathsf{T}}YY^{\mathsf{T}}X$ at the start of the procedure can be used to define another form of the PLS procedure.
- A simpler form of PLS arises if the deflation step is omitted. There is some evidence that this gives reasonable results. This coincides with PLS1 but differs from PLS2.

9.7.3 Properties of PLS

The original solution to the correlation maximization in (9.7.1) was as an eigenvalue problem to be solved sequentially for the t_is. Different from the NIPALS algorithm, a singular value decomposition is used on the matrices X and Y. This can be seen informally by using $t = Xw$ and $u = Yc$ to derive

$$w \propto X^{\mathsf{T}}u \propto X^{\mathsf{T}}Yc \propto X^{\mathsf{T}}YY^{\mathsf{T}}t \propto X^{\mathsf{T}}YY^{\mathsf{T}}Xw,$$

giving

$$X^{\mathsf{T}}YY^{\mathsf{T}}Xw = \lambda w$$

and

$$XX^{\mathsf{T}}YY^{\mathsf{T}}t = \lambda t \text{ with } u = YY^{\mathsf{T}}t$$

for the first eigenvalues and first PLS. Expressions for the cs and ws follow, and subsequent PLSs emerge by appropriately deflating the matrices; see Rosipal (2005) or Abdi (2007) for more details.

A second important property of PLSs is that, like PCs and independent components, they can be used in regression. As noted in Rosipal and Kramer (2006), using (9.7.2) and (9.7.4) gives the regression model

$$Y = TDQ^{\mathsf{T}} + (HQ^{\mathsf{T}} + F) = TC^{\mathsf{T}} + F^*, \tag{9.7.5}$$

where $C^{\mathsf{T}} = DQ^{\mathsf{T}}$ is $\ell \times q$ and F^* is the residual. Expression (9.7.5) is formally the same as the usual least squares regression model since $T = XW(P^{\mathsf{T}}W)^{-1}$ can be derived and plugged into (9.7.5) to give

$$Y = XB + F^*,$$

in which $B = W(P^{\mathsf{T}}W)^{-1}C^{\mathsf{T}} = X^{\mathsf{T}}U(T^{\mathsf{T}}XX^{\mathsf{T}}U)^{-1}T^{\mathsf{T}}Y$ after some derivations.

Third, PLS is similar to PCA and CCA. First, the choice of ℓ in all three techniques is not clear. The same sort of techniques as discussed for PCs can be used with PLS. Second, in contrast with (9.7.1) or (9.7.3), CCA and PCA find

$$\max_{\|\alpha\|=\|\beta\|=1} \text{Corr}(\boldsymbol{X\alpha}, \boldsymbol{Y\beta}) \text{ and } \max_{\|\beta\|=1} \text{Var}(\boldsymbol{X\beta}),$$

respectively. Thus, t is seen that PLS is a weighted form of CCA but that both differ from PCs in that they seek linear combinations highly correlated with the response rather than seeking directions with large variability. Also, as written, CCA and PLS are supervised, while PCA is unsupervised. However, $\boldsymbol{X}$ can be partitioned into $\boldsymbol{X} = (\boldsymbol{X}_1, \boldsymbol{X}_2)$, and both CCA and PLS can be described in terms of two matrices, $\boldsymbol{X} = \boldsymbol{X}_1$ and $\boldsymbol{Y} = \boldsymbol{X}_2$, so that no actual response is used. In this way, PLS or CCA can be made unsupervised, although the interpretation would change.

Finally, like PCs and ICA, PLS can be extended. Nonlinear PLS can be defined by extending the linear model to $\boldsymbol{U} = g(\boldsymbol{T}) + \boldsymbol{h} = g(\boldsymbol{X}, \boldsymbol{w}) + \boldsymbol{h}$. Another sort of nonlinearization of PLSs – and PCs and ICA – is via the kernel trick. As before, this uses a function $\boldsymbol{\Phi}$ to give a transformation of the original space into an RKHS that becomes a higher-dimensional feature space. The NIPALS method extends to this case as well.

9.8 Supervised Dimension Reduction: Sufficient Dimensions in Regression

In regression models for a single response Y as a function of a p-dimensional $\boldsymbol{X}$, the goal of sufficient dimension reduction (SDR) is to replace $\boldsymbol{X}$ with its projection onto a subspace $\mathscr{S}$ of $\mathbb{R}^p$ without loss of information about the conditional distribution $(Y|\boldsymbol{X})$. If the subspace is as small as possible, then optimal dimension reduction can be achieved. When a model with relatively few predictors is true and can be found, the Curse is avoided.

The minimal $\mathscr{S}$ is often denoted $\mathscr{S}(Y|\boldsymbol{X})$ and called the central subspace. It is

$$\mathscr{S}(Y|\boldsymbol{X}) = \bigcap_{\{\mathscr{S}:\, Y \perp (\boldsymbol{X}|P_{\mathscr{S}}\boldsymbol{X})\}} \mathscr{S}, \tag{9.8.1}$$

where $\perp$ indicates stochastic independence and $P_{\mathscr{S}}$ is the projection in the Euclidean inner product of $\mathbb{R}^p$ onto $\mathscr{S}$. In (9.8.1), the idea is that when Y is independent of $(\boldsymbol{X}|P_{\mathscr{S}}\boldsymbol{X})$, then, parallel to the concept of sufficiency, $P_{\mathscr{S}}\boldsymbol{X}$ has all the information about Y that was in $\boldsymbol{X}$. The central subspace (CS) is unique and permits reduction of $Y = \boldsymbol{X\beta} + \varepsilon$ to $Y = \boldsymbol{U}\boldsymbol{\gamma} + \varepsilon^{\mathsf{T}}$, where $\boldsymbol{U}$ can be regarded as $\boldsymbol{RX}$, in which $\boldsymbol{R}$ is $p \times \dim(\mathscr{S}(Y|\boldsymbol{X}))$, whose columns $\boldsymbol{R}_j$ for $j = 1, \ldots, \dim(\mathscr{S}(Y|\boldsymbol{X}))$ form a basis for $\mathscr{S}(Y|\boldsymbol{X})$. The linear combinations $\boldsymbol{R}_j\boldsymbol{X}$ are called the sufficient predictors. In general, linear transformations of $\boldsymbol{X}$ lead to linear transformations of $\mathscr{S}(Y|\boldsymbol{X})$. Indeed, for $\boldsymbol{Z} = \Sigma^{-1/2}(\boldsymbol{X} - \mathbb{E}\boldsymbol{X})$, $S(Y|\boldsymbol{X}) = \Sigma^{-1/2}\mathscr{S}(Y|\boldsymbol{Z})$; see Cook and Ni (2005).

Expression (9.8.1) is an instance of the more general concept of a dimension reduction subspace, which is any $\mathscr{S}$ with the property that

$$Y \perp (\boldsymbol{X}|\boldsymbol{RX}) \tag{9.8.2}$$

for some $p \times K$ matrix $\boldsymbol{R}$. The CS is the dimension reduction subspace of the smallest dimension. Arguably, a more important reduction is to the subspace of the CS that contains the mean as a function of the predictors. Its definition differs from (9.8.1) in only requiring a subspace of the predictor space that can express the mean of Y, not necessarily the whole distribution of Y. Thus, in parallel with (9.8.2), $\mathscr{S}$ is a mean dimension reduction subspace if

$$Y \perp (\mathbb{E}(Y|\boldsymbol{X})|\boldsymbol{R}\boldsymbol{X}), \tag{9.8.3}$$

where $\mathscr{S}$ is spanned by $\boldsymbol{R}\boldsymbol{X}$. Note that (9.8.2) implies (9.8.3), so every dimension reduction subspace is a mean dimension reduction subspace (see Cook and Li (2002)) and leads to a central mean dimension reduction subspace (CMS) by intersection as in (9.8.1). Fortunately, for location regressions, the CMS and CS coincide.

A standard way to estimate the CS is by inverse regression, trying to model $\boldsymbol{X}$ as a function of Y rather than Y as a function of $\boldsymbol{X}$. This approach is presented in detail in Cook and Ni (2005) and is followed closely here. To make a link between the CS and inverse regression, the linearity condition

$$\mathbb{E}(\mathbf{Z}|P_{\mathscr{S}(Y|\mathbf{Z})}\mathbf{Z}) = P_{\mathscr{S}(Y|\mathbf{Z})}\mathbf{Z} \tag{9.8.4}$$

is often imposed. Condition (9.8.4) is equivalent to requiring that $\mathbb{E}(\boldsymbol{X}|\boldsymbol{R}\boldsymbol{X})$ be a linear function of $\boldsymbol{R}\boldsymbol{X}$, where the columns of $\boldsymbol{R}$ form a basis for $\mathscr{S}(Y|\boldsymbol{X})$. When (9.8.4) holds, $E(\mathbf{Z}|Y) \in \mathscr{S}(Y|\mathbf{Z})$ and this implies

$$\mathsf{Span}(\mathrm{Cov}(E(\mathbf{Z}|Y))) \subseteq \mathscr{S}(Y|\mathbf{Z}).$$

When Y is discrete, $\mathbb{E}(\boldsymbol{X}|Y=y)$ can be estimated by averaging the $\boldsymbol{x}_i$s that correspond to y. When Y is continuous, discrete approximations of the same form can be used. Specifically, sliced inverse regression (SIR) as developed by Li (1991), uses a discretized version Y_{disc} of Y by partitioning the range of Y into h slices. In general, $\mathscr{S}(Y_{disc}|\boldsymbol{X}) \subseteq \mathscr{S}(Y|\boldsymbol{X})$ and, when h is large enough, equality is achieved; i.e., $\mathscr{S}(Y_{disc}|\boldsymbol{X}) = \mathscr{S}(Y|\boldsymbol{X})$. Consequently, for the rest of this section, Y is assumed discrete, taking H values $h = 1, \ldots, H$ called slices.

In Chapter 4, the SIR methodology was presented. The core result can be restated here as follows. First, the design condition in Li (1991), which applies to the marginal distribution of the predictors rather than to the conditional distribution $(Y|\boldsymbol{X})$, is also equivalent to (9.8.4). Then, if $\dim(\mathscr{S}(Y|\mathbf{Z}))$ is known, $\mathsf{Span}(\mathrm{Cov}(\mathbb{E}(\mathbf{Z}|Y))) = \mathscr{S}(Y|\mathbf{Z})$, and a consistent estimate $\widehat{\mathrm{Cov}}(\mathbb{E}(\mathbf{Z}|Y))$ of $\mathrm{Cov}(E(\mathbf{Z}|Y))$ is available. Specifically, the span of the eigenvectors from the $\dim(\mathscr{S}(Y|\mathbf{Z}))$ largest eigenvalues of $\widehat{\mathrm{Cov}}(\mathbb{E}(\mathbf{Z}|Y))$ is a consistent estimator for $\mathscr{S}(Y|\mathbf{Z})$. Thus, SIR, like PCs, rests on a spectral decomposition of the covariance matrix. Note that the number of eigenvectors used is the dimension of the subspace to which the p variates in $\boldsymbol{X} = (X_1, \ldots, X_p)$ is reduced and, in principle, one can choose fewer than $\dim(\mathscr{S}(Y|\boldsymbol{X}))$ of them.

To go beyond SIR to get better estimates of the CS, consider the parameter

$$\mathscr{S}_\xi = \mathsf{Span}\{\xi_1, \ldots, \xi_H\},$$

in which

$$\xi_y = \Sigma^{-1/2}\mathbb{E}\left(\Sigma^{-1/2}(\boldsymbol{X} - \mathbb{E}(\boldsymbol{X}))\middle| Y = y\right) = \Sigma^{-1/2}\mathbb{E}(\boldsymbol{Z}|Y = y).$$

Under linearity (i.e., (9.8.4)), $\mathscr{S}_\xi \subseteq \mathscr{S}(Y|\boldsymbol{X})$; sometimes $\mathscr{S}_\xi = \mathscr{S}(Y|\boldsymbol{X})$, a condition called coverage.

Inference about $\mathscr{S}_\xi$ does not depend on linearity or coverage, however. Estimates of a basis for $\mathscr{S}_\xi$ are also estimates of a basis for the CS. For $d = \dim(\mathscr{S}_\xi)$, denote a basis of $\mathscr{S}_\xi$ by the matrix $\boldsymbol{\beta} = (\beta_1, \ldots, \beta_d)$, where $\beta_k \in \mathbb{R}^p$. Then $\boldsymbol{\beta}$ is an estimate of the basis of the CS. In terms of the ξ_js whose span defines $\mathscr{S}_\xi$, one can define a matrix $\boldsymbol{\gamma}$ with columns given by the vectors γ_y by $\xi_y = \boldsymbol{\beta}\boldsymbol{\gamma}_y$ so that

$$\boldsymbol{\xi} = (\xi_1, \ldots, \xi_H) = \boldsymbol{\beta}\boldsymbol{\gamma}.$$

By the definition of conditional expectation, this gives $\boldsymbol{\xi}(P(Y = 1), \ldots, P(Y = h))^{\mathsf{T}} = 0$, the intrinsic location constraint.

The natural way to estimate $\mathscr{S}_\xi$ is to estimate the ξ_ks. Given a sample $(Y_i, \boldsymbol{X}_i)$ for $i = 1, \ldots, n$, Cook and Ni (2005) suggest writing $\boldsymbol{X}_{y,j}$ to mean the jth observation of $\boldsymbol{X}$ in slice y for $j = 1, \ldots, n_y$, where $n_1 + \cdots + n_h = n$. Then $\bar{\boldsymbol{X}}_{\cdot,\cdot}$ is the overall mean and $\bar{\boldsymbol{X}}_{y,\cdot}$ is the mean on the yth slice containing n_y points with $Y = y$. Provided the estimate $\hat{\Sigma}$ of Σ is positive definite, one can set $\hat{\xi}_y = \hat{\Sigma}^{-1}(\bar{\boldsymbol{X}}_{y,\cdot} - \bar{\boldsymbol{X}}_{\cdot,\cdot})$ so that

$$\hat{\boldsymbol{\xi}} = (\hat{\xi}_1, \ldots, \hat{\xi}_H) \in \mathbb{R}^{p \times H}.$$

The problem is that, for large H, the case of greatest importance for continuous Y, this estimate may have more than d dimensions purely by chance. Thus, even though it is reasonable to estimate $\mathscr{S}_\xi$ by the column space of $\hat{\boldsymbol{\xi}}$, it is unclear how to obtain a d-dimensional estimate of $\mathscr{S}_\xi$.

Cook and Ni (2005) resolve this problem by proposing a quadratic discrepancy function. The idea is to define two matrices $\boldsymbol{B} \in \mathbb{R}^{p \times d}$ and $\boldsymbol{C} \in \mathbb{R}^{d \times \ell}$ to express the optimal reduction to d dimensions and then to minimize

$$F_d(\boldsymbol{B}, \boldsymbol{C}) = \|\mathsf{vec}(\hat{\boldsymbol{\xi}}\boldsymbol{Q}) - \mathsf{vec}(\boldsymbol{B}\boldsymbol{C})\|_{\boldsymbol{V}} \tag{9.8.5}$$

over $\boldsymbol{B}$ and $\boldsymbol{C}$ for fixed d. In (9.8.5), the $p\ell \times p\ell$ matrix $\boldsymbol{V}$ defines the inner product giving the sense in which the reduced column space of $\hat{\boldsymbol{\xi}}$ is to be found. The $H \times \ell$ matrix $\boldsymbol{Q}$ can be absorbed by replacing $\boldsymbol{C}$ with $\boldsymbol{C}\boldsymbol{Q}^{-1}$. So, without loss of generality, (9.8.5) only depends on $\boldsymbol{B}$, $\boldsymbol{C}$, $\boldsymbol{V}$, and $\hat{\boldsymbol{\xi}}$. The columns of $\boldsymbol{B}$ represent a basis for $\mathsf{Span}(\boldsymbol{\xi}\boldsymbol{Q})$, and $\boldsymbol{C}$ represents the coordinates of $\boldsymbol{\xi}$ relative to $\boldsymbol{B}$. Note that the choice of $d < H$ is what makes the product $\boldsymbol{B}\boldsymbol{C}$ force a reduction in dimension; the role of ℓ only appears to help indicate how $\boldsymbol{C}$ or $\boldsymbol{Q}$ organizes the columns of $\hat{\boldsymbol{\xi}}$.

One procedure for finding the minimizing values of $\boldsymbol{B}$ and $\boldsymbol{C}$ is an alternating least squares method; see Cook and Ni (2005).

Let $\hat{\boldsymbol{\zeta}} = \hat{\boldsymbol{\xi}}\boldsymbol{D}\boldsymbol{U}$, where $\boldsymbol{U}^\top\boldsymbol{U} = I_{H-1}$ and $\boldsymbol{U}^\top\mathbf{1}_H = 0$ and $\boldsymbol{D}$ is diagonal with n_y/ns on the main diagonal. Then set $\boldsymbol{\zeta} = \boldsymbol{\beta}\boldsymbol{\gamma}\boldsymbol{D}\boldsymbol{U}$.

The idea of the procedure is to fix $\boldsymbol{V}$, absorb $\boldsymbol{Q}$ so it can be ignored, and then start with a guess $\boldsymbol{B}_0$ of $\boldsymbol{B}$, obtaining $\boldsymbol{C}_0$ as the least squares coefficients from the regression of $\hat{\boldsymbol{\zeta}}\boldsymbol{Q}$ on $\boldsymbol{B}$. This gives an initial value $F_d(\boldsymbol{B}_0, \boldsymbol{C}_0)$. Redo the regression of $\boldsymbol{\zeta}$ on $\boldsymbol{B}_0$, but with the kth column of $\boldsymbol{B}_0$ deleted to get the residual (a column vector formed from the matrix of residuals). A new kth column for $\boldsymbol{B}_0$ can be found by minimizing $F_d(\boldsymbol{B}_{0,-k}, \boldsymbol{C}_{0,-k})$ subject to the constraint that the new kth column be orthogonal to all the columns of $\boldsymbol{B}_{0,-k}$. With this new $\boldsymbol{B}_1$, reminimize $F_d(\boldsymbol{B}_1, \boldsymbol{C})$ over $\boldsymbol{C}$ to get the new $\boldsymbol{C}_1$. Then repeat the procedure with $\boldsymbol{B}_1$ and $\boldsymbol{C}_1$ and continue until $F_d(\boldsymbol{B}, \boldsymbol{C})$ no longer decreases. Now, $\mathsf{Span}(\boldsymbol{B}_\infty)$ is the estimate of $\mathscr{S}_\xi$, where $\boldsymbol{B}_\infty$ is the final value of $\boldsymbol{B}$ from the alternating least squares procedure.

Each choice of $\boldsymbol{V}$ and d gives a dimension reduction scheme. However, it is desirable to find an estimate $\hat{\boldsymbol{V}}$ of $\boldsymbol{V}$ that will result in the smallest discrepancy. To develop this, Cook and Ni (2005), Theorem 1 establish that

$$\sqrt{n}(\mathsf{vec}(\hat{\boldsymbol{\xi}}\boldsymbol{D})) - \mathsf{vec}(\boldsymbol{\beta}\boldsymbol{\gamma}\mathrm{diag}(P(Y=1), \ldots, P(Y=H))) \to N(0, \Gamma),$$

where $\Gamma = \mathrm{Cov}(\mathsf{vec}(\Sigma^{-1/2}\boldsymbol{Z}\boldsymbol{\varepsilon}^\top))$. The vector $\boldsymbol{\varepsilon} = (\varepsilon_1, \ldots, \varepsilon_H)^\top$ is a collection of residuals obtained as follows. Consider H indicator functions for the slices of Y. That is, let $I_h(\boldsymbol{X}) = 1$ when $\boldsymbol{X}$ is in the hth slice and zero otherwise. Then $\mathbb{E}(I_h(\boldsymbol{X})) = f_h$, the frequency of the slice, $P(Y = h)$, and for $h = 1, \ldots, H$,

$$\varepsilon_h = I_h - E\, I_h - \boldsymbol{Z}\mathbb{E}(\boldsymbol{Z}I_h)$$

are the population residuals from an OLS fit of I_h on $\boldsymbol{Z}$.

Cook and Ni (2005) also establish that

$$\sqrt{n}(\mathsf{vec}(\hat{\boldsymbol{\zeta}}) - \mathsf{vec}(\boldsymbol{\beta}\boldsymbol{\gamma})) \to N(0, \Gamma_{\hat{\zeta}}),$$

where $\Gamma_{\hat{\zeta}}$ is $p(H-1) \times p(H-1)$ and that a consistent estimate $\hat{\Gamma}_{\hat{\zeta}}$ of $\Gamma_{\hat{\zeta}}$ can be obtained from Γ. (In fact, $\Gamma_{\hat{\zeta}} = (\boldsymbol{U}^\top \otimes \boldsymbol{I})\Gamma(\boldsymbol{U} \otimes \boldsymbol{I})$.)

The best $\boldsymbol{V}$ to use in F_d (i.e., the one giving the smallest variances for estimates of any function of $\mathsf{vec}(\boldsymbol{\zeta})$) is a consistent estimate $\hat{\Gamma}_{\hat{\zeta}}^{-1}$ for $\Gamma_{\hat{\zeta}}^{-1}$, so it remains to provide an estimate of Γ. This can be done by using sample versions of Σ, the z_is, and $\boldsymbol{\varepsilon}$. The standard estimate of Σ and the resulting z_is suffice, and a sample version of $\boldsymbol{\varepsilon}$ can be found by substituting sample means for $\mathbb{E}(I_h)$ and $\mathbb{E}\boldsymbol{Z}I_h$ for each ε_h. This gives the sample version $\hat{\varepsilon}_{h,i} = I_{h,i} - n_h/n - (\boldsymbol{x}_i - \boldsymbol{x}_{\cdot\cdot})^\top \hat{\xi}_h(n_h/n)$ for $i = 1, \ldots, n$. Using these, $\hat{\Gamma}$ can be found and therefore so can $\hat{\gamma}_{\hat{\zeta}}$.

To choose d, a testing procedure is often used. Let $F_d^{ire}(\boldsymbol{B}, \boldsymbol{C})$ denote the optimal version of F_d, using $\hat{\Gamma}_{\hat{\zeta}}^{-1}$ in place of $\boldsymbol{V}$. That is,

$$F_d^{ire}(\boldsymbol{B}, \boldsymbol{C}) = (\mathsf{vec}(\hat{\zeta}) - \mathsf{vec}(\boldsymbol{B}\boldsymbol{C}))^\top \hat{\Gamma}_{\hat{\zeta}}^{-1} (\mathsf{vec}(\hat{\zeta}) - \mathsf{vec}(\boldsymbol{B}\boldsymbol{C})).$$

Also, let $\hat{F}_d = \min F_d(\boldsymbol{B}, \boldsymbol{C})$, where the minimum is over $\boldsymbol{B}$ and $\boldsymbol{C}$ as found by the alternating least squares procedure. Then, $n\hat{F}_d^{ire}$ has a limiting chi-squared distribution with $(p-d)(H-d-1)$ degrees of freedom, where $d = \dim(\mathscr{S}_\xi)$ and $\mathsf{Span}(\hat{\boldsymbol{\beta}})$ is consistent for $\mathscr{S}_\xi$, where $\hat{\boldsymbol{\beta}} = \arg\min F_d^{ire}(\boldsymbol{B}, \boldsymbol{C})$. Given this, for testing nulls of the form $\mathscr{H}_0 : d = k$ vs. $\mathscr{H}_1 : d > k$, one can compare the sample value $n\hat{F}_k$ with the $1-\alpha$ percentile from the chi-squared distribution with $(p-d)(H-d-1)$ degrees of freedom and reject if $n\hat{F}_k$ is too large. If $\mathscr{H}_0$ is rejected, increment the null by 1 and test again, stopping when rejection at level α is no longer possible. This final test gives the appropriate value of d.

An important special case of the quadratic discrepancy function in (9.8.5) sets $\boldsymbol{V} = \mathrm{diag}(n/n_1, \ldots, n/n_H)\mathrm{diag}(\hat{\Sigma})$ which gives the SIR procedure. SIR is therefore potentially suboptimal in the sense that it limits the class of procedures substantially. It is unclear how bad the suboptimality is or when it matters most.

Finally, another variation on SDR is sliced average variance estimation (SAVE). The core idea of SAVE is to apply the SIR procedure to the matrix $\mathbb{E}(\boldsymbol{I} - \Sigma_{\mathbf{Z}|Y})^2$ instead of $\mathrm{Cov}(\mathbb{E}(\mathbf{Z}|Y))$. Under linearity and other conditions, $\mathsf{Span}(\mathbb{E}(\boldsymbol{I} - \Sigma_{\mathbf{Z}|Y})^2) \subseteq \mathscr{S}(Y|\boldsymbol{Z})$, so that if $\dim(\mathscr{S}(Y|\mathbf{Z})$ is known and $\mathsf{Span}(\mathbb{E}(\boldsymbol{I} - \Sigma_{\mathbf{Z}|Y})^2) = \mathscr{S}(Y|\mathbf{Z})$, then the subspace spanned by the eigenvectors from the $\dim(\mathscr{S}(Y|\mathbf{Z}))$ largest eigenvalues of an estimate of $\mathbb{E}(\boldsymbol{I} - \Sigma_{\mathbf{Z}|Y})^2$ is a consistent estimate of $\mathscr{S}(Y|\mathbf{Z})$.

9.9 Visualization I: Basic Plots

Imagine a continuum of variability such that statistical problems can be ordered by increasing uncertainty about the data generator. At the low end are unimodal, univariate, IID data sets, with largish sample sizes, that can be explained by a parametric family with a small number of parameters. Next would come data sets with the number of parameters increasing with n, or data sets with simple dependence structures, more elaborate modal structures, or larger ps. After that would come data generators that required some of the classes studied in Chapter 4 with potentially infinitely many parameters: neural nets, trees, MARS, projection pursuit, and so forth, possibly with large p, small n, and dependence structures. Another more complicated step would be problems in which even these methods were ineffective and one would have to use clustering or dimension reduction just to find enough regularity to model. Finally would come the classes of problems in which only nonparametric classes of models would be effective.

Within the penultimate class in the continuum of variability, there are many ways a data set can be complicated. The dimension p may be large; the joint density of the components of $\boldsymbol{X}$ may be rough, with many local modes giving unusual dependencies and strange marginals; the sample size may be so small relative to the variability in the data that inferences are weak at best; and there may be substructures or subsamples of the data that are so important that they should be modeled separately. It is not hard to imagine other sorts of complexity as well – the data might have been gathered sequentially, with the experiment changing from time to time, as in Internet data. The consequence is that model uncertainty in all its forms dominates so that for many data sources only approximate models can be proposed and used for inference up to the limit of their reliability.

Aside from their obvious role in displaying results, visualization techniques are most important for this penultimate class in which no parametric family can be proposed a priori. Underneath all the variability there may in fact be a very intelligible model; however, teasing it out from the data in the absence of physical knowledge may be exceedingly difficult. In such cases, visualization, like dimension reduction and clustering, can be regarded as a collection of search strategies to find some regularity in the data strong enough that modeling can say something about it. Indeed, there is great subjectivity in how the boundaries among visualization, clustering, and dimension reduction are drawn. This is so because clustering and dimension reduction find properties of the data to visualize; visualization can be used to find effective dimension reductions or to find clusters; and clustering can lead to dimension reduction and conversely.

To see the importance of visualization in a simple context, consider Fig. 9.1, which depicts the pairs (Y_i, X_i) for $n = 97$, with 27 points on a straight line and 70 points generated with noise. It is easy to see that there are two subsets of points, which should be modeled separately. However, any naive modeling attempt, by regression for instance, will miss both subsets, defaulting to an average solution.

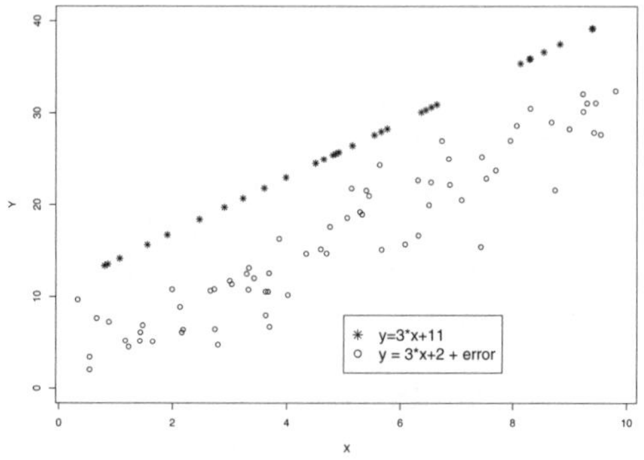

Fig. 9.1 The 97 bivariate data points plotted visually segregate into two clusters, easily seen by the different symbols dots and circles.

For cases such as Fig. 9.1, it would be nice to be able to cherry-pick. In practice, this might mean using a clustering technique to find (hopefully two) clusters. Then, choose the largest cluster and fit a regression model. Provided the model fits well, remove outliers from the cluster according to some reasonable criterion and search the other clusters for points that fit the model well, putting them into the first cluster if they improve the fit or at least do not worsen it much. Reclustering the remaining points and repeating the procedure might reveal the two-cluster structure that is obvious in Fig. 9.1, even though in more complicated settings lower-dimensional substructures could be hard to discern. In general, there are the usual caveats, among which are: (i) The goodness-of-fit measure should not depend on the sample size or p (e.g. adjusted R^2); (ii) the modeling within a cluster should use some trade-off between sample size and number of variables (e.g., Mallows' C_p); and (iii) the number of clusters is subject to uncertainty.

If this kind of procedure were used with the data from Fig. 9.1 and the first cluster with 70 points had been found, the graph of most reasonable assessments of fit would look something like Fig. 9.2 as data points from the smaller cluster were added. This plot is based on using R^2 for fitness with the data from Fig. 9.1. The total sample size used here is 80; the first 70 observations are from the noisy cluster and the last 10 observations were generated exactly on the line, as indicated by the knee in the curve. Although this procedure is not formal, it does accurately reveal the structure of the data in a suggestive way. In principle, one could propose a model for how adding wrong model data affects statistics such as R^2, generate a curve like that in Fig. 9.2 that has an abrupt change in direction at the knee, and test whether the points added beyond the knee were sufficiently different to reflect a different model.

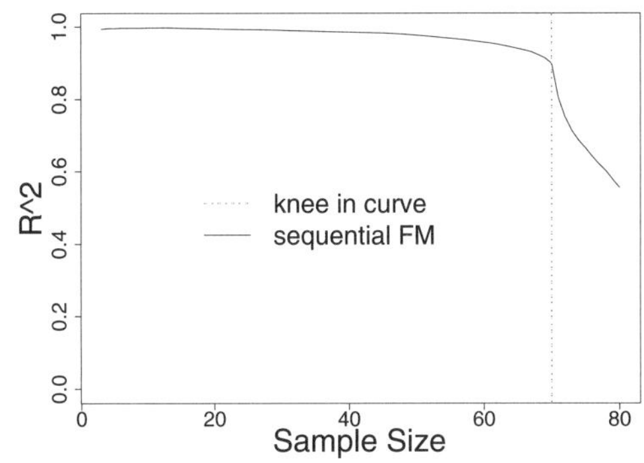

Fig. 9.2 This figure shows how accumulating data within a model affects fit. The first 70, data points fit the model. Past 70 the data points were from the smaller cluster, a different model, and the fit drops off suddenly. FM indicates where the sequentially fitted model begins to include wrong model points.

Just as estimating a variance leaves one fewer degree of freedom for estimating a normal mean or performing a hypothesis test on a parameter "uses" a degree of freedom, so making diagrams from data uses degrees of freedom. This means that the probability of type I error in subsequent testing will be larger than the stated α and the actual standard errors of estimators will be larger than those found by most procedures. In short, inferences after data snooping by visualization, like those after dimension reduction and clustering, will necessarily be much weaker. As tempting as these techniques are, they gobble degrees of freedom faster than calculating individual statistics and really should be done selectively if downstream inference is of great importance and multiple comparison procedures are not available. This degrees of freedom argument does not preclude extensive visualization to present inferences or to search for structure to model, only formal inference procedures. In practice, the cost of these caveats is usually imprudently ignored.

The rest of this section is taken up with a listing of a variety of visualization techniques for revealing the structure within data. It is assumed that the reader is generally familiar with the usual residual plots from regression analyses and exploratory data analysis as presented in, for instance, Neter et al. (1985), Tukey (1977), du Toit et al. (1986), Cleveland (1993), Tufte (2001), and many other texts. So, the first collection of techniques here is easy, based on familiar quantities, but not routine. Accordingly, these are listed and only briefly explained. The challenge is to use them ingeniously in examples – or develop new representations that are better suited to a particular application. While the present list is far from exhaustive – the extensive graphics associated with quality control and time series are omitted – the intent is to focus on graphics that represent the data faithfully so as to bring out features that would not otherwise be obvious. Techniques that include transformations are deferred to the next section.

9.9.1 Elementary Visualization

There is a large and growing body of literature devoted to representations of data with the intent of underscoring their key qualitative features. Key features include whether the data are numeric or categorical; scalar, vector, or more complex; discrete or continuous; ordered or unordered; dependent or independent, and so forth. Such features of the data must have corresponding features in the key aspects of diagrams. Importantly, the physical meaning of geometric aspects of a representation should not mislead a viewer into incorrect inferences. Geometric aspects include position along a scale (common or not, aligned or not); length, direction, and curvature of lines; and angle, area, and volume for shapes. Nongeometric aspects such as coloring, choice of symbol, motion, and adjacency on the page or screen of different components also affect the message a viewer will perceive. Again, the basic rule is that these choices should not lead to incorrect inferences.

The program GGobi, available from `www.ggobi.org`, described in Cook and Swayne (2007), is one important contribution to visualization; another collection of packages is available in R at `seewww.r-project.org`. There are many others.

Below is a listing of several common ways to represent data with little or no processing. These techniques are most useful when the dimension is between 4 and, say, 20 or so. Once the dimension gets too high, visualization methods cannot help understand the patterns very well unless some dimension reduction is used first. In the remainder of this section, descriptions are cursory because the methods are easy to visualize and the figures tell the story better. One technique that is not described here is the parallel coordinate plot; it was already introduced in Chapter 7 to explain polythetic clustering.

9.9.1.1 Profiles and Stars

A profile in p dimensions is a representation of a vector of the form $(x_1,...,x_p)$ in which the values x_j are plotted adjacent to each other. This can be presented as a bar graph with p bars on a common axis or as a polygonal line. Often this is done for all n points in a data set giving n graphs. Profiles can be smoothed to give a polygonal line. Like the parallel coordinate plots used in Fig. 7.4, this can reveal patterns but is highly dependent on the ordering of the variables.

A star in p dimensions is a representation of $(x_1,...,x_p)$ in which the values x_j are plotted on axes drawn from a center point. Any two adjacent axes have the same angle between them. The values x_j are noted on the axes and then connected to form a p-gon. Doing this for a data set gives n p-gons that may exhibit characteristic patterns, depending on the ordering of the x_js.

The typical appearances of stars and profiles can be seen in Fig. 9.3. Using data on 17 classes of household expenditures from nine Canadian cities for 2006, the top panel in Fig. 9.3 shows one star with 17 points for each city. Household expenditures means dollars per year spent on food, shelter, clothing, and so forth. The data, along with precise definitions, can be found at `http://www40.statcan.ca/101/cst01/famil10d.htm`. Note that the absolute number of dollars per year is not shown, although all the dimensions of the stars are comparable. The bottom panel in Figure 9.3 shows the profiles for the cities plotted on the same axis. The four peaks are suggestive, but a little misleading if read too closely: The four largest peaks at 2, 6, 9, and 15 correspond to shelter, transportation, recreation, and taxes. However, the first variable, food, should be a peak: It is higher than all the other variables, except for shelter, transportation, and taxes, which is the biggest expense. The real use of this graph is to see that despite differences suggested by the stars, the distribution of expenditures looks fairly similar across the cities.

The commands used to create the plots were:

```
library(Hmisc)
library(aplpack)
xx<-read.csv(file="canada.csv",head=TRUE,sep=",")
postscript("CanadaStars.eps")
stars(xx, labels = xx[,1], full = FALSE,
main="Canadian Household Expenditures")
postscript("CanadaProfile.eps")
ts.plot(t(xx[,-1]), gpars=list(xlab = "Type of
Expenditure", ylab="Amount Spent", lty=c(1:17)))
```

9.9.1.2 Trees

A tree is a graph that has one root node with branches coming off it. These branches may split into further branches. If each node has either zero or two branches coming

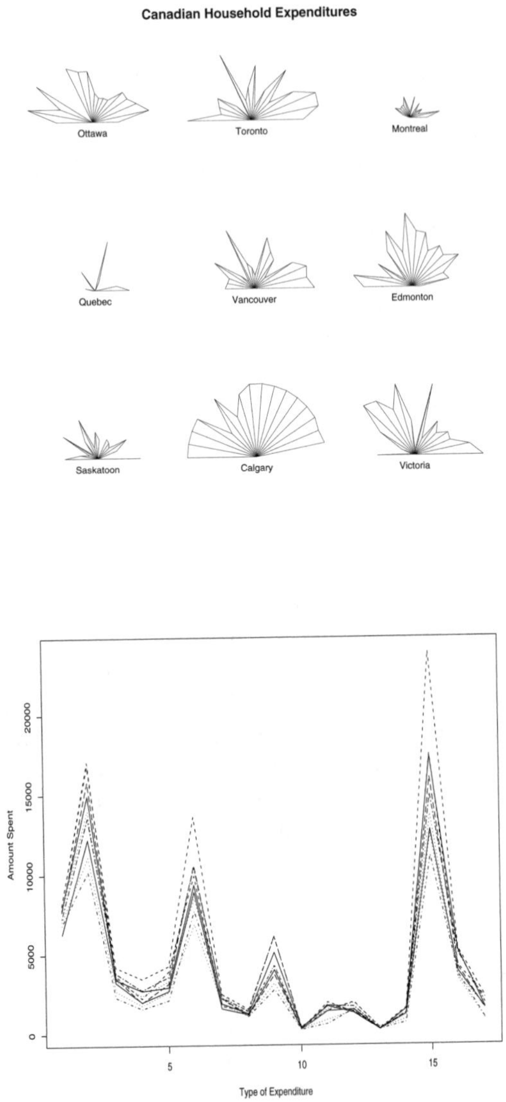

Fig. 9.3 It is seen in the top panel that Calgary's star fans out the most uniformly over all expenditure classes, while Quebec City's star has only five points of any real size; they are food, shelter, transportation, insurance and pensions, and taxes. The lower panel shows the profiles. There are four peaks, but the point is that across Canadian cities the distribution of expenditures is fairly similar.

off it, the tree is binary. If each node has k branches coming off it, the tree is k-ary. Binary trees were seen in a clustering context and called dendrograms; the branching corresponded to splitting or uniting collections of points so the sequence of branches summarizes an inclusion relation. Tree structures were also used in recursive partitioning. In these cases, the branching also reflected an inclusion relation but on the domain of the explanatory variables. The length of the branches sometimes has a meaningful interpretation for clustering. In single linkage, for instance, it is the minimal distance between clusters or the distance between their centroids. However, in recursive partitioning, it is not clear that the length of the branches has any interpretation. On the other hand, when trees are used for regression or classification, the tree structure represents a function whose structure may indicate relationships among the variables.

Often when p is large and n is small it is useful to form a fully connected graph of n points in $\mathbb{R}^p$. If distances can be assigned to the edges, one can delete edges sequentially to minimize the sum of distances while retaining a connected graph. The result is a minimal spanning tree, as discussed in Chapter 8, which may indicate new relationships. The cuts of a spanning tree essentially define graph-based clustering.

9.9.1.3 Graphs

Graphs generalize trees. Formally, a graph is a collection of points, called nodes or vertices, and a collection of links between them, called edges. The structure of feed-forward neural nets is a graph: Each input variable is a vertex, and these p vertices feed forward to the first hidden layer. If the first hidden layer is the only layer, the outputs of that layer are combined to give a model for the response. If there is a second hidden layer, the nodes of the first hidden layer feed forward into it. The connectivity of the NN is therefore a graph structure in which the vertices represent sigmoid functions and the edges indicate how the sigmoid functions are composed.

As has been seen in the clustering context, graphs can be weighted. That is, each edge may have a number assigned to it, intuitively representing its length. Optimizing pathlengths and graph partitions are common ways to represent and solve numerous problems, including clustering.

Graphs also model the dependence structure among variables. This is particularly useful for categorical data. In graphical models, vertices represent variables. Two vertices have no edge between them if the two variables they represent are independent; otherwise there is an edge. In undirected graphical models, two sets of vertices A and B are conditionally independent given a third set C if there is no path between the vertices in A and B that doesn't pass through C. Directed graphical models are more complicated. Techniques for estimating parameters, probabilities, and selecting a graphical model in the first place are specialized and beyond the scope of this text.

Graphs similar to those in Fig. 8.4 can be used to indicate the structure of other models, like Markov chains that represent chemical reactions, networks of genes, population models, random walks, and so forth. The idea is that each node represents a state and each edge coming out of a node represents a move that can be made in one time step with the conditional probability specified by the chain. The lower diagram in Fig. 9.4 is the state diagram for a ten state Markov chain with one source on the left and one sink on the right, with the allowable transitions indicated by edges.

Multidimensional scaling, a visualization technique to be discussed in the next section, takes data points in a high-dimensional space and compresses them down to a lower-dimensional space, representing them as a fully connected weighted graph. Thus, MDS and graphical models are two ways to generate graphs to represent the structure among data points.

9.9.1.4 Heatmaps

A heatmap is a matrix of values that have been color coded, usually so that higher values are brighter and lower values are darker, in analogy with temperature. The columns and rows of the matrix typically have an interpretation. In genomics, the columns and rows may represent genes and subjects so that a bright color in a cell means the given

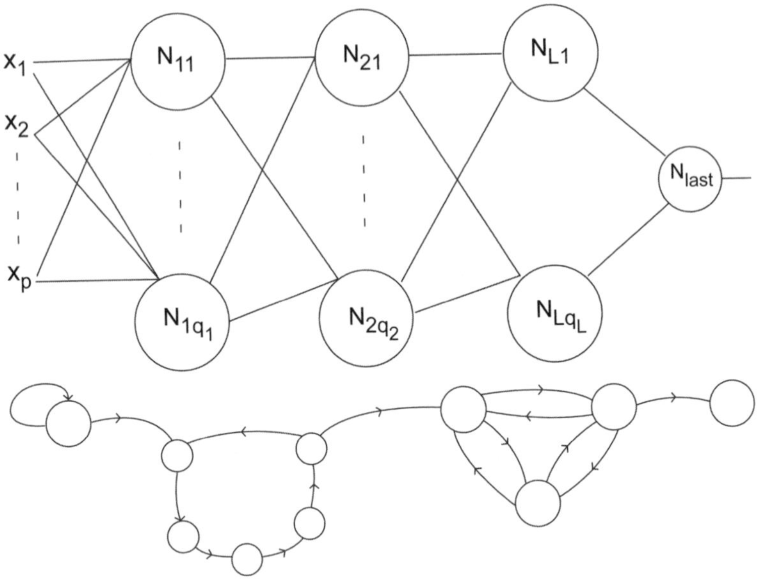

Fig. 9.4 The top panel shows the (directed) graph corresponding to a generic single-output feedforward neural net. The vertices, indicated by circles, indicate functions with arguments from the previous layer and each edge indicates that the output of a vertex feeds into the next layer. The jth layer has q_j vertices, or nodes, for $j = 1,\ldots,L$; all nodes in a layer feed into the next layer. The bottom panel shows a graph for a hypothetical discrete time, discrete space Markov chain: The vertices, indicated by circles, represent states, and the edges represent the allowed transitions between states in a time step. The graph is directed, but if there are two edges between two vertices, with opposite orientations, then both transitions are allowed (though possibly with different probabilities). The leftmost vertex is a source; the next five form a cycle, the next three are fully connected, and the last vertex is a sink.

gene has a high expression in the given subject. Usually the rows and columns are grouped so that those in the same group are next to each other; this leads to figures comprised of homogeneous rectangles. Heatmaps are often good for presenting data once they have been analyzed, but as a practical matter they often do not reveal much initially because the patterns in the data tend to be weak.

A related concept is that color can indicate density of points. In a scatterplot of a large data set, points may accumulate densely in some regions and sparsely in others. In these cases, a local density of data points per unit area can be associated with rectangular regions and the density values discretized and represented by colors, bright for high density and dark for low density. This is much like a contour plot for a distribution but has sides parallel to the axes and usually a lower resolution.

Figure 9.5 depicts a heatmap for data extracted from the 1974 US magazine Motor Trend. It comprises 11 aspects of automobile design and performance for 1973–1974 models. The variables were: mpg, miles/(US) gallon; cyl, number of cylinders; disp, displacement (cu.in.); hp, gross horsepower; drat, rear axle ratio; wt, weight (lb/1000); qsec, 1/4 mile time; vs, the engine's V/S ratio; am, transmission (0 = automatic, 1 = manual); gear, number of forward gears; and carb, number of carburetors.

The commands were

```
x  <- as.matrix(mtcars)
hv <- heatmap(x, col=colorpanel(7, "white", "grey10"),
scale="column", margin=c(5, 10),
xlab="specification variables",
ylab= "Car Models", main="Heatmap for MTcars data")
```

Both the models and the measurements on them have been clustered; this is typical for a heatmap. Roughly, the models of cars are in three clusters: The bottom cluster (Duster to Maserati) consists of cars that are heavy or have big engines; the top cluster (Honda Civic to Toyota Corona) consists of lighter cars with smaller engines; and the middle cluster (Valiant to Mercedes 450SL) is in between. The clustering on the measurements on the cars is less clear: The pair on the right are measures of power and the next two are measures of performance, but it is unclear what the block of seven (cyl to gear) on the left represents.

The heatmap itself shows a clear dividing line between the Honda Civic and the Mercedes 450SL: In each column, if the top part is light, the bottom part is dark, or vice versa. This suggests that the middle cluster has more in common with the lower cluster than the upper and that the small group from Honda Civic to Fiat 128 really belongs to the top half. It is left as an exercise to rearrange the rows and columns (ignoring the clustering) to get two dark blocks on the main diagonal and two lighter blocks off the main diagonal, the ideal way to try to present a heatmap. In fact, typical heatmaps often look much more random than that in Fig. 9.5. In practice, heatmaps are suggestive but little more despite their use (or overuse).

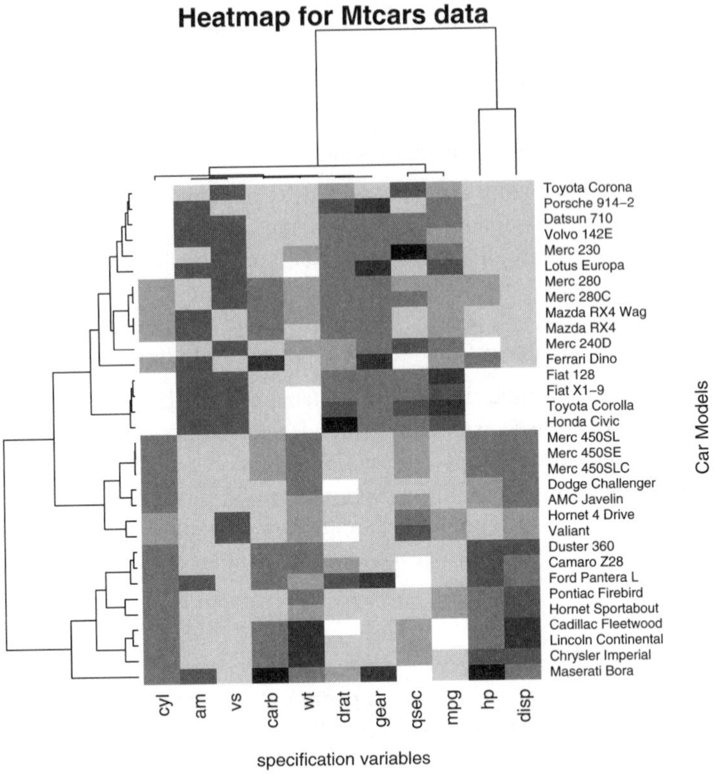

Fig. 9.5 Heatmap for the MTcars data, with clustering on the models and variables done separately. Darker regions correspond to higher values, lighter regions to lower values.

9.9.1.5 Composite Plots

This is a generic term for visualizations that combine two or more representations. For instance, a heatmap may have its columns or row values grouped by the dendrogram of a clustering procedure as in Fig. 9.5. As another instance, multivariate data from distinct locations may be represented as a sequence of stars at those locations as depicted on a map. If each location has measurements over time, the stars for each location form a trajectory with features having some interpretable structure.

Composite plots are efforts to compress huge amounts of data into a single diagram by representing many variables with different geometric or compositional features. Charles Minard's 1869 graphic indicating the fate of Napoleon's Russian campaign remains a classic, see `http://en.wikipedia.org/wiki/Charles_Joseph_Minard`. The flowmap includes a map and scale for physical distances, line thickness for the size of the army, calibrations for time, and other features of that folly.

9.9.2 Projections

Although elementary, it is important to remember that the search for structure in data almost always devolves to representing features of p dimensions on a two dimensional surface. That is, p dimensional data are, in effect understood by inferring structure from their projections. As noted in Chapter 1, projections along many directions will look like an undifferentiated point cloud. So, insight is required to find the directions along which the projection of the data will reveal its features. For instance, consider projecting p-dimensional data points of the form $(x_{i,1},...,x_{i,p})$ for $i = 1,...,n$ using a $(p \times p)$-dimensional idempotent matrix D. It is possible that a value x_{i,j_0} is an outlier in the $(x_i, x_{i'})$ plane but not in the $(x_i, x_{i''})$ plane – think of a curve in the horizontal plane and one point several units above it. Projected to the plane, there are no outliers, but projected into any vertical plane, the extra point is seen to be an outlier. One way to search for outliers is through all pairwise scatterplots from projections into the coordinate planes: $p(p-1)/2 - p$ scatterplots of $(x_j, x_{j'})$ for $j' > j$ can be displayed as an upper triangle of a matrix of scatterplots.

Another way to do this is to spin the data. The idea is to project the data points in a three-dimensional subspace and then rotate the projections. Rotation can be interactive (user controlled) or automated. Systematically doing this so that all representative projections of the data are seen is called a Grand Tour. When the data are sparse for their dimension, such cycling through projections often gives more insight than other methods. In particular, watching the rotations in continuous time reveals the context of informative projections as well as the projected points themselves. So, interesting structure in one projection can be related to the overall structure of the data cloud. Ideally, the shape of clusters or other formations in the data cloud can be identified. In principle, a Grand Tour can be used to help find directions most useful in a projection pursuit modeling sense.

Sometimes a technique called brushing is used in conjunction with projections. The idea is that points conjectured to have some common feature (e.g., forming a cluster or boundary) can be tagged with a color or symbol and tracked through a variety of projections to see if the conjectured feature is retained from other perspectives. Combining brushing with a Grand Tour provides a visual way to explore and test for possible structure of high-dimensional data. From an esthetic standpoint, the resulting movies are an important contribution to modern art.

Although low-dimensional projections from large, high-dimensional data sets often look normal, it is easy to cluster by eye in one, two, or even three dimensions. However, lifting a lower-dimensional clustering of a projection of p-dimensional data back up to its p-dimensional space is problematic. In this context, byplots, where a sequence of scatterplots of two variables is indexed by the values of a third, check that the dependence of a relationship between two variables on a third is not illusory.

Importantly, there are problems in which, even after dimension reduction, one is left with many more than three dimensions. If the statistics cannot be assumed independent, thereby ruling out many procedures, such as multiple comparisons, visualization may be one of the few approaches available.

To see a little of how this works in practice, consider the Australian crab data available from `www.ggobi.org/book/`. The data consist of 200 measurements on a sample of crabs from Australia. Each measurement is five-dimensional: frontal lobe length, FL; rear width, RW; carapace length, CL; carapace width, CW; and body depth, BD. To understand the data, one might start by plotting the projections into the coordinate planes just to see if there is any structure worth exploring. These are shown in Fig. 9.6. It is seen that in some of the projections the data form an elongated vee shape, while in others the data seem to form a line.

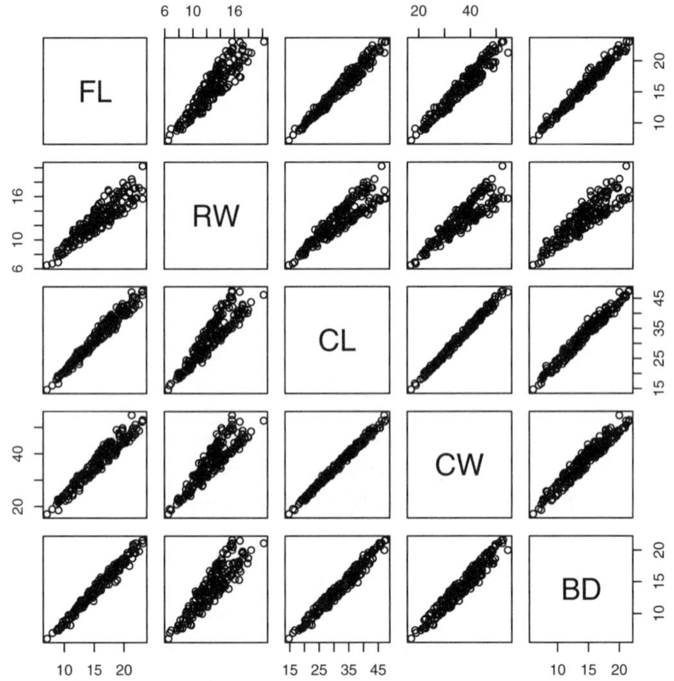

Fig. 9.6 Coordinate plane plots for the five variables in the Australian crab data.

Since the data are five dimensional, one can load them into the ggobi or rggobi visualization system, which can be downloaded freely from `www.ggobi.org`. GGobi can generate Grand Tours and other visualizations. Doing this, one can watch the way the data forms change shape as the perspective is rotated. GGobi can be paused at interesting projections. Since GGobi also gives the unit vectors defining each projection, it is easy to identify the projection matrix and apply it to the data already loaded in R and regenerate the figure found from the Grand Tour.

The three panels on the right in Fig. 9.7 were generated in this way. First, the upper left was found just by watching the Tour and stopping it at a clearly delineated shape. Then the picture was rotated by dragging the cursor over the window with the projection and watching it change appearance. This generated the other two panels on the right. The next step would be to color one of the arms of the vee in the upper left and do the rotation again to see how the points changed their relative positions. Doing this permits separation of each arm of the vee. After some work, it can be determined that the data look roughly like four long thin cones coming out of the origin in four directions that are only seen to separate clearly past about 20 units. Two of the cones that are next to each other are longer than the other two cones.

In fact, this can be confirmed by looking at the right-hand panels in Fig. 9.7. The four cones actually correspond to two species of crabs, one for each sex. The shorter two cones are the females and the longer two are the males. The panels on the right are the same as the panels on the left except that the four classes are labeled (brushed) with different shapes. The top right panel shows that the circles and triangles form the lower arm of the vee and the squares and diamonds form the upper arm. The rotation to the middle panel on the left brings the circles alone to the top, the squares and triangles to the middle region, and the diamonds alone to the bottom. Continuing the rotation, the third panel on the left shows that the circles and squares are on the top and the triangles and diamonds are on the bottom. It is left as an exercise to use GGobi to find a direction (down the center) in which the cones collapse into four blobs, one for each species-sex pair.

9.9.3 Time Dependence

Essentially, the idea is to plot the data points versus their observation number. If the data are believed to be exchangeable, then this plot should have no information and hence reveal no patterns. However, there often is information in the sequence of the data points. While there are extensive graphics associated with time series, the point here is merely to represent the data to suggest its structure. So, with minimal processing of the data, it is natural to look at plots such as (X_i, X_{i+1}), $(X_i, (X_i - X_{i-1})/\hat{\sigma})$ ($\hat{\sigma}$ is a suitable estimate of a scale to make the fraction approximate a derivative), $(i, X_i - X_{i-1})$, or $(X_i, X_i - X_{i-1})$. Sometimes these are called phase portraits in analogy with plots such as $(f(x), f'(x))$ in differential equations. Of course, there are many other similar plots that might be appropriate for particular data sets, especially when searching for structure in multivariate data. One can imagine plotting, for instance, stars or Chernoff faces (see below) from data over time. Indeed, as with projections, one can look at time-dependent plots for statistics obtained from dimension reduction.

Here, Fig. 9.8 shows a series of plots for the sunspot data set for the period 1770–1869. The time period was chosen to include some of the largest changes in amplitudes of the readings since 1700. Sunspots themselves are vortices of gases at or near the surface of the sun that follow a cycle of about 11.1 years and are counted by a technique developed in 1849 by R. Wolf in Zurich. The full data set is available in R.

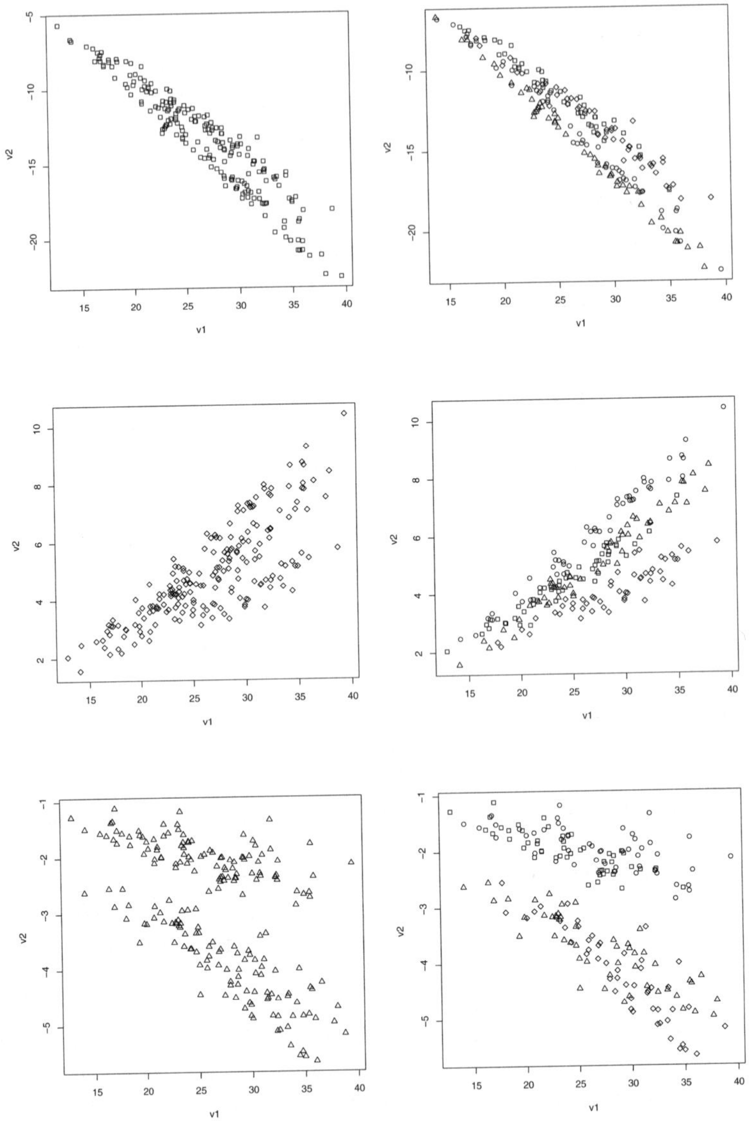

Fig. 9.7 The three left panels show views found from GGobi. The three right panels are the same views but with the cones labeled.

First, the upper left panel in Fig. 9.8 shows the standard visualization of the sunspot time series. Letting X_i denote the sunspot number for year $i = 1, ..., 100$ for 1770–1869, the upper right panel in Fig. 9.8 shows a plot of $(X_i, (X_i - X_{i-1}))$. It is seen that the first difference is a coarse derivative ratio (the denominator is $i - (i-1) = 1$) at X_i. The plot itself is seen to be cyclic over time, rotating around a point near $(30, 0)$, with orbits

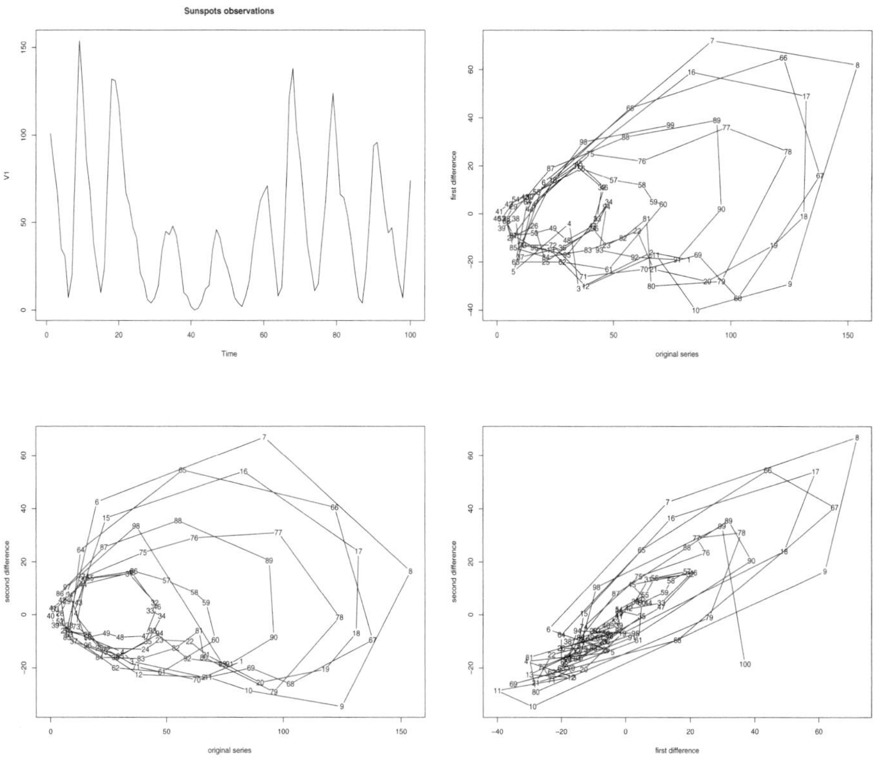

Fig. 9.8 Four plots to show the sunspot time series and the analogs of the (f, f'), (f, f''), and (f', f'') phase portraits. It is seen that the lower left panel is more regular than the other two phase plots.

much tighter on the left than on the right. A similar plot but for the second difference (i.e., $(X_i, X_i - 2X_{i-1} + X_{i-2})$) in the lower left panel of Fig. 9.7, corresponds to (f, f''). It shows the same kind of pattern. Finally, the lower right panel in Fig. 9.7 shows the plot of the first and second differences as i increases. It is seen to be similarly cyclical, though more tilted and flattened.

The commands used to obtain these plots were as follows.

```
x<-as.ts(read.table("sunspots.dat"))
postscript("Sunspots.eps")
plot(x, main="Sunspots observations")
y<-diff(x,1)
postscript("SunspotsPhase1.eps")
plot(x,y, xlab='original series',
ylab='first difference')
# Pre-processing the data
xm1 <- x[-1]
xm2 <- xm1[-1]
```

```
xm2[99]<-0
x11<-diff(x,1)
xp <- x
yp <- 0.5*(x11+(xm2-xm1))
postscript("SunspotsPhase2.eps")
plot(xp,yp, xlab='original series',
ylab='second difference')
postscript("SunspotsPhase12.eps")
plot(diff(x,1),yp, xlab='first difference',
ylab='second difference')
```

9.10 Visualization II: Transformations

In contrast to the techniques of the last section, here the data points are processed in some substantial way. That is, the data are not just represented, functions of them, possibly complex, are chosen and visualized. This is a sort of modeling, but the goal is not to do formal inference, but to search for patterns in the data for which, hopefully, inferential models can be proposed.

Among the numerous ways to do this, three are discussed here. Chernoff faces are an entertaining way to map vectors to facial features. Multidimensional scaling (MDS) is a way to represent a p-dimensional data set as a collection of points in K dimensions, where K is user-chosen. Thus, MDS is a dimension reduction technique as well. It is included here because when $K = 2, 3$, reconfiguration of the p-dimensional points represents the distances between them so the relationships can be seen. Finally, self-organizing maps, (SOMs) are actually a sort of vector quantization. Crudely, they help identify likely modes of a high-dimensional density.

9.10.1 Chernoff Faces

This is the kind of idea that emerges at happy hour as a joke. Except that it works. Chernoff (1973) recognized that people are exquisitely sensitive to small differences in faces and proposed that this be harnessed to visualize high-dimensional data. Thus, under a mapping, a p-dimensional data point is converted to a list of values that specify features of a human face. For instance, the values of $x_{1,i}$ may represent the height of a face, the values of $x_{2,i}$ might represent the width of a face, and so forth. Then, each face generated from the data points has a unique expression. Since people can discriminate very finely across facial expressions, characteristic patterns of the data set can be readily seen. Like profiles and stars, Chernoff faces depend on how the X_js are converted to facial features and are subjective.

Using the Canadian household expenditure data for which stars and profiles were plotted in Fig. 9.3, gives the Chernoff faces in Fig. 9.9. The command is

```
postscript("CanadaChernoff.eps")
faces(xx[,-1], labels = xx[,1],
main = "Canadian Expenditure")
```

The R package automatically transforms the data into 15 facial features. Each variable is scaled so it is between 0 and 1; here this is accomplished by looking at the proportion of income spent on each category. For instance, the first variable, food expenditure as a proportion of total household expenditures, was converted into the height of the face; the second variable, shelter expenditure as a proportion of total household expenditures, was converted into the width of the face; and so on. In fact, there were 17 variables, but R only permits 15, so the last two were omitted.

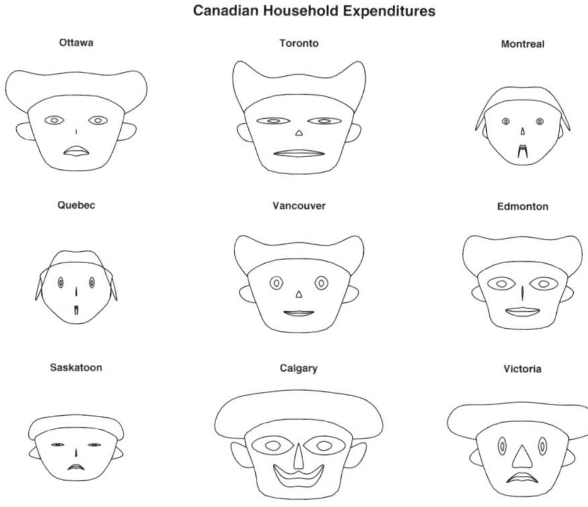

Fig. 9.9 Chernoff faces for household expenditures for nine Canadian cities. Note that food and shelter primarily determine the size of the face. It is great fun to try to match the expression on the face to the character of the city. This shows how exquisitely sensitive we are to facial expressions, underscoring the strength of the method.

9.10.2 Multidimensional Scaling

Multidimensional scaling (MDS) is a class of techniques that converts the set of distances between n given points in $\mathbb{R}^p$ to a collection of n points in $\mathbb{R}^K$ for a prespecified

$K << p$. The conversion is done by minimizing a criterion that measures the distance between the given proximity matrix in p dimensions and the proximity matrices that might arise from a distance in K dimensions. Since the configuration of the points in $\mathbb{R}^p$ tends to be distorted when embedded in $\mathbb{R}^K$, the criterion is called a stress, in analogy with mechanical stress. Nevertheless, the output of an MDS procedure is a fully connected, weighted graph that can be examined.

More formally, let $\boldsymbol{x}_i$, $i = 1, \ldots, n$ be points in $\mathbb{R}^p$ and let $D = D_n = (d_{i,j})_{i,j=1,\ldots,n}$, where $d_{i,j} = d(\boldsymbol{x}_i, \boldsymbol{x}_j)$, be the proximity matrix in p dimensions. It will be seen that the $\boldsymbol{x}$s only affect the minimization through the $n(n-1)/2 - n$ values $d_{i,j}$ for $i < j$. Fix a K and a measure of distance, say δ, in $\mathbb{R}^K$. Parallel to the data $\boldsymbol{X}_i$, let $z_i \in \mathbb{R}^K$ for $i = 1, \ldots, n$, and write $\delta_{i,j} = \delta(z_i, z_j)$ and $\Delta = (\delta_{i,j})_{i,j=1,\ldots,n}$. To be general, let $g : \mathbb{R} \to \mathbb{R}$ be a fixed, nondecreasing function. The MDS task is to minimize a function of the form

$$S = \text{Stress}(z_1, \ldots, z_n) = \sum_{i<j} \frac{w_{i,j}(\delta_{i,j} - g(d_{i,j}))^2}{\text{scale}} \tag{9.10.1}$$

to find the z_is. The denominator in (9.10.1) indicates the scaling of distances and is usually based on either d or δ. Sometimes

$$\text{scale} = \sum_{i<j} w_{i,j}\delta_{i,j}^2 \quad \text{or} \quad \sum_{i<j} w_{i,j}d_{i,j}^2$$

is chosen, in which case Stress is a little like a weighted chi-square distance; i.e., like a goodness-of-fit criterion. Thus, (9.10.1) can be regarded as if it were a norm on real $n \times n$ matrices and MDS is seeking a matrix of distances for points in $\mathbb{R}^K$ as close as possible to the matrix of p-dimensional distances. In particular, this means every reasonable distance on matrix space gives an MDS representation for a data set. In other words, MDS representations depend substantially on the choice of Stress. Often, g is the identity function, or a linear function $g(x) = ax$, in which case Stress is usually minimized over a as well; it may be important to allow g to be nonlinear in cases where only the ordering relations among the $\boldsymbol{x}_i$s are reliable.

Because of the minimization, the K-dimensional points have a configuration that matches the distances as closely as possible, apart from the dimension. Points that are close together or far apart in p dimensions usually lead to points that are close together or far apart in K dimensions, but the relative positions may change. This arises because MDS depends on distances only and so does not distinguish among the orientations of the angles between line segments joining the points in p dimensions. So, the configuration of points in K dimensions will necessarily be distorted. The benefit is that if $K << p$ it is often easier to recognize patterns in the lower-dimensional points. In this sense, MDS is a (nonlinear) dimension reduction technique. However, the goal is usually limited to visualizing the relationship between high dimensional data points by their lower-dimensional analogs because the lower-dimensional representations are typically only weakly representative of any other information in the data points.

There are two main problems to overcome in using MDS. First is finding the zs and second is assessing them. Assessment includes choosing K, determining how representative of the $\boldsymbol{x}$s the zs are, and identifying the patterns in the zs that lift up to give patterns in the $\boldsymbol{x}$s.

There are numerous methods for finding an MDS representation. Given K, the simplest is to choose d and δ to be Euclidean, g the identity, and scale $= w_{i,j} = 1$, so the criterion becomes

$$S(z_1, \ldots, z_n) = \sum_{i<j} (\delta_{i,j} - d_{i,j})^2. \tag{9.10.2}$$

In this case, the optimization problem reduces to finding the first K eigenvectors of the empirical variance matrix and using the coordinates of the points with respect to them. That is, the p-dimensional data points are projected to the space spanned by the principal components corresponding to the K largest eigenvalues. Sometimes this is called Classical MDS.

One step that is more complicated is Newton-Raphson. In this case, differentiate (9.10.1) to get a Hessian matrix $\boldsymbol{H}$. Then, as in the neural network setting where weights are updated, each z_i can be updated. Essentially this finds directions in which to tweak the entries of the z_is to reduce the stress. Basalaj (2001) explains several more techniques, including genetic algorithms, simulated annealing, and a majorization approach somewhat like EM.

Assuming z_is can be found for a range of K, the usual procedure is to form the scree plot $(K, S_{min}(K))$ over K and look for a knee in the curve. The value of K at the knee is a common choice since it represents a trade-off between keeping K small, which is the point, and ensuring that enough of the information in the data has been included in the z_is that including more would be diminishing returns.

Assessing representativity is harder because it is not easy to obtain a standard error for the z_is even though they are functions of the $\boldsymbol{x}_i$s. In addition, the axes of the z space are meaningless. They are only constructs to find a minimum: The zs are moved around unpredictably to approximate the distances in $\mathbb{R}^p$. The distances in $\mathbb{R}^K$ therefore are distorted representations of the configuration of the data; the greater the stress, the greater the distortion. As a consequence, usually all one seeks in the lower-dimensional representation is the dimension and any clustering. This is reasonable because large distances are typically relatively less distorted than small distances. So, distinct clusters are often valid, but within clusters the relative positioning may be an artifact of the optimization. Given a clustering obtained for an MDS representation, a good check is to rerun MDS for the elements of each cluster individually to see if the apparent clustering is borne out.

Another check on the reasonableness of an MDS is a 2D Shepard plot. The idea is to examine how closely the input and output distances match. The input distances in $\mathbb{R}^p$ are on the horizontal axis and the output distances in $\mathbb{R}^K$ are on the vertical axis. The plot consists of the collection of points $(d(\boldsymbol{x}_i, \boldsymbol{x}_j), \delta(z_i, z_j))$. If the scatter around the $y = x$ line is tight, the plot indicates that over the range of the original dissimilarities, the MDS gave a faithful representation. The ranges where the scatter deviates from the $y = x$ line indicate the regions most subject to distortion. In essence, the Shepard plot examines the size of each possible error, neglecting scale. Often there are patterns in this plot; seeking explanations for them may yield insight. Once structure is found, the corresponding points can be removed and the remaining points searched for secondary structures, or the entire MDS can be rerun. In this way, it may be natural to combine MDS with other procedures, such as clustering, perhaps sequentially.

MDS has limitations aside from possible distortions. MDS depends on the choices of distances and can be highly nonrobust. A small perturbation in p dimensions can easily be in a direction for which the small changes in the p coordinates add up to a large effect in the K coordinates. Also, because outliers give unusually large distances, they can distort the lower-dimensional representation. Taken together, this can reduce the interpretability of an MDS plot. Computationally, MDS may not scale up well to large p or moderate n. When MDS does not reduce to PCs, the computing required may be enormous or have convergence problems.

MDS is implemented in the R package SMACOF (Scaling by MAjorizing a Complicated Function), which implements a procedure based on majorizing a stress. The majorization is accomplished using a complicated version of the majorization principle, which can be generically stated as follows. If the goal is to minimize a function $f(x)$, then it is enough to find a more tractable function $g(x,y)$ for which $g(x,y) \geq f(x)$ for some well-chosen supporting point y and $g(x,x) = f(x)$. The minimizing x_{min} should satisfy

$$f(x_{min}) \leq g(x_{min}, y) \leq g(y,y) = f(y).$$

The usual strategy is to find a starting y_0, find x_k such that $g(x_k, y) \leq g(y,y)$, and iterate until $f(y) - f(x_k)$ falls below a threshold. To see how this procedure is implemented in SMACOF for MDS, see de Leeuw and Mair (2008). The package also gives a Shepard plot for the residuals. The command used in the code below, smacofSym, is the basic version of the procedure for symmetric dissimilarity matrices using Euclidean distance, scale $= 1$ and all $w_{i,j} = 1$ as in (9.10.2); the form of the command used below takes an object of class dist – a distance matrix found using a distance measure to give the distances between the rows of a data matrix. Other forms of the command permit more detailed specification of the arguments directly.

Figure 9.10 shows the results of using MDS on the vertices of cubes and simplices in four and five dimensions to get a two-dimensional visualization. The commands that generated Fig. 9.10 are

```
library(MASS)
library(gtools)
library(smacof)
dc     <- 5         # Dimension of cube
nbp    <- 2^dc      # Number of vertices
dcube <- NULL       # Matrix of generated points
sigma <- 2          # SD of normal distribution
ns     <- 5         # Number obs per component
Sigma <- diag(rep(1,dc))
muse   <- permutations(n=2,r=dc,v=0:1,
repeats.allowed=TRUE)
for(i in 1:nbp){
```

```
dcube <-rbind(dcube,mvrnorm(mu=muse[i,],
Sigma=Sigma,ns)) }
x11()
mds.cube<-smacofSym(dist(muse,method= "manhattan"))
plot(mds.cube, main = paste("MDS for Hypercube in
dimension", dc, sep=""))
x11()
simp.id <- which(apply(muse, 1, sum) <= 1)
mds.simplex <- smacofSym(dist(muse[simp.id,],
method="manhattan"))
plot(mds.simplex, main = paste("MDS for Simplex in
dimension", dc, sep=""))
```

In Fig. 9.10, the top pair of panels shows that groups of four points representing two-dimensional faces form clusters; recall that the cube in four dimensions has 16 vertices and the cube in five dimensions has 32 vertices. On the left, each pair of these four-point sets represents a cube that is the projection of the four dimensional cube into two dimensions. On the right, the four-point sets group into eight-point sets and each pair of them represents the projection of a four dimensional cube into two dimensions. The bottom pair of panels in Fig. 9.10 shows how the five or six vertices of simplices in four or five dimensions, respectively, appear in two dimensions. In both cases, the axial vertices become points equally spaced around the vertex at the origin, which is unchanged.

There are numerous variations on the basic MDS template. First, the optimization in (9.10.1) corresponds to metric MDS. Nonmetric MDS is intended for data where the numerical values are unreliable but the ordering on distances is reliable. This occurs when measurements are preferences of one option over another, for instance. In these cases, only $d_{ij} > d_{k\ell}$ or $d_{ij} < d_{kl}$ can be taken as known. To get useful solutions, the function g must be monotonic, satisfying $g(d_{ij}) > g(d_{k\ell})$ when $d_{ij} > d_{k\ell}$. Minimizing the stress in this case is more difficult; however, the Shepard-Kruskal algorithm generally provides solutions. The basic idea is to start with an initial guess about the δ_{ij}^0s – possibly from metric MDS – and update the guesses to δ_{ij}^1, δ_{ij}^2, and so forth.

The updating finds δ_{ij}^1s by finding a monotone regression relationship between the δ_{ij}^0s and the d_{ij}s under the constraint that $d_{ij} < d_{k\ell} \Leftrightarrow \delta_{ij}^0 < \delta_{k\ell}^0$. Specifically, the δ_{ij}^1s are found as follows. For $i, j = 1, ..., n$, the distances between pairs of $\boldsymbol{x}$s (i, j) can be ranked. Write

$$(i_1, j_1) > (i_2, j_2) > \ldots > (i_u, j_u),$$

where $u = n(n-1)/2$, to mean that the distance between $\boldsymbol{x}_{i_v}$ and $\boldsymbol{x}_{i_v}$ is greater than the distance between $\boldsymbol{x}_{i_w}$ and $\boldsymbol{x}_{i_w}$ when $v > w$; i.e., $d_{i_v,j_v} > d_{i_w,j_w}$ for $v > w$. The idea is to make the ranks of the δ_{ij}s match those of the d_{ij}s. So, violations of the monotonicity can be found by examining a plot of the ordered d_{ij} versus their respective δ_{ij}^0s. The strategy for reducing violations of the ordering is to start at the smallest $d_{i,j}$, say d_{i_u,j_u}, and find the first order violation. For this violation, average the violating d_{ij} with the most recent nonviolating d_{ij}; this gives a δ_{ij}^1. Then, proceed to the next violation, where the same procedure can be used. Doing this for all the violations and leaving the δ_{ij}s

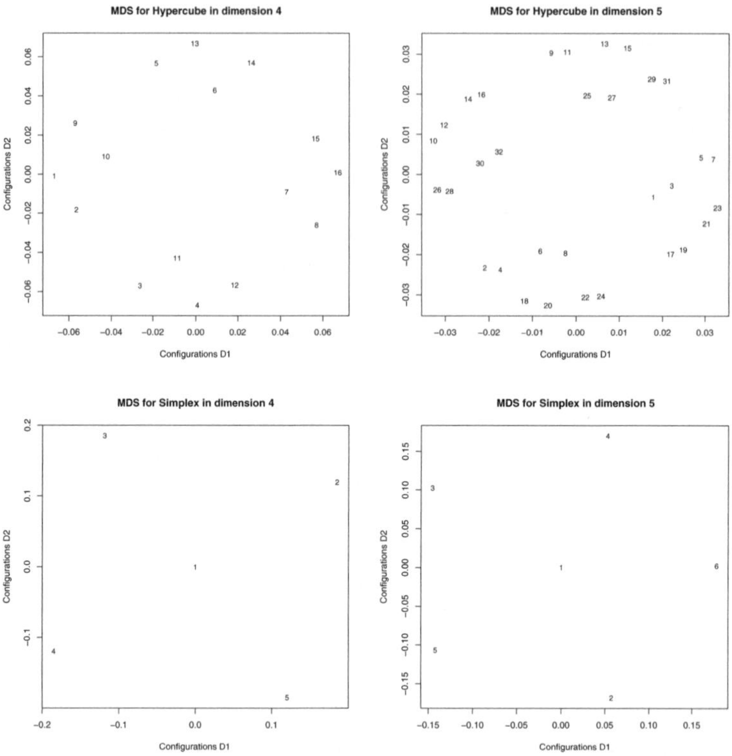

Fig. 9.10 The top pair shows the MDS output for a cube of sidelength one in four and five dimensions, respectively. The bottom pair shows the MDS output from a simplex in dimensions 4 and 5.

that respect the ordering of the d_{ij}s unchanged gives a new collection of distances, δ^1_{ij}, which are closer to the d_{ij}s than the δ_{ij}s were.

Another variation on the MDS procedure is to seek a lower-dimensional geometry as well as finding representations of the data points within it. The idea is that the points $\boldsymbol{x}_i$ might lie in a low-dimensional submanifold in $\mathbb{R}^p$ and that this manifold can be uncovered if its intrinsic geometry is used to give the d_{ij}s instead of deriving them from how the low-dimensional submanifold is embedded in p-dimensional Euclidean space. The procedure called ISOMAP (see Tenenbaum et al. (2000)) uses a sequence of nearest-neighbor regions to assign distances between points that are not neighbors; this represents the intrinsic, within-submanifold distance better than the raw Euclidean distances. With this alternative assignment of the d_{ij}s, the usual MDS optimization may give $\boldsymbol{z}$s that are more reasonable.

A third variation on MDS, invented by Roweis and Saul (2000), is called locally linear embedding. The idea has two steps. First, find weights w_{ij} by minimizing

$$E_1 = \sum_{i=1}^{n} |\boldsymbol{x}_i - \sum_{j \in I_i} w_{ij} \boldsymbol{x}_j|.$$

The index set I_i is usually chosen to be the nearest neighbors of $\boldsymbol{x}_i$ so that each $\boldsymbol{x}_i$ is reconstructed optimally from its neighbors as a convex combination, enforced by requiring $\sum_j w_{ij} = 1$ for each i. Quadratic optimization problems like this can be solved relatively effectively. The second step is to choose a lower dimension K and then use the w_{ij}s to define a new error,

$$E_2 = \sum_{i=1}^{n} |z_i - \sum_{j \in I_i} w_{ij} z_j|,$$

which can be minimized over the z_is to give the lower-dimensional representation. Again, minimizing E_2 is a quadratic optimization that can be done readily. The net effect of these two related optimizations is to transform the nearest-neighbor relations from the p-dimensional space to the K-dimensional space. It is locally linear in the sense that points are represented as a convex combination of their neighbors. Despite the local definition, the lower-dimensional representation often captures important relationships among nonneighboring points as well.

Finally, if the distances in (9.10.1) are Euclidean, then there are variations on MDS in which the Euclidean norms are replaced by (Euclidean) inner products; see Buja et al. (2001) for a discussion and implementation. The properties of this criterion do not seem to have been fully investigated.

9.10.3 Self-Organizing Maps

For electrical engineers, vector quantization is the effort to choose a collection of K canonical representatives, say $z_1,...,z_K$ in $\mathbb{R}^p$, such that any future $\boldsymbol{X} = \boldsymbol{x}$ will be closest to a unique z_j. Then, if the z_js are chosen well (i.e., there is enough of them and they are spread evenly over the range of $\boldsymbol{X}$) one can approximate $\boldsymbol{X}$ as a function by using the z_js. For computer scientists, vector quantization is incorporated in what is called competitive learning: Units within a formal structure compete for the exclusive right to respond to a particular input pattern. For instance, a neuron or small set of neurons in a neural network might specialize in detecting and processing certain classes of inputs. Statisticians would regard vector quantization as a sort of partitional clustering (on the grounds that the input patterns a unit represents optimally are similar to each other and amount to a cluster) followed by choosing a canonical representative for each cluster.

Among the numerous techniques developed for vector quantization, self-organizing maps stand out as a visualization technique, a dimension reduction technique, and a clustering or mode-seeking technique simultaneously. SOMs were introduced in Kohonen (1981) and became popular for visualizing data because the canonical representatives output by the SOM procedure typically exhibited an interpretable structure in low dimensional ($p = 2,3$) examples. See Kohonen (1989) for a full description. SOMs find these canonical representatives by bootstrapping.

The basic structure of an SOM is as follows. Imagine a collection of units, say U_k for $k = 1,...,K$; each unit is located at a point in a lattice of dimension $\ell << p$ that remains fixed throughout the procedure. Usually the lattice points that have units located at them form a square or other regular-shaped grid. To each unit associate an initial vector $\boldsymbol{r}_k(0) \in \mathbb{R}^p$, typically $p >> \ell$. At the tth iteration, the $\boldsymbol{r}_k(0)$s will be updated to $\boldsymbol{r}_k(t)$. For each t, these are called reference vectors. Let $N_k(t)$ be the neighborhood around unit k at iteration t. If the units form a square in a $2D$ lattice, then one choice for the neighborhood around a unit is the unit itself with its eight neighbors (unless the unit is on the edge of the square, in which case its neighborhood has six units, or four if the unit is on a corner). Note that the time dependence in the $N_k(t)$s permits the neighborhoods at different iterations to vary. Usually it is better to choose large neighborhoods for early iterations and let them shrink as t increases.

Given this notation, the SOM procedure is as follows.

Given data $\boldsymbol{x}_1,..., \boldsymbol{x}_n$ in $\mathbb{R}^p$, choose one at random and denote it $\boldsymbol{x}$.

- ☐ Find the unit $U_{k_{opt}}$ with $\boldsymbol{r}_{k_{opt}} = \arg\min_k \|\boldsymbol{x} - \boldsymbol{r}_k\|$.
- ☐ Choose a learning rate, $\alpha(t) \in [0,1]$, to control how much the reference vectors are permitted to change at the $t+1$ iteration.
- ☐ Update all K reference vectors $\boldsymbol{r}_k(t)$ to $\boldsymbol{r}_k(t+1)$ using the neighborhood $N_{k_{opt}}(t)$ by

$$\boldsymbol{r}_k(t+1) = \begin{cases} \boldsymbol{r}_k(t) + \alpha(t)(\boldsymbol{x} - \boldsymbol{r}_k(t)) & \text{if } k \in N_{k_{opt}}(t), \\ \boldsymbol{r}_k(t) & \text{if } k \notin N_{k_{opt}}(t). \end{cases} \tag{9.10.3}$$

- ☐ If the set $\boldsymbol{r}_k(t+1)$ for $k = 1,...,K$ satisfies a convergence criterion stop. Otherwise, draw another $\boldsymbol{x}$ and repeat.

The random reselection of the $\boldsymbol{x}$ in the first step is a draw from the empirical distribution of the data, which is the effective input to an SOM. The idea is to use the $\boldsymbol{x}$ to choose which $\boldsymbol{r}_k$, and hence N_k, to update. The updating moves the reference vectors in N_k closer to the drawn $\boldsymbol{x}$. The output of the whole bootstrapping procedure is a set of K vectors $\boldsymbol{r}_k(\infty)$ of dimension p. These vectors are indexed by the points on the grid, in the lattice of dimension ℓ, and, subject to that, summarize the information in the data. The intent is that if two regions are close in the original high-dimensional space they should map to gridpoints that remain close in the SOM in the low-dimensional space.

The interpretation of an SOM can be given only roughly. As can be guessed from the procedure, the $\boldsymbol{r}_k(\infty)$s will tend to accumulate around modes of the empirical distribution, one or more at or near each mode, and the others spread over regions to reflect their relative probability. The question becomes how to summarize the information in the vectors associated with the units; if two or more reference vectors are quite close to each other, it may be reasonable to rerun the SOM with fewer reference vectors to prevent different units from representing the same information. Thus, the output of an SOM may be interpreted as a collection of canonical representatives from a clustering procedure. As such, the SOM can be regarded, weakly, as putting together modes so as to represent a density from a high-dimensional space in a low-dimensional space. Also, the SOM can be regarded as a submanifold-seeking procedure trying to identify a low-dimensional submanifold embedded in a high-dimensional space and represent the density on it. Neither of the interpretations are entirely accurate, but the SOM output grid can often be interpreted in terms of them, at least locally. Also, trying to determine, simultaneously, a submanifold and a density on it is an extraordinarily difficult problem, not least because estimates of a density and the manifold supporting it likely converge at different rates (probably slow and very slow, respectively).

Since the SOM procedure depends on the choice of so many subjective features – the value of ℓ, the collection of units, the $N_k(t)$s, the $\alpha(t)$, and the norm and the initial reference vectors – it is worthwhile looking at the typical behavior of a few choices and the idealized cases they imply. So, suppose $p = 2$ and ℓ is chosen to be one so the lattice is just the integers. After the SOM procedure converges, the $\boldsymbol{r}_k(\infty)$ are on a line of units that can be identified with points in the $\boldsymbol{X}$ space by the reference vectors. If the distribution of $\boldsymbol{X}$ is uniform over an equilateral triangle, for example, the line of units typically ends up being identified with a space-filling curve in the triangle because no regions of $\boldsymbol{X}$ can be meaningfully distinguished from other regions. In other words, the line of units reconstructs, as best it can, a uniform distribution.

If $p = 2$ and $\ell = 2$, there is no dimension reduction. If the lattice is all pairs of integers and the $\boldsymbol{r}_k(\infty)$ are on a rectangle of units, then the grid itself, not just the reference vectors, can be identified with points in the $\boldsymbol{X}$-space. Now, if the distribution of $\boldsymbol{X}$ is uniform on a rectangle, the grid of units typically ends up matching the rectangle.

If $p = 3$ and $\ell = 2$, there is dimension reduction. The lattice may be regarded again as pairs of integers and the reference vectors $\boldsymbol{r}_k(\infty)$ can be on a rectangle of units. Now, as with the $\ell = 1$ case above, the grid of units can no longer be identified with points in the $\boldsymbol{X}$-space. However, the reference vectors will converge, typically, to the modes of the $p = 3$ dimensional distribution, if there are any, and will become evenly spread out over any uniform regions. In between, the reference vectors will accumulate roughly in line with the density. One example in this context that may indicate the structure is to regard the $\boldsymbol{X}$s as proportions of red, green, and blue light in a beam. Then the reference vectors indicate a color. It is then natural to color the K units with the color identified by their reference vectors. In this way, the modal colors, if any, may be seen. In general, interpreting an SOM requires some way to visualize the reference vectors. Kohonen (1990) gives further examples.

There seems to be little theory for SOMs, partially because they are defined by a class of procedures rather than by an optimality principle (which would be much more stringent and informative). It is well known that SOM procedures need not converge at all; aside from actually diverging, SOM outputs often cycle around a weak representation of the higher-dimensional structure, never settling down to a limit. Much of what is known about SOMs is heuristic, based on many examples or analogies with other methods thought to be similar. For instance, there is the suggestion that there is a sense in which SOMs escape the Curse of Dimensionality. Kohonen (1990) states that the number of iterations typically needed for convergence is independent of p. However, since SOMs are an extremely general procedure, this statement is necessarily tentative, perhaps only holding under restrictive conditions.

SOMs are expected to behave like other similar vector quantization techniques. In particular, as was noted in the three low-dimensional instances above with uniform distributions, vector quantization methods are well known to approximate the distribution of the input $\boldsymbol{X}$. For a class of vector quantization techniques related to neural modeling, Ritter (1991) shows that in one-dimensional cases the asymptotic density q of the quantization in terms of $p(x)$ follows a power law $q(x) = Cp(x)^{\alpha}$, where $\alpha = \alpha_N$ depends on the number N of neighbors on each side of a unit. As N increases, $\alpha \rightarrow 2/3$, which is consistent with uniform $\boldsymbol{X}$s leading to uniform grids, $p = \ell = 2$, or approximations to uniform distributions from the reference vectors, $p = 2, \ell = 1$ and $p = 3, \ell = 2$.

Some argue that SOMs are related to K-means clustering. For instance, the first step in an SOM iteration, finding k_{opt}, is like assigning an outcome to the cluster represented by $\boldsymbol{r}_{k_{opt}}$. More generally, if the partition from a vector quantization procedure can be summarized by the means of its partition elements, it resembles a clustering procedure that is similar to K-means. This kind of reasoning is heuristic but is often borne out in examples and deserves development.

SOMs are also similar in spirit to principal curves. Recall that a principal curve is the conditional expectation over a level set. In SOMs, the reference vectors can also be exhibited as conditional expectations of the data (see the batch map algorithm in Kohonen (1995)). Thus, SOMs can also be regarded as a discrete version of principal curves.

Finally, another analogy made for SOMs is with unsupervised neural networks. However, this is somewhat more distant than analogies with vector quantization, clustering, and principal curves. The reason is that although one can regard the updating from neighbors as a connectivity between nodes, this is to update the grid as opposed to feeding the output of one unit into another unit to get an overall output. Thus, SOMs may have some vague relation to the architecture of a network, but neural nets and SOMs are fundamentally different procedures.

In the absence of good theoretical characterizations for SOMs there are practical defaults in various settings. Kohonen (1990), p. 1469 explains several. The number of iterations, for instance, should be on the order of 500K or more. The choice of α is fairly broad. As long as $\alpha(1) \approx ... \approx \alpha(1000) \approx 1$ its later behavior is not very important provided it eventually is less than .01. Setting $\alpha(t) = .9(1 - t/1000)$ is one recommended choice. Likewise, there are a broad range of neighborhoods that work well. Usually, $N_k(t)$s should be large for small t, up to over half the grid, shrinking possibly linearly down to a single nearest neighbor.

To see what kind of representations SOMs give, it's worth looking at two examples. The first is simulated data from a uniform distribution on the three-dimensional region formed by the union of two unit cubes, one at the origin with edges parallel to the positive axes and the other a copy of it but shifted by adding $(2,2,2)$. The second is the Australian crab data visualized earlier using projections. SOMs can be generated for both of these data sets using the contributed R package som.

The commands used to generate the panels in Figs. 9.11 and 9.2 were

```
gr <- somgrid(topo = "hexagonal")
cube.som <- SOM(dcube, gr,
alpha = list( seq(0.0005, 0, len = 1e4),
seq(0.0002, 0, len = 1e5) ),
radii = list(seq(4, 1, len = 1e4),
seq(2, 1, len = 1e5)))
```

The first command indicates the neighborhood structure N_k and tells som to use its default for the size of the map. Often this is a rectangle with $5\sqrt{n}$ gridpoints for a medium-sized map. The second command generates the SOM for the cube data, called cube.som. The arguments of the SOM command are the data (dcube), the grid (gr), the learning rate α, and the radius of the map, which controls how the neighborhoods of shape hexagonal (in this case) shrink. The default number of iterations is $rlen = 10,000$; the default initial assignments to the gridpoints are random from the data (without replacement); the default SOM representation of the reference vectors uses stars at each of the gridpoints. The learning rate depends on the iteration; i.e., $\alpha = \alpha(t)$ in (9.10.3). The default in som is for $\alpha(t)$ to decline linearly from 0.05 to 0 over *rlen* updates. Even though the form of the neighborhood is chosen, its size must be specified. Thus, radii gives the length in gridpoints of the neighborhood to be used for each iteration. The default is for this to decrease linearly from 4 to 1 over *rlen* iterations. Note that, in the command above, two sequences are used for α and radii. This is to provide a more effective search over SOMs. Roughly, the first sequence is to explore the space and the second is to converge to a good SOM. This helps reduce dependence on initial conditions as well as stabilize the limiting behavior of the SOM as a function of more iterations. In Figs. 9.11 and 9.12, variations on α, radii, and *rlen* were done as part of the tuning.

For the pair of cubes data, Fig. 9.11 shows six SOMs generated by different values of the tuning parameters. The gridpoints are the locations of the stars, giving the reference vector for their gridpoint. In all cases, it is seen that the unit cube at the origin gives the medium stars in the SW corner and the other cube gives the large stars spread over the remainder. The tiny stars seem to represent the intercube region that has no data; they are scattered to a greater or lesser degree throughout the region representing the cube with vertex at $(2,2,2)$. The panels of Fig. 9.11 only hint at the range of SOMs generated by som under a variety of reasonable choices of tuning parameters. Indeed, it is quite easy to get SOMs that do not represent any aspect of this simple data set accurately. With practice, however, one can become more adept at choosing tuning parameters and interpreting the patterns in SOMs in terms of densities and submanifolds.

For the Australian crab data, Fig. 9.12 gives two SOMs. Recall that this data set looks (very roughly) like four cones coming out of the origin in five dimensions. The left-hand SOM barely picks this up, vaguely looking like three regions: a SW region, a vacant NE region, and a string of largish stars separating them. The string of largish

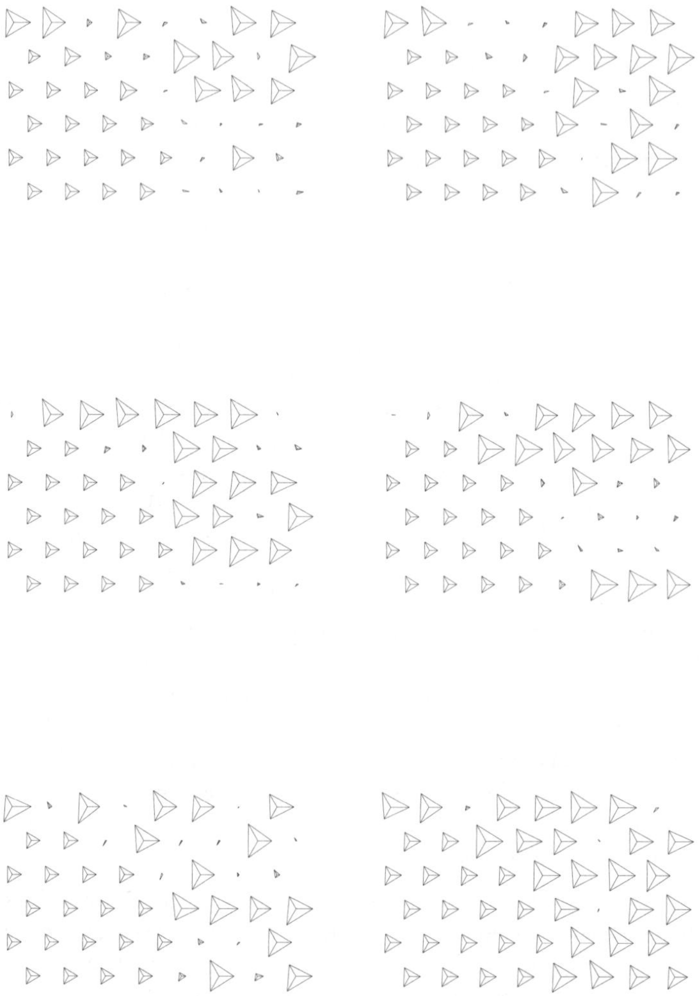

Fig. 9.11 All of these SOMs were formed using 100 data points from each of the cubes, excpt the one at the bottom right, which used 100 data points from the unit cube at the origin and 200 from the other cube.

stars actually has thickness two, which it is tempting to interpret as representing two different regions in the data space. The right-hand SOM correctly suggests four regions: There is a NE cluster of large stars (i), a string of very small stars, suggesting a vacant region, another string of large stars (ii), another string of tiny stars, a string of largish stars (iii), and finally, in the SE corner, a group of medium-sized stars (iv). As with the cube data, it is easy to obtain SOMs that are either uninterpretable or lead to incorrect interpretations (for instance, a smooth gradient across the map).

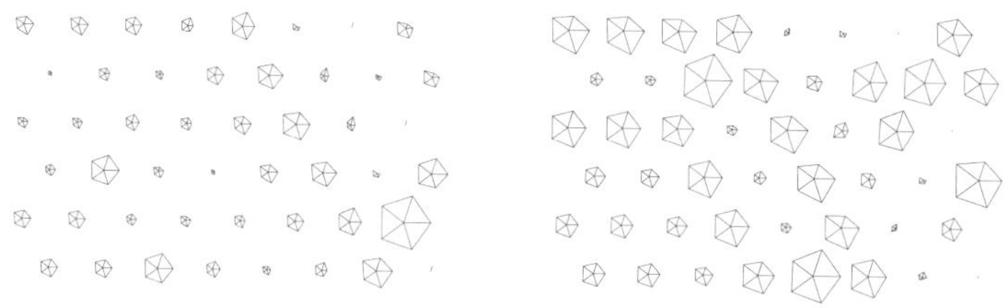

Fig. 9.12 Two SOMs for the Australian crab data. Note that they are quite different. This suggests that, even among SOMs that are reasonably well tuned, a lot of variation is to be expected.

Overall, these examples suggest several caveats when using SOMs. First, SOMs seem to represent disjoint regions separated by edges better than regions where the density is a gradient. This means SOMs may be useful for some image segmentation problems. Second, SOMs are exquisitely sensitive to initial conditions, the number of iterations used, the radius, the neighborhood function, and the weight α. In practice, tuning so many choices is very complex, indeed nearly impossible, unless the expected pattern is at least roughly known in advance or a user can guide the SOM interactively to a solution thought to be reasonable. That is, automatic procedures don't seem to work reliably. Even so, SOMs probably only give interpretable answers that are not far wrong (over repeated usage) more often than truly wrong answers. Perhaps SOMs are best regarded as offering a rough guide in the most complicated problems (although the supervised version of SOMs, not presented here, provides much better results).

Finally, the contrast between SOMs and MDS must be made. Both are techniques defined by procedures, so all the nonrobustness problems (sensitivity to perturbations, outliers, subjective choices of inputs) with MDS carry over to SOMs. However, SOMs are often easier to compute after allowing for a greater number of choices that must be made to specify the algorithm. With MDS, the distortions arise from dimension reduction, whereas with SOMs the question is how well the reference vectors provide an interpretable visualization of the modal structure.

A second difference is that the optimization in MDS tries to preserve interpoint distances and tends to distort smaller distances more than larger distances. SOMs are somewhat the reverse. SOMs tend to be locally valid but otherwise tend not to preserve interpoint distances. The reason may be that it's as if SOMs cluster first and then identify representatives for the clusters, thereby preserving topological properties, whereas MDS tries to preserve metric properties, ignoring clustering.

A final problem with most methods like SOMs and MDS that are defined by a computational procedure is that there is no obvious way to specify uncertainty. While this makes it possible to get answers – indeed many answers, often incompatible – it means there is little basis, apart from robustness or modeling, that can be done to evaluate how good the answers are. In these settings, however, modeling is usually difficult, whence the default to an algorithmic approach, and robustness can only be checked informally, by rerunning the procedure with variants of the inputs. These can be part of an uncertainty analysis, but they often are exceedingly difficult to do well, and it is generally unclear how well they can substitute for a proper uncertainty analysis.

9.11 Exercises

Exercise 9.1 (Multivariate normal warm-up). Consider n independent outcomes, $\boldsymbol{Y}_i$ for $i = 1,...,n$, drawn form $N(\mu_i, \Sigma)$ where the normal is q-variate and $\mu_i = B^{\mathsf{T}}\boldsymbol{x}_i$, in which B is a $q \times p$ matrix and $\boldsymbol{x}$ is p-dimensional. Observe that B is a matrix of regression parameters and that the vector $\boldsymbol{x}_i$ is the i-th design point for the q-variate response. Let X be the design matrix; i.e., the $n \times p$ matrix with rows given by $\boldsymbol{x}_i^{\mathsf{T}}$. Assume X is full-rank.

1. Identify a criterion under which $B^{\mathsf{T}}\boldsymbol{x}_0$ is a good point predictor for $\boldsymbol{Y}_0$, for a given design point $\boldsymbol{x}_0$. Verify the criterion is satisfied.
2. Show that, even in the multivariate setting, $\hat{B} = (X'X)^{-1}X^{\mathsf{T}}Y$, where $Y = (\boldsymbol{y}_1, ..., \boldsymbol{y}_n)$ is an $n \times q$ matrix, remains a good estimator for B.
3. Find the covariance matrix of $\hat{B}^{\mathsf{T}}\boldsymbol{x}_0$.
4. Give a confidence interval for a linear combination of the form $\boldsymbol{c}^{\mathsf{T}}\boldsymbol{y}_0$, where $c = (c_1, ..., c_q)$.

Exercise 9.2 (Special cases for PCs). Here are two simple special cases where the PCs can be readily found.

1. Let X_1 and X_2 be random variables with mean zero, variance 1, and suppose their correlation is ρ. Find the PCs of the vector response $\boldsymbol{X} = (X_1, X_2)$.
2. Let $\boldsymbol{Y}$ be a q-variate normal, $N(0, \Sigma)$, and write $\Sigma = \sigma^2 I + \delta^2 \mathbf{1}\mathbf{1}^{\mathsf{T}}$. What are the PCs of $\boldsymbol{Y}$?

Exercise 9.3 (PCs and other procedures). One of the strengths of PCs is that they are well understood and relatively easy to implement. As a consequence, they can be readily combined with other analytic methods. Recall that the essence of the PC-reduction is to replace the data points $\boldsymbol{x}_i$, for $i = 1, \ldots n$, with new vectors $\boldsymbol{u}_i$, where $\boldsymbol{u}_i = \boldsymbol{A}\boldsymbol{x}_i$, and then project onto the first K entries of the $\boldsymbol{u}_i$s to reduce the dimension.

To see how well this works when combined with other analytic methods, consider the following comparisons that can be done using the contributed R package princomp.

1. Variable selection: Define PC-based variable selection as follows. From a PC analysis, find the eigenvector corresponding to the smallest eigenvalue. Find the entry in this eigenvector that has the largest coefficient and drop the corresponding variable. This can be repeated for the second-largest eigenvalue and so forth. Compare PC-based variable selection with backwards elimination in a standard linear regression problem. Use any regression data set with, say, $p \geq 3$.
2. Clustering: Use one of the data sets described at the end of Section 8.8 – flea beetles, olive oils, or PRIM7 – and apply any of the clustering techniques first to the data set and then to the PC-reduced version of the data set for several choices of K. As K increases to p, the reduced dimension clustering should approach the unreduced clustering. Can you relate the number or location of clusters to the choice of K?
3. Classification: The same type of comparison as above for clustering can be done with classification data. That is, the reduced dimension data can be used to develop a classifier. So, reanalyze the data sets from Section 8.8, or any other data set with large enough dimension (say $p \geq 4$), using, for instance, SVMs. Again, as K increases, the SVM classification for the dimension-reduced data should approach the SVM classification for the unreduced data. Do the support vectors change? If RVMs are used, do the relevant vectors change?
4. Regression: As with clustering and classification, dimension reduction techniques can be used to preprocess data for a regression analysis. Using any data set with a reasonably large p and a univariate Y, compare the pure unreduced regression with the results for $K = p-1, p-2, \ldots$ in terms of, say, MSE or R^2_{adj}.
5. Thresholding: Finally, consider a shrinkage form of PCs. In a regression analysis based on PCs, choose a threshold value $c > 0$ for the components of the matrix $\boldsymbol{A}$. When a component a_{ij} of $\boldsymbol{A}$ has absolute value below c, set it to zero. Otherwise leave it unchanged. Call the resulting matrix $\boldsymbol{A}_s$, as in Section 9.1. How does this shrinkage form $\boldsymbol{A}_s$ of $\boldsymbol{A}$ compare with $\boldsymbol{A}$ itself in the clustering, classification, and regression cases above? In particular, how are the results affected by the choice of c?

Exercise 9.4 (Simple case of classical FA model). Consider n independent outcomes $\boldsymbol{Y}_i$, $i = 1, ..., n$ of the form $\boldsymbol{Y}_i \sim N(\mu_i, \Sigma)$, where the normal is q-variate. Form a factor analysis model for Σ with $K = 1$ by using

$$\Sigma = \text{Var}(\boldsymbol{Y}_i) = \Gamma\Gamma^{\mathsf{T}} + \sigma^2 I_q$$

in place of (9.2.5), where Γ is the vector $(\gamma_1, \ldots, \gamma_q)$.

1. Derive the MLE, $\hat{\Gamma}$, for Γ.
2. Find the first eigenvector of Σ and show that $\widehat{Gamma}$ is proportional to it.

Exercise 9.5 (Latent trait model). In item response theory, a manifest property is a quantity that can be directly measured and called a trait. This is in contrast to a latent trait that cannot be measured directly. An example would be test scores as a way to evaluate mathematics ability. Often, the trait that can be measured is attributed to a latent trait. One way to express this is

$$X_j = C_j F + D_j e_j \tag{9.11.1}$$

for $j = 1, ..., p$, in which the C_js and D_js here can be taken as known constants, the F is the latent variable, and the e_js are error terms. To be definite, fix $p = 3$, group the unobserved quantities into the vector $\boldsymbol{Y} = (F, e_1, e_2, e_3)^{\mathsf{T}}$, and group the observed quantities into the vector $\boldsymbol{X} = (X_1, X_2, X_3)^{\mathsf{T}}$.

1. Find a matrix A so that $\boldsymbol{X} = A\boldsymbol{Y}$.
2. If $\text{Cov}(\boldsymbol{Y}) = I_4$, find $\text{Cov}(\boldsymbol{X})$.
3. An obvious estimator for F is the average $\hat{F} = (X_1 + X_2 + X_3)/3$. Find the $\text{Corr}(\hat{F}, F)$. When $|\text{Corr}(\hat{F}, F)|$ is high, is it reasonable to regard $\hat{F}$ as good? Explain.
4. Suppose that p increases in (9.11.1) but that all the other assumptions are maintained. How is $\text{Corr}(\hat{F}, F)$ affected by increasing p?

Exercise 9.6 (FA with rotation). The point of this exercise is to use FA, with a factor rotation. Consider the following data analyzed in Abdi (2003). There are five wines and seven measurements are made on each by a panel of experts: how pleasing it is, how well it goes with meat or dessert, its price, sweetness, alcohol content and acidity. Thus, much of the information in the data is subjective.

Wine	Hedonic	Meat	Dessert	Price	Sugar	Alcohol	Acid
1	14	7	8	7	7	13	7
2	10	7	6	4	3	14	7
3	8	5	5	10	5	12	5
4	2	4	7	16	7	11	3
5	6	2	4	13	3	10	3

1. First, verify that a PC decomposition gives four PCs (with eigenvalues 4.76, 1.81, .35, and .07).
2. Argue that a factor analytic dimension reduction to two factors is appropriate, and give the 2×7 matrix of factor loadings. The contributed R package factanal is one way to do the computations, as is factor.pa.
3. It was seen in Section 9.2 that FA solutions are only determined up to an orthogonal transformation. So, practitioners often use a "varimax" rotation to maximize the variance of the squared elements in the columns of the matrix of factor loadings. The idea is to find the linear combination of the original factors such that

$$\mathscr{V} = \sum_{j=1,k=1}^{p,K} (\lambda_{j,k}^2 - \bar{\lambda^2})$$

for a given K, where $\lambda_{j,k}^2$ is the squared jth loading for the kth factor with mean (over j and k) denoted $\bar{\lambda}^2$. The linear combination is a rotation and hence is orthogonal. Use the varimax command in R, or any other software that does FA, to find the new 2×7 matrix of factor loadings that maximize $\mathscr{V}$. (The optimal rotation is 15°.)

4. Justify the varimax criterion, and explain its typical effect. (Are there other criteria for choosing an orthogonal transformation that are reasonable?)

Exercise 9.7. Projection pursuit as a technique for dimension reduction has not been as fully explored as other techniques. Its potential is suggested by the following.

1. In what sense can you argue that both SIR from Chapter 4 (and its generalization in Section 9.8) and projection pursuit regression (also from Chapter 4) can be regarded as special cases of the projection pursuit procedure presented in Section 9.3?
2. Consider the model

$$Y = (X_1 + 5X_2)^2 + (X_3 - 5X_4)^4 + \varepsilon$$

where $\varepsilon \sim N(0, \sigma^2)$ and $(X_1, X_2, X_3, X_4) \sim N(0, I_4)$. Simulate data from this model (use $\sigma = .1, .3, .5$) and use the SIR procedure to find the effective dimension reduction dimensions. How does the size of σ affect the results?

3. (Not for the faint of heart.) Estimate a basis for the space $\mathscr{S}_\xi$ defined in Section 9.8 following the Cook and Ni (2005) procedure.
4. Derive $I_{cum}(X) = \kappa_3^2 + (1/4)\kappa_4^2$ for univariate X by approximating $I_{ent}(X)$. Using the Gram-Charlier expansion

$$f(x)\phi(x) = 1 + \kappa_3 H_3(x)/6 + \kappa_4 H_4(x)/24 + \ldots, \tag{9.11.2}$$

where H_k is the kth Hermite polynomial, assume $f(x) = (1 + \varepsilon(x))\phi(x)$ and

$$\int \phi(u)\varepsilon(u)u^k du = 0 for k = 0, 1, 2, \ldots.$$

(See Jones and Sibson (1987) for the argument.)

5. Finally, use a gradient descent approach on I_{cum} in the procedure in Section 9.3 to find good directions for the data simulated from (9.11.2). Compare the results of the three methods (EDR from SIR, the estimated CS from the Cook-Ni technique, and the projections from I_{cum}).

Exercise 9.8 (ICA and the uniform distribution). Here is an example worked out in detail in Hyvarinen and Oja (2000). Consider two independent $Uniform[-\sqrt{3}, \sqrt{3}]$ random variables S_1 and S_2, with mean zero and variance one. Let

$$\begin{pmatrix} X_1 \\ X_2 \end{pmatrix} = \begin{pmatrix} 2 & 3 \\ 2 & 1 \end{pmatrix} \begin{pmatrix} S_1 \\ S_2 \end{pmatrix}.$$

Now suppose that both (S_1, S_2) and $\boldsymbol{A}$ are unknown and only outcomes of (X_1, X_2) are available.

1. It is easy to see that $\boldsymbol{X} = (X_1, X_2)$ is uniform on a parallelogram. How would you estimate the edges of the parallelogram and use them to uncover (S_1, S_2)? What is the limitation of this approach?

2. Consider functions of the form $\boldsymbol{Y} = \boldsymbol{W}\boldsymbol{X}$ and consider the Shannon mutual information of $\boldsymbol{Y}$, $I(Y_1,Y_2) = \mathbb{E}\ln p(Y_1,Y_2)/p(Y_1)p(Y_2)$. Show that if the Y_is are uncorrelated ($\mathrm{Corr}(Y_i,Y_j) = 0$ for $i \neq j$) and have unit variance ($\forall i\ \mathrm{Var}(Y_i) = 1$)then $\det(\boldsymbol{W})$ must be constant. (**Hint**: Observe that

$$I(Y_1,Y_2) = \sum_{i=1}^{n} H(Y_i) - H(\boldsymbol{X}) - \ln|\det \boldsymbol{W}|,$$

where H indicates the entropy of a random variable.)

3. Let the negentropy of a random variable Z be

$$J(Z) = H(Z_G) - H(Z),$$

where Z_G is a Gaussian random variable with the same mean and variance as Z. Now obtain

$$I(Y_1,Y_2) = C - \sum_{i=1}^{2} J(Y_i),$$

where C is a constant.

4. Argue that ICA ,by minimizing mutual information, is formally equivalent to maximizing non-Gaussianity when the estimates are uncorrelated.

5. (Not easy.) Use ICA defined by mutual information to recover S_1 and S_2 from X_1 and X_2. How different is the ICA solution from the uniform inputs?

Exercise 9.9 (Identifiability, uniqueness, and separability in ICA). These examples are taken from Eriksson and Koivunen (2003) to demonstrate that nonidentifiability is nontrivial, identifiability can hold without separability or uniqueness, and that uniqueness can hold without separability. Verify the following for ICA representations:

1. If $\boldsymbol{S}$ is normal, then multiplying it by any orthogonal or diagonal matrices shows it is nonidentifiable.

2. Let S_i for $i = 1,2,3,4$ be independent and nonnormal and Z_1 and Z_2 be standard normal and independent. Show that $\boldsymbol{X} = (X_1, X_2)$ is identifiable but not separable or unique when $X_1 = S_1 + S_3 + S_4 + Z_1 + Z_2$ and $X_2 = S_1 + S_3 - S_4 + Z_1 - Z_2$; i.e., find two choices of $\boldsymbol{S}$ that produce $\boldsymbol{S}$ using the same $\boldsymbol{T}$.

3. Let S_1, S_2, and S_3 be independent with nonvanishing characteristic functions and let $\boldsymbol{X} = (X_1, X_2) = (S_1 + S_3, S_2 + S_3)$. Show that the representation is unique but not separable.

Exercise 9.10 (Principal curve summarization). This example is treated in Hastie (1984), Hastie and Stuetzle (1989), and Delicado (2001). Generate 100 independent data points by setting $S \sim Uniform[0, 2\pi]$, $\varepsilon_j \sim N(01)$ (for $j = 1,2$) and finding outcomes of the form

$$\begin{pmatrix} X_1 \\ X_2 \end{pmatrix} = \begin{pmatrix} 5\sin(S) \\ 5\cos(S) \end{pmatrix} + \begin{pmatrix} \varepsilon_1 \\ \varepsilon_2 \end{pmatrix}. \tag{9.11.3}$$

Apart from noise, this gives points scattered around a circle of radius five.

1. Verify that (9.11.3) uses the arc-length parametrization.
2. Use the contributed R-package pcurve to generate a principal curve to summarize the data.
3. What is the estimated length of the curve, and how does this compare to the population length of the curve?
4. What is $\text{Var}(S)$?
5. What is the estimated total variability along the curve?
6. What is the average residual variance in the orthogonal directions to the curve? Is it valid to compare this to $\text{Var}(\varepsilon)$? Explain.

Exercise 9.11 (Wine data and PLS). Reconsider the wine data from Exercise 9.6, but let $\boldsymbol{Y} = (Y_1, Y_2, Y_3)$ be the subjective measurements Hedonic, Meat, and Dessert, respectively, and let $\boldsymbol{X} = (X_1, X_2, X_3, X_4)$ be the objective measurements Price, Sugar, Alcohol, and Acid. As before, $n = 5$. Using the contributed R package pls, which does partial least squares regression (and principal components regression), analyze the wine data. Find the $\boldsymbol{T}$, $\boldsymbol{U}$, $\boldsymbol{P}$, and $\boldsymbol{Q}$ matrices. How many latent vectors should a good model retain?

An earlier analysis (see Abdi (2007)) used one, two, and three latent vectors and found

Latent vector	PE $\boldsymbol{X}$	CPE $\boldsymbol{X}$	PE $\boldsymbol{Y}$	CPE $\boldsymbol{Y}$
1	70	70	63	63
2	28	98	22	85
3	2	100	10	95

where PE is the percentage of the variance explained and CPE is the cumulative percentage of the variance explained. Thus, two latent variables are reasonable. Further analysis of the regression coefficients shows that Sugar is mainly responsible for choosing a dessert wine, Price is negatively correlated with the perceived quality of the wine, and Alcohol is positively correlated with the perceived quality of the wine. Thus, the latent vectors reflect Price and Sugar. Does your output confirm this?

Exercise 9.12 (PLS with large p, small n). One of the most important features of PLS-1 is that it can be used when $p >> n$. Here is a simple version (see Garthwaite (1994)) that is a variant on the method of Section 9.7. Consider expressing a univariate Y as a function of X_1, ..., X_p. The idea is to do a search over linear functions of the X_js to find a small collection of linear combinations of the X_js to use as in a regression function for Y. That is, the task is to form a predictor $\hat{Y} = b_1 T_1 + ... + b_K T_K$ for some K, in which each $T_j = c_1 X_1 + ... + c_p X_p$. In fact, the regression function in terms of the T_js reduces to the form $\hat{Y} = a_1 x_1 + ... + a_p x_p$ (the data are assumed centered so that $a_0 = 0$) but the goal is to find T_js on the grounds that they might be meaningful in their own right.

Start by finding T_1 (i.e., the c_js for a candidate T_1) by p univariate regressions of the X_js on Y. Write $\hat{X}_j = \hat{\alpha}_j + \hat{\beta}_j Y$, and choose $c_j = \hat{\beta}_j$. Given this T_1, find b_1 by regressing Y on T_1 and take the least squares optimal coefficient of T_1 as b_1. Now examine the residuals from this regression. Call them Y_1. Find T_2 for Y_1 as T_1 was found for the original Y. Then, refind b_1 and b_2 from the regression of Y on both T_1 and T_2. The T_js are the latent factors that can be used to predict Y.

1. Verify that this procedure really is a variant on the technique of Section 9.7. (Recall that least squares coefficients can be expressed in terms of $\mathrm{Cov}(X_j, Y)$.)
2. Consider the $p = 9$, $n = 4$ data set on protein consumption from four European countries from 1973.

Country	RM	WM	Egg	Mk	F	Cer	St	N	FV
France	18.0	9.9	3.3	19.5	5.7	28.1	4.8	2.4	6.5
Italy	9.0	5.1	2.9	13.7	3.4	36.8	2.1	4.3	6.7
USSR	9.3	4.6	2.1	16.6	3.0	43.6	6.4	3.4	2.9
W. Germany	11.4	12.5	4.1	18.8	3.4	18.6	5.2	1.5	3.8

   The coding is Red Meat, White Meat, Eggs, Milk, Fish, Cereal, Starch, Nuts, Fruit and Vegetables; the full data set is available from `http://lib.stat.cmu.edu/DASL/Datafiles/Protein.html` or Hand et al. (1994). Use the simplified procedure above to do a PLS regression analysis on these data. How many latent factors are reasonable? Can you interpret them?
3. Redo the analysis using software such as pls. Do you get the same result?

Exercise 9.13. Since projection matrices arise frequently in the visualization and analysis high dimensional data, here are some facts from linear algebra that are useful to remember.

1. Let X be a matrix. Verify that XX^{T} is symmetric.
2. Show that the eigenvalues of a XX^{T} are the squares of the singular values of X. (It is a fact that symmetric real matrices have a full set of eigenvalues.)
3. Let v_1 and v_2 be two PCs from a variance matrix with different eigenvalues. Show that if you project normal data onto the span of a v_1 and separately onto the span of v_2 then the correlation between the projected data is zero. Explain how to choose the basis vectors of an eigenspace so that even when the eigenvalues are the same, this property holds.
4. How can you use PCs to choose projections to visualize high dimensional data? What are the limitations of PCs for this purpose?
5. What other choices of directions for projection would be reasonable? For each choice of directions, indicate what kind of structure the directions would detect and what type of structure they would miss. (PCs, for instance, would likely detect single modes quite well but would probably break down if there were two cigar-shaped modes parallel to each other.)

Exercise 9.14. MDS can sometimes be used to represent heuristically the vague belief that some objects are more or less similar to other objects. Consider the following data extracted from Kaufman and Rousseeuw (1990) attempting to represent (in a unidimensional summary) how 9 different countries (Belgium, Brazil, China, Cuba, Israel, France, India, USA, and Zaire) are from each other on the basis of their social and political environments.

	BEL	BRA	CHI	CUB	ISR	FRA	IND	USA
BRA	5.58							
CHI	7.00	6.50						
CUB	7.08	7.00	3.83					
ISR	3.42	5.50	6.42	5.83				
FRA	2.17	5.75	6.67	6.42	3.92			
IND	6.42	5.00	5.58	6.00	6.17	6.42		
USA	2.50	4.92	6.25	7.33	2.75	2.25	2.75	
ZAI	4.75	3.00	6.08	6.67	4.83	5.58	6.17	5.67

1. Using software such as GGobi or R (the cmdscale function), produce a two-dimensional (metric) MDS plot for these data. Identify the form of the stress used in the program, and generate a Shepard plot to evaluate the distortion.
2. Cluster the MDS version of the data (in two dimensions). Do the clusters have an obvious interpretation?
3. Repeat items 1 and 2, but use a three-dimensional embedding. How would you validate a clustering in this context?
4. Arguably, robustness is a key desideratum of an MDS visualization. So consider adding two more countries (USSR and Yugoslavia) one at a time after the USA line:

	BEL	BRA	CHI	CUB	ISR	FRA	IND	USA	USS	YUG
USS	6.08	6.67	4.25	2.67	6.92	6.17	6.92	6.17		
YUG	5.25	6.83	4.50	3.75	5.83	5.42	5.83	6.67	3.67	
ZAI	4.75	3.00	6.08	6.67	6.17	5.58	6.17	5.67	6.50	6.92

How does adding each of the extra countries affect the embedding in two or three dimensions and their Shepard plots? Can you construct a scree plot for the distortion as a function of dimension and find the knee?

Exercise 9.15. To see how self-organizing maps behave, consider a cube in four dimensions with edgelength 10 situated in the positive orthant with one corner at the origin. Choose the line of points $\boldsymbol{u} = (u,u,u,u)$ for $u = 0,1,2,...,10$ and for each $\boldsymbol{u}$ assign a four dimensional normal with mean $\boldsymbol{u}$ and variance $(1/16)I_4$. Generate 110 tetradimensional points from the equally weighted mixture of the 11 normals.

1. Consider a one-dimensional grid with points at $-5,-4,\ldots,15$ at which the SOM units will be located. Use the neighborhood function that looks only at the closest neighbor for each unit; i.e., for units between -4 and 14 there are two neighbors,

but for -5 and 15 there is only one neighbor. Try $\alpha(t) = .9(1 - t/1000)$. Run the SOM procedure, for instance in R (command som or use the contributed package kohonen); there are other software packages, too. Does the result look like 11 equal bins as intuition would suggest? What if the variance is increased to, say, I_4?

2. Redo item 1, but use a grid of units in the plane. Try the grid generated by $\{-5,\ldots,15\} \times \{-5,\ldots,15\}$. Do you get a line of equal-sized units on the points $(0,0)$, ..., $(10,10)$?
3. If the original data points are regenerated with unequal weights on the normals, are the modes accurately reflected in the size of the units on the line or plane? Explain. (What would you expect if the weights on the normals are chosen so that the mixture they generate in four dimensions has a single mode?)
4. Now, generate data from a collection of normal distributions with means located at the corners of the 10-cube using variance $(2.5)^2 I_4$ and repeat items 1 and 2. Do your SOMs detect the difference between the data from a line through the center of the cube and the data from the corners of the cube?

Chapter 10
Variable Selection

So far, the focus has been on nonparametric and intermediate tranche model classes, clustering, and dimension reduction. So, the modeling techniques presented have been abstract and general. The perspective has been to search for a reasonable model class via unsupervised learning or dimension reduction or to assume a reasonable model class had been identified. In both cases, the goal was understanding the model class so the problem could become finding elements of the class that fit reasonably well and gave good predictions. The focus was on the model class as a whole more than on the models in the class. In this chapter, the focus is on the models themselves rather than the general properties of the class they came from.

Variable selection is distinct from model selection in that it focuses on searching for the best set of variables to include in a model rather than necessarily finding the best function of them. Best often means parsimonious. Since identifying covariates that predict well may result in a great many serviceable models, variable selection is a key way to winnow down a large number of covariates to relatively few.

Of course, variable selection is usually done in the context of a specific model class. So, finding the best collection of variables can be much at one with finding the right functional form. When this is not the case, finding the best collection of variables typically reduces the range of possible models considerably. The reduced range may be amenable to some of the techniques presented here or in earlier chapters. In addition, sometimes uncertainty about the model can be reduced by asking how variables interact with each other and what they mean physically. However, it is important not to let the desire for interpretability of models reduce the model class so much that it can no longer adequately represent the complexity of the data. This is a common predilection of subject matter specialists.

In general, the sequence of choices an analyst must make are: (i) the model class to use, (ii) the variables to include, (iii) the correct functional form within the class for those variables, and (iv) the correct values for any parameters estimated. Assigning a meaningful variance and bias expression to each of these choices is another layer of complexity. Having looked at a variety of model classes, the main topic of this chapter is (ii). Since (ii) and (iii) are closely related, some aspects of (iii) will necessarily arise here as well. However, (iv) will be neglected because estimation of parameters

B. Clarke et al., *Principles and Theory for Data Mining and Machine Learning*, Springer Series in Statistics, DOI 10.1007/978-0-387-98135-2_10,

is comparatively better understood, usually presenting computational difficulties more than statistical impediments.

As before, assume a data set $\{(\boldsymbol{X}_i, Y_i), i = 1, \cdots, n\}$ is available where $\boldsymbol{X}_i \in R^p$ is the p-dimensional covariate vector and Y_i is the response. To reduce modeling bias, it is common to start with more covariates than are actually believed to be required. This may necessitate a substantial search to determine which variables are most important to include, especially when there is a true model that is relatively sparse. Subject matter specialists prefer parsimonious models on the grounds that they are easier to interpret; statisticians prefer parsimonious models because they give smaller variances.

Most of the early work on variable selection was done in the context of linear models, and many of the core ideas can be seen most easily in this context. Many of the later developments in variable selection owe much to these early contributions. For instance, the idea of ranking variables or subsets of variables in terms of importance of inclusion in a linear model recurs in settings like information methods and shrinkage. For this reason, it is important to begin with some familiar material from linear models before describing the main topics of this chapter.

10.1 Concepts from Linear Regression

Consider the linear regression model

$$Y_i = \boldsymbol{X}_i^{\mathsf{T}} \boldsymbol{\beta} + \varepsilon_i, \quad i = 1, \cdots, n, \tag{10.1.1}$$

where $\boldsymbol{\beta} = (\beta_1, \ldots, \beta_p)^{\mathsf{T}}$ is the vector of linear regression coefficients and $\varepsilon_1, \ldots, \varepsilon_n$ are IID mean-zero errors with constant variance σ^2. Let $\boldsymbol{y} = (y_1, \ldots, y_n)^{\mathsf{T}}$ be the observed response vector and $\mathbf{X} = (\boldsymbol{X}_1, \ldots, \boldsymbol{X}_n) = (x_{ij})$ for $i = 1, \cdots, n$, $j = 1, \ldots, p$ be the design matrix, taken as nonstochastic. Without loss of generality, the response Y is assumed centered (i.e., $\sum_{i=1}^n y_i = 0$) by subtracting $\bar{y}$ and the predictors are similarly standardized so that, for $j = 1, \ldots, p$, $\sum_{i=1}^n x_{ij} = 0$ and $\sum_{i=1}^n x_{ij}^2 = 1$. For the present, also assume that $n > p$. Now, the ordinary least squares (OLS) estimator of $\boldsymbol{\beta}$ can be defined and is

$$\widehat{\boldsymbol{\beta}}^{ols} = \arg\min_{\boldsymbol{\beta}} (\boldsymbol{y} - \mathbf{X}\boldsymbol{\beta})^{\mathsf{T}} (\boldsymbol{y} - \mathbf{X}\boldsymbol{\beta}) = (\mathbf{X}^{\mathsf{T}}\mathbf{X})^{-1}\mathbf{X}^{\mathsf{T}}\boldsymbol{y}. \tag{10.1.2}$$

If OLS is used to fit a linear model, it is well known that as the number of variables included in the model increases, the variance of the predicted values increases monotonically. In parallel, the bias decreases monotonically until the true model is contained in the fitted model.

indexvariable selection To see this in more detail, observe that variable selection in the context of (10.1.1) is equivalent to the problem of selecting a best subset model from a class of models, say $\mathscr{M}$. This can be represented by choosing the correct nonvoid subset of $\mathbf{A} = \{1, \cdots, p\}$. Without loss of generality, let $\mathbf{A}_0 = \{1, \cdots, p_0\}$ be the index set for the true model for some $p_0 \leq p$. Then, the design matrix is

$$\mathbf{X} = \left(\mathbf{X}_{\mathbf{A}_0}, \mathbf{X}_{\mathbf{A}_0^c}\right),$$

and the true coefficients are $\boldsymbol{\beta}^* = (\boldsymbol{\beta}_{\mathbf{A}_0}^{*\top}, \mathbf{0}^\top)^\top$. If the correct choice of variables were known, the OLS estimator of $\boldsymbol{\beta}_{\mathbf{A}_0}^*$ would be

$$\widehat{\boldsymbol{\beta}}_{\mathbf{A}_0}^{ols} = \arg\min_{\boldsymbol{\beta}} (\boldsymbol{y} - \mathbf{X}_{\mathbf{A}_0}\boldsymbol{\beta}_{\mathbf{A}_0})^\top (\boldsymbol{y} - \mathbf{X}_{\mathbf{A}_0}\boldsymbol{\beta}_{\mathbf{A}_0}) = (\mathbf{X}_{\mathbf{A}_0}^\top \mathbf{X}_{\mathbf{A}_0})^{-1}\mathbf{X}_{\mathbf{A}_0}^\top \boldsymbol{y}. \tag{10.1.3}$$

It is not hard to show (see Exercise 10.1) that, for any $\boldsymbol{x} \in \mathbb{R}^p$,

$$\mathbb{E}(\boldsymbol{x}^\top \widehat{\boldsymbol{\beta}}^{ols}) = \boldsymbol{x}^\top \boldsymbol{\beta}^* = \boldsymbol{x}_{\mathbf{A}_0}^\top \boldsymbol{\beta}_{\mathbf{A}_0}^* = \mathbb{E}(\boldsymbol{x}_{\mathbf{A}_0}^\top \widehat{\boldsymbol{\beta}}_{\mathbf{A}_0}^{ols}) \tag{10.1.4}$$

and

$$\mathrm{Var}(\boldsymbol{x}^\top \widehat{\boldsymbol{\beta}}^{ols}) \geq \mathrm{Var}(\boldsymbol{x}_{\mathbf{A}_0}^\top \widehat{\boldsymbol{\beta}}_{\mathbf{A}_0}^{ols}), \tag{10.1.5}$$

where $\boldsymbol{x}_{\mathbf{A}_0}$ are the first p_0 elements of $\boldsymbol{x}$. Thus, the inclusion of extraneous variables can be detrimental, leading to inflated estimates of coefficients and wider confidence and prediction intervals. In practice, a perfectly true model is rarely known, so as more variables are added to the model, reduced bias is traded off against increased variance. Unfortunately, if the added variable is not in the true model, then the increase in prediction variance does not permit a reduction in bias.

Let $\mathscr{S}$ be a variable varying over subsets of $\mathbf{A}$. Then any given $\mathscr{S}$ defines a linear model

$$\mathbb{E}(Y|\boldsymbol{X} = \boldsymbol{x}) = \mathbb{E}_{\mathscr{S}}(Y|\boldsymbol{X} = \boldsymbol{x}) = \sum_{j \in \mathscr{S}} \beta_j x_j. \tag{10.1.6}$$

Classical subset selection seeks an $\mathscr{S}$ of a certain size that achieves a relatively small residual sum of squares or SSE. One important measure of how well a given model defined by $\mathscr{S}$ describes the data is the coefficient of determination R^2,

$$R^2 = R^2(\mathscr{S}) = 1 - \frac{SSE}{TSS}, \tag{10.1.7}$$

where $SSE = SSE(\mathscr{S}) = \sum_{i=1}^n (y_i - \hat{y}_i)^2$ is the sum of squared errors from fitting (10.1.6), and $TSS = \sum_{i=1}^n (y_i - \bar{y})^2$ is the total sum of squares. Sometimes when the permitted sets $\mathscr{S}$ are understood, SSE_k is used to mean the fitted model contains k variables. Usually, R^2 is interpreted as the proportion of variability in the data that is explained by the model. It is seen that values of R^2 close to one indicate a good fit and values close to zero a poor fit.

One main disadvantage of R^2 is that it favors large models unduly because its value decreases monotonically as variables are added to the model. To correct for this but retain the advantages of R^2, the R^2 value can be corrected by dividing the residual sums of squares by their degrees of freedom, giving

$$\left[R^2_{adj}\right]_k = 1 - \frac{SSE/(n-k)}{TSS/(n-1)}, \tag{10.1.8}$$

in which k is the size of the fitted model. For most variable-selection purposes, R^2_{adj} is more effective than R^2.

Mallows (1964, 1973) proposed a useful criterion to determine the optimal number of variables to be retained in a model. The idea is to search for the correct $\mathscr{S}$ by finding the best model of the best size k. For any model of size k, Mallows' criterion is

$$C_k = \frac{SSE_k}{s^2} - (n-2k), \tag{10.1.9}$$

where $s^2 = SSE_p/(n-p)$ is the residual mean square from the full model containing all p covariates. The intuition is that, for the full model of p variables, $SSE_p/(n-p) \approx \sigma^2$, even when the true model only uses a proper subset of the p variables. Given this, if a particular model of size k is correct, then, for this model, $SSE_k/(n-k) \approx \sigma^2$ as well. Consequently, for a good k,

$$C_k \approx \frac{(n-k)\sigma^2}{\sigma^2} - (n-2k) = k.$$

Mallows (1964) recommended plotting a graph of C_k against k for various ks and models of size k and noted that models with small values of C_k are desirable.

Standard variable-selection procedures typically fall into one of two categories: subset selection and variable ranking. It will be seen that in subset selection the best $\mathscr{S}$s are those with relatively small SSEs. At their root, general subset selection methods are just slightly more sophisticated than using R^2_{adj}, MSE, or Mallows' C_k directly. Such methods can be very suggestive, but it must be remembered that while SSE is a good assessment of fit, it says little directly about prediction. By contrast, variable ranking tries to assign a worth to each variable X_j for $j = 1, \ldots, p$, traditionally its association with Y. Although this is a univariate approach, there are some recent extensions that are promising.

10.1.1 Subset Selection

In principle, an exhaustive search could be done by fitting all $2^p - 1$ candidate models and selecting the best one under some criterion, such as the smallest R^2_{adj}, MSE, or Mallows' C_k. However, this can be computationally expensive if p is large. Consequently, a variety of algorithms have been developed to speed the search over sets $\mathscr{S}$. Usually, the main idea is to identify the best subsets by a greedy procedure so large numbers of suboptimal subsets can be ruled out efficiently.

One of the earliest procedures achieving improved computational efficiency is due to Hocking and Leslie (1967). They described a method for identifying parsimonious models by eliminating entire subsets of variables from further consideration when they are demonstrably inferior to other subsets already evaluated. The basic idea is to start by fitting the full regression model and using it to reorder and relabel the variables according to the magnitudes of their t-statistics. This means that variables with larger absolute t-statistics, which are marginally worse than those with smaller absolute t-statistics, will have lower indices.

The efficiency gains are based on the observation that when the SSE due to eliminating a set of variables for which the maximum subscript j is less than the SSE due to eliminating the variable $(j+1)$, then no subset including any variables with subscripts greater than j can result in a smaller reduction. To see how this principle helps, let $j=3$ and consider identifying the best set of three variables to be deleted. Let $SSE_{-(123)}$ denote the SSE for the model in which the (relabeled) variables 1, 2, and 3 are not used and $SSE_{(-4)}$ is the SSE if only variable 4 is deleted. If $SSE_{(-123)} < SSE_{(-4)}$, then the deletion of any other set of three variables will have $SSE > SSE_{-(123)}$, so there is no need to evaluate the other subsets of size three. On the other hand, if $SSE_{-(123)} \geq SSE_{(-4)}$, then it is necessary to examine other subsets of size three taken from the first four variables. However, if the smallest SSE from that group is less than $SSE_{(-5)}$, again one can stop since there is no need to evaluate other sets of size three. Continuing in this way allows us to find the optimal set of three variables to delete. Simulation studies suggest that only 20–30% of the $2^p - 1$ models have to be examined.

Furnival (1971) proposed a computationally efficient implementation of the Hocking and Leslie (1967) procedure called "leaps and bounds". The algorithm, which uses sophisticated matrix routines, is developed in Furnival and Wilson (1974). It is based on the principle that

$$\mathscr{S}_1 \subset \mathscr{S}_2 \Rightarrow SSE(\mathscr{S}_1) \geq SSE(\mathscr{S}_2) \tag{10.1.10}$$

for subsets $\mathscr{S}_1$ and $\mathscr{S}_2$. It is seen that inequality (10.1.10) is a variant on the Hocking and Leslie observation about SSEs. Both the Hocking and Leslie (1967) and Furnival and Wilson (1974) procedures assume that the explanatory variables are independent, even though they are often used in the dependent case as well.

Both procedures are also instances of a class of optimization methods called branch and bound. In these methods, large numbers of candidates are ruled out en masse by using estimated upper and lower bounds on the quantity being optimized, in this case SSE or some variant of it such as C_k or R^2_{adj}. This type of algorithm works well when there are a small number of important variables that dominate the regression and so can easily be found. While branch and bound type procedures scale up better than all subsets procedures, they still have exponential running time. They are really only effective for up to 50 or so variables.

As a consequence, sequential search methods such as forward selection, backward elimination, and stepwise regression are widely used. Forward selection begins with no variables in the model and at each step adds the variable that results in the maximum decrease in SSE to the current model. If there are k variables in the current model, then the new SSE from adding another variable is

$$SSE_{k+1}(j) = SSE_k - \frac{[\mathbf{y}^\top (I - H_k)\boldsymbol{x}_j]^2}{\boldsymbol{x}_j(I - H_k)\boldsymbol{x}_j},$$

where $\boldsymbol{x}_j$ is the vector of inputs for the new variable and $H_k = X_k(X_k^\top X_k)^{-1}X_k^\top$ is the projection matrix for the k-variable model.

Adding the variable that gives the maximum decrease in SSE is equivalent to selecting the variable K_{k+1} whose partial correlation with the response, given the current variables, is maximum. (The partial correlation is the usual correlation but between two sets of residuals from regressing on the same variables. In this case, it is the correlation between the residuals from regressing X_{k+} on $X_1, \ldots, X_k$ and from regressing the response Y on $X_1, \ldots, X_k$.) The method stops when adding the next variable does not give a significant improvement in the fit under some criterion. A common stopping criterion is the critical value of the F-statistic for testing the hypothesis $\mathscr{H} : \beta_{k+1} = 0$ in the $(k+1)$-variable model. Thus, the variable X_{k+1} is added to the current model if

$$F_{k+1} = \max\left(\frac{SSE_k - SSE_{k+1}}{SSE_{k+1}/(n-k-1)}\right) > F_{in}, \tag{10.1.11}$$

where $F_{in} = F(\alpha; 1, n-k-1)$, the "$F$-to-enter" value.

Backward elimination is the reverse of this. It begins with all p variables in the model and at each step removes the variable making the smallest contribution. Suppose there are k variables, $k \leq p$, in the current model, and the corresponding design matrix is X_k. Then the new SSE from deleting the jth $(1 \leq j \leq k)$ variable from the current k-variable model is

$$SSE_{k-1}(j) = SSE_k + \frac{(\hat{\beta}_j)^2}{s^{jj}}, \tag{10.1.12}$$

where $\widehat{\boldsymbol{\beta}}_k = (\hat{\beta}_1, \ldots, \hat{\beta}_k)^\top$ is the vector of current regression coefficients and s^{jj} is the jth diagonal element of $(X_1^\top X_1)^{-1}$. (Equation (10.1.12) is Exercise 10.2 at the end of the Chapter.) Deletion of variables continues until it starts harming the fit. As with forward selection, a common stopping criterion is based on the F-statistic: The variable X_j is deleted from the current model if

$$F_j = \min\left(\frac{SSE_{k-1} - SSE_k}{SSE_k/(n-k)}\right) < F_{out}, \tag{10.1.13}$$

where $F_{out} = F(\alpha; 1, n-k)$, the "$F$-to-delete" value.

One problem with forward selection and backward elimination is that once a decision has been made to include or exclude a variable, it is never reversed. Stepwise selection, proposed by Efroymson (1960), overcomes this drawback – at a cost. Stepwise selection begins like forward selection with no variables in the model. Then variables are added sequentially until a stopping criterion such as (10.1.11) is met. Once there are variables in the model, one or more of them may be deleted if they satisfy (10.1.13). Once this deletion step is complete, the variables not in the model are examined again; the ones that satisfy (10.1.11) are added one at a time to the model. The procedure stops when all variables in the model pass the F-test for staying in and all the other variables fail the test for being added.

Stepwise selection can dramatically reduce the computational cost of the best subset selection. However, it has have two costs: First, it is not guaranteed to find the global optimal set. Almost as bad is its instability. Because variables are either added or dropped, subset selection is a discrete process. As noted in Breiman (1996), a relatively small change in the data may cause a large change in which variables are selected. This is partially due to the fact that the criteria for adding or dropping are based on squared errors, which can be oversensitive to the data. Thorough reviews on subset selection procedures can be found in Linhart and Zucchini (1986), Rao and Wu (2001), and Miller (1990).

10.1.2 Variable Ranking

As p increases, traditional variable selection procedures become more onerous computationally. When $p > n$, they break down completely because OLS estimates are not defined. In these cases, it is helpful to screen variables to eliminate those that are redundant or noisy. Sometimes, this can reduce the dimension of the model from p to $q < n$. If this can be done, then, in principle, traditional selection procedures can be applied. Such screening is often done by ranking the variables on the basis of some criterion and eliminating all variables that do not have a high enough score. For instance, the X_js can be ordered by using some measure of association strength to Y. The hope is that the magnitude of such a measure will reliably identify the most important variables.

To overcome the Curse of Dimensionality, most ranking methods are based on marginal models for the response; that is, on a univariate model for Y using a single X_j. For instance, in linear models, p univariate models may be fitted, each assuming a single K_j is linearly related to the response. In this case, the predictors can be ordered by their individual associations with Y. Various coefficients reflecting association have been proposed as the ranking criteria, including the Pearson correlation coefficient, the t-statistic (proportional to the correlation for individual variables), the p-value, and more generally Kendall's τ (see Kendall (1938)) and Spearman's ρ.

In the classification context, variable ranking is also important and is often done by similar methods. One of these is based on sums of squares familiar from two sample t-tests. It is called the BW ratio – between-class variation versus within-class variation. Consider a binary classification problem with response $Y \in \{+1, -1\}$. Let n_1 and n_2 be the sample sizes for two classes $+1$ and -1, respectively. The BW ratio is

$$\mathrm{BW}(j) = \frac{n_1(\bar{x}_{j,+1} - \bar{x}_j)^2 + n_2(\bar{x}_{j,+1} - \bar{x}_j)^2}{\sum_{y_i=1}(x_{ij} - \bar{x}_{j,+1})^2 + \sum_{y_i=-1}(x_{ij} - \bar{x}_{j,-1})^2},$$

where $\bar{x}_{j,+1}$ and $\bar{x}_{j,-1}$ are the sample means of the jth predictor for classes $+1$ and -1 respectively and $\bar{x}_j$ is the overall sample mean for the jth predictor. Variables with high

BW ratios tend to be more homogeneous within the two classes and heterogeneous between the two classes and so are more useful for discriminating between them.

The ideal ranking technique for dimension reduction would retain all the important variables and discard all the unimportant ones. One way to formalize this is called sure independence screening (SIS; see Fan and Lv (2008)) which, asymptotically in n, identifies the correct variables for a model. Essentially, this is consistent model selection. Since SIS is based on the marginal correlations of individual variables X_j and Y, it is easy to describe. Suppose the $n \times p$ design matrix $\mathbf{X} = (\boldsymbol{X}_1, \ldots, \boldsymbol{X}_n)$ is standardized columnwise; i.e., the explanatory variables have been centered to have mean zero and rescaled to have variance one. The vector of marginal correlations of individual variables X_j and Y scaled by the standard deviation of $\boldsymbol{y}$ is $\boldsymbol{\omega} = \mathbf{X}^{\mathsf{T}}\mathbf{y}$.

For a given $\gamma \in (0,1)$, SIS sorts the p componentwise magnitudes of $\omega = (\omega, \ldots, \omega_p)$ into decreasing order to define submodels of the form

$$\mathscr{S}_\gamma = \{1 \leq j \leq p: \ |\omega_j| \text{ is one of the } \lfloor \gamma n \rfloor \text{ largest entries in } \omega\},$$

where $\lfloor \gamma \rfloor$ is the integer part of γn. Clearly, as γ increases, $\mathscr{S}$ increases, so smaller γs give smaller models. Under several regularity conditions, Fan and Lv (2008) (Theorem 2) showed that if the distribution of $Z = X\mathrm{Cov}(X)^{-1/2}$ is spherically symmetric for normal data having a concentration property (essentially that the eigenvalues of any $n \times q$ submatrix of $(1/q)ZZ'$ are bounded away from zero and infinity with high probability), then SIS really is sure in the sense that it captures all the important variables with probability tending to one as n increases.

Unfortunately, variable ranking based on marginal correlation may not work well in the presence of collinearity, often found in high-dimensional data. It is possible that many unimportant variables are highly correlated with important ones and so may be likely to be selected over important predictors having weaker marginal correlation with the response. Also, marginal correlation ignores the interaction effects among variables: It is possible that an important predictor that is marginally uncorrelated but jointly correlated with the response will be filtered out.

One technique to help get around this is the Dantzig selector (see (Candes and Tao, 2007)) which is a ranking method that tries to find estimators $\boldsymbol{\beta}_{\mathbf{A}}$ with minimal norms subject to constraints on residuals of the form $\mathbf{y} - \mathbf{X}_{\mathbf{A}}\boldsymbol{\beta}_{\mathbf{A}}$. Overall, it is as yet unclear how univariate marginal procedures, usually based on independence, compare with these procedures. More procedures like this will be presented in Section 10.3 on shrinkage methods.

A separate problem is that when the predictors have nonlinear effects on the response, the linear correlation coefficient is not likely to work well. For instance, in an additive model

$$Y_i = \sum_{j=1}^{p} f_j(X_{j,i}) + \varepsilon_i, \quad i = 1, \cdots, n,$$

each f_j may have an arbitrary functional form, making the model highly nonlinear. It seems to be an open question what a proper measure of association between individual X_js and Y should be so that the ranking will reflect the relative importance of individual predictors to the response. One natural solution may be to fit a univariate smoother for each X_i and use a goodness-of-fit statistic or measure of association on it.

10.1.3 Overview

There is a huge body of literature on variable selection, and only a fraction of it can be presented here. This is so because variable selection is extremely broad, going far beyond linear models, which themselves have been extensively studied. The wide range of techniques stems largely from the fact that the various classes of nonlinear models have very different properties. Moreover, transformations of the original variables can be important, and this too is a wide range of possibilities going far beyond using a product of two variables to model an interaction. Indeed, transformation includes general functions of one or more explanatory variables and may require using sets of basis functions for some model space. These cases must be borne in mind even though they are not treated extensively here. However, the focus will be on complex models as much as possible.

To begin, Section 10.2 introduces a few classical and recently developed information criteria. These include Mallows' C_p, Akaike's information criterion (AIC; Akaike (1973, 1974, 1977)), and Schwartz's Bayesian information criterion (BIC; Schwarz (1978); Hannan and Quinn (1979)) widely used to gauge the number of variables to include in a model. Another type of criterion is cross-validatory, evaluating model performance by internal prediction. Section 10.2 also covers CV as a technique for variable selection, model evaluation, and parameter tuning.

In Section 10.3, the wide variety of regularization methods that have been proposed and studied intensely will be surveyed. These methods are based on linear models but impose penalties on regression coefficients to shrink them toward zero. Popular examples include ridge regression, the least absolute selection and shrinkage operator (LASSO; Tibshirani (1996)), and their generalization to bridge regression (Frank and Friedman (1993); Fu (1998)) which uses an L_p penalty on the coefficients with $p \geq 0$. An important penalty term, different from norm-type penalties on the coefficients, is the smoothly clipped absolute deviation (SCAD; Fan and Li (2001)) which is also presented. It can give better fits than bridge regression. Other recent developments discussed here are least angle regression (LARS) which is primarily a computational procedure generalizing LASSO in a different direction from bridge, and boosted LASSO, in which the boosting idea from Chapter 6 is used to improve LASSO fits. Variable selection in the context of other penalization methods such as for trees, generalized linear models, smoothing splines, basis pursuit, and SVMs is also discussed.

For contrast, Section 10.4 presents Bayesian approaches, which treat variable selection as a model selection problem. Assume the entire model space associated with p covariates consists of models indexed by $\gamma = \{1, ..., 2^p\}$, so that each model corresponds to a distinct subset of variables. Bayesian methods take model uncertainty into account by putting a prior distribution on both γ and the associated coefficient βs. The posterior $P(\gamma|Y)$ is then used to identify good models. Prior selection, and the computational methods to find the posterior, are the main tasks that must be accomplished to implement Bayesian variable selection. Prior selection will be discussed here, but the extensive computational techniques required are too specialized for substantial discussion in the present treatment.

10.2 Traditional Criteria

Recall that for any subset index $\mathscr{S} \subseteq \{1,...,p\}$, the linear model

$$E(Y|\boldsymbol{X}) = \sum_{j \in \mathscr{S}} \beta_j X_j$$

can be fit and best subset selection means finding the S to minimize some criterion. The first class of criteria will be informational – based on penalizing a log likelihood by a complexity – rather than based on residual error as in the traditional techniques, although the SSE is much the same as the log likelihood for normal errors. Another type of criterion is cross-validatory, evaluating model performance using internal validation on the data set. Interestingly, the criteria from these two classes are closely related and asymptotically equivalent under some conditions.

Unsurprisingly, numerous information criteria have been proposed for selecting the most parsimonious yet correct model. Mallows' C_p can be regarded as being in this category (Mallows (1973, 1995)). More typical members are Akaike's information criterion (AIC; Akaike (1973)), the Bayes' information criterion (BIC; Schwarz (1978)), and the HQ criterion (Hannan and Quinn (1979)). Likewise, there are many versions of cross validation.

To begin, consider the linear model (10.1.1), and let $g(y|\boldsymbol{x},\boldsymbol{\beta})$ be the density of the response y. For a sample of n observations $(\boldsymbol{x}_i, y_i), i = 1, \cdots, n$, the log-likelihood is

$$\log L = \sum_{i=1}^{n} \log g(y_i|\boldsymbol{x}_i, \boldsymbol{\beta}).$$

Typically, an information criterion is of the form

$$\mathrm{IC}_k = -2[\log(L_k(\hat{\beta}_{MLE})) - \phi(n)k], \tag{10.2.1}$$

where $\log(L_k(\hat{\beta}_{MLE}))$ is the maximized log likelihood of a subset model containing a choice of k variables and $\phi(n)$ is a factor specifying the penalty on the model dimension. Two models of the same size can have different IC_k values because their L_ks may differ. In general, the function $\phi(n)$ is increasing in n and takes different forms for different criteria. The information number IC_k can be interpreted as a combination of goodness of fit and model complexity or bias and variance.

In particular, if Y_is are assumed to be independently distributed with $N(\boldsymbol{X}_i^{\top}\boldsymbol{\beta}, \sigma^2)$,

$$\log L(\boldsymbol{\beta}) = -\frac{n}{2}\log(2\pi\sigma^2) - \sum_{i=1}^{n}(y_i - \boldsymbol{x}_i^{\mathsf{T}}\boldsymbol{\beta})^2/(2\sigma^2) = -\frac{n}{2}\log(2\pi\sigma^2) - SSE/(2\sigma^2). \tag{10.2.2}$$

This leads to

$$\mathrm{IC}_k = \frac{SSE_k}{\sigma^2} + 2\phi(n)k. \tag{10.2.3}$$

When σ^2 in (10.2.3) is unknown, it is often estimated by $s^2 = SSE_p/(n-p)$ under the full model. Recall that Mallows' C_p was defined in (10.1.9) as

$$C_k = \frac{SSE_k}{s^2} - (n - 2k),$$

so C_k is an information criterion with $\phi(n) = 1$.

Under a criterion such as (10.2.1), one chooses the best model having size $\hat{k}$ defined as

$$\hat{k} = \arg\min_{1 \le k \le p} IC_k. \tag{10.2.4}$$

To evaluate how good a choice of k this is, assume the true model has size $p_0 \le p$. Models with $k < p_0$ parameters are misspecified and models with $k > k_0$ parameters are correctly specified (at least for the number of terms) but overparametrized. An information criterion is consistent if

$$\lim_{n\to\infty} \mathbb{P}(\hat{k} = p_0) = 1. \tag{10.2.5}$$

As long as $\lim_{n\to\infty}\phi(n)/n = 0$, the information criterion in (10.2.1) is unlikely to lead to a misspecified model asymptotically. Indeed, as $n \to \infty$ it is not hard to see that $\mathbb{P}(\hat{k} < p_0) \to 0$. This follows by noting that $\limsup_{n\to\infty}\mathbb{P}(\hat{k} < p_0)$ is bounded from above by

$$\begin{aligned}
&\limsup_{n\to\infty}\mathbb{P}(IC_{p_0} > IC_k \text{ for some } k < p_0)\\
&= \limsup_{n\to\infty}\mathbb{P}(-2\log L_{p_0}/n + 2p_0\phi(n)/n > -2\log L_k/n + 2k\log\phi(n)/n\\
&\quad \text{for some } k < p_0)\\
&= \limsup_{n\to\infty}\mathbb{P}(\log L_{p_0}/n - \log L_k/n < (p_0 - k)\phi(n)/n \text{ for some } k < p_0)\\
&= \sum_{k<p_0}\limsup_{n\to\infty}\mathbb{P}(\log L_{p_0}/n - \log L_k/n < 0) = 0,
\end{aligned} \tag{10.2.6}$$

where the third equality holds because $\log L_{p_0}/n - \log L_k/n$ is $O_p(1)$, while $(p_0 - k)\phi(n)/n$ is $o(1)$. AIC, BIC, and HQ all satisfy this one-sided property.

Note that consistency of model selection also requires $P(\hat{k} > p_0) \to 0$ as $n \to \infty$. However, only some choices of $\phi(n)$ satisfy this extra one-sided property; a variety of cases are covered in a unifying theorem stated at the end of this section. Although they will not be developed here, two important points to bear in mind are that (i) asymptotic results do not guarantee a satisfactory performance for finite data and (ii) the standard errors for parameter estimation and prediction following model selection will necessarily be increased due to the sampling distribution of the model selection. This extra variability is usually ignored, even though it can have a big impact on inference.

10.2.1 Akaike Information Criterion (AIC)

Akaike (1973, 1974, 1977) proposed a selection criterion by seeking the model that best explains the data with the fewest variables. The AIC score is

$$\mathrm{AIC}_k = -2[\log(L_k(\hat{\beta}_{MLE})) - k], \tag{10.2.7}$$

where k is the number of variables included in the model and L_k is the maximized value of the likelihood function for the chosen model. It is easy to see that AIC corresponds to setting $\phi(n) = 1$ in (10.2.1) and the preferred model is the one with the lowest AIC. Since the model complexity penalty in AIC is simply the number of parameters, AIC not only rewards goodness of fit, but also discourages overfitting, though not strongly.

The most typical form of (10.2.7) is for normal data. With the normal likelihood from (10.2.2), the maximal log likelihood for fitting a k-variable model occurs when

$$\hat{\sigma}_k^2 = SSE_k/n,$$

where SSE_k is the residual sum of squares. So, the maximum value of the log likelihood for this model is

$$\log(L_k(\hat{\beta}_{MLE})) = -\frac{n}{2}\log(2\pi\hat{\sigma}_k^2) - \frac{n}{2},$$

and the AIC for normal data becomes

$$\mathrm{AIC}_k = n\log(2\pi\hat{\sigma}_k^2) + n + 2k = \text{constant} + n\log(\hat{\sigma}^2) + 2k. \tag{10.2.8}$$

Minimizing (10.2.8) gives the desired value of k.

The AIC procedure and its variations are information-theoretic because they can be derived from the relative entropy or Kullback-Leibler number. It is easy and worthwhile to outline this argument. Suppose $g(\mathbf{y}|\boldsymbol{X}, \boldsymbol{\beta})$ is the true conditional density of $\mathbf{y}$ given $\boldsymbol{X}$, and let h be the density of an approximating model. By definition, the relative entropy between g and h is

$$\mathrm{D}(g \,||\, h) = \int g(\mathbf{y}) \log\left(\frac{g(\mathbf{y})}{h(\mathbf{y})}\right) d\mathbf{y} = \int g(\mathbf{y}) \log(h(\mathbf{y}))d\mathbf{y} - \int g(\mathbf{y}) \log(h(\mathbf{y}))d\mathbf{y}. \tag{10.2.9}$$

When using $\mathrm{D}(g \,||\, h)$ for fixed h and variable g to identify the minimal relative entropy distance model, it should be enough to maximize the comparative term

$$\mathrm{CD}(g,h) = \int g(\mathbf{y}) \log(h(\mathbf{y}))d\mathbf{y}. \tag{10.2.10}$$

Akaike (1974) suggests that the central issue for getting a selection criterion is to estimate the expected form of CD in (10.2.10),

$$\mathbb{E}_{\boldsymbol{Y}}\mathbb{E}_{\boldsymbol{Y}^*}\left[\log(\hat{h}(\boldsymbol{Y}^*))|\hat{h}\right], \tag{10.2.11}$$

for a candidate $\hat{h}$, where $\boldsymbol{Y}$ and $\boldsymbol{Y}^*$ are independent random samples from g and $\hat{h}$ is the MLE. The inner, conditional expectation in (10.2.11) represents the predictive behavior of $\hat{h}$; the outer expectation represents the overall performance. Akaike (1973) argues that

$$\mathbb{E}_{\boldsymbol{Y}}\mathbb{E}_{\boldsymbol{Y}^*}\left[\log(\hat{h}(\boldsymbol{Y}^*))|\hat{h}\right] \approx \log(L(\hat{\theta}|\boldsymbol{Y})) - k,$$

which essentially gives the AIC. The argument is that $(1/n)\log(L(\hat{\theta}|\boldsymbol{Y}))$ using $\hat{h}$ is a reasonable estimator of (10.2.10) but has a bias that is corrected by using

$$\text{bias} = \mathbb{E}_g\left(\frac{1}{n}\sum_{i=1}^{n}\log h(\boldsymbol{x}_i|\hat{\theta})\right) - \mathbb{E}_g \log h(\boldsymbol{x}_i|\hat{\theta}) = \frac{n(k+1)}{n-k-2} \approx k \tag{10.2.12}$$

(see Bozdogan (1987)), in which the exact expression $n(k+1)/(n-k-2)$ is derived by Sugiura (1978). In essence, although maximizing $\log(L(\hat{\theta}|\mathbf{y})$ is natural for estimating (10.2.11), the maximized log nlikelihood is biased upward, and the bias can be approximated by the model size k.

AIC is not consistent, as shown in Shibata (1983), because it often gives models with too many terms. Indeed, if $k > p_0$ the likelihood ratio test for testing the true model against a model with k variables gives

$$2(\log L_k - \log L_{p_0}) \longrightarrow_d \chi^2_{k-p_0} \quad \text{as } n \longrightarrow \infty \tag{10.2.13}$$

by Wilks' theorem, where, henceforth, $\log L_k = \log(L_k(\hat{\beta}_{MLE}))$. Therefore,

$$\text{AIC}_{p_0} - \text{AIC}_k = 2(\log L_k - \log L_{p_0} - (k-p_0)) \longrightarrow_d \chi^2_{k-p_0} - 2(k-p_0),$$

leading to

$$\lim_{n\longrightarrow\infty} \mathbb{P}(\text{AIC}_{p_0} > \text{AIC}_k) = \mathbb{P}(\chi^2_{k-p_0} - 2(k-p_0)) > 0.$$

This implies that the AIC-selected model asymptotically satisfies

$$\lim_{n\longrightarrow\infty} \mathbb{P}(\hat{k} \geq p_0) = 1 \quad \text{but} \quad \lim_{n\longrightarrow\infty} \mathbb{P}(\hat{k} > p_0) > 0,$$

which means that AIC tends to choose models that are too complex as $n \to \infty$.

Note that the penalty in AIC is relatively small – not dependent on n, for instance. This means that the AIC tends to permit larger models and therefore is better for prediction than other variable selection techniques. This is especially true when robust prediction is important but model identification is not. This arises because the *SSE* is roughly the sum of a squared bias and a variance. Since the penalty has the same order as the model dimension, the whole AIC value remains of the same order as the sum of the squared bias and the estimation error (model dimension over the sample size), which is most important when the number of models is bounded. Using this observation, the

AIC has been shown to be minimax-rate optimal over linear regression functions under a squared error type loss, as in Barron et al. (1999) and Yang and Barron (1999).

In particular, Yang (2005) gives a bound on the risk in terms of a bias and variance. To state this result, consider the average squared error over the design points for estimating a function $f \in \mathscr{F}$ using a selection criterion δ giving a model $\hat{f}_{\hat{k}}$. This is

$$ASE(\hat{f}_{\hat{k}}) = \frac{1}{n}\sum_{i=1}^{n}\left(f(\boldsymbol{x}_i) - \hat{f}_{\hat{k}}(\boldsymbol{x}_i;\hat{\beta}_{\hat{k}})\right)^2 = \frac{\|f - \hat{f}(\cdot\,;\hat{\beta}_{\hat{k}})\|_n}{n}, \tag{10.2.14}$$

where $\hat{\beta}_{\hat{k}}$ is the OLS estimate of the parameter in the model, giving the empirical risk

$$R(f;\delta;n) = \frac{1}{n}\sum_{i=1}^{n}\mathbb{E}\left(f(\boldsymbol{x}_i) - \hat{f}_{\hat{k}}(\boldsymbol{x}_i)\right)^2 = \frac{\|f - \hat{f}(\cdot)\|_n}{n}$$

for $\hat{f}_{\hat{k}}(\boldsymbol{x}_i) = \hat{f}_{\hat{k}}(\boldsymbol{x}_i;\hat{\beta}_{\hat{k}})$. A model selection criterion δ is minimax-rate optimal over a class of regression functions $\mathscr{F}$ if

$$\sup_{f\in\mathscr{F}} R(f;\delta;n) \simeq \inf_{\hat{f}}\sup_{f\in\mathscr{F}}\frac{1}{n}\sum_{i=1}^{n}\mathbb{E}\left(f(\boldsymbol{x}_i) - \hat{f}(\boldsymbol{x}_i)\right)^2,$$

meaning that the left and right sides converge at the same rate, where $\hat{f}$ ranges over all estimators based on $y_1,\cdots,y_n$. Now, suppose $\mathscr{M}$ is the collection of all the candidate models at most countable, and let $\mathscr{M}_k \subset \mathscr{M}$ be the collection of models of size k with cardinality N_k assumed to be subexponential in the sense that $N_k \le \exp^{ck}$ for some $c > 0$. Let $\delta_{AIC} = \hat{f}_{\hat{k}}(\boldsymbol{x};\hat{\boldsymbol{\beta}}_{\hat{k}})$, the estimator of f, where $\hat{k}$ is selected by the AIC. Letting P_k denote the projection matrix down to $\mathscr{M}_k$ with $\mathrm{rank}(P_k) = r_k$, Yang (2005) showed the following.

Proposition (Yang, 2005): There exists a constant $C > 0$ depending only on c so that for every $f \in \mathscr{F}$,

$$R(f;\delta_{AIC};n) \le C\inf_{k\in\mathscr{M}}\left(\frac{\|f - P_k f\|_n^2}{n} + \frac{r_k}{n}\right), \tag{10.2.15}$$

where $\|a\|_n^2$ is the Euclidean norm of an n dimensional vector a. Furthermore, if $k^* \in \mathscr{M}$ indicates the correct model of size k^*, then

$$\sup_{f\in\mathscr{M}_{k^*}} R(f;\delta_{AIC};n) \le \frac{CN_{k^*}}{n}. \quad \square$$

From this it is seen that when the true model is in $\mathscr{M}$, the worst-case risk of δ_{AIC} converges at the parametric rate $1/n$, which is minimax-optimal. More generally, if the true regression function is infinite-dimensional, $\|f - P_k f\|_n^2/n$ is nonzero for all k but often decreasing. For smooth classes such as a Sobolev ball, with an appropriate choice of the candidate models, it can be shown that the term on the right in (10.2.15) also converges at the minimax rate. Expressions (10.2.15) and (10.2.16) are Oracle inequalities similar in form to those given in Chapter 6.

10.2.2 Bayesian Information Criterion (BIC)

Recall that in the decision-theoretic Bayes formulation, the mode of the posterior is the point estimate for the parameter under zero-one loss and that, similarly, the Bayes factor is the Bayes action for a binary decision problem such as hypothesis testing under generalized zero-one loss. Putting these together leads to the BIC, which is to minimize

$$\mathrm{BIC}_k = -2[\log L_k - \log(n)k] \tag{10.2.16}$$

over k, where $\log L_k = \log(L_k(\hat{\beta}_{MLE}))$. This corresponds to setting $\phi(n) = \log(n)$ in (10.2.1). Though BIC looks similar to AIC, the difference in their penalty terms has a big effect on the number of variables selected because the BIC punishes high-dimensional models more than AIC does.

The BIC follows from the Bayesian formulation of the model selection problem. Given p covariates, there are $K = 2^p - 1$ candidate models denoted, say, $\mathscr{S}_k$ for $k = 1, \cdots, K$ with corresponding model parameters $\boldsymbol{\beta}_k$. Write the prior over the model space as $W(\mathscr{S}_k)$ and the prior distribution for the parameters within model $\mathscr{S}_k$ as $P(\boldsymbol{\beta}_k|\mathscr{S}_k)$. Then the posterior probability for a given model is

$$\mathbb{P}(\mathscr{S}_k|\mathbf{y}) \propto W(\mathscr{S}_k)\mathbb{P}(\mathbf{y}|\mathscr{S}_k) \tag{10.2.17}$$

$$\propto W(\mathscr{S}_k)\int \mathbb{P}(\mathbf{y}|\boldsymbol{\beta}_k, \mathscr{S}_k)\mathbb{P}(\boldsymbol{\beta}_k|\mathscr{S}_k)d\boldsymbol{\beta}_k. \tag{10.2.18}$$

The model achieving the mode of the posterior is the one best supported by the data; in a zero-one loss sense, this is the natural estimate to take for the true model.

An asymptotic form can be derived for the modal model. With only slight loss of generality, assume the prior on the model space is uniform so $w(\mathscr{S})$ is a constant. A standard Laplace approximation on the integral in (10.2.18) (see Walker (1969) or de Bruijn (1959)) gives that

$$\log \mathbb{P}(\mathbf{y}|\mathscr{S}_k) = \log \mathbb{P}(\mathbf{y}|\widehat{\boldsymbol{\beta}}_k, \mathscr{S}_k) - \frac{k}{2}\log(n) + O(1),$$

where $\widehat{\boldsymbol{\beta}}_k$ is the maximum likelihood estimate. So (10.2.16) follows by taking the leading terms multiplied by two, and choosing the model with a minimum BIC is equivalent to choosing the model with the largest posterior probability.

The BIC is information theoretic in that it has a close relationship to $D(P_\theta||M_n)$, where M_n is the mixture distribution from the n-fold parametric family P_θ and the prior $W(\theta)$. Indeed, the penalty term in BIC comes out of the approximation $D(P_\theta||M_n) \approx (p/2)\log n + \mathsf{constant} + o_p(1)$, and close examination of constant reveals how it comes from the limit of $\log p(x^n|\hat{\theta})$; see Clarke and Barron (1990, 1994).

From (10.2.17), it is also seen that the BIC is also a sort of MLE since it maximizes $\mathbb{P}(\mathbf{y}|\mathscr{S}_k)$, which can be regarded as a sort of likelihood. Indeed, it can be seen that the BIC approximation gives

$$\frac{\exp\{-\mathrm{BIC}_k/2\}}{\sum_{k=1}^{K}\exp\{-\mathrm{BIC}_k/2\}}$$

as an approximation to the posterior distribution across the models. It can be readily verified that the effect of the prior drops out at rate $\mathscr{O}(1/n)$ and so can be safely ignored in many settings.

From a hypothesis testing standpoint, (10.2.16) has been justified by comparing two models $\mathscr{S}_k$ and $\mathscr{S}_{k'}$ via their Bayes factor,

$$\mathrm{BF} = \frac{\mathbb{P}(\mathbf{y}|\mathscr{S}_k)}{\mathbb{P}(\mathbf{y}|\mathscr{S}_{k'})} = \frac{\mathbb{P}(\mathscr{S}_k|\mathbf{y})/\mathbb{P}(\mathscr{S}_{k'}|\mathbf{y})}{w(\mathscr{S}_k)/w(\mathscr{S}_{k'})}.$$

Schwarz (1978) uses this approach to provide a strictly Bayesian derivation of the BIC. In addition, the BIC is a minimum description length, apart from a minus sign (see Barron and Cover (1991)) and has other interpretations in Shannon coding theory (see Clarke and Barron (1990)). However, minimum description length is a more general concept not treated here.

Like most Bayesian procedures, consistency almost always holds when the true model is in the support of the prior. Indeed, a heuristic consistency proof for the BIC can be seen as follows. Since $\phi(n) \to \infty$, (10.2.13) gives that

$$\limsup_{n\to\infty} -2(\log L_{p_0} - \log L_k)/\phi(n) = 0.$$

Therefore,

$$\limsup_{n\to\infty}(\mathrm{BIC}_{p_0} - \mathrm{BIC}_k)/\phi(n) = \limsup_{n\to\infty} -2(\log L_{p_0} - \log L_k)/\phi(n) + 2(p_0 - k) = p_0 - k,$$

which is less than or equal to -1. This gives

$$\lim_{n\to\infty} \mathbb{P}(\mathrm{BIC}_{p_0} \geq \mathrm{BIC}_{k+1}) = 0,$$

implying that $\lim_{n\to\infty}\mathbb{P}(\hat{k} > p_0) = 0$. Coupled with (10.2.6), consistency follows.

In fact, the validity of this argument is limited: Stone (1979) provides examples where the BIC in (10.2.16) is not consistent. Berger et al. (2003) demonstrate that, in certain contexts with many parameters, the hypotheses for the derivation of the BIC are not satisfied. The exact Bayes procedure is consistent; however, the approximations to it are not always accurate enough to reflect this. In contrast to AIC, the BIC is not minimax-rate optimal. Foster and George (1994) show that, in parametric settings, the BIC can converge suboptimally in worst-case risk performance; this may reflect the effect of the prior.

10.2.3 Choices of Information Criteria

Any choice of ϕ in (10.2.1) gives an information criterion; AIC and BIC just reflect two choices. In fact, (10.2.6) appears to hold as long as $\phi(n) = o(n)$, so the open question is how large ϕ must be to allow consistency and some sort of efficiency. This is partially answered by the unifying theorem at the end of this section. Now, for the sake of intuition, some comparative comments are as follows.

10.2.3.1 Between AIC and BIC

General guidelines for when to use AIC, BIC, or other selection criteria are hard to give beyond a few simple considerations. Roughly, if the true model is simple or finite-dimensional, then higher penalties such as BIC should be used. But if the true model is complex or infinitely-dimensional, then smaller penalties such as AIC should be used. In other words, keeping models small with high penalties on complexity helps for identification, but allowing larger models by way of smaller complexity penalties helps for prediction. Most AIC versus BIC comparisons (see Shao (1997), Yang (2005) for asymptotics and Burnham and Anderson (2002) for practical performance) come down to properties of their sampling distributions $\mathbb{P}_A$ and $\mathbb{P}_B$. Typically, $\mathbb{P}_A$ is more spread out than $\mathbb{P}_B$ is. This means the AIC is slower than the BIC for identifying the right model as n increases – when the AIC identifies the right model at all. However, the comparative inefficiency of the AIC allows it to be more robust and hence better for other purposes, such as prediction.

Reinforcing this, Shao (1997) pointed out that the support of $\mathbb{P}_A$ and $\mathbb{P}_B$ is a determining factor for the asymptotic performance of variable selection procedures. In particular, the question is whether or not the model space $\mathscr{M}$ contains a correct model with a fixed dimension (i.e., whether or not the problem is M-closed in the sense of Bernardo and Smith (1994)). If the problem is M-closed, then the consistency of the BIC makes it preferred over the AIC. However, if the true model is not in the model space but the number of models of the same dimension does not grow too fast with the dimension, the average squared error of the model selected by AIC is asymptotically equivalent to the smallest possible offered by candidate models, as shown by Shibata (1983), Li (1987), Polyak and Tsybakov (1990), Zhang (1992), and Shao (1997). The BIC does not have this property.

Interestingly, Yang (2005) showed that there is an unbridgeable gap between AIC and BIC: The strengths of AIC and BIC cannot be shared. That is, if any model selection criterion is consistent like BIC, it cannot be minimax-rate optimal; i.e., it must have a worse mean average squared error than AIC. Consistency in selection and minimax-rate optimality in estimating the regression function are therefore conflicting measures of performance for model selection. However, recently van Erven et al. (2008) provided a reconciliation between the properties of AIC and BIC by showing that switching from AIC (for searching) to BIC (for model identification) at some stage in sequential prediction is better than using either AIC or BIC alone. The switching permits a prediction scheme to take advantage of the fact that simpler models can be better at predicting than complex models when the sample size is preasymptotic.

10.2.3.2 Other Information Criteria

As discussed above, the AIC may perform poorly when the true model is small or when there are too many parameters relative to n (see Sakamoto et al. (1986)). Also, the BIC may perform poorly when the true model is large. So, many variants on the AIC and BIC have been proposed; usually they have penalty terms between those of the AIC and BIC.

For instance, Hurvich and Tsai (1989) studied the small-sample properties of the likelihood function and derived a corrected AIC,

$$\mathrm{AIC}_{c,k} = \mathrm{AIC}_k + \frac{2k(k+1)}{n-k-1} = -2\left[\log L_k - \frac{n}{n-k-1}k\right], \tag{10.2.19}$$

with $\phi(n) = n/(n-k-1)$ (see Sugiura (1978)). The adjustment is based on small-sample properties of the estimates assuming that the models have been chosen a priori. Since $\phi(n) > 1$, AIC_c gives a sharper cutoff and tends to select smaller subsets than AIC when p is large compared with n. As $n \to \infty$, $\phi(n) \to 1$, AIC_c thus asymptotically behaves like AIC. Generally, Burnham and Anderson (2002) advocate the use of AIC_c when the ratio n/p is small, say less than 40.

Hannan and Quinn (1979) considered selecting time series models in which the number of parameters k increases with n. In particular, they studied the selection of the order of an autoregressive model and proposed minimizing

$$\mathrm{HQ}_k = -2[\log L_k - \log\log(n)k]. \tag{10.2.20}$$

The HQ criterion imposes a heavier complexity penalty than the AIC but a lighter penalty than the BIC. HQ is seen to be asymptotically consistent for model selection by an argument similar to that for BIC. Indeed, Hannan and Quinn (1979) establish that, under some regularity conditions, if $\hat{k}$ minimizes (10.2.20) over $k \le K$, for $p_0 \le K$, then $\hat{k}$ converges a.s. to p_0.

The deviance information criterion (DIC) arises from calling $D(\theta) = -2\log f(\mathbf{y}|\theta)$ a deviance and setting $\bar{D} = E_{\theta|\mathbf{Y}}D(\theta)$ and $\bar{\theta} = E_{\theta|\mathbf{Y}}\Theta$ so the criterion is to minimize

$$\mathrm{DIC} = k_D + \bar{D} = 2\bar{D} - D(\bar{\theta}),$$

in which $k_D = \bar{D} - D(\bar{\theta})$ is an effective number of parameters. The DIC also represents a sort of trade-off between fit and complexity.

There are many other information-like criteria, including the risk inflation criterion (RIC; see George and Foster (2000)) which uses $SSE_{\boldsymbol{\gamma}}$ plus a penalty dependent on the number of nonzero parameters in the selection of terms summarized by $\boldsymbol{\gamma}$. The RIC is derived from a ratio of predictive risks of the form $\mathbb{E}(\mathbb{X}\boldsymbol{\beta} - \mathbb{X}\hat{\boldsymbol{\beta}})^2$. The informational complexity (IC; see Bozdogan (1987)) is based on penalizing the log likelihood by functions of the Fisher information to obtain invariance properties. The focused information criterion (FIC; see Clasekens and Hjort (2003)) is an effort to minimize an unbiased limiting form of the risk using the asymptotic normality of MLEs. The FIC also uses a penalty term, on a risk, derived from the Fisher information; the FIC is often similar to AIC but relies on many more limiting quantities. Atkinson (1980) also penalizes a risk, but by a constant times $k\hat{\sigma}^2$. The covariance inflation criterion (CIC; see Tibshirani and Knight (1999)) penalizes the *SSE* by an average of covariances between predictions and outcomes of the response by bootstrapping. Finally, there are many information criteria having the general form $\sum_i \rho(e_i) + Ca(n)b(\beta)$, where ρ is a convex function, e_i is a residual, and a and b are functions of the indicated arguments; see Nishii (1984), Konishi and Kitagawa (1996), and Rao and Wu (2001).

10.2.4 Cross Validation

Cross validation (CV) was introduced in Chapter 1 as a blackbox tool for choosing models based on their predictive ability. Here the focus is on the mechanics of how CV works and what it means. Recall that CV partitions the sample into subsets and that an analysis (e.g., estimating the coefficients in a model) is performed on each subset in turn so that the data not used in the analysis can be used for evaluating the analysis. CV is routinely used to select the important subset of variables in linear models, select and build the architecture of neural networks and trees, choose the regularization parameters for smoothing splines or other penalized methods, and to select the bandwidth for kernel and other estimators.

CV is technically not an information criterion. Unlike AIC and BIC, CV does not arise from a complexity penalty on a log likelihood (or risk). However, at the end of this section, it will be seen that CV, AIC, and C_p are asymptotically equivalent in some important cases. Admittedly, CV and AIC/C_p diverge in some cases as well. Nevertheless, the four techniques AIC, C_p, BIC, and CV can be regarded as similar enough that it makes sense to group them together. Indeed, the omnibus theorem at the end of this section covers those four techniques, and several others, mostly on the basis of norm properties.

The motivation for the CV procedure can be seen easily in the context of a linear model $Y = \boldsymbol{X}^{\top}\boldsymbol{\beta} + \varepsilon$. Given a data set $\mathscr{D} = \{(\boldsymbol{x}_i, y_i), i = 1, \cdots, n\}$, imagine fitting a regression model $\hat{f}$ using $\mathscr{D}$. One common measure for assessing the predictive performance of $\hat{f}$, on average, is the mean squared prediction error (MSPE),

$$\text{MSPE} = \mathbb{E}_{\boldsymbol{X},Y}\left[Y - \hat{f}(\boldsymbol{X})\right]^2, \tag{10.2.21}$$

where $(\boldsymbol{X}, Y)$ follows the same distribution as the original data points $(\boldsymbol{x}_i, y_i)$s. Since the true distribution is unknown in real data analysis, it is necessary to have test data samples independent of the original data and compute

$$\widehat{\text{MSPE}} = \frac{1}{n^*}\sum_{i=1}^{n^*}[y_i^* - \hat{f}(\boldsymbol{x}_i^*)]^2, \tag{10.2.22}$$

where $\mathscr{D}^* = \{(\boldsymbol{x}_i^*, y_i^*), i = 1, \cdots, n^*\}$ are generated from the same distribution as the original data. The quantity (10.2.22) can be interpreted as the average predictor error over the sample $\mathscr{D}^*$. The usual terminology is that the original data set $\mathscr{D}$ used for model fitting is the training set and the set $\mathscr{D}^*$ used for model assessment is the validation set. If the MSPE is used to tune a procedure (e.g., select the best tuning parameter) $\mathscr{D}^*$ is called the tuning set, and if it is used to evaluate the performance of a fitted model or compare the performance of several different procedures, $\mathscr{D}^*$ is called the test set.

In practice, however, independent test data are frequently difficult or expensive to obtain, and it is undesirable to hold back data from the original data to use for a separate test because it weakens the training data. CV is a technique for performing independent tests without requiring separate test sets and without reducing the amount of data used to fit the model. In particular, one can split the data set $\mathscr{D}$ into two complementary parts as $\mathscr{D} = \mathscr{D}_1 \cup \mathscr{D}_2$: The first part $\mathscr{D}_1$ contains n_1 data points used for constructing a model (training subset), whereas the second part $\mathscr{D}_2$ contains the remaining $n_2 = n - n_1$ data points used for assessing the predictive ability of the model (validating subset). This process is then repeated many times, and there are $\binom{n}{n_1}$ ways of partitioning the data. The CV score is the average prediction errors based on the different ways the data were partitioned.

Depending on the nature of the partitioning scheme, there are various types of CV scores. For example, if $\mathscr{D}_1$ is chosen such that $n_2 = 1$ (i.e., $\mathscr{D}_2$ is a singleton set) the result is leave-one-out cross-validation (LOOCV). For the LOOCV, there are n ways to split the data set, and the $\mathscr{D}_2$s corresponding to different splits are all disjoint. K-fold CV is based on another type of scheme. The data are first divided into K roughly equal parts. Then $\mathscr{D}_1$ is formed by claiming $K-1$ parts and $\mathscr{D}_2$ consists of the remaining part. So $n_1 = n(K-1)/K$ and $n_2 = n/K$; there are K ways to form $\mathscr{D}_1$ and $\mathscr{D}_2$. In addition to the LOOCV and K-fold CV, there are other variations on CV including the generalized CV (GCV; see Craven and Wahba (1979)), Monte Carlo CV (MCCV), Bayesian CV, and median CV.

10.2.4.1 Leave-One-Out CV

One of the earliest statements of the idea of LOOCV is in Mosteller and Tukey (1968). They asserted, informally, that LOOCV extracts all the information in a data set. LOOCV is also mentioned as the prediction sum of squares (PRESS) by Allen (1971). As the name suggests, LOOCV involves using a single observation from the original sample as the validation data set, and the remaining observations as the training set. In particular, let $\boldsymbol{y}^{-i}$ be the vector with the ith observation, y_i, removed from the original response vector $\boldsymbol{y}$ and $\hat{f}^{-i}$ be the estimate based on $\boldsymbol{y}^{-i}$. This is repeated for each $i = 1, \cdots, n$. Then the average of n prediction errors is called the LOOCV score

$$LOOCV = \frac{1}{n}\sum_{i=1}^{n}[y_i - \hat{f}^{-i}(\boldsymbol{x}_i)]^2. \tag{10.2.23}$$

To compute the LOOCV score, one needs to fit the model n times, once for each delete-one data set $\boldsymbol{y}^{-i}$. Hence, if the sample size n is large, the computation cost of LOOCV can be expensive. Fortunately, for many situations one can calculate the LOOCV using only one fit for all the data if the model satisfies the following "leave-one-out" property. Let

$$\tilde{\boldsymbol{y}}^i = (y_1, \cdots, y_{i-1}, \hat{f}^{-i}(\boldsymbol{x}_i), y_{i+1}, \cdots, y_n)^\top,$$

the result of replacing the ith element in $\boldsymbol{y}$, y_i, with the evaluation of the delete-i fit $\hat{f}^{-i}$ at $\boldsymbol{x}_i$. Let $\tilde{f}^{-i}$ be the estimate of f with data $\tilde{\boldsymbol{y}}^i$. Then some fits satisfy the following.

LEAVE-ONE-OUT PROPERTY:

$$\tilde{f}^{-i}(\boldsymbol{x}_i) = \hat{f}^{-i}(\boldsymbol{x}_i), \quad i = 1, \cdots, n.$$

The leave-one-out property is that adding a new point that lies exactly on the surface $\hat{f}^{-i}$ does not change the fitted regression.

To see what this means, it helps to recall that a smoothing method $\widehat{\boldsymbol{f}}$ is linear if

$$\widehat{\boldsymbol{f}} = S\boldsymbol{y},$$

where $\widehat{\boldsymbol{f}} = (\hat{f}(\boldsymbol{x}_1), \cdots, \hat{f}(\boldsymbol{x}_n))^\top$, and the $n \times n$ matrix S depends on the input vectors $\boldsymbol{x}_i, i = 1, \cdots, n$ but not on the y_is. For any linear smoother $\hat{f}$,

$$\hat{f}(\boldsymbol{x}_i) - \tilde{f}^{-i}(\boldsymbol{x}_i) = \sum_{j=1}^{n} s_{ij}y_j - \left[\sum_{j\neq i}^{n} s_{ij}y_j + s_{ii}\hat{f}^{-i}(\boldsymbol{x}_i)\right] = s_{ii}(y_i - \hat{f}^{-i}(\boldsymbol{x}_i)).$$

If the leave-one-out property holds for $\hat{f}$, then

$$\hat{f}(\boldsymbol{x}_i) - \hat{f}^{-i}(\boldsymbol{x}_i) = \hat{f}(\boldsymbol{x}_i) - \tilde{f}^{-i}(\boldsymbol{x}_i) = s_{ii}(y_i - \hat{f}^{-i}(\boldsymbol{x}_i)),$$

which implies that

$$\frac{y_i - \hat{f}(\boldsymbol{x}_i)}{y_i - \hat{f}^{-i}(\boldsymbol{x}_i)} = 1 - s_{ii}. \tag{10.2.24}$$

Equation (10.2.24) describes the effect of dropping one individual datum on the final fit, which can be used to adjust the optimized result for the omitted individual.

Using (10.2.24) in (10.2.23) gives that the LOOCV is

$$LOOCV = \frac{1}{n}\sum_{i=1}^{n}\left[\frac{y_i - \hat{f}(\boldsymbol{x}_i)}{1 - s_{ii}}\right]^2. \tag{10.2.25}$$

The computation of (10.2.25) only requires one model fit based on the entire data set and therefore saves computation time. The only extra cost is to save all the entries lying on the diagonal of the smoother matrix S. Linear regression models, Nadaraya-Watson kernel estimators, and cubic smoothing splines all satisfy (10.2.24).

Indeed, it is easy to see (10.2.24) explicitly for a one-dimensional linear model $y_i = \beta_0 + \beta_1 x_i + \varepsilon_i$. The least squares fit is given by $\hat{f}(x_i) = \bar{y} + \hat{\beta}_1(x_i - \bar{x})$, where $\hat{\beta}_1 = \sum_{i=1}^n (x_i - \bar{x}) y_i / \sum_{i=1}^n (x_i - \bar{x})^2$. Then (10.2.24) holds with

$$s_{ii} = \frac{1}{n} + \frac{(x_i - \bar{x})^2}{\sum_{i=1}^n (x_i - \bar{x})^2}.$$

LOOCV is approximately unbiased for the true prediction error but can have high variance because the n training sets are so similar to one another. Discussion and theory on LOOCV under various situations can be found in Allen (1974), Stone (1974, 1977), Geisser (1975), Wahba and Wold (1975), Efron (1983, 1986), Picard and Cook (1984), Herzberg and Tsukanov (1986), and Li (1987). In particular, Stone (1977) studies the asymptotic consistency of CV and its asymptotic efficiency.

10.2.4.2 *K-fold CV*

The idea of multifold CV (MCV) first appeared in Geisser (1975), where instead of deleting one observation as in LOOCV, $d > 1$ observations are deleted. The delete-d MCV criterion is

$$MCV_d = \frac{1}{\binom{n}{d}} \sum_{\mathscr{T}} \sum_{i \in \mathscr{T}} [y_i - \hat{f}^{-\mathscr{T}}(\boldsymbol{x}_i)]^2 / d, \tag{10.2.26}$$

where $\mathscr{T}$ denotes a subset of $\{1, \cdots, n\}$ size d, $\hat{f}^{-\mathscr{T}}$ is the model fit using the remaining data after leaving $\mathscr{T}$ out, and the first summation runs over all possible subsets of size d. Zhang (1993) shows that the delete-d MCV criterion can be asymptotically equivalent to information criteria in some cases where $d \to \infty$. The main disadvantage of MCV is that it demands intensive computation; this motivates many useful alternative CV methods such as K-fold CV.

In K-fold CV (see Breiman et al. (1984)), the observations are removed in groups. Suppose the original sample is partitioned into K roughly equal-sized groups. For each $k = 1, \cdots, K$, one fits the model using all the data except the kth group and then calculates the prediction error of the fitted model using the data in the kth group. This process is repeated K times, with each group used exactly once as the validation set. The K prediction errors are then averaged to produce a single estimation. In particular, define the index function $\kappa : \{1, \cdots, n\} \to \{1, \cdots, K\}$, which assigns the observation i to some group by a random scheme. Let $\hat{f}^{-k}(\boldsymbol{x})$ be the fitted function based on the data excluding the kth group. Then the K-fold CV estimate of the prediction error is

$$CV_K = \frac{1}{n} \sum_{i=1}^n [y_i - \hat{f}^{-\kappa(i)}(\boldsymbol{x}_i)]^2. \tag{10.2.27}$$

If the models are indexed by a parameter $\alpha \in \Lambda$, let $\hat{f}_{-k}(\boldsymbol{x}; \alpha \in \Lambda)$ be the fitted model associated with α and trained using the data excluding the kth group. Then the CV estimate of the prediction error is

$$CV_K(\alpha) = \frac{1}{n}\sum_{i=1}^{n}[y_i - \hat{f}^{-\kappa(i)}(\boldsymbol{x}_i; \alpha)]^2. \tag{10.2.28}$$

The function $CV_K(\alpha)$ provides an estimate of the test error curve, and the optimal tuning parameter $\hat{\alpha}$ is found as the minimizer of the $CV_K(\alpha)$ curve. The final model chosen is $\hat{f}(\boldsymbol{x}; \hat{\alpha})$, fitted with the entire data. In practice, $K = 5$ or $K = 10$ are common choices and give good performance in practice (see Zhang (1993)).

10.2.4.3 Generalized CV

The calculation of LOOCV is often expensive since the whole process requires fitting the model n times. GCV provides a convenient way to approximate the LOOCV for linear fitting methods under the squared-error loss. Indeed, GCV should really be called ACV – approximate CV – because it is obtained from the CV as

$$GCV = \frac{1}{n}\sum_{i=1}^{n}\frac{[y_i - \hat{f}(\boldsymbol{x}_i)]^2}{[1 - \mathrm{tr}(S)/n]^2} \tag{10.2.29}$$

using $s_{ii} \approx \sum_{i=1}^{n} s_{ii}/n = \mathrm{trace}(S)/n$. The GCV score GCV works well if the s_{ii}s are not very different from each other; the quantity trace(S) is the effective number of parameters. GCV is usually easier to compute than the LOOCV since only a single fit based on the entire data is needed.

Interestingly, GCV has many connections with other variable selection criteria. First, GCV can be viewed as a weighted version of LOOCV since

$$GCV = \frac{1}{n}\sum_{i=1}^{n} w_{ii}[y_i - \hat{f}(\boldsymbol{x}_i)]^2 \tag{10.2.30}$$

with

$$w_{ii} = \left[\frac{1 - s_{ii}}{1 - \mathrm{trace}(S)/n}\right]^2.$$

If w_{ii} is independent of i, then $LOOCV = GCV$. For models satisfying the leave-one-out property, GCV may save additional time since the computation of trace(S) is sometimes easier than finding the individual s_{ii}s.

Second, GCV is close to Mallows' C_p and the AIC. In the context of subset selection for linear models, the GCV for a model of size k is

$$GCV_k = \frac{SSE_k}{n}\frac{1}{[1 - k/n]^2}. \tag{10.2.31}$$

Taylor expanding gives $(1-t)^{-2} \approx 1+2t$ for small$t$, so when $n^{-1}\text{trace}(S)$ is small,

$$GCV_k \approx \frac{SSE_k}{n} + 2\frac{SSE_k}{n} \cdot \frac{k}{n} \approx \frac{SSE_k}{n} + 2s^2 \cdot \frac{k}{n}, \tag{10.2.32}$$

where SSE_k/n estimates the error variance. Recall that in C_p the variance estimate is based on a full model, $s^2 = SSE_p/(n-p)$. Because $SSE_k/n \to \sigma^2$ as $n \to \infty$, (10.2.32) suggests that minimizing GCV_k is asymptotically equivalent to minimizing AIC and Mallows' C_p,

In the context of smoothing splines, GCV can be mathematically justified because asymptotically it minimizes the mean squared error for estimation of f (see O'Sullivan (1983), Wahba (1985, 1990)). More generally, GCV is typically used for linear smoothers because GCV has nice theoretical properties and gives good performance for them. For more complicated modeling procedures, such as nonlinear smoothers, GCV is not necessarily a suitable choice and other criteria, such as LOOCV and K-fold CV, are often better.

10.2.4.4 CV and Other Criteria

In linear regression, it has long been known that the LOOCV is asymptotically equivalent to other model selection criteria such as AIC, Mallows' C_p, the bootstrap (Efron (1983, 1986)), and the jackknife. To see the link between LOOCV and either AIC or C_p, let $\gamma \in \mathscr{M}$ be the model index and write

$$AIC = -2[\log L(\hat{\alpha}_\gamma;\gamma) - k_\gamma],$$

where $L(\alpha_\gamma;\gamma)$ is the log likelihood function of the model indexed by γ, $\hat{\alpha}_\gamma$ is the maximum likelihood estimate of the parameter α_γ, and k_γ is the dimension of α_γ. For normal multiple linear regression models with known variance σ^2, Mallows' C_p is given by

$$C_p = SSE_\gamma/\sigma^2 - (n - 2k_\gamma)$$

and is equivalent to AIC.

To see that these are equivalent to LOOCV, follow the notation of Stone (1977) and set

$$g(y|\boldsymbol{x},\gamma,\mathscr{D}) \equiv g_\gamma(y|\boldsymbol{x},\hat{\alpha}_\gamma(\mathscr{D})),$$

where $\{g_\gamma(y|\boldsymbol{x},\alpha_\gamma), \alpha_\gamma \in \Lambda_\gamma\}$ are the densities for a conventional parametric model γ associated with parameter α_γ and $\hat{\alpha}_\gamma(\mathscr{D})$ is the maximum likelihood estimator maximizing $L(\alpha_\gamma;\gamma) = \sum_{i=1}^n g_\gamma(y_i|\boldsymbol{x}_i,\alpha_\gamma)$. The general LOOCV based on the likelihood is now

$$LOOCV(\gamma) = \sum_{i=1}^n \log g_\gamma(y_i|\boldsymbol{x}_i,\mathscr{D}_{-i}), \tag{10.2.33}$$

where $\mathcal{D}_{-i}$ is the data $\mathcal{D}$ omitting the ith observation. Under weak conditions, Stone (1977) shows that (10.2.30) is asymptotically equivalent to AIC, making it equivalent to Mallows' C_p as well.

When $K = n$, CV is approximately unbiased for the true prediction error but can have high variance because the n training sets are so similar to one another. The computational burden is also considerable, requiring the model to be fit n times. On the other hand, if K is chosen small, then CV has a lower variance than LOOCV but possibly a larger bias, depending on how the performance of the fitted model varies with the size of the training set. As noted above, $K = 5$ and $K = 10$ often achieve a satisfactory trade-off between bias and variance.

Careful use of the CV includes two further steps. First, practitioners look at the standard error of the CV score to calibrate the CV. This partially justifies the one-standard-error rule to avoid underfitting in subset selection in linear models. The idea is that since CV strongly favors smaller models, choosing the most parsimonious model whose error is no more than one standard error above the error of the best model is a reasonable fix. Second, it is important to look at a histogram of the errors $y_i - \hat{f}(\boldsymbol{x}_i)$ that give the summands in the CV. If this histogram is roughly normal, the CV is probably more reliable than if it is skewed or multimodel – features that reflect bad fit.

10.2.4.5 CV for Model Selection

The basic model selection problem is to choose a prediction density for Y given $\boldsymbol{x}$ from a class of candidate models indexed by $\gamma \in \mathcal{M}$. Recall the average squared error (ASE) of a fitted model $\hat{f}_\gamma(\boldsymbol{x})$ over the design points given in (10.2.14). A model selection procedure is asymptotically optimal if

$$\frac{ASE_n(\hat{\gamma})}{\inf_{\gamma \in \Gamma} ASE_n(\gamma)} \to 1, \quad \text{in probability}, \ n \to \infty, \tag{10.2.34}$$

where $\hat{\gamma}$ is the model selected by the procedure. Similar to AIC and C_p, when the number of predictors in each model is fixed as n increases, the LOOCV is asymptotically inconsistent in the sense that the probability of selecting the best model does not converge to 1 as $n \to \infty$. If the number of predictors in the models under consideration increases as n increases, Li (1987) shows that the LOOCV can be consistent and asymptotically optimal under some regularity conditions. More recently, see Shao (1997). In other words, the model selected by LOOCV asymptotically approaches the minimum value of ASE among all possible models.

To organize the summary below, let n_1 be the size of the training set so that $n_2 = n - n_1$ is the size of the validation set. One of the key issues for implementing CV is the choice n_2. This question is equivalent to specifying the K in K-fold CV. The following are typical choices.

$n_2 = 1$

This gives the LOOCV, which is approximately unbiased for the true prediction error. The LOOCV can have high variance because the n training sets are so similar to one another. The computational burden is also considerable, requiring the model to be fit n times. When the number of predictors in each model is fixed as n increases , the LOOCV is asymptotically inconsistent in the sense that the probability of selecting the best model does not converge to 1 as $n \to \infty$. In practice, the LOOCV tends to select unnecessarily large models. Shao (1993) shows that the probability of selecting the optimal models by using the LOOCV can be very low; e.g., $\leq 50\%$ in Shao's simulations. The more zero components the $\boldsymbol{\beta}$ has, the worse performance LOOCV has. The behavior of LOOCV is clarified in the theorem of the next section.

$n_2 = 2$

The issue of using more than one observation at time for validation in LOOCV was raised by several researchers, including Geisser (1975), Herzberg and Tsukanov (1986), Burman (1989), Zhang (1992), and Shao (1993). For example, Herzberg and Tsukanov (1986) discovered that the leave-two-out CV is better than the LOOCV in some situations, although these two procedures are asymptotically equivalent in theory.

$n_2 = n/K$

Between $n_2 = 2$ and limiting behavior as n increases, one can choose K for K-fold CV. This is usually done by simulations to find the best variance–bias trade-off. Larger K gives a small bias but larger variance, and the computing time may be high. Smaller K reduces the computing time and variance but may have a large bias. For sparse data, $K = 1$ may work best. For large data sets, $K = 5$ or $K = 10$ is common, even though K as low as three may be adequate.

$n_2/n \to 1$

Shao (1993) showed that the inconsistency of the LOOCV can be fixed by using a large n_2, necessarily depending on n. In particular, an asymptotically correct CV procedure requires an n_2 that has the same rate of divergence to infinity as n; in other words, $n_2/n \to 1$ as $n \to \infty$. As explained in Shao (1993), one does not necessarily need a very accurate model in the fitting step, but one does need an accurate assessment of the prediction error in the validation step because the overall purpose of CV is to select a model, and the selected model will then be refitted using the full data set for prediction. From this standpoint, the value n_1 is not necessarily close to n in the fitting step. Shao (1993) shows that the MCCV is consistent for model selection if $n_2/n \to 1$.

Nonparametric Regression

In the context of nonparametric regression, the LOOCV for smoothing parameter selection leads to consistent regression estimators, as shown in Wong (1983) for kernel regression and Li (1984) for nearest-neighbor estimation. It also leads to an asymptotically optimal, or rate optimal, choice of smoothing parameters and optimal regression estimation, as shown by Speckman (1985) and Burman (1990) for spline regression and Hardle et al. (1988) for kernel estimation. Recently, Yang (2007) established the consistency of CV when used to compare parametric and nonparametric methods or within nonparametric methods. Under some regularity conditions, with an appropriate choice of data-splitting ratio, Yang (2007) shows that CV is consistent in the sense of selecting the better procedure with probability approaching one. Furthermore, when comparing two models converging at the same nonparametric rate, the validation size n_2 does not need to dominate; this contrasts with the parametric case described above.

10.2.4.6 A Unifying Theorem for Consistency

Separate from criteria like (10.2.34) that focus on optimality and hope to give consistency are criteria that focus on consistency, which may give some sort of optimality. Here, a general theorem on forms of consistency will be given to reconcile the inconsistency of AIC or CV with their use, which is extensive and effective. The key issue is that the way some criteria are inconsistent is relatively harmless: They overfit. That is, the models they choose are incorrect in the sense that they have extra terms but those terms help give a better approximation.

Consider a true model of the form $Y = f(\boldsymbol{x}) + \varepsilon$ with $\mathbb{E}(\varepsilon) = 0$ and $\text{Var}(\varepsilon) = \sigma^2$. Suppose that, for n outcomes, a p term linear model $\boldsymbol{Y} = \mathbf{X}\boldsymbol{\beta} + \boldsymbol{\varepsilon}$ is available. Let k_0 denote the correct selection of nonzero β_js and let K_0 denote all the submodels of the p term model that contain k_0. The correct k_0 contains the coefficients in the p term model that satisfy

$$\boldsymbol{\beta}_f(n) = \arg\min_{\boldsymbol{\beta}\in\mathbb{R}^p} \|\mathbf{X}\boldsymbol{\beta} - \boldsymbol{f}\| \quad \text{for finite } n \text{ and } \quad \liminf_{n\to\infty} |(\beta_f(n))_j| > 0$$

for the given f. Now, let $CRIT$ be a criterion such as AIC or CV by which a model $\hat{m}$ is selected from a list of models using the data. Now, define $CRIT$ to be k_0 consistent if $\mathbb{P}(\hat{m} = k_0) \to 1$ and define $CRIT$ to be K_0 consistent if $\mathbb{P}(\hat{m} \supset k_0) \to \infty$. That is, k_0 consistency is the usual form of consistency but K_0 is a weaker form that permits extra terms, beyond those in k_0, because they provide a better, if incorrect, fit to f.

To fix notation, let K be a collection of submodels containing k_0. Write the projection matrix for the full model as $P = \mathbf{X}(\mathbf{X}^\mathsf{T}\mathbf{X})^{-1}\mathbf{X}^\mathsf{T}$, and for a model $k \in K$ write $P_k = \mathbf{X}_k(\mathbf{X}_k^\mathsf{T}\mathbf{X}_k)^{-1}\mathbf{X}_k^\mathsf{T}$. Without loss of generality, suppose $\hat{m} = \arg\min_{k\in K} CRIT(k)$ and write $\|\cdot\|_A$ as the norm with respect to an inner product defined by the matrix A.

Following Muller (1993) (see also Muller (1992)), criteria can be divided into two classes for the purpose of consistency. These are

$$CRIT(k) = \begin{cases} \|\mathbf{y} - P_k\mathbf{y}\|^2 + a_n \mathsf{rank}(P_k)\hat{\sigma}^2_{Q(k)} & \text{Type A} \\ \|\mathbf{y} - P_k\mathbf{y}\|^2_{S(k)} & \text{Type B}, \end{cases} \tag{10.2.35}$$

where $\hat{\sigma}^2_{Q(k)} = (1/n)\|\mathbf{y}\|_{Q(k)}$ and $a_n \geq 0$. The two positive semidefinite matrices $Q(k)$ and $S(k)$ will be discussed below. The table lists nine selection criteria, six of which have been discussed here, and notes their types. Note that type is not strictly exclusive.

To establish a unifying theorem over the nine criteria, Muller (1993) uses five regularity conditions. They are relatively weak: (i) The true model is nontrivial. (ii) The projection matrix is asymptotically full rank; in particular, $(1/n)\mathbf{X}^{\mathsf{T}}\mathbf{X} \to M$, where M is $k \times k$ positive definite. (iii) The average of the squared norm of the vector of values of the true function at the design points is bounded, $(1/n)\|\boldsymbol{f}\|^2 = \mathcal{O}(1)$. (iv) The squared norm of the orthogonal projection is bounded, $\|\boldsymbol{f}\|^2_{P-P_{k_0}} = \mathcal{O}(1)$. And (v) some even moment of ε greater than four is finite, $\mathbb{E}|\varepsilon|^{2s} < \infty$ for some $s \geq 2$.

The matrices $Q(k)$ and $S(k)$ control the sense in which minimizing $CRIT$ leads to a model. In particular, the differences between models must be detectable and estimates of variances must be bounded. There are two conditions that ensure this for $Q(k)$:

$$\forall k: \quad q_{n,k} = \lambda_{max}(Q(k)) = \mathcal{O}(1), \tag{10.2.36}$$

$$\forall k: \quad \exists Q \qquad Q(k) r_{n,k} Q + s_{n,k}(Id - P_k), \tag{10.2.37}$$

where, in the context of (10.2.37), (i) λ_{max} is the maximum eigenvalue and $r_{n,k}, s_{n,k} \geq 0$, (ii) $\forall g \supset k$ we have that

$$r_{n,k} \geq r_{n,g} \quad \text{and} \quad s_{n,k} \geq s_{n,g},$$

and (iii) at least one of

$$\exists r_k > 0: \quad r_{n,k} Q \geq r_k Q_p, \tag{10.2.38}$$

where Q_p is a projection matrix with $\text{rank}(Id - Q_p) = \mathcal{O}(1)$, and

$$\exists s_k: \quad s_{n,k} \geq s_k > 0 \tag{10.2.39}$$

holds. If $a_n \to \infty$, then (10.2.36) is not a restriction. If $Q = Id - P$, then (10.2.37) is satisfied for all the Type A criteria in the table.

CRIT	Expression	Type
Mallows' C_p	$C_p(k) = \|\mathbf{y} - P_k\mathbf{y}\|^2 + 2\text{rank}(P_k)\hat{\sigma}^2$	A
AIC	$AIC(k) = n\log((1/n)\|\mathbf{y} - P_k\mathbf{y}\|^2) + 2\text{rank}(P_k)$	A
Final Pred. Err.	$FPE(k) = (1 + 2\text{rank}(P_k)/n)\|\mathbf{y} - P_k\mathbf{y}\|^2$	A
BIC	$BIC(k) = n\log((1/n)\|\mathbf{y} - P_k\mathbf{y}\|^2) + \text{rank}(P_k)\log n$	A
LOOCV	$LOOCV(k) = (\mathbf{y} - P_k\mathbf{y})^{\mathsf{T}}(Id - T_k)^{-2}(\mathbf{y} - P_k\mathbf{y})$	B
Stand. RSS	$SRSS(k) = (\mathbf{y} - P_k\mathbf{y})^{\mathsf{T}}(Id - T_k)^{-1}(\mathbf{y} - P_k\mathbf{y})$	B
GCV	$GCV(k) = (\|\mathbf{y} - P_k\mathbf{y}\|^2)/(1 - \text{rank}(P_k)/n)^2$	A or B
Res. Mean. Sq.	$RMS(k) = (\|\mathbf{y} - P_k\mathbf{y}\|^2)/(1 - \text{rank}(P_k)/n)$	A or B
SSE	$SSE(k) = \|\mathbf{y} - P_k\mathbf{y}\|^2$	A or B

Table 10.1 For Mallows' C_p, $\hat{\sigma}^2 = (1/(n-k))\|\mathbf{y} - P\mathbf{y}\|$ and for LOOCV, $T_k = \text{diag}((P_k)_{1,1}, \ldots, (P_k)_{n,n})$.

The two assumptions on $S(k)$ are like those in Nishii (1984):

$$\forall k: \quad S(k) \quad \text{is diagonal with} \quad S(k) \geq Id, \tag{10.2.40}$$

$$\forall k: \quad \lim_{n \longrightarrow \infty} \max\{S(k)_{i,i} \mid i = 1, \ldots, n\} = 1. \tag{10.2.41}$$

Expression (10.2.41) is satisfied if the design points are in a circle with radius bounded by $o(n)$. Note that (10.2.40) is satisfied by CV and SRSS.

Now, Muller's theorem can be stated. It has four parts: Two are for k_0 and K_0 consistency. The other two parts give some bounds on the rates of consistency.

Theorem (Muller, 1993): Suppose the first four regularity conditions are satisfied $a_n = o(n)$, and that *CRIT* is a criterion function.

(i) If *CRIT* is of Type A or B as in (10.2.35) and satisfies (10.2.36), (10.2.40), and (10.2.41), then *CRIT* is K_0 consistent.

(ii) If *CRIT* is of Type A as in (10.2.35) and satisfies (10.2.36) and (10.2.37), then *CRIT* is k_0 consistent.

(iii) Under the conditions of (i), if $k \notin K_0$ and ε satisfies regularity condition (v) above, then

$$\mathbb{P}(\hat{m} = k) = \mathcal{O}\left(\frac{1}{(c_k n)^s}\right) + \mathcal{O}\left(\frac{q_{n,k_0} K a_n}{u_k n^{3/2}}\right) + \mathcal{O}\left(\frac{t_{n,k_0} - 1}{v_k \sqrt{n}}\right),$$

where K is the number of models under consideration, t_{n,k_0} is the largest diagonal element of $S(k_0)$, and c_k, u_k, and v_k are positive constants. If, in addition, $\varepsilon \sim N(0, \sigma^2)$, then

$$\mathbb{P}(\hat{m} = k) = \mathcal{O}(e^{-c_k n}) + \mathcal{O}(e^{-u_k n^2 / q_{n,k_0} K a_n}) + \mathcal{O}(e^{-v_k n / (t_{n,k_0} - 1)}).$$

(iv) Under the conditions in (ii), if $k \notin K_0$, the conclusions of (C) continue to hold. Under regularity condition (v) above, for $k \in K_0 \setminus \{k_0\}$,

$$\mathbb{P}(\hat{m} = k) = \mathcal{O}\left(\frac{1}{n^{s/2}}\right) + \mathcal{O}\left(\frac{1}{(z_k a_n)^s}\right),$$

where the z_ks denote positive constants. If $\varepsilon \sim N(0, \sigma^2)$, then for a $w > 0$,

$$\mathbb{P}(\hat{m} = k) = \mathcal{O}(e^{-wn}) + \mathcal{O}(e^{-z_k a_n}).$$

Proof: Omitted. □

It may be that the consistency of K-fold CV follows from this theorem because, when the K is chosen well, K-fold CV gives a lower MSE than LOOCV, which is of Type B. Indeed, if the design points are taken K at a time and the goal is to model KY, then K-fold CV becomes LOOCV. However, the consistency properties of K-fold CV do not seem as extensively studied as for LOOCV.

Overall, as noted in Muller (1993), consistency rates of the criteria can be categorized by the rate of a_n. When $a_n = o(n)$, K_0 consistency holds and $\mathbb{P}(\hat{m} \notin K_0)$ decreases exponentially fast under normal errors but with a rate $o(1/n^{s/2})$ more generally. For Type A criteria, when $a_n = \mathcal{O}(1)$, the rate $\mathbb{P}(\hat{m} \notin K_0) = \mathcal{O}(1/n^s)$ holds. If $a_n \rightarrow \infty$, then Type A criteria are k_0 consistent with rates $\mathbb{P}(\hat{m} \neq k_0) = \mathcal{O}(\exp(-ca_n))$ for some $c > 0$ and $\mathbb{P}(\hat{m} \neq k_0) = \mathcal{O}(1/n^{s/2}) + \mathcal{O}(1/a_n^2)$. Accordingly, from an asymptotic rate of convergence standpoint, choices of $a_n \sim n^\alpha$ for $\alpha \in (1/2, 1)$ are optimal – not AIC, BIC, or most of the other criteria commonly used.

10.2.4.7 Other Variants on Basic CV

There are many variants on the basic CV idea. Here, four are worth describing. First, the number of subsets into which the n data points are partitioned reflects the estimation procedure. Here, a test set and training set were used. However, if the modeling strategy has two steps (e.g., model selection via cost complexity in trees followed by parameter estimation), the data can be partitioned into three subsets: two for inferences and one for testing. This is a straightforward extension.

Second, Picard and Cook (1984) and Shao (1993) suggest an alternative way to implement CV using Monte Carlo. Randomly draw, with or without replacement, a collection of B subsets of $\mathcal{D}$ that have size n_1, denoted $\mathcal{S}_1, \cdots, \mathcal{S}_B$. The Monte Carlo CV (MCCV) is

$$MCCV(\alpha) = \frac{1}{Bn_2} \sum_{b=1}^{B} \sum_{(\mathbf{x}_i, y_i) \in \mathcal{S}_b} [y_i - \hat{f}^{-\mathcal{S}_b}(\mathbf{x}_i; \alpha)]^2. \tag{10.2.42}$$

This method uses B random splits of the original data and averages the squared prediction error over the splits. This provides an estimate of (10.2.21) by sampling from the possible subsets of size n_1.

Third, Bayesian CV is close in spirit to (10.2.30), being based on the predictive density. Start with a full model for $m < n$ data points $w(\beta(\alpha), \sigma) g(\mathbf{x}_1, \ldots, \mathbf{x}_m | \beta(\alpha), \sigma)$, where $\beta(\alpha)$ indicates the dependence of a parameter β on a hyperparameter α. and a posterior density obtained from the remaining $n - m$ points. Then, form the predictive density

$$\begin{aligned} g_\alpha(\mathbf{x}_1, \ldots, \mathbf{x}_m \mid \mathbf{x}_{m+1}, \ldots, \mathbf{x}_n) \\ = \int g(\mathbf{x}_1, \ldots, \mathbf{x}_m | \beta(\alpha), \sigma) w(\beta(\alpha), \sigma | \mathbf{x}_{m+1}, \ldots, \mathbf{x}_n) d\beta(\alpha) d\sigma. \end{aligned}$$

Now, instead of the squared error and the arithmetic mean, Bayesians often use the geometric mean of the logarithms of densities. For B splits this is

$$CV(\alpha) = \frac{1}{B} \sum_{b=1}^{B} \log g_\alpha(\mathcal{S}_b | \mathcal{S}_b^c).$$

It is proved in Samanta and Chakrabarti (2008) that when each $\mathbf{x}_i$ appears the same number of times in the $\mathcal{S}_b$s, $\arg\max_\alpha CV(\alpha)$ is asymptotically consistent and optimal.

Finally, it must be remarked that (10.2.21) and (10.2.22) derive their importance from squared error and the mean, neither of which are necessary. Indeed, any loss function can be used and any function of the data that reflects the location of the loss can be used. An obvious variant is taking the median instead of the mean in (10.2.21) and using it to define a median CV in (10.2.22), again by taking the median instead of the mean. Yu (2009) provide evidence that median CV has a higher probability of choosing the correct model than mean-based CV when the tails of the error distribution are heavy or there are many outliers in the data. Moreover, median CV does not underfit as severely as CV. One can argue that when the squared residuals are far from being normally distributed that the median is a better summary of the variation than the mean is. It is easy to imagine other functions of the data to use besides the mean and the median to define other, possibly more useful, cross-validation functions.

10.3 Shrinkage Methods

In contrast to the model selection methods of the last section, penalized methods, also called shrinkage or regularization, are usually based on subtracting a penalty from the risk rather than from a log likelihood. Usually, the penalty is a function of the parameter times a constant, called a decay parameter. The point of shrinkage methods is to order the variables for inclusion by the size of the decay parameter, thereby reducing the general problem to a nested case. Then, once the decay parameter is estimated, the variable selection is complete. This can be done for a wide variety of model classes, not just linear models.

From an abstract perspective, shrinkage methods aim to make a problem well-posed when it is ill-posed because of nonuniqueness of the solution and are efforts to obtain sparse solutions by automatically shrinking small entries of $\hat{\boldsymbol{\beta}}^{ols}$ to zero to reduce the variance. For instance, $\hat{\boldsymbol{\beta}}^{ols}$ need not be unique when X is not of full rank. Even when $\hat{\boldsymbol{\beta}}^{ols}$ exists and is unbiased, it has a large variance due to nonsparsity when X is nearly collinear. In contrast to OLS, some shrinkage methods, like ridge regression below, make a problem well-posed but not sparse. Others, like LASSO below, make the problem well-posed and achieve sparsity at the same time. Overall, shrinkage estimators generally give better prediction and smaller variances than the OLS estimators when p is large.

From a more operational perspective, a good variable selection procedure aims (i) to filter out unimportant variables from the candidate variables and (ii) to estimate the regression coefficients for important variables consistently, with a high level of efficiency. Such variable selection procedures are described as "oracle", see Donoho and Johnstone (1994). Let model (10.1.1) have true regression coefficients $\boldsymbol{\beta}^*$. Write the index set for the important variables as $\mathbf{A}_0 = \{j = 1, \ldots, p : \beta_j^* \neq 0\}$ and the index set for the unimportant ones as $\mathbf{A}_0^c$. Without loss of generality, suppose $\mathbf{A}_0 = \{1, \cdots, p_0\}$. Then $\boldsymbol{\beta}^* = (\boldsymbol{\beta}^*_{\mathbf{A}_0}, \boldsymbol{\beta}^*_{\mathbf{A}_0^c}) = (\boldsymbol{\beta}^*_{\mathbf{A}_0}, \mathbf{0})$, and the true model is

$$Y_i = \sum_{j \in \mathbf{A}_0} X_{ij}\beta_j + \varepsilon_i.$$

Now, a variable selection procedure for linear models is oracle if, with probability tending to one, the resulting estimator satisfies:

(i) Consistent selection: The procedure is able to identify $\mathbf{A}_0$ correctly; i.e.,

$$\hat{\beta}_j = 0, \;\; \forall j \in \mathbf{A}_0, \;\; \text{and} \;\; \hat{\beta}_j \neq 0, \;\; \forall j \in \mathbf{A}_0^c.$$

(ii)Optimal estimation: The estimates of nonzero parameters are as efficient as if $\mathbf{A}_0$ were known in advance; i.e.,

$$\sqrt{n}(\widehat{\boldsymbol{\beta}}_{\mathbf{A}_0} - \beta^*_{\mathbf{A}_0}) \longrightarrow_d N(\mathbf{0}, \Sigma^*),$$

where Σ^* is the covariance matrix under the true subset model.

Consequently, an oracle procedure performs as well asymptotically as a procedure that actually knew the true model structure. That is, it can uncover the correct model so quickly that the convergence of the parameter estimation is not affected by being based on the wrong, preconvergence model. Although not well studied, there probably are cases where the model selection procedure is so slow that the parameters converge before the model does, leading to poor inference. It will be seen below that several regularization methods satisfy the oracle property.

Note that the use of the term oracle here is in a different sense than in Oracle inequalities. The common feature is that the inference procedures that satisfy them perform as well as if the correct model were known. However, Oracle inequalities are not asymptotic, while oracle estimators are only oracle in a limiting sense as $n \to \infty$. Several oracle penalized methods will be presented below; some nonoracle penalized methods will also be presented since they, too, can give good performance.

Note that the stepwise methods of Section 10.1 are not oracle even though they often give effective performance with finite data. There are two reasons for this. First, being greedy searches, they make the best selection locally, looking only one step ahead, rather than seeking the globally optimal step. Therefore, they tend to produce suboptimal models, calling the consistency into question. Second, the nature of subset selection is discrete; i.e., a coefficient is either forced to zero or retained in the model with possible inflation due to a complete exclusion of other covariates. This calls the efficiency (and consistency) into question, as well as making prediction less accurate.

There is a third reason stepwise regression is not oracle: Subset selection can be unstable with respect to small perturbations in the data, as illustrated in Breiman (1996). For instance, suppose backwards elimination is used to obtain a sequence of subsets $\{\mathscr{S}_k; k = 1, \cdots, p\}$ of size $|\mathscr{S}_k| = k$. Then removing a single point $(\boldsymbol{x}_n, y_n)$ from the data set and using backwards elimination again will often give a different sequence of subsets $\{\mathscr{S}'_k; k = 1, \cdots, p\}$. Thus, a slight perturbation in the data may cause a dramatic change in the prediction equation.

As before, $X = (\boldsymbol{X}_1, ..., \boldsymbol{X}_p)$ is an $n \times p$ design matrix, the jth column indicating the repeated sampling of the jth variable. Let $\mathbf{y}$ be the vector of n responses, and assume all variables are standardized so that $\mathbf{y}^\top \mathbf{1}_n = 0$, $\boldsymbol{X}_j^\top \mathbf{1}_n = 0$, and $\boldsymbol{X}_j^\top \boldsymbol{X}_j = 1$.

10.3.1 Shrinkage Methods for Linear Models

In general, a shrinkage method solves the optimization problem

$$\min_{\boldsymbol{\beta}} (\mathbf{y} - X\boldsymbol{\beta})^{\mathsf{T}} (\mathbf{y} - X\boldsymbol{\beta}) + \lambda \sum_{j=1}^{p} J(|\beta_j|), \tag{10.3.1}$$

where $J(|\beta_j|)$ is the penalty function and $\lambda \geq 0$ is the decay or tuning parameter that controls the amount of shrinkage on the parameters β_j. The larger the λ is, the greater the shrinkage effect imposed on the βs. Different λs can also be used for different coefficients, allowing distinct levels of penalty on coefficients. Also, some methods use a more general form of penalty $J_\lambda(|\beta|)$, not necessarily linear in λ.

Fan and Li (2001) provide a thoughtful discussion on the principles of constructing a good penalty function from the standpoint of unbiased estimation and sparse modeling. They conclude that a good penalty function should produce an estimator $\widehat{\boldsymbol{\beta}}$ satisfying: (i) nonzero coefficients of $\boldsymbol{\beta}^*$ are nearly or asymptotically unbiased; (ii) $\widehat{\boldsymbol{\beta}}$ is sparse (i.e., small coefficients are automatically set to zero); and (iii) $\widehat{\boldsymbol{\beta}}$ is continuous as a function of the data. Fan and Li (2001) also argue that $J'(|\beta|) = 0$ for large $|\beta|$s is a sufficient condition for unbiased estimation of large coefficients, and the singularity of J at 0 is a necessary condition for the sparsity and continuity of the solutions.

An alternative formulation for (10.3.1) is to solve

$$\min_{\boldsymbol{\beta}} (\mathbf{y} - X\boldsymbol{\beta})^{\mathsf{T}} (\mathbf{y} - X\boldsymbol{\beta}), \quad \text{subject to} \quad \sum_{j=1}^{p} J(|\beta_j|) \leq s, \tag{10.3.2}$$

where $s > 0$ is a tuning parameter. For some regularization methods, using (10.3.2) can greatly facilitate computation. For example, as will be seen later, the LASSO (Tibshirani, 1996) solution is piecewise linear in s and its solution path can be efficiently computed using the LARS algorithm proposed by Efron et al. (2004).

Note that (10.3.2) is essentially the constrained optimization problem with Lagrange form (10.3.1), so solving (10.3.1) and (10.3.2) are equivalent. Denote the solution to (10.3.1) as $\widehat{\boldsymbol{\beta}}_\lambda$ and the minimizer of (10.3.2) as $\widehat{\boldsymbol{\beta}}_s$. Then, for any $\lambda_0 > 0$ and the corresponding solution $\widehat{\boldsymbol{\beta}}_{\lambda_0}$, there exists s_{λ_0} such that $\widehat{\boldsymbol{\beta}}_{\lambda_0} = \widehat{\boldsymbol{\beta}}_{s_{\lambda_0}}$. On the other hand, for any given $s_0 > 0$ and the corresponding solution $\widehat{\boldsymbol{\beta}}_{s_0}$, there exists λ_{s_0} such that $\widehat{\boldsymbol{\beta}}_{s_0} = \widehat{\boldsymbol{\beta}}_{\lambda_{s_0}}$. That is, there is a one-to-one correspondence between λ and s.

Interestingly, if the design matrix is orthogonal, the optimization problem (10.3.1) can be decomposed into p componentwise shrinkage problems. To see this, assume that $X^{\mathsf{T}}X = I_n$. The column orthogonality of X implies that the OLS estimator can be computed componentwise as $\hat{\beta}_j^{ols} = \sum_{i=1}^{n} x_{ij} y_i$ for $j = 1, \cdots, p$. Then the objective function in (10.3.1) becomes

$$(\mathbf{y}-X\boldsymbol{\beta})^{\mathsf{T}}(\mathbf{y}-X\boldsymbol{\beta})+\lambda\sum_{j=1}^{p}J(|\beta_j|)=(\mathbf{y}-\hat{\mathbf{y}})^{\mathsf{T}}(\mathbf{y}-\hat{\mathbf{y}})+\sum_{j=1}^{p}(\beta_j-\hat{\beta}_j^{ols})^2+\lambda\sum_{j=1}^{p}J(|\beta_j|),$$

where $\hat{\mathbf{y}} = XX^{\mathsf{T}}\mathbf{y}$. Consequently, the optimization in (10.3.1) is equivalent to solving p one-dimensional shrinkage problems:

$$\min_{\beta_j} \ (\beta_j - \hat{\beta}_j^{ols})^2 + \lambda J(|\beta_j|), \quad j = 1, \cdots, p. \tag{10.3.3}$$

10.3.1.1 Ridge Regression

When the design matrix X is not full rank or $p < n$, the OLS estimate is not unique. The simplest way to make it unique is to invoke an extra criterion as a constraint. Hoerl and Kennard (1970) introduced ridge regression (RR), which modifies OLS by introducing a penalty to shrink βs toward zero. The RR estimator is

$$\widehat{\boldsymbol{\beta}}^{ridge} = \arg\min_{\boldsymbol{\beta}}(\mathbf{y}-X\boldsymbol{\beta})^{\mathsf{T}}(\mathbf{y}-X\boldsymbol{\beta})+\lambda\sum_{j=1}^{p}\beta_j^2 \tag{10.3.4}$$

for $\lambda > 0$. The decay parameter λ should be chosen suitably based on data, often by CV, and is sometimes called the Tikhonov factor. The explicit expression for the minimizer of (10.3.4) is

$$\widehat{\boldsymbol{\beta}}^{ridge} = (X^{\mathsf{T}}X + \lambda I_p)^{-1}X^{\mathsf{T}}\mathbf{y} \equiv R_{\lambda}\mathbf{y}, \tag{10.3.5}$$

where I_p is the $n \times n$ identity matrix and $R_{\lambda} = (X^{\mathsf{T}}X + \lambda I_p)^{-1}X^{\mathsf{T}}$ is called the shrinkage operator. It is easy to derive a standard error for $\widehat{\boldsymbol{\beta}}^{ridge}$ using (10.3.5).

Interest in (10.3.5) centers on small values of λ because that's where the penalty term contributes least; when $\lambda = 0$, (10.3.5) reduces to OLS. When $n > p$, the solution is typically overdetermined and does not exist. Expression (10.3.4) roughly corresponds to the goal of choosing a solution $\boldsymbol{\beta}$ to the original equation that has smallest λ weighted norm. RR is most useful when X is nonsingular but has high collinearity; i.e., is close to singular. In those cases, a small λ controls the variance inflation by enlarging the subspace in which the eigenvalues of the inverse are bounded. The name derives from this stabilization of the inverse matrix by adding the "ridge" of λI in R_{λ}. Note that this increases the trace of the hat matrix, which corresponds to the degrees of freedom used in fitting the model (see Sen and Srivastava, 1990, Chapters 5, 12). So RR acts like adding extra data points, but fewer as λ decreases.

Another way to see the effect of regularization is to use the singular value decomposition (SVD). Write $X_{n\times p} = U_{n\times p}S_{p\times p}V_{p\times p}^{\mathsf{T}}$, where $U = (\boldsymbol{u}_1,...,\boldsymbol{u}_n)$, $S = \text{diag}(s_1,...,s_p)$, and $V = (\mathbf{v}_1,...,\mathbf{v}_p)$, with $X\mathbf{v}_k = s_k\boldsymbol{u}_k$, $X^{\mathsf{T}}\boldsymbol{u}_k = s_k\mathbf{v}_k$, and the $\boldsymbol{u}_k$s and $\mathbf{v}_k$s are orthonormal so $U^{\mathsf{T}}U = I_n$ and $V^{\mathsf{T}}V = I_p$. Now,

$$R_\lambda = \left(VSU^\top USV^\top + \lambda VI_pV^\top\right)^{-1} VSU^\top = V\mathrm{diag}\left(\frac{s_1}{s_1^2+\lambda}, \dots, \frac{s_p}{s_p^2+\lambda}\right)U^\top.$$

It is seen that if λ is large relative to a given s_k, then the component in the kth direction is downweighted, while if s_k is large relative to λ, then the component in the kth direction gets its full weight. Obviously, choosing small λ means R_λ captures more and more of the image of X, providing an inverse on a larger domain.

It is clear from (10.3.4) that $\widehat{\boldsymbol{\beta}}^{ridge}$ is biased, with the bias decreasing as $\lambda \to 0$. On the other side, as λ increases, the $\hat{\beta}_j^{ridge}$s get closer to zero, even though they rarely equal zero. Of course, one could force shrinkage of the coefficients to zero by choosing a cutoff below which coefficients are set to zero. This would not have any obvious meaning but might help smooth out noisy data and produce robust solutions. Overall, despite the bias, the variance will usually be much smaller than that of the OLS estimator. Therefore, RR often gives a smaller MSE and so better prediction.

10.3.1.2 Nonnegative Garrote

Different from the RR modification of the OLS estimator, Breiman (1995) proposed a shrinkage estimator based on OLS to overcome the instability of subset selection methods. The nonnegative garrote (NG) estimator seeks a set of nonnegative scaling factors c_j for $j = 1, \dots, p$ that achieve

$$\min_{\boldsymbol{c}} \sum_{i=1}^{n}\left(y_i - \sum_{j=1}^{p} c_j x_{ij}\hat{\beta}_j^{ols}\right)^2 + \lambda \sum_{j=1}^{p} c_j, \qquad \text{subject to } c_j \geq 0, \ \ j = 1, \dots, p, \tag{10.3.6}$$

where λ is the decay or tuning parameter. Minimizing gives $\widehat{\boldsymbol{\beta}}^{ng} = (\hat{c}_1\hat{\beta}_1^{ols}, \dots, \hat{c}_p\hat{\beta}_p^{ols})^\top$ in which $\hat{c}_j = \hat{c}_j(\lambda)$. Each of the $\hat{c}_j$s go to zero as λ increases.

To see how the NG estimator gives greater stability than the OLS, consider the special case of orthogonal design, $X^\top X = I_n$. The best subset model of size k consists of those X_js with the k largest $|\hat{\beta}_j^{ols}|$s. That is, the coefficients of a best subset regression are given by

$$\hat{\beta}_j^{subset} = \begin{cases} \hat{\beta}_j^{ols}, & \text{if } |\hat{\beta}_j^{ols}| \text{ is among the largest } k \text{ coefficients;} \\ 0, & \text{otherwise,} \quad \text{for } j = 1, \cdots, p. \end{cases}$$

In the NG, $c_j \geq 0$, $j = 1, \cdots, p$ and solves

$$\min_{c_j}(c_j\hat{\beta}_j^{ols} - \hat{\beta}_j^{ols})^2 + \lambda c_j.$$

Basic univariate calculus leads to

$$\hat{c}_j = \left(1 - \frac{\lambda}{2(\hat{\beta}_j^{ols})^2}\right)_+, \quad j = 1, \ldots, p, \tag{10.3.7}$$

where $(t)_+ = \max(t, 0)$. That is, in the NG, large OLS coefficients have c_js close to 1 and small OLS coefficients have c_js close to zero. Finally, the NG solution is

$$\hat{\beta}_j^{ng} = \left(1 - \frac{\lambda}{2(\hat{\beta}_j^{ols})^2}\right)_+ \hat{\beta}_j^{ols}, \quad j = 1, \ldots, p. \tag{10.3.8}$$

By contrast, univariate calculus gives that the solution to the RR form of (10.3.3) is

$$\hat{\beta}_j^{ridge} = \frac{1}{1+\lambda} \hat{\beta}_j^{ols}. \tag{10.3.9}$$

It is seen in (10.3.8) that the NG estimator uses the magnitude of the c_js to retain important variables and shrink the coefficients of $p-k$ redundant variables to zero. It is this property of (10.3.8) that stabilizes the subset selection of NG compared with the instability of RR (or OLS) as can be inferred from (10.3.9).

A drawback of the original NG is its dependence on the OLS, which can perform poorly when the design matrix is ill-posed and fail to exist uniquely when $p > n$. However, the idea of the NG can be generalized to take any initial estimate $\hat{\beta}_j^{init}$ in (10.3.6) in place of $\hat{\beta}_j^{ols}$. The NG can be implemented by standard quadratic programming problems, and recently Yuan and Lin (2007) showed that the NG has a piecewise linear solution path and derived an efficient algorithm for computing the whole path. In addition, Zou (2006) and Yuan and Lin (2007) show that NG is consistent for variable selection and estimation.

10.3.1.3 Least Absolute Selection and Shrinkage Operator (LASSO)

Tibshirani (1996) imposes an L_1 penalty on the least squares error and solves the convex optimization problem

$$\widehat{\boldsymbol{\beta}}^{lasso} = \arg\min_{\boldsymbol{\beta}} (\mathbf{y} - X\boldsymbol{\beta})^{\mathsf{T}} (\mathbf{y} - X\boldsymbol{\beta}) + \lambda \sum_{j=1}^{p} |\beta_j|, \tag{10.3.10}$$

where $\lambda > 0$ is the decay parameter. The important insight embedded in this functional is that the contours of the squared error loss are ellipses but the contours of absolute value are lozenge-shaped, with corners on the coordinate axes, meaning some coefficient is zero. Thus, viewed as a Lagrange multiplier problem, some of the LASSO estimates $\hat{\beta}_j^{lasso}$ will actually equal zero rather than just move closer to zero as with RR. This happens because the intersection of the elliptical and lozenge contours will often be at the corner of the lozenge. As $\lambda \to \infty$, LASSO forces more and more $\hat{\beta}_j$s to be zero; as $\lambda \to 0$, LASSO becomes closer and closer to OLS. The sequence in which variables leave the model as λ increases indicates their relative importance. An optimal value of λ is therefore needed to balance model fit, via RSS, and model complexity.

The penalty term in (10.3.10) is equivalent to putting a double exponential prior with parameter λ on $\boldsymbol{\beta}$. As λ increases, the double exponential puts ever more of its mass near 0, shrinking the $\boldsymbol{\beta}$ to a point mass at zero. Since many of the $\hat{\beta}_j$s end up being set to zero for reasonable values of λ, LASSO combines variable selection with shrinkage on the regression function in one minimization. This can be seen from the form of the LASSO solution in the orthogonal design case $X^{\mathsf{T}}X = I_n$, which is

$$\hat{\beta}_j^{lasso} = \text{sign}(\hat{\beta}_j^{ols})(|\hat{\beta}_j^{ols}| - 2\lambda)_+ \tag{10.3.11}$$

for $j = 1, \ldots, p$ analogous to (10.3.8) and (10.3.9). The operation defined in (10.3.11) is also known as soft thresholding.

Unlike OLS and RR, where there are straightforward estimates of the standard error of $\widehat{\boldsymbol{\beta}}^{ols}$ and $\widehat{\boldsymbol{\beta}}^{ridge}$, the absolute error means there is no model-derived estimate for $SE(\widehat{\boldsymbol{\beta}}^{lasso})$, even for normal errors. One way to get an approximate estimated standard error is by the bootstrap: Fix λ, and generate n bootstrap samples and their corresponding $\widehat{\boldsymbol{\beta}}_i^{lasso}(\lambda)$s for $i = 1, \ldots, n$. From these an SE can be found. If λ is not fixed, it can be estimated by CV. Use, say, four-fifths of the data to get $\hat{\boldsymbol{\beta}} = \hat{\boldsymbol{\beta}}(\lambda)$ as a function of λ and then predict the remaining fifth of the data. Rotate through the other four-fifths of the data to get a full set of predictions dependent on λ. Find λ by minimizing the sum of the predictive errors over the whole data set. Thus $\hat{\lambda}$ could be found first and then the bootstrap applied to get an SE for the $\widehat{\boldsymbol{\beta}}_i^{lasso}(\hat{\lambda})$s. Alternatively, one could write the penalty as $\lambda \sum_{j=1}^{p} \beta_j^2/|\beta_j|$ and argue that the SE for RR is not too far wrong (see Tibshirani, 1996, Section 3).

Theoretical properties of LASSO have been studied. The L_1 penalty approach was first proposed in Donoho and Johnstone (1994) and called basis pursuit. They proved the near-minimax optimality of LASSO for orthogonal predictors. Later, Knight and Fu (2000) showed that if $\lambda = O(\sqrt{n})$ then $\widehat{\boldsymbol{\beta}}_n^{lasso}$ is root-n consistent and has a well-defined asymptotic distribution, that can have positive probability mass at zero for $j \in \mathbf{A}_0^c$; details will be given in the next subsection. Greenshtein and Ritov (2004) also show that, for suitable λ, LASSO does give asymptotically (in n) consistent prediction.

However, the LASSO is not an oracle procedure. Fan and Li (2001) show that LASSO shrinkage can produce biased estimates for large coefficients, and therefore it can be suboptimal in terms of estimation risk. Meinshausen and Buhlmann (2008) show that the optimal λ for prediction gives inconsistent variable selection outside a condition on the design matrix and a sort of neighborhood stability requirement. Recently, Zou (2006) proved that if $\lambda = O(\sqrt{n})$, then $\limsup_{n\to\infty} P(\widehat{\mathbf{A}}^{lasso} = \mathbf{A}_0) \le \eta < 1$,where η is some constant depending on the true model. Furthermore, Zou (2006) gave a necessary condition for LASSO to be consistent for variable selection. Improving this, Zhao and Yu (2006) identify an irrepresentable condition, based on covariances, which is nearly necessary and sufficient for selection consistency of the LASSO. In summary, LASSO is not always a consistent variable selection procedure, and it tends to select over-parametrized models with positive probability asymptotically.

LASSO has shown good empirical performance in various contexts. When a true model really is sparse (i.e., has a few terms that greatly dominate) and p is large relative to n, LASSO often outperforms AIC, BIC, stepwise methods and RR in predictive senses like MSE. The reasoning may be that the smooth constraint tends to give more stable performance than stepwise or best subset selection, and the trade-off of permitting a little bias to get a large reduction in variance for LASSO may typically be better than for other regularization methods.

On the other hand no method will perform best under all circumstances; see No Free Lunch in Chapter 7. LASSO does poorly compared with other methods when the true model is not sparse – i.e., there really are a large number of variables, each influencing Y a little. In these cases, LASSO often gives too many large, wrong models. In large p, small n problems, LASSO often gives n variables. Also, if some variables are highly correlated, LASSO tends to pick one randomly, and if the design matrix is too correlated, LASSO can readily fail to be consistent. In such cases, RR tends to perform better. LASSO may not be robust to outliers in the response; however, replacing the squared error loss with absolute error often corrects this (see Wang et al. (2007)). Like other regularization methods, LASSO is sensitive to λ; if CV is used, too many variables may result and bias can be high if LASSO overshrinks large $\hat{\beta}_j$s .

One great advantage of LASSO is its convenient computation. For each fixed λ, the computation of the LASSO solution can be formulated as a quadratic programming problem and solved by standard numerical analysis algorithms (Tibshirani (1996)). Even so, it would be ideal if one could compute the entire solution path $\widehat{\boldsymbol{\beta}}^{lasso}(\lambda)$ all at once (i.e., calculate the LASSO solutions simultaneously for all values of λ). Recently, Efron et al. (2004) showed that the LASSO has a piecewise linear solution path; i.e., for some m, there exist $\lambda_0 = 0 < \lambda_1 < \ldots < \lambda_m = \infty$ and $\boldsymbol{\xi}_0, \cdots, \boldsymbol{\xi}_{m-1} \in R^p$ such that

$$\widehat{\boldsymbol{\beta}}^{lasso}(\lambda) = \widehat{\boldsymbol{\beta}}^{lasso}(\lambda_k) + (\lambda - \lambda_k)\boldsymbol{\xi}_k \quad \text{for } \lambda_k \le \lambda \le \lambda_{k+1}, \tag{10.3.12}$$

for $k = 0, \ldots, m-1$. This means one can efficiently generate the whole regularized path $\widehat{\boldsymbol{\beta}}^{lasso}(\lambda), 0 \le \lambda \le \infty$, by sequentially calculating the step size between λ values and the directions $\boldsymbol{\xi}_1, \cdots, \boldsymbol{\xi}_{m-1}$. Efron et al. (2004) propose the least angle regression (LARS) algorithm to find the entire path of the LASSO solution. They show that the number of linear pieces in the LASSO path is approximately p, and the complexity of getting the whole LASSO path is $O(np^2)$, the same as the cost of computing a single OLS fit. LARS will be presented in the next section; similar algorithms are suggested in Osborne et al. (2000).

10.3.1.4 Bridge Estimators

The LASSO and RR differ only in the penalty form imposed on the regression coefficients: RR uses the squared L_2 norm of parameters, while LASSO uses the L_1 norm.

So, obviously, they can be embedded in a larger class. For $r \geq 0$, let

$$\widehat{\boldsymbol{\beta}}^{bridge} = \arg\min_{\boldsymbol{\beta}} (\boldsymbol{y} - X\boldsymbol{\beta})^{\mathsf{T}}(\boldsymbol{y} - X\boldsymbol{\beta}) + \lambda \sum_{j=1}^{p} |\beta_j|^r. \tag{10.3.13}$$

Frank and Friedman (1993) suggested using (10.3.13) and called it bridge regularization, adding that it was desirable to estimate r as a parameter rather than choosing it for convenience. Nevertheless, r is often fixed in practice. The bridge estimator includes some interesting special cases:

$$\begin{array}{ll} L_0(\boldsymbol{\beta}) = \sum_{j=1}^{p} I(\beta_j \neq 0), & \\ L_1(\boldsymbol{\beta}) = \sum_{j=1}^{p} |\beta_j| & \text{LASSO}, \\ L_2(\boldsymbol{\beta}) = \sum_{j=1}^{p} \beta_j^2 & \text{ridge regression}, \\ L_\infty(\boldsymbol{\beta}) = \max_j |\beta_j|. & \end{array}$$

The penalty functions $L_0, L_{0.5}, L_1$, and L_2 are plotted in Figure 10.1.

When the L_0 penalty is used and the design matrix is orthogonal, the optimization becomes

$$\min_{\beta_j} (\beta_j - \hat{\beta}_j^{ols})^2 + \lambda I(|\beta_j| \neq 0),$$

which has the solution

$$\hat{\beta}_j = \text{sign}(\hat{\beta}_j^{ols}) I(|\hat{\beta}_j^{ols}| > \sqrt{\lambda}). \tag{10.3.14}$$

The operator (10.3.14) is also known as the hard-thresholding rule, which shrinks small coefficients to zeros while keeping large coefficients intact. Theoretically speaking, the L_0 penalty is the ideal choice for variable selection since it directly penalizes the size of model. However, it is discontinuous and nonconvex so the optimization is numerically hard and unstable.

The cases $r \in (0,1]$ are known as soft-thresholding penalties, having contours that are not convex. Choosing an $r \in (0,1]$ regularly gives estimates that are 0 for some of the β_js. Model selection methods that penalize the number of parameters can be seen as limiting cases of bridge estimation as $r \to 0$ because $|\beta_j|^r \to 0, 1$ accordingly as $\beta_j = 0$ or $\beta_j \neq 0$ when $r \to 0$.

The most important results for the general class of regularizations giving (10.3.13) are that the estimators are consistent and have asymptotic distributions. These results are due to Knight and Fu (2000) and are worth seeing in detail. Denote the functional by

$$\Delta_n = \Delta_{n,p}(\boldsymbol{\beta}, y_1, \ldots, y_n, \boldsymbol{x}_1, \ldots, \boldsymbol{x}_n, \lambda_n, r) = \frac{1}{n} \sum_{i=1}^{n} (y_i - \boldsymbol{x}_i^{\mathsf{T}} \boldsymbol{\beta})^2 + \frac{\lambda_n}{n} \sum_{j=1}^{p} |\beta_j|^r. \tag{10.3.15}$$

Assume

$$\sum_{i=1}^{n} \boldsymbol{x}_i \boldsymbol{x}_i^{\mathsf{T}} \to C \quad \text{and} \quad (1/n) \max_{1 \leq i \leq n} \boldsymbol{x}_i^{\mathsf{T}} \boldsymbol{x}_i \to 0,$$

where C is a strictly positive definite $p \times p$ matrix, and define the limiting form

$$\Delta(\boldsymbol{\beta}) = (\boldsymbol{\beta} - \boldsymbol{\beta}^*)^{\mathsf{T}} C(\boldsymbol{\beta} - \boldsymbol{\beta}^*) + \lambda_0 \sum_{j=1}^{p} |\beta_j|^r,$$

where $\boldsymbol{\beta}^*$ are the true regression coefficients.

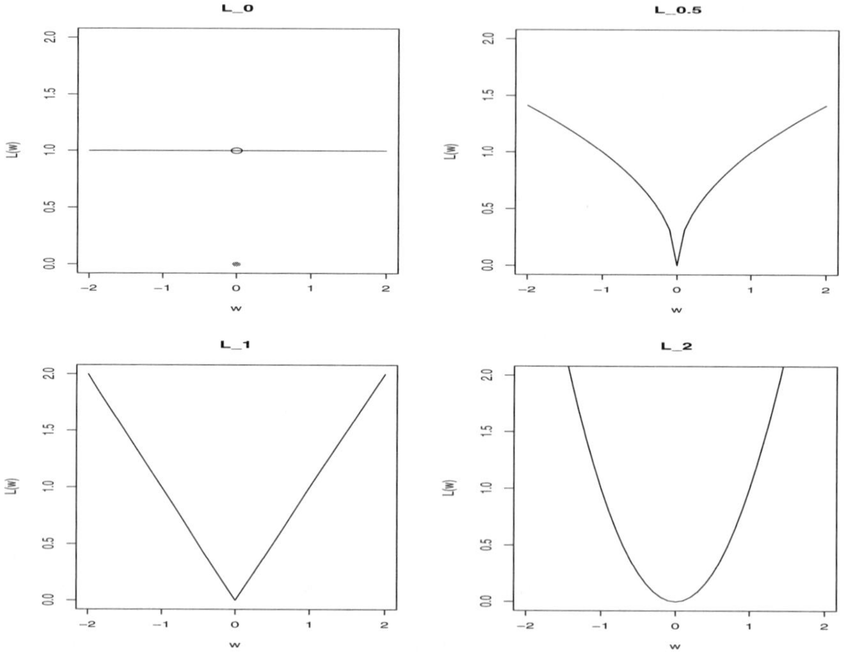

Fig. 10.1 Various shrinkage-type penalty functions.

Theorem (Knight and Fu, 2000): Suppose $\lambda_n/n \to \lambda_0 \geq 0$. Then,

$$\hat{\boldsymbol{\beta}}^{bridge} \to_p \arg \min_{\boldsymbol{\beta}} \Delta.$$

So, if $\lambda_n = o(n)$, arg $\min_{\boldsymbol{\beta}} \Delta = \boldsymbol{\beta}^*$ and $\hat{\boldsymbol{\beta}}^{bridge}$ is consistent.

Proof: It is enough to show that, for any compact set K of $\boldsymbol{\beta}$s,

$$\sup_{\boldsymbol{\beta} \in K} |\Delta_n - \Delta| \to_p 0 \tag{10.3.16}$$

and

$$\hat{\boldsymbol{\beta}} = \mathcal{O}_p(1). \tag{10.3.17}$$

If that is done, then (10.3.16) and (10.3.17) imply arg $\min_{\boldsymbol{\beta}} \Delta_n \to_p$ arg $\min_{\boldsymbol{\beta}} \Delta$, as desired.

Let $r \geq 1$ so that Δ_n is convex. Then (10.3.16) and (10.3.17) follow from the pointwise convergence of Δ_n to $\Delta + \sigma^2$ by directly taking the limit of Δ_n under the convergence assumptions on $\sum_{i=1}^n \boldsymbol{x}_i \boldsymbol{x}_i^\top$.

Let $r < 1$; now Δ_n is not convex. However, (10.3.16) still holds as before. For (10.3.17),

$$\Delta_n(\boldsymbol{\beta}) \geq \frac{1}{n} \sum_{i=1}^n (y_i - \boldsymbol{x}_i^\top \boldsymbol{\beta})^2 \equiv \Delta_0(\boldsymbol{\beta}).$$

It is clear that arg $\min_{\boldsymbol{\beta}} \Delta_0(\boldsymbol{\beta}) = \mathcal{O}_p(1)$, so arg $\min_{\boldsymbol{\beta}} \Delta(\boldsymbol{\beta}) = \mathcal{O}_p(1)$, too. □

The distributions of bridge estimators converge to limits with various locations at rates different from the usual $\sqrt{n}$. In fact, there are three possible limiting distributions, depending on the value of r. Accordingly, consider the following three functions of U, a generic $N(0, \sigma^2 C)$ random variable:

$r > 1$:

$$V_2(\boldsymbol{u}) = -2\boldsymbol{u}^\top W \boldsymbol{u}^\top C \boldsymbol{u} + \lambda_0 \sum_{j=1}^p u_j \text{sign}(\beta_j) |\beta_j|^{r-1};$$

$r = 1$:

$$V_1(\boldsymbol{u}) = -2\boldsymbol{u}^\top W \boldsymbol{u}^\top C \boldsymbol{u} + \lambda_0 \sum_{j=1}^p u_j \text{sign}(\beta_j) I(\beta_j \neq 0) + |u_j| I(\beta_j = 0);$$

$r < 1$:

$$V_0(\boldsymbol{u}) = -2\boldsymbol{u}^\top W \boldsymbol{u}^\top C \boldsymbol{u} + \lambda_0 \sum_{j=1}^p |u_j|^r I(\beta_j = 0).$$

As with consistency, the technique is to identify a finite-sample version V_n of the V_js that converges and can be optimized.

Theorem (Knight and Fu, 2000):

(i) Suppose $r > 1$: If $\lambda_n / n \to \lambda_0 \geq 0$, then

$$\sqrt{n}(\hat{\boldsymbol{\beta}}_n - \boldsymbol{\beta}^*) \to \lambda_0 \arg \min_{\boldsymbol{\beta}} V_2.$$

(ii) Suppose $r = 1$: If $\lambda_n / n \to \lambda_0 \geq 0$, then

$$\sqrt{n}(\hat{\boldsymbol{\beta}}_n - \boldsymbol{\beta}^*) \to \lambda_0 \arg \min_{\boldsymbol{\beta}} V_1.$$

(iii) Suppose $r < 1$: If $\lambda_n / n^{r/2} \to \lambda_0 \geq 0$, then

$$\sqrt{n}(\hat{\boldsymbol{\beta}}_n - \boldsymbol{\beta}^*) \to \lambda_0 \arg \min_{\boldsymbol{\beta}} V_0.$$

Proof: It will be enough to prove (i) and (ii); (iii) is similar, and details can be found in Knight and Fu (2000).

Consider the finite-sample version of the V_is denoted

$$V_n(\boldsymbol{u}) = \sum_{i=1}^{n}\left[(\varepsilon_i - \boldsymbol{u}^\top \boldsymbol{x}_i/\sqrt{n})^2 - \varepsilon_i^2\right] + \lambda_0 \sum_{j=1}^{p}\left[|\beta_j + u_j/\sqrt{n}|^r - |\beta_j|^r\right].$$

It is seen that V_n is minimized at $\sqrt{n}(\hat{\boldsymbol{\beta}} - \boldsymbol{\beta})$. To see how V_n converges, note that the first term

$$\sum_{i=1}^{n}\left[(\varepsilon_i - \boldsymbol{u}^\top \boldsymbol{x}_i/\sqrt{n})^2 - \varepsilon_i^2\right] \to_d -2\boldsymbol{u}^\top W \boldsymbol{u}^\top C \boldsymbol{u}.$$

When $r > 1$, the second term satisfies

$$\lambda_0 \sum_{j=1}^{p}\left[|\beta_j + u_j/\sqrt{n}|^r - |\beta_j|^r\right] \to \lambda_0 \sum_{j=1}^{p} u_j \text{sign}(\beta_j)|\beta_j|^{r-1}.$$

And, when $r = 1$, the second term satisfies

$$\lambda_0 \sum_{j=1}^{p}\left[|\beta_j + u_j/\sqrt{n}|^r - |\beta_j|^r\right] \to \lambda_0 \sum_{j=1}^{p} u_j \text{sign}(\beta_j) I(\beta_j \neq 0) + |u_j| I(\beta_j = 0).$$

Thus, $V_n(\boldsymbol{u})$ converges to $V_2(\boldsymbol{u})$ or $V_1(\boldsymbol{u})$ in distribution accordingly as $r > 1$ or $r = 1$. Since V_n is convex and V_i for $i = 1, 2$ has a unique minimum,

$$\arg \min_{\boldsymbol{\beta}} V_n = \sqrt{n}(\widehat{\boldsymbol{\beta}} - \boldsymbol{\beta}^*) \to_d \arg \min_{\boldsymbol{\beta}} V_i. \ \square$$

Note that the limiting forms reveal the degree of bias that may be associated with bridge regression. Although the bias is $\mathcal{O}_p(1/\sqrt{n})$, it may be significant for large $\boldsymbol{\beta}$s especially when $r > 1$.

10.3.1.5 Adaptive LASSO

Zou (2006) proposed a generalization of LASSO, adaptive LASSO, that permits different weights for different parameters. The optimization problem is to find

$$\widehat{\boldsymbol{\beta}}^{alasso} = \arg\min_{\boldsymbol{\beta}} (\boldsymbol{y} - X\boldsymbol{\beta})^\top (\boldsymbol{y} - X\boldsymbol{\beta}) + \lambda \sum_{j=1}^{p} w_j |\beta_j|, \tag{10.3.18}$$

where the w_js are positive weights to ensure good sampling performance of the ALASSO estimator. This adaptivity permits larger penalties to be imposed on unimportant covariates and smaller penalties on important ones. In the special case orthogonal design $X^\top X = I_n$, the ALASSO solution is

$$\hat{\beta}_j^{alasso} = \text{sign}(\hat{\beta}_j^{ols})(|\hat{\beta}_j^{ols}| - 2\lambda w_j)_+$$

for $j = 1, \ldots, p$. It will be seen that the ALASSO estimates can be obtained from the LARS algorithm.

One reasonable choice for the weights was suggested by Zou (2006),

$$w_j = \frac{1}{|\tilde{\beta}_j|^\gamma}, \quad j = 1, \cdots, p,$$

where $\widetilde{\boldsymbol{\beta}} = (\tilde{\beta}_1, \ldots, \tilde{\beta}_p)$ is a root-n consistent estimate of $\boldsymbol{\beta}$ and $\gamma > 0$ is a fixed constant. For instance, OLS estimates can be used when they exist.

ALASSO has two important optimality properties. First, it is near-minimax optimal. Let $\{y_1, \cdots, y_n\}$ be n independent observations from

$$Y_i = \mu_i + Z_i, \quad i = 1, \cdots, n,$$

where the Z_i are IID mean zero and known variance σ^2, and suppose the goal is to estimate $\boldsymbol{\mu} = (\mu_1, \cdots, \mu_n)$ by an estimator $\widehat{\boldsymbol{\mu}}$ with risk given by $R(\hat{\mu}) = \mathbb{E}\left[\sum_{i=1}^n (\mu_i - \hat{\mu}_i)^2\right]$. To identify the ideal estimator, consider the family of diagonal linear projections $\{\delta_i y_i\}_{i=1}^n$ with $\delta \in \{0, 1\}$. If there were an oracle who could tell us which μ_i were larger than the noise σ, then the ideal coefficients would be $\delta_i = I(|\mu_i| > \sigma)$ and the ideal risk would be $R(\text{ideal}) = \sum_{i=1}^n \min(\mu_i^2, \sigma^2)$, see Donoho and Johnstone (1994).

However, in general, the ideal risk cannot be achieved for all μ by any estimator. Next best is to say an estimator $\widehat{\boldsymbol{\mu}}$ is near-minimax optimal if its performance differs from the ideal performance by a factor; $2\log n$ has been found reasonable. Under several regularity conditions, Zou (2006) generalizes Donoho and Johnstone (1994) to show that if $w_j = 1/|y_i|^\gamma$ for some $\gamma > 0$, then the ALASSO estimate $\hat{\mu}_i^{alasso} = \text{sign}(y_i)(|y_i| - 2\lambda/|y_i|^\gamma)$ is near-minimax optimal, as in the following.

Theorem (Zou 2006, Theorem 3): Let $\lambda = 2(2\log n)^{\frac{1+\gamma}{2}}$. Then

$$\inf_{\hat{\mu}} \sup_{\mu} \frac{R(\widehat{\boldsymbol{\mu}}^{alasso})}{\sigma^2 + R(\text{ideal})} \sim 2\log n. \quad \square$$

Second, ALASSO is oracle. Zou (2006) Theorem 2 gives conditions to ensure that if $\lambda/\sqrt{n} \to 0$ and $\lambda n^{\frac{\gamma-1}{2}} \to \infty$, then ALASSO has the oracle property for estimation and selection. Other researchers have also proved ALASSO is oracle in quantile regression (see Wu and Liu (2009)) and in Cox's proportional hazard models (Zhang and Lu (2007)). The proof that ALASSO is oracle is inspired by the work of Knight and Fu (2000), but is more complex.

10.3.1.6 Smoothly Clipped Absolute Deviation (SCAD)

Although difficult to ensure, a good penalty function should lead to a sparse, nearly unbiased estimator that is continuous in the data and converges rapidly. Unfortunately, RR is not sparse and LASSO can be biased. To remedy this, Fan and Li (2001) propose a penalty term called the smoothly clipped absolute deviation (SCAD), that satisfies all the desired conditions. Mathematically, the SCAD penalty function is

$$q_\lambda(|\beta|) = \begin{cases} \lambda|\beta| & \text{if } |\beta| \le \lambda, \\ -\frac{(|\beta|^2 - 2a\lambda|\beta| + \lambda^2)}{2(a-1)} & \text{if } \lambda < |\beta| \le a\lambda, \\ \frac{(a+1)\lambda^2}{2} & \text{if } |\beta| > a\lambda, \end{cases} \tag{10.3.19}$$

where $a > 2$ and $\lambda > 0$ are tuning parameters. The function q_λ is a symmetric quadratic spline with knots at λ and $a\lambda$. Except for its singularity at the origin, $q_\lambda(|\beta|)$ has a continuous first-order derivative. In Fig. 10.2, the SCAD function with $a = 3$ and $\lambda = 0.4$ is plotted.

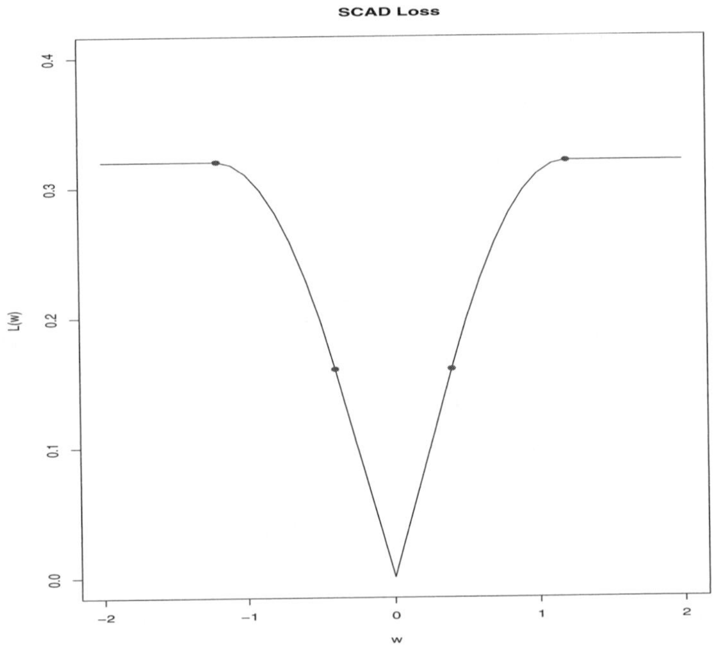

Fig. 10.2 The SCAD penalty function with $\lambda = 0.4, a = 3$.

Note that the SCAD function is the same as the L_1 penalty for small $|\beta|$s. However, for large coefficients, the SCAD penalty is constant while the L_1 penalty is linear. The SCAD estimator solves

$$\widehat{\boldsymbol{\beta}}^{scad} = \arg\min_{\boldsymbol{\beta}} (\mathbf{y} - X\boldsymbol{\beta})^\top (\mathbf{y} - X\boldsymbol{\beta}) + \sum_{j=1}^{p} q_\lambda(|\beta_j|). \tag{10.3.20}$$

The main reason that SCAD avoids causing bias for estimating large coefficients is that $q_\lambda(|\beta|)$ does not increase in $|\beta|$. Indeed, under regularity conditions, Fan and Li

(2001) establish that the SCAD estimator is oracle when the parameter λ is tuned properly. ALASSO is oracle, too, but the bias may not decrease as rapidly as with the SCAD penalty in many cases.

Unfortunately, the SCAD penalty can be hard to implement because (10.3.19) is a non-convex optimization problem. Also, the SCAD penalty is not differentiable at zero, the most important point, causing extra difficulty for gradient-based methods like Newton-Raphson. Nevertheless, there are four algorithms to solve (10.3.19).

First, despite the nondifferentiability at zero of SCAD, Fan and Li (2001) propose to solve a local quadratic approximation (LQA) of (10.3.20), iteratively. Let $\widehat{\boldsymbol{\beta}}_{(0)}$ be an initial value. If $\hat{\boldsymbol{\beta}}_j^{(0)}$ is close to zero, set $\hat{\beta}_j = 0$; otherwise, approximate the penalty function locally by the quadratic function

$$q_\lambda(|\beta_j|) \approx q_\lambda(|\hat{beta}_j^{(0)}|) + \frac{\{\beta_j^2 - \hat{\beta}_j^{(0)}\}^2 q'_\lambda(|\hat{\beta}_j^{(0)}|+)}{2|\hat{\beta}_j^{(0)}|},$$

where $q'_\lambda(|\beta|+)$ denotes the right limit of $q'_\lambda(|\beta|)$. Using this LQA, a Newton-Raphson algorithm can be used to solve

$$\min_{\boldsymbol{\beta}}(\mathbf{y} - X\boldsymbol{\beta})^{\mathsf{T}}(\mathbf{y} - X\boldsymbol{\beta}) + \frac{1}{2}\boldsymbol{\beta}^{\mathsf{T}}\Sigma_\lambda(\boldsymbol{\beta}_1)\boldsymbol{\beta}, \tag{10.3.21}$$

where $\Sigma_\lambda(\boldsymbol{\beta}^{(0)}) = \mathrm{diag}\{q'_\lambda(|\hat{\beta}_1^{(0)}|)/|\hat{\beta}_1^{(0)}|, \ldots, q'_\lambda(|\hat{\beta}_p^{(0)}|)/|\hat{\beta}_p^{(0)}|\}$. Since (10.3.21) is a quadratic minimization, it can be iteratively solved like the ridge regression estimator. In general, the solutions to (10.3.21) are not exactly zero, so a thresholding rule is needed to decide whether to set a coefficient to zero or not. One drawback of the algorithm (see Hunter and Li (2005)) is that once a covariate is deleted at any step in the procedure, it is permanently removed from the final model.

Second, Hunter and Li (2005) suggested a majorize-minimize (MM) procedure to solve (10.3.20). In principle, the MM algorithm substitutes a simple minimization problem for a difficult minimization problem and solves the simple problem iteratively until convergence is obtained.

Let θ_0 be the current iterate. A function $\Phi_{\theta_0}(\theta)$ is said to majorize $q_\lambda(|\theta|)$ at θ_0 if it satisfies

$$\Phi_{\theta_0}(\theta) \geq q_\lambda(|\theta|), \ \forall\theta; \quad \Phi_{\theta_0}(\theta) = q_\lambda(|\theta_0|).$$

This condition implies

$$\Phi_{\theta_0}(\theta) - \Phi_{\theta_0}(\theta_0) \geq q_\lambda(|\theta|) - q_\lambda(|\theta_0|),$$

which leads to the descent property

$$\Phi_{\theta_0}(\theta) < \Phi_{\theta_0}(\theta_0) \quad \text{implies} \ \ q_\lambda(|\theta|) < q_\lambda(|\theta_0|). \tag{10.3.22}$$

Thus (10.3.22) implies that any decrease in the value of $\Phi_{\theta_0}(\theta)$ guarantees a decrease in $q_\lambda(|\theta|)$.

For the SCAD penalty, if the current iterate $\hat{\beta}_j^{(m)}$ is nonzero for all j, then Hunter and Li (2005) suggested using

$$S_m(\boldsymbol{\beta}) \equiv (\mathbf{y} - X\boldsymbol{\beta})^\top (\mathbf{y} - X\boldsymbol{\beta}) + \sum_{j=1}^{p} \Phi_{\beta_j^{(m)}}(\beta_j)$$

as a majorizing function for (10.3.20), where

$$\Phi_{\hat{\beta}_j^{(m)}}(\beta_j) \approx q_\lambda(|\hat{\beta}_j^{(m)}|) + \frac{\{\beta_j^2 - (\hat{\beta}_j^{(m)})^2\} q'_\lambda(|\hat{\beta}_j^{(m)}|+)}{2|\hat{\beta}_j^{(m)}|}. \tag{10.3.23}$$

If some $\hat{\beta}_j^{(m)}$ is zero, Hunter and Li (2005) suggested replacing $2|\hat{\beta}_j^{(m)}|$ by $2(\varepsilon + |\hat{\beta}_j^{(m)}|)$ for some $\varepsilon > 0$, leading to a perturbed objective function:

$$S_{m,\varepsilon} \equiv (\mathbf{y} - X\boldsymbol{\beta})^\top (\mathbf{y} - X\boldsymbol{\beta}) + \Phi_{\hat{\beta}_j^{(m)}}(\beta_j) + \frac{\{\beta_j^2 - (\hat{\beta}_j^{(m)})^2\} q'_\lambda(|\hat{\beta}_j^{(m)}|+)}{2|\hat{\beta}_j^{(m)} + \varepsilon|}.$$

They show that, as long as ε is small, the minimizer of $S_{m,\varepsilon}$ approximately solves the original SCAD problem (10.3.20). Since the MM procedure optimizes a perturbed version of LQA, it is less likely than LQA to exclude variables inappropriately.

Third, nonconvex penalties can sometimes be decomposed as the difference of two convex functions, leading to a difference convex algorithm (DCA; An and Tao (1997)) that can be solved as a series of convex problems. For the SCAD case, Wu and Liu (2009) note that, for (10.3.19),

$$q_\lambda(|\beta|) = q_{\lambda,1}(|\beta|) - q_{\lambda,2}(|\beta|), \tag{10.3.24}$$

where

$$q'_{\lambda,1}(|\beta|) = \lambda, \quad q'_{\lambda,2}(|\beta|) = \lambda \left(1 - \frac{(a\lambda - |\beta|)_+}{(a-1)\lambda}\right) I(|\beta| > \lambda).$$

In this case, Wu and Liu (2009) observe that the DCA is an instance of the MM since at each step the DCA majorizes a nonconvex objective function and then performs a minimization.

With the DC decomposition, (10.3.20) can be decomposed as $Q_1(\boldsymbol{\beta}) - Q_2(\boldsymbol{\beta})$, where

$$Q_1(\boldsymbol{\beta}) = (\mathbf{y} - X\boldsymbol{\beta})^\top (\mathbf{y} - X\boldsymbol{\beta}) + \sum_{j=1}^{p} q_{\lambda,1}(|\beta_j|),$$

$$Q_2(\boldsymbol{\beta}) = n \sum_{j=1}^{p} q_{\lambda,2}(|\beta_j|).$$

The algorithm suggested by Wu and Liu (2009) uses an initializing $\boldsymbol{\beta}^{(0)}$ and then cycles through

$$\widehat{\boldsymbol{\beta}}^{(m+1)} = \arg\min_{\boldsymbol{\beta}} \; Q_1(\boldsymbol{\beta}) + \left\langle -Q'_2(\widehat{\boldsymbol{\beta}}^{(m)}), \boldsymbol{\beta} - \widehat{\boldsymbol{\beta}}^{(m)} \right\rangle,$$

where $\langle \boldsymbol{u}, \boldsymbol{v} \rangle$ is the inner product of $\boldsymbol{u}$ and $\boldsymbol{v}$, until convergence is observed.

One difference between DCA and MM is that, at each iteration, DCA majorizes the nonconvex function using a linear approximation, while the MM algorithm uses a quadratic approximation. Indeed,

$$-Q_2'(\widehat{\boldsymbol{\beta}}^{(m)}) = -n\left(q_{\lambda,2}'(|\hat{\beta}_1^{(m)}|)\text{sign}(\hat{\beta}_1^{(m)}), \ldots, q_{\lambda,2}'(|\hat{\beta}_p^{(m)}|)\text{sign}(\hat{\beta}_p^{(m)})\right)^{\mathsf{T}}.$$

So, at the $(m+1)$th step, DCA approximates the second function linearly and essentially solves

$$\min_{\boldsymbol{\beta}}(\boldsymbol{y} - X\boldsymbol{\beta})^{\mathsf{T}}(\boldsymbol{y} - X\boldsymbol{\beta}) + \sum_{j=1}^{p} q_{\lambda,1}(|\beta_j|) - n\sum_{j=1}^{p} q_{\lambda,2}'(|\hat{\beta}_j^{(m)}|)\text{sgn}(\hat{\beta}_j^{(m)})(\beta_j - \hat{\beta}_j^{(m)}).$$

Fourth, Zou and Li (2007) proposed computing $\widehat{\boldsymbol{\beta}}^{scad}$ based on local linear approximation (LLA). A distinctive feature of this is that, at each step, the LLA estimator has a sparse representation. Let $\widehat{\boldsymbol{\beta}}^{(0)}$ be an initial value. Then, the penalty function can be locally approximated by

$$q_\lambda(|\beta_j|) \approx q_\lambda(|\hat{\beta}_j^{(0)}|) + (\beta_j - \hat{\beta}_j^{(0)})q_\lambda'(|\hat{\beta}_j^{(0)}|).$$

Now, Zou and Li (2007) suggest cycling via

$$\widehat{\boldsymbol{\beta}}^{(m+1)} = \arg\min_{\boldsymbol{\beta}} \; (\boldsymbol{y} - X\boldsymbol{\beta})^{\mathsf{T}}(\boldsymbol{y} - X\boldsymbol{\beta}) + n\sum_{j=1}^{p} q_\lambda'(|\hat{\beta}_j^{(m)}|)|\beta_j|$$

until convergence is observed. It is seen that, at each step, the LLA algorithm must solve a LASSO-type problem. This can be done efficiently by LARS (Efron et al. (2004)).

Compared with LQA, LLA is more stable numerically and does not have the problem of deleting variables in the middle of iteration. The LLA algorithm is an instance of MM and hence has a descent property. Also, Zou and Li (2007) prove that the SCAD estimate from the LLA algorithm, with $\widehat{\boldsymbol{\beta}}^{ols}$ as the initial estimator, has the oracle property if the decay parameter is appropriately chosen.

10.3.2 Grouping in Variable Selection

Even without leaving the $p \le n$ paradigm, when p is too large for an exhaustive search of the model space to be feasible, it may make sense to partition the set of variables into groups and then do exhaustive searches within the groups (Gabriel and Pun (1979)). Selecting groups of variables arises naturally in several scenarios. For instance, if there are groups of highly correlated variables and each group works together to influence the response, it may be desirable to include or exclude the whole group. Two popular techniques for examining groups of variables in the regression setting are the grouped LASSO and the elastic net (EN).

10.3.2.1 Grouped LASSO

Assume the set of p variables can be partitioned into J groups, denoted $Z_1, \cdots, Z_J$, in which each Z_j consists of n measurements of p_j variables so that $p_1 + \ldots + p_J = p$. Now, the linear model can be expressed as

$$Y = \sum_{j=1}^{J} Z_j \boldsymbol{\beta}_j + \boldsymbol{\varepsilon}, \tag{10.3.25}$$

where $\mathbf{y} = (y_1, \ldots, y_n)$ is the response, Z_j is an $n \times p_j$ matrix corresponding to the jth group of variables, and $\boldsymbol{\beta}_j$ is a vector of length p_j, $j = 1, \cdots, J$. Thus, the entire design matrix is $\mathbf{X} = (Z_1, \cdots, Z_J)$ and the coefficient vector $\boldsymbol{\beta} = (\boldsymbol{\beta}_1^{\mathsf{T}}, \cdots, \boldsymbol{\beta}_J^{\mathsf{T}})^{\mathsf{T}}$. The group LASSO estimator (Yuan and Lin (2007)) is

$$\widehat{\boldsymbol{\beta}}^{glasso} = \arg\min_{\boldsymbol{\beta}} \left[\left(\mathbf{y} - \sum_{j=1}^{J} Z_j \boldsymbol{\beta}_j \right)^{\mathsf{T}} \left(\mathbf{y} - \sum_{j=1}^{J} Z_j \boldsymbol{\beta}_j \right) + \lambda \sum_{j=1}^{J} p_j (\boldsymbol{\beta}_j^{\mathsf{T}} \boldsymbol{\beta}_j)^{1/2} \right]. \tag{10.3.26}$$

Note that the groupwise L_2 norm $\|\boldsymbol{\beta}_j\| = (\boldsymbol{\beta}_j^{\mathsf{T}} \boldsymbol{\beta}_j)^{1/2}$ permits the selection or removal of groups of variables simultaneously, and this selection is invariant under groupwise orthogonal transformations.

The grouped LASSO gives better performance than traditional stepwise selection procedures in factor selection problems (Yuan and Lin (2007)). A grouped NG and a grouped LARS (see below) can also be defined; they can be efficiently solved by using the solution path algorithm. However, the grouped LASSO does not have a piecewise linear solution path.

10.3.2.2 Elastic Net

As a variant on single penalty terms, Zou and Hastie (2005) propose the elastic net (EN) as a way to combine the benefits of the L_1 and L_2 regularizations and simultaneously ensure that related X_js get comparably sized coefficients. First, the naive elastic net estimator is

$$\widehat{\boldsymbol{\beta}}^{enet} = \arg\min_{\boldsymbol{\beta}} (\mathbf{y} - X\boldsymbol{\beta})^{\mathsf{T}} (\mathbf{y} - X\boldsymbol{\beta}) + \lambda_1 \sum_{j=1}^{p} |\beta_j| + \lambda_2 \sum_{j=1}^{p} \beta_j^2, \tag{10.3.27}$$

where λ_1 and λ_2 are decay parameters. The penalty in (10.3.27) causes a double shrinkage, which can introduce bias to the estimation. To correct this, the elastic net coefficient is defined as a rescaled naive estimate $(1+\lambda_2)\widehat{\boldsymbol{\beta}}^{enet}$. In contrast to the LASSO, the EN penalty is strictly convex and so $\widehat{\boldsymbol{\beta}}^{enet}$ is particularly appropriate when p is much larger than n. The EN estimator combines the strength of LASSO and the ridge regression: The L_1 penalty encourages the model to be parsimonious, and the L_2 penalty encourages highly correlated predictors to have similar coefficient estimates.

An important property of EN estimators that makes them appropriate for assessing groups of variables is that they are provably close to each other when their associated variables are correlated. In particular, define the difference between the coefficients of predictors X_j and X_k as

$$D(j,k) = \frac{1}{\sum_{i=1}^{n} |y_i|} |\hat{\beta}_j^{enet} - \hat{\beta}_k^{enet}|.$$

The grouping effect of EN is illustrated by the following theorem.

Theorem (Zou and Hastie, 2005): Assume $\boldsymbol{y}$ is centered and the predictors X are standardized. Define the sample correlation $\rho_{jk} = \boldsymbol{x}_j^{\mathsf{T}} \boldsymbol{x}_k$ for each $j \neq k$. Then

$$D(j,k) \leq \frac{1}{\lambda_2} \sqrt{2(1-\rho_{jk})}. \quad \square$$

This theorem implies that, if X_j and X_k are almost perfectly correlated (i.e. $|\rho_{jk}| \approx 1$), then the difference between $\hat{\beta}_j^{enet}$ and $\hat{\beta}_k^{enet}$ is nearly zero.

Zou and Hastie (2005) proposed a LARS-EN algorithm to compute the entire EN path efficiently. Recently, Zou and Zhang (2008) extended the EN to an adaptive EN by solving

$$\widehat{\boldsymbol{\beta}}^{aenet} = \arg\min_{\boldsymbol{\beta}} (\boldsymbol{y} - X\boldsymbol{\beta})^{\mathsf{T}} (\boldsymbol{y} - X\boldsymbol{\beta}) + \lambda_1 \sum_{j=1}^{p} w_j |\beta_j| + \lambda_2 \sum_{j=1}^{p} \beta_j^2, \tag{10.3.28}$$

and they proved that the adaptive EN estimator is oracle under regularity conditions.

10.3.3 Least Angle Regression

The computational technique called least angle regression (Efron et al. (2004)) aimed at getting a parsimonious model for predictive purposes, is a less greedy version of traditional forward selection methods. The intuitive idea of LARS is that rather than adding explanatory variables to a submodel one at a time, look at the column space of the design matrix, identify the equiangular vector that forms the same size angle with each column vector, add a fraction of it to the model, and form a new set of predictions. This is somewhat abstract, so it is worthwhile giving some details of the LARS procedure (see Efron et al. (2004)).

Suppose the data are standardized. Then,

- ☐ Initialize: Let $\hat{\boldsymbol{\mu}}_0$ be a first choice for the fitted values, possibly $\mathbf{0}$, possibly based on a choice of explanatory variables already in the model.
- ☐ Compute inputs: Form the n vector of "current correlations"

$$\boldsymbol{c} = X^{\mathsf{T}}(\boldsymbol{y} - \hat{\boldsymbol{\mu}}_0),$$

and use this to find

$$\mathscr{A} = \{j : |c_j| = \max_j |c_j|\}.$$

For $j \in \mathscr{A}$, set

$$s_j = \text{sign}(c_j) = \pm 1$$

and let

$$X_{\mathscr{A}} = (\ldots, s_j \boldsymbol{x}_j, \ldots)_{j \in \mathscr{A}}, \quad K_{\mathscr{A}} = (\mathbf{1}_{\mathscr{A}}^{\mathsf{T}}(X_{\mathscr{A}}^{\mathsf{T}} X_{\mathscr{A}})^{-1} \mathbf{1}_{\mathscr{A}})^{-1/2}.$$

Now, the equiangular vector is

$$\boldsymbol{u}_{\mathscr{A}} = K_{\mathscr{A}} X_{\mathscr{A}} (X_{\mathscr{A}}^{\mathsf{T}} X_{\mathscr{A}})^{-1} \mathbf{1}_{\mathscr{A}}.$$

(Note: $\boldsymbol{u}_{\mathscr{A}}$ is the unit vector making equal angles with the columns of $X_{\mathscr{A}}$: $X_{\mathscr{A}}^{\mathsf{T}} u_{\mathscr{A}} = K_{\mathscr{A}} \mathbf{1}_{\mathscr{A}}$ and $\|u_{\mathscr{A}}\| = 1$.) The inner product vector is

$$\mathbf{a} = X^{\mathsf{T}} \boldsymbol{u}_{\mathscr{A}}.$$

- ☐ Update $\widehat{\boldsymbol{\mu}}_0$: Let

$$\widehat{\boldsymbol{\mu}}_1 = \widehat{\boldsymbol{\mu}}_0 + \gamma \boldsymbol{u}_{\mathscr{A}}$$

where

$$\gamma = \min_{j \in \mathscr{A}^c} \left[\frac{\max_j |c_j| - c_j}{K_{\mathscr{A}} - a_j}, \frac{\max_j |c_j| + c_j}{K_{\mathscr{A}} + a_j} \right]^+$$

and the $+$ means the positive part; i.e., negative values are excluded.

- ☐ Iterate to find $\hat{\boldsymbol{\mu}}_2$ and so forth until a stopping criterion is met.

Thus, LARS augments a set of fitted values $\widehat{\boldsymbol{\mu}}_k$ by finding a set of variables that are maximally correlated with its residual and using their central (equiangular) vector to shift $\widehat{\boldsymbol{\mu}}_k$ to $\widehat{\boldsymbol{\mu}}_{k+1}$. It's as if $\boldsymbol{y}$ is regressed on the equiangular vector with coefficient γ. That is, the fitted values are sequentially improved by finding explanatory (equiangular) vectors from the explanatory variables to reduce residuals.

LARS is closely related to another iterative algorithm, called forward stagewise linear regression, which begins with $\hat{\boldsymbol{\mu}}_0 = \mathbf{0}$ and builds up the regression function in successive small steps. In particular, if $\hat{\boldsymbol{\mu}}$ is the current stagewise estimate, let $\boldsymbol{c}(\hat{\boldsymbol{\mu}})$ be the vector of current correlations (i.e., $\boldsymbol{c}(\hat{\boldsymbol{\mu}}) = X^{\mathsf{T}}(\boldsymbol{y} - \hat{\boldsymbol{\mu}})$), so that the jth component of $\boldsymbol{c}(\hat{\boldsymbol{\mu}})$, denoted as $\hat{c}_j$, is proportional to the correlation between the covariate X_j and the current residual vector. At the next step, the stagewise algorithm takes the direction of the greatest current correlation,

$$\hat{j} = \arg\max |\hat{c}_j|$$

and updates the estimate as

$$\hat{\boldsymbol{\mu}} \to \hat{\boldsymbol{\mu}} + \varepsilon \cdot \text{sign}(\hat{c}_j) \cdot \boldsymbol{x}_{\hat{j}},$$

where ε is some small constant.

There are several stopping criteria and variants of this procedure that have been studied, but the details are beyond the present scope. For instance, although not obvious, LARS can be modified to give LASSO: If the procedure is stopped when γ falls below a threshold and $\mathscr{A}$ is sequentially modified only by adding or removing one variable at a time from $\mathscr{A}$, LARS recovers all LASSO solutions (see Efron et al. (2004)) with the same order of magnitude of computational effort as one OLS fit using the full set of covariates. LARS is implemented in R with the function lars and is designed so that other estimates such as LASSO and forward stagewise regression can be obtained from it as options.

As an instance of how LARS performs, consider the Boston housing data set concerning housing values in the suburbs of Boston, D. Harrison and Rubinfeld (1978). There are 13 predictors to predict the median value of owner-occupied homes and $n = 506$. The predictors consist of a variety of social variables (e.g., crime rate) and property location variables (e.g., proportion of residential land zoned for lots over 25,000 square feet). The full data set is available in the R package MASS.

Figure 10.3 plots the solution paths (see (10.3.12)) of the LASSO, LARS, and forward stagewise regression estimates obtained by the LARS algorithm for the Boston housing data. It is seen that there is high similarity among the solution paths of these three methods. In all three cases, the variables come into the regression in essentially the same order. For instance, the first three are variable 13 (status of the population), variable 9 (access to highways), and variable 10 (property taxes). Also, the numerical values for the coefficients from the three methods are seen to be similar, too. As noted in Efron et al. (2004), all three procedures can be viewed as moderately greedy forward stepwise procedures, which progress according to the compromise among the currently most correlated covariates. LARS moves along the most obvious compromise direction, the equiangular vector, while LASSO and forward stagewise regression put some restrictions on the equiangular strategy.

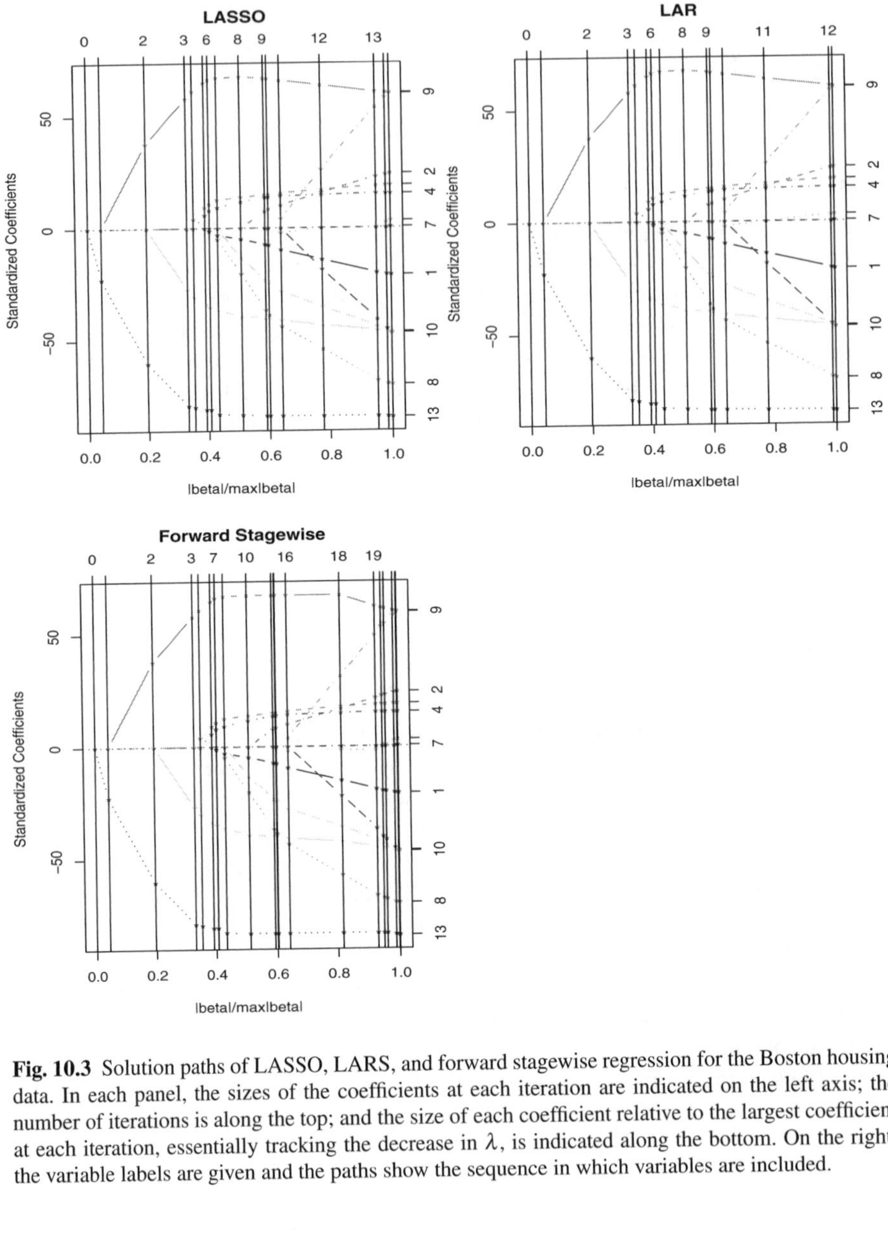

Fig. 10.3 Solution paths of LASSO, LARS, and forward stagewise regression for the Boston housing data. In each panel, the sizes of the coefficients at each iteration are indicated on the left axis; the number of iterations is along the top; and the size of each coefficient relative to the largest coefficient at each iteration, essentially tracking the decrease in λ, is indicated along the bottom. On the right, the variable labels are given and the paths show the sequence in which variables are included.

10.3.4 Shrinkage Methods for Model Classes

So far, the idea of penalization has been seen only in linear models. However, penalization is much more general and can be applied to both regression and classification problems with a wide variety of model classes, including ensembles, GLM, trees, splines, and SVMs. It is worth looking at these in turn.

10.3.4.1 Boosted LASSO

Recall that ensemble methods were discussed as a general class in Chapter 6 and several specific ensemble methods – boosting in particular – were presented. It is desirable to merge the predictive performance of ensemble methods with the parsimony of penalized methods in the hope of getting better small-sample behavior. This could mean combining several of the estimators from several regularization methods with weights chosen by some ancillary optimization. Or it could mean finding a penalty term on ensembles that would enforce parsimony. The first might be suboptimal because of the use of the data in each regularization; the second might lead to bias because pruning an ensemble weakens its predictive power. A more sophisticated approach is due to Zhao and Yu (2007), where boosting and LASSO are combined on a methodological level rather than just using them in sequence. Of course, any ensemble method and any penalty would lead to another instance of this template.

Recall that boosting can be regarded as minimizing a risk, $\mathbb{E}L(Y, F(\boldsymbol{x}))$ where F varies over a class of ensembles of "base learners", say

$$\mathscr{F} = \{F \mid F(\boldsymbol{x}) = \sum_{j=1}^{\infty} \beta_j h_j(\boldsymbol{x})\}.$$

Using $\boldsymbol{\beta} = (\beta_1, \ldots)$, write the minimization of $\mathbb{E}(L)$ over $\mathscr{F}$ as

$$\boldsymbol{\beta} = \arg\min_{\boldsymbol{\beta}} \sum_{i=1}^{n} L(Z_i, \boldsymbol{\beta}), \tag{10.3.29}$$

where $Z_i = (X_i, Y_i)$. One can attempt to solve (10.3.29) directly, but this can easily lead to overfit solutions when many β_js cannot be taken as zero. Boosting builds up a solution iteratively – and can be stopped at any time to avoid overfitting.

Let $\boldsymbol{e}_j$ be the vector of length n, which is 1 in the jth entry and zero everywhere else. Zhao and Yu (2007) write the boosting procedure algorithmically as a step from iteration t to $t+1$. For a given $\widehat{\boldsymbol{\beta}}^{(t)}$ at step t, set

$$(\hat{j}, \hat{s}) = \arg\min_{j,s} \sum_{i=1}^{n} L(Z_i, \widehat{\boldsymbol{\beta}}^{(t)} + s\boldsymbol{e}_j).$$

Then, find

$$\widehat{\boldsymbol{\beta}}^{(t+1)} = \widehat{\boldsymbol{\beta}}^{(t)} + \hat{s}_{\hat{j}} \boldsymbol{e}_{\hat{j}}.$$

LASSO is seen to be a regularized version of this when L is squared error. The minimization finds j and s such that $\boldsymbol{e}_j$ is added to the emerging $\hat{\beta}^t$ with a factor of $\hat{s}_j$. If $s = \pm\varepsilon$ instead of being continuous, the procedure is called forward stagewise fitting.

Roughly, the regularization in LASSO is used in the boosted LASSO procedure as part of a decision criterion. Denote the functional defining LASSO in (10.3.10) by $\Gamma = \Gamma(\mathbf{Z}, \boldsymbol{\beta}, \lambda) = \sum_{i=1}^{n} L(Z_i, \boldsymbol{\beta}) + \lambda \|\boldsymbol{\beta}\|_1$. In boosted LASSO, if $\Gamma(\mathbf{Z}, \boldsymbol{\beta}, \lambda)$ is reduced enough by an iteration on $\boldsymbol{\beta}^{(t)}$ for fixed λ, there is no need to reduce λ; otherwise a new search over j is done and typically a reduced λ is found. This can be seen in the shrinkage step in the formal statement of the procedure.

Given data $\mathbf{Z} = (Z_1, ..., Z_n)$, step size $\varepsilon > 0$, and threshold $\xi > 0$:

☐ Initialize: Start with $\widehat{\boldsymbol{\beta}}^{(-1)} = \mathbf{0}$ and find

$$(\hat{j}, \hat{s}) = \arg \min_{j, s = \pm\varepsilon} \sum_{i=1}^{n} L(Z_i, s\mathbf{e}_j)$$

and

$$\widehat{\boldsymbol{\beta}}^{(0)} = \hat{s}_{\hat{j}} \mathbf{e}_{\hat{j}}.$$

Let $\lambda = \lambda_0$ be the initial decrease in empirical risk at the first step, scaled by ε:

$$\lambda_0 = \frac{1}{\varepsilon} \left(\sum_{i=1}^{n} L(Z_i, \mathbf{0}) - \sum_{i=1}^{n} L\left(Z_i, \widehat{\boldsymbol{\beta}}^{(0)} \right) \right),$$

and let the initial active set $\mathscr{A}_0 = \{\hat{j}\}$. This completes the $t = 0$ step.

☐ Shrinkage: If shrinking the coefficient of an element of $\mathscr{A}$ also improves the LASSO criterion enough, then use it to update the estimates. That is, find

$$\hat{j} = \arg \min_{j \in \mathscr{A}_t} \sum_{i=1}^{n} L\left(Z_i, \widehat{\boldsymbol{\beta}}^{(t)} + s_j \mathbf{e}_j \right)$$

with $s_j = \text{sign}(\hat{\beta}_j)\varepsilon$, and if

$$\Gamma\left(\mathbf{Z}, \widehat{\boldsymbol{\beta}}^{(t)} + s_{\hat{j}} \mathbf{e}_{\hat{j}}, \lambda_t \right) < \Gamma\left(\mathbf{Z}, \widehat{\boldsymbol{\beta}}^{(t)}, \lambda_t \right) + \xi, \tag{10.3.30}$$

then shrink the $\hat{j}$th coefficient and update:

$$\widehat{\boldsymbol{\beta}}^{(t+1)} = \widehat{\boldsymbol{\beta}}^{(t)} + s_{\hat{j}} \mathbf{e}_{\hat{j}}; \quad \lambda_{t+1} = \lambda_t; \quad \mathscr{A}_{t+1} = \mathscr{A}_t - \{\hat{j}\}.$$

☐ Search: If (10.3.30) fails, then search $\mathscr{A}^c$ for a new j. Find

$$(\hat{j}, \hat{s}) = \arg \min_{j, s = \pm\varepsilon} \sum_{i=1}^{n} L\left(Z_i, \widehat{\boldsymbol{\beta}}^{(t)} + s\mathbf{e}_j \right)$$

and update. Again, set $\widehat{\boldsymbol{\beta}}^{(t+1)} = \widehat{\boldsymbol{\beta}}^{(t)} + \hat{s}\mathbf{e}_{\hat{j}}$ and then set

$$\lambda_{t+1} = \min\left[\lambda_t, \frac{1}{\varepsilon}\left(\sum_{i=1}^{n} L\left(Z_i, \widehat{\boldsymbol{\beta}}^{(t)}\right) - \sum_{i=1}^{n} L\left(Z_i, \widehat{\boldsymbol{\beta}}^{(t+1)}\right)\right)\right]; \quad \mathscr{A}_{t+1} = \mathscr{A}_t \cup \{\hat{j}\}.$$

☐ Stopping: Increase t by 1 and repeat the steps until $\lambda_t \leq 0$.

The shrinkage is a sort of backwards step to refine a coefficient; the search is a forward step to find a new coefficient. The role of λ in the boosted LASSO is not quite the same as in (10.3.10), but its role is similar since it tracks the decrease in empirical risk. The choice of ε and ξ can vary widely; ε can easily range from .05 to 50, for instance. Smaller values tend to give smoother curves for the quantities over time. Note that the procedure is written so that the penalty term is generic. This means that the L_1 penalty can be used but so can L_2, SCAD or any other.

10.3.4.2 GLM and Trees

Recall that the GLM was discussed in Section 4.2 and Trees were discussed in Sections 4.5 and 5.3. Here, both model classes can be situated in a penalization context. Recall that in GLMs there is a link function $g^{-1}(\cdot)$ that transforms the expected value of Y so it can be expressed in terms of the regression function, not necessarily on the whole real line. the correct choice of link function is particularly important for binomial, count, and categorical data, but not usually as important for continuous data. Thus, there is no reason not to try the same penalty term as in (10.3.13) and define

$$\widehat{\boldsymbol{\beta}} = \arg\min_{\boldsymbol{\beta}} \left\{\sum |y_i - g(\boldsymbol{x}_i^{\mathsf{T}}\boldsymbol{\beta})|^2 + \lambda \sum |\beta_j|^r\right\}. \tag{10.3.31}$$

Indeed, any of the penalties discussed could be used in place of the bridge penalty. Note that g is applied to the population mean in the definition of a GLM model, so in (10.3.31), it is applied to the expression in $\boldsymbol{x}$. When g has local properties that make it linearly approximable, the results of penalizing in GLMs should be similar to the linear model case, though transformed, although this does not seem to have been investigated. Likewise, properties such as being oracle or near-minimax optimal should hold for GLMs if they hold for the linear model case.

Recall that tree-based regression and classification partitions the space of explanatory variables into, say, p regions $R_1, R_2, \ldots, R_p$ by a binary tree structure and then defines an estimator for Y on each region. The response for $\boldsymbol{x}$ in the region at the terminal node can be modeled as a linear regression, often just the average of the y_is on it. More formally, this is

$$y_i = \sum_{j=1}^{p} \text{ave}(y_i|\boldsymbol{x}_i \in R_j) I(\boldsymbol{x} \in R_j) + \varepsilon_i.$$

The trade-off is that a very large tree risks overfitting the data while a small tree might not capture important structure. As before, minimizing a cost-complexity

$$C(T) = \sum_{j=1}^{|T|} \sum_{x_i \in R_j} (y_i - \hat{c}_j)^2 + \lambda |\mathcal{T}| \tag{10.3.32}$$

helps find an optimally sized tree, where $\hat{c}_j = \text{ave}(y_i | \boldsymbol{x}_i \in R_j)$ and $|\mathcal{T}|$ denotes the number of terminal nodes in tree $\mathcal{T}$.

It can be seen in retrospect that cost-complexity optimization was a penalization method. Moreover, (10.3.32) is a regularized (empirical) risk similar to those used before. Indeed, the number of parameters in $\mathcal{T}$ is an affine function of $|\mathcal{T}|$, suggesting L_0-like properties. Other assessments of the size of $\mathcal{T}$ can be used and may be more appropriate depending on the class of trees and data. It is tempting to use $|\mathcal{T}|^r$ hoping that it corresponds to bridge. However, only penalizing the number of nodes in will likely not be enough in general; the actual norms of the parameters in the $\hat{c}_j$s must be penalized if the node function can be more complicated than the average. Penalized trees in this sense do not seem to have been extensively studied.

10.3.4.3 Multidimensional Smoothing Splines

Chapter 3 gave a detailed treatment on smoothing splines. There, the focus was on univariate function estimation; i.e., x was one-dimensional. Here, function estimation with smoothing splines will use a multivariate argument $\boldsymbol{x}$ from a product domain $\prod_{j=1}^{p} \mathcal{X}_j$. A convenient approach to constructing the space of multidimensional splines is by taking the tensor product of one-dimensional spaces of polynomial splines; a tensor product of vector spaces is a way to generate a new vector space using elements formed from combining the basis elements of the individual spaces. This is the main idea of smoothing splines ANOVA and SS-ANOVA (see Wahba (1990), Gu (2002)).

Similar to the standard analysis of variance, in the SS-ANOVA framework, a multidimensional function $f(\boldsymbol{x}) = f(x_1, ..., x_p)$ is decomposed as a sum of main effects, two-way interaction terms, and higher-order interaction terms:

$$f(\boldsymbol{x}) = b_0 + \sum_{j=1}^{p} f_j(x_j) + \sum_{j<l} f_{jl}(x_j, x_l) + \cdots. \tag{10.3.33}$$

Certain side conditions can be imposed on the f_js to guarantee the uniqueness of this ANOVA-like decomposition. The modeling is accomplished by saying that each component function in (10.3.33) is in an RKHS and that the RKHSs for bivariate and higher-order terms are derived by taking tensor products of the RKHS of the univariate functions. That is, each f_j varies over an RKHS $\mathcal{H}^{(j)}$ of univariate functions associated to it, and each $\mathcal{H}^{(j)}$ has its own kernel. Then, the bivariate function f_{jl} of $x_j$and x_l for $j < l$ in (10.3.33) is assumed to be in the tensor product space $\mathcal{H}^{(j)} \otimes \mathcal{H}^{(l)}$. The trivariate and later terms are similar, depending on tensor products of more and more spaces. Thus, each component function in (10.3.33) is in a subspace of the form $\otimes_{j=1}^{p} \mathcal{H}^{(j)}$ with its own reproducing kernel derived from the reproducing kernels of the individual spaces.

Since tensor products of different numbers of spaces are independent, the overall pth order function space $\mathscr{H}$ of which $f(\boldsymbol{x})$ is an element can be written as a direct sum,

$$\mathscr{H} = \otimes_{j=1}^{p} \mathscr{H}^{(j)} = [1] \oplus \sum_{j=1}^{p} \mathscr{H}_1^{(j)} \oplus \sum_{j<l} [\mathscr{H}_1^{(j)} \otimes \mathscr{H}_1^{(l)}] \oplus \cdots,$$

where $\mathscr{H}^{(j)} = [1] \oplus \mathscr{H}_1^{(j)}$ for each j, $[1]$ is the constant subspace, and $\mathscr{H}_1^{(j)}$ is its complement subspace. It can be shown that $\mathscr{H}$ is an RKHS with a reproducing kernel derived from the reproducing kernels of its constituent spaces. An example of this $\mathscr{H}^{(j)}$ is the second-order Sobolev space $W_2[0,1]$,

$$W_2[0,1] = \{g : g(t), g'(t) \text{ absolutely continuous, } g''(t) \in \mathscr{L}_2[0,1]\},$$

treated in Exercise 10.14.

Typically only low-order interactions in (10.3.33) are considered. The simplest is the additive model

$$f(\boldsymbol{x}) = b_0 + \sum_{j=1}^{p} f_j(x_j). \tag{10.3.34}$$

In this case, the selection of functional components is equivalent to variable selection and (10.3.33) and (10.3.34) reduce to the setting of Chapter 3. That is, the f to be estimated is in an RKHS $\mathscr{H}$ equipped with the norm $\|\cdot\|_{\mathscr{H}}$, and the decomposition (10.3.34) corresponds to the function space $\mathscr{H}$ in SS-ANOVA, which can be written

$$\mathscr{H} = [1] \oplus \bigoplus_{j=1}^{p} \mathscr{H}_j. \tag{10.3.35}$$

In (10.3.35), $\mathscr{H}_1$, ..., $\mathscr{H}_p$ are p orthogonal subspaces of $\mathscr{H}$ for the main effects f_js. A traditional smoothing spline type method finds $f \in \mathscr{H}$ to minimize

$$\frac{1}{n}\sum_{i=1}^{n} (y_i - f(\mathbf{x}_i))^2 + \lambda \sum_{j=1}^{p} \theta_j^{-1} \|P^j f\|_{\mathscr{H}}^2, \tag{10.3.36}$$

where $P^j f$ is the orthogonal projection of f onto $\mathscr{H}_j$ and the $\theta_j \geq 0$ and λ are multiple tuning parameters. In more complex SS-ANOVA models not explicitly studied here, model selection requires the selection of interaction terms in the SS-ANOVA decomposition.

As seen earlier, shrinkage methods in the linear model context devolve to coefficient shrinkage because X_j is regarded as unimportant when $\hat{\beta}_j = 0$. For nonparametric models, the parallel property is that a component f_j is excluded from the model when it is estimated by the zero function; i.e., $\hat{f}_j(X_j) \equiv 0$, a much stronger condition. Consequently, shrinkage in general nonlinear models is much harder than in linear models. Nevertheless, two classes of methods have been developed to implement function shrinkage for (10.3.33): basis pursuit (BP) methods and direct function shrinkage methods. The idea of BP is to make the nonparametric problem parametric: Represent each component function in terms of a basis and then shrink the coefficients of basis elements toward zero. Direct methods, by contrast, define a soft-thresholding operator to shrink the actual function components to the zero function. For ease of presentation, the focus here will be on additive models as in (10.3.34), thereby assuming there are no interaction effects between the covariates used. The methods presented here extend to the general case (10.3.33); i.e., to functions f that include nonunivariate terms in (10.3.33).

Basis Pursuit

BP originated as a principle for decomposing a signal into an optimal superposition of prechosen elements, where the sense of optimality is that the superposition has the smallest possible l_1 norm of coefficients among all such decompositions; see Chen et al. (1998) for BP in a wavelet regression context. Given the finite representation of smoothing splines, Zhang et al. (2004) use enough basis functions to span the model space and estimate the target function in that space. In particular, they assume for each j that each component function can be approximated by a linear combination of N basis functions $B_{jl}, l = 1,\ldots,N$; i.e.,

$$f_j(x_j) = \sum_{l=1}^{N} c_{jl} B_{jl}(x_j).$$

The BP solution is

$$\hat{f}^{bp} = \arg\min_{f} \frac{1}{n} \sum_{i=1}^{n} (y_i - f(\boldsymbol{x}_i))^2 + \sum_{j=1}^{p} \lambda_j \sum_{l=1}^{N} |c_{jl}|, \tag{10.3.37}$$

where $f(\boldsymbol{x}) = b_0 + \sum_{j=1}^{p} f_j(x_j) = b_0 + \sum_{j=1}^{p} \sum_{l=1}^{N} c_{jl} B_{jl}(x_j)$ and the λ_js are the decay parameters, which are allowed to take distinct values for different components.

It can be seen that (10.3.37) is a direct generalization of the LASSO penalty to the context of nonparametric models. Moreover, the l_1 penalty can still produce basis coefficients that are exactly zero. Therefore, BP gives a sparse representation for each component. Computing BP solutions can be done through quadratic programming or LARS. Instead of the L_1 penalty, any other penalty, such as SCAD, group LASSO, or adaptive LASSO, can be imposed. Although not shown here, if a representation of f that includes bivariate terms as in (10.3.33) is adopted, then an extra penalty term appears. The penalty term ensures the norm of the coefficients of the basis elements for each copy of $\mathscr{H}_1^{(j)} \otimes \mathscr{H}_1^{(l)}$ is small enough.

A Direct Method: COSSO

The component selection and smoothing operator (COSSO; see Lin and Zhang (2006)) is a direct shrinkage method for variable selection in nonparametric regression. Specifically, COSSO solves

$$\hat{f}^{cosso} = \arg\min_{f \in \mathscr{H}} \sum_{i=1}^{n} [y_i - f(\boldsymbol{x}_i)]^2 + \lambda \sum_{j=1}^{p} \|P^j f\|_{\mathscr{H}}, \tag{10.3.38}$$

where $\|\cdot\|_{\mathscr{H}}$ is the norm defined in the RKHS $\mathscr{H}$ and $\lambda \geq 0$ is the smoothing parameter. The penalty term $\sum_{j=1}^{p} \|P_j f\|_{\mathscr{H}}$ in (10.3.38) is a sum of reproducing kernel Hilbert space norms, not the squared norm penalty used in traditional smoothing splines in (10.3.36). Note that the objective function in (10.3.38) is continuous and convex in $f \in \mathscr{H}$, so its minimizer exists. In addition, the finite representer theorem holds for the estimator $\hat{f}$. This is an important result for computational purposes since it ensures $\hat{f}$ can be characterized by a finite number of parameters even for $\mathscr{H}$ infinite-dimensional.

As shown in Lin and Zhang (2006), COSSO applies soft thresholding to the function components to achieve sparse function estimation in SS-ANOVA models. Instead of solving (10.3.38) directly, consider the equivalent formulation

$$\min_{f \in \mathscr{H}, \boldsymbol{\theta}} \sum_{i=1}^{n} [y_i - f(\boldsymbol{x}_i)]^2 + \lambda_0 \sum_{j=1}^{p} \theta_j^{-1} \|P^j f\|_{\mathscr{H}}^2 + \lambda_1 \sum_{j=1}^{p} \theta_j, \text{ subject to } \theta_j \geq 0 \, \forall j, \tag{10.3.39}$$

where $\lambda_0 \geq 0$ is constant and λ_1 is the smoothing parameter. The form of (10.3.39) is similar to the common smoothing spline (10.3.36) with multiple smoothing parameters, except that there is an additional penalty on the θ_js, although there is still only one smoothing parameter λ. This means that the θs are part of the estimate rather than free smoothing parameters. The additional penalty on the θ_js means some θ_js can be zero, thereby giving zero function components in the COSSO estimate.

For additive models, Lin and Zhang (2006) show that COSSO converges at rate $\mathscr{O} n^{-m/(2m+1)}$, where m is the order of function smoothness. For the case of a tensor product design with periodic functions, COSSO selects the correct model structure with probability tending to one. However, Storlie et al. (2007) observe that COSSO tends to oversmooth nonzero functional components in the process of setting unimportant function components to zero. This parallels the way LASSO can overshrink nonzero coefficients in when doing variable selection. COSSO is not oracle for nonparametric model selection although a variant of it, presented next, can be.

Adaptive COSSO

Recently, Storlie et al. (2007) extended COSSO by including adaptive weights in the COSSO penalty, ACOSSO. This allows for more flexibility in estimating important functional components. It also penalizes unimportant functional components more heavily because a rescaled norm is used to smooth each of the components. Formally, the ACOSSO estimator for $f \in \mathscr{F}$ is

$$\hat{f}^{acosso} = \arg\min \sum_{i=1}^{n} [y_i - f(\boldsymbol{x}_i)]^2 + \lambda \sum_{j=1}^{p} w_j \|f_j\|_{\mathscr{H}}, \tag{10.3.40}$$

where the $0 \leq w_j \leq \infty$ are prechosen weights and $\lambda > 0$ is a smoothing parameter.

As with COSSO, ACOSSO has an equivalent formulation that is better for computation. Storlie et al. (2007) show that the optimization in (10.3.40) is equivalent to the problem of finding $\boldsymbol{\theta} = (\theta_1, \ldots, \theta_p)^{\mathsf{T}}$ and $f \in \mathscr{H}$ to minimize

$$\frac{1}{n}\sum_{i=1}^{n}[y_i - f(\boldsymbol{x}_i)]^2 + \lambda_0 \sum_{j=1}^{p} \theta_j^{-1} w_j^{2-\vartheta} \|P^j f\|_{\mathscr{H}}^2 + \lambda_1 \sum_{j=1}^{p} w_j^{\vartheta} \theta_j \text{ subject to } \theta_j \geq 0 \, \forall j, \tag{10.3.41}$$

where $0 \leq \vartheta \leq 2$, $\lambda_0 \geq 0$ is a fixed constant, and $\lambda_1 \geq 0$ is a smoothing parameter. In (10.3.41), the $0 \leq w_j < \infty$ are usually chosen to depend on an initial f, say $\tilde{f}$.

Ideally the w_js should be chosen so that more important components have a smaller penalty relative to less important components. Unfortunately, in contrast to the linear model case, in nonparametric settings there is no single coefficient, or set of coefficients, to measure the importance of a component f_j infallibly. However, the L_2 norm of the projection of a reasonable estimator may be adequate in many cases. That is, if $\tilde{f}$ is a reasonable initial estimator, then $\|P^j \tilde{f}\|_{L_2}^2 = (\int_{\mathscr{X}_j} (P^j \tilde{f}(x))^2 dx)^{1/2}$ should reflect the importance of the component f_j. If so, then the weights might be chosen as

$$w_j = \|P^j \tilde{f}\|_{L_2}^{-\gamma}, \quad j = 1, \ldots, p; \tag{10.3.42}$$

setting $\gamma = 1$ appears to work well for many purposes. Storlie et al. (2007) suggested choosing the traditional smoothing spline estimate as $\tilde{f}$ and showed that using the weights in (10.3.42) had good theoretical properties. Indeed, if $\|f\|_n^2 = 1/n \sum_{i=1}^n f^2(\boldsymbol{x}_i)$ is the squared norm of f at the design points, then, under some regularity conditions, the ACOSSO estimator is asymptotically nonparametric oracle; i.e., $\|\hat{f} - f\|_n \to 0$ at the optimal rate and $P^j \hat{f} \equiv 0$ with probability tending to one for all the unimportant components f_j.

10.3.4.4 Shrinkage Methods for Support Vector Machines

Recall that SVMs were studied in detail in Chapter 5; they were seen to give sparse solutions in the sense of the number of terms because the classifiers they gave only depended on the support vectors, typically far fewer than n. SVMs are used extensively with large p, small n data because of this sparsity and good performance. However, SVMs are not sparse in the number of variables. This means that, even when there are relatively few terms, the prediction accuracy of SVMs may suffer from the presence of redundant variables and that the SVM solution often will not provide any insight on the effects of individual variables. Thus, it is desirable to achieve sparsity over p as well as over the number of terms.

In more precise form, this abstract goal can be described as follows. Let $\{\boldsymbol{x}_1, \cdots, \boldsymbol{x}_n\}$ be n points in p-dimensional space. Then, an SVM is sparse over the points in terms of selecting support vectors. Variable selection, by contrast, is sparse over the coordinates of the vectors. If only the first two vectors $\boldsymbol{x}_1$ and $\boldsymbol{x}_2$ are support vectors and only the first three variables are selected, then the classification boundary depends on $2 \times 3 = 6$ coordinates rather than np, giving a sort of double sparsity.

Many algorithms have been proposed for selecting variables prior to classification. Ranking methods are the most popular; these rank individual variables according to some predetermined criteria and then use top-ranked variables for classification. Two commonly-used ranking criteria are correlation coefficients and hypothesis testing statistics such as two-sample t-tests. Although useful in practice, these do selection only over individual variables, ignoring correlation information. In addition, there are kernel scaling methods; see Weston et al. (2000). Here, the focus will be on shrinkage to ensure sparsity; the core idea is to replace the L_2 norm $||\boldsymbol{w}||^2$ with a shrinkage penalty on $\boldsymbol{w}$.

Variable Selection for Binary SVMs

The L_1 SVM solves

$$(\hat{b}^{l1}, \widehat{\boldsymbol{w}}^{l1}) = \arg\min_{b,\boldsymbol{w}} \frac{1}{n} \sum_{i=1}^{n} [1 - y_i(b + \boldsymbol{w} \cdot \boldsymbol{x}_i)]_+ + \lambda \sum_{j=1}^{p} |w_j|. \tag{10.3.43}$$

Different from the standard SVM that solves a linearly constrained quadratic optimization, the L_1 SVM in (10.3.43) leads to a linear programming problem,

$$\begin{aligned} \min_{\boldsymbol{\xi},\boldsymbol{u},\boldsymbol{v}} \quad & \mathbf{1}^{\mathsf{T}}\boldsymbol{\xi} + \lambda \mathbf{1}^{\mathsf{T}}(\boldsymbol{u} + \boldsymbol{v}) \\ \text{s.t. } & Y^{\mathsf{T}}(b\mathbf{1} + X\boldsymbol{w}) + \boldsymbol{\xi} \geq \mathbf{1}, \\ & \boldsymbol{u}, \boldsymbol{v}, \boldsymbol{\xi} \geq \mathbf{0}, \end{aligned} \tag{10.3.44}$$

where $Y = \text{diag}[y_1, ..., y_n]$, $X_{n\times p}$ is the design matrix with the ith row being the input vector $\boldsymbol{x}_i$. Zhu et al. (2003) studied the solution properties of the L_1 SVM by reformulating (10.3.44) into a Lagrange version,

$$\begin{aligned} \min_{b,\boldsymbol{w}} \quad & \tfrac{1}{n}\textstyle\sum_{i=1}^{n}[1 - y_i(b + \boldsymbol{w} \cdot \boldsymbol{x}_i)]_+ \\ s.t. \quad & ||\boldsymbol{w}|| = |w_1| + \cdots + |w_p| \leq s, \end{aligned} \tag{10.3.45}$$

where $s > 0$ is the tuning parameter, which has the same role as λ. They showed the solution of (10.3.45), say $\boldsymbol{w}(s)$, is piecewise linear in s and gave an efficient algorithm to compute the whole solution path. This facilitates adaptive selection of s. Fung and Mangasarian (2004) developed a fast Newton algorithm to solve the dual problem of (10.3.44), which scales up to large p, including $p >> n$.

The SCAD SVM (Zhang et al., 2006) solves

$$(\hat{b}^{scad}, \widehat{\boldsymbol{w}}^{scad}) = \arg\min_{b,\boldsymbol{w}} \frac{1}{n} \sum_{i=1}^{n} [1 - y_i(b + \boldsymbol{w} \cdot \boldsymbol{x}_i)]_+ + \sum_{j=1}^{p} p_\lambda(|w_j|), \tag{10.3.46}$$

where p_λ is the nonconvex penalty defined in (10.3.19). Compared with the L_1 SVM, the SCAD SVM often gives a more compact classifier and achieves higher classification accuracy. A sequential quadratic programming algorithm can be used to optimize in (10.3.46) by solving a series of linear equation systems.

Variable Selection for Multiclass SVM

The linear multicategory SVM (MSVM) estimates K discriminating functions

$$f_k(\boldsymbol{x}) = b_k + \sum_{j=1}^{p} w_{kj}x_j, \quad k = 1,\ldots,K,$$

in which each f_j is associated with one class, so that any $\boldsymbol{x}_{new}$ is assigned to the class

$$\hat{y}_{new} = \arg\max_k f_k(\boldsymbol{x}_{new}).$$

One version of MSVM finds f_k for $k = 1,\ldots,K$ by solving

$$\min_{f_k:k=1,\ldots,K} \quad \frac{1}{n}\sum_{i=1}^{n}\sum_{k=1}^{K} I(y_i \neq k)[f_k(\boldsymbol{x}_i)+1]_+ + \lambda \sum_{k=1}^{K}\sum_{j=1}^{p} w_{kj}^2 \tag{10.3.47}$$

under the sum-to-zero constraint, $f_1(x)+\cdots+f_K(x) = 0$ (see Lee et al. (2004)); however, other variants are possible. To force sparsity on variable selection, Wang and Shen (2007) impose an L_1 penalty and solve

$$(\widehat{\boldsymbol{b}}^{l1}, \widehat{\boldsymbol{w}}^{l1}) = \arg\min_{\boldsymbol{b},\boldsymbol{w}} \quad \frac{1}{n}\sum_{i=1}^{n}\sum_{k=1}^{K} I(y_i \neq k)[b_k + \mathbf{w}_k^{\mathsf{T}}\boldsymbol{x}_i + 1]_+ + \lambda \sum_{k=1}^{K}\sum_{j=1}^{p} |w_{kj}| \tag{10.3.48}$$

under the sum-to-zero constraint (see Section 5.4.10).

The problem with (10.3.48) is that the L_1 penalty treats all the coefficients equally, no matter whether they correspond to the same or different variables. Intuitively, if a variable is not important, all coefficients associated with it should be shrunk to zeros simultaneously. To correct this, Zhang et al. (2008) proposed penalizing the supnorm of all the coefficients associated with a given variable. For each X_j, let the collection of coefficients associated with it be $\mathbf{w}_{(j)} = (w_{1j},\cdots,w_{Kj})^{\mathsf{T}}$, with supnorm $\|\mathbf{w}_{(j)}\|_\infty = \max_{k=1,\cdots,K}|w_{kj}|$. This means the importance of X_j is measured by its largest absolute coefficient.

The supnorm MSVM solves

$$(\widehat{\boldsymbol{b}}^{sup}, \widehat{\boldsymbol{w}}^{sup}) = \arg\min_{\boldsymbol{w},\boldsymbol{b}} \quad \sum_{i=1}^{n}\sum_{k=1}^{K} I(y_i \neq k)[f_k(\boldsymbol{x}_i)+1]_+ + \lambda \sum_{j=1}^{p} \|\mathbf{w}_{(j)}\|_\infty,$$

$$\text{subject to} \quad \sum_{l=1}^{K} b_k = 0, \ \sum_{k=1}^{K} \mathbf{w}_k = \mathbf{0}. \tag{10.3.49}$$

For three-class problems, the supnorm MSVM is equivalent to the L_1 MSVM after adjusting the tuning parameters. Empirical studies showed that the Supnorm MSVM tends to achieve a higher degree of model parsimony than the L_1 MSVM without compromising the classification accuracy.

Note that in (10.3.49) the same tuning parameter λ is used for all the terms $\|\mathbf{w}_{(j)}\|_\infty$ in the penalty. To make the sparsity more adaptive (i.e., like ACOSSO as compared with COSSO), different variables can be penalized according to their relative importance. Ideally, larger penalties should be imposed on redundant variables to eliminate them, while smaller penalties should be used on important variables to retain them in the fitted classifier. The adaptive supnorm MSVM achieves

$$(\widehat{\boldsymbol{b}}^{asup}, \widehat{\boldsymbol{w}}^{asup}) = \arg\min_{\boldsymbol{w},\boldsymbol{b}} \sum_{i=1}^{n}\sum_{k=1}^{K} I(y_i \neq k)[f_k(\boldsymbol{x}_i)+1]_+ + \lambda \sum_{j=1}^{p} \tau_j \|\mathbf{w}_{(j)}\|_\infty,$$

$$\text{subject to} \quad \sum_{k=1}^{K} b_k = 0, \; \sum_{k=1}^{K} \mathbf{w}_k = \mathbf{0}, \tag{10.3.50}$$

where the weights $\tau_j \geq 0$ are adaptively chosen. Similar to ACOSSO, let $(\tilde{\mathbf{w}}_1, \cdots, \tilde{\mathbf{w}}_d)$ be from the MSVM solution to (10.3.47). A natural choice is

$$\tau_j = \frac{1}{\|\tilde{\mathbf{w}}_{(j)}\|_\infty}, \quad j = 1, \cdots, p,$$

which often performs well in numerical examples. The case where $\|\tilde{\mathbf{w}}_{(j)}\|_\infty = 0$ implies an infinite penalty is imposed on w_{kj}s in which case all the coefficients $\hat{w}_{kj}, k = 1, \cdots, K$ associated with X_j are taken as zero.

10.3.5 Cautionary Notes

As successful and widespread as shrinkage methods have been, three sorts of criticisms of them remain largely unanswered. First, as a class, they seem to be unstable in the sense that they rest delicately on the correct choice of λ – a problem that can be even more serious for ALASSO and ACOSSO, which have p decay parameters. Poor choice of decay can lead to bias and increased variance. Separate from this, sensitivity to dependence structures in the explanatory variables has been reported. This is an analog to collinearity but seems to be a more serious problem for shrinkage methods than conventional regression.

A related concern is that although shrinkage methods are intended for $p < n$, as suggested by the definition of oracle, which is asymptotic in n, there is an irresistible temptation to use them even when $p > n$. The hope is that the good performance of shrinkage for $p < n$ will extend to the more complicated scenario. While this hope is reasonable, the existing theory, by and large, does not justify it as yet. In particular, estimating a parameter by setting it to zero and not considering an SE for it is as unstatistical as ignoring the variability in model selection but may be worse in effect. However, the Bayesian interpretation of shrinkage, developed at the end of the next section, may provide a satisfactory justification.

Second, the actual choice of distances in the objective functions is arbitrary because so many of them give the oracle property. For instance, using L^1 error in both the risk and the penalty (see Wang et al. (2007)) is also oracle. Indeed, the class of objective functions for which the oracle property (or near-minimaxity) holds remains to be characterized. So far, the oracle property holds for some adaptive penalties on squared error, ALASSO and ACOSSO, for some bounded penalties like SCAD, and for the fully L^1 case. However, the "oracle" class is much more general than these four cases. One way to proceed is to survey the proofs of the existing theorems and identify the class of objective functions to which their core arguments can be applied. In practice, a partial solution to the arbitrariness of the objective function is to use several penalized methods and combine them.

Third, separate from these two concerns is the argument advanced in a series of papers culminating in Leeb and Potscher (2008). They argue that shrinkage estimators have counterintuitive asymptotic behavior and that the oracle property is a consequence of the sparsity of the estimator. That is, any estimator satisfying a sparsity property, such as being oracle, has maximal risk converging to the supremum of the loss function, even when the loss is unbounded. They further argue that when the penalty is bounded, as with SCAD, the performance of shrinkage estimators can be poor. Overall, they liken the effectiveness of shrinkage methods to a phenomenon such as superefficiency known to hold only for a set of parameters with Lebesgue measure zero.

While this list of criticisms does not invalidate shrinkage methods, it is serious and may motivate further elucidation of when shrinkage methods work well and why.

10.4 Bayes Variable Selection

Bayesian techniques for variable selection are a useful way to evaluate models in a model space $\mathcal{M} = \{M_1, \ldots, M_K\}$ because a notion of model uncertainty (conditional on the model list) follows immediately from the posterior on $\mathcal{M}$. Indeed, as a generality, it is more natural in the Bayes formulation to examine whole models rather than individual terms. The strictest of axiomatic Bayesians disapprove of comparing individual terms across different models on the grounds that the same term in two different models has two different interpretations. Sometimes called the "fallacy of Greek letters", this follows from the containment principle that inference is only legitimate if done on a single measure space. While this restriction does afford insight, it also reveals a limitation of orthodox Bayes methods in that reasonable questions cannot be answered.

Before turning to a detailed discussion of Bayesian model or variable selection, it is important to note that there are a variety of settings in which Bayesian variable or model selection is done. This is important because the status of the priors and the models changes from setting to setting. In the subjective Bayesian terminology of Bernardo and Smith (1994), the first scenario, called $\mathcal{M}$-closed, corresponds to the knowledge that one of models in $\mathcal{M}$ is true, without knowing which. That is, the real-world data generator of the data is in $\mathcal{M}$. The second scenario, called $\mathcal{M}$-complete, assumes that

$\mathcal{M}$ contains a range of models available for comparison to be evaluated in relation to the experimenter's actual belief model M_{true}, which is not necessarily known. It is understood that $M_{true} \notin \mathcal{M}$, perhaps because it is more complex than anything in $\mathcal{M}$. Intuitively, this is the case where $\mathcal{M}$ only provides really good approximations to the data generator.

The third scenario, called $\mathcal{M}$-open, also assumes $\mathcal{M}$ is only a range of models available for comparison. However, in the $\mathcal{M}$-open case, there need not be any meaningful model believed true; in this case, the status of the prior changes because it no longer represents a degree of belief in a model, only its utility for modeling a response. Also, the models are no longer descriptions of reality so much as actions one might use to predict outcomes or estimate parameters. Many model selection procedures assume the $\mathcal{M}$-closed perspective; however, the other two perspectives are usually more realistic.

So, recall that the input vector is $\boldsymbol{X} = (X_1, \cdots, X_p)$ and the response is Y. Then, the natural choice for $\mathcal{M}$ in a linear model context is the collection of all possible subsets of $X_1, \cdots, X_p$. Outside nonparametric contexts, it is necessary to assume a specific parametric form for the density of the response vector $\boldsymbol{y} = (y_1, \cdots, y_n)$. Usually, $\boldsymbol{Y}$ is assumed drawn from a multivariate normal distribution $\boldsymbol{Y} \sim N_n(X\boldsymbol{\beta}, \sigma^2 I_n)$, where I_n is the identity matrix of size n and the parameters taken together as $\boldsymbol{\theta} = (\boldsymbol{\beta}, \sigma)$.

To express the uncertainty of the models in $\mathcal{M}$ and specify the notation for the Bayesian hierarchical formulation, let γ_j be a latent variable for each predictor X_j, taking 1 or 0 to indicate whether X_j is in the model or not. Now, each model $M \in \mathcal{M}$ is indexed by a binary vector

$$\boldsymbol{\gamma} = (\gamma_1, \cdots, \gamma_p), \quad \text{where } \gamma_j = 0 \text{ or } 1, \quad \text{for } j = 1, \ldots, p.$$

Correspondingly, let $\boldsymbol{X}_{\boldsymbol{\gamma}}$ denote the design matrix consisting of only variables with $\gamma_j = 1$ and let $\boldsymbol{\beta}_{\boldsymbol{\gamma}}$ be the regression coefficients under the design matrix $\boldsymbol{X}_{\boldsymbol{\gamma}}$. Define $|\boldsymbol{\gamma}| = \sum_{j=1}^{p} \gamma_j$. So, $\boldsymbol{X}_{\boldsymbol{\gamma}}$ is of dimension $n \times |\boldsymbol{\gamma}|$ and $\boldsymbol{\beta}_{\boldsymbol{\gamma}}$ is of length $|\boldsymbol{\gamma}|$.

A Bayesian hierarchical model formulation has three main components:

- ☐ a prior distribution, $w(\boldsymbol{\gamma})$, for the candidate models $\boldsymbol{\gamma}$,
- ☐ a prior density, $w(\boldsymbol{\theta}_{\boldsymbol{\gamma}}|\boldsymbol{\gamma})$, for the parameter $\boldsymbol{\theta}_{\boldsymbol{\gamma}}$ associated with the model $\boldsymbol{\gamma}$,
- ☐ a data-generating mechanism conditional on $(\boldsymbol{\gamma}, \boldsymbol{\theta}_{\boldsymbol{\gamma}})$, $P(\boldsymbol{y}|\boldsymbol{\theta}_{\boldsymbol{\gamma}}, \boldsymbol{\gamma})$.

Once these are specified, obtaining the model posterior probabilities is mathematically well defined and can be used to identify the most promising models. Note that $w(\boldsymbol{\gamma})$ is a density with respect to counting measure not Lebesgue measure; this is more convenient than distinguishing between the discrete $W(\cdot)$ for $\boldsymbol{\gamma}$ and the continuous $w(\cdot|\boldsymbol{\gamma})$ for $\boldsymbol{\theta}$.

To see how this works in practice, assume that for any model $\boldsymbol{\gamma}_k$ and its associated parameter $\boldsymbol{\theta}_k$ the response vector has density $f_k(\boldsymbol{y}|\boldsymbol{\theta}_k, \boldsymbol{\gamma}_k)$. For linear regression models,

$$\boldsymbol{Y}|(\boldsymbol{\beta}, \sigma, \boldsymbol{\gamma}_k) \sim N_n(\boldsymbol{X}_{\boldsymbol{\gamma}}\boldsymbol{\beta}_{\boldsymbol{\gamma}}, \sigma^2 I_n).$$

The marginal likelihood of the data under $\boldsymbol{\gamma}_k$ can be obtained by integrating with respect to the prior distribution for model-specific parameters $\boldsymbol{\theta}_k = (\boldsymbol{\beta}_{\gamma_k}, \sigma^2)$,

$$p(\mathbf{y}|\boldsymbol{\gamma}_k) = \int f_k(\mathbf{y}|\boldsymbol{\theta}_k, \boldsymbol{\gamma}_k) w(\boldsymbol{\theta}_k|\boldsymbol{\gamma}_k) d\boldsymbol{\theta}_k.$$

The posterior probability for the model indexed by $\boldsymbol{\gamma}_k$ is

$$\begin{aligned} \mathbb{P}(\boldsymbol{\gamma}_k|\mathbf{y}) &= \frac{w(\boldsymbol{\gamma}_k)p(\mathbf{y}|\boldsymbol{\gamma}_k)}{m(\mathbf{y})} = \frac{w(\boldsymbol{\gamma}_k)p(\mathbf{y}|\boldsymbol{\gamma}_k)}{\sum_{l=1}^K w(\boldsymbol{\gamma}_l)p(\mathbf{y}|\boldsymbol{\gamma}_l)} \\ &= \frac{w(\boldsymbol{\gamma}_k)\int f_k(\mathbf{y}|\boldsymbol{\theta}_k, \boldsymbol{\gamma}_k) w(\boldsymbol{\theta}_k|\boldsymbol{\gamma}_k) d\boldsymbol{\theta}_k}{\sum_{l=1}^K w(\boldsymbol{\gamma}_l)\int f_l(\mathbf{y}|\boldsymbol{\theta}_l, \boldsymbol{\gamma}_l) w(\boldsymbol{\theta}_l|\boldsymbol{\gamma}_l) d\boldsymbol{\theta}_l}, \end{aligned} \quad (10.4.1)$$

where $m(\mathbf{y})$ is the marginal density of $\mathbf{y}$. The posterior distribution (10.4.1) is the fundamental quantity in Bayesian model selection since it summarizes all the relevant information in data about the model and provides the post-data representation of model uncertainty.

A common Bayesian procedure is to choose the model with the largest $\mathbb{P}(\boldsymbol{\gamma}_k|\mathbf{y})$. This is the Bayes action under a generalized 0-1 loss and is essentially the BIC. Generalizing this slightly, one can identify a set of models with high posterior probability and use the average of these models for future prediction. Using all the models would correspond to Bayes model averaging (which is optimal under squared error loss). In any case, search algorithms are needed to identify "promising" regions in $\mathscr{M}$.

Before delving into Bayes variable, or model, selection, some facts about the Bayes methods must be recalled. First, Bayes methods are consistent: If the true model is among candidate models, has positive prior probability, and enough data are observed, then Bayesian methods uncover the true model under very mild conditions. Even when the true model is not in the support of the prior, Berk (1966) and Dmochowski (1996) show that Bayesian model selection will asymptotically choose the model that is closest to the true model in terms of Kullback-Leibler divergence. Second, as will be seen more formally at the end of this section, Bayes model selection procedures are automatic Ockham's razors (see Jefferys and Berger (1992)) typically penalizing complex models and favoring simple models that provide comparable fits.

Though the concept of Bayes model selection is conceptually straightforward, there are many challenging issues in practical implementation. Since the model list is assumed given, arguably the two biggest challenges are (i) choosing proper priors for models and parameters and (ii) exploiting posterior information. In the hierarchical framework, two priors must be specified: the prior $w(\boldsymbol{\gamma})$ over the $\boldsymbol{\gamma}_k$s and the priors $w(\boldsymbol{\theta}_k|\boldsymbol{\gamma}_k)$ within each model $\boldsymbol{\gamma}_k$. From (10.4.1), it is easy to see that $\mathbb{P}(\boldsymbol{\gamma}_k|\mathbf{y})$ can be small for a good model if $w(\boldsymbol{\gamma}_j, \boldsymbol{\theta}_j)$ is chosen unwisely. For example, if too much mass is put at the wrong model, or if the prior mass is spread out too much, or if the prior probability is divided too finely among a large collection of broadly adequate models, then very little weight may be on the region of the model space where the residual sum of squares is smallest. A separate problem is that when the number of models under consideration is enormous, calculating all the Bayes factors and posterior probabilities can be very time-consuming. Although it is not the focus here, it is important to note that recent developments in numerical and Markov chain Monte Carlo (MCMC) methods have led to many algorithms to identify high-probability regions in the model space efficiently.

10.4.1 Prior Specification

The first step in a Bayesian approach is to specify the model fully, the prior structure in particular. At their root, Bayes procedures are hierarchical and Bayes variable selection represents the data generating mechanism as a three-stage hierarchical mixture. That is, the data $\mathbf{y}$ were generated as follows:

- ☐ Generate the model $\boldsymbol{\gamma}_k$ from the model prior distribution $w(\boldsymbol{\gamma})$.
- ☐ Generate the parameter $\boldsymbol{\theta}_k$ from the parameter prior distribution $w(\boldsymbol{\theta}_k|\boldsymbol{\gamma}_k)$.
- ☐ Generate the data $\mathbf{y}$ from $f_k(\mathbf{y}|\boldsymbol{\theta}_k, \boldsymbol{\gamma}_k)$.

It is seen from (10.4.1) that the model posterior probabilities depend heavily on both priors. One task in prior specification is to find priors that are not overly influential as a way to ensure the posterior distribution on the model list puts a relatively high probability on the underlying true model and neighborhoods of it.

There has been a long debate in the Bayes community as to the roles of subjective and objective priors (see Casella and Moreno (2006), for instance). Subjective Bayes analysis is psychologically attractive (to some) but intellectually dishonest. The purely subjective approach uses priors chosen to formalize the statistician's pre-experimental feelings and preferences about the unknowns but does not necessarily evaluate whether these feelings or preferences mimic any reality. Sincere implementation of this approach for model-selection problems when the model list is very large is nearly impossible because providing careful subjective prior specification for all the parameters in all the models requires an exquisite sensitivity to small differences between similar models. In particular, subjective elicitation of priors for model-specific coefficients is not recommended, particularly in high-dimensional model spaces, because experimenters tend to be overoptimistic as to the efficacy of their treatments. Consequently, objective priors are more commonly used in model selection contexts, and the role of subjective priors is exploratory, not inferential. Subjective priors may be used for inference, of course, if they have been validated or tested in some way by data.

A related prior selection issue is the role of conjugate priors. Conjugate priors have several advantages: (i) If p is moderate (less than 20), they allow exhaustive posterior evaluation; (ii) if p is large, they allow analytic computations of relative posterior probabilities and estimates of total visited probability; and (iii) they allow more efficient MCMC posterior exploration. However, the class of conjugate priors is typically too small to permit good prior selection. Unless further arguments can be made to validate conjugate priors, they are best regarded as a particularly mathematically tractable collection of subjective priors. In addition, apart from any role in inference directly, conjugate priors can be a reasonable class for evaluating prior robustness.

10.4.1.1 Priors on the Model Space

Recall that the total number of candidate models in $\mathcal{M}$ is $K = 2^p$. If p is large, $\mathcal{M}$ is high-dimensional, making it hard to specify $w(\boldsymbol{\gamma})$. In practice, to reduce the complexity, the independence assumption is often invoked (i.e., the presence or absence of one variable is independent of the presence or absence of other variables). When the covariates are highly correlated, independent priors do not provide the proper "dilution" of posterior mass over similar models. Priors with better dilution properties are also presented below.

Independence Priors

The simplification used in this class of priors is that each X_j enters the model independently of the other coefficients and does so with probability

$$P(\gamma_j = 1) = 1 - P(\gamma_j = 0) = w_j,$$

in which the w_js are p hyperparameters. This leads to the prior on $\mathcal{M}$,

$$w(\boldsymbol{\gamma}) = \prod_{j=1}^{p} w_j^{\gamma_j}(1 - w_j)^{1-\gamma_j}. \tag{10.4.2}$$

Chipman et al. (2001) note that this prior is easy to specify, substantially reduces computational requirements, and often yields sensible results; see Clyde et al. (1996) and George and McCulloch (1993, 1997).

Choices of w_js are flexible and may be problem dependent. If some predictors are not favored, say due to high cost or low interest, the corresponding w_js can be made smaller. The Principle of Insufficient Reason suggests $w_1 = \cdots = w_p = w$ be chosen giving the model prior

$$w(\boldsymbol{\gamma}) = w^{|\boldsymbol{\gamma}|}(1 - w)^{p-|\boldsymbol{\gamma}|},$$

and the single hyperparameter w is the a priori expected proportion of X_js in the model. In particular, w can be chosen small if a sparse model is desired, as is often the case for high-dimensional problems. If $w = 1/2$, the uniform prior results; it assigns all the models the same probability,

$$w(\boldsymbol{\gamma}_k) = \frac{1}{2^p}, \quad k = 1, \cdots, 2^p.$$

The uniform prior can be regarded as noninformative and the model posterior is proportional to the marginal likelihood under this prior,

$$\mathrm{P}(\boldsymbol{\gamma}|\mathbf{y}) \propto \mathrm{P}(\mathbf{y}|\boldsymbol{\gamma}).$$

This is appealing because the posterior odds is equivalent to the BF comparison.

One problem of the uniform prior is that, despite being uniform over all models, it need not be uniform over model neighborhoods, thereby biasing the posterior away from good models. For instance, the uniform prior puts most of its weight near models with size close to $|\boldsymbol{\gamma}| = p/2$ because there are more of them in the model space. So the uniform prior does not provide the proper "dilution" of posterior mass over similar models. This makes the uniform prior unreasonable for model averaging when there are groups of similar models (Clyde (1999); Hoeting et al. (1997); and George (2000)).

To overcome this problem, w can be chosen small; this tends to increase the relative weight on parsimonious models. Or, one can specify a hierarchical model over the model space by assigning a prior to w as a random variable and take a fully Bayesian or empirical Bayes approach. Cui and George (2007) use a uniform prior on w that induces a uniform prior over the model size and therefore increases the chance of models with small or large sizes.

Dilution Model Priors

If there is a dependence relation among covariates, say interaction or polynomial terms of predictors are included in the model, then independence priors (such as the uniform) are less satisfactory. They tend to ignore differences and similarities between the models. Instead, priors that can capture the dependence relation between the predictors are more desired. Motivated by this, George (1999) proposed dilution priors; these take into account covariate dependence and assign probabilities to neighborhoods of the models. Dilution priors are designed to avoid placing too little probability on good but unique models as a consequence of massing excess probability on large sets of nearby similar models. That is, as in George (2000), it is important to ensure that a true model surrounded by many nearly equally good models will not be erroneously seen as having low posterior probability.

The following example from Chipman et al. (2001) illustrates how a dilution prior can be constructed.

In the context of linear regression, suppose there are three independent main effects X_1, X_2, X_3 and three two-factor interactions X_1X_2, X_1X_3, X_2X_3. Common practice to avoid dilution problems is to impose a hierarchical structure on the modeling: Interaction terms such as X_1X_2 are only added to the model when their main effects X_1 and X_2 are already included. A prior for $\boldsymbol{\gamma} = (\gamma_1, \gamma_2, \gamma_3, \gamma_{12}, \gamma_{13}, \gamma_{23})$ that reflects this might satisfy

$$w(\boldsymbol{\gamma}) = w(\gamma_1)w(\gamma_2)w(\gamma_3)w(\boldsymbol{\gamma}_{12}|\gamma_1,\gamma_2)w(\boldsymbol{\gamma}_{23}|\gamma_2,\gamma_3)w(\boldsymbol{\gamma}_{13}|\gamma_1,\gamma_3),$$

typically with

$$w(\boldsymbol{\gamma}_{12}|0,0) < \{w(\boldsymbol{\gamma}_{12}|0,1), w(\boldsymbol{\gamma}_{12}|1,0)\} < w(\boldsymbol{\gamma}_{12}|1,1).$$

Similar strategies can be used to downweight or eliminate models with only isolated high-order terms or isolated interaction terms. Different from the independence priors in (10.4.2), dilution priors concentrate more on plausible models. This is essential in applications, especially when $\mathscr{M}$ is large.

Conventional independent priors can also be modified to dilution priors. Let R_γ be the correlation matrix so that $R_\gamma \propto X_\gamma^T X_\gamma$. When the columns of X_γ are orthogonal, $|R^\gamma| = 1$ and as the columns of X_γ become more redundant, $|R^\gamma|$ decreases to 0. Define

$$w_h \propto h(|R_{\boldsymbol{\gamma}}|) \prod_{j=1}^{p} w_j^{\gamma_j} (1-w_j)^{1-\gamma_j},$$

where h is a monotone function satisfying that $h(0) = 0$ and $h(1) = 1$. It is seen that w_D is a dilution prior because it downweights models with redundant components. The simplest choice for h is the identity function.

Dilution priors are particularly desirable for model averaging using the entire posterior because they avoid biasing the average away from good models. These priors are also desirable for MCMC sampling of the posterior because Markov chains sample more heavily from regions of high probability. In general, failure to dilute posterior probability across clusters of similar models biases model search, model averaging, and inference more broadly.

10.4.1.2 Priors for Parameters

In the normal error case, variable selection is equivalent to selecting a submodel of the form

$$p(\mathbf{y}|\boldsymbol{\beta}, \sigma^2, \boldsymbol{\gamma}) = N_n(X_{\boldsymbol{\gamma}} \boldsymbol{\beta}_{\boldsymbol{\gamma}}, \sigma^2 I_n), \tag{10.4.3}$$

where $X_{\boldsymbol{\gamma}}$ is the $n \times |\boldsymbol{\gamma}|$ matrix whose columns consist of the subset of $X_1, \cdots, X_p$ corresponding to 1s of $\boldsymbol{\gamma}$, and $\boldsymbol{\beta}_{\boldsymbol{\gamma}}$ is the vector of regression coefficients. In order to select variables, one needs to zero out those coefficients that are truly zero by making their posterior mean values very small. In general, there are two ways of specifying the prior to remove a predictor X_j from the model: (i) Assign an atom of probability to the event $\beta_j = 0$; and (ii) use a continuous distribution on β_j with high concentration at 0. Therefore, the data must strongly indicate a β_j is nonzero for X_j to be included.

The use of improper priors for model-specific parameters is not recommended for model selection because improper priors are determined only up to an arbitrary multiplicative constant. Although constants can cancel out in the posterior distribution of the model-specific parameters, they remain in the marginal likelihoods. There, they can lead to indeterminate posterior model probabilities and Bayes factors. To avoid indeterminacies in posterior model probabilities, and other problems such as excess dispersion and marginalization paradoxes, proper priors for $\boldsymbol{\theta}_{\boldsymbol{\gamma}}$ under each model are often required. Below, a collection of commonly occurring proper priors is given.

Spike and Slab Priors

Lempers (1971) and Mitchell and Beauchamp (1988) proposed spike and slab priors for $\boldsymbol{\beta}$. For each variable X_j, the regression coefficient β_j is assigned a two-point mixture distribution made up of a uniform flat distribution (the slab) and a degenerate distribution at zero (the spike) given by

$$\beta_j \sim (1-h_{j0}) U(-a_j, +a_j) + h_{j0}\delta(0), \quad j = 1, \cdots, p,$$

where $\delta(0)$ is a point mass at zero, the a_js are large positive numbers, and $U(-a,a)$ is the uniform distribution. This prior has an atom of probability for the event $\beta_j = 0$. If the prior on σ is chosen to be $\log(\sigma) \sim U(-\ln(\sigma_0), +\ln(\sigma_0))$, the prior for $\boldsymbol{\gamma}_k$ is

$$w(\boldsymbol{\gamma}_k) = \prod_{j\in\boldsymbol{\gamma}_k}(1-w_j)\prod_{j\notin\boldsymbol{\gamma}_k} w_j,$$

where $j \in \boldsymbol{\gamma}_k$ denotes the variable X_j in the model M_k indexed by $\boldsymbol{\gamma}_k$. Note the specification of these priors is not the same as the hierarchical formulation, which first sets the prior $w(\boldsymbol{\gamma})$ and then the parameter prior $w(\boldsymbol{\theta}|\boldsymbol{\gamma})$. Using some approximations for integrals, Mitchell and Beauchamp (1988) express the model posterior probability as

$$P(\boldsymbol{\gamma}_k|\mathbf{y}) = g \cdot \frac{\prod_{j\notin\boldsymbol{\gamma}_k}[2h_{j0}a_j/(1-h_{j0})]\cdot w^{(n-|\boldsymbol{\gamma}_k|)/2}}{|X_{\boldsymbol{\gamma}_k}^{\top}X_{\boldsymbol{\gamma}_k}|^{1/2}RSS_{\boldsymbol{\gamma}_k}^{(n-|\boldsymbol{\gamma}_k|)/2}},$$

where g is a normalizing constant and $RSS_{\boldsymbol{\gamma}_k}$ is the residual sum of squares for model $\boldsymbol{\gamma}_k$. Clearly the posterior probabilities above are highly dependent on the choice of h_{j0} and a_j for each variable.

George and McCulloch (1993) proposed another spike and slab prior using zero-one latent variables, each β_j having a scale mixture of two normal distributions,

$$\beta_j|\gamma_j \sim (1-w(\gamma_j))N(0,\tau_j^2) + w(\gamma_j)N(0,c_j\tau_j^2), \quad j = 1,\cdots,p,$$

where the value for τ_j is chosen to be small and c_j is chosen to be large. As a consequence, coefficients that are promising have posterior latent variables $\gamma_j = 1$ and hence large posterior hypervariances and large posterior β_js. Coefficients that are not important have posterior latent variables $\gamma_j = 0$ and hence have small posterior hypervariances and small posterior β_js. In this formulation, each β_j has a continuous distribution but with high concentration at 0 if $\gamma_j = 0$. It is common to assign a prior for γ_j derived from independent $Bernoulli(w_j)$ distributions with $w_j = 1/2$ as a popular choice.

Point-Normal Prior

The conventional conjugate prior for $(\boldsymbol{\beta}, \sigma^2)$ is a normal-inverse-gamma,

$$w(\boldsymbol{\beta}_{\boldsymbol{\gamma}}|\sigma^2,\boldsymbol{\gamma}) = N_{|\boldsymbol{\gamma}|}(\mathbf{0},\sigma^2\Sigma_{\boldsymbol{\gamma}}), \quad (10.4.4)$$

$$w(\sigma^2|\boldsymbol{\gamma}) = IG(\nu/2, \nu\lambda/2), \quad (10.4.5)$$

where λ, $\Sigma_{\boldsymbol{\gamma}}$, and ν are hyperparameters that must be specified for implementations. Note that the prior on σ^2 is equivalent to assigning $\nu\lambda/\sigma^2 \sim \chi^2_\nu$. When coupled with the prior $w(\boldsymbol{\gamma})$, the prior in (10.4.4) implicitly assigns a point mass at zero for coefficients that are not contained in $\boldsymbol{\beta}_{\boldsymbol{\gamma}}$. If σ^2 is integrated out in (10.4.4), the prior on $\boldsymbol{\beta}_{\boldsymbol{\gamma}}$, conditional only on $\boldsymbol{\gamma}$, is $w(\boldsymbol{\beta}_{\boldsymbol{\gamma}}|\boldsymbol{\gamma}) = T_{|\boldsymbol{\gamma}|}(\nu, \mathbf{0}, \lambda\Sigma_{\boldsymbol{\gamma}})$, which is the multivariate T distribution centered at $\mathbf{0}$ with ν degrees of freedom and scale $\lambda\Sigma_{\boldsymbol{\gamma}}$.

What makes (10.4.4) appealing is its analytical tractability: It has closed-form expressions for all marginal likelihoods. This greatly speeds up posterior evaluation and

MCMC exploration. Note that the conditional distribution of $\boldsymbol{\theta}_\gamma = (\boldsymbol{\beta}_\gamma, \sigma^2)$ given $\boldsymbol{\gamma}$ is conjugate for (10.4.3), so that $(\boldsymbol{\beta}_\gamma, \sigma^2)$ can be eliminated by routine integration from

$$p(\mathbf{y}, \boldsymbol{\beta}_\gamma, \sigma^2|\boldsymbol{\gamma}) = p(\mathbf{y}|\boldsymbol{\beta}_\gamma, \sigma^2, \boldsymbol{\gamma})p(\boldsymbol{\beta}_\gamma|\sigma^2, \boldsymbol{\gamma})w(\sigma^2|\boldsymbol{\gamma}),$$

leading to

$$p(\mathbf{y}|\boldsymbol{\gamma}) \propto |X_\gamma^\mathsf{T} X_\gamma + \Sigma_\gamma^{-1}|^{-1/2}|\Sigma_\gamma|^{-1/2}(\nu\lambda + S_\gamma^2)^{-(n+\nu)/2}, \tag{10.4.6}$$

where

$$S_\gamma^2 = \mathbf{y}^\mathsf{T}\mathbf{y} - \mathbf{y}^\mathsf{T} X_\gamma (X_\gamma^\mathsf{T} X_\gamma + \Sigma_\gamma^{-1})^{-1} X_\gamma^\mathsf{T} \mathbf{y}.$$

These priors have been extensively used.

Since it is not easy to make a good subjective choice of hyperparameters $(\lambda, \Sigma_\gamma, \nu)$, they are often chosen to "minimize" the prior influence. How to do this for ν and λ is treated in Clyde et al. (1996) and Raftery et al. (1997). Indeed, the variance is often reexpressed as $\Sigma_\gamma = gV_\gamma$, where g is a scalar and V_γ is a preset form. Common choices include $V_\gamma = (X_\gamma^\mathsf{T} X_\gamma)^{-1}$, $V = I_{|\gamma|}$ and combinations of them. The first of these gives Zellner's g-prior, Zellner (1986).

Zellner's g-prior

For Bayes variable selection, Zellner (1986) proposed a class of priors defined by

$$w(\sigma) = \frac{1}{\sigma}, \quad w(\boldsymbol{\beta}|\sigma, \boldsymbol{\gamma}) \sim N_{|\gamma|}(\mathbf{0}, g\sigma^2(X_\gamma^\mathsf{T} X_\gamma)^{-1}), \tag{10.4.7}$$

where g is a hyperparameter interpreted as the amount of information in the prior relative to the sample. Under a uniform prior on the model space, g controls the model complexity: Large values of g tend to concentrate the prior on parsimonious models with a few large coefficients, while small values of g typically concentrate the prior on large models with small coefficients (George and Foster (2000)). When the explanatory variables are orthogonal, the g-prior reduces to a standard normal, and outside of this case, dependent explanatory variables tend to correspond to higher marginal variances for the β_js.

One big advantage of g-priors is that the marginal density, $p(\mathbf{y})$, has a closed-form

$$p(\mathbf{y}) = \frac{\Gamma(n/2)}{2\pi^{n/2}(1+g)^{|\gamma|/2}} \left(\mathbf{y}^\mathsf{T}\mathbf{y} - \frac{g}{1+g}\mathbf{y}^\mathsf{T} X_\gamma (X_\gamma^\mathsf{T} X_\gamma)^{-1} X_\gamma^\mathsf{T} \mathbf{y}\right)^{-n/2}. \tag{10.4.8}$$

Likewise, the Bayes factors and posterior model probabilities also have closed-form expressions. This resulting computational efficiency for evaluating marginal likelihoods and doing model searches makes g-priors popular for Bayes variable selection.

Heuristically, g must be chosen large enough that $w(\boldsymbol{\beta}_\gamma|\boldsymbol{\gamma})$ is relatively flat over the region of plausible values of $\boldsymbol{\beta}_\gamma$. There are typically three ways to do this: (i) Deterministically preselect a value of g, (2) estimate g using the empirical Bayes (EB) method, and (3) be fully Bayes and assign a prior to g. Shively et al. (1999) suggest

$g = n$. Foster and George (1994) calibrated priors for model selection based on the risk information criterion (RIC) and recommended the use of $g = p^2$ from a minimax perspective. Fernandez et al. (2001) recommended $g = \max(n, p^2)$. Hansen and Yu (2001) developed a local EB approach to estimate a separate g for each model, deriving

$$\hat{g}_{\boldsymbol{\gamma}}^{LEB} = \max\{F_{\boldsymbol{\gamma}} - 1, 0\},$$

where $F_{\boldsymbol{\gamma}}$ is the F-test statistic for $\mathscr{H}_0 : \boldsymbol{\beta}_{\boldsymbol{\gamma}} = \mathbf{0}$,

$$F_{\boldsymbol{\gamma}} = \frac{R_{\boldsymbol{\gamma}}^2/|\boldsymbol{\gamma}|}{(1 - R_{\boldsymbol{\gamma}}^2)/(n - |\boldsymbol{\gamma}|)},$$

and $R_{\boldsymbol{\gamma}}^2$ is the ordinary coefficient of determination of the model $\boldsymbol{\gamma}$. George and Foster (2000) and Clyde and George (2000) suggest estimating g by EB methods based on its marginal likelihood.

Recently, Liang et al. (2008) proposed the mixture of g-priors using a prior $w(g)$ on g. This includes the Zellner-Siow Cauchy prior (see Zellner and Siow (1980)) as a special case. They show that the fixed g-prior imposes a fixed shrinkage factor $g/(1+g)$ on the posterior mean of $\boldsymbol{\beta}_{\boldsymbol{\gamma}}$, while the mixture of g-priors allows adaptive data-dependent shrinkage on $\boldsymbol{\theta}_{\boldsymbol{\gamma}}$. This adaptivity makes the mixture of g-prior procedure robust to misspecification of g and consistent for model selection.

It can be shown that Bayes selection with fixed choices of g may suffer some paradoxes in terms of model selection consistency. Here are two examples. Suppose one compares the linear model $\boldsymbol{\gamma} : \boldsymbol{Y} = X_{\boldsymbol{\gamma}}\boldsymbol{\beta}_{\boldsymbol{\gamma}} + \boldsymbol{\varepsilon}$ versus the null model $\boldsymbol{\gamma}_0 : \boldsymbol{\beta} = \mathbf{0}$. It can be shown that, as the least squares estimate $\widehat{\boldsymbol{\beta}}_{\boldsymbol{\gamma}}$ goes to infinity, so that the evidence is overwhelmingly against $\boldsymbol{\gamma}_0$, the BF of $\boldsymbol{\gamma}_0$ to $\boldsymbol{\gamma}$ will go to $(1+g)^{(|\boldsymbol{\gamma}|-n)/2}$, a nonzero constant. Another undesired feature of the g-prior is Bartlett's paradox: It can be shown that as $g \to \infty$, where n and $|\boldsymbol{\gamma}|$ are fixed, the Bayes factor of $\boldsymbol{\gamma}_0$ and $\boldsymbol{\gamma}$ will go to zero. This means the null model is favored by the BF, regardless of the data, an unintended consequence of the noninformative choice of g. Liang et al. (2008) have shown that, in some cases, the mixture of g-priors with an empirical Bayes estimate of g resolves these Bayes factor paradoxes.

Normal-Normal Prior

A drawback for the normal-inverse-gamma prior is that when n is large enough, the posterior tends to retain unimportant β_js; i.e., the posterior favors retention of X_j as long as $|\beta_j| \neq 0$, no matter how small. To overcome this, George and McCulloch (1993, 1997) propose a normal-normal prior that excludes X_j whenever $|\beta_j|$ is below a preassigned threshold. Under this prior, X_j is removed from the model if $|\beta_j| < \delta_j$ for a given $\delta_j > 0$.

Under the normal-normal formulation, the data follow the full model

$$p(\mathbf{y}|\beta, \sigma^2, \boldsymbol{\gamma}) = N_n(X\boldsymbol{\beta}, \sigma^2 I_n) \tag{10.4.9}$$

for all $\boldsymbol{\gamma}$, and different values of $\boldsymbol{\gamma}$ index different priors on the βs so that submodels of (10.4.9) can be chosen. For each $\boldsymbol{\gamma}$, the corresponding coefficients have

$$w(\boldsymbol{\beta}|\sigma^2,\boldsymbol{\gamma}) = N_p(0, D_{\boldsymbol{\gamma}}R_{\boldsymbol{\gamma}}D_{\boldsymbol{\gamma}}) \tag{10.4.10}$$

as a prior, where $R_{\boldsymbol{\gamma}}$ is a correlation matrix and $D_{\boldsymbol{\gamma}}$ is a diagonal matrix with the jth element being $\sqrt{v_{0j}}$ if $\gamma_j = 0$ and $\sqrt{v_{1j}}$ if $\gamma_j = 1$, for $j = 1, \cdots, p$. Here v_{0j} and v_{1j} are hyperparameters that must be specified. Note that $\boldsymbol{\beta}$ is independent of σ^2 in (10.4.10), and it is convenient to choose an inverse Gamma prior for σ^2. Obvious choices of $R_{\boldsymbol{\gamma}}$ include $R_{\boldsymbol{\gamma}} \propto (X^\top X)^{-1}$ and $R = I_p$.

Under the model space prior $w(\boldsymbol{\gamma})$, the marginal prior distribution of each component β_j is a scale mixture of two normal distributions:

$$w(\beta_j) = (1 - w(\gamma_j))N(0, v_{0j}) + w(\gamma_j)N(0, v_{1j}). \tag{10.4.11}$$

George and McCulloch (1993) suggest that the hyperparameters v_{0j} is set small while v_{1j} be set large, so that $N(0, v_{0j})$ is concentrated and $N(0, v_{1j})$ is diffuse. In this way, a small coefficient β_j is more likely to be removed from the model if $\gamma_j = 0$. Given a threshold δ_j, higher posterior weighting of those $\boldsymbol{\gamma}$ values for which $|\beta_j| > \delta_j$ when $\gamma_j = 1$ can be achieved by choosing v_{0j} and v_{1j} such that $p(\beta_j|\gamma_j = 0) = N(0, v_{0j}) > p(\beta_j|\gamma_j = 1) = N(0, v_{1j})$ precisely on the interval $(-\delta_j, \delta_j)$. In turn, this can be achieved by choosing v_{0j} and v_{1j} to satisfy

$$\log(v_{1j}/v_{0j})/(v_{0j}^{-1} - v_{1j}^{-1}) = \delta_{j\gamma}^2. \tag{10.4.12}$$

Under (10.4.10), the joint distribution of $(\boldsymbol{\beta}, \sigma^2)$ given $\boldsymbol{\gamma}$ is not conjugate for the likelihood of the data; this can substantially increase the cost of posterior computations. To address this, George and McCulloch (1993) modify (10.4.10) and (10.4.11) to propose a normal prior,

$$w(\boldsymbol{\beta}|\sigma^2,\boldsymbol{\gamma}) = N_p(0, \sigma^2 D_{\boldsymbol{\gamma}}R_{\boldsymbol{\gamma}}D_{\boldsymbol{\gamma}}), \tag{10.4.13}$$

for $\boldsymbol{\beta}$ and an inverse gamma prior for σ^2,

$$w(\sigma^2|\boldsymbol{\gamma}) = IG(\nu/2, \nu\lambda/2).$$

It can be shown that the conditional distribution of $(\boldsymbol{\beta}, \sigma^2)$ given $\boldsymbol{\gamma}$ is conjugate. This allows $(\boldsymbol{\beta}, \sigma^2)$ to be integrated out to give

$$p(\mathbf{y}|\boldsymbol{\gamma}) \propto |X^\top X + (D_{\boldsymbol{\gamma}}R_{\boldsymbol{\gamma}}D_{\boldsymbol{\gamma}})^{-1}|^{-1/2}|D_{\boldsymbol{\gamma}}R_{\boldsymbol{\gamma}}D_{\boldsymbol{\gamma}}|^{-1/2}(\nu\lambda + S_{\boldsymbol{\gamma}}^2)^{-(n+\nu)/2}, \tag{10.4.14}$$

where $S_{\boldsymbol{\gamma}}^2 = \mathbf{y}^\top\mathbf{y} - \mathbf{y}^\top X(X^\top X + (D_{\boldsymbol{\gamma}}RD_{\boldsymbol{\gamma}})^{-1})^{-1}X^\top\mathbf{y}$. This dramatically simplifies the computational burden of posterior calculation and exploration.

Under (10.4.13), the inverse gamma prior for σ^2, and a model space prior $w(\boldsymbol{\gamma})$, the marginal distribution of each β_j is a scale mixture of t-distributions,

$$w(\beta_j|\boldsymbol{\gamma}) = (1 - \gamma_j)t(\nu, 0, \lambda v_{0j}) + \gamma_j t(\nu, 0, \lambda v_{1j}), \tag{10.4.15}$$

where $t(\nu,0,\lambda v)$ is a one-dimensional t-distribution centered at 0 with ν degrees of freedom and scale v. Note that (10.4.15) is different from the normal mixture of (10.4.11). Similar to the nonconjugate prior, v_{0j} and v_{1j} are to be chosen small and large, respectively, so that a small coefficient β_j is more likely to be removed from the model if $\gamma_j = 0$. Given a threshold δ_j, the pdf $p(\beta_j|\gamma_j = 0) = t(\nu,0,\lambda v_{0j}) > p(\beta_j|\gamma_j = 1) = t(\nu,0,\lambda v_{1j})$ precisely on the interval $(-\delta_j, \delta_j)$, resulting in

$$(v_{0j}/v_{1j})^{\nu/(\nu+1)} = [(v_{0j} + \delta_j^2)/(\nu\lambda)]/[v_{1j} + \delta_j^2/(\nu\gamma)],$$

parallel to (10.4.12).

10.4.2 *Posterior Calculation and Exploration*

In order to do Bayes variable selection, it is enough to find the model posterior probability (10.4.1). For moderately sized $\mathcal{M}$, when a closed-form expression for $w(\boldsymbol{\gamma}|\mathbf{y})$ is available, exhaustive calculation is feasible. However, when a closed form for $w(\boldsymbol{\gamma}|\mathbf{y})$ is unavailable or if p is large, it is practically impossible to calculate the entire posterior model distribution. In such cases, inference about posterior characteristics ultimately relies on a sequence like

$$\boldsymbol{\gamma}^{(1)}, \boldsymbol{\gamma}^{(2)}, \ldots \tag{10.4.16}$$

whose empirical distribution converges (in distribution) to $w(\boldsymbol{\gamma}|\mathbf{y})$. In particular, the empirical frequency estimates of the visited $\boldsymbol{\gamma}$ values are intended to provide consistent estimates for posterior characteristics. Even when the length of (10.4.16) is much smaller than 2^p, it may be possible to identify regions of $\mathcal{M}$ containing high probability models $\boldsymbol{\gamma}$ because they appear more frequently.

In practice, Markov chain Monte Carlo (MCMC) methods are the main technique for simulating approximate samples from the posterior. These samples can be used to explore the posterior distribution, estimate model posterior characteristics, and search for models with high posterior probability over the model space. Since the focus here is on Bayes variable selection, only the two most important MCMC methods are described here: the Gibbs sampler (Geman and Geman (1984); Gelfand and Smith (1990)) and the Metropolis-Hastings algorithm (Metropolis et al. (1953); Hastings (1970)). Other general MCMC posterior exploration techniques, such as reversible jump and particle methods, are more sophisticated and are not covered here.

10.4.2.1 Closed Form for $w(\mathbf{y}|\boldsymbol{\gamma})$

One great advantage of conjugate priors is that they lead to closed-form expressions, for instance in (10.4.6) and (10.4.14), that are proportional to the marginal likelihood of the data $p(\mathbf{y}|\boldsymbol{\gamma})$ for each model $\boldsymbol{\gamma}$. This facilitates posterior calculation and estimation enormously. Indeed, if the model prior $w(\boldsymbol{\gamma})$ is computable, conjugate priors lead to closed-form expressions $g(\boldsymbol{\gamma})$ satisfying

$$g(\boldsymbol{\gamma}) \propto p(\mathbf{y}|\boldsymbol{\gamma})w(\boldsymbol{\gamma}) \propto p(\boldsymbol{\gamma}|\mathbf{y}).$$

The availability of a computable $g(\boldsymbol{\gamma})$ enables exhaustive calculation of $p(\boldsymbol{\gamma}|\mathbf{y})$ when p is small or moderate. This is done by calculating the value $g(\boldsymbol{\gamma})$ for each $\boldsymbol{\gamma}$ and then summing over the $\boldsymbol{\gamma}$s to obtain the normalization constant. In many situations, the value of $g(\boldsymbol{\gamma})$ can also be updated rapidly when one of the components in $\boldsymbol{\gamma}$ is changed. This also speeds posterior evaluation and exploitation.

As shown by George and McCulloch (1997), the availability of $g(\boldsymbol{\gamma})$ can also be used to obtain estimators of the normalizing constant C for $p(\boldsymbol{\gamma}|\mathbf{y})$. That is, an MCMC sequence $\boldsymbol{\gamma}^{(1)}, \cdots, \boldsymbol{\gamma}^{(L)}$ from (10.4.16) can be used to find

$$C = C(\mathbf{y}) = g(\boldsymbol{\gamma})/p(\boldsymbol{\gamma}|\mathbf{y}).$$

The idea is to choose a set A of $\boldsymbol{\gamma}$ values and write $g(A) = \sum_{\boldsymbol{\gamma} \in A} g(\boldsymbol{\gamma})$ so that $P(A|\mathbf{y}) = Cg(A)$. Then a consistent estimator of C is

$$\hat{C} = \frac{1}{g(A)L} \sum_{l=1}^{L} I_A(\boldsymbol{\gamma}^{(l)}),$$

where $I_A()$ is the indicator of the set A.

The availability of $g(\boldsymbol{\gamma})$ also allows for the flexible construction of MCMC algorithms that simulate (10.4.16) directly as a Markov chain. Such chains are very useful in terms of both computational and convergence speeds. Numerous MCMC algorithms have been proposed to generate the sequences like (10.4.16) based on Gibbs sampler and Metropolis-Hastings algorithms. Detailed introductions on these algorithms are given by Casella and George (1992), Liu et al. (1994), Chib and Greenberg (1995), and Chipman et al. (2001), among others.

10.4.2.2 Stochastic Variable Search Algorithms

George and McCulloch (1993) proposed the stochastic search variable selection (SSVS) algorithm. Built in the framework of hierarchical Bayesian formulation, the SSVS indirectly samples from the posterior model distribution, identifies subsets that appear more frequently in the sample, and therefore avoids the problem of calculating the posterior probabilities of all 2^p subsets of $\mathcal{M}$. That is, instead of calculating the posterior model distribution for all possible models, SSVS uses a sampling procedure to identify promising models associated with high posterior probabilities in the model space. As effective as this procedure is, it does not seem to scale up to large p as well as others do, for instance Hans et al. (2007), discussed below.

To understand the basics behind these procedures, suppose that analytical simplification of $p(\boldsymbol{\beta}, \sigma^2, \boldsymbol{\gamma}|\mathbf{y})$ is unavailable. MCMC methods first simulate a Markov chain

$$\boldsymbol{\beta}^{(1)}, \sigma^{(1)}, \boldsymbol{\gamma}^{(1)}, \boldsymbol{\beta}^{(2)}, \sigma^{(2)}, \boldsymbol{\gamma}^{(2)}, \cdots, \tag{10.4.17}$$

that converges to $p(\boldsymbol{\beta}, \sigma^2, \boldsymbol{\gamma}|\mathbf{y})$ and take out the subsequence $\boldsymbol{\gamma}^{(1)}, \boldsymbol{\gamma}^{(2)}, \cdots$. Now, the two most fundamental general procedures can be described. However,note that verifying convergence of the estimates from the chains remains a topic of controversy despite extensive study.

Gibbs Samplers

In practice, Gibbs sampling (GS) is often used to identify promising models with high posterior probability. In GS, the parameter sequence (10.4.17) is obtained by successive simulations from the distribution conditioned on the most recently generated parameters. When conjugate priors are used, the simplest strategy is to generate each component of $\boldsymbol{\gamma} = (\gamma_1, \cdots, \gamma_p)$ from the full conditionals

$$\gamma_j | \boldsymbol{\gamma}_{-j}, \mathbf{y}, \quad j = 1, \cdots, p,$$

where $\boldsymbol{\gamma}_{(-j)} = (\gamma_1, \cdots, \gamma_{j-1}, \gamma_{j+1}, \cdots, \gamma_p)$. This decomposes the p-dimensional data simulation problem into one-dimensional simulations, and the generation of each component can be obtained as a sequence of Bernoulli draws.

When nonconjugate priors are used, the full parameter sequence (10.4.17) can be successively simulated using GS from the full conditionals in sequence:

$$p(\boldsymbol{\beta}|\sigma^2, \boldsymbol{\gamma}, \mathbf{y}),$$

$$p(\sigma^2|\boldsymbol{\beta}, \boldsymbol{\gamma}, \mathbf{y}),$$

$$p(\gamma_j|\boldsymbol{\beta}, \sigma^2, \boldsymbol{\gamma}_{(-j)}, \mathbf{y}), \quad j = 1, \cdots, p.$$

Metropolis-Hastings Algorithm

The Metropolis-Hastings (MH) algorithm is a rejection sampling procedure to generate a sequence of samples from a probability distribution that is difficult to sample directly. In this sense MH, generalizes GS. (The availability of $g(\boldsymbol{\gamma}) \propto p(\boldsymbol{\gamma}|\mathbf{y})$ also facilitates the use of the MH algorithm for direct simulation of (10.4.16).) The MH works by successive sampling from an essentially arbitrary probability transition kernel $q(\boldsymbol{\gamma}|\boldsymbol{\gamma}')$ and imposing a random rejection step at each transition. Because $g(\boldsymbol{\gamma})/g(\boldsymbol{\gamma}') = p(\boldsymbol{\gamma}|\mathbf{y})/p(\boldsymbol{\gamma}/|\mathbf{y})$, the general MH algorithm has the following form.

At the ith step, for $i = 0, 1, ...,$:

- ☐ simulate a candidate $\boldsymbol{\gamma}^*$ from a transition kernel $q(\boldsymbol{\gamma}^*|\boldsymbol{\gamma}^{(i)})$.
- ☐ Accept the candidate sample $\boldsymbol{\gamma}^*$; i.e. set $\boldsymbol{\gamma}^{(i+1)} = \boldsymbol{\gamma}^*$, with probability

$$\alpha(\boldsymbol{\gamma}^*|\boldsymbol{\gamma}^{(i)}) = \min\left\{\frac{q(\boldsymbol{\gamma}^{(i)}|\boldsymbol{\gamma}^*)}{q(\boldsymbol{\gamma}^*|\boldsymbol{\gamma}^{(i)})}\frac{g(\boldsymbol{\gamma}^*)}{g(\boldsymbol{\gamma}^{(i)})}, 1\right\}.$$

Otherwise, reject the candidate sample; i.e., set $\boldsymbol{\gamma}^{(i+1)} = \boldsymbol{\gamma}^{(i)}$.

The idea is that only new γs representative of g are likely to be retained.

A special case of the MH algorithm is the Metropolis algorithm, which is obtained using symmetric transition kernels q. The acceptance probability then simplifies to

$$\alpha^M(\boldsymbol{\gamma}^*|\boldsymbol{\gamma}^{(i)}) = \min\left\{\frac{g(\boldsymbol{\gamma}^*)}{g(\boldsymbol{\gamma}^{(i)})}, 1\right\}.$$

One choice of symmetric transition kernel is $q(\boldsymbol{\gamma}_k|\boldsymbol{\gamma}_l) = 1/p$ if $\sum_{j=1}^p |\gamma_{kj} - \gamma_{lj}| = 1$. This gives the Metropolis algorithm:

- ☐ Simulate a candidate $\boldsymbol{\gamma}^*$ by randomly changing one component of $\boldsymbol{\gamma}^{(i)}$.
- ☐ Set $\boldsymbol{\gamma}^{(i+1)} = \boldsymbol{\gamma}^*$ with probability $\alpha^M(\boldsymbol{\gamma}^*|\boldsymbol{\gamma}^{(i)})$. Otherwise, reject the candidate sample and set $\boldsymbol{\gamma}^{(i+1)} = \boldsymbol{\gamma}^{(i)}$.

High-Dimensional Model Search

Standard MCMC methods usually perform well when p is small. However, when p is high, standard MCMC is often ineffective due to slow convergence because MCMC chains tend to get trapped near local maxima of the model space. To speed the search for "interesting" regions of the model space when p is high, some strategies exploit local collinearity structures, for example shotgun stochastic search (SSS; Hans et al. (2007)). Compared with standard MCMC methods, SSS often identifies probable models rapidly and moves swiftly around in the model spaces when p is large.

The key idea of the SSS is that, for any current model, there are many similar models that contain either overlapping or collinear predictors and these models form a neighborhood of the current model. The neighborhood makes it possible to consider each possible variable at each step, allowing the search to move freely among models of various dimensions. Therefore, quickly identifying these neighborhoods generates multiple candidate models so the procedure can "shoot out" proposed moves in various directions in the model space.

In Hans et al. (2007), for a current model $\boldsymbol{\gamma}$ of dimension $|\boldsymbol{\gamma}| = k$, the neighborhood $N(\gamma)$ can be given as three sets $N(\gamma) = \{\boldsymbol{\gamma}^+, \boldsymbol{\gamma}^0, \boldsymbol{\gamma}^-\}$, where $\boldsymbol{\gamma}^+$ is a set containing the models obtained by adding one of the remaining variables to the current model γ, the addition moves; $\boldsymbol{\gamma}^0$ is a set containing all the models obtained by replacing any one variable in γ with one not in γ, the replacement moves; and $\boldsymbol{\gamma}^-$ is a set containing the models obtained by deleting one variable from γ, the deletion moves. For large p problems, typically $|\boldsymbol{\gamma}^0| \gg |\boldsymbol{\gamma}^+| \gg |\boldsymbol{\gamma}^-|$, making it hard to examine models of different dimensions. To correct this, Hans et al. (2007) suggested a two-stage sampling: First sample three models $\boldsymbol{\gamma}_*^+, \boldsymbol{\gamma}_*^0, \boldsymbol{\gamma}_*^-$ from $\boldsymbol{\gamma}^+, \boldsymbol{\gamma}^0, \boldsymbol{\gamma}^-$, respectively, and then select one of the three. Let $\boldsymbol{\gamma}$ be a regression model and $S(\boldsymbol{\gamma})$ be some (unnormalized) score that can be normalized within a set of scores to become a probability. The following is the detailed SSS sampling scheme:

Given a starting model $\boldsymbol{\gamma}^{(0)}$, iterate in $t = 1, \cdots, T$ the following steps:

□ In parallel, compute $S(\boldsymbol{\gamma})$ for all $\boldsymbol{\gamma} \in \text{nbd}(\boldsymbol{\gamma}^{(t)})$, constructing $\boldsymbol{\gamma}_*^+, \boldsymbol{\gamma}_*^0, \boldsymbol{\gamma}_*^-$. Update the list of the overall best models evaluated.

□ Sample three models $\boldsymbol{\gamma}_*^+, \boldsymbol{\gamma}_*^0, \boldsymbol{\gamma}_*^-$ from $\boldsymbol{\gamma}^+, \boldsymbol{\gamma}^0, \boldsymbol{\gamma}^-$ respectively, with probabilities proportional to $S(\boldsymbol{\gamma})^{\alpha_1}$, normalized within each set.

□ Sample $\boldsymbol{\gamma}^{(t+1)}$ from $\{\boldsymbol{\gamma}_*^+, \boldsymbol{\gamma}_*^0, \boldsymbol{\gamma}_*^-\}$ with probability proportional to $S(\boldsymbol{\gamma})^{\alpha_2}$, normalized within this set.

The positive annealing parameters α_1 and α_2 control how greedy the search is: Values less than one flatten out the proposal distribution, whereas very large values lead to a hill-climbing search.

10.4.2.3 Bayes Prediction

A typical Bayes selection approach chooses the single best model and then makes inferences as if the selected model were true. However, this ignores the uncertainty about the model itself and the uncertainty due to the choice of $\mathcal{M}$. This leads to overconfident inferences and risky decisions. A better Bayes solution is Bayes model averaging (BMA), presented in Chapter 6, which involves averaging with respect to the posterior for the models over all elements of $\mathcal{M}$ to make decisions, and especially predictions, about quantities of interest.

Berger and Mortera (1999) show that the largest posterior probability model is optimal if only two models are being entertained and is often optimal for variable selection in linear models having orthogonal design matrices. For other cases, the largest posterior model is not in general optimal. For example, in nested linear models, the optimal single model for prediction is the median probability model (Barbieri and Berger (2004)). The median model is the model consisting of only those variables which have posterior inclusion probabilities greater than or equal to one-half. In this case, only the posterior inclusion probabilities for the variables must be found, not the whole posterior as required for BMA.

10.4.3 Evaluating Evidence

There are many ways to extract information from the posterior, but the most popular is the Bayes factor, which is the Bayes action under generalized zero-one loss, the natural loss function for binary decision problems such as hypothesis testing. Obviously, using a different loss function would lead to a different Bayes action. Moreover, there are modifications to the Bayes factor that are efforts to correct some of its deficiencies.

10.4.3.1 Bayes Factors

Let the prior over the model space be $w(\boldsymbol{\gamma})$ and the prior for the parameters $\boldsymbol{\theta}$ in $\boldsymbol{\gamma}$ be $w(\boldsymbol{\theta}|\boldsymbol{\gamma})$. Then the posterior probability of the model $\boldsymbol{\gamma}$ is given by (10.4.1) as

$$\mathbb{P}(\boldsymbol{\gamma}|\mathbf{y}) = \frac{w(\boldsymbol{\gamma})}{m(\mathbf{y})}\int f(\mathbf{y}|\boldsymbol{\theta},\boldsymbol{\gamma})w(\boldsymbol{\theta}|\boldsymbol{\gamma})d\boldsymbol{\theta},$$

where m is the mixture of distributions that makes the right-hand side integrate to one. The posterior odds in favor of model $\boldsymbol{\gamma}_1$ over an alternative model $\boldsymbol{\gamma}_2$ are

$$\begin{aligned}\frac{\mathbb{P}(\boldsymbol{\gamma}_1|\mathbf{y})}{\mathbb{P}(\boldsymbol{\gamma}_2|\mathbf{y})} &= \frac{p(\mathbf{y}|\boldsymbol{\gamma}_1)}{p(\mathbf{y}|\boldsymbol{\gamma}_2)}\cdot\frac{w(\boldsymbol{\gamma}_1)}{w(\boldsymbol{\gamma}_2)}\\ &= \frac{\int f_1(\mathbf{y}|\boldsymbol{\theta}_1,\boldsymbol{\gamma}_1)w(\boldsymbol{\theta}_1|\boldsymbol{\gamma}_1)d\boldsymbol{\theta}_1}{\int f_2(\mathbf{y}|\boldsymbol{\theta}_2,\boldsymbol{\gamma}_2)w(\boldsymbol{\theta}_2|\boldsymbol{\gamma}_2)d\boldsymbol{\theta}_2}\cdot\frac{w(\boldsymbol{\gamma}_1)}{w(\boldsymbol{\gamma}_2)}.\end{aligned}$$

The Bayes factor (BF) of the model $\boldsymbol{\gamma}_1$ to the model $\boldsymbol{\gamma}_2$ is defined as the ratio

$$\mathrm{BF}_{12} = \frac{\mathbb{P}(\mathbf{y}|\boldsymbol{\gamma}_1)}{\mathbb{P}(\mathbf{y}|\boldsymbol{\gamma}_2)} = \frac{\int f_1(\mathbf{y}|\boldsymbol{\theta}_1,\boldsymbol{\gamma}_1)w(\boldsymbol{\theta}_1|\boldsymbol{\gamma}_1)d\boldsymbol{\theta}_1}{\int f_2(\mathbf{y}|\boldsymbol{\theta}_2,\boldsymbol{\gamma}_2)w(\boldsymbol{\theta}_2|\boldsymbol{\gamma}_2)d\boldsymbol{\theta}_2}. \tag{10.4.18}$$

The BF is the weighted likelihood ratio of $\boldsymbol{\gamma}_1$ and $\boldsymbol{\gamma}_2$; it represents the comparative support for one model versus the other provided by the data. That is, through the Bayes factor, the data updates the prior odds to the posterior odds. Computing BF_{12} requires both priors $w(\boldsymbol{\theta}_k|\boldsymbol{\gamma}_k)$ for $j = 1,2$ be specified.

Posterior model probabilities can also be obtained from BFs. If $w(\boldsymbol{\gamma}_k)$s are available for $k = 1,\cdots,K$, then the posterior probability of $\boldsymbol{\gamma}_k$ is

$$\mathbb{P}(\boldsymbol{\gamma}_k|\mathbf{y}) = \frac{w(\boldsymbol{\gamma}_k)\mathbb{P}(\mathbf{y}|\boldsymbol{\gamma}_k)}{\sum_{l=1}^{K} w(\boldsymbol{\gamma}_l)\mathbb{P}(\mathbf{y}|\boldsymbol{\gamma}_l)} = \left[\sum_{l=1}^{K}\frac{w(\boldsymbol{\gamma}_l)}{w(\boldsymbol{\gamma}_k)}B_{lk}\right]^{-1}. \tag{10.4.19}$$

A special case for the prior over models is uniform, $w(\boldsymbol{\gamma}_k) = 1/K$ for $k = 1,\cdots,K$. For this, the posterior model probabilities are the same as the renormalized marginal probabilities,

$$\mathbb{P}^*(\mathbf{y}|\boldsymbol{\gamma}_k) = \frac{\mathbb{P}(\mathbf{y}|\boldsymbol{\gamma}_k)}{\sum_{l=1}^{K}\mathbb{P}(\mathbf{y}|\boldsymbol{\gamma}_l)}, \tag{10.4.20}$$

so $B_{lk} = \mathbb{P}^*(\mathbf{y}|\boldsymbol{\gamma}_l)/\mathbb{P}^*(\mathbf{y}|\boldsymbol{\gamma}_k)$.

Strictly, BFs just give the Bayes action from a particular decision problem. Nevertheless, they are used more generally as an assessment of evidence. However, in the model selection context, the general use of BFs has several limitations. First, when the models have parameter spaces of different dimensions, use of improper noninformative priors for model-specific parameters can make the BF become indeterminate. For instance, suppose $w(\boldsymbol{\theta}_1|\boldsymbol{\gamma}_1)$ and $w(\boldsymbol{\theta}_2|\boldsymbol{\gamma}_2)$ are improper noninformative priors for model-specific parameters. Then the BF is BF_{12} from (10.4.18). However, because the priors are improper, the noninformative priors $c_1 w(\boldsymbol{\theta}_1|\boldsymbol{\gamma}_1)$ and $c_2 w(\boldsymbol{\theta}_2|\boldsymbol{\gamma}_2)$ are equally valid. They would give $(c_1/c_2)\mathrm{BF}_{12}$ as the BF, but, since c_1/c_2 is arbitrary, the BF is indeterminate. When the parameter spaces for $\boldsymbol{\gamma}_1$ and $\boldsymbol{\gamma}_2$ are the same, it is usually reasonable to choose $c_1 = c_2$, but when the parameter spaces have different dimensions, $c_1 = c_2$ can give bad answers (Spiegelhalter and Smith (1982), Ghosh and Samanta (1999)). Second, the use of vague proper priors usually gives unreliable answers in Bayes model selection, partially because the dispersion of the prior can overwhelm the information in the data. Berger and Pericchi (2001) argue that one should never use arbitrary vague proper priors for model selection, but improper noninformative priors may give reasonable results. Jeffreys (1961) also dealt with the issue of indeterminacy of noninformative priors by using noninformative priors only for common parameters in the models and using default proper priors for parameters that would appear in one model but not the other.

10.4.3.2 Other Bayes Factors

The dependence of BFs on the priors matters most when the prior is weak (i.e., too spread out) because as the tail behavior becomes too influential the BFs become unstable. To address this issue, Berger and Pericchi (1996) and O'Hagan (1995, 1997) suggested the use of partial Bayes factors: Some data points, say m, are used as a training sample to update the prior distribution, effectively making it more informative, and the remaining $n-m$ data points are used to form the BF from the updated prior.

To see this, let $\boldsymbol{y} = (\tilde{\boldsymbol{y}}_{(m)}^{\mathsf{T}}, \tilde{\boldsymbol{y}}_{(n-m)}^{\mathsf{T}})^{\mathsf{T}}$, where $\tilde{\boldsymbol{y}}_{(m)}$ are the m training points, and let $w(\boldsymbol{\theta}_j|\boldsymbol{\gamma}_j, \tilde{\boldsymbol{y}}_{(m)})$ be the posterior distribution of the parameter $\theta_j, j = 1, 2$ given $\tilde{\boldsymbol{y}}_{(m)}$. The point is to use $\tilde{\boldsymbol{y}}_{(m)}$ to convert improper priors $w(\theta_k|\boldsymbol{\gamma}_k)$ to proper posteriors $w(\boldsymbol{\theta}_j|\boldsymbol{\gamma}_j, \tilde{\boldsymbol{y}}_{(m)})$. Now, the partial BF for model $\boldsymbol{\gamma}_1$ against model $\boldsymbol{\gamma}_2$ is

$$\mathrm{BF}_{12}^{part} = \frac{\int p(\tilde{\boldsymbol{y}}_{(n-m)}|\boldsymbol{\theta}_1, \boldsymbol{\gamma}_1) w(\boldsymbol{\theta}_1|\boldsymbol{\gamma}_1, \tilde{\boldsymbol{y}}_{(m)}) d\boldsymbol{\theta}_1}{\int p(\tilde{\boldsymbol{y}}_{(n-m)}|\boldsymbol{\theta}_2, \boldsymbol{\gamma}_2) w(\boldsymbol{\theta}_2|\boldsymbol{\gamma}_2, \tilde{\boldsymbol{y}}_{(m)}) d\boldsymbol{\theta}_2}. \tag{10.4.21}$$

Compared with BF_{12}, the partial BF is less sensitive to the priors. The BF^{part} does not depend on the absolute values of prior distributions, and instead it depends on the relative values of priors, the training sample $\tilde{\boldsymbol{y}}_{(m)}$, and the training sample size m. When the training size m increases, the sensitivity of the partial BF to prior distributions decreases, but at the cost of less discriminatory power. Also, BF_{12}^{part} depends on the arbitrary choice of the training sample $\tilde{\boldsymbol{y}}_{(m)}$.

To eliminate this dependence and to increase the stability, Berger and Pericchi (1996) propose an intrinsic Bayes factor (IBF), which averages the partial BF over all possible training samples $\tilde{\boldsymbol{y}}_{(m)}$. Depending on how the averaging is done, there are different versions of IBFs. Commonly used are the arithmetic IBF (IBF^a), the encompassing IBF (IBF^{en}), the expected IBF (IBF^e), and the median IBF (IBF^m); see Berger and Pericchi (1996) for their exact definitions. Berger and Pericchi (1996, 1997, 2001) do extensive evaluations and comparisons of different BFs under various settings. They conclude that different IBFs are optimal in different situations and that IBFs based on training samples can be used with considerable confidence as long as the sample size is not too small. In particular, they suggest that the expected IBF should be used if the sample size is small, the arithmetic IBF be used for comparing nested models, the encompassing IBF be used for multiple linear models, and the median IBF be used for other problems including nonnested models. Computational issues are also addressed in Varshavsky (1995). Typically IBFs are the most difficult to compute among default Bayes factors since most of them involve training sample computations. When the sample size n is large, computation of IBFs is only possible by using suitable schemes for sampling from the training samples.

To reduce the computational burden of averaging needed in the IBF, O'Hagan (1995) suggests the fractional Bayes factor (FBF) based on much the same intuition as for the IBF. Instead of using part of the data to turn noninformative priors into proper priors, the FBF uses a fraction, b, of each likelihood function, $p(\mathbf{y}|\boldsymbol{\theta}_k, \boldsymbol{\gamma}_k)$, with the remaining $1-b$ fraction of the likelihood used for model discrimination. The FBF of $\boldsymbol{\gamma}_1$ to $\boldsymbol{\gamma}_2$ is

$$\mathrm{FBF}_{12} = \mathrm{BF}_{12} \cdot \frac{\int [p(\mathbf{y}|\boldsymbol{\theta}_2, \boldsymbol{\gamma}_2)]^b w(\boldsymbol{\theta}_2|\boldsymbol{\gamma}_2) d\boldsymbol{\theta}_2}{\int [p(\mathbf{y}|\boldsymbol{\theta}_1, \boldsymbol{\gamma}_1)]^b w(\boldsymbol{\theta}_1|\boldsymbol{\gamma}_1) d\boldsymbol{\theta}_1}. \tag{10.4.22}$$

One common choice is $b = m/n$, where m is the minimal training sample size, as in O'Hagan (1995) and Berger and Mortera (1999). The asymptotic motivation for (10.4.22) is that if m and n are both large, the likelihood based on $\tilde{\mathbf{y}}_{(m)}$ is approximated by the one based on $\mathbf{y}$, raised on the power m/n. The FBF is in general easier to compute than the IBF.

10.4.4 Connections Between Bayesian and Frequentist Methods

To conclude this section, it is worthwhile to see that even though Bayesian and frequentist variable selection methods have different formulations, they are actually closely related. Indeed, Bayes selection can be seen as a generalization of penalization methods, and posterior maximization subsumes several information criteria.

10.4.4.1 Bayes and Penalization

Many shrinkage estimators introduced in Section 10.3, such as ridge, LASSO, SCAD, ALASSO, have Bayesian interpretations. Assume the prior $w(\boldsymbol{\beta})$ on $\boldsymbol{\beta}$ and an independent prior $w(\sigma^2)$ on $\sigma^2 > 0$. Then, the posterior for $(\boldsymbol{\beta}, \sigma^2)$, given $\mathbf{y}$, is

$$w(\boldsymbol{\beta}, \sigma^2|\mathbf{y}) \propto w(\sigma^2)(\sigma^2)^{-(n-1)/2} \exp\left\{ -\frac{1}{2\sigma^2}(\mathbf{y} - X\boldsymbol{\beta})^{\mathrm{T}}(\mathbf{y} - X\beta) + \log w(\boldsymbol{\beta}) \right\}.$$

Shrinkage procedures can now be seen to correspond to different choices of $w(\boldsymbol{\beta})$.

First, assume the $w(\beta_j)$s are independent normal $N(0, 2\lambda)$s; i.e.,

$$w(\boldsymbol{\beta}) = \prod_{j=1}^{p} \frac{1}{\sqrt{2\pi}(2\lambda)^{-1/2}} e^{-\lambda \beta_j^2}.$$

Then the posterior for $(\boldsymbol{\beta}, \sigma^2)$, given $\mathbf{y}$, becomes

$$w(\boldsymbol{\beta}, \sigma^2|\mathbf{y}) \propto w(\sigma^2)(\sigma^2)^{-(n-1)/2} \exp\left\{ -\frac{1}{2\sigma^2}(\mathbf{y} - X\boldsymbol{\beta})^{\mathsf{T}}(\mathbf{y} - X\beta) - \lambda \sum_{j=1}^{p} \beta_j^2 \right\}.$$

Now, for any fixed value of σ^2, the maximizing $\boldsymbol{\beta}$ is the RR estimate, which is the posterior mode.

Next, when the priors on parameters are independent double-exponential (Laplace) distributions (i.e., $w(\boldsymbol{\beta}) = \prod_{j=1}^{p} \frac{\lambda}{2} e^{-|\beta_j|\lambda}$) the posterior for $(\boldsymbol{\beta}, \sigma^2)$, given $\mathbf{y}$, is

$$w(\boldsymbol{\beta}, \sigma^2|\mathbf{y}) \propto w(\sigma^2)(\sigma^2)^{-(n-1)/2} \exp\left\{ -\frac{1}{2\sigma^2}(\mathbf{y} - X\boldsymbol{\beta})^{\mathsf{T}}(\mathbf{y} - X\beta) - \lambda \sum_{j=1}^{p} |\beta_j| \right\}.$$

So, again, for any fixed values of σ^2, the maximizing $\boldsymbol{\beta}$ is the LASSO estimate, a posterior mode, as noted in Tibshirani (1996). If different scaling parameters are allowed in the prior (i.e., $w(\boldsymbol{\beta}) = \prod_{j=1}^{p} \frac{\lambda_j}{2} \exp\{-|\beta_j|\lambda_j\}$), the posterior distribution becomes

$$w(\boldsymbol{\beta}, \sigma^2|\mathbf{y}) \propto w(\sigma^2)(\sigma^2)^{-(n-1)/2} \exp\left\{ -\frac{1}{2\sigma^2}(\mathbf{y} - X\boldsymbol{\beta})^{\mathsf{T}}(\mathbf{y} - X\beta) - \sum_{j=1}^{p} \lambda_j |\beta_j| \right\},$$

and the posterior mode is the ALASSO estimate.

More generally, using the prior

$$w(\boldsymbol{\beta}) = C(\lambda, q) \exp\left\{ -\sum_{j=1}^{p} \beta_j^q \right\}, \quad q > 0,$$

on a normal likelihood leads to the bridge estimator, which can likewise be seen as a posterior mode. Also, the elastic net penalty corresponds to using the prior

$$w(\boldsymbol{\beta}) = C(\lambda, q) \exp\left\{ -\lambda \left[\alpha \sum_{j=1}^{p} \beta_j^2 + (1 - \alpha) \sum_{j=1}^{p} |\beta_j| \right] \right\},$$

which is a compromise between the Gaussian and Laplacian priors. The Bayes procedures, unlike the frequentist versions, also give an immediate notion of uncertainty different from bootstrapping as might be used in LASSO, for instance.

Note that this conversion from penalties to priors is mathematical. A Bayesian would be more likely to try to think carefully about the topology of the space of regression functions and the meaning of the penalty in that context. That is, the use of penalties for their mathematical properties of getting good behavior would probably not satisfy a Bayesian on the grounds that the class of such priors was large and would not lead to correct assignments of probabilities more generally.

10.4.4.2 Bayes and Information Criteria

To see how several information-based criteria can be regarded as instances of Bayes selection, consider the simple independence prior for model space combined with the point-normal prior on the parameters,

$$\begin{aligned} w(\boldsymbol{\beta}_\gamma|\sigma^2,\gamma) &= N_{|\gamma|}(\mathbf{0}, c\sigma^2 X_\gamma^\mathsf{T} X_\gamma)^{-1}, \\ w(\boldsymbol{\gamma}) &= w^{|\gamma|}(1-w)^{p-|\gamma|}, \end{aligned}$$

and assume σ^2 is known. George and Foster (2000) show that

$$\begin{aligned} w(\boldsymbol{\gamma}|\mathbf{y}) &\propto w^{|\gamma|}(1-w)^{p-|\gamma|}(1+c)^{-|\gamma|/2}\exp\left\{-\frac{\mathbf{y}^\mathsf{T}\mathbf{y}-SS_\gamma}{2\sigma^2}-\frac{SS_\gamma}{2\sigma^2(1+c)}\right\} \\ &\propto \exp\left[\frac{c}{2(1+c)}\{SS_\gamma/\sigma^2 - F(c,w)|\boldsymbol{\gamma}|\}\right], \end{aligned} \qquad (10.4.23)$$

where

$$F(c,w) = \frac{1+c}{c}\{2\log\frac{1-w}{w} + \log(1+c)\},$$

$SS_\gamma = \widehat{\boldsymbol{\beta}}_\gamma^\mathsf{T} X_\gamma^\mathsf{T} X_\gamma \widehat{\boldsymbol{\beta}}_\gamma$, and $\widehat{\boldsymbol{\beta}}_\gamma = (X_\gamma^\mathsf{T} X_\gamma)^{-1} X_\gamma^\mathsf{T}\mathbf{y}$.

It can be seen from (10.4.23) that, for fixed c and w, $w(\boldsymbol{\gamma}|\mathbf{y})$ is monotonic in

$$SS_\gamma/\sigma^2 - F(c,w)|\boldsymbol{\gamma}|.$$

Therefore, the $\boldsymbol{\gamma}$ maximizing the posterior model probability $w(\boldsymbol{\gamma}|\mathbf{y})$ is equivalent to model selection based on the penalized sum of squares criterion. As pointed out by Chipman et al. (2001), many frequentist model selection criteria can now be obtained by choosing a particular value of c, w, and $F(c,w)$. For example, if c and w are chosen so that $F(c,w) = 2$, this yields Mallows C_p and approximately the AIC. Likewise, choice of $F(c,w) = \log n$ leads to the BIC, and $F(c,w) = 2\log p$ yields the RIC (Donoho and Johnstone (1994), Foster and George (1994)). In other words, depending on c, w, and F, selecting the highest-posterior model is equivalent to selecting the best AIC/C_p, BIC, and RIC models, respectively.

Furthermore, since c and w control the expected size and the proportion of the nonzero components of $\boldsymbol{\beta}$, the dependence of $F(c,w)$ on c and w provides a connection between the penalty F and the models it will favor. For example, large c will favor the models with large regression coefficients, and small w will favor models where the proportion of nonzero coefficients is small. To avoid fixing c and w, they can be treated as parameters and estimated by empirical Bayes by maximizing the marginal likelihood

$$\begin{aligned} L(c,w|\mathbf{y}) &\propto \sum_\gamma p(\boldsymbol{\gamma}|w)p(\mathbf{y}|\boldsymbol{\gamma},c) \\ &\propto \sum_\gamma w^{|\gamma|}(1-w)^{p-|\gamma|}(1+c)^{-|\gamma|/2}\exp\left\{\frac{cSS_\gamma}{2\sigma^2(1+c)}\right\}. \end{aligned}$$

However, this can be computationally overwhelming when p is large and X is not orthogonal.

10.5 Computational Comparisons

To conclude this chapter, it is important to present some comparisons of variable selection methods to show how well they work and suggest when to use them. The setting for examining some traditional methods, shrinkage methods, and Bayes methods is a standard linear regression model of the form

$$Y = X\boldsymbol{\beta} + \varepsilon, \quad \varepsilon \sim N(0, \sigma). \tag{10.5.1}$$

The true model used here is a four-term submodel of (10.5.1). When $n > p$, this mimics the usual linear regression setting. Here, however, in some simulations, the explanatory variables are assigned nontrivial dependence structures.

The second subsection here permits $p > n$ in (10.5.1). The same true model is used as a data generator but now the task is to find it from a vastly larger overall model. To do this, sure independence screening (SIS) is applied first and then various shrinkage methods are used on the output to identify a final model. In this way, the oracle property of some of the shrinkage methods may be retained. Note that, in this section, all data sets are assumed standardized.

10.5.1 The $n > p$ Case

To present simulations in the context of (10.5.1), several more specifications must be given. First, the true value of $\boldsymbol{\beta}$ was taken as $\boldsymbol{\beta} = (2.5, 3, 0, 0, 0, 1.5, 0, 0, 0, 4, 0)$. So, the correct number of important covariates is $q = 4$, and these have indices $\{1, 2, 6, 9\}$. The covariates were generated in several different ways, usually with an autoregressive structure of order one, $AR(1)$. That is, the X_js were generated as $X_j = \rho X_{j-1} + u_j$ with u_j taken as $N(0, 1)$. In this case, the covariance structure is $\text{Corr}(X_j, X_k) = \rho^{|j-k|}$. The other covariance structure simply took all the pairwise correlations among the X_js as the same, $\text{corr}(X_j, X_k) = \rho$ for $j \neq k$. In the cases examined, the two covariance structures often gave broadly similar results, so only the $AR(1)$ computations are presented in this section. The values chosen for the correlation were $\rho = 0, .5, .9$, corresponding to no dependence, moderate dependence, and high dependence.

The error term was assigned three values σ: $\sigma = 1$, to match the variance of the $\rho = 0$ variance of the X_js, and $\sigma = 2, 3$ to see how inference behaves when the noise is stronger. Computations with $n = 50, 100$ were done, and unless otherwise specified, the number of iterations was $N = 500$; this was found to be sufficient to get good approximations. The code for generating the data can be found in the Notes at the end of this chapter.

For each of the traditional and shrinkage methods for variable selection presented here, five numerical summaries of the sampling distribution are given. First is the average MSE,

$$MSE = \mathbb{E}[(X\widehat{\boldsymbol{\beta}} - X\boldsymbol{\beta})^{\mathsf{T}}(X\widehat{\boldsymbol{\beta}} - X\boldsymbol{\beta})] = (\widehat{\boldsymbol{\beta}} - \boldsymbol{\beta})^{\mathsf{T}}\mathbb{E}(X^{\mathsf{T}}X)(\widehat{\boldsymbol{\beta}} - \boldsymbol{\beta}).$$

Second, the number of explanatory variables correctly found to be zero is given. The true value is six. Third is the number of explanatory variables incorrectly found to be zero. The correct value is zero; the worst value is four. If $\mathscr{H} : \beta_j = 0$ is taken as a null hypothesis, then the second summary counts the number of false rejections and the third counts the number of false acceptances. Thus, the second and third numerical summaries correspond roughly to Type I and Type II errors. Fourth is the probability that the method selected the correct model; this is the fraction of times the correct model was chosen over the 500 iterations. Fifth are the inclusion probabilities of each of the explanatory variables.

For Bayes methods, frequentist evaluations are inappropriate since it is the posterior distribution that provides inference, not the sampling distribution. For these cases, in place of the first three numerical summaries, a graphical summary of the posterior distribution and its properties over the class of models can be provided. Parallel to the fourth and fifth summaries, the posterior probabilities of selecting the correct model and variable inclusion probabilities are given.

10.5.1.1 Traditional Methods

Consider using AIC, BIC, and GCV to find good submodels of (10.5.1). Since there are 2^{10} models to compare, it is common to reduce the problem by looking only at a sequence of p models formed by adding one variable at a time to a growing model. For instance, here, rather than evaluating AIC on all 2^{10} models, an initial good model with one variable is found and its AIC computed. Then, forward selection is used to find the next variable it is best to include, and its AIC is found. Continuing until all ten variables have been included and ten AIC scores have been found gives a list of ten good submodels from which the model with the highest AIC score can be chosen. This can be done for BIC and GCV as well. Although forward selection is used here, backward selection and stepwise selection could have been used instead and often are.

To get the results, the leaps package was used to get the whole sequence for the forward model selection. In the same notation as the Notes at the end of this chapter, the commands used here were:

```
library(leaps)
  forward_fit <- regsubsets(Xtr,ytr,method="forward")
  aic <- which.min(2*(2:(p+2))+n*log(forward_fit$rss/n)
  +n+n*log(2*pi))
```

```
bic <- which.min(log(n)*(2:(p+2))
+n*log(forward_fit$rss/n)+n+n*log(2*pi))
gcv <- which.min(forward_fit$rss
/(n*(1-(2:(p+2))/n)^2))
```

The command library() is used to include special packages in R, such as leaps, that are not in the base library.

To begin, Table 10.2 shows the results of using AIC, BIC, and GCV when $\rho = 0$. The first column indicates the sample size. The second column indicates the SD of the error term. The third column indicates the criterion used. The next four columns contain the first four of the numerical summaries; the numbers in parentheses give the SE of the MSE.

				Average # of Zero Coef.		
n	σ	IC	MSE	Corr. Zero (6)	Inc. Zero (0)	PCS
50	1	AIC	0.186 (0.006)	5.060	0	0.366
		GCV	0.187 (0.006)	5.036	0	0.356
		BIC	0.139 (0.006)	5.722	0	0.754
	2	AIC	0.747 (0.025)	5.060	0.002	0.366
		GCV	0.751 (0.024)	5.036	0.002	0.356
		BIC	0.568 (0.024)	5.712	0.008	0.750
	3	AIC	1.722 (0.057)	5.022	0.040	0.354
		GCV	1.731 (0.057)	5.000	0.040	0.344
		BIC	1.476 (0.061)	5.598	0.120	0.660
100	1	AIC	0.082 (0.002)	5.044	0	0.364
		GCV	0.082 (0.002)	5.030	0	0.360
		BIC	0.055 (0.002)	5.816	0	0.832
	2	AIC	0.328 (0.010)	5.044	0	0.364
		GCV	0.329 (0.009)	5.030	0	0.360
		BIC	0.222 (0.008)	5.816	0	0.832
	3	AIC	0.739 (0.021)	5.042	0.002	0.362
		GCV	0.742 (0.022)	5.028	0.002	0.358
		BIC	0.501 (0.018)	5.814	0.002	0.830

Table 10.2 Model selection and fitting results when the explanatory variables are $AR(\rho = 0)$. When $n = 50$, it is seen that, as σ increases, the MSE for each method increases. Likewise, the other columns increase or decrease as anticipated with increasing noise. The same holds for $n = 100$. However, a few small reversals from what would be expected can be seen across sample sizes. For instance, the PCS for AIC with $\sigma = 2$ drops from .366 when $n = 50$ to .364 when $n = 100$.

It can be seen that AIC and GCV give comparable performances; perhaps AIC does slightly better. This is no surprise since AIC, GCV, and cross-validation are all fairly similar in many cases. Also, BIC performs meaningfully better, especially in terms of probability of correct selection (PCS) nearly doubling the values of AIC or GCV.

The fifth collection of numerical summaries consists of the variable inclusion probabilities. These are the proportion of times a variable was included in the estimated model out of the 500 iterations. Table 10.3 gives the inclusion probabilities for AIC,

BIC, and GCV in the context of (10.5.1) for $\rho = 0$. Recall that the data generator used X_1, X_2, X_6, and X_9.

n	σ	IC	X_1	X_2	X_3	X_4	X_5	X_6	X_7	X_8	X_9	X_{10}
50	1	AIC	1	1	0.152	0.184	0.154	1	0.146	0.164	1	0.140
		GCV	1	1	0.152	0.194	0.164	1	0.146	0.168	1	0.140
		BIC	1	1	0.052	0.054	0.048	1	0.034	0.052	1	0.038
	2	AIC	1	1	0.152	0.184	0.154	0.998	0.144	0.162	1	0.142
		GCV	1	1	0.152	0.194	0.164	0.998	0.144	0.166	1	0.142
		BIC	1	1	0.054	0.052	0.048	0.992	0.034	0.052	1	0.040
	3	AIC	1	1	0.148	0.184	0.156	0.960	0.146	0.164	1	0.140
		GCV	1	1	0.148	0.194	0.164	0.960	0.146	0.168	1	0.140
		BIC	0.998	1	0.054	0.054	0.048	0.882	0.034	0.054	1	0.038
100	1	AIC	1	1	0.176	0.162	0.144	1	0.158	0.156	1	0.160
		GCV	1	1	0.178	0.166	0.144	1	0.158	0.156	1	0.168
		BIC	1	1	0.036	0.034	0.024	1	0.036	0.022	1	0.032
	2	AIC	1	1	0.176	0.162	0.144	1	0.158	0.156	1	0.160
		GCV	1	1	0.178	0.166	0.144	1	0.158	0.156	1	0.168
		BIC	1	1	0.036	0.034	0.024	1	0.036	0.022	1	0.032
	3	AIC	1	1	0.176	0.162	0.144	0.998	0.158	0.156	1	0.160
		GCV	1	1	0.178	0.166	0.144	0.998	0.158	0.156	1	0.168
		BIC	1	1	0.036	0.034	0.024	0.998	0.036	0.022	1	0.032

Table 10.3 Selection probabilities of the explanatory variables for $AR(\rho = 0)$. When $n = 50$, as σ increases, the selection probabilities of correct variables generally increase and the selection probabilities of incorrect variables generally decrease. However, there are some reversals: For X_7, both the AIC and GCV inclusion probabilities decrease as σ increases from one to two. When $n = 100$, the entries are preternaturally stable but consistent with intuition as σ increases. However, again the comparison between sample sizes yields inconsistencies. For instance, for X_3, the inclusion probabilities for AIC and GCV rise as sample size rises for fixed σ. The same is seen for X_7 and X_{10}.

Tables 10.4 and 10.5 parallel Tables 10.2 and 10.3 but have $\rho = .5$. As ρ increases, the dependence among the explanatory variables increases. Moreover, the dependence structure of X_j involves the previous $j-1$ variables, so the complexity of the dependence within the explanatory variables increases with the index. Note that the PCS for each method at each σ and n does not change much but that the drop in performance of AIC and GCV for fixed σ as n increases is much larger than for Table 10.2.

Similarly, in comparing Table 10.3 with Table 10.5, the extent of the reversals from what one would expect increases. For instance, with GCV, X_{10}, and $\sigma = 1$, the difference in inclusion probabilities from $n = 50$ to $n = 100$ is .036 in Table 10.5 but is .02 in Table 10.3. Overall, BIC seems to be least affected by the dependence, although the performance of BIC also deteriorates. It is tempting to conjecture that the forward selection is breaking down because of the dependence; however, the good performance of BIC seems to militate against this.

n	σ	IC	MSE	Average # of Zero Coef. Corr. Zero (6)	Inc. Zero (0)	PCS
50	1	AIC	0.187 (0.006)	5.030	0	0.392
		GCV	0.189 (0.006)	4.996	0	0.374
		BIC	0.135 (0.005)	5.718	0	0.748
	2	AIC	0.757 (0.025)	5.018	0.012	0.394
		GCV	0.763 (0.024)	4.984	0.012	0.376
		BIC	0.553 (0.024)	5.698	0.012	0.742
	3	AIC	1.786 (0.060)	4.910	0.096	0.364
		GCV	1.793 (0.059)	4.886	0.090	0.352
		BIC	1.589 (0.063)	5.458	0.206	0.594
100	1	AIC	0.081 (0.002)	5.018	0	0.362
		GCV	0.081 (0.002)	5.016	0	0.358
		BIC	0.055 (0.002)	5.808	0	0.820
	2	AIC	0.326 (0.009)	5.018	0	0.362
		GCV	0.326 (0.009)	5.016	0	0.358
		BIC	0.220 (0.008)	5.808	0	0.820
	3	AIC	0.734 (0.021)	5.016	0.002	0.360
		GCV	0.735 (0.022)	5.014	0.002	0.356
		BIC	0.511 (0.020)	5.794	0.012	0.812

Table 10.4 Model selection and fitting results for $AR(\rho = .5)$.

n	σ	IC	X_1	X_2	X_3	X_4	X_5	X_6	X_7	X_8	X_9	X_{10}
50	1	AIC	1	1	0.172	0.186	0.164	1	0.158	0.160	1	0.130
		GCV	1	1	0.178	0.190	0.170	1	0.164	0.166	1	0.136
		BIC	1	1	0.046	0.048	0.046	1	0.044	0.054	1	0.044
	2	AIC	1	1	0.170	0.186	0.168	0.988	0.156	0.160	1	0.130
		GCV	1	1	0.176	0.190	0.174	0.988	0.162	0.166	1	0.136
		BIC	1	1	0.048	0.048	0.048	0.988	0.044	0.056	1	0.046
	3	AIC	0.998	1	0.170	0.184	0.172	0.906	0.168	0.166	1	0.134
		GCV	0.998	1	0.172	0.188	0.178	0.912	0.176	0.172	1	0.138
		BIC	0.988	1	0.050	0.050	0.064	0.806	0.062	0.066	1	0.044
100	1	AIC	1	1	0.172	0.170	0.144	1	0.172	0.158	1	0.166
		GCV	1	1	0.172	0.166	0.146	1	0.178	0.158	1	0.164
		BIC	1	1	0.040	0.030	0.032	1	0.032	0.016	1	0.042
	2	AIC	1	1	0.172	0.170	0.144	1	0.172	0.158	1	0.166
		GCV	1	1	0.172	0.166	0.146	1	0.178	0.158	1	0.164
		BIC	1	1	0.040	0.030	0.032	1	0.032	0.016	1	0.042
	3	AIC	1	1	0.172	0.170	0.144	0.998	0.172	0.158	1	0.166
		GCV	1	1	0.172	0.166	0.146	0.998	0.178	0.158	1	0.164
		BIC	1	1	0.040	0.030	0.032	0.988	0.034	0.016	1	0.042

Table 10.5 Selection frequencies of variables for $AR(\rho = 0.5)$.

The corresponding tables for $\rho = .9$ differed from those for $\rho = .5$, in much the same way as the tables for $\rho = .5$ differed from those for $\rho = 0$. All the methods deteriorated in all scenarios; their relative performance remained much the same.

10.5.1.2 Shrinkage Methods

Consider using shrinkage methods to find good submodels of (10.5.1). A good range of methods is provided by using the elastic net (Enet), LASSO, a variant on Enet called the adaptive Enet (AEnet), Adaptive LASSO (ALASSO), and SCAD penalties. These methods automatically reduce the number of models to be considered from 2^{10} to 10 since the size of the decay parameter orders the variables for inclusion.

The computations for Enet, LASSO, AEnet, and ALASSO can be done with existing R packages so it is worth giving some explanation of this. However, the optimization for SCAD (see (10.3.20)) is not yet available in established packages. So, it is enough to comment that the results here are obtained using private code to implement the local quadratic approximation (LQA) theorem suggested by Fan and Li (2001). Also, while the discussion here uses a fixed value of the decay parameter λ, in practice λ is chosen using the BIC for all the methods in this section.

To describe the four procedures amenable to R, assume X is the matrix of standardized predictors, y is the vector of responses, and a fixed value of the decay parameter λ is chosen. To begin the description, it is easiest to start with LASSO and ALASSO. Recall the LARS algorithm was described in Section 10.3.3 and can provide an entire sequence of coefficients and fits, starting from zero, for least squares fitting. For LASSO, the optimization in (10.3.10) can be done in R using the following:

```
> library(lars)
> lasso_fit <- lars(X,y,type="lasso")
> lasso_coef <- coef(lasso_fit, type="coef",
mode="lambda")
```

Next, ALASSO solutions can be derived from lars by using a transformation. Recall that ALASSO solves (10.3.18), in which the w_js are prespecified weights. Let $w = (w_1, \cdots, w_d)^{\mathrm{T}}$ be the weight vector. Zou (2006) constructs a transformed design matrix X^* of dimension $n \times p$ as

$$X^*_{ij} = X_{ij}/w_j \quad \text{for } i = 1, \cdots, n \text{ and } j = 1, \cdots, d.$$

Then lasso is used in lars to obtain

$$\widehat{\boldsymbol{\beta}}^* = \arg\min(\boldsymbol{y} - X^*\boldsymbol{\beta})^{\mathrm{T}}(\boldsymbol{y} - X^*\boldsymbol{\beta}) + \lambda \sum_{j=1}^{p} |\beta_j|,$$

so that the ALASSO solution is found as $\widehat{\boldsymbol{\beta}} * alasso = (\hat{\beta}^*_1/w_1, \cdots, \hat{\beta}^*_d/w_d)^{\mathrm{T}}$.

For Enet and AEnet, recall that the elastic net estimator achieves (10.3.27). Within R, Enet solutions can be found using

```
> library(elasticnet)
> enet_fit <- enet(X,y,lambda)
> enet_coef <-coef(enet_fit, type="coef",
mode="penalty")
```

The AEnet is an adaptive form of the Enet derived by using more weights in the penalty, much like ALASSO is obtained from LASSO and ACOSSO is obtained from COSSO. That is, the adaptive elastic net estimate is

$$\widehat{\boldsymbol{\beta}}^{aenet} = (1+\lambda_2)\left\{\arg\min_{\boldsymbol{\beta}}(\boldsymbol{y}-X\boldsymbol{\beta})^{\mathrm{T}}(\boldsymbol{y}-X\boldsymbol{\beta})+\lambda_1\sum_{j=1}^{p} w_j|\beta_j|+\lambda\sum_{j=1}^{p}\beta_j^2\right\}.$$

The AEnet estimate can be obtained by solving a transformed Enet problem, much like ALASSO can be obtained from LASSO. Moreover, like the other adaptive versions of shrinkage methods, the AEnet is oracle; see Ghosh (2007).

Parallel to Tables 10.2–10.5, Tables 10.6–10.9 show the performance of Enet, LASSO, AEnet, ALASSO, and SCAD for (10.5.1), but here $\rho = 0, .9$; the $\rho = .5$ case is omitted since its results are intermediate. For comparison, the same numerical summaries are provided on data generated the same way. Note that the ordering on the methods is roughly in terms of performance for $\rho = 0$; in some cases below, however, ALASSO or AEnet outperforms SCAD.

Tables 10.6 and 10.7 match intuition for each method as σ and n vary. Overall, it can be seen that the three methods that are oracle (AEnet, ALASSO, SCAD) give better results than the two that are not (LASSO, Enet). However, there is a slight tendency for ALASSO to give better variable selection probabilities for incorrect variables than for SCAD. This is seen in Table 10.7 for X_{10} when $n = 50$, $\sigma = 3$. However, more typically for this case, ALASSO is intermediate between SCAD and AEnet.

When $\rho = .9$, Tables 10.8 and 10.9 show results that can be compared with those in Tables 10.6 and 10.7. Again, for each method, performance increases with n and decreases with σ. Also, the three oracle methods outperform the two nonoracle methods. However, the relative performance changes in particular among AEnet, ALASSO, and SCAD. Note that except for the case where $n = 50$ and $\sigma = 1$, ALASSO has a higher PCS than AEnet or SCAD. AEnet also appears to outperform SCAD when n and σ are larger. This suggests, but does not establish, that SCAD may be more sensitive to dependence structure than ALASSO or AEnet.

Table 10.9 exhibits the same trends as Table 10.8: In the presence of dependence, ALASSO tends to give better inclusion probabilities than SCAD and AEnet for both the variables in the data generator and for the noise variables. Likewise, AEnet outperforms SCAD. Interestingly, the inclusion probabilities do not seem to depend on the index of the explanatory variables apart from being closer to zero for noise variables and closer to one for correct variables.

In contrast to Tables 10.6 and 10.7, Tables 10. 8 and 10.9 show that all five methods deteriorate substantially, in all five summaries, in the presence of strong dependence.

10.5.1.3 Bayes Methods

As noted in Section 10.3, Bayes inference rests on the posterior. The posterior is formed from the prior and likelihood. In (10.5.1), the likelihood is given by the

n	σ	IC	MSE	Average # of Zero Coef. Corr. Zero (6)	Inc. Zero (0)	PCS
50	1	Enet	0.233 (0.007)	4.042	0	0.152
		LASSO	0.221 (0.006)	4.506	0	0.250
		AEnet	0.151 (0.007)	5.658	0	0.734
		Alasso	0.128 (0.005)	5.758	0	0.796
		SCAD	0.112 (0.005)	5.866	0	0.920
	2	Enet	0.893 (0.026)	4.194	0.002	0.182
		LASSO	0.889 (0.027)	4.510	0.004	0.252
		AEnet	0.611 (0.025)	5.610	0.010	0.704
		Alasso	0.574 (0.023)	5.688	0.010	0.736
		SCAD	0.517 (0.022)	5.760	0.006	0.824
	3	Enet	2.001 (0.059)	4.324	0.046	0.194
		LASSO	2.011 (0.059)	4.530	0.046	0.250
		AEnet	1.533 (0.057)	5.562	0.108	0.588
		Alasso	1.522 (0.057)	5.612	0.114	0.622
		SCAD	1.564 (0.061)	5.592	0.156	0.622
100	1	Enet	0.105 (0.003)	4.404	0	0.206
		LASSO	0.101 (0.003)	4.804	0	0.328
		AEnet	0.058 (0.002)	5.792	0	0.838
		Alasso	0.052 (0.002)	5.826	0	0.860
		SCAD	0.047 (0.002)	5.924	0	0.964
	2	Enet	0.400 (0.011)	4.532	0.00	0.232
		LASSO	0.403 (0.011)	4.804	0.00	0.328
		AEnet	0.243 (0.009)	5.748	0.00	0.804
		Alasso	0.220 (0.008)	5.788	0.00	0.834
		SCAD	0.203 (0.007)	5.884	0.00	0.916
	3	Enet	0.890 (0.024)	4.626	0.002	0.260
		LASSO	0.905 (0.024)	4.804	0.002	0.326
		AEnet	0.549 (0.019)	5.716	0.002	0.778
		Alasso	0.530 (0.018)	5.748	0.002	0.800
		SCAD	0.513 (0.021)	5.778	0.010	0.876

Table 10.6 Model selection and fitting results for $AR(\rho = 0)$. As n increases or σ decreases, the performance of each method increases.

distribution on ε but also depends on the sampling properties of the explanatory variables. In effect, Bayesians condition on the explanatory variables and a careful Bayesian would build their distribution into the posterior to reflect the dependence. Here, however, this will be neglected. Thus, even though there is a dependence structure on $(X_1, \ldots, X_{10})$, they will be treated as deterministic and independent in the underlying mathematics. While incorrect, it will be seen that the results will still be interpretable and reflect the distribution of the X_js.

So, to specify the posterior, it remains to specify the priors. There are two priors, one across models and one within models. First recall that there are $2^{10} = 1024$ models and that within each model there are up to 11 parameters: up to 10 for the coefficients of the variables and one more for σ. For simplicity, the prior over models will be uniform; i.e., each model gets probability $1/2^{10}$. The prior on the coefficients within models will be either $N(0,1)$ or given by Zellner's g-prior; here $\sigma = 1,2,3$ will be taken as fixed so

n	σ	IC	X_1	X_2	X_3	X_4	X_5	X_6	X_7	X_8	X_9	X_{10}
50	1	Enet	1	1	0.294	0.296	0.310	1	0.284	0.356	1	0.302
		LASSO	1	1	0.218	0.250	0.246	1	0.238	0.268	1	0.224
		AEnet	1	1	0.056	0.054	0.054	1	0.046	0.070	1	0.046
		Alasso	1	1	0.042	0.046	0.040	1	0.028	0.036	1	0.036
		SCAD	1	1	0.026	0.024	0.018	1	0.016	0.032	1	0.016
	2	Enet	1	1	0.280	0.284	0.302	0.994	0.280	0.336	1	0.294
		LASSO	1	1	0.218	0.250	0.246	0.994	0.238	0.268	1	0.224
		AEnet	1	1	0.066	0.070	0.070	0.990	0.060	0.088	1	0.066
		ALASSO	1	1	0.052	0.060	0.056	0.990	0.042	0.060	1	0.048
		SCAD	1	1	0.044	0.052	0.048	0.988	0.036	0.052	1	0.040
	3	Enet	1.000	1	0.262	0.276	0.270	0.946	0.272	0.324	1	0.260
		LASSO	1.000	1	0.214	0.244	0.238	0.942	0.232	0.260	1	0.220
		AEnet	1.000	1	0.072	0.084	0.072	0.878	0.070	0.088	1	0.070
		ALASSO	1.000	1	0.066	0.082	0.070	0.870	0.062	0.058	1	0.052
		SCAD	0.996	1	0.076	0.072	0.072	0.832	0.066	0.084	1	0.068
100	1	Enet	1	1	0.316	0.228	0.250	1	0.292	0.314	1	0.306
		LASSO	1	1	0.224	0.198	0.198	1	0.228	0.222	1	0.202
		AEnet	1	1	0.044	0.036	0.038	1	0.046	0.044	1	0.056
		Alasso	1	1	0.046	0.030	0.028	1	0.040	0.012	1	0.036
		SCAD	1	1	0.004	0.014	0.004	1	0.014	0.006	1	0.008
	2	Enet	1	1	0.294	0.222	0.242	1	0.284	0.308	1	0.286
		LASSO	1	1	0.224	0.198	0.198	1	0.228	0.222	1	0.202
		AEnet	1	1	0.054	0.046	0.038	1	0.056	0.050	1	0.068
		Alasso	1	1	0.032	0.032	0.028	1	0.044	0.016	1	0.040
		SCAD	1	1	0.016	0.020	0.020	1	0.026	0.008	1	0.030
	3	Enet	1	1	0.280	0.216	0.228	0.998	0.274	0.284	1	0.258
		LASSO	1	1	0.224	0.196	0.198	0.998	0.228	0.222	1	0.202
		AEnet	1	1	0.052	0.056	0.042	0.994	0.048	0.048	1	0.066
		Alasso	1	1	0.036	0.044	0.038	0.996	0.034	0.034	1	0.042
		SCAD	1	1	0.034	0.044	0.028	0.990	0.018	0.018	1	0.042

Table 10.7 Selection frequencies of variables for $AR(\rho = 0)$. As n increases or σ decreases, the inclusion probabilities of incorrect variables decrease and the inclusion probabilities of correct variables increase.

no prior on it needs to be specified. The g in the g-prior is chosen by a cross-validation (empirical Bayes) criterion, so no prior needs to be specified on it either.

It is important to remember that Bayes evaluations are done conditionally on the data via the posterior. As before, sample sizes of $n = 50, 100$ were used; only results for $n = 100$ will be shown. In the key inferential calculations, therefore, the conditional probabilities are shown for exactly one set of data. For evaluating methods overall by inclusion probabilities, one can find the actual inclusion probability from the posterior or, as done here, average over N iterations. When this is done, $N = 50$ is used rather than $N = 500$, strictly for convenience.

For the uniform prior on models and the normal prior on coefficients, the computations can be done in R. The BMA package can be loaded and called by

```
library(BMA)
bma.result <- bicreg(X,y, strict=FALSE, OR = 20)
```

				Average # of Zero Coef.		
n	σ	IC	MSE	Corr. Zero (6)	Inc. Zero (0)	PCS
50	1	Enet	0.177 (0.005)	3.508	0.002	0.040
		LASSO	0.176 (0.005)	3.822	0.002	0.056
		AEnet	0.144 (0.006)	5.572	0.020	0.706
		Alasso	0.141 (0.006)	5.658	0.020	0.766
		SCAD	0.133 (0.006)	5.626	0.018	0.778
	2	Enet	0.664 (0.020)	3.570	0.112	0.054
		LASSO	0.683 (0.021)	3.846	0.120	0.072
		AEnet	0.762 (0.027)	5.324	0.414	0.560
		Alasso	0.811 (0.029)	5.470	0.474	0.588
		SCAD	0.770 (0.026)	5.166	0.402	0.482
	3	Enet	1.371 (0.039)	3.628	0.332	0.088
		LASSO	1.453 (0.044)	3.900	0.374	0.112
		AEnet	1.773 (0.049)	5.268	0.974	0.358
		Alasso	2.016 (0.058)	5.434	1.182	0.334
		SCAD	1.642 (0.048)	5.078	0.828	0.400
100	1	Enet	0.076 (0.002)	3.594	0.000	0.050
		LASSO	0.075 (0.003)	3.852	0.000	0.080
		AEnet	0.055 (0.002)	5.760	0.000	0.820
		Alasso	0.050 (0.002)	5.886	0.000	0.896
		SCAD	0.051 (0.002)	5.830	0.000	0.894
	2	Enet	0.299 (0.008)	3.572	0.016	0.044
		LASSO	0.301 (0.010)	3.850	0.020	0.082
		AEnet	0.270 (0.010)	5.554	0.110	0.734
		Alasso	0.273 (0.010)	5.674	0.128	0.790
		SCAD	0.293 (0.011)	5.358	0.114	0.656
	3	Enet	0.646 (0.018)	3.622	0.112	0.054
		LASSO	0.661 (0.018)	3.860	0.112	0.084
		AEnet	0.729 (0.023)	5.384	0.422	0.602
		Alasso	0.795 (0.026)	5.560	0.504	0.620
		SCAD	0.714 (0.022)	5.204	0.358	0.552

Table 10.8 Model selection and fitting results for $AR(\rho = .9)$. Note that in contrast to Table 10.6, ALASSO usually gives better values than SCAD or AEnet in the four columns indicating performance of the methods. AEnet also does better than SCAD for those cases.

(The notation is as in the Notes at the end of this chapter.) This searches the model space and computes all that is needed for Bayes model averaging. As a consequence, it provides the posterior weights (probabilities) of all the models. The command help(bicreg) can be used to find all the outputs of the BMA package.

For Zellner's g-prior, the function bayes.model.selection by Jim Albert, was taken from `http://learnbayes.blogspot.com/2007/11/` and go to the section `bayesian-model-selection.html`. It is freely available.

Turning now to the actual results, for the uniform-normal case, Fig. 10.4 shows the posterior probabilities for $\rho = 0$ for one set of generated data. It is seen that all the top models had X_1, X_2, and X_9. All but one of the top ten models had X_6. Other variables occasionally appeared, but not with any obvious pattern.

n	σ	IC	X_1	X_2	X_3	X_4	X_5	X_6	X_7	X_8	X_9	X_{10}
50	1	Enet	1	1	0.400	0.358	0.422	0.998	0.408	0.448	1	0.456
		LASSO	1	1	0.360	0.324	0.384	0.998	0.364	0.400	1	0.346
		AEnet	1	1	0.068	0.058	0.082	0.980	0.070	0.082	1	0.0648
		Alasso	1	1	0.052	0.058	0.070	0.980	0.058	0.060	1	0.044
		SCAD	1	1	0.056	0.044	0.068	0.982	0.082	0.070	1	0.054
	2	Enet	0.998	0.998	0.404	0.352	0.414	0.892	0.394	0.442	1	0.424
		LASSO	0.998	0.998	0.354	0.318	0.380	0.886	0.360	0.398	0.998	0.344
		AEnet	0.960	0.970	0.118	0.098	0.118	0.658	0.128	0.134	0.998	0.080
		ALASSO	0.942	0.962	0.092	0.086	0.100	0.624	0.112	0.088	0.998	0.052
		SCAD	0.968	0.980	0.142	0.118	0.142	0.652	0.150	0.152	0.998	0.130
	3	Enet	0.968	0.974	0.392	0.346	0.406	0.734	0.384	0.430	0.992	0.414
		LASSO	0.958	0.970	0.348	0.316	0.368	0.706	0.344	0.388	0.992	0.336
		AEnet	0.824	0.852	0.124	0.112	0.108	0.382	0.134	0.144	0.968	0.110
		ALASSO	0.734	0.806	0.098	0.098	0.086	0.328	0.116	0.100	0.950	0.068
		SCAD	0.882	0.900	0.152	0.124	0.140	0.420	0.178	0.160	0.970	0.168
100	1	Enet	1	1	0.412	0.320	0.360	1	0.448	0.454	1	0.412
		LASSO	1	1	0.384	0.292	0.332	1	0.406	0.404	1	0.330
		AEnet	1	1	0.046	0.022	0.024	1	0.044	0.042	1	0.062
		Alasso	1	1	0.022	0.016	0.018	1	0.026	0.010	1	0.022
		SCAD	1	1	0.038	0.016	0.020	1	0.040	0.022	1	0.034
	2	Enet	1	1	0.420	0.334	0.358	0.984	0.446	0.446	1	0.424
		LASSO	1	1	0.384	0.292	0.332	0.980	0.406	0.404	1	0.332
		AEnet	1	1	0.072	0.044	0.072	0.890	0.114	0.078	1	0.066
		Alasso	1	1	0.104	0.032	0.068	0.874	0.092	0.040	1	0.042
		SCAD	1	1	0.034	0.076	0.098	0.886	0.132	0.122	1	0.110
	3	Enet	1	1	0.414	0.318	0.362	0.888	0.444	0.446	1	0.394
		LASSO	1	1	0.384	0.290	0.328	0.888	0.404	0.402	1	0.332
		AEnet	0.972	0.972	0.102	0.056	0.106	0.634	0.136	0.122	1	0.094
		Alasso	0.938	0.956	0.062	0.044	0.098	0.602	0.118	0.066	1	0.052
		SCAD	0.988	0.980	0.138	0.092	0.122	0.674	0.164	0.148	1	0.132

Table 10.9 Selection frequencies of variables for $AR(\rho = .9)$. The conclusions suggested by Table 10.8 are generally supported.

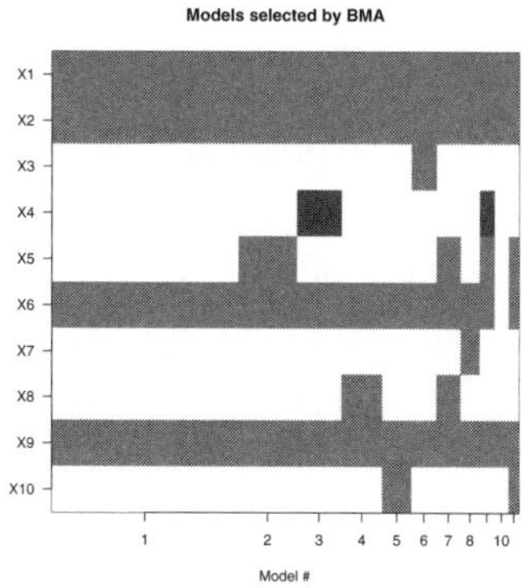

Fig. 10.4 Posterior probabilities for the uniform-normal case. The ten models (out of the 1024 models) with the highest posterior probabilities are indicated along the horizontal axis. The ten explanatory variables are noted on the vertical axis. A solid square on the grid means that the variable was included in the model. A white square means the variable was not included. The few indeterminate cases are also indicated.

As ρ increases, Fig. 10.5 shows results similar to those in Fig. 10.4. Note that larger values of ρ are associated with the posterior distribution being less concentrated so that, rather than ten models, 15 or 30 are required, as indicated on the horizontal axis. Thus, the figures have many smaller rectangles to indicate which variables are present, and many of these are incorrect. Note that Figs. 10.4 and 10.5 are not really analogous to any of the earlier tables because they only represent a single data set.

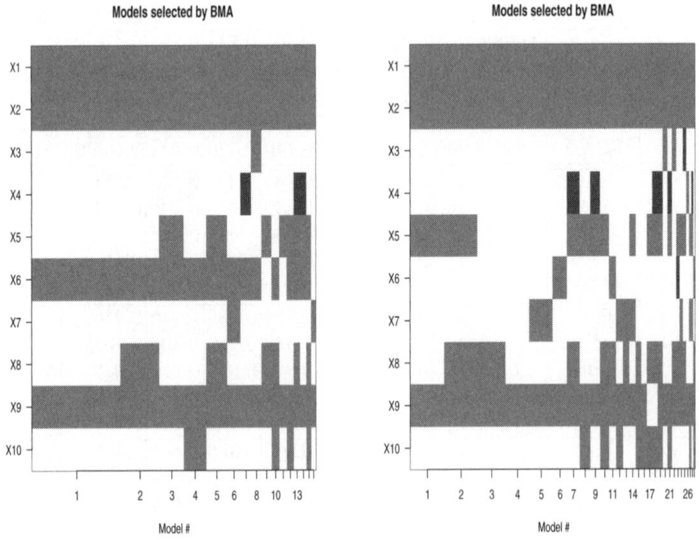

Fig. 10.5 The two panels here are the same as in Fig. 10.4 but for $\rho = .5, .9$.

If the analysis for a single data set given in Fig. 10.4 or 10.5 is repeated and the variable inclusions counted over 50 iterations, average inclusion probabilities in a frequentist sense can be computed and compared with the corresponding results for traditional and shrinkage methods. These are recorded in Table 10.10.

ρ	X_1	X_2	X_3	X_4	X_5	X_6	X_7	X_8	X_9	X_{10}
0	1.00	1.00	.139	.139	.124	.994	.912	.130	1.00	.122
0.5	1.00	1.00	.129	.147	.142	.974	.963	.137	1.00	.119
0.9	.912	.951	.119	.141	.159	.559	.129	. 146	.996	.109

Table 10.10 Here the inclusion probabilities for the ten variables are given for the three correlations among the explanatory variables in the uniform-normal case.

It is seen that the inclusion probabilities deteriorate as ρ increases. However, they are broadly comparable to the results in Tables 10.3 and 10.7 for $\rho = 0$, to Table 10.5 for $\rho = .5$, and to Table 10.9 for $\rho = .9$.

For the g-prior, code to generate visual summaries like Figs. 10.4 and 10.5 does not exist. Albert's code does, however, provide equivalent information, and this is presented next. First, parallel to Figs. 10.4 and 10.5, the posterior probabilities of the 1024 models can be output. For $\rho = 0$, one run of the simulation gave that the true model was the model of the posterior and had probability .527. For $\rho = .5$, again the true model was the mode of the posterior but had lower posterior probability, .459. When $\rho = .9$, however, the modal model had explanatory variables 1, 2, and 9, omitting 6, with probability .228. The correct model had only the fifth highest posterior probability, .058. The intervening models all had X_1, X_2, and X_9; however, they included a variable other than X_6 and had probabilities .134, .111, and .093.

Finally, parallel to Table 10.10, a table shows of inclusion probabilities can be compiled as in Table 10.11. Again, performance deteriorates as ρ increases, and the values are broadly comparable to the earlier cases.

ρ	X_1	X_2	X_3	X_4	X_5	X_6	X_7	X_8	X_9	X_{10}
0	1	1	0.093	0.094	0.085	0.999	0.064	0.090	1	0.084
.5	1	1	0.084	0.101	0.105	0.936	0.073	0.095	1	0.083
.9	0.821	0.901	0.094	0.112	0.128	0.470	0.113	0.116	0.990	0.086

Table 10.11 Here the inclusion probabilities for the ten variables are given for the three correlations among the explanatory variables for the uniform-g-prior case.

10.5.2 When $p > n$

In many problems, the number of potentially important variables is so large that the first task is to reduce them to a manageable number. In these cases, practitioners sometimes believe that the true model is quite small (i.e., sparse), so that little is lost and much is gained by the reduction.

To mimic this procedure, consider the case where (10.5.1) has $p = 200$ but the data-generating model is the same (i.e., has X_1, X_2, X_6, and X_{10}, with the same coefficients as before). Suppose the sample size is $n = 50, 100$ so that $p > n$ and that $\sigma = 1, 2$. Instead of $AR(1)$ generation of the explanatory variables, suppose they exhibit compound symmetry (CS). That is, all the pairwise correlations among the X_js are the same, $\text{Corr}(X_j, X_k) = \rho$ for $j \neq k$. This sort of dependence is more extensive, and typical for sparse models, than serial dependence is.

To search for a sparse model, recall that sure independence screening (SIS) is one way to reduce the available p variables to a number below n. Essentially, SIS ranks the explanatory variables according to the absolute values of their correlation with the response. Since the variance of the X_js is one and the variance of Y is common across all X_js, it is enough to look at the absolute covariance between Y and each X_j. When these are sorted by size, only those X_js with a high enough absolute covariance are retained. This is a simple procedure that can be coded easily. Here, for a variety of reasons, the SIS threshold was chosen to be 37. Fan and Lv (2008) suggest $n/\ln n$; however, here, comparisons over different sample sizes are made. Thresholds involving p may seem more reasonable; however, they also have drawbacks.

Once SIS is used to eliminate enough explanatory variables, shrinkage methods, for instance, may be applied to select an even smaller model; this second step is the same as in the last section. Combining shrinkage with SIS vastly expands the scope of variable selection problems that can be resolved effectively.

Below, the output of the combined SIS and shrinkage procedures is reported. The summaries roughly parallel those given earlier. However, there are three noteworthy differences. First, MSE was calculated using the model chosen from SIS and a shrinkage method. Thus, different iterations gave models with different numbers of explanatory variables. Second, the case $\rho = 0$ was used as a baseline. However, only computations with $\rho = .3$ were done because once ρ was much greater than .3, the performance was terrible. (The MSEs were around 14 for $\rho = .5$.) Third, in Table 10.12, the expected number of explanatory variables is given instead of the probability of correct selection. This is easier to compare with the other columns within the table, and the numbers would otherwise be quite low.

In Table 10.12, it is seen that the performance indices improve with larger n or smaller σ. Also, the numbers are much better behaved than the earlier ones where SIS did not need to be used. Not only are there few reversals, but the jumps from entry to entry and the ranges of performance within the cells are smaller. This may be due to the fact that none of the variables SIS selects are bad. They may be irrelevant, but they are less obviously irrelevant than in earlier cases. So, the numbers smooth out more. One consequence is that the difference in performance between shrinkage methods that have oracle properties and those that don't is much less than when SIS is not used. On the other hand, the Inc. Zero column is a little bit higher than for the earlier cases, too, partially because the combined procedure is more complex. Note that the average model sizes are much larger than four, often by a factor of 2, and that the sum of the last three columns is roughly 200.

Table 10.13 parallels Tables 10.3, 10.5, 10.7, and 10.11. However, it is just the first ten columns in a 200 column table where the columns correspond to the explanatory variables. The correct variables are 1, 2, 6, and 9, so all 190 later variables do not matter except in that they make the correct model hard to find. The entries in the irrelevant explanatory variables among the first ten give a good picture of the general behavior. The values are seen to be reasonable but generally worse for the correct variables (especially for X_1 and X_6) than in the earlier cases, where SIS was not used. This only means that the problem is more complicated, not that SIS is ineffective. The values for the irrelevant variables are lower than in the earlier cases; however, the sheer number of explanatory variables means that even though the chances of including any one of them are lower, the chances of including some of them are higher.

For a last comparison, the problem was rerun with $\rho = .3$. The results are seen in Tables 10.14 and 10.15. By and large, the deterioration is more visible in the worsening selection of correct variables. However, an overall increase in selection of incorrect variables is seen, too. Since the model sizes increase as well, there is a greater chance

n	σ	IC	MSE	Average # of Zero Coef. Corr. Zero (196)	Inc. Zero (0)	Model Size (4)
50	1	SIS+Enet	3.576 (0.218)	183.864	0.406	15.730
		SIS+LASSO	3.187 (0.210)	185.610	0.406	13.984
		SIS+AEnet	3.322 (0.217)	189.870	0.406	9.724
		SIS+Alasso	2.757 (0.206)	191.138	0.406	8.456
		SIS+SCAD	3.606 (0.236)	190.014	0.410	9.576
	2	SIS+Enet	4.970 (0.170)	186.732	0.454	12.814
		SIS+LASSO	4.665 (0.161)	187.868	0.454	11.678
		SIS+AEnet	4.083 (0.187)	192.074	0.458	7.468
		SIS+Alasso	3.351 (0.181)	193.074	0.458	6.648
		SIS+SCAD	5.530 (0.237)	190.542	0.468	5.530
100	1	SIS+Enet	0.733 (0.059)	189.242	0.126	10.632
		SIS+LASSO	0.669 (0.054)	190.576	0.126	9.298
		SIS+AEnet	0.660 (0.062)	193.194	0.126	6.680
		SIS+Alasso	0.526 (0.057)	194.162	0.126	5.712
		SIS+SCAD	0.693 (0.064)	192.306	0.126	7.568
	2	SIS+Enet	1.590 (0.056)	190.306	0.154	9.540
		SIS+LASSO	1.538 (0.056)	191.208	0.154	8.638
		SIS+AEnet	1.220 (0.063)	193.616	0.154	6.230
		SIS+Alasso	0.908 (0.058)	194.436	0.154	5.410
		SIS+SCAD	1.580 (0.079)	192.014	0.154	7.832

Table 10.12 Model selection and fitting results for $AR(\rho = 0)$. The values are less spread out than in Tables 10.2, 10.4, 10.6, and 10.8. Also, it is seen that SIS with ALASSO routinely gives better performance than SIS composed with any of the other four methods.

that wrong models with only one or two correct variables and a dozen or so incorrect variables will result. The effect will be high variance and high bias.

10.6 Notes

10.6.1 Code for Generating Data in Section 10.5

The code for the bayes.model.selection function was taken from `http://learnbayes.blogspot.com/2007/11/bayesian-model-selection.html`, and is due to Jim Albert.

For the frequentist results, the code is below, followed by the code used to call Albert's code.

```
library(BMA)
library(MASS)
library(MCMCpack)
library(LearnBayes)
```

n	σ	IC	X_1	X_2	X_3	X_4	X_5	X_6	X_7	X_8	X_9	X_{10}
50	1	SIS+Enet	0.952	0.986	0.070	0.068	0.070	0.656	0.066	0.072	1	0.054
		SIS+LASSO	0.952	0.986	0.064	0.060	0.060	0.656	0.062	0.058	1	0.044
		SIS+AEnet	0.952	0.986	0.036	0.046	0.028	0.656	0.036	0.036	1	0.020
		SIS+Alasso	0.952	0.986	0.026	0.028	0.026	0.656	0.036	0.034	1	0.016
		SIS+SCAD	0.952	0.986	0.038	0.040	0.022	0.652	0.040	0.036	1	0.016
	2	SIS+Enet	0.946	0.986	0.048	0.054	0.054	0.614	0.046	0.050	1	0.052
		SIS+LASSO	0.946	0.986	0.042	0.048	0.048	0.614	0.038	0.050	1	0.040
		SIS+AEnet	0.946	0.986	0.014	0.022	0.020	0.610	0.016	0.030	1	0.022
		SIS+Alasso	0.946	0.986	0.010	0.016	0.020	0.610	0.018	0.022	1	0.016
		SIS+SCAD	0.946	0.986	0.030	0.036	0.028	0.600	0.026	0.026	1	0.020
100	1	SIS+Enet	0.994	1	0.036	0.040	0.036	0.88	0.030	0.046	1	0.032
		SIS+LASSO	0.994	1	0.030	0.026	0.028	0.88	0.020	0.040	1	0.028
		SIS+AEnet	0.994	1	0.014	0.016	0.014	0.88	0.012	0.026	1	0.014
		SIS+Alasso	0.994	1	0.010	0.008	0.004	0.88	0.006	0.014	1	0.012
		SIS+SCAD	0.994	1	0.020	0.014	0.014	0.88	0.010	0.018	1	0.024
	2	SIS+Enet	0.994	1	0.024	0.018	0.028	0.852	0.026	0.036	1	0.032
		SIS+LASSO	0.994	1	0.020	0.014	0.024	0.852	0.024	0.028	1	0.030
		SIS+AEnet	0.994	1	0.010	0.010	0.010	0.852	0.012	0.014	1	0.014
		SIS+Alasso	0.994	1	0.006	0.004	0.002	0.852	0.008	0.012	1	0.008
		SIS+SCAD	0.994	1	0.016	0.014	0.014	0.852	0.024	0.026	1	0.032

Table 10.13 Selection frequencies of variables for $AR(\rho = 0)$.

```
##INPUT parameters
N = 500    #number of simulations
n = 100    #sample size
p = 10     #total number of covariates
p0 = 5     #number of nonzero regression coefficients
sigma = 3 #the standard deviation of noise
rho = 0.0 #the correlation coefficient in AR or CS
truebeta = c(2.5,3,0,0,0,1.5,0,0,4,0)
c =  500
##OUTPUT results
betahat = matrix(0,N,p)
beta0hat = rep(0,N)
modelerr = rep(0,N)
varprob = matrix(0,N,p)
gVarprob = matrix(0,N,p)
##specify the AR(rho) covariance matrix for X
Xcov<-matrix(0,p,p)
for (i in 1:p)
  {for (j in 1:p)
    Xcov[i,j]<-rho^(abs(i-j))
  }
##specify the CS(rho) covariance matrix for X
#Xcov<-rho*matrix(1,p,p)
#diag(Xcov) <- 1
```

n	σ	IC	MSE	Average # of Zero Coef. Corr. Zero (196)	Inc. Zero (0)	Model Size (4)
50	1	SIS+Enet	6.866 (0.460)	182.746	0.672	16.582
		SIS+LASSO	6.730 (0.465)	183.540	0.674	15.786
		SIS+AEnet	6.597 (0.440)	188.948	0.684	10.368
		SIS+Alasso	6.576 (0.440)	189.148	0.684	10.168
		SIS+SCAD	7.612 (0.427)	187.562	0.684	11.754
	2	SIS+Enet	7.047 (0.279)	186.168	0.734	13.098
		SIS+LASSO	6.691 (0.275)	187.376	0.734	11.890
		SIS+AEnet	6.199 (0.287)	192.526	0.758	6.716
		SIS+Alasso	5.870 (0.292)	193.240	0.776	5.984
		SIS+SCAD	8.781 (0.375)	191.062	0.772	8.166
100	1	SIS+Enet	1.553 (0.119)	186.938	0.332	12.730
		SIS+LASSO	1.518 (0.118)	187.758	0.332	11.910
		SIS+AEnet	1.444 (0.121)	192.840	0.332	6.828
		SIS+Alasso	1.394 (0.119)	193.202	0.332	6.466
		SIS+SCAD	1.756 (0.133)	190.412	0.332	9.256
	2	SIS+Enet	2.483 (0.120)	187.160	0.36	12.480
		SIS+LASSO	2.296 (0.118)	188.472	0.36	11.168
		SIS+AEnet	1.911 (0.125)	193.896	0.36	5.744
		SIS+Alasso	1.708 (0.120)	194.610	0.36	5.030
		SIS+SCAD	2.615 (0.143)	191.814	0.36	7.826

Table 10.14 Model selection and fitting results for $AR(\rho = .3)$. It is seen that the output here is nearly uniformly worse than in Table 10.12.

```
svd.Xcov<-svd(Xcov)
v<-svd.Xcov$v
d<-svd.Xcov$d
D<-diag(sqrt(d))
S<-(v)%*%D%*%t(v)
## LOOP starts here
for (i in 1:N)
{set.seed(2009+i)
 ##generate X
 Ztr<-matrix(rnorm(n*p),n,p)
 X<-Ztr%*%S
 #generate y
 ymean<-X%*%truebeta
 y<-ymean+sigma*rnorm(n)

 #data fit (here we apply the model fitting method)

 #-----------------------------------------#
 # BMA with normal on beta and uniform on models #
 #-----------------------------------------#
 # lma <- bicreg(X,y, strict=FALSE, OR = 20)
 # varprob[i,] <- lma$probne0
```

n	σ	IC	X_1	X_2	X_3	X_4	X_5	X_6	X_7	X_8	X_9	X_{10}
50	1	SIS+Enet	0.832	0.918	0.064	0.070	0.070	0.590	0.080	0.078	0.988	0.058
		SIS+LASSO	0.832	0.918	0.064	0.066	0.062	0.588	0.072	0.068	0.988	0.054
		SIS+AEnet	0.830	0.918	0.034	0.032	0.036	0.580	0.046	0.046	0.988	0.036
		SIS+Alasso	0.830	0.918	0.034	0.032	0.038	0.580	0.042	0.044	0.988	0.034
		SIS+SCAD	0.832	0.918	0.042	0.042	0.034	0.578	0.042	0.042	0.988	0.032
	2	SIS+Enet	0.814	0.908	0.048	0.046	0.052	0.564	0.052	0.052	0.980	0.052
		SIS+LASSO	0.814	0.908	0.042	0.038	0.048	0.564	0.048	0.050	0.980	0.042
		SIS+AEnet	0.814	0.906	0.012	0.016	0.026	0.542	0.018	0.028	0.980	0.018
		SIS+Alasso	0.814	0.904	0.006	0.016	0.022	0.526	0.008	0.024	0.980	0.018
		SIS+SCAD	0.814	0.904	0.020	0.038	0.026	0.530	0.034	0.032	0.980	0.024
100	1	SIS+Enet	0.956	0.988	0.036	0.058	0.056	0.724	0.040	0.034	1	0.032
		SIS+LASSO	0.956	0.988	0.030	0.044	0.050	0.724	0.034	0.030	1	0.028
		SIS+AEnet	0.956	0.988	0.006	0.022	0.016	0.724	0.012	0.014	1	0.016
		SIS+Alasso	0.956	0.988	0.006	0.022	0.014	0.724	0.008	0.014	1	0.016
		SIS+SCAD	0.956	0.988	0.016	0.038	0.028	0.724	0.024	0.028	1	0.032
	2	SIS+Enet	0.946	0.984	0.042	0.052	0.052	0.71	0.036	0.040	1	0.036
		SIS+LASSO	0.946	0.984	0.036	0.042	0.046	0.71	0.036	0.034	1	0.034
		SIS+AEnet	0.946	0.984	0.006	0.010	0.004	0.71	0.008	0.010	1	0.014
		SIS+Alasso	0.946	0.984	0.004	0.006	0.004	0.71	0.006	0.004	1	0.012
		SIS+SCAD	0.946	0.984	0.016	0.026	0.016	0.71	0.028	0.028	1	0.032

Table 10.15 Selection frequencies of variables for $AR(\rho = .3)$. It is seen that the output here is nearly uniformly worse than in Table 10.13.

```
#-------------------------------------------------#
# Bayes Model Selection with g-prior on beta     #
# and uniform on models                           #
#-------------------------------------------------#

X    <- data.frame(X)
prob <- matrix(0,1,p)

gPriorBayes    <- bayes.model.selection(y, X, c,
constant = FALSE) # Call to Jim Albert's code
post.prob.col  <- dim(gPriorBayes$mod.prob)[2]
sorted.gPrior  <- sort(gPriorBayes$mod.prob
[,post.prob.col], index.return = TRUE,
decreasing = TRUE)

# Type "gPriorBayes$mod.prob[sorted.gPrior$ix,]"
to see list of models for this run

# Extract post prob of var inclusion
post.prob <- gPriorBayes$mod.prob[,post.prob.col]
for(j in 1:p){
   prob[j] <- sum(post.prob[which
   (gPriorBayes$mod.prob[,j] == TRUE)]) }
```

```
gVarprob[i,] <- prob # Store post prob of var
                     # inclus for run i }
```

10.7 Exercises

Exercise 10.1 (Overfitting and underfitting). Variable selection is important for regression since there are problems in either using too many (irrelevant) or too few (omitted) variables in a regression model. Consider the linear regression model $y_i = \boldsymbol{x}_i^\top \boldsymbol{\beta}_0 + \varepsilon_i$, where the input vector is $\boldsymbol{x}_i \in R^p$ and the errors are IID, satisfying $E(\varepsilon_i) = 0$ and $\mathrm{Var}(\varepsilon_i) = \sigma^2$. Let $\boldsymbol{y} = (y_1, \ldots, y_n)^\top$ be the response vector and $X = (x_{ij}), i = 1, \cdots, n; j = 1, \ldots, p$ be the design matrix.

Assume that only the first p_0 variables are important. Let $\mathbf{A} = \{1, \ldots, p\}$ be the index set for the full model and $\mathbf{A}_0 = \{1, \ldots, p_0\}$ be the index set for the true model. The true regression coefficients can be denoted as $\boldsymbol{\beta}^* = (\boldsymbol{\beta}_{\mathbf{A}_0}^{*\top}, \mathbf{0}^\top)^\top$. Now consider three different modeling strategies:

- *Strategy I:* Fit the full model. Denote the full design matrix as $X_{\mathbf{A}}$ and the corresponding OLS estimator by $\widehat{\boldsymbol{\beta}}_{\mathbf{A}}^{ols}$.
- *Strategy II:* Fit the true model using the first p_0 covariates. Denote the corresponding design matrix by $X_{\mathbf{A}_0}$ and the OLS estimator by $\widehat{\boldsymbol{\beta}}_{\mathbf{A}_0}^{ols}$.
- *Strategy III:* Fit a subset model using only the first q covariates for some $q < p_0$. Denote the corresponding design matrix by $X_{\mathbf{A}_1}$ and the OLS estimator by $\widehat{\boldsymbol{\beta}}_{\mathbf{A}_1}^{ols}$.

1. As noted in Section 10.1, one possible consequence of including irrelevant variables in a regression model is that the predictions are not efficient (i.e., have larger variances) though they are unbiased. For any $\boldsymbol{x} \in R^p$, show that

$$\mathbb{E}(\boldsymbol{x}_{\mathbf{A}}^\top \widehat{\boldsymbol{\beta}}_{\mathbf{A}}^{ols}) = \boldsymbol{x}_{\mathbf{A}_0}^\top \boldsymbol{\beta}_{\mathbf{A}_0}^*, \quad \mathrm{Var}(\boldsymbol{x}_{\mathbf{A}}^\top \widehat{\boldsymbol{\beta}}_{\mathbf{A}}^{ols}) \geq \mathrm{Var}(\boldsymbol{x}_{\mathbf{A}_0}^\top \widehat{\boldsymbol{\beta}}_{\mathbf{A}_0}^{ols}),$$

   where $\boldsymbol{x}_{\mathbf{A}_0}$ consists of the first p_0 elements of $\boldsymbol{x}$.

2. One consequence of excluding important variables in a linear model is that the predictions are biased, though they have smaller variances. For any $\boldsymbol{x} \in R^p$, show that

$$\mathbb{E}(\boldsymbol{x}_{\mathbf{A}_1}^\top \widehat{\boldsymbol{\beta}}_{\mathbf{A}_1}^{ols}) \neq \boldsymbol{x}_{\mathbf{A}_0}^\top \boldsymbol{\beta}_{\mathbf{A}_0}^*, \quad \mathrm{Var}(\boldsymbol{x}_{\mathbf{A}_1}^\top \widehat{\boldsymbol{\beta}}_{\mathbf{A}_1}^{ols}) \leq \mathrm{Var}(\boldsymbol{x}_{\mathbf{A}_0}^\top \widehat{\boldsymbol{\beta}}_{\mathbf{A}_0}^{ols}),$$

   where $\boldsymbol{x}_{\mathbf{A}_1}$ consists of the first q elements of $\boldsymbol{x}$.

Exercise 10.2 (SSE from backward elimination).] Backward elimination starts with the model that includes all the variables. At each step, it removes the variable making the smallest contribution. Assume that there are currently k variables in the model, and the corresponding design matrix is X_1. Verify that the equation (10.1.12) holds; i.e., the new sum of squared errors (SSE) from fitting resulting from deleting the jth $(1 \le j \le k)$ variable from the current k-variable model is

$$SSE_{k-1} = SSE_k + \frac{(\hat{\beta}_{1j})^2}{s^{jj}},$$

where $\widehat{\boldsymbol{\beta}}_1 = (\hat{\beta}_{11}, \ldots, \hat{\beta}_{1k})^\mathsf{T}$ is the vector of current regression coefficients and s^{jj} is the jth diagonal element of $(X_1^\mathsf{T} X_1)^{-1}$.

Exercise 10.3 (Forward, backward, stepwise selection). Compare the performance of forward selection, backward elimination, and stepwise selection on the Boston housing data, which are available in the R package MASS.

1. What are the final selected covariates for each of the three methods?
2. What are the five-fold CV errors for each of the three methods?

Exercise 10.4 (Linearity of regression). As seen in Chapter 2, for a general regression model $y_i = f(\boldsymbol{x}_i) + \varepsilon_i$, for $i = 1, \cdots, n$, an estimator $\hat{f}$ of f is linear if the vector of predicted values $\hat{\boldsymbol{y}} = (\hat{f}(\boldsymbol{x}_1), \cdots, \hat{f}(\boldsymbol{x}_n))^\mathsf{T}$ satisfies

$$\hat{\boldsymbol{y}} = S\boldsymbol{y},$$

where S is the smoother matrix depending on $\boldsymbol{x}_i, i = 1, \cdots, n$ only but not on y_is. Here, apply this to the linear model $y_i = \beta_0 + \boldsymbol{\beta}\boldsymbol{x}_i + \varepsilon_i$, $i = 1, \cdots, n$.

1. Show that the least squares estimator is a linear smoother, and find its smoother matrix.
2. Show that the ridge estimator is a linear smoother, and find its smoother matrix.

Exercise 10.5 (Leave-one-out Property). Consider the regression model $y_i = f(\boldsymbol{x}_i) + \varepsilon_i$ for $i = 1, \cdots, n$. Denote by $\hat{f}(\boldsymbol{x})$ the fitted function based on the full data set and by $\hat{f}^{-i}$ the function fitted with the ith data deleted. Let

$$\tilde{\boldsymbol{y}}^i = (y_1, \cdots, y_{i-1}, \hat{f}^{-i}(\boldsymbol{x}_i), y_{i+1}, \cdots, y_n)^\mathsf{T},$$

be the perturbed data, by replacing the ith element in $\boldsymbol{y}$, y_i, with the evaluation of the delete-i fit $\hat{f}^{-i}$ at $\boldsymbol{x}_i$. Let $\tilde{f}^{-i}$ be the estimate of f with data $\tilde{\boldsymbol{y}}^i$.

1. Prove that the leave-one-out property holds for the least squares estimate; i.e.,

$$\tilde{f}^{-i}(\boldsymbol{x}_i) = \hat{f}^{-i}(\boldsymbol{x}_i), \quad i = 1, \cdots, n,$$

   if $\hat{f}(\boldsymbol{x}) = \boldsymbol{x}^\mathsf{T}\widehat{\boldsymbol{\beta}}^{ols}$.
2. Based on item 1, show that

$$|y_i - \hat{f}^{-i}(\boldsymbol{x}_i)| \geq |y_i - \hat{f}(\boldsymbol{x}_i)|, \quad \forall i.$$

Exercise 10.6 (Penalized least squares). The penalized least squares problem can be reformulated as a quadratic optimization problem subject to constraints. Take the example of LASSO regression. Consider the two problems

$$\widehat{\boldsymbol{\beta}}_{\lambda}^{lasso} = \arg\min_{\boldsymbol{\beta}} (\boldsymbol{y} - X\boldsymbol{\beta})^{\mathsf{T}}(\boldsymbol{y} - X\boldsymbol{\beta}) + \lambda \sum_{j=1}^{p} |\beta_j| \tag{10.7.1}$$

and

$$\widehat{\boldsymbol{\beta}}_{s}^{lasso} = \arg\min_{\boldsymbol{\beta}} (\boldsymbol{y} - X\boldsymbol{\beta})^{\mathsf{T}}(\boldsymbol{y} - X\boldsymbol{\beta}), \quad \text{subject to} \quad \sum_{j=1}^{p} |\beta_j| \leq s. \tag{10.7.2}$$

Prove that solving these two problems are equivalent. In other words, do the following:

1. Given any $\lambda_0 > 0$ and its corresponding solution $\widehat{\boldsymbol{\beta}}_{\lambda_0}$ to (10.7.1), find a unique value s_{λ_0} and its corresponding solution to (10.7.2) such that $\widehat{\boldsymbol{\beta}}_{\lambda_0}^{lasso} = \widehat{\boldsymbol{\beta}}_{s_{\lambda_0}}^{lasso}$.
2. Given any $s_0 > 0$ and its corresponding solution $\widehat{\boldsymbol{\beta}}_{s_0}$ to (10.7.2), find a unique value λ_{s_0} and its corresponding solution to (10.7.1) such that $\widehat{\boldsymbol{\beta}}_{s_0}^{lasso} = \widehat{\boldsymbol{\beta}}_{\lambda_{s_0}}^{lasso}$.

Therefore, there is a one-to-one correspondence relationship between λ and s.

Exercise 10.7 (Penalized least squares and orthogonal design). In the special case of an orthogonal design matrix (i.e., $X^{\mathsf{T}}X = I_n$) the penalized least squares problem

$$\min_{\boldsymbol{\beta} \in R^p} (\boldsymbol{y} - X\boldsymbol{\beta})^{\mathsf{T}}(\boldsymbol{y} - X\boldsymbol{\beta}) + \lambda \sum_{j=1}^{p} J(|\beta_j|)$$

becomes solving p one-dimensional shrinkage problems:

$$\min_{\beta_j} (\beta_j - \hat{\beta}_j^{ols})^2 + \lambda J(|\beta_j|), \quad j = 1, \cdots, p.$$

Now, consider four types of penalized least squares problems.

1. When X is orthogonal, show that the ridge estimates are given by

$$\hat{\beta}_j^{ridge} = \frac{1}{1+\lambda} \hat{\beta}_j^{ols}, \quad j = 1, \ldots, p.$$

2. When X is orthogonal, the nonnegative garrote (NG) estimator seeks a set of nonnegative scaling factors c_j for $j = 1, \ldots, p$ by solving

$$\min_{c_j} (c_j \hat{\beta}_j^{ols} - \hat{\beta}_j^{ols})^2 + \lambda c_j, \quad \text{s.t. } c_j \geq 0.$$

Show that the solution has the expression

$$\hat{c}_j = \left(1 - \frac{\lambda}{2(\hat{\beta}_j^{ols})^2}\right)_+, \quad j = 1, \ldots, p,$$

where $(t)_+ = \max(t, 0)$. Therefore, the final NG estimator for the βs is

$$\hat{\beta}_j^{ng} = \left(1 - \frac{\lambda}{2(\hat{\beta}_j^{ols})^2}\right)_+ \hat{\beta}_j^{ols}, \quad j = 1, \ldots, p. \tag{10.7.3}$$

3. When X is orthogonal, show that the lasso solution is given by

$$\hat{\beta}_j^{lasso} = \text{sign}(\hat{\beta}_j^{ols})(|\hat{\beta}_j^{ols}| - 2\lambda)_+, \quad j = 1, \ldots, p.$$

4. When X is orthogonal, the L_0-penalized least squares problems solve

$$\min_{\beta_j} (\beta_j - \hat{\beta}_j^{ols})^2 + \lambda I(|\beta_j| \neq 0), \quad j = 1, \ldots, p.$$

Show that the solution is given by

$$\hat{\beta}_j = \text{sign}(\hat{\beta}_j^{ols}) I(|\hat{\beta}_j^{ols}| > \sqrt{\lambda}).$$

This is often called the hard thresholding rule.

5. Comment on the difference in shrinkage effects achieved in items 1–4.

Exercise 10.8 (SCAD penalty and convexity). Fan and Li (2001) proposed a penalized least squares using a smoothly clipped absolute deviation (SCAD) penalty of the form

$$q_\lambda(|w|) = \begin{cases} \lambda|w| & \text{if} \quad |w| \leq \lambda, \\ -\frac{(|w|^2 - 2a\lambda|w| + \lambda^2)}{2(a-1)} & \text{if} \quad \lambda < |w| \leq a\lambda, \\ \frac{(a+1)\lambda^2}{2} & \text{if} \quad |w| > a\lambda, \end{cases}$$

where $a > 2$ and $\lambda > 0$ are tuning parameters.

1. Show that q_λ has a continuous first-order derivative everywhere except at the origin.
2. Show that q_λ is a symmetric quadratic spline with knots at λ and $a\lambda$.
3. Show that q_λ is not convex.
4. Show that q_λ can be decomposed as the difference of two convex functions,

$$q_\lambda(|w|) = q_{\lambda,1}(|w|) - q_{\lambda,2}(|w|),$$

where q_1 and q_2 are convex and satisfy

$$q'_{\lambda,1}(|w|) = \lambda, \quad q'_{\lambda,2}(|w|) = \lambda\left(1 - \frac{(a\lambda - |w|)_+}{(a-1)\lambda}\right) I(|w| > \lambda).$$

This result was first proved in Wu and Liu (2009).

Exercise 10.9 (Bayes factors). Here are two examples of Bayes factors (BF).

1. Suppose that n samples are chosen from a population of size N. The number of samples X found to have a certain trait (e.g., satisfactory quality) is distributed as $Binomial(n,\theta)$, and $\theta \sim Beta(1,9)$ is thought reasonable. If $X=0$ is observed and $N >> n$, consider $\mathscr{H}_0 : \theta \leq .1$ vs. $\mathscr{H}_1 : \theta > .1$. Find the posterior probabilities of the two hypotheses, the posterior odds ratio, and the BF.
2. Now do a simple robustness analysis: Vary the 1 and the 9 in the Beta, the value of X, the sample size n, and the threshold of the test. Do the results change much? What happens if N is not enormously large compared with n?
3. The waiting time for a meal in a restaurant at a certain time of day is $Uniform(0,\theta)$ minutes. An ardent Bayesian is a patron of the restaurant and wants to test $\mathscr{H}_0 : 0 \leq \theta \leq 15$ vs. $\mathscr{H}_0 : \theta > 15$ to see if there is enough time for her guests to have dinner there before they go to the Bayesian dance party. She assigns a Pareto(5,3) prior and from personal experience has observed wait times of 5, 8, 10, 15, and 20 minutes. (Recall that if $X \sim Pareto(x_0,\alpha)$, then

$$f(x|x_0,\alpha) = \frac{\alpha}{x_0}\left(\frac{x_0}{x}\right)^{\alpha+1} \quad \text{for } x \geq x_0,$$

for which $\mathbb{E}(X) = (\alpha x_0)/(\alpha+1)$ when $\alpha > 1$.) Find her posterior probability, her posterior odds ratio, and her Bayes factor.
4. Again do a simple robustness analysis: Vary the prior, the data, the thresholds of the test, and the sample size to see how the conclusions change. Can you vary the likelihood as well? Which of these inputs seems to have the largest effect?

Exercise 10.10 (BFs and dimension). Steve MacEachern observes that Bayes factors do not always work well in general when the dimension of the two hypotheses is different. To see this, consider two models. Model I is one-dimensional: Let $\theta \sim N(0,\sigma^2)$, and suppose $(X_i|\theta) \sim N(\theta,1)$ are independent for $i=1,2$. Model II is two-dimensional: Let $(\theta_1,\theta_2)^t \sim N(\mathbf{0},\sigma^2 I)$, and suppose that $(X_i|\theta) \sim N(\theta_i,1)$ for $i=1,2$. Thus, in Model I, the two variables are tied together by a parameter while in Model II they are not.

1. Find the form of the BF and verify that it will essentially always choose Model I.
2. What happens as $\sigma \to \infty$?
3. Conclude that prior selection for one dimension is not compatible in general with prior selection for other dimensions and that some technique other than naive use of BFs may be required for comparing models of different dimensions.

Exercise 10.11 (Zellner g-prior). Consider a linear model of the form

$$M_\gamma : \mathbf{Y} = \mathbf{1}\beta_0 + \boldsymbol{X}_\gamma \boldsymbol{\beta}_\gamma + \boldsymbol{\varepsilon},$$

for n outcomes, where $\boldsymbol{X}_\gamma$ is the design matrix with columns corresponding to the γth subset of $(X_1, ..., X_p)$, the vector $\boldsymbol{\beta}$ contains the regression coefficients, and $\boldsymbol{\varepsilon} \sim N(0, \sigma^2 I_p)$. (The notation bf 1 means the vector of p ones.) Letting $p_\gamma = \dim(\boldsymbol{\beta}_\gamma)$, assign the Zellner g-prior by choosing priors

$$w(\boldsymbol{\beta}_\gamma|\gamma, g) = N_{p_\gamma}(0, g\sigma^2(\boldsymbol{X}_\gamma^t \boldsymbol{X}_\gamma)^{-1}) \text{ and } w(\beta_0, \sigma^2|\gamma) = \frac{1}{\sigma^2}.$$

Without loss of generality, assume the predictors are centered at zero so that dependence on β_0 is removed.

1. Verify that, for any γ,

$$BF(\gamma, 0) = (1+g)^{(n-p_\gamma-1)/2}(1 - g(1-R_\gamma^2))^{-(n-1)/2},$$

where R_γ^2 is the usual coefficient of determination.

2. Suppose the data support γ strongly so that $R_\gamma^2 \to 1$ for fixed values of n and g. What is the limiting value of $BF(\gamma, 0)$? What does this mean about the use of BFs?

Exercise 10.12 (LARS and LASSO). The LARS procedure described in Section 10.3.3 gives an efficient algorithm to solve the lasso. In particular, the fit can be obtained with the R commands

```
load(lars)
fit <- lars(x,y, type="lasso")
```

where $\boldsymbol{x}$ is the vector of covariates and y is the response.

1. Show that LARS can be used to solve the adaptive LASSO optimization,

$$\min_{\boldsymbol{\beta}} (\boldsymbol{y} - X\boldsymbol{\beta})^\top (\boldsymbol{y} - X\boldsymbol{\beta}) + \lambda \sum_{j=1}^p w_j |\beta_j|.$$

2. Modify the above code to get the entire solution path for the adaptive LASSO, given a set of prespecified weights $w_1, \cdots, w_p$.

Exercise 10.13 (Comparing shrinkage estimators). Compare the performance of the LASSO, the elastic net and the adaptive LASSO on the Boston housing data, which are available in the R package MASS.

1. What are the final selected covariates for each of the three methods?
2. What are the five-fold CV errors for each of the three methods?
3. Compare the results in items 1 and 2 with those from Exercise 10.3.

Exercise 10.14 (Example of an RKHS). In this exercise, you can show that the second-order Sobolev space $W_2[0,1]$,

$$W_2[0,1] = \{g : g(t), g'(t) \text{ absolutely continuous}, \ g''(t) \in \mathscr{L}_2[0,1]\},$$

equipped with the norm

$$(f,g) = f(0)g(0) + f'(0)g'(0) + \int_0^1 f''(t)g''(t)dt,$$

is a reproducing kernel Hilbert space. For any $t \in [0,1]$, define the evaluation functional $[t](\cdot)$

$$[t](f) = f(t) \quad \forall f \in W_2[0,1].$$

For each $t \in [0,1]$, define the function

$$R_t(s) = 1 + st + \int_0^1 (s-u)_+(t-u)_+du.$$

1. Show that $W_2[0,1]$ is a Hilbert space.
2. For each $t \in [0,1]$, show that the function R_t satisfies:
   i. $R_t \in W_2[0,1]$.
   ii. R_t has the reproducing property; i.e.,
   $$(R_t, f) = f(t), \quad \forall f \in W_2[0,1].$$
3. Using item 2, show that the evaluation functional $[t](\cdot)$ is bounded for any $t \in [0,1]$.

Taken together, these three items imply that $W_2[0,1]$ is an RKHS associated with the reproducing kernel $R(t,s) = R_t(s)$.

Exercise 10.15 (COSSO optimization). Lin and Zhang (2006) proposed the component selection and smoothing operator (COSSO) as a direct shrinkage method to do variable selection in nonparametric regression,

$$\min_{f \in \mathscr{H}} \sum_{i=1}^n [y_i - f(\mathbf{x}_i)]^2 + \lambda \sum_{j=1}^p \|P^j f\|_{\mathscr{H}}, \tag{10.7.4}$$

where $\|\cdot\|_{\mathscr{H}}$ is the norm defined in the RKHS $\mathscr{H}$ and $\lambda \geq 0$ is the smoothing parameter. Lin and Zhang (2006) further showed that solving (10.7.4) is equivalent to solve the problem

$$\min_{f \in \mathscr{H}, \boldsymbol{\theta}} \sum_{i=1}^n [y_i - f(\boldsymbol{x}_i)]^2 + \lambda_0 \sum_{j=1}^p \theta_j^{-1} \|P^j f\|_{\mathscr{H}}^2 + \lambda_1 \sum_{j=1}^p \theta_j, \ \text{subject to } \theta_j \geq 0 \ \forall j, \tag{10.7.5}$$

where λ_0 and λ_1 are chosen to satisfy $\lambda_1 = \lambda^4/(4\lambda_0)$.

1. For any $\lambda > 0$, if $\hat{f}$ minimizes (10.7.4), set $\hat{\theta}_j = \lambda_0^{1/2}\lambda_1^{-1/2}\|P^j \hat{f}\|$ and show that the pair $(\hat{\theta}, \hat{f})$ minimizes (10.7.5).

2. On the other hand, if a pair $(\hat{\theta}, \hat{f})$ minimizes (10.7.5), show that $\hat{f}$ also minimizes (10.7.4).

Exercise 10.16 (Bayes factors in a linear model). Suppose that the $n \times p$ design matrix $\boldsymbol{X}$ can be partitioned into $\boldsymbol{X}_1$, $n \times p_1$, containing the variables that do not belong in a model for Y and $\boldsymbol{X}_2$, $n \times p_2$, containing the variables that do belong in a model for Y, and suppose their coefficients are $\boldsymbol{\beta}_1$ and $\boldsymbol{\beta}_2$ with dimensions p_1 and p_2, respectively.

1. Verify that the linear model for $\boldsymbol{Y}$ can be written as

$$\boldsymbol{Y} = \mathbf{1}\beta_0 + \boldsymbol{X}_1\boldsymbol{\eta}_\gamma + \boldsymbol{V}\boldsymbol{\beta}_2 + \boldsymbol{\varepsilon},$$

where $\boldsymbol{V} = (I - \boldsymbol{X}_1(\boldsymbol{X}_1^t\boldsymbol{X}_1)^{-1}\boldsymbol{X}_1^t)\boldsymbol{X}_2$, $\boldsymbol{\eta} = \boldsymbol{\beta}_1 + (\boldsymbol{X}_1^t\boldsymbol{X}_1)^{-1}\boldsymbol{X}_1^t\boldsymbol{X}_2\boldsymbol{\beta}_2$, and $\boldsymbol{X}_1^t\boldsymbol{V} = 0$ (see Zellner and Siow (1980)).

2. Consider testing $\mathscr{H}_0 : \boldsymbol{\beta}_2 = 0$ vs. $\mathscr{H}_1 : \boldsymbol{\beta}_2 \neq 0$ assuming the prior probabilities of $\mathscr{H}_0$ and $\mathscr{H}_1$ are equal and the priors on the β_js are noninformative (for instance, normal with variance tending to infinity). Verify that the Bayes factor can be expressed as

$$BF(0,1) = b\left[\frac{\nu_1}{2}\right]^{p_2/2}\left[1 + \frac{p_2}{\nu_1}F_{p_2,\nu_1}\right]^{(\nu_1-1)/2},$$

where $\nu_1 = n - p_1 - p_2 - 1$, $b = \sqrt{\pi}/\Gamma((p_2+1)/2)$, and $F_{p_2,\nu_1} = \hat{\boldsymbol{\beta}}^t\boldsymbol{V}^t\boldsymbol{V}\hat{\boldsymbol{\beta}}/(p_2 s_1^2)$. In the last expression, $s_1^2 = (1/\nu_1)(\boldsymbol{Y} - \bar{y}\mathbf{1} - \boldsymbol{X}_1\hat{\boldsymbol{\eta}} - \boldsymbol{V}\hat{\boldsymbol{\beta}}_2)^t(\boldsymbol{Y} - \bar{y}\mathbf{1} - \boldsymbol{X}_1\hat{\boldsymbol{\eta}} - \boldsymbol{V}\hat{\boldsymbol{\beta}}_2)$, $\hat{\boldsymbol{\beta}}_2 = (\boldsymbol{V}^t\boldsymbol{V})^{-1}\boldsymbol{V}^t\boldsymbol{Y}$, and $\hat{\boldsymbol{\eta}} = (\boldsymbol{X}_1^t\boldsymbol{X}_1)^{-1}\boldsymbol{X}_1^t\boldsymbol{Y}$ (see Moulton (1991), Yardimci and Erar (2002)).

Exercise 10.17 (Exploration with complex sequential data). Generate IID data from a complex function or use a complex real data set, and treat the data as if it were arriving sequentially. Next, choose three of the shrinkage methods from the chapter, e.g., LASSO, Ridge regression, SCAD, and use the default values of λ when fitting each of them.

Now, using different ensembles of terms – polynomial, spline, trigonometric – compare the sequential predictive performance of each of the three shrinkage methods with the sequential predictive performance of their stacking average. Iterate the computation enough times that a variance can be assigned to the predictions and the residuals at each time step.

Do you see any characteristic patterns in the variation in the residuals over time? Is the improvement in bias from stacking enough to warrant the increase in variance from estimating parameters in three models along with two stacking coefficients? Explain.

Chapter 11
Multiple Testing

The Neyman-Pearson formulation of the hypothesis testing problem comes down to maximizing power, usually denoted $1-\beta$, where β is the probability of making a Type II error, subject to a level constraint, usually denoted α. That is, for testing

$$H : \mu \in \Omega_H \quad \text{vs.} \quad K : \mu \in \Omega_K, \tag{11.0.1}$$

where μ is the parameter and Ω_H and Ω_K denote sets in the parameter space, the optimal rejection region R_n for a sample of size n maximizes the power

$$P_\mu(R_n) = 1 - \beta_{R_n}(\mu) \text{ over } R_n \text{ satisfying } \sup_{\mu \in \Omega_H} P_\mu(R_n) \le \alpha \tag{11.0.2}$$

over $\mu \in \Omega_K$. It is not generally possible to maximize P_μ uniformly over Ω_K, but that is the ideal, which is sometimes achieved. In practice, R_n is typically identified by thresholding a statistic.

Variants on the basic Neyman-Pearson formulation include the search for rejection regions for H that permit $\alpha = \alpha_n$ and $\beta = \beta_n$ to decrease simultaneously, preferably both at an exponential rate of the form $\mathscr{O}(e^{-\gamma n})$ for some $\gamma > 0$. (The Stein test based on the empirical relative entropy is one test that achieves this in some cases and $\gamma = D(P||Q)$ for good choices of P and Q in the null and alternative hypotheses.) By contrast, the Bayes approach treats hypothesis testing as a decision problem with action space $\{\text{Reject H}, \text{Accept H}\}$ and minimizes the posterior risk under generalized zero-one loss. This gives the posterior probabilities of the hypotheses as the basis for the test; in many cases the correct hypothesis has posterior probability tending to one at an exponential rate (i.e., $W(\Omega_{correct}|X^n) \ge 1 - e^{-\gamma n}$ for some $\gamma > 0$).

All of this is reasonable for testing individual hypotheses; however, it is rare that only one hypothesis is of interest. What's even more serious is that practitioners often do data snooping to see what hypotheses are suggested by the data with a view toward testing only them. This amounts to selecting test statistics more likely to give a desired conclusion. That is, many hypothesis tests are entertained initially and only some are carried out. Done naively, this violates the assumption that the origin of the test and the generation of the data are independent of each other. The effect of this is to overstate

B. Clarke et al., *Principles and Theory for Data Mining and Machine Learning*, Springer Series in Statistics, DOI 10.1007/978-0-387-98135-2_11, © Springer Science+Business Media, LLC 2009

the power and level by understating the probability of a Type II error. Specifically, letting the data motivate a hypothesis test that will give a rejection at nominal level α has actual level $\alpha' < \alpha$. This means the nominal power $1-\beta$ is greater than the actual power $1-\beta'$ and therefore the nominal probability of Type II error is $\beta < \beta'$.

The way this is overcome classically is through multiple comparison procedures; i.e., procedures specifically derived to be valid for a large collection of tests regardless of which elements of the collection are actually done. Arguably, the best-known instances of this occur in analysis of variance, ANOVA.

Consider a balanced one-way ANOVA model with k cells, $Y_{i,j} = \mu_i + \varepsilon_{i,j}$ with $j = 1,...,n$ with independent $N(0,\sigma^2)$ error terms $\varepsilon_{i,j}$. Tukey's method for multiple comparisons is a way to perform all pairwise tests $H_{i,j} : \mu_i = \mu_j$ vs. $K_{i,j} : \mu_i \neq \mu_j$ simultaneously. It is based on the following reasoning. Let $Y_1,...,Y_r\tilde{N}(\mu,\sigma^2)$ be IID, and suppose an estimate of σ^2, say s^2, based on ν degrees of freedom, is available. Then, the studentized range statistic

$$q(\nu,r) = \frac{\max_i Y_i - \min_i Y_i}{s}$$

has percentiles $q_{r,\nu}(1-\alpha)$ that can be tabulated for each pair (r,ν). The percentiles can be used for testing any two means and also give confidence intervals for any difference of two means. Naturally, Tukey's procedure gives tests and confidence intervals that are weaker and wider, respectively, than would be the case for individual comparisons. However, Tukey's method is valid for all pairwise tests, regardless of which ones were deemed most important before the data were collected. Scheffe's method has a similar interpretation but for a more general class of tests.

One general bound comes from the Bonferroni method, which rests on the union of events bound. Let $E_1, E_2, \cdots, E_k$ be a collection of events. Then, if each $\{E_i\}$ is the rejection region for a test with level α/k, the overall level of the test is

$$\Pr\left[\bigcup_{i=1}^{k} E_i\right] \leq \sum_{i=1}^{k} \Pr(E_i) \leq \alpha.$$

In essence, to guarantee that the inferences are correct simultaneously with an overall probability $1-\alpha$, one must adjust for the multiplicity, ensuring each individual inference on μ_i is correct with a probability somewhat higher than $1-\alpha$. Failure to adjust for multiplicity makes the rate of incorrect decisions unacceptably high.

The limitation of these classical procedures is the sense in which they control errors. As the number of tests increases, the probability that at least one error will be made increases no matter what level is chosen. Even worse, if the error is not controlled correctly, most methods will have very small levels and hence very low power, as can be seen easily for Bonferroni. This problem is compounded by the fact that the test statistics defining the individual rejection regions are typically not independent. Since quantifying dependence is difficult, it can be hard to improve on Bonferroni bounds.

This chapter begins by analyzing the hypothesis testing problem to see how controlling errors, in a sense different from the level and power, might be valuable. Then, several definitions of errors to be controlled can be listed. These include the family-wise error rate, the per comparison error rate, the per family error rate, the false discovery rate and the positive false discovery rate. In succeeding sections, controlling them is discussed. Finally, conditional assessments such as q-values and the Bayesian formulation are presented. It is important to bear in mind that the techniques of this chapter are intended for large numbers of hypothesis tests where classical methods break down. Where classical methods are effective, it may not make sense to use the techniques here.

11.1 Analyzing the Hypothesis Testing Problem

There are several analyses of the hypothesis testing problem; the classical Neyman-Pearson framework is merely one of them that is particularly well known. In this section, an analysis of Miller (1981) is presented first and then several alternative senses of error are introduced.

11.1.1 A Paradigmatic Setting

To illustrate the principles underlying simultaneous statistical inference, Miller (1981) proposes the problem of comparing the means of two normals with common variance: Let Y_1 and Y_2 be independent random variables with $Y_1 \sim N(\mu_1, 1)$ and $Y_2 \sim N(\mu_2, 1)$, and consider simultaneous inferences on μ_1 and μ_2. It will be seen that this problem already contains the core elements of general multiple-comparisons problems.

11.1.1.1 Analyzing Miller's Problem

From a significance test perspective, a natural null hypothesis would be

$$H_0: \quad \mu_1 = 0 \qquad \mu_2 = 0,$$

and the natural alternative is the negation of H_0, without further consideration. However, from a multiple-comparisons perspective, the alternative matters and corresponds to multiple nulls. For instance, exploring both H_0 and *not* H_0 means comparing all the following hypotheses:

$$\begin{array}{lll} H_1: & \mu_1 = 0 & \mu_2 = 0, \\ H_2: & \mu_1 = 0 & \mu_2 \neq 0, \\ H_3: & \mu_1 \neq 0 & \mu_2 = 0, \\ H_4: & \mu_1 \neq 0 & \mu_2 \neq 0. \end{array}$$

In other words, in simultaneous inference, it is not enough to stop with a decision against H_0 since rejection of H_0 does not indicate which clause in H_0 was wrong. Moreover, among the alternatives, the ones corresponding to rejected nulls need to be known, as they are likely to be of great interest to the investigator.

In the present case, making multiple comparisons between H_0 and its alternatives requires making statement S_1 about μ_1 and statement S_2 about μ_2; for instance $S_1 = \{\mu_1 = 0\}$ and $S_2 = \{\mu_2 \neq 0\}$. Thus, to examine two means, one needs to make two statements. So, the investigator can make zero, one, or two errors, and one measure of adequacy is the frequency of errors made, or more typically the error rate.

11.1.1.2 The General Structure

For a given problem, there may be a family of possible statements, say $\mathscr{F}$. In the two-means example, the family is $\mathscr{F} = \{S_1, S_2\}$. More generally, the family is defined as $\mathscr{F} = \{S_i : i \in \mathscr{I}\}$, where S_i relates to parameter i. The number of statements in the family $\mathscr{F}$ is $N(\mathscr{F})$; for the two-means case, $N(\mathscr{F}) = 2$. A variety of statements about both μ_1 and μ_2 can be made, and the question arises as to how to group statements together to form a family. Deferring this central question but assuming a family has been selected, define $N_w(\mathscr{F})$ to be the number of incorrect statements made from $\mathscr{F}$. Now, the family error rate is

$$Err\{\mathscr{F}\} = \frac{N_w(\mathscr{F})}{N(\mathscr{F})}.$$

Given data, the numerator in $Err\{\mathscr{F}\}$ is a random variable and so is $Err\{\mathscr{F}\}$. Inference about the procedure generating $Err\{\mathscr{F}\}$ therefore requires the computation of some nonrandom characteristic of the distribution of $Err\{\mathscr{F}\}$. The two natural candidates are:

- the probability of a nonzero family error rate, defined as

$$P\{N_w(\mathscr{F}) > 0\} = P\left[\frac{N_w(\mathscr{F})}{N(\mathscr{F})} > 0\right],$$

- and the expected family error rate, defined as

$$\mathbb{E}\left[\frac{N_w(\mathscr{F})}{N(\mathscr{F})}\right].$$

Using the indicator function

$$I(S_i) = \begin{cases} 1 \text{ if } S_i \text{ is incorrect,} \\ 0 \text{ if } S_i \text{ is correct,} \end{cases}$$

it is seen that

$$\mathbb{E}\{N_w(\mathscr{F})\} = \mathbb{E}\{I(S_1)\} + \mathbb{E}\{I(S_2)\} + \cdots + \mathbb{E}\left\{I(S_{N(\mathscr{F})})\right\},$$

giving an obvious expression for the expected family error rate. In principle, it can be based on the rejection regions of a collection of test statistics. In addition,

$$\begin{aligned}\mathbb{P}\{N_w(\mathscr{F}) > 0\} &= P\left\{\bigcup_i [I(S_i) = 1]\right\} \\ &= \sum_i \mathbb{P}\{I(S_i) = 1\} - \sum_{i_1 < i_2} \mathbb{P}\{I(S_{i_1})I(S_{i_2}) = 1\} + \cdot \\ &+ (-1)^{N(\mathscr{F})-1}\mathbb{P}\{I(S_1)I(S_2)\cdots I(S_{N(\mathscr{F})})\}. \end{aligned} \tag{11.1.1}$$

11.1.1.3 The $N(\mathscr{F}) = 2$ Case

Expression (11.1.1) is usually analytically intractable because the probabilities of intersections cannot be found. The only case that allows an easy expression for $\mathbb{P}\{\mathscr{F}\}$ is the case where the statements in the family $\mathscr{F}$ are independent, in which case it is straightforward to show that

$$1 - \mathbb{P}\{N_w(\mathscr{F}) > 0\} = \prod_{i=1}^{N(\mathscr{F})} (1 - \mathbb{P}\{I(S_i) = 1\}). \tag{11.1.2}$$

While (11.1.2) does provide an analytical expression for $P\{N_w(\mathscr{F}) > 0\}$, the independence assumption is rarely satisfied.

Without the assumption of independence, it is hard to relate $P\{N_w(\mathscr{F}) > 0\}$ to the individual probabilities for the statements in the family $\mathscr{F}$. Usually, only approximations are available, for instance from the Bonferroni inequality. Write

$$1 - \mathbb{P}\{N_w(\mathscr{F}) > 0\} \geq 1 - \sum_{i=1}^{N(\mathscr{F})} \alpha_i, \tag{11.1.3}$$

where $\alpha_i = \mathbb{P}\{I(S_i) = 1\}$ for $i = 1, 2, \cdots, N(\mathscr{F})$.

The expected family error has a slightly more tractable interpretation. Setting

$$\alpha_i = P\{I(S_i) = 1\} = \mathbb{E}\{I(S_i)\}$$

for $i = 1, 2, \cdots, N(\mathscr{F})$, the expected family error is

$$\frac{1}{N(\mathscr{F})}(\alpha_1 + \alpha_2 + \cdots + \alpha_{N(\mathscr{F})}). \tag{11.1.4}$$

The importance of the probability of a nonzero family error rate and the expected family error rate is that they are conceptually different from the Neyman-Pearson framework and do not focus on ensuring that each individual test gives the right answer.

Indeed, the count $N_w(\mathscr{F})$, as defined above, does not even distinguish between null and alternative hypotheses. Below, the notions of null and alternative will be imposed on testing problems, but it is not clear that this is logically necessary in general.

Arguably, the main limitation of classical testing is its emphasis on controlling the per-test probability of making Type I errors because this causes a severe reduction in the power of testing procedures when the number of tests grows larger. That is, as the number of tests increases, the power of each individual test is substantially reduced and may become unacceptably low. The effort to minimize the probability of making even a single Type I error substantially reduces the overall coverage. In other words, no matter what the α, the probability of at least one false positive goes to one as the number of tests increases.

In effect, recent multiple testing procedures rest on the assumption that mistakes are inevitable. So, it doesn't make sense to seek rejection regions that correspond to essentially no mistakes, but it does make sense to model the mistake-making process. Hence, multiple testing often uses a relative measure in place of an absolute one that tries to minimize or control mistakes rather than eliminate them.

11.1.2 Counts for Multiple Tests

Let H_i denote the null hypothesis for the test of the ith parameter. Then $H_1, H_2, ..., H_m$ are the null hypotheses of interest with alternatives taken as the negations of the H_is. For each individual test above, the classical approach to testing is:

1. Prespecify H_i and the alternative.
2. Prespecify an acceptable significance level (Type I error rate) α.
3. Seek the most powerful test – the test with the smallest Type II error (β), or highest power $(1-\beta)$, among all the tests with the same α.
4. Collect the data.
5. See if the test statistic lands in the rejection region or not.

The possible outcomes of each test can be summarized in Table 11.1. However, when m is large, it is likely that there will be many false positives (Type II errors), especially when the null hypotheses are all true. So, rather than trying to eliminate them, consider counting them. In fact, note that conditional on H_i being true, the outcome of a test, reject H_i or do not reject H_i, amounts to a binomial. Similarly, if H_i is false, the decision from a test is also a binomial. Combining the outcomes from these two possibilities gives a contingency table that summarizes the counts from the individual tests. Let V denote the number of true null hypotheses rejected (false positives), and let S be the number of alternative hypotheses that are correctly declared significant (true positives). The new summary is given in Table 11.2.

The sum $R = V + S$ is the random variable that counts the number of rejected null hypotheses, of which V is the number of false rejections and S is the number of correct

Truth	Decision	
	Do not Reject H_i	Reject H_i
H_i is true	Correct $(1-\alpha)$	Type I Error (α)
H_i is false	Type II Error (β)	Correct $(1-\beta)$

Table 11.1 Summary of the possible outcomes of an individual hypothesis test.

Truth	Decision		Total
	# Non Discoveries	# Discoveries	
	(Null Not Rejected)	*(Null Rejected)*	
# True Null	U	V	m_0
# True Alternative	T	S	$m-m_0$
Total	W	R	m

Table 11.2 Summary of the counts of outcomes for the m tests with m_0 true nulls.

rejections. Likewise, $W = U+T$ is the number of nonrejected nulls, of which U are correct and T are incorrect. The number of true nulls is m_0, so ideally there would be $m-m_0$ rejections and m_0 nonrejections. However, S, T, U, V, and m_0 are unobservable; only m and R can be observed.

11.1.3 Measures of Error in Multiple Testing

As in classical testing of individual hypotheses, a reasonable multiple testing procedure should keep the number of false positives (V) and the number of false negatives (T) as small as possible in some sense. Also as in classical testing, it is typical to focus on controlling the number of false positives, whether in absolute or relative terms. Thus, most of the commonly used measures of error are generalizations of Type I error, as can be seen from the following list:

- Family-wise error rate, FWER: For m simultaneous tests for which a joint decision is sought with an overall significance level α, the most widely used technique in the classical approach to multiple testing has been to control

$$FWER = \mathbb{P}[V \geq 1]. \tag{11.1.5}$$

  The FWER is the probability of declaring at least one false positive out of the m hypotheses under consideration. It is the classical Type I error rate extended to multiple hypotheses. So, for a given α, the classical criterion is expressible as

$$FWER \leq \alpha.$$

- Per-comparison error rate, PCER: This is a Type I error rate defined as

$$PCER = \frac{\mathbb{E}[V]}{m}.$$

The PCER is the expected number of Type I errors per hypothesis test.

- **Per-family error rate, PFER**: This is not really a rate, but the expected number of Type I errors,

$$PFER = \mathbb{E}[V],$$

the numerator of the PCER. Really, the $PFER$ is a frequency or count.

- **False discovery rate, FDR**: This is the expected proportion of falsely rejected null hypotheses among the rejected null hypotheses. More formally, the FDR is

$$FDR = \mathbb{E}\left[\frac{V}{R}|R > 0\right] \mathbb{P}[R > 0].$$

If all the null hypotheses are true (i.e., $m = m_0$) then controlling FDR coincides with controlling $FWER$. One appeal of the FDR approach comes from the fact that if $m = m_0$ is not very likely, then the $FWER$ criterion can be too stringent and somewhat unrealistic, thereby reducing in test power.

- **Positive false discovery rate, pFDR**: This is yet another measure of Type I error rate that addresses some limitations of the FDR by conditioning. It is defined as

$$pFDR = \mathbb{E}\left[\frac{V}{R}|R > 0\right].$$

The $pFDR$ is the rate at which discoveries are false, presuming a nonzero probability that there are "discoveries" to start with, hence the term positive. It is easy to see that if $\Pr(R > 0) \to 1$ as $m \to \infty$, then $pFDR$ and FDR become identical.

Bayesian forms of the error in multiple testing will also be presented in Section 11.6.

Similar lists of types of errors and some of their properties can be found in Dudoit et al. (2003) and in Ge et al. (2003), who also present many of their standard properties.

It has become customary to refer to rejected hypotheses as "discoveries", although any new, conclusive inference can claim to be a discovery. This terminology suggests that only the probability of false rejection of the null has been controlled. Presumably, if the probability of false rejection of the alternative were controlled, not rejecting the null would also be a "discovery".

Given the variety of measures of error in testing, it is of interest to compare them theoretically and in practice. First, asymptotically in m, $FDR \approx pFDR \approx \mathbb{E}(V)/\mathbb{E}(R)$, the proportion of false positives. Another comparison is also nearly immediate: It can be easily shown that

$$PCER \leq FDR \leq pFDR \leq FWER \leq PFER. \tag{11.1.6}$$

However, it must be noted that these five measures may also differ in how much power the testing procedures based on them have. Some of the measures tend to be more stringent than others, leading to more conservative test procedures.

11.1.4 Aspects of Error Control

In general, the concepts of a conservative test and p-value carry over to the multiple-comparisons context. A test is conservative if it protects primarily against false rejection of the null, in the belief that the cost of Type I error is greater than that of Type II error. It is conservatism that breaks the symmetry of the testing procedure because a conservative test requires the Type I error rate to be less than the overall significance level α, ignoring the Type II error rate. In practice, this leads to a reduction of test power. Nonconservative tests, for instance tests that minimize the sum of the probability of false rejection of the null and false rejection of the alternative, typically allow for an increase in power in exchange for less control over the type I error rate.

The p-value is defined relative to a test that amounts to a collection of nested rejection regions, $\{\Gamma\}$. Given a value t of a statistic T, the p-value for $T = t$ is

$$p - \text{value}(t) = \min_{\{\Gamma : t \in \Gamma\}} \mathbb{P}[T \in \Gamma | H = 0],$$

where the notation $H = 0$ means that the hypothesis H is true. Typically, the regions Γ are defined in terms of T. That is, the p-value is the lowest significance level for which H would be falsely rejected for a future set of data using the regions $\{\Gamma\}$ and a threshold from the data gathered. Informally, it is the probability of seeing something more discordant with the null than the data already gathered. In principle, it can be computed for any test regions.

To extend these classical ideas to the multiple testing context, consider one hypothesis H_i per parameter, with m parameters. Then, define random variables H_i for $i = 1, ..., m$ corresponding to the hypotheses by

$$H_i = \begin{cases} 0, \text{ if the } i\text{th null hypothesis is true,} \\ 1, \text{ if the } i\text{th null hypothesis is false.} \end{cases}$$

Now, let the set of indices of all the m hypotheses under consideration be denoted by $\mathscr{C} = \{1, 2, \cdots, m\}$, and set $\mathscr{N} = \{i : H_i = 0\}$ and $\mathscr{A} = \{i : H_i = 1\}$. These are the indices corresponding to the sets of true null and true alternative hypotheses, respectively. So, $\mathscr{C} = \mathscr{N} \cup \mathscr{A}$. Note that $m_0 = |\mathscr{N}|$ and $m_1 = m - m_0 = |\mathscr{A}|$ are unknown, but m is known. Also, the set $\mathscr{C}$ is known, while the sets $\mathscr{N}$ and $\mathscr{A}$ are unknown and must be estimated from the data. The set corresponding to the complete null hypothesis is $H_{\mathscr{C}}$,

$$H_{\mathscr{C}} = \bigcap_{i=1}^{m} \{H_i = 0\},$$

which means that all m nulls are true. Parallel to this, $H_{\mathscr{N}}$ is the collection of the m_0 true nulls out of the $m \geq m_0$ hypotheses being tested. Thus, $H_{\mathscr{N}}$ is

$$H_{\mathscr{N}} = \bigcap_{i \in \mathscr{N}} \{H_i = 0\}.$$

The importance of $H_{\mathscr{N}}$ stems from the fact that the type of control exerted on the Type I error rate will depend on the truth value of $H_{\mathscr{N}}$.

In fact, the set of rejected hypotheses is an estimate of $\mathscr{A}$ just as the set of hypotheses not rejected is an estimate of $\mathscr{N}$. In this sense, multiple testing is a model selection procedure. For instance, if the rejection of nulls is decided by a vector of statistics $T_n = (T_{n,1}, ..., T_{n,m})$, where $T_{n,i}$ corresponds to H_i, then the null distribution for T is determined under the complete (joint) null $\mathscr{C}$ and $\mathscr{N}$ is estimated by $\hat{\mathscr{N}} = \{i : H_i \text{ is rejected}\} = \{i : |T_{n,i}| \geq \tau_i\}$ for thresholds τ_i.

11.1.4.1 Kinds of Control

There are several ways to control hypotheses under a given choice of error assessment. Here, these ways are presented in terms of the FWER, but the ideas apply to the other error assessments, in particular FDR and pFDR.

For practical purposes, there are two kinds of control of the Type I error rate, weak control and strong control, both of which are defined in terms of how the collection of true null hypotheses $H_{\mathscr{N}}$ is handled.

First, weak control for the FWER comes from a naive generalization of the p-value to m simultaneous tests. So, consider the rate of falsely rejected null hypothesis conditional on the complete null hypothesis $H_{\mathscr{C}}$ being true. If FWER is the Type I error rate, this translates into finding

$$\mathbb{P}[V > 0 | H_{\mathscr{C}}].$$

With m simultaneous tests, $H_{\mathscr{C}}$ is just one out of the 2^m possible configurations of the null hypotheses. Clearly, controlling 1 out of the 2^m is very little control since the other $2^m - 1$ possibilities are unaccounted for, hence the expression weak control for characterizing strategies that only account for the complete null hypothesis $H_{\mathscr{C}}$.

By contrast, strong control is when the Type I error rate is controlled under any combination of true and false null hypotheses. The intuition here is that since $H_{\mathscr{N}}$ is unknown, it makes sense to include all 2^m possible configurations ranging from the ones with $m_0 = 1$ to the one with $m_0 = m$. Under strong control, one does indeed perform complete multiple testing and multiple comparisons. This is clearly the most thorough way to perform a realistic multiple testing procedure and can be done in many ways; e.g., Bonferroni and Sidák, among others. However, in many other cases, strong control can require intensive computations for even moderately large m.

11.1.4.2 Terminology for Multiple Testing

If a sequence of hypothesis tests is to be performed, each individual test has a null, a marginal p-value, and a marginal threshold for the p-value. The point of multiple testing is to combine the marginal tests into one joint procedure to provide control of the overall error. Once the joint procedure has been specified, the m tests are often done stepwise (i.e., one at a time). However, the marginal thresholds are no longer valid. One can either change the thresholds or transform the raw p-values, a process called adjustment. It will be seen that most of the popular techniques for multiple testing are based on some form of adjustment. Although the adjustment may involve the joint distribution, these methods are often called marginal because they combine results from the m tests. (Truly joint methods combine across tests by using a high-dimensional statistic and comparing it with a null distribution but are beyond the present scope.)

There are two kinds of adjustments that are commonly applied to raw p-values in stepwise testing. The simpler ones are called single-step: In a single step adjustment, all the p-values from the m tests are subject to the same adjustment independent of the data. Bonferroni and Sidák in the next section are of this type. Thus, stepwise testing may use a single-step adjustment of the raw p-values. The more complicated adjustments are called stepwise: In these, the adjustments of the p-values depend on the data. Thus, stepwise testing may also use a stepwise adjustment of the raw p-values. Note that the term stepwise has a different meaning when used to describe a testing procedure than it does when describing an adjustment procedure.

Stepwise adjustments themselves are usually one of two types. Step-down methods start with the most significant p-values (i.e., the smallest) test sequentially, reducing the adjustment at each step, and stop at the first null not rejected. Step-down methods do the reverse: They start with the least significant p-values (i.e., the largest) test sequentially, increasing the adjustment at each step, and stop at the first rejected null. Step-down and step-up adjustments tend to be less conservative than single-step adjustments.

The notion of an adjusted p-value for a fixed testing procedure can be made more precise. Fix n and consider a testing procedure based on statistics T_i for testing hypotheses H_i. If the tests are two-sided, the raw p-values are $p_i = \mathbb{P}(|T_i| \geq t_i | H_i = 0)$. The adjusted p-value $\tilde{p}_i$ for testing H_i is the level of the entire procedure at which H_i would just be rejected when $T_i = t_i$, holding all the other statistics T_j, for $j \neq i$, fixed. More explicitly, the adjusted p-values based on the optimality criteria in conventional significance testing are

$$\tilde{p}_i = \inf_{\alpha \in [0,1]} \{\alpha \mid H_i \text{ is rejected at level } \alpha \text{ given } t_1, \ldots, t_m\}. \tag{11.1.7}$$

These $\tilde{p}_i$s can be used to give an estimate $\hat{\mathcal{N}}$ of $\mathcal{N}$. Note that the adjustment uses all the values of t_i for $i = 1, \ldots, m$, but the adjusted p-values still have a marginal interpretation. As will be seen below, the decision rule based on $\tilde{p}_i$s can sometimes be expressed in terms of threshold functions $T_i = T_i(p_1, \ldots, p_m) \in [0,1]$, where the ith hypothesis is rejected if $p_i \leq T_i(p_1, \ldots, p_m)$.

Expressions like (11.1.7) can be given for other optimality criteria in testing. For the FWER, for instance, the adjusted p-values would be

$$\tilde{p}_i = \inf_{\alpha \in [0,1]} \{\alpha \mid H_i \text{ is rejected at } FWER = \alpha \text{ given } t_1, \ldots, t_m\}.$$

Other measures such as FDR or q-values have similar expressions. In the FWER situation, H_i would be rejected if $\tilde{p}_i \leq \alpha$.

11.2 Controlling the Familywise Error Rate

The vast majority of early multiple-comparisons procedures from ANOVA, such as Tukey, Scheffe, studentized maximum modulus, and so forth, are single-step adjustments of p-values to provide strong control of the FWER. By contrast, a goodness-of-fit test in ANOVA that all the μ_is are zero, would be an instance of weak control with FWER. Obviously, weak control is not enough for large studies involving thousands of tests. The big problem with weak control is that as the number m of tests grows larger, the probability of declaring false positives gets closer to 1 very fast even if all the null hypotheses are true.

To dramatize this, if an experimenter performs one individual test at the $\alpha = 0.05$ significance level, then the probability of declaring the test significant under the null hypothesis is 0.05. In this case, $FWER \leq \alpha$. However, if the same experimenter performs $m = 20$ independent tests, each at level 0.05, then the probability of declaring at least one of the tests significant is

$$FWER = 1 - (1-\alpha)^2 = 1 - 0.95^2 = 0.0975 > 0.05 = \alpha,$$

nearly twice the level, even though all the null hypotheses are assumed true. It gets worse as m gets larger. In fact, with $m = 20$ tests, this probability becomes $1-(1-\alpha)^{20} = 0.642$ for the experimenter to reject at least one correct null. Thus, the probability of declaring at least one of the tests significant under the null converges quickly to one as m increases. This means that if each individual test is required to have the same significance level α, the overall joint test procedure cannot have $FWER \leq \alpha$.

In practice, this means that tests have to be adjusted to control $FWER$ to the desired size. One way to do this is to adjust the threshold α_i for the p-value of test i to ensure that the entire study has a false positive rate no larger than the prespecified overall acceptable Type I error rate α. Unfortunately, all the classical techniques aimed at achieving strong control of the FWER that achieve $FWER \leq \alpha$ turn out to be quite conservative for large m.

11.2.1 One-Step Adjustments

As noted, the Bonferroni correction is the simplest and most widely used technique for implementing a strong control of FWER. It uses the threshold $\alpha_i = \alpha/m$ for each test, thereby guaranteeing that the overall test procedure has $FWER \leq \alpha$.

BONFERRONI CORRECTION FOR m TESTS: For each null hypothesis H_i out of the m hypotheses under consideration:

- ☐ Compute the unadjusted p-value p_i.
- ☐ Compute the adjusted p-value $\tilde{p}_i$ with

$$\tilde{p}_i = \min(mp_i, 1).$$

☐ Reject H_i if $\tilde{p}_i \leq \alpha$ or equivalently $p_i \leq \alpha/m$.

The Bonferroni correction on m independent tests always achieves $FWER < \alpha$ under the null hypotheses. For instance, let $\alpha = 0.05$. With $m = 2$, one has $FWER = 1 - (1-0.05/2)^2 = 0.0494 < 0.05 = \alpha$. For $m = 10$, the control is still achieved since one has $FWER = 1-(1-0.05/10)^{10} = 0.0489 < 0.05 = \alpha$. For $m = 20$, the same applies; i.e., $FWER = 1-(1-0.05/20)^{20} \approx e^{-.05} \approx 0.0488 < 0.05 = \alpha$. Bonferroni, like Sidák below, is conservative with respect to individual hypotheses but not very conservative overall in the independent case. The practical appeal of Bonferroni lies in its simplicity and its symmetric treatment of the null hypotheses.

Another adjustment technique that allows control of the FWER within a prespecified level is the Sidák adjustment:

SIDÁK ADJUSTMENT FOR m TESTS: For each null hypothesis H_i out of the m hypotheses under consideration:

☐ Compute the unadjusted p-value p_i.

☐ Compute the adjusted p-value $\tilde{p}_i$ with

$$\tilde{p}_i = \min(1-(1-p_i)^m, 1).$$

☐ Reject H_i if $\tilde{p}_i \leq \alpha$ or equivalently $p_i \leq 1-(1-\alpha)^{1/m}$.

Unfortunately, with $\alpha_i = 1-(1-\alpha)^{1/m}$ decreasing as m gets larger, Sidák is just as conservative as Bonferroni. Essentially, the only difference between Bonferroni and Sidák is that the first is additive and the second multiplicative in their single-step slicing of the overall significance level. The price paid for guaranteeing $FWER < \alpha$ in the case of strategies like Bonferroni and Sidák is a substantial reduction in the ability to reject any null hypothesis, as $\alpha_i = \alpha/m$ becomes ever smaller as m grows. In other words, the power of the overall test is dramatically reduced for this type of single-step adjustment.

11.2.1.1 Two More One-Step Adjustments

Westfall and Young (1993) propose two, more elaborate, single-step p-value adjustment procedures that are less conservative and also take into account the dependence structure among the tests. Let P_i denote the p-value from test i as a random variable so that, under the null H_i, P_i is $Uniform[0,1]$ when H_i is exact. Then, it makes sense to compare P_i with p_i, the p-value obtained from the specific data set. Their first adjustment is known as the single-step minP, which computes the adjusted p-values as

$$\tilde{p}_i = \mathbb{P}\left[\min_{l=1,\cdots,m} P_l \leq p_i \middle| H_{\mathscr{C}}\right]. \tag{11.2.1}$$

Their second single-step adjustment is known as the single-step maxT and computes the adjusted p-values as

$$\tilde{p}_i = \mathbb{P}\left[\max_{l=1,\cdots,m} |T_l| \geq |t_i| \middle| H_{\mathscr{C}}\right]. \tag{11.2.2}$$

Both of these provide weak control under the complete null; under extra conditions, they give strong control (see Westfall and Young (1993), Section 2.8). Comparisons among Bonferroni, Sidák, minP, and maxT tend to be very detailed and situation-specific; see Ge et al. (2003).

11.2.1.2 Permutation Technique for Multiple Testing

Often minP and maxT are used in a permutation test context in which the distribution of the test statistic under the null hypothesis is obtained by permuting the sample labels. To see how this can be done in a simple case, recall that the data form an $m \times n$ matrix. Of the n subjects in the sample, suppose that n_1 are treatments and n_2 are controls. The gist of the permutation approach in this context is to create permutations of the control/treatment allocation in the original sample. With that, there are

$$B = \binom{n}{n_1} = \binom{n_1+n_2}{n_1} = \frac{n!}{n_1!n_2!}$$

permutations of the labels, so that in principle B different samples have been generated from the data. If the statistic for test i is T_i with realized values t_i, then the following table shows all the sample statistics that can be computed from all the permutations, denoted Perm 1,...,B. The notation var i is meant to indicate the variable from the i-th measurement in a sample in the experiment. For instance in genomics, this would be the i-th gene and the outcome t_i of T_i could be the usual t-test for testing the difference of means in two independent normal samples with common variance.

	Perm 1	Perm 2	$\cdots$	Perm b	$\cdots$	Perm B
var 1	t_{11}	t_{12}	$\cdots$	t_{1b}	$\cdots$	t_{1B}
var 2	t_{21}	t_{22}	$\cdots$	t_{2b}	$\cdots$	t_{2B}
$\vdots$	$\vdots$	$\vdots$	$\cdots$	$\vdots$	$\ddots$	$\vdots$
var i	t_{i1}	t_{i2}	$\cdots$	t_{ib}	$\cdots$	t_{iB}
$\vdots$	$\vdots$	$\vdots$	$\cdots$	$\vdots$	$\ddots$	$\vdots$
var m	t_{m1}	t_{m2}	$\cdots$	t_{mb}	$\cdots$	t_{mB}

In a typical multiple testing setting, the following template describes how the adjusted p-values are computed for the single-step maxT procedure.

Create B permutations of labels for $b = 1, \cdots, B$.

- ☐ Pick the bth permutation out of B possibilities; for $i = 1, \cdots, m$, compute the test statistics t_{ib}.
- ☐ For $i = 1, \cdots, m$, set $p_i^* = \dfrac{\#\{b : \max_i |t_{ib}| \geq |t_i|\}}{B}$.

The p_i^* from the pseudocode is the estimate of $\tilde{p}_i$ in (11.2.2). A similar procedure for (11.2.1) can be used; just compute p_i^* as $p_i^* = \#\{b : \min_i p_{ib} \leq p_i\}/B$. For instance, if the minimal unadjusted p-value is $p_{min} = .003$, then count how many times the minimal p-value from the permuted sample label pseudodata sets is smaller than .003. If this occurs in 8% of these B data sets, then $\tilde{p}_{min} = .08$.

In general, even though minP and maxT are less conservative (outside of special cases) than Bonferroni or Sidák, they remain overly conservative. This seems to be typical for single-step adjustment procedures, but see Dudoit et al. (2003) and Pollard and van der Laan (2004) for further results. Improving on single-step procedures in the sense of finding related procedures that are less conservative but still effective seems to require stepwise adjustment procedures.

11.2.2 Stepwise p-Value Adjustments

To illustrate the idea behind step-down procedures, let $p_{(i)}$ for $i = 1, \ldots, m$ be the order statistics from the p-values from m tests. The procedure in Holm (1979) is essentially a step-down version of Bonferroni that is as follows. For weak control of the FWER at level α, start with $i = 1$ and compare $p_{(1)}$ with $\alpha/(m-1+1)$, $p_{(2)}$ with $\alpha/(m-2+1)$, and so forth. Thus, the thresholds for the ordered p_is increase as the $p_{(i)}$s do. Identify the first value i_0 for which the i_0th order statistic exceeds its threshold, $p_{(i_0)} > \alpha/(m - i_0 + 1)$, indicating nonrejection of $H_{(i)}$, where the ordering on the hypotheses matches that of the p-values. Then, reject $H_{(1)}, \ldots, H_{(i_0-1)}$ but do not reject $H_{(i_0)}, \ldots, H_{(m)}$. If there is no i_0, then all the H_is can be rejected. If $i_0 = m$, then no hypotheses can be rejected. It can be verified that the Holm step-down adjusted p-values are

$$\tilde{p}_{(i)} = \max_{k=1,\ldots,i} \min((m-k+1)p_{(k)}, 1), \tag{11.2.3}$$

which shows that the coefficient on the ordered p-values increases rather than being constant, m at each step, as in Bonferroni. Note that, as i increases, the maximum in (11.2.3) is taken over a larger and larger set so that $\tilde{p}_i \leq \tilde{p}_{i+1}$. This means that rejection of a hypothesis necessitates the rejection of the hypotheses corresponding to the earlier order statistics.

A similar extension to a step-down procedure can be given for the Sidák adjustment. The comparable expression to (11.2.3) is

$$\tilde{p}_{(i)} = \max_{k=1,\ldots,i} 1 - (1 - p_{(i)})^{m-k+1}, \tag{11.2.4}$$

in which the factor becomes an exponent, analogous to the difference between the single-step versions of Bonferroni and Sidák. Indeed, in applications, for both the single step (see the two boxes) and step-down, (11.2.3) and (11.2.4), versions of the Bonferroni and Sidák testing procedures, the raw p-values denoted p_i can be replaced with the p_i^* versions obtained from the permutation technique.

Following Ge et al. (2003) or Dudoit et al. (2003), the minP and maxT single-step procedures also extend to step-down procedures, giving adjusted p-values analogous to those in (11.2.1) and (11.2.2). These guarantee weak control of the FWER in all cases and strong control under additional conditions. The maxT step-down adjustment for the p-values is

$$\tilde{p}_{(i)} = \max_{k=1,\cdots,i} \left\{ \mathbb{P}\left[\max_{u=k,\cdots,m} |T_{(u)}| \geq |t_{(k)}| \middle| H_{\mathscr{C}} \right] \right\}, \tag{11.2.5}$$

and the minP step-down adjustment for the p-values is

$$\tilde{p}_{(i)} = \max_{k=1,\cdots,i} \left\{ \mathbb{P}\left[\min_{u=k,\cdots,m} P_{(u)} \leq p_{(k)} \middle| H_{\mathscr{C}} \right] \right\}. \tag{11.2.6}$$

Under some conditions, the minP procedure reduces to Holm's procedure, but more generally Holm's procedure is more conservative than minP, as one would expect by analogy with Bonferroni.

11.2.2.1 Permutation Techniques for Step-down minP and maxT

The computational procedure given for the single-step adjusted p-values for maxT and minP is too simple for the step-down method because there is an optimization inside the probability as well as outside the probability in (11.2.5) and (11.2.6). Consequently, there is an extra procedure to evaluate the maximum and the minimum, respectively. Since the two cases are analogous, it is enough to describe minP; see Ge et al. (2003).

WESTFALL AND YOUNG MINP STEP-DOWN PROCEDURE: For each null hypothesis H_i out of the m hypotheses under consideration:

☐ Compute the unadjusted p-values p_i from the data.

☐ Generate all the B permutations of the original sample of the n data points.

☐ For each permutation, compute the order statistic from the raw p-values,

$$p_{(1),b},\ldots,p_{(m),b}, \quad b = 1,\ldots,B.$$

☐ Find all m values of the successive minima, $q_{i,b} = \min_{k=i,\ldots,m} p_{(k),b}$, based on the raw p-values from the bth permutation:

For $i = m$, set $q_{m,b} = p_{(m),b}$ and, for $i = m-1, ..., 1$, recursively set $q_{i,b} = \min\left(q_{i+1,b}, p_{(i),b}\right)$.

- ☐ From the B repetitions of this, find

$$\tilde{p}_{(i)} = \#(\{b : q_{i,b} \leq p_{(i)}\})/B$$

for $i = 1, ..., m$.

- ☐ Enforce monotonicity on the $\tilde{p}_{(i)}$s by using $\tilde{p}^*_{(i)} = \max\{\tilde{p}_{(1)}, ..., \tilde{p}_{(i)}\}$.

The maxT case is the same apart from using the appropriate maximization over sets defined by permutations of the statistics in place of the p-values; the maxT and minP are known to be essentially equivalent when the test statistics are identically distributed. In practice, maxT is more computationally tractable because typically estimating probabilities is computationally more demanding than estimating test statistics. Ge et al. (2003), however, give improved forms of these algorithms that are more computationally tractable; see also Tsai and Chen (2007).

In general, the permutation distribution of Westfall and Young (1993) gives valid adjusted p-values in any setting where a condition called subset pivotality holds. The distribution of $P = (P_1, ..., P_m)$ has the subset pivotality property if and only if the joint distribution of any subvector $(P_{i_1}, ..., P_{i_K})$ of P is the same under the complete null $H_{\mathscr{C}}$ as it is under $\cap_{j=1}^K H_{i_j}$. Essentially, this means that the subvector distribution is unaffected by the truth or falsehood of hypotheses not included. When subset pivotality is satisfied, it implies that upper bounds on conditional probabilities of events defined in terms of subvectors of P, given $H_{\mathscr{C}}$, give upper bounds on the same events conditional only on the hypotheses in the subvector; see Westfall and Young (1993), Section 2.3. Consequently, whether the test is exact and has $P_i \sim Uniform[0,1]$, or is conservative in which P_i is stochastically larger than $Uniform[0,1]$ is not really the issue for validity of the permutation distribution even though it is valid. In fact, the permutation distribution is valid even for tests that have P_i stochastically smaller than $Uniform[0,1]$.

The benefit of step-down procedures, and their step-up counterparts, is that they are a little less conservative and have more power than single-step procedures. This arises from the way in which the adjusted p-values tie the m tests together. There is also some evidence that this holds in large m, small n contexts. It will be seen that the FDR paradigm achieves the same goal as Westfall and Young (1993) but for a different measure of Type I error. Step-down testing for FWER is studied in Dudoit et al. (2003).

11.3 PCER and PFER

The central idea of PCER or PFER is to apportion the error level α for m tests to the tests individually. Informally, roughly it's as if each test is allowed to have α_i error associated with falsely rejecting the null in the ith test. In other words, the value of V

is composed of the errors of the individual tests, and the task is to choose the α_is so they add up to α overall.

The techniques for how to do this for PCER or PFER given in this section are from the important contribution by Dudoit et al. (2003). Although the development here is for PFER and PCER, the techniques apply to any measure of Type I error rate that can be expressed in terms of the distribution of V, including FWER, the median-based PFER, and the generalized FWER; the last two measures of error rate are not explicitly studied here. On the other hand, the false discovery rate to be presented in the next section cannot be expressed as an operator on V because it involves R as well.

PCER can be expressed in terms of the distribution of V because if F_n is the distribution function of V, having support on $\{1,...,m\}$, then $PCER = \int v dF_n(v)/m$; PCER is similar. Observe that for, one-sided tests, the set $\mathscr{N}$ is estimated by

$$S_n = \{i : T_{n,i} > \tau_i\},$$

where the threshold $\tau_i = \tau_i(T_n, Q_0, \alpha)$. Note that $S_n = S(T_n, Q_0, \alpha)$, where Q_0 is the distribution assigned to T under the null. In terms of Table 1.2, the number of hypotheses rejected is $R = R_n = \#(S_n)$, and the number not rejected is $\#(S_n^c) = m - \#(S_n)$. The key variable is $V = V_n = \#(S_n \cap S_0)$, where $S_0 = S_0(\mathbb{P})$ is the number of nulls that are true when $\mathbb{P}$ is the true distribution, so $m_0 = \#(S_0)$ and $m_1 = m - m_0 = \#(S_0^c)$.

11.3.1 Null Domination

Write $\mathbb{P}$ to mean a candidate probability for the data in a testing problem. The distribution of a test statistic $T = (T_1, \ldots, T_m)$ based on n data points can be denoted $Q = Q_n(\mathbb{P})$ and contrasted with the null Q_0 used to get level α cutoffs for T. Note that Q_0 need not be $Q(\mathbb{P})$ for any $\mathbb{P}$ and that testing procedures are very sensitive to the choice of Q_0; see Pollard and van der Laan (2004).

To be explicit, note that there is a big conceptual distinction between

$$R_n = R(S(T_n, Q_0, \alpha)|Q_n(\mathbb{P})); \qquad V_n = V(S(T_n, Q_0, \alpha)|Q_n(\mathbb{P})),$$

the number of rejected and the number of falsely rejected hypotheses when $\mathbb{P}$ is true and

$$R_0 = R(S(T_n, Q_0, \alpha)|Q_0); \qquad V_0 = V(S(T_n, Q_0, \alpha)|Q_0),$$

the same quantities when Q_0 is taken as true.

If $\mathbb{P}_0$ is the null distribution in an exact test then it may make sense to set $Q_0 = Q_n(\mathbb{P}_0)$. However, more generally it is difficult to obtain reliable testing procedures unless it can be ensured that controlling the Type I error under Q_0 implies that the Type I error under the $\mathbb{P}$s in the null is controlled at the same level. One way to do this is to require that

$$R_0 \geq V_0 \geq V_n; \tag{11.3.1}$$

i.e., the number of rejected hypotheses under Q_0 is bounded below by the number of falsely rejected hypotheses under Q_0 (a trivial bound), which in turn is bounded below by the number of falsely rejected hypotheses under the true distribution. Note that the random variables in (11.3.1) only assume values 0, 1,.., m.

Expression (11.3.1) can be expressed in terms of distribution functions denoted F_X for random variable X as

$$\forall s: \ F_{R_0}(s) \leq F_{V_0}(s) \leq F_{V_n}(s), \tag{11.3.2}$$

which is the usual criterion for "stochastically larger than". Expression (11.3.2) is the null domination condition for the number of Type I errors.

Recall that $PFER$ and $PCER$ are error criteria that can be regarded as parameters of the distribution function. Indeed, $PCER$ and $PFER$ can be expressed as increasing functions of the distribution function. This ensures that bounding the error of V_n also provides a bound on the error under R_0. That is, for the case of $PFER$,

$$PFER(F_{V_n}) \leq PFER(F_{V_0}) \leq PFER(F_{R_0}),$$

and if the right is bounded by α, so is the left.

The null domination condition can also be expressed in terms of statistics. Form the subvector $T_{n,S}$ of T_n consisting of those $T_{n,j}$ for which $j \in S_0$ and consider two distributions for T. First, $T_n \sim Q_0$, a null chosen to give useful cutoffs. Second, if $\mathbb{P}_0$ is the true distribution, then $T_n \sim Q_n = Q_n(\mathbb{P}_0)$. Now, the domination condition relates Q_0 and Q_n by the inequality

$$Q_n(T_{n,j} \leq \tau_j, \ j \in S_0) \geq Q_0(T_{n,j} \leq \tau_j, \ j \in S_0).$$

That is, if the left-hand side is small (indicating rejection), then the right-hand side is small also.

11.3.2 Two Procedures

For single-step procedures, which are the main ones studied to date for PFER and PCER, Dudoit et al. (2003) propose a generic strategy that can be implemented in two ways. It is the following.

DUDOIT, VAN DER LAAN, AND POLLARD GENERIC STRATEGY:

☐ To control the Type I error rate $PCER(F_{V_n})$ for $T_n \sim Q_n(\mathbb{P})$, find a testing null Q_0 that satisfies

$$PCER(F_{V_n}) \leq PCER(F_{V_0}).$$

- ☐ By the monotonicity of PCER as a parameter of a distribution, since $V_0 \leq R_0$, $F_{V_0} \geq F_{R_0}$, so $PCER(F_{V_0}) \leq PCER(F_{R_0})$.
- ☐ Control $PCER(F_{R_0})$, which correspond to the observed number of rejections under Q_0. That is, assume $T_n \sim Q_0$, and ensure $PCER(F_{R_0}) \leq \alpha$.

Note that the first two steps are conservative; often the bound will not be tight in the sense that $PCER(F_{R_0}) \leq \alpha$ may mean $PCER(F_{V_n}) < \alpha$.

This generic strategy can be implemented in two different ways, the common quantile and common cutoff versions. They correspond to choosing rejection regions for the H_is with thresholds representing the same percentiles with different numerical cutoffs or rejection regions with the same cutoffs but different percentiles.

11.3.2.1 Common Quantiles: Single Step

The common quantile version of the Dudoit et al. (2003) generic strategy is to define a function $\delta = \delta(\alpha)$ that will be the common quantile for the m hypotheses. Then, the null H_j for $j = 1, ..., m$ is rejected when $T_{n,j}$ exceeds the $\delta(\alpha)$ quantile of the testing null Q_0, say $\tau_j(Q_0, \delta(\alpha))$. The function δ translates between the overall level α and the common level of the m tests. Therefore, to control the Type I error rate $PCER(F_{V_n})$ at level α, $\delta(\alpha)$ is chosen so that $PCER(F_{R_0})$ is bounded by α. Note that F_{R_0} is the distribution for the observed number of rejections R_0 when Q_0 is assumed true. That is, Q_0 is used to set the thresholds for testing H_i.

Suppose this procedure is used and the null H_j is rejected when $T_{n,j} > \tau_j$, where the $\tau_j = \tau_j(Q_0, \alpha)$ are quantiles of the marginals $Q_{0,j}$ from Q_0 for T_n. Gathered into a single vector, this is $\tau = (\tau_1, ..., \tau_m)$. So, the number of rejected hypotheses is

$$R(\tau|Q) = \sum_{j=1}^{m} \mathbf{1}_{\{T_{n,j} > \tau_j\}}, \tag{11.3.3}$$

and the number of Type I errors among the $R(\tau|Q)$ rejections is

$$V(\tau|Q) = \sum_{j \in S_0} \mathbf{1}_{\{T_{n,j} > \tau_j\}}, \tag{11.3.4}$$

although S_0 is not known. As before, following Dudoit et al. (2003), R_n, V_n, R_0, and V_0 are the versions of R and V with the true (but unknown) distribution $Q_n = Q_n(\mathbb{P})$ and the testing null Q_0 in place of Q in (11.3.3) and (11.3.4). That is,

$$R_n = R(\tau|Q_n),\ V_n = V(\tau|Q_n),\ R_0 = R(\tau|Q_0),\ V_0 = R(\tau|Q_0). \tag{11.3.5}$$

Now, the Dudoit et al. (2003) single-step common quantile control of the Type I error rate $PCER(F_{V_n})$ at level α is defined in terms of the common quantile cutoffs $\tau(Q_0, \alpha) = (\tau_1(Q_0, \delta(\alpha)), ..., \tau_m(Q_0, \delta(\alpha)))$, where $\delta(\alpha)$ is found under Q_0. Their procedure is the following.

☐ Let Q_0 be an m-variate testing null, let $\delta \in [0,1]$, and write

$$\tau = (\tau_1(Q_0,\delta),\ldots,\tau_m(Q_0,\delta))$$

to mean a vector of δ-quantiles for the m marginals of Q_0. Formally, if $Q_{0,j}$ is the distribution function of the jth marginal from Q_0, this means

$$\tau_j(Q_0,\delta) = Q_{0,j}^{-1}(\delta) = \inf\{x|Q_{0,j}(x) \geq \delta\}.$$

☐ Given the desired level α, set

$$\delta = \delta(\alpha) = \inf\{\delta|PCER(F_{R(\tau(Q_0,\delta)|Q_0)}) \leq \alpha\},$$

where $R(\tau(Q_0,\delta)|Q_0)$ is the number of rejected hypotheses when the same quantile is used for each marginal from Q_0.

☐ The rejection rule is: For $j = 1,\ldots,m$,

$$\text{Reject the null } H_j \text{ when } T_{n,j} > \tau_j(Q_0,\delta(\alpha)).$$

Equivalently, estimate the set of false hypotheses by the set

$$S(T_n,Q_0,\alpha) = \{j|T_{n,j} > \tau(Q_0,\delta(\alpha))\}.$$

This procedure is based on the marginals, largely ignoring dependence among the $T_{n,j}$s. Moreover, its validity rests on the fact that $PCER(F)$ is a continuous function of distributions F and that if $F \geq G$ then, $PCER(F) \leq PCER(G)$.

11.3.2.2 Common Quantiles: Adjusted p-Values

Single-step common quantile p-values can be converted into adjusted p-values. The conversion presented in this subsection assumes Q_0 is continuous and the marginals $Q_{0,j}$ have strictly increasing distribution functions. While these conditions are not necessary, they do make the basic result easier to express. Recall that the raw p-value $p_j = p_j(Q_0)$ for testing the null H_i under Q_0 can be represented as

$$p_j = 1 - Q_{0,j}(T_{n,j}) = \bar{Q}_{0,j}(T_{n,j})$$

for $j = 1,\ldots,n$. Thus, the common quantile method uses thresholds

$$\tau_j(Q_0, 1-p_j) = Q_{0,j}^{-1}(1-p_j) = \bar{Q}_{0,j}^{-1}(p_j) = \bar{Q}_{0,j}^{-1}(\bar{Q}_{0,j}(T_{n,j})).$$

Indeed, $\tau_j(Q_0, 1-p_j) = T_{n,j}$. Now, the adjusted p-values for the common quantile procedure can be stated explicitly.

Proposition (Dudoit et al. 2003): The adjusted p-values for the single-step common quantile procedure for controlling the Type I error rate under $PCER$ using Q_0 are

$$\tilde{p}_j = PCER(F_{R(\tau_j(Q_0,1-p_j)|Q_0)}).$$

Equivalently, the set of false hypotheses is estimated by the set

$$S(T_n, Q_0, \alpha) = \{j : \tilde{p}_j \leq \alpha\}. \tag{11.3.6}$$

Proof: Recall that, for fixed Q_0, the function $\phi : \delta \to \phi(\delta) = PCER(F_{R(\tau(Q_0,\delta)|Q_0)})$ is monotonically increasing so its inverse ϕ^{-1} exists and

$$\phi^{-1}(\alpha) = \inf\{\delta | \phi(\delta) \leq \alpha\}.$$

So, the common quantile cutoffs can be written

$$\tau_j(Q_0, \delta(\alpha)) = Q_{0,j}^{-1}(\delta(\alpha)) = Q_{0,j}^{-1}(\phi^{-1}(\alpha)).$$

Now, the adjusted p-values are

$$\begin{aligned}
\tilde{p}_j &= \inf\{\alpha \in [0,1] : \tau_j(Q_0, \delta(\alpha)) < T_{n,j}\} \\
&= \sup\{\alpha \in [0,1] : Q_{0,j}^{-1}(\phi^{-1}(\alpha)) < T_{n,j}\} \\
&= \sup\{\alpha \in [0,1] : \phi^{-1}(\alpha) < Q_{0,j}(T_{n,j})\} \\
&= \sup\{\alpha \in [0,1] : \alpha < \phi(Q_{0,j}(T_{n,j}))\} \\
&= \phi(Q_{0,j}(T_{n,j}) = \phi(1 - p_j) \\
&= PCER(F_{R(\tau_j(Q_0,1-p_j)|Q_0)}),
\end{aligned} \tag{11.3.7}$$

as claimed. The second part follows from the definitions. □

11.3.2.3 Common Cutoffs: Single Step

The Dudoit et al. (2003) generic strategy can be implemented as a common cutoff procedure as well. The common cutoff procedure is simpler: Reject the null H_j when $T_{m,j} > c(Q_0, \alpha)$, where α satisfies $PCER(F_{R_0}) \leq \alpha$.

As before, the single-step common cutoff procedure for controlling the Type I error rate $PCER(F_{V_n})$ at level α is defined in terms of the common cutoffs $c(Q_0, \alpha)$ and can be given as follows.

Let Q_0 be an m-variate testing null and let $\alpha \in (0,1)$ be the desired level.

- □ Define the common cutoff $c(Q_0, \alpha)$ by

$$c(Q_0, \alpha) = \inf\{c | PCER(F_{R((c,...,c)|Q_0)}) \leq \alpha\},$$

  where $R((c,...,c)|Q_0)$ is the number of rejected hypotheses for the common cutoff c under Q_0 for T_n.
- □ The rejection rule is: For $j = 1, ..., m$,

$$\text{Reject the null } H_j \text{ when } T_{n,j} > c(Q_0, \alpha).$$

☐ The set of false hypotheses is estimated by

$$S(T_n, Q_0, \alpha) = \{j | T_{n,j} > c(Q_0, \alpha)\}.$$

The single-step common cutoff and common quantile procedures here reduce to the single-step minP and maxT procedures based on ordering the raw p-values or test statistics; see Dudoit et al. (2003).

11.3.2.4 Common Cutoffs: Adjusted p-Values

The single-step common cutoff p-values can be converted into adjusted p-values as was the case with single-step common quantiles. Again, the conversion assumes Q_0 is continuous and the marginals $Q_{0,j}$ have strictly increasing distribution functions. These conditions are not necessary, but they do make the basic result easier to express.

Proposition (Dudoit et al. 2003): The adjusted p-values for the single-step common cutoff procedure for controlling the Type I error rate under $PCER$ using Q_0 are

$$\tilde{p}_j = PCER(F_{R((T_{n,j},\ldots,T_{n,j})||Q_0)}) \tag{11.3.8}$$

for $j = 1, \ldots, m$. Equivalently, the set of false hypotheses is estimated by the set

$$S(T_n, Q_0, \alpha) = \{j : \tilde{p}_j \leq \alpha\}. \quad \square$$

The proof is omitted; it is similar to that of the common quantile case. Indeed, much of the difference is a reinterpretation of the notation. For instance, note that in (11.3.8), the expression for $\tilde{p}_j$, the common cutoff $T_{n,j}$ appears m times in the argument of R and the set estimation expression is the same as in (11.3.6), although the adjusted p-values are from (11.3.8) rather than from (11.3.6).

Overall, choosing between common quantile and common cutoff procedures is a matter of modeling, outside of special cases. For instance, the two procedures are equivalent when the test statistics $T_{n,j}$ are identically distributed. More generally, the procedures give different results because the m tests are either done at different levels (and hence weighted in importance) or are done at the same level (implying all tests are equally important in terms of the consequences of errors). As a generality, common-quantile based procedures seem to require more computation than common cutoff based procedures; this may make common quantile methods more sensitive to the choice of Q_0. This may force common quantile procedures to be more conservative than cutoff-based methods.

11.3.3 Controlling the Type I Error Rate

It can be proved that the single-step common quantile and common cutoff tests are asymptotically level α; this is done in the first theorem below. There are several assumptions; the most important one is that V_n is stochastically smaller than V_0. Verifying that this condition is satisfied is not trivial: It rests on constructing a satisfactory testing null Q_0. Accordingly, it is important to identify sufficient conditions for a satisfactory Q_0 to exist. This is the point of the second theorem in this subsection.

Unfortunately, while these results enable identification of a *PCER* level α test, they do not say anything about whether the *PCER* for any element in the alternative is small. That is, the analog of power for Neyman-Pearson testing, which leads to unbiasedness of tests, has not been examined for *PCER*. Nevertheless, if the dependence of the behavior on the test statistics $T_{n,j}$ depends on $\mathbb{P}$ strongly enough, it is reasonable to conjecture that the analogs of power and unbiasedness from Neyman-Pearson testing can be established for the Dudoit et al. (2003) methods.

11.3.3.1 Asymptotic *PCER* Level

Recall that the generic procedure compares V_n with V_0 and then compares V_0 with R_0. The second of these is trivial because $V_0 \le R_0$ by definition. So, it is enough to focus on the first. Restated, this is the requirement that the number of Type I errors, V_n under the true m-dimensional distribution $Q_n = Q_n(\mathbb{P})$ for the test statistics $T_{n,j}$ be stochastically smaller, at least asymptotically, than the number of Type I errors V_0 under the testing null Q_0. Formally, this means

$$\forall x \ \liminf_{n\to\infty} F_{V_n}(x) \ge F_{V_0}(x).$$

In the present setting, this can be written in terms of events of indicator functions. The criterion becomes that the joint distribution $Q_n = Q_n(\mathbb{P})$ of the test statistics T_n satisfies an asymptotic null domination property when compared with Q_0,

$$\liminf_{n\to\infty} P_{Q_n}\left(\sum_{j\in S_0} \mathbf{1}_{T_{n,j}>c_j} \le x\right) \ge P_{Q_0}\left(\sum_{j\in S_0} \mathbf{1}_{Z_j>c_j} \le x\right), \tag{11.3.9}$$

for all $x = 0,\ldots,m$ and all $\mathbf{c} = (c_1,\ldots,c_m)$, where $Z \sim Q_0 = Q_0(\mathbb{P})$.

The proof that the single-step common quantile procedure is level α also requires the monotonicity of *PCER*; i.e., given two distribution functions F_1 and F_2,

$$F_1 \ge F_2 \Rightarrow PCER(F_1) \le PCER(F_2), \tag{11.3.10}$$

where the $\ge$ on the left holds in the sense of stochastic ordering and the representation of the *PCER* as a functional with a distribution function argument is continuous. The continuity can be formalized as requiring that, for any two sequences of distribution functions F_k and G_k,

$$\lim_{k\to\infty} d(F_k, G_k) = 0 \Rightarrow \lim_{k\to\infty}\Big(PCER(F_k) - PCER(G_k)\Big) = 0 \qquad (11.3.11)$$

for some metric d. One natural choice is the Kolmogorov-Smirnov metric, $d(F,G) = \sup_x |F(x) - G(x)|$. Since the distribution functions of concern here only assign mass at $x = 0, 1, ..., m$, any metric that ensures the two distribution functions match at those points will make the representation of $PCER$ continuous as a functional.

The level α property of the single-step common quantile procedure can now be stated and proved.

Theorem (Dudoit et al., 2003): Suppose there is a random variable Z so that (11.3.9) holds. Also, assume that (11.3.10) and (11.3.11) hold. Then the single-step procedure with common quantile cutoffs given by $c(Q_0,\alpha) = \tau(Q_0, \delta(\alpha))$ gives asymptotic level α control over the $PCER$ Type I error rate,

$$\limsup_{n\to\infty} PCER(F_{V_n}) \le \alpha. \qquad (11.3.12)$$

The number of Type I errors for $T_n \sim Q_n(P)$ is

$$V_n = V(c(Q_0,\alpha)|Q_n) = \sum_{j\in S_0} \mathbf{1}_{T_{n,j} > c_j(Q_0,\alpha)}.$$

Proof: By construction, $V_0 \le R_0$. So, $F_{V_0}(x) \ge F_{R_0}(x)$, and so

$$PCER(F_{V_0}) \le PCER(F_{R_0}) \le \alpha \qquad (11.3.13)$$

when the cutoffs $c(Q_0,\alpha) = \tau(Q_0, \delta(\alpha))$ are used to ensure $PCER(F_{R_0}) \le \alpha$. The theorem will follow if

$$\limsup_{n\to\infty} PCER(F_{V_n}) \le PCER(F_{V_0}). \qquad (11.3.14)$$

To see (11.3.14), write

$$F_{V_n} = F_{V_0} + (F_{V_n} - F_{V_0}) \ge F_{V_0} + \min(0, F_{V_n} - F_{V_0}).$$

By (11.3.9), $\liminf F_{V_n} \ge F_{V_0}$, so

$$\lim_{n\to\infty} (F_{V_0}(x) + \min(0, F_{V_n} - F_{V_0})(x)) = F_{V_0}(x)$$

since the limit exists. Using (11.3.11) gives

$$\lim_{n\to\infty} PCER(F_{V_0} + \min(0, F_{V_n} - F_{V_0})) = PCER(F_{V_0}).$$

By (11.3.10),

$$PCER(F_{V_n}) \le PCER(F_{V_0} + \min(0, F_{V_n} - F_{V_0})),$$

and so

$$\limsup_{n\to\infty} PCER(F_{V_n}) \leq \lim_{n\to\infty} PCER(F_{V_0} + \min(0, F_{V_n} - F_{V_0})) = PCER(F_{V_0}). \quad \square$$

It is not hard to see that, under the same conditions, the single-step common cutoff procedure is also level α. Moreover, it is straightforward to see that the key assumptions satisfied by $PCER$ or $PFER$ that make the proof possible are also satisfied by $FWER$ and other criteria. Therefore, level α tests for common quantile procedures for those criteria can also be found.

11.3.3.2 Constructing the Null

The remaining gap to be filled for the common quantile and common cutoff procedures for $PCER$ and $PFER$ is the identification of Q_0. Essentially, asymptotic normality can be invoked provided a suitable shift is made to ensure V_n is stochastically smaller than V_0 and hence R_0. Once this is verified, a bootstrap procedure can be used to give an estimate of Q_0. The central result is the identification of a limiting procedure that relates the statistics $T_{n,j}$ to the criterion (11.3.9).

To state the theorem, suppose there are vectors $\lambda = (\lambda_1, ..., \lambda_m)$ and $\gamma = (\gamma_1, ..., \gamma_m)$ with $\gamma_j \geq 0$ that bound the first two moments of T_n for $j \in S_0$. That is, when j indexes a true null hypothesis H_j,

$$\limsup_{n\to\infty} \mathbb{E}\, T_{n,j} \leq \lambda_j \tag{11.3.15}$$

and

$$\limsup_{n\to\infty} \mathrm{Var}(T_{n,j}) \leq \gamma_j. \tag{11.3.16}$$

The λ_js will be used to relocate the $T_{n,j}$s into random variables $Z_{n,j}$ that are stochastically larger. The γ_js will be used to rescale the relocated $T_{n,j}$s so their standardized form will have a limiting distribution that does not degenerate to a single value.

To do this, let

$$\nu_j = \nu_{n,j} = \sqrt{\min\left(1, \frac{\gamma_j}{\mathrm{Var}(T_{n,j})}\right)}, \tag{11.3.17}$$

and set

$$Z_{n,j} = Z_j = \nu_j\,(T_{n,j} + \lambda_j - \mathbb{E}(T_{n,j})) \tag{11.3.18}$$

for $j = 1, ..., m$. The key assumption for the theorem will be that $Z_n = (Z_1, ..., Z_m)$ has a well-defined limiting distribution. Although (11.3.17) supposes that the $T_{n,j}$s are scaled appropriately to make their variances converge to a (usually nonzero) constant, expression (11.3.17) also ensures that a gap between the limit superior and the upper bound on the variance will not generally affect the limiting distribution of Z_n. The following theorem establishes that the assumption (11.3.9) holds in general.

Theorem (Dudoit et al., 2003): Suppose that $\mathbb{P}$ is the true probability measure and that, for some m-dimensional random variable Z,

$$Z_n \Rightarrow Z \sim Q_0(\mathbb{P}). \tag{11.3.19}$$

Then, for $Q_0 = Q_0(\mathbb{P})$, (11.3.9) is satisfied for any x and c_js. That is, for $Q_n = Q_n(\mathbb{P})$,

$$\liminf_{n\to\infty} Q_n\left(\sum_{j\in S_0} \mathbf{1}_{T_{n,j}>c_j} \le x\right) \ge Q_0\left(\sum_{j\in S_0} \mathbf{1}_{Z_j>c_j} \le x\right). \tag{11.3.20}$$

Proof: Consider a vector $\bar{Z}_n = (\bar{Z}_{n,j} : j \in S_0)$, with entries corresponding to the true hypotheses S_0 in which

$$\bar{Z}_{n,j} = \bar{Z}_j = T_{n,j} + \max(0, \lambda_j - ET_{n,j}).$$

By construction, $T_{n,j} \le \bar{Z}_j$.

From (11.3.15) and (11.3.16) for $j \in S_0$, it is seen that $\lim_n \nu_{n,j} = 1$ and that $\bar{Z}_n$ and Z_n have the same limiting distribution Z. That is,

$$\bar{Z} \Rightarrow Z \sim Q_{0,S_0},$$

where Q_{0,S_0} indicates the marginal joint from Q_0 corresponding to S_0. Letting P denote the probability of $\bar{Z}$, the limiting property of $\bar{Z}$ and the upper bound on T_n give

$$\begin{aligned}\liminf_{n\to\infty} Q_n\left(\sum_{j\in S_0} \mathbf{1}_{T_{n,j}>c_j} \le x\right) &\ge \liminf_{n\to\infty} P\left(\sum_{j\in S_0} \mathbf{1}_{\bar{Z}_j>c_j} \le x\right)\\ &= Q_{0,S_0}\left(\sum_{j\in S_0} \mathbf{1}_{Z_j>c_j} \le x\right)\end{aligned} \tag{11.3.21}$$

for any vector of c_js and any x. □

In some cases, Q_0 can be a mean-zero normal. However, the scaling per se is not needed to get level α so much as the relocating to ensure the stochastic ordering. A consequence of this theorem is that the λ_js and γ_js only depend on the marginals for the $T_{n,j}$s under the true hypothesis; in many cases, they can be taken as known from univariate problems. Dudoit et al. (2003) Section 5 use t-statistics and F-statistics as examples and replace the λ_js and γ_js by estimators.

Even given this theorem, it remains to get a serviceable version of Q_0 and derive the cutoffs from it. This can be done by bootstrapping. A thorough analysis of this is given in Dudoit et al. (2003) Section 4, ensuring that the bootstrapped quantities are consistent for the population quantities. This analysis largely boils down to making sure that the bootstrapped version of Q_0 converges to $Q(\mathbb{P})$ for the true distribution

$\mathbb{P}$, essentially a careful verification that empirical distributions converge to their limits uniformly on compact sets. As a consequence, the following procedure gives the desired construction for the estimate of a testing null Q_0:

- ☐ Generate B bootstrap samples $X_{1,b},...,X_{n,b}$. For fixed b, the $X_{i,b}$s are n IID realizations.
- ☐ From each of the B samples, find the test statistics $T_{n,b} = (T_{n,b,1},...,T_{n,b,m})$. This gives an $m \times B$ matrix $\mathbf{T}$ as in the permutation technique in Section 2.1.
- ☐ The row means and variances of $\mathbf{T}$ give m estimates of $ET_{n,j}$ and $\text{Var}(T_{n,j})$ for $j = 1,...,m$.
- ☐ Use the means and variances from the last step, together with user-chosen values λ_j and γ_j for $j = 1,...,m$, to relocate and rescale the entries in $\mathbf{T}$ by (11.3.18). Call the resulting matrix $\mathbf{M}$.
- ☐ The empirical distribution from the columns $M_{b,n}$ of $\mathbf{M}$ is the bootstrap estimate $Q_{0,B}$ for Q_0 from the last theorem.
- ☐ The bootstrap common quantiles or common cutoffs are row quantities of $\mathbf{M}$.

Note that this procedure for estimating Q_0 is quite general and can be adapted, if desired, to other testing criteria.

11.3.4 Adjusted p-Values for PFER/PCER

To conclude this section, it is revealing to give expressions for the adjusted p-values for the common quantile and common cutoff procedures. In this context, the notion of adjusted does not correspond to step-down or step-up procedures but only to p-values for H_j that take into account the values of $T_{n,j'}$ for $j' \neq j$. The essence of the result is that adjustment does not make any difference for the PCER in the common quantile case, whereas adjustment amounts to taking averages in the marginal distribution for the common cutoff procedure.

Proposition (Dudoit et al., 2003): Suppose the null distribution Q_0 is continuous with strictly monotone marginal distributions. For control of the PCER, the adjusted p-values for the single-step procedures are as follows:

(i) For common quantiles, $\tilde{p}_j = \bar{Q}_{0,j}(T_{n,j}) = p_j$, i.e., they reduce to the unadjusted, raw p-values for $j = 1,...,m$.

(ii) For common cutoffs,

$$\tilde{p}_j = \frac{1}{m}\sum_{k=1}^{m} \bar{Q}_{0,k}(T_{n,j});$$

i.e., they become identical, with common a value given by the average of the p-values from the m tests.

Proof: Let $Z \sim Q_0$ and write the PCER as an operator on a distribution, $PCER(F) = \int x dF(x)/m$.

For Part (i), the adjusted p-value for testing the null H_j is

$$\tilde{p}_j = PCER(F_{R(\tau(Q_0,1-p_j)|Q_0)}) = \frac{1}{m}\sum_{k=1}^{m} Q_0(Z_k > \bar{Q}_0^{-1}(p_j))$$
$$= \frac{1}{m}\sum_{k=1}^{m} \bar{Q}_0(\bar{Q}_0^{-1}(p_j)) = p_j.$$

For part (ii), the adjusted p-value for testing the null H_j is

$$\tilde{p}_j = PCER(F_{R((T_{n,j},\ldots,T_{n,j})|Q_0)}) = \frac{1}{m}\sum_{k=1}^{m} Q_0(Z_k > T_{n,j}) = \frac{1}{m}\sum_{k=1}^{m} \bar{Q}_{o,k}(T_{n,j}). \quad \square$$

It should be remembered that all the results in this section apply not just to PCER but have analogs for FWER and any other measure of Type I error that can be represented as a monotone, continuous functional of distribution functions. Indeed, the only place that specific properties of PCER were used was in the last proposition. However, even it has an analog for other Type I error measures, including generalized FWER, gFWER, which is defined by $P(V_n \geq k+1)$ so the FWER is gFWER for $k = 0$. The result is the interesting fact that the single-step adjusted p-values for the common quantile and common cutoff for gFWER control are expressible in terms of the ordered raw p-values similar to the step-down procedures for FWER; see Dudoit et al. (2003), Section 3.3 for details.

11.4 Controlling the False Discovery Rate

While controlling FWER is appealing, the resulting procedures often have low power; the effect of low power is to make it hard to reject the null when it's false so the interesting effects (usually the alternatives) become hard to find. Indeed, although minP and maxT improve power and account for some dependence among the test statistics, the fact that they control the FWER criterion can make them ill-suited for situations where the number of tests m is large: FWER-based tests typically have low power for high values of m. The properties of PCER are not as well studied but overall seem to suggest that PCER is at the other extreme in that the control it imposes on Type I error is not strong enough: It doesn't force the number of false positives low enough. Thus the question of whether there is a measure that can achieve an acceptable control of the Type I error while at the same time maintaining a usefully high power for the overall testing procedure remains.

One answer to this question is the FDR, introduced by Benjamini and Hochberg (1995), often written simply as $\mathbb{E}(V/R)$. Of course, when the number of rejections is $R = 0$, the number of false rejections $V = 0$, too, so if $0/0 \equiv 0$, then $\mathbb{E}(V/R)$ is the same as the formal definition in Section 11.1. The integrand V/R is sometimes called the false discovery proportion. In some cases, the slight variant on V/R is used and is written explicitly as the random variable

$$FDP(t) = \frac{\sum_{j=1}^{m}(1-H_j)\mathbf{1}_{p_j \leq t}}{\sum_{j=1}^{m}\mathbf{1}_{p_j \leq t}} + \mathbf{1}_{\text{all } p_j > t}, \tag{11.4.1}$$

where H_j is an indicator function, $H_j = 0, 1$, according to whether the jth null hypothesis is true or false, and p_j is the jth p-value (which should be written as P_j since it is being treated as a random quantity). The ratio amounts to the number of false discoveries over the total number of discoveries. Thus, if T is a multiple testing threshold, the FDR can be regarded as

$$FDR = FDR_T = EFDP(T).$$

The net effect of using a relative measure of error like FDR, which doesn't directly control the absolute number of rejections in any sense, is that testing essentially becomes just a search procedure. Rather than relying on testing to say conclusively which H_is are true and which are not, the goal is that the rejected H_is hopefully reduce the number of hypotheses that need to be investigated further. For instance, if each H_j pertains to a gene and $H_j = 0$ means that the gene is not relevant to some biochemical process, then the rejected H_js indicate the gene that might be investigated further in subsequent experiments. Thus, the point of rejecting nulls is not to conclude alternatives per se but to identify a reduced number of cases, the discoveries. The hope is that these statistical discoveries will be a small subset of cases, that hopefully contain the even fewer actual cases being sought. Then, studying the discoveries is an effective step between the set of m possibilities and scientific truth. Of course, false acceptances of nulls may be present, but controlling the level is a way to minimize these in practice to a point where they can be ignored.

While the FDR and pFDR are used ever more commonly, it is important to recall they they are just one choice of criterion to impose on a testing procedure. So, for background, it will first be important to present some variants on them that may be reasonable. Then, the Benjamini-Hochberg (BH) procedure will be presented. In fact, BH goes back to Eklund and Seeger (1965) and Simes (1986) but was recently rediscovered and studied more thoroughly. For instance, there have been a wide variety of extensions to the BH method, including using dependent p-values, and studies to develop an intuition for how BH performs and how to estimate and use both the FDR and pFDR. Finally, there is an interesting Bayesian perspective afforded from comparing the FDR to the pFDR leading to the q-value, a Bayesian analog of the p-value. Some material on the FDR is deferred to Section 11.5 on the pFDR since the two criteria can be treated together for some purposes.

11.4.1 FDR and other Measures of Error

The FDR and pFDR are two possible ratios involving false discoveries that have been studied and found to be well justified. One argument favoring the FDR is that when some of the alternatives are true (i.e., $m_0 < m$) the FDR is smaller than the FWER. Indeed, if $V = v \geq 1$ and $R = r$, then $v/r \leq 1$ so $\mathbb{E}(V/R) \leq \mathbf{1}_{\{V \geq 1\}}$. Taking expectations gives $P(V \geq 1) \geq FDR$, so any test controlling the FWER also controls the FDR. Since the inequality is far from tight, a procedure that only bounds the FDR will be less stringent and hence may have higher power. As the number of false nulls, $m - m_0$, increases, S tends to increase and the difference between the FWER and FDR tends to increase as well. Thus, the power of an FDR-based testing scheme tends to increase as the number of false nulls does. Indeed, the intuition behind the FDR approach is that tolerating a few false positives can give a testing procedure with a much higher overall power.

Indeed, even controlling the FDR or the pFDR has unexpected subtleties: If $m = m_0$ and a single hypothesis is rejected, then $v/r = 1$. Thus, V/R cannot be forced to be small and neither can $(V/R|R > 0)$ under the same conditions. Their means can be forced to be small, and this is the point, but the random variables themselves seem hard to control, even though that would be the ideal.

Variants on the FDR or pFDR are numerous. The proportion of false discoveries as a proportion of all discoveries would be $\mathbb{E}(V)/R$ – a mix of a mean and a random variable. This unusual criterion is similar to the conditional expectation of V/R, $\mathbb{E}(V/R|R = r) = \mathbb{E}(V|R = r)/r$, but $\mathbb{E}(V) \neq \mathbb{E}(V|R = r)$. The ratio $\mathbb{E}(V)/R$ is also impossible to control as before when $m_0 = m$ because of the random part.

The proportion of false positives would be $\mathbb{E}(V)/\mathbb{E}(R)$. Like the random variable ratios, when $m_0 = m$, $\mathbb{E}(V)/\mathbb{E}(R) = 1$, making it impossible to control. In principle, one could use $\mathbb{E}(R|R > 0)$ in the denominator, but using $\mathbb{E}(V|R > 0)$ in the numerator for symmetry again cannot be controlled when $m = m_0$.

In the present context, the usual Type I error is a false positive. To see how false positives and the FDR are related, consider a single test of H_j. Walsh (2004) notes that the FDR is

$$FDR = \mathbb{P}(H_j \text{ is truly null}|\, j \text{ is significant}).$$

The false positive rate is the reverse,

$$FPR = \mathbb{P}(j \text{ is significant}|H_j \text{ is truly null}),$$

controlled at level α. The conditioning in the FDR includes both false positives and true positives, and the relative fraction depends on what proportion of the H_js are truly null. That is, the FDR is heuristically like a Bayesian's posterior probability in contrast to the usual frequentist conditioning.

The posterior error rate is the probability that a single rejection is a false positive,

$$PER = \mathbb{P}(V = 1|R = m = 1).$$

If $FDR = \delta$ then the PER for a randomly drawn significant test is also δ. Now let α be the level and β be the probability of a Type II error, and suppose π is the fraction of true nulls, $\pi = m_0/m$. Then,

$$PER = \frac{\mathbb{P}(\text{false positive}|\text{null true})\mathbb{P}(\text{null})}{\mathbb{P}(R = m = 1)} = \frac{\alpha\pi}{\mathbb{P}(R = m = 1)}.$$

Again, follow Walsh (2004) and consider the event where a single randomly chosen hypothesis is rejected; i.e., discovered or declared significant. This event can happen if the hypothesis is true and a Type I error is made or if an alternative is true and a Type II error is avoided. In the second case, the power is $S/(m - m_0)$, the fraction of alternatives that are found significant. Since the power is $1 - \beta$, the probability that a single randomly drawn test is significant is

$$\mathbb{P}(R = m = 1) = \alpha\pi + (1 - \beta)(1 - \pi).$$

So, as a function of α and β,

$$PER = \frac{1}{1 + \frac{(1-\beta)(1-\pi)}{\alpha\pi}}.$$

It can be seen that the Type I error rate and the PER for a significant test are very different because the PER depends on the power and the fraction of true nulls, as well as α. To get a satisfactorily low PER usually requires $1 - \pi > \alpha$.

11.4.2 The Benjamini-Hochberg Procedure

Although there are precedents for what is often called the Benjamini-Hochberg or BH procedure (see Simes (1986) for instance) the earlier versions generally did not focus on the procedure itself as a multiple-comparisons technique for general use in its own right. Benjamini and Hochberg (1995) not only proposed the procedure as a general solution to the multiple comparisons problem, they established it has level α.

First, the BH procedure is the following. Fix m null hypotheses $H_1,...,H_m$. Rank the m p-values p_i in increasing order $p_{(1)}, \cdots, p_{(m)}$ and then find

$$K_{(BH)} = \arg \max_{1 \le i \le m} \left\{ p_{(i)} \le \frac{i}{m} \cdot \alpha \right\}. \tag{11.4.2}$$

The rule becomes: Reject all the hypotheses corresponding to $P_{(1)}, P_{(2)}, \cdots, P_{(K_{BH})}$. It can be seen that BH corresponds to adjusting the p-values as

$$\tilde{p}_{r_i} = \min_{k=i,\cdots,m} \left\{ \min \left\{ \left(\frac{m}{k} \right) p_{r_k}, 1 \right\} \right\}.$$

Establishing that this procedure has level α is nontrivial, even with the added assumption that the p-values are independent as random variables. Indeed, the proof is unusual in testing because of how it uses induction and partitions on the conditioning p-values as random variables.

Theorem (Benjamini and Hochberg, 1995): Let $\alpha > 0$, and suppose all the p_is are independent as random variables. Then, for any choice of m_0 false nulls, the BH procedure given in (11.4.2) is level α; that is,

$$\mathbb{E}(V/R) \leq \frac{m_0}{m}\alpha. \tag{11.4.3}$$

Proof: See the Notes at the end of this chapter. □

To conclude this subsection, it is worthwhile stating formally a result proved at the beginning, comparing FDR and FWER.

Proposition: For all combinations of null and alternative hypotheses,

$$\mathbb{E}[FDR] \leq P(\text{Reject at least one true null}) = FWER$$

with equality if all the nulls are true. □

11.4.3 A BH Theorem for a Dependent Setting

Benjamini and Yakutieli (2001) establish a more general form of the BH theorem by introducing a dependence concept for m-dimensional statistics called positive regression dependency on a subset (PRDS) where the subset I of indices must be specified. If I is specified, it is assumed to be the full set of indices. To state PRDS, define a set $D \subset \mathbb{R}^m$ to be increasing if, given $\boldsymbol{x} \in D$ and $\boldsymbol{y} \geq \boldsymbol{x}$, then $\boldsymbol{y} \in D$. An increasing set is more general than a cone (which contains all positive multiples of its elements) and very roughly corresponds to an orthant with origin located at the smallest element of D. Letting $\boldsymbol{X} = (X_1, ..., X_m)$ and $I = \{1, \ldots, m\}$,

$$\boldsymbol{X} \text{ satisfies } PRDS \Leftrightarrow \mathbb{P}(\boldsymbol{X} \in D | X_i = x_i) \text{ nondecreasing in } x_i, \tag{11.4.4}$$

with corresponding versions if $I \subset \{1, \ldots, m\}$.

The right-hand side of (11.4.4) is implied by multivariate total positivity of order 2 (roughly $f(\boldsymbol{x})f(\boldsymbol{y}) \leq f(\min(\boldsymbol{x},\boldsymbol{y}))f(\max(\boldsymbol{x},\boldsymbol{y}))$, the max and min taken coordinatewise) and implies positive association (in the sense that $\text{cov}(f(\boldsymbol{X}), g(\boldsymbol{x})) \geq 0$ for increasing f, g). PRDS differs from positive regression dependency, which is that $P(\boldsymbol{X} \in D | X_1 = x_1, \ldots, X_m = x_m)$ be nondecreasing in the m arguments x_i, even though the two are clearly similar. The PRDS assumption will be applied to the distribution of m test statistics T_1, ..., T_m, giving p-values p_1,..., p_m for m hypotheses H_1,..., H_m. It will turn out that choosing I to be the indices of the hypotheses that are true permits a new BH-style theorem and it is only the T_js for $j \in I$ that must be PRDS.

Benjamini and Yakutieli (2001) verify that multivariate normal test statistics satisfy the PRDS criterion provided the elements of the covariance matrix are positive. Also, the absolute values of a multivariate normal and the studentized multivariate normal (divided by the sample standard deviation) are PRDS. Accordingly, asymptotically normal test statistics are very likely to be PRDS. In addition, there are latent variable models that satisfy the PRDS property.

Essentially, under PRDS, Benjamini and Yakutieli (2001) refine the BH procedure so it will be level α for FDR for PRDS dependent tests. In other words, this is a new way to estimate $K_{(BH)}$. In this case, the procedure is based on

$$K_{BH} = \arg \max_{1 \le i \le m} \left\{ P_{(i)} \le \frac{i}{m \cdot c(m)} \cdot \alpha \right\},$$

where $c(m) = 1$ for independent tests and more generally $c(m) = \sum_{i=1}^{m} 1/i$. This corresponds to adjusting the p-values as

$$\tilde{p}_{(i)} = \min_{k=i,\cdots,m} \left\{ \min \left\{ \left(m \sum_{j=1}^{k} \frac{1}{j} \right) p_{(k)}, 1 \right\} \right\}.$$

Formally, the Type I error of FDR can be controlled as follows. It can be seen in Step 4 of the proof that one of the main events whose probability must be bounded is an increasing set. The event is defined in terms of p-values, and its conditional probability, given a p-value, is amenable to the PRDS assumption. It is this step that generalizes the control of (11.7.6) (in which an extra conditioning on a p-value is introduced) which is central to the independent case.

Theorem (Benajamini and Yekutieli, 2001): Suppose the distribution of the test statistic $\boldsymbol{T} = (T_1, ..., T_m)$ is PRDS on the subset of T_js corresponding to the true H_is. Then, the BH procedure controls the FDR at level no more than $(m_0/m)\alpha$; i.e.,

$$\mathbb{E}\left(\frac{V}{R}\right) \le \frac{m_0}{m}\alpha. \tag{11.4.5}$$

Proof: See the Notes at the end of this chapter. □

The hypotheses of this theorem are stronger than necessary; however, it is unclear how to weaken them effectively. Indeed, the next theorem has this same problem: The use of $\sum_{i=1}^{m} 1/i$ gives a test that may be excessively conservative.

Theorem (Benjamini and Yekutieli, 2001): Let $\alpha' = \alpha / \sum_{i=1}^{m} 1/i$. If the BH procedure is used with α' in place of α, then

$$\mathbb{E}\left(\frac{V}{R}\right) \le \frac{m_0}{m}\alpha. \tag{11.4.6}$$

Proof: It is enough to use α and show that the increase in FDR is bounded by $\sum_{j=1}^{m}(1/j)$.

Let $C_k(i) = \underset{v+s=k}{\cup} C_{v,s}(i)$, where $C_{v,s}(i)$ is the event that H_i is rejected along with $v-1$ true nulls and s falls nulls; i.e., there are k rejections total. Define

$$p_{ijk} = \mathbb{P}\left(\left\{P_i \in \left[\frac{j-1}{m}\alpha, \frac{j}{m}\alpha\right]\right\} \cap C_k(\hat{\imath})\right),$$

so that

$$\sum_{k=1}^{m} p_{ijk} = \mathbb{P}\left(\left\{P_i \in \left[\frac{j-1}{m}\alpha, \frac{j}{m}\alpha\right]\right\} \cap \left(\cup_{k=1}^{m} C_k(\hat{\imath})\right)\right) \leq \frac{\alpha}{m}.$$

Using this, the FDR is

$$\begin{aligned} \mathbb{E}\left(\frac{V}{R}\right) &= \sum_{i=1}^{m_0}\sum_{k=1}^{m}\frac{1}{k}\sum_{j=1}^{k} p_{ijk} = \sum_{i=1}^{m_0}\sum_{j=1}^{k}\sum_{k=j}^{m}\frac{1}{k}p_{ijk} \\ &\leq \sum_{i=1}^{m_0}\sum_{j=1}^{k}\sum_{k=j}^{m}\frac{1}{j}p_{ijk} \leq \sum_{i=1}^{m_0}\sum_{j=1}^{k}\sum_{k=1}^{m}\frac{1}{j}p_{ijk} = m_0\sum_{j=1}^{m}\frac{1}{j}\frac{\alpha}{m}. \quad \square \end{aligned}$$

11.4.4 Variations on BH

As noted earlier, there are a large number of choices of error criteria for testing. Since this is an area of ongoing research and the procedures are being continually improved, it is important to explain several related directions.

11.4.4.1 Simes Inequality

A simpler form of the procedure in (11.4.2) is called Simes' procedure; see Simes (1986). Suppose a global test of $H_1,\ldots, H_m$ is to be done with p-values $p_1,\ldots,p_m$ so that the null is $H_0 = \cap H_i$ and the level of significance is α. Then, Simes' procedure is a restricted case of BH. That is, Simes' method is to reject H_0 if $p_{(i)} \leq i\alpha/n$ for at least one i. This can be regarded as a more powerful Bonferroni correction because Bonferroni would reject H_0 if any $p_{(i)} \leq \alpha/n$, which is easier to satisfy. However, unlike Bonferroni, Simes' procedure does not really allow investigation of which of the H_i's are rejected or not.

Simes' procedure is in the fixed α framework in which test power is the optimality criterion. Simes (1986) showed that for continuous independent test statistics his method was level α; i.e., the probability of correct acceptance of H_0 is

$$\mathbb{P}_{H_0}(p_{(i)} \geq i\alpha/n : i = 1,\ldots,n) \geq 1-\alpha. \tag{11.4.7}$$

Thus, Simes' procedure is the same as the BH procedure in the case that $m_0 = m$, i.e., all m hypotheses are true. In fact, although (11.4.7) is attributed to Simes (1986), it was first studied by Seeger (1968).

Although the BH theorem permits an arbitrary number m_0 of the m hypotheses to be true, the BH theorem is still limited by the assumption that the p-values are independent; this assumption is used in the proof to ensure that the ith p-value under H_i is $Uniform(0,1)$. In practice, p-values are often dependent and, in many cases, even the Benjamini-Yekutieli theorem will not be entirely sufficient. Like BH, Simes' inequality also holds for some dependent cases, but the focus here is on multiple testing, not global tests.

11.4.4.2 Combining FDR and FNR:

A dual quantity to the FDR is the false nondiscovery rate or FNR. Whereas the FDR is the proportion of incorrect rejections, the FNR is the proportion of incorrect acceptances, or nonrejections. Thus,

$$FNR = \begin{cases} \frac{T}{W} & R < m \\ 0 & R = m. \end{cases}$$

Obviously, it is not enough to control the FDR at level α without ensuring some quantity like the FNR is not too big. Otherwise it would be like ensuring a test is level α without investigating its power on the alternative.

To set this up, Genovese and Wasserman (2002) look at the asymptotics of the BH procedure as $m \to \infty$ in the special case where all the alternatives are the same fixed simple hypothesis with the same distribution. The nulls are also all the same, so their asymptotic analysis is the generalization of the simple versus simple case on which the Neyman-Pearson theory is built. To formalize the limiting behavior of the testing procedure requires that the proportions of true and false hypotheses be constant. Thus, write $a_0 = m_0/m$ and $a_1 = m_1/m$ and assume these are constants bounded away from zero as m increases. Now, each test can be written $H_{0,i} : \theta = \theta_0$ vs. $H_{1,i} : \theta = \theta_1$ for some θ_0, θ_1 taken to index the distributions specified under the nulls and alternatives. For simplicity, write $F_0 = F_{\theta_0}$ and $F_1 = F_{\theta_1}$.

The main random quantity in the BH procedure is D, say, the largest value of i for which $p_{(i)} < \alpha i/m$, because this is the last rejected hypothesis. So, let $D = \max\{i : p_{(i)} < \alpha i/m\}$. Now, the limiting behavior of D/m from the BH procedure can be expressed in terms of u^*, a solution to an equation involving F_1 and a constant $\beta = (1/\alpha - a_0)/(1 - a_0)$.

Theorem (Genovese and Wasserman, 2002): Suppose that F_1 is strictly concave, that $F'(0) > \beta$, and that u^* is the solution to $F_1(u) = \beta u$. Then, $D/m \to u^*/\alpha$, in probability, and hence the BH threshold $D\alpha/m \to u^*$ in probability.

Proof: Omitted. □

This means that $\alpha/m \leq u^* \leq \alpha$, so the BH procedure is between Bonferroni and uncorrected testing. Moreover, the BH threshold at which rejection stops can, in principle, be replaced for large m by the constant u^*. That is, there is a fixed, asymptotically valid threshold that one could use to approximate the BH procedure so that in the limit step-up (or step-down) procedures are not really needed. It would be enough, in principle, to use the right correction to the p-values and compare them to a single fixed threshold.

Two consequences of this result use the characterization of the limit to express other properties of the BH procedure. First, let $\delta = (\delta_1, ..., \delta_m)$ be indicator functions for the truth of the ith hypothesis, $\delta_i = 0$ if $H_{0,i}$ is true and $\delta_i = 1$ if $H_{i,i}$ is true. The empirical version of δ is $\hat{\delta} = (\hat{\delta}_1, ..., \hat{\delta}_m)$ where $\hat{\delta}_i = 0$ if $H_{0,i}$ is accepted by a procedure and $\hat{\delta}_i = 1$ if $H_{0,i}$ is rejected in level α testing. The difference between δ and $\hat{\delta}$ is summarized by

$$R_m = \frac{1}{m}\mathbb{E}\left(\sum_{i=1}^{m} |\delta_i - \hat{\delta}_i|\right),$$

which combines false positives and false negatives. This is a classification risk, and its limiting behavior is summarized by the following.

Theorem (Genovese and Wasserman, 2002): As $m \to \infty$,

$$R_m \to R_F = a_0 u^* + a_1(1 - F_1(u^*)) = a_0 u^* + a_1(1 - \beta u^*).$$

Proof (sketch): The BH procedure is

$$\text{reject } H_i \text{ with } p - \text{value } P_i \Leftrightarrow P_i \leq P_D.$$

Also, $P_i \leq P_{(D)} \Leftrightarrow P_i \leq D\alpha/m$. Using this,

$$\begin{aligned} R_m &= \frac{1}{m}\mathbb{E}\left[\sum_{i=1}^{m_0} \chi_{P_i < D\alpha/m} + \sum_{i=1}^{m_1} \chi_{P_i > D\alpha/m}\right] \\ &= a_0 \mathbb{P}(P_0 < D\alpha/m) + a_1 \mathbb{P}(P_1 > D\alpha/m), \end{aligned}$$

where P_0 is a p-value under H_0 and P_1 is a p-value under H_1 since they are all the same. The last theorem gives the result. □

With this in hand, one can verify that the risk of uncorrected testing is $R_U = a_0\alpha + a_1(1 - F_1(\alpha))$ and the risk of the Bonferroni method is $R_B = a_0\alpha/m + a_1(1 - F_1(\alpha/m))$. Genovese and Wasserman (2002) verify that BH dominates Bonferroni but examining when $R_U - R_F > 0$ reveals that neither method dominates the other.

Second, Genovese and Wasserman (2002) also characterize the limiting behavior of the expected FNR under the BH procedure in the generalized simple versus simple context.

Theorem (Genovese and Wasserman, 2002): Suppose that F_1 is strictly concave, that $F_1'(0) > \beta$, and that u^* is the solution to $F_1(u) = \beta u$. Then,

$$\mathbb{E}(FNR) \to a_1(1-\beta u^*).$$

Proof: Similar to that of the last theorem. □

Equipped with these results, $\mathbb{E}(FNR)$ can be minimized subject to $\mathbb{E}(FDR) \le \alpha$ asymptotically as $m \to \infty$. Since the BH procedure rejects hypotheses with p-values under u^*, consider $P_i < c$ in general. Following Genovese and Wasserman (2002), the FDR is

$$FDR = \frac{\sum_{i=1}^{m_0} \chi_{P_i<c}}{\sum_{i=1}^{m_0} \chi_{P_i<c} + \sum_{i=1}^{m_1} \chi_{P_i<c}}.$$

The sums are binomial random variables, $\sum_{i=1}^{m_0} \chi_{P_i<c} \sim Binomial(m_0, c)$ and $\sum_{i=1}^{m_0} \chi_{P_i<c} \sim Binpmial(m_1, F_1(c))$. So,

$$\mathbb{E}(FDR) = \frac{a_0}{a_0 c + a_1 F_1(c)} + \mathscr{O}\left(\frac{1}{\sqrt{m}}\right).$$

Neglecting the big-O term, c satisfies $E(FDR) \le \alpha$ when

$$\frac{F_1(c)}{c} \ge \beta - \frac{1}{\alpha},$$

which is satisfied for $c = u^*$ for any F_1 and a_0. This verifies that the BH procedure satisfies the $\mathbb{E}(FDR)$ level constraint for any distributions. That is,

$$\frac{a_0 u^*}{a_0 u^* + a_1 F_1(u^*)} = \frac{m_0}{m}.$$

The asymptotics in this case for the FNR are based on

$$FNR = \frac{\sum_{i=1}^{m_0} \chi_{P_i>c}}{\sum_{i=1}^{m_0} \chi_{P_i>c} + \sum_{i=1}^{m_1} \chi_{P_i>c}}.$$

Recognizing the binomials gives the approximation

$$\mathbb{E}(FNR) = \frac{a_1(1-F_1(c))}{a_1(1-F_1(c)) + a_0(1-c)} + \mathscr{O}\left(\frac{1}{\sqrt{m}}\right).$$

Now, the assumed concavity of F_1 gives that the FDR is increasing in c, so minimizing $\mathbb{E}(FNR)$ forces the choice $\mathbb{E}(FDR) = \alpha$, implying that the optimal c is the c^* satisfying $F_1(c^*)/c^* = \beta - 1/\alpha$ so that

$$\frac{F_1(u^*)}{u^*} - \frac{F_1(c^*)}{c^*} = \frac{1}{\alpha}.$$

The difference in effect from c^* to u^* is the cost BH pays for being effective independently of the distributions in the nulls and alternatives. Consequently, the BH procedure does not achieve the minimal $\mathbb{E}(FNR)$.

However, there is a sense in which the BH procedure is $\mathbb{E}(FNR)$ optimal. Write the threshold in the BH procedure in general as $\ell(t) = r(t)/m$ so the BH procedure corresponds to the choice $r(t) = \alpha t$ with $D = \max\{i : P_{(i)} < \ell(i)\}$. It can be verified (see Genovese and Wasserman (2002)) that, since BH uses the last upcrossing from the left of the p-values, above a threshold minimizing the $\mathbb{E}(FNR)$ corresponds to choosing the D as big as possible, which leads to $r(t) = \alpha t$ as in BH. Thus, the BH procedure is optimal in the class of last upcrossing procedures.

Related work by Sarkar (2004) uses step-up/step-down procedures to control the FDR and FNR by comparing FDR procedures in terms of their FNRs.

11.4.4.3 False Discovery Proportion

Another variant on the BH method is to examine V/R as a random variable rather than taking its expectation as in (11.4.1). This is called the false discovery proportion (FDP) and is the number of false rejections divided by the total number of rejections. If there are no rejections, then $V/R = 0/0$ is taken as zero. The goal is to preset $\gamma > 0$ to be the tolerable proportion of false discoveries and devise a testing procedure subject to

$$\mathbb{P}(FDP > \gamma) \leq \alpha. \tag{11.4.8}$$

In this case, setting $\gamma = 0$ gives back the FWER since the probability of no false rejections is bounded.

A simple relationship between the FDP and FDR follows from Markov's inequality. For any random variable X,

$$\mathbb{E}(X) = \mathbb{E}(X|X \leq \gamma)\mathbb{P}(X \leq \gamma) + \mathbb{E}(X|X > \gamma)\mathbb{P}(X > \gamma) \leq \gamma \mathbb{P}(X \leq \gamma) + \mathbb{P}(X > \gamma).$$

As noted in Romano and Shaikh (2006), this gives

$$\frac{\mathbb{E}(X) - \gamma}{1 - \gamma} \leq \mathbb{P}(X > \gamma) \leq \frac{\mathbb{E}(X)}{\gamma}. \tag{11.4.9}$$

Setting $X = (V/R)$ implies that if a method gives $FDR \leq \alpha$, then the same method controls the FDP by $\mathbb{P}(FDP > \gamma) \leq \alpha/\gamma$. From the other direction, the first inequality in (11.4.9) gives that if (11.4.8) holds, then $FDR \leq \alpha(1-\gamma) + \gamma \geq \alpha$, but not by much. Crudely, controlling one of FDR and FDP controls the other.

However, since it is a mean, directly controlling FDR seems to require stronger distributional assumptions on the p-values than directly controlling the FDP, which is a random variable. Indeed, Romano and Shaikh (2006) establish that the level of an FDP controlling procedure can be enforced when the p-values from the hypothesis tests are bounded by a uniform distribution, as would be obtained for a single hypothesis test. That is,

$$\forall u \in (0,1) \quad \forall \ H_{0,i} \ \ P_{0,i}(p_i \leq u) \leq u, \tag{11.4.10}$$

where $P_{0,i}$ is any element in the ith null. When the null is simple, $P_{0,i}$ is unique. No further distributional requirements such as independence, asymptotic properties, or dependence structures such as PRDS need be imposed. In fact, (11.4.10) generally holds for any sequence of nested rejection regions: If $S_i(\alpha)$ is a level α rejection region for H_i and $S_i(\alpha) \subset S_i(\alpha')$ when $\alpha < \alpha'$, then p-values defined by $p_i(X) = \inf\{\alpha : X \in S_i(\alpha)$ satisfy (11.4.10).

To find a variant on BH that satisfies (11.4.8), consider the following line of reasoning. Recall that V is the number of false rejections. At stage i in the BH procedure, having rejected $i-1$ hypotheses, the false rejection rate should be $V/i \leq \gamma$ (i.e., $V \leq \lfloor \gamma i \rfloor$), where $\lfloor x \rfloor$ is the first integer less than or equal to x. If $k = \lfloor \gamma i \rfloor + 1$, then $\mathbb{P}(V \geq k) \leq \alpha$ so that the number of false rejections is bounded by k. Therefore, in a step-down procedure, the BH thresholds $i\alpha/m$ should be replaced by the step-down thresholds

$$\alpha_i = \frac{(\lfloor \gamma i \rfloor + 1)\alpha}{m - i + \lfloor \gamma i \rfloor + 1}, \tag{11.4.11}$$

in which $m-i$ is the number of tests yet to be performed and $(\lfloor \gamma i \rfloor + 1)$ is the tolerable number of false rejections in the first i tests expressed as a fraction of i. Unfortunately, Romano and Shaikh (2006) show that this procedure is level α for the FDP when a dependence criterion like PRDS is satisfied. (Their condition uses conditioning on the p-values from the false nulls.) Thus, the test depends on the dependence structure of the p-values.

In general, increasing the α_is makes it easier to reject hypotheses, thereby increasing power. So, the challenge is to maintain a level α while increasing the α_is and enlarging the collection of distributions the p-values are allowed to have (to require only (11.4.10) for instance). One choice is to use $\alpha' = \alpha_i / C_{\lfloor \gamma m \rfloor + 1}$, where $C_{\lfloor \gamma m \rfloor + 1} = \sum_{i=1}^{\lfloor \gamma m \rfloor + 1} (1/i)$. However, it is possible to do better.

Romano and Shaikh (2006) propose a step-down method that replaces α_is with $\alpha_i'' = \alpha_i / D$ where $D = D(\gamma, m)$ is much smaller than $C_{\lfloor \gamma m \rfloor + 1}$. This procedure controls the FDP at level α but for all $i = 1, \ldots, m$, $\alpha_i'' > \alpha_i'$, so that the test will reject more hypotheses and have higher power.

The test itself is as follows: For $k = 1, \ldots, \lfloor \gamma m \rfloor$, set

$$\beta_k = k / \max(m + k - \lceil k/\gamma \rceil + 1, m_0) \text{ and } \beta_{\lfloor \gamma m \rfloor + 1} = \frac{\lfloor \gamma m \rfloor + 1}{m_0}, \beta_0 = 0.$$

Then let

$$N = N(\gamma, m, m_0) = \min\left(\lfloor \gamma m \rfloor + 1, m_0, \lfloor \gamma \left(\frac{m - m_0}{1 - \gamma} + 1 \right) \rfloor + 1 \right),$$

$$S = S(\gamma, m, m_0) = m_0 \sum_{i=1}^{N} \frac{\beta_i - \beta_{i-1}}{i},$$

and finally

$$D(\gamma, m) = \max\{S(\gamma, m, k) : k = 1, \ldots, m_0\}.$$

Despite the odd appearance of this test, it is level α for FDP.

Theorem (Romano and Shaikh, 2006): Suppose the p_is satisfy (11.4.10). The step-down testing procedure with thresholds $\alpha_i'' = \alpha_i/D$ satisfies $\mathbb{P}(FDP > \gamma) \leq \alpha$.

Proof: Omitted. □

Note that $D(\gamma, m)$ is a maximum and so does not depend on m_0, the unknown number of true hypotheses. If m_0 were known, or could be estimated extremely well (see below), $S(\gamma, m, m_0)$ could be used in place of D.

11.5 Controlling the Positive False Discovery Rate

The pFDR is a variant on the FDR resulting from conditioning on there being at least one rejected null. That is, the pFDR is

$$pFDR = \mathbb{E}\left[\frac{V}{R}|R > 0\right].$$

The motivation for this criterion is that the event $\{R = 0\}$ makes the testing pointless so there is no reason to consider it. Indeed, this is why the factor $\mathbb{P}(R > 0)$ is not usually part of the definition. Although similar to the FDR, the pFDR and FDR have surprisingly different properties. These are studied in detail in Storey (2002, 2003).

The pFDR is relatively tractable. It will first be seen that the pFDR has a natural Bayesian interpretation. Then several theoretical properties of pFDRs can be given. In many settings, however, it is not the theoretical properties that are most important to investigate further. Rather, it is the implementation that bears development.

11.5.1 Bayesian Interpretations

Unexpectedly, it is the Bayesian formulation that makes the pFDR tractable. The pure Bayes treatment will be seen in the last section of this chapter; for the present, the central quantity in Bayes testing, the posterior probability of the null, appears naturally as an expression for the pFDR that can be related to frequentist testing with p-values. As a consequence, a Bayesian analog to the p-value, the q-value, can be defined from the rejection regions of the frequentist tests.

11.5.1.1 pFDR and the Posterior

Recall that each test H_i can be regarded as a random variable so that, for $i = 1, 2, \cdots, m$,

$$H_i = \begin{cases} 0 & i\text{th null hypothesis is true,} \\ 1 & i\text{th null hypothesis is false.} \end{cases}$$

Now, suppose the m hypotheses are identical and are performed with independent, identical test statistics $T_1, T_2, \cdots, T_m$. Then, the H_is can be regarded as independent Bernoullis with $\Pr[H_i = 0] = \pi_0$ and $\mathbb{P}[H_i = 1] = \pi_1 = 1 - \pi_1$. In other words, π_0 and π_1 are the prior probabilities of the regions meant by H_i and H_i^c, $(T_i|H_i) \sim (1 - H_i)F_0 + H_iF_1$, where F_0 and F_1 are the distributions of T_i under H_i and H_i^c, and $H_i \sim Bernoulli(\pi_1)$; implicitly this treats H_i and H_i^c as a simple versus simple test. Let Γ_i denote a fixed test region under T_i for H_i. By the identicality, the Γ_is are all the same and can be denoted generically as Γ. The first important result is that the pFDR is a posterior probability.

Theorem (Storey, 2003): Let m identical hypothesis tests be performed with independent test statistics T_i with common rejection region Γ. If the prior probabilities of the hypotheses are all $P(H = 0) = \pi_0$ and $P(H = 1) = \pi_1$, then

$$pFDR(\Gamma) = \frac{\pi_0 \mathbb{P}[T \in \Gamma | H = 0]}{\mathbb{P}[T \in \Gamma]} = \mathbb{P}[H = 0 | T \in \Gamma],$$

where $\mathbb{P}[T \in \Gamma] = \pi_0 \mathbb{P}[T \in \Gamma | H = 0] + (1 - \pi_0)\mathbb{P}[T \in \Gamma | H = 1]$.

Proof: First, let $\theta = \mathbb{P}[H = 0 | T \in \Gamma]$ be the probability that the null hypothesis is true given that the test statistic led to rejection. If there are k rejections (discoveries) among the m null hypotheses, they can be regarded as k independent Bernoulli trials, with success being the true positives and failure being the false positives. Let $V(\Gamma)$ denote the number of false positives and $R(\Gamma)$ be the total number of positives. Conditioning on the total number of discoveries being k, this formulation implies that the expected number of false positives is

$$\mathbb{E}[V(\Gamma)|R(\Gamma) = k] = k\theta = k\mathbb{P}[H = 0 | T \in \Gamma].$$

Returning to the statement of the theorem, it is easy to see that

$$\begin{aligned} pFDR(\Gamma) &= \mathbb{E}\left[\frac{V(\Gamma)}{R(\Gamma)} | R(\Gamma) > 0\right] \\ &= \sum_{k=1}^{m} \mathbb{E}\left[\frac{V(\Gamma)}{k} | R(\Gamma) = k\right] \mathbb{P}[R(\Gamma) = k | R(\Gamma) > 0] \\ &= \sum_{k=1}^{m} \frac{k\mathbb{P}[H = 0 | T \in \Gamma]}{k} \mathbb{P}[R(\Gamma) = k | R(\Gamma) > 0] \\ &= \mathbb{P}[H = 0 | T \in \Gamma] \sum_{k=1}^{m} \mathbb{P}[R(\Gamma) = k | R(\Gamma) > 0] \\ &= \mathbb{P}[H = 0 | T \in \Gamma]. \quad \square \end{aligned}$$

This result shows that the pFDR is the posterior probability from a Bayesian test; later it will be seen that the pFDR also has an interpretation in terms of p-values because it can be defined by rejection regions rather than by specifying levels α.

Note that this result is independent of m and m_0 and that from Bayes rule

$$pFDR(\Gamma) = \mathbb{P}(H=0|T \in \Gamma) = \frac{\pi_0 \mathbb{P}(\text{Type I error of } \Gamma)}{\pi_0 \mathbb{P}(\text{Type I error of } \Gamma) + \pi_1 \mathbb{P}(\text{power of } \Gamma)},$$

so that pFDR is seen to increase with increasing Type I error and decrease with increasing power.

11.5.1.2 The q-Value

The pFDR analog of the p-value is called the q-value; roughly the event in a p-value becomes the conditioning in the Bayesian formulation. For tests based on T_i, the reasoning from p-values suggests rejecting H_i when

$$pFDR(\{T \geq t\}) = \frac{\pi_0 \mathbb{P}(T \geq t|H=0)}{\pi_0 \mathbb{P}(T \geq t|H=0) + \pi_1 \mathbb{P}(T \geq t|H=1)} \tag{11.5.1}$$

$$= \frac{\pi_0 \mathbb{P}(U_0 \geq)}{\pi_0 \mathbb{P}(U_0 \geq) + \pi_1 \mathbb{P}(U_1 \geq)} \tag{11.5.2}$$

$$= \mathbb{P}(H=0|T \geq t),$$

is small enough, where U_0 and U_1 are random variables with the distribution specified under H_0 and H_1 and $T=t$ is the value of T obtained for the data at hand.

The left-hand side is the q-value and is seen to be a function of a region defined by T; however, this is not necessary. It is enough to have a nested set of regions Γ_α with $\alpha \leq \alpha' \Rightarrow \Gamma_\alpha \subset \Gamma_{\alpha'}$. Indeed, the index α is not strictly needed, although it may provide an interpretation for the rejection region if $\alpha = t$ is the value of the statistic, for instance. In general, however, no index is needed: For a nested set of rejection regions denoted $\langle C \rangle = \langle C_i \rangle|_{i \in I}$, the p-value of an observed statistic $T=t$ is

$$q\text{-value}(t) = \min_{\{C: t \in C\}} \mathbb{P}[T \in C|H=0],$$

and the q-value is

$$p\text{-value}(t) = \min_{\{C: t \in C\}} \mathbb{P}[H=0|T \in C] = \min_{\{C: t \in C\}} pFDR(C).$$

Thus, the q-value is a measure of the strength of $T=t$ under pFDR rather than probability. It is the minimum pFDR that permits rejection of H_0 for $T=t$. Now, the last theorem can be restated as follows.

Corollary (Storey, 2003): If Γ is replaced by a set of nested regions Γ_α parametrized by test levels α, then

$$q\text{-value}(t) = \inf_{\{\Gamma_\alpha : t \in \Gamma_\alpha\}} \mathbb{P}(H=0|T \in \Gamma_\alpha). \quad \square$$

As a consequence, the rejection region determining a q-value can be related to the ratio of Type I error to power. By Bayes rule and rearranging,

$$\arg \inf_{\{\Gamma_\alpha : t \in \Gamma_\alpha\}} \mathbb{P}(H=0|T \in \Gamma_\alpha) = \arg \inf_{\{\Gamma_\alpha : t \in \Gamma_\alpha\}} \frac{\mathbb{P}(T \in \Gamma_\alpha | H=0)}{\mathbb{P}(T \in \Gamma_\alpha | H=1)}. \tag{11.5.3}$$

It is intuitively reasonable that the q-value minimizes the ratio because the pFDR measures how frequently false positives occur relative to true positives.

Moreover, it is seen that there is a close relationship between the rejection regions for p-values and q-values. This extends beyond (11.5.3) because the p-values can be used directly to give the q-value. Recall that the $T_i \sim \pi_0 F_0 + \pi_1 F_1$ are IID for $i = 1, ..., m$, where F_i is the distribution specified under H_i. Let

$$G_i(\alpha) = \mathbb{P}(T \in \Gamma_\alpha | H = i),$$

for $i = 0, 1$, be the distribution functions of T under the null and alternative hypotheses. Clearly, $G_0(\alpha)$ represents the level and $G_1(\alpha)$ represents the power. The ratio $\alpha/G_1(\alpha)$, from (11.5.3), is minimized over α to find the smallest α satisfying $\alpha = G_1(\alpha)/G_1'(\alpha)$. Thus, as noted in Storey (2003), $\alpha/G_1(\alpha)$ can be minimized graphically by looking for the line from the origin that is tangent to a concave portion of the function $G_1(\alpha)$ for $\alpha \in (0,1)$ and has the highest slope. This line is tangent to the point on the curve where $\alpha/G_1(\alpha)$ is minimized. In particular, if $G_1(\cdot)$ is strictly concave on $[0,1]$, then the ratio of power to level increases as $\alpha \to 0$ (i.e., as the regions Γ_α get smaller), and therefore in (11.5.1) the minimal pFDR corresponding to the minimal level-to-power ratio is found for small α. More formally, we have the following.

Proposition (Storey, 2003): If $G_1(\alpha)/\alpha$ is decreasing on $[0,1]$, the q-value is based on the same significance region as the p-value,

$$\arg \min_{\{\Gamma_\alpha : t \in \Gamma_\alpha\}} \mathbb{P}(H=0|T \in \Gamma_\alpha) = \arg \min_{\{\Gamma_\alpha : t \in \Gamma_\alpha\}} \mathbb{P}(T \in \Gamma_\alpha | H=0).$$

Proof: Since $G_1(\alpha)/\alpha$ decreasing implies G_1 is concave, the Γ_α that contains t and minimizes $pFDR(\Gamma_\alpha)$ also minimizes $P(T \in \Gamma_\alpha | H = 0)$ because one would take the same significance region with the smallest α with $tin\Gamma_\alpha$ to minimize $\alpha/G_1(\alpha)$. Consequently, the same significance region is used to define both the p-value and the q-value. □

Not only can p-values and q-values be related as in the corollary, defining the q-value in terms of regions specified by statistics is equivalent to defining the q-value in terms of the p-values from those statistics. Let $pFDR^T(\Gamma_\alpha)$ be the pFDR based on the original statistics and let $pFDR^P(\Gamma_\alpha)$ be the pFDR based on the p-values; that is, $pFDR^P(\{p \le \alpha\})$. The relationship between these quantities is summarized as the first part means that the q-value can be found from the raw statistics or their p-values.

Proposition (Storey, 2003): For m identical hypothesis tests,

$$pFDR^T(\Gamma_\alpha) = pFDR^P(\{p \le \alpha\}).$$

Moreover, when the statistics are independent and follow a mixture distribution,

$$q\text{-value}(t) = pFDR^P(\{p : p \leq p\text{-value}(t)\}) \tag{11.5.4}$$

if and only if $G_1(\alpha)/\alpha$ is decreasing in u.

Proof: Since the Γ_αs are nested, it is trivial to see that p-value$(t) \leq \alpha \Leftrightarrow t \in \Gamma_\alpha$. This implies the first statement.

For the second, for any $T = t$, let

$$\Gamma_{\alpha'} = \arg \min_{\{\Gamma_\alpha : t \in \Gamma_\alpha\}} pFDR^T(\Gamma_\alpha),$$

so that

$$q\text{-value}(t) = pFDR^T(\Gamma_{\alpha'}) = pFDR^P(\{p : p \leq \alpha'\}).$$

Since it is also true that

$$\Gamma_{\alpha'} = \arg \min_{\{\Gamma_\alpha : t \in \Gamma_\alpha\}} \mathbb{P}(T \in \Gamma_\alpha H = 0),$$

it is seen that p-value $= \alpha'$.

For the converse, suppose (11.5.4) holds for each t. By the definition of the q-value, q-value(t) is increasing in p-value(t), so $G_1(\alpha)/\alpha$ is decreasing in α. □

11.5.2 Aspects of Implementation

Continuing to assume that the p-values are independent, rejection regions expressed in terms of p-values are always of the form $[0, \gamma]$, where γ represents the p-value. It is usually convenient for implementation purposes to replace the abstract Γ with such intervals, often just using $\gamma > 0$ to mean the whole interval $[0, \gamma]$. Now the theorem can be stated as

$$pFDR(\gamma) = \frac{\pi_o \mathbb{P}(P \leq \gamma | H = 0)}{\mathbb{P}(P \leq \gamma)} = \frac{\pi_0 \gamma}{\mathbb{P}(P \leq \gamma)},$$

in which, under the null, the p-value P is distributed as $Uniform[0, 1]$ and $\mathbb{P}$ in the denominator is the mixture probability over the null and alternative.

Now, if good estimators for π_0 and $\mathbb{P}(P \leq \gamma)$ can be given, the pFDR can be estimated. Let $\lambda > 0$. Following Storey (2002), since the p-values are $Uniform[0, 1]$, the numerator can be estimated by

$$\hat{\pi}_0(\lambda) = \frac{\#\{p_i \geq \lambda\}}{(1-\lambda)m} = \frac{W(\lambda)}{(1-\lambda)m}, \tag{11.5.5}$$

where $W = \#\{p_i \geq \lambda\}$ and λ remains to be chosen. (Note that $\mathbb{E}W(\gamma) = m\mathbb{P}(P_i > \gamma)$ and $\pi_0(1-\gamma) = \mathbb{P}(P_i > \gamma, H_i = 0) \leq \mathbb{P}(P_i > \gamma) \leq m(1-\gamma)$.) This estimator is reasonable since the largest p-values are likely to come from the null and π_0 is the prior probability of the null. Similarly, the denominator can be estimated for any γ empirically by

$$\hat{\mathbb{P}}(P \leq \gamma) = \frac{\#\{p_i \leq \gamma\}}{m} = \frac{R(\lambda)}{m}, \tag{11.5.6}$$

where $R(\lambda) = \#\{p_i \leq \lambda\}$. The ratio of (11.5.5) over (11.5.6) is an estimator for the pFDR. However, it can be improved in two ways. First, if $R(\gamma) = 0$, the estimate is undefined, so replace $R(\gamma)$ with $\max(R(\gamma), 1)$. Second, $R(\gamma) \geq 1 - (1-\gamma)^m$ because $\mathbb{P}(R(\gamma) > 0) \geq 1 - (1-\gamma)^m$. So, since the pFDR is conditioned on $R > 0$, divide by $1 - (1-\gamma)^m$ as a conservative estimate of the probability of Type I error. Taken together,

$$\widehat{pFDR}_\lambda(\gamma) = \frac{\hat{\pi}_0(\lambda)\gamma}{\hat{P}(P \leq \gamma)(1-(1-\gamma)^m)} = \frac{W(\lambda)\gamma}{(1-\lambda)\max(R(\gamma), 1)(1-(1-\gamma)^m)}$$

is a natural estimator for pFDR, apart from choosing λ.

The same reasoning leads to

$$\widehat{FDR}_\lambda(\gamma) = \frac{\hat{\pi}_0(\lambda)\gamma}{\hat{P}(P \leq \gamma)} = \frac{W(\lambda)\gamma}{(1-\lambda)\max(R(\gamma), 1)}$$

as a natural estimator for FDR, apart from choosing λ which will be done shortly.

11.5.2.1 Estimating $FDR(\gamma)$ and $pFDR(\gamma)$

Since the procedures for obtaining useful estimates of pFDR and FDR are so similar, the treatment here will apply to both but focus on pFDR. As with the PCER, the main computational procedures devolve to bootstrapping. First, recall that by assumption all the tests are identical, and suppose the abstract region Γ_α is replaced by the region defined by ranges of p-values as in the last proposition. Now, the following algorithm, from Storey (2002), results in an estimate $\widehat{pFDR}^b_\lambda(\gamma)$ of $pFDR_\lambda(\gamma)$, where the choice of λ is described in the next subsection and γ is the upper bound on the value of a p-value, usually chosen using the actual p-value from the data for the test.

ALGORITHM FOR ESTIMATING $FDR(\gamma)$ AND $pFDR(\gamma)$:

☐ For the m hypothesis tests, calculate their respective realized p-values $p_1, p_2, \cdots, p_m$. Then, estimate π_0 and $\mathbb{P}[P \leq \gamma]$ by

$$\widehat{\pi}_0(\lambda) = \frac{W(\lambda)}{(1-\lambda)m}$$

and

$$\widehat{P}[P \leq \gamma] = \frac{R(\gamma) \vee 1}{m},$$

where $R(\gamma) = \#\{p_i \leq \gamma\}$ and $W(\lambda) = \#\{p_i > \lambda\}$.

- ☐ For any fixed rejection region of interest $[0, \gamma]$, estimate $pFDR(\gamma)$ using

$$\widehat{pFDR}_\lambda(\gamma) = \frac{\widehat{\pi}_0(\lambda)\gamma}{\hat{P}[P \leq \gamma]\{1 - (1-\gamma)^m\}}$$

  for some well-chosen λ.

- ☐ For B bootstrap samples from $p_1, p_2, \cdots, p_m$, calculate the bootstrap estimates $\widehat{pFDR}^b_\lambda(\gamma)$ for $b = 1, ..., B$.

- ☐ Take the $1-\alpha$ quantile of $\widehat{pFDR}^b_\lambda(\gamma)$ for $b = 1, ..., B$ as an upper confidence bound. This gives a $1-\alpha$ upper confidence interval for $pFDR(\gamma)$.

  For $FDR(\gamma)$, use the same procedure apart from choosing

$$\widehat{FDR}_\lambda(\gamma) = \frac{\widehat{\pi}_0(\lambda)\gamma}{\widehat{\mathbb{P}}[P \leq \gamma]}.$$

It is seen from this procedure that pFDR and FDR procedures are somewhat the reverse of the usual Neyman-Pearson procedure. That is, rather than fixing a level and then finding a region that satisfies it, one fixes a procedure (based on taking γ as a p-value, say) and finds its "level" by bootstrapping. Nevertheless, iterating this process can result in a region with a prespecified level as in the traditional theory.

11.5.2.2 Choosing λ

To complete the last procedure, specification of λ is necessary, and the procedure is straightforward. As suggested in Storey (2002), since $\lambda \in [0, 1]$, start with a grid of values such as $G = \{\lambda = .05u \mid u = 0, 1, ..., 19\}$ and find both

$$\widehat{pFDR}_\lambda(\gamma) = \frac{\widehat{\pi}_0(\lambda)\gamma}{\hat{\mathbb{P}}[P \leq \gamma]\{1 - (1-\gamma)^m\}}$$

and $\widehat{pFDR}^b_\lambda(\gamma)$ for $b = 1, ..., B$, for each λ in the grid as in the last algorithm. These can be used to form a mean squared error,

$$\widehat{MSE}(\lambda) = \frac{1}{B}\sum_{b=1}^{B}\left(\widehat{pFDR}^b_\lambda(\gamma) - \min_{\lambda' \in G}\widehat{pFDR}_{\lambda'}(\gamma)\right)^2.$$

Now, choose $\hat{\lambda} = \arg\min \widehat{MSE}(\lambda)$ to form the estimate $\widehat{pFDR}(\gamma) = \widehat{pFDR}_{\lambda'}(\gamma)$.

Again, the $1-\alpha$ quantile of the bootstrap estimates $\widehat{pFDR}_{\lambda'}^{b}$ for $b = 1,...,B$ gives a $1-\alpha$ upper confidence bound on $\widehat{pFDR}(\gamma)$. For FDR, the same procedure can be applied to $\widehat{FDR}(\gamma)$.

11.5.2.3 Calculating the q-Value

Recall that for an observed statistic $T = t$ the q-value is

$$q(t) = \inf_{\Gamma : t \in \Gamma} pFDR(\Gamma).$$

So, in terms of p-values, it is

$$q(t) = \inf_{\gamma \geq p} pFDR(\gamma) = \inf_{\gamma \geq p} \frac{\pi_0 \gamma}{\mathbb{P}(P \leq \gamma)},$$

in which the second equality only holds when the H_i are independent Bernoulli variables with $\mathbb{P}(H_i = 0) = \pi_0$.

If the p-values from the m tests are $p_1,...,p_m$, with order statistic $p_{(1)},...,p_{(m)}$ then the corresponding q-values are $q_i = q_1(p_{(i)})$, with $q_i \leq q_{i+1}$, for $i = 1,...,m$. Denote the estimates of the q-values by $\hat{q}_i = \hat{q}_i(p_{(i)})$. Then, $\hat{q}(p_{(i)})$ gives the minimum pFDR for rejection regions containing $[0, p_{(i)}]$. That is, for each p_i, there is a rejection region with $pFDR = q(p_{(i)})$ so that at least $H_{(1)},...,H_{(i)}$ are rejected.

ALGORITHM FOR CALCULATING THE q-VALUE:

- ☐ For the m hypothesis tests, calculate the p-values and order them to get $p_{(1)} \leq p_{(2)} \leq \cdots \leq p_{(m)}$.
- ☐ Set $\widehat{q}(p_{(m)}) = \widehat{pFDR}(p_{(m)})$.
- ☐ For $i = m-1, m-2, \cdots, 2, 1$, set $\widehat{q}(p_{(i)}) = \min\left[\widehat{pFDR}(p_{(i)}), \widehat{q}(p_{(i+1)})\right]$.

11.5.2.4 Estimating m_0

The number of true hypotheses m_0 is unknown but, surprisingly, can be estimated, although the techniques can be elaborate. Among other authors, Storey (2002) suggests estimating m_0 using

$$\hat{m}_0(\lambda) = \frac{\sum_{i=1}^{m} I(p_i \geq \lambda)}{1-\lambda},$$

where $\lambda \in (0,1)$ can be estimated by cross-validation. Meinshausen and Buhlmann (2008) give a $1-\alpha$ upper confidence bound on m_0 by way of bounding functions through a more complicated technique.

As a separate issue, controlling sample size as a way to ensure a minimal pFDR has been studied in Ferreira and Zwinderman (2006) and Chi (2007). While this sort of control is possible and important, effective techniques are not yet fully established, even though they can be based substantially on the classical theory of p-values.

11.6 Bayesian Multiple Testing

To a conventional Bayesian, multiple testing is just another problem that fits comfortably into the existing Bayesian paradigm. The usual questions of what loss function to choose, what reasonable priors would be, and how to do the computing still occur and demand resolution. However, the formulation is not otherwise conceptually difficult. The usual hierarchical models suffice even when m is large. As a consequence, the Bayesian treatment is much easier to present and understand than the frequentist.

An orthodox Bayesian would not only argue that the unified Bayesian treatment is more straightforward and scientifically accurate, but would also criticize the other methods. The Bayesian would be especially critical of the use of p-values on the grounds that they are frequentist probabilities of regions that did not occur (and hence are irrelevant) and are thoroughly ad hoc, having no inferential interpretation.

More specifically, Bayesians would ignore the BH procedure, and others like it, for at least four reasons: First, it has no apparent decision theoretic justification. The procedure is not a Bayes rule, or necessarily optimal, under any meaningful criterion. Thus, there is no reason to use it. Second, it is wide open to cheating: If you want to accept a hypothesis, just find enough hypotheses with really small p-values so that the BH threshold is reduced enough that you can accept the hypotheses you want. (The Bayesian does not accuse the frequentist of cheating, just that the frequentist method is so lacking in justification that it is likely to be misused inadvertently.) Third, the central ingredients, whether p-values or other quantities, including the FDR, PCER and so forth, are just the wrong quantities to use because they involve post-data use of the entire sample space, not conditioning on the realized outcomes. Indeed, even the Bayesian interpretation of the q-value is not really Bayesian because the conditioning is on a set representing tail behavior rather than on the data. Fourth, a more pragmatic Bayesian criticism is that p-values generally overreject so that rejection at the commonly used thresholds .05 or .01 is far too easy. This is sometimes argued from a prior robustness standpoint.

Naturally, frequentists have spirited responses to these criticisms. The point for the present is not to examine the Bayes-frequentist divide so much as to explain the motivations Bayesians have for developing alternative techniques. These techniques begin with a hierarchical formulation of the multiple testing problem, which leads to more complicated decision-theoretic Bayes formulations.

11.6.1 Fully Bayes: Hierarchical

Following the critical survey of Bayarri and Berger (2006), on which much of this section is based, the Bayes multiple testing problem can be formulated as follows. Assume observables $\boldsymbol{X} = (X_1, ..., X_m)$ and consider m tests

$$H_{0,i} : X_i \sim f_{0,i} \text{ vs. } H_{1,i} : X_i \sim f_{1,i}. \tag{11.6.1}$$

The X_is may be outcomes, statistics T_i, or any other data-driven quantities. The hypothesized distributions $f_{0,i}$ and $f_{1,i}$ may have parameters. If the parameters are fixed by the hypotheses, the tests are simple versus simple. Let $\boldsymbol{\gamma} = (\gamma_1, ..., \gamma_m)$ in which the γ_is are indicator variables for the truth of $H_{0,i}$, $\gamma_i = 0, 1$ according to whether $H_{0,i}$ is true or false. There are 2^m models $\boldsymbol{\gamma}$.

As usual in Bayesian analyses, inference is from the posterior probabilities. In this case, it is not the model selection problem (i.e., $\boldsymbol{\gamma}$) that is of interest but the posterior probabilities that each $\gamma_i = 0, 1$. Whichever of 0, 1 has higher posterior probability is the favored inference. If the m tests are independent and $\boldsymbol{\gamma} = 0$ (i.e., all the nulls are true), then overall there should be αm rejections where α is the threshold for the posterior probabilities. The goal is to do better than this since excessive false rejections mask the detection of incorrect nulls by rejecting true nulls.

11.6.1.1 Paradigmatic Example

Suppose that each $X_i \sim N(\mu_i, \sigma^2)$ with σ unknown and the task is to tell which of the μ_is are nonzero. Write $\boldsymbol{\mu} = (\mu_1, ..., \mu_m)$. Then $\gamma_i = 0, 1$ according to whether $\mu_i = 0$ or $\mu_i \neq 0$. The conditional density for $\boldsymbol{X} = (X_1, \ldots, X_m)$ with $\boldsymbol{x} = (x_1, \ldots, x_m)$ is

$$p(\boldsymbol{x}|\sigma^2, \gamma, \boldsymbol{\mu}) = \Pi_{i=1}^m \frac{e^{-(x_j - \gamma_j \mu_j)^2/2\sigma^2}}{\sqrt{2\pi}\sigma}.$$

To determine which entries in $\boldsymbol{\mu}$ are nonzero, there are m conditionally independent tests of the form

$$H_{0,i} : \mu_i = 0 \text{ vs. } H_{1,i} : \mu_i \neq 0.$$

Fix the same prior for each γ_i and for each μ_i. Thus, using w and W to denote prior densities and probabilities generically, set $W(\gamma_i = 0) = p_0$ to represent the proportion of nulls thought to be true. Write $W(\gamma_i = 0|\boldsymbol{x}) = p_i$ for the posterior for γ_i and set $\mu_i \sim N(0, \tau^2)$. Now, the density $w(\mu_i|\gamma_i, \boldsymbol{x})$ is well defined. If a hyperprior for p_0 is given as well, the joint posterior density for all the hyperparameters $w(p_0, \sigma^2, \tau^2|\boldsymbol{x})$ is well defined and these three are the main quantities of interest.

The hierarchy can be explicitly given as:

- ☐ $(X_i|\mu_i, \sigma^2, \gamma_i) \sim N(\gamma_i \mu_i, \sigma^2)$, IID,
- ☐ $(\mu_i|\tau^2) \sim N(0, \tau^2)$, $(\gamma_i|p_0) \sim Bernoulli(1 - p_0)$, and

□ $(\tau^2, \sigma^2) \sim w(\tau^2, \sigma^2)$, $p_0 \sim w(p_0)$.

The joint density for the parameters and data is

$$w(p_0, \tau^2, \sigma^2, \boldsymbol{\gamma}, \boldsymbol{\mu}) p(\boldsymbol{x}|\boldsymbol{\mu}, \boldsymbol{\gamma}, \sigma^2)$$
$$= w(p_0) w(\tau^2|\sigma^2) w(\sigma^2) (\Pi_{i=1}^m w(\gamma_i|p_0)) (\Pi_{i=1}^m w(\mu_i|\tau^2)) p(\boldsymbol{x}|\boldsymbol{\mu}, \boldsymbol{\gamma}, \sigma^2),$$

with the specifications indicated in the hierarchy. It is also reasonable to take $w(p_0) \sim Uniform[0,1]$, although this may be generalized to a $Beta(a,b)$ since $a = 1, b = 1$ gives the uniform and allows mass to be put in the part of $[0,1]$ thought to contain the proportion of true nulls.

Now, the posterior probability that $\mu_i \neq 0$ is $W(\mu_i \neq 0|\boldsymbol{x}) = \pi_i$, where

$$1 - \pi_i = \frac{\int_0^1 \int_0^1 \Pi_{j \neq i} (p_0 + (1 - p_0)\sqrt{1 - u} e^{ux_i^2/2\sigma^2}) dp_0 du}{\int_0^1 \int_0^1 \Pi_{j=1}^m (p_0 + (1 - p_0)\sqrt{1 - u} e^{ux_i^2/2\sigma^2}) dp_0 du},$$

which can be computed numerically or by importance sampling.

The prior inclusion probabilities are $W(\gamma_i = 1)$, leading to posterior inclusion probabilities $W(\gamma_i = 1|\boldsymbol{x})$. These quantities are usually the most interesting since they determine which of the μ_is really are nonzero. Of course, if $\mu_i \neq 0$, the distribution of the μ_i remains interesting. However, the distribution $W(\boldsymbol{\gamma}|\boldsymbol{x})$ is not particularly interesting because the problem is hypothesis testing, not model selection.

Even so, Ghosh et al. (2004) use a hierarchical Bayes model to show the pFDR in a variable selection context. Briefly, if $Y_i \sim N(\boldsymbol{X}_i\beta, \sigma^2)$ and $\beta_i|\gamma_i \sim (1-\gamma_i)N(0, \tau_i^2) + \gamma_i N(0, c_i^2\tau_i^2)$ with $\gamma_i \sim Bernoulli(p_i)$ and $\sigma^2 \sim IG(\nu/2, \nu/2)$, then

$$\mathbb{P}(\gamma_i = 0|\hat{\beta}_i = 0) = 1 - \left[\frac{(\sigma_i^2/\tau_i^2 + c_i^2)}{(\sigma_i^2/\tau_i^2 + 1)}\right]^{1/2}.$$

This posterior probability is closely related to the pFDR, and the calculation generalizes to conditioning on other regions.

One benefit of the Bayes approach is that $W(p_0|\boldsymbol{x})$ can be defined and found readily. This is the proportion of true nulls, and its distribution comes for free from the Bayesian formulation. By contrast, the estimator $\hat{m}_0$ from, say, the pFDR is a point estimator. While a standard error for it can doubtless be given, it is not as good a summary as the full distribution. Bayarri and Berger (2006) give examples of computations for all the relevant posteriors. The effect of prior selection clearly matters, as does the computing, which can be demanding. See Scott and Berger (2006) for a discussion of prior selection and techniques for using importance sampling.

11.6.1.2 A Bayesian Stepdown Procedure

Arguably, there is a parallel between p-values and Bayes factors in that both are interpreted to mean the strength of the support of the data for the hypothesis. Recall that the Bayes factor is the ratio of the posterior odds to the prior odds and is the Bayes action under generalized 0-1 loss. In the present case, the marginal Bayes factors for the m hypotheses are

$$B_i = \frac{W(H_i|\boldsymbol{x})}{1 - W(H_i|\boldsymbol{x})} \frac{1 - \pi_{i,0}}{\pi_{i,0}}. \tag{11.6.2}$$

In the Scott and Berger (2006) example above, all the $\pi_{i,0}$s were the same value p_0 which could therefore be treated as a hyperparameter. The joint Bayes factor for testing $H_0 : \cap_{i=1}^m H_i$ vs. $H_1 : \cup_{i=1}^m H_i^c$ is

$$B = \frac{\int_{H_0} w(\boldsymbol{\theta}|\boldsymbol{x})d\boldsymbol{\theta}}{1 - \int_{H_0} w(\boldsymbol{\theta}|\boldsymbol{x})d\boldsymbol{\theta}} \frac{1 - \int_{H_0} w(\boldsymbol{\theta})d\boldsymbol{\theta}}{\int_{H_0} w(\boldsymbol{\theta})d\boldsymbol{\theta}}. \tag{11.6.3}$$

The difference between B_i and B is analogous to the difference between tests based on the marginal distributions from the statistics T_i from m experiments and tests based on the joint distribution from the vector $(T_1, ..., T_m)$ in the frequentist case.

Now let $\theta = (\theta_1, ..., \theta_m)$ and let $\Omega \subset \mathbb{R}^m$ be the whole parameter space. Suppose the marginal testing problems are

$$H_{0,i} : \theta \in \Theta_i \text{ vs } H_{1,i} : \theta \in \Theta_i^c$$

rather than simple versus simple, where $\Theta_i \cap \Theta_i^c$ is void and $\Theta_i \cup \Theta_i^c = \Omega$. This means Θ_i, Θ_i^c do not constrain the $m-1$ testing problems for θ_j, $j \neq i$. Therefore, neither $H_{0,i}$ nor $H_{0,1}$ constrain θ_j for $j \neq i$ so the hypotheses only constrain the ith component θ_i. Thus,

$$\pi_{i,0} = \int_{H_{0,i}} \pi(\theta)d\theta \text{ and } P(H_{0,i}|\boldsymbol{x}) = \int_{H_{0,i}} \pi(\theta|\boldsymbol{x})d\theta,$$

and similarly for the $H_{1,i}$s. If Ω is a Cartesian product of intervals so the hypotheses can be written as $H_{0,i} : \theta_i \leq \theta_{0,i}$ for all i and the data are IID, then

$$\int_{H_0} \pi(\theta)d\theta = \Pi_{i=1}^m \int_{H_{0,i}} \pi(\theta)d\theta \text{ and } \int_{H_0} \pi(\theta)f(\boldsymbol{x}|\theta)\pi(\theta)d\theta = \Pi_{i=1}^m \int_{H_{0,i}} \pi(\theta_i|\boldsymbol{x})d\theta_i,$$

with similar expressions for H_1.

If the data are truly from unrelated sources, there is no reason to combine the m tests; nothing can be gained in the Bayesian paradigm. However, even if the data are related, so it may be useful to combine the tests, it may be easier to use the B_is instead of investigating B itself. In these cases, to account for the effect of the joint distribution, it may make sense to use a stepwise procedure on the marginal Bayes factors. This has been proposed heuristically (i.e., without decision-theoretic or other formal justification) by Chen and Sarkar (2004). In essence, their method is the BH procedure applied to marginal Bayes factors.

BAYES VERSION OF THE BH PROCEDURE:

Find $B_{(1)} \leq B_{(2)}, \leq \ldots B_{(m)}$ the order statistic from the marginal Bayes factors, with $B_{(i)}$ corresponding to $H_{(i)} = H_{0,(i)}$.

For $r = 0, \ldots, m$, construct composite hypotheses

$$H^{(r)} : \left[\cap_{i=1}^{r} H_{1,(i)}\right] \cap \left[\cap_{i=r+1}^{m} H_{0,(i)}\right],$$

so the number of nulls is decremented by one with each increment in r. (When $r = 0$, the first intersection does not appear.)

For each r, the stepwise Bayes factor $B^{(r)}$ for testing $H^{(r)}$ vs. $H^{(r+1)}, \ldots, H^{(m)}$ is

$$B^{(r)} = \frac{W(H^{(r)}|\boldsymbol{x})}{\sum_{i=r+1}^{m} W(H^{(i)}|\boldsymbol{x})} \frac{\sum_{i=r+1}^{m} W(H^{(i)})}{W(H^{(r)})}.$$

- ☐ Start with $r = 0$, the intersection of all m hypotheses. If $B^{(0)} > 1$, then accept $H^{(0)} = \cap_{i=1}^{m} H_{(i)}$ and stop. If $B^{(0)} \leq 1$, then proceed.
- ☐ For $r = 1, \ldots m-1$, find $B^{(r)}$. If $B^{(r)} > 1$, then accept $H^{(r)}$ and stop. Otherwise, $B^{(r)} \leq 1$, so reject all $H_{(i)}$s for $i \leq r+1$ and proceed.
- ☐ For $r = m$, find $B^{(m)}$. If $B^{(m)} > 1$, then accept $H^{(m)}$ and stop. Otherwise, $B^{(m)} \leq 1$, so reject all the $H_{(i)}$s.

Note that the threshold used for the Bayes factors is denoted as 1; this means that acceptance and rejection are a priori equally acceptable. A larger threshold would give a more stringent criterion for rejection.

Chen and Sarkar (2004) give the formulas for testing a point null versus a continuous alternative by using a point mass prior on the null. They report that the procedure works well in examples and may be easier to implement.

11.6.2 Fully Bayes: Decision theory

Recall that the pFDR has a Bayesian interpretation in that the pFDR for a rejection region in terms of a statistic T corresponds to a q-value, which in turn can be regarded as a conditional probability. That is, because

$$pFDR(\Gamma) = P(\gamma_i = 0 | T_i \in \Gamma),$$

for a region Γ usually taken to be defined as the rejection region from a statistic T, one can abuse notation and write

$$pFDR = P(H_0 \text{ true}| \text{ reject} H_0).$$

However, as noted, this is not really Bayesian because the conditioning is on a region not the data. Nevertheless, Bayarri and Berger (2006) observe that a Bayesian version of this is finding

$$1 - \pi_i = P(H_0 \text{ true}|t_i),$$

and then integrating it over the rejection region based on T. To turn this into a test, one can try to control the pFDR at level α and reject H_i if $\pi_i > p^*$, where

$$p^* = \arg\min_c \left\{ \frac{\sum_{i=1}^m \mathbf{1}_{\pi > c}(1 - \pi_i)}{\sum_{i=1}^m bf 1_{\pi > c}} \leq \alpha \right\}.$$

While this procedure is intuitively sensible, it rests on assuming that some version of the FDR is the right criterion in the first place. While this may be true, it does not seem to correspond to a decision-theoretic framework. That is, the FDR does not obviously correspond to risks under a loss, Bayes or otherwise, which can be minimized to give an optimal action. (BH is known to be inadmissible Cohen and Sackrowitz (2005) and so cannot be Bayes either.)

However, other quantities do have a decision-theoretic interpretation. Under variations on the zero-one loss, the most popular choices lead to thresholding rules for π_i as the optimal strategy for making decisions under posterior risk. More generally, zero-one loss, and its variants, amount to being entirely right or entirely wrong. This is unrealistic since an alternative that is closer to the null will usually be nowhere near as suboptimal in practice as one that is far from the null even though both have the same risk. Accordingly, linear or other losses are often more reasonable in general, even though they are more difficult to use.

11.6.2.1 Proportion of False Positives

A quantity similar to the pFDR called the Proportion of False Positives, $PFP = \mathbb{E}(V)/\mathbb{E}(R)$, may have a decision theory justification of sorts (see Bickel (2004)) which extends to the pFDR. Even though $FDR \approx pFDR \approx \mathbb{E}(V)/\mathbb{E}(R)$, asymptotically in m, so the differences among them only matter in finite samples, it does not appear that Bickel's argument applies to the FDR.

Bickel (2004) regards Table 11.2 as a summary of the results from the m tests individually. Thus, the number of rejections is $R = \sum_{i=1}^m R_i$, where each R_i is 1 if the ith null is rejected and 0 otherwise. If $H_i = 0, 1$, as before, to indicate the i-th null is true or false, respectively, then for each $i = 1, ..., m$, $V_i = (1 - H_i)R_i$ so that $V = \sum_{i=1}^m V_i$ is the number of false discoveries. The other entries in Table 11.2 can be treated similarly. So, $PFP = (\mathbb{E}\sum_{i=1}^m V_i)/(\mathbb{E}\sum_{i=1}^m R_i)$.

The PFP can be derived from a sort of cost–benefit structure: see Bickel (2004). If c_i is the cost of false rejection of H_i and b_i is the benefit of rejecting H_i when it's false, then the overall desirability of any pattern of rejections can be written as

$$d_m(\boldsymbol{b},\boldsymbol{c}) = \sum_{i=1}^{m} b_i R_i - \sum_{i=1}^{m} (b_i + c_i) V_i.$$

If all the b_is and c_is are the same, say b_1 and c_1, then the expectation simplifies to

$$\mathbb{E}d_m = b_1 \mathbb{E}\left(\sum_{i=1}^{m} R_i - \left(1+\frac{c_1}{b_1}\right)\sum_{i=1}^{m} V_i\right) = b_1\left(1-\left(1+\frac{c_1}{b_1}\right)\frac{\mathbb{E}(V)}{\mathbb{E}(R)}\right)\mathbb{E}(R),$$

provided $\mathbb{E}(R) \neq 0$. Bickel (2004) calls $\nabla = \mathbb{E}(V)/\mathbb{E}(R)$ the decisive FDR, dFDR, on the grounds that it has a decision-motivated interpretation.

Storey (2003) establishes that PFP and dFDR amount to the same criterion. Indeed, a Corollary to Storey's theorem is the following.

Corollary (Storey, 2003): Under the hypotheses of Storey's theorem,

$$\mathbb{E}\left[\frac{V(\Gamma)}{R(\Gamma)}|R(\Gamma)>0\right] = \frac{\mathbb{E}(V(\Gamma))}{\mathbb{E}(R(\Gamma))}.$$

Proof: This follows from the theorem because $\mathbb{E}(V(\Gamma)) = m\pi_0\mathbb{P}(T\in\Gamma|H=0)$ and $\mathbb{E}(R(\Gamma)) = m\mathbb{P}(T\in\Gamma)$. □

A consequence is that the argument leading to dFDR also applies, in many cases, to the pFDR.

Bickel (2004) uses Storey's corollary for each of the m tests to see that

$$\begin{aligned}\nabla = \frac{\mathbb{E}(\sum_{i=1}^m (1-H_i)R_i)}{\mathbb{E}(\sum_{i=1}^m R_i)} &= \frac{\pi_0 \sum_{i=1}^m \mathbb{P}(t_i\in\Gamma|H_i=0)}{\sum_{i=1}^m \mathbb{P}(t_i\in\Gamma)}\\ &= \frac{\pi_0 m\mathbb{P}(t\geq\tau|H=0)}{m\mathbb{P}(t\geq\tau)} = \frac{\pi_0(1-F_0(\tau))}{1-F(\tau)},\end{aligned}$$

where $\Gamma=[\tau,\infty)$ is the rejection region, F_0 is the distribution of T under H_0, and F denotes the distribution of T under an alternative. In the case of a simple alternative F_1, $F=F_1$. More generally, the choice of F depends on which element of H_1 is under consideration.

11.6.2.2 Zero-One Loss and Thresholds

Instead of proposing a multiple testing criterion and seeing if it corresponds to a loss, one can propose a loss and see what kind of criteria it can motivate. Following Bayarri and Berger (2006), consider the zero-one loss commonly used for Bayes testing. Let $\boldsymbol{d}=(d_1,\ldots,d_m)$ be a decision rule for m hypothesis tests; $d_i=0$ if the ith null is accepted and $d_i=1$ if the ith null is rejected. As before, $\boldsymbol{\gamma}=(\gamma_1,\ldots,\gamma_m)$ and $\gamma_i=0,1$

according to whether the ith null is true or false. Under the generalized zero-one loss, for each test i there are four possibilities:

	$d_i = 0$	$d_i = 1$
$\gamma_i = 0$	0	c_1
$\gamma_i = 1$	c_0	0

The zeros indicate where a correct choice was made, and the c_0, c_1 are the costs for false acceptance and false rejection of the null. The standard theory now says that, given a prior on γ_i, it is posterior-risk optimal to reject the null; i.e., $d_i = 1$ if and only if $W(\gamma_i = 1|\boldsymbol{x}) = \pi > c_1/(c_0 + c_1)$.

Recalling that in Table 11.2 it is only V and T that correspond to the number of errors and that for m tests $V = \sum_{i=1}^m V_i$ and $T = \sum_{i=1}^m T_i$, the natural global loss for m independent tests is

$$L(\boldsymbol{d}, \boldsymbol{\gamma}) = \sum_{i=1}^{m} L(d_i(x_i), \gamma_i) = \sum_{i=1}^{m} c_{1,i} V_i + \sum_{i=1}^{m} c_{0,i} T_i = c_1 V + c_0 T$$

if the costs $c_{0,i}$ and $c_{1,i}$ are the same c_0 and c_1 for all tests. The consequence is that the posterior risk is necessarily of the form

$$\mathbb{E}_{\boldsymbol{\gamma}|\boldsymbol{x}} L(\boldsymbol{d}, \boldsymbol{\gamma}) = c_1 E_{\boldsymbol{\gamma}|\boldsymbol{x}}(V) + c_0 E_{\boldsymbol{\gamma}|\boldsymbol{x}}(T).$$

This seems to imply that under a zero-one loss regime, only functions of $\mathbb{E}(V)$ and $\mathbb{E}(T)$ can be justified decision-theoretically, and the least surprising optimal strategies d_i would involve thresholding the posterior probabilities as above.

11.6.2.3 Alternative Loss Functions

Muller et al. (2004) study the performance of four objective functions. One is

$$L_1(\boldsymbol{d}, \boldsymbol{\gamma}) = c\mathbb{E}_{\boldsymbol{\gamma}|\boldsymbol{x}}(V) + \mathbb{E}_{\boldsymbol{\gamma}|\boldsymbol{x}}(T),$$

which results from setting $c_0 = 1$. Setting

$$FDR(\boldsymbol{d}, \boldsymbol{\gamma}) = \frac{\sum_{i=1}^m d_i(1-\gamma_i)}{R} \quad \text{and} \quad FNR(\boldsymbol{d}, \boldsymbol{\gamma}) = \frac{\sum_{i=1}^m \gamma_i(1-d_i)}{n-R},$$

and ignoring the possibility of zeros in the denominator, the posterior means are

$$\mathbb{E}_{\boldsymbol{\gamma}|\boldsymbol{x}}(FDR) = \frac{\sum_{i=1}^m d_i(1-\pi_i)}{R},$$

$$\mathbb{E}_{\boldsymbol{\gamma}|\boldsymbol{x}}(FNR) = \frac{\sum_{i=1}^m \pi_i(1-d_i)}{m-R}.$$

The behavior of L_1 can be contrasted with

$$\begin{aligned} L_2(\boldsymbol{d},\boldsymbol{x}) &= c\mathbb{E}_{\boldsymbol{\gamma}|\boldsymbol{x}}(FDR) + \mathbb{E}_{\boldsymbol{\gamma}|\boldsymbol{x}}(FNR), \\ L_3(\boldsymbol{d},\boldsymbol{x}) &= (\mathbb{E}_{\boldsymbol{\gamma}|\boldsymbol{x}}(FDR), \mathbb{E}_{\boldsymbol{\gamma}|\boldsymbol{x}}(FNR)), \\ L_4(\boldsymbol{d},\boldsymbol{x}) &= (\mathbb{E}_{\boldsymbol{\gamma}|\boldsymbol{x}}(V), \mathbb{E}_{\boldsymbol{\gamma}|\boldsymbol{x}}(T)), \end{aligned}$$

in which L_3 and L_4 are two-dimensional objective functions that must be reduced to one dimension; L_1 and L_2 are just two possibilities. In effect, Muller et al. (2004) treat the FDR or FNR as if it were the loss function itself.

First, note that the optimal strategy under L_2 is similar to the thresholding derived for L_1; the main difference is in the thresholding value. Writing $d_i = I_{\gamma_i > t}$, the optimal t for L_1 was the ratio of costs. The optimal t for L_2 can be derived as follows.

Since L_2, like L_3 and L_4, depends only on $\boldsymbol{d}$ (and $\boldsymbol{x}$), direct substitution gives

$$L_2(\boldsymbol{d},\boldsymbol{x}) = c - \left(\frac{c}{R} + \frac{1}{m-R}\right)\sum_{i=1}^{m} d_i \pi_i + \frac{1}{m-R}\sum_{i=1}^{m} \pi_i.$$

Only the second term depends on the d_is, so for fixed R the minimum occurs by setting $d_i = 1$ for the R largest posterior probabilities π_i, $\pi_{(m)}$, ..., $\pi_{(m-R+1)}$. Using this gives

$$\min_r L_2(\boldsymbol{d},\boldsymbol{x}|R=r) = \min_r \left[c - \left(\frac{c}{r} + \frac{1}{m-r}\right)\sum_{i=m-r+1}^{m} \pi_{(r)} + \frac{1}{m-R}\sum_{i=1}^{m} \pi_i\right].$$

So, the optimal thresholding is $d_i = I_{\pi_i > t}$, where

$$t = t(\boldsymbol{x}) = \pi_{(m-r^*)} \quad \text{and} \quad r^* = \arg\min_r L_2(\boldsymbol{d},\boldsymbol{x}|R=r).$$

The optimal strategies for L_3 and L_4 can also be thresholding rules. Often, the two-dimensional objective function is reduced by minimization. In this case, it is natural to minimize $\mathbb{E}_{\boldsymbol{\gamma}|\boldsymbol{x}}(FNR)$ subject to $\mathbb{E}_{\boldsymbol{\gamma}|\boldsymbol{x}}(FDR) \le \alpha$ and $\mathbb{E}_{\boldsymbol{\gamma}|\boldsymbol{x}}(T)$ subject to $\mathbb{E}_{\boldsymbol{\gamma}|\boldsymbol{x}}(V) \le \alpha m$, respectively. These minimizations can be done by a Lagrange-style argument on

$$f_\lambda(\boldsymbol{d}) = \mathbb{E}_{\boldsymbol{\gamma}|\boldsymbol{x}}(FNR) - \lambda(\alpha - \mathbb{E}_{\boldsymbol{\gamma}|\boldsymbol{x}}(FNR))$$

and the corresponding expression for L_4. The thresholding rules for L_2, L_3, and L_4 are data-dependent, unlike that for L_1, which is genuinely Bayesian.

Unfortunately, all of these approaches have deficiencies. For instance, since $\mathbb{E}_{\boldsymbol{\gamma}|\boldsymbol{x}}(FDR)$ is bounded, even as m increases, it is possible that some hypotheses with $\pi_i \approx 0$ will end up being rejected, so L_3 may give anomalous results; L_4 may have the same property (although slower as m increases). Under L_1, $\mathbb{E}_{\boldsymbol{\gamma}|\boldsymbol{x}}(FDR) \to 0$ as m increases, so it may end up being trivial. Finally, L_2 appears to lead to jumps in $\mathbb{E}_{\boldsymbol{\gamma}|\boldsymbol{x}}(FDR)$, which seems anomalous as well.

11.6.2.4 Linear Loss

To demonstrate how loss functions that are explicitly sensitive to the distance between a point null and elements of the alternative behave, Scott and Berger (2006) (see also Bayarri and Berger (2006)) develop the appropriate expressions for a linear loss. Let

$$L(d_i = 0, \mu_i) = \begin{cases} 0 & \text{if } \mu_i = 0, \\ c|\mu_i| & \text{if } \mu_i \neq 0 \end{cases}$$

$$L(d_i = 1, \mu_i) = \begin{cases} 1 & \text{if } \mu_i = 0, \\ 0 & \text{if } \mu_i \neq 0 \end{cases}$$

where c indicates the relative costs of the two types of errors. Letting π denote a prior on μ_i, the posterior risks are given by

$$\mathbb{E}(L(d_i = 0, \mu_i)|\boldsymbol{x}) = \int L(d_i, \mu_i)\pi(\mu_i|\boldsymbol{x})d\mu_i = \pi_i,$$

$$\mathbb{E}(L(d_i = 1, \mu_i)|\boldsymbol{x}) = c(1 - \pi_i)\int |\mu_i|\pi(\mu_i|\gamma_i = 1, \boldsymbol{x})d\mu_i.$$

Consequently, the posterior expected loss is minimized by rejecting $H_{0,i}$ (i.e., setting $d_i = 1$) when

$$1 - \pi_i < \frac{c\int_{\mathbb{R}} |\mu_i|\pi(\mu_i|\gamma_i = 1, \boldsymbol{x})d\mu_i}{1 + c\int_{\mathbb{R}} |\mu_i|\pi(\mu_i|\gamma_i = 1, \boldsymbol{x})d\mu_i}. \tag{11.6.4}$$

Note that this, too, is a thresholding rule for π_i, and the larger $E(|\mu_i||\gamma_i = 1, \boldsymbol{x})$ is, the smaller the threshold. There is some evidence that, for appropriate prior selection, the posterior expectations of the μ_is would be large enough that $H_{0,i}$ would be rejected for extreme observations even when the posterior odds against an outcome x_i representing a nonzero μ_i are large. This appears to happen in cases where the number of observations that are noise (i.e., come from μ_is best taken as 0) is large.

11.7 Notes

11.7.1 Proof of the Benjamini-Hochberg Theorem

For the sake of clarity, use uppercase and lowercase to distinguish between realized p-values and p-values as random variables. Also, let P_i for $i = 1, \ldots, m_1 = m - m_0$ be the p-values for the false hypotheses and let P_i' for $i = 1, \ldots, m_0$ be the p-values for the true null hypotheses. For convenience, write $P_{(m_0)}$ for the largest p-value of the true nulls.

Assuming that the first m_0 hypotheses (i.e., $H_1,...,H_{m_0}$) are true, the theorem follows by taking expectations on both sides of

$$\mathbb{E}(V/R\,|P_{m_0+1}=p_{m_0+1},...P_m=p_{m_0+m_1})\leq\frac{m_0}{m}\alpha, \tag{11.7.1}$$

so it is enough to prove (11.7.1) by induction on m.

The case $m=1$ is immediate since there is one hypothesis, which is either true or false. The induction hypothesis is that (11.7.1) is true for any $m'\leq m$. So, it is enough to verify (11.7.1) for $m+1$. In this case, m_0 ranges from 0 to $m+1$ and m_1 ranges from $m+1$ to 0, correspondingly.

Now, for $m+1$, if $m_0=0$, all the nulls are false, $V/R=0$, and

$$\mathbb{E}(V/R\,|P_1=p_1,...,P_m=p_m)=0\leq\frac{m_0}{m+1}\alpha. \tag{11.7.2}$$

So, to control the level at the $m+1$ stage, it is enough to look at $m_0\geq 1$.

To do this, fix a nonzero value of m_0 and consider P_i for $i=1,...,m_0$, the m_0 p-values corresponding to the true nulls. These can be regarded as m_0 independent outcomes from a $Uniform(0,1)$ distribution. Write the largest order statistic from these p-values as $P_{(m_0)}$. Without loss of generality, order the p-values corresponding to the false nulls so that the $P_{m_0+i}=p_{m_0+i}$ for $i=1,...,m_1$ satisfy $p_{m_0+1}\leq p_{m_0+2}\leq\cdots\leq p_{m_0+m_1}$.

Now, consistent with the BH method, let j_0 be the largest value of j in $\{0,...,m_1\}$ for which the BH method rejects H_j and necessarily all hypotheses with smaller p-values. That is, set

$$j_0=\max\left\{j|p_{m_0+j}\leq\frac{m_0+j}{m+1}\alpha, j=1,...,m_1\right\}, \tag{11.7.3}$$

and write p^* to mean the maximum value of the threshold,

$$p^*=\frac{m_0+j_0}{m+1}\alpha.$$

Next represent the conditional expectation given the p-values from the false nulls as the integral in which the expectation has been further conditioned on the largest p-value $P_{(m_0)}$ from the true nulls,

$$\begin{aligned}&\mathbb{E}(V/R\,|P_{m_0+1},\ldots,P_{m_0+m_1}=p_{m_0+m+1})\\&\quad=\int_{p^*}^{1}\mathbb{E}(V/R\,|P_{(m_0)}=p,P_{m_0+1},...,P_{m_0+m_1}=p_{m_0+m+1})f_{P_{(m_0)}}(p)dp\\&\quad+\int_0^{p^*}\mathbb{E}(V/R\,|P_{(m_0)}=p,P_{m_0+1},...,P_{m_0+m_1}=p_{m_0+m+1})f_{P_{(m_0)}}(p)dp,\end{aligned} \tag{11.7.4}$$

in which the density $f_{P_{(m_0)}}(p) = m_0 p^{(m_0-1)}$ for $P_{(m_0)}$ comes from the fact that it is the largest order statistic from a $Uniform(0,1)$ sample.

In the second term on the right in (11.7.4), $p \le p^*$, so all m_0 true hypotheses are rejected, as are the first j_0 false hypotheses. That is, on this domain of integration, $m_0 + j_0$ hypotheses are rejected and $(V/R) = m_0/(m_0 + j_0)$. Now, the integral in the second term of (11.7.4) is

$$\frac{m_0}{m_0+j_0}(p^*)^{m_0} \le \frac{m_0}{m_0+j_0}\frac{m_0+j_0}{m+1}\alpha(p^*)^{m_0-1} = \frac{m_0}{m+1}\alpha(p^*)^{m_0-1}, \quad (11.7.5)$$

in which the inequality follows from the bound in (11.7.3).

To deal with the first term in (11.7.4), write it as the sum

$$\int_{p^*}^{p_{m_0+j_0+1}} \mathbb{E}(V/R\,|P_{(m_0)} = p, P_{m_0+1} = p_{m_0+1}, \ldots, P_{m_0+m_1} = p_{m_0+m+1}) f_{P'_{(m_0)}}(p)dp$$
$$+ \sum_{i=1}^{m_1-j_0-1} \int_{p_{m_0+j_0+i}}^{p_{m_0+j_0+i+1}} \mathbb{E}(V/R\,|P_{(m_0)} = p, P_{m_0+1} = p_{m_0+1}, \ldots$$
$$\ldots, P_{m_0+m_1} = p_{m_0+m+1}) f_{P_{(m_0)}}(p)dp$$
$$+ \int_{p_{m_0+m_1}}^{1} \mathbb{E}(V/R\,|P_{(m_0)} = p, P_{m_0+1} = p_{m_0+1}, \ldots$$
$$\ldots, P_{m_0+m_1} = p_{m_0+m+1}) f_{P_{(m_0)}}(p)dp, \quad (11.7.6)$$

in which $p_{m_0+j_0} \le p^* < P_{(m_0)} = p < p_{m_0+j_0+1}$ for the first term, $p_{m_0+j_0} < p_{m_0+j} \le P_{(m_0)} = p < p_{m_0+j+1}$ for the terms in the summation, and the last term is just the truncation at 1.

To control (11.7.6), it is enough to get an upper bound on the integrands that depends on p (but not on the domain of integration) so as to set up an application of the induction hypothesis.

So, fix one of the terms in (11.7.6) and observe that, because of the careful way j_0 and p^* have been defined, no hypothesis can be rejected because of the values of p, $p_{m_0+j_0+1}$, ..., $p_{m_0+m_1}$, because they are bigger than the cutoff p^*. Therefore, when all $m_0 + m_1$ hypotheses are considered together and their p-values ordered from 1 to m, a hypothesis $H_{(i)}$ corresponding to $p_{(i)}$ is rejected if and only if

$$\exists k \in \{i, \ldots, m_0 + j_0 - 1\} : p_{(k)} \le \frac{k}{m+1}\alpha$$
$$\Leftrightarrow \frac{p_{(k)}}{p} \le \frac{k}{m_0+j_0-1}\frac{m_0+j_0-1}{(m+1)p}\alpha. \quad (11.7.7)$$

When conditioning on $P_{(m_0)} = p$, the $m_0 + j_0 - 1$ p-values, the $p_{(k)}$s, on the right-hand side of (11.7.4) have two forms. Some, m_0 of them, correspond to true H_is. Of these, the largest is the condition $P_{(m_0)} = p$. For the other $m_0 - 1$ true hypotheses, $p_{(k)}/p$ really is of the form P_i/p for some $i = 1, \ldots, m_0 - 1$, which are independent $Uniform(0,1)$ variates. The rest, $j_0 - 1$ of them, correspond to false H_is. In these cases, $p_{(k)}/p$ corresponds to p_{m_0+i}/p for $i = 1, \ldots, j_0 - 1$.

Using the criterion (11.7.7) to test the $m_0 + j_0 - 1 \leq m$ hypotheses is equivalent to using the BH method with α chosen to be

$$\alpha' = \frac{m_0 + j_0 - 1}{(m+1)p}\alpha.$$

Therefore, the induction hypothesis (11.7.1) for this choice of α' and the extra conditioning on p, with m_0 replaced by $m_0 - 1$, can be applied. The result is

$$\begin{aligned} &\mathbb{E}(V/R\,|P_{(m_0)} = p, P_{m_0+1} = p_{m_0+1}, ..., P_{m_0+m_1} = p_{m_0+m+1}) \\ &\leq \frac{m_0 - 1}{m_0 + j_0 - 1} \times \frac{m_0 + j_0 - 1}{(m+1)p}\alpha = \frac{m_0 - 1}{(m+1)p}\alpha, \end{aligned} \tag{11.7.8}$$

which depends on p but not on the i in (11.7.6) for which it was derived. That is, the bound in (11.7.8) is independent of the segment $p_{m_0+i} \leq p \leq p_{m_0+i+1}$, the initial segment bounded by p^*, and the terminal segment bounded by 1.

Using (11.7.8) as a bound on the integrand in the first term in (11.7.4) gives that it is

$$\begin{aligned} &\int_{p^*}^{1} \mathbb{E}(V/R\,|P_{(m_0)} = p, P_{m_0+1}, ..., P_{m_0+m_1} = p_{m_0+m+1}) f_{P_{(m_0)}}(p)dp \\ &\leq \int_{p^*}^{1} \frac{m_0 - 1}{(m+1)p}\alpha \times m_0 p^{m_0-1}dp = \int_{p^*}^{1} (m_0 - 1)p^{m_0-2}dp \\ &= \frac{m_0}{m+1}\alpha(1 - (p^*)^{m_0-1}). \end{aligned} \tag{11.7.9}$$

Finally, adding the bounds on the two terms in (11.7.4) from (11.7.5) and (11.7.9) gives (11.7.4) for $m+1$, so the induction is complete. □

11.7.2 Proof of the Benjamini-Yekutieli Theorem

Let $\alpha_i = (1/m)\alpha$ be the threshold for the ith p-value for $i = 1, ..., m$. Now partition the sample space into sets

$$A_{v,s} = \{\boldsymbol{x} : \text{under } \boldsymbol{T},\ \text{BH rejects exactly } v \text{ true and } s \text{ false hypotheses}\},$$

so that the FDR is

$$\mathbb{E}\left(\frac{V}{R}\right) = \sum_{s=0}^{m_1}\sum_{v=1}^{m_0} \frac{v}{v+s}\mathbb{P}(A_{v,s}), \tag{11.7.10}$$

and let P_i be the p-value for the ith true hypothesis, as a random variable, from T_i for $i = 1, ..., m_0$.

Step 1: For any fixed v, s,

$$\mathbb{P}(A_{v,s}) = \frac{1}{v}\sum_{i=1}^{m_0}\mathbb{P}(\{P_i \le \alpha_{v+s}\} \cap A_{v,s}).$$

Let w be a subset of $\{1,...,m_0\}$ of size v and define the event

$$A^w_{v,s} = \{\text{the } v \text{ true nulls rejected are in } w\}. \tag{11.7.11}$$

Now, $A_{v,s}$ is the disjoint union of the $A^w_{v,s}$s over all possible distinct ws.

Now, ignoring the v, consider the sum on the right-hand side. It is

$$\sum_{i=1}^{m_0}\mathbb{P}(\{P_i \le \alpha_{v+s}\} \cap A_{v,s}) = \sum_w\sum_{i=1}^{m_0}\mathbb{P}(\{P_i \le \alpha_{v+s}\} \cap A^w_{v,s}). \tag{11.7.12}$$

There are two cases: $i \in w$ and $i \notin w$. If $i \in w$, (i.e., H_i is rejected), then, by construction, $P_i \le \alpha_{v+s}$ and conversely, if $P_i \le \alpha_{v+s}$, then H_i must be rejected and so corresponds to an outcome in $A^w_{v,s}$. Thus, for $i \in w$, $\mathbb{P}(P_i \le \alpha_{v+s} \cap A^w_{v,s}) = \mathbb{P}(A^w_{v,s})$. If $i \notin w$, then $\mathbb{P}(P_i \le \alpha_{v+s} \cap A^w_{v,s}) = 0$ because the two events are disjoint.

So, the right-hand side of (11.7.12) is

$$\sum_w\sum_{i=1}^{m_0}\chi_{i\in w}\mathbb{P}(A^w_{v,s}) = \sum_w v\mathbb{P}(A^w_{v,s}) = v\mathbb{P}(A_{v,s}).$$

To state Step 2, two classes of sets must be defined. Let

$$C_{v,s}(\hat{i}) = \{\text{ if } H_i \text{ is rejected, then so are } v-1 \text{ other true nulls and } v \text{ false nulls}\},$$

and denote unions over these sets by

$$C_k(\hat{i}) = \cup_{v,s:v+s=k}C_{v,s}(\hat{i}).$$

Roughly, $C_{v,s}(\hat{i})$ is the event that H_i is one of the v rejected hypotheses and $C_k(\hat{i})$ is the event that, out of all the ways to reject exactly k hypotheses, one of them is H_i.

Step 2: The FDR can be written as

$$\mathbb{E}\left(\frac{V}{R}\right) = \sum_{i=1}^{m_0}\sum_{k=1}^{m}\frac{1}{k}\mathbb{P}(\{P_i \le \alpha_k\} \cap C_k(\hat{i})).$$

Start by using Step 1 in (11.7.11) to get

$$\mathbb{E}\left(\frac{V}{R}\right) = \sum_{s=0}^{m_1}\sum_{v=1}^{m_0}\sum_{i=0}^{m_0}\frac{1}{v+s}\mathbb{P}(P_i \le \alpha_{v+s} \cap A_{v,s}). \tag{11.7.13}$$

It is the intersected events in the probability that can be simplified. Note that $C_{v,s}(\hat{i}) \subset A_{v,s}$ since the event that one rejected hypothesis out of $v+s$ rejections is H_i is a subset of there being v_s rejected hypotheses in total. So,

$$P_i \leq \alpha_{v+s} \cap A_{v,s} = \left[P_i \leq \alpha_{v+s} \cap C_{v,s}(\hat{i})\right] \cup \left[P_i \leq \alpha_{v+s} \cap (A_{v,s} \setminus C_{v,s}(\hat{i}))\right],$$

in which the second intersection is void because $P_i \leq \alpha_{v+s}$ means that H_i is rejected and $A_{v,s} \setminus C_{v,s}(\hat{i})$ means H_i is not rejected. Using this substitution in (11.7.13) and noting that for each i the events $C_k(\hat{i})$ are mutually disjoint (for k and $k' \neq k$, different numbers of H_is are rejected) and so their probabilities can be summed, gives Step 2.

To state Step 3, define

$$D_k(\hat{i}) = \cup_{j: j \leq k} C_j(\hat{i})$$

for $k = 1, ..., m$. This is the set on which k or fewer of the m hypotheses are rejected, one of them being H_i, regardless of whether they are true or not. This will set up an application of the PRDS property to bound the inner sum in (11.7.13) by α/m.

Step 3: The set $D_k(\hat{i})$ is nondecreasing.

To see Step 3, it is enough to reexpress $D_k(\hat{i})$ in terms of inequalities on p-values. First, let $\mathbf{P}(\hat{i})$ be the ordered vector of $m-1$ p-values formed by leaving out the ith p-value corresponding to H_i. Now, $\mathbf{P}(\hat{i})$ has $m-1$ entries,

$$\mathbf{P}(\hat{i}) = (P_{(1)}(\hat{i}), ..., P_{(m-1)}(\hat{i})).$$

On the set $D_k(\hat{i})$, H_i is rejected, so its p-value must be below its BH threshold. Also, $k-1$ other hypotheses must be rejected so the smallest BH threshold that a p-value can be above is α_{k+1} and the smallest p-value must be the kth entry of $\mathbf{P}(\hat{i})$. That is, on $D_k(\hat{i})$, $\alpha_{k+1} \leq P_{(k)}(\hat{i})$. The next smallest BH threshold is α_{k+2}, and the next smallest p-value must be the $k+1$ entry of $\mathbf{P}(\hat{i})$ and on $D_k(\hat{i})$ they must satisfy $\alpha_{k+2} \leq P_{(k+1)}(\hat{i})$. Proceeding in this way gives that

$$D_k(\hat{i}) = \{\boldsymbol{p} = \boldsymbol{p}_k(\hat{i}) | \alpha_{k+1} \leq p_{(k)}(\hat{i}), ..., \alpha_m \leq p_{(m-1)}(\hat{i})\},$$

from which it is easily seen that $D_k(\hat{i})$ is nondecreasing.

Step 4: For $i = 1, ..., m-1$,

$$\sum_{k=1}^{m} \frac{\mathbb{P}(\{P_i \leq \alpha_k\} \cap C_k(\hat{i}))}{\mathbb{P}(P_i \leq \alpha_k)} \leq 1.$$

This is where the PRDS property is used. For any nondecreasing set D, $p \leq q$ implies $\mathbb{P}(D|P_i = p) \leq \mathbb{P}(D|P_i = q)$. So, $\mathbb{P}(D|P_i \leq p) \leq \mathbb{P}(D|P_i \leq q)$, see Lehmann (1966) (which could be used as the definition of PRDS). Setting $D = \{P_i \leq q_k\} \cap D_k(\hat{i})$, $p = \alpha_k$, and $q = \alpha_{k+1}$ gives

$$\frac{\mathbb{P}(\{P_i \leq \alpha_k\} \cap D_k(\hat{i}))}{\mathbb{P}(P_i \leq \alpha_k)} \leq \frac{\mathbb{P}(\{P_i \leq \alpha_{k+1}\} \cap D_k(\hat{i}))}{\mathbb{P}(P_i \leq \alpha_{k+1})}. \tag{11.7.14}$$

Using $D_{j+1}(\hat{i}) = D_j(\hat{i}) \cup C_{j+1}(\hat{i})$ in (11.7.14) gives

$$\frac{\mathbb{P}(\{P_i \le \alpha_k\} \cap D_k(\hat{\imath}))}{\mathbb{P}(P_i \le \alpha_k)} + \frac{\mathbb{P}(\{P_i \le \alpha_{k+1}\} \cap D_k(\hat{\imath}))}{\mathbb{P}(P_i \le \alpha_{k+1})}$$

$$\le \frac{\mathbb{P}(\{P_i \le \alpha_{k+1}\} \cap D_k(\hat{\imath}))}{\mathbb{P}(P_i \le \alpha_{k+1})} + \frac{\mathbb{P}(\{P_i \le \alpha_{k+1}\} \cap D_k(\hat{\imath}))}{\mathbb{P}(P_i \le \alpha_{k+1})}$$

$$= \frac{\mathbb{P}(\{P_i \le \alpha_{k+1}\} \cap D_{k+1}(\hat{\imath}))}{\mathbb{P}(P_i \le \alpha_{k+1})}$$

for $k = 1, ..., m-1$.

To complete Step 4, take the sum over $k = 1, ..., m-1$: Since $D_1(\hat{\imath}) = C_1(\hat{\imath})$ for each i, once $k = 2$ on the left the first term on the left cancels with the term on the right for $k = 1$ and so forth for $k = 3, 4, ...m-1$ until the last uncanceled term on the right is

$$\frac{\mathbb{P}(\{P_i \le \alpha_m\} \cap D_m(\hat{\imath}))}{\mathbb{P}(P_i \le \alpha_m)} \le 1 \tag{11.7.15}$$

since $D_m(\hat{\imath})$ is the whole sample space.

Step 5: To complete the proof, note that, under H_i, $\mathbb{P}(P_i \le \alpha_k) \le \alpha_k = (k/m)\alpha$, so that from Step 2

$$\mathbb{E}\left(\frac{V}{R}\right) \le \sum_{i=1}^{m_0} \sum_{k=1}^{m} \frac{\alpha}{m} \frac{\mathbb{P}(\{P_i \le \alpha_k\} \cap C_k(\hat{\imath}))}{\mathbb{P}(P_i \le \alpha_k)}. \tag{11.7.16}$$

Now, Step 4 completes the proof. □

References

Abdi, H. (2003). Factor rotations in factor analysis. In B. Lewis-Beck and Futing (Eds.), *Encyclopedia of Social Sciences Research Methods*, pp. 978–982. Thousand Oaks, Sage.

Abdi, H. (2007). Partial least squares regression. In N. Salkind. (Ed.), *Encyclopedia of Measurement and Statistics*, pp. 740–744. Thousand Oaks, Sage.

Akaike, H. (1973). Maximum likelihood identification of gaussian autoregressive moving average models. *Biometrika 60*, 255–265.

Akaike, H. (1974). A new look at statistical model identification. *IEEE Trans. Auto. Control 19*, 716–723.

Akaike, H. (1977). On entropy maximization principle. In P. R. Krishnaiah (Ed.), *Proceedings of the Symposium on Application of Statistics*, pp. 27–41. North Holland, Amsterdam.

Aksoy, S. and R. Haralick (1999). Graph-theoretic clustering for image grouping and retrieval. In D. Huijsmans and A. Smeulders (Eds.), *Proceedings of the Third International Conference in Visual Information and Information Systems*, Number 1614 in Lecture Notes in Computer Science, pp. 341–348. New York: Springer.

Allen, D. M. (1971). Mean square error of prediction as a criterion for selecting variables. *Technometrics 13*, 469–475.

Allen, D. M. (1974). The relationship between variable selection and data agumentation and a method for prediction. *Technometrics 16*, 125–127.

Amato, R., C. Del Mondo, L. De Vinco, C. Donalek, G. Longo, and G. Miele (2004). *Ensembles of Probabilistic Principal Surface and Competitive Evolution on Data.* British Computer Society: Electronic Workshops in Computing.

An, L. and P. Tao (1997). Solving a class of linearly constrained indefinite quadratic problems by d.c. algorithm. *J. Global Opt. 11*, 253–285.

Anderson, T. W. (1984). *An Introduction to Multivariate Statistical Analysis* (2nd ed.). New York: Wiley and Sons.

B. Clarke et al., *Principles and Theory for Data Mining and Machine Learning*, Springer Series in Statistics, DOI 10.1007/978-0-387-98135-2_BM2,

Andrieu, C., L. Breyer, and A. Doucet (2001). Convergence of simulated annealing using foster-lyapunov criteria. *J. Appl. Probab. 38*(4), 975–994.

Aronszajn, N. (1950). Theory of reporducing kernels. *Trans. Amer. MAth. Soc. 68*, 522–527.

Atkinson, A. (1980). A note on the generalized information criterion for choice of a model. *Biometrika 67*, 413–418.

Banks, D., R. Olszewski, and R. Maxion (2003). Comparing methods for multivariate nonparametric regression. *Comm. Statist. Sim. Comp. 32*(2), 541–571.

Banks, D. L. and R. Olszewski (2004). Data mining in federal agencies. In H. Bozdogan (Ed.), *Statistical Data Mining and Knowledge Discovery*, pp. 529–548. New York: Chapman & Hall.

Barbieri, M. and J. O. Berger (2004). Optimal predictive model selection. *Ann. Statist. 32*, 870–897.

Barbour, A. D., L. Holst, and S. Janson (1992). *Poisson Approximation*. Gloucestershire: Clarendon Press.

Barron, A. (1991). Approximation bounds for superpositions of a sigmoidal function. In *1991 IEEE International Symposium on Information Theory*, pp. 67–82.

Barron, A. (1993). Universal approximation bounds for superpositions of a sigmoidal function. *IEEE Trans. Inform. Theory 39*(3), 930–945.

Barron, A., L. Birge, and P. Massart (1999). Risk bounds of model selection via penalization. *Probability Theory and Related Fields 113*, 301–413.

Barron, A. and T. M. Cover (1991). Minimum complexity density estimation. *IEEE Trans. Inform. Theory 37*, 1034–1054.

Barron, A. R. and X. Xiao (1991). Discussion of multivariate adaptive regression splines. *Ann. Statist. 19*(1), 6–82.

Basalaj, W. (2001). Proximity visualization of abstract data. Tech report, `http://www.pavis.org/essay`.

Baum, L. E. and T. Petrie (1966). Statistical inference for probabilistic functions of finite state markov chains. *Ann. Math. Statist. 37*, 1554–1563.

Bayarri, S. and J. Berger (2006). Multiple testing: Some contrasts between bayesian and frequentist approaches. slides. Personal communication.

Bellman, R. E. (1961). *Adaptive Control Processes*. Princeton University Press.

Benjamini, Y. and T. Hochberg (1995). Controlling the false discovery rate: A practical cand powerful approach to multiple testing. *J. Roy. Statist. Soc. Ser. B 85*, 289–300.

Benjamini, Y. and D. Yakutieli (2001). Control of the false discovery rate in multple testing under dependency. *Ann. Statist. 29*(4), 1165–1188.

Berger, J., J. K. Ghosh, and N. Mukhopadhyay (2003). Approximation and consistency of bayes factors as model dimension grows. *J. Statist. Planning and Inference 112*, 241–258.

Berger, J. and J. Mortera (1999). Default Bayes factors for one-sided hypothesis testing. *J. Amer. Statist. Assoc. 94*, 542–554.

Berger, J. and L. Pericchi (1996). The intrinsic Bayes factor for model selection and prediction. *J. Amer. Statist. Assoc. 91*, 109–122.

Berger, J. and L. Pericchi (1997). On the justification of default and intrinsic Bayes factors. In e. a. J. C. Lee (Ed.), *Modeling and Prediction*, New York, pp. 276–293. Springer-Verlag.

Berger, J. and L. Pericchi (2001). Objective Bayesian methods for model selection: introduction and comparison (with discussion). In P. Lahiri (Ed.), *Model Selection, Institute of Mathematical Statistics Lecture Notes*, pp. 135–207. Beachwood Ohio.

Berk, R. (1966). Limiting behavior of posterior distributions when the model is incorrect. *Ann. Math. Statist. 37*(1), 51–58.

Bernardo, J. M. and A. Smith (1994). *Bayesian Theory*. Wiley & Sons, Chichester.

Bhattacharya, R. N. and R. Ranga Rao (1976). *Normal Approximation and Asymptotic Expansions*. Malabar FL: Robert E. Krieger Publishing Company.

Bickel, D. (2004). *Error Rate and Decision Theoretic Methods of Multiple Testing: Which Genes Have High Objective Probabilities of Differential Expression?*, Volume 5. Art. 8, `http://www.bepress.com/sagmb`: Berkeley Electronic Press.

Bickel, P. and E. Levina (2004). Some theory of fisher's linear discriminant function, 'naive bayes', and some alternatives when there are many more variables than observations. *Bernoulli 10*(6), 989–1010.

Billingsley, P. (1968). *Convergence of Probability Measures*. Inc., New York, NY: John Wiley & Sons.

Bilmes, J. (1998). A gentle tutorial of the em algorithm and its application to parameter estimation for gaussian mixture and hidden markov models. Tech. Rep. 97-021, Dept of EE and CS, U.C. Berkeley.

Boley, D. (1998). Principal direction divisive partitioning. *Data Mining and Knowledge Discovery, Vol 2*(4), 325–344.

Borman, S. (2004). *The Expectation-Maximization Algorithm: A short tutorial*. See `www.seanborman.com/publications/EM\_algorithm.pdf`.

Box, G. E. P., W. G. Hunter, and J. S. Hunter (1978). *Statistics for Experimenters: An Introduction to Design, Data Analysis, and Model Building*. New York, NY, USA: John Wiley and Sons.

Bozdogan, H. (1987). Model selection and akaike's information criterion (aic): the general theory and its analytical extensions. *Psychometrika 52*, 345–370.

Breiman, L. (1994). Bagging predictors. Technical Report 421, Dept. of Statistics, U.C. Berkeley.

Breiman, L. (1995). Better subset regression using the nonnegative garrote. *Technometrics 37*, 373–384.

Breiman, L. (1996). Stacked regressions. *Machine Learning 24*, 49–64.

Breiman, L. (2001). Random forests. *Machine Learning 45*, 5–32.

Breiman, L. and A. Cutler (2004). *Random Forests*. `http://www.stat.berkeley.edu/breiman/RandomForests/cc-home.htm`.

Breiman, L. and J. Friedman (1985). Estimating optimal transformations for regression and correlation (with discussion). *J. Amer. Statist. Assoc. 80*, 580–619.

Breiman, L., J. Friedman, R. Olshen, and C. Stone (1984). *Classification and Regression Trees*. Wadsworth: Belmont, CA.

Buhlman, P. and B. Yu (2002). Analyzing bagging. *Ann. Statist. 30*, 927–961.

Buja, A. and Y.-S. Lee (2001). Data mining criteria for tree-based regression and classification. *Proceedings of KDD 2001.*, 27–36.

Buja, A. and W. Stuetzle (2000a). The effect of bagging on variance, bias, and mean squared error. Preprint, AT&T Labs-Research.

Buja, A. and W. Stuetzle (2000b). Smoothing effects of bagging. Preprint, AT&T LabsResearch.

Buja, A., D. Swayne, M. Littman, N. Dean, and H. Hofmann (2001). Interactive data visualization with multidimensional scaling. Technical report, `http://www.research.att.com/areas/stat/xgobi/papers/xgvis.pdf`.

Buja, A., H. T., and T. R. (1989). Linear smoothers and additive models. *Ann. Statist. 17*, 453–555.

Bullinaria, J. (2004). Introduction to neural networks course. Lecture notes, `http://www.cs.bham.ac.uk/~jxb/inn.html`.

Burges, C. (1998). A tutorial on support vector machines for pattern recognition. *Data mining and Knowledge Discovery 2*, 121–167.

Burman, P. (1989). A comparative study of ordinary cross-validation, v-fold cross-validation, and the repeated learning-testing methods. *Biometrika 76*, 503–514.

Burman, P. (1990). Estimation of optimal transformation using v-fold cross validation and repeated learning-testing methods. *Sankhya 52*, 314–345.

Burnham, K. P. and D. R. Anderson (2002). *Model Selection and Multimodel Inference: A Practical Information-Theoretical Approach*. New York: Springer-Verlag.

Candes, E. and T. Tao (2007). The dantzig selector: Statistical estimation when p is much larger than n. *Ann. Statist. 35*, 2313–2351.

Canu, S., C. S. Ong, and X. Mary (2005). *Splines with Non-Positive Kernels*. World Scientific. See: `http://eprints.pascal-networks.org`.

Cappe, O., E. Moulines, and T. Ryden (2005). *Inference in Hidden Markov Models*. Berlin: Springer.

Carreira-Perpinan, M. (1997). A review of dimension reduction techniques. Tech. Rep. CS-96-09, Dept. of Computer Science, University of Sheffield.

Casella, G. and E. I. George (1992). Explaining the Gibbs sampler. *The American Statistician 46*, 167–174.

Casella, G. and E. Moreno (2006). Objective Bayes variable selection. *J. Amer. Statist. Assoc. 101*, 157–167.

Chang, K. and J. Ghosh (2001). A unified model for probabilistic principal surfaces. *IEEE Trans. Pattern Anal. Mach. Int 23*(1), 22–41.

Chen, H. (1991). Estimation of a projection-pursuit type regression model. *Ann. Statist. 19*, 142–157.

Chen, J. and S. Sarkar (2004). Multiple testing of response rates with a control: a bayesian stepwise approach. *J. Statist. Plann. inf 125*, 3–16.

Chen, S., D. Donoho, and M. Saunders (1998). Atomic decomposition by basis pursuit. *SIAM J. Sci. Comput. 20*, 33–61.

Chernoff, H. (1956). Large sample theory: Parametric case. *Ann. Math. Statist. 27*(1), 1–22.

Chernoff, H. (1973). The use of faces to represent points in k-dimensional space graphically. *J. Amer. Statist. Soc 68*(342), 361–368.

Chernoff, H. (1999). Gustav elfving's impact on experimental design. *Statist. Sci. 14*(2), 201–205.

Chi, Z. (2007). Sample size and positive false discovery rate control for multiple testing. *Elec. J. Statist. 1*, 77–118.

Chib, S. and E. Greenberg (1995). Understanding the metropolis-hastings algorithm. *The American Statistician 49*, 327–335.

Chipman, H., E. George, , and McCulloch (2005). Bayesian additive regression trees. Preprint.

Chipman, H., E. George, and R. McCulloch (1996). Bayesian cart model search. *J. Amer. Statist. Assoc. 93*(443), 935–948.

Chipman, H., E. I. George, and R. E. McCulloch (2001). The practical implementation of Bayesian model selection. In P. Lahiri (Ed.), *Model Selection, Institute of Mathematical Statistics Lecture Notes*, pp. 65–116. Beachwood Ohio.

Chipman, H. and H. Gu (2001). Interpretable dimension reduction. Research Report 01-01, `http://www.bisrg.uwaterloo.ca/archive/`.

Clarke, B. (2004). Comparing stacking and bma when model mis-specification cannot be ignored. *J. Mach. Learning Res. 4*(4), 683–712.

Clarke, B. (2007). Information optimality and bayesian models. *J. Econ. 138*(2), 405–429.

Clarke, B. and A. R. Barron (1990). Information-theoretic asymptotics of bayes' methods. *IEEE Trans. Inform. Theory 36*(3), 453–471.

Clarke, B. and A. R. Barron (1994). Jeffreys prior is asymptotically least favorable under entropy risk. *J. Statist. Planning Inference 41*, 37–60.

Clasekens, G. and N. Hjort (2003). Focused information criterion. *J. Amer. Statist. Assoc. 98*, 879–899.

Clemen, R. T. (1989). Combining forecasts: A review and annotated bibliography. *Int. J. Forecasting 5*, 559–583.

Cleveland, W. (1993). *Visualizing Data*. Summit, NJ: Hobart Press.

Cleveland, W. S. (1979). Robust locally weighted regression and smoothing scatterplots. *J. Amer. Statist. Assoc. 74*, 829–836.

Cleveland, W. S. and S. J. Devlin (1988). Locally weighted regression: An approach to regression by local fitting. *J. Amer. Statist. Assoc. 83*, 596–610.

Clyde, M. (1999). Bayesian model averaging and model search strategies. In J. M. Bernardo, J. O. Berger, A. P. Dawid, and A. F. M. Smith (Eds.), *Bayesian Statistics*, Number 6, pp. 157–185. Oxford University Press, Oxford.

Clyde, M., H. DeSimone, and G. Parmigiani (1996). Prediction vis orthogonalized model mixing. *J. Amer. Statist. Assoc. 91*, 1197–1208.

Clyde, M. and E. I. George (2000). Flexible empirical Bayes estimation for wavelets. *J. Roy. Statist. Soc. Ser. B 62*, 681–698.

Cohen, A. and H. Sackrowitz (2005). Characterization of bayes procedures for multiple endpoint problems and inadmissibility of the step-up procedure. *Ann. Statist. 33*, 145–158.

Comon, P. (1994). Independent component analysis: A new concept? *Signal Processing 36*, 287–314.

Cook, D. and B. Li (2002). Dimension reduction for conditional mean in regression. *Ann. Statist. 30*, 455–474.

Cook, D. and L. Ni (2005). Sufficient dimension reduction via inverse regression: A minimum discrepancy approach. *J. Amer. Statist. Assoc. 100*(470), 410–428.

Cook, D. and D. Swayne (2007). *Interactive and Dynamic Graphics for Data Analysis*. New York: Springer.

Cook, R. D. and S. Weisberg (1991). Comment on: Sliced inverse regression for dimension reduction. *J. Amer. Statist. Assoc. 86*, 328–332.

Cover, T. (1965). Geometrical and statistical properties of systems of linear inequalities with applications to pattern recognition. *IEEE Trans. Elec. Comp. 14*, 326–334.

Craven, P. and G. Wahba (1979). Smoothing noisy data with spline functions: Estimating the correct degree of smoothing by the method of gcv. *Numer. Math. 31*, 377–403.

Cui, W. and E. I. George (2007). Empirical Bayes vs. fully Bayes variable selection. *J. Statist. Planning and Inference 138*, 888–900.

Cutting, D., D. Karger, J. Pedersen, and J. Tukey (1992). Scatter/gather: A cluster based approach to browsing large document collections. *In: Proceedings of the 15th annual international ACM SIGIR conference on Research and development in information retrieval*, 318–329.

D. Harrison, D. and D. Rubinfeld (1978). Hedonic housing prices and the demand for clean air. *J. Env. Econ. Management 5*, 81–102.

Daniel, W. (1990). *Applied Nonparametric Statistics*. Boston: PWS-Kent Publishing.

Dawid, A. P. (1984). Present position and potential developments: some personal views. statistical theory. the prequential approach (with discussion). *J. Roy. Statist. Soc. Ser. B 147*, 278–292.

de Bruijn, N. G. (1959). Pairs of slowly oscillating functions occurring in asymptotic problems concerning the laplace transform. *Nieuw Archief Wiskunde 7*, 20–26.

de Leeuw, J. (1988). Multivariate analysis with optimal scaling. In S. DasGupta and J. Ghosh (Eds.), *Proceedings of the International Conference on Advances in Multivariate Statistical Analysis*, pp. 127–160. Indian Statistical Institute, Calcutta.

de Leeuw, J. (2005). *Nonlinear Principal Component Analysis*. U. of California: eScholarship Repository.

de Leeuw, J. and P. Mair (2008). Multidimensional scaling using majorization: Smacof in r. Technical report, `http://preprints.stat.ucla.edu/`.

Delicado, P. (2001). Another look at principal curves and surfaces. *J. Mult. Anal. 77*, 84–116.

Dempster, A., N. Laird, and D. Rubin (1977). Maximum likelihood from incomplete data via the em algorithm. *J. Roy. Statist. Soc. Ser. B 39*(1), 1–38.

Devroye, L., G. L., A. Krzyzak, and G. Lugosi (1994). On the strong universal consistency of nearest neighbor regression function estimates. *Ann. Statist. 22*(3), 1371–1385.

Devroye, L. and T. J. Wagner (1980). Distribution free consistency results in nonparametric discrimination and regression function estimation. *Ann. Statist. 8*, 231–9.

Diaconis, P. and D. Freedman (1984). Asymptotics of graphical projection pursuit. *Ann. Statist. 12*(3).

Dietterich, T. G. (1999). Machine learning research: Four current directions. *AI Magazine 18*(4), 97–136.

Ding, C. and X. He (2002). Cluster merging and splitting in hierarchical clustering algorithms. *Proceedings of the IEEE International Conference on Data Mining*, 139–146.

Dmochowski, J. (1996). Intrinsic priors via kullback-leibler geometry. In J. M. Bernado (Ed.), *Bayesian Statistics 5*, London, pp. 543–549. Oxford University Press.

Dodge, Y., V. Fedorov, and H. P. Wynn (1988). *Optimal Design and Analysis of Experiments*. Saint Louis: Elsevier Science Ltd.

Donoho, D. L. and I. M. Johnstone (1994). Ideal spatial adaptation by wavelet shrinkage. *Biometrika 81*, 425–455.

Doob, J. (1953). *Stochastic Processes*. New York: John Wiley.

Dreiseitl, S. and L. Ohno-Machado (2002). Logistic regression and artificial neural network models: a methodology review. *J.Biomedical Informatics 35*(5-6), 352–359.

du Toit, Steyn, and Stumpf (1986). *Graphical Exploratory Data Analysis*. New York: Springer-Verlag.

Duan, N. and K.-C. Li (1991). Slicing regression: A link free regression method. *Ann. Statist. 19*(2), 505–530.

Duda, R., P. Hart, and D. Stork (2000). *Pattern Classification* (2nd ed.). Wiley.

Dudoit, A., M. van der Laan, and K. Pollard (2003). Multiple testing i: Procedures for control of general type i error rates. Tech. Rep. 138, Div. Biostatistics, UC Berkeley. See www.bepress.com/ucbiostat/paper138.

Efron, B. (1979). Bootstrap methods: Another look at the jackknife. *Ann. Statist. 7*, 1–26.

Efron, B. (1983). Estimating the error rate of a prediction rule: improvement on cross validation. *J. Amer. Statist. Assoc. 78*, 316–331.

Efron, B. (1986). How biased is the apparent error rate of a prediction rule? *J. Amer. Statist. Assoc. 81*, 461–470.

Efron, B., T. Hastie, I. Johnstone, and R. Tibshirani (2004). Least angle regression. *Ann. Statist. 32*, 407–451.

Efron, B. and R. Tibshirani (1994). *An Introduction to the Bootstrap*. New York, NY: Chapman and Hall.

Efroymson, M. A. (1960). Multiple regression analysis. In *Mathematical Methods for Digital Computers*, pp. 191–203. Wiley: New York.

Eklund, G. and P. Seeger (1965). Massignifikansanalys. *Statistisk Tidskrift Stockholm 3*(4), 355–365.

Eriksson, J. (2004). *Contributions to Theory and Algorithms of ICA and Signal Separation.* Ph. D. thesis, Signal Processing Laboratory, Helsinki University of Technology.

Eriksson, J. and V. Koivunen (2003). Identifiability, separability, and uniqueness of linear ica models revisited. *Proceedings of the 4th International Symposium on ICA and Blind Signal Separation*, 23–27.

Eriksson, J. and V. Koivunen (2004). Identifiability, separability, and uniqueness of linear ica models. *IEEE Signal Proc. Letters 11*(2), 601–604.

Eubank, R. (1988). *Spline Smoothing and Nonparametric Regression*. New York: Marcel Dekker.

Faber, P. and R. Fisher (2001a). Euclidean fitting revisted. *Proc. 4th Intl. Workshop on Visual Form*, 165–175.

Faber, P. and R. Fisher (2001b). Pros and cons of euclidean fitting. Technical report, `http://citeseer.ist.psu.edu/735323.html`.

Fan, J. and I. Gijbels (1996). *Local Polynomial Modeling and Its Application*. London: Chapman & Hall.

Fan, J. and J. Jiang (2005). Nonparametric inferences for additive models. *J. Amer. Statist. Assoc. 100*, 890–907.

Fan, J. and R. Z. Li (2001). Variable selection via penalized likelihood. *J. Amer. Statist. Assoc. 96*, 1348–1360.

Fan, J. and J. Lv (2008). Sure independence screening for ultra-high dimensional feature space. *J. Roy. Statist. Soc. Ser. B 70*, 849–911.

Feraud, R. and F. Clerot (2002). A methodology to explain neural network classification. *Neural Networks 15*, 237–246.

Fernandez, C., E. Ley, and M. F. Steel (2001). Benchmark priors for Bayesian model averaging. *J. Econ. 100*, 381–427.

Ferreira, J. and A. Zwinderman (2006). Approximate power and sample size calculations for the benjamini and hochberg method. *Int'l J. Biostat. 2*(1). Art. 8.

Fisher, L. and J. V. Ness (1971). Admissible clustering procedures. *Biometrika 58*, 91–104.

Fix, E. and J. L. Hodges (1951). Discriminatory analysis – nonparametric discrimination: consistency properties. Tech. Rep. Project No. 21-49-004, USAF School of Aviation Medicine, Randolph Field, TX.

Flake, G., R. Tarjan, and K. Tsioutsiouliklis (2004). Graph clustering and minimum cut trees. *Internet Math. 1*, 385–408.

Fodor, I. (2002). Survey of dimension reduction techniques. Technical report, `http://www.llnl.gov/CASC/Sapphire/pubs/148494.pdf`.

Fokoue, E. and P. Goel (2006). The relevance vector machine: An interesting statistical perspective. Technical Report EPF-06-10-1, Department of Mathematics, Kettering University, Flint, Michigan, USA.

Foster, D. and E. George (1994). The risk inflation criterion for multiple regression. *Ann. Statist. 22*, 1947–1975.

Fraley, C. and A. Raftery (2002). Model-based clustering, discriminant analysis, and density estimation. *J. Amer. Statist. Assoc. 97*(458), 611–631.

Frank, I. E. and J. H. Friedman (1993). A statistical view of some chemometrics regression tools. *Technometrics 35*, 109–148.

Freedman, D. and R. Purves (1969). Bayes method for bookies. *Ann. Math. Statist. 40*, 1177–1186.

Freund, Y. and R. E. Schapire (1999). A short introduction to boosting. *J. of the Japanese Soc. Art. Intelligence 14*(5), 771–780.

Friedman, J. (1987). Exploratory projection pursuit. *J. Amer. Statist. Assoc. 82*(397), 249–266.

Friedman, J. and P. Hall (2000). On bagging and nonlinear estimation. `http://www.-stat.stanford.edu/~jhf/`.

Friedman, J., T. Hastie, and R. Tibshirani (2000). A statistical view of boosting. *Ann. Statist. 28*, 337–407.

Friedman, J. and B. E. Popescu (2005). Predictive learning via rule ensembles. Technical Report, Department of Statistics, Stanford University.

Friedman, J. and Stuetzle (1981). Projection pursuit regression. *J. Amer. Statist. Assoc. 76*, 817–823.

Friedman, J. H. (1984). Smart user's guide. Technical Report 1, Laboratory for Computational Statistics, Stanford University.

Friedman, J. H. (1991). Multivariate adaptive regression splines. *Ann. Statist. 19*(1), 1–67.

Friedman, J. H. and J. J. Meulman (2004). Clustering objects on subsets of attributes (with discussion). *J. Roy. Statist. Soc. Ser. B 66*(4), 815–849.

Fu, W. J. (1998). Penalized regression: the bridge versus the LASSO. *J. Comp. Graph. Statist. 7*, 397–416.

Fukunaga, K. (1990). *Introduction to Statistical Pattern Recognition.* New York: Academic Press.

Fung, G. and O. L. Mangasarian (2004). A feature selection newton method for support vector machine classification. *Comp. Opt. Appl. J. 28(2)*, 185–202.

Furnival, G. (1971). All possible regressions with less computations. *Technometrics 13*, 403–408.

Furnival, G. and R. Wilson (1974). Regression by leaps and bounds. *Technometrics 16*, 499–511.

Gabriel, K. R. and F. C. Pun (1979). Binary prediction of weather event with several predictors. In *6th Conference on Prob. and Statist. in Atmos. Sci., Amer. Meteor. Soc.*, pp. 248–253.

Gallant, R. (1987). *Nonlinear Statistical Models.* New York: J. Wiley and Sons.

Garey, M. and D. Johnson (1979). *Computers and Intractability - A Guide to the Theory of NP-completeness.* New York: Freeman.

Garthwaite, P. (1994). An interpretation of partial least squares. *J. Amer. Statist. Assoc. 89*, 122–127.

Ge, Y., S. Dudoit, and T. Speed (2003). Resampling-based multiple testing for microarray data analysis. *Test 12*(1), 1–77.

Geiringer, H. (1937). On the probability theory of arbitrarily linked events. *Ann. Math. Statist. 9*(4), 260–271.

Geisser, S. (1975). The predictive sample reuse method with applications. *J. Amer. Statist. Assoc. 70*, 320–328.

Gelfand, A. E. and A. F. M. Smith (1990). Sampling-based approaches to calculating marginal densities. *J. Amer. Statist. Assoc. 85*, 398–409.

Geman, S. and D. Geman (1984). Stochastic relaxation, Gibbs distributions, and the Bayesian restoration of images. *IEEE Trans. Pattern Anal. Mach. Int. 6*, 721–741.

Genovese, C. and L. Wasserman (2002). Operating characteristics and extensions of the false discovery rate procedure. *J. Roy. Statist. Soc. Ser. B 64*(3), 499–517.

Genton, M. (2001). Classes of kernels for machine learning: A statistics perspective. *J. Mach. Learning Res. 2*, 299–312.

George, E. and D. Foster (2000). Calibration and empirical bayes variable selection. *Biometrika 87*, 731–747.

George, E. I. (1999). Discussion of "Bayesian model averaging and model selection strategies" by M. Clyde. In J. M. Bernado, A. P. Dawid, J. O. Berger, and A. F. M. Smith (Eds.), *Bayesian Statistics 6*, pp. 157–185. Oxford University Press, London.

George, E. I. (2000). The variable selection problem. *J. Amer. Statist. Assoc. 95*, 1304–1308.

George, E. I. and R. E. McCulloch (1993). Variable selection via Gibbs sampling. *J. Amer. Statist. Assoc. 88*, 881–889.

George, E. I. and R. E. McCulloch (1997). Approaches for Bayesian variable selection. *Statistica Sinica 7*, 339–373.

Gey, S. and E. Nedelec (2005). Model selection and cart regression trees. *Trans. Inform. Theory 51*(2), 658–670.

Ghosh, D., W. Chen, and T. Raghuanthan (2004). The false discovery rate: A variable selection perspective. Working Paper 41, Univ. Mich. SPH.

Ghosh, J. and Ramamoorthi (2003). Bayesian nonparametrics. New York: Springer.

Ghosh, J. K. and T. Samanta (1999). Nonsubjective bayesian testing - an overview. Technical report, Indian Statistical Institute, Calcutta.

Ghosh, S. (2007). Adaptive elastic net: An improvement of elastic net to achieve oracle properties. Technical report, Indiana University-Purdue Univeraity at Indianapolis.

Globerson, A. and N. Tishby (2003). Sufficient dimensionality reduction. *J. Mach. Learning Res. 3*, 1307–1331.

Goldberger, J. and S. Roweis (2004). Hierarchical clustering of a mixture model. *Advances in Neural Information Processing Systems 17*, 505–512.

Green, P. and B. Silverman (1994). *Nonparametric Regression and Generalized Linear Models: A Roughness Penalty Approach*. London.

Greenshtein, E. and Y. Ritov (2004). Persistence in high-dimensional linear predictor selection and the virtue of overparametrization. *Bernoulli 10*, 971988.

Gu, C. (2002). *Smoothing Spline ANOVA Models*. New York: Springer.

Gu, C. and G. Wahba (1991). Minimizing gcv/gml scores with multiple smoothing parameters via the newton method. *SIAM J. Sci. Statist. Comput. 12*, 383–398.

Hall, P. (1989). On projection pursuit regression. *Ann. Statist. 17*(2), 573–588.

Hall, P. (1992). *The Bootstrap and Edgeworth Expansion.* New York: Springer-Verlag.

Hall, P., B. Park, and R. Samworth (2008). Choice of Neighbor Order in Nearest-neighbor Classification. *Ann. Statist. 36*, 2135–2152.

Hand, D., F. Daly, A. Lunn, K. McConway, and E. Ostrowski (Eds.) (1994). *A Handbook of Small Data Sets.* London: Chapman & Hall.

Hannan, E. J. and B. G. Quinn (1979). The determination of the order of an autoregression. *J. Roy. Statist. Soc. Ser. B 41*, 190–195.

Hans, C., A. Dobra, and M. West (2007). Shotgun stochastic search for "large p" regression. *J. Amer. Statist. Assoc. 102*, 507–516.

Hansen, M. H. and B. Yu (2001). Model selection and the principle of minimum description length. *J. Amer. Statist. Assoc. 96*, 746–774.

Hardle, W. (1990). *Applied Nonparametric Regression.* Cambridge: Cambridge Univ. Press.

Hardle, W., P. Hall, and J. S. Marron (1988). How far are automatically chosen regression smoothing parameters from their optimum? (with discussion). *J. Amer. Statist. Assoc. 83*, 86–95.

Hardle, W. and L. Simar (2003). *Applied Multivariate Statistical Analysis.* See: `http://www.xplore-stat.de/tutorials/`.

Harris, B. (1966). *Theory of Probability.* Reading, MA: Addison-Wesley, Statistics Series.

Hartigan, J. (1985). Statistical theory in clustering. *J. Class. 2*, 63–76.

Hartley, H. (1958). Maximum likelihood estimation from incomplete data. *Biometrics 14*, 174–194.

Hartuv, E. and R. Shamir (2000). A clustering algorithm based on graph connectivity. *Information Processing Letters 76*, 175–181.

Hastie, T. (1984). *Pincipal Curves and Surfaces.* Ph. D. thesis, Stanford.

Hastie, T. and W. Stuetzle (1989). Principal curves. *J. Amer. Statist. Assoc. 84*(406), 502–516.

Hastie, T. and R. Tibshirani (1990). *Generalized Additive Models.* New York: Chapman & Hall.

Hastie, T. and R. Tibshirani (1996). Generalized additive models. In S. Kotz and N. Johnson (Eds.), *Encyclopedia of Statistical Sciences.* New York: Wiley and Sons, Inc.

Hastie, T., R. Tibshirani, and J. Friedman (2001). *Elements of Statistical Learning.* New York: Springer.

Hastings, W. (1970). Monte carlo sampling methods using Markov chains and their applications. *Biometrika 57*, 97–109.

Hebb, D. (1949). *The Organization of Behavior: A Neuropsychological Theory*. New York: Wiley.

Heckman, N. E. (1997). The theory and application of penalized least squares methods or reproducing kernel hilbert spaces made easy. Technical Report 216, Statistics Dept., Univ. Brit. Columbia.

Heller, K. A. and Z. Ghahramani (2005). Bayesian hierarchical clustering. In L. Raedt and S. Wrobel (Eds.), *ACM International Conference Proceeding Series 119*, pp. 297–304. Proceedings of the Twenty-Second International Conference Machine Learning.

Helzer, A., M. Barzohar, and D. Malah (2004). Stable fitting of 2d curves and 3d surfaces by implicit polynomials. *IEEE Trans. Pattern Anal. and Mach. Intelligence 26*(10), 1283–1294.

Herzberg, A. M. and A. V. Tsukanov (1986). A note on modifications of the jackknife criterion for model selection. *Utilitas Math. 29*, 209–216.

Hinkle, J. and W. Rayens (1994). Partial least squares and compositional data: Problems an alternatives. Lecture notes, `http://www.ms.uky.edu/jhinkle/JEHdoc/JEHdoc.html`.

Ho, Y. C. and D. L. Pepyne (2002). Simple explanation of the no-free-lunch theorem and its implications. *J. Opt. Theory and Appl. 115*, 549.

Hocking, R. R. and R. N. Leslie (1967). Selection of the best subset in regression analysis. *Technometrics 9*, 531–540.

Hoerl, A. E. and R. W. Kennard (1970). Ridge regression: Biased estimation for nonorthogonal problems. *Technometrics 42*, 80–86.

Hoeting, J. A., D. Madigan, A. E. Raftery, and C. T. Volinsky (1997). Bayesian model averaging: A tutorial. *Statistical Science 14*, 382–417.

Holm, S. (1979). A simple sequentially rejective multiple test procedure. *Scand. J. Statist. 6*, 65–70.

Hotelling, H. (1933). Analysis of complex statistical variables into principal components. *J. Educ. Psych 24*(417-441), 498–520.

House, L. and D. Banks (2004). Cherrypicking as a robustness tool. In Banks, H. D., M. L., A. F, and G. P. (Eds.), *Classification, Clustering, and Data Mining Applcations*, pp. 197–206. Berlin: Springer.

Huang, L. (2001). A roughness penalty view of kernel smoothing. *Statist. Prob. Letters 52*, 85–89.

Huber, P. (1967). Behavior of maximum likelihood estimates under nonstandard conditions. *In Proc. 5th Berkeley Symp. on Math. Statist. and Prob. 1*, 221–233.

Huber, P. (1985). Projection pursuit. *Ann. Statist. 13*, 435–475.

Hunter, D. and R. Li (2005). Variable selection using mm algorithm. *Ann. Statist. 33*, 1617–1642.

Hurvich, C. M. and C. Tsai (1989). Regression and time series model selection in small samples. *Biometrika 76*, 297–307.

Hwang, Y. T. (2001). Edegeworth expansions for the product limit estimator under left-truncation and right censoring with the bootstrap. *Statist. Sinica 11*, 1069–1079.

Hyvarinen, A. (1999). Survey on independent component analysis. *Neural Computing Surveys 2*, 94–128.

Hyvarinen, A. and E. Oja (2000). Independent component analysis: Algorithms and applications. *Neural Networks 13*(4-5), 411–430.

Ibragimov, I. and R. Hasminksy (1980). On nonparametric estimation of regression. *Soviet Math. Dokl 21*, 810–814.

Inselberg, A. (1985). The plane with parallel coordinates. *Special Issue on Computational Geometry, The Visual Computer 1*, 69–91.

Jain, A., M. Topchy, A. Law, and J. Buhmann (2004). Landscape of clustering algorithms. *Proc. 17-th Int'l conference on Pattern Recognition*, 260–263.

Janowitz, M. (2002). The controversy about continuity in clustering algorithms. Tech. Report DIMACS 2002-04, Rutgers University. See: `http://citeseer.ist.psu.edu/549596.html`.

Jefferys, W. and J. O. Berger (1992). Ockham's razor and bayesian analysis. *American Scientist 80*, 64–72.

Jeffreys, H. (1961). *Theory of Probability*. London: Oxford University Press.

Jennings, E., L. Motyckova, and D. Carr (2000). Evaluating graph theoretic clustering algorithms for reliable multicasting. Proceedings of IEEE GLOBECOM 2001. See `http://citeseer.ist.psu.edu/631430.html`,.

Johnson, N. and S. Kotz (1977). *Urn Models and Their Applications*. New York: Wiley.

Johnson, R. and D. Wichern (1998). *Applied Multivariate Statistical Analysis* (4th ed.). Upper Saddle River, NJ: Prentice-Hall.

Jones, L. (1987). On a conjecture of huber concerning the convergence of projection pursuit regression. *Ann. Statist. 15*(2), 880–882.

Jones, L. (1992). A simple lemma on greedy approximation in hilbert space and convergence rates for projection pursuit regression and neural network training. *Ann. Statist. 20*(1), 608–613.

Jones, L. (2000). Local greedy approximation for nonlinear regression and neural network training. *Ann. Statist. 28*(5), 1379–1389.

Jones, M. and R. Sibson (1987). What is projection pursuit? *J. Roy. Statist. Soc. Ser. B 150*(1), 1–37.

Juditsky, A. and A. Nemirovskiii (2000). Functional aggregation for nonparametric regression. *Ann. Statist. 28*(3), 681–712.

Kagan, A., Y. Linnik, and C. Rao (1973). *Characterization Problems in Mathematical Statistics*. Probability and Mathematical Statistics.

Karatzoglou, A., A. J. Smola, K. Hornik, and A. Zeileis (2004). Kernlab: An s4 Package for Kernel Methods in R. *Journal Statistical Software. 11*, 1–20.

Karhunen, J. (2001). Nonlinear independent component analysis. In S. Roberts and R. Everson (Eds.), *Indpendent Component Analysis: Principles and Practice*, pp. 113–134. Cambridge Univ. Press.

Karhunen, J., P. Pajunen, and E. Oja (1998). The nonlinear pca criterion in blind source separation: Relations with other approaches. *Neurocomputing 22*, 5–20.

Kaufman, L. and P. J. Rousseeuw (1990). *Finding groups in data an introduction to cluster analysis.* Wiley Series in Probability and Mathematical Statistics. New York: Wiley.

Kendall, M. (1938). A new measure of rank correlation. *Biometrika 30*, 81–89.

Kiefer, J. and J. Wolfowitz (1960). The equivalence of two extremum problems. *Can. J. Statist. 12*, 363–366.

Kimeldorf, G. and G. Wahba (1971). Some results on Tchebycheffian spline functions. *J. Math. Anal. Applic. 33*, 82–85.

Kimmeldorf and G. Wahba (1971). Correspondence between bayesian estimation of stochastic process and smoothing by splines. *Ann. Math. Statist. 41*, 495–502.

Kirkpatrick, S., C. D. J. Gerlatt, and M. P. Vecchi (1983). Optimization by simulated annealing. *Science 220*, 671–680.

Kleinberg, J. (2003). An impossibility theorem for clustering. *Advances in Neural Information Processing Systems 15.*

Knight, K. and W. J. Fu (2000). Asymptotics for Lasso-type estimators. *Ann. Statist. 28*, 1356–1378.

Knuth, D. E. (1988). Fibonacci multiplication. *AML 1*, 57–60.

Koepke, H. (2008a). Bayesian cluster validation. Master's thesis, Department of Computer Science, University of British Columbia.

Koepke, H. (2008b). Personal communication. *University of British Columbia..*

Kohonen, T. (1981). Automatic formation of topological maps of patterns in a self-organizing system. *Proc. 2nd Scandinavian Conference on Image Analysis. Espoo, Finland.*, 214–220.

Kohonen, T. (1989). *Self-Organization and Associative Memory* (3rd ed.). New York: Springer-Verlag.

Kohonen, T. (1990). The self-organizing map. *Proc. of IEEE 78*(9), 1464–1479.

Kohonen, T. (1995). *Self-Organizing Maps.* Berlin: Springer.

Konishi, S. and G. Kitagawa (1996). Generalised information criteria in model selection. *Biometrika 83*(4), 875–890.

Kruskal, J. (1969). Toward a practical method which helps uncover the structure of a set of multivariate observations by finding the linear transformation which optimizes a new index of condensation. In R. C. Milton and J. A. Nelder (Eds.), *Statistical Computation*. New York: Academic Press.

LeBlanc, M. and R. Tibshirani (1994). Adaptive principal surfaces. *J. Amer. Statist. Assoc. 89*(425), 53–64.

LeCué, G. (2006). Optimal oracle inequality for aggregation of classifiers under low noise condition. In G. Lugosi and H. U. Simon (Eds.), *Proceedings of the 19th Annual Conference On Learning Theory, COLT06*, Volume 32, Berlin, pp. 364–378. Springer. LNAI.

Lee, H. (2000). Model selection for neural network classification. Working Paper 18, ISDS - Duke University.

Lee, H. (2004). *Bayesian Nonparametric via Neural Networks*. ASA-SIAM series on Statistics and Applied Probability.

Lee, W. S., P. Bartlet, and R. Williamson (1996). Efficient agnostic learning of neural networks with bounded fan-in. *IEEE Trans. Inform. Theory 42*(6), 2118–2132.

Lee, Y., Y. Lin, and G. Wahba (2004). Multicategory support vector machines, theory, and application to the classification of microarray data and satellite ra diance data. *J. Amer. Statist. Assoc. 99*, 67–81.

Leeb, H. and B. Potscher (2008). Sparse estimators and the oracle property, or the return of hodges' estimator. *J of Econometrics 142*, 201–211.

Lehmann, E. (1966). Some concepts of dependence. *Ann. Math. Statist. 37*, 1137–1153.

Lempers, F. B. (1971). *Posterior Probabilities of Alternative Linear Models*. Rotterdam: Rotterdam University Press.

Li, K. C. (1984). Consistency for cross-validated nearest neighbor estimates in nonparametric regression. *Ann. Statist. 12*, 230–240.

Li, K. C. (1986). Asymptotic optimality of c_l and gcv in ridge regression with application to spline smoothing. *Ann. Statist. 14*(3), 1101–1112.

Li, K. C. (1987). Asymptotic optimality for c_p, c_l, cv and gcv: Discrete index set. *Ann. Statist. 15*(3), 958–975.

Li, K.-C. (1991). Sliced inverse regression for dimension reduction. *J. Amer. Statist. Assoc. 86*, 316–327.

Liang, F., R. Paulo, G. Molina, M. Clyde, and J. O. Berger (2008). Mixtures of g-priors for Bayesian variable selection. *J. Amer. Statist. Assoc. 103*, 410–423.

Lin, Y. and H. H. Zhang (2006). Component selection and smoothing in smoothing spline analysis of variance models. *Ann. Statist. 34*, 2272–2297.

Lindsay, B. (1995). *Mixture Models: Geometry, Theory, and Applications*. Hayward, CA: IMS Lecture Notes in Statistics.

Linhart, H. and W. Zucchini (1986). *Model Selection*. New York: Wiley.

Little, R. and D. Rubin (2002). *Statistical Analysis with Missing Data*. New Jersey: Wiley. Hoboken.

Liu, J., W. Wong, and A. Kong (1994). Covariance structure of the Gibbs sampler with applications to the comparisons of estimators and augmentation schemes. *Biometrika 81*, 27–40.

Liu, Y. and X. Shen (2006). Multicategory psi-learning and support vector machine: computational tools. *J. Amer. Statist. Assoc. 99*, 219–236.

Ljung, L. (1977). Analysis of recursive stochastic algorithms. *Trans. Autom. Control 22*, 551–575.

Luo, Z.-Q. and J. Tsitsiklis (1994). Data fusion with minima communication. *IEEE Trans. Inform. Theory 40*(5), 1551–1563.

Luxburg, U. (2007). A tutorial on spectral clustering. *Statistics and Computing 17*(4), 395–416.

Luxburg, U., M. Belkin, and O. Bousquet (2008). Consistency of spectral clustering. *Ann. Statist. 36*(2), 555–586.

Luxburg, U. v. and S. Ben-David (2005). Towards a statistical theory of clustering. Tech report, `http://www.cs.uwaterloo.ca/shai/LuxburgBendavid05.pdf`.

MacQueen, J. (1967). Some methods for classification and analysis of multivariate observations. *Proceedings of the Fifth Berkeley Symposium on Mathematical Statistics and Probability*, 281–297.

Madigan, D. and A. Raftery (1984). Model selection and accounting for model uncertainty in graphical models using occam's window. *J. Amer Statist. Assoc. 89*, 1535–1546.

Makato, I. and T. Tokunaga (1995). Hierarchical bayesian clustering for automatic text classification. Tech. Rep. 15, Dept. of Computer Science, Tokyo Institute of Technology.

Mallows, C. L. (1964). Choosing a subset regression. In *Central Regional Meeting of the Institute of Mathematical Statistics*.

Mallows, C. L. (1973). Some comments on c_p. *Technometrics 15*, 661–675.

Mallows, C. L. (1995). More comments on c_p. *Technometrics 37*, 362–372.

Mammen, E., O. Linton, and J. P. Nielsen (1999). The existence and asymptotic properties of a backfitting projection algorithm under weak conditions. *Ann. Statist. 27*, 1443–1490.

Mammen, E., J. Marron, B. Turlach, and M. Wand (2001). A general projection framework for constrained smoothing. *Statist. Sci 16*(3), 232–248.

Mammen, E. and B. Park (2005). Bandwidth selection for smooth backfitting in additive models. *Ann. Statist. 33*(3), 1260–1294.

Maron, O. and A. Moore (1997). The racing algorithm: Model selection for lazy learners. *Artificial Intelligence Review 11*, 193–225.

Marron, J. and W. Härdle (1986). Random approximations to an error criterion of nonparametric statistics. *J. Mult. Anal. 20*, 91–113.

Marron, J. and D. Nolan (1988). Canonical kernels for density estimation. *Statist. Prob. Letters 7*, 195–199.

McCullagh, P. and J. Nelder (1989). *Generalized Linear Models* (second ed.). Chapman & Hall.

McLachlan, G. and T. Krishnan (1997). *EM Algorithm and Extensions*. New York: Wiley.

Meila, M. and L. Xu (2002). Multiway cuts and spectral clustering. Tech report, `www.stat.washington.edu/mmp/Papers/nips03-multicut-tr.ps`.

Meinshausen, N. and P. Buhlmann (2008). Variable selection and high-dimensional graphs with the lasso. *Ann. Statist. 34*(3), 1436–1462.

Meir, R. and G. Ratsch (2003). An introduction to boosting and leveraging. In *Lecture Notes in Computer Science Vol. 2600*, pp. 118–183. Berlin: Springer.

Meng, X. (1994). On the rate of convergence of the ecm algorithm. *Ann. Statist. 22*(1), 326–339.

Meng, X. and D. Rubin (1991). Using em to obtain asymptotic variance covariance matrices: The sem algorithm. *J. Amer. Statist. Assoc. 86*, 899–909.

Metropolis, A. W., M. N. Rosenbluth, A. H. Rosenbluth, and E. Teller (1953). Equations of state calculations by fast computing machines. *Journal of Chemical Physics 21*, 1087–1092.

Miller, A. J. (1990). *Subset Selection in Regression*. London: Chapman & Hall.

Miller, R. (1981). *Simultaneous Statistical Inference*. Springer Series in Statistics. New York: Springer-Verlag.

Mitchell, T. J. and J. J. Beauchamp (1988). Bayesian variable selection in linear regression (with discussion). *J. Amer. Statist. Assoc. 83*, 1023–1036.

Mizuta, M. (1983). Generalized pca invariant under rotations of a coordinate system. *J. Jap. Statist. Soc 14*, 1–9.

Mizuta, M. (2004). Dimension reduction methods. Technical report, `http://www.xplore-stat.de/ebooks/scripts/csa/html/node158.html`.

Mosteller, F. and J. W. Tukey (1968). Data analysis including statistics. In G. Lindzey and E. Aronson (Eds.), *Handbook of Social Psychology*, Volume 2, pp. 80–203. Reading, Mass.: Addison-Wesley.

Moulton, B. (1991). A bayesian approach to regression, selection, and estimation with application to price index for radio services. *J. Econometrics 49*, 169–193.

Muller, M. (1992). *Asymptotic properties of model selection procedures in regression analysis (in German)*. Ph. D. thesis, Humboldt-Universitat zu Berlin.

Muller, M. (1993). Consistency properties of model selection criteria in multiple linear regression. Technical report, `citeseer.comp.nus.edu.sg/184201.html`.

Muller, P., G. Parmigiani, C. Robert, and J. Rousseau (2004). Optimal samples size for multiple testing: The case of gene expression microarrays. *J. Amer. Statist. Assoc. 99*, 990–1001.

Nelder, J. A. and R. Mead (1965). A simplex method for function minimization. *Computer Journal 7*(4), 308–313.

Nemirovsky, A., P. B., and A. Tsybakov (1985). Rate of convergence of nonparametric estimates of maximum likelihood type. *Problemy peredachi informatisii 21*, 258–72.

Neter, J., W. Wasserman, and M. Kutner (1985). *Applied Linear Statistical Models*. Homewood: Irwin.

Ng, A., M. Jordan, and Y. Weiss (2002). On spectral clustering: Analysis and a review. In Z. G. T.G. Dietterich, S. Becker (Ed.), *Advances in Neural Information Processing Systems*, Volume 14, Cambridge, MA, USA, pp. 849–856.

Ng, S., T. Krishnan, and G. McLachlan (2004). The em algorithm. In W. H. J. Gentle and Y. Mori (Eds.), *Handbook of Computational Statistics*, Volume 1, pp. 137–168. New York: Springer-Verlag.

Niederreiter, H. (1992a). *Random Number Generation and Quasi-Monte Carlo Methods*. Philadelphia, Pennsylvania: Society for Industrial and Applied Mathematics.

Niederreiter, H. (1992b). *Random number generation and quasi-Monte Carlo methods*. Philadelphia, PA: SIAM, CBMS-NSF.

Nishii, R. (1984). Asymptotic properties of criteria for selection of variables in multiple regression. *Ann. Statist. 12*, 758–765.

Nussbaum, M. (1985). Spline smoothing in regression models and asymptotic efficiency in l^2. *Ann. Statist. 13*, 984–997.

O'Hagan, A. (1995). Fractional Bayes factors for model comparisons. *J. Roy. Statist. Soc. Ser. B 57*, 99–138.

O'Hagan, A. (1997). Properties of intrinsic and fractional Bayes factors. *Test 6*, 101–118.

Opsomer, J. D. (2000). Asymptotic properties of backfitting estimators. *J. Mult. Anal. 73*, 166–179.

Osborne, M. R., B. Presnell, and B. A. Turlach (2000). On the lasso and its dual. *J. Comp. Graph. Statist. 9*, 319–337.

O'Sullivan, F. (1983). The analysis of some penalized likelihood estimation schemes. Technical Report 726, Statistics Department, UW-Madison.

Pearce, N. D. and M. P. Wand (2005). Penalized splines and reproducing kernel methods. Tech. rep., Dept of Statistics, School of Mathematics, Univ. of New South Wales, Sydney 2052, Australia. `http://www.maths.unsw.edu.au/`.

Pelletier, M. (1998). Weak convergence rates for stochastic approximation with application to multiple targets and simulated annealing. *Ann. Appl. Probab 8*(1), 10–44.

Petrov, V. V. (1975). *Sums of Independent Random Variables*. New York: Springer.

Picard, R. and R. Cook (1984). Cross-validation of regression models. *J. Amer. Statist. Assoc. 79*, 575–583.

Pollard, K. and M. van der Laan (2004). Choice of null distribution in resampling based multiple testing. *J. Statist. Planning and Inference 125*, 85–100.

Polyak, B. T. and A. B. Tsybakov (1990). Asymptotic optimality of thec_p-test for the orthogonal series estimation of regression. *Theory Probab. Appl. 35*, 293–306.

Pontil, M., S. Mukherjee, and F. Girosi (1998). On the noise model of support vector machine regression. Tech Report A.I. memo-1651, MIT Artificial Intelligence Lab.

Priestley, M. B. and M. T. Chao (1972). Nonparametric function fitting. *J. Roy. Statist. Soc. Ser. B 34*, 385–92.

Pukelsheim, F. (1993). *Optimal Design of Experiments*. New York: John Wiley and Sons.

Quintana, F. (2006). A predictive view of bayesian clustering. *J. Statist. Planning Inference 136*, 2407–2429.

Raftery, A. E., D. Madigan, and J. A. Hoeting (1997). Bayesian model averaging for linear regression models. *J. Amer. Statist. Assoc. 92*, 179–191.

Rahmann, S. and E. Rivals (2000). Exact and efficient computation of the expected number of missing and common words in random texts. In R. Giancarlo and D. Sankoff (Eds.), *Combinatorial Pattern Matching*, Volume 1848 of *Lecture Notes in Computer Science*, pp. 375–387. Springer, Berlin.

Rao, C. and Y. Wu (2001). On model selection (with discussion). In *Institute of Mathematical Statistical Lecture Notes - Monograph Series, P. Lahiri (ed.)*, Volume 38, pp. 1–64.

Raudys, S. and D. Young (2004). Results in statistical discriminant analysis: A review of the former soviet union literature. *J. Mult. Analysis 89*, 1–35.

Reiss, R. D. (1989). *Approximate Distributions of Order Statistics*. New York: Springer-Verlag.

Ritter, H. (1991). Asymptotic level density for a class of vector quantization processes. *IEEE Trans. Neural Net. 2*(1), 173–175.

Romano, J. and A. Shaikh (2006). On stepdown control of the false discovery proportion. *IMS Lecture Notes Monograph Series, Second Lehmann Symposium – Optimality 49*, 33–50.

Rosenblatt, F. (1958). The perceptron: A probabilistic model for information and storage in the brain. *Cornell Aeronautical Library, Psychological Review Vol 65*(6), 386–408.

Rosipal, R. (2005). Overview and some aspects of partial least squares. Technical report, `http://www.ofai.at/roman.rosipal/Papers/pascal05.pdf`.

Rosipal, R. and N. Kramer (2006). *Overview and Recent Advances in Partial Least Squares*. `http://www.ofai.at/~roman.rosipal/Papers/pls\_book06.pdf`.

Rosipal, R. and L. Trejo (2001). Kernel partial least squares regression in reproducing kernel hilbert space. *J. Mach. Learning Res. 2*, 97–123.

Roweis, S. and L. Saul (2000). Nonlinear dimensionality reduction by locally linear embedding. *Science 290*, 2323–2326.

Royden, H. L. (1968). *Real Analysis*. New York: MacMillan Publishing Co.

Ruczinski, I., C. Kooperberg, and M. Leblanc (2003). Logic regression. *J. Comp. Graph. Statist. 12*(2), 475–511.

Safavian, S. R. and D. Landgrebe (1991). A survey of decision tree classifier methodology. *IEEE Trans. Sys. Man. Cyber. 21*(3), 660–674.

Sakamoto, Y., M. Ishiguro, and G. Kitagawa (1986). *Akaike Information Criterion Statistics*. Tokyo: KTK Scientific Publishers.

Samanta, T. and A. Chakrabarti (2008). Asymptotic optimality of cross-validatory predictive approach to linear model selection. In B. Clarke and S. Ghosal (Eds.), *Pushing the Limits of Contemporary Statistics: Contributions in Honor of J. K. Ghosh.*, pp. 138–154. Beachwood, OH: Institute of Mathematical Statistics.

Sarkar, S. (2004). Fdr-controlling stepwise procedures and their false negatives rates. *J. Statist. Planning and Inference 125*, 119–137.

Sarle, W. (1996). The number of clusters. Technical report, `http://www.pitt.edu/~wplib/clusfaq.html`. *Adapted from: SAS/STAT User's Guide 1990, Sarle, W. and Kuo, A-H. (1993) The MODECLUS procedure. SAS Technical Report P-256, Cary NC: SAS Institute Inc.*

Savaresi, S., D. Boley, S. Bittanti, and G. Gazzaniga (2002). Choosing the cluster to split in bisecting divisive clustering algorithms. *Proceedings Second SIAM International Conference on Data Mining*, 299–314.

Schapire, R. (1990). The strength of weak learnability. *Machine Learning 5*(2), 197–227.

Schapire, R. E., Y. B. P. Freund, and W. Lee (1998). Boosting the margin: A new explanation for the effectiveness of voting methods. *Ann. Statist. 26*(5), 1651–1686.

Scholkopf, B., A. Smola, and K. Muller (1998). Nonlinear component analysis and a kernel eigenvalue problem. *Neural Comp. 10*, 1299–1319.

Scholkopf, B. and A. J. Smola (2002). *Learning with Kernels: Support Vector Machines, Regularization, Optimization and Beyond.* Cambridge. MA, USA: MIT Press.

Schwarz, G. (1978). Estimating the dimension of a model. *Ann. Statist. 6*, 461–464.

Scott, D. W. (1992). *Multivariate Density Estimation, Theory, Practice and Visualization*. New York: John Wiley and Sons.

Scott, D. W. and M. P. Wand (1991). Feasibility of multivariate density estimates. *Biometrika 78*, 197–205.

Scott, G. and J. Berger (2006). An exploration of bayesian multiple testing. *J. Statist. Planning and Inference 136*(7), 2144–2162.

Seeger, M. (2004). Gaussian processes for machine learning. *Int. J. Neural Sys. 14*(2), 69–106.

Seeger, P. (1968). A note on a method for analysis of significance en mass. *Technometrics 10*, 586–593.

Sen, A. and M. Srivastava (1990). *Regression Analysis: Theory, Methods, and Applications*. New York: Springer.

Shao, J. (1993). Linear model selection by cross-validation. *J. Amer. Statist. Assoc. 88*, 486–494.

Shao, J. (1997). An asymptotic theory for linear model selection, with discussion. *Statist. Sinica 7*, 221–264.

Shi, J. and J. Malik (2000). Normalized cuts and image segmentation. *IEEE Trans. Pattern Anal. Mach. Int. 22*(8), 888–905.

Shibata, R. (1983). Asymptotic mean efficiency of a selection of regression variables. *Ann. Inst. Statist. Math. 35*, 415–423.

Shively, T., R. Kohn, and S. Wood (1999). Variable selection and function estimation in additive nonparametric regression using a data-based prior. *J. Amer. Statist. Assoc. 94*, 777–806.

Shtarkov, Y. (1987). Universal sequential coding of single messages. *Problems of Information Transmission 23*, 175–186.

Sibson, R. and N. Jardine (1971). *Mathematical Taxonomy*. London: Wiley.

Silverman, B. (1984). *Density Estimation for Statistics and Data Analysis*. London: Chapman & Hall.

Simes, J. (1986). An improved bonferroni procedure for multiple tests of significance. *Biometrika 73*, 75–754.

Smola, A. and B. Scholkopf (2003). *A Tutorial on Support Vector Regression*. `citeseer.ist.psu.edu/smola98tutorial.html`.

Smola, A. J. and B. Scholkopf (1998). On a kernel-based method for pattern recognition, regression, approximation and operator inversion. *Algorithmica 22*, 211–231.

Speckman, P. (1985). Spline smoothing and optimal rates of convergence in nonparametric regression models. *Ann. Statist. 8*, 1348–1360.

Spiegelhalter, D. and A. F. Smith (1982). Bayesian factors and choice criteria for linear models. *J. Roy. Statist. Soc. Ser. B 42*, 213–220.

Steinbach, M., G. Karypis, and V. Kumar (2000). A comparison of document clustering techniques. In *Proceedings of the Workshop on Text Mining, 6th ACM SIGKDD International Conference on Knowledge Discovery and Data Mining*, Boston. `http://www.cs.umn.edu/tech_reports_upload/tr2000/00-034.pdf`.

Steinwart, I. (2001). On the influence of the kernel on the consistency of support vector machines. *J. Mach. Learning Res. 2*, 67–93.

Stone, C. (1974). Cross-validatory choice and assessment of statistical predictions. *J. Roy. Statist. Soc. Ser. B 36*, 111–147.

Stone, C. (1977). Asymptotics for and against cross validation. *Biometrika 64*, 29–38.

Stone, C. (1979). Comments on model selection criteria of akaike and schwarz. *J. Roy. Statist. Soc. Ser. B 41*, 276–278.

Stone, C. (1980). Optimal rate of convergence for nonparametric estimators. *Ann. Statist. 8*(6), 1348–1360.

Stone, C. (1982). Optimal global rates of convergence of nonparametric regression. *Ann. Statist. 10*, 1040–1053.

Stone, C. J. (1985). Additive regression and other nonparametric models. *Ann. Statist. 13*, 689–705.

Stone, M. (1959). Application of a measure of information to the design and comparison of regression experiments. *Ann. Math. Statist. 30*, 55–69.

Storey, J. (2002). A direct approach to false discovery rates. *J. Roy. Statist. Soc Ser. B. 64*, 479–498.

Storey, J. (2003). The positive false discovery rate: A bayesian interpretation and the q value. *Ann. Statist. 31*(6), 2013–2035.

Storlie, C., H. Bondell, B. Reich, and H. Zhang (2007). The adaptive COSSO for nonparametric surface estimation and variable selection. Technical report, North Carolina State University, `www4.stat.ncsu.edu/~bondell/acosso.pdf`.

Strobl, C. (2004). Variable selection bias in classification trees. Master's thesis, University of Munich, `http://epub.ub.uni-muenchen.de/1788/1/paper\_419.pdf`.

Stuetzle, W. (2003). Estimating the cluster tree of a density by analyzing the minimal spanning tree of a sample. *J. Class. 20*(5), 25–47.

Sugiura, N. (1978). Further analysis of the data by akaike's information criterion and the finite corrections. *Comm. Statist., Theory and Methods 7*, 13–26.

Sun, J. and C. Loader (1994). Simultaneous confidence bands for linear regression and smothing. *Ann. Statist. 22*(3), 1328–1345.

Sutton, C. D. (2005). Classification and regression trees, bagging, and boosting. In *Handbook of Statistics*, Volume 24, pp. 303–329. Elsevier.

Taleb, A., , and C. Jutten (1999). (1999) source separation in post-nonlnear mixtures. *IEEE Trans. Signal Processing 47*(10), 2807–2820.

Taubin, G. (1988). Nonplanar curve and surface estimation in 3-space. *Proc. IEEE Conf. Robotics and Automation*, 644–645.

Taubin, G., F. Cukierman, S. Sullivan, J. Ponce, and D. Kriegman (1994). Parametrized families of polynmials for bounded algebraic curve and surface fitting. *IEEE Trans. Pattern Anal. and Mach. Intelligence 16*(3), 287–303.

Tenenbaum, J., de Silva V., and J. Langford (2000). A global geometric framework for nonlinear dimension reduction. *Science 290*, 2319–2323.

Terrell, G. (1990). Linear density estimates. *Amer. Statist. Assoc. Proc. Statist. Comput.*, 297–302.

Tibshirani, R. (1988). Estimating transformations for regression via additivity and variance stabilization. *J. Amer. Statist. Assoc. 83*, 394–405.

Tibshirani, R. J. (1996). Regression shrinkage and selection via the lasso. *J. Roy. Statist. Soc. Ser. B 58*, 267–288.

Tibshirani, R. J. and K. Knight (1999). The covariance inflation criterion for model selection. *J. Roy. Statist. Soc. Ser. B 61*, 529–546.

Tipping, M. E. (2001). Sparse bayesian learning and the relevance vector machine. *J. Mach. Learning Res. 1*, 211–244.

Traub, J. F. and H. Wozniakowski (1992). Perspectives on information based complexity. *Bull. Amer. Math Soc. 26*(1), 29–52.

Tsai, C. and J. Chen (2007). Kernel estimation for adjusted p-values in multiple testing. *Comp. Statist. Data Anal. 51*(8), 3885–3897.

Tufte, E. (2001). *The Visual Display of Quantitative Information.* Graphics Press.

Tukey, J. R. (1961). Curves as parameters and toch estimation. *Proc. 4-th Berkeley Symposium*, 681–694.

Tukey, J. R. (1977). *Exploratory Data Analysis.* Upper Saddle River: Addison-Wesley.

Van de Geer, S. (2007). Oracle inequalities and regularization. In E. Del Barrio, P. Deheuvels, and S. van de Geer (Eds.), *Lectures on Empirical Processes: Theory and Statistical Applications*, pp. 191–249. European Mathematical Society.

van Erven, T., P. Grunwald, and S. de Rooij (2008). Catching up faster by switching sooner: a prequential solution to the aic-bic dilemma. arXiv:0807.1005.

Vanderbei, R. and D. Shanno (1999). An interior point algorithm for nonconvex nonlinear programming. *Comp. Opt. and Appl. 13*, 231–252.

Vapnik, V. N. (1995). *The Nature of Statistical Learning Theory*. Springer-Verlag, New York.

Vapnik, V. N. (1998). *Statistical Learning Theory*. Wiley, New York.

Vapnik, V. N. and A. Y. Chervonenkis (1971). On the uniform convergence of relative frequencies of events to their probabilities. *Theory of Probab. and It Applications 16*(2), 264–280.

Varshavsky, J. (1995). *On the Development of Intrinsic Bayes factors*. Ph. D. thesis, Department of Statsitics, Purdue University.

Verma, D. and M. Meila (2003). A comparison of spectral clustering algorithms. Tech Report 03-05-01, Department of CSE, University of Washington, Seattle, WA, USA.

Voorhees, E. and D. Harman (Eds.) (2005). *Experiment and Evaluation in Information Retrieval*. Boston, MA: The MIT Press.

Wahba, G. (1985). A comparison of gcv and gml for choosing the smoothing parameter in the generalized spline smoothing problem. *Ann. Statist. 13*, 1378–1402.

Wahba, G. (1990). *Spline Models for Observational Data*, Volume 59. Philadelphia: SIAM CBMS-NSF Regional Conference Series.

Wahba, G. (1998). Support vector machines, reproducing kernel hilbert spaces, and the randomized gacv. Tech. Rep. 984, Dept. of Statistics, Univ. Wisconsin, Madison, `http://www.stat.wisc.edu/wahba`.

Wahba, G. (2005). Reproducing kernel hilbert spaces and why they are important. Tech. Rep. xx, Dept. Statistics, Univ. Wisconsin, Madison, `http://www.stat.wisc.edu/wahba`.

Wahba, G. and S. Wold (1975). A completely automatic french curve: Fitting spline functions by cross validation. *Comm. Statist. - Sim. Comp. 4*, 1–17.

Walker, A. M. (1969). On the asymptotic behavior of posterior distributions. *J. Roy. Statist. Soc. Ser. B 31*, 80–88.

Walsh, B. (2004, May). Multiple comparisons:bonferroni corrections and false discovery rates. Lecture notes, University of Arizona, `http://nitro.biosci.arizona.edu/courses/EEB581-2006/handouts/Multiple.pdf`.

Wang, H., G. Li, and G. Jiang (2007). Robust regression shrinkage and consistent variable selection via the lad-lasso. *J. Bus. Econ. Statist. 20*, 347–355.

Wang, L. and X. Shen (2007). On l1-norm multiclass support vector machines: methodology and theory. *J. Amer. Statist. Assoc. 102*, 583–594.

Wasserman, L. (2004). *All of Statistics: A Concise Course in Statistical Inference*. New york: Springer.

Wasserman, L. A. (2000). Asymptotic inference for mixture models by using data-dependent priors. *J. Roy. Statist. Soc. Ser. B 62*(1), 159–180.

Wegelin, J. (2000). A survey of partial least squares methods, with emphasis on the two-block case. Technical report, `http://citeseer.ist.psu.edu/`

wegelin00survey.html.

Weisberg, S. (1985). *Applied Linear Regression*. New York: Wiley.

Weisstein, E. W. (2009). Monte carlo integration. Technical report, MathWorld–A Wolfram Web Resource., http://mathworld.wolfram.com/Quasi-MonteCarloIntegration.html.

Welling, M. (2005). Fisher-lda. Technical report, http://www.ics.uci.edu/welling/classnotes/papers_class/Fisher-LDA.pdf.

Westfall, P. and S. Young (1993). *Resampling Based Methods for Multiple Testing: Examples and Methods for p-value Adjustment*. New York: Wiley.

Weston, J., S. Mukherjee, O. Chapelle, M. Pontil, T. Poggio, and V. Vapnik (2000). Feature selection for SVMs. *Advances in Neural Information Processing Systems 13*.

Weston, J. and C. Watkins (1999). Support vector machines for multiclass pattern recognition. In *Proceedings of the Seventh European Symposium on Artificial Neural Networks*, http://www.dice.ucl.ac.be/Proceedings/esann/esannpdf/es1999-461.pdf.

White, H. (1981). Consequences and detection of misspecified nonlinear regression models. *J. Amer. Statist. Assoc. 76*, 419–433.

White, H. (1989). Some asymptotic results for learning in single hidden layer feedforward network models. *J. Amer. Statist. Assoc. 84*, 1003–1013.

Wilf, H. (1989). *Combinatorial Algorithms: An Update*, Volume 55 of *CBMS-NSF Regional Conference in Applied Mathematics*. philadelphia: SIAM.

Wolfe, J. H. (1963). *Object Cluster Analysis of Social Areas*. Ph. D. thesis, UC Berkeley.

Wolfe, J. H. (1965). A computer program for the maximum likelihood analysis of types. *Technical Bulletin of the US Naval Personnel and Training Research Activity* (SRM 65-15).

Wolfe, J. H. (1967). Normix: Computational methods for estimating the parameters of multivariate normal mixtures of distributions. Research Activity SRM 68-2, San Diego, CA, USA.

Wolfe, J. H. (1970). Pattern clustering by multivariate mixture analysis. *Multivariate Behavioral Research 5*, 329–350.

Wolpert, D. (2001). *The Supervised Learning No Free Lunch Theorems*. http://ic.arc.nasa.gov/ic/projects/bayes-group/people/dhw/.

Wolpert, D. and W. G. Macready (1995). No free lunch theorems for search. Working paper SFI-TR-95-02-010, Santa Fe Institute.

Wolpert, D. and W. G. Macready (1996). Combining stacking with bagging to improve a learning algorithm. Technical report, See: http://citeseer.ist.psu.edu/wolpert96combining.html.

Wolpert, D. and P. Smyth (2004). Stacked density estimation. Technical report.

Wolpert, D. H. (1992). On the connection between in-sample testing and generalization error. *Complex Systems 6*, 47–94.

Wong, D. and M. Murphy (2004). Estimating optimal transformations for multivariate regression using the ace algorithm. *J. Data. Sci. 2*, 329–346.

Wong, H. and B. Clarke (2004). Improvement over bayes prediction in small samples in the presence of model uncertainty. *Can. J. Statist. 32*(3), 269–284.

Wong, W. H. (1983). On the consistency of cross-validation in kernel nonparametric regression. *Ann. Statist. 11*, 1136–1141.

Wozniakowski, H. (1991). Average case complexity of multivariate integration. *Bull. AMS 24*(1), 185–193.

Wu, C. F. J. (1983). On the convergence properties of the em algorithm. *Ann. Statist. 11*(1), 95–103.

Wu, Y. and Y. Liu (2009). Variable selection in quantile regression. *Statistica Sinica*. To Appear.

Wu, Z. and R. Leahy (1993). An optimal graph theoretic approach to data clustering: Theory and its application to image segmentation. *IEEE Pattern Anal. and Mach. Intel. 15*, 1101–1113.

Wyner, A. J. (2003). On boosting and the exponential loss. Tech report, See: `http://citeseer.ist.psu.edu/576079.html`.

Yang, Y. (2001). Adaptive regression through mixing. *J. Amer. Statist. Assoc. 96*, 574–588.

Yang, Y. (2005). Can the strengths of aic and bic be shared? a conflict between model identification and regression estimation. *Biometrika 92*, 937–950.

Yang, Y. (2007). Consistency of cross validation for comparing regression procedures. *Ann. Statist. 35*, 2450–2473.

Yang, Y. and A. Barron (1999). Information-theoretic determination of minimax rates of convergence. *Ann. Statist. 27*, 1564–1599.

Yardimci, A. and A. Erar (2002). Bayesian variable selection in linear regression and a comparison. *Hacettepe J. Math. and Statist. 31*, 63–76.

Ye, J. (2007). Least squares discriminant analysis. In *Proceedings of the 24th international conference on Machine learning*, Volume 227, pp. 1087–1093. ACM International Conference Proceeding Series.

Ye, J., R. Janardan, Q. Li, and H. Park (2004). Feature extraction via generalized uncorrelated linear discriminant analysis. In *Twenty-First International Conference on Machine Learning*, pp. 895902. ICML 2004.

Yu, B. and T. Speed (1992). Data compression and histograms. *Probab. theory Related Fields 92*, 195–229.

Yu, C. W. (2009). *Median Methods in Statistical Analysis with Applications*. Ph. D. thesis, Department of Statsitics, University of British Columbia.

Yuan, M. and Y. Lin (2007). On the non-negative garrote estimator. *J. Roy. Statist. Soc. Ser. B 69*, 143–161.

Zellner, A. (1986). On assessing prior distributions and Bayes regression analysis with *g*-prior distributions. In P. K. Goel and A. Zellner (Eds.), *Bayesian Inference and Decision Techniques: Essays in Honor of Bruno de Finetti*, pp. 233–243. North-Holland/Elsevier.

Zellner, A. and A. Siow (1980). Posterior odd ratios for selected regression hypotheses. In J. M. Bernardo, M. H. DeGroot, D. V. Lindley, and A. F. M. Smith (Eds.), *Bayesian Statistics: Proceedings of the First International Meeting held in Valencia*, pp. 585–603. Valencia University Press.

Zhang, H., J. Ahn, X. Lin, and C. Park (2006). Gene selection using support vector machines with nonconvex penalty. *Bioinformatics 22*, 88–95.

Zhang, H., Y. Liu, Y. Wu, and J. Zhu (2008). Variable selection for multicategory SVM via supnorm regularization. *Elec. J. Statist. 2*, 149–167.

Zhang, H. and W. Lu (2007). Adaptive-lasso for cox's proportional hazards model. *Biometrika 94*, 691–703.

Zhang, H. H., G. Wahba, Y. Lin, M. Voelker, M. Ferris, R. Klein, and B. Klein (2004). Nonparametric variable selection via basis pursuit for non-Gaussian data. *J. Amer. Statist. Assoc. 99*, 659–672.

Zhang, P. (1992). On the distributional properties of model selection criteria rule. *J. Amer. Statist. Assoc. 87*, 732–737.

Zhang, P. (1993). Model selection via multifold cross validation. *Ann. Statist. 21*, 299–313.

Zhang, T. (2004). Statistical behavior and consistency of classification methods based on convex risk minimization. *Ann. Statist. 32*, 56–85.

Zhao, P. and B. Yu (2006). On model selection consistency of lasso. *J. Mach. Learning Res. 7*, 2541–2563.

Zhao, P. and B. Yu (2007). Stagewise lasso. *J. Mach. Learning Res. 8*, 2701–2726.

Zhao, Y. and C. G. Atkeson (1991). Some approximation properties of projection pursuit learning networks. *NIPS*, 936–943.

Zhao, Y. and C. G. Atkeson (1994). Projection pursuit learning: Approximation properties. *ftp://ftp.cis.ohio-state.edu/pub/neuroprose/yzhao.theory-pp.ps.Z*.

Zhao, Y. and G. Karypis (2002). Evaluation of hierarchical clustering algorithms for document data sets. *Proceedings of the 11th ACM Conference on Information and Knowledge Management*, 515–524.

Zhao, Y. and G. Karypis (2005). Hierarchical clustering algorithms for document data sets. *Data Mining and Knowledge Discovery 10*(2), 141–168.

Zhou, S., X. Shen, and D. A. Wolfe (1998). Local asymptotics for regression splines and confidence regions. *Ann. Statist. 26*, 1760–1782.

Zhou, S. K., B. Georgescu, Z. X. S., and D. Comaniciu (2005). Image based regression using boosting method. *International Conference on Computer Vision 1*, 541–548.

Zhu, J. and T. Hastie (2001). Kernel logistic regression and the import vector machine. Tech report, See `http://citeseer.ist.psu.edu/479963.html`.

Zhu, J., S. Rosset, T. Hastie, and R. Tibshirani (2003). 1-norm support vector machines. *17th Annual Conference on Neural Information Processing Systems 16.*

Zou, H. (2006). The adaptive lasso and its oracle properties. *J. Amer. Statist. Assoc. 101*, 1418–1429.

Zou, H. and T. Hastie (2005). Regularization and variable selection via the elastic net. *J. Roy. Statist. Soc. Ser. B 67*, 301–320.

Zou, H. and R. Li (2007). One-step sparse estimates in nonconcave penalized likelihood models. *Ann. Statist. 32*, 1–28.

Zou, H. and H. H. Zhang (2008). On the adaptive elastic-net with a diverging number of parameters. *Ann. Statist.*, to appear.

Index

Printed in the United States of America